32nd Symposium of Malaysian Chemical Engineers (SOMChE2021)

IOP Conference Series: Matarials Science and Engineering
Volume 1195

Kuching, Sarawak
15 - 16 July 2021

Editors:

Jaka Sunarso
Ngu Lock Hei
Chew Jiuan Jing
How Bing Shen
Elaine Yeu Yee Lee
Mahmoud Ahmed Mussad Elhadi
Siti Salwa Hashim
Bazlul Mobin Siddique
Lai Jia Chi
Claudia li Chia Chirng
Megan Soh Zheyin
Lai Jia Yen
Liza Melia Anak Terry
Shirleen Lo Lee Yuen
Karen Kong Gah Hie
Adeline Lim Kher Li
Luke Ngu Wei Wei
Slyvester Chai Yew Wang
Frederick Phang Jit Fook
Jonathan Cheng Lin Yang
Wong Jung Lin
Yeo Lip Siang

ISBN: 978-1-7138-4654-3
ISSN: 1757-8981

Printed from e-media with permission by:

Curran Associates, Inc.
57 Morehouse Lane
Red Hook, NY 12571

Some format issues inherent in the e-media version may also appear in this print version.

Institute of Physics
Dirac House, Temple Back
Bristol BS1 6BE UK

Phone: 44 1 17 929 7481
Fax: 44 1 17 920 0979

techtracking@iop.org

Additional copies of this publication are available from:

Curran Associates, Inc.
57 Morehouse Lane
Red Hook, NY 12571 USA
Phone: 845-758-0400
Fax: 845-758-2633
Email: curran@proceedings.com
Web: www.proceedings.com

TABLE OF CONTENTS

IOP Conf. Series: Materials Science and Engineering 1195 (2021) 011001 doi:10.1088/1757-899X/1195/1/011001

Editorial Note

The 32nd Symposium of Malaysian Chemical Engineers (SOMChE2021) was the first of its kind to be held virtually on 15th and 16th July 2021, which was delayed due to the pandemic from its original event date in March 2020 to July 2021. SOMChE2021 was jointly organised by Swinburne University of Technology Sarawak Campus and the Institution of Chemical Engineers (IChemE) Malaysia. Presently, industrial chemical processes rely heavily on unsustainable fossil-fuel based resources. The theme "Navigating the Frontier of Sustainability" serves to address the ongoing global sustainability issues and present viable long-term solutions. SOMChE2021 has provided a platform for a wide community of young researchers, scholars, engineers, and practitioners to present their latest research findings and innovation in chemical engineering (and related disciplines) and to develop long-term research collaborations. A total of 75 delegates from Malaysia, Philippines, Taiwan, Indonesia, Brunei, Australia, and Japan have attended and presented their research findings at SOMChE2021, encompassing a diverse range of topics including green and sustainable processing, environmental and waste management, bioprocessing and biotechnology, materials science and engineering, and energy in terms of energy conversion and management, and renewable and alternative energy. These proceedings serve as a compilation of publications that showcase the progress in engineering and science to bridge the gap between research and industry in solving sustainability concerns, of which all accepted papers have undergone stringent screening and review processes. Overall, the success of SOMChE2021 was made possible through the collective contribution of the Organising Committee, International Scientific Committee, and Sponsors.

Editors Assoc. Prof. Jaka Sunarso, Ir. Dr. Ngu Lock Hei, Dr. Chew Jiuan Jing, Dr. How Bing Shen, Dr. Elaine Yeu Yee Lee, Dr. Mahmoud Ahmed Mussad Elhadi, Dr. Siti Salwa Hashim, Dr. MD. Bazlul Mobin Siddique, Ms. Lai Jia Chi, Dr. Claudia Li Chia Chirng, Ms. Megan Soh Zheyin, Ms. Lai Jia Yen, Ms. Liza Melia Anak Terry, Ms. Shirleen Lo Lee Yuen, Ms. Karen Kong Gah Hie, Ms. Adeline Lim Kher Li, Mr. Luke Ngu Wei Wei, Mr. Slyvester Chai Yew Wang, Mr. Frederick Phang Jit Fook, Mr. Jonathan Cheng Lin Yang, Mr. Wong Jung Lin, Mr. Yeo Lip Siang
Swinburne University of Technology Sarawak Campus

IOP Conf. Series: Materials Science and Engineering 1195 (2021) 011001 doi:10.1088/1757-899X/1195/1/011001

Organising Committee

Co-chairs Prof. Su Hieng Tiong, Assoc. Prof. Jaka Sunarso

Vice Chair Dr. Chew Jiuan Jing

Secretariat Team Dr. How Bing Shen (Chair), Ms. Shirleen Lo Lee Yuen, Ms. Karen Kong Gah Hie, Mr. Yeo Lip Siang

Finance Ms. Lai Jia Chi (Chair), Ms. Audrey Chong Hui Kian, Mr. Shariman Hoklai

Scientific Program Assoc. Prof. Jaka Sunarso (Chair), Ir. Dr. Ngu Lock Hei, Dr. Claudia Li Chia Chirng, Ms. Megan Soh Zheyin, Ms. Lai Jia Yen, Ms. Liza Melia Anak Terry, Ms. Karen Kong Gah Hie, Ms. Adeline Lim Kher Li, Mr. Luke Ngu Wei Wei, Mr. Slyvester Chai Yew Wang, Mr. Frederick Phang Jit Fook, Mr. Jonathan Cheng Lin Yang, Mr. Wong Jung Lin

Sponsorship Dr. Chew Jiuan Jing (Chair), Dr. Elaine Yeu Yee Lee, Ms. Adeline Lim Kher Li

Public Relations Dr. Elaine Yeu Yee Lee (Chair), Dr. Chua Hong Siang

Post-event Management Team Dr. Xavier Wezen Chee (Chair), Ms. Ee Sia, Ms. Emily George Jimbai, Mr. Jeremiah Idek Angking, Mdm. Nurlida Mohamad Salleh

Virtual Trade Display Dr. Mahmoud Ahmed Mussad Elhadi (Chair), Dr. Daniel Tan Lee Tung, Mr. Luke Ngu Wei Wei

Publicity Dr. Elammaran Jayamani (Chair), Mdm. Ting Lik Fong, Dr. MD. Bazlul Mobin Siddique

Design and Printing Dr. Irine Runnie Henry Ginjom (Chair), Ms. Jenny Liew Ngi San (Designer)

Virtual Registration Dr. Siti Salwa Hashim (Chair), Dr. Elaine Yeu Yee Lee, Ms. Ee Sia, Ms. Emily George Jimbai, Mr. Jeremiah Idek Angking, Mdm. Nurlida Mohamad Salleh

Advisory Panel Assoc. Prof. Dr. Peter Morin Nissom (Sarawak Research Development Council), Prof. Dr. Michael Cloke (IChemE), Assoc. Prof. Dr. Hwang Siaw San (Swinburne Sarawak), Prof. Lau Hieng Ho (Swinburne Sarawak)

IOP Conf. Series: Materials Science and Engineering 1195 (2021) 011001 doi:10.1088/1757-899X/1195/1/011001

International Scientific Committee

Prof. Dr. Chen Xiao Dong (Soochow University, China)
Prof. Dr. Manabu Kano (Kyoto University, Japan)
Prof. Dr. Wuled Lenggoro (Tokyo University of Agriculture and Technology, Japan)
Prof. Michael Kobina Danquah (University of Tennessee, USA)
Assoc. Prof. Kazi Bayzid Kabir (Bangladesh University of Engineering and Technology, Bangladesh)
Ir. Dr. Dave Mangindaan (Bina Nusantara University, Indonesia)
Prof. Ir. Tjandra Setiadi (Institut Teknologi Bandung, Indonesia)
Dr. Yalun Arifin (Prasetiya Mulya Universiti, Indonesia)
Prof. Renanto Handogo (Institut Teknologi Sepuluh Nopember, Indonesia)
Prof. Dr. Ing. Misri Gozan (Universitas Indonesia, Indonesia)
Prof. Ir. Rochmadi (Universitas Gadjah Mada, Indonesia)
Prof. Ir. Suryo Purwono (Universitas Gadjah Mada, Indonesia)
Prof. Kathleen B. Aviso (De La Salle University, Philippines)
Prof. Dr. Maria Natalia R. Dimaano (University of Santo Tomas, Philippines)
Dr. Michelle C. Almendrala (Mapua University, Philippines)
Dr. Hazwani Hj Suhaimi (Universiti Brunei Darussalam, Brunei)
Dr. Muhammad Saifullah Abu Bakar (Universiti Brunei Darussalam, Brunei)
Dr. Rama Rao Karri (Universiti Teknologi Brunei, Brunei)
Assoc. Prof. Dr. Chong Meng Nan (Monash University, Malaysia)
Dr. Irene Chew Mei Leng (Monash University, Malaysia)
Dr. Serene Lock Sow Mun (Monash University, Malaysia)
Assoc. Prof. Poh Phaik Eong (Monash University, Malaysia)
Prof. Ir. Dr. Chai Siang Piao (Monash University, Malaysia)
Prof. Ir. Dr. Haslenda Hashim (UTM, Malaysia)
Prof. Dr. Azman Bin Hassan (UTM, Malaysia)
Ir. Dr. Tan Lian See (UTM, Malaysia)
Dr. Nor Ruwaida binti Jamian (UTM, Malaysia)
Dr. Nurfatehah Wahyuny binti Che Jusoh (UTM, Malaysia)
Dr. Dayang Norulfairuz Abang Zaidel (UTM, Malaysia)
Prof. Datuk Dr. Ahmad Fauzi Ismail (UTM, Malaysia)
Ir. Dr. Kiew Peck Loo (UTM, Malaysia)
Prof. Ir. Ts. Dr. Pau Loke Show (Nottingham University, Malaysia)
Assoc. Prof. Dr. Nishanth Chemmangattuvalappil (Nottingham University, Malaysia)
Assoc. Prof. Dr. Lim Siew Shee (Nottingham University, Malaysia)
Ir. Dr. Leong Wei Dong (Nottingham University, Malaysia)
Dr. Ngan Sue Lin (Nottingham University, Malaysia)
Prof. Ir. Dr. Suzana Yusup (UTP, Malaysia)
Prof. Dr. Thanabalan Murugesan (UTP, Malaysia)

Dr. Chew Thiam Leng (UTP, Malaysia)
Dr. Lam Man Kee (UTP, Malaysia)
Assoc. Prof. Dr. Yeong Yin Fong (UTP, Malaysia)
Mr. Adrian Chun Minh Loy (UTP, Malaysia)
Dr. Chan Yi Herng (UTP, Malaysia)
Dr. Chai Yee Ho (UTP, Malaysia)
Assoc. Prof. Syamsul Rizal Abd Shukor (USM, Malaysia)
Prof. Dr. Bassim H. Hameed (USM, Malaysia)
Assoc. Prof. Dr. Low Siew Chun (USM, Malaysia)
Ir. Dr. Wendy Ng Pei Qin (Curtin University, Malaysia)
Assoc. Prof. Dr. Mubarak Mujawar (Curtin University, Malaysia)
Assoc. Prof. Dr. Agus Saptoro (Curtin University, Malaysia)
Dr. Bridgid Chin Lai Fui (Curtin University, Malaysia)
Dr. Jibrail Kansedo (Curtin University, Malaysia)
Dr. Tan Inn Shi (Curtin University, Malaysia)
Dr. Henry Foo Chee Yew (Curtin University, Malaysia)
Assoc. Prof. Dr. Chong Chien Hwa (Taylor's University, Malaysia)
Dr. Rebecca Lim Xiao Yien (Taylor's University, Malaysia)
Dr. Anis Suhaila binti Shuib (Taylor's University, Malaysia)
Dr. Farizul Hafiz Bin Kasim (UNIMAP, Malaysia)
Assoc. Prof. Dr. Muhammad Syarhabil Bin Ahmad (UNIMAP, Malaysia)
Dr. Siti Kartini binti Enche Ab Rahim (UNIMAP, Malaysia)
Dr. Loh Soh Kheang (MPOB, Malaysia)
Prof. Dr. Ir. Thomas Choong Shean Yaw (UPM, Malaysia)
Dr. Nozieana Khairuddin (UPM, Malaysia)
Dr. Azizul Hakim Lahuri (UPM, Malaysia)
Dr. Tengku Sharifah Marliza Tengku Azmi (UPM, Malaysia)
Assoc. Prof. Ir. Dr. Wan Azlina bt. Wan Abdul Karim Ghani (UPM, Malaysia)
Assoc. Prof. Jaka Sunarso (Swinburne University, Malaysia)
Dr. Chew Jiuan Jing (Swinburne University, Malaysia)
Ir. Dr. Ngu Lock Hei (Swinburne University, Malaysia)
Dr. How Bing Shen (Swinburne University, Malaysia)
Dr. Mahmoud Ahmed Mussad Elhadi (Swinburne University, Malaysia)
Dr. Siti Salwa Hashim (Swinburne University, Malaysia)
Dr. Bazlul Mobin Siddique (Swinburne University, Malaysia)
Dr. Elaine Yeu Yee Lee (Swinburne University, Malaysia)
Ms. Lai Jia Chi (Swinburne University, Malaysia)
Assoc. Prof. Moritz Müller (Swinburne University, Malaysia)
Assoc. Prof. Hwang Siaw San (Swinburne University, Malaysia)
Dr. Irine Runnie Henry Ginjom (Swinburne University, Malaysia)
Dr. Daniel Tan Lee Tung (Swinburne University, Malaysia)

IOP Conf. Series: Materials Science and Engineering 1195 (2021) 011001 doi:10.1088/1757-899X/1195/1/011001

Dr. Xavier Chee Wezen (Swinburne University, Malaysia)
Ms. Ting Lik Fong (Swinburne University, Malaysia)
Dr. Elammaran Jayamani (Swinburne University, Malaysia)
Prof. Dr. Duduku Krishnaiah (UMS, Malaysia)
Assoc. Prof. Dr. Chu Chi Ming (UMS, Malaysia)
Dr. Tham Heng Jin (UMS, Malaysia)
Assoc. Prof. Dr. Rubiyah bt Haji Baini (UNIMAS, Malaysia)
Prof. Dr. Mohammad Omar Abdullah (UNIMAS, Malaysia)
Dr. Yiin Chung Loong (UNIMAS, Malaysia)
Prof. Datuk Ir. Dr. Abdul Wahab Mohammad (UKM, Malaysia)
Prof. Datuk Ir. Dr. Wan Ramli Wan Daud (UKM, Malaysia)
Assoc. Prof. Dr. Masli Irwan Rosli (UKM, Malaysia)
Prof. Ir. Dr. Ong Kok Seng (UTAR, Malaysia)
Dr. Mah Shee Keat (UTAR, Malaysia)
Dr. Pang Yean Ling (UTAR, Malaysia)
Dr. Steven Lim (UTAR, Malaysia)
Dr. Lee Kiat Moon (UCSI University, Malaysia)
Dr. Chai Pui Vun (UCSI University, Malaysia)
Assoc. Prof. Dr. Yeap Swee Pin (UCSI University, Malaysia)
Dr. Sum Jing Yao (UCSI University, Malaysia)
Prof. Zulkafli bin Hassan (UMP, Malaysia)
Assoc. Prof. Ts. Dr. Sumaiya binti Zainal Abidin @ Murad (UMP, Malaysia)
Prof. Ir. Dr. Abdul Aziz bin Abdul Rahman (UM, Malaysia)
Assoc. Prof. Dr. Ong Wee Jun (Xiamen University, Malaysia)
Assoc. Prof. Ir. Dr. Parthiban Siwayanan (Xiamen University, Malaysia)
Dr. Viknesh Andiappan (Heriot-Watt University, Malaysia)
Prof. Ir. Dr. Denny Ng Kok Sum (Heriot-Watt University, Malaysia)

IOP Conf. Series: Materials Science and Engineering 1195 (2021) 011001 doi:10.1088/1757-899X/1195/1/011001

Sponsors

IOP Conf. Series: Materials Science and Engineering 1195 (2021) 011002 doi:10.1088/1757-899X/1195/1/011002

Peer review declaration

All papers published in this volume of *IOP Conference Series: Materials Science and Engineering* have been peer reviewed through processes administered by the Editors. Reviews were conducted by expert referees to the professional and scientific standards expected of a proceedings journal published by IOP Publishing

- **Type of peer review:** Single-blind (authors' identities are known to the reviewers, reviewers' identities are hidden from authors)

- **Describe criteria used by Reviewers when accepting/declining papers. Was there the opportunity to resubmit articles after revisions?**
  Reviewers were requested to review the manuscripts by pointing out the strengths, weaknesses, errors, omissions, and inconsistencies in the papers. Authors were encouraged to revise the papers based on the reviewer's comments. The review process was then repeated until the reviewers were satisfied with the quality of the submissions.

- **Conference submission management system:** Committee-managed system using Google's suite of products. Submissions were received through Google forms and managed within Google drive. Correspondence with authors was conducted using Gmail.

- **Number of submissions received:** 118

- **Number of submissions sent for review:** 118

- **Number of submissions accepted:** 62

- **Acceptance Rate (Number of Submissions Accepted / Number of Submissions Received X 100):** 52.5%

- **Average number of reviews per paper:** 2

- **Total number of reviewers involved:** 106

- **Any additional info on review process (ie plagiarism check system):** Plagiarism was checked using Turnitin, which is embedded in Canvas - Swinburne Sarawak's course management system.

- **Contact person for queries:** Jaka Sunarso (jsunarso@swinburne.edu.my)

Composting of food waste in passive aerated bioreactor with turning mode

A Y Zahrim, M Darwis, D Samantha, A Z Siti Hasanah, S A Nur Aqeela, L Junidah, S Sariah and R Mariani

Faculty of Engineering, Universiti Malaysia Sabah, Jalan UMS, 88400 Kota Kinabalu, Sabah, Malaysia

Email: zahrim@ums.edu.my

Abstract. Almost 45% of municipal solid waste in Malaysia consist of food waste. Composting is one of the sustainable ways to manage food waste compared to incineration and landfilling. This paper investigates the physicochemical and phytotoxicity characteristics during food waste composting in passive aerated bioreactor assisted with compost turning. The initial compost mixture consists of 124 kg of food waste mixed with 62 kg of dry leaves. The composting process was conducted for 40 days, and physicochemical characteristics i.e., temperature, moisture content, total organic carbon, pH and conductivity were monitored. Seed germination test was conducted with cabbage seeds (Brassica oleracea). The highest temperature and final moisture content obtained were 42 °C and 78%, respectively. The seed germination index value was 127%, indicating that the compost is suitable for plant growth.

1. Introduction

Food waste disposal worldwide, including in Malaysia is getting more challenging due to land shortage for landfilling. Malaysia generates about 38,000 tonnes of waste per day and around 17,000 tonnes is food waste [1]. Owing to the high moisture content of food waste, the incineration process is expensive [2]. In addition to that, dioxins and greenhouse gases may release during the process [3, 4]. Besides incineration, composting could be an alternative for food waste disposal. Composting is an aerobic microbiological process where organic waste is converted into compost, a humus-like structure. Various bulking agents have been used to enhance the degradation process with minimum odor generation, such as sawdust [5] and mushroom waste [6]. Nevertheless, the awareness among the community on sustainable food waste disposal is still lacking.

Higher learning institutions should play a vital role in their local community for sustainable food waste processing via composting. At Kean University, Mu *et al.* [7] composting food waste in an in-vessel composter. Their finding stated that the compost produced can be used to grow vegetables and able to generate profit from the activities. Recently, Saalah *et al.* [8] successfully produced compost from food waste collected at the Universiti Malaysia Sabah cafeteria. The initial compost mixture contains 1:1 (dry leaves:food waste) in a passive aerated bioreactor with turning mode. In this study, a similar process will be carried out with a different ratio of food waste-dry leaves i.e., 1:2 (dry leaves: food waste) to increase the final compost nutrients content.

The composting process can be conducted in passive (natural) or forced aeration mode. Bhave and Kulkarni [9] found that the quality of compost concerning Germination Index (GI), root length index and NPK value was better in an active aerated reactor compared to passive aerated reactor. Similarly, Sánchez *et al.* [10] also stated that composting process using forced aerated reduces methane and nitrous oxide emission. However, in another study, Varma *et al.* [11] found that passive aeration was better than force aeration in water hyacinth composting. Zahrim *et al.* [12] investigated composting of anaerobically treated palm oil mill effluent, leaves, and pre-digested papers using a passive aerated reactor. The authors found that the highest temperature achieved was 43 °C, 66% overall mass reduction, the organic degradation was 72% and the moisture content, total organic carbon and carbon to nitrogen (C/N) ratio decreased during the composting process. Karnchanawong and Nissaikla [13] have successfully carried out food waste and dry leaves composting using a passive aerated bin. In another study, Zahrim *et al.* [14] also applied passive aeration for co-composting of empty fruit bunches (EFB) and palm oil mill effluent (POME). With proper passive aeration arrangements, composting of paper and grass clippings with anaerobically treated palm oil mill effluent (AnPOME) can produce acceptable quality compost that is suitable for soil conditioner Zahrim *et al.* [15]. Thus, this study aims to investigate the physicochemical and phytotoxicity characteristics during food waste composting in passive aerated bioreactor assisted with compost turning.

2. Material and methods

2.1. Feedstocks

The feedstocks for the composting process consist of food waste and dry leaves with a ratio of 1:2 (dry leaves: food waste). Food waste was collected from the Faculty of Engineering cafeteria and dry leaves were gathered around the Faculty of Engineering area. The food waste in this study consisted of leftover cooked rice, noodle, vegetable, chicken, meat, fish and fruit peelings. Both food waste and dry leaves were segregated from not compostable materials such as plastic before feeding it into the composter.

2.2. Composting reactor

The composting process was carried out using Active Zone-Yield composter (AZY) located at the Faculty of Engineering, University Malaysia Sabah. The AZY composter consists of three-compartment which are A, B and C, as shown in figure 1. Compartment A and B are where the turning process was carried out and food waste degradation occurred. Compartment C is the storage of yield. The design of the front and the back view of the composter are shown in figure 2 and figure 3, respectively. Each compartment has 15 PVC pipe holes at the back wall with a diameter of 2.6 cm and 10 cm length to maintained natural air convection.

Figure 1. Active Zone-Yield composter.

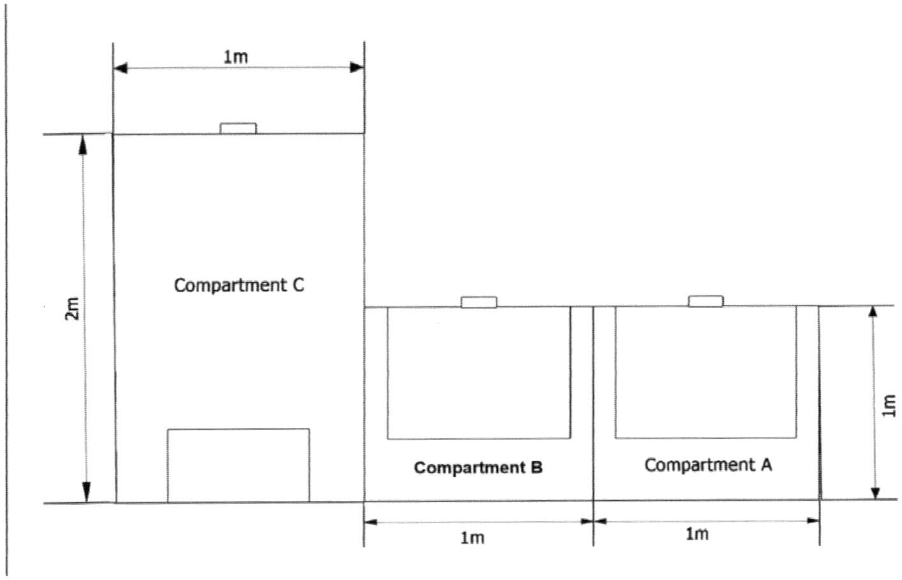

Figure 2. Front view of Active Zone-Yield composter.

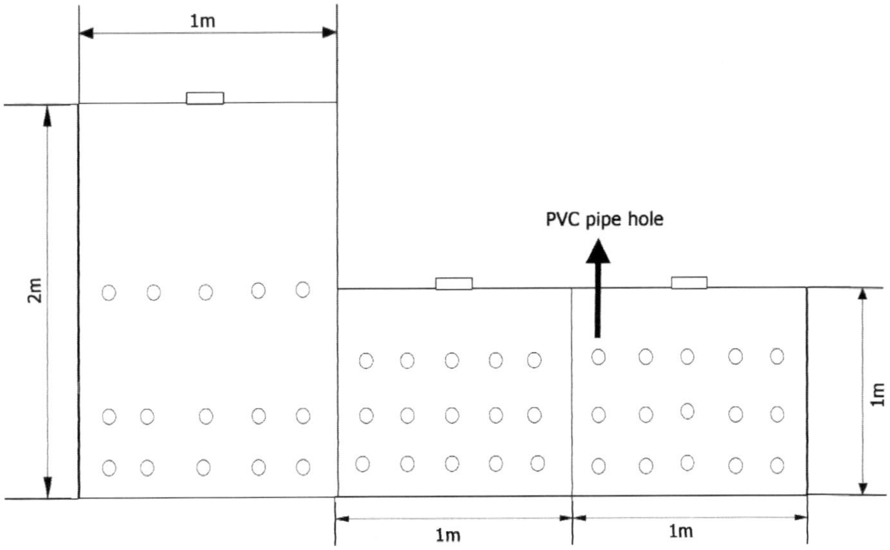

Figure 3. Back view of Active Zone-Yield composter.

2.3. Composting experiment

The composter was filled with food waste and dry leaves up to 90% of the volume and it was considered day 0. Around 62 kg of dry leaves and 124 kg of food waste were properly mixed (1:2 by weight). The composting process was carried out for 40 days. The turning process was carried out once a week on days 3, 11, 20, 25, and 38.

2.4. Analytical measurements
Laboratory analyses included moisture content measurements, total organic carbon (TOC), temperature, volume, pH and conductivity. About 0.4 kg of the compost was collected from five different places. The sample was dried at 103 °C in an incubator (Binder) for 24 hours; the weight loss was taken as the moisture content. The percentage of moisture content was calculated using equation (1). The oven-dried samples were further heated using a high temperature furnace (Thermolyne 46100) at 550 °C for 4 hours. The TOC was calculated using the equation (2) [16]:

$$Moisture\ content\ (\%) = \frac{weight\ of\ wet\ sample - weight\ of\ dried\ sample}{weight\ of\ wet\ sample} \quad (1)$$

$$TOC\ (\%) = \frac{100 - ash\ (\%)}{1.8} \quad (2)$$

The temperature was recorded every day at 12:30 pm. There are four data for temperature, including ambient temperature T0, upper-level T1, middle-level T2, and bottom level T3. For measuring pH and conductivity of compost, 5 g sample of the compost was added to 50 mL of distilled water. The mixture was mixed using a magnetic stirrer for 20 minutes and left for 24 hours. After the mixture was filtered, the pH and conductivity were measured using a pH/EC/TSD/°C portable meter (Hanna Hi 9811-5).

The germination test was performed three days in the dark cupboard with 10 cabbage seeds *(Brassica oleracea).* The seed was placed on a filter paper and soaked in petri dish filled with 5 mL of compost extract [16]. The germination test was repeated with another three replicates for compost extract and distilled water as a control. The seed germination percentage and root length of the cabbage seeds in the extract were determined. The following equation (3), equation (4), and equation (5) were used to calculate the percentage of the relative seed germination (RSG), relative root growth (RRG), and germination index (GI), respectively [15, 17].

$$RSG\ (\%) = \frac{number\ of\ seeds\ germinated\ in\ sample\ extract}{number\ of\ seeds\ germinated\ in\ control}\ x\ 100 \quad (3)$$

$$RRG\ (\%) = \frac{root\ length\ in\ sample\ extract}{root\ length\ in\ control}\ x\ 100 \quad (4)$$

$$GI\ (\%) = \frac{RSG\ x\ RRG}{100} \quad (5)$$

3. Results and discussion

3.1. Temperature
Figure 4 illustrated the changes in ambient and compost temperature. Temperature is the critical factor influencing composting efficiency, compost quality, and microbial activity [18]. The temperature variation during composting in this study went through three typical degradation phases which are mesophilic, thermophilic, and curing. The ambient temperature ranged from 28 °C to 35 °C throughout the composting period. Initially, the temperature was 31 °C, which in the mesophilic phase. The temperature increased quickly at the beginning of the composting process and reached the thermophilic phase (>40°C) on day 13 due to heat accumulation from microbial respiration. The temperature in the compost mixture is maintained above 40 °C for only three days might be due to an increase in moisture content that enhances the cooling effect. From day 23 onwards, the temperature decreased rapidly to the ambient temperature (30 ± 3 °C). The temperature started to drop due to the loss of readily biodegradable substances and reduced microorganism activities [19]. After that, the composting was in the curing phase, where the stabilization of product occurred.

Every time the turning (day 3, 11, 20, 25, 38) was carried out, the temperature shows some increment. This was because it loosens the organic compost material and the aeration was improved [20]. Hence, the organic matter and microorganisms could be reallocated, and as a result, it will increase the temperature [19].

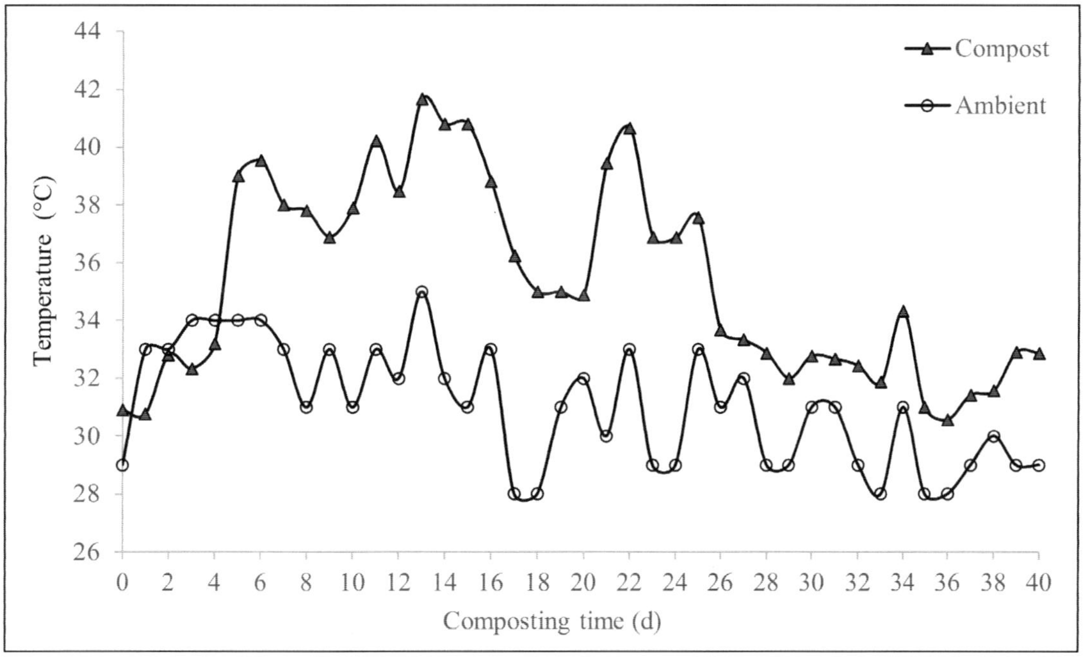

Figure 4. Compost and ambient temperature during composting.

3.2. pH and conductivity

The pH and the conductivity profile during composting are shown in figure 5. The initial pH was 3.2. The pH was low during the initial phase of composting due to the presence of organic acid in food waste [21]. As composting proceeds, the pH increases rapidly up to 7.5 in the first week. The increase in pH was potentially related with releasing of NH_3 [21]. According to Jain and Kalamdhad [22], the optimum pH range for composting is 7-8.

Compost conductivity indicates the salinity of the compost mixture matrix and it can be a limiting factor for seed germination and plant growth [5, 23]. Figure 5 showed that the initial electrical conductivity (EC) was 350 µS/cm and it gradually increased to 900 µS/cm. The increase in conductivity attributed to the released of soluble components, such as ammonium, volatile fatty acids, phosphate and potassium during decomposition and mineralization of organic substances [5, 20, 24]. Then it started to be maintained where the final EC is 870 µS/cm. Wang *et al.* [25] reported a similar result that the EC increased rapidly at the initial stage of the composting process. Table 1 shows the optimum range of EC for certain crops. According to ASCP Guidelines 2001, the conductivity in this study was at an acceptable level (<2500 µS/cm) for plant growth [26].

Table 1. Optimum range of electrical conductivity for certain crops.

Crops	Electrical Conductivity (µS/cm)	References
Cabbage	1800	
Carrot	1000	
Lettuce	1300	[27]
Spinach	2000	
Tomato	2500	

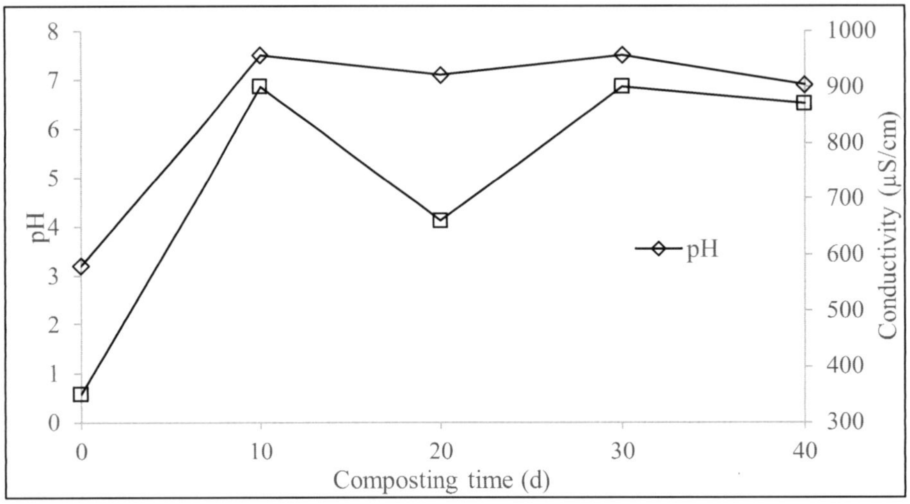

Figure 5. Profile of pH and conductivity of compost.

3.3. Moisture content and total organic carbon

The profile of moisture content (MC) and total organic carbon (TOC) during the composting process is shown in figure 6. Based on the figure, the initial MC of the compost was found to be 57%. Compared to other studies, this study has shown an increment trend in MC. The MC was decreased during composting of municipal solid waste [28], and a study by [29] also demonstrated a decreasing pattern in MC for food waste composting. However, in this study, at the end of the experiment, the MC value was 78% due to water production during degradation and the weak effect of vaporisation. The MC affects the temperature of the compost and the composting process [21, 30]. If the compost has a low value of MC, it can cause dehydration and premature compost. Simultaneously, an excess of MC will restrict the airflow and the composting process will become anaerobic [31] and thus lower the compost mixture temperature. The result indicates that the composting process had maintained the process since the acceptable range of moisture content as the recommended MC for composting process is 50-70% [30]. The initial TOC content is 48%, and it declined to 35% on day 40 of the composting process. The decreasing of TOC may attribute to the continuous degradation of organic matter [19]. TOC minimum value of 16% was recommended by the Government of India [32]. The TOC content in this study was slightly higher than the FCO standard. Nevertheless, high TOC value (40%) has been reported in a study on food waste composting in a passive aerated reactor [9].

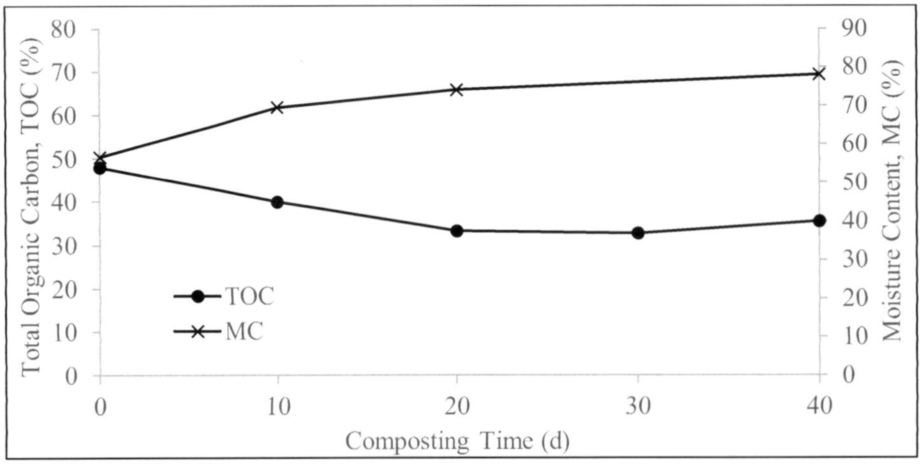

Figure 6. Profile of moisture content and total organic carbon during composting.

3.4. Germination index

Figure 7 and figure 8 illustrate the variation of GI during the composting process. The GI at day 0 was 66%. Bhave and Kulkarni [9] using wheat seed and Lee *et al.* [33] using cabbage seed for GI test also reported low initial GI value in their study on food waste composting, which were 45% and 21%, respectively. The NH_4^+-N and volatile fatty acids were released in high concentration during the initial of composting causing a low GI value [34]. At the end of the experiment, the GI value was 127%. The increase in the GI value shows that the decomposition of toxic and/or inhibitory substances [19]. The maturity and stability of compost are vital issues for land application as it can affect plant growth and the soil environment.

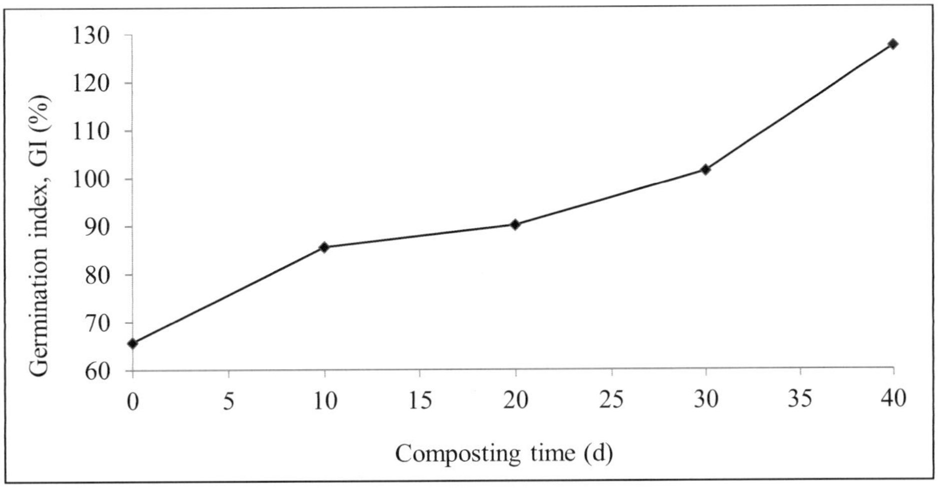

Figure 7. Germination index profile during composting process.

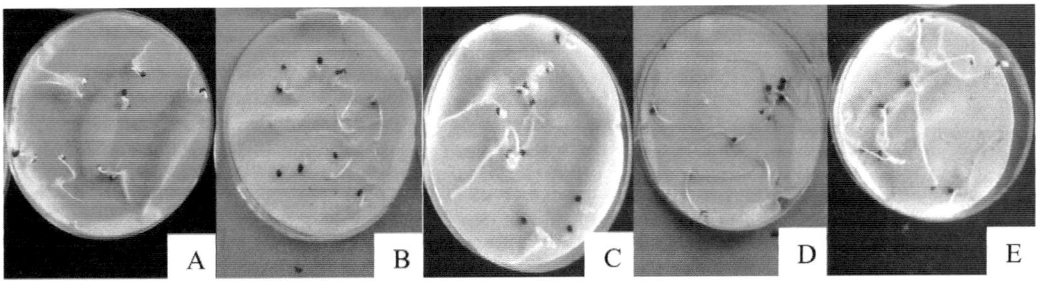

Figure 8. Profile of seed germination index during composting: (A) 0 day; (B) 10 days; (C) 20 days; (D) 30 days; (E) 40 days.

Previous studies stated that a GI value of more than 80% indicates the compost is mature and phytotoxic-free [5, 23, 35]. In this study, mature compost was found can be produced within 10 days. Based on compost maturity standards in the US, the germination index 80 – 90% are considered mature [36]. Table 2 shows the comparison of composting duration, GI and nutrients contents for passive aeration composting.

N, P, and K are important nutrients for the plant growth. The root growth and plant metabolism affected by the phosphorus content, while potassium acts as a regulator of the water content of the plant cell and is important for proteins and carbohydrate production [37]. The N, P, and K for this study were 2.50%, 0.003% and 0.0053%, respectively. Van Fan *et al.* [38] stated that the recommended range for compost is N $\geq$ 1%, P: 0.6 - 1.7% and K:0.4 - 1.11%. From table 2, the value of most of the studies was within the recommended range. It should be noted that the value of P and K in this study was lower. This might be due to the lower initial compost P and K than other studies. The initial value of both P and K were 0.001% and much lower than other studies where P: 0.61% K: 0.61 [9], P: 0.29% K: 1.19 [20] and P: 3.26% K: 0.92% [39]. Jamaludin *et al.* [40] reported a low value of N, P, and K in food waste, coconut fiber, salt, and breadfruit peels composting. In their study the N, P, and K value increase from 0.008%, 0.0002% and 0.013% to 0.224%, 0.001% and 0.069%, respectively. Gautam *et al.* [41] and Pathak *et al.* [42] also reported low values of N, P and K in municipal solid waste (mainly kitchen waste, vegetable waste, fruit waste) composting.

Table 2. Comparison of type of aeration, composting duration, germination index (GI) and nutrients for food waste composting.

Feedstocks	Aeration	Composting duration (days)	Germination Index, GI (%)	Nutrients (%)	References
Food waste, dry leaves	Passive with turning	40	127	N: 2.50 P: 0.003 K: 0.005	This study
Food waste	Passive aeration	110	71	N: 0.90 P: 0.91 K: 0.72	[9]
Food waste, dry leaves, rice bran	Passive with turning	56	402	N: 2.1 P: 0.4 K: 1.37	[37]
Food waste, garden and yard waste	Passive aeration	82	25	N: NA P: NA K: NA	[43]
Food waste, garden waste	Passive with turning	60	85	N: NA P: 0.15 K: 0.65	[29]
Food waste, coconut fiber, salt and breadfruit peels	Passive with turning	140	NA	N: 0.22 P: 0.001 K: 0.07	[40]
Food waste, dry leaves	Passive aeration	184	139	N: 2.52 P: 0.87 K: 2.26	[13]
Kitchen waste	Passive with turning	135	NA	N: 1.17 P: 0.04 K: 0.38	[42]
Kitchen waste, vegetable waste, fruit waste	Passive with turning	90	NA	N: 0.05 P: 0.003 K: 0.34	[41]
Pineapple leaves, chicken manure slurry	Passive with turning	57	84	N: 2.31 P: 0.47 K: 2.67	[20]
Vegetable waste, cow manure, sawdust, dry leaves	Passive with turning	20	NA	N: 3.01 P: 3.27 K: 1.70	[39]
Vegetable waste, cow dung, saw dust	Passive with turning	30	110	N: 2.40 P: NA K: NA	[44]
Vegetable waste, yard waste	Passive with turning	105	>100	N: 2.04 P: 0.18 K: 0.11	[45]

*NA: not available

3.5. Other chemical properties

Table 3 shows the C/N ratio and nutrients content at the end of composting. The C/N ratio in this study was ~9. The quality and maturity of the compost can also be assessed using the C/N ratio and the C/N ratio less than 20 was reported as matured compost [30]. The nutrients (Ca, Mg, Na, Fe, Al, C, and H) in this study were lower compared to others study and the standard range. The availability of the nutrients depends on the feedstock composition of the compost. The magnesium (Mg) in compost will protect the plants from different scarcity. In contrast, the metabolism of plants and the chlorophyll will be helped by sodium (Na) in compost [46].

Table 3. Carbon to nitrogen ratio and nutrients content in compost.

References	[42]	[47]	[31]	[48]	[29]	Standard range [49]	This Study
Feedstock	Kitchen waste	Food waste, yard waste	Food waste, wood chips	Food waste, leaves, grass clipping	Food waste, green waste		Food waste, dry leaves
C/N	21.75	23.0	20.00	11.7	12.1	10.00 – 15.00	9.14
Ca (%)	NA	0.02	1.47	3.6 4	0.87	1.50 – 3.50	0.008
Mg (%)	NA	0.005	0.4	0.30	0.12	0.25 - 0.70	0.001
Na (%)	2.80	NA	1.97	0.23	0.45	< 0.60	0.003
Fe (%)	0.12	NA	NA	0.22	0.015	NA	0.001
Al (%)	NA	NA	NA	0.00003	0.003	NA	0.001
C (%)	NA	0.004	NA	34.1	NA	NA	22.85
H (%)	NA	NA	NA	NA	NA	NA	4.12

*NA: not available

4. Conclusions

This study evaluated the performance of food waste composting in a passive aerated bioreactor with food waste to dry leaves ratio of 2:1. The temperature reached above 40 °C on day 13 of composting. At the end of the composting process, the pH and conductivity were 6.9 and 870 µS/cm, respectively. TOC reduction of 26% was observed and the final value of moisture content was 78%. The final compost was matured within 10 days, with a GI value of more than 90 % on day 40 indicating very mature compost. The N, P and K value were 2.5%, 0.003% and 0.005%, respectively. Based on the result reported in this study, it is best to conclude that the compost produced is suitable for plant growth.

Acknowledgement

The authors would like to acknowledge Universiti Malaysia Sabah for the financial aids (SDN0042-2019, SDK0102-2019 & SDG03-2020).

References

[1] SWCorp 2018 Info Sisa Makanan SwCorp News

[2] Van Fan Y, Lee C T, Leow C W, Chua L S and Sarmidi M R 2016 Physico-chemical and biological changes during co-composting of model kitchen waste, rice bran and dry leaves with different microbial inoculants *Malays. J. Anal. Sci.* **20** 1447–57

[3] Adhikari B K, Barrington S, Martinez J and King S 2009 Effectiveness of three bulking agents for food waste composting *Waste Manage. (Oxford)* **29** 197–203

[4] Paritosh K, Kushwaha S K, Yadav M, Pareek N, Chawade A and Vivekanand V 2017 Food waste to energy: An overview of sustainable approaches for food waste management and nutrient

recycling *Biomed Res. Int.* **2017** 23709–27

[5] Zhou Y, Selvam A and Wong J W C 2018 Chinese medicinal herbal residues as a bulking agent for food waste composting *Bioresour. Technol.* **249** 182–8

[6] Lee J-H *et al.* 2019 Comparing the composting characteristics of food waste supplemented with various bulking agents *Korean J. Agric. Sci.* **46** 897–905

[7] Mu D, Horowitz N, Casey M and Jones K 2017 Environmental and economic analysis of an in-vessel food waste composting system at Kean University in the U.S *Waste Manage. (Oxford)* **59** 476–86

[8] Saalah S, Rajin M, Yaser A Z, Azmi N A S and Mohammad A F F 2020 *Green Engineering for Campus Sustainability,* ed A Z Yaser (Singapore: Springer Singapore) pp 173–91

[9] Bhave P P and Kulkarni B N 2019 Effect of active and passive aeration on composting of household biodegradable wastes: A decentralized approach *Int. J. Recycl. Org. Waste Agric.* **8** 335–44

[10] Sánchez A *et al.* 2015 Greenhouse gas emissions from organic waste composting *Environ. Chem. Lett.* **13** 223–38

[11] Varma V S, Prasad R, Deb S and Kalamdhad A S 2018 Effects of aeration during pile composting of water hyacinth operated at agitated, passive and forced aerated condition *Waste Biomass Valorization* **9** 1339–47

[12] Zahrim A, Sariah S, Mariani R, Azreen I, Zulkiflee Y and Fazlin A 2019 *Research Methods and Applications in Chemical and Biological Engineering,* ed A Pourhashemi et al (: Apple Academic Press Inc) p 217

[13] Karnchanawong S and Nissaikla S 2014 Effects of microbial inoculation on composting of household organic waste using passive aeration bin *Int. J. Recycl. Org. Waste Agric.* **3** 113-9

[14] Zahrim A *et al.* 2018 Effect of pre-treatment and inoculant during composting of palm oil empty fruit bunches *ASEAN J. Chem. Eng.* **2** 1–16

[15] Zahrim A Y, Leong P S, Ayisah S R, Janaun J, Chong K P, Cooke F M and Haywood S K 2016 Composting paper and grass clippings with anaerobically treated palm oil mill effluent *Int. J. Recycl. Org. Waste Agric.* **5** 221–30

[16] Yaser A Z, Rahman R A and Kalil M S 2007 Co-composting of palm oil mill sludge-sawdust *Pak. J. Biol. Sci.* **10** 4473–8

[17] Tiquia S M, Tam N F Y and Hodgkiss I J 1996 Effects of composting on phytotoxicity of spent pig-manure sawdust litter *Environ. Pollut.* **93** 249–56

[18] Guo W, Zhou Y, Zhu N, Hu H, Shen W, Huang X, Zhang T, Wu P and Li Z 2018 On site composting of food waste: A pilot scale case study in China *Resour. Conserv. Recycl.* **132** 130–8

[19] Yang F, Li Y, Han Y, Qian W, Li G and Luo W 2019 Performance of mature compost to control gaseous emissions in kitchen waste composting *Sci. Total Environ.* **657** 262–9

[20] Ch'ng H Y, Ahmed O H, Kassim S and Majid N M A 2013 Co-composting of pineapple leaves and chicken manure slurry *Int. J. Recycl. Org. Waste Agric.* **2** 23

[21] Kumar M, Ou Y-L and Lin J-G 2010 Co-composting of green waste and food waste at low C/N ratio *Waste Manage.* **30** 602–9

[22] Jain M S and Kalamdhad A S 2019 Drum composting of nitrogen-rich Hydrilla Verticillata with carbon-rich agents: Effects on composting physics and kinetics *J. Environ. Manage.* **231** 770–9

[23] Neves L, Ferreira V and Oliveira R 2009 Co-composting cow manure with food waste: The influence of lipids content *World Acad. Sci. Eng. Technol.* **58** 986–91

[24] He Z, Lin H, Hao J, Kong X, Tian K, Bei Z and Tian X 2018 Impact of vermiculite on ammonia emissions and organic matter decomposition of food waste during composting *Bioresour. Technol.* **263** 548–54

[25] Wang X, Selvam A, Chan M and Wong J W C 2013 Nitrogen conservation and acidity control during food wastes composting through struvite formation *Bioresour. Technol.* **147** 17–22

[26] Fuchs J, Galli U, Schleiss K and Wellinger A 2001 ASCP Guidelines 2001: Quality criteria for composts and digestates from biodegradable waste management *The Association of Swiss Compost Plants (ASCP) in Collaboration with the Swiss Biogas Forum*

[27] Hanlon E A, McNeal B L and Kidder G 1993 *Soil and container media electrical conductivity interpretations* (University of Florida: Florida Cooperative Extension Service, Institute of Food and Agricultural Science)
[28] Elango D, Thinakaran N, Panneerselvam P and Sivanesan S 2009 Thermophilic composting of municipal solid waste *Appl. Energy* **86** 663–8
[29] Manu M K, Kumar R and Garg A 2017 Performance assessment of improved composting system for food waste with varying aeration and use of microbial inoculum *Bioresour. Technol.* **234** 167–77
[30] Pandey P K, Cao W, Biswas S and Vaddella V 2016 A new closed loop heating system for composting of green and food wastes *J. Clean. Prod.* **133** 1252–9
[31] Margaritis M, Psarras K, Panaretou V, Thanos A G, Malamis D and Sotiropoulos A 2018 Improvement of home composting process of food waste using different minerals *Waste Manage.* **73** 87–100
[32] Mandal P, Chaturvedi M, Bassin J, Vaidya A and Gupta R 2014 Qualitative assessment of municipal solid waste compost by indexing method *Int. J. Recycl. Org. Waste Agricult.* **3** 133–9
[33] Lee I B, Kim P J and Chang K W 2002 Evaluation of stability of compost prepared with Korean food wastes *Soil Sci.Plant. Nutr.* **48** 1–8
[34] Chan M T, Selvam A and Wong J W C 2016 Reducing nitrogen loss and salinity during 'struvite' food waste composting by zeolite amendment *Bioresour. Technol.* **200** 838–44
[35] Guo R, Li G, Jiang T, Schuchardt F, Chen T, Zhao Y and Shen Y 2012 Effect of aeration rate, C/N ratio and moisture content on the stability and maturity of compost *Bioresour. Technol.* **112** 171–8
[36] Brinton W F 2000 Compost quality standards and guidelines. In: *Final Report by Woods End Research Laboratories for the New York State Association of Recyclers*
[37] Van Fan Y, Lee C T, Klemeš J J, Chua L S, Sarmidi M R and Leow C W 2018 Evaluation of Effective Microorganisms on home scale organic waste composting *J. Environ. Manage.* **216** 41–8
[38] Van Fan Y, Lee C T, Klemeš J J, Bong C P C and Ho W S 2016 Economic assessment system towards sustainable composting quality in the developing countries *Clean Technol. Envir.* **18** 2479–91
[39] Varma V S, Mayur C and Kalamdhad A 2014 Effects of bulking agent in composting of vegetable waste and leachate control using rotary drum composter *Sustain. Environ. Res.* **24** 245–256
[40] Jamaludin S N, Abdul Kadir A and Azhari N W 2017 Study on NPK performance in food waste composting by using agricultural fermentation *MATEC Web Conf.* **103** 05015
[41] Gautam S, Bundela P, Pandey A, Awasthi M and Sarsaiya S 2010 Composting of municipal solid waste of Jabalpur City *Global J. Environ. Res.* **4** 43–6
[42] Pathak A, Singh M, Kumara V, Arya S and Trivedi A 2012 Assessment of physico-chemical properties and microbial community during composting of municipal solid waste (viz. kitchen waste) at Jhansi City, UP (India) *Recent Res. Sci. Technol.* **4** 10–14
[43] Kamaruddin M A, Norashiddin F A, Idrus A F M, Zawawi M H and Alrozi R 2018 A study on the effects of different microbial inoculants on the decomposition of organic waste by using semi passive aerated reactor *AIP Conf. Proc.* **2030** 020199
[44] Rich N, Bharti A and Kumar S 2018 Effect of bulking agents and cow dung as inoculant on vegetable waste compost quality *Bioresour. Technol.* **252** 83–90
[45] Sangamithirai K M, Jayapriya J, Hema J and Manoj R 2015 Evaluation of in-vessel co-composting of yard waste and development of kinetic models for co-composting *Int. J. Recycl. Org. Waste Agric.* **4** 157–65
[46] Sharma D, Yadav K D and Kumar S 2018 Role of sawdust and cow dung on compost maturity during rotary drum composting of flower waste *Bioresour. Technol.* **264** 285–9
[47] Faucette L, Das K and Risse L 2000 Evaluation of aerated container composting of university preconsumer and postconsumer foodwaste. In: *Composting in the Southeast: 2000 Conf. Proc., October,* pp 9–11
[48] Storino F, Arizmendiarrieta J S, Irigoyen I, Muro J and Aparicio-Tejo P M 2016 Meat waste as

feedstock for home composting: Effects on the process and quality of compost *Waste Manage.* **56** 53–62

[49] Sullivan D M, Bary A I, Miller R O and Brewer L J 2018 Interpreting compost analyses. In: *Proc. of the International Conf. on Soils Across latitudes. San Diego, California,*

Lipid yields of algae dried in an enhanced solar chimney

L O Morris, C M Chu and E Suali

Chemical Engineering Programme, Faculty of Engineering
Universiti Malaysia Sabah
88400 Kota Kinabalu, Sabah, Malaysia

Email: chrischu@ums.edu.my

Abstract. This work focuses on the lipid yield of algae dried using an enhanced solar chimney dryer with the aim towards high quality biodiesel production and low energy consumption. Jaworski's medium was chosen for the cultivation in this research project as it can provide enough nutrients to green algae of type *Chlorella. Sp.* Centrifugation is an effective method to harvest the algae from its medium prior to a drying process. In this project, the methods used for drying were oven drying, open-sun drying and an enhanced solar chimney drying. The moisture content was determined where the average moisture content were 82.5% for oven drying method, 81.6% for open sun drying and 82.2% for solar chimney drying. Methods were found to affect differently the algae properties in terms of the lipid yield and mineral content. The lipid yields were 23.7%, for oven drying, 20.6% for open sun drying and 24.4% for enhanced solar chimney drying. While the oven drying was found to be the fastest way to dry the algae, the solar chimney drying proved best in energy saving while producing the same amount or more algae lipid within reasonable drying times.

1. Introduction

Algae based fuels are considered to be the most sustainable, renewable, effective and environment-friendly response to climate change and food security threats, as well as the only renewable energy resource that has the capacity to meet the global demand for fuels in the long term. A roadblock to harvesting of algae is the high concentration of water and it can reduce the extraction efficiency. Drying is a critical step in the production of biodiesel from algae, as it removes the water content in algae before the lipid extraction. There are many ways of drying microalgae such as oven drying, drum drying, freeze drying, spray drying, open sun drying, solar chimney drying and vacuum drying. Each method affects the properties of the algae at varying degrees such as the volume of lipid extraction yield as well as the fatty acids content in lipid, mineral content and other significant properties in algae for conversion to biodiesel.

Generally, the production of microalgae biodiesel is no different from the production of first generation biodiesel. The biomass is cultivated, then harvested for lipid extraction to undergo transesterification for conversion into FAMES (fatty acid methyl ester), or biodiesel. Unlike the oil seed plants, harvesting microalgae cells can be quite challenging. The tiny cells floating in water cannot be collected as easily as plant seeds or leaves, and accordingly oil extraction is more complicated than the cold-press procedure traditionally used for oil seeds extraction. Furthermore, algal cultures are often dilute in nature, around 1% through autotrophic growth and up to 10% by heterotrophic growth, and dewatering is necessary prior to biomass use. Thus, harvesting stages can be varied from one to several different stages, where each stage's operation can vary according to the final desire solid concentration. Generally, algae cultivation results in nothing more than a green slurry. In order to increase its calorific value, the slurry needs to be concentrated through the process of drying [1].

Depending on the different usage, the harvesting methods may vary. For instance, nutraceutical products require algae free from chemical contamination and still maintaining its natural characteristics, therefore physical process for harvesting is much preferred [2]. For fish feeding products, a longer shelf-life is prioritised, thus the continuous centrifugation of algae slurry is preferred. This method is highly effective and it is still the most widely used due to its efficiency and well documentation of the techniques.

One major challenge in developing a viable biodiesel production process from microalgae is finding an effectively harvest the biomass in a cost-effective manner [3]. Until recently technology development for downstream processing with low energy consumption while producing high yield is still advocated as a priority [4, 5]. The aim of this research project is to evaluate a strategy for its efficiency and cost effectiveness in harvesting microalgae. One of the factors to consider when choosing a harvesting method is the size of particles of the algae produced.

To produce biodiesel from algae economically is the primary challenge to date. After harvesting, the wet slurry of the microalgae undergoes dewatering for further processing. Drying of the harvested algal biomass which is 90% water presents an obstacle to making the algal biodiesel commercially viable, as it takes up vast amount of energy and time in order to gain dried biomass for lipid extraction. It was reported that drying of microalgae biomass by centrifugation and tangential filtration can account for 20% to 57% of the final total production costs of algae biomass [4, 6–7] which makes up the overall net energy ratio [8]. Though many researchers use freeze drying technique to dry the wet algal biomass [9], the drying method is believed to affect the phytochemical properties in algae such as the fatty acids as well as the anti-oxidant content in algae.

Sun drying has long been traditionally employed to dry crops, thus, it is not new to the area of drying research. Drying is a continuous process with simultaneous changes in the moisture content of the crop being dried and in the air humidity [10].

Bagchi *et al.* [9] wrote about an oven drying protocol to improve biodiesel production for an indigenous chlorophycean microalga *Scenedesmus sp.* They concluded that spray and freeze drying technique are not economically feasible for microalgae biofuel production. Oven drying has become one of the common drying methods other than sun drying. Usually for research purpose, the temperatures are set to be constant at 3 different levels, namely 60, 80 and 105 °C, as recommended by the standards [11].

The objective of this research is to examine the parameters that could affect the drying process and this paper will report the best possible drying method for high lipid yield. A research at Faculty of Engineering of Universiti Malaysia Sabah on a modified solar chimney dryer [12] has found an increase of 13% in the evaporation of water, supporting the earlier results of natural convection air flowrate enhancement by the same type of chimney [13], and the solar chimney drying system was therefore adopted for this project.

2. Material and Methods

The project experiments involved 5 stages which were cultivation of algae, harvesting, drying, lipid extraction and analysis and profiling (figure 1). The species of algae used was *Chlorella sp.* It was decided to cultivate the algae locally using Jaworski's medium to ensure the harvest yield is of sufficient quality and lessen quantity lost. It is usually the medium used to cultivate freshwater algae or usually for green freshwater algae.

Figure 1. Flow chart of the project.

A pyranometer was used to measure solar irradiation rate and thermocouples at various places to measure the air temperature in the ambient, in the oven and at the chimney outlet. Humidity was also measured digitally together with the air temperature.

2.1. *Drying rate and Moisture Content*

The weight of sample was measured according to the interval chosen. The average weight of each sample was approximately 0.5 g, in thin slices. The drying rate is usually calculated by observing the mass loss during the drying using the following equation [14–16]:

$$Drying\ rate = \frac{Initial\ weight - Final\ weight}{Time\ Interval\ (h)} \tag{1}$$

The moisture or water content of algae is the difference of the initial weight of sample and the weight of sample after being dried in oven when all the water has been removed. Moisture content of algae is calculated using the formula given below [11]:

$$MC = \frac{Initial\ Weight - Oven\ dry\ Weight}{Initial\ Weight} \times 100\% \tag{2}$$

Centrifugation is the fastest harvesting method but is also the most expensive due to the high consumption of energy. To achieve high harvesting efficiencies, longer retention times are required for sedimentation due to the relatively low density of the cells. In these experiments the centrifuge was operated at a low flow rate of 3000 rpm for 5 minutes.

2.2. *Drying*

In this project the methods used were oven drying, open sun drying and solar chimney drying. Drying of sample was done for 4 runs on different days with 1 sample for each different drying method. The experiments were commenced at 8.00 a.m. and ended at about 6.00 p.m. depending on the weight measured.

Thick microalgae slurry was spread uniformly in glass petri dishes in identical manner and placed in a laboratory-scale hot air oven, in the open sun (figure 2) and in the solar chimney dryer where the water pool had been removed and replaced with the dishes (figure 3). The transparent sheet shown was made of polyethylene capable of blocking UV radiation. The oven drying temperature usually would be set at 60 °C, 80 °C and 105 °C but for this research the temperature was kept constant at 60 °C to conserve the composition and the sample was weighed at 2-hour intervals until the reading of the weight reached a constant value [8].

2.3. *Lipid Extraction*

There are several ways of lipid extraction, where some extraction methods do not require drying step and can be performed in the presence of water and algae. The lipid extraction method used was Bligh and Dyer method, considered to be more effective and reliable than the classical Folch's method [16]. The Bligh and Dyer extraction method requires drying steps to remove the water. Each sample of dried microalgae was mixed with a solution of (2:1) chloroform and methanol. The ratio of extraction solvent was 2:1 of Chloroform-methanol and the total solvent was 20 mL per gram. Since only 0.5 g of sample was used for lipid extraction so the total volume of solvent used was only 10 mL. samples and solvent undergo homogenizing. Next, solution was filtrated to remove the non-lipid substance and any sizeable biomass debris. Instead of using water on its own, sodium chloride solution was used as the washer and centrifuged at 3000 rpm for about 5 minutes to form two layers. The upper layer was removed while the lower layer (chloroform) containing lipid was recovered.

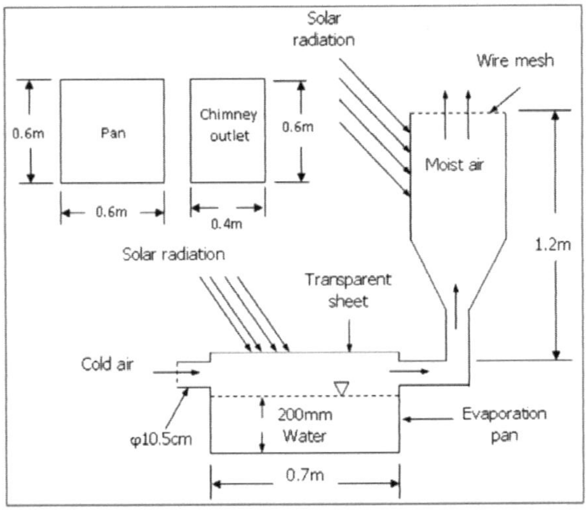

Figure 2. Open sun and solar chimney dryer [12]. **Figure 3.** Solar chimney dryer dimension [12].

3. Results Analysis and Profiling

Table 1 shows the lipid yield extracted from dried chlorella sp. using the Bligh and Dyer extraction method.

Each sample yielded different amounts of lipid but the trend of yield for each drying method is consistent for each sample. Table 1 displays that the average lipid yields for the three different drying methods are 0.119 g for oven drying, 0.103 g for sun drying and 0.122 g for solar chimney drying. This places the solar chimney drying method at the highest position of lipid yield and shows that this drying method has negligible damaging effect on the phytochemicals content in algae. The lipid yield in terms of percentage is also tabulated where the average lipid yield for oven drying is 23.65%, sun drying is 20.6% and solar chimney drying is 24.4%. The highest yield from solar chimney drying is probably because of the lower temperature and lower exposure to solar radiation. While the oven was not exposed to the solar radiation, 60 °C is still considered a high temperature compared to the temperature in the chimney solar dryer and high temperature can cause composition deterioration. Sun drying had been exposed to an ambient temperature of about 35 °C to 45 °C, lower than that of the oven drying method, but the samples dried by sun drying were directly exposed to the UV-B radiation which could alter the DNA, proteins, lipids as well as the membranes in the algae [17] while solar chimney drying with the polyethylene plastic sheet appeared to be less affected by the UV-B radiation and also the operating temperature was lower than the oven drying.

Enamala *et al.* [18] compiled a large database of algae strain and the vital and required growth parameters. There were nine sources for *Chlorella sp.* Covering marine and fresh water habitat, and various nutrients. The maximum yield as lipid wt% of biomass, was 48%, and the minimum 4.9%, where the average yield was 22.4%, which is of the same order as the results here. Aguoru and Okibe [19] reported on the content and composition of lipid produced by *Chlorella Vulgaris* for biodiesel production with the average lipid production at about 28% to 32% while the research on lipid production carried out by Hussain *et al.* [20] by employing different drying methods and different extraction methods resulted in lipid yield in the range of 14% to 23% by using Soxhlet extraction, Halim Extraction and Bligh and Dyer methods. It would appear that the final yield depends on both the drying and the extraction methods since different lipid yields were obtained.

Table 1. Lipid yield in algae *Chlorella sp.*

No. Sample	Oven Drying Lipid Yield, g / %		Open Sun Drying Lipid Yield, g / %		Solar Chimney Drying Lipid Yield, g / %	
Sample 1	0.132	26.0	0.095	19.0	0.112	22.4
Sample 2	0.153	30.6	0.119	23.8	0.099	19.8
Sample 3	0.078	15.6	0.064	12.8	0.137	27.4
Sample 4	0.112	22.4	0.134	26.8	0.139	27.8
Average	0.119	23.65	0.103	20.6	0.122	24.4
Error	±0.0319	±6.3	±0.0306	±6.1	±0.0195	±3.9

A research on the effect of ultraviolet on plant cells mentioned that the isolated or conjugated double bonds in lipids can be photo-chemically modified by UV absorbance [17]. Phospholipid and glycolipids, the main components of plant cell membranes, contain unsaturated fatty acids which are destroyed by UV-B radiation in the presence of oxygen. This research supports the common findings concerning the effect of sun drying on the lipid yield of the algae where the sun drying lipid yield was the lowest amongst the other two methods, while solar chimney dryer was better and it was close to the oven drying method in terms of production of lipid for biodiesel. Until recently oven drying has been preferred as the best method in drying applications due to its speed while solar chimney drying is still a technology under development in harnessing a natural energy source (figure 4).

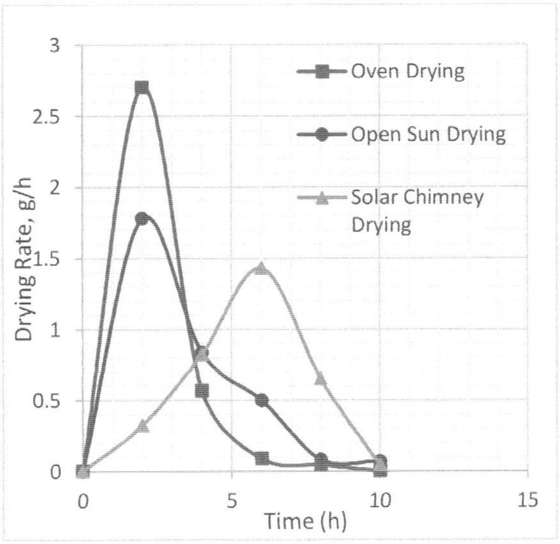

Figure 4. Graph of drying rates versus time for different drying methods.

4. Conclusion and Recommendation

In this paper the objective of determining and comparing lipid yield extracted from the microalgae *Chlorella sp.* via three different drying methods has been achieved. The three drying methods are oven drying, open sun drying and solar chimney drying. The solar chimney drying method could yield a relatively high lipid content to that of the oven drying method and the lowest was the sun drying. The open sun drying was found to be significantly affected by the UV-B radiation since other researchers on the effect of ultraviolet have mentioned that UV-B light can affect the lipid content in the algae, where high temperature can be one of the factors that reduce the lipid production. Therefore, in terms of lipid yield, solar chimney drying with UV protection might be the best choice since it appeared to preserve the quality of the algae and can save the usage of energy where solar energy can be the source of renewable energy to generate hot air through the application of a stable ventilation. For future study, research on the solar chimney drying method is recommended to be carried out on a pilot-scale to replace the conventional ways of drying.

Acknowledgments
The authors would like to thank the Ministry of Higher Education of Malaysia for providing a fundamental research grant under the number FRG0352-TK-1/2013 and Universiti Malaysia Sabah for the facility and laboratory support to carry out the experimental research.

References
[1] Aziz M, Oda T and Kashiwagi T 2013 Enhanced high energy efficient steam drying of algae *Appl. Energy* **109** 163–170
[2] Leite G B, Abdelaziz A E M and Hallenbeck P C 2013 Algal biofuels: Challenges and opportunities *Bioresour. Technol.* **145** 134–41
[3] Uduman N, Qi Y, Danquah M K, Forde G M and Hoadley A 2010 Dewatering of microalgal cultures: A major bottleneck to algae-based fuels *J. Renew. Sust. Energy* **2** 012701
[4] Zhu L D, Hu T Y, Li S X, Nugroho Y K, Li B S, Cao J, Show P L and Hiltunen E 2020 Effects of operating parameters on algae Chlorella vulgaris biomass harvesting and lipid extraction using metal sulfates as flocculants, *Biomass and Bioenergy* **132** 1–7
[5] Yew G Y, Lee S Y, Show P L, Tao Y, Law C L, Nguyen T T C and Chang J S 2019 Recent advances in algae biodiesel production: From upstream cultivation to downstream processing *Bioresource Technology Reports* **7** 100227
[6] Molina Grima E, Belarbi E-H, Acién Fernández F G, Robles Medina A and Chisti Y 2003 Recovery of microalgal biomass and metabolites: Process options and economics *Biotechnol. Adv.* **20** 491–515
[7] Van Den Hende S, Vervaeren H, Desmet S and Boon N 2011 Bioflocculation of microalgae and bacteria combined with flue gas to improve sewage treatment *New Biotechnol.* **29** 23–31
[8] Sander K and Murthy G S 2010 Life cycle analysis of algae biodiesel *Int. J. Life Cycle Assess* **15** 704–14
[9] Bagchi S K, Rao P S and Mallick N 2015 Development of an oven drying protocol to improve biodiesel production for an indigenous chlorophycean microalga Scenedesmus sp. *Bioresource Technology* **180** 207–13
[10] Toğrul I T T 2005 Convective heat tranfer coefficient of some fruits under open sun drying conditions *J. Food Technol.* **3** 10–4
[11] Reeb J E and Milota M R 1999 *Moisture content by the oven-dry method for industrial testing* 66–74 May WDKA Oregon State University USA
[12] Rahman M M, Rahman A M, Noor Ajian M L and Chu C M 2019 Analysis of evaporation data from an enclosed pond ventilated by solar chimney *IOP Conf. Series: Materials Science and Engineering* **606** 012013
[13] Chu C M and Rahman M M 2009 A method to achieve robust aerodynamics and enhancement of updraft in natural draft dry cooling towers HT2009-88289 19-23 July *ASME Summer Heat Transfer Conference* San Francisco Calif. USA
[14] Gupta S, Cox S and Abu-Ghannam N 2011 Effect of different drying temperatures on the moisture and phytochemical constituents of edible Irish brown seaweed *LWT - Food Science and Technology* **44** 1266–72
[15] Özcan M, Arslan D and Unever A 2005 Effect of drying methods on the mineral content of basil

(Ocimum basilicum L.) Effect of drying methods on the mineral content *J. Food Eng.* **69** 375–9

[16] Ayanwale B, Ocheme O and Oloyede O 2007 The effect of sun-drying and oven-drying on the nutritive value of meat pieces in hot humid environment *Pak. J. Nutr.* **6** 370–4

[17] Hollósy F 2002 Effects of ultraviolet radiation on plant cells *Micron* **33** 179–97

[18] Enamala M K, Enamala S, Chavali M, Donepudi J, Yadavalli R, Kolapalli B, Aradhyula T V, Velpuri J and Kuppam C 2018 Production of biofuels from microalgae — A review on cultivation, harvesting, lipid extraction, and numerous applications of microalgae *Renewable Sustainable Energy Rev.* **94** 49–68

[19] Aguoru C U and Okibe P O 2015 Content and composition of lipid produced by Chlorella Vulgaris for Biodiesel Production *Advances in Life Science and Technology* **36** 96–101

[20] Hussain J, Liu Y, Lopes W A, Druzian J I, Souza C O, Carvalho G C, Nascimento I A and Liao W 2015 Effects of different biomass drying and lipid extraction methods on algal lipid yield, fatty acid profile, and biodiesel quality *Appl. Biochem. Biotechnol.* **175** 3048–57

IOP Conf. Series: Materials Science and Engineering 1195 (2021) 012003 doi:10.1088/1757-899X/1195/1/012003

Investigation on the use of aluminium rich sandy soil as natural adsorbent in the removal of lead from water

Y S Ng, Y J Tan, K J Heng and Y H Ong

Lee Kong Chian Faculty of Engineering and Science, Universiti Tunku Abdul Rahman, Jalan Sungai Long, Bandar Sungai Long, Cheras, 43000 Kajang, Selangor, Malaysia

Email: ngys@utar.edu.my

Abstract. The feasibility of aluminium rich sandy soil collected from Jeram as natural adsorbent in removing lead (Pb) from water was investigated without any surface modification. The investigation on the effect of initial concentration, solution pH, and soil:solution ratio was carried out using response surface methodology. The adsorption efficiency was increased at higher pH and soil:solution ratio, as well as lower initial concentration, as a result of higher availability of adsorption sites and less adsorbate competition. The adsorption followed Langmuir isotherm and monolayer chemisorption with an adsorption capacity of 10.64 mg g^{-1}. The process followed pseudo-second order kinetic model, with a rate constant of 0.011 g mg^{-1} min^{-1} at optimum adsorption pH of 4-5.

1. Introduction

Water is an irreplaceable natural resource that is essential for the survival of humanity and all species on the earth. However, one of the major environmental issues that the world is currently facing is the insufficient availability of clean water. The reported data shows that there are million of people have no access to clean drinking water in this world [1], as a result of water pollution.

Heavy metal contamination is one of the main contribution factors for water pollution problem. Among many types of heavy metals, lead (Pb) has been reported as a toxic metal element [2] whereby the maximum allowable concentration in wastewater has been set at 0.1-0.5 mg L^{-1} in Malaysia [3]. Pb contamination can be attributed to different sources such as pipes and fittings, mining, smelting and milling processes [4–6], leading to high exposure risk of Pb for human. The overexposure and intake of Pb can cause several negative impacts to human such as damage to some elemental cellular processes, central nervous system, reproductive system, liver and kidney [4, 5]. Therefore, an effective counter measure is required urgently in order to ensure Pb is removed from water effectively before it is being consumed.

Many types of water treatment technologies have been reported for removing Pb from water such as adsorption, chemical precipitation, membrane filtration and etc. Among these water treatment technologies, adsorption emerges as a promising treatment method due to the advantages of resilience in design/operation, simple, low maintenance cost, ability to produce greater treated water quality and its ability to regenerate the adsorbent easily using proper desorption processes for future use [7, 8]. Recently, different types of natural adsorbents have been investigated to replace conventional activated carbons such as biosorbents [9–13], drinking water treatment sludge [14], mineral coated peats/sands/clays [15–17] and foams [18]. Positive results and significant Pb uptake capacity have been reported.

In this study, the ability of aluminium rich sandy soil as natural adsorbent in the adsorption of Pb from water was investigated. The effect of initial Pb concentration, solution pH and soil:solution ratio on the Pb removal efficiency were investigated. The adsorption isotherms and kinetic of the process were also evaluated.

2. Methodology

2.1. Materials

The Pb source in this study was lead (II) nitrate ($Pb(NO_3)_2$), which was purchased from R&M Chemicals (Malaysia). The pH adjustment for the Pb solution was carried out using hydrochloric acid and sodium hydroxide, which were purchased by R&M Chemicals (Malaysia). The soil samples, which were rich in aluminium content (11894 mg kg^{-1}) were originated from Jeram, Kuala Selangor. The soil was pre-treated by removing huge rocks and branches. Then, the soil was dried, grinded and sieved through 2 mm size mesh. The soil samples were then stored in desiccator prior to use for the experiments. The general characteristics study of the soil has been investigated by Mukhopadhyay *et al.* [19] and they are as shown in table 1.

Table 1. Soil characteristics [19].

USDA soil classification		
Soil distribution of soil particles	**Size distribution, mm**	
Sand (Mainly medium and find sand)	0.045 < x < 2	97.45 %
Silt + clay	<0.045	2.40 %
Physical characteristic of soil		
Moisture content (% wt)		3.60
Density (kg L^{-1})		2.52
Organic matter (Loss by ignition method, %wt)		1.21
pH (USEPA SW-846 Method 9045D)		3.45
Oxidation Reduction Potential (ASTM Method D 1498-93, mV)		333
Electrical Conductivity (USEPA SW-846 Method 9045D, mS cm^{-1})		8.25
Metal content of the soil, mg kg^{-1} (USEPA Method 3050B)		
Al		11894
Ca		221
Fe		1000
Mg		464
Mn		42
Pb		218
Zn		10303

2.2. Experimental procedures

2.2.1. Effect of operating parameters. Response surface methodology (RSM) is a mathematical and statistical technique that is useful for the analysis and modelling of problems into a polynomial equation in which the response/outcome is influenced by several variables [20]. Three operating parameters were investigated using RSM in this study, namely: i) initial concentration (500 ppm to 1900 ppm), ii) solution pH (1 to 5), and iii) soil:solution ratio (0.067 to 0.2). The number and sequence of the experiments are planned using Box-Behnken Design, as shown in Table 2. The experiment was conducted by contacting a volume of 100 mL Pb solution at pre-determined concentration and pH with a mass of soil. The mixture was homogenized at 300 rpm and 25°C for four hours. After the experiment, the aqueous sample was taken, filtered, diluted and analysed for Pb concentration using Inductively Coupled Plasma – Optical Emission Spectroscopy (ICP-OES).

Table 2. Parameters studied and the results in Box-Behnken Design.

Run	Initial concentration, ppm	Solution pH	Soil:solution ratio, g mL^{-1}	Pb uptake, mg g^{-1}	Removal efficiency, %	Ln (Removal efficiency)	
						Experimental	Predicted
1	1225	3	0.135	8.11	88.42	4.48	4.48
2	1225	3	0.135	8.10	88.32	4.48	4.48
3	1900	3	0.067	13.27	46.79	3.85	3.85
4	1900	5	0.135	10.68	75.06	4.32	4.32
5	1225	3	0.135	8.10	88.3	4.48	4.48
6	1900	3	0.2	8.47	89.21	4.49	4.49
7	550	3	0.067	7.41	90.31	4.50	4.50
8	1225	1	0.067	5.39	29.48	3.38	3.38
9	550	3	0.2	2.75	99.84	4.60	4.60
10	1225	3	0.135	8.14	88.67	4.48	4.48
11	1225	3	0.135	8.13	88.62	4.48	4.48
12	1225	5	0.067	11.22	61.37	4.12	4.12
13	1225	5	0.2	5.96	97.35	4.58	4.58
14	1225	1	0.2	4.72	76.99	4.34	4.34
15	1900	1	0.135	5.20	36.53	3.60	3.60
16	550	5	0.135	4.04	98.15	4.59	4.59
17	550	1	0.135	3.32	80.67	4.39	4.39

2.2.2. Adsorption isotherms. Synthetic Pb contaminated water was prepared by dissolving $Pb(NO_3)_2$ into deionzied water at different concentrations of 550 ppm to 1900 ppm at pH 5. A volume of 100 mL of the Pb solution was contacted with 10 g of soil and was homogenized using a shaker at 300 rpm and 25°C for four hours. Aqueous samples were taken after the experiment and they were filtered, diluted, and analysed using ICP-OES. The final Pb concentration in the aqueous phase and the Pb uptake were evaluated and were fitted into Langmuir isotherm and Freundlich isotherm.

2.2.3. Adsorption kinetics. The study was investigated by mixing 250 mL of 200 ppm Pb solution with 25 g of soil. The mixture was homogenized at 200 rpm and three hours contact time. Aqueous samples were taken from time to time and they were filtered, diluted and analyzed using ICP-OES. The experimental results were evaluated by different kinetic models.

2.3. Analytical methods

The removal efficiency can be calculated based on Pb concentration obtained from the sample, as shown in equation (1) where C_f is the Pb concentration (mg L^{-1}) after the experiment and C_i is the initial Pb concentration, (mg L^{-1}). The Pb uptake, q (mg g^{-1}) can be determined using equation (2), where m_s is the mass of soil used (g) and V is the volume of Pb solution (L).

$$\text{Removal Efficiency, \%} = \frac{C_i - C_f}{C_i} \times 100 \tag{1}$$

$$\text{Pb uptake, q, mg/g} = \frac{(C_i - C_f)V}{m_s} \tag{2}$$

3. Results and discussion

3.1. Soil characterization

According to the results as shown in table 1, the soil is acidic and is rich in aluminium and has significant magnesium and iron content. The X-Ray Diffraction (XRD) analysis on its mineralogical structure by Mukhopadhyay *et al.* [21] revealed that the the minerals that are presence in the soil are silicon oxide (SiO_2), aluminium silicate hydroxide ($Al_2Si_2O_5(OH)_4$), potassium magnesium iron aluminum silicate hydrate (K-Mg-Fe-Al-Si-O-H_2O), sodium aluminium silicate hydroxide hydrate ($Na_{0.3}Al_6(Si, Al)_8O_{20}(OH)_{10}$•$14H_2O$), zeolite NdY ($H_{42}Al_{52.45}Nd_{18.258}O_{405}Se_{32}Si_{139.55}$), and magnesium silicate hydroxide hydrate ($Mg_4Si_6O_{15}(OH)_2$•$16H_2O$). The presence of these minerals and organic matter content of 1.21% suggested that the soil has ability to retain Pb cation from water, as heavy metals are reported to be adsorbed on iron and aluminium based minerals [22, 23].

3.2 Effect of operating parameters

3.2.1. Analysis of Variance (ANOVA). In this study, the parameters are denoted as (A) for initial concentration, (B) for pH, and (C) for soil:solution ratio. The analysis of variance (ANOVA) as shown in table 3 shows that the model is significant with insignificant lack of fit. This is desirable and this shows that the experimental data fits well with the model.

Table 3. Analysis of Variance (ANOVA) for the data.

Source	Sum of Squares	DF	Mean Square	F-Value	Prob> F	
Model	2.078	11	0.19	29170.585	< 0.0001	significant
A	0.148	1	0.15	22892.639	< 0.0001	
B	0.234	1	0.23	36153.715	< 0.0001	
C	0.505	1	0.51	77973.203	< 0.0001	
B^2	0.280	1	0.28	43174.919	< 0.0001	
C^2	0.061	1	0.06	9357.397	< 0.0001	
AB	0.069	1	0.07	10598.446	< 0.0001	
AC	0.074	1	0.07	11464.045	< 0.0001	
BC	0.062	1	0.06	9593.912	< 0.0001	
A^2B	3.32×10^{-4}	1	3.32×10^{-4}	51.306	0.0008	
A^2C	0.057	1	0.057	8811.135	< 0.0001	
AB^2	0.011	1	0.011	1626.174	< 0.0001	
Residual	3.24×10^{-5}	5	6.48×10^{-6}			
Lack of Fit	1.75×10^{-5}	1	1.75×10^{-5}	4.720	0.0955	not significant
Pure Error	1.49×10^{-5}	4	3.71×10^{-6}			
Cor Total	2.078	16				
Std. Dev.	2.545×10^{-3}			R-Squared	1.0000	
Mean	4.30			Adj R-Squared	1.0000	
C.V.	0.059			Pred R-Squared	0.9984	
PRESS	3.224×10^{-3}			Adeq Precision	571.488	

The model shows that the terms A, B, C and the interaction effect between AB, AC and BC have significant impact on the removal efficiency. The high value of R^2s (0.9984-1) further confirm the model fits the data well. An adequate precision of 571.488, which is >4 further justify the adequacy of the model. The coded mathematical equation is as shown in equation (3) where the initial concentration, pH and soil:solution ratio, are represented by A, B, and C, respectively, and the range of the parameters are from -1 (lowest level) to 1 (highest level).

$$\text{Ln (Removal efficiency)} = 4.48 - 0.19A + 0.24B + 0.36C - 0.26B^2 - 0.12C^2 + 0.13AB + 0.14AC - 0.12BC - 0.013A^2B - 0.17A^2C - 0.073AB^2 \quad (3)$$

3.2.2. Interaction effect between pH and initial concentration. Figure 1 shows that, when the pH is increased, the removal efficiency is increased, regardless of initial concentration used. This was in line with the results for heavy metal adsorption in other reported literature [10, 15, 24]. The pH condition not only caused significant influence on the surface characteristics of the adsorbents, but also on the chemical speciation of Pb ion [13, 25]. Low adsorption rate occurring at low pH could be described by the competition between the adsorption of H^+ and Pb^{2+} ions on the sorption sites [10]. Furthermore, the excessively acidic environment may also cause the protonation of the surface adsorption sites from negative to positive charge, which further induced electrostatic repulsion that restricted the adsorption of Pb ions on the soil surface [9, 11, 14–16, 24]. The concentration of H^+ was decreased when the pH was increased to 5, and this reduced the competition for adsorption sites as well as the electrostatic repulsion mechanism via deprotonation of the adsorption sites for electrostatic interaction between Pb ions and the soil [11, 15–16, 24]. Thus, better Pb uptake was facilitated.

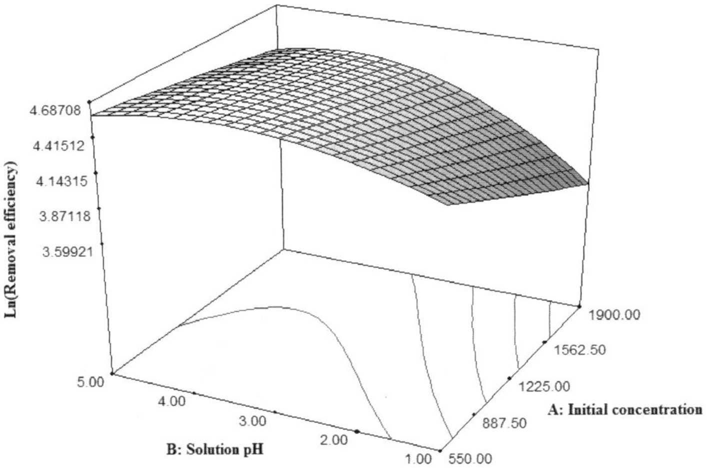

Figure 1. 3D response surface on the effect of initial concentration and solution pH on the removal efficiency (soil:solution ratio: 0.135).

In contrast, the removal efficiency is inversely proportional to the initial concentration, as shown in figure 1. From the view of removal efficiency, the effect given by initial concentration has higher impact at low pH condition, as shown in figure 1 and table 2. When the initial concentration was increased from 550 ppm to 1900 ppm, the decrement in the removal efficiency at pH 1 was 44.15% (80.67% to 36.53%) in comparison to pH 5 at 23.10% (98.15% to 75.06%). This trend was also in line with the work of Karnib *et al.* [26] on the use of activated carbon for heavy metals adsorption. In general, high pH condition and low initial concentration will give higher removal efficiency, as a result of the abundance of adsorption sites due to: i) less competition between H^+ and Pb^{2+}, and ii) higher ratio of adsorption site to Pb. At low concentration, there was abundance of adsorption sites on the soil surface for Pb adsorption. With the increase of Pb concentration, the adsorption sites were filled and saturated [16, 26], resulting inadequacy of the number of adsorption sites. Thus, the removal efficiency was decreased. It was worth noting that the adsorption site was utilised effectively for Pb solution at higher concentration. In this study, Pb uptake by the soil was found to improve at higher initial concentration as more Pb ions were available for the adsorption. The increase in the initial concentration from 550 ppm to 1900 ppm increased the Pb uptake from 3.32 mg g^{-1} to 5.2 mg g^{-1} and 4.04 mg g^{-1} to 10.68 mg g^{-1} when the system was operated at pH 1 and pH 5, respectively.

3.2.3. Interaction effect between pH and soil:solution ratio. Figure 2 illustrates that the use of higher soil:solution ratio increases the removal efficiency and this applies to all pH conditions in this study. This was due to the fact that at higher ratio, the number of available and accessible adsorption sites for Pb-adsorbent interaction was increased and this consequently caused easier access of Pb ion to the sorption sites [24]. Therefore, higher removal efficiency and adsorption rate could be achieved, as reported in the works of Hafsejani *et al.* [10], Chaudhry *et al.* [16] and Wen *et al.* [13]. Nevertheless, the enhancement given by higher soil:solution ratio was highly dependent on the pH condition. As shown in figure 2 and table 2, the enhancement in removal efficiency of 47.51% is observed at pH 1 when the soil:solution ratio is increased from 0.067 to 0.2. This was because at low pH condition, higher number of adsorption sites improved effective Pb adsorption, thereby reducing the effect of competition between Pb^{2+} and H^{+}. In contrast, the increase in soil:solution ratio from 0.067 to 0.2 only increases the removal efficiency by 35.97% when the system was operated at pH 5, as shown in figure 2 and table 2. In this case, the competition between H^{+} and Pb^{2+} was less significant and thus, the requirement on the number of adsorption sites was less critical. However, the use of high soil:solution ratio was not feasible in terms of economical point of view. This argument was deduced from the experimental results which showed that the increase in soil:solution ratio from 0.067 to 0.2 reduced Pb uptake from 5.39 mg g^{-1} to 4.72 mg g^{-1} and from 11.22 mg g^{-1} to 5.96 mg g^{-1} when the system was operated at pH 1 and pH 5, respectively. The decrease in adsorption equilibrium at high soil:solution ratio was reported to be mainly caused by two mechanisms, namely i) unsaturation of adsorption sites [11, 13, 15], and ii) particles aggregation [11]. In addition, the higher decrement in Pb uptake per mass of soil at pH 5 can be attributed to the presence of excess adsorption sites due to the absence of competitive adsorption between hydrogen and Pb ions.

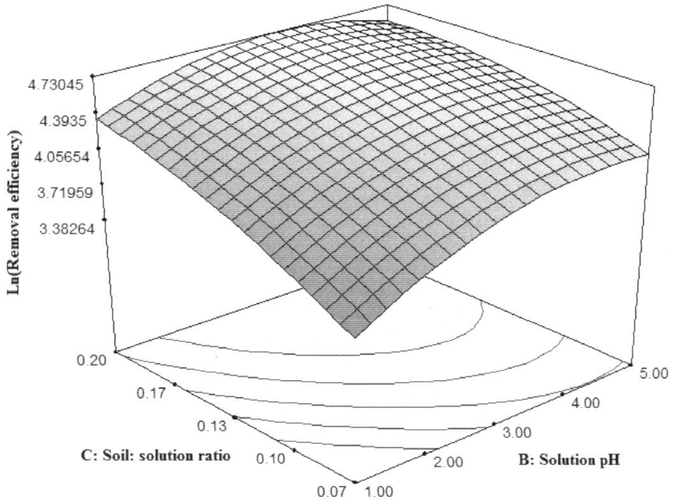

Figure 2. 3D response surface on the effect of solution pH and soil:solution ratio on the removal efficiency (initial concentration: 1225 ppm).

3.3 Adsorption isotherms

The adsorption isotherm was investigated using Langmuir isotherm and Freundlich isotherm. Table 4 shows that the adsorption of Pb on soil is likely to follow Langmuir isotherm, as the curve fitting has highest R^2 of 0.993 in comparison to Freundlich isotherm which only shows R^2 of 0.867. Therefore, it was suggested that the adsorption was monolayer chemisorption [11] where the adsorption process occurred due to the chemical bond between Pb and the finite number of adsorption sites which was mainly consisted of aluminium and iron as well as organic matter on soil surface and there was no transmigration of adsorbate on the soil surface [9, 27]. This adsorption isotherm was also observed for heavy metals adsorption on different types of adsorbent such as manganese coated sand [16], porous hydrogel adsorbents [25], carbonaceous based adsorbent [28], alginate-based attapulgite foams [18], ash biosorbents [9, 11, 13], suggesting that the adsorption process using the soil as adsorbent has similar mechanisms as other types of surface modified adsorbents. The maximum Pb adsorption capacity for the soil, Q_{max} was 10.64 mg g^{-1}.

Table 4. Adsorption isotherms analysis at solution pH of 5.

Isotherm	**Constant**				**R^2**
	Q_{max}, mg g^{-1}	**K_L**	**n**	**K_f**	
Langmuir	10.64	0.03	-	-	0.993
Freundlich	-	-	4.93	2.74	0.867

3.4 Adsorption kinetic study

Figure 3 illustrates the changes of Pb concentration in the aqueous solution with time at different pH conditions, in which equilibrium removal was achieved at 150 min. The remaining Pb concentration in the solution was found to be higher at pH 2 (51.87 ppm) and the concentration was stable at (21.22 ppm – 24.61 ppm) when the pH was increased from 3 to 5, suggesting that better Pb uptake by the soil. This revealed that at low pH, Pb^{2+} ion is competing with H^+ for the adsorption site. Thus, the adsorption was lower in comparison to high pH condition, as discussed in Section 3.2.

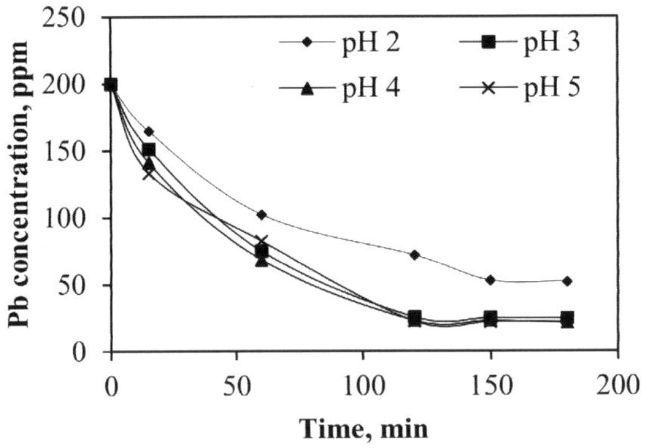

Figure 3. Changes of Pb concentration in aqueous solution with time.

The adsorption kinetic study shows that Pb adsorption to the soil follows pseudo second order kinetic model as the R^2 obtained are the highest in comparison to other models for all pH conditions, as shown in table 5. The pseudo second order rate constant k_{p2} and initial adsorption rate, V_0 are found to stabilize at 0.0110 g mg^{-1} min^{-1} and 0.056 mg g^{-1} min^{-1}, respectively, at pH 4-5, as shown in table 6. The results are in line with the works of Bozbas and Boz [9], Lim and Lee [29] and Chaudhry *et al.* [16] which reported that the adsorption of heavy metals on *anadara inaequivalvis* shells, soil and manganese coated sand, respectively, followed pseudo second order kinetic model [24].

Table 5. R^2 for the Pb adsorption on soil at different pHs using different kinetic models.

Kinetic model	Linearized equation	R^2			
		pH 2	pH 3	pH 4	pH 5
Pseudo-first order kinetic model	$\log(q_e - q_t) = \log q_e - \frac{k_{p1}}{2.303} t$	0.836	0.96	0.956	0.947
Pseudo-second order kinetic model	$\frac{t}{q_t} = \frac{1}{V_0} + \frac{1}{q_e} t;\ V_0 = k_{p2} q_e^2$	0.994	0.987	0.994	0.981
Elovich model	$q_t = \alpha \ln(a\alpha) + \alpha \ln t$	0.994	0.981	0.982	0.96
Film diffusion mass transfer kinetic model	$\ln\left(1 - \frac{q_t}{q_e}\right) = -R^1 t$	0.836	0.956	0.956	0.947

Table 6. Equilibrium Pb uptake, pseudo second order rate constant, and initial adsorption rate at different pH conditions.

pH	Equilibrium Pb uptake q_e, mg g^{-1}	Pseudo second order rate constant k_{p2}, g mg^{-1} min^{-1} ($V_0 = k_{p2}q_e^2$)	Initial adsorption rate V_0, mg g^{-1} min^{-1}
2	2.110	0.0065	0.029
3	2.353	0.0080	0.044
4	2.242	0.0111	0.056
5	2.237	0.0109	0.055

3.5 Comparison with other types of adsorbents

Table 7. Comparison on Pb adsorption capacity between different types of adsorbents.

Type of adsorbents	Adsorbents	Fe or Al content mg kg^{-1}	Adsorption capacity, mg g^{-1}	References
Activated carbons	• Activated carbon	-	294.11	[26]
	• Carbonaceous adsorbent	-	130.54	[28]
Biosorbents	• Tomato waste and apple juice residue	-	108-152	[24]
	• Nanostructured cedar leaf ash	-	7.23-8.05	[10]
	• Exhausted coffee grounds	-	61.6	[9]
	• *anadara inaequivalvis* shells	-	621.1	[11]
Soil	• Manganese coated sand	-	0.15	[16]
	• Natural iron rich sandy soil	3719 (Fe) 2400 (Al)	0.9-1.0	[30]
	• Natural aluminium rich sandy soil	11894 (Al) 1000 (Fe)	10.64	Present study

A comparison on the adsorption capacity between the soil used in this study and other adsorbents was conducted. The summary as shown in table 7 shows that in general, the activated carbons and biosorbents show relatively higher Pb adsorption capacity than the soil based adsorbents. Nevertheless, it is worth noting that the soil undergoes less surface modification or treatment for the adsorption process. Thus, it is expected that the adsorption capacity is lower. From the results, it was found that the natural aluminium rich sandy soil has a much higher Pb adsorption capacity (10.64 mg g^{-1}) than natural iron rich sandy soil (0.9-1.0 mg g^{-1}), as a result of its relatively high total concentration in aluminium and iron content, as the aluminium and iron oxides serve as the main adsorption sites for removing Pb from water [22, 23].

4. Conclusions

The study investigated the use of aluminium rich sandy soil from Jeram as natural adsorbent in Pb removal from water. The soil was found to show promising ability as the adsorbent for removing Pb from water. The removal of Pb was mainly through the adsorption on the surface minerals of aluminium and ferric oxides and this mechanism was highly affected by the solution pH, initial Pb concentration, and soil: solution ratio. The adsorption efficiency was found to be increased at higher pH (4-5), higher soil:solution ratio and lower initial concentration, as a result of Pb exposure to higher number of adsorption surface area/sites at low competition with H^+ ion.

The Pb adsorption process followed Langmuir isotherm and pseudo-second order kinetic model, with the adsorption capacity of 10.64 mg g^{-1} at optimum pH of 4-5. Even though the adsorption capacity is lower than typical commercial adsorbents, its highly abundance in Jeram area and its natural properties that does not require additional activation steps or surface modification make it suitable to be potential candidate as the adsorbent for water treatment in removing Pb.

Acknowledgment

This work was supported by Universiti Tunku Abdul Rahman.

References

[1] Nahal S, 2014 Blue Revolution – global water primer. Retrieved 13 October 2019, from https://www.longfinance.net/media/documents/baml_bluerevolution_2014.pdf
[2] McCartor A and Becker D, 2010 *Blacksmith institute's world worst pollution problem report 2010.* New York: Blacksmith Institute
[3] DoE-Malaysia, 2009 Contaminated land management and control guidelines No. 1: Malaysian recommended site screening levels for contaminated land. Retrieved 30 August 2019, from http://www.doe.gov.my/portal/wp-content/uploads/Contaminated-Land-Management-and-Control-Guidelines-No-1_Malaysian-Recommended-Site-Screening-Levels-for-Contaminated-Land.pdf
[4] Duruibe J O, Ogwuegbu M O C and Egwurugwu J N 2007 Heavy metal pollution and human biotoxic effects *Int. J. Phys. Sci.* **2** 112–8
[5] Wuana R A and Okieimen F E 2011 Heavy metals in contaminated soils: A review of sources, chemistry, risks and best available strategies for remediation. *ISRN Ecology* **2011** 1–20
[6] Harvey P J, Handley H K and Taylor M P 2016 Widespread copper and lead contamination of household drinking water, New South Wales, Australia. *Environ. Res.* **151** 275–85
[7] Mishra S P, Singh V K and Tiwari D 1996 Radiotracer technique in adsorption study part XV. Removal behaviour of barium titanate for mercury ions *Radiochim. Acta* **73** 49–54
[8] Pan B, Pan B, Zhang W, Lv L, Zhang Q and Zheng S 2009 Development of polymeric and polymer-based hybrid adsorbents for pollutants removal from waters *Chem. Eng. J.* **151** 19–29
[9] Babu A N, Reddy D S, Kumar G S, Ravindhranath K and Mohan G V K 2012 Removal of lead and fluoride from contaminated water using exhausted coffee grounds based bio-sorbent *J. Environ. Manage.* **218** 602-12
[10] Hafshejani L D, Nasab S B, Gholami R M, Moradzadeh M, Izadpanah Z, Hafshejani S B, and Bhatnagar A 2015 Removal of zinc and lead from aqueous solution by nanostructured cedar leaf ash as biosorbent *J. Mol. Liq.* **211** 448–56

[11] Bozbas S K and Boz Y 2016 Low-cost biosorbent: Anadara inaequivalvis shells for removal of Pb(II) and Cu(II) from aqueous solution *Process Saf. Environ. Prot.* **103** 144–52
[12] Noor N M, Othman R, Mubarak N M and Abdullah E C 2017 Agricultural biomass-derived magnetic adsorbents: Preparation and application for heavy metals removal *J. Taiwan Inst. Chem. Eng.* **78** 168–77
[13] Wen X, Du C, Zeng G, Huang D, Zhang J, Yin L, Tan S, Huang L, Chen H, Yu G, Hu X, Lai C, Xu P and Wan J 2018 A novel biosorbent prepared by immobilized bacillus licheniformis for lead removal from wastewater *Chemosphere* **200** 173–9
[14] Abo-El-Enein S A, Shebl A and Abo El-Dahab S A 2017 Drinking water treatment sludge as an efficient adsorbent for heavy metals removal *Appl. Clay Sci.* **146** 343–9
[15] Han R, Zou W, Zhang Z, Shi J and Yang J 2006 Removal of copper(II) and lead(II) from aqueous solution by manganese oxide coated sand I. Characterization and kinetic study *J. Hazard. Mater.* **B137** 384–95
[16] Chaudhry S A, Khan T. A and Ali I 2016 Adsorptive removal of Pb(II) and Zn(II) from water onto manganese oxide-coated sand: Isotherm, thermodynamic and kinetic studies *Egypt. J. Basic Appl. Sci.* **3** 287–300
[17] Kasiuliene A, Carabante I, Bhattacharya P, Caporale A G, Adamo P and Kumpiene J 2018 Removal of metal(oid)s from contaminated water using iron-coated peat sorbent, *Chemosphere* **198** 290–6
[18] Wang Y, Feng Y, Zhang X-F, Zhang X, Jiang J and Yao J 2018 Alginate-based attapulgite foams as efficient and recyclable adsorbents for the removal of heavy metals *J. Colloid Interface Sci.* **514** 190–8
[19] Mukhopadhyay S, Mukherjee S, Adnan N F, Hayyan A, Hayyan M, Hashim M A and Sen Gupta B 2016 Ammonium-based deep eutectic solvents as novel soil washing agent for lead removal *Chem. Eng. J.* **294** 316–22
[20] Montgomery C D 2001 *Design and analysis of experiments (5th ed.)* (New York: John Wiley & Sons, Inc)
[21] Mukhopadhyay S, Mukherjee S, Hayyan A, Hayyan M, Hashim M A and Sen Gupta B 2016 Enhanced removal of lead from contaminated soil by polyol-based deep eutectic solvents and saponin *J. Contam. Hydrol.* **194** 17–23
[22] Strawn D G, Scheidegger A M and Sparks D L 1998 Kinetics and mechanisms of Pb(II) sorption and desorption at the aluminum oxide-water interface *Environ. Sci. Technol.* **32** 2596–601
[23] Jacukowicz-Sobala I, Ociński D and Kociołek-Balawejder E 2015 Iron and aluminium oxides containing industrial wastes as adsorbents of heavy metals: Application possibilities and limitations *Waste Manag. Res.* **33** 612–29
[24] Heraldy E, Lestari W W, Permatasari D and Arimurti D D 2018 Biosorbent from tomato waste and apple juice residue for lead removal *J. Environ. Chem. Eng.* **6** 1201–8
[25] Zhou G, Luo J, Liu C, Chu L and Crittenden J 2018 Efficient heavy metal removal from industrial melting effluent using fixed-bed process based on porous hydrogel adsorbents *Water Res.* **131** 246–54
[26] Karnib M, Kabbani A, Holail H and Olama Z 2014 Heavy metals removal using activated carbon, silica and silica activated carbon composite *Energy Procedia* **50** 113–20
[27] Dada A O, Olalekan A P, Olatunya A M and Dada O 2012 Langmuir, Freundlich, Temkin and Dubinin-Radushkevich isotherms studies of equilibrium sorption of Zn2+ unto phosphoric acid modified rice husk *IOSR J. Appl. Chem.* **3** 38–45
[28] Wajima T 2017 A new carbonaceous adsorbent for heavy metal removal from aqueous solution prepared from paper sludge by sulfur-impregnation and pyrolysis *Process Saf. Environ. Prot.* **112** 342–52
[29] Lim S-F and Lee A Y W 2015 Kinetic study on removal of heavy metal ions from aqueous solution by using soil *Environ. Sci. Pollut. R.* **22** 10144–58

IOP Conf. Series: Materials Science and Engineering 1195 (2021) 012003 doi:10.1088/1757-899X/1195/1/012003

[30] Ng Y S, Sen Gupta B and Hashim M A 2016 Performance evaluation of natural iron-rich sandysoil as a low-cost adsorbent for removal of lead from water *Desalin. Water Treat.* **57** 5013–24

IOP Conf. Series: Materials Science and Engineering 1195 (2021) 012004 doi:10.1088/1757-899X/1195/1/012004

Reduction of tar from biomass gasification using a dielectric barrier discharge reactor

M Lim[1] and Z Alimuddin[2]

[1] Biomass & Plasma Technologies, Renewable Energy and Green Technology TNB Research Sdn. Bhd., Research Institution Area, 43000 Kajang, Selangor, Malaysia
[2] School of Mechanical Eng., Universiti Sains Malaysia, Engineering Campus, 13600 Nibong Tebal, Penang, Malaysia

Email: mook.tzeng@tnb.com.my; mooktzeng.lim@live.com

Abstract. A non-thermal plasma reactor was used to investigate its effectiveness in reducing the by-products from biomass gasification. Biomass is used for generating heat and power through gasification, which is a process of converting solid fuel to gaseous fuel at temperatures of 700 to 900 °C by operating a reactor in sub-stoichiometric conditions. This gas mixture can be utilized for liquid fuel synthesis or for fuel cells. However, the by-product of gasification consists of tar, which consists of oxygenates, ringed-aromatics, phenolic compounds, and polyaromatic hydrocarbons (PAH). Depending on the composition, the condensation temperature can be as high as 450 °C, fouling downstream equipment. In this study, a dielectric barrier discharge (DBD) reactor with a coil as the inner electrode was used to reduce toluene, a model tar compound. Toluene was injected into a mixing chamber that was heated to 900 °C, evaporating the toluene, and is entrained by nitrogen into the DBD reactor. High voltage is injected into the DBD reactor to initiate ionization, decomposing the toluene into lighter hydrocarbons. A sampling bottle submerged in an ice bath collects the residual toluene, and the resulting decomposition rate is as high as 70%.

1. Introduction

Biomass is increasingly being used for generating heat and power through gasification, which is a process of converting solid fuel to gaseous fuel at temperatures of 700 to 900 °C by operating a reactor in sub-stoichiometric conditions [1]. In stoichiometric conditions, sufficient (or excess) amount of air is supplied for reaction of oxygen with solid fuels, and combustion is the main reaction where the products are mostly carbon dioxide and water. In sub-stoichiometric conditions, the amount of air into the reactor is less than required for the stoichiometric conditions, such that the there is a partial oxidation of the solid fuels (as opposed to combustion) produces syngas that consists of carbon monoxide, methane, hydrogen, and other hydrocarbons [1, 2]. This gas mixture can be utilized for liquid fuel synthesis [3, 4] or for fuel cells [5]. The scale and type of the gasifier affects the concentration of the by-product, also known as tar, which can be substantial and causes operational issues by fouling downstream equipment. Tar consists of compounds including oxygenates, ringed-aromatics, phenolic compounds, and polyaromatic hydrocarbons (PAH) [6–8]. The condensation temperature of tar depends on its' composition, which can be as high as 450 °C [9, 10], making prevention measures difficult.

There are several methods to reduce tar, which include the use of low temperature pre-treatment methods, catalysts and scrubbers [7, 11]. Low temperature pre-treatments involve the carbonization of biomass at temperatures of 300 to 400 °C before gasification. Results showed that the syngas from the gasification of carbonized biomass contains 58% less tar than biomass which were not pre-treated [12]. Several catalysts are commonly used to reduce tar, such as nickel, cobalt, alumina, platinum, rubidium and etc. However, the catalysts are known to deactivate during the reduction of the tar [13]. Recently, dielectric barrier discharge (DBD) reactors initiate non-thermal plasma (NTP) to reduce tar content. NTP has been known to convert toluene (C_7H_8, a model tar compound [14, 15]) by 90%, producing H_2, CO, and CO_2 at a power consumption of 90 W and a temperature of 300 °C [16]. Naphthalene was also used as a tar model compound in a reforming experiment, using a gliding arc discharge reactor (a form of NTP) with the addition of steam. Results showed that the naphthalene conversion rate was 85%, producing acetylene yield of 33%, with a total gas yield of 72% with a steam-to-carbon ratio of 2. The advantage of NTP is its' lower reaction temperature, allowing process and material requirements to be less stringent. In this study, a dielectric barrier discharge (DBD) reactor was used to investigate its effectiveness in reducing toluene, a compound that is classified as a Class 3 tar [17].

2. Methodology

Figure 1 shows a schematic of the experimental set-up. Toluene (5 to 7 g) was injected using a syringe into a mixing chamber of 0.2 m in diameter and 0.1 m in height. The injection rate of toluene is timed with a stopwatch, to ensure similar amount of toluene is injected into the mixing chamber within a period of time for each experimental run. The mixing chamber is heated up to temperatures of 900 °C such that the injected toluene evaporates. The toluene vapor is entrained by nitrogen (as a carrier gas) at a flow rate of 10 LPM, giving a toluene concentration of 17%. The gas mixture of toluene and nitrogen then flows into the DBD reactor.

The detailed dimensions and construction of the DBD reactor is described in [18], and is connected to a high voltage (HV) supply shown in figure 2. A detailed description of the HV supply and set-up is available in literature [19, 20]. The discharge power of the DBD is determined via a correlation with the power consumption, recorded by a power meter, and is varied between 1 to 56 W to determine its effect on the reduction rate of toluene.

When the voltage is high enough and the resonance frequency is achieved, ionization is initiated in the gap between the dielectric barrier and the ground electrode. Electrons begin to accelerate to the ground electrode, colliding with the toluene and nitrogen gas mixture.

The ionisation intensity of the DBD reactor is varied according to discharge power of between. A sampling bottle filled with silica is submerged in an ice bath to collect the by-products from the reaction with the DBD. The sampling bottle is weighed (in grams) before and after the experimental run to determine the rate of decomposition or reduction. The results show that the higher the discharge power (from 0 to 56 W), the higher the decomposition rate of toluene (from 28 to 55%).

M_o is the initial weight of the toluene collected (in grams) after 20 s of the experiment, while M_i is the weight of toluene collected (in grams) after each subsequent 20 s (e.g., t_o = 20 s; t_1 =40 s; t_3 =60 s…) The reduction rate of toluene is thus determined by:

$$\text{Reduction rate of toluene (\%)} = 1 - \left({M_i}/{M_o}\right) \quad (1)$$

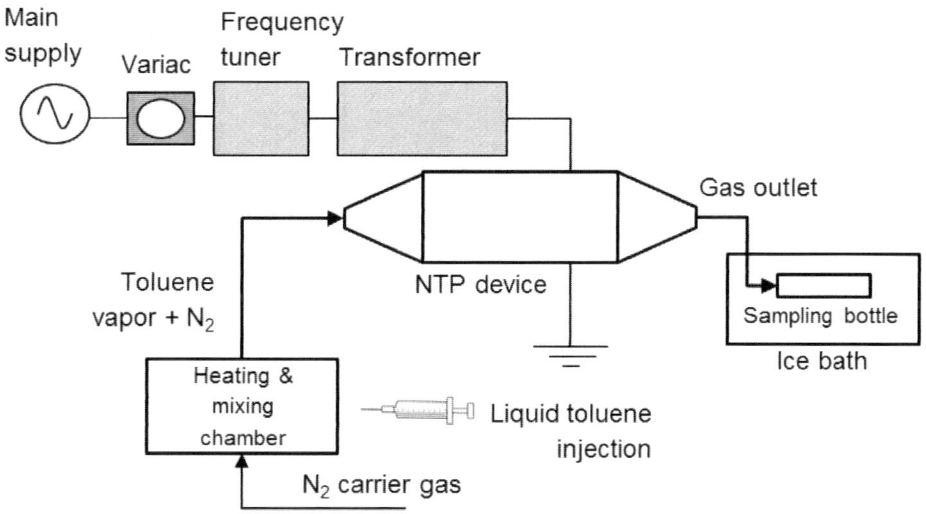

Figure 1. Schematic for non-thermal plasma reduction of toluene.

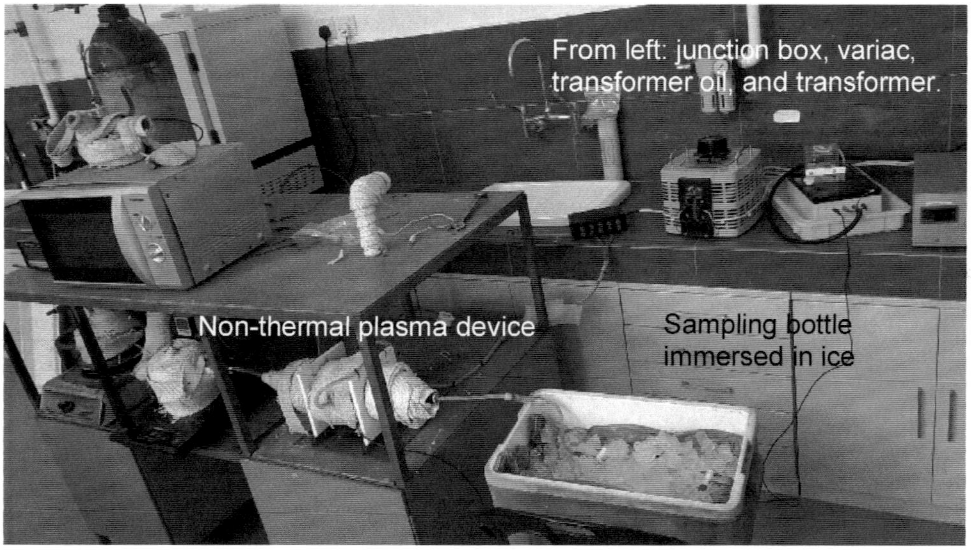

Figure 2. High voltage supply for dielectric barrier discharge reactor and experimental set-up for reduction of toluene.

3. Results and Discussion

Figure 3 shows the reduction rate of toluene with respect to time and discharge power of DBD reactor. After t = 100s, t little or no discharge power (~1W), the decomposition rate of toluene in the sampling bottle is at 32%. When the discharge power of the DBD increases to 26 W, the reduction rate increases to 50%, and is at its highest at 41 or 56 W, with a reduction rate of 62%. After 200 s, the reduction rate reaches 70% at a discharge power of 56 W.

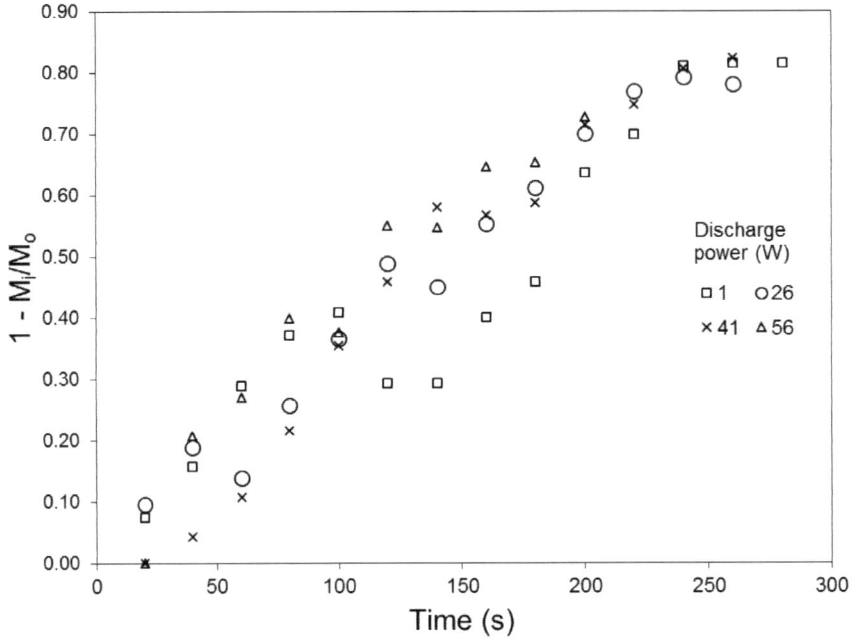

Figure 3. Decomposition rate of toluene with respect to time and discharge power of DBD reactor.

The results show that with increasing discharge power, there is an increasing number of excited electrons due to the higher ionization intensity. The higher number of electrons collide to initiate the breakdown of the toluene into lighter hydrocarbons. Reports have shown that the decomposition of toluene produces benzoic acid, benzaldehyde, benzyl alcohol and nitrophenol [21]. Real time monitoring of the lighter hydrocarbons showed that initially acetone was produced, and was decomposed into other organic byproducts such as methanol and and acetic acid [22]. Other results suggest that when toluene is decomposed, the major gases that are produced are H_2, CO and C_2H_2, yields of 34.9%, 39% and 73.9% respectively [23].

The aforementioned elements and reactions can corrode the electrodes in the DBD reactor (figure 4), hence causing the surface to become uneven and the non-thermal plasma to be unstable, with sharp edges on the electrodes promoting sparks instead of the non-thermal plasma that is more effective in reducing tar formation.

A non-concentric DBD reactor was used in this study, with the advantage of a longer operating lifetime. When the surface of the electrode is corroded, the system could be tuned off temporarily so that the electrode is rotated to other new surface areas (figure 4). Operation of the non-thermal plasma is resumed for decomposition of the pollutants. Thus, the operating lifetime of the tar reduction device is extended.

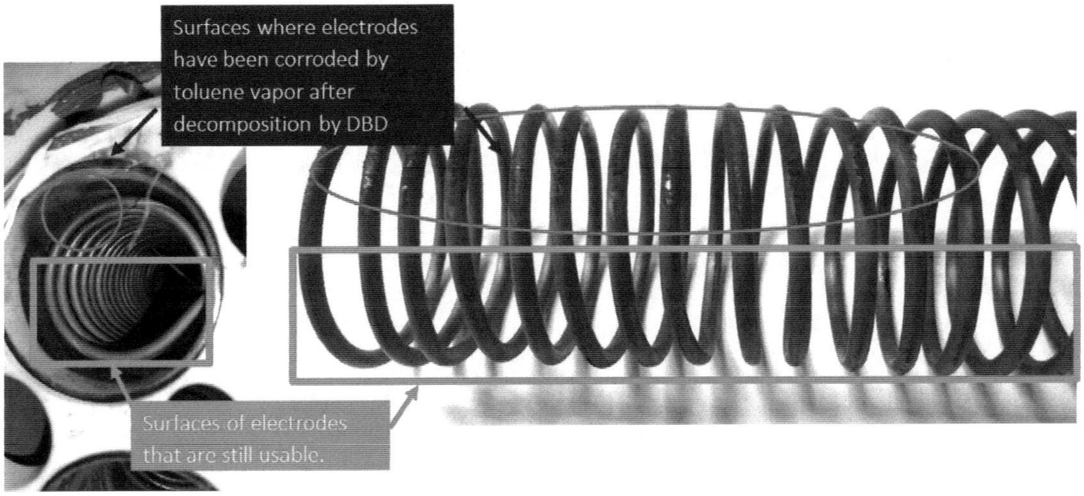

Figure 4. Corroded surfaces of electrodes and other surfaces which could be rotated for continued use.

4. Conclusions

A non-thermal plasma system in the form of a dielectric barrier discharge (DBD) reactor has been shown to reduce the toluene content by as much as 70%. The non-concentric DBD used in this study has the advantage of a longer operating lifetime. If a surface of the electrode is corroded due to reactions with the toluene or other tar compounds, a new surface for reducing tar is available by extracting and rotating the electrode. A stable non-thermal plasma is now able to be generated for the decomposition of tar. The operating lifetime of the device is prolonged, since the reduction of tar continues until the entire surfaces of the electrodes have been corroded and exhausted. Future work includes the use of the device for the decomposition of other gases, such as biogas, and also pre-commercial trials in pilot scale gasifiers.

Acknowledgments

The authors wish to acknowledge the funding for this work from Tenaga Nasional Berhad (TNBR/SF240/2016).

References

[1] Pereira E G, da Silva J N, de Oliveira J L and Machado C S 2012 Sustainable energy: A review of gasification technologies *Renewable Sustainable Energy Rev.* **16** 4753–62

[2] Wang L, Weller C L, Jones D D and Hanna M A 2008 Contemporary issues in thermal gasification of biomass and its application to electricity and fuel production *Biomass Bioenergy* **32** 573–81

[3] Wang Z *et al.* 2016 Design and operation of a pilot plant for biomass to liquid fuels by integrating gasification, DME synthesis and DME to gasoline *Fuel* **186** 587–96

[4] Luque R and Speight J G, R. Luque and J.G. Speight, Editors. 2015, Woodhead Publishing. p. 3

[5] Subotić V, Baldinelli A, Barelli L, Scharler R, Hochenauer C and Anca-Couce A 2019 Optimization of an integrated biomass gasifier-fuel cell system: An experimental study on the cell response to process variations *Energy Procedia* **158** 2052–7

[6] Kiel J, Van Paasen S, Neeft J, Devi L, Ptasinski K and Janssen F 2004 Primary measures to reduce tar formation in fluidised-bed biomass gasifiers *ECN* https://pure.tue.nl/ws/portalfiles/portal/4367413/576697.pdf Accessed on 7 Dec 2020

[7] Abdoulmoumine N, Adhikari S, Kulkarni A and Chattanathan S 2015 A review on biomass gasification syngas cleanup *Appl. Energy* **155** 294–307

[8] Zhou B, Dichiara A, Zhang Y, Zhang Q and Zhou J 2018 Tar formation and evolution during biomass gasification: An experimental and theoretical study *Fuel* **234** 944–53
[9] Edinger P, Schneebeli J, Struis R P W J, Biollaz S M A and Ludwig C 2016 On-line liquid quench sampling and UV-Vis spectroscopy for tar measurements in wood gasification process gases *Fuel* **184** 59–68
[10] Grootjes A J 2012 Solid Phase Adsorption (SPA) method for tar and sulphur compounds *ECN (funded by EDGaR program and the BRISK project)* https://publicaties.ecn.nl/PdfFetch.aspx?nr=ECN-L--12-043 Accessed on 7 Dec 2020
[11] Shen Y, Wang J, Ge X and Chen M 2016 By-products recycling for syngas cleanup in biomass pyrolysis – An overview *Renewable Sustainable Energy Rev.* **59** 1246–68
[12] Umeda K, Nakamura S, Lu D and Yoshikawa K 2019 Biomass gasification employing low-temperature carbonization pretreatment for tar reduction *Biomass Bioenergy* **126** 142–9
[13] Liu L, Zhang Z, Das S and Kawi S 2019 Reforming of tar from biomass gasification in a hybrid catalysis-plasma system: A review *Appl. Catal., B* **250** 250–72
[14] Huang Z *et al.* 2020 In-situ removal of toluene as a biomass tar model compound using $NiFe_2O_4$ for application in chemical looping gasification oxygen carrier *Energy* **190** 116360
[15] Chen Y-H, Schmid M, Kertthong T and Scheffknecht G 2020 Reforming of toluene as a tar model compound over straw char containing fly ash *Biomass Bioenergy* **141** 105657
[16] Liu L, Wang Q, Song J, Yang X and Sun Y 2018 Dry reforming of model biomass pyrolysis products to syngas by dielectric barrier discharge plasma *Int. J. Hydrogen Energy* **43** 10281–93
[17] ECN. *Thersites - the ECN tar dew point site: Classification*. 2009
[18] Mook Tzeng L, Ahmad Zulazlan S Z, Kanesh Kumar J and Oihoong C 2017 Development of a dimensionless parameter for characterization of dielectric barrier discharge devices with respect to geometrical features *Plasma Sci. Technol* **19** 095402
[19] Lim M and Zulkifli A Z S 2018 Investigation of biomass surface modification using non-thermal plasma treatment *Plasma Sci. Technol.* **20** 115502
[20] Lim M and Lea-Langton A R 2020 Investigation of non-thermal plasma assisted combustion of solid biomass fuels: Effects on flue gas composition and efficiency *Plasma Chem. Plasma Process.* **40** 1465–83
[21] Xu X, Wang P, Xu W, Wu J, Chen L, Fu M and Ye D 2016 Plasma-catalysis of metal loaded SBA-15 for toluene removal: Comparison of continuously introduced and adsorption-discharge plasma system *Chem. Eng. J.* **283** 276–84
[22] Guo T, Li X, Li J, Peng Z, Xu L, Dong J, Cheng P and Zhou Z 2018 On-line quantification and human health risk assessment of organic by-products from the removal of toluene in air using non-thermal plasma *Chemosphere* **194** 139–46
[23] Liu S, Mei D, Wang L and Tu X 2017 Steam reforming of toluene as biomass tar model compound in a gliding arc discharge reactor *Chem. Eng. J.* **307** 793–802

Combustion temperature analysis in a fluidized-bed reactor by utilizing palm oil biomass for a renewable energy

Erdiwansyah[1,6], Mahidin[2], Husni Husin[2], Nasaruddin[3], Muhibbuddin[4], M Faisal[5] and Muhtadin[5]

[1] Doctoral Program, School of Engineering, Universitas Syiah Kuala, Darussalam Banda Aceh, 23111 Indonesia

[2] Department of Chemical Engineering, Syiah Kuala University, Banda Aceh 23111, Indonesia

[3] Department of Electrical and Computer Engineering, Syiah Kuala University, Banda Aceh 23111, Indonesia

[4] Department of Mechanical Engineering, Syiah Kuala University, Banda Aceh 23111, Indonesia

[5] Department of Mechanical Engineering, Universitas Abulyatama Aceh, Aceh Besar 23372, Indonesia

[6] Faculty of Engineering, Universitas Serambi Mekkah, Banda Aceh 23245, Indonesia

Email: mahidin@unsyiah.ac.id

Abstract. Biomass from palm oil is a renewable energy source that can be utilized and has very promising availability. Biomass energy is a renewable and sustainable energy that can replace conventional (fossil) fuels. The main objective of the experiment in this article is to analyze the combustion temperature, emissions, and efficiency of palm oil biomass fuel to use and applied in rural/remote areas. The palm oil biomass used in this study is palm kernel shells, empty fruit bunches, oil palm midrib, and oil palm fibers. The experiments in the research carried out in a fluidized-bed combustion chamber designed explicitly with capacities of up to 5 kg of biomass. The results of operations on fluidized-bed when the valve is open 100%, 75%, and 50% with overall palm oil biomass show a high combustion temperature. The highest combustion temperature was recorded in the TC test for 100% open valves with 3 kg biomass of 943 °C. While the minimum combustion temperature obtained on TF2 at 50% open valve with 1 kg biomass of 619 °C, overall combustion temperatures in this experiment showed high results. The maximum emission for O_2 is 20.4% which is obtained at 50% open valve, while for CO_2 the maximum emission is produced when 100% open valve is 19.9% with a biomass weight of 1 kg and 3 kg, respectively. The yield for maximum combustion efficiency when using 1 kg of biomass recorded at 50% open valve was 94.9%. While the minimum efficiency of 87.7% is obtained when the valve is 100% open with biomass of 2 kg. As the biomass fuel used in fluidized-bed increases, the combustion temperature also increases significantly.

1. Introduction

The climate that is increasingly erratic throughout the world today has triggered impacts on the environment. Most of these environmental impacts are caused by air pollution generated from fuel. Air pollution is a major environmental problem that poses various health risks for humans. These areas are reported by the World Health Organization (WHO), where every year, outdoor air pollution and those caused by household appliances cause 7 million deaths shown in figure 1 [1, 2]. In recent decades, many researchers have researched in the field of renewable and sustainable energy intending to statement the problem of air pollution that continues to increase throughout the world. Renewable energy sources in Southeast Asia are currently abundant enough to be produced to replace conventional fuels [3–5]. This renewable energy source can be obtained very easily and cheaply, such as biomass from palm oil mills. Biomass from palm oil is a renewable energy that can use for electricity generation [6–11].

Indonesia and Malaysia are the most significant producing countries in supplying oil palm in the Southeast Asia region [12, 13]. The palm oil biomass can be presented as a power plant to replace conventional fuels. Power plants with biomass are also environmentally friendly if the management is appropriately done. Some researchers have previously conducted studies on palm oil biomass which use as a power plant [12–14]. The investigation and optimizing the process of electricity generation and heat by utilizing biomass waste in the palm oil industry has been carried out by [15]. In their research, they made three combinations of biomass, EFB, PKS, and Biogas with a preheater, which was used in the simulation. The results of the simulation of palm oil and biogas solid biomass waste from factory waste can provide sustainable and renewable fuels for palm oil production. Besides, the availability of additional electricity is possible because it is one of the best choices for the utilization of palm oil biomass waste. The investigation of various wastes used for electricity generation has also been studied comprehensively by [16]. Research to improve the physical and chemical properties originating from biomass and garbage to increase the thermal efficiency of power plants has been studied [17]. This biomass use to mix renewable energy in reducing dependence on fossil fuels, especially coal [17]. Sustainable energy supply from palm oil biomass such as empty fruit bunches, mesocarp fiber, oil palm shells, oil palm leaves, oil palm trunks obtained from oil palm plantations, and also wood biomass used as fuel have been investigated in Malaysia [18]. Palm oil biomass fuel can save energy up to 10 MW for biomass power plants. Based on the results of the analysis that generating 10 MW of biomass can reduce CO_2 emissions (50,130 t), SO_2 (750 t), NOx (218.65 t), and CO (22.83 t) in the environment compared to the energy mix. Evaluation of upstream and downstream oil palm farming activities is critical to carry out so that the goal of making palm oil waste a sustainable energy industry can achieve. The industrial roadmap representations and policies are designed to provide a unique improvement to the FiT system. Besides, the right strategies and frameworks can provide biomass as a sustainable energy industry [19, 20].

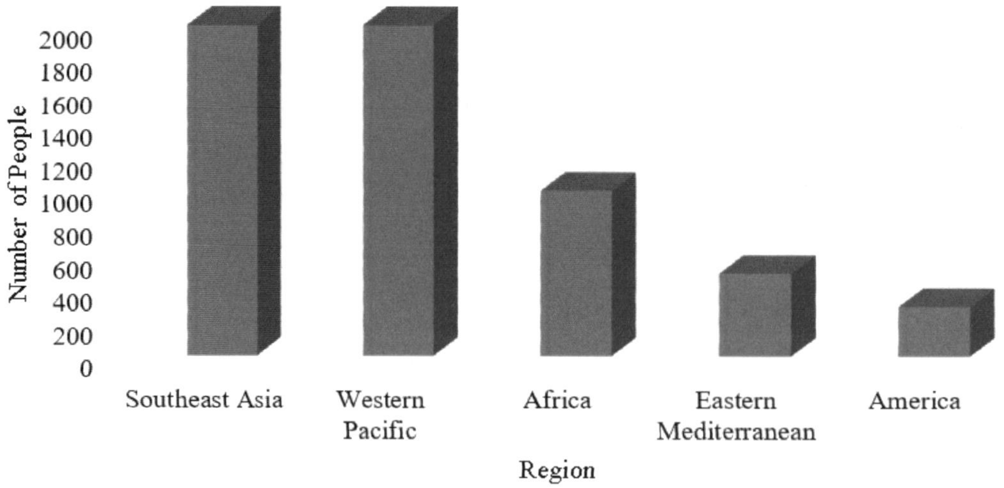

Figure 1. Number of deaths due to air pollution by region.

Indonesia has abundant biomass resources which are a source of renewable and cost-effective energy to be made as renewable and sustainable energy and environmentally friendly [21–23]. Forest biomass resources and from palm oil mills can be converted into bioenergy through various technologies which are one of the sources in Indonesia's energy mix in the future. In 2013 Indonesia produced biomass of around 50.4% of the remaining harvest and residual yields from wood processing reached 49.6% [22]. The amount of biomass found throughout Indonesia as a whole can contribute to the bioenergy potential of 87% of the total. Besides, there are three large islands in Indonesia, which provide 95% of the total bioenergy potential. The study of decentralized palm oil biomass gasification in Indonesia, which aims to reduce dependence on fossil fuel-based power plants in facilitating electrification in rural areas or remote islands has been studied [24]. Technically the palm oil biomass of gasification method also has been evaluated by reviewing various existing literature. Besides, proposals from two scenarios (V1 and V2, and M1 and M2) are each explicitly proposed for the use of power plants in villages and factories [24]. Indonesia has several regions and provinces that have a sufficient amount of biomass for the needs of biomass power plants.

One area that has the most palm oil biomass is the Sumatran region, specifically the Aceh Province. The city of oil palm plantations in Aceh in 2018 reached 26,660 hectares, with a production capacity volume of Certified Sustainable Palm Oil (CSPO) (RSPO certified sustainable palm oil) of 95,234 metric tons [25]. The amount of oil palm can produce palm oil mill waste that is large enough every year that can deliver into electricity generation. PKS with a capacity of 100 thousand tons of fresh fruit bunches annually can produce shells of around 6 thousand tons, fibers of 12 thousand tons, and empty fruit bunches of 23 thousand tons [26]. Whereas the generating efficiency of 25%, can generate electricity around each for the shells from 7.2 to 8.4 GW (e) h, fibers 9.2-15.9 GW (e) h, and 30 GW empty fruit bunches (e) h. With the availability of biomass, the biomass power plant from the palm oil mill can generate electricity around 1.4-1.6 GW (e) h if it is processed and produced correctly. However, the management of electricity from biomass in Aceh Province has not been adequately utilized. These are due to the lack of existing human resources and government policies on renewable energy not in favor of local investors and outside investors. Thus, to be able to utilize this abundant renewable energy source, it is very much needed support from various parties, especially from stakeholders or the government. With better support and policies, the development of renewable energy can be carried out well so that renewable energy can be applied as soon as possible, especially in remote/outermost regions [27, 28].

Power generation from biomass has been widely studied and applied in various countries in several applications both for industry and for domestic. Overall testing in several studies of biomass from palm oil mills for power plants shows promising results. The biomass power plant is renewable and sustainable energy because its availability in the plant is also very adequate to meet the energy in the future. The focus of this research is to apply and utilize renewable electricity generation energy from palm oil biomass for rural/remote areas in Aceh Province. Besides, analyses were also carried out on the temperature, emissions, and efficiency of the palm oil biomass combustion. This study, conducted experiments using four types of palm oil biomass such as empty fruit bunches (EFB), oil palm midrib (OPM), oil palm fiber (OPF), and palm kernel shell (PKS). This experiment was carried out in three stages, namely; the first stage is burning with 1 kg, 2 kg, and 3 kg of biomass. Each operation was carried out with 100% open valve, 75% and 50% with 4 different parameters.

2. Material and Setup

This study conducted experiments with biomass material from palm oil mills such as EFB, OPM, OPF, and PKS shown in figure 2. The biomass used in this experiment was 1 kg, 2 kg, and 3.5 kg, respectively. Each biomass material was treated four times with three different valve openings (50%, 75%, and 100%). While the chemical composition of each palm oil biomass is shown in table 1. Schematic diagrams in this experiment shown in figure 3. While the analysis tools used in this study include Thermocouple, Flue Gas DWYER 1207A, and Atomic Absorption Spectroscopy.

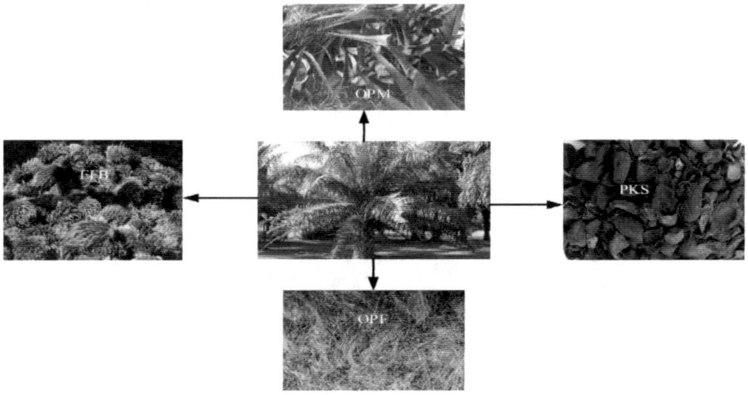

Figure 2. Material for palm oil biomass.

Table 1. Composition of palm oil biomass compounds [29].

Categories	EFB	OPM	PKS	OPF
Carbon (C)	46.50 %	44.97 %	45.74 %	50,27 %
Oxygen (O)	39.89 %	43.58 %	47.19 %	38,28 %
Nitrogen (N)	0.89 %	0.45 %	0.25 %	0,42 %
Hydrogen (H)	7.13 %	6.99 %	5.54 %	7,07 %
Total Sulphur (S)	0.21 %	0.14 %	0.09 %	0,63 %
Calorific Value (CV)	4,469 Cal/gr	4,278 Cal/gr	4,515 Cal/gr	-
Volatile Matter (VM)	68.47 %	71.47 %	69.95 %	75,99 %
Ash (A)	5.38 %	3.87 %	1.19 %	5,33 %

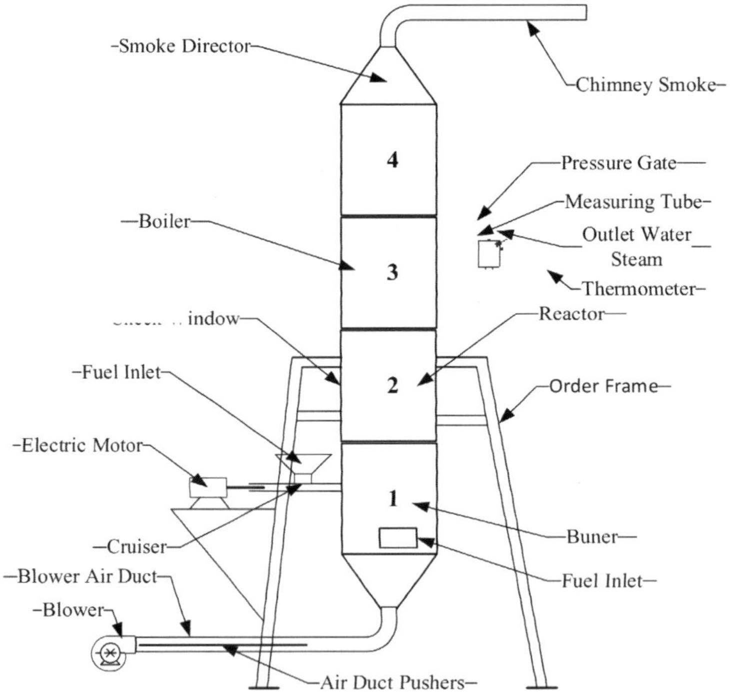

Figure 3. A fluidized-bed combustor (FBC) for biomass combustion.

3. Result and discussion

3.1. Profile of temperature

This study focuses on the burning characteristics of palm oil biomass in a fluidized-bed combustion chamber. Four types of biomass weighing 1 kg, 2 kg and 3 kg respectively tested on valves 50%, 75%, and 100% with four trials on each biomass used. The focus of this experiment is to analyze the combustion temperature, efficiency, O_2, and CO_2 coming out of the fluidized-bed combustion chamber. Airflow rates varying from 0.625, 0.9375, and 1.25 m^3/s are used as variations of the combustion time. Temperature for four experiments when the valve opened 100% ranged from 722 °C to 943 °C. A series of tests and temperatures in four operations with 100% open valves and the use of biomass to investigate combustion behavior shown in table 2.

Table 2. Operating conditions and combustion behavior of palm oil biomass at valve opening of 100%.

Fuel Type	Total of using Biomass (kg)	Maximum of Temperature (°C)	Time of Burning (s)
	1	893	53
EFB	2	905	42
	3	943	53
	1	838	66
OPM	2	873	40
	3	830	56
	1	766	52
PKS	2	804	40
	3	808	51
	1	722	49
OPF	2	762	38
	3	851	48

Figure 4 (a) - (d) is a temperature profile shown from measurements taken from four experiments when the valve is open 100%, (a) temperature combustor (TC), (b) freeboard temperature 1 (TF1), (c) freeboard temperature 2 (TF2) and (c) freeboard temperature 3 (TF3). This observation was carried out for all four cases with similar biomass. The combustor temperature gradually decreased by 1 kg of biomass from the four trials conducted. As for biomass 2 kg and 3 kg at 200 seconds of combustion time until the end of combustion shows normal temperature. The best temperature produced 3 kg of biomass for all four trials with 100% open valve. The maximum temperature of the four tests with 100% open valve recorded at TC was 893 °C, 905 °C, and 943 °C for 1 kg, 2 kg, and 3 kg biomass, respectively. The minimum temperature recorded on TF3 is 722 °C and 762 °C. However, the 3 kg biomass at TF3 851 °C showed higher results than TF2 and TF 1 for biomass 1 and 23 kg, which were 838 °C and 830 °C, respectively. The results of testing with biomass 1 kg, 2 kg, and 3 kg with four trials on open valves 100% higher than the results of tests conducted by [29]. The maximum temperature achieved in their study was between 650 °C and 680 °C with a burning time of 120 minutes. As for combustion in the test, they did use palm oil stone material in a pilot-scale bed reactor.

Subsequent experiments were carried out on 75% open valves with similar fuel in testing for 100% open valves. The results of trials conducted at this stage shown in table 3. Figure 5 (a) - (d) shows the temperature profile from the measurement results with a 75% open valve in the fluidized-bed combustion chamber. This observation was carried out from the first to the fourth experiment with a time between 10 to 60 seconds. The 75% open valve shows that the temperature gradually decreases for all biomass used. These were very different in testing with 100% open valves. However, a more significant decrease was recorded in 1 kg biomass compared to 2 and 3 kg biomass at the same time. Besides, the time taken for the 75% valve is also longer than the test when the valve is 100% open.

The reduction in combustion temperature in this test arranged from 250 seconds to the end of combustion, while for the open valve 100% the decline began at 200 seconds until the end of burning for 1 kg of biomass. Maximum temperature in the experiment was recorded in biomass 2 kg and 3 kg of TC 899 °C and 878 °C, respectively.

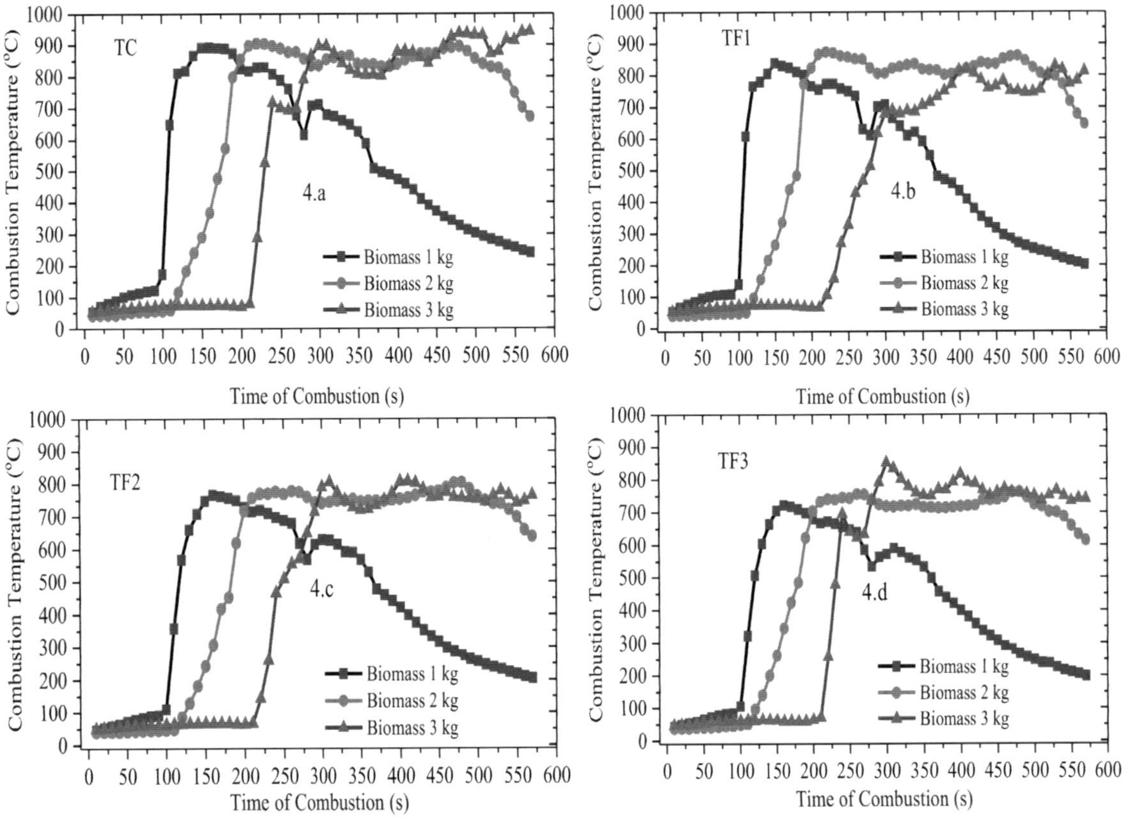

Figure 4. Combustion temperature for different biomass at valve opening of 100% (4.a) Measurement point one (4.b) Measurement point two (4.c) Measurement point tree (4.d) Measurement point four in the FBC.

Meanwhile, for 1 kg of biomass on TC, the temperature slightly decreased compared to 2 kg and 3 kg of biomass. The lowest maximum value for the four trials in 75% open valve recorded 1 kg of biomass for TF2 and TF3 of 685 °C and 689 °C, respectively. The results of testing by burning biomass 1 kg, 2 kg, and 3 kg at 75% open valve maximum temperature showed lower by 6 °C and 65 °C for biomass 2 and 3 kg at TC. The results shown in figure 4 are still higher than the results of tests conducted by [30]. The maximum temperature recorded in their trials between 650 °C and 680 °C with a burning time of 120 minutes.

Table 3. Operating conditions and combustion behavior of palm oil biomass at valve opening of 75%.

Fuel Type	Total of using Biomass (kg)	Maximum of Temperature (°C)	Time of Burning (s)
	1	764	74
EFB	2	899	50
	3	878	33
	1	748	52
OPM	2	844	49
	3	844	30
	1	685	51
PKS	2	783	46
	3	784	33
	1	689	69
OPF	2	739	49
	3	741	33

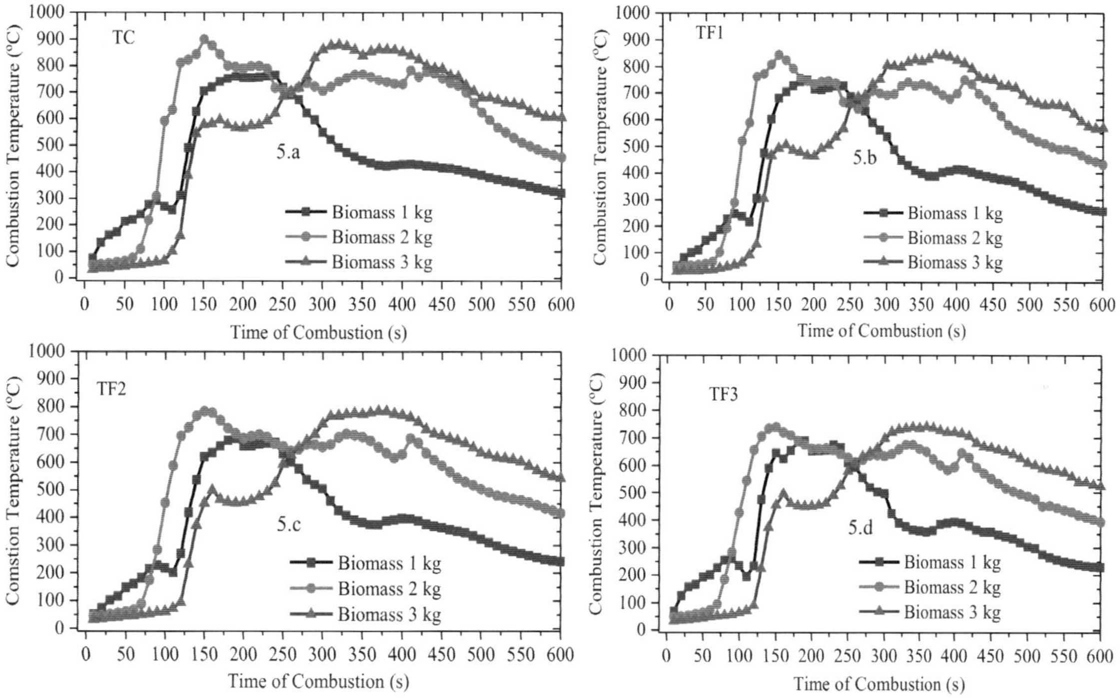

Figure 5. Combustion temperature for different biomass at valve opening of 75% (5.a) Measurement point one (5.b) Measurement point two (5.c) Measurement point tree (5.d) Measurement point four in the FBC.

Table 4. Operating conditions and combustion behavior of palm oil biomass at valve opening of 50%.

Fuel Type	Total of using Biomass (kg)	Maximum of Temperature (°C)	Time of Burning (s)
	1	682	44
EFB	2	840	43
	3	780	48
	1	649	44
OPM	2	799	44
	3	738	45
	1	619	40
PKS	2	718	40
	3	680	42
	1	604	39
OPF	2	675	38
	3	665	44

Figure 6 (a) - (d) combustion temperature profile in a 50% open valve with similar biomass as 100% and 75% open valve. However, the combustion time carried out in this trial is longer than in the two previous tests. The combustion time on the valve opens 50% for 830 seconds compared to 570 and 600 seconds in the first and second tests. As in the first and second tests, in this experiment, the combustion temperature was analyzed based on the time and use of biomass 1 kg, 2 kg, and 3 kg. The highest temperature recorded on the 2 kg biomass for all experiments carried out with detailed results shown in figure 5. The maximum temperature for four experiments was found at 2 kg biomass with each of 840 °C (TC), 799 °C (TF1), 718 °C (TF2) and 675 °C (TF3). While the minimum temperature was recorded at 1 kg of biomass. The combustion temperature profile on the 50% open valve shows that biomass for 1 kg and 2 kg decreases gradually at an average time of 300 minutes compared to the 100% and 75% open valve. While 3 kg of biomass shows a better trend until the burning time is over. These show that the smaller the valve is open during combustion, the more time needed. Besides, the resulting temperature decreases compared to the wider open valve. The results of testing four experiments with similar biomass on the open valve 50% of the combustion temperature produced higher test results by [30]. While in their study, trials were carried out eight times compared to four experiments in this work. While a series of experimental results on 50% open valves with four experiments performed shown in table 4. Burning biomass in a fluidized bed to study the temperature and emissions of combustion has also been studied by several previous researchers [31, 32]. Their results show that NOx emissions were reduced by 30% and significant reductions were noted in CO emissions. However, their research combustion using biomass from non-wood like (straw, miscanthus, and peanuts).

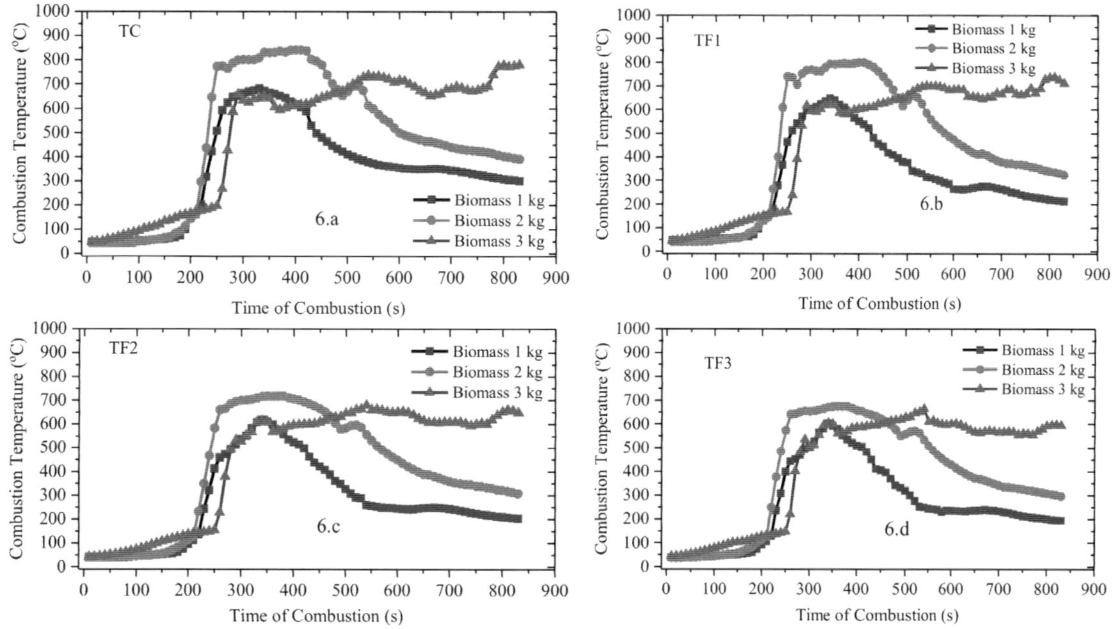

Figure 6. Combustion temperature for different biomass at valve opening of 50% (6.a) Measurement point one (6.b) Measurement point two (6.c) Measurement point tree (6.d) Measurement point four in the FBC.

3.2. Gas emissions

Unreacted O_2 concentrations show significant differences between 1 kg, 2 kg, and 3 kg palm oil biomass for 50%, 75%, and 100% open valves. The highest O_2 release recorded at 50% open valve is 20.4%, 20.3%, and 18.9% respectively shown in figure 7. While the lowest maximum value is observed at 2 kg of biomass at 100% open valve at 16.1%. Furthermore, followed by 3 kg of biomass on 75% 0.4% open valve and 0.7% for 50% open valve also obtained 3 kg of biomass. The experiments conducted in this article are part of a trial for electricity generation in rural/outside areas in Aceh Province. Where the design of tools made for this experiment can apply and utilize. A series of trial results show that the biomass power plant is very suitable for use in remote areas with adequate biomass availability. This biomass electricity is also beneficial to the environment and the cost required is not too expensive.

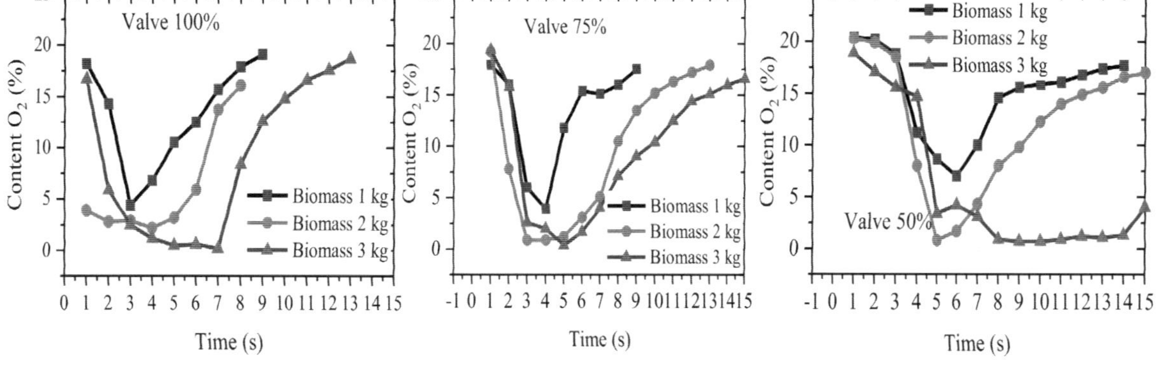

Figure 7. O_2 emissions for different valve opening and materials of biomass.

Figure 8 is the CO_2 emissions from the combustion analysis results with biomass from a coconut factory tested with three trials and three different types of biomass. CO_2 emissions show a significant reduction at 4 and 5 minutes. However, the difference indicated for 3 kg biomass at 50% open valve. The results obtained increased significantly from the beginning to the end of combustion compared with biomass 1 kg and 2 kg. The minimum value of CO_2 from the three parameters with different biomass recorded in the starting valve 100% for biomass 1 kg and biomass 1 kg and 2 kg respectively 2.9% for 75% open valve. Whereas the maximum CO_2 emissions for the three parameters found at 3 kg of biomass 19.9% in the 100% open valve compared to the biomass and other open valves in this test.

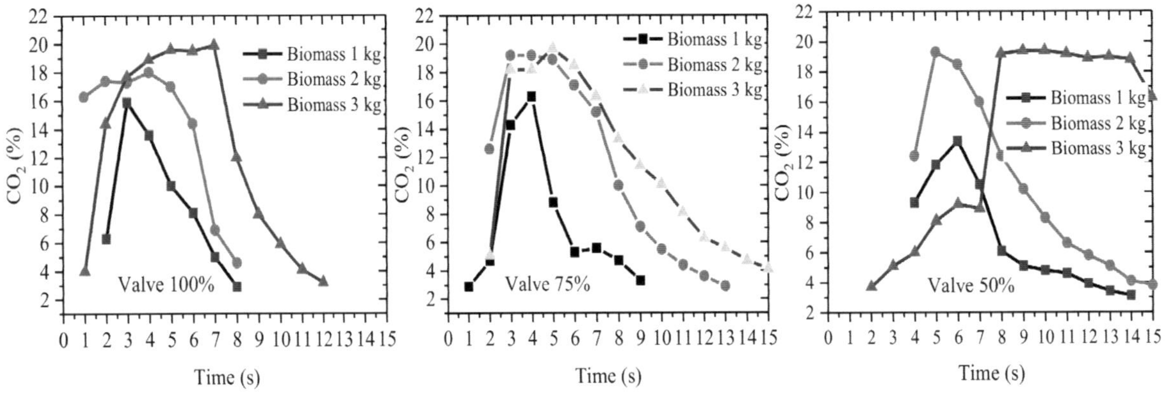

Figure 8. CO_2 emissions for different valve opening and materials of biomass.

3.3. Combustion efficiency

Finally, the analysis in this experiment carried out on the efficiency of combustion of biomass in open valves 100%, 75%, and 50%. Where the results of combustion efficiency in this test, as shown in figure 9. The maximum combustion efficiency in this trial was recorded at 1 kg biomass for 50% open valve compared to 100% 91.7% open valve and 75% for 94.5% for 3 kg biomass. Combustion efficiency shows results following the temperature of the burner. Where the increased combustion temperature shows better efficiency, as shown in figure 8.

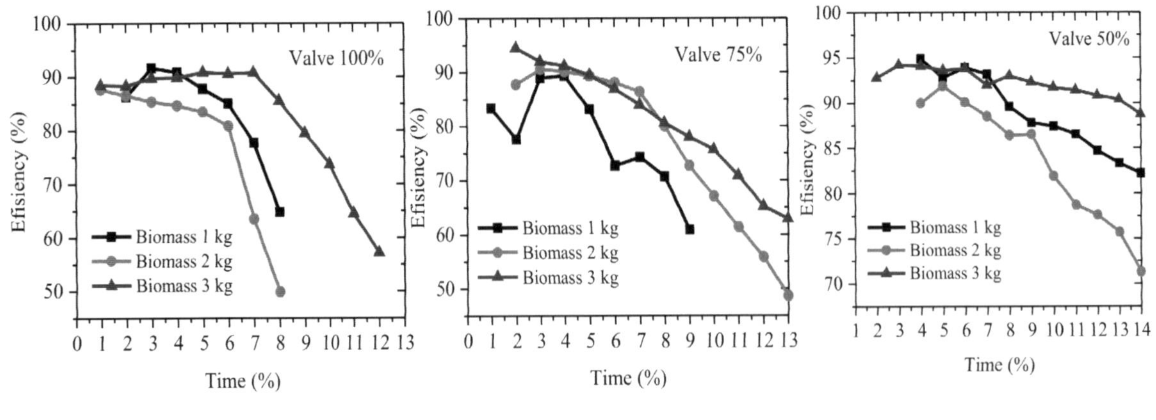

Figure 9. Efficiency for different valve opening and materials of biomass.

3.4. Ash Deposit

Ash deposit is the residual ash from the burning of palm oil biomass contained in this study is a mixture of EFB, OPM, PKS, and OPF. Ash deposit generated in this study taken in the combustor fluidized-bed combustor. The burning ash of palm oil biomass is one of the biomasses from the plantation sector with an abundant amount. This palm oil waste can be obtained at a little cost in a sustainable and environmentally friendly manner. Oil palm waste has been thrown away for nothing and without proper management so that its existence has caused various environmental problems and endangered human health risks [33]. The ash deposit investigation aims to determine the composition of chemical compounds found in the residual combustion of palm oil biomass and then compared them with chemical compounds in ash from the burning of coal biomass. The structure of the ash deposit chemical compounds from burning palm oil biomass in this study is shown in table 5.

Table 5. Composition of chemical compounds in palm biomass ash deposits.

No	Chemical Compounds	Total	Unit
1	Al_2SO_3	4,710	%
2	SiO_2	30,860	%
3	K_2O	1,910	%
4	MgO	2,54	%
5	CaO	10,55	%
6	Iron (Fe)	0,63	%
7	Natrium (Na)	0,11	%
8	Cadmium (Cd)	< 0,0004	mg/kg
9	Copper (Cu)	151,5964	mg/kg
10	Nickel (Ni)	15,1967	mg/kg

The combustion results of palm biomass contain enough chemical compounds that can use for various other functions. One of them is as an inexpensive alternative fertilizer, this is because the ash from the burning of palm oil biomass contains nutrients needed by plants such as K, Ca, and Mg [33]. The ash deposit composition obtained from palm oil biomass contained in chemical compounds is similar to the results of coal combustion. The utilization of the content used as a parameter for additives in mixing for the manufacture of concrete such as SiO_2 content. The ash results from burning palm oil biomass. The SiO_2 content contained in coal reaches 30-50%. The residual ash from the combustion of palm oil biomass produced in this study contained 30.86% SiO_2. This content can use as a mixture for making concrete because SiO_2 contained in the ash from the burning of palm oil biomass is the primary raw material for cement [34]. Besides, ash from the burning of palm oil biomass can use as an air filter in motor vehicles. Because with this ash content of palm oil biomass can save fuel consumption up to 10% [36]. Combustion ash from palm oil biomass can also use as an additional composition for the manufacture of polystyrene polymers. This is because it has better power than polystyrene in general [33]. Besides, the ash from combustion can also be used as a mixture in the mortar (cement, sand, and water) to strengthen the compressed concrete [37].

4. Conclusion

The thermochemical characteristics of biomass fuels from oil palm mills were evaluated using a fluidized bed combined with various testing parameters, including the effects of temperature, emissions, and combustion efficiency. The main experimental results from this study are Combustion of palm oil biomass in a variety of open valves against the effects of temperature, emissions and combustion efficiency in fluidized-bed has been successfully carried out. The combustion temperature for all types of biomass used decreases when the valve is open 50% and vice versa has an increase in the 100% open valve. The maximum O_2 emission gas measured in this study reached 19.9% at 3 kg biomass for 100% open valves while the value of the minimum O_2 emissions gas recorded at 1 kg of biomass 2.9% for 100% open valves and biomass 1 and 2 kilograms of 75% open valves respectively at 2.9%. The maximum CO_2 emissions were recorded at 1 kg of biomass for 50% open valves by 94.9% and a minimum of 87.7% for 2 kg of biomass at 100% open valves. Combustion temperature maximum found in 3 kg biomass for open valves 100% 943 °C, while the minimum combustion temperature of 604 °C on the open valve is 50% for 1 kg of biomass.

Acknowledgments
This research is supported by PNBP Universitas Syiah Kuala, Research institutions, and community service with the contract number of (20/UN11.2.1/PT.01.03/PNBP/2021).

References

[1] WHO 2019 Air pollution causes 7 million deaths per year in the world World Health Organization.
[2] WHO 2019 [Online] Available: https://databoks.katadata.co.id/datapublish/2019/06/07/polusi-udara-sebabkan-7-juta-kematian-per-tahun-di-dunia.
[3] Erdiwansyah, Mamat R, Sani M S M and Sudhakar K 2019 Renewable energy in Southeast Asia: Policies and recommendations *Sci. Total Environ* **670** 1095–102
[4] IRENA and ACE 2016 Renewable Energy Outlook for ASEAN *Renewable Energy Outlook for ASEAN*
[5] Erdiwansyah, Mahidin, Mamat R, Sani M S M, Khoerunnisa F and Kadarohman A 2019 Target and demand for renewable energy across 10 ASEAN countries by 2040 *Electr. J* **10** 106670
[6] Asfar J Al, AlShwawra A, Shaban N A, Alrbai M, Qawasmeh B R, Sakhrieh A, Hamdan M A and Odeh O 2020 Thermodynamic analysis of a biomass-fired lab-scale power plant *Energy* **194** 116843
[7] Hani M R, Mahidin, Husin H, Khairil, Hamdani, Erdiwansyah and Mel M 2020 Experimental Studies on Combustion Characteristics of OilPalm Biomass in Fluidized-Bed: A Heat Energy Alternative *Journal of Advanced Research in Fluid Mechanics and Thermal Sciences* **68** 9–28
[8] Cheng W, Zhang Y and Wang P 2020 Effect of spatial distribution and number of raw material collection locations on the transportation costs of biomass thermal power plants *Sustain. Cities Soc* **55** 102040
[9] Suresh N S, Thirumalai N C and Dasappa S 2019 Modeling and analysis of solar thermal and biomass hybrid power plants *Appl. Therm. Eng* **160** 114–21
[10] Yan L, Wang Z, Cao Y and He B 2020 Comparative evaluation of two biomass direct-fired power plants with carbon capture and sequestration *Renew. Energy* **147** 1188–98
[11] Smith J S, Safferman S I and Saffron C M 2019 Development and application of a decision support tool for biomass co-firing in existing coal-fired power plants *J. Clean. Prod* **236** 117–375
[12] Othman M R, Martunus, Zakaria R and Fernando W J N 2009 Strategic planning on carbon capture from coal fired plants in Malaysia and Indonesia: A review *Energy Policy* **37** 1718–35
[13] Mahidin, Saifullah, Erdiwansyah, Hamdani, Hisbullah, Hayati A P, M.Zhafran, Sidiq M A, Rinaldi A, Fitria B, Tarisma R and Y.Bindar 2020 Analysis of power from palm oil solid waste for biomass power plants: A case study in Aceh Province *Chemosphere* **253** 126714
[14] Jayed M H, Masjuki H H, Kalam M A, Mahlia T M I, Husnawan M and Liaquat A M 2011 Prospects of dedicated biodiesel engine vehicles in Malaysia and Indonesia *Renew Sustain. Energy Rev* **15** 220–35
[15] Wu Q, Qiang T C, Zeng G, Zhang H, Huang Y and Wang Y 2017 Sustainable and renewable energy from biomass wastes in palm oil industry: A case study in Malaysia *Int. J. Hydrogen Energy* **37** 23871–7
[16] Khalil M, Berawi M A, Heryanto R and Rizalie A 2019 Waste to energy technology: The potential of sustainable biogas production from animal waste in Indonesia *Renew. Sustain. Energy Rev* **105** 323–31

[17] Hamzah N, Tokimatsu K and Yoshikawa K 2017 Prospective for power generation of solid fuel from hydrothermal treatment of biomass and waste in Malaysia *Energy Procedia* **142** 369–73
[18] Malek A B M A, Hasanuzzaman M, Rahim N A and Turki Y A Al 2017 Techno-economic analysis and environmental impact assessment of a 10 MW biomass-based power plant in Malaysia *J. Clean. Prod* **141** 502–13
[19] Umar M S, Urmee T and Jennings P 2018 A policy framework and industry roadmap model for sustainable oil palm biomass electricity generation in Malaysia *Renew. Energy* **128** 275–84
[20] Sukiran M A, Abnisa F, Wan Daud W M A, Abu Bakar N and Loh S K 2017 A review of torrefaction of oil palm solid wastes for biofuel production *Energy Convers. Manag* **149** 101–20
[21] Singh R and Setiawan A D 2013 Biomass energy policies and strategies: Harvesting potential in India and Indonesia *Renew. Sustain. Energy Rev* **22** 332–45
[22] Simangunsong B C H, Sitanggang V J, Manurung E G T, Rahmadi A, Moore G A, Aye L and Tambunan A H 2017 Potential forest biomass resource as feedstock for bioenergy and its economic value in Indonesia *For. Policy Econ* **81** 10–7
[23] Putrasari, Praptijanto Y, Santoso A, Lim W B and Ocktaeck 2016 Resources, policy, and research activities of biofuel in Indonesia: A review *Energy Reports* **2** 237–45
[24] You S, Tong H, Armin-Hoiland J, Tong Y W and Wang C H 2017 Techno-economic and greenhouse gas savings assessment of decentralized biomass gasification for electrifying the rural areas of Indonesia *Appl. Energy* **208** 495–510
[25] "Palm Oil Dominates Plantation Sector in Aceh," 2019 [Online] Available: https://minanews.net/kelapa-sawit-dominasi-sektor-perkebunan-di-aceh/
[26] Budiarto R 2009 Potensi Energi Limbah Pabrik Kelapa Sawit *Jur. Tek. Fis. Fak. Tek. Gadjah Mada* **325** 1–6.
[27] Kraal D 2019 Petroleum industry tax incentives and energy policy implications: A comparison between Australia, Malaysia, Indonesia and Papua New Guinea *Energy Policy* **126** 212–22
[28] Braadbaart O and Braadbaart F 1997 Policing the urban pumping race: Industrial groundwater overexploitation in Indonesia *World Dev* **25** 199–210
[29] Hani M H, Mahidin, Husin H, Hamdani and Khairil 2018 Oil palm biomass utilization as an energy source and its possibility use for polygeneration scenarios in Langsa City, Aceh Province, Indonesia *IOP Conf. Ser. Mater. Sci. Eng* **334** 12003
[30] Razuan R, Chen Q, Finney K N, Russell N V, Sharifi V N and Swithenbank J 2011 Combustion of oil palm stone in a pilot-scale fluidised bed reactor *Fuel Process. Technol* **12** 2219–25
[31] Sher F, Pans M A, Afilaka D T, Sun C and Liu H 2017 Experimental investigation of woody and non-woody biomass combustion in a bubbling fluidised bed combustor focusing on gaseous emissions and temperature profiles *Energy* **141** 2069–80
[32] Ahmad F B, Zhang Z, Doherty W O S and O'Hara I M 2019 The outlook of the production of advanced fuels and chemicals from integrated oil palm biomass biorefinery *Renew. Sustain. Energy Rev* **109** 386–411
[33] Asyri F, Hafni K N and Simamora A H 2015 Pengaruh limbah abu pembakaran biomassa kelapa sawit terhadap sifat-sifat fisika dan mekanik high impact polystyrene *J. Tek. Kim. USU* **4** 23–8
[34] Ekawati I and Purwanto Z 2012 Potensi abu limbah pertanian sebagai sumber alternatif unsur hara kalium, kalsium, dan magnesium untuk menunjang kelestarian produksi tanaman *Semin. Nas. Kedaulatan Pangan dan Energi* **27** 135–9
[35] Mulyati S, Dahlan D and Adril E 2012 Pengaruh persen massa hasil pembakaran serbuk kayu dan ampas tebu pada mortar terhadap sifat mekanik dan sifat fisisnya *J. Ilmu Fis* **4** 31–9
[36] Prayitno H and Wardono H 2014 Pengaruh variasi massa dan temperatur aktivasi fly ash pelet dari cangkang dan serabut kelapa sawit terhadap konsumsi bahan bakar sepeda motor bensin 4 langkah *J. Ilm. Tek. Mesin* **2** 57–63

[37] Umboh A H 2014 Pengaruh pemanfaatan abu terbang (fly ash) dari PLTU II sulawesi utara sebagai substitusi parsial semen terhadap kuat tekan beton *J. Sipil Statik* **7** 352–8

IOP Conf. Series: Materials Science and Engineering 1195 (2021) 012006 doi:10.1088/1757-899X/1195/1/012006

Prediction of solvation properties of low transition temperature mixtures (LTTMs) using COSMO-RS and NMR approach

N R Yusuf[1], S Yusup[1,2], C L Yiin[3], P J Ratri[4], A A Halim[1] and N A Razak[1]

[1] HICoE- Centre for Biofuel and Biochemical Research (CBBR), Institute of Self-Sustainable Building, Universiti Teknologi PETRONAS (UTP), 32610, Bandar Seri Iskandar, Perak, Malaysia
[2] Department of Chemical Engineering, Universiti Teknologi PETRONAS (UTP), 32610, Bandar Seri Iskandar, Perak, Malaysia
[3] Department of Chemical Engineering and Energy Sustainability, Faculty of Engineering, Universiti Malaysia Sarawak (UNIMAS), Kota Samarahan 94300, Sarawak, Malaysia
[4] Chemistry Department, Universitas Pertamina, Jl Teuku Nyak Arief, Simprug, Jakarta Selatan, 12220, Indonesia

Email: drsuzana_yusuf@utp.edu.my

Abstract. The concept of sustainable and green solvent has always highlighted in the field of energy and environmental science. The synthesis and application of natural-based Low Transition Temperature Mixture (LTTM) as a novel and green solvent for the lignocellulose biomass pre-treatment such as delignification of Oil-Palm Empty Fruit Bunch (EFB) have been greatly emphasized. In this present work, the investigation of LTTM efficiency as green solvent in delignification process was conducted using both theoretical and experimental studies. Initially, screening of solvation properties of different types of hydrogen bond acceptor (HBA) and predicted hydrogen bond donor (HBD) for synthesis of LTTMs was conducted using conductor-like screening model (COSMO-RS) software and formation of hydrogen bonding was evidenced using NMR spectroscopy analysis. Three types of HBA namely sucrose, choline chloride and monosodium glutamate were mixed with malic acids as HBD and their charge density distribution on the surface was determined through sigma profile (σ). The COSMO-RS results determined the σ profile of pure component malic acid to be 11.42, sucrose to be 25.37 and the total value of σ profile for mixtures is 14.19 as the best combination of LTTM composition compared to LTTM from choline chloride and monosodium glutamate (MSG). The reliability of the COSMO-RS predictions data was correlated with Nuclear Magnetic Resonance (NMR) analysis through determination of peaks with chemical shifts hydrogen bonding that suggested existence of potential interaction between malic acids and sucrose has occurred.

1. Introduction

The dependency of fossil fuels leads to the environment problems such as environmental contamination and global heating. Thus, most researchers are looking for bio-based materials that can offer renewable, biodegradable, and biocompatible sources as alternative to fossil based materials. According to researchers Ong & Wu [1] Zhang *et al.* [2], Zakzeski *et al.* [3] and others, lignocellulosic biomass as feedstock for bio-chemicals and fuels conversion as the alternative to fossil feedstocks increasing yearly. Malaysia is one of the countries that can capitalize the renewable energy resources based on resources availability of the biomass generated from the plantations and the mills. Malaysia generated at least 168 million tons yearly from their biomass, including timber and oil palm waste, rice husk, coconut trunk fibers, municipal waste and sugarcane bagasse [4–6]. Among all this lignocellulosic biomass, oil palm has the biggest contribution towards Malaysia's economy. Only 10% of the total palm oil can be recovered. Another 90% biomass such as oil palm trunk (OPT), oil palm fronds (OPF), mesocarp fiber (MF), palm oil mill effluent (POME) and empty fruit bunch (EFB) is considered as wastes. Among this palm oil biomass, based on researchers data [7,8], empty fruit bunch (EFB) is one of the most abundant lignocellulosic biomass.

Lignocellulosic biomass attracted wood industry as an alternative fiber as it is most abundant feedstock worldwide and they contain of 35-50% cellulose, 35% hemicelluloses, 5-30% lignin [9,10]. Researchers also showed growing interest in lignocellulosic biomass conversion which contributed to successful production in renewable chemicals and biorefining technologies. Hence, an efficient conversion technology is crucial to produce high conversion bio-chemicals. The technologies involved chemical or enzymatic treatments which required high temperature and pressure are judged as low efficiency and expensive. Delignification process is one of conversion method that hydrolyzed the hardwood. Delignification efficiency depends on variables namely reaction temperature and time. The bonding of lignin usually broken down under lower temperature. Higher temperature should be avoided as it will reduce cellulose purity through the byproduct formation. There are few solvent used in the delignification process such as ionic liquid, deep eutectic solvent (DES) and Natural deep eutectic solvents (NADES) [11,12]. Natural deep eutectic solvent (NADES) is bio-derived deep eutectic solvent which comprised of two or more compounds such as organic acid, sugar, alcohols, amines, and amino acid. The deep eutectic identified as a mixture of solid compound transformed to liquids that have lower melting point than room temperature. These unique properties resulted NADES as alternative solvent in many applications as conventional solvents is generally avoided because of toxicity issue [13]. On the other hand, a new approach was introduced which is low-transition-temperature-mixtures (LTTMs) solvent. LTTMs helps to break lignocellulosic structure and lead to the dissolution effectively. Compared to NADES, LTTMs showed higher biocompatibility, low toxicity, lower cost, and it requires less time to prepare. Other than that, the delignification process is viable with LTTMs as its existing hydrogen-bond can disrupt the lignocellulose chain of lignin which contributed by formation hydrogen bond between LTTMs and biomass [14,15].

To access the solvation properties and potential of hydrogen bond of LTTMs, the effect of LTTMs components in prediction potential hydrogen bond donor (HBD) and hydrogen bond acceptor (HBA) was evaluated using solvation-thermodynamics molecular simulation. COSMO-RS is one of solvation simulation model which can be used as screening model of molecular interaction among pure components and components in mixtures [16–18]. Thus, in this present work, COSMO-RS simulation was employed to compute the value of the σ profile of monosodium glutamate, sucrose, and choline chloride with malic acids. Reliability of COSMO-RS predictions data was correlated with Nuclear Magnetic Resonance (NMR) analysis to determine the formation of hydrogen bonding in LTTM as well as to prove rationalization of solvation behavior. In this sense, screening of the most ideal HBD and HBA, for synthesis of LTTMs through COSMO-RS simulation and correlation with NMR analysis are important to discover the efficiency of natural LTTMs for biomass delignification.

2. Methodology

2.1. COSMO-RS computation model

COSMO-RS computational (Amsterdam Modeling Suite 2019.3) was calculated the chemical potentials of the molecules within liquid solutions or to predict the solvation properties of pure compound and mixed fluids. The COSMO-RS data estimates other important thermodynamic parameters, such as solubility or activity coefficient. In this study, sets of LTTMs (malic acid:sucrose:water, choline chloride:sucrose:water and monosodium glutamate:sucrose:water) were used for validation of the model. Compositions and weight ratio of the LTTMs used in the study are shown in table 1.

Table 1. Ratio of the LTTMs component used in the study [19].

HBA	HBD	HBD/HBA Solvent	Weight Ratio
Sucrose	Malic Acid	Water	2:1:1
Choline Chloride	Malic Acid	Water	2:1:1
Monosodium Glutamate	Malic Acid	Water	2:1:1

2.2. Theory and Background
The molecular interaction between pure compound of HBA, HBD and their solvent in table 1 is based on interaction surface segments of the molecules. This interaction surface segment is determined by the surface charge density distribution of the molecules which calculated from Quantum mechanical calculation in the COSMO-RS simulation [20,21]. The created surface segments from the each molecules is indicated by their screening density charge and area. The main parameters in calculating the total interaction energy between the compounds are hydrogen bonding, the electrostatic misfit, and van der Waals interactions. Sigma profile (σ-profile) is surface composition function unit for calculation of screening charge density distribution on the surface [20]. The whole σ- profile system namely $P_S(\sigma)$ can be then described as the sum of the σ-profiles of its individual components weighted by their molar fractions. These calculated by chemical potential of the surface (μ_s) and expressed as equations below.

$$\mu_s = -\frac{RT}{a_{eff}} ln\left[\int P_s(\sigma')exp\left(\frac{a_{eff}}{RT}(\mu_s(\sigma') - e(\sigma,\sigma'))\right)d\sigma'\right] \quad (1)$$

Where,

$$e(\sigma,\sigma') = E_{VdW}(\sigma,\sigma') + E_{HB}(\sigma,\sigma') + \frac{E_{MF}(\sigma,\sigma')}{a_{eff}} \quad (2)$$

Where, E for Elastic modulus, *intramolecular forces and also* dispersive, van der Waals (vdW), d is diameter of material, a_{eff} for effective segment radius, E_{MF} for misfit energy, E_{HB} for hydrogen bonds energy, T for temperature.

2.3. Computational Details
In this work, molecular structure of the pure compounds that include malic acids and sucrose was obtained from the chemical structure in Amsterdam Data Function (ADF) database. However, chemical structure of choline chloride and monosodium glutamate are not available in the ADF library thus musvt be generated using ADF input program. Fast σ profile can be determined for the component that existed in the library by using the chemical abstract service (CAS) and simplified molecular-input line entry (SMILE). These information and structure of the compounds were summarized in table 2.

COSMO-RS estimation were set by Amsterdam Data Function (ADF)-COSMO result files and generated by following computational program. The ADF-Graphical User Interface (ADF-GUI) function was applied for the quantum-chemical calculation to initiate the COSMO files for choline chloride and monosodium glutamate compound. Density functional theory (DFT) is important to optimize the compound molecular geometry at the DF module. Therefore, this optimization applied Beck-Perdew (GGA:BP86) density functional theory and related to triplezeta polarized (TZP). Klamt radii, and Dellysurface area for atoms were used in the calculations until the compound converged [16,17,22]. Subsequently, the obtained geometries of these compounds were further analysed. The molecular surface polarity distributions (σ-profiles), and other thermodynamical properties behaviour can be calculated and obtained at given temperature with reference to the retrieved from COSMO output files.

2.4. LTTM Experimental Methodology

LTTM preparation followed the optimum condition and procedure by Yiin *et al.* [19]. Malic acid, sucrose and water were mixed in a ratio of 1:2:1, respectively in a beaker and stirred. The temperature was set to 60°C while stirring speed was set to 300 rpm and the process continues until mixture become a clear liquid at around 30 minutes. Another two combination of HBDs and HBAs were used in preparation of LTTMs solvent followed same optimum condition. The HBD and HBA compounds used were summarized in table 2.

Table 2. Compositions and molar ratio of the LTTMs used in the study.

Component	CAS	SMILE	Structure
Malic Acid	6915-15-7	C(C(C(=O)O)O)C(=O)O	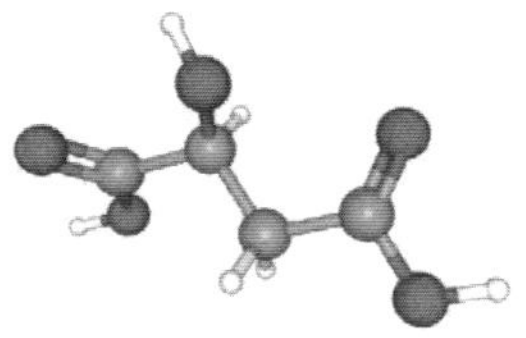 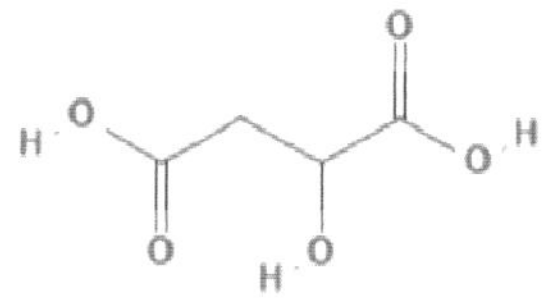
Sucrose	57-50-1	C(C1C(C(C(C(O1)OC2(C(C(C(O2)CO)O)O)CO)O)O)O)O	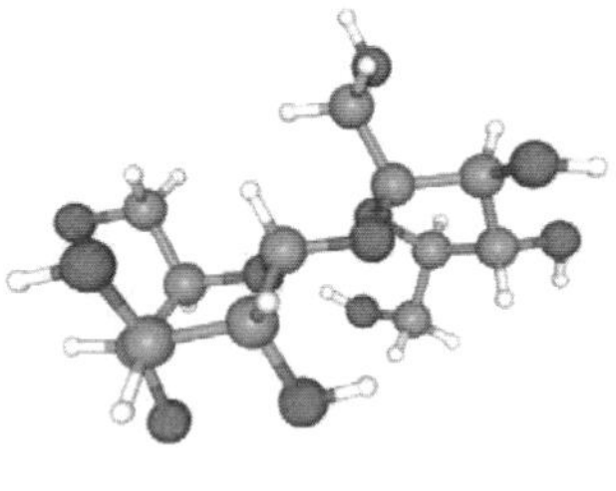 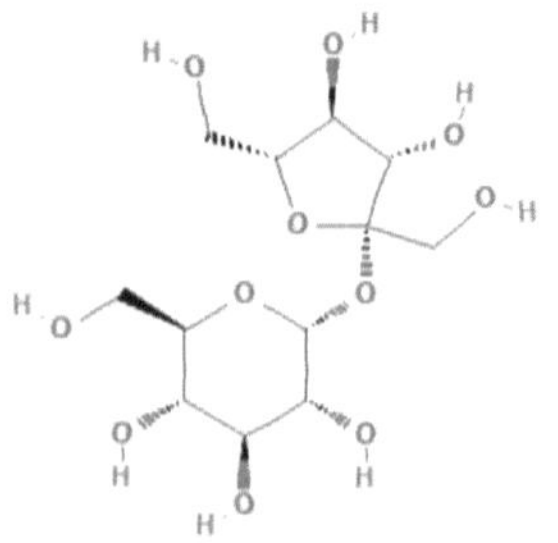

Choline Chloride	67-48-1	C[N+](C)(C)CCO.[Cl-]	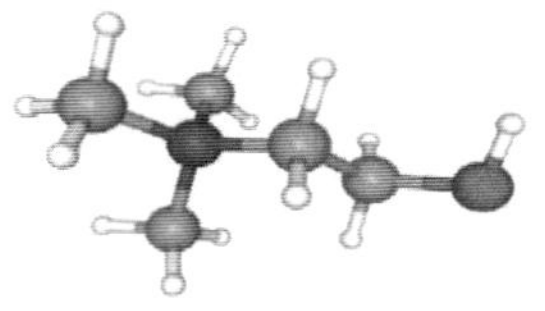 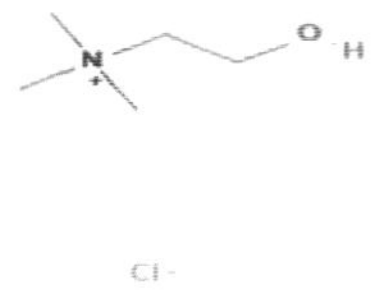
Monosodium Glutamate	142-47-2	C(CC(=O)O)[C@@H](C(=O)[O-])N.[Na+]	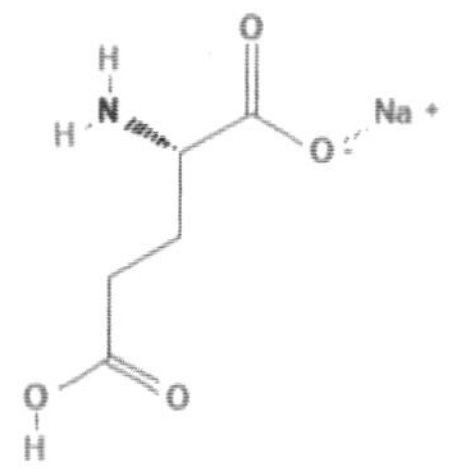 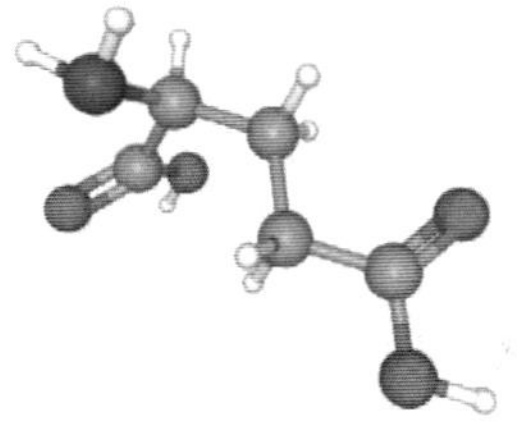

2.5. NMR Spectroscopy
H NMR spectroscopy analysis was performed using a Bruker DRX-500 spectrometer (500 MHz) and TMS as a calibration standard at room temperature. All spectra were recorded using Nuclear Magnetic Resonance Bruker AVANCE 400 MHz III operating at 400.17 MHz. Approximately of the sample is dissolved in 0.6 mL of Toluene-d8 and the dissolved solution was put into a NMR tube with 5-mm diameter. 1H NMR spectra were taken using 7.5-degree pulse angle, 2.5 s recovery delay, and 32 scan times. Measurement conditions were 90-degree pulse angle and 200 scan times.

3. Result and Discussion
In this work, the pure compound and mixture of LTTM was validated through COSMO-RS simulation and NMR analysis. The screening of pure compound and the mixture of LTTMs were performed via thermodynamic properties (i.e., infinite dilution and activity coefficients) to identify the most suitable interaction for solvating. Based on the theory and computational details, the compounds were obtained using Turbomole BP/TZVP. The COSMO-RS charge density was calculated by averaging the results for all cases. Calculation for the charge distribution showed that using different models resulted in different mixture behaviour restive of the activity coefficient at infinite dilution and sigma profiles.

3.1. COSMO-RS Simulation

3.1.1. Activity Coefficients of Pure Compounds. The selection of solutes for synthesis LTTM was based on Activity coefficients unit at infinite dilution. They reflected the interaction between LTTMs materials because the solute–solute interaction can be negligible at infinite dilution. Thus, in this work, activity coefficients of solutes in LTTMs at infinite dilution collected from experimental data from previous study by Yiin [14] is referred to optimize the COSMO-RS model parameters. Thus, regression of the interaction parameters of the component was done. Their precision were based on interaction of computed thermodynamic quantities with measured solubility data. Three different types of mixtures were suggested in this work which involved pure compound of malic acid, choline chloride, monosodium glutamate and sucrose. Table 3 showed the activity coefficient of these compound and the values are close to ideal behaviours in liquid phase [16]. Furthermore, the mixture's behaviours and performance of all compounds of each model was validated against the experimental data through sigma profile and hydrogen bonding of the components.

Table 3. The activity coefficient of LTTM compound from COSMO-RS.

Component	Activity Coefficient, ɣ
Malic Acid	1.340
Sucrose	6.945
Choline Chloride	10.692
Monosodium Glutamate	25.210

3.1.2. Sigma Profile and Hydrogen Bonding of Pure Compounds Coefficients of Pure Compounds. Sigma profile described the charge density distribution (σ-profile) which is a mathematical calculation of molecule's polarity. Thin peaks in the σ-profiles illustrated to low polarity, whereas wide peaks indicate high polarity [16,17,22] as displayed in figure 1. figures 2 (a), (b) and (c) showed σ-profiles of the HBAs and HBDs used in this work. The σ-profile can be segregated into three main regions: (i) the HBD region, positive charge densities (blue surface colour) in the range of $-\infty < \sigma < -0.0082$ [eA^{-2}] correspond to the hydrogen donating surface. (ii) the nonpolar region described almost neutral charge densities and (iii) the HBA region, negative charge densities (red in colour), in the range of $-0.0082 < \sigma < 0.0082$ [eA^{-2}]: correspond to polar regions. Figure 2, 3 and 4 showed the sigma profile of choline chloride, monosodium glutamate and sucrose. Generally, based on the figures shown, all of compound indicated in the HBA region and sucrose showed highest value of sigma profile of 25.373, followed by choline chloride at 20.134 and monosodium glutamate of 12.223. The hydrogen bonding parts of the components are also illustrated in figure 2, 3 and 4. The hydrogen peaks of choline chloride, sucrose and monosodium glutamate also fall in the hydrogen bond acceptor (HBA) region which shows the highest peak is on the right side of the graph. The hydrogen bond part sigma profile for all compounds also determined that sucrose showed the highest value of 11.095 compared to monosodium glutamate of 10.124 and choline chloride of 1.665. Thus, this result indicated sucrose has the higher tendency to accept more hydrogen compared to others. The screening was repeated for the mixtures.

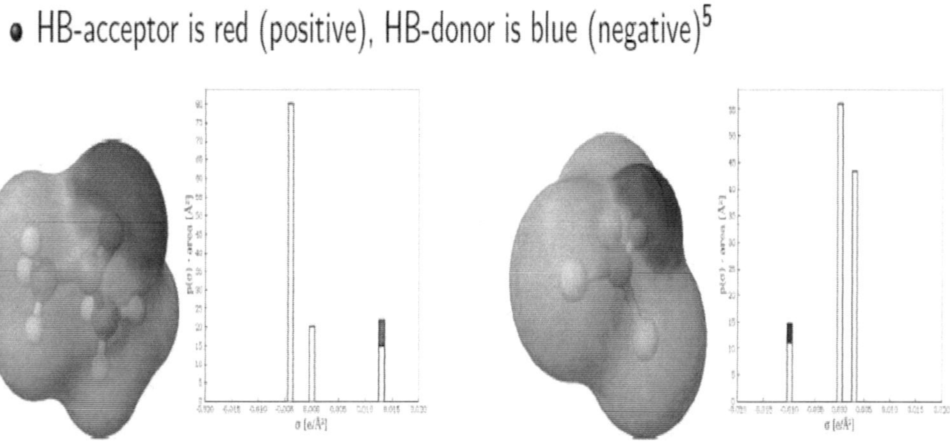

Figure 1. Chemical information processed by the COSMO-RS method [23].

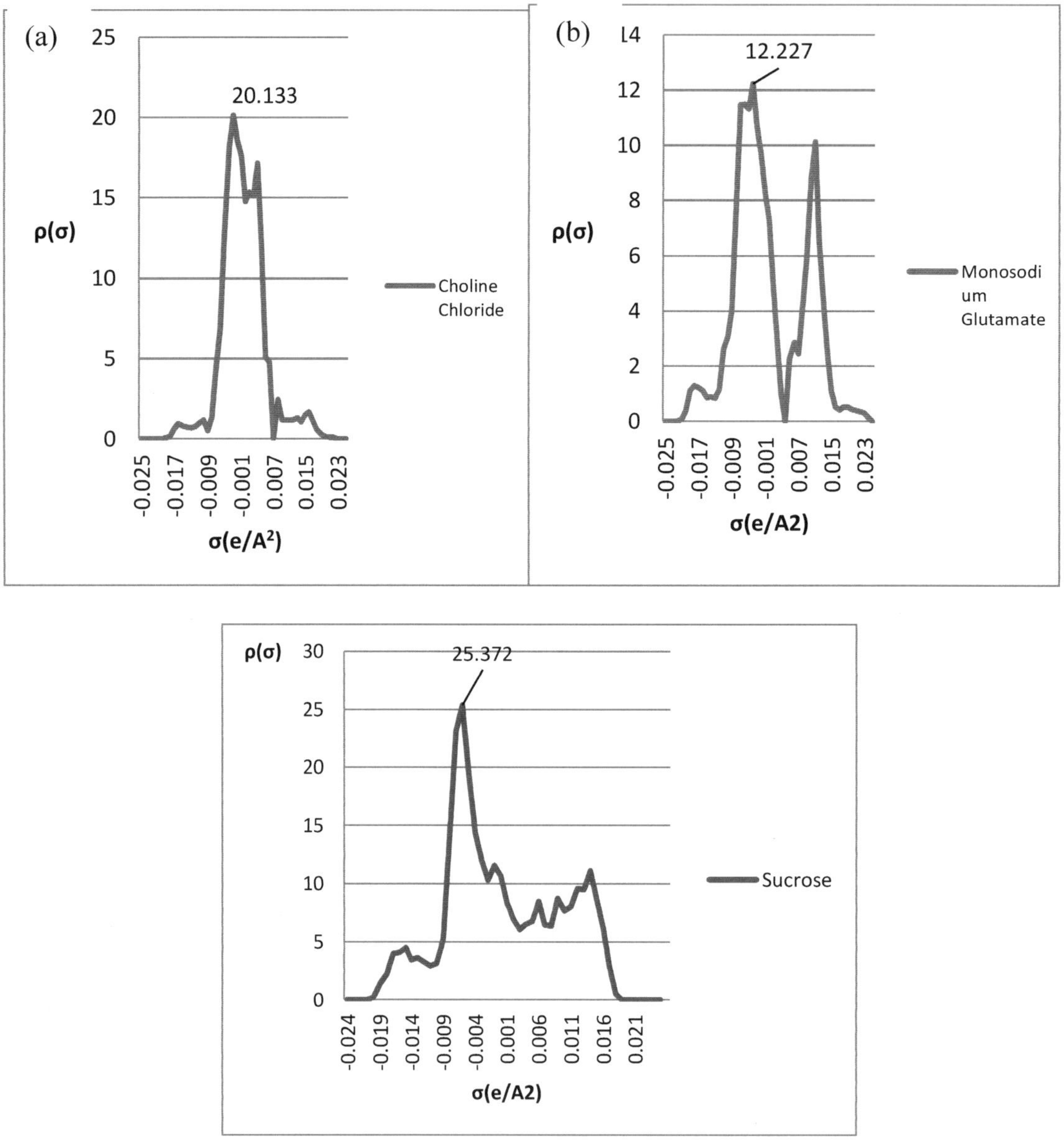

Figure 2. (a) Sigma profile for choline chloride pure component and (b) Sigma profile for monosodium glutamate pure component and (c) Sigma profile for sucrose pure component.

Figure 3 (a), (b) and (c) showed the sigma profile for the HBA of sucrose, monosodium glutamate and and choline chloride with malic acids. Generally, all mixtures showed high sigma profile values which indicated high attraction to H-bond acceptor group and H-bond donor in mixtures. Figure 3 (a) showed positive values of the σ-profiles between the regions σ 0.002 and σ 0.02 e/Å2, indicating the presence of the oxygen atoms within the molecules that responsible for accepting the hydrogen atoms and forming the hydrogen bond of choline chloride. The H-bond donor character which appears from the hydroxyl group of the malic acids was presented by a low intensity peak at −0.01 and −0.019 on their sigma profile. The results correspond with as corresponds with another researcher's finding [24]. Figure 3 (b) showed σ-profiles of monosodium glutamate at region 0.005 and 0.02 e/Å2 and HBD σ-profiles of malic acids as described at -0.01 and -0.022. Figure 3 (c) determined high intensity sigma profile for HBA

(Sucrose) at σ-profiles region of 0.007 and 0.0019 and HBD (malic acids) at region -0.001 and -0.021. This described the high interactions between malic acids and sucrose through formation of hydrogen bonds as compared to others HBA (choline chloride and monosodium glutamate) [16,19,25].

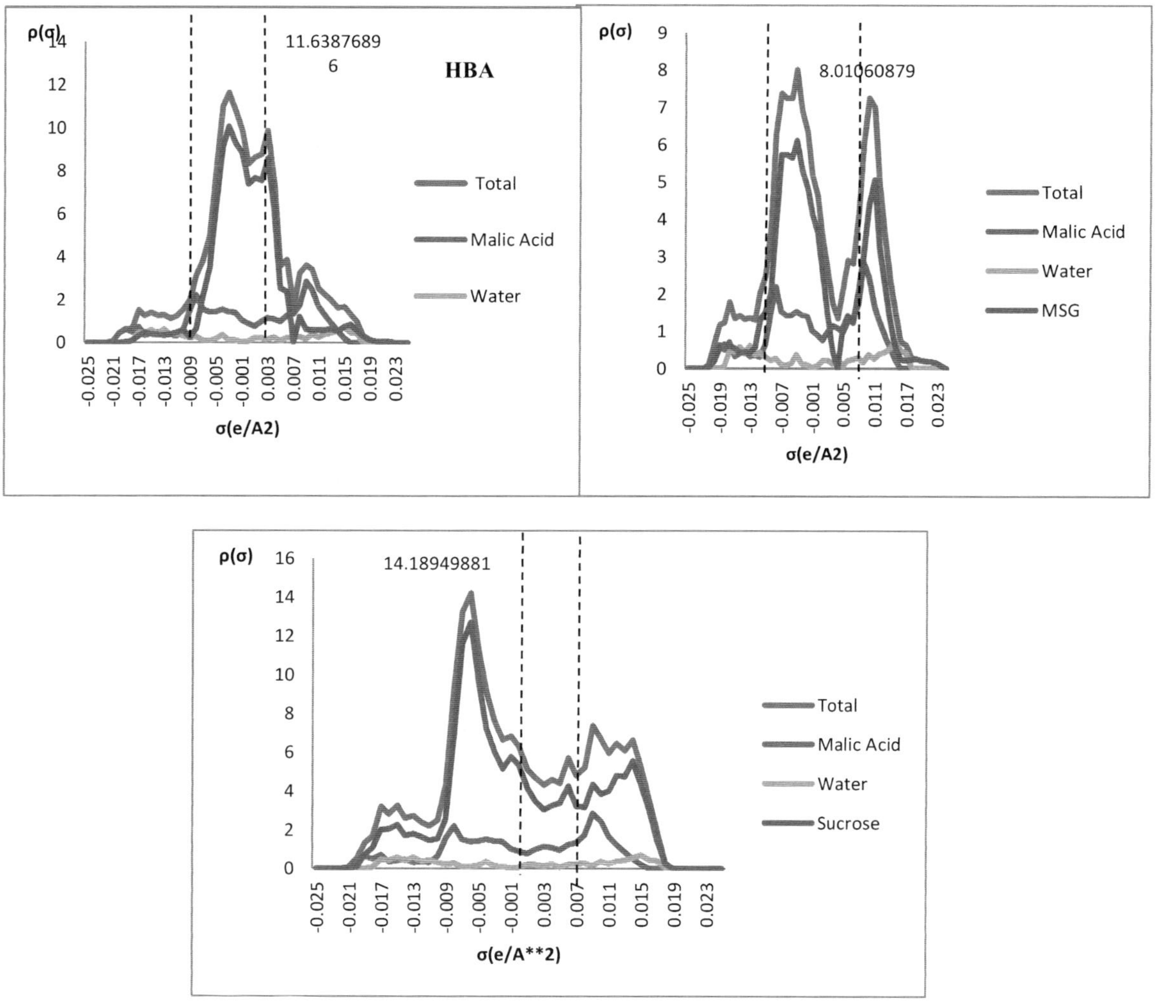

Figure 3. (a) Sigma Profile for Mixture of Malic Acid and Choline Chloride, (b) Sigma Profile for Mixture of Malic Acid and Monosodium Glutamate and (c) Sigma Profile for Mixture of Malic Acid and Sucrose.

3.2. Determination of Hydrogen bonds via NMR Detection

The solvation and interaction between solute and solvent study are a complex process, as they be influenced by several factors. In the view of thermodynamic terms, increasing dissolution tendency illustrated in previous COSMO-RS sigma profile showed stronger hydrogen bonding among the HBA and HDA constituents. Furthermore, NMR spectroscopy was applied to verify hydrogen bonding which involved in this interaction [26,27]. Thus, determination of hydrogen bond detection of H bond donor and acceptor were carried out and utilized the same frequencies as for hydroxyl proton detection. Table 4 and figure 4 showed chemical shift component and H-NMR spectroscopy of sucrose compound. Table 5 and figure 5 showed chemical shift component and H-NMR spectroscopy of malic acids compound. The up field in the signals of characteristic hydroxyl protons in sucrose and malic acids reflect the hydrogen bond structures existence. The downshift and intensity of characteristic group were indicative of many hydrogen bonding emerged between HBA and HBD [27]. This indicated that there were large hydrogen bond structures comes from (OH$^-$ ions in sucrose as HBA and the hydrogen malate of acids malic as the HBD. The high intensity of peak in the sucrose spectra indicated that a part of the protons of D_2O.

Similar observation has been explained by Christofides *et al.* [28], whereby peaks of methyl and methylene groups was observed in downfield shift spectra, and the relative number of protons detection on the hydroxyl groups which relative to the protons on the methyl group. Therefore, it is indicated the hydrogen bonding donor in malic acids, and oxygen atom from hydroxyl groups for accepts a hydrogen bond. Both group are important to made stronger bonding as compare to the compound that consist oxygen atom as donor group only. These observations provide suggestion of interaction of the COOH bonds of malic acids will form a hydrogen bond with hydroxyl group of sucrose compound. The result are in agreement with review paper by Espino *et al.* [11]. They concluded that although LTTM NADES can be derived from prepared from different routes, the same chemical shift will be showed in ^{1}H NMR spectra but with same HBA/ HBD components.

Table 4. Summary of H-NMR with possible sucrose component.

Chemical shift	Splitting	Integral	No of H	(refer to H-NMR chemical shift diagram)
5.2-5.3	2	1.00	1	(-OH), (C=C-H)
4.0-4.1	2	1.02	1	(-OH), (-CH2-O)
3.9	3	1.03	1	(-OH), (-CH2-O)
3.7-3.8	5	1.05	1	(-OH), (-CH2-O)
3.6-3.7	3	4.30	4	(-OH), (-CH2-O)
3.6	3	1.10	1	(-OH), (-CH2-O)
3.5	1	1.97	1	(-OH), (-CH2-O)
3.4	4	1.00	1	(-OH), (-CH2-O)
3.3	3	0.99	0	(-OH), (-CH2-O)

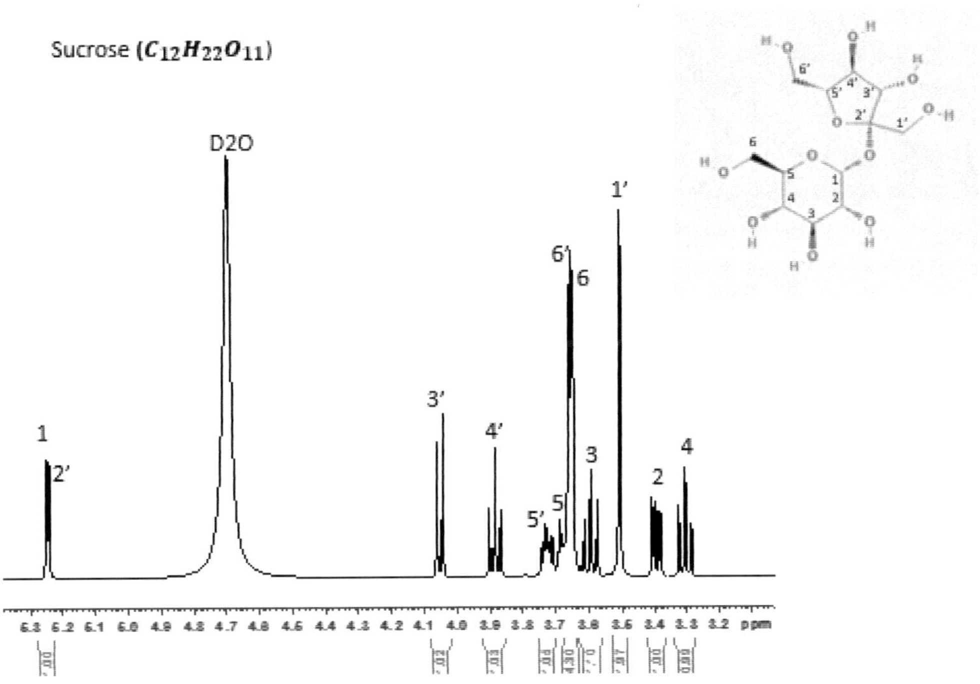

Figure 4. H-NMR spectrum for sucrose structure.

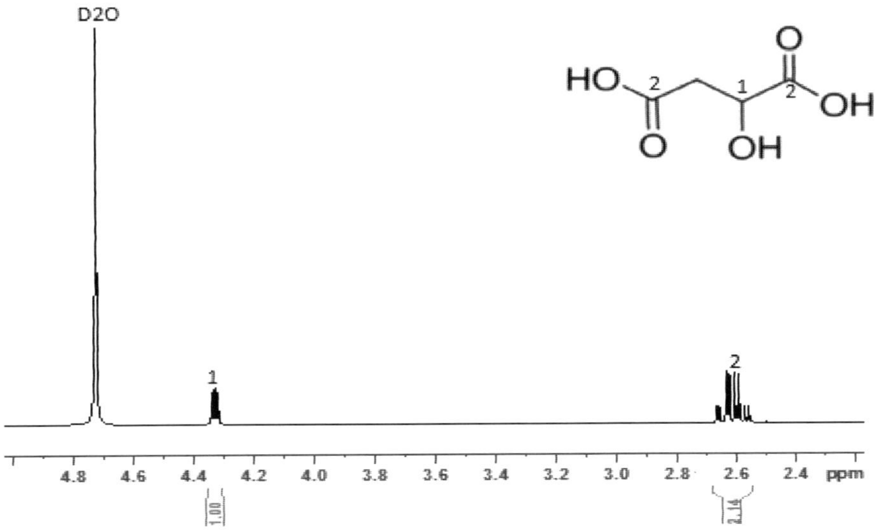

Figure 5. H-NMR spectrum for malic acids structure.

Table 5. Summary of H-NMR with possible malic acids component.

Chemical shift	Splitting	Integral	No of H	Proposed Structures
4.3-4.4	3	1.00	1	-OH
2.5	8	2.14	2	2(-OH) 2(H-C-C=O)

4. Conclusion

In this study, COSMO-RS simulation was explored to forecast the solvation properties of LTTM compounds. It is possible to screen for most effective LTTMs based on the solvation power and ^{1}H nuclear magnetic resonance (H-NMR) analysis. The H-NMR validated the hydrogen bonding presence. Generally, COSMO-RS software determined sigma profile of choline chloride, monosodium glutamate and sucrose in the hydrogen bond acceptor (HBA) region and sucrose showed highest value of sigma profile of 25.373, followed by choline chloride at 20.134 and monosodium glutamate of 12.223. The same results were indicated by the mixture compound of HBA in the presence of malic acids (HBD). High intensity sigma profile for HBA (Sucrose) at σ-profiles region of 0.007 and 0.0019 e/Å2 and HBD (Malic acids) at region -0.001 and -0.021 e/Å2 indicated great repulsion to both H-bond donor and H-bond acceptor group in the mixture. ^{1}H nuclear magnetic resonance spectroscopy demonstrated that there was intensive hydrogen-bonding interactions between sucrose and malic acids based on signals reflected by the hydroxyl protons. The 'up field' region showed the hydroxyl protons signals in sucrose and malic acids in the hydrogen bond structure. The down field showed representative group were resulted of many hydrogen bonding formed between the components.

Acknowledgments

The authors gratefully acknowledge support from Ministry of Education Malaysia through HICoE award to CBBR centre, financial support for International Collaborative Research Funding (ICRF) grant and Universiti Teknologi PETRONAS for a URIF grant. The authors would also like to thank the laboratory technical assistants, and other staff, at the Biomass Processing Laboratory for their assistance.

References

[1] Ong V Z and Wu T Y 2020 An application of ultrasonication in lignocellulosic biomass valorisation into bio-energy and bio-based products *Renew. Sustain. Energy Rev.* **132** 109924
[2] Zhang Y, Huang M, Su J, Hu H, Yang M, Huang Z, Chen D, Wu J and Feng Z 2019 Overcoming biomass recalcitrance by synergistic pretreatment of mechanical activation and metal salt for enhancing enzymatic conversion of lignocellulose *Biotechnol. Biofuels* **12** 1–15
[3] Zakzeski J, Bruijnincx P C A, Jongerius A L and Weckhuysen B M 2010 The catalytic valorization of lignin for the production of renewable chemicals *Chem. Rev.* **110** 3552–99
[4] Chan Y H, Cheah K W, How B S, Loy A C M, Shahbaz M, Singh H K G, Yusuf N R, Shuhaili A F A, Yusup S, Ghani W A W A K, Rambli J, Kansha Y, Lam H L, Hong B H and Ngan S L 2019 An overview of biomass thermochemical conversion technologies in Malaysia *Sci. Total Environ.* **680** 105–23
[5] Ozturk M, Saba N, Altay V, Iqbal R, Hakeem K R, Jawaid M and Ibrahim F H 2017 Biomass and bioenergy: An overview of the development potential in Turkey and Malaysia *Renew. Sustain. Energy Rev.* **79** 1285–302
[6] Mohamed M, Yusup S and Loy A C M 2019 Effect of Empty Fruit Bunch in Calcium Oxide for Cyclic CO2 Capture *Chem. Eng. Technol.* **42** 1840–51

[7] Yiin C L, Yusup S, Quitain A T, Uemura Y, Sasaki M and Kida T 2018 Delignification kinetics of empty fruit bunch (EFB): a sustainable and green pretreatment approach using malic acid-based solvents *Clean Technol. Environ. Policy* **20** 1987–2000

[8] Mohtar S S, Tengku Malim Busu T N Z, Md Noor A M, Shaari N and Mat H 2017 An ionic liquid treatment and fractionation of cellulose, hemicellulose and lignin from oil palm empty fruit bunch *Carbohydr. Polym.* **166** 291–9

[9] Zoghlami A and Paës G 2019 Lignocellulosic Biomass: Understanding recalcitrance and predicting hydrolysis *Front. Chem.* **7** 874

[10] Francisco M, Van Den Bruinhorst A and Kroon M C 2012 New natural and renewable low transition temperature mixtures (LTTMs): Screening as solvents for lignocellulosic biomass processing *Green Chem.* **14** 2153–7

[11] Espino M, de los Ángeles Fernández M, Gomez F J V and Silva M F 2016 Natural designer solvents for greening analytical chemistry TrAC - Trends *Anal. Chem.* **76** 126–36

[12] Craveiro R, Aroso I, Flammia V, Carvalho T, Viciosa M T, Dionísio M, Barreiros S, Reis R L, Duarte A R C and Paiva A 2016 Properties and thermal behavior of natural deep eutectic solvents *J. Mol. Liq.* **215** 534–40

[13] Wagle D V., Adhikari L and Baker G A 2017 Computational perspectives on structure, dynamics, gas sorption, and bio-interactions in deep eutectic solvents *Fluid Phase Equilib.* **448** 50–8

[14] Yiin C L, Quitain A T, Yusup S, Sasaki M, Uemura Y and Kida T 2016 Characterization of natural low transition temperature mixtures (LTTMs): Green solvents for biomass delignification *Bioresour. Technol.* **199** 258–64

[15] Rodríguez N R, González A S B, Tijssen P M A and Kroon M C 2015 Low transition temperature mixtures (LTTMs) as novel entrainers in extractive distillation *Fluid Phase Equilib.* **385** 72–8

[16] Silva L P, Fernandez L, Conceiçao J H F, Martins M A R, Sosa A, Ortega J, Pinho S P and Coutinho J A P 2018 Design and Characterization of Sugar-Based Deep Eutectic Solvents Using Conductor-like Screening Model for Real Solvents *ACS Sustain. Chem. Eng.* **6** 10724–34

[17] Lemaoui T, Darwish A S, Hammoudi N E H, Abu Hatab F, Attoui A, Alnashef I M and Benguerba Y 2020 Prediction of Electrical Conductivity of Deep Eutectic Solvents Using COSMO-RS Sigma Profiles as Molecular Descriptors: A Quantitative Structure-Property Relationship Study *Ind. Eng. Chem. Res.* **59** 13343–54

[18] Zurob E, Cabezas R, Villarroel E, Rosas N, Merlet G, Quijada-Maldonado E, Romero J and Plaza A 2020 Design of natural deep eutectic solvents for the ultrasound-assisted extraction of hydroxytyrosol from olive leaves supported by COSMO-RS Sep. *Purif. Technol.* **248** 117054

[19] Yiin C L, Ho S, Yusup S, Quitain A T, Chan Y H, Loy A C M and Gwee Y L 2019 Recovery of cellulose fibers from oil palm empty fruit bunch for pulp and paper using green delignification approach *Bioresour. Technol.* **290** 121797

[20] Paduszyński K 2017 An overview of the performance of the COSMO-RS approach in predicting the activity coefficients of molecular solutes in ionic liquids and derived properties at infinite dilution *Phys. Chem. Chem. Phys.* **19** 11835–50

[21] Mu T, Rarey J and Gmehling J 2007 Performance of COSMO-RS with sigma profiles from different model chemistries *Ind. Eng. Chem. Res.* **46** 6612–29

[22] Ruiz E, Ferro V R, Palomar J, Ortega J and Rodriguez J J 2013 Interactions of ionic liquids and acetone: Thermodynamic properties, quantum-chemical calculations, and NMR analysis *J. Phys. Chem. B* **117** 7388–98

[23] Gerber R P and Soares R P 2013 Assessing the reliability of predictive activity coefficient models for molecules consisting of several functional groups Brazilian *J. Chem. Eng.* **30** 1–11

[24] Hayyan M, Mbous Y P, Looi C Y, Wong W F, Hayyan A, Salleh Z and Mohd-Ali O 2016 Natural deep eutectic solvents: cytotoxic profile *Springerplus* **5** 1–12

[25] Wojeicchowski J P, Ferreira A M, Abranches D O, Mafra M R and Coutinho J A P 2020 Using

COSMO-RS in the Design of Deep Eutectic Solvents for the Extraction of Antioxidants from Rosemary *ACS Sustain. Chem. Eng.* **8** 12132–41

[26] Kalhor P and Ghandi K 2019 Deep eutectic solvents for pretreatment, extraction, and catalysis of biomass and food waste *Molecules* **24**

[27] Xin R, Qi S, Zeng C, Khan F I, Yang B and Wang Y 2017 A functional natural deep eutectic solvent based on trehalose: Structural and physicochemical properties *Food Chem.* **217** 560–7

[28] Christofides J C, Davies D B, Martin J A and Rathbone E B 1986 Intramolecular Hydrogen Bonding in 1'-Sucrose Derivatives Determined by SIMPLE 1H NMR Spectroscopy *J. Am. Chem. Soc.* **108** 5738–43

Impact of heavy hydrocarbon impurities on polytetrafluoroethylene (PTFE) membrane stability

S M Saleh, Z P Chan and F F A C Chang

PETRONAS Research Sdn Bhd, Lot 3288 & 3289, Off Jalan Ayer Itam, Kawasan Institusi Bangi, 43000 Kajang, Selangor, Malaysia

Email: syafiqa_saleh@petronas.com.my

Abstract. Membrane contactor technology has attained considerable attention as a promising technology to reduce CO_2 content in natural gas. In this study, the main objective is to investigate the effect of heavy hydrocarbons impurities, often present in natural gas, on polytetrafluoroethylene (PTFE) hollow fibre membrane. The membranes were immersed for months in n-heptane, 1-decene, benzene and toluene, and analysed periodically through its surface morphology, composition, functional groups, hydrophobicity, and thermal stability. The characteristics of PTFE fibres remained unchanged even after long term exposure with heavy hydrocarbons. This study provides a better understanding of the robustness of using PTFE membrane fibre for CO_2 removal in membrane contactor system.

Keywords: Polytetrafluoroethylene (PTFE), membrane, heavy hydrocarbon

1. Introduction

In the recent decades, removal of impurities, especially CO_2, from natural gas has become a concerning issue. High CO_2 content would cause severe corrosion issues, decrease the value of natural gas as a fossil fuel and reduces pipeline capacity [1]. Membrane contactor is a potential technology that can be developed further to reduce CO_2 content in natural gas. Membrane contactor uses microporous hollow fiber membrane (HFM) as phase barrier allowing gas and liquid to flow on different sides of the membrane; hence more efficient mass transfer without two phases dispersing into each other. It offers modularity, larger contact area and advantageous as it has no flooding, channelling, or foaming issues. Gas selectivity is dependent on the absorbent of choice.

As in other membrane-based CO_2 removal technologies, membrane material selection is crucial in determining the stability of membrane from plasticization and wetting effect during CO_2 removal process. Polyimide, polyethylene (PE), polypropylene (PP), polytetrafluoroethylene (PTFE) and polyvinylidene fluoride (PVDF) are the common polymeric membrane materials for CO_2 capture. Previous studies reported the challenges in membrane based gas separation includes plasticization, conditioning and physical aging due to the interaction between impurities from feed stream and membrane [2-5]. A review from Suleman, Lau and Yeong indicated that destruction of the polymer structure could occur due to plasticization from dissolution of different penetrants in the polymer matrix [6].

Fluoropolymer such as polytetrafluoroethylene (PTFE) has drawn attention due to its high physical and chemical stabilities [7]. Its outstanding chemical stability and resistance to plasticization, in presence of condensable organics, is due to its carbon backbone, and strong C-F and C-C bonds. It is protected by the electron cloud of the large fluorine atoms which are tightly bonded with monomer unit [8], shown in figure 1. For long-term operation of over 180 days, it is reported that PTFE HFM indicated no apparent deterioration of CO_2 removal performance during operation. PTFE polymer also maintained good stability in absorbent solutions over 24 weeks [3][6]. In a study by deMontigny *et al.*, membrane contactor with PTFE showed better resistance towards wetting from aminomethyl propanol (AMP) and monoethanolamine (MEA), and higher CO_2 transfer coefficient; whilst polypropylene (PP) was easily wetted [9].

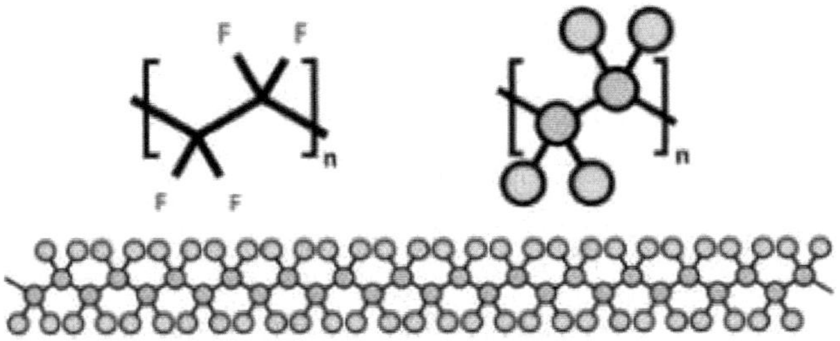

Figure 1. Molecular arrangement of PTFE polymer [10].

Natural gas commonly contains heavy hydrocarbons which includes benzene, toluene, ethylbenzene and xylene (BTEX) ranged in 200 – 3000 ppm [11]. These heavy hydrocarbons may act as plasticizer which can cause swelling of most polymeric membranes. The larger pores, caused by swelling, will then allow other gas species to pass through and result in significant drop of CO_2 permeability [12-13]. For membrane contactor, enlargement of membrane pores would result in dramatic reduction of critical entry pressure and eventual membrane wetting [13].

Encouragingly, Scholes *et al.* showed that CO_2 and CH_4 separation with Teflon AF1600 and Hyflon AD60 membrane films in the presence of hexane and toluene showed minor performance reduction [12]. There have been several research studies carried out to determine the effect of heavy hydrocarbons such as pentane, heptane, and toluene on membrane materials such as PP, PE, polyimide and polyamide in CO_2 removal. However, research studies on PTFE HFM for longer duration remain limited due to the toxicity and volatility of liquid heavy hydrocarbons, and high cost of PTFE membrane fabrication compared to other polymeric membranes [9].

Therefore, the purpose of this study is to examine the impact of long-term exposure of heavy hydrocarbons on the characteristics of PTFE HFM. Often lab test on CO_2 performance is limited to using binary gas for testing and are of short duration; hence this study shall provide indication on what to be expected for continuous exposure to heavy hydrocarbons present in actual natural gas. This study is conducted through immersion of PTFE HFM into heavy hydrocarbons and analysing its surface morphology, hydrophobicity, chemical and thermal stability. The result from this study shall serve as valuable reference on the long-term stability of PTFE membrane for application in actual natural gas fields.

2. Experimental Methodology

2.1. Materials

The PTFE hollow fibre membranes were supplied by Dalian Institute of Chemical Physics (DICP) for testing. The characteristics of PTFE membrane used in this study are listed in table 1. The heavy hydrocarbons used for the immersion of membrane were n-heptane, 1-decene, benzene and toluene, and were all purchased from Merck KGaA, Germany. Distilled water and amine solvent (mixture of methyldiethanolamine, MDEA and piperazine at concentration of 35 wt. %) were used for the contact angle measurement; the amines were purchased from Sigma-Aldrich.

Table 1. Characteristics of the PTFE HFM.

Characteristics	Value
Outer diameter (cm)	0.08949
Inner diameter (cm)	0.04344
Wall thickness (cm)	0.02225
Porosity (%)	47.3092

2.2. Membrane Immersion

As shown in figure 2, roughly about 10 cm of PTFE HFMs were immersed in vials containing n-heptane, 1-decene, benzene and toluene, respectively. The PTFE HFMs were fully immersed in every vial at ambient condition over 4 months. At the end of each month, samples were taken, rinsed with deionised water and vacuum dried at 900 mbar and 70 °C for 24 hours to remove traces of heavy hydrocarbon from membrane surfaces, before proceeding with further characterisation.

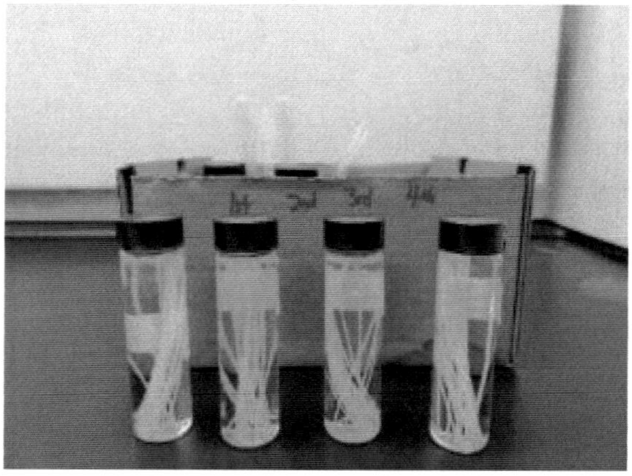

Figure 2. Immersion of PTFE HFM in heavy hydrocarbons.

2.3. Characterisation

The characterisation of the initial and immersed membrane samples includes the analyses on surface morphology, porosity, hydrophobicity, and chemical and thermal stability.

2.3.1. Field-Emission Scanning Electron Microscope (FESEM). The membrane fibres were fractured using liquid nitrogen to obtain a smooth cross-sectional area. The fractures were then coated with gold using ion sputtering device (Quorum, Q150RS). The morphologies of inner, outer, and cross-sectional surfaces of PTFE HFM were examined by FESEM (EVO (Hitachi, Model SU7000). The membrane thickness, inner and outer diameters were also measured through FESEM.

2.3.2. Porosity Test. The effect of heavy hydrocarbons towards the overall porosity, ε of the PTFE HFM was determined from the average outer diameter (OD) (in mm), and inner diameter (ID)(in mm) obtained from FESEM analysis, and the weight of 1 metre long of PTFE fiber (W) (in gram) using analytical balance. This measures on the porosity for CO_2 permeability. The porosity, ε of the PTFE HFM is determined by the following equation:

$$\varepsilon\,(\%) = [1 - W/2.17/\pi*(OD*OD - ID*ID)/4*100] \tag{1}$$

2.3.3. Attenuated Total Reflection Fourier Transform Infrared (ATR-FTIR) Spectroscopy. ATR-FTIR was used to determine chemical stability of the immersed membrane fibres. The spectra were obtained using spectrophotometer (Perkin Elmer, Spectrum 100 FTIR). The fibre was pressed against ATR crystal in transmittance mode from 450 to 4000 cm^{-1} using 4 scans at ambient temperature to determine the composition, structure, and the functional groups in the sample. Additionally, the heavy hydrocarbons that were used for immersion were also measured with ATR-FTIR as to compare their spectra with the immersed PTFE HFM spectra respectively to identify the presence of hydrocarbon in the membrane matrix during the immersion. This is in case traces are still present even after high temperature (70 °C) vacuum drying method has been applied on the membrane.

2.3.4. Contact Angle Measurement. Tensiometer (Model T200, Biolin Scientific) was used to perform contact angle measurements. Sessile drop method was done, using 35 wt. % amine solvent to conduct the analysis, to determine the variation of hydrophobicity of PTFE HFM after the immersion. The volume of liquid drop was controlled manually at the range of 0.5 – 2.0 μL through a syringe and measured by tensiometer software. Minimum of eight readings were taken and averaged, similar to other researchers in membrane study [7][14-15].

2.3.5. Thermogravimetric Analysis (TGA). Thermogravimetric analysis (TGA) of the membrane fibres was completed using a Mettler Toledo Thermogravimetric Analyzer. 10 mg of PTFE HFM is analysed at the heating rate of 10 °C/min, from 30 to 800 °C and a nitrogen flow of 50 mL/min. Thermogravimetric loss of the PTFE HFMs were compared to determine the effect of heavy hydrocarbons immersion on the membrane thermal stability. The decomposition temperature of PTFE HFM and existence of heavy hydrocarbon in membrane after the immersion were determined through the thermogravimetric curve.

3. Results and Discussions

3.1. Membrane Surface Morphology and Porosity

figure 3(a) shows the surface morphologies of the initial membrane fibres. The structure of PTFE HFM has microstructures of nodes interconnected by fibrils; where node is the solid agglomerates of particles and fibrils bridged the two adjoining nodes together. The size of pores on the inner surface of PTFE membrane are commonly larger than those on the outer surface, due to higher reduction ratio and higher extrusion pressure on the outer surface [14]. The cross-section of the non-immersed PTFE HFM (figure 3(a)iii) shows a continuous porous layer from inner to outer surface.

After immersion, the morphology of PTFE HFM has no significant change. The structures at 5000 of magnification (in figure 3(b) to 3(e)) show no displacement of fibrils and the surface area of nodes remained at almost the same even after long term in contact with heavy hydrocarbons. PTFE HFM does not suffer plasticization as its microstructure of C-F bonds in the polymer matrix and extra-electronegative protective layer from fluorine atoms inhibited the penetration of heavy hydrocarbons molecules into the membrane matrix [16–18]. Other polymer, such as polyimide membrane has its small pores closed and large pores enlarged just after 4 days in contact with gas plasticizing agent [5].

Porosity, ε of the PTFE HFM is determined from measurement of the inner and outer diameters, and weight of the membrane, and through calculation using equation (1). Result in figure 4 shows the porosity of the PTFE HFM throughout the four months' immersion. Based on t-test done, there is no significant change observed where the overall porosity calculated, for before and after immersion, is about 47 %. Membrane porosity is important to maintain CO_2 flux whilst preventing membrane wetting phenomenon; higher porosity would lead to lower breakthrough pressure and ease the amine solvent to enter the membrane pores. Consistent porosity of membrane in long term membrane contactor operation shall allow stable CO_2 absorption performance and prevent unnecessary wetting [19-20].

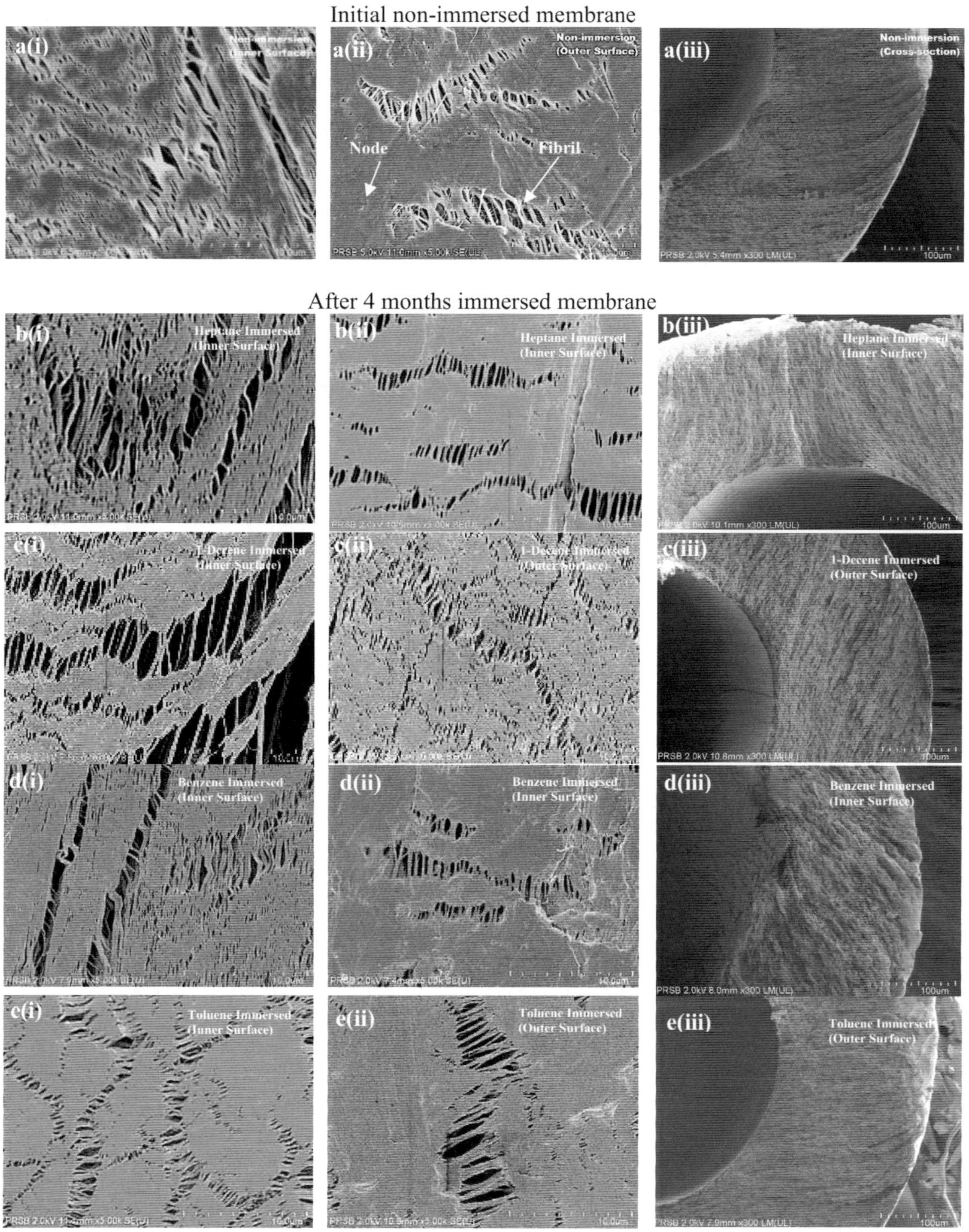

Figure 3. FESEM images of PTFE HFMs. (a) Initial; (b) to (e): Immersed PTFE HFM. (i): outer surface, (ii): inner surface, (iii): cross-sectional surface.

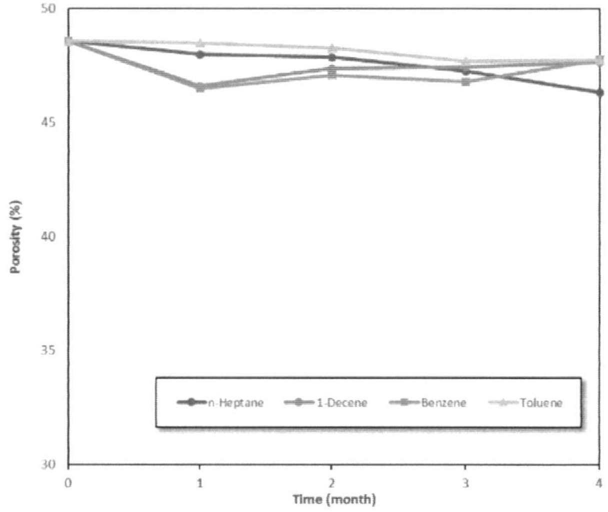

Figure 4. Porosity of immersed and non-immersed PTFE HFM.

3.2. Compositional and Functional Group Analysis

ATR - FTIR spectroscopy is used to identify the composition and functional groups of the sample. figure 5(a) shows the IR spectra of results of initial membrane; whilst figure 5(b) shows the overlapped IR bands for the immersed PTFE HFM samples.

Peaks of C–F bonds could be detected within range of 1000-1400 cm^{-1}. In figure 5, C-F bonds are observed clearly at 1204 cm^{-1} and 1149 cm^{-1}. These peaks remain practically unchanged in their respective intensities over the periods of immersion as shown in figure 5(b), since there is no active reaction happening during the immersion that may deteriorate the composition of the PTFE functional groups. The consistent intensity of functional peaks indicate that PTFE polymer is inert towards organic solvent [3][12].

If there was any degradation and defluorination were to occur, it would have been detected by presence of C=C bond within the range of 665 – 995 cm^{-1}, and presence of any C-H and O-H bonds at the range 2840 – 3000 cm^{-1} and 1330 – 1420 cm^{-1}, respectively [21-22]. The absence of those peaks indicate again that no reaction has taken place on the PTFE even after 4 months of immersion.

This stability is due to PTFE polymer microstructure where the ($-CF_2-CF_2-$) has the spiral molecular structure. Its carbon backbone is totally covered and protected by extremely high electronegative F atoms which are tightly connected. The size of the F atoms resulted in close contact among each other and allowed continuous coverage around the C–C bonds which protects them from attack. Both C–C bonds and C–F bonds are extremely strong (C–C, 607 kJ/mol; C–F, 552 kJ/mol). Hence this lead to chemical resistance and stability to the molecule [3]. In addition, the PTFE polymer is highly crystalline which contributes to its insolubility in most solvents, even at higher temperature [23]. Only a few chemicals, such as molten alkali metals, can attack PTFE molecules [3]. Therefore, it is very difficult for the heavy hydrocarbon molecules to disrupt the PTFE membrane matrix and cause swelling.

Hence, the IR analysis indicated that PTFE polymer exhibited strong chemical stability for long term contact with the heavy hydrocarbons. This is important for minimising membrane damage and the need to put the membrane offline for rejuvenation. The good chemical stability also helps to lengthen the membrane module lifespan and to cut cost from having to change membrane module too frequently.

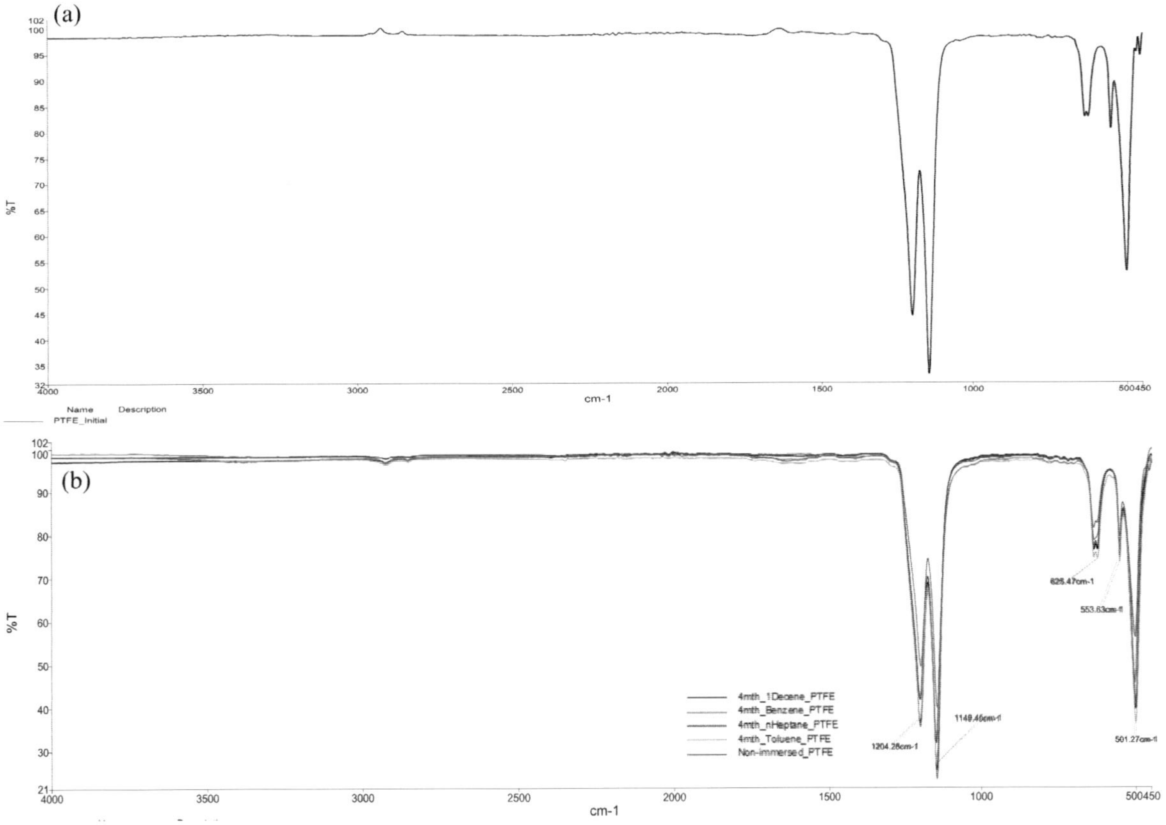

Figure 5. Comparison of IR spectra between (a) initial fresh membrane and (b) the immersed PTFE HFM.

3.3. Contact Angle Measurement

After the PTFE membrane fibres have undergone 4 months of immersion in heavy hydrocarbons, contact angle measurements for each of the PTFE HFM (figure 6) were still able to maintain its high value. Contact angle readings of immersed PTFE HFM remained in non-wetting mode (i.e. >90°) with minimum value averaged at 96.97° after 4 months of immersion. Furthermore, based on t-test between the initial and immersed PTFE HFM in table 2, there was no significant difference between the results.

Hence, it can be concluded that the heavy hydrocarbons immersion does not cause deterioration on the hydrophobicity of the PTFE HFM. This is important because it is critical for the membrane to maintain high hydrophobicity in membrane contactor operation, in order to prevent the liquid phase from easily going into the membrane. Reduction in hydrophobicity could result in increase of membrane wetting, which would lead to reduction in performance.

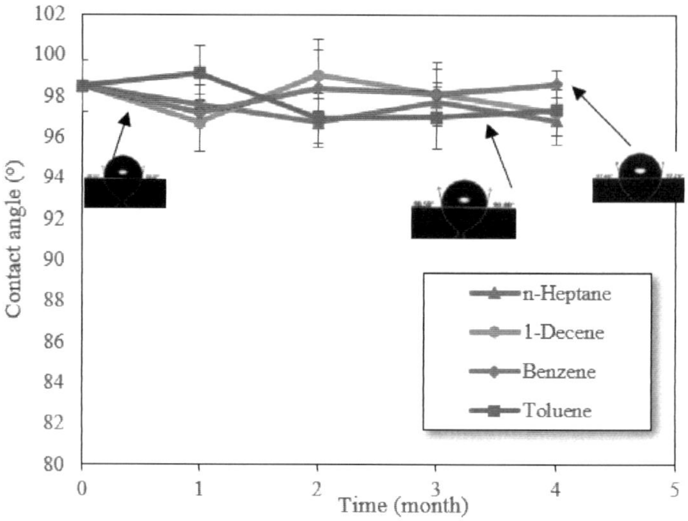

Figure 6. CA of PTFE HFM with amine after immersion.

Table 2. t-test of CA measurement.

	Fresh PTFE HFM	Immersed PTFE HFM
Mean	104.19	103.5168
Variance	37.26183	38.62356
Observations	8	8
Pooled Variance	37.94269	
Hypothesized Mean Difference	0	
df	14	
t Stat	0.218567	
P(T<=t) one-tail	0.415071	
t Critical one-tail	1.76131	
P(T<=t) two-tail	0.830141	
t Critical two-tail	2.144787	

3.4. Thermal Stability Analysis

The thermal stability of membrane was investigated by TGA to determine the decomposition temperature and weight loss of the PTFE HFM samples as a function of temperature. figure 7(a) displays the TGA curves of the initial PTFE membrane fibres, whilst figure 7(b) shows those after 4-months immersed in heavy hydrocarbons.

The results indicate no weight loss identified below 400 °C for all the samples. Boiling temperature of heavy hydrocarbons and water are all below 400 °C. This indicates that there were no heavy hydrocarbons trapped in the matrix and pores of membranes [24].

As temperature increased, the weight loss transition of PTFE HFM samples began after 460 °C and was noticeably complete after 610°C. Decomposition of the immersed PTFE HFM remained the same as the initial PTFE, where the membrane began to lose weight at temperature of 460 °C. This indicated that tetrafluoroethylene (TFE) began to dissociate from the membrane matrix and decomposed. Most of the weight loss occurred at around 600 °C, and this is observed from the steep slope in the TGA graph. The result is consistent with the works by R. Wang, Xu & He [22] and Zhu *et al.* [7].

Therefore, the membrane exhibited good thermal stability even after long exposure to heavy hydrocarbon.

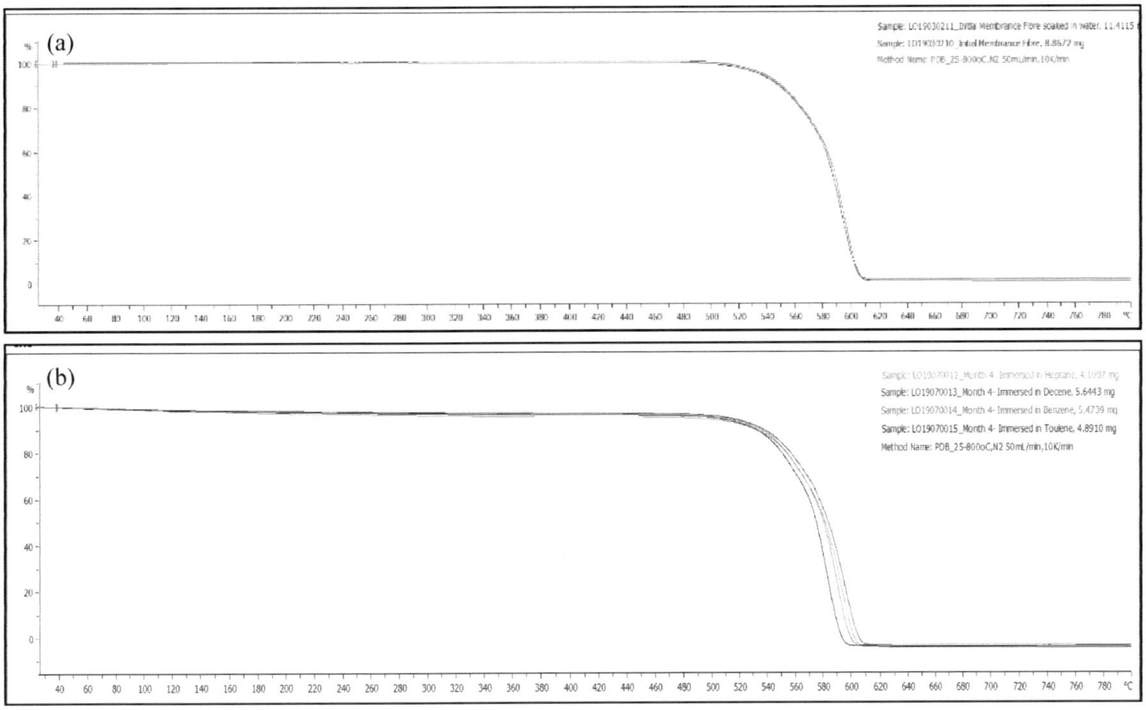

Figure 7. Thermogravimetric curve of (a) initial membrane and (b) immersed PTFE HFM.

4. Conclusion

In this study, the impact of heavy hydrocarbons towards PTFE HFM was investigated. PTFE hollow fibre membranes were immersed for months in n-heptane, 1-decene, benzene and toluene, and analysed periodically through its surface morphology, composition, functional groups, hydrophobicity, and thermal stability. The results indicate that there is no significant change in the characteristics of PTFE HFM. This study serves as a good reference on the robustness of PTFE membrane for application in membrane contactor process for natural gas sweetening.

References

[1] Tan L S, Lau K K, Bustam M A and Shariff A M 2012 Removal of high concentration CO_2 from natural gas at elevated pressure via absorption process in packed column *J. Nat. Gas Chem.* **21** 7–10

[2] Adewole J K, Ahmad A L, Ismail S and Leo C P 2013 Current challenges in membrane separation of CO_2 from natural gas: A review Int. *J. Greenhouse Gas Control* **17** 46–65

[3] Tang H, Zhang Y, Wang F, Zhang H, and Guo Y 2016 Long-Term Stability of Polytetrafluoroethylene (PTFE) Hollow Fiber Membranes for CO_2 Capture *Energy and Fuels* **30** 492–503

[4] Franco J A, deMontigny D, Kentish S E, Perera J M and Stevens G W 2009 Effect of amine degradation products on the membrane gas absorption process *Chem. Eng. Sci.* **64** 4016–23

[5] Zhang L, Li J, Zhou L, Liu R, Wang X and Yang L 2016 Fouling of Impurities in Desulfurized Flue Gas on Hollow Fiber Membrane Absorption for CO_2 *Capture Ind. Eng. Chem. Res.* **55** 8002–10

[6] Suleman M S, Lau K K and Yeong Y F 2016 Plasticization and Swelling in Polymeric Membranes in CO_2 Removal from Natural Gas *Chem. Eng. Technol.* **39** 1604–16

[7] Zhu H, Wang H, Wang F, Guo Y, Zhang H and Chen J 2013 Preparation and properties of PTFE hollow fiber membranes for desalination through vacuum membrane distillation *J. Memb. Sci.* **446** 145–53

[8] Feng S, Zhong Z, Wang Y, W. Xing W and Drioli E 2018 Progress and perspectives in PTFE membrane: Preparation, modification, and applications *J. Memb. Sci.* **549** 332–49

[9] deMontigny D, Tontiwachwuthikul P and Chakma A 2006 Using polypropylene and polytetrafluoroethylene membranes in a membrane contactor for CO_2 absorption *J. Memb. Sci.* **277** 99–107

[10] Crawford C B and Quinn B 2017 Physiochemical properties and degradation *Microplastic Pollutants* 57–100.

[11] Baker R W and Lokhandwala K 2008 Natural Gas Processing with Membranes: An Overview *Ind. Eng. Chem. Res* **47** 2109–21

[12] Scholes C A, Kentish S E and Stevens G W 2009 Effects of minor components in carbon dioxide capture using polymeric gas separation membranes *Sep. Purif. Rev.* **38** 1–44

[13] Hasan R, Scholes C A, Stevens G W and Kentish S E 2009 Effect of Hydrocarbons on the separation of carbon dioxide from methane through a polyimide gas separation membrane *Ind. Eng. Chem. Res.* **48** 5415–19

[14] Wang H, Ding S, Zhu H, Wang F, Guo Y, Zhang H and Chen J 2014 Effect of stretching ratio and heating temperature on structure and performance of PTFE hollow fiber membrane in VMD for RO brine *Sep. Purif. Technol.* **126** 82–94

[15] Naim R, Ismail A F, Cheer N B and Abdullah M S 2014 Polyvinylidene fluoride and polyetherimide hollow fiber membranes for CO_2 stripping in membrane contactor *Chem. Eng. Res. Des.* **92** 1391–8

[16] Scholes C A, Stevens G W and Kentish S E 2016 Impact of Heavy Hydrocarbons on Natural Gas Sweetening Using Perfluorinated Polymeric Membranes *Ind. Eng. Chem. Res.* **55** 7696–703

[17] Al-Juaied M and Koros W J 2006 Performance of natural gas membranes in the presence of heavy hydrocarbons *J. Memb. Sci.* **274** 227–43

[18] Chenar M P, Savoji H, Soltanieh M, Matsuura T and Tabe S 2010 The effect of ethane on the performance of commercial polyphenylene oxide and cardo-type polyimide hollow fiber membranes in CO_2/ CH_4 separation applications *Korean J. Chem. Eng.* **27** 1876–81

[19] Huang F Y C and Arning A 2019 Performance Comparison between Polyvinylidene Fluoride and Polytetrafluoroethylene Hollow Fiber Membranes for Direct Contact *Membrane Distillation Membranes* **9** 52

[20] Mosadegh-Sedghi S, Rodrigue D, Brisson J and Iliuta M C 2014 Wetting phenomenon in membrane contactors – Causes and prevention *J. Memb. Sci.* **452** 332–53

[21] Marshall J E, Zhenova A, Roberts S, Petchey T, Zhu P, Dancer C E J, McElroy C R, Kendrick E and Goodship V 2021 On the Solubility and Stability of Polyvinylidene Fluoride *Polymers* **13** 1–31

[23] Feng S, Zhong Z, Wang Y, Xing W and Drioli E 2018 Progress and perspectives in PTFE membrane: Preparation, modification, and applications *J. Memb. Sci.* **549** 332–49

[24] Jusoh N, Lau K K, Shariff A M and Yeong Y F 2014 Capture of bulk CO_2 from methane with the presence of heavy hydrocarbon using membrane process Int. *J. Greenh. Gas Control* **22** 213–22

[25] Wang R, Xu G and He Y 2017 Structure and properties of polytetrafluoroethylene (PTFE) fibers *e-Polymers* **17** 215–20

Microwave-assisted synthesis of photoluminescent carbon dots from palm fronds biomass wastes

W L Ang[1,2], Q A Alqasem[1] and A W Mohammad[1,2]

[1] Department of Chemical and Process Engineering, Faculty of Engineering and Built Environment, Universiti Kebangsaan Malaysia, 43600 Bangi, Selangor, Malaysia.
[2] Centre for Sustainable Process Technology (CESPRO), Faculty of Engineering and Built Environment, Universiti Kebangsaan Malaysia, 43600 Bangi, Selangor, Malaysia.

Email: wl_ang@ukm.edu.my

Abstract. Oil palm fronds (OPF) is one of the largest biomass sources in Malaysia that has been underutilized. In this work, OPF has been used as a precursor to synthesize carbon dots (CDs) via microwave irradiation method. The impacts of irradiation power and duration and the reacting solution have been investigated. It was discovered that CDs with the highest photoluminescence intensity was obtained at microwave irradiation power of 385 W for 30 s. Irradiation at lower or higher power resulted in incomplete or over carbonization that reduced the fluorescence property. In addition, CDs synthesized with diethylene glycol (DEG) as reacting solution possessed higher photoluminescence intensity as compared to ultrapure water solution. This could be attributed to more complete CDs formation that happened at higher temperature, which could only be achieved by DEG solution (higher boiling point). The CDs were then tested as a sensor for lead (II) ions. The UV-vis absorbance was found to be reduced with the presence of lead (II) ions. This indicated that the lead (II) ions might interact with CDs and disrupted with the absorbance of UV light. Overall, OPF could be a potential precursor for the synthesis of low-cost and easily available CDs for environmental applications.

1. Introduction

Quantum dots are nanoparticles that have gained significant fame in diagnosis and labeling applications due to its intriguing optical (fluorescence) and electronic properties [1]. However, the synthesis process of quantum dots is not environmentally friendly as it involves the use of cadmium and arsenic heavy metals which makes them potentially hostile towards the biological applications [2,3]. In this scenario, carbon dots (CDs) synthesized from carbon precursors have emerged as an attractive alternative to quantum dots. CDs are oxygenous carbon nanoparticles with a size of less than 10 nm [4]. Owing to their low cytotoxicity and great water solubility, biocompatibility, and fluorescence properties, CDs have found promising application potential in various fields such as optical, optoelectronic, bioimaging, biomarking, solar cells, drug delivery, and photodynamic therapy [1,5–7].

The synthesis of CDs can be generally categorized into top-down (e.g. laser ablation, electro deposition, arc discharge, etc.) and bottom-up (microwave, hydro and solvothermal, etc.) methods [8]. Top-down synthesis approaches break down larger carbon structures such as graphite and graphene into CDs, whereas bottom-up synthesis approaches generate CDs by decomposition, fusion or carbonization of small or large carbon precursors including both natural and synthetic polymers [8]. As compared to top-down synthesis approaches, the bottom-up processes are generally considered facile, cost effective, and environmentally friendly without the need of harsh reagents or conditions [9–14]. Among these, the microwave synthesis method has gained increasing attention due to its fast reaction rates, milder reaction conditions, higher quantum yield, and lower energy usage [15].

Various studies have shown that CDs could be successfully produced from microwave synthesis method. For instance, Jusuf *et al.* [16] managed to obtain CDs from eggshell membrane ashes by using microwave irradiation technique within a short duration (60 s). The CDs was capable to degrade methylene blue (pollutant) under sunlight irradiation, indicating the photodegradation property of the sample [16]. In another study, Hinterberger *et al.* (2018) synthesized CDs with pH-dependent fluorescence color property from citric acid and o-phenylene-diamine as precursors via microwave-assisted hydrothermal approach [17]. The fluorescence emission color of the CDs solution could be tuned precisely and reproducibly from blue to white to yellow by controlling the pH. The 8aforementioned cases indicate that the precursors of CDs synthesis can come from a wide range of materials. However, the use of organic chemicals as precursors could potentially heighten the overall synthesis cost [9]. Hence, the use of natural materials and biomass waste (e.g. sugarcane bagasse pulp, carrot roots, tamarind, citrus peels, and used green tea) as precursors of CDs synthesis has received extensive interest recently, in line with the concept of green chemistry [18–22].

The use of non-toxic, easily available, and low-cost precursor for the synthesis of CDs could fulfill the green chemistry approach since the material is obtained from nature and thus the CDs produced are biocompatible in nature [23]. In this context, oil palm fronds (OPF) could be a promising carbon precursor for the synthesis of CDs. In Malaysia, more than 50 million ton of OPF biomass waste has been generated yearly from the palm oil industry, the most important agriculture sector in the country [24]. The use of OPF as carbon precursor for the synthesis of CDs for environmental application is feasible since it is available throughout the year and can be obtained at low to no cost. Furthermore, the conversion of OPF into CDs could help to valorize the biomass waste. Hence, this study aimed to utilize OPF as carbon precursor for the synthesis of CDs via microwave-assisted method. The impacts of the synthesis conditions (microwave irradiation power and period and synthesis solution) on the property and performance of CDs for heavy metal sensing were also investigated.

2. Methodology

2.1. Materials

The OPF (consist of leaflets and rachis) used as the precursor for CDs synthesis were collected from the oil palm in Universiti Kebangsaan Malaysia. Diethylene glycol (DEG) with 99.9 % purity was purchased from Sigma-Aldrich (USA) and ammonia solution 28 % was obtained from J.K. Baker (USA). Ultrapure water (UPW) with a quality of ~18.2 MΩ.cm at 25 °C produced from Sartorius Arium PRO (Fisher Scientific, USA) water purification system was used as solution in the synthesis and sample preparation. Lead (II) nitrate $Pb(NO_3)_2$ as the heavy metal was purchased from Merck, Malaysia.

2.2. Carbon Dots Synthesis

The collected OPF were dried under the sun for approximately 5 days to remove the moisture content of the sample. Afterwards, the OPF were cut into smaller pieces before ground and sieved to obtain the sample in fine powder form of 0.2 mm. The ground OPF samples were then mixed with 38 ml of DEG and 2 ml of ammonia solution to form a mixture with concentration of OPF powder of 10 g/L. The sample mixtures were sonicated for 30 min to homogenize the dispersion of OPF powder in the solution. The samples were subsequently heated in the microwave oven (Midea MM720CXM, Malaysia) at different heating power (low: 119 W, medium: 285 W, and high: 700 W) and duration ranging within 30-120 s. Table 1 summarized the synthesis condition for this experiment. After undergoing microwave irradiation process, the samples were cooled to room temperature before centrifuged at 10000 rpm for 20 min to remove bulk particles or agglomerates. The supernatant was collected and kept for further analysis and application testing.

Table 1. Synthesis conditions of carbon dots from OPF via microwave-assisted approach.

Set	Microwave irradiation power (W)	Microwave irradiation time (s)
A1	119	30
A2		60
A3		90
A4		120
B1	385	30
B2		60
B3		90
B4		120
C1	700	30
C2		60
C3		90
C4		120

2.3. Characterization of CDs
Fourier Transformed Infrared (FTIR) (Nicolet, 6700, United States) was used to characterize the functional groups present on the OPF powder and synthesized CDs. The absorption spectra of the CDs were determined by Lambda 950 UV-visible spectrophotometer (Perkin Elmer, United States). Fluorescence spectrophotometer (FLS920, Edinburgh Instrument, United Kingdom) was used to determine the photoluminescence spectra and fluorescence intensity of the synthesized CDs. Transmission Electron Microscope (TEM) (Thermo Fisher, Talos 120 C, United States) was employed to obtain the high-resolution images of the CDs.

2.4. Detection of Lead (II) Nitrate
Testing solutions at different concentrations of lead (II) nitrate (20 ppm, 50 ppm, 100 ppm, and 150 ppm) were prepared for the evaluation of fluorescence quenching in the presence of heavy metal. Next, 5 ml of the as-synthesized CDs was mixed with the lead (II) nitrate solutions and the effectiveness of metal ions sensing was determined through the quenching of photoluminescence spectrum of CDs.

3. Results and Discussion

3.1. Characterization of CDs
Figure 1 illustrates the appearance of CDs synthesized under different conditions viewed under normal light and UV light (365 nm). As all the CDs solutions look the same under normal light (pale brownish), only one representative was included in figure 1. Since the CDs emitted green-blue fluorescence upon irradiated with UV light, it can be said that the OPF has been successfully converted to CDs. For benchmarking purpose, CDs synthesized with water and DEG medium (without ammonia solution) at microwave irradiation power of 385 W and time of 90 s were included too. It can be seen that CDs with medium irradiation power possessed the brightest green-blue fluorescence as compared to the rest. This could be attributed to the degree of carbonization where insufficient irradiation power (119 W) results in incomplete carbonization while excessive irradiation (700 W) causes over carbonization, where both scenarios lead to the formation of defect or degraded CDs. This observation aligned with the reported study where both incomplete carbonization and over carbonization (decomposition of precursor into

ashes) would lead to the loss of the fluorescence property [25]. The effect of irradiation duration was uncertain and would be discussed based on other analysis. Meanwhile, the use of DEG as solution medium and ammonia solution as doping agent seemed to enhance the fluorescence of CDs as brighter light was observed for the sample set A, B, and C. Further discussion will be provided below by referring to other analysis results.

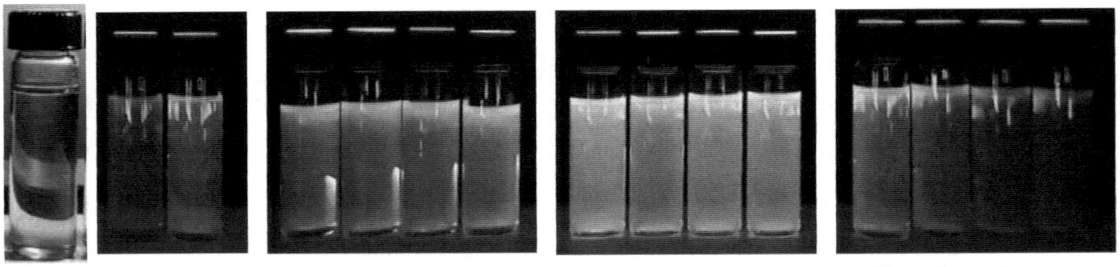

Figure 1. CDs viewed under normal light and UV (365 nm) viewer chamber.

To gain insight into the surface and structural information that contributed to the fluorescence property of CDs, FTIR, and TEM have been conducted. Considering that the FTIR results of all CDs sample were almost the same, only one spectra was displayed as in figure 2. The peak at 3386 cm^{-1} could be assigned to O-H stretching while the presence of carbonyl functional group was revealed with C=O stretching at 1658 cm^{-1}. The peak at 1058 cm^{-1} indicated the presence of C-O-C or C-O stretching. All these functional groups were normally known to contribute to the fluorescence property of CDs as reported in past literature [13,14,16,26]. Hence, the CDs synthesized from OPF possessed the typical functional groups responsible for its luminescent property.

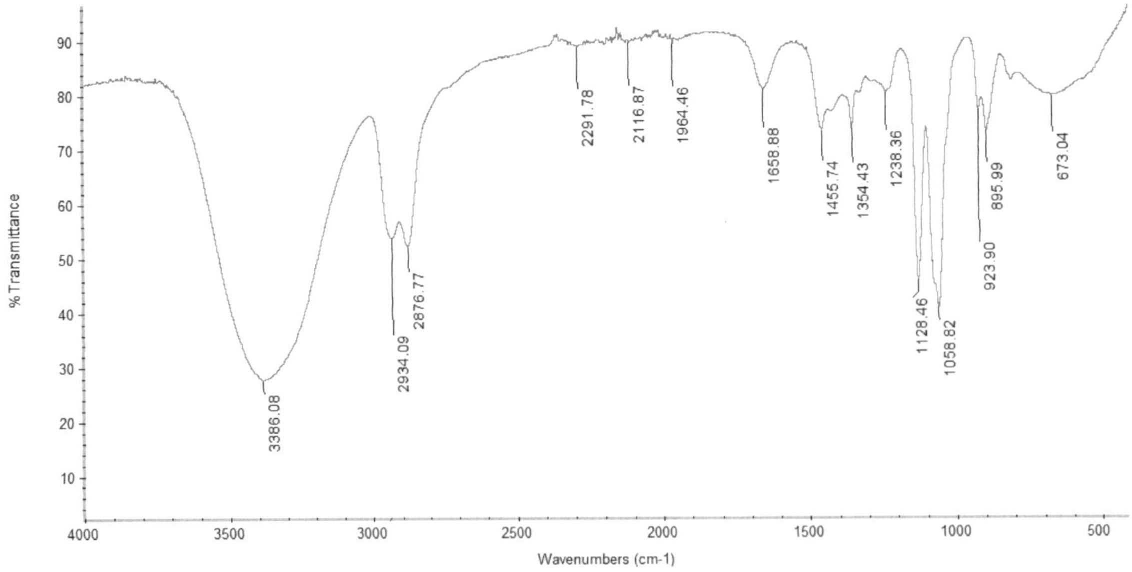

Figure 2. FTIR spectra of CDs synthesized at microwave irradiation power of 119 W and duration of 120 s.

The size and shape of CDs were determined through TEM images as shown in figure 3. In general, all CDs were spherical in shape and with a size less than 10 nm, regardless of the synthesis conditions. This morphology was quite similar with the features reported in other studies [26,27]. The influence of irradiation power on the size of CDs was more apparent as compared to irradiation duration. As can be seen from figure 3, the sizes of CDs were between 3.0-5.0 nm, 2.0-3.5 nm, and 3.0-7.0 nm for microwave irradiation power of 117 W, 385 W, and 700 W, respectively. This could be an indication that the carbonization process was more appropriate for microwave irradiation power of 385 W, while under or over carbonization would lead to the formation of less perfect CDs. The size observation supported the fluorescence intensity as shown in figure 1.

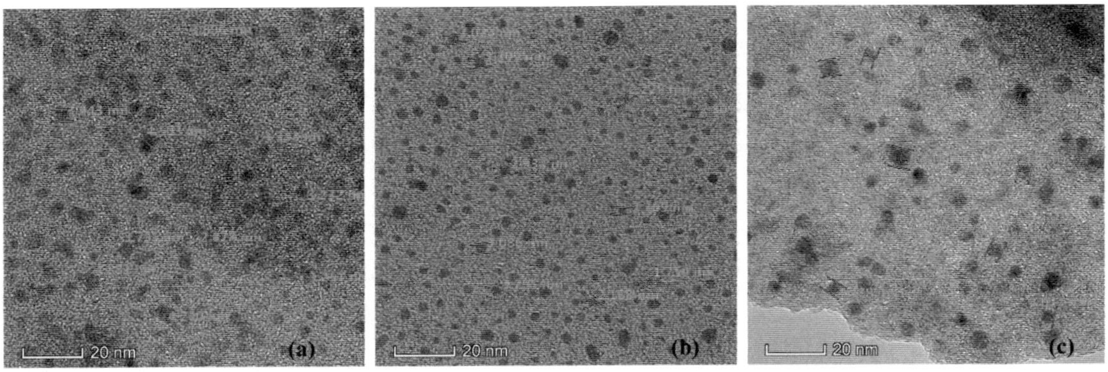

Figure 3. TEM images of CDs samples synthesized under different conditions (a) A3; (b) B4; (c) C1.

3.2. Photoluminescence Property of CDs

Figure 4 (a) illustrates the photoluminescence spectra of CDs for each microwave irradiation power with the highest intensity and control sets of using ultrapure water and DEG as reacting medium (CDs synthesized with microwave irradiation power of 385 W and duration of 90 s). In general, the highest photoluminescence intensity was obtained within the shortest irradiation period of 30 s, regardless of microwave irradiation power (as can be seen in figure 4(b-d)). Nonetheless, prolonging the microwave irradiation duration was found to reduce the photoluminescence intensity for each set of CDs, with more drastic reduction recorded for high microwave irradiation power (700 W). This could be due to phenomena of over carbonization where under long period of exposure to high irradiation power destroyed the structure of CDs and thus resulted in the lower photoluminescence intensity [25]. Subsequently, the best CDs sample in set C (C1) recorded lower photoluminescence intensity as compared to set A and B (figure 4 (a)).

The CDs sample with the highest photoluminescence intensity was obtained at 385 W microwave irradiation power and duration of 30 s. The finding further strengthened the observation where set B CDs emitted the brightest luminescence light when viewed under UV light (figure 1). As mentioned previously, microwave irradiation at lower power could not lead to complete carbonization while higher irradiation power could damage to structure of CDs. Subsequently, the photoluminescence intensity was also affected and appeared to be weaker compared to microwave irradiation power of 385 W. Figure 4 also shows an interesting phenomenon where the photoluminescence intensity of CDs with DEG as the heating solution and ammonia as the doping solution was higher compared to sample with ultrapure water. This could be attributed to the higher boiling point of DEG, which allowed a more complete reaction for the formation of CDs through several possible mechanisms such as intermolecular dehydration and polymerisation that only happened at higher temperature [28]. In addition, the doping agent (ammonia solution) could enhance the photoluminescence property of the CDs.

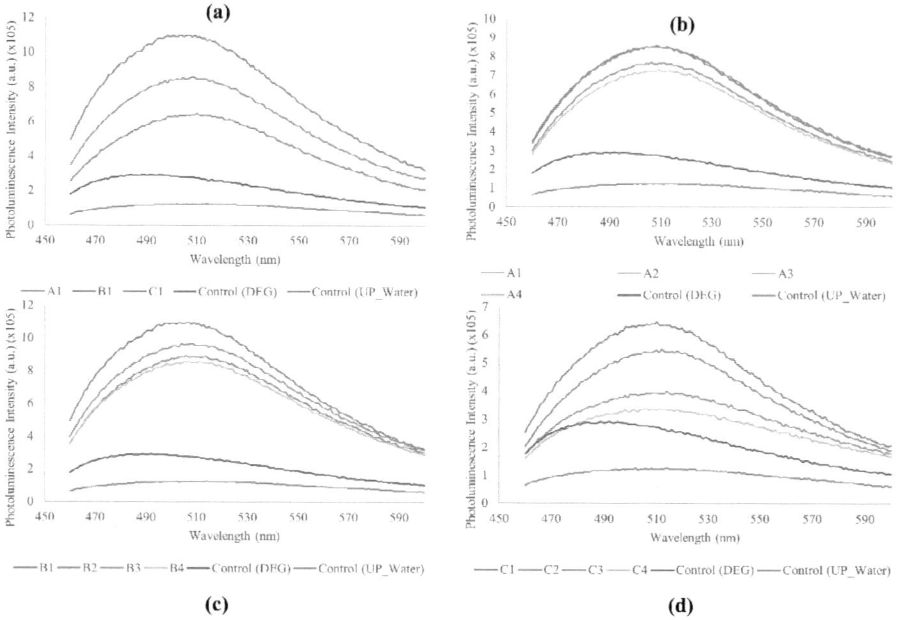

Figure 4. Photoluminescence spectra of CDs at excitation wavelength of 370 nm.

3.3. Detection of Copper (II) Ions

The luminescent property of CDs is of great interest since the quenching of the absorbance could provide an indication of the presence of targeted compounds, e.g. pollutants [29]. This confers the CDs the potential to be used as a sensing material for environmental pollution detection purpose. In this study, the capability of CDs as heavy metal pollutant-copper (II) ions sensing has been evaluated by observing the changes of absorbance spectra with the addition of metal ions at different concentration. It can be seen from figure 5 that the UV-vis absorbance declined with the increase of copper (II) ions concentration, with a reduction range of 7-58 %. This indicated that the heavy metal ions could react with the oxygeneous functional groups present on the CDs surface through chelation for metal ion complexation [30,31]. The formation of complex between CDs and copper (II) ions would facilitate the electron/hole recombination annihilation, which subsequently lead to the reduction of absorbance [32,33]. The decline pattern of absorbance for other CDs (A1 and C1) followed the same trend as in figure 5 and hence was not included here. The range of absorbance decline of A1 CDs and C1 CDs was within 4-28 % and 13-43 %, respectively. The lower reduction of absorbance could be attributed to the under and over carbonization situation, where the amount of CDs was lesser or the property of CDs was damaged during the microwave irradiation process. Hence, it can be deduced that the CDs synthesized from low-cost and easily available OPF could be used as a quick pollutant detection material to cope with environmental pollution issues.

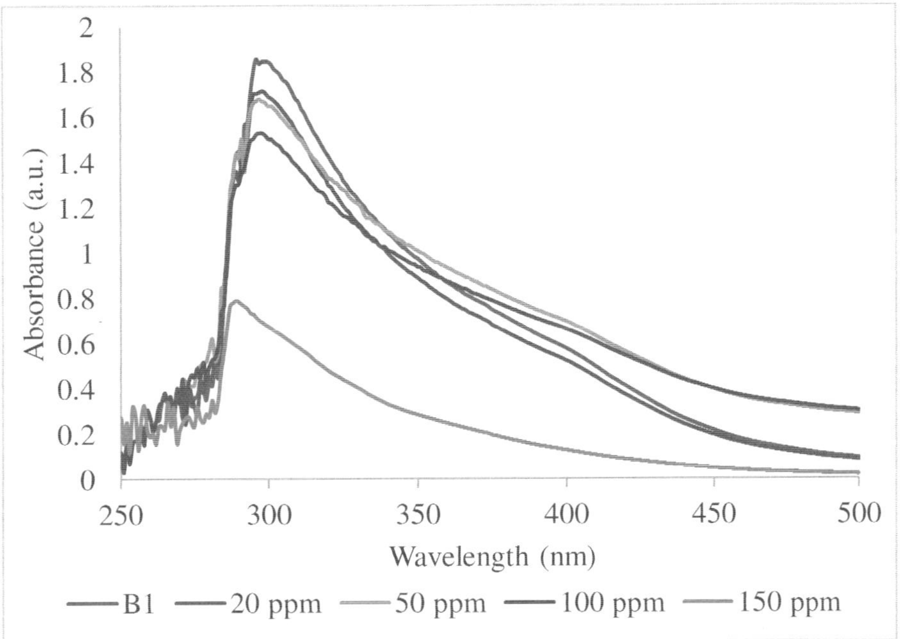

Figure 5. UV-vis absorbance spectrum of B1 on different concentration of lead (II) ions.

4. Conclusion

The OPF has been successfully converted into CDs via microwave-assisted synthesis approach with fluorescence property. It was discovered that microwave irradiation power has a more dominant effect on the fluorescence property as compared to microwave irradiation duration. Microwave irradiation at low (119 W) and high (700 W) power led to incomplete and over carbonization, respectively. This resulted in lower photoluminescence intensity as compared to CDs synthesized from microwave irradiation power of 385 W. On top of that, DEG as the reacting solution enabled a more complete formation of CDs and ammonia as the doping solution further enhanced the fluorescence property of the CDs. In terms of lead (II) ions detection, it was observed that the increase of lead (II) ion concentration reduced the UV-vis absorbance of the CDs solution, indicating there was interaction between CDs and lead (II) ions. Overall, the decline of absorbance values indicated the potential of the CDs synthesized from OPF to be used for lead (II) sensor and detection.

Reference

[1] Baker S N and Baker G A 2010 Luminescent carbon nanodots: Emergent nanolights *Angew. Chemie - Int. Ed.* **49** 6726–44
[2] Tan W B, Huang N and Zhang Y 2007 Ultrafine biocompatible chitosan nanoparticles encapsulating multi-coloured quantum dots for bioapplications *J. Colloid Interface Sci.* **310** 464–70
[3] Cho S J, Maysinger D, Jain M, Röder B, Hackbarth S and Winnik F M 2007 Long-term exposure to CdTe quantum dots causes functional impairments in live cells *Langmuir* **23** 1974–80
[4] Jia X, Li J and Wang E 2012 One-pot green synthesis of optically pH-sensitive carbon dots with upconversion luminescence *Nanoscale* **4** 5572–5
[5] Hu L, Sun Y, Li S, Wang X, Hu K, Wang L, Liang X J and Wu Y 2014 Multifunctional carbon dots with high quantum yield for imaging and gene delivery *Carbon N. Y.* **67** 508–13
[6] Shen J, Zhu Y, Yang X and Li C 2012 Graphene quantum dots: Emergent nanolights for bioimaging, sensors, catalysis and photovoltaic devices *Chem. Commun.* **48** 3686–99
[7] Liu H, Ye T and Mao C 2007 Fluorescent carbon nanoparticles derived from candle soot *Angew.*

Chemie - Int. Ed. **46** 6473–5

[8] Chu K W, Lee S L, Chang C J and Liu L 2019 Recent progress of carbon dot precursors and photocatalysis applications *Polymers (Basel).* **11** 689

[9] Liang Q, Ma W, Shi Y, Li Z and Yang X 2013 Easy synthesis of highly fluorescent carbon quantum dots from gelatin and their luminescent properties and applications *Carbon N. Y.* **60** 421–8

[10] Bhunia S K, Saha A, Maity A R, Ray S C and Jana N R 2013 Carbon nanoparticle-based fluorescent bioimaging probes *Sci. Rep.* **3** 1473

[11] Zhang Z, Sun W and Wu P 2015 Highly photoluminescent carbon dots derived from egg white: Facile and green synthesis, photoluminescence properties, and multiple applications *ACS Sustain. Chem. Eng.* **3** 1412–8

[12] Kumawat M K, Srivastava R, Thakur M and Gurung R B 2017 Graphene quantum dots from mangifera indica: Application in near-infrared bioimaging and intracellular nanothermometry *ACS Sustain. Chem. Eng.* **5** 1382–91

[13] Liao J, Cheng Z and Zhou L 2016 Nitrogen-doping enhanced fluorescent carbon dots: Green synthesis and their applications for bioimaging and label-free detection of Au3+ ions *ACS Sustain. Chem. Eng.* **4** 3053–61

[14] Sahu S, Behera B, Maiti T K and Mohapatra S 2012 Simple one-step synthesis of highly luminescent carbon dots from orange juice: Application as excellent bio-imaging agents *Chem. Commun.* **48** 8835–7

[15] Xu M, He G, Li Z, He F, Gao F, Su Y, Zhang L, Yang Z and Zhang Y 2014 A green heterogeneous synthesis of N-doped carbon dots and their photoluminescence applications in solid and aqueous states *Nanoscale* **6** 10307–15

[16] Jusuf B N, Sambudi N S, Isnaeni I and Samsuri S 2018 Microwave-assisted synthesis of carbon dots from eggshell membrane ashes by using sodium hydroxide and their usage for degradation of methylene blue *J. Environ. Chem. Eng.* **6** 7426–33

[17] Vanessa H, Wenshuo W, Cornelia D, Simon W, Martin T and Wolfgang P 2018 Microwave-assisted one-step synthesis of white light-emitting carbon dot suspensions *Opt. Mater. (Amst).* **80** 110–9

[18] D'souza S L, Chettiar S S, Koduru J R and Kailasa S K 2018 Synthesis of fluorescent carbon dots using Daucus carota subsp. sativus roots for mitomycin drug delivery *Optik (Stuttg).* **158** 893–900

[19] S. T and D. R S 2016 Green synthesis of highly fluorescent carbon quantum dots from sugarcane bagasse pulp *Appl. Surf. Sci.* **390** 435–43

[20] Hsu P C, Chen P C, Ou C M, Chang H Y and Chang H T 2013 Extremely high inhibition activity of photoluminescent carbon nanodots toward cancer cells *J. Mater. Chem. B* **1** 1774–81

[21] Asha Jhonsi M and Thulasi S 2016 A novel fluorescent carbon dots derived from tamarind *Chem. Phys. Lett.* **661** 179–84

[22] Chatzimitakos T, Kasouni A, Sygellou L, Avgeropoulos A, Troganis A and Stalikas C 2017 Two of a kind but different: Luminescent carbon quantum dots from Citrus peels for iron and tartrazine sensing and cell imaging *Talanta* **175** 305–12

[23] Himaja A L, Karthik P S, Sreedhar B and Singh S P 2014 Synthesis of carbon dots from kitchen waste: Conversion of waste to value added product *J. Fluoresc.* **24** 1767–73

[24] Ooi Z X, Teoh Y P, Kunasundari B and Shui S H 2017 Oil palm frond as a sustainable and promising biomass source in Malaysia: A review *Environ. Prog. Sustain. Energy* **36** 1864–74

[25] Tan X W, Romainor A N B, Chin S F and Ng S M 2014 Carbon dots production via pyrolysis of sago waste as potential probe for metal ions sensing *J. Anal. Appl. Pyrolysis* **105** 157–65

[26] Mewada A, Pandey S, Shinde S, Mishra N, Oza G, Thakur M, Sharon M and Sharon M 2013 Green synthesis of biocompatible carbon dots using aqueous extract of Trapa bispinosa peel *Mater. Sci. Eng. C* **33** 2914–7

[27] Yang Y, Cui J, Zheng M, Hu C, Tan S, Xiao Y, Yang Q and Liu Y 2012 One-step synthesis of

amino-functionalized fluorescent carbon nanoparticles by hydrothermal carbonization of chitosan *Chem. Commun.* **48** 380–2

[28] He G, Shu M, Yang Z, Ma Y, Huang D, Xu S, Wang Y, Hu N, Zhang Y and Xu L 2017 Microwave formation and photoluminescence mechanisms of multi-states nitrogen doped carbon dots *Appl. Surf. Sci.* **422** 257–65

[29] Ang W L, Boon Mee C A L, Sambudi N S, Mohammad A W, Leo C P, Mahmoudi E, Ba-Abbad M and Benamor A 2020 Microwave-assisted conversion of palm kernel shell biomass waste to photoluminescent carbon dots *Sci. Rep.* **10** 1–15

[30] Praneerad J, Thongsai N, Supchocksoonthorn P, Kladsomboon S and Paoprasert P 2019 Multipurpose sensing applications of biocompatible radish-derived carbon dots as Cu 2+ and acetic acid vapor sensors *Spectrochim. Acta - Part A Mol. Biomol. Spectrosc.* **211** 59–70

[31] Su Q, Wei X, Mao J and Yang X 2019 Carbon nanopowder directed synthesis of carbon dots for sensing multiple targets *Colloids Surfaces A Physicochem. Eng. Asp.* **562** 86–92

[32] Sun S, Guan Q, Liu Y, Wei B, Yang Y and Yu Z 2019 Highly luminescence manganese doped carbon dots *Chinese Chem. Lett.* **30** 1051–4

[33] Tabaraki R and Sadeghinejad N 2018 Microwave assisted synthesis of doped carbon dots and their application as green and simple turn off–on fluorescent sensor for mercury (II) and iodide in environmental samples *Ecotoxicol. Environ. Saf.* **153** 101–6

Acknowledgments

The authors would like to thank the Ministry of Education Malaysia for funding this work through Fundamental Research Grant Scheme (FRGS) under the grant code FRGS/1/2018/TK10/UKM/02/5.

Hydrogenation of CO_2 to methanol using Cu-based catalyst supported on oxide pellets

N H Berahim[1], A Abu Seman[1], N H Yasin[1], N S Abd Halim[2] and N A Mohd Zabidi[2]

[1] Gas Sustainability Technology Department, PETRONAS Research Sdn Bhd, Lot 3288 & 3289, Off Jalan Ayer Itam, Kawasan Institusi Bangi, 43000 Kajang, Selangor.
[2] Department of Fundamental and Applied Sciences, Institute of Contaminant Management, Universiti Teknologi PETRONAS, 32610 Bandar Seri Iskandar, Perak, Malaysia.

Email: hafizah.berahimjusoh@petronas.com

Abstract. Hydrogenation of CO_2 into methanol is one of the most economical process to reduce CO_2 concentration in the atmosphere. Since methanol is an industrial commodity used in chemical products as well as transportation fuel, this process has gained considerable interest, which enables the effective utilization of CO_2. Nevertheless, the efficiency of direct CO_2 hydrogenation to produce methanol is strongly reliant on the activity of the catalyst. In this regard, the present work highlights the synthesis of methanol, catalytic evaluation and characterization of catalysts Cu/ZnO supported on Al_2O_3 and SBA-15 pellets with the addition of group IV, V and VII metal oxides mixture as promoters. The catalysts were systematically prepared *via* impregnation technique with fixed Cu:Zn and promoter ratio from group VII:V:IV. The synthesized catalysts were characterized by H_2- temperature-programmed reduction (H_2-TPR), field emission scanning electron microscopy (FESEM), X-ray fluorescence (XRF), N_2 adsorption-desorption and N_2O pulse chemisorption method. The crushing strength of the pellets were also tested. Catalytic performances were evaluated for methanol synthesis from CO_2 hydrogenation in a tubular, stainless steel fixed-bed reactor at 250 °C, 2 MPa, gas hourly space velocity (GHSV) 4000 ml/g.h and H_2/CO_2 ratio of 3:1. The tri-promoted Cu/ZnO supported on Al_2O_3 pellet resulted in CO_2 conversion of 13.3 % compared to 11.61 % from that of SBA-15-supported catalyst. However, the catalyst supported on SBA-15 pellet exhibited 54.59% methanol selectivity, whereas Al_2O_3-supported catalyst only resulted in 46.73 % methanol selectivity.

1. Introduction

When fossil fuels such as coal, oil and natural gas are used for power generation, massive volumes of greenhouse gas (GHG) are released into the atmosphere, often in the form of carbon dioxide (CO_2) [1, 2]. Emission of high CO_2 to the environment has caused the mother earth to suffer from greenhouse effect. This later causing the earth climate to rise in temperature by ~1.5 °C in average on month-to-month basis [3]. As a requirement for future consumption, the production of sustainable and clean energy necessitates the use of energy-efficient and modern technology [4, 5]. The use of renewable H_2 to convert CO_2 into value-added products has proven to be an emerging approach for achieving this goal

[6]. At present, there has been an immense effort in converting CO_2 directly to methanol. However, due to the limited equilibrium conversion of CO_2 to methanol at high temperatures, developing effective catalysts to break the thermodynamic barriers is critical. To improve methanol synthesis, a selective catalyst is necessary [7]. Hence, catalyst development for methanol synthesis from CO_2 hydrogenation is still considered as a research gap to be filled.

Since the beginning of the century, there has been a lot of interest in methanol synthesis. This is because of its importance in industrial processes as well as its prospects as a cleaner energy bearer [8]. Being a useful intermediate chemical, methanol yields product such as formaldehyde, dimethyl ether (DME), acetic acid in addition to its use as a primary fuel [9]. Traditionally, methanol has been produced catalytically using various oxide catalyst such as mixed Cr_2O_3-ZnO material and the process took place at high temperature and pressure using natural gas as the feed stock. Methanol synthesis was then innovated by adapting Cu/Zn/Al_2O_3 catalyst and operated at a much lower temperature and pressure [9]. Since then, catalysts for methanol synthesis are being developed industrially by adding other additives as well in the effort to control the sintering effect of Cu and deactivation of catalyst.

According to S.H Kang *et al.*, addition of Al_2O_3 as the third component of Cu-based catalyst helps to disperse and stabilize the Cu/ZnO structure and also prevents the Cu from agglomeration due to the long term process [10]. Although Al_2O_3 has been widely used throughout the industries as it is rather non-expensive and easy to use, the acidic sites of Al_2O_3 can lead to the formation of dimethyl ether (DME) which is non-desirable for methanol synthesis reaction [11]. Apart from that, Al_2O_3 has a low surface area (350 m^2/g), which leads to activity loss throughout catalytic reactions. One of its major drawbacks is its wide pore size distribution (3-15 nm) [12].

An ordered mesoporous structure such as SBA-15 has shown significant potential for material production especially in response to a vast range of environmental and technical demanding [13]. The advancement of SBA-15 ordered mesoporous structure, which is optimized with normal channel systems, well-defined pore size distribution, long-range ordering, and good thermal stability, has also led to advancements in the chemistry and catalysis fields [14]–[16]. The aim to synthesize and apply this solid have gained much attention in order to understand the templating mechanism and exploring its potential.

Metals from group IV, V and VII were incorporated together into the catalyst formulation as they act as structural stabilizer that prevents Cu nanoparticles from sintering. Addition of the mixture of metal promoter increase the basic property of the catalyst that later helps to increase the CO_2 adsorption throughout the reaction [17]. Addition of the metals also led to the formation of new active sites as a result of the increased in metal-support interaction [18]. In this work, combination of metal precursors, Cu/ZnO with promoters from transition metal groups were impregnated on Al_2O_3 and SBA-15 pellet to study effect of each pellet as support to the performance of methanol synthesis from CO_2 hydrogenation reaction.

2. Methodology

2.1. Preparation of Catalyst Pellet Support (Al_2O_3 and SBA-15)
All catalyst supports were pretreated in the tube furnace under Argon flow at temperature 400 °C for 5 hours to remove moisture and impurities.

2.2. Preparation of Cu/ZnO based Catalyst with the Addition of Tri Promoters
15 wt.% of Cu/ZnO with ratio of (70:30) and 3 wt.% of promoters (GVII/GV/GIV) (the specific salts for each group can't be revealed at this stage of study) with ratio of (1:1:1) were synthesized using wetness impregnation method. The amount of each precursor and promoter added were calculated based on the amount of total catalyst mass being prepared. Copper (II) nitrate trihydrate ($Cu(NO_3)_2.3H_2O$), zinc nitrate hexahydrate ($Zn(NO_3)_2.6H_2O$), Group VII salt, Group V salt, and Group IV salt were dissolved in deionized water to produce an aqueous solution. The solutions were stirred for 2 hours at room temperature. Then, the prepared aqueous precursor solutions were added dropwise into a beaker

containing the pellet supports (Al_2O_3 and SBA-15). No further stirring is required to avoid breakage of the pellet. The impregnated samples were dried at room temperature for 24 hours and calcined at 350 °C for 4 hours. The catalyst samples were denoted as CZ(M)A for Cu/ZnO/GVII/GV/GIV/Al_2O_3 and CZ(M)S for Cu/ZnO/GVII/GV/GIV/SBA-15. M represents the tri-promoters (GVII/GV/GIV).

2.3. *Catalyst Characterization*

Reduction study was conducted by Temperature Programmed Reduction (TPR) using a Thermo Finnigan TPD/R/O 110 CE equipped with a thermal conductivity detector (TCD). The catalyst was put in the quartz tube and pre-treated with pure N_2 at 250 °C with a holding time of 1 hr for moisture and impurities removal. The analysis subsequently continued by switching the gas flow to 5 % H_2/Ar (20 mL/min) to a maximum temperature of 950 °C with ramping rate of 10 °C/min and held for an hour. The hydrogen consumption is monitored by a thermal conductivity detector (TCD).

The morphology of the synthesized catalyst was observed using Hitachi SU8000 field emission scanning electron microscope (FESEM).

Textural analysis was performed on a Micromeritics ASAP 2020 analyzer by measuring the nitrogen adsorption/desorption isotherms at -196 °C. Prior to analysis, the sample was degassed at 350 °C with heating rate 10 °C/min. The adsorption data were used to calculate the Brunauer-Emmett-Teller (BET) basic surface area (S_{BET}) and pore volume (V_P) while the Barrett-Joyner-Halenda (BJH) model was used to calculate the mean pore diameter (D_{BJH}) of the isotherm desorption branch [19].

The bulk compositions of active metals impregnated onto the pellet support was determined using XRF method. 3.0 g of calcined catalyst sample was put into the sample holder and the analysis were carried out based on calibration curves obtained using certified method (ASTM E1621).

The surface area of Cu and the degree of dispersion were determined using pulse chemisorption on a Micromeritics Autochem II 2920. The process involved two steps where the first step was reduction of the active metal to Cu^0. Second step involved the injection of the known and constant doses of the reactive gas (N_2O). Injection process continued until it became saturated. With an appropriate calibration, amount of the exact reactive gas used to oxidize the copper surface can be determined which eventually lead to the quantification of the Cu dispersion and surface area.

2.4. *Catalytic Performance Evaluation*

CO_2 hydrogenation reaction was performed in a stainless steel fixed-bed reactor (PID Eng & Tech). The catalysts were reduced in-situ at atmospheric pressure with 5 vol% H_2 in He for 2 hours. After the reduction, the reactor tube was pressurized with He to 2 MPa, then switched to feed gas containing CO_2 and H_2 at a molar ratio of 1:3. The reaction temperature was set at 250 °C for 5 hours with GHSV 4000 ml/g.h [20] and the reactor effluents were analyzed using an on-line gas chromatograph (GC) outfitted with a thermal conductivity detector (TCD) and flame ionization detector (FID). The CO_2 conversion and MeOH selectivity were used to assess the catalytic activity.

3. Results and Discussion

3.1. *The Reduction of Catalysts*

The reduction of the modified catalysts was determined using H_2-TPR profiles, as shown in figure 1. CZ(M)A showed two reduction peaks; a low temperature peak at 366 °C (α peak) due to the reduction of widely distributed CuO surface species, and a higher temperature peak at 566 °C (β peak) owing to the reduction of bulk-like CuO [21, 22]. While for CZ(M)S, the catalyst showed a large reduction peak in the 200 °C to 500 °C region that could originate from the Cu species and other precursor salt species. Table 1 results for CZ(M)S indicates a main reduction peak at 383 °C and a shoulder at 631 °C. The CZ(M)S exhibited an intense and broader peak compared to CZ(M)A. This phenomenon might be due to the support difference and the uniform crystallite size of Cu dispersed on SBA-15 support [23]. Furthermore, the broader peak could also relates to the higher active site exposed to the H_2 where the

IOP Conf. Series: Materials Science and Engineering 1195 (2021) 012009 doi:10.1088/1757-899X/1195/1/012009

surface area of the CZ(M)S is higher than that of CZ(M)A as presented in table 2. The catalysts were reduced at 400 °C before the CO_2 hydrogenation reaction.

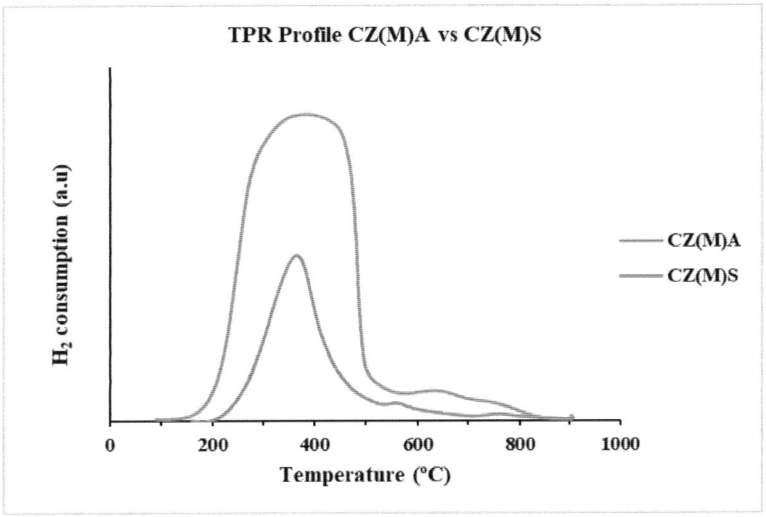

Figure 1. H_2-TPR profiles of CZ(M)A and CZ(M)S.

Table 1. TPR quantification data for CZ(M)A and CZ(M)S.

Catalyst	H_2 Consumption (µmol/g)	Reduction Temperature Peak (°C)	
		T1	*T2*
CZ(M)A	411.12	366	566
CZ(M)S	1258.75	383	631

3.2. Morphology

FESEM images of Al_2O_3, CZ(M)A, SBA-15 and CZ(M)S are shown in figure 2. All samples had an irregular shape, however for the impregnated samples, agglomeration can be observed. The catalysts, CZ(M)A and CZ(M)S contained particles which were distributed onto the supports and filled the pores, which led to the reduction in the surface area of the samples, as reported in table 2 [24].

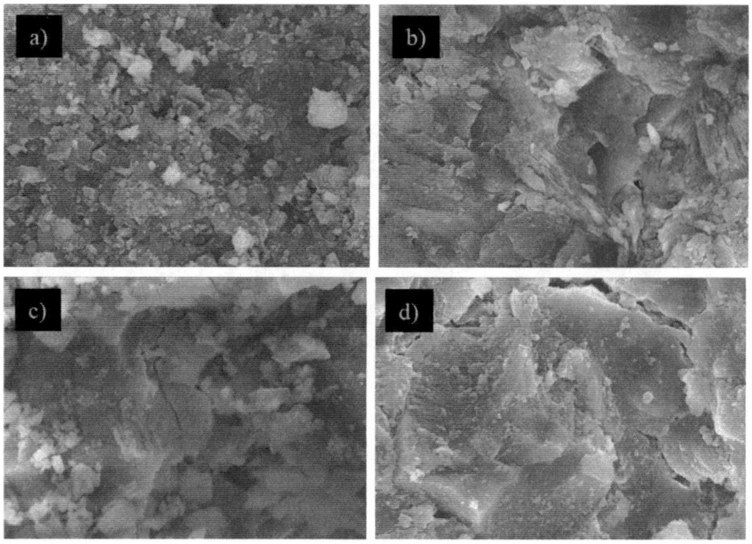

Figure 2. FESEM images of a) Al_2O_3 b) CZ(M)A c) SBA-15 d) CZ(M)S at 5000 magnifications.

3.3. Textural and Physical Properties

Table 2 displays textural properties, namely catalyst surface area (S_{BET}), pore volume (V_P) and pore diameter (D_{BJH}). SBA-15 and loaded catalyst CZ(M) supported on SBA-15 have the highest BET surface area of 447.01 and 339.17 m^2/g, respectively compared to those of Al_2O_3 and its CZ(M) supported catalyst. The loaded samples show slightly lower values of BET surface area and pore volume compared to those of bare supports which indicated that the impregnated metals have occupied the pores of the bare support. On the other hand, an opposite trend was observed on the diameter of the pores. The incorporation of active metals precursor into the pores of the support decreased the surface area and pore volume, however this incorporation leads to the internal pore strain thus increasing the pore diameter [25].

Table 2. Textural properties of the catalysts.

Sample	S_{BET} (m^2/g)	V_P ($cm^{3/}g$)	D_{BJH} (nm)
Al_2O_3	341.52	0.46	4.25
CZ(M)A	302.96	0.43	4.32
SBA-15	447.01	0.49	3.77
CZ(M)S	339.17	0.41	3.83

Table 3 displays the XRF analysis for the bulk compositions of the impregnated active metals. The prepared catalysts at Cu to Zn ratio of 7:3 were impregnated on the pellets with 15 wt.%, while all the tri promoters (M), were loaded with 1 wt.% each. The outcome of XRF analysis were different from the expected value by which theoretically, the amount of Cu should be about 10.5 wt.%, Zn 4.5 wt.% and tri promoters (M) at 1 wt.% each, indicating that the impregnation method did not result in homogeneous metal deposition on the pellet surface.

Table 3. Bulk composition of the catalysts.

Catalyst	Composition of support (wt.%)		Composition of active metals (wt. %)		Composition of promoters (wt. %) Tri-promoters (M)		
	Al	*Si*	*Cu*	*Zn*	*GVII*	*GV*	*GIV*
CZ(M)A	82	-	11.16	3.84	1.67	0.33	1
CZ(M)S	-	82	11.65	3.35	0.52	1.15	1.33

3.4. Chemisorption Analysis

Table 4 shows the findings of N_2O pulse chemisorption. The Cu surface area as well as the dispersion of the active sites over the support were affected significantly by the type of the oxide carriers. The CZ(M)A catalyst exhibits higher Cu metal dispersion (0.50 %) and Cu metal surface area (0.34 m^2/g) as compared to those of CZ(M)S catalyst which indicated a more homogeneous distribution of copper particles on the Al_2O_3 pellet compared to that of SBA-15 support. In addition, the average particle size of Cu, d_{Cu} for CZ(M)A sample was found to be 209 nm, which is significantly smaller than the average particle size of 321 nm for the CZ(M)S sample. Cu particle size is a significant parameter of the catalyst in methanol hydrogenation reaction as smaller Cu particles increase the interfacial area with the adjacent metal oxide, thus enhancing the synergistic effect and increasing the catalyst's efficiency [26].

Table 4. N_2O pulse chemisorption.

Catalyst	Cu metal dispersion, D_{Cu} (%)	Cu surface area, S_{Cu} (m^2/g)	Cu particle size, d_{Cu} (nm)
CZ(M)A	0.50	0.34	209
CZ(M)S	0.32	0.22	321

3.5. Catalytic Performance

Table 5 shows the catalytic performance of CZ(M)A and CZ(M)S catalysts in terms of CO_2 conversion (X_{CO2}) and methanol selectivity (S_{CH3OH}). Under the conditions of the catalytic test, it shows that, Al_2O_3-supported catalyst, CZ(M)A resulted in higher CO_2 conversion of 13.30 % as compared to CZ(M)S at 11.61 %. Even so, an inverse trend was observed for methanol selectivity where CZ(M)S exhibited higher value than that of CZ(M)A at 54.59 % and 46.73 %, respectively. As reported by previous findings [27, 28], the determination of catalytic activity for Cu-based catalysts is heavily dependent on CuO particles, Cu surface area, and small crystallite size, which explains the higher activity exhibited by the CZ(M)A catalyst in terms of CO_2 conversion. Although the CZ(M)S has lower Cu surface area, lower Cu metal dispersion and large Cu particle size as compared to CZ(M)A, it is selective towards methanol formation. It is also well known that catalytic activity is heavily dependent on the interaction among Cu and oxide components for the Cu-based catalysts for methanol synthesis. Even though the CZ(M)S has higher surface area and pore volume, the CO_2 conversion is lower than that CZ(M)A. There is no clear connection between the accessibility factors (S_{BET} and V_p) and catalyst efficiency, which should be governed by other material properties [29, 30]. The methanol yield for both CZ(M)A and CZ(M)S are comparable at 6.22 % and 6.34 % respectively. It is hard to conclude at this preliminary stage, however, methanol selectivity is an important aspect to look at for this reaction as methanol is the targeted product which need to be maximized. In this case, CZ(M)S can be considered as the most efficient catalyst due to its high methanol selectivity. Even though CZ(M)A has higher CO_2 conversion, the amount of CO_2 converted to other products also high which concluded its lower methanol selectivity compared to CZ(M)S. Figure 3 and figure 4 show the time on stream of CO_2 conversion and methanol selectivity for both catalysts. It was found that the reactions for both catalysts were stable over the time-on-stream of 5 hours.

Table 5. Catalyst performance evaluation.

Catalyst	X_{CO2} (%)	S_{CH3OH} (%)
CZ(M)A	13.30	46.73
CZ(M)S	11.61	54.59

reaction conditions: T=250°C, P=2 MPa, H_2/CO_2 (volume ratio) = 3:1, and GHSV=4000 mL/(g·h)

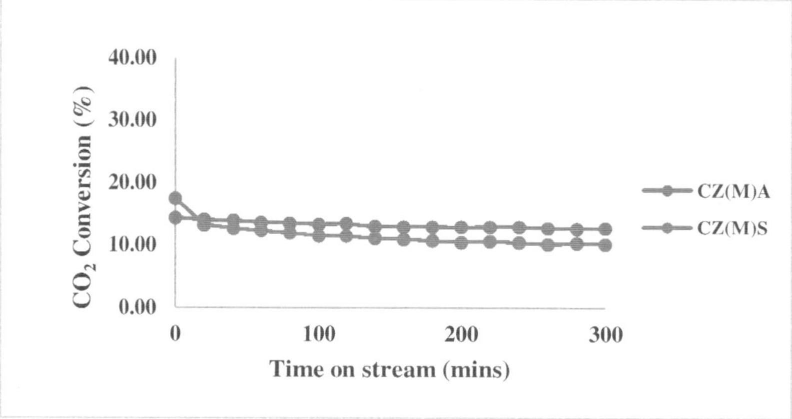

Figure 3. CO_2 conversion time on stream profiles.

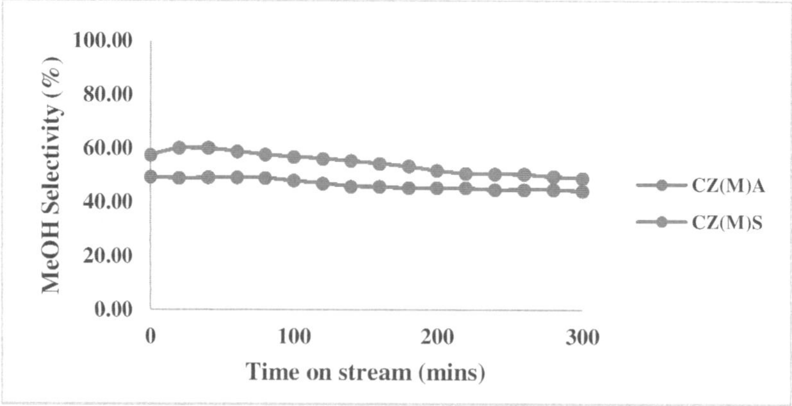

Figure 4. MeOH selectivity time on stream profiles.

4. Conclusion

In this study, Al_2O_3 and SBA-15 pellets were used as catalyst carriers for the tri-promoted Cu/ZnO catalyst. The catalyst reduction, its morphology, textural and physical properties were investigated. The catalytic performance for both CZ(M)A and CZ(M)S catalysts in a CO_2 hydrogenation reaction were evaluated under the same reaction conditions. The results show that CZ(M)A exhibited CO_2 conversion of 13.30 %, which was higher compared to that of CZ(M)S catalyst. However, the CZ(M)S catalyst was more selective towards methanol formation compared to the CZ(M)A catalyst. The methanol selectivity obtained over the CZ(M)S and CZ(M)A catalysts were 54.59 % and 46.73 %, respectively. It is recommended that future work to focus on the optimization of the synthesis technique for the pellet catalyst which will influence its' physicochemical properties. Process reaction optimization such as temperature, pressure, feed ratio and GHSV will also provide a larger picture on the catalytic performance evaluation. Catalyst deactivation study can also be part of the future work to determine the longevity of the catalyst in the reaction.

References

[1] Broumand M, Albert-Green S, Yun S, Hong Z and Thomson M J 2020 Spray combustion of fast pyrolysis bio-oils: Applications, challenges, and potential solutions *Prog. Energy Combust. Sci.* **79** 100834

[2] Watson J, Wang T, Si B, Chen W-T , Aierzhati A and Zhang Y 2020 Valorizationof hydrothermal liquefaction aqueous phase: Pathways towards commercial viability *Prog. Energy Combust. Sci.* **77** 100819

[3] Hawkins E, FæHn T, Fuglestvedt J 2019 The climate spiral demonstrates the power of sharing creative ideas *Bul.l Am. Meteorol. Soc.* **100** 753–6

[4] Gielen D , Boshell F , Saygin D , Bazilian M D , Wagner N and Gorini R 2019 The role of renewable energy in the global energy transformation *Energy Strateg. Rev.* **24** 38–50

[5] Zhang S, Shen Y, Wang L, Chen J and Lu Y 2019 Phase change solvents for post–combustion CO_2 capture: Principle, advances, and challenges *Appl. Energy* **239** 876–97

[6] Saeidi S, Najari S, Hessel V, Wilson K, Keil F J, Concepcion P, Suib L S and Rodrigues E A 2021 Recent advances in CO_2 hydrogenation to value-added products — Current challenges and future directions *Prog. Energy Combust. Sci.* **85** 100905

[7] Etim U J, Song Y and Zhong Z 2020 Improving the Cu/ZnO-based catalysts for carbon dioxide hydrogenation to methanol as a renewable energy storage media *Frontiers in Energy Res.* **8** 545431

[8] Liu X M, Lu G Q, Yan Z F and Beltramini J 2003 Recent advances in catalysts for methanol synthesis via hydrogenation of CO and CO_2 *Ind. Eng. Chem. Res.* **42** 6518–30

[9] Bowker M 2019 Methanol synthesis from CO_2 hydrogenation *ChemCatChem.* 4238–46.

[10] Kang S H, Bae J W, Prasad P S S, Oh J H, Jun K W, Song S L and Min K S 2009 Influence of Ga addition on the methanol synthesis activity of Cu/ZnO catalyst in the presence and absence of alumina *J. Ind. Eng. Chem.* **15** 665–9

[11] Toyir J, Ramírez D L P P, Fierro J L G and Homs N 2001 Highly effective conversion of CO_2 to methanol over supported and promoted copper-based catalysts: Influence of support and promoter *Appl. Catal. B Environ.* **29** 207–15

[12] Said S, Mikhail S and Riad M 2019 Recent progress in preparations and applications of meso-porous alumina *Mater. Sci. Energy Technol.* **2** 288–97

[13] Zholobenko V L, Khodakov A Y, Impéror-Clerc M, Durand D and Grillo I 2008 Initial stages of SBA-15 synthesis: An overview *Adv. Colloid Interface Sci.* **142** 67–74

[14] Beck J S, Vartuli J C, Roth W J, Leonowicz M E, Kresge C T, Schmitt K D, Chu T W, Olson D H, Sheppard E W, McCullen S B *et al.* 1992 A new family of mesoporous molecular sieves prepared with liquid crystal templates *J. Am. Chem Soc.* **114** 10834–43

[15] Zhuang J, Lu B, Gu F, Zhong Z and Su F 2018 Ordered mesoporous Cu–Ca–Zr: A superior catalyst for direct synthesis of methyl formate from syngas *Carbon Resour. Convers.* **1** 174–82

[16] Zhao D, Feng J, Huo Q, Melosh N, Fredrickson G H, Chmelka B F and Stucky G D 1998 Triblock copolymer syntheses of mesoporous silica with periodic 50 to 300 angstrom pores *Science* **279** 548–52

[17] Karelovic A, Bargibant A, Fernández C and Ruiz P 2012 Effect of the structural and morphological properties of Cu/ZnO catalysts prepared by citrate method on their activity toward methanol synthesis from CO_2 and H_2 under mild reaction conditions *Catal. Today.* **197**

109–18

[18] Da Silva R J, Pimentel A F, Monteiro R S and Mota C J A 2016 Synthesis of methanol and dimethyl ether from the CO_2 hydrogenation over Cu·ZnO supported on Al_2 and Nb_2 *J. CO2 Util.* **15** 83–8

[19] Barrett E P, Joyner L G and Halenda P P 1951 The determination of pore volume and area distributions in porous substances I. computations from nitrogen isotherms *J. Am. Chem. Soc.* **73** 373–80

[20] Tasfy S F H, Zabidi N A M, Shaharun M S, Subbarao D and Elbagir A 2017 Carbon dioxide hydrogenation to methanol over Cu/ZnO-SBA-15 catalyst: Effect of metal loading *Defect and Diff. Forum.* **380** 151–160

[21] Zhang Y, Zhong L, Wang H, Gao P, Li X and Xiao S 2016 Catalytic performance of spray-dried $Cu/ZnO/Al_2O_3/ZrO_2$ catalysts for slurry methanol synthesis from CO_2 hydrogenation. Biochem. Pharmacol. 15 72–82

[22] Słoczy J 2003 Effect of Mg and Mn oxide additions on structural and adsorptive properties of $Cu/ZnO/ZrO_2$ catalysts for the methanol synthesis from CO_2 *Appl. Catal. A :General* **249** 129–38

[23] Kruatim J, Jantasee S and Jongsomjit B 2016 Improvement of cobalt dispersion on Co/SBA-15 and Co/SBA-16 catalysts by ultrasounds and vacuum treatments during post-impregnation step *Eng. J.* 21

[24] Tasfy S F H, Zabidi N A M, Shaharun M S and Subbarao D 2015 The role of support morphology on the performance of Cu/ZnO-catalyst for hydrogenation of CO_2 to methanol *AIP Conf. Proc.* 1669

[25] Nguyen L P, Tri V T, Phuong N T, Luong H N and Thuat T T 2018 Correlation between the porosity of γ-Al_2O_3 and the performance of CuO–ZnO–Al_2O_3 catalysts for CO_2 hydrogenation into methanol *Reac. Kinet. Mech. Cat.* **124** 171–185

[26] Natesakhawat S, Lekse J W, Baltrus J P, Ohodnicki P R, Howard B H, Deng X and Matranga C 2012 Active sites and structure-activity relationships of copper-based catalysts for carbon dioxide hydrogenation to methanol *ACS Catal.* **2** 1667–1676

[27] Guo X, Mao D, Lu G, Wang S and Wu G 2010 Glycine–nitrate combustion synthesis of CuO–ZnO–ZrO_2 catalysts for methanol synthesis from CO_2 hydrogenation *J. of Catal.* **271** 178–185

[28] Avgouropoulos G, Ioannides T and Matralis H 2005 Influence of the preparation method on the performance of CuO–CeO_2 catalysts for the selective oxidation of CO *Appl. Catal. B Environ.* **56** 87–93

[29] Witoon T, Permsirivanich T, Donphai W, Jaree A and Chareonpanich M 2013 CO_2 hydrogenation to methanol over Cu/ZnO nanocatalysts prepared *via* a chitosan-assisted co-precipitation method *Fuel Process. Tech.* **116** 72–78

[30] Sun Q, Zhang Y L, Chen H Y, Deng J F, Wu D and Chen S Y 1997 A Novel process for the preparation of Cu/ZnO and $Cu/ZnO/Al_2O_3$ ultrafine catalyst: Structure, surface properties, and activity for methanol synthesis from CO_2+H_2 *J. of Catal.* **105** 92–105

Acknowledgements
The authors acknowledge the support provided by PETRONAS Research Sdn.Bhd. and Universiti Teknologi Petronas (UTP).

Thermogravimetric kinetic analysis of in-situ catalytic pyrolysis of palm oil wastes with the presence of palm oil wastes ash catalyst

S W Hii[1], B L F Chin[1], C L Yiin[2], F R S Majing[1], Z A Jawad[3], H Y Lim[4], S Yusup[4], M N Acda[5], P Unrean[6] and E Rianawati[7]

[1] Department of Chemical Engineering, Faculty of Engineering and Science, Curtin University Malaysia, CDT 250, 98009, Miri Sarawak
[2] Department of Chemical Engineering and Energy Sustainability, Faculty of Engineering, Universiti Malaysia Sarawak (UNIMAS), Kota Samarahan 94300, Sarawak, Malaysia
[3] Department of Chemical Engineering, College of Engineering, Qatar University, P.O. Box: 2713, Doha, Qatar
[4] National HiCoE Thermochemical Conversion of Biomass, Centre for Biofuel and Biochemical Research, Institute of Sustainable Building, Chemical Engineering Department, Universiti Teknologi PETRONAS, Seri Iskandar, Perak 32610, Malaysia
[5] College of Forest Products and Paper Science, University of the Philippines Los Baños, Laguna 4031, Philippines
[6] National Center for Genetic Engineering and Biotechnology (BIOTEC), 113 Thailand Science Park Paholyothin Road, Klong 1, Klong Luang, Pathumthani 12120, Thailand
[7] Resilience Development Initiative, Jl. Imperial Imperial 2, No. 52, Bandung 40135, Indonesia

Email: bridgidchin@curtin.edu.my

Abstract. The thermal degradation and kinetic analysis for oil palm frond (OPF) and oil palm trunk (OPT) with its ashes were investigated using thermogravimetric approach (TGA). OPF ash, OPT ash and its mixtures are used as a natural source of catalysts in the pyrolytic conversion of the palm oil wastes to bioenergy. The TGA experiments were conducted in a range of heating rates of 10-100 K/min from the temperature of 323 K to 1173 K. Coats-Redfern model is applied in this study to evaluate the activation energy (E_A) and pre-exponential factor (A). The average E_A values ranged 24.27-32.36 kJ.mol^{-1} and 41.42-46.10 kJ.mol^{-1} for pyrolysis of OPF and OPT respectively. Meanwhile, the average E_A values ranged 24.27-31.06 kJ.mol^{-1} and 31.77-43.45 kJ.mol^{-1} for catalytic pyrolysis of OPF and OPT respectively.

1. Introduction

Renewable energy is an alternative energy to reduce environmental impacts produced on this earth. From the various types of renewable energy resources, biomass energy is chosen in this present study as it is said to be easily available, does not rely with the climate change and the topography of the country as compared to wind, solar and hydroelectric energy [1]. Besides that, biomass does not released carbon dioxide (CO_2) to the environment as it will absorbs the CO_2 during the process for electricity generation. Furthermore, biomass is considered as the clean energy and heating sources. The production rate of biomass is high and can be easily be obtained to form as one of the energy resources compared to other renewable energy resources.

In this work, palm oil wastes such as oil palm frond (OPF) and oil palm trunk (OPT) were selected as the biomass for pyrolysis process due to its abundant availability in ASEAN countries particularly in Malaysia and Indonesia. It was reported an increasing trend of palm oil production in Indonesia these recent years and successfully achieved an estimated production of 34.5 million tons in year 2016 [2,3]. Meanwhile, an estimated of 19.16 million tons of palm oil production were recorded in Malaysia for that same year. As a result, around 43-45 % of biomass was generated from the palm oil plantation [4].

Palm oil biomass can potentially be used as a biofuel to replace fossil fuels production. It is reported that palm oil produced the highest yield oil when compared with other vegetable oils [5]. There are many industries built that utilise palm oil biomass as a feedstock in the aim of biofuel production and electricity generation [6]. In this work, OPF and OPT are selected as the oil palm biomass in the pyrolysis process. It is reported that the pre-exponential factor (A) of 1.2 x 10^{13} - 1.1 x 10^{17} min^{-1} and activation energy (E_A) of 160 to 199 $kJ.mol^{-1}$ can be achieved in OPF and OPT in pyrolysis process [7, 8]. The presence of chemical composition such as hemicellulose, cellulose, and lignin content provides a potentiality as a source of fuel for thermochemical and biochemical conversion. The chemical composition for OPF in terms of hemicellulose, cellulose and lignin content are reported to be in 31.93 %, 42.12 % and 26.05 % respectively [9]. Meanwhile, the hemicellulose, cellulose, and lignin in OPT are 30.36 %, 50.78 % and 17.87 % respectively [9].

There are many studies found in literature using palm oil wastes in pyrolysis process with the presence of commercial catalyst. Cheah *et al.* [10] had investigated the catalytic pyrolysis for OPF with the presence of commercial HZSM-5 zeolite and graphite nanofiber (GNF). The authors had discovered that GNF catalyst performed better than HZSM-5 zeolite in the OPF pyrolysis by achieving a conversion performance between 66.57 wt% and 72.40 wt% [10]. Lim and Andrésen [11] had studied on the catalytic effect of boric oxide on empty fruit bunches (EFB) and OPF in a fixed pyrolysis bed reactor. It is reported that boric oxide has the ability to remove the hydroxyl and methoxy groups in the bio-oil with an efficiency of 50 %-80 % [11]. Kabir *et al.* [12] conducted a comparative study of palm oil mesocarp fiber and OPF in a slow-heating fixed reactor (2.2 cm × 35 cm) with alkali metal as catalyst. It is reported that the highest bio-oil yield produced is 50 wt% and 47 wt% for palm oil mesocarp fiber and OPF, respectively at temperature 550 °C and 600 °C [12]. The role of the catalyst enables the reaction temperature to be lowered, quality of pyrolysis products can be improved in terms of stability and reduction of the oxygen compounds, and increased of product selectivity [13]. Few examples of different catalysts utilised in pyrolysis process are molecular sieve, metal oxides, transition metals had been evaluated on the bio-oil yields and properties [14].

To the authors best knowledge, it is discovered that the kinetic analysis for catalytic of OPF and OPT pyrolysis utilising its ashes as a natural catalyst had not been conducted before. However, there are studies conducted using rice hull ash in pyrolysis of rice husk [15] and coal bottom ash in pyrolysis of rice husk [16]. Therefore, in this work, the kinetic analysis for pyrolysis of OPF and OPT are investigated with the absence and presence of OPF ash, OPT ash, and OPF/OPT ash using thermogravimetric approach (TGA). The catalysts such as OPF ash, OPT ash, and OPF/OPT ash are utilised in the pyrolysis process of palm oil wastes to enhance the biomass to biofuel production. Coats-Redfern model (iso-conversional) is used to evaluate the kinetic parameters.

2. Experimental materials and methods

2.1. Biomass preparation

Both OPF and OPT are obtained from a local palm oil mill, Woodman, Miri Sarawak. Both of these biomass are dried using an oven until the moisture content are less than 10 wt% and grounded into powder form in a particle size less than 500 μm. The ultimate analysis for OPF are 40.5 wt% carbon, 6,8 wt% hydrogen, 3.5 wt% nitrogen, 0.3 wt% sulphur, and 48.9 wt% oxygen using LECO CHNS-932 elemental analyzer. The ultimate analysis for OPT are 42.4 wt% carbon, 7.3 wt% hydrogen, 0.3 wt% nitrogen, 0.7 wt% sulphur, and 49.3 wt% oxygen.

2.2. *Catalyst preparation*

The catalysts that are used for pyrolysis process are from the ashes produced from the combustion process of OPT and OPF through continuous heating process with temperature of 973 K for 4 hours using Laboratory Muffle Furnace, Carbolite ELF 11/14B. The ashes of OPT and OPF are grounded to fine particle sizes less than 500 μm. Thereafter, the ashes are then sent to oven for 12 hours at 373 K for drying purpose and left to cool down to 25 °C prior for TGA analysis.

2.3. *Thermogravimetric analysis (TGA) approach*

Pyrolysis experiments are conducted using a thermogravimetric analyser (TGA-DSC 1, Mettler Toledo). An approximate of 10 mg of samples are placed in the ceramic crucible under non-isothermal condition. The samples are pyrolysed from room temperature to 1173 K at different five heating rates of 10 K.min^{-1}, 20 K.min^{-1}, 30 K.min^{-1}, 50 K.min^{-1}, and 100 K.min^{-1}. Thereafter, the temperature of 1173 K is maintained for 10 minutes when it reached 1173 K. The biomass are mixed with the ashes in the weight ratio of 1:1 meanwhile the biomass/OPF ash/OPT ash samples are blended in the weight ratio of 1:0.05:0.05.

2.4. *Kinetic theory*

The kinetic principles that used in pyrolysis process are based on iso-conversional principle. This principle indicates the rate of the reaction will be affected by the temperature only with a basis of constant conversion rate [17]. Thus, thermal degradation of the lignocellulose component in biomass will occurred simultaneously within a conversion rate of the time, proposed a reaction mechanism of the pyrolysis process as shown as below equation (1).

$$Biomass\ (Palm\ oil\ waste) \xrightarrow{k} Volatiles + Char \qquad (1)$$

where k refers to pyrolysis conversion rate constant in transforming the biomass (palm oil waste) into volatiles and char. According to Sadhukhan *et al.* [18], the conversion rate of the pyrolysis process is assumed as single step process which follows Arrhenius and show in equation (2) as shown below.

$$k = Ae^{\left(-\frac{E_A}{RT}\right)} \qquad (2)$$

where A represents pre-exponential factor (s^{-1}), E_A represents activation energy (kJ.mol^{-1}), R represents the universal gas constant (8.314 J.(K.mol)$^{-1}$) and T represents the absolute temperature in Kelvin (K). The degradation rate of the reaction can be further expressed in equation (3) by showing the rate conversion of the biomass from solid to volatile state.

$$\frac{d\alpha}{dt} = k(T) \cdot f(\alpha) \qquad (3)$$

The time rate of degradation or conversion, $\frac{d\alpha}{dt}$ represents the conversion rate which is temperature-dependant and α is the degree of conversion fraction of the sample in the respective time. The degree of conversion fraction is defined as the decomposition rate of the biomass sample at any range of the temperature in thermogravimetric analysis which is expressed in equation (4) as shown below.

$$\alpha = \frac{m_i - m_t}{m_i - m_f} \tag{4}$$

where m_i represents the initial mass of biomass sample, m_t represents the mass of the sample at the given time, t, and m_f represents the final mass of sample

By substituting equation (2) into equation (3) to obtain equation (5) as depicts below.

$$\frac{d\alpha}{dt} = Ae^{\left(-\frac{E_A}{RT}\right)} \cdot f(\alpha) \tag{5}$$

Non-isothermal method is commonly used in a solid-state kinetics where the heating rate, β is remain constant and defined in equation (6).

$$\beta = \frac{dT}{dt} \tag{6}$$

By combining both equations (5) and (6), which gives equation (7) as follows:

$$\frac{d\alpha}{dT} = \frac{A}{\beta} e^{\left(-\frac{E_A}{RT}\right)} \cdot f(\alpha) \tag{7}$$

By integrating equation (7) above with the function of $g(\alpha) = \int_0^\alpha \frac{d\alpha}{f(\alpha)}$, which simplifies the equation for numerical approximation solving and gives equation (8)

$$g(\alpha) = \int_0^\alpha \frac{d\alpha}{f(\alpha)} = \frac{A}{\beta} \int_{T_0}^{T} e^{\left(-\frac{E_A}{RT}\right)} dT \tag{8}$$

In iso-conversional kinetics, the pyrolysis kinetic order is specified under first order reaction, which gives equation (9) as shown below.

$$g(\alpha) = -\ln(1 - \alpha) \tag{9}$$

2.5. Kinetic models

Coats-Redfern model is the selected kinetic iso-conversional model to evaluate the kinetic parameters of the biomass pyrolysis in this study. The equation of the models will be linearised and plotted in a linear graph to verify the values of *A* and the E_A of the reaction. Table 1 shows the linear expression of

Coats-Redfern models with the respective x-axis, y-axis, y-intercept and slope. By plotting y-axis versus x-axis of each model, the E_A can be obtained from the slope of the graph while the pre-exponential can be obtained from the y-intercept of the graph.

Table 1. Linear expression of three different models

Model	Equation	y-axis	x-axis	Equation
Coats-Redfern	$\ln\left(\frac{-\ln(1-\alpha)}{T^2}\right) = \ln\left(\frac{AR}{\beta E_A}\right) - \frac{E_A}{RT}$	$\ln\left(\frac{-\ln(1-\alpha)}{T^2}\right)$	$\frac{1}{T}$	(10)

3. Results and discussion

3.1. Thermal degradation behaviour of OPF and OPT in pyrolysis process

In the present study, the thermal degradation effect of the OPF and OPT with or without the presence of catalyst were analyzed by using thermogravimetric analysis (TGA) equipment. Figure 1 shows the curves of the thermogravimetric (TG) and derivative thermogravimetric (DTG) of OPF and OPT in non-catalytic and catalytic pyrolysis reaction with five different types of heating rates 10 to 100 K.min^{-1}. The TG curve is used to determine the weight loss of the sample in weight percentage with the increased of temperature meanwhile the DTG curve is used to analyse the initial, maximum and final biomass degradation temperature. Based on this figure, the trend of the curves can be categorized into three main stages of thermal degradation namely, Stage I: dehydration, Stage II: main devolatilisation and Stage III: decomposition of carbonaceous components.

Stage I of thermal degradation for OPF and OPT pyrolysis started from room temperature to 495 K and 500 K respectively with a slight reduction of weight fraction of OPF and OPT. This is where the dehydration process takes place to reduce the moisture content and light volatiles of the component [7]. The surface tension will tend to cause evaporation to both moisture content and the water bounded of the OPF and OPT biomass surface. When the temperature reached the end of Stage I decomposition, the internal forces such as intermolecular and intramolecular forces start to break the hydrogen bond in lignocellulose component and form water molecules to generate the char. When most of the moisture content in the OPF and OPT have been evaporated, the weight of the OPF and OPT decreases dramatically in all five different heating rates. This indicated that the biomass had experienced Stage II decomposition in temperature range from 495 to 750 K and 500 to 710 K for OPF and OPT respectively. This is the phase where the organic compound of OPF and OPT is undergoing main devolatilisation process for carbon and volatile component extraction. In this stage, transglucosidation process is happened where the intermediate cellulose component starts to degrade to levoglucosan and oligosaccharides [19]. In stage III, a slight weight loss of OPF and OPT with the temperature range from 750 to 1173 K and 500 to 1173 K respectively due to occurrence of passive devolatilisation process. Passive devolatilisation process is where the carbonaceous component started to decompose and both CO and CO_2 started to form by vaporisation of non-volatile carbon compounds that contained in the solid residue of the biomass [20]. Besides that, the aromatisation and bond cracking happens to convert the cellulose and carbonaceous material to char and volatile matter.

Figure 1 shows that there are two exothermic peaks in between the temperature range from 500 to 820 K for both OPF and OPT occurred in this curve for the five different heating rates. A lateral shift to the right hand side in the thermogravimetric curve was observed as the heating rates increases. The first peak occurred due to the thermal decomposition of protein and carbohydrates component while the decomposition of lignin happened in the second peak. The second peak for both OPF and OPT are relatively higher than the first peak due to high content of lignin content in oil palm wastes. It is where the main decomposition stage happened in the second decomposition stage as mentioned in the description for the TG curve. In last stage (stage III), a small endothermic peak indicates that the remaining residue undergoes final decomposition stage after the temperature reached 820 K. Based on figure 1, the optimum mass loss was observed to be in the stage II where the temperature ranged from 495 to 750 K. The mass loss was observed in the average range of the residues weight (%) from the initial temperature, $T_{initial}$ until the final temperature of stage II, T_{final} for the main pyrolysis stage (stage II).

3.2. Effect of heating rates on OPT and OPT thermal decomposition
It is said that high heating rate of the process will favors the depolymerisation of hemicellulose and cellulose and volatile cracking in secondary reactions. Additionally, the volatile compound in biomass will decreased due to the generation of polyhydroxy aromatics component [21]. Fast pyrolysis usually used high temperature for the process in order to generate high yield of bio-char, oil and non-condensable gases. From figure 1, it is seen that both TG and DTG curves increased with heating rates. The trend is in similar behavior with other biomass as reported by Soon et al. [7], Khan et. al [8], Fong et al. [22], Liew et al. [23], Loy et al. [24], Majid et al. [25] and Teng et al. [26].

3.3. Effect of catalyst on OPF and OPT thermal decomposition
This study enhances the important of catalytic pyrolysis process using renewable resources such as OPF ash and OPT ash catalyst. Chen et al. [20] reported that the EA of the palm oil pyrolysis was decreased significantly using rice husk ash as the natural catalyst in representing CaO catalyst. The main purpose of the study is to enhance the reusability of the oil palm wastes and improved the bio-fuel production [5].

In this present study, the EA of OPF was reduced from 28.49 kJ.mol-1 to 23.66 kJ.mol-1 with the addition OPF/OPT ash. Furthermore, the DTGmax reduced significantly from 5.74 to 4.43×10-2 %.min-1 at the heating rate of 100 K.min-1. A similar behavior is observed for OPT pyrolysis when comparing with literature findings by Liu et al. [21] and Chen et al. [27]. This can be further explained that the OPF/OPT ash is able to transform the complicated reaction into a semi simple-complex reaction involving less EA. The semi-complex reaction refers to the reagent involved in the system which can move freely meanwhile the activated complex have restricted movement.

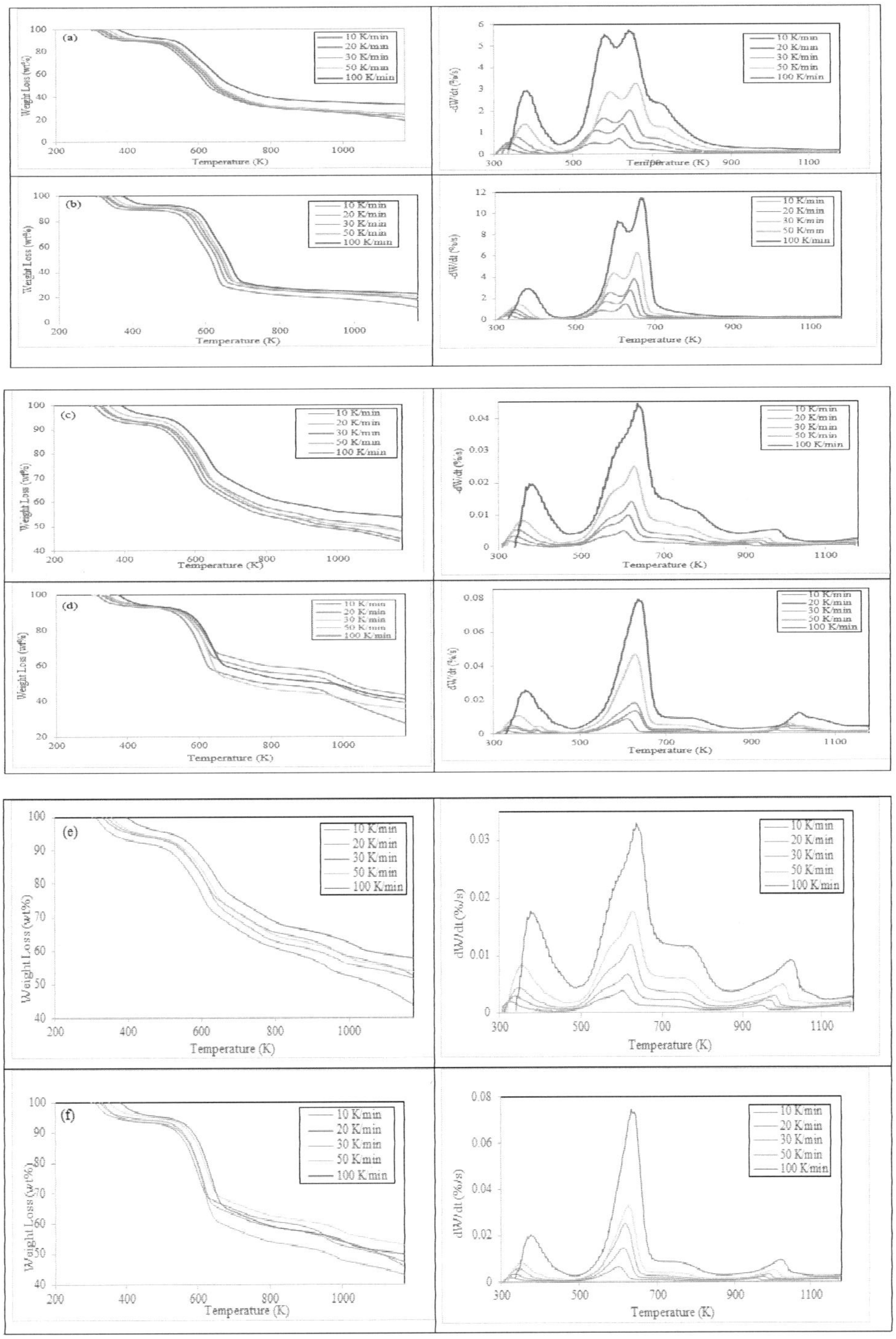

Figure 1. TG and DTG graphs of (a) non-catalytic pyrolysis of OPF, (b) non-catalytic pyrolysis of OPT, (c) pyrolysis of OPF using OPF ash catalyst, (d) pyrolysis of OPT using OPT ash catalyst, (e) pyrolysis of OPF using OPF/OPT ash catalysts (f) pyrolysis of OPT using OPF/OPT ash catalysts.

3.4. Kinetic study

Coats-Redfern, an is conversional method was selected to identify the kinetic mechanism involved in the OPF and OPT pyrolysis with the presence of OPF ash and OPT ash as catalyst in this study. Based on the TG curve, the main devolatilization stage (Stage II) was selected as the main pyrolysis stage for both OPF and OPT in five different heating rates respectively. This model is applied to determine the kinetic parameters such as E_A and A for the pyrolysis of OPF and OPT in the absence and presence of catalyst. The role of E_A determines the amount of energy that reactants need to overcome the pathway before the chemical reaction meanwhile A represents the frequency of the collisions of the chemical reactants molecules during pyrolysis reaction [21]. The values of E_A and A were calculated based on the slope and intercept point of the kinetic plot. It was found that the average of E_A and A for pure OPF were 28.49 kJ.mol^{-1} and 1.12×10^6 min^{-1}. With the presence of OPF ash, the values of E_A and A are slightly reduced to 28.02 kJ/mol and 1.15×10^6 min^{-1}. With the presence of OPF/OPT ash catalyst, the values of E_A and A obtained have been reduced to 23.66 kJ.mol^{-1} l and 1.16×10^6 min^{-1} respectively when compared without catalyst. On the other hand, the average of E_A and A for pure OPT were 44 kJ.mol^{-1} and 5.78×10^6 min^{-1} respectively. With the presence of OPT and OPF/OPT ash, the values of E_A and A reduced to the range of 36.86 – 38.86 kJ.mol^{-1} and 1.14×10^6 - 1.16×10^6 min^{-1}. The presence of these biomass ash catalysts (OPF ash, OPT ash, and OPF/OPT ash) in the pyrolysis process enhances the rate of secondary reaction. The diffusivity from the OPT and OPF molecules will takes place on the porous structure of these biomass ash catalysts, followed by the metal oxides and char components existed in these biomass ash catalysts which allow the continuation of the degradation reaction. According to Xu *et al.* [20], they mentioned that good catalytic effect reduces the values of E_A with increasing rate of reaction in the pyrolysis process.

Table 2 shows the average correlation coefficients (R^2) values for Coats-Redfern model for five different heating rates. It was observed that the R^2 values were above 95 % for the studied heating rates and conclude that this model is suitable for kinetic analysis with high accuracy of the results. As seen in table 2, the average values of the E_A and A for OPF pyrolysis are observed to be 28.49 kJ.mol^{-1} and 1.12×10^6 min^{-1}. Besides that, it was observed that the values of E_A and A increased with increasing heating rates from 10 to 100 K.min^{-1} which is in good agreement with the findings found by Soon *et al.* [7]. It is found that the presence of OPF/OPT ash in OPF pyrolysis had reduced both the E_A and A values. The E_A values had been reduced from 28.02 to 23.66 kJ.mol^{-1} and A values had been reduced from 1.15×10^6 to 1.11×10^6 min^{-1} when OPF/OPT ash is added in OPF pyrolysis. On the other hand, the average values of the E_A and A for OPT pyrolysis are observed to be 44 kJ.mol^{-1} and 5.78×10^6 min^{-1} respectively. With the presence of OPF/OPT ash catalyst in the OPT pyrolysis, the values of E_A and A were increased from 36.86 to 38.86 kJ.mol^{-1} and 1.14×10^6-1.16×10^6 min^{-1} respectively. The observation of the results found in this study is in accordance with the literature as reported by Gan *et al.* [28] where the A were range from 1.1×10^6 min^{-1} to 189.4×10^9 min^{-1} using Coats-Redfern model. It was observed that the pyrolysis of OPF with OPF/OPT ash catalyst gave the lowest E_A and A for all samples using Coats-Redfern model.

Table 2. Kinetic evaluation for the catalytic and non-catalytic pyrolysis of OPF and OPT using Coats-Redfern model.

		Kinetic analysis		
	β (K.min^{-1})	E_A (KJ.mol^{-1})	A (min^{-1})	R^2
OPF	10	24.27	2.43×10^5	0.9881
	20	26.79	5.09×10^5	0.9955
	30	28.81	7.85×10^5	0.9976
	50	30.22	1.34×10^6	0.9980
	100	32.36	2.74×10^6	0.9961
	Average	28.49	1.12×10^6	0.9951
OPF - OPF ash	10	24.27	2.24×10^5	0.9950
	20	28.14	5.26×10^5	0.9936
	30	26.77	7.84×10^5	0.9901
	50	31.06	1.38×10^6	0.9920
	100	29.89	2.82×10^6	0.9792
	Average	28.02	1.15×10^6	0.9700
OPF - OPF/OPT ash	10	18.66	2.19×10^5	0.9672
	20	22.13	4.87×10^5	0.9871
	30	26.05	7.88×10^5	0.9911
	50	24.44	1.29×10^6	0.9906
	100	27.02	2.75×10^6	0.9887
	Average	23.66	1.11×10^6	0.9849
OPT	10	41.42	2.44×10^5	0.9878
	20	46.10	4.57×10^5	0.9871
	30	43.97	7.48×10^5	0.9762
	50	44.48	1.26×10^6	0.9636
	100	44.00	2.67×10^6	0.9504
	Average	44.00	5.78×10^6	0.9730
OPT - OPT ash	10	38.89	2.68×10^5	0.9795
	20	34.60	5.42×10^5	0.9715
	30	33.61	8.64×10^5	0.9612
	50	43.45	1.29×10^6	0.9614
	100	33.73	2.87×10^6	0.9160
	Average	36.86	1.16×10^6	0.9579
OPT - OPT/OPF ash	10	31.77	2.72×10^5	0.9430
	20	39.71	5.34×10^5	0.9664
	30	40.77	7.89×10^5	0.9976
	50	40.13	1.37×10^6	0.9590
	100	41.92	2.76×10^6	0.9417
	Average	38.86	1.14×10^6	0.9615

4. Conclusions

A kinetic analysis for pyrolysis of various palm oil wastes such as oil palm frond (OPF), and oil palm trunk (OPT) was successfully conducted with the absence and presence of OPF ash, OPT ash, and OPF/OPT ash via thermogravimetric approach. Coats-Redfern model had successfully achieved the regression coefficients above 95 % in pyrolysis process utilizing these feedstock. The degradation results were found in this sequence: OPF-OPF/OPT ash > OPF-OPF ash > OPF > OPT-OPT ash > OPT-OPF/OPT ash > OPT.

Acknowledgments
We would like to acknowledge ASEAN Science Technology and Innovation Fund (ASTIF) on the project grant which entitled, "Sustainable Production of Biofuels in the ASEAN Region from Oil Palm Residues", and JASTIP-Net 2019 Grant Award under the category of Energy and Environment Joint Laboratory for their financial support.

References
[1] Chin B L F, Gorin A, Chua H B and Twaiq F 2016 *J. Energy Inst.* **89** 713
[2] Nambiappan B, Ismail A, Hashim N, Ismail N, Shahari D N, Nik Idris N A, Omar N, Salleh K
[3] M, Mohd Hassan N A and Kushairi A 2018 *J. Oil Palm Res.* **30** 13
[4] Rafiie S A K 2018 Oil Palm *Ind. Eco. J.* **18** 25
[5] Abdullah N, Sulaiman F and Aliasak Z 2013 *AIP Conf. Proc.* **1528** 75
[6] Kurnia J C, Jangam S V, Akhtar S, Sasmito A P and Mujumdar A S 2016 *Biofuel Res. J.* **9** 332
[7] Yusniati L P and Sulaiman O K 2018 *J. Phy. Conf. Series* **1007** 012053
[8] Soon V S Y, Chin B L F and Lim A C R 2016 *IOP Conf. Sci. Eng.* **121** 012004
[9] Khan Z, Yusup S and Murni M A 2011 *Proc. First Conf. Clean Energy Tech.* 27
[10] Lai L and Ani I 2013 *BioResources* **8** 2792
[11] Cheah K W, Yusup S, Muda N S and Abdul Kapor N Z 2016 *Int. J. Biomass Renew.* **5** 14
[12] Lim X Y and Andrésen J M 2011 *Fuel Proc. Technol.* **92** 1796
[13] Kabir G, Mohd D A T, Hameed BH 2017 *Bioresour. Tech.* **241** 563
[14] Johansson A N, Sandström L, Olov G W Ö, Jilbero H 2018 *J. Anal. Appl. Pyro.l* **134** 102
[15] Zhang L, Bao Z, Xia S, Lu Q, Walters K B 2018 *Catalysts* **8** 659
[16] Loy A C M, Gan D K W, Yusup S, Chin B L F, Lam M K, Shahbaz M, Unrean P, Acda M N and Rianawati E 2018 *Bioresour Technol.* **261** 213
[17] Loy A C M, Yusup S, Lam M K, Chin B L F, Shahbaz M, Yamamoto A and Acda M N 2018 *Energy Conserv. Manage.* **165** 541
[18] Kan T, Strezov V and Evans T 2016 *Energy Fuel* **30** 1564
[19] Sadhukhan A K, Gupta P and Saha R K 2009 *Bioresour. Technol.* **100** 3134
[20] Urych B 2014 *J. Sustain. Mining* **13** 3
[21] Xu Y, Liu Z, Peng Y, You T and Hu T 2017 *J. Renew. Sustain. Energy* **9** 063105
[22] Liu Y Q, Lingo R X, Lim J, Wang R Y and Aparat M 2012 *Ind. Eng. Chem. Res.* **51** 10320
[23] Fong M J B, Loy A C M, Chin B L F, Lam M K, Yusup S, Jawad Z A 2019 *Bioresour. Technol.* **289** 121689
[24] Liew J X, Loy A C M, Chin B L F, AlNouss A, Shahbaz M, Al-Ansari T, Govindan R and Chai Y H 2021 *Renew. Energy* **170** 948
[25] Loy A C M, Yusup S, How B S, Yiin C L, Chin B L F, Muhammad M, Gwee Y L 2021 *Bioresour Technol.* **294** 122089
[26] Majid M, Chin B L F, Jawad Z A, Chai Y H, Lam M K, Yusup S and Cheah K W 2021 *Bioresour. Technol.* **329** 124874
[27] Teng S Y, Loy A C M, Leong W D, How B S, Chin B L F and Máša V 2019 *Bioresour. Technol.* **292** 121971
[28] Chen G Y, Shan R, Shi J F and Yan B B 2015 *Fuel Process. Technol.* **133** 8
[29] Gan, D K W, Loy A C M, Chin B L F, Yusup S, Unrean P, Rianawati E and Acda M N 2018 *Bioresour. Technol.* **265** 180

Inter-entities symbiotic relationships with the use of multi-period methodology in energy planning

K G H Kong[1], S L Y Lo[1], B S How[1], W D Leong[2] and J Sunarso[1]

[1] Research Centre for Sustainable Technologies, Faculty of Engineering, Computing and Science, Swinburne University of Technology Sarawak Campus, Jalan Simpang Tiga, Kuching, Sarawak 93350, Malaysia
[2] School of Engineering, Chemical Engineering Discipline, Monash University Malaysia, Jalan Lagoon Selatan, Bandar Sunway, Selangor Darul Ehsan 47500, Malaysia

Email: bshow@swinburne.edu.my

Abstract. The deficiency of natural resources and serious climate change have driven the global community to optimize energy planning using various process integration approaches. The inter-entities energy planning that allows internal sharing of resources poses a great potential to enhance energy planning. It is believed that the effective management of such relationships is crucial to gaining collaborative synergies, which provide economic benefits and minimize environmental impact. The developed inter-collaborated energy sharing model gives a handy lens to evaluate the effectiveness of the suggested inter-entities collaboration and how it provides economic benefits for the involved "players". To demonstrate the economic viability of the inter-entities' energy planning, an energy sharing model is developed and applied to an illustrative case study that involved two entities. The results show that, when energy sharing is enabled, the involved entities can reduce their monthly electricity bill by 16.72 % (MYR 14940.73) for entity 1 and 14.29 % (MYR 14218.50) for entity 2, with a 20 % carbon emission constraint limit.

1. Introduction and Literature Review

Energy has always been the main fuel for the nation's economy as well as mankind's development worldwide. However, the exploitation of energy comes assuredly with climate change due to the increasing consumption of fossil-based energies for industrialization and urbanization [1,2]. The impacts of climate change have been experienced drastically in every nook and cranny, urging the need in limiting climate change and adapting the inevitable [3-4]. Numerous actions have been implemented to focus on climate change issues, but yet, they have been insufficient [5]. International Panel on Climate Change (IPCC) [4] stated that greenhouse gases (GHG) emissions need to be reduced by 50-85 % by 2050 compared with those at 2000 to limit global warming to 1.5 °C.

Nonetheless, the global GHG emissions are not decreasing at a desired rate [5]. Given the global energy-related emissions in 2020 is about 31.5 Gt, it is still far to achieve a "net-zero" emission goal, which is planned to be achieved by 2050 [5,6]. To that extent, efforts are tremendously required especially in developing carbon-constrained energy planning to reduce GHG emissions and mitigate climate change. Many developing countries have already taken their actions by focusing on renewable energy as an alternative for fossil-based energies in adapting to climate change [7]. Nearly 70 % of worldwide power capacities are generated from renewable energy sources in 2017 [8], showing that the evolution of renewables has gained massive attention from various nations.

Numerous process integration (PI) techniques and mathematical optimization models have been developed in handling carbon-constrained energy planning problems (e.g., pinch analysis [9], mixed-integer linear programming (MILP) [10], Automated Targeting Model (ATM) [11], and etc.) with the use of renewables. For example, Tan and Foo [9] solved the carbon-constrained energy planning using Carbon Emission Pinch Analysis (CEPA) and Ooi *et al.* [11] developed an Automated Targeting Model (ATM) for the carbon-constrained energy planning with the consideration of carbon capture and storage (CCS) systems. Thereafter, the provision research areas of PI techniques for energy planning-related problems have been critically reviewed by Klemeš *et al.* [12] and lately, by Kong et.al [13]. Both works outlined the need for accurate time-sliced based model for more reliable energy planning.

Recently, Kong *et al.* [14] developed a two-step multi-period optimization model that incorporated billing system in a proposed carbon-constrained energy planning. The minimum amount of self-generated renewables required to attain the carbon emissions target is determined in the first step (Targeting step). They will be optimally distributed in the second step (Scheduling step) with the aim of minimizing the total electricity bill of a given entity. The self-generated and consumed renewables made the owners active prosumers in the case study. Given this context, Tushar *et al.* [15] presented a comprehensive peer-to-peer (P2P) energy sharing concept in which the prosumers can independently decide their energy sharing parameters (e.g., who and when to share the energy, unit price of the shared energy, and etc.) to ensure the "players" involved gain economic benefits. This concept has been subsequently extended by Zhou *et al.* [16], where various P2P energy sharing models are developed to effectively reduce the energy bill of the prosumers.

It has been shown that idea of P2P energy sharing is one of the effective mechanisms for future smart grid, which enables direct energy trading between energy prosumers and consumers [15]. The P2P energy sharing concept can help in empowering the prosumers and give them flexibility in managing their own generated energy. To the best of the authors' knowledge, there are no specific P2P energy-sharing models developed that incorporate the actual billing system with micro time-basis (minutes) for carbon-constrained energy planning to-date.

To address this research gap, a modified version of multi-period P2P energy sharing model that incorporates an actual billing system is developed in this paper. A two-entity case based in Malaysia is used to demonstrate the effectiveness of the proposed model. Thus, the billing scheme from Tenaga Nasional Berhad (TNB), one of the largest electricity retailers in Malaysia, is adopted in this case study. Note that the offered P2P energy sharing is supposed to be flexible between peers, in which the excess energy can be traded or sold locally to one another as long as the scheme poses an overall positive impact to the "players" involved. The developed model is then benchmarked with the normal scenario where energy sharing is not considered, so that the feasibility of having collaborative involvement among entities (or "players") can be demonstrated.

2. Problem Statement

The multi-period pinch-based framework, which was proposed by Kong *et al.* [14], is adopted in this work and has been further modified to enable inter-entities collaborations among entities. Two different scenarios are carried out for both entities. The first scenario is assumed wherein both entities do not consider energy sharing whereas the sharing of energy between entities is performed in the latter scenario. The first two steps of the framework (targeting and scheduling) proposed by Kong *et al.* [14] is performed individually for both entities that allow the entities to meet the overall emission limit and minimize the total electricity bill in the first scenario. Finally, the extended step, energy sharing between entities is used to determine how the different energy resources should be allocated in order to further reduce the electricity bill for both entities. Therefore, in this work, the multi-period energy plannings for two entities, with and without consideration of inter-entity energy sharing are performed. The objective of this work is to evaluate the feasibility and the effectiveness of P2P energy sharing between entities with the utilization of multi-period carbon-constrained energy planning framework. The daily energy usage of these two entities is presented in figure 1.

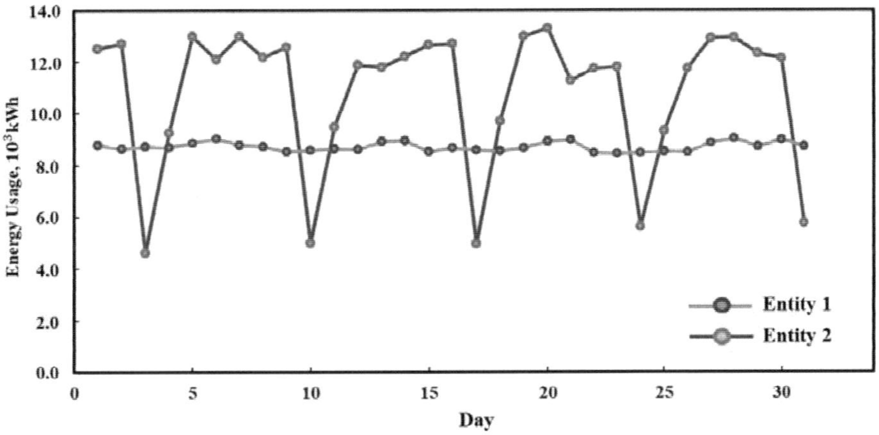

Figure 1. Daily energy usage of two entities.

3. Model Formulation

The multi-period energy planning model is an extension of the two-step optimization model developed by Kong *et al.* [14], which incorporates energy sharing possibilities between entities. The time intervals of 30 min in a month is assigned as T = {1, 2.... N}, while the total energy demand in each interval is represented as DEMAND = $\{D_{T1}, D_{T2}...D_{TN}\}$. The energy sources, i.e. renewables energies, fossil-based energies (derived from TNB), energy from the energy storage system, and shared renewables from the partner entity are assigned as SOURCES= $\{S_F, S_R, S_{ER}, S_{SR}\}$, which are used to fulfill the total energy demand. Each energy supply has its emission factor (denoted as CI_F and CI_R for fossil and renewable energy sources). Energy storage system with a capacity of S_{ER} is allocated to allow energy to be stored during a given time. The stored energy can either be self-consumed or sold to the partner entity. Note that the shared renewables and the energy from the storage system utilized the same emission factor with renewables as they are homogeneous. The renewable source considered in this work is assumed to be solar energy. Furthermore, the operation of the case study is also assumed to operate at steady-state condition. The following subsection presents the formulation of the proposed model.

3.1. Carbon Emission Pinch Analysis (CEPA) -- Targeting

In this targeting step, Carbon Emission Pinch Analysis (CEPA), which was developed by Tan and Foo [9], is used to determine the minimum amount of renewables required (so that the pre-defined carbon emission limit can be achieved) for both entities. The minimum amount of renewables is identified in this step as the deployment of renewable energy is still more cost-intensive as compared to the use of conventional energy sources (oil, natural gas, and coal). As shown in equation (1), the objective of this work is to determine the minimum amount of renewables needed for each entity, so that the additional investment for renewable energy deployment can be kept minimal while ensuring the targeted carbon emission limit is achieved. Note that $S_{Total\ Renewables}$ (kWh) indicates the total renewables required across the one-month period while $S_{R,T}$ (kWh) denotes the amount of renewable energy that is applied to the system at time interval T.

$$Min\ S_{Total\ Renewables} = \sum_T S_{R,T} + \sum_T S_{ER,T} \quad (1)$$

The carbon emission of each time interval, E_T (kg) is the product of energy demand and their respective carbon intensity, CI_F and CI_R (kg/kWh), by using the equation (2). Note that the renewables and energy storage utilizes the same carbon intensity, CI_R in this case, as they are derived from the same energy supply.

$$E_T = S_{F,T}CI_F + S_{R,T}CI_R + S_{ER,T}CI_R \quad \forall T \tag{2}$$

Equation (3) represents the energy demand at each time interval T, (D_T (kWh)), which can be fulfilled by either fossil-based ($S_{F,T}$), renewables energies ($S_{R,T}$) or energy storage for renewables ($S_{ER,T}$) during day time (7 a.m. to 7 p.m.)

$$D_T = S_{F,T} + S_{R,T} + S_{ER,T} \quad \forall T \tag{3}$$

Whereas on the other hand, equation (4) ensures that the renewable energy supply is only derived from energy storage from 12 a.m. to 7 a.m. and 7 p.m. to 12 a.m.

$$D_T = S_{F,T} + S_{ER,T} \quad \forall T \tag{4}$$

The initial carbon emission limit, E_{in} (kg) is presumed to be caused by fossil-based energy sources, which are derived from TNB, so it can be expressed as equation (5):

$$E_{in} = \sum_T (S_{F,T} \times CI_F) \tag{5}$$

As for the carbon emission limit E_L (kg), it can be represented as equation (6) where β indicates the emission reduction ratio. Note that in this work, the emission reduction ratio is set as 0.2 (20 % lower than the initial emission), but the emission reduction ratio can vary based on the preferences of the decision-makers and policymakers.

$$E_L = E_{in} \times (1 - \beta) \tag{6}$$

The above equations are incorporated into the first linear programming model, which is solved using LINGO 18.0 [17].

3.2. Automated Targeting Model (ATM) – Scheduling

As for the scheduling step, optimal energy allocations of renewable energy for both entities are performed. The optimized energy allocation can be determined *via* the developed mixed-integer linear programming (MILP) model using LINGO 18.0 [17]. Note that the total amount of renewables distributed in this step is equivalent to the targeted amount of renewables in the first step. The objective of the scheduling step is to minimize the total electricity bill, $Cost_{Total,Without\ Sharing}$, across all time intervals, which is expressed as equation (7). The indicated C_T denotes the electricity cost at time interval T while C_{MD} is the MD charges:

$$Min\ Cost_{Total,Without\ Sharing} = \sum_T C_T + C_{MD} \tag{7}$$

The electricity cost at time interval T, C_T (MYR), can be mathematically expressed as equation (8), whereas CF_F (MYR/kWh) and CF_R (MYR/kWh) are the cost factor of fossil-based energy and renewable energy at different time intervals (indicated in table 1).

$$C_T = (S_{F,T} * CF_F) + (S_{R,T} * CF_R) \quad \forall T \tag{8}$$

For the MD charges, the highest MD value is selected between mid-peak, $MDV_{Mid-Peak}$ (kW) or peak period MDV_{Peak} (kW), and multiply it with their respective MD cost factor, $MD_{Mid-Peak}$ and MD_{Peak} (MYR/kW). The $MDV_{Mid-Peak}$ can be obtained by dividing the summation of fossil-based energies in every mid-peak period ($\sum S_{F,T=Mid-peak}$) to the number of mid-peak time intervals ($\sum_T T = Mid - peak$). Whereas MDV_{Peak} can be obtained by dividing the summation of fossil-based energies in every peak period ($\sum S_{F,T=Peak}$) to the number of peak-time intervals ($\sum_T T = peak$). The selection of the MD value of mid-peak and peak period can be expressed as equation (9) and (10).

$$MDV_{Mid-Peak} = \frac{\sum S_{F,T=Mid-peak}}{\sum_T T=Mid-peak} \quad (9)$$

$$MDV_{Peak} = \frac{\sum S_{F,T=Peak}}{\sum_T T=Peak} \quad (10)$$

Whereas the MD charges can be expressed as equation (11):

$$C_{MD} = \max(MDV_{Mid-Peak}, MDV_{Peak}) \times \left\{ \frac{\begin{matrix} MD_{Mid-Peak} | MDV_{Mid-Peak} = max \\ MD_{Peak} | MDV_{Peak} = max \end{matrix}}{0.5h} \right. \quad (11)$$

Note that the original electricity bill ($Cost_{Total,Original}$) can be calculated using equation (8) to equation (11), the calculation can be formally represented as equation (12).

$$Cost_{Total,Original} = \sum_T C_T + C_{MD} \quad (12)$$

In this step, a modified version of ATM is presented. Three different cascades (energy cascade, emission cascade, and cost cascade) are performed simultaneously to determine the optimal scheduling of renewables at different time intervals. Figure 2 shows the generic Automated Targeting Model (ATM); while the cost factors of different time classifications are tabulated in Table 1. Note that, the models formulated in Section 3.1 and Section 3.2 are modified from Kong *et al.* [14], where they are used to determine the optimal renewable energy allocation without the consideration of energy sharing.

Table 1. Cost factor and MD charge for each time interval.

Time	Time Interval (30 min/interval/day)	Classification	Cost Factor[a], $CF_{F,T}$ (MYR/kWh)	Maximum Demand Charge[a], MD_T (MYR/kW)	Cost Factor[b], $CF_{R,T}$ (MYR/kWh)
00:00-08:30	1-16	Off-Peak	0.202	0.00	
08:30-11:30	17-22	Mid-Peak	0.327	35.00	
11:30-12:30	23-24	Peak	0.576	38.30	
12:30-14:30	25-28	Mid-Peak	0.327	35.00	0.276
14:30-17:30	29-34	Peak	0.576	38.30	
17:30-21:00	30-43	Mid-Peak	0.327	35.00	
22:00-24:00	44-48	Off-Peak	0.202	0.00	

[a]Data obtained from Tenaga Nasional Berhad [18]
[b]Data obtained from Renewable Power Generation Costs in 2019, IRENA [19].

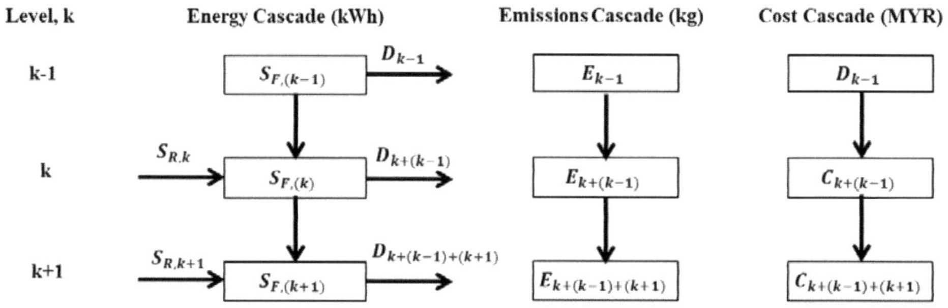

Figure 2. Generic Automated Targeting Model (ATM) (adapted from Kong *et al.* [13]).

3.3. *Energy Sharing*

This subsection presents the mathematical model, which incorporates energy sharing. In this illustration, the developed model only enables energy sharing during the off period of the participating entities. As shown in figure 1, less energy is consumed on the off-period (i.e., day 3, 10, 17, 24, and 31) for entity 2. The allocated renewables for entity 2 during the off period can, therefore, be sold to entity 1. The optimal allocation of the shared renewables that allows both entities to achieve their minimum electricity bill can be determined *via* the model developed below.

The objective of the energy sharing model is expressed as equation (13), which is to maximize the weighted-sum satisfaction of both entities (S_E), where ω refers to the assigned weighting coefficients (both entities are assumed equally important, thus 0.5 is taken for each entity). The index E represents the type of entity indicated (i.e., entity 1 and entity 2). The objective function ensures both entities can further minimize their electricity bills compared to the one without energy sharing.

$$Max \quad \sum_E S_E \times \omega \tag{13}$$

The calculation of total electricity bill normalization of each entity can be expressed as equation (14), whereas the $Cost_{Total,Sharing,E}$ represents the total electricity bill after energy sharing where sharing cost, $C_{SR,E}$ (MYR) (positive sign is taken when energy is sold; negative sign is used when energy is imported from the partner entity), is incorporated.

$$S_E = \frac{Cost_{Total,Without\ Sharing,E} - Cost_{Total,Sharing,E}}{Cost_{Total,Original,E} - Cost_{Total,Sharing,E}} \tag{14}$$

$$Cost_{Total,Sharing,E} = \sum_T C_{T,E} + C_{MD} \pm C_{SR,E} \tag{15}$$

The sharing cost can be expressed as equation (16), in which CF_{SR} represents the unit cost of each kWh of renewables sold and $S_{SR,T,E}$ denotes the amount of shared renewable energy that is applied to the system at time interval T. It is set as MYR 0.45/kWh in this work (assumed to be 21.88 % lower than the cost factor in the peak period).

$$C_{SR,E} = \sum_T S_{SR,T,E} \times CF_{SR} \tag{16}$$

It should be noted that the carbon emission and energy demand at each time interval of entity 1 during entity 2's off days are then revised as equations (17) to (18).

$$E_T = S_{F,T}CI_F + S_{R,T}CI_R + S_{ER,T}CI_R + S_{SR,T}CI_R \quad \forall T \tag{17}$$

$$D_T = \begin{cases} S_{F,T} + S_{R,T} + S_{ER,T} + S_{SR,T}\big|_{7\text{ a.m.to 7p.m.}} \\ S_{F,T} + S_{ER,T} + S_{SR,T}\big|_{7\text{ p.m.to 7a.m.}} \end{cases} \forall T \tag{18}$$

The carbon emission and energy demand of entity 2 during its' off days are then revised as equations (19) and (20):

$$E_T = S_{F,T}CI_F \quad \forall T \tag{19}$$

$$D_T = S_{F,T} \quad \forall T \tag{20}$$

4. Results and Discussion

The original electricity bill for entity 1 and entity 2 is tabulated as in Table 2. Both entities do not utilize any renewables and the original electricity bill serves as the benchmark of the study.

Table 2. Original Electricity bill for entity 1 and 2 (without energy planning)

	Entity 1	Entity 2
Mid peak value (kWh)	209.9268	213.9410
Peak value (kWh)	237.7551	212.0308
Maximum Demand Cost (MYR)	9106.0222	7487.9367
Normal Electricity Cost (MYR)	95059.336	105683.002
Total Electricity Bill (MYR)	104165.40	113170.90
Carbon Emission (kg)	210998.9	262465.20

The minimum renewable energy required can be determined using the equations formulated in Section 3.1. The results are illustrated in figure 3.

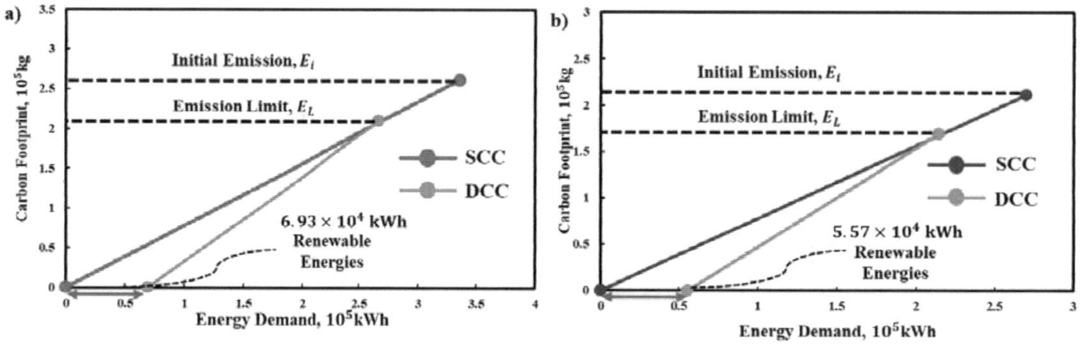

Figure 3. Carbon Emission Pinch Analysis of a) entity 1; and b) entity 2.

Figure 3 provides the visualization results of carbon emission pinch analysis for entity 1 and entity 2. The graph encompasses two composite curves, i.e., (i) source composite curve (SCC) and (ii) demand composite curve (DCC). SCC is the total energy supply (input) while the DCC is the emission limit of the problem. The intersection point is known as the pinch point. The horizontal distance (right side) between the two composite curves denotes the minimum amount of renewable energy required to meet the emission limit. In this work, 20 % of carbon emission reduction is set as the constraint for both entities and the minimum renewables needed for entity 1 and entity 2 are 6.93 x 10^4 kWh and 5.57 x 10^4 kWh, respectively.

The scheduling part, on the other hand, is designed to obtain the optimal allocation of the renewables targeted from the targeting step for both entities. Similar to the targeting step, the minimum electricity bill and optimal allocation of renewables can be determined through MILP model developed in Section 3.2. As for the energy sharing part, it can be solved through the model developed in Section 3.3 by using LINGO V18 [17]. The sharing of renewables allows both entities to benefit from trading the excess renewables to another, which leads to lower net cost. Table 3 shows the total electricity bill for both entities after energy sharing. Note that the total renewables used in both entities are still the same, therefore, the sharing of energy does not affect the total carbon emission. In other words, the net carbon emissions still fulfil the total emissions reduction target. The respective optimal energy allocation for both entities (before and after energy sharing) are illustrated in figure 4 to figure 7. The complete ATMs (1st to 31st day) for both entities are compiled in Appendix A-Appendix D.

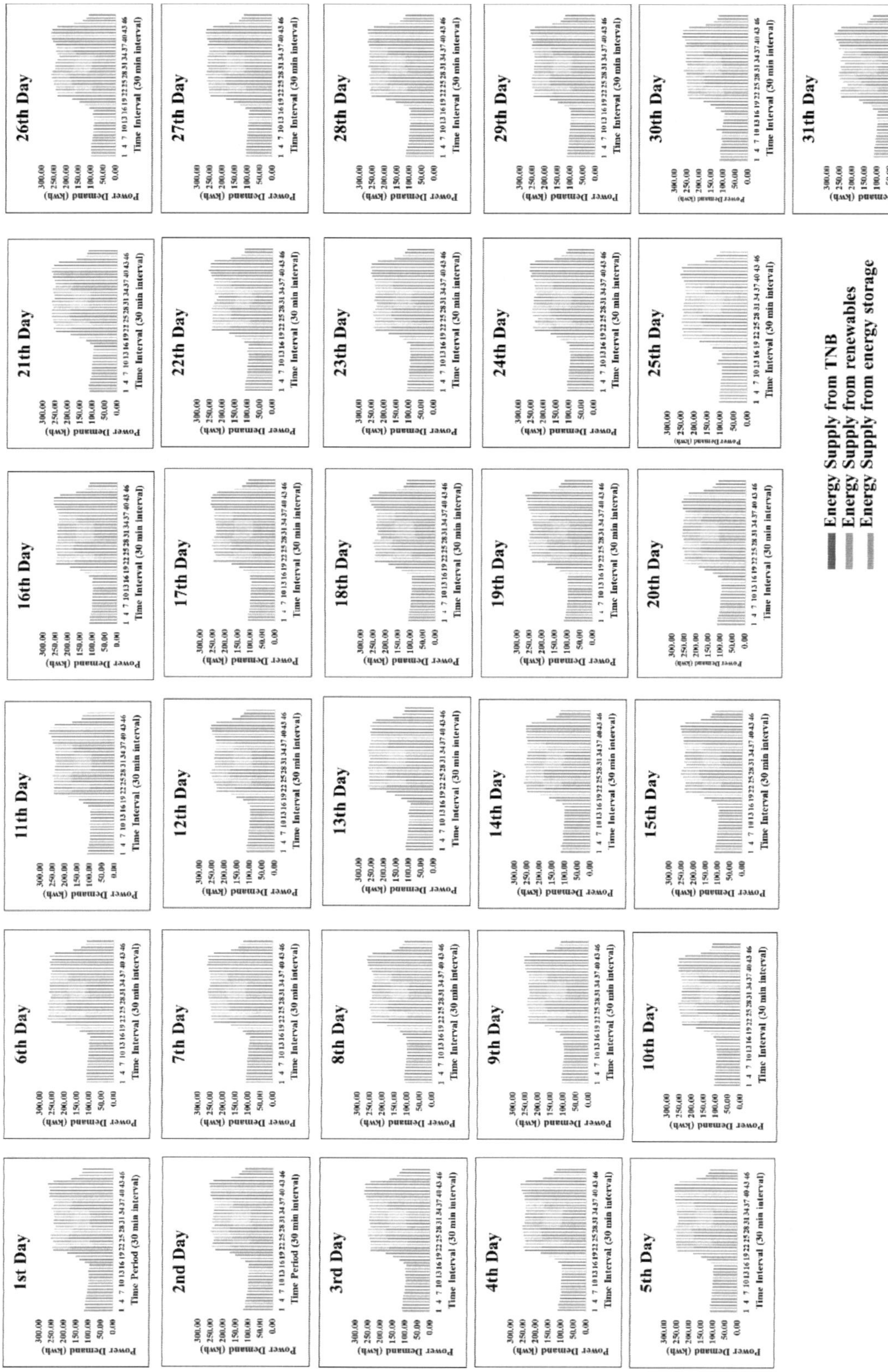

Figure 4. Energy allocation of entity 1 before energy sharing.

Figure 5. Energy allocation of entity 1 after energy sharing.

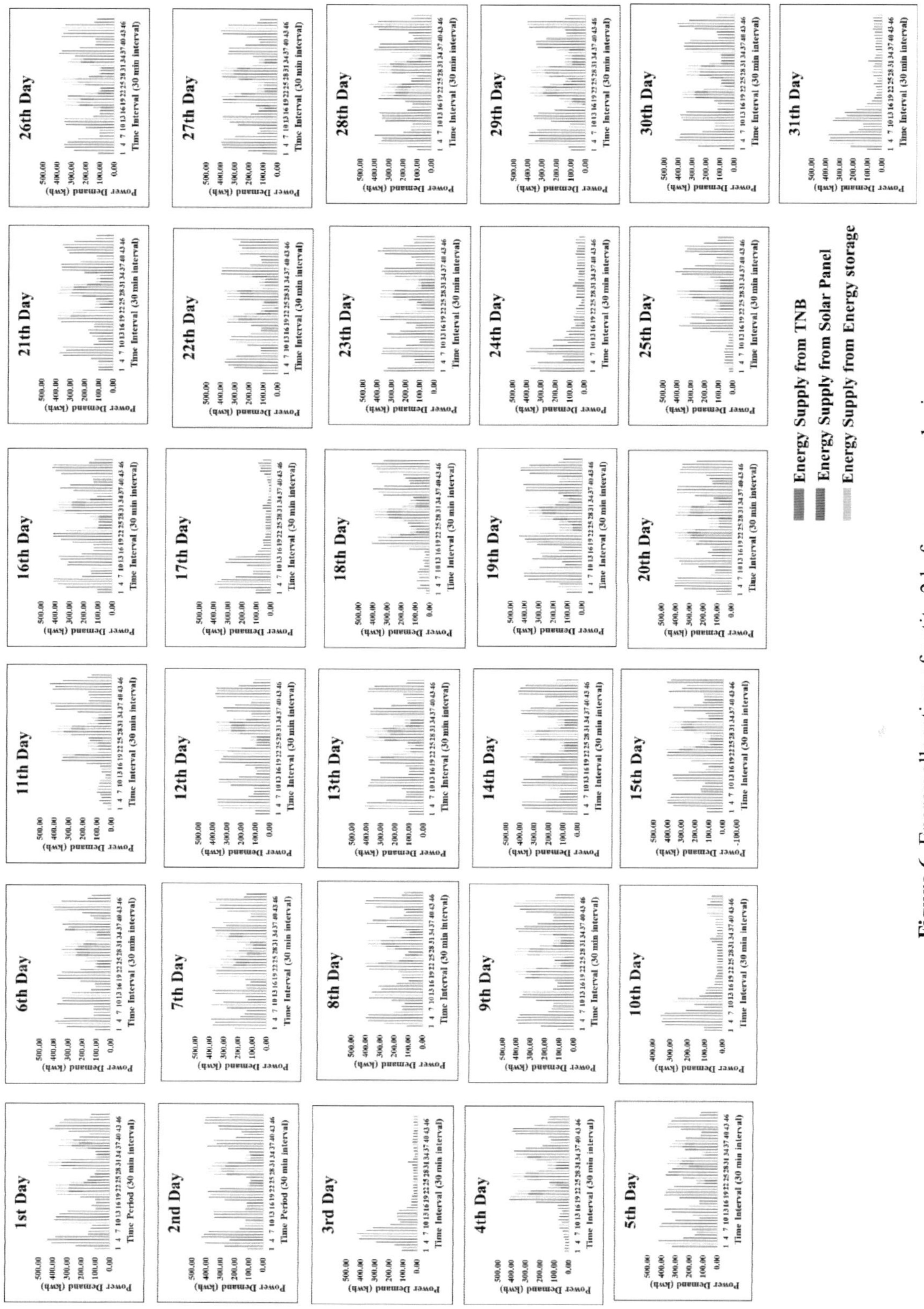

Figure 6. Energy allocation of entity 2 before energy sharing.

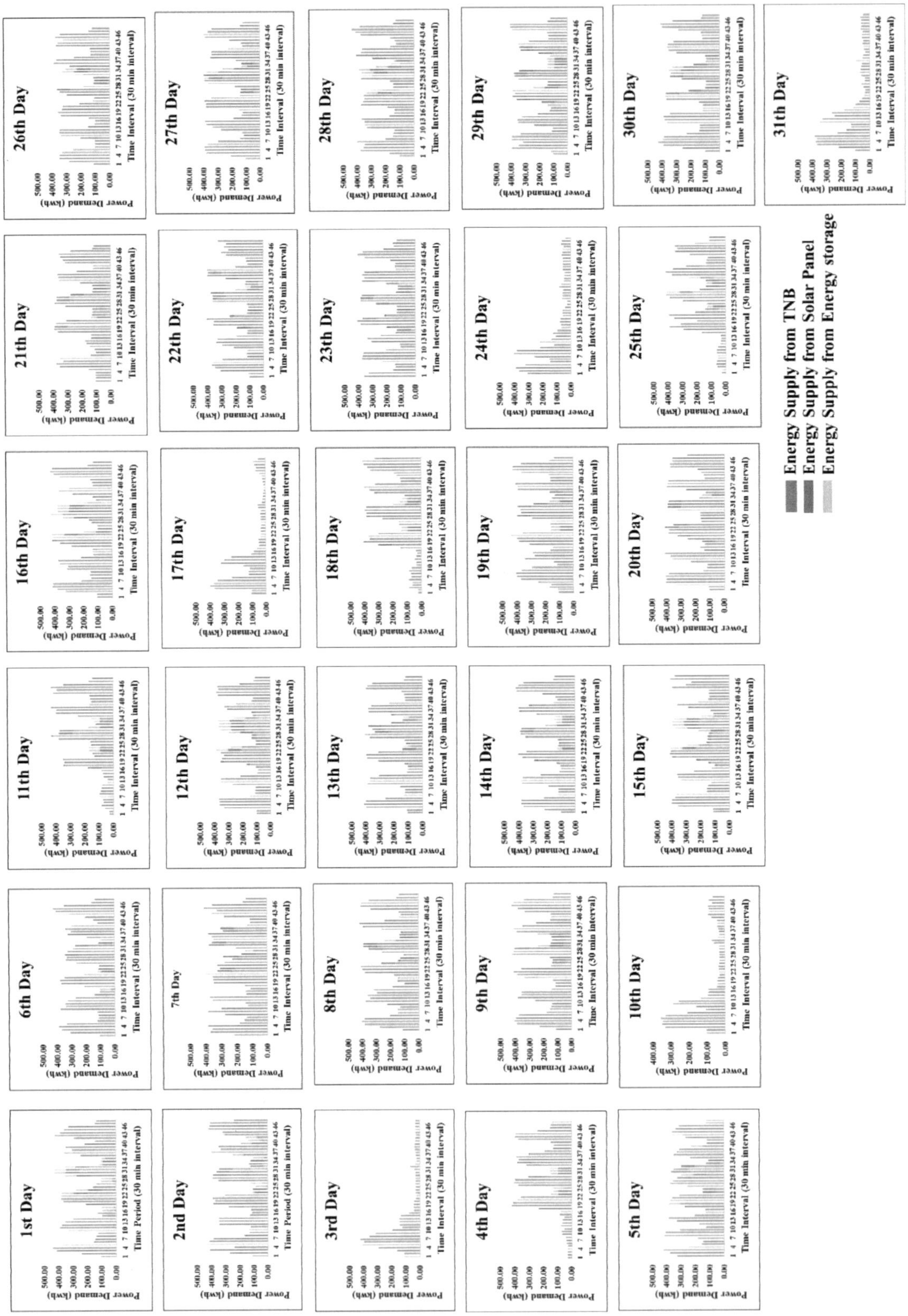

Figure 7. Energy allocation of entity 2 after energy sharing.

Table 3. Electricity bill for entity 1 and 2 (before and after energy sharing)

	Before energy sharing		After energy sharing	
	Entity 1	Entity 2	Entity 1	Entity 2
Mid peak value (kWh)	180.9352	177.8171	160.3396	217.8014
Peak value (kWh)	92.7970	69.6838	45.9429	109.38
Maximum Demand Cost (MYR)	6332.7316	6223.5980	5611.884	7623.049
Normal Electricity Cost (MYR)	67588.449	74086.268	52633.75	87863.20
Renewables Cost (MYR)	15421.60	19183.20	15421.59	19183.2
Sharing Renewables Cost (MYR)	-	-	+15557.40	-15557.40
Total Electricity Bill (MYR)	89342.80	99493.04	89224.63	99112.05
Carbon Emission (kg)	168799.20	209972.1	236166.87	142604.43

The results show that with the aid of the targeting and scheduling steps, entity 1 and 2 can reduce their electricity bill by 14.23 % and 12.08 %, respectively. In the first scenario (before energy sharing), the utilization of renewables and the optimal scheduling strategy enables both entities to drastically reduce their MD charges by 30.4 % (MYR 2773.29) and 16.95 % (MYR1269.21), respectively. The targeted renewables are distributed in mid-peak and peak periods accordingly. As for the second scenario (energy sharing scenario), 34572.01 kWh of renewables generated from entity 2 are proposed to be sold to entity 1 with a unit cost of MYR 0.45 per kWh. It can be seen from the ATMs (Appendix B) and energy allocation models (figure 5) that the sold renewables are mainly distributed to the mid-peak and peak-period for entity 1, which effectively further reduces the MD charges (MYR 720.85) and the normal electricity cost (MYR 14954.69). As for the peak periods for entity 1, more renewables are utilized (from the solar panel, energy storage, and renewables from entity 2) as the MD charges are higher than that of the mid-peak periods. Thus, with the aid of energy sharing, entity 1 can further reduce its electricity bill by MYR 118.17 (0.13 %) even though 34742.66 kWh of renewables are purchased from entity 2. On the other hand, entity 2 can mitigate its total electricity bill by MYR 380.98 (0.38 %) by exporting its generated renewables to the partner entity (figure 7 and Appendix D). Note that even though the MD cost and normal electricity cost of entity 2 have been increased by MYR 1399.45 and MYR 13776.93, respectively due to lower utilization of renewables, the income generated from the energy sharing (MYR 15557.40) is high enough to compensate for the losses. By looking at the comparative results illustrated in Table 3, the proposed multi-period energy sharing framework that incorporates the actual billing system of TNB has shown the positive economic benefits of energy sharing for both entities in energy planning.

5. Conclusions

This work evaluates the potential of P2P energy sharing between two entities using the proposed multi-period optimisation model. This work benchmarks the two energy planning scenarios, i.e., with and without the consideration of energy sharing between entities, while ensuring the overall electricity bills are kept at minimal. The results generated from the developed model proved that the sharing of the renewables can reduce the electricity bill by 14.34 % (MYR 14940.77) and 12.42 % (MYR 14058.90) as compared to the original scenario where the energy planning is not optimised. It is worth emphasizing that the P2P energy sharing concept is capable of offering positive impacts on carbon-constrained energy planning, particularly in enhancing the economic viability without the need of compromising the environmental benefit. However, the P2P energy sharing concept is often limited by a series of barriers, which include environmental regulation constraints, lack of information sharing, and trust issues between the players in the industries [20]. Therefore, more effort is needed in exploring game-theoretic solutions to rationalize the energy sharing scheme [13]. In addition, the model can be extended to determine the optimal carbon reduction ratio, without the compromising the profitability of each entity.

References

[1] Cao X 2003 Climate change and energy development: Implications for developing countries *Resour. Pol.* **29** 61–67

[2] IRENA 2018 Global energy transformation: A roadmap to 2050 International Renewable Energy, Abu Dhabi *International Renewable Energy Agency*

[3] IPCC 2014 Climate Change 2014: Synthesis Report Contribution of Working Groups I, II and III to the Fifth Assessment Report of the Intergovernmental Panel on Climate Change [Core Writing Team, R.K. Pachauri and L.A. Meyer (eds.)] *Intergovernmental panel on climate change*

[4] IPCC 2018 Summary for Policymakers In: Global warming of 1.5°C, 2018: An IPCC Special Report on the impacts of global warming of 1.5°C above pre-industrial levels and related global greenhouse gas emission pathways, in the context of strengthening the global response to the threat of climate change, sustainable development, and efforts to eradicate poverty [V. Masson-Delmotte, P. Zhai, H. O. Pörtner, D. Roberts, J. Skea, P.R. Shukla, A. Pirani, W. Moufouma-Okia, C. Péan, R. Pidcock, S. Connors, J. B. R. Matthews, Y. Chen, X. Zhou, M. I. Gomis, E. Lonnoy, T. Maycock, M. Tignor, T. Waterfield (eds.)] *Intergovernmental panel on climate change*

[5] Hurlimann A C, Moosavi S and Browne G R 2021 Climate change transformation: A definition and typology to guide decision making in urban environments *Sustain. Cities. Soc.* **70** 102890.

[6] IEA 2021 Global Energy Review 2021 IEA Paris *International Energy Agency*

[7] Tan X, Zhu, K, Meng X, Gu B, Wang Y, Meng F, Liu G, Tu T and Li H 2021 Research on the status and priority needs of developing countries to address climate change *J. Clean. Prod.* **289** 125669

[8] REN21 2018 Renewables 2018 Global Status Report (Paris: REN21 Secretariat) *Renewables now*

[9] Tan R R and Foo D C Y 2007 Pinch analysis approach to carbon-constrained energy sector planning *Energy* **32(8)** 1422–1429

[10] Hashim H, Douglas P, Elkamel A and Croiset E 2005 Optimization model for energy planning with CO_2 emission considerations *Ind. Eng. Chem. Res.* **44(4)** 879–890

[11] Ooi R E H, Foo D C Y and Tan R R 2014 Targeting for carbon sequestration retrofit planning in the power generation sector for multi-period problems *Appl. Energ.* **113** 477–487

[12] Klemeš J J, Varbanov P, Walmsley T S and Jia X 2018 New directions in the implementation of Pinch Methodology (PM) *Renew. Suste. Energ. Rev.* **98** 439–468

[13] Kong K G H, How B S, Teng S Y, Leong W D, Foo D C Y, Tan R R and Sunarso J 2021 Towards data-driven process integration for renewable energy planning *Curr. Opin. Chem. Eng.* **31** 100665

[14] Kong K G H, Lo S L Y, How B S, Leong W D, Teng S Y, Ng W P and Sunarso J 2020 Enhanced Automated Targeting Model for Multi-Period Energy Planning *Chem. Eng. Trans.* **81** 607–612

[15] Tushar W, Yuen C, Saha TK, Morstyn T, Chapman A C, Alam M J E, Hanif S and Vincent P H 2021 Peer-to-peer energy systems for connected communities: A review of recent advances and emerging challenges *Appl. Energ.* **282**, Part A 116131

[16] Zhou Y, Wu J, Long C, Cheng M and Zhang C 2017 Performance evaluation of peer-to-peer energy sharing models *Energy Procedia* **143** 817–822

[17] Lindo system INC 2020 LINGO 19.0 - Optimization Modeling Software for Linear, Nonlinear, and Integer Programming *Lindo system INC*

[18] Tenaga Nasional Berhad 2014 Tenaga Nasional Berhad Enhanced Time of Use (ETOU) *Tenaga Nasional Berhad*

[19] IRENA 2019 Renewable Power Generation Costs in 2019 *International Renewable Energy*

Agency
[20] Golev A, Corder G A and Giuro D P 2014 Barriers to industrial symbiosis: Insights from the use of a maturity grid *J. Ind. Ecol.* **19(1)** 141–153

Acknowledgments
The authors would like to acknowledge financial support from Swinburne University of Technology (Sarawak Campus) via Full Fee-Waiver Studentship and the sponsorship from Swinburne University of Technology (Sarawak Campus) School of Research.

IOP Conf. Series: Materials Science and Engineering 1195 (2021) 012011 doi:10.1088/1757-899X/1195/1/012011

Article I. Appendix

Appendix A: Automated Targeting Model for entity 1 (Before Energy sharing) 1st -30th day

1st

Classification	Time Interval (30 min)	Energy Cascade (kWh)			Emission Cascade (kg)	Cost Cascade (MYR)
Off-Peak	1-16		1682.98	1682.98	1314.958	377.65
Mid-Peak	17-22	74.895	1007.80	2765.67	2104.177	739.409
Peak	23-24	299.58	180.983	3246.23	2252.775	935.863
Mid-Peak	25-28	299.58	685.653	4231.46	2795.686	1234.01
Peak	29-34	898.74	547.07	5677.27	3244.697	1805.75
Mid-Peak	35-44	224.685	2308.21	8210.16	5053.56	2565.33
Off-Peak	45-48		562.623	8772.79	5493.154	2678.98

2nd

Classification	Time Interval (30 min)	Energy Cascade (kWh)			Emission Cascade (kg)	Cost Cascade (MYR)
Off-Peak	1-16		1672.16	1672.16	1306.512	374.164
Mid-Peak	17-22	74.895	970.811	2717.87	2066.83	729.986
Peak	23-24	299.58	184.472	3246.23	2218.156	918.383
Mid-Peak	25-28	299.58	678.924	4231.46	2755.81	1218.97
Peak	29-34	898.74	539.837	5677.27	3199.17	1782.34
Mid-Peak	35-44	224.685	2241.59	8210.16	4995.98	2524.91
Off-Peak	45-48		546.197	8631.47	5382.74	2635.24

3rd

Classification	Time Interval (30 min)	Energy Cascade (kWh)			Emission Cascade (kg)	Cost Cascade (MYR)
Off-Peak	1-16		1652.75	1652.75	1291.34	369.129
Mid-Peak	17-22	74.895	966.184	2693.83	2048.047	723.079
Peak	23-24	299.58	181.197	3174.6	2196.811	913.17
Mid-Peak	25-28	299.58	699.349	4173.53	2750.42	1222.17
Peak	29-34	898.74	562.709	5364.98	3211.65	1796.78
Mid-Peak	35-44	224.685	2325.71	8185.38	5034.19	2564.95
Off-Peak	45-48		539.265	8724.64	5455.54	2673.88

4th

Classification	Time Interval (30 min)	Energy Cascade (kWh)			Emission Cascade (kg)	Cost Cascade (MYR)
Off-Peak	1-16		1631.49	1631.49	1274.73	365.554
Mid-Peak	17-22	74.895	980.409	2686.79	2042.553	724.99
Peak	23-24	299.58	186.26	3172.63	2195.273	916.549
Mid-Peak	25-28	299.58	700.198	4172.41	2749.548	1232.35
Peak	29-34	898.74	568.814	5639.97	3215.55	18033.67
Mid-Peak	35-44	224.685	2269.13	8133.79	4993.885	2553.52
Off-Peak	45-48		557.704	8691.49	5429.636	2666.18

5th

Classification	Time Interval (30 min)	Energy Cascade (kWh)			Emission Cascade (kg)	Cost Cascade (MYR)
Off-Peak	1-16		1666.75	1666.75	1301.49	372.584
Mid-Peak	17-22	74.895	993.844	2734.48	2079.814	740.683
Peak	23-24	299.58	205.972	3240.04	2247.936	943.214
Mid-Peak	25-28	299.58	727.351	4266.97	2823.427	1266
Peak	29-34	898.74	595.224	5760.93	3310.063	1853.16
Mid-Peak	35-44	224.685	2330.57	8316.18	5136.397	2620.37
Off-Peak	45-48		541.762	8857.94	5559.691	2729.80

6th

Classification	Time Interval (30 min)	Energy Cascade (kWh)			Emission Cascade (kg)	Cost Cascade (MYR)
Off-Peak	1-16		1679.67	1679.67	1313.37	374.392
Mid-Peak	17-22	74.895	1050.63	2805.2	2135.06	770.251
Peak	23-24	299.58	228.274	3333.05	2320.61	981.06
Mid-Peak	25-28	299.58	761.453	4394.08	2922.74	1311.46
Peak	29-34	898.74	656.801	5949.62	3457.49	1937.44
Mid-Peak	35-44	224.685	2334.91	8509.22	5287.226	2702.64
Off-Peak	45-48		521.967	9031.19	5695.05	2808.07

7th

Classification	Time Interval (30 min)	Energy Cascade (kWh)			Emission Cascade (kg)	Cost Cascade (MYR)
Off-Peak	1-16		1618.33	1618.33	1264.425	361.948
Mid-Peak	17-22	74.895	946.898	2640.13	2006.09	712.657
Peak	23-24	299.58	190.016	3129.72	2161.745	906.505
Mid-Peak	25-28	299.58	705.627	4134.93	2720.26	1224.93
Peak	29-34	898.74	637.956	5671.63	3240.28	1837.80
Mid-Peak	35-44	224.685	2324.65	8220.96	5061.99	2601.82
Off-Peak	45-48		555.995	8776.96	5496.41	2714.13

8th

Classification	Time Interval (30 min)	Energy Cascade (kWh)			Emission Cascade (kg)	Cost Cascade (MYR)
Off-Peak	1-16		1675.3	1675.3	1308.959	376.587
Mid-Peak	17-22	74.895	1002.97	2753.16	2094.407	742.345
Peak	23-24	299.58	193.738	3246.48	2252.97	938.47
Mid-Peak	25-28	299.58	704.356	4250.42	2810.49	1248.64
Peak	29-34	898.74	560.416	5709.57	3269.934	1819.79
Mid-Peak	35-44	224.685	2265.81	8200.06	5045.66	2569.49
Off-Peak	45-48		525.315	8725.38	5456.11	2675.6

9th

Classification	Time Interval (30 min)	Energy Cascade (kWh)			Emission Cascade (kg)	Cost Cascade (MYR)
Off-Peak	1-16		1617.63	1617.63	1263.905	363.36
Mid-Peak	17-22	74.895	977.374	2669.90	2029.355	718.384
Peak	23-24	299.58	173.129	3142.61	2171.815	903.16
Mid-Peak	25-28	299.58	671.491	4113.68	2703.66	1205.06
Peak	29-34	898.74	528.51	5540.93	3138.171	1758.26
Mid-Peak	35-44	224.685	2239.70	8005.32	4893.511	2499.94
Off-Peak	45-48		528.646	8533.97	5306.557	2606.72

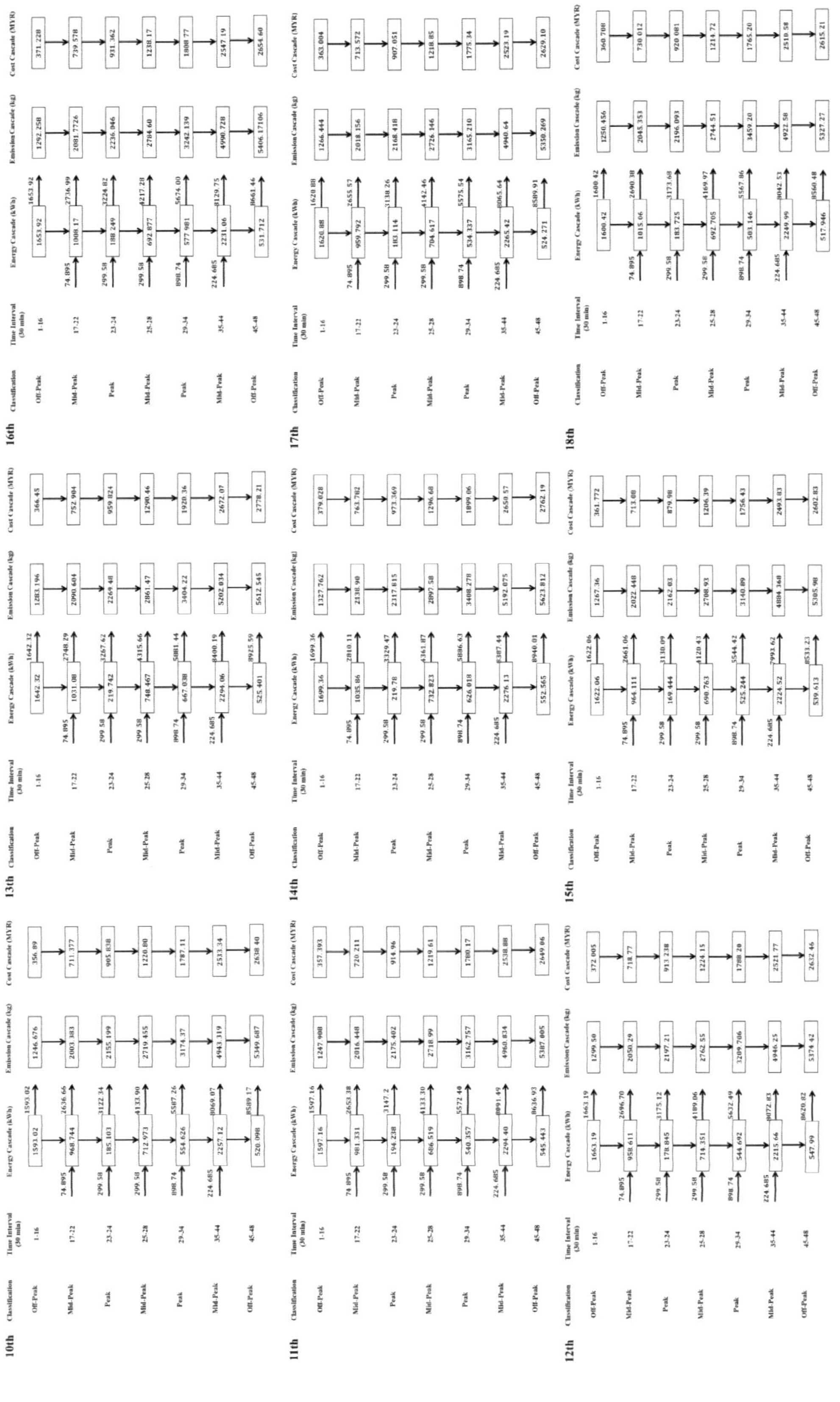
10th
11th
12th
13th
14th
15th
16th
17th
18th
Classification
Time Interval (30 min)
Energy Cascade (kWh)
Emission Cascade (kg)
Cost Cascade (MYR)

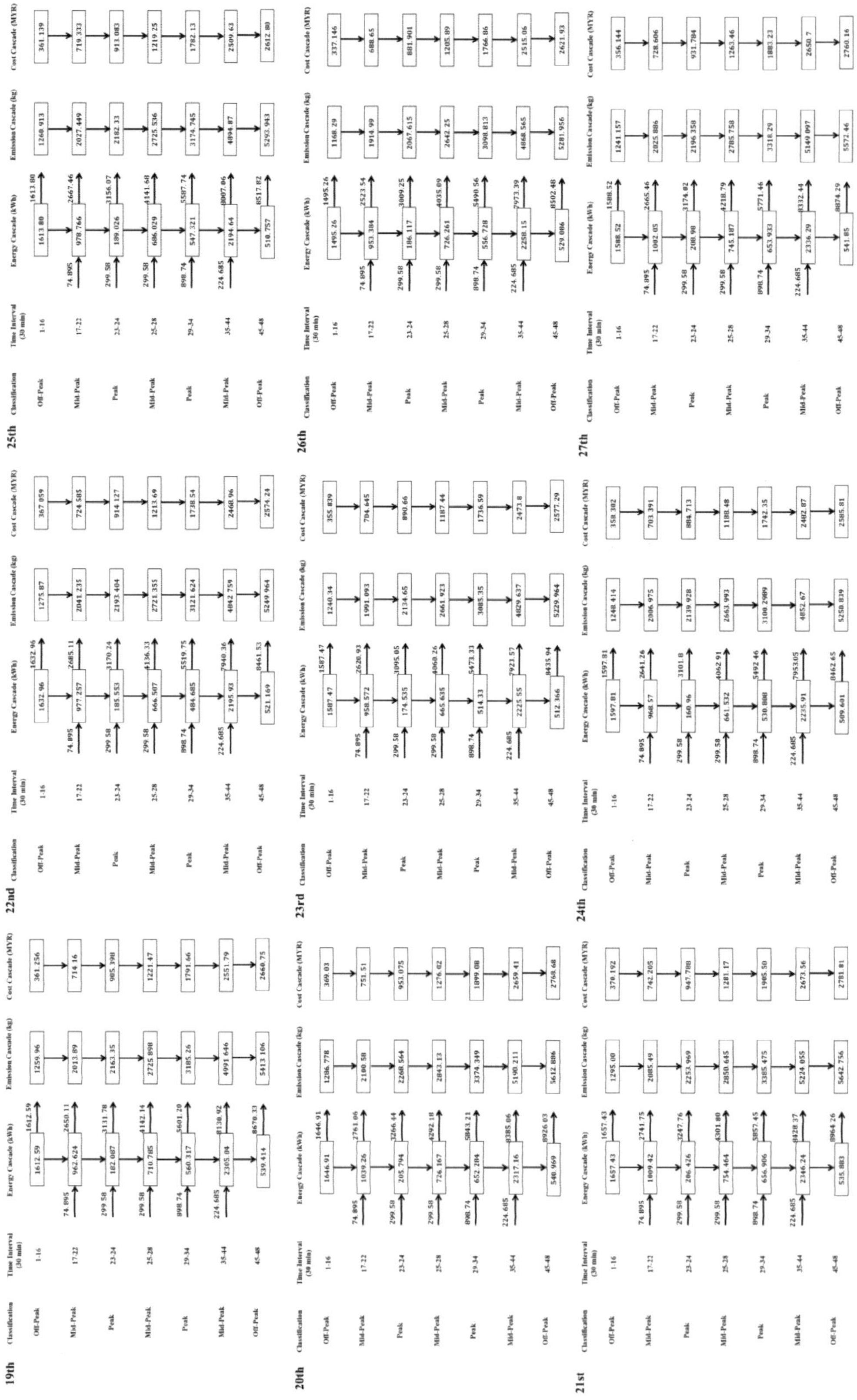
19th
20th
21st
22nd
23rd
24th
25th
26th
27th
Classification
Time Interval (30 min)
Energy Cascade (kWh)
Emission Cascade (kg)
Cost Cascade (MYR)

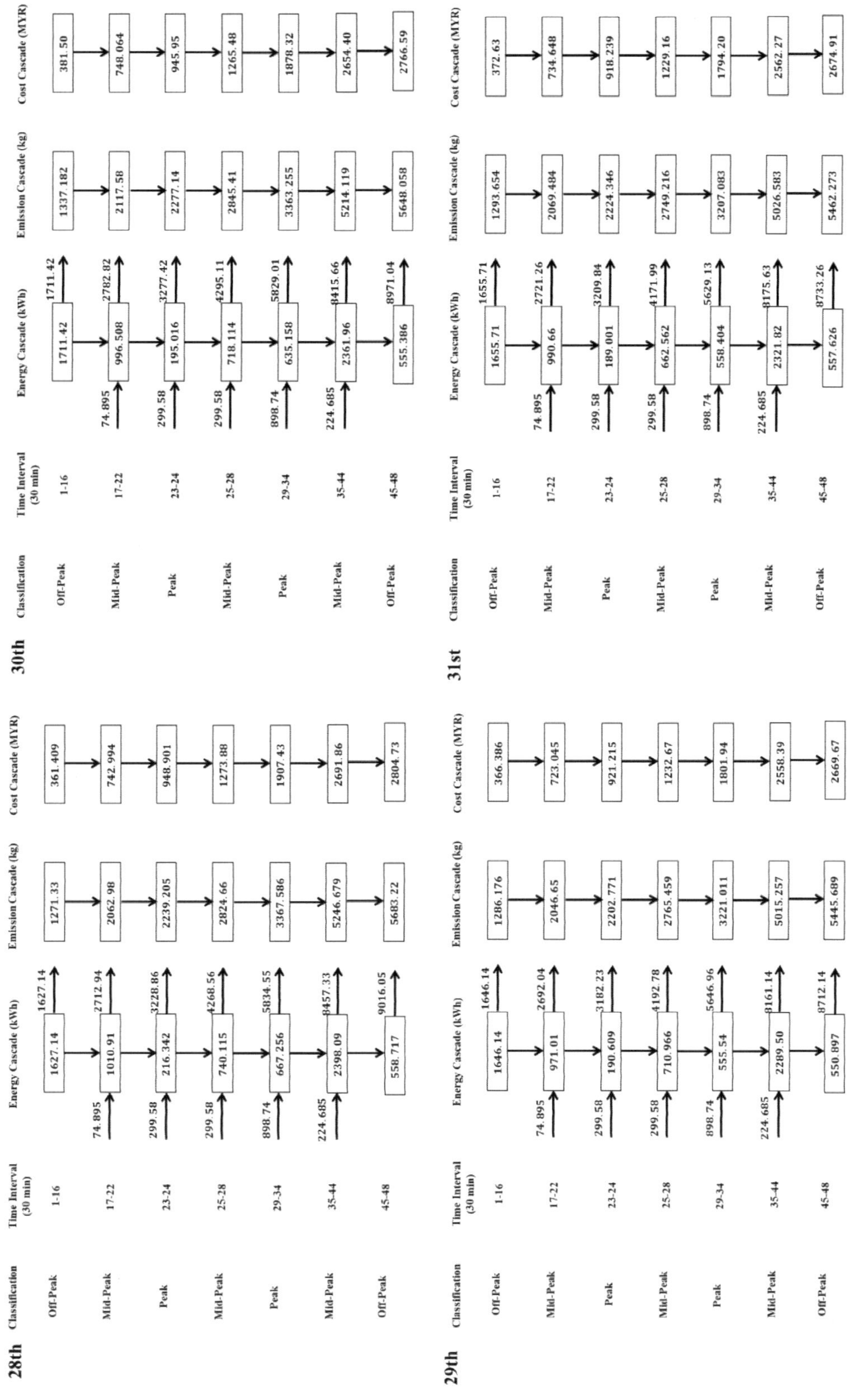
28th
Classification
Time Interval (30 min)
Energy Cascade (kWh)
Emission Cascade (kg)
Cost Cascade (MYR)
Off-Peak 1-16 1627.14 1627.14 1271.33 361.409
Mid-Peak 17-22 74.895 1010.91 2712.94 2062.98 742.994
Peak 23-24 299.58 216.342 3228.86 2239.205 948.901
Mid-Peak 25-28 299.58 740.115 4268.56 2824.66 1273.88
Peak 29-34 898.74 667.256 5834.55 3367.586 1907.43
Mid-Peak 35-44 224.685 2398.09 8457.33 5246.679 2691.86
Off-Peak 45-48 558.717 9016.05 5683.22 2804.73
29th
Classification
Time Interval (30 min)
Energy Cascade (kWh)
Emission Cascade (kg)
Cost Cascade (MYR)
Off-Peak 1-16 1646.14 1646.14 1286.176 366.386
Mid-Peak 17-22 74.895 971.01 2692.04 2046.65 723.045
Peak 23-24 299.58 190.609 3182.23 2202.771 921.215
Mid-Peak 25-28 299.58 710.966 4192.78 2765.459 1232.67
Peak 29-34 898.74 555.54 5646.96 3221.011 1801.94
Mid-Peak 35-44 224.685 2289.50 8161.14 5015.257 2558.39
Off-Peak 45-48 550.897 8712.14 5445.689 2669.67
30th
Classification
Time Interval (30 min)
Energy Cascade (kWh)
Emission Cascade (kg)
Cost Cascade (MYR)
Off-Peak 1-16 1711.42 1711.42 1337.182 381.50
Mid-Peak 17-22 74.895 996.508 2782.82 2117.58 748.064
Peak 23-24 299.58 195.016 3277.42 2277.14 945.95
Mid-Peak 25-28 299.58 718.114 4295.11 2845.41 1265.48
Peak 29-34 898.74 635.158 5829.01 3363.255 1878.32
Mid-Peak 35-44 224.685 2361.96 8415.66 5214.119 2654.40
Off-Peak 45-48 555.386 8971.04 5648.058 2766.59
31st
Classification
Time Interval (30 min)
Energy Cascade (kWh)
Emission Cascade (kg)
Cost Cascade (MYR)
Off-Peak 1-16 1655.71 1655.71 1293.654 372.63
Mid-Peak 17-22 74.895 990.66 2721.26 2069.484 734.648
Peak 23-24 299.58 189.001 3209.84 2224.346 918.239
Mid-Peak 25-28 299.58 662.562 4171.99 2749.216 1229.16
Peak 29-34 898.74 558.404 5629.13 3207.083 1794.20
Mid-Peak 35-44 224.685 2321.82 8175.63 5026.583 2562.27
Off-Peak 45-48 557.626 8733.26 5462.273 2674.91

Appendix B: Automated Targeting Model for entity 1 (After Energy sharing) 1st -30th day

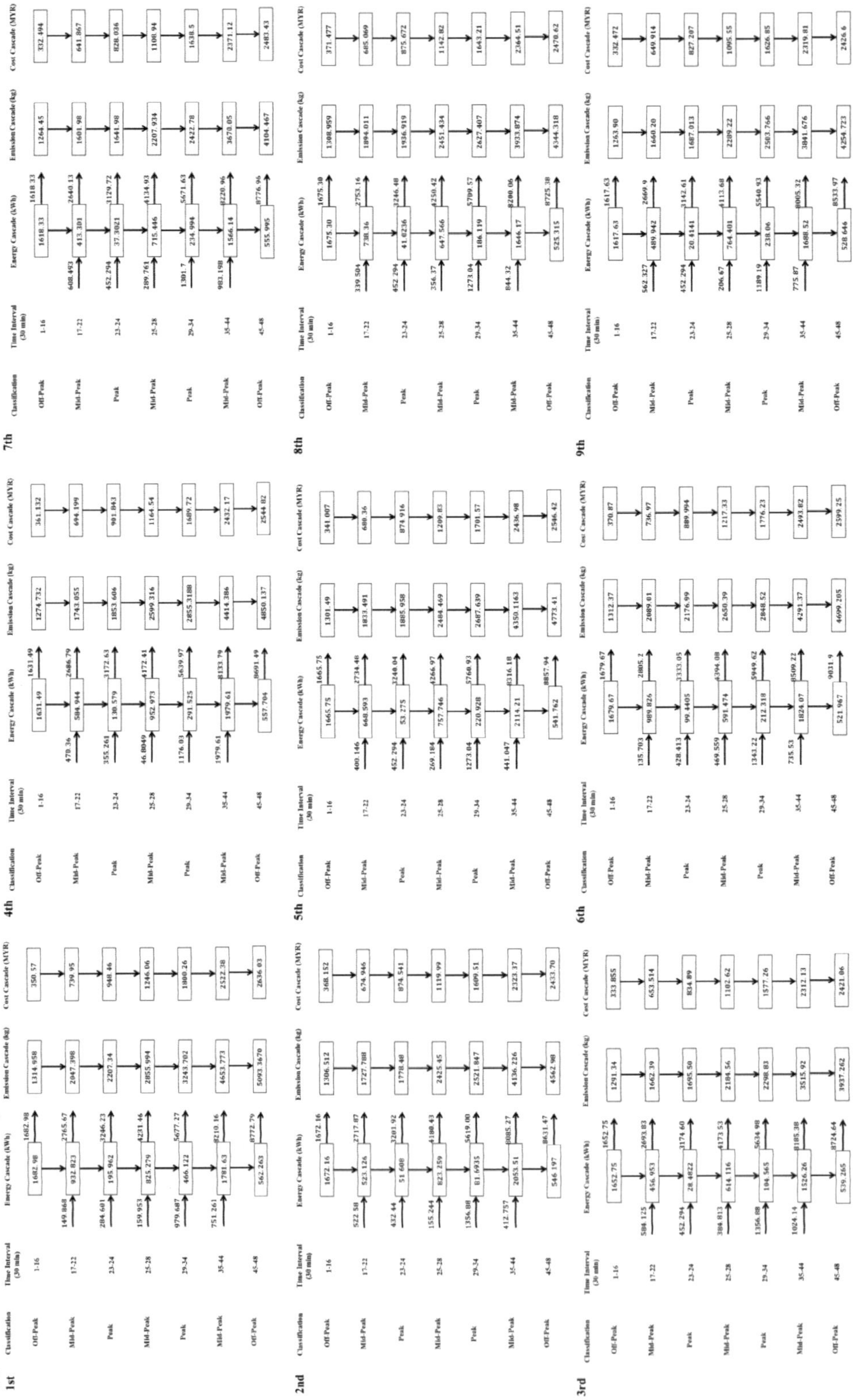

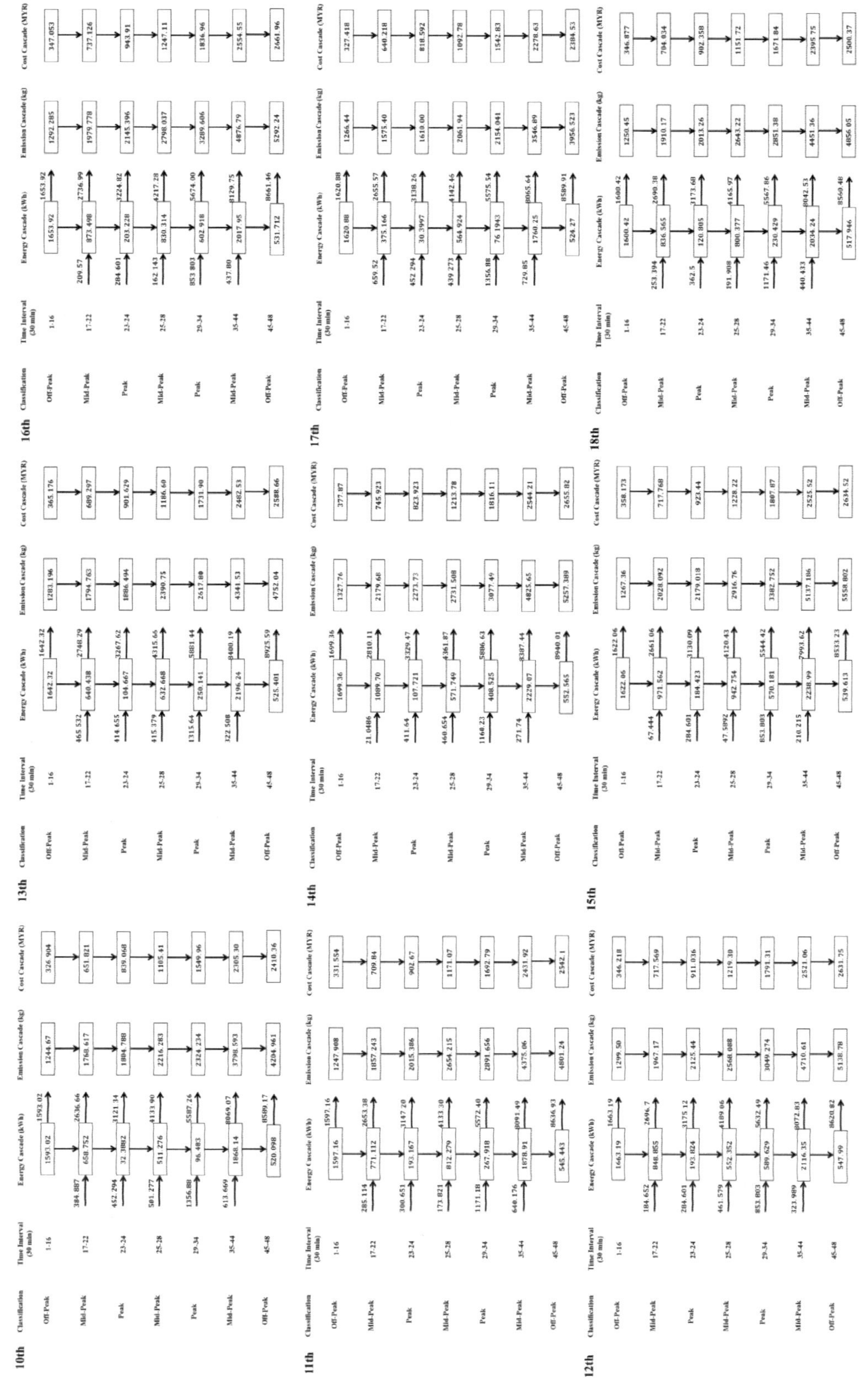

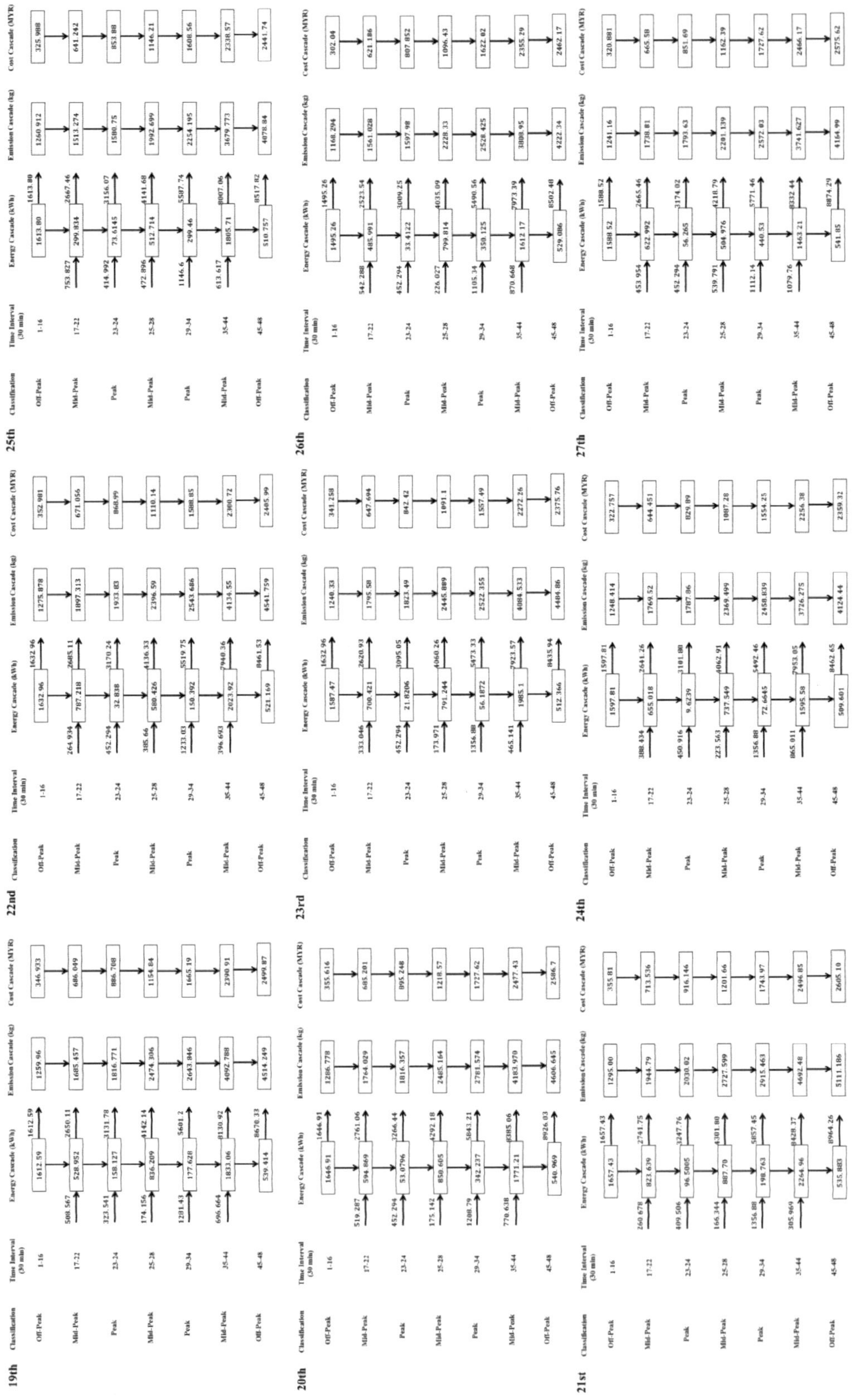

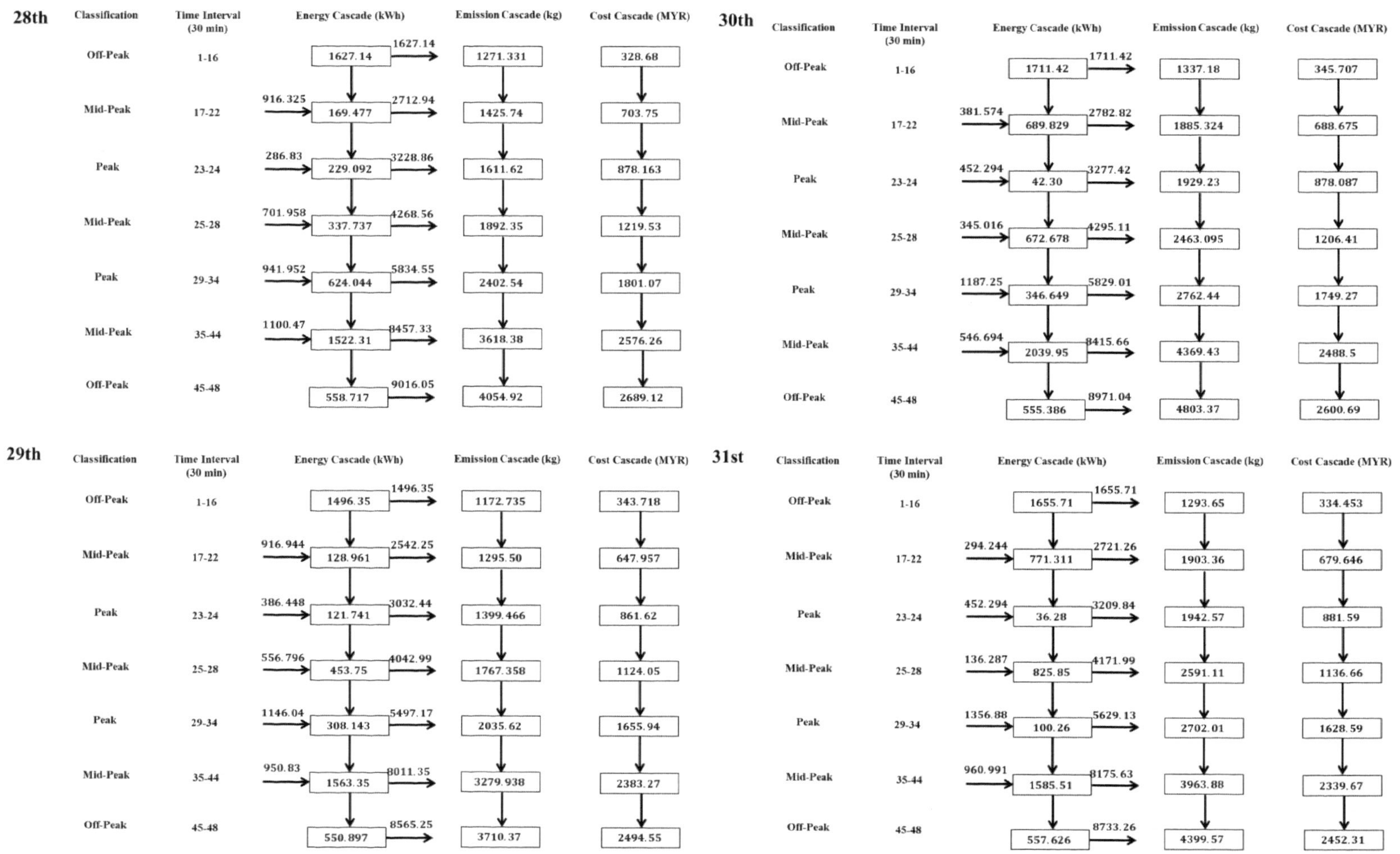
28th
Classification
Time Interval (30 min)
Energy Cascade (kWh)
Emission Cascade (kg)
Cost Cascade (MYR)
Off-Peak 1-16 1627.14 1627.14 1271.331 328.68
Mid-Peak 17-22 916.325 169.477 2712.94 1425.74 703.75
Peak 23-24 286.83 229.092 3228.86 1611.62 878.163
Mid-Peak 25-28 701.958 337.737 4268.56 1892.35 1219.53
Peak 29-34 941.952 624.044 5834.55 2402.54 1801.07
Mid-Peak 35-44 1100.47 1522.31 8457.33 3618.38 2576.26
Off-Peak 45-48 558.717 9016.05 4054.92 2689.12
30th
Classification
Time Interval (30 min)
Energy Cascade (kWh)
Emission Cascade (kg)
Cost Cascade (MYR)
Off-Peak 1-16 1711.42 1711.42 1337.18 345.707
Mid-Peak 17-22 381.574 689.829 2782.82 1885.324 688.675
Peak 23-24 452.294 42.30 3277.42 1929.23 878.087
Mid-Peak 25-28 345.016 672.678 4295.11 2463.095 1206.41
Peak 29-34 1187.25 346.649 5829.01 2762.44 1749.27
Mid-Peak 35-44 546.694 2039.95 8415.66 4369.43 2488.5
Off-Peak 45-48 555.386 8971.04 4803.37 2600.69
29th
Classification
Time Interval (30 min)
Energy Cascade (kWh)
Emission Cascade (kg)
Cost Cascade (MYR)
Off-Peak 1-16 1496.35 1496.35 1172.735 343.718
Mid-Peak 17-22 916.944 128.961 2542.25 1295.50 647.957
Peak 23-24 386.448 121.741 3032.44 1399.466 861.62
Mid-Peak 25-28 556.796 453.75 4042.99 1767.358 1124.05
Peak 29-34 1146.04 308.143 5497.17 2035.62 1655.94
Mid-Peak 35-44 950.83 1563.35 8011.35 3279.938 2383.27
Off-Peak 45-48 550.897 8565.25 3710.37 2494.55
31st
Classification
Time Interval (30 min)
Energy Cascade (kWh)
Emission Cascade (kg)
Cost Cascade (MYR)
Off-Peak 1-16 1655.71 1655.71 1293.65 334.453
Mid-Peak 17-22 294.244 771.311 2721.26 1903.36 679.646
Peak 23-24 452.294 36.28 3209.84 1942.57 881.59
Mid-Peak 25-28 136.287 825.85 4171.99 2591.11 1136.66
Peak 29-34 1356.88 100.26 5629.13 2702.01 1628.59
Mid-Peak 35-44 960.991 1585.51 8175.63 3963.88 2339.67
Off-Peak 45-48 557.626 8733.26 4399.57 2452.31

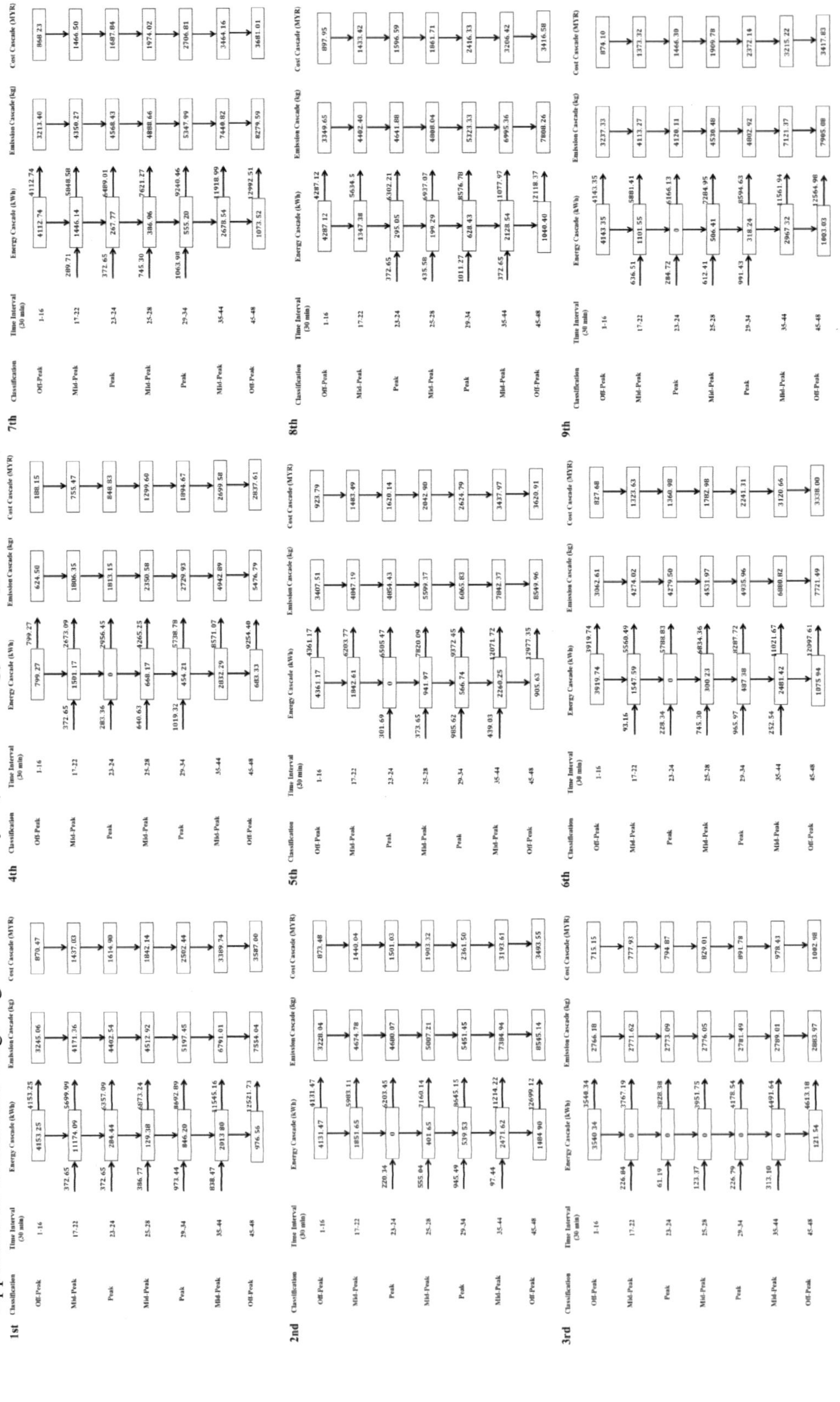

Appendix C: Automated Targeting Model for entity 2 (Before Energy sharing) 1st -30th day

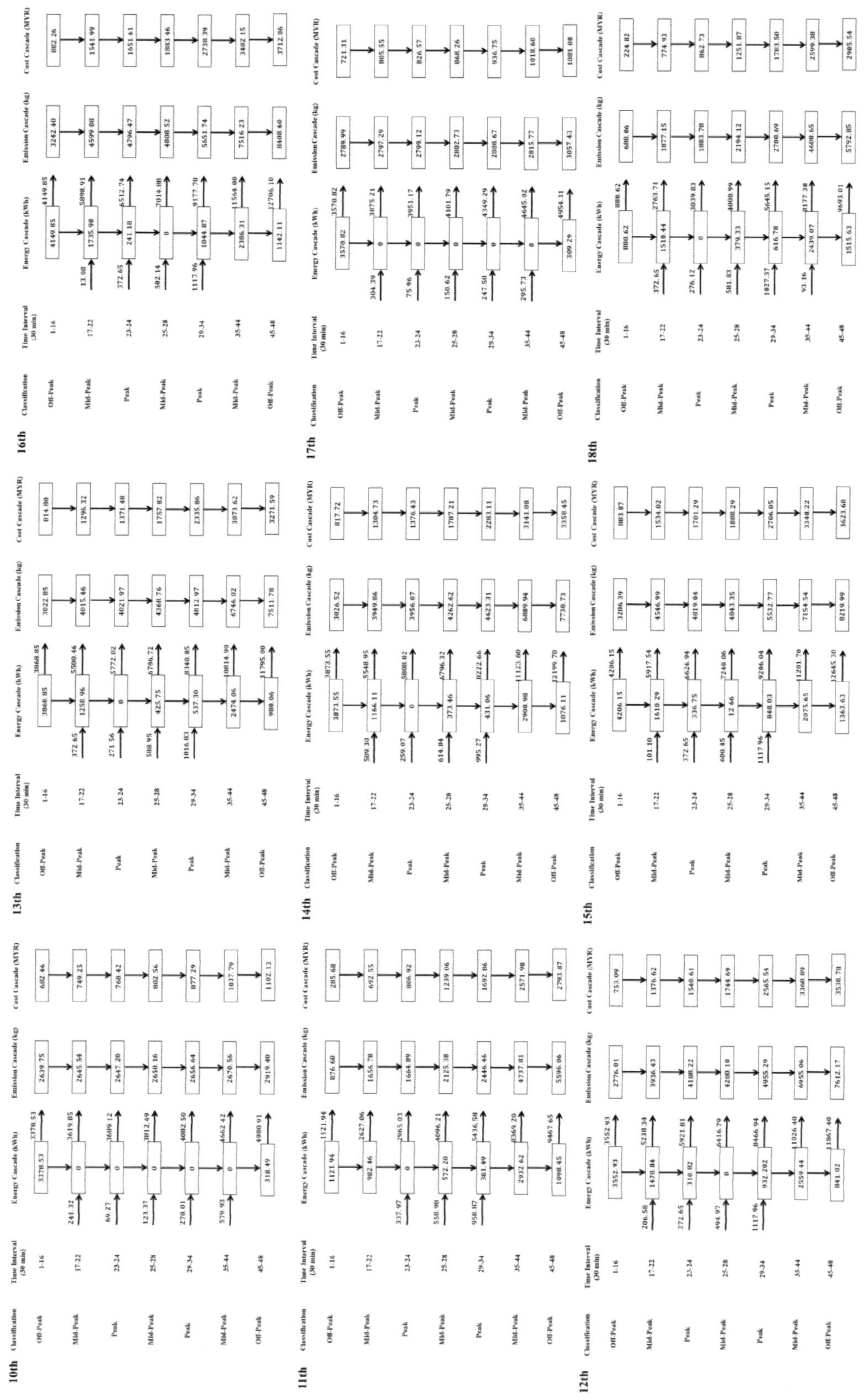
10th
11th
12th
13th
14th
15th
16th
17th
18th
Classification
Time Interval (30 min)
Energy Cascade (kWh)
Emission Cascade (kg)
Cost Cascade (MYR)

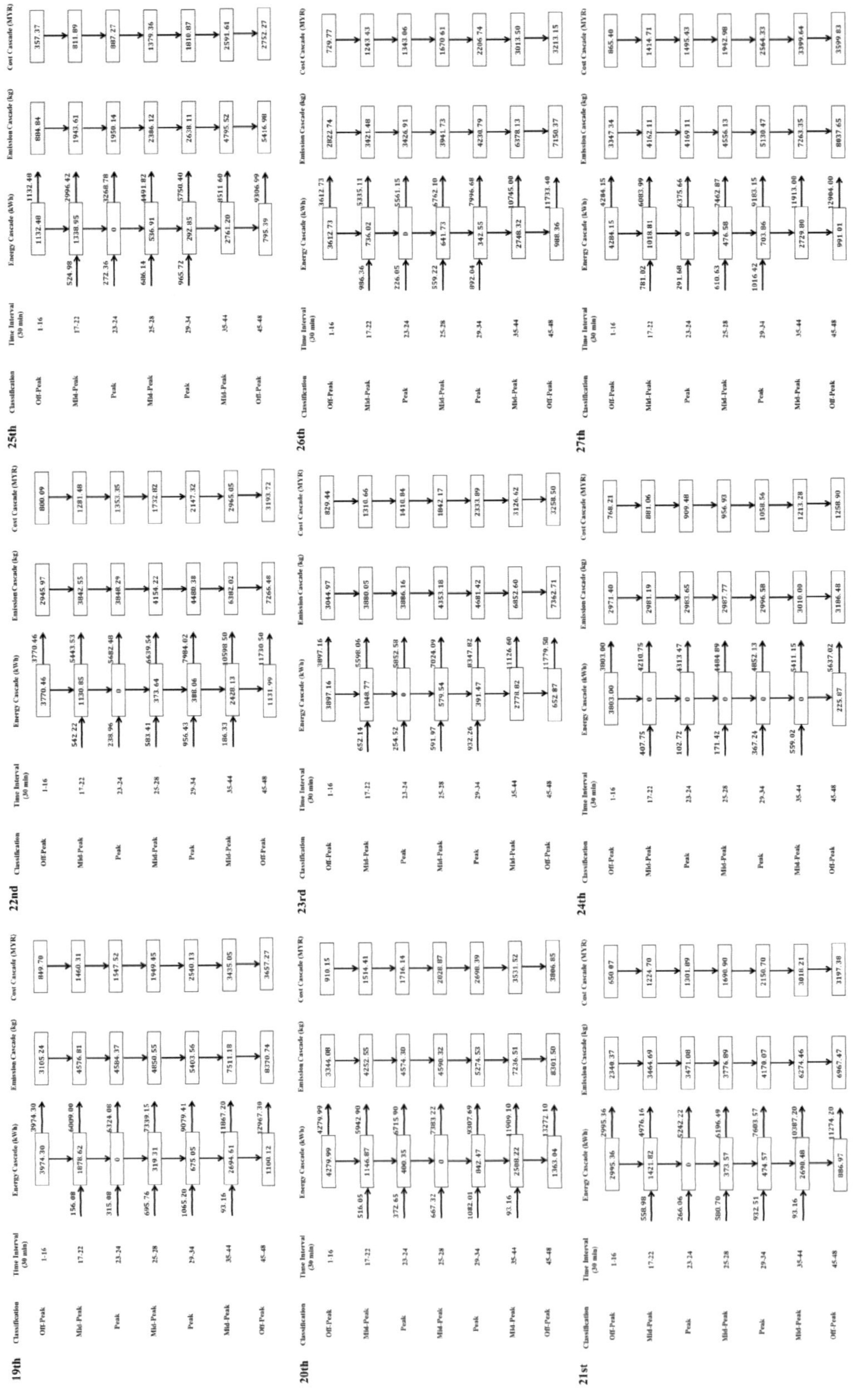
19th
20th
21st
22nd
23rd
24th
25th
26th
27th

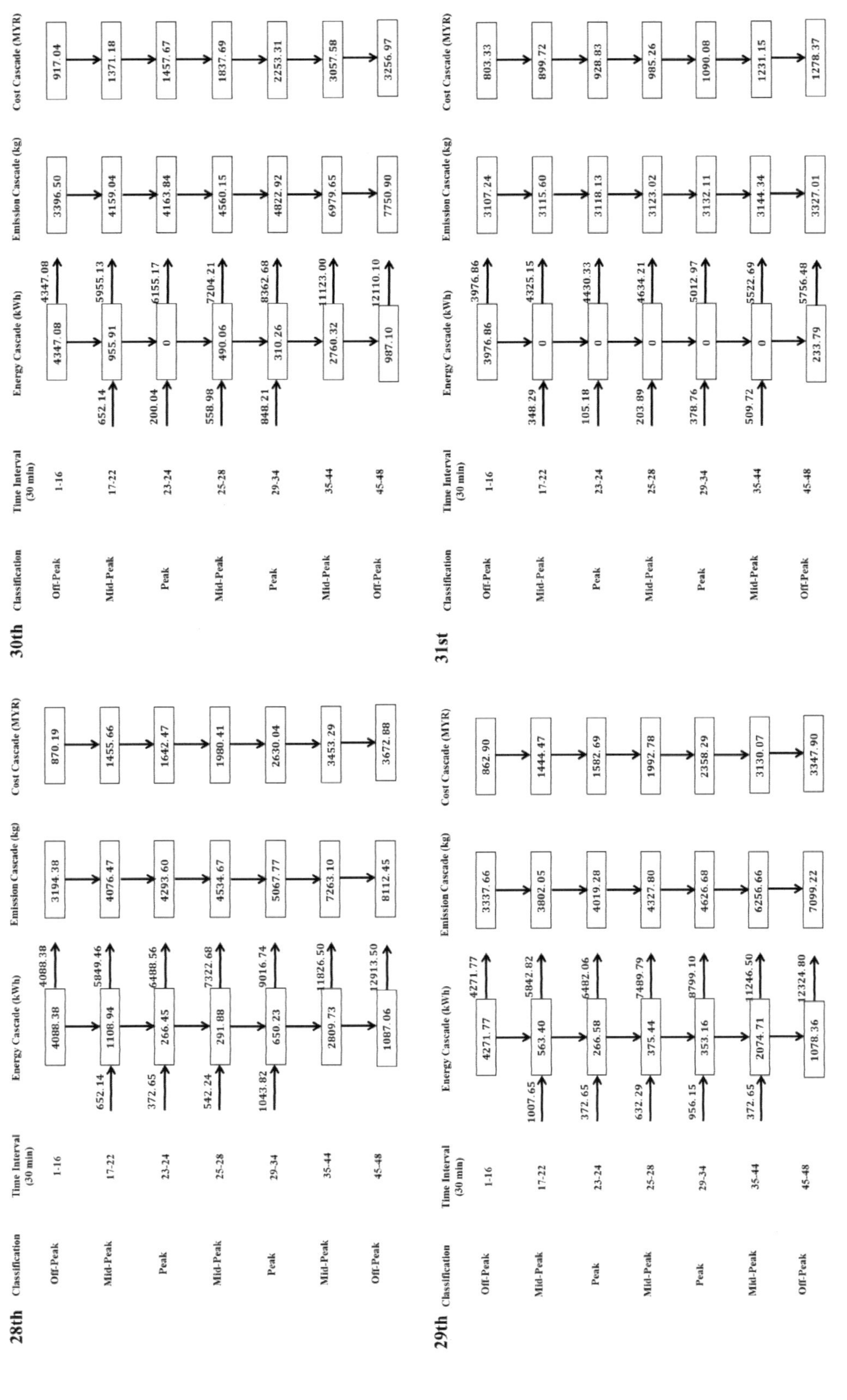
28th
Classification
Time Interval (30 min)
Energy Cascade (kWh)
Emission Cascade (kg)
Cost Cascade (MYR)
Off-Peak 1-16 4088.38 4088.38 3194.38 870.19
Mid-Peak 17-22 652.14 1108.94 5849.46 4076.47 1455.66
Peak 23-24 372.65 266.45 6488.56 4293.60 1642.47
Mid-Peak 25-28 542.24 291.88 7322.68 4534.67 1980.41
Peak 29-34 1043.82 650.23 9016.74 5067.77 2630.04
Mid-Peak 35-44 2809.73 11826.50 7263.10 3453.29
Off-Peak 45-48 1087.06 12913.50 8112.45 3672.88
29th
Classification
Time Interval (30 min)
Energy Cascade (kWh)
Emission Cascade (kg)
Cost Cascade (MYR)
Off-Peak 1-16 4271.77 4271.77 3337.66 862.90
Mid-Peak 17-22 1007.65 563.40 5842.82 3802.05 1444.47
Peak 23-24 372.65 266.58 6482.06 4019.28 1582.69
Mid-Peak 25-28 632.29 375.44 7489.79 4327.80 1992.78
Peak 29-34 956.15 353.16 8799.10 4626.68 2358.29
Mid-Peak 35-44 372.65 2074.71 11246.50 6256.66 3130.07
Off-Peak 45-48 1078.36 12324.80 7099.22 3347.90
30th
Classification
Time Interval (30 min)
Energy Cascade (kWh)
Emission Cascade (kg)
Cost Cascade (MYR)
Off-Peak 1-16 4347.08 4347.08 3396.50 917.04
Mid-Peak 17-22 652.14 955.91 5955.13 4159.04 1371.18
Peak 23-24 200.04 0 6155.17 4163.84 1457.67
Mid-Peak 25-28 558.98 490.06 7204.21 4560.15 1837.69
Peak 29-34 848.21 310.26 8362.68 4822.92 2253.31
Mid-Peak 35-44 2760.32 11123.00 6979.65 3057.58
Off-Peak 45-48 987.10 12110.10 7750.90 3256.97
31st
Classification
Time Interval (30 min)
Energy Cascade (kWh)
Emission Cascade (kg)
Cost Cascade (MYR)
Off-Peak 1-16 3976.86 3976.86 3107.24 803.33
Mid-Peak 17-22 348.29 0 4325.15 3115.60 899.72
Peak 23-24 105.18 0 4430.33 3118.13 928.83
Mid-Peak 25-28 203.89 0 4634.21 3123.02 985.26
Peak 29-34 378.76 0 5012.97 3132.11 1090.08
Mid-Peak 35-44 509.72 0 5522.69 3144.34 1231.15
Off-Peak 45-48 233.79 5756.48 3327.01 1278.37

Appendix D: Automated Targeting Model for entity 2 (After Energy sharing) 1st -30th day

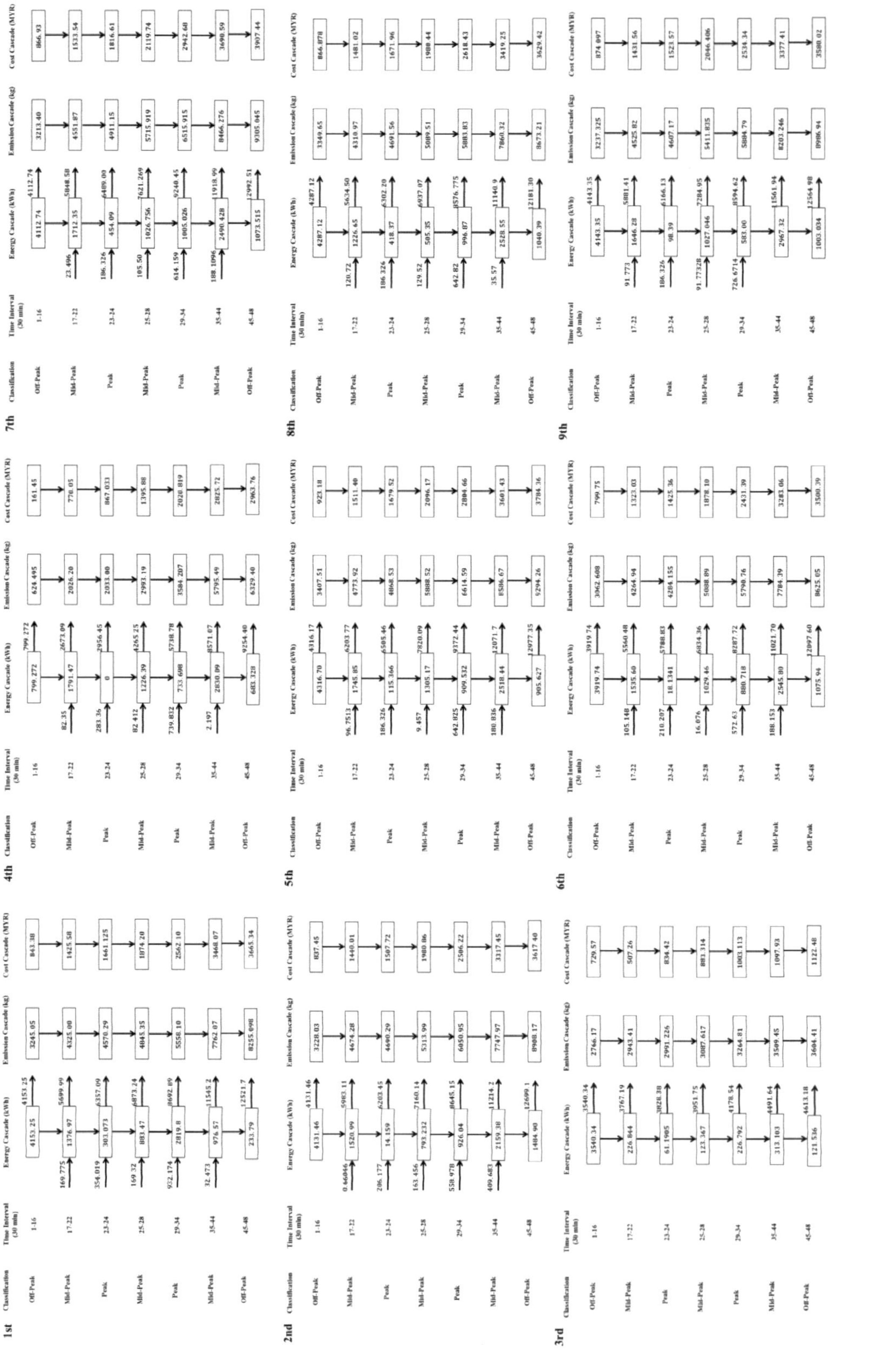

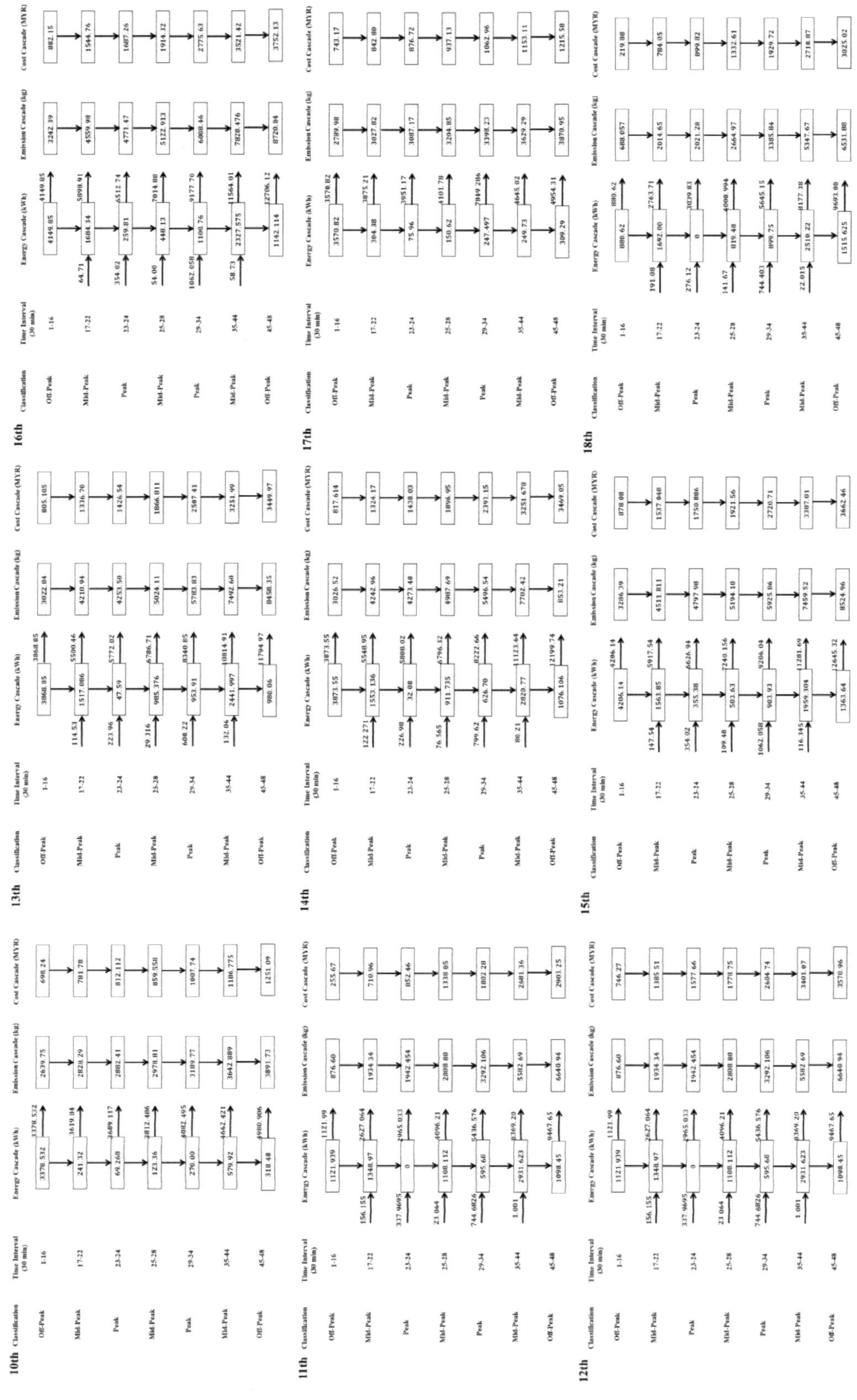
10th
11th
12th
13th
14th
15th
16th
17th
18th
Classification
Time Interval (30 min)
Energy Cascade (kWh)
Emission Cascade (kg)
Cost Cascade (MYR)

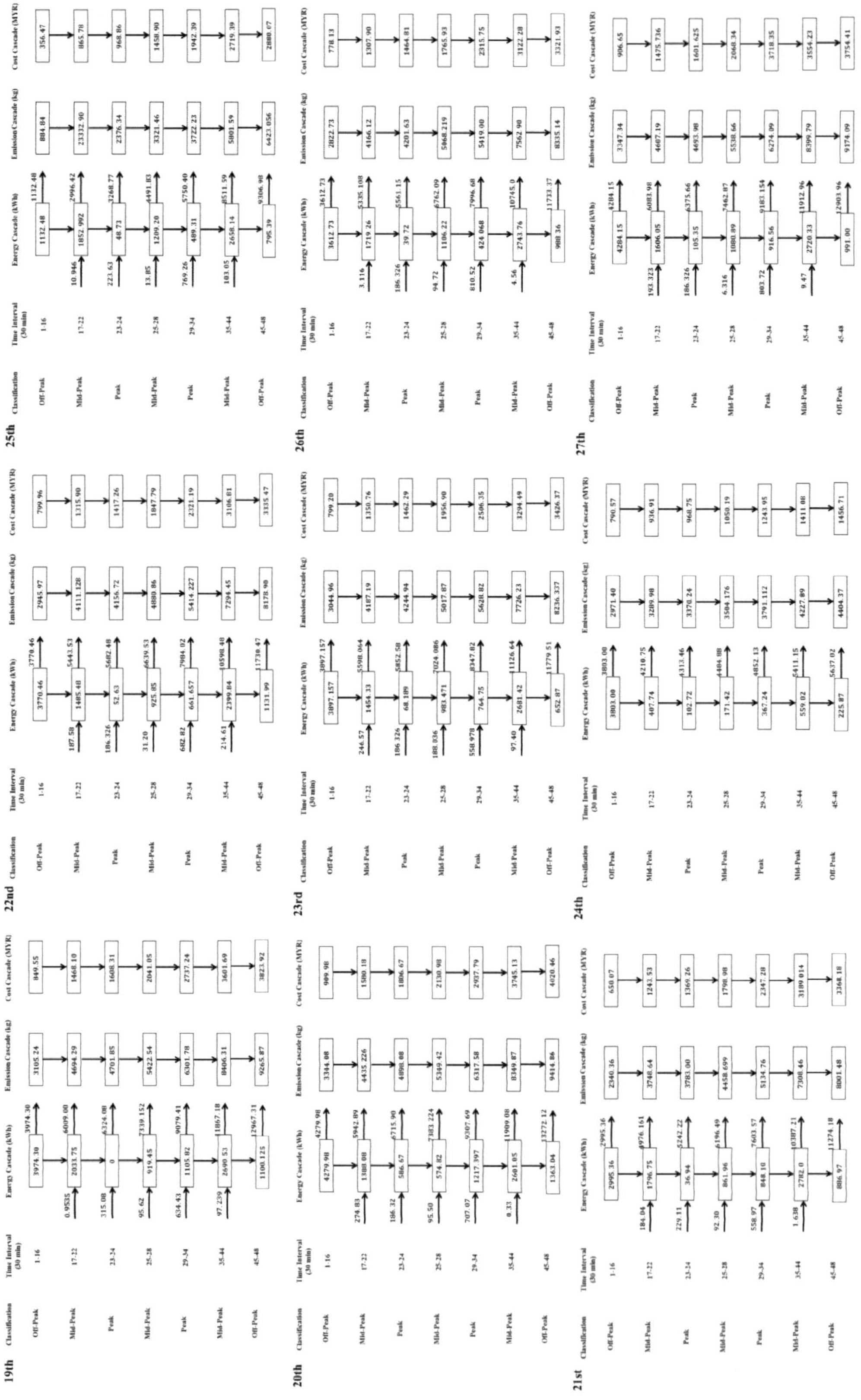

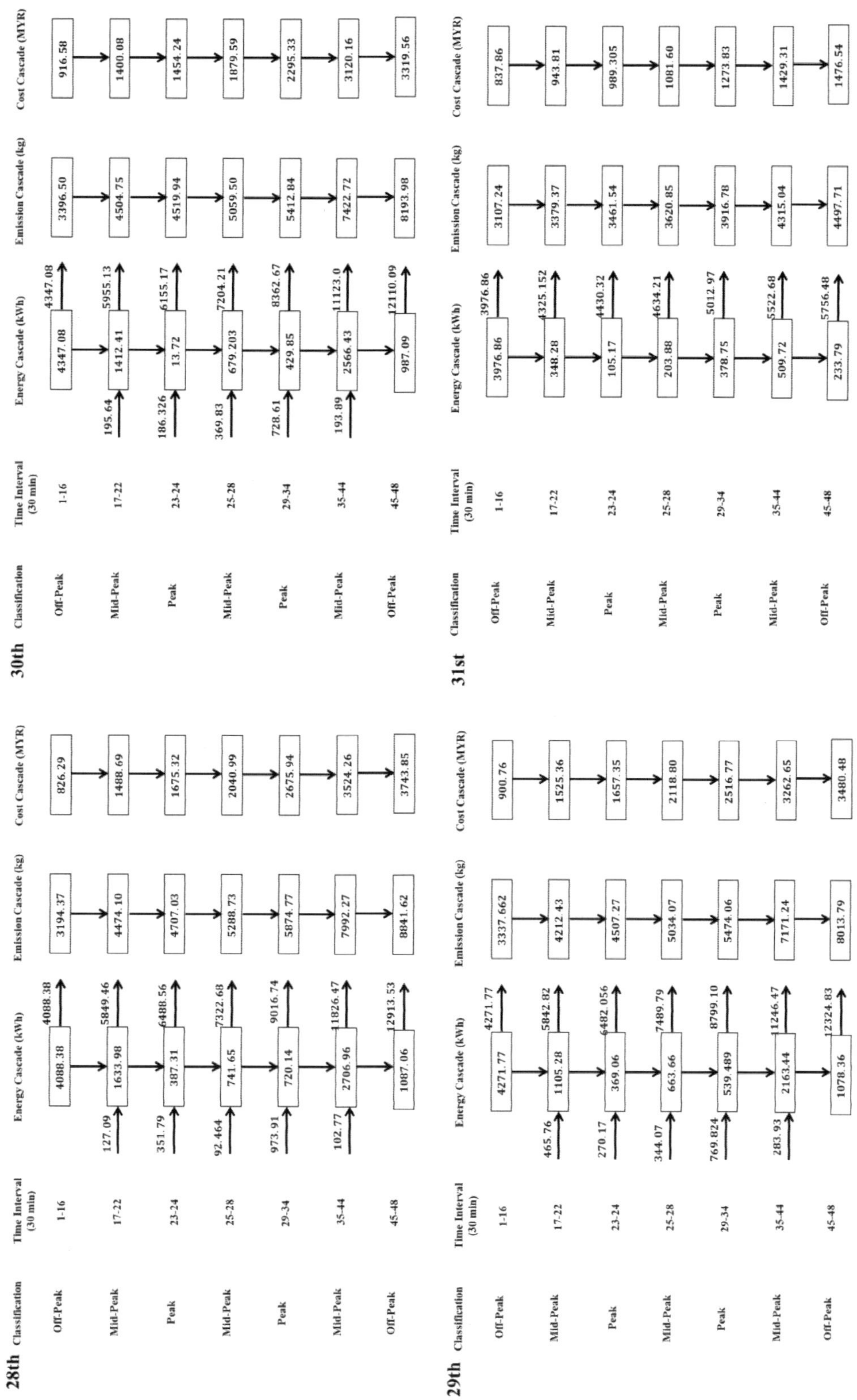
28th
29th
30th
31st
Classification
Time Interval (30 min)
Energy Cascade (kWh)
Emission Cascade (kg)
Cost Cascade (MYR)
Off-Peak
Mid-Peak
Peak
1-16
17-22
23-24
25-28
29-34
35-44
45-48

IOP Conf. Series: Materials Science and Engineering 1195 (2021) 012012 doi:10.1088/1757-899X/1195/1/012012

Simulation of ozonolysis of volatile organic compounds: Effect on flue gas composition

M Lim[1*] **and A R Lea-Langton**[2]
[1] TNB Research Sdn. Bhd., Research Institution Area, 43000 Kajang, Selangor, Malaysia.
[2] Tyndall Centre for Climate Change Research, Department of Mechanical, Aerospace & Civil Engineering, University of Manchester, Oxford Road, Manchester, M13 9PL, United Kingdom

Email: mook.tzeng@tnb.com.my; mooktzeng.lim@live.com

Abstract. This study shows that the reaction of ozone with various volatile organic compounds (VOC) yields different flue gas composition in terms of the carbon dioxide, oxygen and moisture contents. Steam production and thermal output requirements from a combustion system (i.e., a boiler) may dictate the range of operating conditions, such as the air to fuel mass flow rates. To improve the combustion efficiency in these operating conditions, low temperature plasmas have been used to ionize air and generate ozone as an oxidant for ozonolysis with the VOC. Therefore, this study simulates the reaction mechanism of the ozonolysis of VOC and the effect on the flue gas composition, which affects the combustion efficiency. Simulation results show that residual oxygen in the flue gas reduces, reducing the excess air. Thus, the corresponding efficiency loss through dry flue gas would be reduced. Literature data shows that emissions of alkanes, alkynes and alkenes per unit mass of solid fuel is evident for both coal and biomass, and thus ozonolysis of these VOC would reduce the excess air, improving the combustion efficiency.

1. Introduction

Current power generation technologies rely on the combustion of solid fuels such as coal and biomass. The combustion of both coals and biomass emits volatile organic compounds (VOC) which include alkanes, alkenes, alkynes and aromatics. The composition profiles of VOC differ for different fuels, and the concentration is related to the fuel volatile matter content.

Emission factors of VOCs for coal and biomass range from 0.04 to 4.5 g/kg and 0.06 to 19.69 g/kg [1, 2]. The variation in the emission factors is due to the type of fuel and combustion system. Open burning of crops emits higher amounts of VOCs (10 g/kg) compared to residential stoves burning wood (1.92 g/kg) and biomass boilers with wood pellets as fuel (0.15 g/kg) [1, 2]. The amount and reaction of the VOCs with the oxidant determines the flue gas composition. Of particular concern is the residual oxygen and carbon dioxide (CO_2) content in the flue gas, which determine the excess air and the dry flue gas losses, and hence the combustion efficiency. The higher the residual oxygen content, the higher is the excess air, which results in a higher dry flue gas loss and lower combustion efficiency.

Steam production and thermal output requirements from a combustion system (i.e., a boiler) demands a certain amount of airflow rate, resulting in a smaller window of operating conditions for optimum excess air and low dry flue gas loss. Various studies used low temperature plasmas to generate ozone as the oxidant for reaction with the VOC (i.e. ozonolysis of VOC) to improve the combustion performance [3, 4]. Ehn *et al.* [3] reported that addition of low concentrations of O_3 into low-swirl methane (CH_4)-air flame at lean conditions increased formaldehyde (CH_2O), determined from Planar Laser Induced Fluorescence (PLIF) imaging of hydroxyl (OH) and CH_2O. The increase in CH_2O was well predicted by Large Eddy Simulations (LES) with CH_4-air-O_3 reaction mechanisms. The sub-mechanism for O_3 included its presence in the flame and its interaction with all other species. Simulation results also showed that the addition of O_3 increased the laminar flame speed, allowing the flame to burn under more turbulent conditions [3] .

Similar results are shown in the work by Gao et. al.[5], who investigated the effect of ozone addition on laminar flame speeds of alkenes and alkanes. Ozonolysis of alkanes (CH_4 and C_3H_8) have shown to increase the flame speed over lean to rich equivalence ratios, whereas for alkenes, the flame speed was reduced [5]. A photocatalytic and ozone oxidation (UV/TiO_2+O_3) system was combined to decompose VOCs from the fumes of cooking stoves. Fiberglass filter (FGF) was coated with TiO_2, and a continuous-flow reactor combined photocatalysis with ozone oxidation. The VOCs in the simulated cooking fume decomposed by 64 %, with ozonolysis only contributing 34 % [6].

Ozonolysis of VOCs was also performed to understand the effects of a precursor hydrocarbon on the formation of secondary organic aerosols (SOA), and to predict the SOA formation potential [7]. A model was developed to predict the VOC (mainly alkenes) conversion rates by O_3, OH, and NO_3 in residential spaces [8].

The above studies did not present the final flue gas composition as a result of the ozonolysis of the VOC. The flue gas composition, specifically, the carbon dioxide (CO_2), residual oxygen (O_2) and moisture content (H_2O), are often used to determine combustion efficiencies. The carbon dioxide and residual oxygen content are used to calculate efficiency losses due to dry flue gas exiting the stack, and the moisture content determines the wet flue gas losses. Thus, this study investigates the reaction mechanisms of the ozonolysis of VOCs and the effect on the flue gas composition by thermodynamic equilibrium simulations.

2. Methodology

2.1. Volatile Organic Compound Composition of Biomass and Coal

Coal and biomass have a volatile matter (VM) content of approximately 40% to 60 % and 80% respectively [9, 10]. Table 1 shows the properties of coal and wood that were obtained using a CHNS analyzer (LECO, model 628) for ultimate analysis and a thermogravimetric analyzer (LECO) for proximate analysis. Under optimal boiler conditions, flaming combustion is dominant, and the residual pyrolysis products are emitted as VOCs. For the purpose of the simulation, 0.5 kg and 0.8 kg of volatile matter are estimated to be emitted from 1 kg of coal and biomass [11]. The corresponding VOC compositions can be found in [11]. For alkenes/alkynes, the amount is 0.35 and 0.41 kg/kg of coal and biomass; for alkanes, the content is 0.030 and 0.14 kg per kg of coal and biomass; for aromatics, the content is 0.08 and 0.064 to 0.312 kg/kg of coal and biomass respectively.

2.2. Simulation of ozonolysis of volatile organic compounds

The simulation was performed with FactSage 7.2, a commercially available software that calculates Gibbs free energy to simulate various chemical reactions. For the sake of brevity, the methodology is simplified, and readers are referred to the FactSage website for more details about the equations involved to simulate the reactions [12]. Ozone reacts with alkane, alkene, and alkyne in 10 kg of air at 600 °C according to the following reaction:

$$aO_3 + bC_xH_y + cN_2 + dO_2 \qquad (1)$$

IOP Conf. Series: Materials Science and Engineering 1195 (2021) 012012 doi:10.1088/1757-899X/1195/1/012012

In this simulation, C_2 based hydrocarbons are simulated, therefore x is 2, and y corresponds respectively for ethyne, ethene and ethane. The amount of ozone produced from the low temperature reactor, a *is 0.01 and 0.02 kg. This amount is subtracted from the mass of oxygen,* d *is thus (2.1 –* a*) kg. The amount of mass for nitrogen,* c *is 7.9 kg. The amount of VOC,* b *is estimated at 0.2 kg. More details of the simulation are found in [11].*

3. Results and Discussion

3.1. Effect of ozonolysis of volatile organic compounds on the O_2 and CO_2 contents

For flue gas from C_2 hydrocarbons, Table 1 shows that the O_2 and CO_2 are the lowest for C_2H_6 with no ozonolysis. With increasing O_3 to 20 g, the residual O_2 decreases by 0.002 %. Similar results are observed for C_2H_2 and C_2H_4.

Table 1. Simulated composition of emissions for the ozonolysis of volatile organic compounds

Mass of O_3 (g)	CH_4		C_2H_2		C_2H_4		C_2H_6	
	O_2 (%)	CO_2 (%)	O_2 (%)	CO_2 (%)	O_2 (%)	CO_2 (%)	O_2 (%)	CO_2 (%)
0	11.203	3.4615	13.104	4.3700	12.379	4.0188	11.754	3.7195
10	11.202	3.4615	13.103	4.3703	12.378	4.0188	11.753	3.7195
20	11.201	3.4615	13.102	4.3703	12.377	4.0187	11.752	3.7195

Based on literature, the aromatic emissions from biomass are higher on average than coal [11]. Aromatics were not included in the simulation work, since the least complex form of aromatics has six carbons (C_6H_6), and the ozonolysis of a C_6 hydrocarbon would yield a higher amount of CO_2 compared to C_2 hydrocarbons. The ozonolysis of CH_4 and C_2H_6 show a significant difference in the CO_2 emissions (3.4615 % and 3.7195 % respectively).

3.2. Effect of ozonolysis of volatile organic compounds on H_2O, NO and OH contents and temperature

The other main constituents of the flue gas from ozonolysis of C_2H_6 by 20 g of O_3 are N_2 (78.7700 %), H_2O (5.5726 %), NO (0.1704 %), and intermediates such as OH (0.0137), O and HO_2 (<0.001 %). The residual O_3 is less than 1 ppm. Table 2 shows the composition of H_2O, NO and OH in the flue gas after ozonolysis of the VOC. With increase in O_3, the OH concentration increases for the ozonolysis of C_2H_2 by 5.2 %; C_2H_4 by 4.3 %; and C_2H_6 by 3.3 %. The increase in OH concentration is an increase in O_3 concentration [13]:

$$O_3 \rightarrow O_2 + O \quad (2)$$

$$O + CH_4 \rightarrow CH_3 + OH \quad (3)$$

$$H + O_2 \rightarrow OH + O \quad (4)$$

$$H + O_3 \rightarrow OH + O_2 \quad (5)$$

The chain termination reaction of $OH + H_2 \rightarrow H_2O$ [13] suggests that an increase in OH should increase H_2O or moisture content in the flue gas. Instead, an increase in OH corresponds with a decrease of H_2O, by 0.0092 % for the ozonolysis of C_2H_2; 0.0049 % for C_2H_4; and 0.0036 % for C_2H_6. This is because the increase in O_3 increases the reaction temperature as shown in Figure 1, and the increase in temperature reduces the amount of H_2O in the flue gas, more so for C_2H_2 (alkynes), followed by C_2H_4 (alkenes) and C_2H_6 (alkanes).

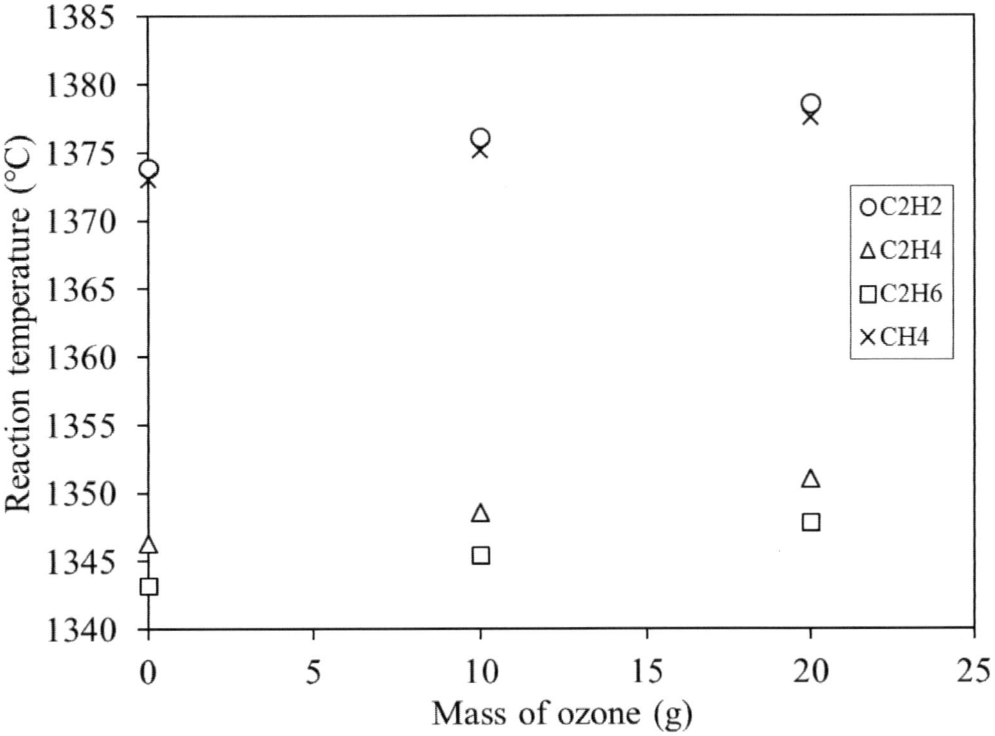

Figure 1. Variation of reaction temperature with ozonolysis of volatile organic compounds

Table 2. Composition of H_2O, NO and OH after ozonolysis of respective volatile organic compounds

	CH_4			C_2H_2		
Mass of O_3 (g)	H_2O (%)	NO (%)	OH (%)	H_2O (%)	NO (%)	OH (%)
0	6.9144	0.1837	0.01813	2.1800	0.2012	0.01065
10	6.9143	0.1853	0.01843	2.1799	0.2036	0.01083
20	6.9141	0.1870	0.01873	2.1798	0.2056	0.01121

	C_2H_4			C_2H_6		
Mass of O_3 (g)	H_2O (%)	NO (%)	OH (%)	H_2O (%)	NO (%)	OH (%)
0	4.0131	0.1744	0.0116	5.5728	0.1671	0.01324
10	4.0130	0.1761	0.0118	5.5727	0.1687	0.01346
20	4.0129	0.1778	0.0121	5.5726	0.1704	0.01370

4. Conclusions

This study shows that the reaction of ozone with volatile organic compounds reduces efficiency losses associated with excess air in combustion systems. Increasing the amount of ozone decreases the amount of residual O_2 from by 0.002 %. The effect of ozonolysis on the dry flue gas loss is more pronounced for alkanes (0.0019 %), followed by alkynes (0.0016 %) and alkenes (0.0009 %). The increase in ozone increases the reaction temperature, and the increase in temperature reduces the amount of moisture in the flue gas, more so for alkynes, followed by alkenes and alkanes. An increase in ozonolysis of the VOC increases the OH concentration and decreases the moisture content in the flue gas. The moisture content decreases because the increase in O_3 increases the reaction temperature, and the increase in temperature reduces the amount of moisture in the flue gas. Overall, the simulation suggests that reaction of ozone with volatile organic compounds increases the efficiency of combustion. The simulation suggests that the residual oxygen and moisture content in the flue gas decreases, increasing the reaction temperature. The experimental investigation of the effect of the ozone addition on the combustion efficiency is presented in an upcoming article.

Acknowledgments
Tenaga Nasional Berhad funded this work (TNBR/SF195/2015 & TNBR/SF240/2016).

References
[1] Geng C, Yang W, Sun X, Wang X, Bai Z and Zhang X 2019 Emission factors, ozone and secondary organic aerosol formation potential of volatile organic compounds emitted from industrial biomass boilers *J. Environ. Sci.* **83** 64
[2] Yan Y, Yang C, Peng L, Li R and Bai H 2016 Emission characteristics of volatile organic compounds from coal-, coal gangue-, and biomass-fired power plants in China *Atmos. Environ.* **143** 261
[3] Ehn A, *et al.* 2015 Plasma assisted combustion: Effects of O_3 on large scale turbulent combustion studied with laser diagnostics and Large Eddy Simulations *Proc. Combust. Inst.* **35** 3487
[4] Sun W, Gao X, Wu B and Ombrello T 2019 The effect of ozone addition on combustion: Kinetics and dynamics *Prog. Energy Combust. Sci.* **73** 1
[5] Gao X, Zhang Y, Adusumilli S, Seitzman J, Sun W, Ombrello T and Carter C 2015 The effect of ozone addition on laminar flame speed *Combust. Flame* **162** 3914.
[6] Li Y H, Cheng S W, Yuan C S, Lai T F and Hung C H 2018 Removing volatile organic compounds in cooking fume by nano-sized TiO_2 photocatalytic reaction combined with ozone oxidation technique *Chemosphere* **208** 808.
[7] Keywood M D, Varutbangkul V, Bahreini R, Flagan R C and Seinfeld J H 2004 Secondary organic aerosol formation from the ozonolysis of cycloalkenes and related compounds *Environ. Sci. Technol.* **38** 4157
[8] Waring M S and Wells J R 2015 Volatile organic compound conversion by ozone, hydroxyl radicals, and nitrate radicals in residential indoor air: Magnitudes and impacts of oxidant sources *Atmos. Environ.* **106** 382
[9] Zhou C, Liu G, Wang X and Qi C 2016.Co-combustion of bituminous coal and biomass fuel blends: Thermochemical characterization, potential utilization and environmental advantage *Bioresour. Technol.* **218** 418
[10] Kijo-Kleczkowska A, Środa K, Kosowska-Golachowska M, Musiał T and Wolski K 2016 Combustion of pelleted sewage sludge with reference to coal and biomass *Fuel* **170** 141
[11] Lim M and Lea-Langton A R 2020 Investigation of non-thermal plasma assisted combustion of solid biomass fuels: Effects on flue gas composition and efficiency *Plasma Chem. Plasma Process* **40** 1465
[12] FactSage Module 2020 *The Equilibrium Module*
[13] Liang X, Wang Z, Weng W, Zhou Z, Huang Z, Zhou J and Cen K 2013 Study of ozone-enhanced combustion in $H_2/CO/N_2$/air premixed flames by laminar burning velocity measurements and kinetic modeling *International J. Hydrogen Energy.* **38** 1177

Continuous desalting concept on ionic liquid-mediated de-acidification process of crude oil: A pilot study

A Hussain and J Basar

PETRONAS Research Sdn. Bhd., Lot 3288 & 3289, Off Jalan Ayer Itam, Kawasan Institusi Bangi, 43000 Kajang, Selangor Darul Ehsan, MALAYSIA

Email: azlanshah_hh@petronas.com

Abstract. Desalting process concept was tested using methyltrimethylammonium methylcarbonate $[N_{4441}][MeCO_3]$ treated Pyrenees crude oil (initial Total Acid Number (TAN) of 1.6 mg KOH/g oil) with the aim to gain empirical evidences on the effectiveness of in-line water washing and electrostatic aided phase separation as mean to recover the naphthenic acid derivatives for recycling. The treated crude oil (final TAN value of less than 0.3 mg KOH/g oil) was subjected to typical operating scheme such as single stage desalting and effects of water wash volumes. The novelty of the work comes from the utilisation of ionic liquids to neutralise acid components of the crude oil. Furthermore, the work is also able to test the hypothesis of whether naphthenate salts behave as is its inorganic counterpart and quantify the solubility behaviour in water as extraction medium. The effectiveness of such scheme will be measured against naphthenic acids derivative percent recovery in the wash water. The results indicate the electrostatic conditions can facilitate the recovery of the naphthenate salts post neutralization with high recovery rate of average of 70.6 % with 30 % water wash volume in a single-stage contact, observed over 12 hours steady-state operation. The water wash weight was observed to increase post separation which indicate hydrocarbon carry-over in the heavy phase due to formation of tight water – oil emulsion. The technique is viable should the amount of water required is available and the process water can be recycled safely into the desalter again without causing tripping to the desalter. Ionic liquid can be used in conjunction with desalter and the presence of electrostatic field did hasten the separation of the phases, however the amount of water used may hinder the viability of the solution.

1. Introduction

Desalting process is a common feature of a refinery and is designed to remove inorganic salts and dehydrate crude oil through water-oil contact and hasten phase separation through coalescence of water droplets in a vessel that has electrostatic field generated in it [1-3]. The removal of salts and water is vital to the downstream process as the absence of the two will reduce fouling and corrosion tendencies in the hot heat exchanger train, boiler tube protection and water load reduction at overhead section of crude distillation unit [4]. Typical desalter package will have an electrostatic generator, separation vessel, pressure drop mixer and mud wash. There are two types of desalters which are Alternating

Current (AC) and Dual Polarity. The main difference is the type of electrostatic field use as the latter employs both AC and DC field to induce coalescence of water droplets.

Recent reviews provide an overview of techniques used to reduce or eliminate naphthenic acids (NAs) in crude oils which include esterification, decarboxylation and neutralisation with metal hydroxides [5, 6]. Naphthenic acids are carboxylic acids with one or more carboxylic acid functionalities attached to one or more saturated cyclic structure, in general. Naphthenic acids are commonly written as $C_nH_{2n+z}O_2$ with z values corresponds to the degree of hydrogen deficiency that starts from 0 to negative even integer. They can also refer be referred to other chemical structures containing carboxylic acids such as heteroatom (S, N, B) or aromatics derived structures [7-12]. Ionic liquids or ionic liquid-based chemistry were explored as a method to reduce or eliminate NAs due to its tuneable properties i.e. solubility, low volatility and basicity [13-16]. Anderson *et al.* [17, 18] demonstrated the usage of two different classes of ionic liquids i.e., amino acid based ionic liquids (AAILs) and carbonate based ionic liquids on solvent-assisted removal of naphthenic acids from crude oil matrix. The work also demonstrated that upon reaction with ionic liquid, the solubility of the resulting naphthenate salts are more in the methanol phase as compared with metal naphthenates form, suggesting good recovery of the ionic liquid for regeneration effort. Hydroxide based ionic liquids were found to be effective with tetralkylammonium cation being more efficient than tetralakylphosponium cation and can be regenerated using ion-exchange method [19]. There are few examples in the open literature that exemplified the synergy between desalter and ioic liquid technology. Vadaraj *et al.* [20, 21] utilised aqueous alkoxylated tertiary amine and contacted it with crude oil containing naphthenic acids in the normal desalter setup. Similarly, like BASIL™ process, the amine shall form quarternary amine salts with the naphthenic acids and partitioned in the aqueous phase [22]. However, there was evidence of 4% oil loss to the aqueous phase and no information on alkoxylated amine salt cross solubility into the oil phase was offered. Simple organic carboxylic acids were shown to have different solubility behaviour between organic phase and ionic liquid. Manan *et al.* demonstrated the phase behaviour of cyclohexanoic acid, a model naphthenic acid with different ionic liquids of which the solubility of cyclohexanoic acid is dependent on lipophilicity of the cation and coordinating capability of the anion such as triluoroacetate and triflate [23]. However, with increasing alkyl chain of the cation, so does the solubility of dodecane indicating higher loss of hydrocarbon. Similar observations were also made with phenolate based ionic liquids [24]. Coto *et al.* recently disclosed 1-ethyl-3-methylimidazolium ethyl sulfate ([EMIM][$EtSO_4$]) achieved 25% TAN reduction post contact with model oil system [25]. Functionalization of the 1-butylimidazolium tetrafluoroborate ([Bmim][BF4]) cation moiety was found to increase the interaction between naphthenic acid and substitutional groups such as carboxylic acid (-COOH) and methyl ester (-$COOCH_3$) through formation of hydrogen bond [26]. This paper shall discuss in detail the work described by Anderson *et al.* [27]. The aim of this work is to examine and quantify process stability, mass balance generation and the behaviour of ionic liquid treated crude oil in a continuous desalting process. Desalter's operating parameter of water wash volume and one versus two stage desalting will be benchmarked against the recovery percentage of the cation. Ultimately, the findings from this work is to aid engineering design and operating philosophy of a facile process that utilises ionic liquid and desalter as means to reduce TAN value of crude oils of which shall increase flexibility of refinery, widen crude oil diet for higher refining margin and mitigate adverse risks to the vacuum distillation units due to high temperature corrosion.

2. Materials and methodologies

Pyrenees crude oil was supplied by PETRONAS from Australia (TAN of 1.63 mg KOH/g oil). Tributylamine (99 %), n-hexane (HPLC), methanol (HPLC), deuterated benzene (C_6D_6) and deuterated dimethyl sulfoxide (DMSO) were supplied by Sigma-Aldrich. Dimethylcarbonate (99 %) was supplied by Alfa Aesar. Purified water with conductivity of 18 mS/cm was used throughout. All chemicals are used as received unless stated otherwise.

2.1. *Synthesis of methyltributylammonium methylcarbonate [N_{4441}][CH_3CO_3]*

The preparation of methyltributylammonium methylcarbonate $[N_{4441}][CH_3CO_3]$ was performed according to published method [28]. Typically, the ionic liquid obtained by reacting tributylamine (37.07 g, 0.2 mol) with dimethyl carbonate (36.03 g, 0.4 mol) in 50 cm^3 of methanol in an autoclave at 6 bar and 393.15 K for 12 hours. After the reaction the excess methanol was removed under reduced pressure and was washed with n-hexane (3 x 100 cm^3) to remove excess dimethylcarbonate affording a light-yellow liquid. The yield (calculated using NMR) was 78 % for the $[N_{4441}]^+$ based ionic liquid. The ionic liquid was then further dried under high vacuum (1.0 x 10^{-3} Pa) at 338.15 K for 24 hours. The resultant ionic liquid crystallised upon standing at room temperature.

2.2. Total acid number (TAN) measurement
Typical procedure used for TAN determination using an automatic potentiometer with combined glass electrode (Metrohm Titrino Plus model 785), in accordance of ASTM D664, is by weighing a known amount of oil in 125 ml beaker which is then diluted with titration solvent. The titration solvent is a mixture of methanol/isopropanol, toluene and water in the ratio of 45:50:5 by volume. The electrode needs to be washed with titration solvent before usage. The homogenous solution is then titrated using 0.1 N of methanolic potassium hydroxide solution (Sigma Aldrich) until it reaches equilibrium point. The equilibrium point is then calculated using equation (1) below to get the final TAN value.

The Total Acid Number (TAN) is then calculated based on the equation (1).

$$Acid\ Number = \frac{(A - B) \times M \times 56.1}{w} \tag{1}$$

where A is the volume of KOH titrated to reach the equilibrium point, B is the volume of KOH for blank titration, M is the concentration of the KOH solution and w is the sample mass. The accuracy of the method is dependant of mass of sample with reproducibility around 2 %. The electrode is conditioned by immersing in pH 4 buffer solution (Metrohm UK) for a minimum of 10 minutes before reusing it again for titration. After several titrations, it is recommended for the electrode to be lightly wiped with soapy water and reconditioned in its electrolyte solution (3.0 M potassium chloride in ethanol, Methrohm UK) for overnight.

2.3. Desalter setup and crude oil neutralisation
The high throughput desalter pilot unit is located at Intertek PARC at Pittsburgh, USA and was manufactured by Howe Baker Engineers (Serial No.: 1498; Year built: 1980). The unit consists of two process stages unit, shown in figure 1. The pilot unit has capability of 53 litres/hr (0.053 m^3/hr), injection pump facility of maximum 0.5 litres/hr, max temperature of 230 °C and working pressure of 10 bar. The setup is equipped with mixing valve with max pressure drop of 2 barg and flexibility of running the experiment in one or two-stages.

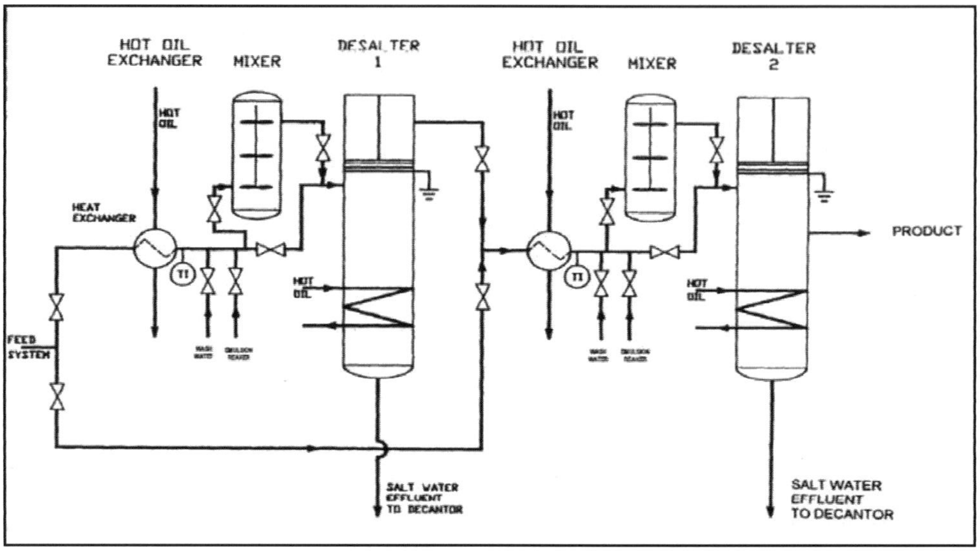

Figure 1. Intertek desalter pilot unit process flow diagram [38].

Ionic liquid and crude oil were pre-mixed in a holding tank with recirculation pump using a predetermined ratio between 0.7 wt% to 1.0 wt% at constant stirring of 500 rpm at 80 °C for 24 hours. The neutralisation effectiveness was monitored using ASTM D664 method for determining total acid number [29]. The crude oil is considered completely neutralised if the value reached <0.3 mg KOH/g of oil. The neutralised crude oil was pumped according to the conditions listed in table 1 and pre-heated using a series of heat exchanger. Wash water of pre-determined ratio between 10 – 30v = vol% , was contacted with the oil phase using static mixer and pressure drop mixer of 10 barg to induce mixing and formation of droplets. The mixed phase was then subjected to the electrostatic field and allows to phase separate based on the vessel designed residence time of 15 mins. All the incoming and outgoing flows were stored and measured using weight balance recorder. Samples of both water and oil phase were taken from the vessels at predetermined 30 mins period for analysis. The process was allowed to reached equilibrium when the mass of wash water drawn became stable. As for this work, a minimum of 15 hours of continuous flow was set or until the crude oil runs out, whichever comes first. All experiments were performed in duplicate unless stated otherwise.

2.4. *Quantification of naphthenate salts in water phase*

The method is described in detail elsewhere [28, 30]. The ionic liquid recovery is expressed in mol% based on ^{1}H NMR spectroscopy using DMSO as an internal standard and the methylene protons adjacent to the ammonium centre (N-CH_3) as the cation signal. The molar amount of the recovered ionic liquid, n_2 is calculated by integrating the signal attributed to the cation in reference to the internal standard signal, expressed by equation (2).

$$\text{Recovery of naphthenate salts } n_2 = \alpha \times n_{std} \times \frac{P_1}{P_2} \quad (2)$$

$$\text{Recovery percentage, } \% = \frac{n_2}{n_1} \times 100 \quad (3)$$

where α is the total mass of extracted material/mass used for NMR analysis, n_1 is the amount of IL used and the expression P_1/P_2 is the ratio of the amount of proton of the cation signal over the amount of proton in the internal standard and n_{std} is the mole amount of the internal standard used. The percentage IL recovery is then calculated from equation (2).

3. Discussions

The operation conditions of the desalter pilot plant are summarised in table 1 below. The operation conditions were chosen based on typical operating conditions used for this type of crude oil and hydraulic limitation in PETRONAS refineries [31]. Table 2 below is the properties of the crude oil used in the pilot plant and commonly related to the operation of a desalter such as relative density, salt content and water content. All the properties tested and found it to be consistent with its crude assay [32]. Consistency of the sample and the published crude assay is important because of two reasons which are it is a new crude oil produced in Australia and is being considered to be a good crude oil to be part of the refinery crude oil diet that will use this technology. Any data generated here will be used in the refinery as base line for refinery crude oil and yield analysis. Secondly, it is important that the variation of the acid content to be at the minimum as this would confirm that the oil was drawn from the same hydrocarbon field and co-mingled in the same vessel prior to sampling. Sampling was done at the charter crude oil carrier off the concerned refinery. As such the heterogeneity of the naphthenic acids present in the oil was kept at minimum, which was gauged from the density, API value and ASTM D664.

Table 1. Operation conditions for desalter pilot.

Parameter	Unit	Value
Flowrate	m^3/hr	30
Temperature	°C	90
Pressure	barg	10
Water flowrate	vol%	10 and 30
Electrostatic field strength	kV	6.0
Pressure drop across the mixing valve	barg	0.5

Table 2. Raw pyrenees crude oil properties.

Property	Method	Unit	Value
Density	ASTM D6822	g/cm^3	0.932
API	ASTM D287	n/a	20.3
Viscosity	ASTM D341	cP at 30 °C	68.0
		cP at 40 °C	41.0
		cP at 50 °C	26.0
		cP at 60 °C	7.0
Total Acid Number	ASTM D664	mgKOH/g	1.63
Basic Sediment & Water	ASTM D4007	%	0.2

3.1. Process mass balance and stability

One of the key questions in adapting the desalter process unit with the current application is how stable the extraction of ionic liquid and its derivatives under electrostatic field. The consensus coming off the performed series of experiments with varying conditions is there is no observable adverse effect of electrostatic field and presence of ionic liquid and its derivatives (naphthenate salts) on process parameters namely, temperature and pressure, voltage stability, resistance or current value and chemical structure of ionic liquids. Crude oil here shall refers to the de-acidified crude oil or crude oil that has a

IOP Conf. Series: Materials Science and Engineering 1195 (2021) 012013 doi:10.1088/1757-899X/1195/1/012013

TAN value of < 0.3 mg KOH/g post mixing with $[N_{4441}][CH_3CO_3]$. Figure 2 below illustrates the amount of total feed to a process in which water is used in an amount of 30 % by volume of the mixture, as well as the amounts of crude oil product and water product from the process, as a function of process operating time. Followed by figure 3 which is a graph showing the amounts of feed mixture and feed water to a process in which water is used in an amount of 30 % by volume of the mixture, as well as the amounts of crude oil product and water product from the process, as a function of process operating time.

As mentioned before, ionic liquids' properties can be tuned to have differ from the metal salt analogues in this case would solubility of the resultant organic salts or naphthenate salts. Again, by taking desalter operation as a model, water would be the best solvent as it is available in refinery system. Therefore, it is best to empirically determined the weight of the outgoing streams. The masses of incoming streams were determined using density and flow rate. Both figure 2 and figure 3 indicate that all the mass i.e., de-acidified crude oil and wash water, going into the desalter are accounted for which includes the ionic liquid. The outgoing stream of wash water post-contact with electrostatic field was found to increase in weight, attributed by the dissolved naphthenate salts and possibly small amount of oil carry-over. Whenever the flowrate was stable, the proportionate mass is consistent with every stream recorded. For example, the flow of each stream was compared to what phases they were. The observed weight in outgoing water streams increased by 0.9 kg (1.97 lb) of which similar loss in weight recorded in the product oil stream for recorded flow of 0800 hrs in Figure 3. The dips observed at 1400 hours and 0200 hours were due to crude oil feed tank change out.

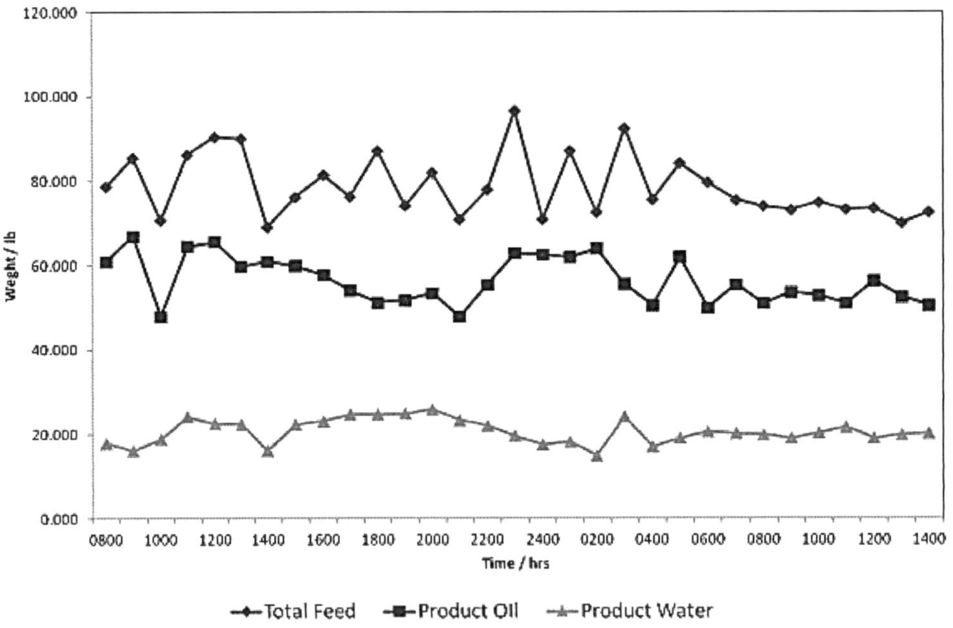

Figure 2. Flow of total feed and product streams over time.

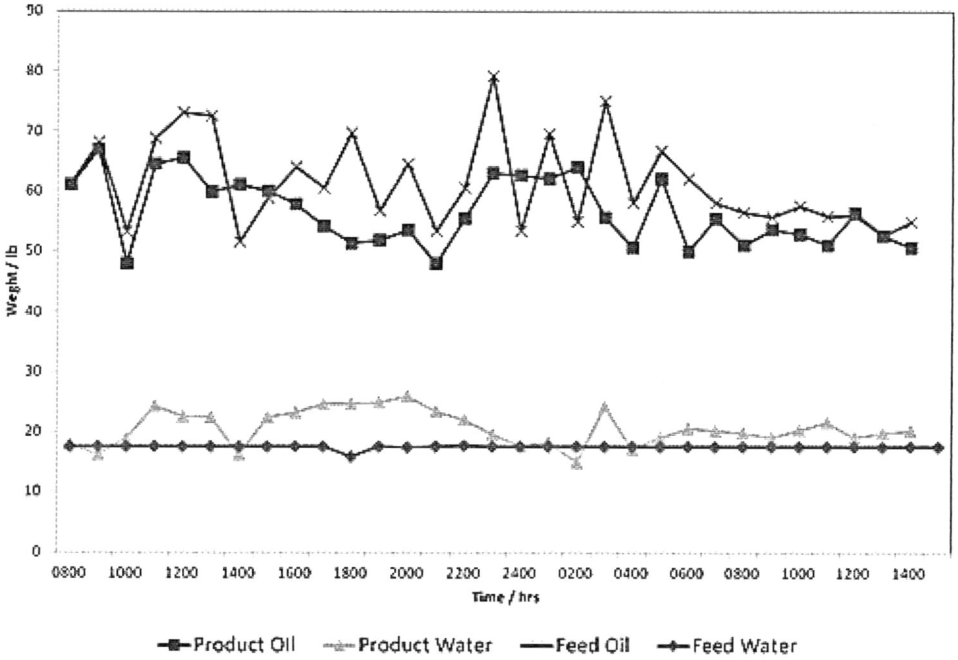

Figure 3. Flow of total feed and product streams over time.

3.2. *Effect of wash water volume on salt recovery*

The amount of water typically used in refinery for desalting process ranges from 5 – 10 vol% and is a major factor contributing to salt removal efficiency (SRE) [33-35]. The standard operating conditions of desalting conditions in PETRONAS refineries was applied in the pilot trial [31]. The measure of success is analogous to SRE which is the maximization of mass recovery of naphthenate salts in the water phase.

10 vol% water wash relative to the volume of the treated crude oil were performed on the pilot plant as a baseline. The trial was monitored on hourly basis as to determine the hour of which the process has reached steady state. Figure 4 below summarizes the three observations made during the trial. The first observation is the maximum recovery of the ionic liquid – product or naphthenate salts using 10 vol% water wash is at average of 0.2 mol per unit mass of naphthenate salts recovered as compared of incoming stream of 0.7 mol/unit mass. This translates to 29 % recovery. The final observation is the amount of time to reach steady state is between 8 to 10 hours. The third observation which is the utilization of desalter as potential process unit to recover naphthenate salt and ionic liquid post neutralization is viable. The primary function of desalter in conventional refinery setting is to recover wash water containing inorganics from the crude oil prior to distillation process. The inorganics may cause scaling and fouling in the hot train and furnace potentially causing unscheduled shutdowns. And the electrostatic field help to hasten the process of water recovery by encouraging contact between water droplet and allows gravity to take over and form the continuous aqueous phase [37]. Clear separation between the oil phase and aqueous phase was observed and maintained throughout the run. As comparison, more than six (6) hours is required for a clear separation to takes place in absence of electrostatic field [36].

The low recovery rate of the ionic liquid derivatives in the water phase was discussed in detail elsewhere [18,30]. Factors cited were hydro- and lipophilicity behavior of the resultant ammonium cation and naphthenate anions, molecular structure of the acids themselves which some may exhibits surfactant-like propensity and formation of tight water – oil emulsion. Similar observation was made by

Hussain [36] whereby limited by solubility of naphthenate salts in simple alcohols around 0.26 mol/unit mass would require significant amount of alcohol to increase the recovery up to 80 %. This is further supported by analysis done by Anderson *et al.* [28] that shows the range of NA's molecular weight can be up to carbon number of 34. The formation of tight emulsion of oil-water-oil was promoted by large surfactant – like naphthenic acid derivatives and get carried over by the aqueous phase.

A trial of 30 vol% water wash was performed and the results is presented in figure 5. Steady state was achieved after the 4th hour with maximum recovery of 0.49 mol of naphthenate salts/unit mass, which is equivalent to ~70 % mass recovery. Higher amount of water wash was not considered due to diminishing economic return on ionic liquid recovery. The additional mass recovery was in the single digit range against incremental volume of 10 % [36].

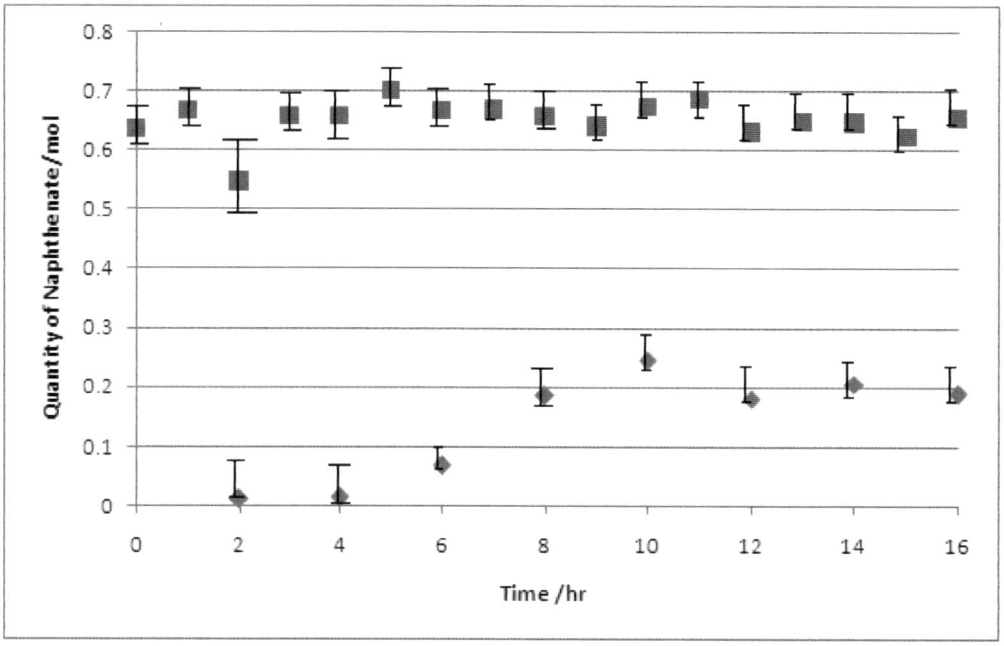

Figure 4. Quantity of naphthenate salt extracted, utilising water in an amount of 10 % by volume of the mixture, as a function of process operating time. (◊ - naphthenate salts in recovered water phase; € - Calculated amount of naphthenate salts in the incoming stream).

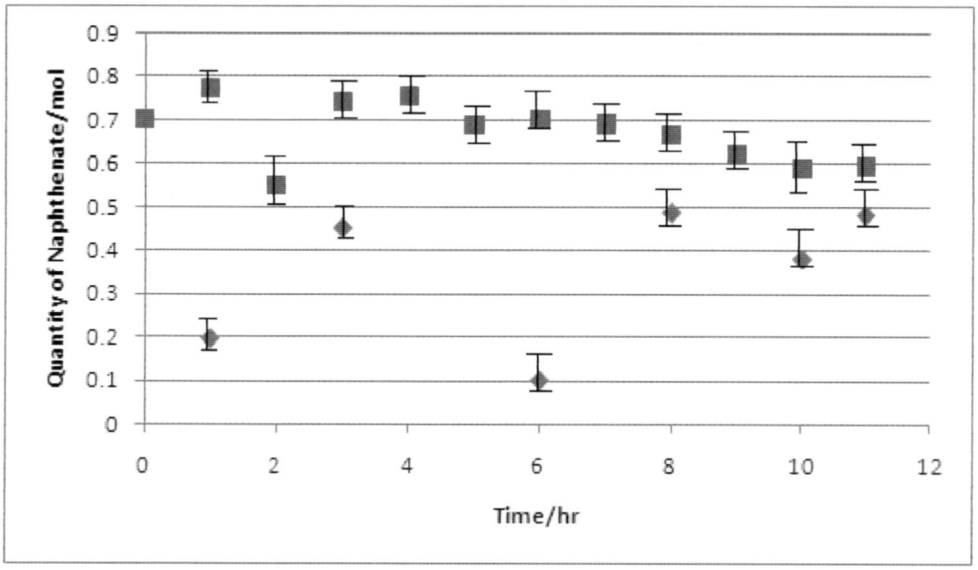

Figure 5. Quantity of naphthenate salt extracted, utilising water in an amount of 30% by volume of the mixture, as a function of process operating time. (◊ - naphthenate salts in recovered water phase; € - Calculated amount of naphthenate salts in the incoming stream).

Preliminary investigation on the feasibility of two-stage desalting with consecutive two (2) cycle of 30 vol% water wash, each consists of freshwater makeup, was performed. The second stage desalter only able to achieve additional 3.5 % recovery. Considering that, no more test was pursued on that idea. This confirmed the observation made by previous studies [18,30,36] on the solubility limitation posed by the anion of the salt which is naphthenate itself. The electrostatic field itself works as design which is to increase the movement of the water droplet and increased collision until the water droplet becomes big enough to be influenced by Stokes Law and merge with the separated phase [2]. Figure 6 illustrates the changing conditions of the recovered wash water containing the ionic liquid as the extraction process reaching steady state, suggesting more hydrocarbon being co-extracted in the water phase as a result of emulsion formation. The follow-up tests in Malaysia revealed that non-naphthenate based hydrocarbon was within 1 – 2 wt% using the method described in [28]. This is rather consistent with work done by Anderson *et al.* [28] whereby as the emulsion become tighter because of the lipophilicity of the naphthenate salt, more oil become entrapped as oil-water-oil formation which could not be easily broken up by mere droplet to droplet collision caused by electrophoretic forces [37]. To break such emulsion, chemical intervention is needed such as pH control and emulsion breaker to ensure lower hydrocarbon losses. Hydrocarbon losses would translate into higher end-of-pipe treatment cost and opportunity cost to the refinery.

Figure 7 compares the extraction efficiency of 10 vol% versus 30 vol% water wash with maximum recovery at 68 % - 72 %, with extra cautions to note on mixing and sample homogeneity. Compared to gravity-assisted settlement performed by PRSB of which requires visual inspection, bigger density difference between phases and long-standing time (more than one hour) for separation of two or more non-miscible phases, faster separation can be performed adequately with electrophoretic force induced by oscillating current and bigger droplets via gravity-assisted settlement (Stokes' Law) of which is found adequate for desalter's residence time of 15 mins and less.

Figure 6. Quantity of naphthenate salt extracted, utilising water in an amount of 30% by volume of the mixture, as a function of process operating time.

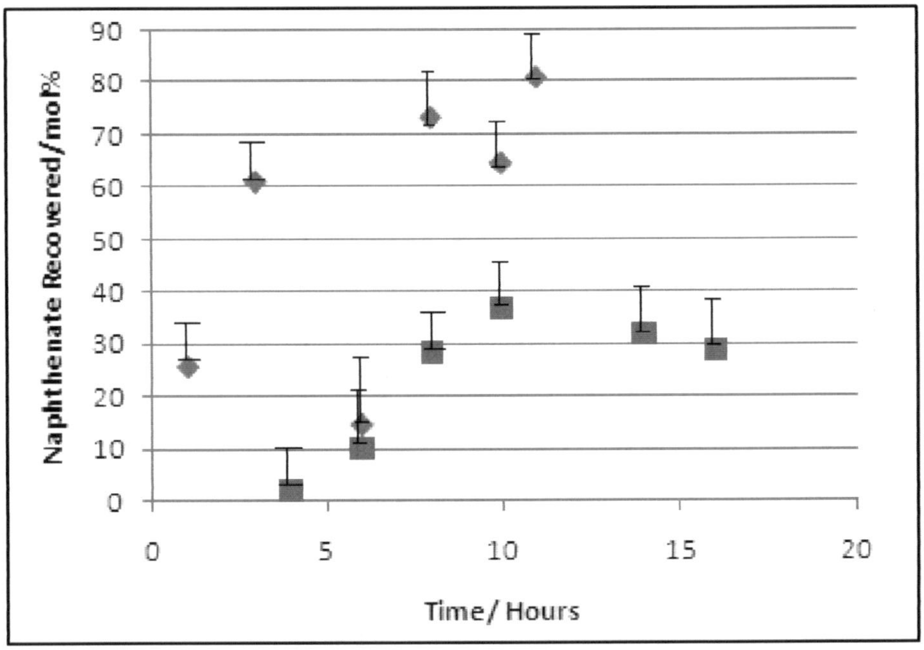

Figure 7. Quantity of naphthenate salt extracted, utilising water in an amount of 10% and of 30% by volume of the mixture, as a function of process operating time; (◊ -30 vol% water wash; € - 10 vol% water wash).

3.3. Preliminary finding on in-line mixing for [N_{4441}][CH_3CO_3] and pyrenees oil

The Intertek desalting pilot plant is equipped with a ¾ inch piping with optional heat tracing that can be used for injecting chemicals without the need to pre-mixed it with water or hydrocarbon itself (no figure available to illustrate this). The aim is to seek evidence that the residence time and mixing power required for fresh crude oil and [N_{4441}][CH_3CO_3] to get to low TAN value can be achieved via pipeline volume/length and pressure drop mixing valve.

The mixing point (which is a three-way pipe connection) is located upstream of the mixing valve and based on the volume and flowrate used, the length of pipeline would allow seven (7) mins residence time. Samples were taken out from a sampling point made available at the desalter itself as no other sampling port are available. The result indicates that the concept is feasible as the resultant crude oil met the low TAN value of <0.3 mg KOH/g of oil. The reaction itself is quite fast but the mechanics of diffusivity of ionic liquid into crude oil bulk phase to achieve 'deep de-acidification' can be an issue.

The combination of pipeline volume and pressure drop mixing valve provide adequate mixing and time, in addition to that the hold-up volume of the desalter would provide additional 15–20 mins.

4. Conclusions

All objectives outlined for this work were achieved. The technical performance criteria were met as the desalter allows the recovery of the ionic liquid derivatives or naphthenate salts post contact with the aid of 30 vol% water wash, at temperature and pressure of 90 °C and 10 barg and within the residence time of the vessel. The novelty of this work which is the continuous extraction of the ionic liquid was demonstrated with good process stability and exemplified chemical stability pre and post extraction, a huge improvement in term design challenge as this will eliminate the need for huge buffer vessel to accommodate long standing period to aid phase separation. Furthermore, the ionic liquid is now can be regenerated with minimum fresh make-up needed.

Unconventional water amount needed may pose significant barrier for commercialization effort due to availability and finite nature of the resource for most refineries. New techniques and process strategy are being explored such as designing smaller cation as to induce better solubility in the water. Tight water-hydrocarbon emulsions were observed in the water phase and it will present itself as opportunity cost if it is not recovered.

References

[1] Zeidani K and Bahadori A 2006 *J. Canadian Pet. Tech.* **44** 5
[2] Goode M 2001 *5th Intl. Conf. Refining Processes* AIChE
[3] Pereira J, Velasquez I, Blanco R, Sanchez M, Pernalete C and Canelon C 2015 *Advances in Petrochemicals* (IntechOpen)
[4] Pruneda E F, Escobedo E R B and Vazques F J G 2005 *J. Mex. Chem. Soc* **49** 4–19
[5] Adams F V, Chukwuneke C E and Agboola B O 2019 *Processing of Heavy Crude Oils: Challenges and Opportunities* (IntechOpen)
[6] Mandal P C and Sasaki M 2018 *Recent Insights in Petroleum Science & Engineering* (IntechOpen)
[7] Brown L D and Ulrich A C 2015 *Chemosphere* **127** 276–90
[8] Clemente J and Fedorak P A 2005 *Chemosphere* **5** 585–600
[9] Fan T 1991 *Energy Fuels* **3** 371–5
[10] Othmer K 2007 *Kirk-Othmer Encyclopedia of Chemical Technology* (New Jersey: John Wiley & Sons)
[11] Qian K, Robbins W, Hughey C, Cooper H, Rodgers R and Marshall A 2001 *Energy Fuels* **6** 1505–11
[12] Tomczyk N, Winans R, Shinn J and Robinson R 2001 *Energy Fuels* **6** 1498–1504
[13] Wasserscheid P and Welton T 2007 *Ionic Liquids in Synthesis* (Germany: Wiley – VCH)
[14] Deetlefs M and Seddon K R 2006 *Chim. Oggi* **2**
[15] Greaves T and Drummond C 2008 *Chem. Rev* **1** 206–37
[16] Luo H and Luo H 2009 *J Phys. Chem. B* **13** 4181
[17] Anderson K, Goodrich P, Hardacre C, Hussain A and Rooney D W 2013 *Fuel* **108** 715–22
[18] Anderson K, Atkins M P, Goodrich P, Hardacre C, Hussain A S, Pilus R and Rooney D W 2015 *Fuel* **146** 60–8
[19] Shah S N, Mutalib M I A, Pilus R and Lethesh K C 2015 *Energy Fuels* **29** 106–11
[20] Varadaraj R, Savage D W and Wales W E 1998 US5961821A
[21] Varadaraj R, Pugel T M and Savage D W 2000 US6096196A
[22] Maase M and Massonne K 2005 Ionic liquids IIIB: fundamentals, progress, oppurtinities & challenges *ACS Symp. Ser.*
[23] Hart P R and Cunningham T L 2008 WO137523A1
[24] Manan N A, Atkins M P, Jacquemin J, Hardacre C and Rooney D W 2012 *Sep. Sci. Tech.* **47** 312–24

[25] Shah S N, Mutalib M I A, Ismail M F, Suleman H, Lethesh K C and Pilus R M 2016 *J. Mol. Liquid* **219** 513–25
[26] Coto B, Suarez I, Chirita M, Conde J, Gimenez R, Rodrigues N, Alvarez N and Pena J L 2019 *Sep. Purification Tech.* **223**(15) 234–42
[27] Anderson K, Hussain A S, Atkins M P and Basar J 2014 WO2014016425A1
[28] Anderson K, Atkins M P, Goodrich P, Hardacre C, Hussain A S, Pilus R and Rooney D W 2015 *Fuel* **146** 60–8
[29] American Association for Testing and Materials International (ASTM) 2006 ASTM D664 *Standard Test Method for Acid Number of Petroleum Products by Potentiometric Titrations*
[30] Hussain A and Basar J 2021 *Proc. Int. Conf. Process Eng. Adv. Mater. (ICPEAM)* **287** 04002
[31] Personal communications between project team and PETRONAS Penapisan Melaka
[32] Pyrenees Crude Oil Overview 2009 BHP
[33] Sellami M H, Naam R and Temmar M 2016 *J. Pet. Env. Biotech.* **271** 71–6
[34] Al-Otaibi M, Elkamel A, Al-Sahhaf T and Ahmed A S 2003 *Chem. Eng. Comm.* **190** 65–82
[35] Abdul-Wahab S, Elkamel A, Madhuranthakam C R and Al-Otaibi M 2006 *Chem. Eng. Procs.* 45568–577
[36] Hussain A S 2010 Extraction of acids using ionic liquids *PhD Thesis* Queens University of Belfast
[37] Sams G 2010 Challenges of desalting canadian crudes *Presentation* PETRONAS
[38] Hughes H B 1981 *Operating Manual for Two Stages Single Phase 60kV Desalter* (USA: Intertek PARC)

Acknowledgments
The authors would like to acknowledge PETRONAS and Malaysian Refining Corporation Sdn. Bhd (MRCSB) for funding this work.

IOP Conf. Series: Materials Science and Engineering 1195 (2021) 012014 doi:10.1088/1757-899X/1195/1/012014

Grease formulation and characterization from waste automotive engine oil with the use of complex thickener

N W A Rahman, S S Y Ying and M A A Aziz

Faculty of Chemical & Process Engineering Technology, Universiti Malaysia Pahang, Kuantan, Pahang, Malaysia

Email: maizudin@ump.edu.my

Abstract. Waste engine oil one of the most abundant wastes in Malaysia, and through the reutilization of waste automotive engine oil helps to create a sustainable environment. The objective of this research is to develop the best formulation of lithium complex grease derived from waste automotive engine oil as base oil. The main focused parameter in this study is the different formulation ratio of base oil, thickener and co-thickener. Lithium 12-hydroxystearate is mixed with azelaic acid to produce lithium complex 12-hydroxystearate. Two different type of base oils, i.e. fresh automotive engine oil (FAO) and waste automotive engine oil (WEO) are used to formulate Li-complex grease. The grease derived from FAO is used to compare the physical properties derived from WEO. The texture of the formulation of base oil higher than 82 weight percentage was very fluid. The formulation of grease is carried out by differencing the ratio of the waste automotive engine oil, lithium complex 12-hydroxystearate and azelaic acid, which are 82:18, 80:20 and 70:30. The properties of the grease formulated is conducted through several tests, such as ASTM approach, Fourier Transform Infrared Spectroscopy (FTIR) characterization, oil separation and thermogravimetric analysis (TGA). Such interesting properties included consistency, chemical compound of the grease, oil separation and thermal stability. Based on the finding, the best formulated Li-complex grease is WG_3, classified NLGI 3. The significant peak derived from FAO and WEO to observe is 1710 cm^{-1} as this peak indicated the oxidation stability. From the result, the intensity of carboxylic acid is weak that ranged 1709 – 1711 cm^{-1}. Hence, this indicated the grease formulated exhibited better oxidation stability. Furthermore, the formulated grease was thermally stable as the onset temperature was 250.09 °C. In conclusion, the formulation of Li-complex from WEO can be used as an alternative source of base oil in the grease industry, due to the good properties exhibition and preserving the environment as well as the increment of fossil fuel's demand and cost.

1. Introduction

The depletion of natural resources and environmental issue have putting pressure on industries to find the natural resources scarcity as well as reducing the impact to the environment caused by industrial waste materials. Due to the effect of lubricants exert on the environment, every year, it is expected more than 10 million tons of waste oil such as engine, hydraulic and industrial oils are dumped to the environment [1]. To solve the problem, more research on biodegradable grease using vegetable oil are being actively researched at present day. Several studies have been done by Universiti Malaysia Pahang

(UMP) researchers related to reutilizing waste oil such as waste cooking oil and waste transformer oil [2, 3].

Greases mostly acted to reduce wear and friction in machinery as it is an important machine element making the demand for their improvement in performance to be ongoing in the industry. Thickener, oil and additives is the most common element in grease [4]. The compositions for grease is commonly include base oils at 70-95%, thickeners at 3-30%, and additives at 0-10%. The thickener forms an entangled network which holds or traps the oil and contribute to the appropriate rheological and tribology behaviour of the grease [5]. The most commonly used of thickening agent are soaps, such as calcium, aluminium, lithium and sodium.

Lithium complex grease is commonly formulated from 12-hydroxystearate and azelaic acid [6, 7]. 12-hydroxystearic acid is a saturated long chain fatty acid with one hydroxyl group on aliphatic chain and is described to be a good gelator [8]. Greases produced from lithium 12-hydroxystearate have higher dropping point when compared to lithium stearate. The improvements in properties basically connected to the capability of the 12-hydroxystearate anion to the hydrogen bond. Lithium 12-hydrostearate has been described as the "foundation stone of the current generation of lubricating grease" [9].

Two problems are taken in consideration in this field of study. Firstly, it is the improper handling of WEO towards the environment. Improper handling of waste oil can cause waste emulsion. Secondly, it is the cost effective of crude oil. Due to its higher demand, this has also indirectly threatened public health and pose a serious threat to the environment at the same time. Consequently, the increase in prices of crude oil and environmental concerns have induced the idea of finding alternatives to replace mineral oils [10]. Since base oil accounts for the highest composition for about 70-95 weight percentage, there is a great need for the reutilization of waste automotive engine oil into recycled products. Recycling is beneficial to the environment and also to economic development as it prevents resource scarcity, decreases the demand of landfilling and saves energy too [11]. Hence, it is wise to reutilize WEO as base oil in grease production for its economic value and environmentally friendly aspect.

In this research, the main objective is to develop the best formulation of lithium complex grease from WEO which is something new to be discovered as the base oil for this research. Grease formulated from FAO was used as reference to compare the grease derived from WEO. This new formulated grease derived from WEO was analysed for its exhibited properties through several tests such as consistency test, oil separation, FTIR characterization and TGA analysis.

2. Experimental work

2.1. Material
Waste automotive engine oil (WEO) and fresh automotive engine oil (FAO) were collected from automotive shop located in Kuantan. Lithium 12-hydroxystearate with industrial grade from China was purchased with Orioner Hightech Sdn. Bhd. Azelaic acid (solid form and 98% purity) was purchased from Revlogi Materials (RM).

2.2. Pre-treatment of waste engine oil (WEO)
The pre-treatment of WEO is needed to remove any contaminants such as suspended solid in the waste oil. WEO is filtered using vacuum filter pump. Glass microfiber filer paper with diameter of 47mm was used to remove the impurities contained in WEO. Next, in order to remove any volatile component and water content in the WEO by evaporation process, the oil was heated at temperature ranging 90 °C-120 °C.

2.3. Formulation of Lithium 12-HSA grease from FAO (FG_i) and WEO (WG_i)
To compare the properties of formulated grease by WEO, grease containing FAO was also formulated. The formulated grease consists of two type, with and without the addition of complexing agent which is azelaic acid.

Lithium 12-hydroxystearate, WEO and azelaic acid were weighed according ratio, i.e., 70%-82% (w/w). First, WEO was preheated at temperature ranged 90°C - 120°C. Azelaic acid is melted at a higher temperature until the acids were completely melted. Mixture of azelaic acid and WEO is homogenized for about 15 to 20 minutes. Lithium 12-hydroxystearate was added portion wise into the mixture. The mixture was homogenized for thickening process to disperse the thickener [3]. The remaining WEO was added to control the texture of the grease and the homogenization process was continued. After that, the grease was allowed to cool down. The steps were repeated to formulate with different ratio of base oil, thickener and azelaic acid. Then, the same ratios were repeated to formulate Li-complex from FAO

2.4. *Grease Analysis*

The formulated grease was analysed through several tests to determine their properties. The tests that were carried out were consistency test, oil separation test, Fourier Transform Infrared Spectroscopy (FTIR) characterization and Thermogravimetric Analysis (TGA).

2.4.1. Consistency Test. Consistency test was carried out to decide where the consistency of the grease belong to according to NLGI (National Lubricating Grease Institute). The NLGI ranged from 000 to 6 (table 1). According to Japae *et al.* (2019) the test was in accordance of ISO 2137 which determined the consistency of the lubricating grease when only small samples of greases were available [3]. The NLGI obtained indicated the consistency level of the grease. Greases of higher NLGI grade are firm, tend to stay in place and are better option when leakage is a concern [2].

Table 1. NLGI classification system based on Consistency [2].

NLGI Grade	Consistency
000 00	Very soft
0	Soft
1	Soft
2	Creamy texture (buttery)
3	Semi-solid
4	Stiff
5	Stiff
6	Hard solid

2.4.2. Oil separation test (ASTM D 1742). ASTM D-1742 is used to examine the separation of oil at room temperature condition of grease after storing it for some period of time. The grease started to lose its original consistency and ability when the oil begins to separate in grease. The test is done by storing the grease in a partially filled or normally filled container to test the tendency of the grease to separate oil. According to Suhaila *et al.* (2018), the sample is considered stable if the oil separation is less than 4%. The grease exhibits better stability when lesser oil separation is observed [12]. However, this method is not suitable for grease with NLGI No.1 grade [13].

2.4.3. Fourier transform infrared spectroscopy (FTIR) characterization. Fourier transform infrared spectroscopy is functioning to monitor the additive degradation, builds up of contaminant and the base stock ageing in oil lubricants. Besides, the ability of FTIR to detect a functional-groups by wide range correlated to the chemical changes in lubricants. This will allow the assessment of the overall quality of lubricants and decided whether a lubricant is required to be replaced or not [14]. The FTIR method is also carried out on lubricating grease and its extracted base oil or its separated thickener to detect the difference in the level of the components of the grease during chemical deterioration and base oil evaporation [15]. In short, FTIR was carried out to identify the functional group of the components, the presence of contamination and the oxidation stability of the grease samples. The grease samples formulated were characterized at the wavenumber of 500 – 4000 cm^{-1}.

2.4.4. Thermogravimetric (TGA) analysis. The thermal stability and phase transitions of the lubricating grease have attracted several researchers' attention [16]. It was therefore the analysis was carried out to determine the thermal behaviour of the grease samples formulated. TGA can provide qualitative information from a simple experiment. The test was conducted under an inert air flow via nitrogen gas or oxygen gas at a certain heating rate within a range of temperature [17].

3. Result and Discussion

3.1. Consistency of the formulated grease

The results of the grease samples from the testing were in accordance to the NLGI consistency grade. Table 2 shows the results obtained on the consistency number of each grease sample. Sample of greases that fell under the NLGI grades of 00 to 0 were FG_6 and WG_5. Their applications are suitable to be used in a centralized lubrication system and enclosed gears. These greases were categorized into the fluid to semi-fluid range. Sample greases of FG_5, WG_3, WG_4 and WG_6 were recorded from NLGI of 1 to 3. Greases for bearings are typically found in NLGI grade 1, 2 and 3 [18]. NLGI grade of 4 to 5 were FG_1, FG_2, FG_3, FG_4, WG_1 and WG_2. The most common of NLGI grade used in the automotive bearings would be NLGI grade 2 [19]. Grease of NLGI 2 have a superior properties balance making it easy to pump through the dispensing systems [20]. NLGI grade 4 is suitable to be used for a device that move at high speed which is like greater than 15,000 rotations per minute. It is because these devices cause more friction and produce higher heat, hence, a stiffer and channelling grease is a better choice [21].

Table 2. Properties of each grease sample.

Sample Grease	NLGI	Description	FAO	WEO	Li-12 HSA	Azelaic Acid	Oil Separation
FG_1	4-5	Hard -Very hard	70	-	27	3	-
FG_2	3-4	Medium Hard - Hard	80	-	18	2	-
FG_3	4	Hard	82	-	16.2	1.8	-
FG_4	5	Very Hard	70	-	30	-	-
FG_5	1-2	Soft – Medium	80	-	20	-	-
FG_6	00	Semifluid	82	-	18	-	-
WG_1	4-5	Hard – Very Hard	-	70	27	3	-
WG_2	4	Hard	-	80	18	2	-
WG_3	3	Medium-Hard	-	82	16.2	1.8	-
WG_4	3	Medium-Hard	-	70	30	-	-
WG_5	0-00	Very fluid – Semifluid	-	80	20	-	-
WG_6	2-3	Medium – Medium Hard	-	82	18	-	-

3.2. *Oil separation*

Oil separation occurs when the grease is being kept for a longer period of time. All the grease samples were observed. Table 1 showed the results obtained for the oil separation of grease after being kept in a container for one month. Based on the observation, no oil separation happened for all the grease formulations. Lithium 12-hydroxystearate and azelaic acid will acts as thickener to house the oil in the thickener system. In short, the complexing process makes strong adsorption interactions is developed between the soap, lithium 12-hydroxystearate and complexing agent, azelaic acid during the preparation of grease [22]. According to Suhaila *et al.* (2018), the sample is acceptable if the oil separation is less than 4% [12].

3.3. *FTIR characterization*

Table 3 shows the characteristic peaks and corresponding molecular groups and bonds of the grease sample in the spectral range of 4000 to 500 cm^{-1}. Figure 1 and figure 2 show the FTIR peaks of grease samples from FAO and WEO respectively whereas figure 3 presented the peaks of all grease formulations. The two figures showed that very strong absorbance usually in the range of 2849 – 2954 $^{-1}$. The range of these are the characteristics of the asymmetrical and symmetrical vibrations of hydrophobic groups, i.e., methyl, CH_3 and methylene groups, CH_2. These spectrums originated from lithium 12-hydroxystearate molecules [23]. According to Abdul and Bilal (2012), their research revealed the spectra of azelaic acid was at 1695 cm^{-1} [24]. The main peak observed was the dicarboxylic acid peak at wavenumber range of 1709 – 1711 cm^{-1}. The absorbance of peak at 1709 – 1711 cm^{-1} is weak. In FTIR analysis, it is essential to observe the peak at 1710 cm^{-1} as it indicated the oxidation of hydrocarbon with the formation of carboxylic acid. The higher the intensity of the peak, the less oxidation stability of the lubricant. However, the intensity of this peak is weak hence it indicated better oxidation stability of the formulated grease [25]. The mode of -COO- carbonyl stretching modes is seen at 1559, 1578 and 1579 cm^{-1}. The -COO- carbonyl groups were found in all formulation as lithium 12-hydroxystearate were used as thickener. Base oil components were also found at the wavenumber 1376 cm^{-1}. The absorbance of 720 – 721 cm^{-1} might be related to the 30 % of the oil bound by the molecular attraction forces between the oil and thickener components [26].

Table 3. Characteristic peaks and corresponding molecular groups and bonds of the grease.

Characteristic peak/cm^{-1}	Functional group assignment	Band	Source
2849	Symmetrical stretching vibration of CH_2 group	C-H	
2954	Vibration of CH_3 group	C-H	
1709 - 1711	Dicarboxylic acid	COOH	Azelaic acid
1578	Asymmetric stretching vibration of COO group	COO	Thickener
1579	Asymmetric stretching vibration of COO group	COO	Thickener
1559	Symmetrical stretching vibration of COO group	COO	Thickener
1452	Asymmetrical deformation vibration of CH_3 group	C-H	Base oil
1376	Bending vibrations of CH_2 groups	C-H	Base oil
720 - 721	CH_2 rocking vibration	C-H	

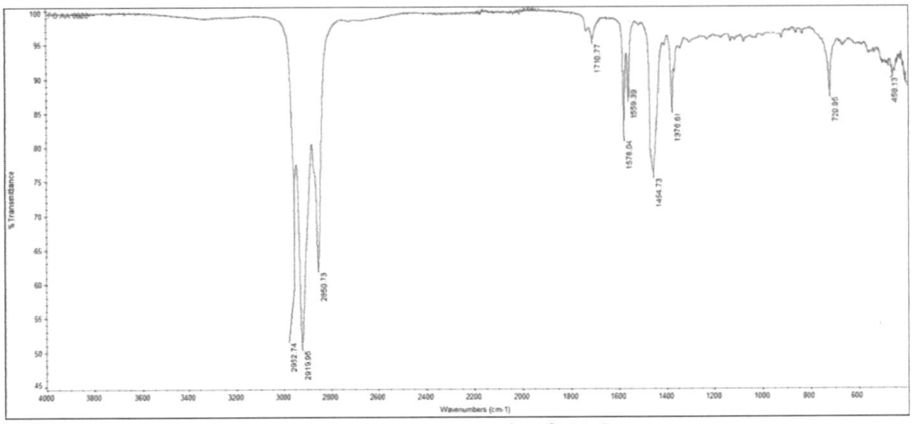

Figure 1. FTIR peak of FAO.

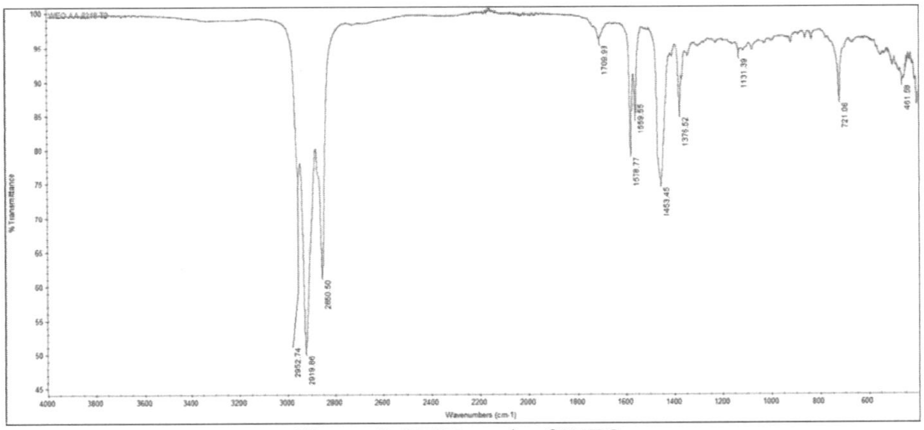

Figure 2. FTIR peak of WEO.

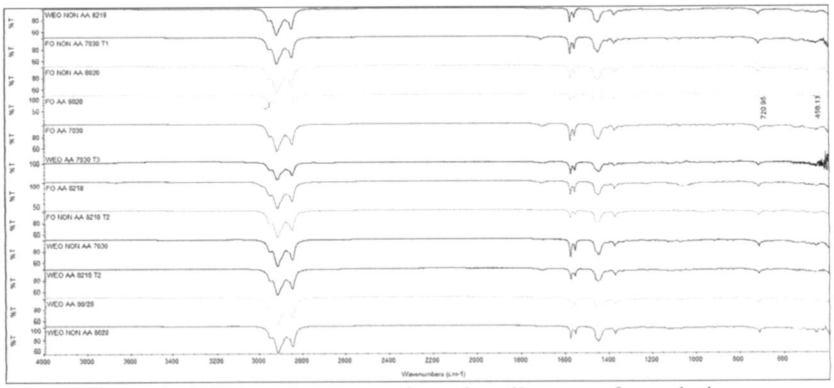

Figure 3. FTIR peaks comparison for all grease formulations.

3.4. TGA analysis

Based on figure 4, the figure depicted the sample grease have high thermal stability until about 250 °C where the start of decreasing weight of the sample grease remarkably. In the temperature range 190 – 540 °C, the TG curve displayed two weight loss steps. At the first onset temperature of 250.09 °C It displayed that the grease undergoes thermal degradation at that temperature with a total mass loss of

86.06 %. °C. In the present study, the first one which range 190 – 360 °C was most probably happens due to the elimination of low molecular weight products. According to Tripathi and Vinu [17], temperature ranged 150 – 350 °C the evaporation hydrocarbons presented in the low end of molecular weight distribution, and degradation of base oil components occur. Based on Santos, Santos [27], this step is the most crucial as it determines the thermal stability of the lubricant. Next, the second weight loss step which ranged 448 – 540 °C was related to the decomposition of higher molecular weight compounds. According to El-Adly and Turky [16], the commercial of Li-complex grease undergone thermal degradation and decomposition during the temperature range 250 – 500 °C. When comparing the result with the literature, the result of thermal degradation for grease sample, WG_3 is the same. This meant that the sample grease was intrinsically stable at higher temperature

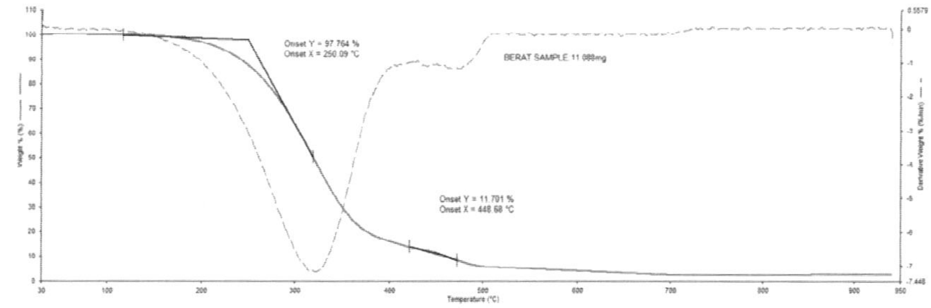

Figure 4. TGA analysis of WG3.

4. Conclusion

In conclusion, the idea of utilizing treated waste engine automotive oil is for the formulation of lithium 12-hydroxystearate complex grease is viable. The best formulation of lithium 12-hydroxystearate from waste engine automotive oil grease is WG3. The consistency number obtained is 3, which falls under common grease grade. The FTIR results obtained of utilizing WEO do not show significant difference with the one using FAO except for the presence of dicarboxylic peak which indicated the oxidation stability of the grease. The grease samples formulated can be concluded good as there was no sign of oil separation happened. The formulated grease is also thermally stable and has high heat resistance as the degradation happens at temperature at 250 °C. Thus, waste automotive engine oil can be a promising base oil in the grease production industry. It can be an alternative source of base oil for grease production to replace the mineral oil derived from crude oil.

Further experimental work related to tribology especially on the viscosity is required to compare with the commercial grease available in the market. The addition of solid additives can be suggested to improve the tribology of the grease. Lastly, more parameters can be study such as temperature and the speed of homogenizer during the process of grease formulation.

Acknowledgments

The researchers would like to extend their gratitude to Universiti Malaysia Pahang (UMP) for providing the grants for this study under the grant number RDU190398.

References

[1] Madanhire, I and Mbohwa C 2016 *Mitigating environmental impact of petroleum lubricants* (Springer International Publishing)

[2] Razali M N *et al.* 2017 Synthesis of Grease from Waste Oils and Red Gypsum *Aust. J. Basic Appl. Sci.* **11**154–9

[3] Japar A N S, Aziz A M A and Razali M N 2019 Formulation of fumed silica grease from waste transformer oil as base oil *Egyptian J. of Pet.* **28** 91–6

[4] Shu J, Harris K, Munavirov B, Westbroek R, Leckner J and Glavatskih S 2017 Tribology of polypropylene and Li-complex greases with ZDDP and MoDTC additives *Tribol. Int.* **118** 189–95

[5] Lugt P M 2012 Grease lubrication in rolling bearings, John Wiley & Sons
[6] Dresel W and Heckler R 2001 Lubricating Greases *Lubricants and Lubrications*, ed W Dresel and T Mang (Wiley-VCH) chapter 16 pp 784–841
[7] Beret S, Boersig T J, Loh W and Wong P 1999 Continuous lithium complex grease formation using methyl esters of 12-hydroxystearic acid and azelaic acid *NLGI spokesman* **62** 14–7
[8] Lipowski B M 2014 Phase Behavior of 12-Hydroxystearic Acid Gels *University of Akron*
[9] Donahue C J 2006 Lubricating grease: A chemical primer *J. Chem. Educ.* **83** 862
[10] Battersby N 2005 Biodegradable lubricants: What does 'biodegradable'really mean? *J. Synth. Lubr* **22** 3–18
[11] Rajvanshi A and Pandey P K 2016 Lubricating grease from waste cooking oil and waste motor sludge *Int. J. Chem. Biol. Eng.* **10** 1220–3
[12] Suhaila N, Japar A, Aizudin M, Aziz A and Razali M N 2018 Formulation of lubricating grease using beeswax thickener *Mater. Sci Eng.* **342** 012007
[13] Clark. *ASTM D1742*. 2015
[14] Voort F R, Sedman J, and Pinchuk D 2011 An overview of progress and new developments in FTIR Lubricant condition monitoring methodology *J. ASTM Int.* **8** 103344
[15] Asghar R and Khonsari M M 2016 On monitoring physical and chemical degradation and life estimation models for lubricating greases *Lubricants* **4** 34
[16] El-Adly R A and Turky G M 2018 Comparative study between prepared electrical grease and the imported one *Egyptian J. of Pet.* **27** 209–13
[17] Tripathi A and Vinu R 2015 Characterization of thermal stability of synthetic and semi-synthetic engine oils *Lubricants* **3** 54–79
[18] Nicole S P 2019 Need to Know: Grease Consistency *Nye Lubricants*
[19] Suhaila N, Japar A, Aizudin M, Aziz A and Razali M N 2018 Formulation of lubricating grease using beeswax thickener *Mater. Sci Eng.* **342** 012007
[20] Neale M J 1995 *The Tribology Handbook* (Oxford: Butterworth-Heinemann)
[21] Cash W 2019 Choosing a High-speed Grease *Machinery Lubrication*
[22] Lorimor J J 2009 An investigation into the use of boron esters to improve the high-temperature capability of lithium 12-hydroxystearate soap thickened grease *76th Annual NLGI Meeting* (Tucson, Arizona) pp 27–37
[23] Bao P J, Hai C Y, and Yong Y J 2015 Effect of heat treatment on the lubricating properties of lithium lubricating grease *RSC Adv.* **5** 58686–93
[24] Kadhum A A H, Wasmi B A, Mohamad A B, Al-Amiery A A and Takriff M S 2012 Preparation, characterization, and theoretical studies of azelaic acid derived from oleic acid by use of a novel ozonolysis method *Res Chem Intermed* **38** 659–68
[25] El-Naggar A Y, El-adly R, Altalhi T, Alhadhrami A, Farouk M, Ebiad M and Salem A 2018 Oxidation stability of lubricating base oils *Pet. Sci. Technol.* **36** 1–7
[26] Paszkowski M, Olsztynska J S and Wilk I 2014 Studies of the kinetics of lithium grease microstructure regeneration by means of dynamic oscillatory rheological tests and FTIR–ATR spectroscopy *Tribol. Lett.* **56** 107–17
[27] Santos J C O, Santos I M G and Souza A G 2017 Thermal degradation of synthetic lubricating oils: Part III – TG and DSC studies *Pet. Sci. and Technol.* **35** 540–6

Studies on the modification of fly ash structure with alkaline pre-treatment as a green composite flame retardant filler

N A M Radzi, A H Sofian and S S Jamari

Faculty of Chemical and Process Engineering Technology, Universiti Malaysia Pahang,26300, Gambang, Kuantan, Pahang, Malaysia

Email: azizulh@ump.edu.my

Abstract. A green composite made up of renewable and recyclable materials has become one of the advanced material's attractive topics. The smooth fly ash surface used in the green composite for flame retardancy enhancement are hard to bind with hydrophobic polymer. Thus, the surface modification of this filler is needed to increase its surface roughness and pore size to be more compatible with its polymer matrix. In this research study, the alkaline pre-treatment of fly ash has been performed by using sodium hydroxide solution (NaOH) with various concentrations (5 w/w%, 10 w/w%, 15 w/w%, 20 w/w%). For pore size and morphological of the filler's evaluation, few analyses such as Scanning Electron Microscopy-Energy Dispersive X-Ray (SEM-EDX), Barret-Joyner-Halenda (BJH) and Brunauer-Emmett-Teller (BET) pore size and volume analysis were conducted. Treated fly ash with 20 w/w% sodium hydroxide concentration gives the better morphological structure in terms of pore diameter, volume, area and high composition of aluminium, silicon with lower calcium and sulphur contents compared to others. Hence, the potential of the physiochemical properties of the green composite produced by using this modified filler will be improved as the adhesiveness of the filler with its matrix increased.

1. Introduction

As for thermal stability enhancement, the flame retardants (FRs) will be used as it gives a significant impact on suppressing flame to prevent fire spread. Flame retardants have been used in a wide range of applications such as in the manufacturing process [1]. Fly ash is an example of an inorganic material which is suitable to be used as a filler flame retardant material in bio-composite. Fly ash is a waste generated by electric power plant through the coal combustion process. In this process, the pulverized solid fuels consist of combustible organic matter with different composition of minerals will be consumed [2–8].

Around 1.5 billion tonnes of fly ash have been produced worldwide and the percentage of fly ash utilization is approximately 40% [3]. According to [4], the estimated fly ash generated for every year is approximately 600 million tonnes and only 110 million tonnes in India. However, the huge amount of non-utilized fly ash waste is classified as a hazardous material and becomes a serious issue towards the environmental problem if the fly ashes are directly dumped as a landfill material [3,8]. The fly ash waste will also cause air and water pollution containing dust [9]. Thus, utilizing the fly ash in a good way will bring more benefits as nowadays many researchers are interested in exploring the high filler application in polymeric materials to enhance its thermal stability and any useful applications [2,3].

The presence of –OH group in fly ash's surface will enhance the compatibility between the inorganic material and polymer substrate as the functional group of -OH becomes the active sites for functional chain polymer structure to be bonded chemically [1,8]. The addition of fly ash is one of the effective additives for flame retardancy to improve the efficiency of the composite [1]. There is also a study conducted by (Porąbka *et al.* 2015) regarding the fly ash that is used as a reinforcement filler for low-density polyethylene [9]. The research on nanostructured fly ash as a flame retardancy enhancement for polyvinyl chloride (PVC) is also reported by (Patil *et al.* 2016) [6]. In addition, the fly ash is reported to be used as a flame retardant to enhance the flame retardancy of several polymers such as polycarbonate, polyvinyl chloride, epoxy resins, epoxy, polypropylene blends, rigid polyurethane foam (RPUF) synergetic with ammonium phosphate (APP) and pentaerythritol (PER) is also reported by (Dong *et al.* 2015)[3]. The modification of low-density polyethylene by adding fly ash is a suitable and effective solution [9]. Therefore, in this research study, the fly ash treated with various concentration of an alkaline solution (NaOH) (5,10,15 and 20 wt/wt%) with a ratio of fly ash/NaOH 1:1.2 for fly ash activation [5] have been performed to study the morphological structure, pore diameter, area and volume of the fly ash produced.

2. Experimental

2.1. Materials

The fly ash was collected from Jimah power station located at Port Dickson, Negeri Sembilan, Malaysia. Sodium hydroxide, pallets for analysis was purchased from Emsure.

2.2. Methods

2.2.1. Treated Fly Ash (Alkaline pre-treatment). The raw fly ash will be treated by using NaOH solution within the range concentration of 5 to 20% (by weight). The fly ash will be immersed in NaOH solution at room temperature and stirred for approximately 30 minutes with the speed of 750 rpm by using FAVORITE Stirring Hotplate (HS0707V2). The ratio of fly ash to sodium hydroxide used is 1:1.2 (for fly ash activation). Then, the fly ash will be filtered by using a vacuum pump and dried at oven by using a temperature of 115°C for 5 hours.

2.3. Characterization process for the structural of fly ash

Characterization process by using Scanning Electron Microscopy – Energy Dispersive X-Ray (SEM-EDX) (Cambridge S200, microscope), Brunauer-Emmett-teller (BET) and Barret-Joyner-Halenda (BJH) (TrisStat 3000 analyzer) analysis were conducted to determine the best the morphological structure, pore diameter, area and volume distribution of the fly ash.

3. Results and discussion

3.1. Structural surface of untreated and treated fly ash by using Scanning Electron Analysis (SEM)

The surface morphology of raw/untreated fly ash in figure 1 (a) shows the smooth surface which leads to difficulty for a binding process with the polymer substrate. Thus, modifying the raw fly ash surface with alkaline treatment will eventually increase the roughness and forming an active group of the surface to increase the adhesiveness of the fly ash with its polymer substrate [1,2]

As shown in figure 1 (b), (c), (d) and (e), the morphological surface of the treated fly ash by using sodium hydroxide with various concentration which is 5, 10, 15 and 20 (wt/wt%) seems to be rougher and smaller compared with raw or untreated fly ash. This research work is aligned with (Anh *et al.* 2019; Nguyen *et al.* 2019), according to them, the rougher and smaller particle size presented on the treated fly ash surface is due to the abrasive condition during the treatment process [1,2]. This reaction of sodium hydroxide with oxide of fly ash makes the surface of fly ash cracked and aluminium silicate formed. Besides, according to (Mukherjee & Borthakur, 2004), the dissolution of sulphate and other water-soluble components in the fly ash sample is one of the reasons for fly ash size reduction with alkaline treatment [10]. The alkaline solution will react with the alumina and silica presented in the fly ash hence forming the soluble silicate and aluminates compounds. The simple reactions of the fly ash with sodium hydroxide described by (Mukherjee & Borthakur, 2004) are as equations (1) and (2) below [10]. The sodium silicate and aluminium silicate are indeed formed during the alkaline treatment of fly ash by using sodium hydroxide solution [1,2,10]. A high concentration of sodium hydroxide used will form a formation of hydroxysodalite as described in equation (3) [10].

$$SiO_2 + 2NaOH \rightarrow Na_2SiO_3 + H_2O \quad (1)$$

$$Al_2O_3 + 2NaOH \rightarrow 2NaAlO_2 + H_2O \quad (2)$$

$$Na_2(SiO_3)_6 + 2[Na_2(AlO_2)_3] + 2(NaOH) + 2(H_2O) \rightarrow Na_8[AlSiO_4]_6(OH)_2 + 2(H_2O) \quad (3)$$

Based on the structural surface of the fly ash in figure 1, the treated fly ash with 20 wt/wt% of sodium hydroxide gives the best images compared to untreated fly ash and treated fly ash with various concentration of sodium hydroxide 5, 10 and 15 (wt/wt%) in term of having more smaller particles and even agglomeration of the particles in the fly ash. The smallest particles detected on the treated fly ash with 20 wt/wt% of sodium hydroxide surface by using Scanning Electron Analysis (SEM) is between 1.0 to 1.4 μm. Similar morphology of the treated fly ash with alkaline treatment is also found in (Yang, Gai, Cai, & Chen, 2006) research work [11]. As the concentration of sodium hydroxide used increased, the amount of silicate and aluminate will also be increased and resulted in producing aluminosilicates and hydroxysodalite [10,12]. The treated fly ash is also expected to have a better adhesive process with the polymer substrate when the fly ash is used as a filler for the polypropylene polymer matrix [11]. Therefore, having a rougher surface and smaller particles with even agglomeration such as a structural image of treated fly ash with 20 wt/wt% of sodium hydroxide will be more compatible with its polymer matrix.

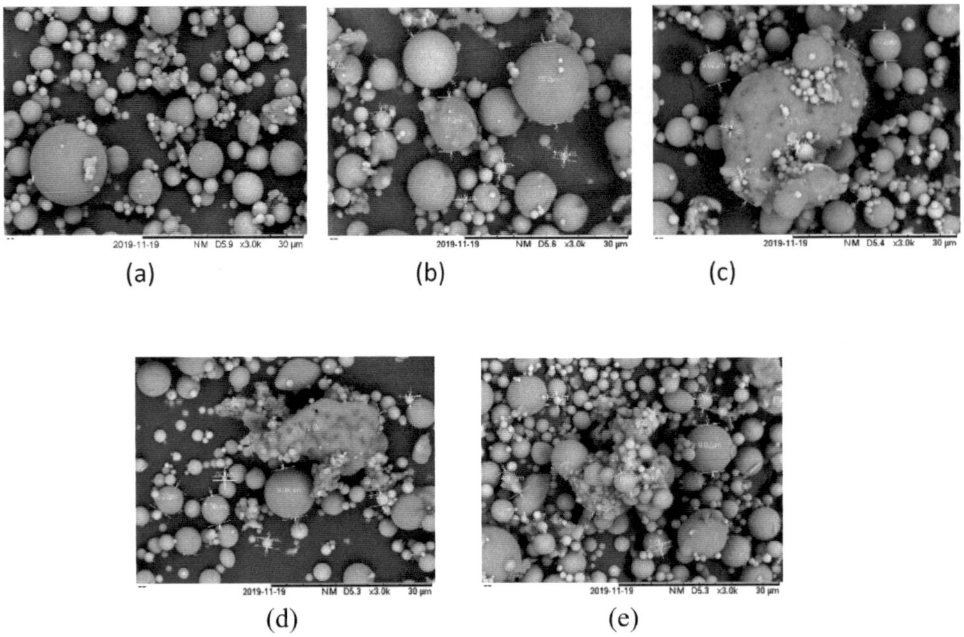

Figure 1. Surface morphology by SEM by 3.0kx magnification scale for (a) Untreated fly ash (b) Treated 5wt/wt% NaOH (c) Treated 10wt/wt% NaOH (d) Treated 15wt/wt% NaOH (e) Treated 20wt/wt% NaOH.

3.2. Element quantification of fly ash by using Energy Dispersive X-ray analysis (EDX)

Figure 2 shows the electron images for the element quantification by using Energy Dispersive X-Ray (EDX) method. The list of elements quantities such as carbon, oxygen, sodium, magnesium, aluminium, silicon, potassium, phosphorus, sulphur, calcium, titanium, iron and platinum presented in the fly ash is tabulated in table 1. The platinum detected in both untreated fly ash and treated fly ash due to the coating method for SEM analysis purposed. Hence, the quantity of platinum detected for both samples can be neglected.

Figure 2. Electron images for Energy Dispersive X-Ray (EDX) quantification (a) Untreated fly ash (b) Treated 20wt/wt% NaOH region A (c) Treated 20wt/wt% NaOH region B.

Table 1. Weight percentages of elements detected on untreated fly ash and treated fly ash with 20 wt/wt% of sodium hydroxide.

Element	Carbon	Oxygen	Sodium	Magnesium	Aluminium	Silicon	Potassium
Weight (wt/wt%) for untreated fly ash	9.315	39.197	-	0.261	1.525	4.689	-
Weight (wt/wt%) for treated fly ash with 20 wt/w% NaOH (region A)	3.547	51.351	1.063	0.314	5.588	26.041	1.410
Weight (wt/wt%) for treated fly ash with 20 wt/w% NaOH (region B)	3.124	55.537	3.716	0.972	5.848	14.974	0.330

Element	Phosphorus	Sulphur	Calcium	Titanium	Iron	Platinum
Weight (wt/wt%) for untreated fly ash	10.250	0.819	25.670	-	2.460	5.814
Weight (wt/wt%) for treated fly ash with 20 wt/w% NaOH (region A)	-	-	0.570	0.721	2.129	7.265
Weight (wt/wt%) for treated fly ash with 20 wt/w% NaOH (region B)	-	-	5.882	0.472	5.196	3.949

Based on the elements presented in table 1 above, the carbon, oxygen, sodium, magnesium, aluminium, silicon, potassium, phosphorus, sulphur, calcium, titanium, iron elements are detected in the fly ash composition. However, the most important elements to be discussed for flame retardancy enhancement filler are silicon and aluminium. The treated fly ash with 20 wt/wt% of NaOH gives the highest silicon and aluminium contents in the fly ash. It is perhaps due to the effect of the alkaline treatment conducted to change the morphological surface of the fly ash. Based on the equation (1) and (2), the aluminosilicate is formed after performing the alkalinisation process by using sodium hydroxide for the fly ash. Excessive sodium hydroxide concentration will make the hydroxysodalite being formed too as described in equation (3) [1,2,10]. Hence, this formation makes the aluminium and silicon content increased for treated fly ash with 20 wt/wt% NaOH. According to (Provis & Deventer, 2002), the best flame retardance filler should be contained with the high composition of aluminosilicate [13]. Furthermore, having the high content of silicon in the polypropylene polymer will reduce the heat release, burning rate and mass loss rate as measured by using nitrogen gasification devices and increase the polymer melt viscosity [14]. Thus, based on the result obtained, the treated fly ash with 20 wt/wt% NaOH gives the best silicon and aluminium composition to act as a flame retardant filler for green composite.

In addition, the treated fly ash with 20 wt/wt% NaOH also gives the lowest value of calcium content compared with the untreated fly ash. According to (Provis & Deventer, 2002), the low calcium content of the fly ash is suitable to be used as a flame retardance filler [13]. Having lower calcium content is more preferable as the high calcium content will make difficulties in the polymerisation process [15]. Low calcium content in the fly ash will also be a good trait for the fly ash to resist the acid attack [15]. Hence, based on the result obtained, the treated fly ash with 20 wt/wt% NaOH gives the best calcium content composition for the better flame retardancy enhancement filler for the composite.

As for the content of sulphur, the sulphur content for treated fly ash with 20 wt/wt% NaOH is not detected in that specific region. The alkaline treatment by using sodium hydroxide has made the sulphur content in the fly ash is being removed. The desulphurization process is increased with an increase in the concentration of the alkaline solution [10]. The sulphur content is good for water and acid resistance but it will not perform well when the fire contact is involved [16]. Thus, zero content of the sulphur element is important for the flame retardance filler. Also, the exposure of sulphur dioxide will cause severe injuries including respiration, irritation, inflammation and eyes burning injuries [17]. Although, the sulphur content for untreated fly ash is less than 1% and might not give a significant impact on the environment. However, having zero content of sulphur in the flame retardant filler is better to prevent environmental pollution. Based on the result obtained, the treated fly ash with 20 wt/wt% NaOH seems suitable to be used as a flame retardancy filler for composite.

3.3. The pore size distribution of untreated fly ash and treated fly ash with 20 wt/wt% filler as NaOH with BJH Analysis

The comparison data for untreated and treated fly ash with 20wt/wt% NaOH data for the particle size distribution including the pore diameter, area and volume by using BJH Analysis is illustrated in figure 3 below. Based on that figure, the range of pore diameter of untreated and treated fly ash with 20 wt/wt% NaOH are between 1.7142 to 177.1200 nm. The overall pattern for both types of fly ash shows that the pore volume keeps decreasing with increment in pore size diameter. The pore volume of untreated fly ash is higher in the range of 1.7142 to 2.6116 nm compared to treated fly ash with 20 wt/wt% NaOH. The depth of the pore for untreated fly ash is deeper compared to treated fly ash in the range of 1.7142 to 2.6116 nm.

The BJH model that has been developed by Barret-Joyner-Halenda in 1951 is based on the Kelvin equation and multilayer adsorption. This method is widely used to calculate the pore size distribution within the mesoporous and microporous range [18]. Hence, the pore volume is determined by using the BJH method is by the multilayer of N_2 gas filling inside the pore. Thus, the N_2 gas filled up inside the pore indicates the pore depth. However, for the pore volume for the particle diameter size of fly ash between 3.3274 to 177.1200 nm, the pore volume of treated fly ash with 20 wt/wt% NaOH has a higher volume compared to untreated fly ash. This data shows that the pore size diameter for treated fly ash with 20 wt/wt% NaOH is deeper compared to untreated fly ash.

As for the pore area perspective, the overall pattern for both types of fly ash shows that the pore area is kept decreasing with increasing pore size diameter. The pore area for untreated fly ash is higher compared to treated fly ash in the range of 1.7142 to 12.4934 nm. However, the pore area for the treated fly ash is relatively higher compared to untreated fly ash in a range pore diameter of 14.2805 to 177.1200 nm.

In short, based on the results obtained for the pore volume, area and diameter for both types of the pore. The untreated fly ash will have bowl shape pore meanwhile the treated fly ash is needle shape type for diameter pore size of 1.7142 to 12.4934 nm and 14.2805 to 177.1200 nm. The images of the pore for both types of fly ash are illustrated in figure 4 (a), (b), (c) and (d) based on the pore volume, area and diameter data obtained in the BJH Analysis.

Therefore, based on the overall data pattern, the treated fly ash with 20 wt/wt% NaOH has a better structural image in terms of the pore size distribution, volume and also area compared to untreated fly ash. The pore area of the larger particle size of treated fly ash is larger compared to untreated fly ash. Thus, treated fly ash with sodium hydroxide indeed improved the structural image for the fly ash to get better adhesiveness with its polymer matrix.

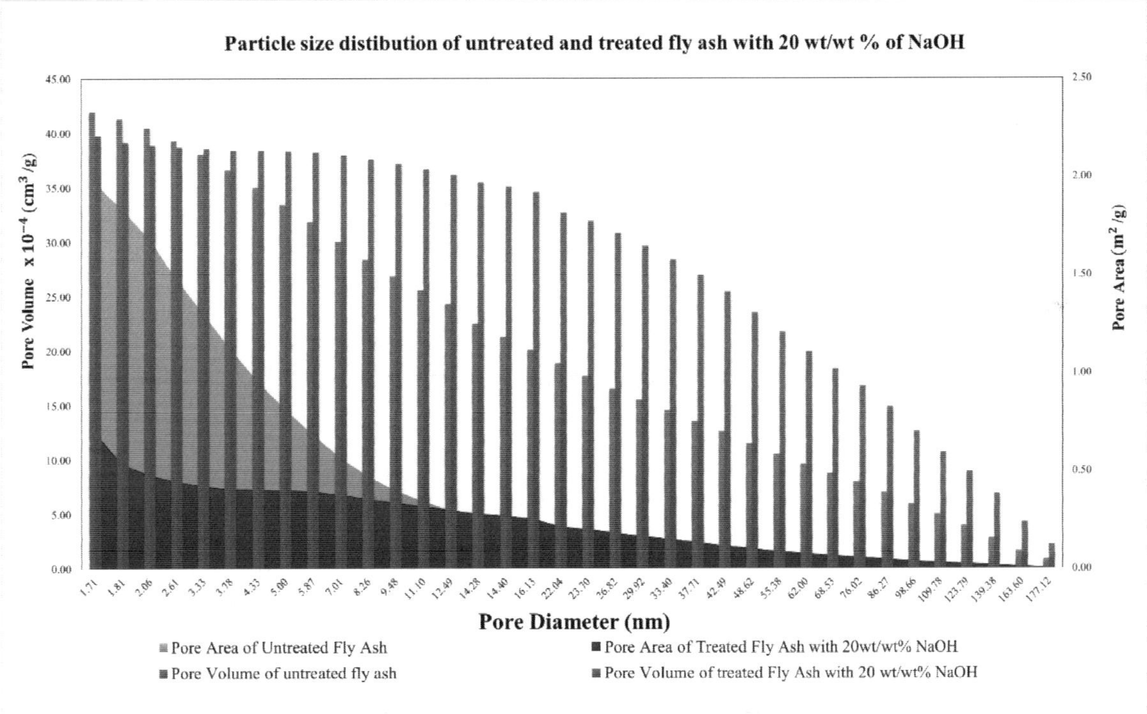

Figure 3. The distribution comparison of pore volume, pore diameter and pore area of untreated and treated fly ash with 20 wt/wt% of NaOH.

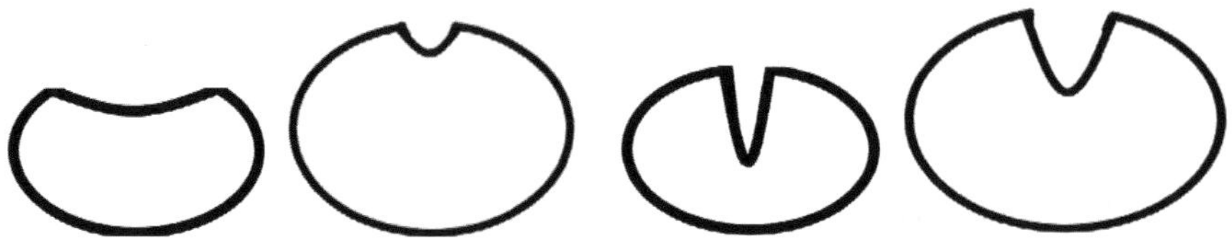

Figure 4. Illustrated image for untreated fly ash for pore diameter of (a) 1.7142 to 12.4934 nm, (b) 14.2805 to 177.1200 nm, Illustrated image for treated fly ash with 20 wt/wt% NaOH for pore diameter of (c) 1.7142 to 12.4934 nm, (d)14.2805 to 177.1200 nm.

3.4. The specific surface area of untreated and treated fly ash by using BET Analysis

Table 2. BET specific surface area (m^2/g).

Sample	Specific surface area (m^2/g)
Untreated fly ash	7.0331
Treated 20wt/wt% NaOH	1.0181

Based on the data tabulated in table 2 above, the specific surface area of untreated fly ash is larger compared to treated 20 wt/wt% of fly ash with NaOH. The treated fly ash with 20 wt/wt% NaOH is having a small BET specific surface area perhaps due to an increase of hydroxysodalite crystallinity. According to (Astuti, Wahyuni, Prasetya, & Bendiyasa, 2012), the small pore size of hydroxysodalite will cover the treated fly ash particle partially including the amorphous component [19]. This statement is related to the relationship of the pore diameter of the fly ash particle with its specific surface area. Increasing hydroxysodalite particles will decrease the specific surface area of the fly ash [19]. The increasing in hydroxysodalite particle will be a good influence in the compressive strength of the geopolymer composite [20]. The optimal thermo-mechanical properties will also be achieved in a high amount of hydroxysodalite produced [13]. Therefore, using treated fly ash with 20 wt/wt% NaOH as a flame retardant filler for a composite might give a good result in thermal and mechanical properties as small BET specific surface area is influenced by the hydroxysodalite component.

4. Conclusion

In a conclusion, the treated fly ash with 20 wt/wt% NaOH gives a better structural and morphological image in terms of pore diameter, volume, area and composition compared to untreated fly ash. The pore area for the large pore diameter size of the particles in treated fly ash is higher compared to untreated fly ash. Having a high pore area at a large particle size of the fly ash will be benefited to enhance the compatibility of filler with its polymeric matrix. In addition, the high composition of the aluminium, silicon and low calcium and sulphur contents in the treated fly ash also gives the positive traits for better flame retardancy enhancement filler with its composite.

Acknowledgments

Authors wishing to acknowledge the financial support from the Ministry of Higher Education, Malaysia through FRGS/1/2019/TK05/UMP/02/18 (RDU1901212), internal grant (RDU1803122) and Master Research Scheme (MRSSM), Universiti Malaysia Pahang.

References

[1] Nguyen, T A, Nguyen Q T, Nguyen X C and Nguyen V H 2019 Study on fire resistance ability and mechanical properties of composites based on Epikote 240 epoxy resin and thermoelectric fly ash: An ecofriendly additive *J. Chem* **2019** 1–8

[2] Anh N T, Tung, N Q, Canh N X and Hoan N V 2019 Eco-friendly flame retardant additives for

epoxy resin nanocomposites *EurAsian J. BioSci.* **13** 323–32
[3] Dong Y, Jow J and Lai S 2015 Fly ash based fillers and flame retardants *2015 World of Coal Ash Conf.* (Nashille, Tennessee)
[4] Katara S 2014 Surface modification of fly ash by thermal activation: A DR/FTIR study *Int. Res. J. Pure Appl. Chem.* **3** 299–307
[5] Nath D C D, Bandyopadhyay S, Yu A, Blackburn D and White C 2010 High strength bio-composite films of poly(vinyl alcohol) reinforced with chemically modified-fly ash *J. Mater Sci.* **45** 1354–60
[6] Patil A G, Mahendran A, Selvakumar M and Anandhan S 2016 Ductility and flame retardancy enhancement of PVC by nanostructured fly ash *Silicon* **11** 2241–51
[7] Sejnoha M and Broucek M 2014 Fire resistance of concrete with fly ash content – Experimental analysis *Eng. Mech.* **21** 159–65
[8] Sugiman S and Setyawan P 2013 Surface treatment of fly ash for improving the tensile strength of fly ash/unsaturated polyester composites *Makara J. Sci.* **17** 128–32
[9] Porąbka A, Jurkowski K and Laska J 2015 Fly ash used as a reinforcing and flame-retardant filler in low-density polyethylene *Polimery/Polymers* **60** 251–7
[10] Mukherjee S and Borthakur P C 2004 Effects of alkali treatment on ash and sulphur removal from Assam coal *Fuel Process. Technol.* **85** 93–101
[11] Yang Y F, Gai, G S, Cai Z F and Chen Q R 2006 Surface modification of purified fly ash and application in polymer *J. Hazard Mater* **133** 276–82
[12] Fukui K, Nishimoto T, Takiguchi M and Yoshida, H 2006 Effects of NaOH concentration on zeolite synthesis from fly ash with a hydrothermal treatment method *KONA PowderPart. J.* **24** 183–91
[13] Provis J L and van Deventer J S J 2002 Introduction to geopolymers *Geopolymers*, ed J L Provis and J S L van Deventer (Woodhead publishing) chapter 1 pp 1–11
[14] Kashiwagi T, Gilman J W, Butler K M, Harris R H, Shields J R and Asano A 2000 Flame retardant mechanism of silica gel / silica *Fire Mater.* **24** 277–89
[15] Kupaei R H, Alengaram U J and Jumaat M Z 2014 The effect of different parameters on the development of compressive strength of oil palm shell geopolymer concrete *Sci. World J.* **2014** 898536
[16] Al-Mehthel M H, Maslehuddin M, Al-di S H and Ali M R 2017 *Patent No. 15/433,708*. Saudi Arabia: US 9,796,629 B1
[17] Grahn H, Schoenberg P, Liljedahl B and Waleij A 2017 The 2016 Al-Mishraq sulphur plant fire : Source and health risk area estimation *Atmos. Environ.* ***169*** 287–96
[18] Musa M A A, Yin C Y and Savory R M 2011 Analysis of the textural characteristics and pore size distribution of a commercial zeolite using various adsorption *J. Appl. Sci.* **11** 3650–4
[19] Astuti W, Wahyuni E T, Prasetya A and Bendiyasa I M 2012 The effect of coal fly ash treatment with NaOH on the characters and adsorption mechanism toward methyl violet in the solution *3rd Int. Conf. on Chemistry and ChemicalEngineering (IPCBEE)* vol 38 (Singapore: LACSIT Press) pp 155–60
[20] Thakur R N and Ghosh S 2009 Effect of mix composition on compressive strength and microstructure of fly ash based geopolymer composites *J. Eng. Appl. Sci.* **4** 68–74

The impact of the mulberry (*Morus nigra* L.) leaf extract on the physicochemical properties of poly(vinyl alcohol) blend films

N S Rafflisman[1], S K Mah[1], S Y Lee[1], K S P Yee[1] and S Chowdhury[2]

[1] Department of Chemical Engineering, Lee Kong Chian Faculty of Engineering and Science, Universiti Tunku Abdul Rahman, Sungai Long Campus, Kajang 43000, Selangor, Malaysia

[2] Department of Chemical Engineering, Jashore University of Science and Technology, 1 Churamonkathi, Chaugachha Road, 7408, Jashore, Bangladesh

Email: mahsk@utar.edu.my

Abstract. In this work, mulberry (*Morus nigra* L.) leaf extract was added in poly(vinyl alcohol) (PVA)-based films and its impact on the film's properties was evaluated. In addition, HCl and glycerol were studied for their use as additives to prepare PVA-based films. The results showed that HCl and glycerol have minimal impacts on the films' appearance, while mulberry leaf extract imparted green colour to the films produced, mainly due to the presence of green pigments. Moreover, the results suggested that a significant interaction has occurred between the polymer matrix and leaf extract, contributing to a more compact and uniform film morphology. The tensile strengths of the films increased from 21.38 to 28.28 MPa after the addition of mulberry leaf extract. Additionally, the films were tested for their application as food wrapping films. Overall, the results showed that PVA-based films incorporated with mulberry leaf extract have higher capability to preserve the freshness of food when compared to commercial cling wraps from brands such as Diamond and Glad. Appropriate proportions of additives (mulberry leaf extract, HCl and glycerol) used in the formulation of P-GH-M20 films showed improvement in its mechanical properties and food preservation capability.

1. Introduction

Food packaging plays an essential role in preserving the quality of food by providing a barrier that avoids excessive contact between food and external environmental factors such as microorganism, moisture, and water vapor. Following the industrial revolution, plastics made from petroleum-based polymer have been introduced into the food packaging industry. However, from environmental perspective, biopolymer is gaining more attention as a sustainable alternative to replace traditional plastic produced from petroleum derivatives. Poly(vinyl alcohol) (PVA) is found to be a promising material owing to its biodegradability, marginal toxicity and outstanding flexibility. Despite its remarkable characteristics, several weaknesses of PVA exist and thus hindering it from being commercialized for industry use. PVA is primarily comprised of hydroxyl bonds with a general formula of $[CH_2CH\ (OH)]_n$, and therefore PVA-based film is highly hydrophilic. Besides, PVA-based film generally has weaker mechanical properties compared to plastic polymers. In view of these constraints, there were numerous studies carried out to improve the properties of PVA-based film by incorporating potential additives into its polymer matrices. The natural extracts obtained from various terrestrial sources such as fruits, leaves and roots of plants and herbs could be a good option. Examples of the several extracts from plants and fruit such as, mint, pomegranate [1], seaweed [2], *Syzygium cumini* [3] and *Azadirachta indica* (neem) extract [4] possess excellent antioxidant activities, good antimicrobial properties, and improve mechanical properties.

In this study, mulberry leaf extract was chosen to be incorporated into PVA-based film. To the best knowledge of authors, there is still limited study on the use of mulberry leaf extract in preparing PVA-based film. Mulberry leaf, or known as *Morus nigra* L., is highly recognized to be used as a herb in traditional Chinese medicine (TCM) to treat several diseases [5]. It was reported that mulberry leaf extract consists of substantial amount of total carotenoids (e.g. β-carotene and lutein) and polyunsaturated fatty acids (e.g. α-linolenic acid and linoleic acid), and therefore contributing to its outstanding antioxidant and cytotoxic activities [6–9]. In this work, mulberry leaf extract was used to prepare PVA-based films, and its impact on the film's appearance, morphology, and mechanical properties, was evaluated. The produced films were also applied as food wrapping films, and their food preservation capability was studied and compared to commercial cling wraps, namely Diamond and Glad cling wraps, which were purchased from the market.

2. Materials and Methods

2.1. *Materials*

PVA (99.8% - 100% hydrolysed) was purchased from Sigma-Aldrich. Glycerol, methanol and HCl were obtained from Merck. All chemicals and solvents were of analytical grade and used without further purification. Commercial food wrapping films, namely Diamond Cling Wrap (Reynolds Consumers Products LLC, USA) and Glad Cling Wrap (The Glad Product Company, USA) were purchased from a local distributor and used directly. The main purpose of using low density polyethylene (LDPE) cling wrap plastic as a reference is to compare the characteristic of both polymer materials in plastic wrapping. The chosen conventional cling wraps are made of Polyethylene (PE). Mulberry leaves were picked from the garden in Putrajaya, Malaysia.

2.2. *Preparation of mulberry leaf extract*

The extraction process was performed following the protocols described by Lee and Choi [9]. Fresh mulberry leaves were picked from plant and washed to remove impurities. Subsequently, the mulberry leaves were dried in an oven at 80°C for 1 h. The oven-dried leaves were grinded using a medicine blender (Golden Bull SY-04, Malaysia). Fine leaf powder which passed through a 100-mesh sieve was used for extraction. 20 g of fine leaf powder was mixed with a solvent which was comprised of 80:20 (v/v) of methanol and distilled water, respectively. The mixture was shaken at 150 rpm for 8 h at room temperature in an orbital shaker. Next, the mixture was filtered using a Whatman No.1 filter paper. The solid residue obtained from the filtration process was mixed with fresh solvent, and the extraction process was repeated for two times. The filtrate obtained was concentrated using a reduced pressure rotary evaporator (Heidolph Hei-VAP Value, Germany) operating at 45 °C. The concentrated extract was stored at 4°C until further use. 5 mL of the concentrated extract was dried at 100 °C until a constant weight was obtainedto calculate the extract concentration. The mulberry leaves extract concentration obtained was 25.92 ± 2.1 mg/mL.

2.3. *Preparation of films*

The polymer films were prepared through the casting method. 10 wt% of PVA solution was prepared by dissolving PVA in water and stirring the mixture at 6000 rpm by using a homogeniser (IKA ULTRA-TURRAX T 18 digital) at 80 °C for 2 h. The PVA-based films with varying compositions, as described in table 1, were prepared. For some films, 15 % glycerol (w/w, based on PVA), was added into the film forming solution, aiming to study the influence of glycerol as a plasticizer on the film's mechanical properties. Besides, the impact of HCl on the film's characteristics was investigated by adding 10 % HCl (v/v, based on the final volume of film making solution) into the film forming solution. Each addition of substance was followed by a constant stirring of the mixture at 6000 rpm for 5 min before the subsequent step.

Table 1. Compositions of PVA-based films studied in this work.

Film sample	10 wt% PVA solution (mL)	Glycerol (mL)	HCl (mL)	Concentrated leaf extract (mL)
P-0	50	-	-	-
P-G	50	0.5952	-	-
P-H	50	-	0.5	-
P-GH	50	0.5952	0.5	-
P-M10	45	-	-	5
P-M20	40	-	-	10
P-M30	35	-	-	15
P-M40	30	-	-	20
P-GH-M10	45	0.5952	0.5	5
P-GH-M20	40	0.5952	0.5	10
P-GH-M30	35	0.5952	0.5	15
P-GH-M40	30	0.5952	0.5	20

All film forming solutions were left overnight so that they were completely degassed before film casting. The film forming solution was casted on a glass plate with a wet thickness of 0.25 mm [10]. The film sample was then dried for 24 h and was peeled and kept in a desiccator for further characterization and application tests.

2.4. Characterization of films

2.4.1. Morphology. Morphology of films was studied using scanning electron microscopy (SEM) (Hitachi, S-3400N, Japan) [11]. All films were sputter-coated with gold before SEM imaging.

2.4.2. Mechanical properties. Mechanical properties of films were acquired by a universal testing machine (Instron 5582Q4970, USA) with a grip separation of 15 mm and a cross-head speed of 20 mm/s. The specimen strips were prepared with a dimension of 25.4 mm × 100 mm and the thickness of specimen strips were measured first before mechanical measurement using a hand-held micrometre (Mitutoyo, Japan).

2.5. Evaluation of film antifungal activity on pumpkin slice. In this test, 40 mm × 20 mm of pumpkin cut was wrapped using the prepared sample film and stored in a refrigerator at 6 °C. The food sample was monitored daily for any growth of mould. The time it took before the pumpkin cut wrapped using the respective film started to grow mould was recorded for each film.

3. Results and discussion

3.1. Physical Appearance

Figure 1 displays the physical appearance of all films studied in this work. Diamond and Glad cling wraps were included for comparison purpose. From figure 1, all PVA-based films (P-0, P-G, P-H and P-GH films) without mulberry leaf extract appeared clear and transparent. This indicates that both glycerol and HCl in P-G, P-H and P-GH did not affect the films' appearance. Whereas, for films which uses mulberry leaf extract as an additive, it was observed that the greenish hue turned darker (from pale to olive green) with increasing amount of mulberry leaf extract. Besides, a brownish green colour was observed for films containing the leaf extract, glycerol and HCl, namely P-GH-M10, P-GH-M20, P-GH-M30 and P-GH-M40. This might be attributed to the colour change of green pigments in the leaf extract under acidic environment when HCl was added in the film making solution. It was frequently reported that chlorophyll molecules change colour from bright green to a dull brown colour in response to the acidity of the water [12]. The colour change of chlorophyll is due to denaturation which takes place when the magnesium ion in the centre of the chlorophyll molecule is replaced with a hydrogen atom [13,14]. On the other hand, both commercial cling wraps by Diamond and Glad showed clear transparency.

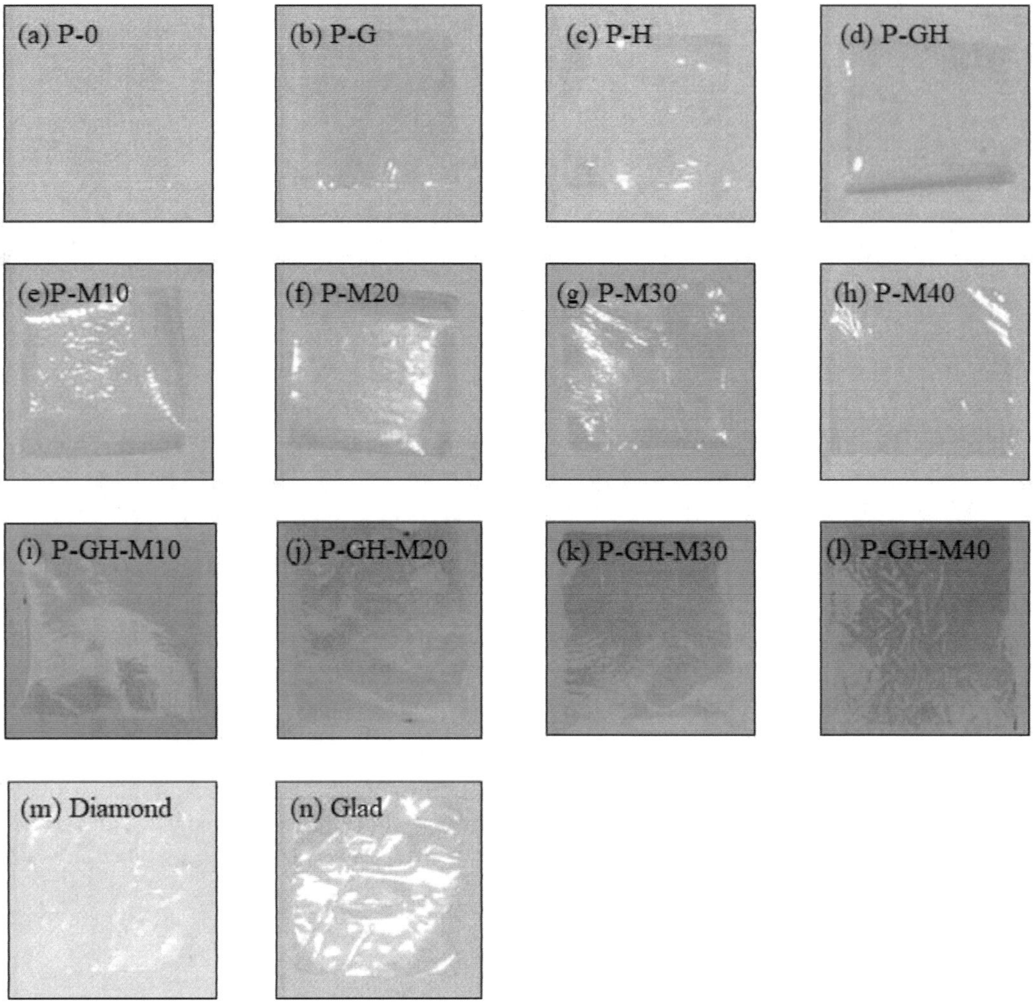

Figure 1. Films studied in this work, (a) P-0; (b) P-G; (c) P-H; (d) P-GH; (e) P-M10; (f) P-M20; (g) P-M30; (h) P-M40; (i) P-GH-M10; (j) P-GH-M20; (k) P-GH-M30; (l) P-GH-M40; (m) Diamond cling wrap; and (n) Glad cling wrap.

3.2. Morphology of films

Figure 2 shows the SEM results of the commercial grade cling wraps and films prepared using different compositions. From figure 2, P-0 surface morphology was found to be more homogeneous than other films with additives. Some droplets and embedded particles were found in the films containing leaf extract, particularly P-M10, P-M20, P-M30 and P-M40. Similar findings were obtained by Ma *et al.* [15] in their study on an intelligent film composed of PVA, chitosan nanoparticle and mulberry extract. The authors observed the presence and formation of droplets in the cross-sections of films incorporated with mulberry extract. This phenomenon was a result of a phase separation or chemical segregation in the chlorophyll pigment found in the leaf extract, where polysaccharide molecules were isolated from the film forming mixture resulting in the formation of droplets [15]. Likewise, Bashir *et al.* [16] reported that many uniformly distributed pores were found in blended tetraethoxysilane-crosslinked films with the addition of mint and grapefruit peel extracts.

On the other hand, there were fewer white particles observed in films incorporated with leaf extract, glycerol and HCl. This might be due to the action of HCl which alters the solubility and nature of the bio-compounds in leaf extract, resulting in smoother cross sections compared to other films with only leaf extract. Commercial Diamond and Glad cling wraps showed slightly non-homogeneous surface morphology compared to P-0, however neither pores nor embedded particle was observed. The non-uniform morphology could be caused by a certain degree of phase separation in films.

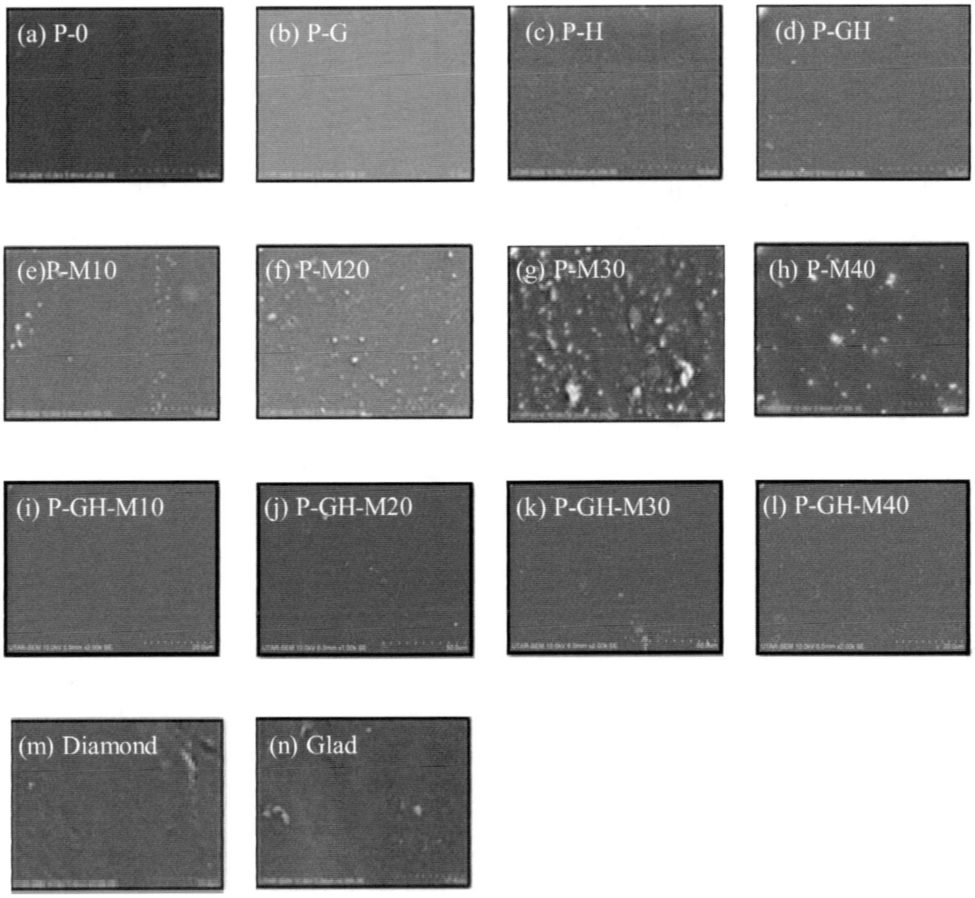

Figure 2. Morphology of films studied in this work, (a) P-0; (b) P-G; (c) P-H;(d) P-GH; (e) P-M10; (f) P-M20; (g) P-M30; (h) P-M40; (i) P-GH-M10;(j) P-GH-M20; (k) P-GH-M30; (l) P-GH-M40; (m) Diamond cling wrap; and (n) Glad cling wrap.

3.3. Mechanical properties of films

It is important to determine the film's mechanical properties since a packaging material will be exposed to various kind of stress during its use. A good food packaging film should not only possess high tensile strength (TS) but also good ductility. Ductility is measured based on fracture strain or also known as elongation at break (E%). It is closely related to the ability of a plastic specimen to resist change in shape without cracking or breaking [17,18]. In this study, the thickness of film strips was measured first before mechanical measurement. P-0, P-G and P-H films' thickness were 0.0250±0.0029 mm, 0.0233±0.0029 mm and 0.015±0 mm, respectively. On the other hand, the thickness of P-GH and P-M_X resulted in stable thickness of 0.0200±0 mm. The thickness of sample PVA added with glycerol, HCl and leaf extract (P-GH-M_X) varied from 0.0130±0.0029 mm to 0.0220±0.0058 mm with increasing among of leaf extract. The conventional cling wrap thickness was 0.0117±0.0029 mm for Glad and 0.0100±0 mm for diamond. The mechanical properties of the films studied were presented in figure 3.

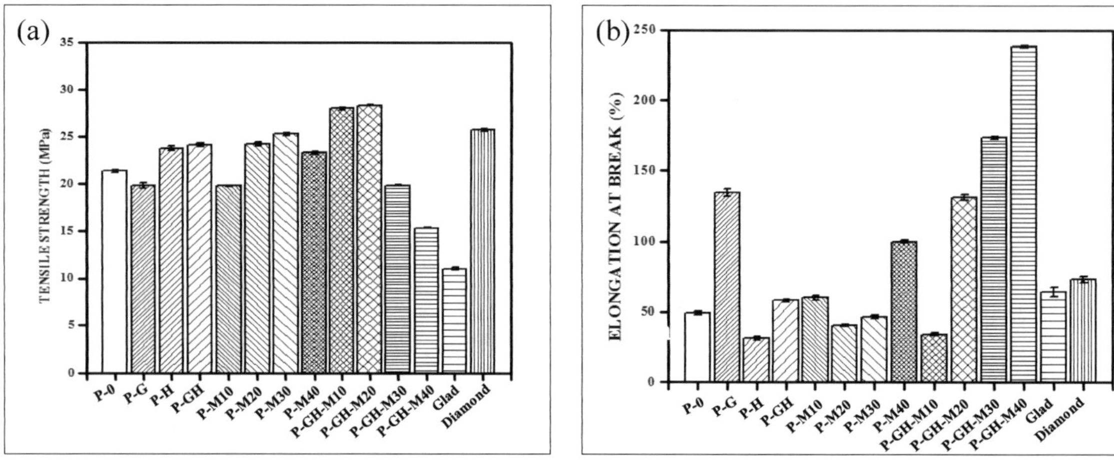

Figure 3. Mechanical properties, including (a) tensile strength and (b) elongation at break of films studied in this work.

In general, all films showed improved in tensile strength when additive were added in the films, except for P-GH-M40 films exhibited exceptionally poor tensile strength. Besides, it was observed that most of the films with glycerol as a plasticizer, namely P-G, P-GH-M20, P-GH-M30 and P-GH-M40, revealed better ductility (higher E%). Several works have highlighted that the addition of glycerol in PVA polymer matrices promotes great flexibility in films by attracting more water into the structure of films [19–21]. On the other hand, the dispersion of lipids glycerol might introduce discontinuities in the polymer matrices which contribute to a reduction in the polymer cohesion forces and thus, the films' have low resistance to break.

On the flip side, the addition of HCl in the PVA polymer film improved TS of films by 10.11% when comparing TS values of P-H and P-0. This might be due to the action of HCl which facilitates the crosslinking process of PVA polymer [22] . However, it was found that the presence of mulberry leaf extract in high amounts hindered the effect of HCl in enhancing the TS of films, as displayed by P-GH-M30 and P-GH-M40 films. An appropriate amount of mulberry leaf extract together with HCl and glycerol could be a good formulation for film composition to produce a film with high TS. Both P-GH-MN10 and P-GH-MN20 have relatively high TS than the Diamond cling wrap. Nevertheless, all films still showed lower TS compared to those reported earlier by Kanatt *et al.* and Qianyun Ma *et al.*[1,15].

3.4. Evaluation of film antifungal activity on pumpkin slice

Food packaging analysis is necessary to evaluate the preservation capability of films towards the freshness of food. The results were expressed in the terms of the time it took for mold to grow on the wrapped pumpkin slice, as summarised in figure 4.

The final results proved that the films with additives such as HCl, glycerol and/or mulberry leaf extract have better food preservation power to resist the growth of mold on the pumpkin compared to pure PVA film, which is P-0. Several films showed promising results, particularly, P-GH-M20 and P-GH-M40. These films were able to preserve the pumpkin cut for at least 29 days before mold started to grow, compared to the commercial Glad cling wrap which was able to preserve the pumpkin for up to 21 days. This might be due to the presence of additives, which consist of antimicrobial behaviour, which helped to extend the freshness of the pumpkin. It was frequently reported that mulberry leaf extract has an antimicrobial effect to inhibit the growth of microorganisms such as *Streptococcus mutans* [23] and *Helicobacter pylori* [24]. P-GH-M20 displayed the highest food preservation power by keeping the food sample away from mould for up to 34 days. This indicated that the optimum amount of mulberry leaf extract to be incorporated into the PVA-based film was 20 % (v/v).

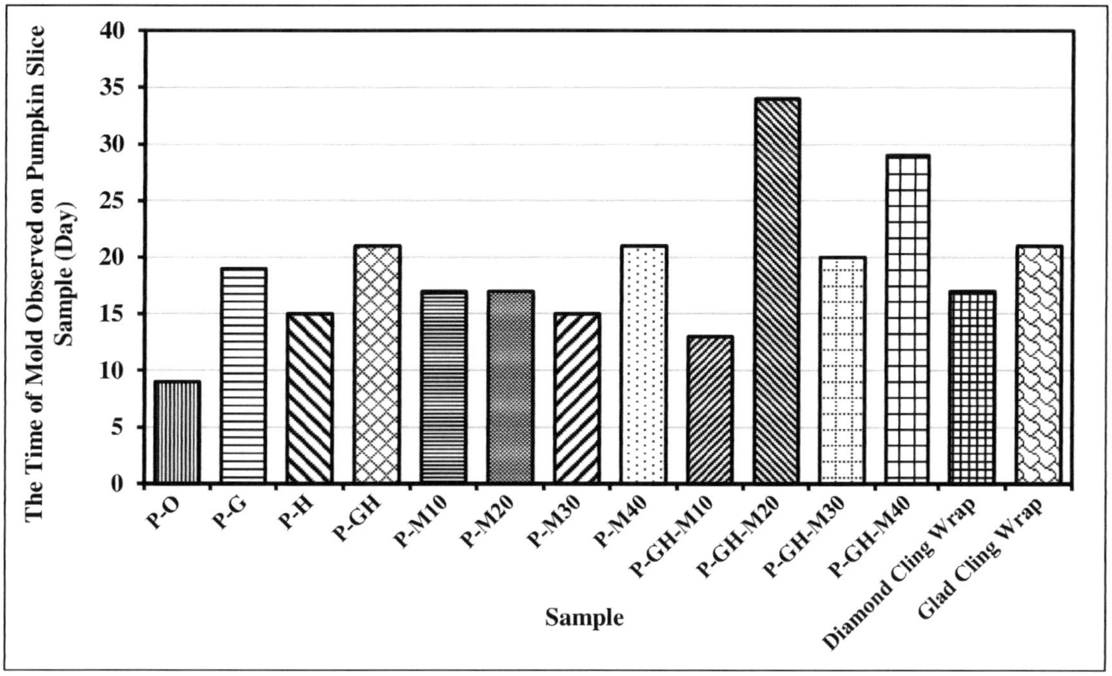

Figure 4. Evaluation of film antifungal activity on pumpkin slice

4. Conclusion

By blending mulberry leaf extract into PVA-based film, developed films have enhanced mechanical properties and their food preservation. Besides, the addition of glycerol and HCl in PVA-based films might affect the film properties. With optimum formulation, it is possible to produce a film with advantageous properties for food packaging application. Among all films, P-GH-M20 demonstrated the highest TS of 28.38 ± 0.07 MPa with a thickness of 0.0130 mm and food freshness preservation of up to 34 days. Besides, a compact and smooth surface section was observed compared to conventional and other developed films. All the films were cast at 0.0250 mm wet thickness and result showed that the dry thickness of sample films varied between 0.0250 and 0.0130 mm, depending on the addition of glycerol, HCL and amount of leaf extract in the PVA solution. Soil biodegradation test will be required in future studies to determine the degradation time of produced films in soil while an antimicrobial study can be done to determine the antimicrobial performance of leaf extract towards *E. coli* bacteria.

References

[1] Kanatt S R, Rao M S, Chawla S P and Sharma A 2012 Active chitosan–polyvinyl alcohol films with natural extracts *Food Hydrocoll* 290–7

[2] Singh S, Gaikwad K K and Lee Y S 2018 Antimicrobial and antioxidant properties of polyvinyl alcohol bio composite films containing seaweed extracted cellulose nano-crystal and basil leaves extract *Int. J. Biol. Macromol* **107** 1879–87

[3] Kasai D, Chougale R, Masti S, Chalannavar R, Malabadi R B and Gani R 2018 Influence of

Syzygium cumini leaves extract on morphological, thermal, mechanical, and antimicrobial properties of PVA and PVA/chitosan blend film. *J. Appl. Polym. Sci.* 46188

[4] Ali A, Shahid M d A, Hossain M d D and Islam M d N 2019 Antibacterial bi-layered polyvinyl alcohol (PVA)-chitosan blend nanofibrous mat loaded with Azadirachta indica (neem) extract *Int. J. Biol. Macromol* **138** 13–20

[5] El-Sayyad H I H, El-Sherbiny M A, Sobh M A, Abou-El-Naga A M, Ibrahim M A N and Mousa S A 2011 Protective effects of Morus alba leaves extract on ocular functions of pups from diabetic and hypercholesterolemic mother rats *Int. J. Biol. Sci.* 715–28

[6] Memon A, Memon N, Luthria D, Bhanger M and Pitafi A 2010 Phenolic acids profiling and antioxidant potential of mulberry (Morus laevigata W., Morus nigra L., Morus alba L.) Leaves and fruits grown in Pakistan *Polish J. Food Nutr. Sci.* 25–32

[7] Anwar F, Kanwal S, Shabir G, Alkharfy K and Gilani A 2015 Antioxidant and Antimicrobial Attributes of Different Solvent Extracts from Leaves of Four Species of Mulberry *Int. J. Pharmacol.* **11** 757–65

[8] Doi K, Kojimata T, Makino M, Kimura Y and Fujimoto Y 2001 Studies on the Constituents of the Leaves of Morus alba L *Chem. Pharm. Bull.* 151–3

[9] Lee W J and Choi S W 2012 Quantitative Changes of Polyphenolic Compounds in Mulberry (Morus alba L.) Leaves in Relation to Varieties, Harvest Period, and Heat Processing *Prev. Nutr. Food Sci* 280–5

[10] Wang E, Batra S and Cakmak M 2015 A real time study on drying and the mechano-optical behavior of polyvinyl alcohol films in solid and swollen state *Polym. J.* **67** 200–7

[11] Jeyaraman P, Jeyabala D, Meenarathi B, Tung K-L and Anbarasan R 2018 Synthesis, characterization, and drug release activity of structurally modified poly(vinyl alcohol) *J. Appl. Polym. Sci* 46620

[12] Koca N, Karadeniz F and Burdurlu H 2007 Effect of pH on chlorophyll degradation and colour loss in blanched green peas *Food Chem* **100** 609–15

[13] Heaton J W and Marangoni A G 1996 Chlorophyll degradation in processed foods and senescent plant tissues *Trends Food Sci. Technol.* 8–15

[14] Isnaeni, Masni and Shiddiq M 2019 Rapid detection of chlorophyll-a reduction in tropical fruit leaves using violet laser *IOP Conf. Ser. Earth Environ. Sci.* **251** 012004

[15] Ma Q, Liang T, Cao L and Wang L 2018 Intelligent poly (vinyl alcohol)-chitosan nanoparticles-mulberry extracts films capable of monitoring pH variations *Int. J. Biol. Macromol.* **108** 576–84

[16] Bashir A, Jabeen S, Gull N, Islam A, Sultan M and Ghaffar A 2018 Co-concentration effect of silane with natural extract on biodegradable polymeric films for food packaging *Int. J. Biol. Macromol.* **106** 351–9

[17] Dean K M, Do M D, Petinakis E and Yu L 2008 Key interactions in biodegradable thermoplastic starch/poly(vinyl alcohol)/montmorillonite micro- and nanocomposites *Compos Sci Technol* 1453–62

[18] Hejri Z, Ahmadpour A, Seifkordi A A and Zebarjad SM 2012 Role of Nano-Sized TiO2 on Mechanical and Thermal Behavior of Starch/Poly (vinyl alcohol) Blend Films *J. Nanosci. Nanotechnol.* 215–26

[19] Andreuccetti C, Carvalho R and Grosso C 2009 Effect of hydrophobic plasticizers on functional properties of gelatin-based films *Food Res. Int.* **42** 1113–21.

[20] Beatty E, Roos Y, Morris M and Kerry J 2013 Development and Characterization of Biodegradable Composite Films Based on Gelatin Derived from Beef, Pork and Fish Sources *Foods* **2** 1–17

[21] Shuaiyang W, Huiling L, Junli R, Chuanfu L, Feng P and Runcang S 2013 Preparation of xylan citrate—A potential adsorbent for industrial wastewater treatment *Carbohydr. Polym.* 1960–5

[22] Harun-or-Rashid M D, Saifur R M D, Enamul K S and Khan M A 2016 Effect of

hydrochloric acid on the properties of biodegradable packaging materials of carboxymethylcellulose/poly(vinyl alcohol) blends *J. Appl. Polym. Sci.*

[23] Islam B, Khan S N, Haque I, Alam M, Mushfiq M and Khan AU 2008 Novel anti-adherence activity of mulberry leaves: inhibition of Streptococcus mutans biofilm by 1-deoxynojirimycin isolated from Morus alba *J Antimicrob. Chemother.* 751–7

[24] Cho Y-J, Ju I, Kim B-O, Kim J-H, Lee B, An B and Choo J 2007 The Antimicrobial Activity against Helicobacter pylori and Antioxidant effect from the extracts of Mulberry leaves (Morus Alba L.) *J. Appl. Biol. Chem.* **50** 334–43

Acknowledgments

This work was supported by Universiti Tunku Abdul Rahman Grants (6200/S74 and 6200/M26).

Synthesis of symmetrical structured triglycerides via a bottom-up process

R C Agapay[1,2], A W Go[3] and Y H Ju[1,3,4]

[1] Department of Chemical Engineering, National Taiwan University of Science and Technology, No. 43, Keelung Road, Section 4, Taipei 10607, Taiwan
[2] Department of Chemical Engineering, University of San Carlos, Nasipit, Talamban, Cebu City 6000, Philippines
[3] Graduate Institute of Applied Science and Technology, National Taiwan University of Science and Technology, No. 43, Keelung Road, Section 4, Taipei 10607, Taiwan
[4] Taiwan Building Technology Center, National Taiwan University of Science and Technology, No. 43, Keelung Road, Section 4, Taipei 10607, Taiwan

Email: ramelitoagapay@yahoo.com

Abstract. The synthesis of symmetrical structured triglycerides (STG) through a bottom-up approach was previously shown to produce 1,3-dioleoyl-2-palmitoyl glycerol in significant quantities. This solvent-free lipase-catalyzed process, consisting of a low-temperature (40 °C) esterification step with glycerol dosing followed by a high-temperature (60 °C) esterification step, was further investigated in the production of symmetrical medium-and-long-chain triglycerides (MLCT). By replacing oleic acid with capric acid in the first step or the palmitic acid by either capric acid or lauric acid in the second step, the effects of free fatty chain length and sequence of fatty acid addition on STG production were established. These produced 1,3-dicaproyl-2-oleoyl glycerol, 1,3-dioleoyl-2-caproyl glycerol, and 1,3-dioleoyl-2-lauroyl glycerol at concentrations of 36.98 g, 36.77 g, and 37.08 g per 100 g of triglycerides respectively after 72 h at an overall FFA1:FFA2:Glycerol of 2:1:1 and 4 g Novozyme 435 per 100 g reactants, without the purification of intermediates and products. The sequence of fatty acid addition had the most significant effect as purer STG products can be obtained when the medium chain fatty acid is introduced in the first step. As the process was carried out without solvents, the STG produced are appropriate for functional food or nutraceutical applications.

1. Introduction

Lipids with structures that have been altered in terms of the fatty acid sequence or position in the glycerol backbone to enhance their nutritional or functional value and/or render a unique property are called structured lipids [1]. The term is commonly associated with triglycerides (TG), but generally includes partial glycerides (PG) composed of monoglycerides (MG) and diglycerides (DG), as well as phospholipids [2]. Structured lipids have several applications, especially the symmetrical or ABA-type where A and B are two different fatty acid ligands attached to the *sn*-1(3) and *sn*-2 positions of the glycerol backbone, respectively. Cocoa butter substitutes (e.g. 1,3-dipalmitoyl-2-oleoyl glycerol or POP, and 1,3-distearoyl-2-oleoyl glycerol or SOS) [3,4], human milk fat substitutes (e.g. 1,3-oleoyl-2-palmitoyl glycerol or OPO) [5], and medium-and-long-chain triglycerides or MLCT (e.g. MLM [6] or LML [7] types) are among the common examples of this type. Having been identified as important ingredients in emerging applications such as drug or food delivery systems in the form of microemulsions [8] and nanostructured lipids [9], these compounds remain to be materials of scientific and commercial interest.

The modification in lipid structure can be accomplished through chemical or enzymatic reactions, e.g. interesterification [10] or acidolysis [11] of oils, (trans)esterification of PG with triglycerides [12], and even genetic engineering of oilseed crops. Each method has its advantages and disadvantages, and a process scheme can be created consisting of one or multiple reaction steps basing on desired specifications, practical considerations, or constraints. Lipid production is not novel as several processes have been explored to synthesize them. However, through innovation or modification of existing processes, product quality, and, more importantly, productivity can still be improved. It is also important to note that the inherently complex reaction mechanism of lipid synthesis produces intermediates or by-products, which oftentimes are as valuable as the main (TG) product. Thus, the knowledge on how product distribution, quality, and yield of the system change when a synthesis or process parameter is changed, and using it to implement simple changes in the process to gain significant results is valuable. Ultimately, tailoring a process to make it adaptable to a wide range of substrates and flexible enough to produce different and valuable products at substantial amounts and qualities is, therefore, the goal.

Bottom-up approaches to structured triglycerides (STG) have been seldom investigated even though these processes, based primarily on esterification, have potentially high reactant mass efficiencies (RME), i.e. the product obtained relative to the reactants used. The few related works reported, usually involve several intermediate purification steps [13] or multiple catalysts [14] which increase process complexity and decrease reaction mass efficiency. Other processes are performed in tandem with top-down reactions [15] and sometimes require special equipment [16]. Some esterification-based processes are focused on the synthesis of STG precursors, specifically DG [17,18], without further demonstration of their use in STG synthesis. Recently, the same group of authors as this study pursued and published a series of works on solvent-free lipase-catalyzed esterification of STG precursors from oleic acid and glycerol [19] and the subsequent production of C52 TG (including OPO) [20], which proved that STG can be adequately produced solely based on a bottom-up process. It was further shown that yields can be improved using a single lipase via reactant feeding strategies without intermediate purification steps. This served as the baseline process on which other types of symmetrical STG, particularly MLCT, were produced and is the main goal of the current work. In particular, the following objectives were pursued: (a) to the determine the effect of free fatty acid chain length, and (b) the effect of the dosing sequence of FFA on the production of STG and its precursors. To address the first objective, palmitic acid (PA) was substituted with medium-chain FFA (MCFA), namely capric acid (CPA) or lauric acid (LA) in the synthesis of MLCT, basing on the optimum conditions identified for OPO. In other trials, CPA was substituted for oleic acid (OA) in the production of the STG precursors or the diglycerides followed by the addition of OA in the second step, to establish the effect of modifying the dosing sequence of FFA on the MLCT production and to ascertain whether reactant feeding is a critical strategy in manipulating the synthesis of specific types of MLCT.

2. Materials and Methods

2.1. Reactants and catalyst for esterification

The free fatty acids and glycerol were secured from various suppliers, as follows: capric acid (≥99 %, Acros Organics, New Jersey, USA), lauric acid ("Baker-analyzed" reagent, J. T. Baker Inc., New Jersey, USA), oleic acid (99 %, Showa Chemical Co. Ltd., Japan), palmitic acid (98 %, Acros Organics, New Jersey), and glycerol (≥99.5 %, J. T. Baker Inc., New Jersey, USA).

The catalyst used, Novozyme 435 (CALB immobilized in acrylic resin, 10,000 PLU/g) was from Novo-Nordisk, Denmark.

2.2. *Solvents, standards and materials for analysis*

The separation and sample analysis were performed using the following solvents: acetic acid (99%, Scharlab, S. L., Barcelona, Spain), acetone (technical grade, Echo Chemical Co. Ltd., Miaoli, Taiwan), ethyl acetate (99.9%, Echo Chemical Co. Ltd., Miaoli, Taiwan), ethyl ether (99%, Echo Chemical Co. Ltd., Miaoli, Taiwan), and n-hexane (>95%, Tedia Company/Tedia High Purity Solvents, Ohio, USA). Nitrogen gas (>99%, Tong Sheng Technic Company, Taipei, Taiwan) was used in purging and removal of residual solvents. The plates used in thin layer chromatography were Silica gel 60 (PF254) containing gypsum (Merck & Co. Inc., Honenbrunn, Germany).

The standards for calibration and identification of lipids were as follows: mixed lipids/lipid mix standard (1787-1AMP, mono-, di- & triglyceride mix), 1,3-diolein (≥99% GC, D3627-10MG), 1,3-dioleoyl-2-palmitoylglycerol (OPO, ≥99%, D1657-25MG), and 1,2-dioleoyl-3-palmitoylglycerol (OOP, ≥99% TLC, D1782-10MG), all from Sigma-Aldrich Corporation.

2.3. *Esterification of fatty acids and glycerol for DG and STG production*

The preparation of precursors and structured triglycerides was patterned after the works of Agapay, *et al.* [19,20]. About 7-10 g of free fatty acids (FFA) and glycerol mixture with the desired FFA:G (2.0-4.4) was prepared in a 50-mL Erlenmeyer flask and incubated at 40°C in an orbital shaker incubator at 150 rpm. A catalyst loading of 4 g of Novozyme 435 per 100 g of reactants was added to start the reaction. Monitoring of DG concentration as the reaction progressed was done by sampling (~30-μL aliquots) at predetermined times. Glycerol dosing was implemented twice at 24 h and 36 h, each time at 25 μL per mL of the original reaction mixture to an overall FFA:G of ~2.0.

The second esterification step was started at 48 h, with the addition of a medium-chain FFA, either capric acid (CPA) or lauric acid (LA) to a molar ratio (CPA:G or LA:G) of ~1.0 and increasing the temperature to 60 °C. The process was terminated after a total reaction time of 72 h. The production of STG was also monitored through sampling at predetermined times. The sequence of the synthesis steps was also reversed in selected trials, starting with the medium-chain FFA, followed by oleic acid (OA) at an OA:G of 1.0. These two processes were hypothesized to produce LML- and MLM-types of MLCT, respectively.

2.4. *Analysis of lipid composition*

Each sample of ~30 μL was dissolved in 1 mL of ethyl acetate, mixed thoroughly, and then filtered using a 13-mm syringe filter with 0.20-μm PTFE membrane (Acrodisc syringe filters, Pall Corporation, Life Sciences). The resulting sample solution was injected in a gas chromatography unit (Shimadzu GC-2010 Plus) equipped with a split injector and FID, and installed with a ZB-5HT Inferno column (15m x 0.32 mm x 0.1μm). The GC program for the analysis of glycerides followed that described in Ting, *et al.* [21]. A total time of 29 minutes was enough to resolve the specific glyceride products and unreacted FFA at these conditions: 80 °C (initial column temperature) ramped to 370 °C (final column temperature), 370 °C (injector and detector temperature), 1.4 mL/min (nitrogen gas flow), and 50 (injection split ratio).

The separation of crude product into lipid classes was performed whenever it is necessary using the procedure based on the principles described in [22]. A sample (25 mg) was mixed with a minimum volume of n-hexane that is enough to dissolve it. An aliquot (~10 μL) of the resulting mixture was blotted on a TLC Silica gel 60 F254 plate (Merck), which was activated before use by drying in an oven at 100 °C for 1 h. The plate was placed in a glass chamber and the glycerides were separated using a pre-mixed solvent of 70:30:1 v/v of n-hexane/ethyl ether/acetic acid. The plate was air-dried and developed using iodine vapor. The bands were scraped, the glycerides extracted using ethyl acetate, and then filtered before injection into the GC unit.

The peaks from the chromatograms obtained were identified individually or classified in groups using lipid standards composed of mono-, di- and triolein and other pure lipids including OPO and OOP. The peak areas were then converted to weight percentages using calibration curves established with the appropriate standards.

2.5. Reaction and process parameters

The effects of manipulated variables were viewed and compared through the following characteristics or responses: product composition, glyceride distribution, FFA conversion, and selectivity.

Product composition (C_i) and glyceride distribution (C'_i) are related quantities that refer to the ratio of a specific component (m_i) in g contained in 100 g of reference material, which are the total product (m_p) and the total glycerides produced m_{Gtotal}, respectively, as expressed in equations (1) and (2).

$$C_i = (m_i/m_p) \times 100 \tag{1}$$

$$C_i' = (m_i/m_{Gtotal}) \times 100 = (m_i \times 100)/(m_{MG} + m_{DG} + m_{TG}) \tag{2}$$

where m_{MG}, m_{DG}, and m_{TG} are the amounts of MG, DG, and TG, respectively.

The FFA conversion (ξ_{FFA}) in % refers to the amount of FFA in moles converted into different products referenced to 100 moles of the total FFA ($n_{FFAtotal} = n_{FFA0} + n_{FFAadded}$) in the mixture, either present at the beginning (n_{FAA0}) or added in between process steps ($n_{FFAadded}$).

$$\xi_{FFA} = [(n_{FFAtotal} - n_{FFA1})/n_{FFAtotal}] \times 100 \tag{3}$$

The selectivity (S_i) is the ratio of the amount of the desired component in the product over that of the undesired components or by-products. For example, DG selectivity (S_{DG}) is expressed as:

$$S_{DG} = m_{DG}/(m_{MG} + m_{TG}) \tag{4}$$

Reliability of the results was ensured by performing the trials at least in duplicates.

3. Results and Discussion

Bottom-up approaches in synthesizing DG [19,21] and STG [20] were previously investigated where a set of conditions for operating a sequential or two-step esterification process was identified (table 1). Two different methods of synthesizing the precursors were initially explored, namely a non-catalyzed high-temperature, nitrogen-purged esterification of OA and glycerol (process A) and a lipase-catalyzed, moderate-temperature version (process B). Both were solvent-free, which were appropriate in the functional food application that was originally projected for their use. The lipase-catalyzed product had a slightly higher DG concentration but attained a lower FFA conversion. The non-catalyzed process, on the other hand, was faster in achieving the maximum DG selectivity but had by-products that contributed to a dark-colored product that was minimized with nitrogen purging. Process A, perceived to be simpler in operation, was therefore adopted as the first step in process C, with process B reserved as a quicker alternative when the need arises.

Table 1. The process conditions and product characteristics of the solvent-free esterification of oleic acid, palmitic acid, and glycerol to synthesize precursors and C52-triglycerides.

Process		*Reaction Composition & Characteristics*	*Operating Conditions*	*FFA Conversion, Product Composition, RME* [a]	*Reference*
Code	*Description*				
A	Non-catalyzed DG synthesis	2:1 (OA:G)	175 °C, 0.1 MPa, 200 rpm, 4 h, with N_2 purging	ξ_{FFA}: 79.5 % C_{DG}: 42.6 g/100 g *1,3-DG*: -nr- *RME*: -nr-	[21]
B	Lipase-catalyzed DG synthesis	2:1 (Overall OA:G), 4.0 wt.% Novozyme 435	40 °C, 0.1 MPa, 150 rpm, 48 h, with glycerol dosing	ξ_{FFA}: 69.5 % C_{DG}: 46.9 g/100 g *1,3-DG*: 70.69 % *RME*: 31.8 g/100 g	[19]
C	2-Step Lipase-catalyzed STG synthesis	2:1:1 (Overall OA:PA:G), 2.88 wt.% Novozyme 435	40→60 °C, 0.1 MPa, 150 rpm, 72 h, with glycerol dosing	ξ_{FFA}: 76.7 % C_{TG}: 58.8 g/100 g *sn-2 PA*: 44.4 % *RME*: 30.6 g/100 g	[20]

[a] -nr- = not reported; C_i [=] g DG or TG per 100 g of glycerol-free product; *1,3-DG* and *sn-2 PA* are both expressed in percentage relative to the total DG and sn-2 FFA, respectively; *RME* [=] g 1,3-DG or OPO per 100 g reactants.

The established baseline process (process C) consists of two esterification steps, with low-temperature esterification at 40 °C to produce DG, and a high-temperature step for STG. Glycerol dosing in the first step ensured high selectivity when initially operated at a high free fatty acid to glycerol molar ratio. The best condition for DG production is at an OA:G of 4.4, 4 wt.% Novozyme 435, with glycerol dosing at 24 h and 36 h carried out at 40 °C until 48 h, which resulted in a DG selectivity of 1.12, and 70.69 % 1,3-DG in the DG fraction of the product. If the precursors were the desired product, consequent purification of the crude DG product could yield a DG-rich fraction with > 89 % purity and > 36 % recovery. The operation of the second step at 60 °C, a PA:G of 1:1 for 24 h with the crude DG intermediates produced a substantial amount of structured lipids containing ~50 % TG of the total glycerides produced and approximately 19 % of which is OPO (~38 % in TG) [20]. As the process was easy to manipulate in terms of the addition of a second FFA to incorporate in the precursors produced or the sequence of the FFA addition, it was used to investigate the synthesis of other symmetrical STG, especially MLCT.

3.1. Effect of Free Fatty Acid to Glycerol Ratio and Fatty Acid Length on Diglyceride Distribution and Selectivity

The esterification of a medium chain FFA, specifically CPA, and glycerol was performed at the same conditions where high DG selectivity was observed for OA and glycerol to establish whether fatty acid length would have significant effects on product composition. Comparing the two systems, it can be observed that the initial rates of formation of glycerides, especially of DG, were higher with CPA than OA (figure 1). Also, the maximum distribution of each component in the CPA system was generally larger and was achieved faster than the OA system within the period investigated as follows: 12.41±0.01 g MG (4 h), 32.09 ± 0.21 g DG (12 h), and 31.42 ± 0.43 g TG (48 h) for the former, and 12.10 ± 0.80 g MG (36 h), 25.80±0.02 g DG (48 h), and 19.59±0.00 g TG (96 h) for the latter, all based on 100 g of glycerol-free product. This also goes for DG selectivity where a maximum of 1.92 ± 0.01 g DG/g (MG+TG) was reached within 8 h for CPA and 1.10±0.03 g DG/g (MG+TG) within 36 h for OA.

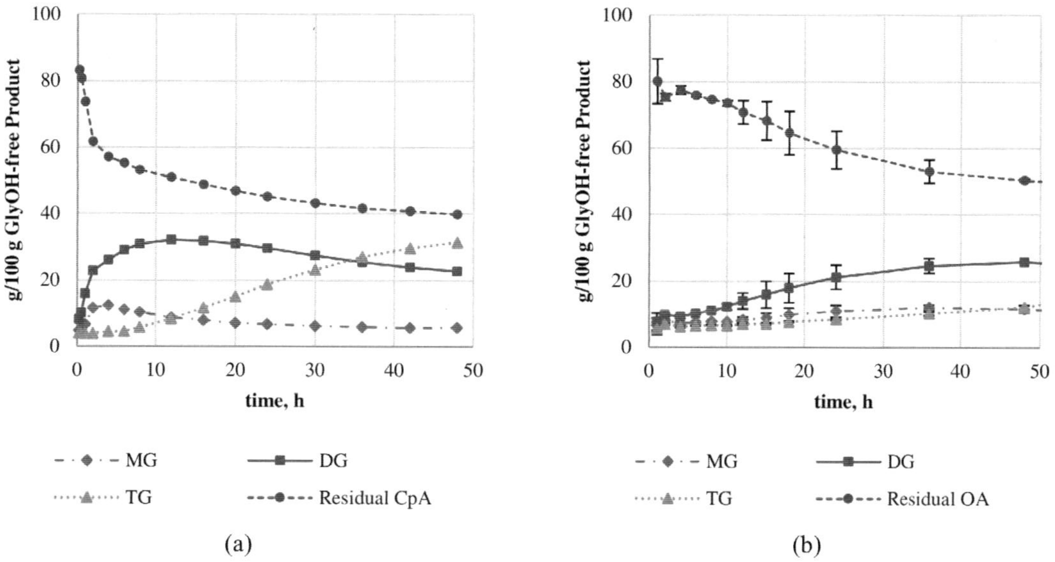

Figure 1. The product composition during the esterification of (a) capric acid and glycerol at CPA:G of 4.4 and (b) oleic acid and glycerol at OA:G of 4.4 catalyzed by 4 g Novozyme 435/100 g reactants carried out at 40 °C.

The esterification reactions were further performed at 40 °C with different CPA:G to gain further information regarding its influence on product distribution and DG (and PG) selectivity. The changes in these responses with time are illustrated in figure 2 and the values for CPA conversion and glyceride distribution at maximum S_{DG} are listed in table 2.

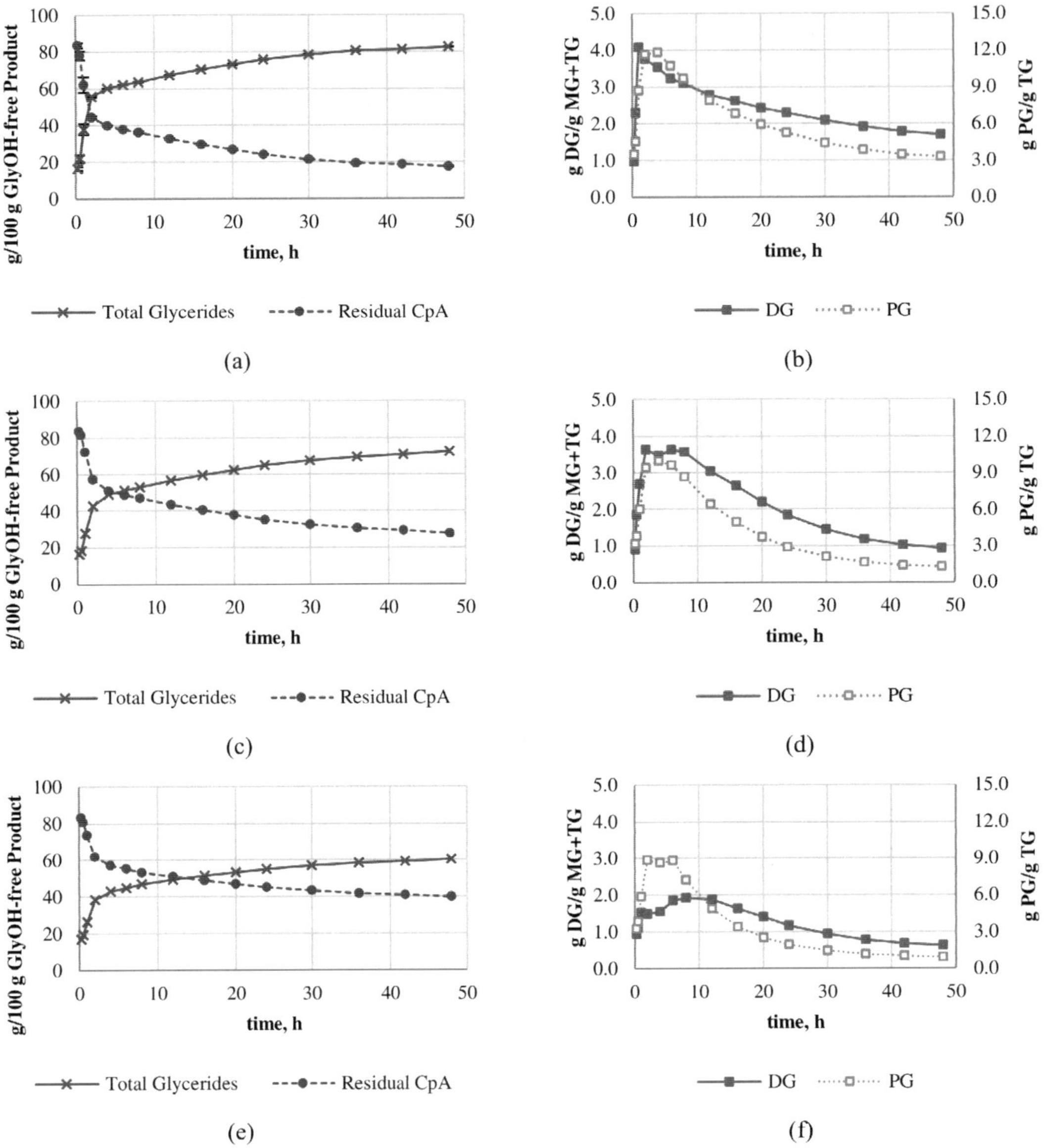

Figure 2. The product composition and glyceride selectivity during the esterification of capric acid and glycerol at an initial CPA:G of 2.0 (a & b), 3.2 (c & d), and 4.4 (e & f) catalyzed by 4 g Novozyme 435/100 g reactants carried out at 40 °C.

Table 2. The glyceride distribution and conversions at maximum DG selectivity during the esterification of capric acid and glycerol at different CPA:G[a].

CPA:G, mol/mol	*Reaction Time, h*	*FFA Conversion, %*	*Glyceride Distribution, g/100 g Total Glycerides*			$S_{DG}(S_{PG})^b$
			C'_{MG}	C'_{DG}	C'_{TG}	
2.0	1	33.41	27.19	62.37	10.44	4.08 (8.70)
3.2	2	37.89	30.71	59.64	9.65	3.63 (9.38)
4.4	8	42.35	22.04	65.77	12.19	1.92 (7.21)

[a] Reaction carried out at 40 °C; [b] S_{DG} is in g DG/g MG+TG and S_{PG} is in g PG/g TG.

From figure 2, it is observed that the lower the CPA:G, the faster the maximum DG selectivity was reached, but at lower conversions of both CPA and glycerol. Also, the point at which maximum DG selectivity was achieved did not coincide with the time to reach maximum PG selectivity, which became shorter but maintained within a wider time window as CPA:G was increased. If the maximum S_{DG} value at CPA:G of 4.4 was referred, then the time window where $S_{DG} \geq 1.92$ g DG/g (MG+TG) increased as CPA:G was decreased, which was about 6 h at a CPA:G of 4.4, 23 h at 3.2, and 35 h at 2.0. From table 2, the maximum S_{DG} at all CPA:G was attained in less than 8 h, much shorter than that for OA:G of 4.4 at 40 °C which is 18 h. This would consequently make the synthesis of MLCT based on CPA and OA faster if the first step was the production of 1,3-dicaprin followed by esterification with OA. The composition of the final product, however, is anticipated to be different if the sequence were reversed, as will be discussed in the following section.

Considering all reaction responses tabulated in table 2, different trends with the response values were observed when a change in CPA:G was implemented. An increase in CPA conversion while a decrease in DG selectivity occurred when the CPA:G was increased. Maximum values for both C'_{MG} and S_{PG} but with corresponding minimum values for both C'_{DG} and C'_{TG} were obtained at a CPA:G of 3.2. Thus, it is obvious that reaction time would differ if the desired response is changed. For example, if CPA conversion was prioritized, then the esterification is preferably run for 8 h at a CPA:G of 4.4, instead of 1 h at a CPA:G of 2.0 for maximum S_{DG}. If all responses were considered at the same degree of importance, then the optimum condition will be a compromise of all the values. In this particular study, a CPA:G of 4.4 was implemented in trials where the first step was the synthesis of 1,3-dicaprin because, at this molar ratio, the reactant conversion was highest. It was previously demonstrated with OA and glycerol systems that the presence of excess glycerol would affect STG formation adversely as PG were formed instead of the desired TG after the second esterification step [19,20].

3.2. *Effect of Sequence of Fatty Acid Addition and Free Fatty Acid Length on MCLT Distribution*

Under the conditions tested, it is apparent that Novozyme 435 had a higher affinity towards MCFA over acids with longer chains. This is supported by a study on the preference of different lipases with various FFA and alcohols in esterification reactions performed in organic media. It was reported that *Candida antarctica* lipase B (CALB) had the following order of FFA selectivity in the esterification with glycerol: C8 > C10, C6, C12 > other fatty acids including C18 [23]. This was carried out with tert-butyl methyl ether as the solvent, mixed fatty acids, and glycerol adsorbed in silica gel. It was also disclosed that using a different alcohol or another catalyst for esterification changes the order of selectivity of some fatty acids. A different case involving n-butanol showed no significant differences in the activities of zirconium sulfate as the catalyst with various FFA [24]. A similar finding was reported in non-catalyzed high-temperature esterification of various FFA with glycerol [25]. These observations indicated two important points: firstly, selectivity is a more prominent feature of enzymes over "chemical" catalysts and non-catalyzed reactions; and secondly, certain structured lipids can be formed predominantly over others when a mixture of fatty acids is used as the raw material in enzyme-catalyzed esterification. These will have to be considered in deciding whether a sequential addition of fatty acid is necessary or a mixture of fatty acids is adequate in the synthesis of MLCT via esterification reactions. As the study by Lee & Parkin, [23] did not report on the selective formation of products, such strategies would have to be verified in terms of their effects on product composition and glyceride selectivity. Considering these notions, three sets of experiments were carried out based on the following 2-step MLCT processes: 1,3-diolein formation followed by esterification with CPA (process D, LML-type MLCT); 1,3-diolein formation, followed by esterification with LA (process E, LML-type MLCT); and 1,3-dicaprin formation followed by esterification with OA (process F, MLM-type MLCT). The first two aimed to show the effect of the length of the MCFA, while the first and the third, the effect of the sequence of fatty acid addition on the composition of the MLCT formed.

IOP Conf. Series: Materials Science and Engineering 1195 (2021) 012017 doi:10.1088/1757-899X/1195/1/012017

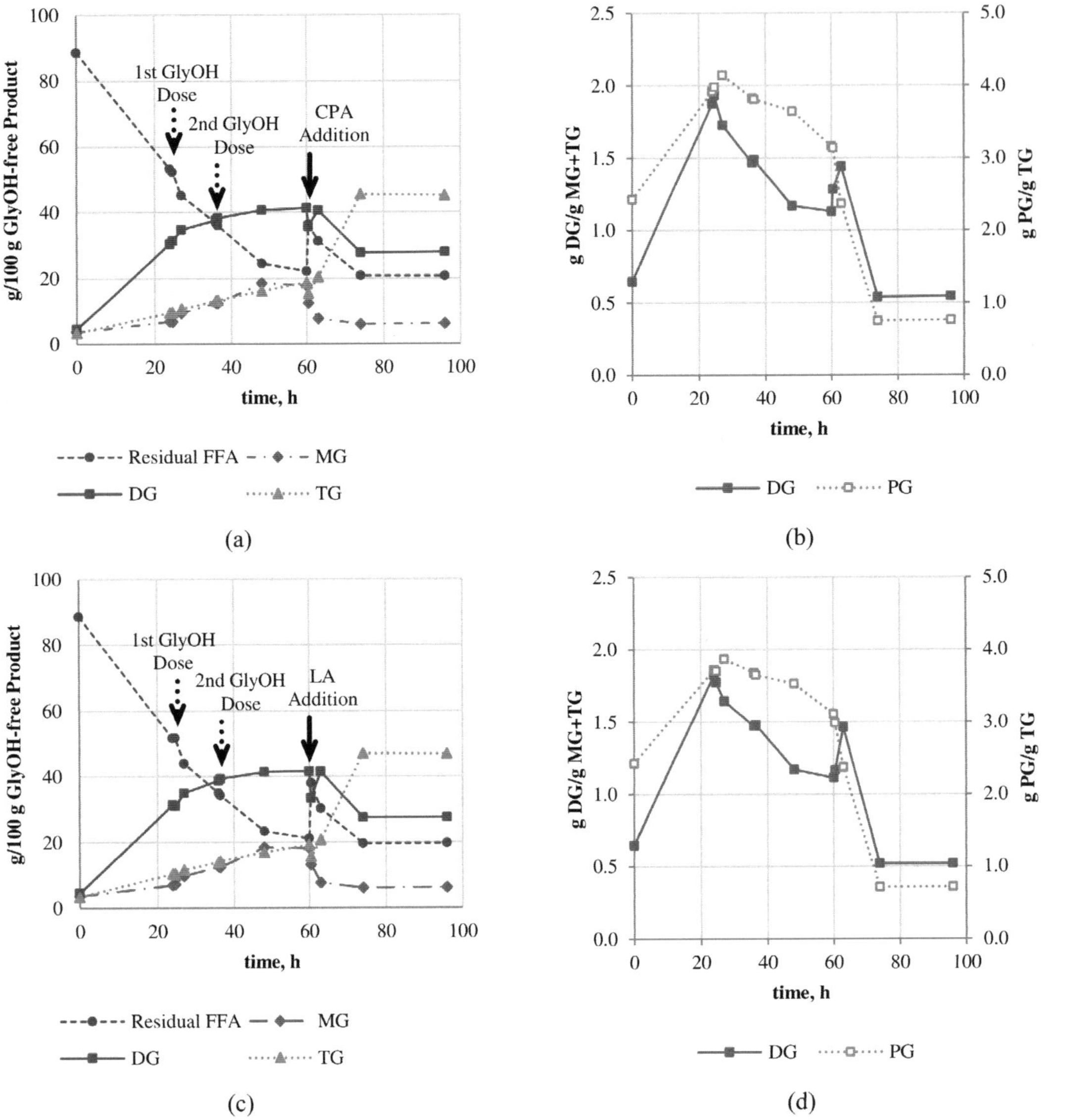

Figure 3. The product composition and glyceride selectivity in the esterification of capric acid (a & b) and lauric acid (c & d) with crude diolein at an overall OA:MCFA:G of 2:1:1 catalyzed by 4 g Novozyme 435/100 g of initial reactants (OA + glycerol) carried out at 40 °C.

The changes in product composition and selectivity in the 2-step MLCT synthesis with CPA and LA as the MCFA are illustrated in figure 3. The figures show similar profiles especially in the first step because they are based on the same strategy. Differences should be visible with the second step, right after the addition of the MCFA, but are in fact, very minimal. The small difference in their chain lengths, C10 for CPA and C12 for LA, may not have strongly affected the reaction systems for their profiles to vary substantially. Moreover, at 72 h of reaction, or 12 h after the addition of either CPA or LA, the product distribution had already reached equilibrium.

The effect may be minimal given that CALB has similar FFA selectivity for CPA and LA when reacted with glycerol as reflected by their α-values of 0.77 and 0.70, respectively [23]. These α-values incorporate the effects of factors that influence reaction rates such as substrate concentration and specificity constants and serve as an index for comparing the affinity of a specific enzyme towards different substrates. It is interesting to note that caprylic acid (CYA, C8:0) was reported to be significantly preferred over the other two fatty acids (i.e. CPA and LA), as well as, over caproic acid (COA, C6:0) when their chain lengths are close to each other [23]. This preference for CYA may be associated with the physiological substrate of the said enzyme. Regarding compositions, minimal differences can be seen with the total glycerides, TG, and LML-type MLCT contents of the two products (table 3), which support the similarity in the selectivity of both acids. The largest difference was obtained with the MLM-type MLCT formed together with the other type, with LA tending to form higher MLM-MLCT as seen by its higher MLM/LML of 0.66 g/g over 0.55 g/g with CPA.

Table 3. The MLM and LML content of crude product from the 2-step esterification process to synthesize MLCT.

Process		*Component*				
Code	*Description*	g Total Glycerides/ 100 g GFP [a]	g TG/ 100 g Total Glycerides	g LML/ 100 g TG	g MLM/ 100 g TG	g MLCT/ 100 g GFP [a]
D	Synthesis of diolein (1st Step) followed by esterification with CPA (2nd Step) for OCPO production	80.27	54.80	36.77	20.07	29.49
E	Synthesis of diolein (1st Step) followed by esterification with LA (2nd Step) for OLO production	80.30	58.35	37.08	24.31	32.94
F	Synthesis of dicaprin (1st Step) followed by esterification with OA (2nd Step) for CPOCP production	67.43	72.32	7.89	36.98	22.85

[a] GFP – glycerol-free product.

The sequence of fatty acid addition in MLCT synthesis had a more prominent effect on product composition as can be seen in the larger differences of product distribution between the first and the third processes. The latter process which started with 1,3-dicaprin formation produced less total glyceride per 100 g glycerol-free product, having a percentage difference of about 22 with the first one although this can be mainly attributed to the difference in molecular weights of OCPO and CPOCP. The proportion of TG in the total glycerides, however, was higher in the third process compared to the first, with a percentage difference close to 28. The most remarkable difference is with the LML-type MLCT produced in the third process resulting in an MLM/LML of 4.69 g/g against 0.55 in the first. These results can be explained by the fact that the enzyme is more selective towards CPA over OA. Because it has a higher affinity towards the MCFA, once the 1,3-dicaprin is formed, it would be more difficult to replace the attached CPA with OA as compared to the reverse sequence where the attached OA in 1,3-diolein would be easier to replace with CPA. This means that following a certain sequence of fatty acid addition, starting with the MCFA followed by the LCFA, ensures a product with significantly higher MLM-type MLCT concentration over the LML-type and a corresponding higher purity with respect to the former type. The reverse sequence, however, is not as effective in producing the desired LML-type MLCT in high purity because the MLM-type MLCT is also produced in the same order of magnitude. Nevertheless, both processes yielded substantial amounts of MCLT of mixed types.

3.3. Prospects of Product Application

In several articles dealing with MLCT production, minimal attention was given to the purity of MLCT produced. Synthesis was often based on interesterification reactions using raw materials that contain mixed lipids such as single-cell oils [26] or microbial oil [27], which are high in PUFA, and blends of oils rich in medium-chain TG [28], that MLCT of pure forms cannot be obtained. The lipid composition in commercial products containing MLCT is also not disclosed and in some cases are essentially blends of MCT added with oils containing long-chain TG or TG with essential fatty acid ligands, e.g. Lipidem® and Smoflipid®. The products obtained from the processes above are of a more predictable composition, and by employing an appropriate separation method, the residual FFA and PG can be removed to yield lipids with MLCT that can maximally reach a concentration of 60 g/100 g TG. Moreover, given that OA can be replaced with essential fatty acids or fatty acids with antioxidant properties, the MLCT products will have diverse functions and broad applications especially in functional food industries.

4. Conclusion

The two-step solvent-free lipase-catalyzed esterification process established to synthesize OPO based on OA, PA, and glycerol is demonstrated to apply well with other FFA, especially with MCFA for the production of MLCT. Replacing PA with CPA or LA in the second step produced MLCT with 29.49 g per 100 g glycerol-free product (36.37 g OCPO per 100 g of triglycerides) and 32.04 g per 100 g glycerol-free product (37.08 g OLO per 100 g of triglycerides), respectively. The sequence of fatty acid addition, particularly in MLCT production, significantly influences the product composition, effectively producing more MLM-type MLCT when the MCFA is reacted first during the low temperature esterification step followed by the LCFA. This produced 22.85 g per 100 g glycerol-free product (36.98 g CPOCP per 100 g triglycerides). The reverse sequence, which originally was the case in this study, is not as effective in producing purer LML-type MLCT as comparable amounts of MLM are simultaneously synthesized. The best feature of the process is that modifications can be implemented easily, which can consequently improve product composition and process productivity.

Acknowledgments

This work was supported by grants from Taiwan Ministry of Science and Technology (MOST 108-2221-E-011-106), Taiwan Building Technology Center, and from the National Taiwan University of Science and Technology for the teaching and research start-up awarded to A.W.G (Grant/Project Number – 1090410305) provided for 2019-2021 in the acquisition of materials used in the study. Authors would also like to thank the Ministry of Science and Technology, Taiwan, for the research support provided through the grant MOST 108-2218-E-011-032-MY3 to join the conference.

References

[1] Guo Y, Cai Z, Xie Y, Ma A, Zhang H, Rao P and Wang Q 2020 Synthesis, physicochemical properties, and health aspects of structured lipids: A review *Compr. Rev. Food Sci. Food Saf.* **19** 759–800
[2] Kim B H and Akoh C C 2015 Recent research trends on the enzymatic synthesis of structured lipids *J. Food Sci.* **80** C1713–24
[3] Naik B and Kumar V 2014 Cocoa butter and its alternatives: A review *J. Bioresour. Eng. Technol.* **1** 7–17
[4] Biswas N, Cheow Y L, Tan C P and Siow LF 2018 Physicochemical properties of enzymatically produced palm-oil-based cocoa butter substitute (cbs) with cocoa butter mixture *Eur. J. Lipid Sci. Technol.* **120** 1–9
[5] Wei W, Jin Q and Wang X 2019 Human milk fat substitutes: past achievements and current trends *Prog. Lipid Res.* **74** 69–86
[6] Utama Q D, Sitanggang A B, Adawiyah D R and Hariyadi P 2019 Lipase-catalyzed interesterification for the synthesis of medium-long-medium (mlm) structured lipids - a review *Food Technol. Biotechnol.* **57** 305–18
[7] Gudmundsdottir A V, Hansen K A, Magnusson CD and Haraldsson GG 2015 Synthesis of reversed structured triacylglycerols possessing epa and dha at their terminal positions *Tetrahedron* **71** 8544–50
[8] Barroso L, Viegas C, Vieira J, Ferreira-Pêgo C, Costa J and Fonte P 2021 Lipid-based carriers

for food ingredients delivery *J. Food Eng.* **295** 110451
[9] Paliwal R, Paliwal S R, Kenwat R, Kurmi B Das and Sahu MK 2020 Solid lipid nanoparticles: a review on recent perspectives and patents *Expert Opin. Ther. Pat.* **30** 179–94
[10] Sivakanthan S and Madhujith T 2020 Current trends in applications of enzymatic interesterification of fats and oils: a review *LWT* **132** 109880
[11] Akil E, Pereira A da S, El-Bacha T, Amaral PFF and Torres AG 2020 Efficient production of bioactive structured lipids by fast acidolysis catalyzed by yarrowia lipolytica lipase, free and immobilized in chitosan-alginate beads, in solvent-free medium *Int. J. Biol. Macromol.* **163** 910–8
[12] Sun S and Hu B 2017 Enzymatic preparation of novel caffeoyl structured lipids using monoacylglycerols as caffeoyl acceptor and transesterification mechanism *Biochem. Eng. J.* **124** 78–87
[13] Morales-Medina R, Munio M, Guadix A and Guadix EM 2017 Development of an up-grading process to produce MLM structured lipids from sardine discards *Food Chem.* **228** 634–42
[14] Sánchez D A, Tonetto G M and Ferreira ML 2020 Valorization of glycerol through the enzymatic synthesis of acylglycerides with high nutritional value *Catalysts* **10** 116
[15] He Y, Li J, Guo Z and Chen B 2018 Synthesis of novel medium-long-medium type structured lipids from microalgae oil via two-step enzymatic reactions *Process Biochem.* **68** 108–16
[16] More S B, Waghmare J T and Gogate PR 2017 Ultrasound pretreatment as a novel approach for intensification of lipase catalyzed esterification of tricaprylin *Ultrason. Sonochem.* **36** 253–61
[17] Duan Z Q, Du W and Liu D H 2012 Rational synthesis of 1,3-diolein by enzymatic esterification *J. Biotechnol.* **159** 44–9
[18] Duan Z Q, Du W and Liu D H 2013 Improved synthesis of 1,3-diolein by novozym 435-mediated esterification of monoolein with oleic acid *J. Mol. Catal.* B Enzym. **89** 1–5
[19] Agapay R C, Ju Y H, Tran-Nguyen P L, Angkawijaya A E, Santoso S P and Go A W 2020 Synthesizing precursors for functional food structured lipids from soybean oil deodorized distillates *Waste Biomass Valori*
[20] Agapay R C, Ju Y H, Tran-Nguyen P L, Ismadji S, Angkawijaya A E and Go AW 2021 Process evaluation of solvent-free lipase-catalyzed esterification schemes in the synthesis of structured triglycerides from oleic and palmitic acids *Asia-Pacific J. Chem. Eng.* **16** 1–12
[21] Ting R R, Agapay R, Angkawijaya A E, Tran Nguyen P L, Truong C T and Ju Y 2019 Diglyceride production via noncatalyzed esterification of glycerol and oleic acid *Asia-Pacific J. Chem. Eng.* **14** 1–11
[22] Touchstone J C 1995 Thin-layer chromatographic procedures for lipid separation *J. Chromatogr.* B Biomed. Sci. Appl. **671** 169–95
[23] Lee C and Parkin K L 2000 Comparative fatty acid selectivity of lipases in esterification reactions with glycerol and diol analogues in organic media *Biotechnol. Prog.* **16** 372–7
[24] Juan J C, Zhang J and Yarmo M A 2008 Efficient esterification of fatty acids with alcohols catalyzed by zr(so4)2·4h2o under solvent-free condition *Catal. Letters* **126** 319–24
[25] Hartman L 1966 Esterification rates of some saturated and unsaturated fatty acids with glycerol *J. Am. Oil Chem. Soc.* **43** 536–8
[26] Korma S A, Zou X, Ali AH, Abed SM, Jin Q and Wang X 2017 Preparation of structured lipids enriched with medium- and long-chain triacylglycerols by enzymatic interesterification for infant formula *Food Bioprod. Process* **107** 121–30
[27] Abed S M, Wei W, Ali A H, Korma S A, Mousa A H, Hassan H M, Jin Q and Wang X 2018 Synthesis of structured lipids enriched with medium-chain fatty acids via solvent-free acidolysis of microbial oil catalyzed by rhizomucor miehei lipase *LWT* **93** 306–15
[28] Aftar Mizan A B, Ayob M K, Ma'aruf A G, Mohamad Yusof M and Izzreen I 2016 Optimization of the synthesis of structured lipids using medium chain fatty acids through acidolysis *Am. J. Agric. Environ. Sci.* **16** 698–706

Comparative laboratory cost analysis of various activated carbon activation process

J Y Lai and L H Ngu

Faculty of Engineering, Computing and Science, Swinburne University of Technology, Jalan Simpang Tiga, 93350 Kuching, Sarawak, Malaysia

Email: lngu@swinburne.edu.my

Abstract. Activated carbon (AC) is an established adsorbent for organic pollutants reduction, metal removal, and liquid and gas adsorption. Cost analysis corresponds to determining the best approach for AC production depending on activation techniques with different degrees of activation is still minimal in literature. A cost estimation of AC production in laboratory scale using different conventional activation and post-activation surface modification process is performed in this study. This study attempts to develop a cost-friendly selection of activation process from laboratory scale prices. Chemicals and utility costs were acquired from vendor quotes (i.e., Sigma-Aldrich and Fisher Scientific) and Sarawak industrial electricity tariffs based on 100 g production. Oil palm-based ACs produced from five different activation or surface modification methods were compared to ascertain the least expensive production approach in terms of estimated production cost. Of the five methods investigated, method that quoted the least expensive production cost is chemical activation using potassium hydroxide (KOH) with minimum estimated cost of \$7.30 whereas the most expensive production cost involves surface modification by polyethyleneimine (PEI) impregnation with cost of \$873.00. Therefore, the estimated production cost for KOH activation is the minimum at \$0.073 g^{-1} while the maximum is \$8.73 g^{-1} for PEI impregnation.

Keywords: Oil palm-based activated carbon (AC); Laboratory scale; Activated carbon production; Production cost; Cost-friendly activation process

1. Introduction

Activated carbon (AC) is an established adsorbent for many different sectors of applications such as reduction of organic pollutants in water, metal removal, and liquid and gas adsorption [1–8]. Recently, oil palm has gain prominence as starting precursor for AC synthesis because it is abundantly available due to its substantial waste generated after harvesting, replanting, and crude oil production. It comprises of prominent compositions such as high carbon, high cellulose, high lignin, high nitrogen content, significant volatile material amount, and low ash content for further porous structure development [9–18]. Oil palm derived AC is extensively used in the applications of volatile organic compounds (VOC) removal, heavy metal reduction, wastewater treatment, flue gas purification, and the more recent carbon dioxide (CO_2) adsorption [6, 19–27].

ACs are synthesized through three types of activation process which are physical, chemical, and physiochemical activation. Physical activation is a method that uses an oxidizing gas (i.e., CO_2, steam, or a binary mixture of gases) to activate the carbon enriched material (char) under elevated temperature (800-1100 °C) after the raw precursor is carbonized at mild temperature (400-600 °C) in an inert nitrogen (N_2) atmosphere [16, 19, 28–31]. In chemical activation, the raw precursor undergoes a chemical impregnation by soaking in a chemical solution which can be prepared from different medium (i.e., acidic (phosphoric acid (H_3PO_4) and sulphuric acid (H_2SO_4)), alkaline (potassium hydroxide (KOH) and sodium hydroxide (NaOH)), and transition metal salt, zinc chloride ($ZnCl_2$)), prior to the simultaneous carbonization and activation at temperature of 400-800 °C [32–37]. Physiochemical activation is a multi-sequences process which combines both chemical and physical activation to initially impregnate the raw precursor in the chemical solution, followed by high temperature (800-1000 °C) gasification by the oxidizing gas [19, 38–45].

In some works [46–52], to improve the surface chemistry of the resulting AC, the AC then further undergoes surface modification by nitrogen incorporation (i.e., ammonia (NH_3) treatment or amine incorporation) or metal salts functionalization, subsequently after the activation process. For NH_3 treatment, the AC is subjected to thermal treatment with the presence of NH_3 gas at an elevated temperature of 400-800 °C [49, 50]. The latter amine functionalization method is initially impregnating the AC in an amine-containing chemical solvent, followed by a mild temperature drying process [25, 48, 53]. In the metal salts functionalization method, the AC is soaked in a metal-containing chemical solvent, prior to high temperature heat treatment at 800 °C [46, 54].

There are some publications on the process scale-up for the AC manufacturing, leading to an estimation of pilot scale production costs. The scale-up of the process helps to identify and estimate the production cost for manufacturing the ACs at a commercial scale, which is used to determine their marketability. Among these are Toles *et al.* [55] which investigated the physical, chemical, and adsorption properties of a series of unoxidized/oxidized, steam or CO_2-activated almond shell carbons via six different activation or activation/oxidation conditions for each series and their estimated production cost, followed by a comparison with commercial carbons to determine best physical activation method and the least expensive for manufacturing the almond shell derived GAC. Lima *et al.* [56] developed an estimation of conceptual capital and operating cost using Superpro Designer process simulation.

A series of H_3PO_4-activated almond shell carbons prepared from six different conditions of activation or activation/oxidation methods were investigated and compared to two commercial carbons by Toles *et al.* [57] to ascertain their relative value in terms of yield, physical, chemical, and adsorption properties, as well as the estimated production cost based on scale-up of processes to determine their commercial viability. Ng *et al.* [58] evaluated a cost estimation for manufacturing pecan shell-based carbons using steam or H_3PO_4 activation at a large-scale production, based on the developed process flow diagrams. Influence of different types of activating agent (i.e., H_3PO_4 and KOH), impregnation ratios, and carbonization temperatures on the properties of the synthesized AC which prepared from Persian Ironwood biomass, their CO_2 adsorption performance and the capital cost estimation of the final production were investigated by Nowrouzi *et al.* [59]. Other than the above-mentioned research works, comprehensive research work of the estimation of production cost for producing AC from oil palm biomass is not yet available.

Therefore, this present study aims to evaluate the production cost at laboratory scale of oil palm-based AC prepared from five different activation or post-activation surface modification methods. A comparison is made between these five activation or surface modification methods to determine the least expensive process to produce the oil palm-based ACs at the same basis of yield (100 g).

2. Methodology

Costs for the five types of activation and post-activation surface modification methods are developed taking on a basis of 100 g production of oil palm-based AC, based on the process conditions presented in earlier research works. A comparative assessment was performed to identify and select the least expensive process from each activation/modification method that is presented with the lowest production/modification cost.

2.1. Materials

Raw oil palm wastes (i.e., oil palm shell, oil palm empty fruit bunch, oil palm frond, oil palm trunk, oil palm fiber, and oil palm seed) available from a local oil palm plantation and palm oil mill as the precursor. Carbon dioxide (CO_2, 0.01-99.80%, Fisher Scientific) and steam (H_2O, Fisher Scientific) are the oxidizing gas for preparation of AC *via* physical and physiochemical activation. Dehydrating agent such as phosphoric acid (H_3PO_4, ≥ 85 wt. % in water, Sigma-Aldrich), sulphuric acid (H_2SO_4, 95.0-98.0%, Sigma-Aldrich), zinc chloride ($ZnCl_2$, ≥ 98%, Sigma-Aldrich), potassium hydroxide (KOH, 0.1 N or ≥ 85%, Sigma-Aldrich), potassium carbonate (K_2CO_3, ≥ 99.0%, Sigma-Aldrich), and sodium hydroxide (NaOH, 0.1 N, 50% in water, 97%, or ≥ 98%, Sigma-Aldrich) are the activating agent for AC preparation by chemical and physiochemical activation. The modifying agent are polyethyleneimine (PEI, 50 wt% in water, Sigma-Aldrich), 2-amino-2-methyl-1-propanol (AMP, 90% with ~5% water, Sigma-Aldrich), 2-amino-2-methyl-1,3-propanediol (AMPD, ≥ 99%, Sigma-Aldrich), 2-(methylamino)ethanol (MMEA, ≥ 98%, Sigma-Aldrich), mono-ethanolamine (MEA, ≥ 99.0%, Sigma-Aldrich), diethanolamine (DEA, ≥ 98.0%, Sigma-Aldrich), ethylenediamine (EDA, ≥ 99.0%, Sigma-Aldrich), diethylenetriamine (DETA, 99%, Sigma-Aldrich), nickel nitrate hexahydrate ($Ni(NO_3)_2{\cdot}6H_2O$, Sigma-Aldrich), ammonia metavanadate (NH_4VO_3, 99%, Sigma-Aldrich), ferum nitrate ($Fe(NO_3)_3{\cdot}9H_2O$, ≥ 98%, Sigma-Aldrich), cerium nitrate ($Ce(NO_3)_3{\cdot}6H_2O$, 99%, Sigma-Aldrich), barium oxide (BaO, 97%, Sigma-Aldrich), magnesium oxide (MgO, 97%, Sigma-Aldrich), copper oxide (CuO, ≥ 99.0%, Sigma-Aldrich), titanium oxide (TiO_2, ≥ 99.0%, Sigma-Aldrich), and cerium oxide (CeO_2, 90% or ≥ 99.0%, Sigma-Aldrich) for surface chemistry modification of the AC. The commercially available methanol (CH_3OH, 99.80%, Sigma-Aldrich) and toluene (C_7H_8, 99.80%, Sigma-Aldrich) as solvent for the modifying agent.

2.2. Activation or post-activation surface modification methods

The procedure and process of oil palm wastes ACs through activation or post-activation surface modification are described by earlier research works [5, 6, 8, 19, 23, 25, 29, 41, 46–48, 52–54, 60–72]. These works shown that AC are produced *via* physical activation using CO_2 and steam, chemical activation using H_3PO_4, H_2SO_4, $ZnCl_2$, KOH, K_2CO_3, and NaOH or physiochemical activation through the combination of chemical and physical activation, using either one-stage or two-stage process by furnace-heating or microwave-heating. The one-stage process carried out simultaneously carbonization and activation process at temperature of 500-1000 °C whereas two-stage process initially carbonized the starting precursor at temperature ranging between 100 °C and 1200 °C to produce carbon-enriched material (char), followed by heat treatment at temperature of 500-1100 °C. The final activated materials then undergo subsequent surface modification through amine (i.e., PEI solution, AMP, AMPD, MMEA, MEA, DEA, EDA, and DETA) or metal (i.e., $Ni(NO_3)_2{\cdot}6H_2O$, NH_4VO_3, $Fe(NO_3)_3{\cdot}9H_2O$, $Ce(NO_3)_3{\cdot}6H_2O$, BaO, MgO, CuO, TiO_2, and CeO_2) impregnation to alter and improve their surface functional groups. The various activation and subsequent surface modification processes for the oil palm-based AC are illustrated in figure 1.

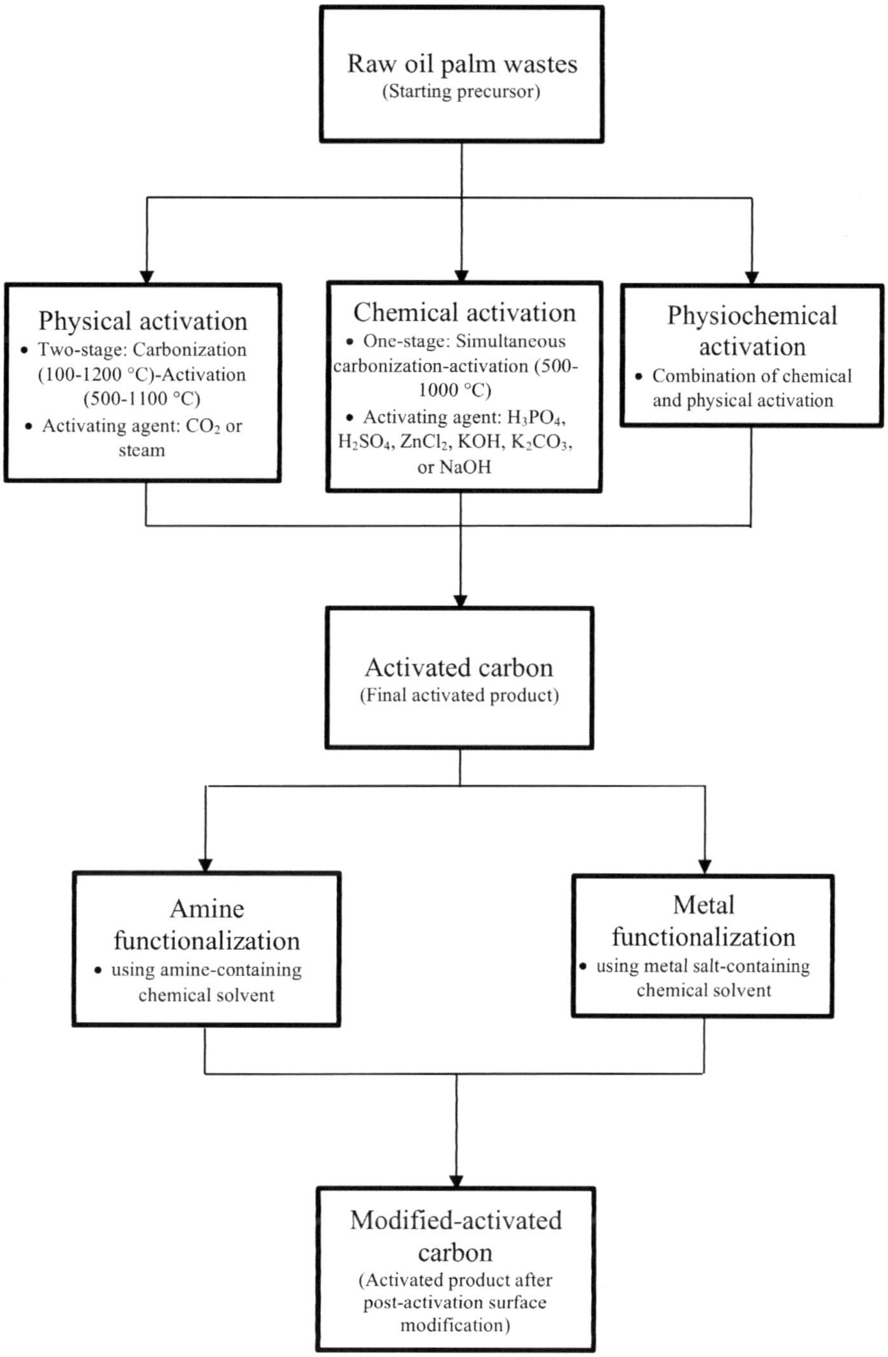

Figure 1. Schematic diagram of a complete activation and subsequent surface modification procedure for oil palm-based AC.

2.3. Determination of cost-effective activation process
Preparation conditions and activities performed in earlier research works [5, 6, 8, 19, 23, 25, 29, 41, 46–48, 52–54, 60–72] without reference to laboratory scale activation carbon production cost estimation method are referred in the evaluation for oil palm-derived AC production cost. The estimation of cost is evaluated for the activation of char after carbonization or starting precursor through physical (CO_2 and

steam), chemical (H_3PO_4, H_2SO_4, $ZnCl_2$, KOH, K_2CO_3, and NaOH), and physiochemical activation, as well as post-activation surface modification of final activated materials by nitrogen incorporation using amine-based chemical solvent and metal salts impregnation. The laboratory scale production cost estimates were developed on a basis of 100 g ACs produced from each activation/surface modification method. Consumable pricing came from budgetary consumable quotations from vendors such as Sigma-Aldrich and Fisher Scientific. Utility requirements were calculated from local electricity provider (Sarawak Energy Berhad) electricity tariffs for industrial. The power consumptions for operating equipment: orbital shaker, drying oven, Soxhlet extraction, electric furnace, and microwave oven used for the production processes for agitation, drying, solvent reflux, and activation are 5.7, 2400, 200, 1000, and 1400 W, respectively. Each activation or post-activation surface modification process estimated production cost was then compared to select the more cost-friendly method. There are 14 methods of activation and post-activation surface modification investigated in this study with the respective commodity chemical (oxidizing gas), chemical agent or modifying agent used and operational time of each process presented in table 1.

IOP Conf. Series: Materials Science and Engineering 1195 (2021) 012018 doi:10.1088/1757-899X/1195/1/012018

Table 1. Activation and post-activation surface modification methods process parameters for oil palm waste-based ACs production used to estimate the laboratory scale production cost (Reproduced from [5, 6, 8, 19, 23, 25, 29, 41, 46–48, 52–54, 60–70]).

Method	Heat treatment conditions			Pre-activation activity	Post-activation treatment				
	Carbonization time (h)	Gas/ Chemical	Activation time (h)	Drying time (h)	Agitation time (h)	Agitation with heating (h)	Drying time (h)	Solvent reflux (h)	Thermal treatment time (h)
(I) Physical activation	0.5-3	CO_2	1-3	NA[a]	NA[a]	NA[a]	NA[a]	NA[a]	NA[a]
	0.5-2	Steam	0.5-7	NA[a]	NA[a]	NA[a]	NA[a]	NA[a]	NA[a]
(II) Chemical activation	NA[a]	H_3PO_4	2	24	NA[a]	NA[a]	NA[a]	NA[a]	NA[a]
	NA[a]	H_2SO_4	2	24	NA[a]	NA[a]	NA[a]	NA[a]	NA[a]
	NA[a]	$ZnCl_2$	1-2	24	NA[a]	NA[a]	NA[a]	NA[a]	NA[a]
	0.33-5	KOH	0.083-2	24	NA[a]	NA[a]	NA[a]	NA[a]	NA[a]
	NA[a]	K_2CO_3	1-2	24	NA[a]	NA[a]	NA[a]	NA[a]	NA[a]
	0.5	NaOH	1	24	NA[a]	NA[a]	NA[a]	NA[a]	NA[a]
(III) Physiochemical activation	2	H_3PO_4 and CO_2	1-7	24	NA[a]	NA[a]	NA[a]	NA[a]	NA[a]
	1	H_2SO_4 and CO_2	1	24	NA[a]	NA[a]	NA[a]	NA[a]	NA[a]
	2	$ZnCl_2$ and CO_2	1-7	24	NA[a]	NA[a]	NA[a]	NA[a]	NA[a]
	2-5	KOH and CO_2	1-2	24	NA[a]	NA[a]	NA[a]	NA[a]	NA[a]
	1-2	NaOH and CO_2	1.5-3	24	NA[a]	NA[a]	NA[a]	NA[a]	NA[a]

Method	Heat treatment conditions			Pre-activation activity	Post-activation treatment				
	Carbonization time (h)	Gas/ Chemical	Activation time (h)	Drying time (h)	Agitation time (h)	Agitation with heating (h)	Drying time (h)	Solvent reflux (h)	Thermal treatment time (h)
(IV) Post-activation surface modification by amine incorporation	NA[a]	PEI solution	NA[a]	NA[a]	72	NA[a]	24	NA[a]	NA[a]
	NA[a]	AMPD	NA[a]	NA[a]	24	NA[a]	24	NA[a]	NA[a]
	NA[a]	AMP	NA[a]	NA[a]	24	NA[a]	24	NA[a]	NA[a]
	NA[a]	MMEA	NA[a]	NA[a]	24	NA[a]	24	NA[a]	NA[a]
	NA[a]	MEA	NA[a]	NA[a]	24	NA[a]	24	NA[a]	NA[a]
	NA[a]	MEA in methanol	NA[a]	NA[a]	24	NA[a]	24	NA[a]	NA[a]
	NA[a]	DEA in Methanol	NA[a]	NA[a]	24	NA[a]	24	NA[a]	NA[a]
	NA[a]	EDA solution in toluene	NA[a]	NA[a]	24	NA[a]	24	4	NA[a]
	NA[a]	DETA solution in toluene	NA[a]	NA[a]	24	NA[a]	24	4	NA[a]
	NA[a]	Pure EDA	NA[a]	NA[a]	24	NA[a]	24	4	NA[a]
	NA[a]	Pure DETA	NA[a]	NA[a]	24	NA[a]	24	4	NA[a]

Method	Heat treatment conditions			Pre-activation activity	Post-activation treatment				
	Carbonization time (h)	Gas/ Chemical	Activation time (h)	Drying time (h)	Agitation time (h)	Agitation with heating (h)	Drying time (h)	Solvent reflux (h)	Thermal treatment time (h)
(V) Post-activation surface modification by metal impregnation	NA[a]	Nickel nitrate hexahydrate ($Ni(NO_3)_2 \cdot 6H_2O$)	NA[a]	NA[a]	NA[a]	5.33	12	NA[a]	4
	NA[a]	Ammonia metavanadate (NH_4VO_3)	NA[a]	NA[a]	NA[a]	5.33	12	NA[a]	4
	NA[a]	Ferum nitrate ($Fe(NO_3)_3 \cdot 9H_2O$)	NA[a]	NA[a]	NA[a]	5.33	12	NA[a]	4
	NA[a]	Cerium nitrate ($Ce(NO_3)_3 \cdot 6H_2O$)	NA[a]	NA[a]	NA[a]	5.33	12	NA[a]	4
	NA[a]	Barium oxide (BaO)	NA[a]	NA[a]	NA[a]	NA[a]	24	NA[a]	1
	NA[a]	Magnesium oxide (MgO)	NA[a]	NA[a]	NA[a]	NA[a]	24	NA[a]	1
	NA[a]	Copper oxide (CuO)	NA[a]	NA[a]	NA[a]	NA[a]	24	NA[a]	1
	NA[a]	Titanium oxide (TiO_2)	NA[a]	NA[a]	NA[a]	NA[a]	24	NA[a]	1
	NA[a]	Cerium oxide (CeO_2)	NA[a]	NA[a]	NA[a]	NA[a]	24	NA[a]	1

[a] NA = Not available in the specific activation/post-activation surface modification process.

3. Results and Discussion

3.1. Activation and surface modification process cost analysis

3.1.1. Physical activation of oil palm waste. A cost analysis of physically AC production *via* CO_2 and steam activation using different percent concentration of CO_2 is summarized in table 2. A comparison of estimated production cost among these physically ACs produced using two different types of activating agents were studied.

Table 2. Operating costs for laboratory physical activation of oil palm waste [73–79].

Item	Unit Cost ($)	Amount	Total cost ($)
Consumables			
CO_2 99.80% (48 L)	826.00	3-15 L	**51.60-258.10**[a]
CO_2 0.01% in N_2 (48 L)	1257.20	3-15 L	78.60-392.90
CO_2 0.1% in N_2 (48 L)	1446.40	3-15 L	90.40-452.00
CO_2 0.5% in carbon monoxide (CO) 0.5%, hydrogen (H_2) 0.5%, oxygen (O_2) 0.5% with N_2 balanced (48 L)	1589.50	3-15 L	99.40-496.70
CO_2 1% in CO 1%, H_2 1%, O_2 1% with N_2 balanced (48 L)	1762.50	3-15 L	110.20-550.80
CO_2 15% in CO 7%, O_2 4%, CH_4 4.5% with N_2 balanced (48 L)	1517.90	3-15 L	94.90-474.40
Steam (CO 7%, O_2 4%, CO_2 15%, CH_4 4.5% in N_2, 48 L)	1517.90	4.5-24 L	142.30-759.00
Utilities cost			
Electricity	$0.056 kWh^{-1}	1-11.2 kWh	0.056-0.63

[a] Value in **bold** indicates minimum value for physical activation of AC production cost.

The raw oil palm biomass is available at no cost and utilized as starting materials for AC production. Hence, the purchased price of the raw oil palm biomass is not included in the cost study. Based on table 2, physical activation carried out using CO_2 in pure form with high purity of 99.80% has the minimum estimated production cost range of $51.70-258.50. Compared to CO_2-AC production, production of steam-activated oil palm carbon gave the maximum estimated production cost range of $142.40-759.60 among the physical activation productions. This manifests that steam activation has higher cost for AC production compared to CO_2 activation, given its higher activating agent consumption required in the activation. The estimated product cost is $0.52-2.59 g^{-1} for CO_2-activated oil palm carbon using pure CO_2 whereas $1.42-7.60 g^{-1} for steam-AC.

3.1.2. Chemical activation of oil palm waste. Cost analysis for chemical activation of oil palm waste in laboratory scale using various dehydrating agent (i.e., H_3PO_4, H_2SO_4, $ZnCl_2$, KOH, K_2CO_3, and NaOH) as activating agent is summarized in table 3. Similar to the cost analysis for physical activation of oil palm wastes, the purchased price for the raw oil palm biomass is not included in the cost study given the raw oil palm biomass is obtained from plantation without charge. Similar assignment is taken in the cost study for physiochemical activation of oil palm waste and post-activation surface modification of oil palm-based ACs. Comparison of the estimated production cost among the oil palm-based ACs produced *via* chemical activation is studied.

Table 3. Operating costs for laboratory chemical activation of oil palm waste [79–89].

Item	Unit Cost ($)	Amount	Total cost ($)
Consumables			
$H_3PO_4 \geq 85$ wt.% in water (2.5 L)	153.00	0.20 L	12.20
H_2SO_4 95.0-98.0% (2.5 L)	85.70	0.20 L	6.90
$ZnCl_2 \geq 98\%$ (500 g)	94.00	30 g	5.60
KOH 0.1 M in water (0.1 N), solution (1 L)	37.90	0.20 L	7.60
KOH ≥ 85%, pellets (500 g)	43.20	30 g	**2.60**[a]
$K_2CO_3 \geq$ 99.0% (500 g)	48.80	30 g	2.90
NaOH ≥ 98%, pellets, anhydrous (500 g)	60.70	30 g	3.60
NaOH 97%, powder (500 g)	262.00	30 g	15.70
NaOH 0.1 M in water (0.1 N) (1 L)	38.40	0.20 L	7.70
NaOH 50% in water (1 L)	55.50	0.20 L	11.10
Utilities cost			
Electricity	$0.056 kWh^{-1}	84.4-98.8 kWh	4.70-5.50

[a] Value in **bold** indicates minimum value for chemical activation of AC production cost.

Based on table 3, minimum production cost of $7.30-8.10 was estimated for chemical activation of oil palm wastes using KOH activating agent with high purity of ≥ 85%, given its low chemical price required for activation. While, conversely, chemical activation with NaOH (97% purity) offers the highest production cost ranging from $20.60 to $21.20. High purchased chemical price for NaOH attributes to a slightly larger amount of cost for production. The estimated product cost for KOH-AC with the use of KOH, ≥ 85% purity is \$0.073-0.081 g^{-1} and NaOH-AC by NaOH, 97% purity is estimated up to \$0.21 g^{-1}.

3.1.3. Physiochemical activation of oil palm waste. Physiochemical activation is performed through different combinations of chemical and physical activation. Different combination presents different amount of cost for production. Types of chemical agent, types of oxidizing gas, and their respective purchased price are significant factors that affect the estimation on production cost. Various physiochemical combination and their respective estimated production cost are summarized in table 4 and compared based on their maximum production cost to determine the physiochemical combination which gives minimum production cost.

Based on table 4, physiochemical combination of chemical activation using various chemical agents (i.e., H_3PO_4, H_2SO_4, $ZnCl_2$, KOH or NaOH) with physical activation using CO_2 in gas mixture with high concentration from 0.5% to 15% cost higher with estimation at \$102.60-132.10 compare to combination with low concentration CO_2 (0.01% and 0.1%) in N_2 at \$86.30-112.30 and pure CO_2 (99.80% purity) at \$59.40-73.50. The estimated production cost for physiochemical activation with acidic dehydrating agents (H_3PO_4 and H_2SO_4) and CO_2 in gas mixture with high concentration range from \$106.70 up to \$129.70. Conversely, AC production through physiochemical activation using combination of alkaline chemicals (KOH and NaOH) and CO_2 in gas mixture with high concentration shows an estimated production cost that ranges from \$102.60 up to \$132.10. Activation with both $ZnCl_2$ and high concentration CO_2 in gas mixture requires a production cost of \$105.70-123.10. The high purchased chemical price of NaOH (\$38.40-262.00) is the principal contribution to the highest production cost in the physiochemical combination of alkaline chemicals and high concentration of CO_2 in gas mixture than the other physiochemical combinations. Therefore, the overall range of the estimated cost for physiochemically AC would be \$0.59-1.32 g^{-1}.

Table 4. Summary of operating costs for laboratory physiochemical activation of oil palm biomass [73–89].

Utility	Unit cost ($)	Amount	Total cost ($)
Electricity	$0.056-0.058 kWh^{-1}	88.4-125.4 kWh	5.00-7.30

	Combination	Total production cost ($)				
		Dehydrating agent				
		(1) H_3PO_4	(2) H_2SO_4	(3) $ZnCl_2$	(4) KOH	(5) NaOH
Activating agent	(1) CO_2 99.80% (48 L)	69.00-71.10	63.40-64.00	**62.40-64.50**[a]	59.40-65.40	60.50-73.50
	(2) CO_2 0.01% in N_2 (48 L)	96.00-98.10	90.40-91.00	89.40-91.50	86.30-92.30	87.50-100.50
	(3) CO_2 0.1% in N_2 (48 L)	107.80-109.90	102.20-102.80	101.20-103.30	98.10-104.20	99.30-112.30
	(4) CO_2 0.5% in carbon monoxide (CO) 0.5%, hydrogen (H_2) 0.5%, oxygen (O_2) 0.5% with N_2 balanced (48 L)	116.70-118.90	111.20-111.70	110.10-112.30	107.10-113.10	108.20-121.20
	(5) CO_2 1% in CO 1%, H_2 1%, O_2 1% with N_2 balanced (48 L)	127.50-129.70	122.00-122.50	120.90-123.10	117.90-123.90	119.00-132.10
	(6) CO_2 15% in CO 7%, O_2 4%, CH_4 4.5% with N_2 balanced (48 L)	112.30-114.40	106.70-107.30	105.70-107.80	102.60-108.60	103.80-116.80

[a] Value in **bold** indicates minimum value for physiochemical activation of AC production cost.

3.1.4. Surface modification of oil palm waste-based AC. Surface modification of oil palm waste-based AC is usually performed through the incorporation of amine or metal on the resulting AC obtained from the activation process. A cost study is developed at a given purchased chemical prices of 100 g physically, chemically and physiochemically ACs (table 5), the purchased chemical prices for chemical solvents (methanol (CH_3OH) and toluene (C_7H_8)), amine-functionalizing doping agents (i.e., PEI solution, AMP, AMPD, MMEA, MEA, DEA, EDA, and DETA), and metal-functionalizing doping agents (i.e., $Ni(NO_3)_2{\cdot}6H_2O$, NH_4VO_3, $Fe(NO_3)_3{\cdot}9H_2O$, $Ce(NO_3)_3{\cdot}6H_2O$, BaO, MgO, CuO, TiO_2, and CeO_2) and the utility costs (table 6 and 7) to estimate the modification cost of amine/metal-functionalized ACs. A summary of modification costs for various modification incorporation are tabulated in table 6 and 7 and compared based on their maximum value in the range of total estimated modification cost to determine the least expensive modification process.

Table 5. Summary of chemical costs for laboratory surface modification of oil palm waste-based AC [90–109].

Item	Unit Cost ($)	Amount	Total cost ($)
Consumables			
PEI solution 50 wt% in H_2O (1 L)	200.00	0.50 L	100.00
AMP 90% with ~5% water (1 L)	57.70	0.50 L	28.90
AMPD ≥ 99% (500 g)	259.00	2 g	1.00
MMEA ≥ 98% (2 L)	79.40	0.50 L	19.90
MEA ≥ 99.0% (2.5 L)	175.00	0.50 L	35.00
CH_3OH ≥ 99.80% (1 L)	67.10	0.005 L	0.30
DEA ≥ 98.0% (500 g)	40.50	2 L	0.20
EDA ≥ 99.0% (2.5 L)	127.00	0.50 L	25.40
C_7H_8 99.80% (1 L)	65.50	0.005 L	0.30
DETA 99% (2.5 L)	124.00	0.50 L	24.80
$Ni(NO_3)_2{\cdot}6H_2O$ (500 g)	104.00	2 g	0.40
NH_4VO_3 99% (1000 g)	304.00	2 g	0.60
$Fe(NO_3)_3{\cdot}9H_2O$ ≥ 98% (500 g)	83.90	2 g	0.30

Item	Unit Cost ($)	Amount	Total cost ($)
$Ce(NO_3)_3 \cdot 6H_2O$ 99% (500 g)	267.00	2 g	1.10
BaO 97% (250 g)	396.00	25 g	39.60
MgO 97% (500 g)	484.00	50 g	48.40
CuO ≥ 99.0% (100 g)	66.10	10 g	6.60
TiO_2 ≥ 99.0% (1000 g)	69.30	100 g	6.90
CeO_2 ≥ 99.0% (100 g)	51.80	10 g	5.20
CeO_2, 99.995% trace metals basis, powder (50 g)	468.00	5 g	46.80
Materials			
CO_2-AC (100 g)	51.70-551.20	100 g	51.70-551.20
Steam-AC (100 g)	142.40-759.60	100 g	142.40-759.60
H_3PO_4-AC (100 g)	17.20-17.80	100 g	17.20-17.80
H_2SO_4-AC (100 g)	11.80-12.40	100 g	11.80-12.40
$ZnCl_2$-AC (100 g)	10.40-11.20	100 g	10.40-11.20
KOH-AC (100 g)	7.30-13.10	100 g	7.30-13.10
K_2CO_3-AC (100 g)	7.70-8.50	100 g	7.70-8.50
NaOH-AC (100 g)	8.50-21.20	100 g	8.50-21.20
H_3PO_4, CO_2-AC (100 g)	69.00-129.70	100 g	69.00-129.70
H_2SO_4, CO_2-AC (100 g)	63.40-122.50	100 g	63.40-122.50
$ZnCl_2$, CO_2-AC (100 g)	62.40-123.10	100 g	62.40-123.10
KOH, CO_2-AC (100 g)	59.40-123.90	100 g	59.40-123.90

Item	Unit Cost ($)	Amount	Total cost ($)
NaOH, CO_2-AC (100 g)	60.50-132.10	100 g	60.50-132.10
Range			571.70-2026.30

Table 6 and 7 present the estimated surface modification cost for the various amine/metal incorporation combination of AC. It can be seen from table 6 that PEI functionalization of steam-activated oil palm carbon requires the highest modification cost of $255.80-873.00, while DEA functionalization of K_2CO_3-activated oil palm carbon in methanol (CH_3OH), conversely, is estimated at lower modification cost of $14.90-15.70. High purchased chemical prices for steam-activated oil palm carbon ($142.40-759.60) and high utility cost ($13.40) for prolonged process operation attributes the PEI functionalization steam-activated oil palm carbon the highest estimated modification cost. The estimated product cost for PEI-functionalized steam-activated oil palm carbon and DEA-functionalized K_2CO_3-activated oil palm carbon is $2.56-8.73 g^{-1} and $0.15-0.16 g^{-1}, respectively.

Comparing to the metal incorporation of oil palm-based AC, as seen in table 7, a modification cost is estimated at $195.70-812.80 for MgO functionalization steam-activated oil palm carbon whereas a lower modification cost of $11.90-12.50 is required for ferum nitrate ($Fe(NO_3)_3 \cdot 9H_2O$) functionalization K_2CO_3-activated oil palm carbon. The MgO functionalization steam-activated oil palm carbon gives the highest modification cost due to high purchased chemical prices for the metal-functionalizing doping agent, MgO and the steam-AC. The estimated product cost for MgO-functionalized steam-activated oil palm carbon is $1.96-8.13 g^{-1} while $0.12-0.13 g^{-1} for $Fe(NO_3)_3 \cdot 9H_2O$-functionalized K_2CO_3-activated oil palm carbon.

IOP Conf. Series: Materials Science and Engineering 1195 (2021) 012018 doi:10.1088/1757-899X/1195/1/012018

Table 6. Summary of operating costs for laboratory amine incorporation of oil palm biomass-derived AC [79, 90–99].

a.

Utility	Unit cost ($)	Amount	Total cost ($)
Electricity	\$0.056-0.058 kWh^{-1}	72.8-231.4 kWh	4.10-13.40

	Incorporation	Total surface modification cost ($) Modifying agent (1) PEI solution	(2) AMP	(3) AMPD	(4) MMEA	(5) MEA
Activated carbon (AC)	(1) CO_2-AC	165.10-664.60	87.20-586.70	59.40-558.90	78.20-577.70	93.40-592.90
	(2) Steam-AC	255.80-873.00	178.00-795.20	150.10-767.30	169.00-786.20	184.10-801.30
	(3) H_3PO_4-AC	130.60-131.20	52.80-53.30	24.90-25.50	43.80-44.30	58.90-59.50
	(4) H_2SO_4-AC	125.20-125.80	47.40-48.00	19.60-20.10	38.40-39.00	53.50-54.10
	(5) $ZnCl_2$-AC	123.80-124.60	46.00-46.70	18.20-18.90	37.00-37.70	52.10-52.90
	(6) KOH-AC	120.70-126.50	42.90-48.70	15.10-20.90	33.90-39.70	49.00-54.80
	(7) K_2CO_3-AC	121.10-121.90	43.30-44.00	15.40-16.20	34.30-35.00	49.40-50.20
	(8) NaOH-AC	121.90-134.60	44.10-56.70	16.20-28.90	35.10-47.70	50.20-62.90
	(9) H_3PO_4, CO_2-AC	182.40-243.10	104.60-165.20	76.80-137.40	95.60-156.20	110.70-171.40
	(10) H_2SO_4, CO_2-AC	176.90-236.00	99.00-158.10	71.20-130.30	90.00-149.10	105.10-164.30
	(11) $ZnCl_2$, CO_2-AC	175.80-236.50	98.00-158.60	70.20-130.80	89.00-149.60	104.10-164.80
	(12) KOH, CO_2-AC	172.80-237.30	94.90-159.50	67.10-131.70	85.90-150.50	101.10-165.60
	(13) NaOH, CO_2-AC	173.90-245.50	96.10-167.60	68.30-139.80	87.10-158.60	102.20-173.80

IOP Conf. Series: Materials Science and Engineering 1195 (2021) 012018 doi:10.1088/1757-899X/1195/1/012018

Table 6 (continued).

b.

Utility	Unit cost ($)	Amount	Total cost ($)
Electricity	$0.056-0.058 kWh^{-1}	72.8-231.4 kWh	4.10-13.40

	Incorporation	Total surface modification cost ($)					
		Modifying agent					
		(6) MEA in CH_3OH	(7) DEA in CH_3OH	(8) EDA	(9) EDA in C_7H_8	(10) DETA	(11) DETA in C_7H_8
AC	(1) CO_2-AC	93.70-593.20	58.90-558.40	81.20-580.60	81.50-584.10	80.60-580.00	80.90-583.50
	(2) Steam-AC	184.40-801.60	149.60-766.80	171.90-789.10	175.40-792.60	171.30-788.50	174.80-792.00
	(3) H_3PO_4-AC	59.20-59.80	24.40-25.00	46.70-47.30	50.20-50.70	46.10-46.70	49.60-50.10
	(4) H_2SO_4-AC	53.90-54.40	19.00-19.60	41.30-41.90	44.80-45.40	40.70-41.30	44.20-44.80
	(5) $ZnCl_2$-AC	52.40-53.20	17.60-18.40	39.90-40.70	43.40-44.10	39.30-40.10	42.80-43.50
	(6) KOH-AC	49.40-55.20	14.50-20.30	36.80-42.60	40.30-46.10	36.20-42.00	39.70-45.50
	(7) K_2CO_3-AC	49.70-50.50	**14.90-15.70**[a]	37.20-37.90	40.70-41.40	36.60-37.30	40.10-40.80
	(8) NaOH-AC	50.50-63.20	15.70-28.40	38.00-50.60	41.50-54.10	37.40-50.00	40.90-53.50
	(9) H_3PO_4, CO_2-AC	111.10-171.70	76.20-136.90	98.50-159.20	102.00-162.60	97.90-158.60	101.40-162.00
	(10) H_2SO_4, CO_2-AC	105.50-164.60	70.60-129.80	92.90-152.00	96.40-155.50	92.30-151.40	95.80-154.90
	(11) $ZnCl_2$, CO_2-AC	104.50-165.10	69.60-130.30	91.90-152.60	95.40-156.00	91.30-152.00	94.80-155.40
	(12) KOH, CO_2-AC	101.40-166.00	66.60-131.10	88.80-153.40	92.30-156.90	88.20-152.80	91.70-156.30
	(13) NaOH, CO_2-AC	102.50-174.10	67.70-139.30	90.00-161.50	93.50-165.00	89.40-160.90	92.90-164.40

[a] Value in **bold** indicates minimum value for surface modification of AC cost.

Table 7. Summary of operating costs for laboratory metal incorporation of oil palm biomass-derived AC [79, 100–109].

a.

Utility	Unit cost ($)	Amount	Total cost ($)
Electricity	$0.058 kWh^{-1}	85-86.40 kWh	4.90-5.00

Total surface modification cost ($)

	Incorporation	Modifying agent			
		(1) $Ni(NO_3)_2{\cdot}6H_2O$	(2) NH_4VO_3	(3) $Fe(NO_3)_3{\cdot}9H_2O$	(4) $Ce(NO_3)_3{\cdot}6H_2O$
Activated carbon (AC)	(1) CO_2-AC	57.10-556.50	57.30-556.70	57.00-556.40	57.80-557.20
	(2) Steam-AC	147.80-764.90	148.00-765.10	147.70-764.80	148.50-765.50
	(3) H_3PO_4-AC	21.50-21.90	21.70-22.10	21.40-21.80	22.10-22.60
	(4) H_2SO_4-AC	16.10-16.50	16.30-16.70	16.00-16.50	16.70-17.20
	(5) $ZnCl_2$-AC	14.70-15.30	14.90-15.50	14.60-15.20	15.40-16.00
	(6) KOH-AC	11.80-17.30	12.00-17.50	11.70-17.20	12.50-17.90
	(7) K_2CO_3-AC	12.00-12.60	12.20-12.80	**11.90-12.50**[a]	12.70-13.30
	(8) NaOH-AC	12.80-25.30	13.00-25.50	12.70-25.20	13.50-26.00
	(9) H_3PO_4, CO_2-AC	73.20-133.20	73.40-133.40	73.20-133.10	73.90-133.90
	(10) H_2SO_4, CO_2-AC	67.70-126.70	67.90-126.90	67.60-126.60	68.40-127.30
	(11) $ZnCl_2$, CO_2-AC	66.60-126.60	66.80-126.80	66.60-126.50	67.30-127.30
	(12) KOH, CO_2-AC	63.60-127.70	63.80-127.90	63.50-127.70	64.20-128.40
	(13) NaOH, CO_2-AC	64.70-135.90	64.90-136.10	64.60-135.80	65.40-136.50

Table 7 (continued).

b.

Utility	Unit cost ($)	Amount	Total cost ($)
Electricity	$0.058 kWh^{-1}	85-86.40 kWh	4.90-5.00

	Total surface modification cost ($)					
		Modifying agent				
	Incorporation	(5) BaO	(6) MgO	(7) CuO	(8) TiO_2	(9) CeO_2
AC	(1) CO_2-AC	96.20-595.60	105.00-604.40	63.20-562.60	63.50-562.90	61.80-602.80
	(2) Steam-AC	186.90-804.00	195.70-812.80	153.90-771.00	154.20-771.30	152.50-811.20
	(3) H_3PO_4-AC	60.60-61.00	69.40-69.80	27.60-28.00	27.90-28.40	26.20-68.20
	(4) H_2SO_4-AC	55.20-55.60	64.00-64.40	22.20-22.60	22.50-23.00	20.80-62.80
	(5) $ZnCl_2$-AC	53.80-54.40	62.60-63.20	20.80-21.40	21.20-21.80	19.40-61.60
	(6) KOH-AC	50.90-56.40	59.70-65.20	17.90-23.40	18.30-23.70	16.50-63.60
	(7) K_2CO_3-AC	51.10-51.70	59.90-60.50	18.10-18.70	18.40-19.00	16.70-58.90
	(8) NaOH-AC	51.90-64.40	60.70-73.20	18.90-31.40	19.20-31.80	17.50-71.60
	(9) H_3PO_4, CO_2-AC	112.30-172.30	121.10-181.10	79.40-139.30	79.70-139.70	77.90-179.50
	(10) H_2SO_4, CO_2-AC	106.80-165.80	115.60-174.60	73.80-132.80	74.10-133.10	72.40-173.00
	(11) $ZnCl_2$, CO_2-AC	105.70-165.70	114.50-174.50	72.80-132.70	73.10-133.10	71.30-172.90
	(12) KOH, CO_2-AC	102.70-166.80	111.50-175.60	69.70-133.90	70.00-134.20	68.30-174.00
	(13) NaOH, CO_2-AC	103.80-175.00	112.60-183.80	70.80-142.00	71.20-142.30	69.40-182.20

[a] Value in **bold** indicates minimum value for surface modification of AC cost.

3.2. Selection of least expensive AC production process

A comparison of the minimum estimated production cost from physical, chemical and physiochemical activation and amine and metal post-activation surface modification process for 100 g AC production is summarized in table 8. Based on table 8, oil palm-based ACs produced from various activation and surface modification approaches minimum estimated cost can be rank as below:

Minimum production/modification cost: KOH activation (chemical activation) < $Fe(NO_3)_3{\cdot}9H_2O$ functionalization (metal surface modification of AC) < DEA functionalization (amine surface modification of AC) < Physiochemical activation by $ZnCl_2$ and CO_2 < CO_2 activation

CO_2 activation with 99.80% CO_2 and the multi-sequences physiochemical activation by $ZnCl_2$ and 99.80% CO_2 estimated respectively 86-97% and 87-88% higher production cost than KOH activation. For post-activation surface modification method, the K_2CO_3 activation with added respective DEA and $Fe(NO_3)_3{\cdot}9H_2O$ functionalization have 48-51% and 54-63% higher estimated modification cost than the sole KOH activation, while 71-94% and 77-95% lower than the CO_2 activation. The main factor which causes the high production cost for CO_2 activation is the high CO_2 amount used for prolonged duration of activation process.

Overall, chemical activation with KOH is the least expensive method among the activation and surface modification methods, in producing the same yield of oil palm-based ACs with the lowest estimated production cost.

Table 8. Comparison of activation/post-activation surface modification process towards AC production with minimum estimated production cost.

Types of process	Estimated production/modification cost ($)	Estimated product cost ($)	
Physical activation: CO_2 activation with 99.80% CO_2	51.70-258.50	0.52-2.59	Highest
Physiochemical activation: $ZnCl_2$ and 99.80% CO_2	62.40-64.50	0.62-0.65	↓
Amine surface modification of AC: DEA functionalization of K_2CO_3-activated oil palm carbon in CH_3OH	14.90-15.70	0.15-0.16	↓
Metal surface modification of AC: $Fe(NO_3)_3{\cdot}9H_2O$ functionalization of K_2CO_3-activated oil palm carbon	11.90-12.50	0.12-0.13	↓
Chemical activation: KOH with ≥ 85% purity KOH pellets	7.30-8.10	0.073-0.081	Lowest

4. Conclusion

A laboratory scale cost estimation of AC production was developed for five activation (i.e., physical, chemical and physiochemical activation) and post-activation surface modification (i.e., amine and metal functionalization) methods. A comparative assessment on the production/modification costs of these activation/modification methods are studied to identify and select a cost-friendly method with the lowest production cost. The production cost is estimated based on the purchased chemical prices and the utility costs quoted from the combination of vendors and official industrial tariffs. Of the five methods investigated for the production cost evaluation, chemical activation with KOH has quoted the lowest production cost with minimum estimated production cost of \$0.073 g^{-1}, while PEI functionalization of steam-activated oil palm carbon presented the highest modification cost of up to \$8.73 g^{-1}. High purchased chemical prices for the steam-activated oil palm carbon and high utility cost due to prolonged process operation are the significant factors that cause the highest estimated modification cost for PEI functionalization. Nonetheless, the estimation of this laboratory scale production cost analysis cannot be considered a promising baseline for potential scaling up to estimate the pilot scale production cost, due to lack of estimation for the purchased equipment cost, capital cost, fixed capital investment cost, and annual operating costs for labor, supplies and general works required for production. Therefore, future studies could include any laboratory capital and operating expenditures associated with the purchase, building, construction, installation, and the operation services required in the production to demonstrate a small pilot scale production cost estimation.

References

[1] Ahmad A L, Loh M M and Aziz J A 2007 Preparation and characterization of activated carbon from oil palm wood and its evaluation on methylene blue adsorption *Dyes Pigm.* **75** 263–72

[2] Arami-Niya A, Daud W M A W and Mjalli F S 2010 Using granular activated carbon prepared from oil palm shell by $ZnCl_2$ and physical activation for methane adsorption *J. Anal. Appl. Pyrolysis* **89** 197–203

[3] Baroutian S, Aroua M K, Raman A A A and Sulaiman N M N 2010 Potassium hydroxide catalyst supported on palm shell activated carbon for transesterification of palm oil *Fuel Process. Technol.* **91** 1378–85

[4] González-Navarro M F, Giraldo L and Moreno-Piraján J C 2014 Preparation and characterization of activated carbon for hydrogen storage from waste African oil-palm by microwave-induced LiOH basic activation *J. Anal. Appl. Pyrolysis* **107** 82–6

[5] Guo J and Lua A C 2003 Adsorption of sulphur dioxide onto activated carbon prepared from oil-palm shells with and without pre-impregnation *Sep. Purif. Technol.* **30** 265–73

[6] Guo J, Xu W S, Chen Y L and Lua A C 2005 Adsorption of NH_3 onto activated carbon prepared from palm shells impregnated with H_2SO_4 *J. Colloid Interface Sci.* **281** 285–90

[7] Hamad B K, Noor A M, Afida A R and Mohd Asri M N 2010 High removal of 4-chloroguaiacol by high surface area of oil palm shell-activated carbon activated with NaOH from aqueous solution *Desalination* **257** 1–7

[8] Wahi R, Ngaini Z and Usun Jok V 2009 Removal of mercury, lead and copper from aqueous solution by activated carbon of palm oil empty fruit bunch *World Appl. Sci. J.* **5** 84–91

[9] Abdul Khalil H, Siti Alwani M, Ridzuan R, Kamarudin H and Khairul A 2008 Chemical composition, morphological characteristics, and cell wall structure of Malaysian oil palm fibers *Polym. Plast. Technol. Eng.* **47** 273–80

[10] Daud W M A W and Ali W S W 2004 Comparison on pore development of activated carbon produced from palm shell and coconut shell *Bioresour Technol.* **93** 63–9

[11] Ilomuanya M O, Nashiru B, Ifudu N D and Igwilo C I 2017 Effect of pore size and morphology of activated charcoal prepared from midribs of Elaeis guineensis on adsorption of poisons using metronidazole and Escherichia coli O157:H7 as a case study *J. Microsc. Ultrastruct.* **5** 32–8

[12] Kabir G, Mohd Din A T and Hameed B H 2017 Pyrolysis of oil palm mesocarp fiber and palm

frond in a slow-heating fixed-bed reactor: a comparative study *Bioresour Technol.* **241** 563–72
[13] Loh S K 2017 The potential of the Malaysian oil palm biomass as a renewable energy source *Energy Convers. Manage.* **141** 285–98
[14] Manocha S M 2003 Porous carbons *Sadhana* **28** 335–48
[15] Okahisa Y, Furukawa Y, Ishimoto K, Narita C, Intharapichai K and Ohara H 2018 Comparison of cellulose nanofiber properties produced from different parts of the oil palm tree *Carbohydr. Polym.* **198** 313–9
[16] Rafatullah M, Ahmad T, Ghazali A, Sulaiman O, Danish M and Hashim R 2013 Oil palm biomass as a precursor of activated carbons: a review *Crit. Rev. Env. Sci. Technol.* **43** 1117–61
[17] Sohni S, Norulaini N A N, Hashim R, Khan S B, Fadhullah W and Mohd Omar A K 2018 Physicochemical characterization of Malaysian crop and agro-industrial biomass residues as renewable energy resources *Ind. Crop. Prod.* **111** 642–50
[18] Tiong Y W, Yap C L, Gan S Y and Yap W S P 2017 One-pot conversion of oil palm empty fruit bunch and mesocarp fiber biomass to levulinic acid and upgrading to ethyl levulinate via indium trichloride-ionic liquids *J. Clean. Prod.* **168** 1251–61
[19] Arami-Niya A, Daud W M A W and Mjalli F S 2011 Comparative study of the textural characteristics of oil palm shell activated carbon produced by chemical and physical activation for methane adsorption *Chem. Eng. Res. Des.* **89** 657–64
[20] Daud W, Ahmad M and Aroua M 2007 Carbon molecular sieves from palm shell: effect of the benzene deposition times on gas separation properties *Sep. Purif. Technol.* **57** 289–93
[21] Foo K Y and Hameed B H 2011 Preparation of oil palm (Elaeis) empty fruit bunch activated carbon by microwave-assisted KOH activation for the adsorption of methylene blue *Desalination* **275** 302–5
[22] Guo J and Lua A C 2002 Microporous activated carbons prepared from palm shell by thermal activation and their application to sulfur dioxide adsorption *J. Colloid Interface Sci.* **251** 242–7
[23] Guo J, Luo Y, Lua A C, Chi R A, Chen Y L, Bao X T and Xiang S X 2007 Adsorption of hydrogen sulphide (H_2S) by activated carbons derived from oil-palm shell *Carbon* **45** 330–6
[24] Jia Q P and Lua A C 2008 Effects of pyrolysis conditions on the physical characteristics of oil-palm-shell activated carbons used in aqueous phase phenol adsorption *J. Anal. Appl. Pyrolysis* **83** 175–9
[25] Khalil S H, Aroua M K and Daud W M A W 2012 Study on the improvement of the capacity of amine-impregnated commercial activated carbon beds for CO_2 adsorbing *Chem. Eng. J.* **183** 15–20
[26] Klose W and Rincón S 2007 Adsorption and reaction of NO on activated carbon in the presence of oxygen and water vapour *Fuel* **86** 203–9
[27] Tan I A W, Ahmad A L and Hameed B H 2008 Adsorption of basic dye using activated carbon prepared from oil palm shell: batch and fixed bed studies *Desalination* **225** 13–28
[28] Hoseinzadeh Hesas R, Arami-Niya A, Wan Daud W M A and Sahu J 2015 Microwave-assisted production of activated carbons from oil palm shell in the presence of CO_2 or N_2 for CO_2 adsorption *J. Ind. Eng. Chem.* **24** 196–205
[29] Hoseinzadeh Hesas R, Arami-Niya A, Wan Daud W M A and Sahu J N 2013 Comparison of oil palm shell-based activated carbons produced by microwave and conventional heating methods using zinc chloride activation *J. Anal. Appl. Pyrolysis* **104** 176–84
[30] Nasri N S, Hamza U D, Ismail S N, Ahmed M M and Mohsin R 2014 Assessment of porous carbons derived from sustainable palm solid waste for carbon dioxide capture *J. Clean. Prod.* **71** 148–57
[31] Rodríguez-Reinoso F and Molina-Sabio M 1992 Activated carbons from lignocellulosic materials by chemical and/or physical activation: An overview *Carbon* **30** 1111–8
[32] Al Bahri M, Calvo L, Gilarranz M A and Rodriguez J J 2012 Activated carbon from grape seeds

upon chemical activation with phosphoric acid: application to the adsorption of diuron from water *Chem. Eng. J.* **203** 348–56

[33] Katesa J, Junpiromand S and Tangsathitkulchai C 2013 Effect of carbonization temperature on properties of char and activated carbon from coconut shell *Suranaree J. Sci. Technol.* **20** 269–78

[34] Loredo-Cancino M, Soto-Regalado E, Cerino-Córdova F J, García-Reyes R B, García-León A M and Garza-González M T 2013 Determining optimal conditions to produce activated carbon from barley husks using single or dual optimization *J. Environ. Manage.* **125** 117–25

[35] Sangchoom W and Mokaya R 2015 Valorization of lignin waste: carbons from hydrothermal carbonization of renewable lignin as superior sorbents for CO_2 and hydrogen storage *ACS Sustain Chem Eng* **3** 1658–67

[36] Sevilla M and Mokaya R 2014 Energy storage applications of activated carbons: Supercapacitors and hydrogen storage *Energy Environ. Sci.* **7** 1250–80

[37] Tang S H and Zaini M A A 2015 Potassium hydroxide activation of activated carbon: A commentary *Carbon Lett.* **16** 275–80

[38] Arami-Niya A, Daud W M A W, Mjalli F S, Abnisa F and Shafeeyan M S 2012 Production of microporous palm shell based activated carbon for methane adsorption: modeling and optimization using response surface methodology *Chem. Eng. Res. Des.* **90** 776–84

[39] Hamad B K 2015 Preparation and characterization of activated carbon from oil palm shell activated by KOH *ZANCO J. Pure Appl. Sci.* **27** 27–38

[40] Hernández-Montoya V, García-Servin J and Bueno-López J I 2012 *Lignocellulosic precursors used in the synthesis of activated carbon-characterization techniques and applications in the wastewater treatment*: InTech)

[41] Lee T, Zubir Z A, Jamil F M, Matsumoto A and Yeoh F Y 2014 Combustion and pyrolysis of activated carbon fibre from oil palm empty fruit bunch fibre assisted through chemical activation with acid treatment *J. Anal. Appl. Pyrolysis* **110** 408–18

[42] Nasri N, Zain H, Usman H, Zulkifli A, Zalilah S, Nurul A S and Nur L A 2015 CO_2 adsorption-breakthrough study on activated carbon derived from renewable oil palm empty fruit bunch *Aust. J. Basic Appl. Sci.* **9** 67–71

[43] Ooi C H, Ang C L and Yeoh F Y 2013 The properties of activated carbon fiber derived from direct activation from oil palm empty fruit bunch fiber *Adv. Mater. Res.* **686** 109–17

[44] Salman J M 2014 Optimization of preparation conditions for activated carbon from palm oil fronds using response surface methodology on removal of pesticides from aqueous solution *Arab J Chem* **7** 101–8

[45] Salman J M, Njoku V O and Hameed B H 2011 Batch and fixed-bed adsorption of 2,4-dichlorophenoxyacetic acid onto oil palm frond activated carbon *Chem. Eng. J.* **174** 33–40

[46] Hidayu A R N and Muda N 2017 Impregnated palm kernel shell activated carbon for CO_2 adsorption by pressure swing adsorption *Indian J. Sci. Technol.* **10** 1–6

[47] Houshmand A, Daud W M A W, Lee M G and Shafeeyan M S 2012 Carbon dioxide capture with amine-grafted activated carbon *Water, Air, Soil Pollut.* **223** 827–35

[48] Lee C S, Ong Y L, Aroua M K and Daud W M A W 2013 Impregnation of palm shell-based activated carbon with sterically hindered amines for CO_2 adsorption *Chem. Eng. J.* **219** 558–64

[49] Shafeeyan M S, Daud W M A W, Houshmand A and Arami-Niya A 2011 Ammonia modification of activated carbon to enhance carbon dioxide adsorption: effect of pre-oxidation *Appl. Surf. Sci.* **257** 3936–42

[50] Shafeeyan M S, Daud W M A W, Shamiri A and Aghamohammadi N 2015 Adsorption equilibrium of carbon dioxide on ammonia-modified activated carbon *Chem. Eng. Res. Des.* **104** 42–52

[51] Sumathi S, Bhatia S, Lee K T and Mohamed A R 2010 Selection of best impregnated palm shell activated carbon (PSAC) for simultaneous removal of SO_2 and NO_x *J. Hazard. Mater.* **176**

1093–6
[52] Yin C Y, Aroua M K and Daud W M A W 2007 Impregnation of palm shell activated carbon with polyethyleneimine and its effects on Cd^{2+} adsorption *Colloids Surf., A* **307** 128–36
[53] Aroua M, Daud W, Yin C and Adinata D 2008 Adsorption capacities of carbon dioxide, oxygen, nitrogen and methane on carbon molecular basket derived from polyethyleneimine impregnation on microporous palm shell activated carbon *Sep. Purif. Technol.* **62** 609–13
[54] Younas M, Leong L K, Mohamed A R and Sethupathi S 2016 CO_2 adsorption by modified palm shell activated carbon (PSAC) via chemical and physical activation and metal impregnation *Chem. Eng. Commun.* **203** 1455-63
[55] Toles C A, Marshall W E, Wartelle L H and McAloon A 2000 Steam- or carbon dioxide-activated carbons from almond shells: physical, chemical and adsorptive properties and estimated cost of production *Bioresour Technol.* **75** 197-203
[56] Lima I M, McAloon A and Boateng A A 2008 Activated carbon from broiler litter: process description and cost of production *Biomass Bioenergy* **32** 568-72
[57] Toles C A, Marshall W E, Johns M M, Wartelle L H and McAloon A 2000 Acid-activated carbons from almond shells: physical, chemical and adsorptive properties and estimated cost of production *Bioresour Technol.* **71** 87–92
[58] Ng C, Marshall W E, Rao R M, Bansode R R and Losso J N 2003 Activated carbon from pecan shell: process description and economic analysis *Ind. Crop. Prod.* **17** 209–17
[59] Nowrouzi M, Younesi H and Bahramifar N 2017 High efficient carbon dioxide capture onto as-synthesized activated carbon by chemical activation of Persian Ironwood biomass and the economic pre-feasibility study for scale-up *J. Clean. Prod.* **168** 499–509
[60] Arami-Niya A, Daud W W and Mjalli F 2010 Production of palm shell-based activated carbon with more homogeniouse pore size distribution *J. Appl. Sci.* **10** 3361–6
[61] Guo J, Gui B, Xiang S X, Bao X T, Zhang H J and Lua A C 2008 Preparation of activated carbons by utilizing solid wastes from palm oil processing mills *J. Porous Mat.* **15** 535–40
[62] Guo J and Lua A C 2003 Textural and chemical properties of adsorbent prepared from palm shell by phosphoric acid activation *Mater. Chem. Phys.* **80** 114–9
[63] Hayashi J i, Horikawa T, Takeda I, Muroyama K and Ani F N 2002 Preparing activated carbon from various nutshells by chemical activation with K_2CO_3 *Carbon* **40** 2381–6
[64] Herawan S G, Hadi M S, Ayob M R and Putra A 2013 Characterization of activated carbons from oil-palm shell by CO_2 activation with no holding carbonization temperature *Sci. World J.* **2013** 1–6
[65] Kongnoo A, Intharapat P, Worathanakul P and Phalakornkule C 2016 Diethanolamine impregnated palm shell activated carbon for CO_2 adsorption at elevated temperatures *J. Environ. Chem. Eng.* **4** 73–81
[66] Lua A C and Guo J 2001 Microporous oil-palm-shell activated carbon prepared by physical activation for gas-phase adsorption *Langmuir* **17** 7112–7
[67] Mohammadi M, Hassani A J, Mohamed A R and Najafpour G D 2010 Removal of rhodamine B from aqueous solution using palm shell-based activated carbon: adsorption and kinetic studies *J. Chem. Eng. Data* **55** 5777–85
[68] Owlad M, Aroua M K and Daud W M A W 2010 Hexavalent chromium adsorption on impregnated palm shell activated carbon with polyethyleneimine *Bioresour Technol.* **101** 5098–103
[69] Parshetti G K, Chowdhury S and Balasubramanian R 2015 Biomass derived low-cost microporous adsorbents for efficient CO_2 capture *Fuel* **148** 246–54
[70] Yin C Y, Aroua M K and Daud W M A W 2009 Fixed-bed adsorption of metal ions from aqueous solution on polyethyleneimine-impregnated palm shell activated carbon *Chem. Eng. J.* **148** 8–14
[71] Sivakumar V M, Kok K L and Mohamed A 2010 Synthesis of carbon molecular sieve from palm shell using deposition of polyfurfuryl alcohol *J. Korean Chem. Soc.* **54** 323–8

[72] Abechi S, Gimba C, Uzairu A and Dallatu Y 2013 Preparation and characterization of activated carbon from palm kernel shell by chemical activation *Res. J. Chem. Sci.* **3** 54–61
[73] Fisher Scientific 2020 Carbon dioxide in nitrogen (100ppm, 48L)-06-802-429. In: *Airgas™ Transportable Pure Gases and Mixtures (48L Size) from Restek*
[74] Fisher Scientific 2020 Carbon dioxide (99.80%, 48L)-06-802-417. In: *Airgas™ Transportable Pure Gases and Mixtures (48L Size) from Restek*
[75] Fisher Scientific 2020 Carbon dioxide in nitrogen (1000ppm, 48L)-06-802-431. In: *Airgas™ Transportable Pure Gases and Mixtures (48L Size) from Restek*
[76] Fisher Scientific 2020 Carbon monoxide 0.5%, carbon dioxide 0.5%, hydrogen 0.5%, oxygen 0.5% in nitrogen (48L)-06-802-470. In: *Airgas™ Transportable Pure Gases and Mixtures (48L Size) from Restek*
[77] Fisher Scientific 2020 Carbon monoxide 1%, carbon dioxide 1%, hydrogen 1%, oxygen 1% in nitrogen (48L)-06-802-473. In: *Airgas™ Transportable Pure Gases and Mixtures (48L Size) from Restek*
[78] Fisher Scientific 2020 Carbon monoxide 7%, oxygen 4%, carbon dioxide 15%, methane 4.5% in nitrogen (48L)-06-802-481. In: *Airgas™ Transportable Pure Gases and Mixtures (48L Size) from Restek*
[79] Sarawak Energy 2020 Tariffs-industrial pricing & tariff. Sarawak Energy)
[80] Sigma-Aldrich 2020 Phosphoric acid (ACS reagent, ⩾85 wt. % in H_2O)-695017
[81] Sigma-Aldrich 2020 Sulfuric acid (ACS reagent, 95.0-98.0%)-258105
[82] Sigma-Aldrich 2020 Zinc chloride (reagent grade, ⩾98%)-208086
[83] Sigma-Aldrich 2020 Potassium hydroxide concentrate (0.1 M KOH in water (0.1N), eluent concentrate for IC)-61699
[84] Sigma-Aldrich 2020 Potassium hydroxide (⩾85% KOH basis, pellets, white)-P1767
[85] Sigma-Aldrich 2020 Potassium carbonate (ACS reagent, ⩾99.0%)-209619
[86] Sigma-Aldrich 2020 Sodium hydroxide (reagent grade, ⩾98%, pellets (anhydrous))-S5881
[87] Sigma-Aldrich 2020 Sodium hydroxide (reagent grade, 97%, powder)-655104
[88] Sigma-Aldrich 2020 Sodium hydroxide concentrate (0.1 M NaOH in water (0.1N), eluent concentrate for IC)-43617
[89] Sigma-Aldrich 2020 Sodium hydroxide solution (50% in H_2O)-415413
[90] Sigma-Aldrich 2020 Poly(ethyleneimine) solution (analytical standard, 50 % (w/v) in H_2O)-P3143
[91] Sigma-Aldrich 2020 2-amino-2-methyl-1-propanol (~5% water, technical grade, 90%)-A65182
[92] Sigma-Aldrich 2020 2-amino-2-methyl-1,3-propanediol (⩾99%)-A9754
[93] Sigma-Aldrich 2020 2-(methylamino)ethanol (⩾98%)-471445
[94] Sigma-Aldrich 2020 Ethanolamine (ACS reagent, ⩾99.0%)-398136
[95] Sigma-Aldrich 2020 Methanol (anhydrous, 99.8%)-322415.
[96] Sigma-Aldrich 2020 Diethanolamine (reagent grade, ⩾98.0%)-D8885
[97] Sigma-Aldrich 2020 Ethylenediamine (ReagentPlus®, ⩾99%)-E26266
[98] Sigma-Aldrich 2020 Toluene (anhydrous, 99.8%)-244511
[99] Sigma-Aldrich 2020 Diethylenetriamine (ReagentPlus®, 99%)-D93856
[100] Sigma-Aldrich 2020 Nickel(II) nitrate hexahydrate (crystals or chunks)-244074
[101] Sigma-Aldrich 2020 Ammonium metavanadate (99%)-205559
[102] Sigma-Aldrich 2020 Iron(III) nitrate nonahydrate (ACS reagent, ⩾98%)-216828
[103] Sigma-Aldrich 2020 Cerium(III) nitrate hexahydrate (99% trace metals basis)-238538
[104] Sigma-Aldrich 2020 Barium oxide (97%)-288497
[105] Sigma-Aldrich 2020 Magnesium oxide (ACS reagent, 97%)-243388
[106] Sigma-Aldrich 2020 Copper(II) oxide (ACS reagent, ⩾99.0%)-241741
[107] Sigma-Aldrich 2020 Titanium(IV) oxide (ReagentPlus®, ⩾99%)-1402

[108] Sigma-Aldrich 2020 Cerium(IV) oxide (⩾99.0%)-22390
[109] Sigma-Aldrich 2020 Cerium(IV) oxide (powder, 99.995% trace metals basis)-202975

Development of poly(ethylene glycol) diacrylate membrane for facilitated CO_2/N_2 separation

T P Kim[1], **Z A Jawad**[2] **and B L F Chin**[1]

[1] Department of Chemical Engineering, Faculty of Engineering and Science, Curtin University Malaysia, CDT 250, 98009 Miri, Sarawak, Malaysia

[2] Department of Chemical Engineering, College of Engineering, Qatar University, P.O.Box: 2713, Doha, Qatar

Email: zjawad@qu.edu.qa

Abstract. Carbon dioxide (CO_2) is responsible for approximately 80% of greenhouse gases emission that is the main source to global climate change causing notable environmental impacts. Poly (ethylene glycol) diacrylate (PEGDA) have polar PEG repeating units, which provide a strong affinity towards carbon dioxide (CO_2) molecules has been blended with 3-aminopropyltrimethoxysilane (APTMS) to synthesize membrane for CO_2 /nitrogen (N_2) separation. The new synthesized membrane is studied for potential applications in gas separation and to be implemented in control CO_2 emission. APTMS is also used to delay the diffusion between polymer and solvent. In this study, concentration of polymer of PEGDA and casting solvent of APTMS in terms of mol ratio from a range of 0.9:1.1 to 1.3:0.7 is discussed. Based on the results, PEGDA membrane shows best gas separation performance at mol ratio of PEGDA to APTMS of 1:1 where the permeance for both CO_2/N_2, and CO_2/N_2 selectivity are 75.21±0.15 GPU, 22.95±0.05 GPU and 3.28±0.12, respectively. An optimal aminosilane/polymer reaction ratio benefits the gas separation performance of the membrane due to the affinity of the membrane towards CO_2 and formation of different membrane surface morphology.

Keywords: Membrane Technology; poly (ethylene glycol) diacrylate; 3-aminopropyltrimethoxysilane; CO_2/N_2 Separation

1. Introduction

Global climate change has been an arising issue in the recent decade prompting organizations such as Intergovernmental Panel on Climate Change into taking proper course of resolving the issue or mitigating the effects of climate change [1]. CO_2 as the primary anthropogenic GHGs emanated from power plants and industrial processes, requires proper constraint or mitigation method to control its emission. Identifying the potential of CCS, this method is executed as a major strategy to resolve the issue [2]. Among three main approaches of CCS, post-combustion capture is favourably practiced given its cost-effective and compatibility for retrofitting applications compared to oxy-fuel combustion and pre-combustion capture [3]. In comparison between conventional CO_2 separation technologies, membrane separation remains favourable in industrial application [4] because of its simple modular design allows process arrangement for retrofitting purposes in industrial and cost optimization in future applications. Membrane technology also requires less maintenance; thus, proving it a cost-effective and energy-efficient technology [5]. However, the energy requirement for this technology depends on the flue gases' partial pressure while the performance of membrane is dependent on multiple parameters such as polymer concentration and casting solution concentration [6].

The principle characteristic of membranes in gas separation applications is to control the permeation of different species limited by their performance, known as Robeson limit (permeability must be sacrificed for selectivity and vice versa) [7]. It is always challenging to have a reproducible and up-scalable method to fulfill the needs. Recently, novel and straightforward strategies have been developed to fabricate high-performance membranes [8–11].

PEGDA polymer have the potential of good mechanical and diffusive properties to be fabricated as gas separation membrane [12]. Increasing PEGDA content in the fabrication of membrane had increased the permeability of CO_2 compared to N_2 due to the strong affinity existing between polar EG repeating units and CO_2 molecules [13]. The enhancement in the mechanical properties of this membrane had significantly increased the CO_2/N_2 selectivity due to the increase in cross-linking density that reduces the pore formation and affects gas separation performance [14].

APTMS are used as crosslinking agents for membrane curing where the synthesized membrane displayed higher crosslinking density, anti-swelling ability and excellent combination between the separation layer and support layer [15]. The increase in APTMS concentration during membrane synthesis had been reduced the formation of macrovoids in the phase inversion process by delaying the mutual diffusion of polymer and solvent [16]. More resistant towards pore formation; thus, smaller pore size of membrane, and higher permeability of CO_2 compared to N_2 based on their respective activation energy required [17].

Therefore, the main aim of this research is to fabricate PEGDA membrane for CO_2/N_2 separation using different polymer and solution concentration to optimize membrane's performance. Up to date, many works have been conducted to identify the effects of polymer concentration of various polymers on the gas separation performance. However, no research on the effects of APTMS concentration has been conducted. Therefore, this research is conduct to understand the relationship between the effects of both polymer (PEGDA) and solvent (APTMS) concentration on the structure and properties of membrane. Besides that, the suitability of APTMS as a casting solvent is to be identified through this experimental work by comparing its advantages and disadvantages exhibited by the membrane synthesized.

2. Materials and Experimental Methodology

2.1. Materials

PEGDA (Purity > 97 %) and APTMS (Purity > 97 %) were obtained from Sigma-Aldrich, Malaysia, for membrane fabrication. Chloroform and acetone, ACS reagent > 99.99 %, were purchased from Merck Chemical, Malaysia.

2.2. Fabrication of PEGDA Membrane

Membranes were synthesized by dry-phase inversion. The casting solution of PEGDA membranes were prepared by mixing a mole ratio of 2 of PEGDA and APTMS in 50 wt % chloroform solution for 24 hours at 25 °C. The casting solution was then casted onto a glass plate at a thickness of 150 μm and allowed to evaporate slowly at 25 °C for 2 days in ambient temperature. Next, the neat membrane was placed in a vacuum oven at 40 °C for 24 hours as the final drying stage. Finally, the resulting membrane was peeled off the glass plate and kept in a desiccator for experiment and performance testing [18]. The concentrations of PEGDA and APTMS are tabulated in Table 1.

Table 1. Mole ratios of PEGDA and APTSM.

Membrane	PEGDA mole ratio	APTMS mole ratio
M1	0.9	1.1
M2	1.0	1.0
M3	1.1	0.9
M4	1.2	0.8
M5	1.3	0.7

2.3. Gas Separation Performance Test
To study the gas separation performance, a gas permeation tests was conducted based on the previous published work [19]. Both flow rates of the permeate and retentate streams in and out of the permeation cell were measured using the bubble gas flow meters during the gas permeation test. Three samples for each fabricated membrane were collected and tested. In this work, standard error of the readings were reported to ensure reproducibility of the data.

2.4. Characterization of Membrane Synthesized

2.4.1. Scanning Electron Microscopy (SEM). Hitachi S3500N scanning electron microscope with a Schottky type thermal field emitter (SU-70 model) was used to scan the microstructure of the membranes surfaces. The analysis was carried out in vacuum conditions ($1x10^{-8}$ Pa) at an accelerating voltage of 15 kV and a secondary electron reception angle of 30.4°. The membrane samples were coated with a layer of gold approx. 2 nm thick using BAL-TEC SCD 050 Sputter Coater from Quorum Technologies to increase conductivity and prevent charge build-up [20]. Analyses of the membranes' surface morphology were carried out at different zooming rate and higher resolution electrographs to obtain clear images. Three samples were carried out for the SEM characterization to ensure reproducibility of the experimental data.

2.4.2. Attenuated Total Reflection- Fourier Transform Infrared Spectroscopy (ATR-FTIR). A schematic of multiple reflections of internal reflection element (IRE) are illustrated in Figure 1 [21, 22]. The spectra of the membranes were recorded by a Thermo Scientific FTIR model (NICOLET iS10, USA) using diamond crystal over the wavenumber range of 4000–400 cm^{-1} with 32 scans.

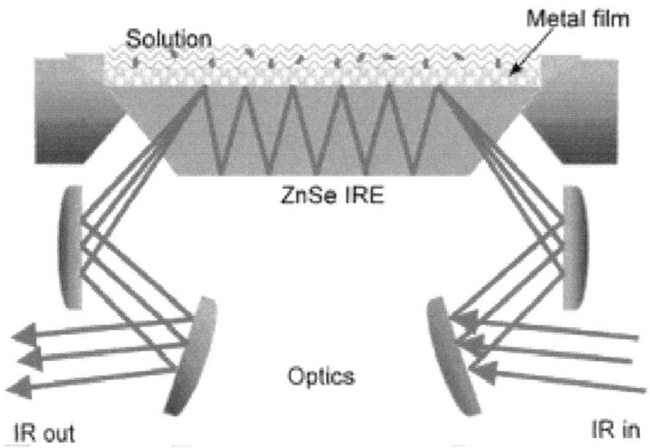

Figure 1. Schematic diagram of ATR-FTIR (I =incident radiation ; R = reflected radiation) [22].

3. Results and Discussion

3.1. The Effects of PEGDA and APTMS Concentration
The main aim of this study is to investigate the effect of the concentration of PEGDA and APTMS on the membrane structure and gas separation performance of PEGDA membrane. According to Ghadimi *et al.* (2014), the increase in the PEGDA content allows higher permeance of CO_2 through the PEGDA membrane, attributed to the dipolar-quadrupolar interactions induced by the strong affinity between the polar ethylene glycol repeating units on the membrane's structure and CO_2 molecules. The chemical affinity increases the gas permeance of CO_2 through the membrane dramatically compared to N_2 gases thus the membrane presents with a relatively high CO_2/N_2 selectivity [23].

According to Ko *et al.* (2011), the increment in the concentration of APTMS increases the affinity of the membrane towards CO_2 which confirmed through the reaction between CO_2 and primary amine, APTMS to form carbamate through a two-step zwitterion mechanism and then regeneration via zwitterion to produce APTMS, as illustrated in Figure 2 [24]. In addition, based on Ilyas *et al.* (2018), the methoxy groups of APTMS has high affinity for CO_2 to provide sites of interaction for CO_2 solvation; thus, improving CO_2 gas permeance [25].

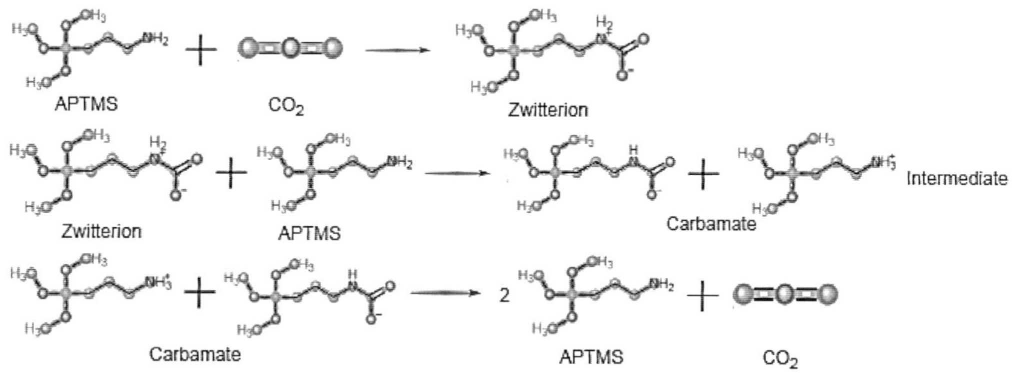

Figure 2. The expected mechanism reaction between CO_2 and primary amines [24].

3.2. ATR-FTIR Analysis

ATR-FTIR analysis was conducted to confirm the presence of organic functional groups in the PEGDA membrane fabricated with different concentration of PEGDA and APTMS of 0.9:1.1 (M1), 1.0:1.0 (M2), 1.1:0.9 (M3), 1.2:0.8 (M4) and 1.3:0.7 (M5) at a casting thickness of 150 μm. As illustrated in Figure 3, the carbonyl group of the carboxylic acid (C=O) functional group is represented by the strong absorption peaks measured at 1713.97 cm^{-1} while the peaks at 1190 cm^{-1} which justified the presence of acrylic bond (C-O) attributed to the stretching of acrylate groups [26]. Both functional groups (C=O and C-O) are ester groups. Interestingly, the disappearance of characteristic peaks of twisting vibration of the acrylic bond (CH_2=CH_2) and the deformation of CH_2=CH bond (i.e., particularly the peaks at 812 cm^{-1} and 1410 cm^{-1}) as presented in the spectra practically indicated the complete reaction conversion between PEGDA and APTMS for membranes [26].

M2 depicted the highest absorbance in relative to the absorption peaks of the PEGDA membrane compared to M1, M3, M4 and M5, as presented in table 2. The decrease in absorbance from M2 to M1 is due to the low mechanical stability of APTMS as polymer concentration of APTMS increases. This causes a lower absorption in IR radiation; thus, a lower absorption peak has been demonstrated in figure 3. Additionally, the reduction in the absorbance peaks from M2 to M5 have been explained by the SEM analysis conducted where the wrinkle formation showed in M3, M4 and M5 resulted lower absorption peaks, and reflection of the evanescent wave [27].

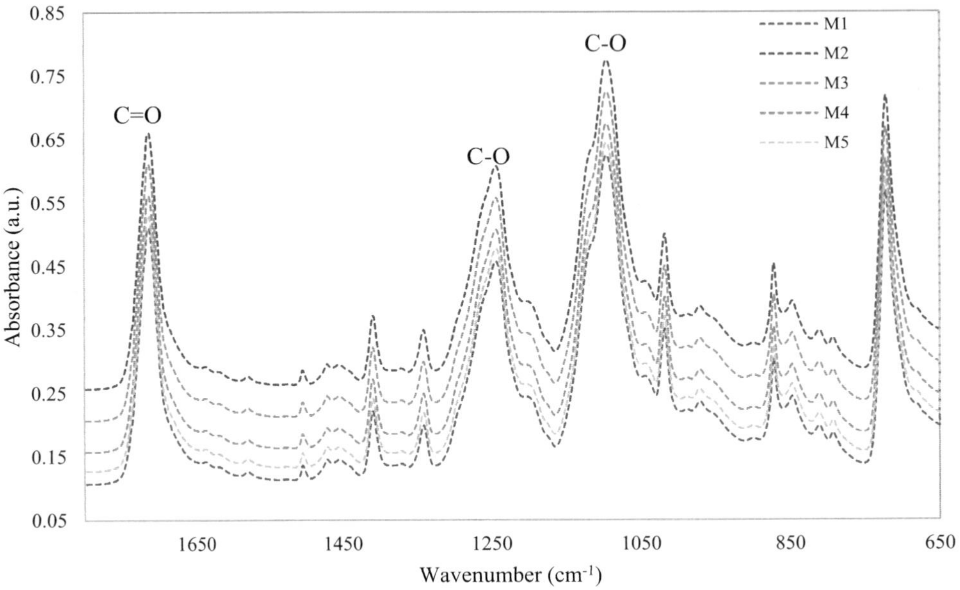

Figure 3. ATR-FTIR spectra of of membrane with different concentration of PEGDA:APTMS for M1 (0.9:1.1); M2 (1:1); M3 (1.1:0.9); M4 (1.2:0.8); M5 (1.3:0.7) at a casting thickness of 150 μm.

Table 2. Assignment of the characteristic FTIR bands of PEGDA membrane.

Vibration Stretching	Wavenumber (cm^{-1})	Absorbance				
		M1	M2	M3	M4	M5
C=O	1713.97	0.51	0.66	0.61	0.56	0.53
C-O	1240.0	0.22	0.37	0.32	0.27	0.24
C-O	1095.9	0.62	0.77	0.72	0.67	0.64

3.3. SEM Analysis

SEM is used for characterization of the physical structure and morphology of the PEGDA membrane fabricated at different concentration of PEGDA and APTMS of M1 (0.9:1.1), M2 (1.0:1.0), M3 (1.1:0.9), M4 (1.2:0.8) and M5 (1.3:0.7) at a casting thickness of 150 μm.

M1 shows a rough surface layer with formation of small molecular size defects, as observed in Figure 4a. Higher concentration of APTMS than PEGDA in M1 exhibits low mechanical stability where the evaporation of solvent forms the molecular size defects at the membrane's surface as the structural integrity of the membrane surface is weaken [28].

The SEM analysis of M2 (Figure 4b) shows a smooth surface layer. The polymer support provided by the concentration of PEGDA in M2 is able to prevent formation of molecular size defects on the membrane surface compared to M1. The hydrophilic properties exhibited by PEGDA is not significant to form wrinkles on the membrane surface as concentration of PEGDA of M2 is lower as compared to M3, M4 and M5 [29].

M3 presented with wrinkle formations on the membrane's top surface as observed in Figure 4c. The hydrophilic properties exhibited by PEGDA causes the diffusion of water vapour during air-drying of membrane to induce interfacial tension with the PEGDA at the membrane surface. The increase in the PEGDA concentration from M2 to M3 results in a stronger diffusion of water vapour through the membrane that caused folding of the membrane surface structure into a wrinkle formation [30]. Therefore, significant wrinkle formations are also observed on the membrane surface of both M4 and

M5 in Figure 4d and Figure 4e, respectively, as the PEGDA concentration increased from M3 to M4 to M5 [27].

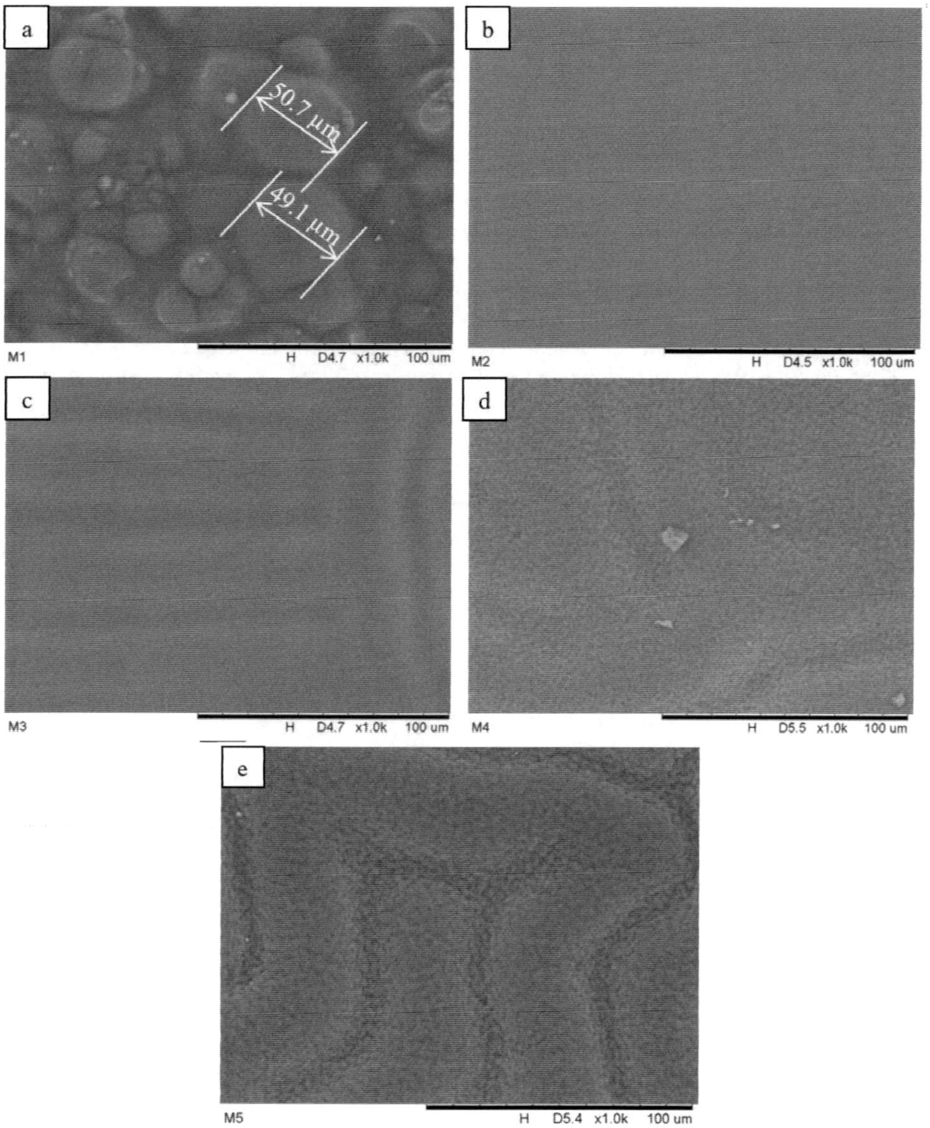

Figure 4. SEM surface morphologies of membrane with different concentration of PEGDA:APTMS for (a) M1 (0.9:1.1); (b) M2 (1:1); (c) M3 (1.1:0.9); (d) M4 (1.2:0.8); (e) M5 (1.3:0.7) at casting thickness of 150 μm.

3.4. Gas Permeation Test

Gas permeance towards CO_2 and N_2 of PEGDA membrane fabricated at different concentration of PEGDA and APTMS of M1 (0.9:1.1), M2 (1.0:1.0), M3 (1.1:0.9), M4 (1.2:0.8) and M5 (1.3:0.7) at a casting thickness of 150 μm is analysed through gas permeation test conducted using a gas permeation test rig.

3.4.1. Membrane Performance towards CO_2. As observed in Figure 5, the gas permeance of CO_2 is the highest for M2 (75.21±0.15 GPU) when the ratio of PEGDA to APTMS is 1:1 compared to M1

(69.42±0.08 GPU), M3 (72.07±0.20 GPU), M4 (70.25±0.23 GPU) and M5 (68.04±0.22 GPU). The CO_2 gas permeance had reduced from M2 to M1 as a result of the increase of APTMS concentration from M2 to M1 during the intermediate reaction between CO_2 and APTMS. Saturated adsorbed CO_2 in the form of carbamate species produced during the reaction between CO_2 and APTMS that is present in the membrane affected the gas permeance of CO_2; therefore, the gas separation performance towards CO_2 of M2 is lower than of M1 [31].

However, the decrease in CO_2 gas permeance from M2 to M5 can be interpreted as the inferior effect of the dipolar-quadrupolar interactions induced between CO_2 and PEGDA compared to the affinity exhibited by the intermediate reaction between CO_2 and APTMS to form carbamate [32]. The methoxy groups present in APTMS also verified the high affinity between CO_2 and APTMS to provide sites of CO_2 interaction; hence, increasing the gas permeance of CO_2 of the membrane as APTMS concentration increases from M5 to M2 [33]. Besides that, based on the FTIR analysis conducted, the absorption peaks recorded for C=O and C-O stretch are the highest in M2 compared to M1, M3, M4 and M5 due to the low mechanical stability in M1 and the wrinkle formations in M3, M4 and M5. Therefore, the higher absorption peaks proves the stronger dipole moment between M2 and CO_2 which is attributed to the increase in gas permeance.

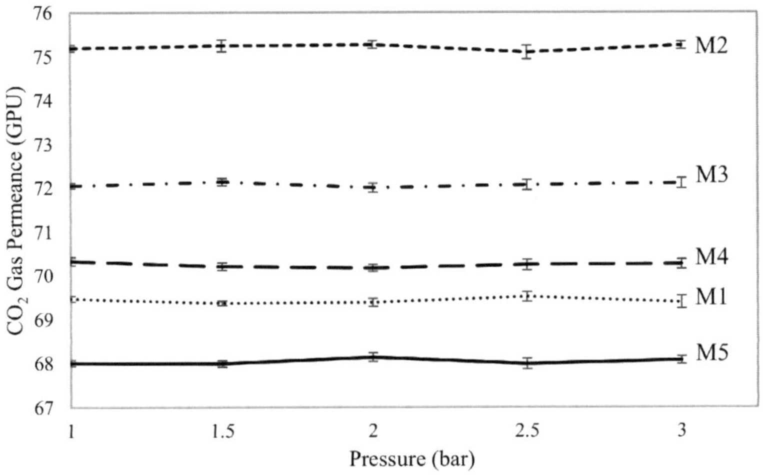

Figure 5. CO_2 permeance of membrane with different concentration of PEGDA:APTMS for M1 (0.9:1.1); M2 (1:1); M3 (1.1:0.9); M4 (1.2:0.8); M5 (1.3:0.7) at a casting thickness of 150 µm.

3.4.2. Membrane Performance towards N_2. Based on Figure 6, the N_2 permeance is the lowest for M2 (22.95±0.05 GPU) when the ratio of PEGDA to APTMS is 1:1 as compared to M1 (23.47±0.11 GPU), M3 (23.36±0.210 GPU), M4 (24.19±0.21 GPU) and M5 (24.79±0.18 GPU). The increment of N_2 permeance from M2 to M1 is a result of the formation of molecular size defects of both 49.1 µm and 50.7 µm that provided sites for gas permeation thus presented with higher permeation of N_2 gases [23]. As for membrane M3, M4 and M5 have higher gas permeance compared to M1 due to their surface morphology where the wrinkles formed on the top surface provides more interlayer is involved in the N_2 gas permeation across the membrane thus increasing the N_2 gas permeance [34].

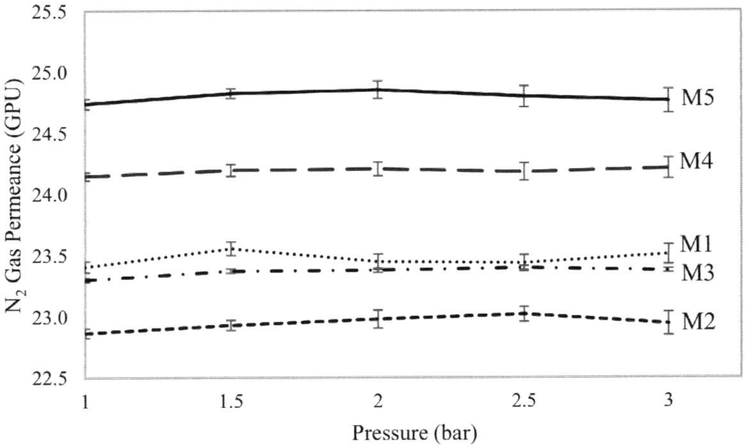

Figure 6. N_2 permeance of membrane with different concentration of PEGDA:APTMS for M1 (0.9:1.1); M2 (1:1); M3 (1.1:0.9); M4 (1.2:0.8); M5 (1.3:0.7) at a casting thickness of 150 µm.

3.4.3. Membrane Performance towards CO_2/N_2 Selectivity. According to Figure 7, the CO_2/N_2 selectivity is the highest for M2 (3.28±0.12) when the ratio of PEGDA to APTMS is 1:1 as compared to M1 (2.96±0.08), M3 (3.08±0.11), M4 (2.90±0.03) and M5 (2.74±0.13). The increasing of APTMS concentration from M2 to M1 resulted in a decrease of CO_2/N_2 selectivity from 3.28±0.12 to 2.96±0.08. This is due to the saturated carbamate species formed by the intermediate reaction between APTMS and CO_2. The saturated carbamate slows the intermediate reaction which affects the CO_2/N_2 selectivity of M1 as compared to M2 [35]. This increasing in CO_2/N_2 selectivity from M5 to M2 is attributed to the affinity between CO_2 and CO_2-affinitive moieties in APTMS (methoxy and amine groups), which is relatively stronger than the affinity between PEGDA and CO_2 as the concentration of APTMS increases from M5 to M2; thus, M5 exhibits a lower CO_2/N_2 selectivity as compared to M2 [36].

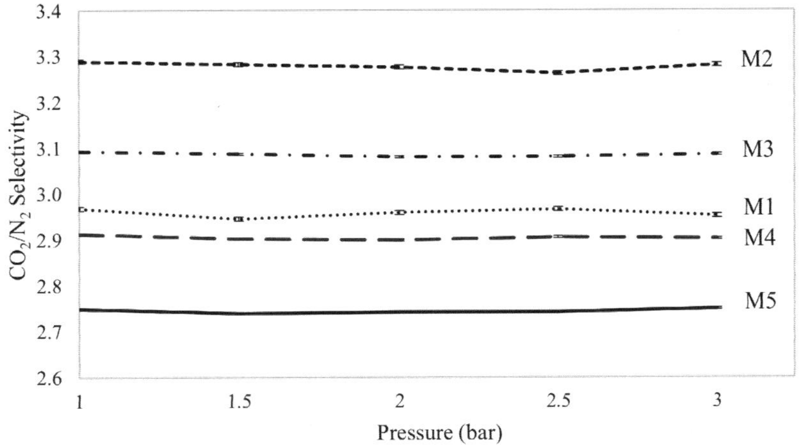

Figure 7. CO_2/N_2 selectivity of membrane with different concentration of PEGDA:APTMS for M1 (0.9:1.1); M2 (1:1); M3 (1.1:0.9); M4 (1.2:0.8); M5 (1.3:0.7) at a casting thickness of 150 µm.

In summary, M2 that has the highest CO_2 permeance (75.21±0.15 GPU) is presented with the highest CO_2/N_2 selectivity (3.278±0.12) and also the lowest N_2 gas permeance (22.95±0.05 GPU) as compared to M1, M3, M4 and M5, as illustrated in figures 5 and 6. This is due to the highest absorption peaks for both C=O and C-O stretch that represents the carbonyl group, as shown in table 2. Therefore, the higher reactivity of the membrane towards CO_2 is attributed to the high CO_2 gas permeance of M2. Besides that, the concentration of PEGDA to APTMS of M2 presents a smooth membrane surface which minimizes the gas permeance of N_2 as compared to M1 that have molecular size defects formation and M3, M4, M5 that have wrinkle formation on the membrane surface, which subsequently increases the N_2 gas permeance across the membrane [30]. The increase in APTMS concentration from M5 to M2 increases the affinity of the membrane towards CO_2 due to the higher presence of amine and methoxy group which has a stronger effect compared to the dipolar-quadrupolar interactions between PEG and CO_2. However, the increment of concentration is limited by the adsorption of CO_2 that affects the gas permeance of CO_2 [28]. Therefore, the optimized concentration of PEGDA to APTMS is determined at a mol ratio of 1:1 of M2.

Furthermore, the performance of the present work has been compared to the high performance membranes, as tabulated in table 3. It is clearly showed that the new synthesized PEGDA membrane need a further modification suggested by developing hollow fiber membrane similar to Liang *et al.* (2017) and Yong *et al.* (2013). Similarly, Naderi *et al.* (2018) proposed the addition of ethylene oxide group that could affect the overall gas permeability and selectivity because it has a high affinity towards CO_2 molecules through quadrupolar interactions [11].

Table 3. Summary of CO_2/N_2 permeation properties achieved from present work compared to previous study.

References	**Types of membrane gas separation**	***P* (CO_2)**	***P* (N_2)**	**CO_2/N_2 Ideal selectivity**
Present work	Poly (ethylene glycol) diacrylate flat sheet membrane	75.21±0.15*	22.95±0.05*	3.28
Liang *et al.* (2017)	Hollow fiber crosslinked polydimethylsiloxane membrane	3195 **	278 **	11.1
Yong *et al.* (2013)	Polymers of intrinsic microporosity /Matrimid Hollow fiber membrane	177.3*	6.7*	27.8
Naderi *et al.* (2018)	Hydrophilic polyethersulfone Flat sheet Membrane	3.74 **	0.11**	34.9

* GPU
** barrer

4. Conclusion

The main objective of this study is to synthesize effective gas separation membranes for CO_2 /N_2 gas separation. Firstly, the neat polymeric membranes PEGDA were fabricated. The performance in regards of permeance towards CO_2 and N_2, and CO_2/N_2 selectivity confirmed that the concentration of PEGDA and APTMS correlated to the membranes' structure and surface morphology as well as the Lewis acid-base interaction towards CO_2. In term of membrane separation performances, the optimal mol ratio of PEGDA and APTMS at 1:1 (M2) exhibited the highest CO_2/N_2 selectivity of (3.278±0.12) with CO_2 permeance (75.21±0.15 GPU) and N_2 permeance (22.95±0.05 GPU). Therefore, PEGDA to APTMS concentration of 1:1 mol ratio is recommended for the fabrication of PEGDA membrane to be utilized in the industrial application for CO_2/N_2 separation and efficiently reduced CO_2 emission into the atmosphere thus mitigate the effect of global warming.

References

[1] Intergovernmental Panel on Climate C 2015 Climate Change 2014: Mitigation of Climate Change: Working Group III Contribution to the IPCC Fifth Assessment Report (Cambridge: Cambridge University Press)

[2] Gale J, Abanades J, Bachu S and Jenkins C 2015 Special Issue commemorating the 10th year anniversary of the publication of the Intergovernmental Panel on Climate Change Special Report on CO_2 Capture and Storage vol 40

[3] Bui M, Adjiman C S, Bardow A, Anthony E J, Boston A, Brown S, Fennell P S, Fuss S, Galindo A and Hackett L A 2018 Carbon capture and storage (CCS): the way forward *Energy & Environ. Sci.* **11** 1062–176

[4] Kim J, Fu Q, Xie K, Scofield J M P, Kentish S E and Qiao G G 2016 CO2 separation using surface-functionalized SiO_2 nanoparticles incorporated ultra-thin film composite mixed matrix membranes for post-combustion carbon capture *J. Membr. Sci.* **515** 54–62

[5] Scofield J M P, Gurr P A, Kim J, Fu Q, Kentish S E and Qiao G G 2016 Blends of fluorinated additives with highly selective thin-film composite membranes to increase CO_2 permeability for CO_2/N_2 gas separation applications *Ind. Eng. Chem. Res.* **55** 8364–72

[6] Sani N A A, Lau W J and Ismail A F 2015 Morphologies and separation characteristics of polyphenylsulfone-based solvent resistant nanofiltration membranes: Effect of polymer concentration in casting solution and membrane pretreatment condition *Korean J. Chem. Eng.* **32** 743–52

[7] Robeson L M 2008 The upper bound revisited *J. Membr. Sci.* **320** 390–400

[8] Liang C Z, Yong W F and Chung T S 2017 High-performance composite hollow fiber membrane for flue gas and air separations *J. Membr. Sci.* **541** 367–77

[9] Yong W F, Li F Y, Xiao Y C, Chung T S and Tong Y W 2013 High performance PIM-1/Matrimid hollow fiber membranes for CO2/CH4, O2/N2 and CO2/N2 separation *J. Membr. Sci.* **443** 156–69

[10] Yong W F, Kwek K H A, Liao K S and Chung T S 2015 Suppression of aging and plasticization in highly permeable polymers *Polymer* **77** 377–86

[11] Naderi A, Yong W F, Xiao Y, Chung T S, Weber M and Maletzko C 2018 Effects of chemical structure on gas transport properties of polyethersulfone polymers *Polymer* **135** 76–84

[12] Anna C, Marta M, Ugo M, Giulia L M and Alessandro S 2017 Photo-crosslinked poly(ethylene glycol) diacrylate (PEGDA) hydrogels from low molecular weight prepolymer: Swelling and permeation studies *J. Appl. Polym. Sci.* **134** 1–9

[13] Teramoto N, Wakayama K, Harima M, Shimasaki T and Shibata M 2017 *Advances in Bioinspired and Biomedical Materials Volume 2*: American Chemical Society) pp 79–91

[14] Saimani S, Dal-Cin M M, Kumar A and Kingston D M 2010 Separation performance of asymmetric membranes based on PEGDA/PEI semi-interpenetrating polymer network in pure and binary gas mixtures of CO_2, N_2 and CH_4 *J. Membr. Sci.* **362** 353–9

[15] Wu H, Zhang X, Xu D, Li B and Jiang Z 2009 Enhancing the interfacial stability and solvent-resistant property of PDMS/PES composite membrane by introducing a bifunctional aminosilane *J. Membr. Sci.* **337** 9

[16] Hisayuki N, Keisuke I, Kensuke M, Noriyasu O and Minoru T 2011 Effect of chemical structure of silane coupling agent on interface adhesion properties of syndiotactic polypropylene/cellulose composite *J. Appl. Polym. Sci.* **119** 1732–41

[17] Madaeni S S and Taheri A H 2011 Effect of Casting Solution on Morphology and Performance of PVDF Microfiltration Membranes *Chem Eng Technol* **34** 1328–34

[18] Mazzoccoli J P 2008 Properties of Poly(ethylene glycol) Diacrylate Blends and Acoustically Focused Multilayered Biocomposites Developed for Tissue Engineering Applications. Case Western Reserve University)

[19] Jawad Z, Ahmad A, Low S and Zein S 2015 Incorporation of inorganic carbon nanotubes fillers into the CA polymeric matrix for improvement in CO_2/N_2 separation *Curr. Nanosci.* **11** 69–79

[20] Kowalik-Klimczak A, Bednarska A, Grądkowski M and Gierycz P 2016 Scanning Electron Microscopy (SEM) in the analysis of the structure of polymeric nanofiltration membranes *Problemy Eksploatacji* **100** 119–28

[21] Georg R and Bernhard L 2013 *Encyclopedia of Analytical Chemistry,* pp

[22] Andrew Chan K L and Kazarian S G 2016 Attenuated total reflection Fourier-transform infrared (ATR-FTIR) imaging of tissues and live cells *Chem. Soc. Rev.* **45** 1850–64

[23] Ghadimi A, Amirilargani M, Mohammadi T, Kasiri N and Sadatnia B 2014 Preparation of alloyed poly (ether block amide)/poly (ethylene glycol diacrylate) membranes for separation of CO2/H2 (syngas application) *J. Membr. Sci.* **458** 14–26

[24] Ko Y G, Shin S S and Choi U S 2011 Primary, secondary, and tertiary amines for CO_2 capture: Designing for mesoporous CO_2 adsorbents *J. Colloid Interface Sci.* **361** 594–602

[25] Ilyas A, Muhammad N, Gilani M A, Vankelecom I F and Khan A L 2018 Effect of zeolite surface modification with ionic liquid [APTMS][Ac] on gas separation performance of mixed matrix membranes *Sep. Purif. Technol.* **205** 176–83

[26] Lin H, Kai T, Freeman B D, Kalakkunnath S and Kalika D S 2005 The Effect of cross-linking on gas permeability in cross-linked poly(ethylene glycol diacrylate) *Macromolecules* **38** 8381–93

[27] Zhang J, Wang N, Liu W, Zhao X and Lu W 2013 Intermolecular hydrogen bonding strategy to fabricate mechanically strong hydrogels with high elasticity and fatigue resistance *Soft Matter* **9** 6331–7

[28] Ismail A F, Khulbe K C and Matsuura T 2015 *Gas Separation Membranes: Polymeric and Inorganic*: Springer International Publishing, Cham) pp 37–192

[29] Gray M L, Soong Y, Champagne K J, Pennline H, Baltrus J P, Stevens R W, Khatri R, Chuang S S C and Filburn T 2005 Improved immobilized carbon dioxide capture sorbents *Fuel Process. Technol.* **86** 1449–55

[30] Yu X, Wang Y, Li L, Li H and Shang Y 2017 Soft and wrinkled carbon membranes derived from petals for flexible supercapacitors *Sci. Rep.* **7** 45378

[31] Fu Y J, Hu C C, Lee K R and Lai J Y 2006 Separation of ethanol/water mixtures by pervaporation through zeolite-filled polysulfone membrane containing 3-aminopropyltrimethoxysilane *Desalination* **193** 119–28

[32] Liu L, Han X, Hu W, Zhao B and Fan A 2017 Desulfurization performance of polydimethylsiloxane membranes by pervaporation: Effect of crosslinking agents *Polym. Eng. Sci.* **57** 1127–35

[33] Ilyas A, Muhammad N, Gilani M A, Ayub K, Vankelecom I F J and Khan A L 2017 Supported protic ionic liquid membrane based on 3-(trimethoxysilyl)propan-1-aminium acetate for the highly selective separation of CO_2 *J. Membr. Sci.* **543** 301–9

[34] Ismail A F, Kusworo T, Mustafa A and Hasbulla H 2005 Understanding the solution-diffusion mechanism in gas separation membrane for engineering students. In: *Regional Conference on Engineering Education RCEE,*

[35] Ashjari H R, Ahmadi A and Dorraji M S S 2018 Synthesis and employment of PEGDA for fabrication of superhydrophilic PVDF/PEGDA electrospun nanofibrous membranes by in-situ visible photopolymerization *Korean J Chem Eng* **35** 289–97

[36] Panwar K, Jassal M and Agrawal A K 2015 In situ synthesis of Ag–SiO_2 Janus particles with epoxy functionality for textile applications *Particuology* **19** 107–12

Optimization of lipids extraction from Moringa oleifera seeds for biodiesel feedstock

A N S Fizal[1], N H Zainordin[1], M I M Jamil[1], H A Hamid[1], N A Khalil[1], M D S Hossain[2], M Zulkifli[1] and A N A Yahaya[1]

[1] Universiti Kuala Lumpur, Malaysian Institute of Chemical and Bioengineering Technology (MICET), Melaka, Malaysia
[2] Division of Environmental Technology, School of Industrial Technology, Universiti Sains Malaysia (USM) Penang, Malaysia

Email: ahmadnaim@unikl.edu.my

Abstract. This paper explore the mechanism of lipid extraction efficiency on Moringa oleifera seeds using Soxhlet extraction method. This present study essential to determine the effect of particle size of the sample, extraction time and type of solvent applied towards the efficiency in extracting the lipids from the material. Soxhlet extraction method utilizing Buchi B118 was use in this study and response surface method was applied to analysed the data and determine the optimum parameter condition to obtain the highest yield of Moringa oil extraction. Moringa oil derived from Moringa oleifera seeds was converted into biodiesel (FAME) via Transesterification process. Conversion of Moringa FAME was observed using three different alcohol oil to molar ratio by based-catalysed. This study shows significant strong correlations between particle size of the sample, extraction time and type of solvent use towards extraction yield. The Response surface analysis shows that 1.3611 mm particle size of sample, 3 hours of extraction time and hexane as extraction solvent was the optimum condition in order to get the highest yield of lipids extract from both Moringa oleifera samples. Authentication extraction based on RSM recommendation showed an average Moringa oil yield of 39.75 % by weight.

Keywords: Moringa oleifera, Soxhlet Extraction, Lipids, FAME

1. Introduction

Biodiesel or also known as fatty acid methyl ester (FAME) is an alternative to conventional diesel or petroleum diesel used in compression ignition engines [1]. It is derived from different types of feedstock such as vegetable oils, animal fats, waste oils or microalgae [2]. Basically, biodiesel or FAME is obtained when feedstock such as vegetable oil or animal fat is reacted chemically with alcohol such as methanol with the presence of a catalyst to produce [3] . The use of biodiesel is important as it lower the dependency on fossil fuels, which will deplete in the future and pose harmful threats towards the environment. Biodiesel can be used to improve diesel in terms of lubricity, performance of fuel and lengthen engine life due to its properties which is similar to conventional diesel [4] .This is achieved by blending the biodiesel with conventional diesel based on different ratio depending on the usage [5]. The advantage of biodiesel is its high cetane number which gives better quality in terms of fuel ignition [6]. The disadvantages of biodiesel are its lower heating value, higher density and higher viscosity compared to conventional diesel which makes it unsuitable for usage in cold climate region [3]. Biodiesel quality is highly dependent on the quality of feedstock utilized and parameters involved during production [4].

1.1. Moringa oleifera

Moringa oleifera commonly known as drumstick tree (English), horseradish tree, ben oil tree, benzoil tree, benzolive (French), kelor, marango, mlonge, mulangay, saijhan, sajna (India) and munggai in Malaysia is a species of oilseed tree from Moringaceae family [7]. The tree is native to northwest India, Africa, Arabia, Southeast Asia, the Pacific and Caribbean Island and South America [1]. Moringa oleifera is regarded as an important commodity due to its nutritional value and medicinal uses [1]. The leaves are used traditionally to treat fever, indigestion and eyes problems [8]. All parts of Moringa oleifera which includes the leaves, flowers immature pods and fruit are used to produce dehydrated Moringa, Moringa powder, Moringa pulp and Moringa seed oil [9]. Moringa fruit has the length range of 25-40 cm. It contains about 20 seeds which are globular in three-angled shape with average weight of 0.3 g and has a diameter of about 1 cm where the seed account to 70-75 % of the weight of the seeds [10]. The seeds contain about 19-47% of oil and the oil is rich in palmitic, oleic, stearic and behenic acids [11]. The seed is rich in oil and protein and can be used in various applications from cosmetic product to environmental water treatment [12]. several researchers have proposed Moringa Oleifera as a potential feedstock in biodiesel production [1, 6, 13]. The focus of this study is to identify the optimize condition for the extraction of Moringa oil based on particle size of seeds, the extraction time and the type of solvent.

Figure 1. Moringa Oleifera seed.

2. Material and Methods

2.1. Sample

Moringa oleifera seeds were obtained from a local supplier. Good seeds were selected and separated from the damaged seeds. The shells were cracked, and the seeds are removed carefully from the shells. The seeds were dried in a drying oven to lower the moisture content to below 4 %. The seeds were then grinded and sieved to selected particle sizes which were 0.85 mm, 1.0 mm and 2.0 mm using plates sieve shaker [14].

2.2. Soxhlet Extraction

Oil was extracted using Buchi B811 Soxhlet extractor. The oil was extracted based on different solvent, particle size and extraction time [14, 15]. The solvents for the extraction process are n-hexane and ethanol. Extraction time was set for 1, 2, and 3 hours. Three selected particle sizes was used in this study are between 2.0 mm, 1.0 mm and 0.85 mm. 15 g of ground samples were placed into a thimble and put into the extraction chamber. 200 mL of n-hexane/ ethanol was poured into the solvent flask. The ratio of solvent to samples was kept constant for 13.33:1 w/w. Soxhlet warm was chosen as the extraction method and suitable process parameter was set. The upper heating is set at 4 and lower heating is set at 10. Extraction time was set at 1 to 3 hours. The extraction process for three consecutive variables was repeated for three times. The extraction process then continued with the other three consecutive variables until all extraction process variables are complete. The oil will be separated from the solvent using Buchi rotary evaporator. The pressure of rotary evaporator was set to 200 millibar to provide a condition under vacuum pressure. The temperature of the water bath was set above the boiling point of solvent under vacuum pressure. The water bath temperature was set to 60 °C [16]. The oil yield percentage was calculated using the formula shown below.

$$Yield\ of\ moringa\ oil\ \left(\frac{W}{W}\%\right) = \frac{Mass\ of\ moringa\ oil\ (g)}{Mass\ of\ moringa\ seeds\ used\ (g)} x\ 100\% \quad (1)$$

3. Results & Discussion

3.1. Soxhlet Extraction

Soxhlet extraction was performed utilizing three different particle sizes of the samples with three different extraction time were done to find out the effect of particle size of the sample and extraction time of Soxhlet extraction method towards the yield production of extracted oil. These two variables were extracted using two different solvents: hexane and ethanol. Figure 2 shows the data from the extraction process using hexane. The highest oil yield recorded for 1 hour extraction time was 35.47 % from 1.0 mm samples, followed by 28.34 % from 0.85 mm samples and 25.37 % from 2.0 mm samples. The highest oil yield recorded for 2 hours extraction time was 39.12 % from 1.0 mm samples, followed by 30.98 % from 0.85 mm samples and 29.63% from 2.0 mm samples. The highest oil yield recorded for 3 hours extraction time was 40.85 % from 1.0 mm samples, followed by 35.65 % from 0.85 mm samples and 33.62 % from 2.0 mm samples. The 1.0 mm samples yielded the highest oil percentage for each extraction time. 2.0 mm size samples yielded the lowest oil percentage for each extraction time.

Figure 3 shows the data from the extraction process using ethanol. The highest oil yield recorded for 1 hour extraction time was 16.40 % from 0.85 mm samples, followed by 15.15 % from 1.0 mm samples and 10.15 % from 2.0 mm samples. The highest oil yield recorded for 2 hours extraction time was 20.40 % from 0.85 mm samples, followed by 18.40 % from 1.0 mm samples and 17.85 % from 2.0 mm samples. The highest oil yield recorded for 3 hours extraction time was 27.20 % from 0.85 mm samples, followed by 26.50 % from 1.0 mm samples and 21.00 % from 2.0 mm samples. The 0.85 mm samples yielded the highest oil percentage for each extraction time. The 2.0 mm samples yielded the lowest oil percentage for each extraction time. N-hexane is known as a non-polar solvent meanwhile, ethanol is a polar solvent. The general rule of solvent is "like dissolves like" which means a polar solvent will more be attracted to polar molecules and vice versa [17, 18]. Since the candlenut oil consists of five major fatty acids in its oil composition (palmitic acid, stearic acid, oleic acid, linoleic and linolenic acid), it is likely to be dissolved in n-hexane compared to ethanol. Previous study also shown that palmitic acid is more soluble in hexane compared to ethanol [19].

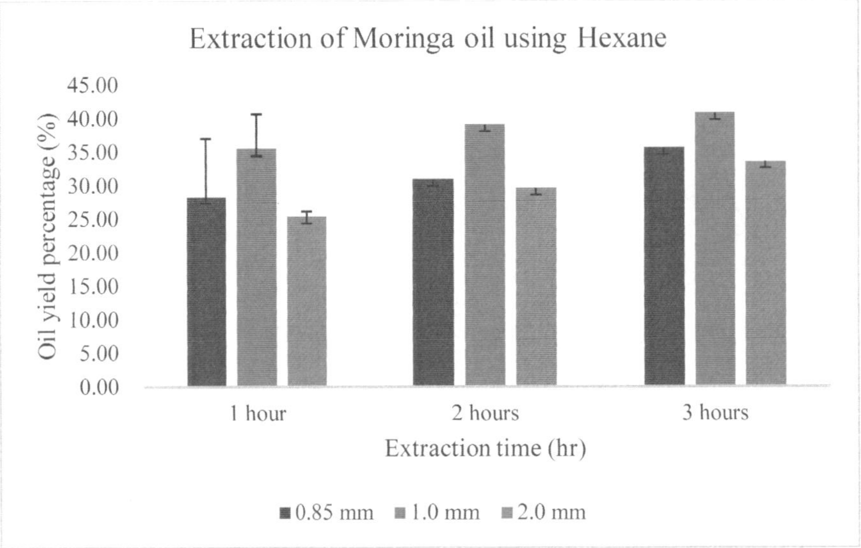

Figure 2. Percentage of Moringa oil extracted using hexane as solvent.

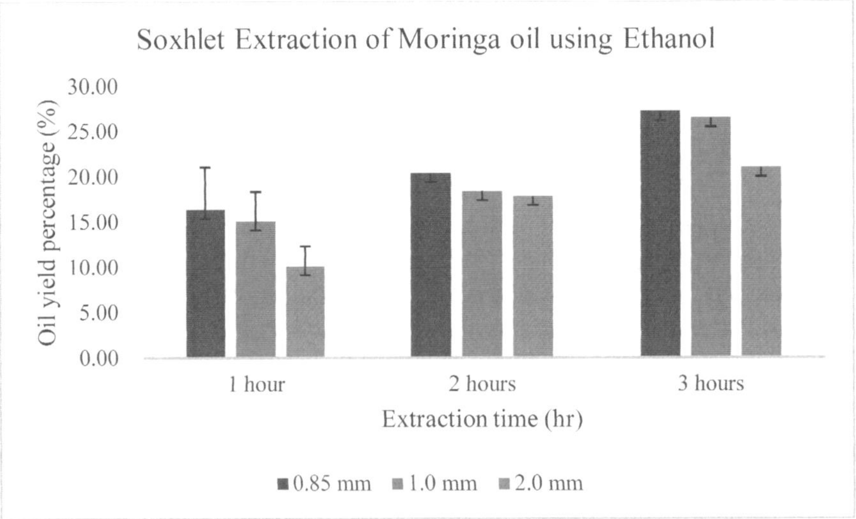

Figure 3. Percentage of Moringa oil extracted using ethanol as solvent.

3.2. Relations between Particle Size, Extraction Time and Solvent in Soxhlet Extraction

Figure 4 shows the data obtained from the first parameter of the extraction process, which was particle size, and its effect in terms of average oil yield obtained for two different solvent. The highest average oil yield obtained when hexane was used as extraction solvent was 38.48 % from 1.0 mm samples followed by 31.65 % from 0.85 mm samples and 29.54 % from 2.0 mm samples. The oil yield increased when the particle size of the samples was increased from 0.85 mm to 1.0 mm and decreased back when the particle size of the samples increased from 1.0 mm to 2.0 mm. The maximum oil yield obtained when hexane was used was not from the lowest particle size samples. The highest average oil yield obtained when ethanol was used as extraction solvent was 21.33 % from 0.85 mm samples followed by

20.02 % from 1.0 mm samples and 16.33 % from 2.0 mm samples. The oil yield continued to decrease when the particle size of the samples was increased from 0.85 mm to 1.0 mm until 2.0 mm. The maximum oil yield obtained when ethanol was used was from the lowest particle size samples.

Small particle size have a wider surface area in contact with the solvent compared to large particle size which increases the oil yield [20]. The data obtained from extraction using hexane showed otherwise which could be explained from the small particle size samples which sometimes has high chance of floating in the extraction solvent which reduce the contact with the solvent therefore affecting the extraction efficiency [21]. Masime *et al.* [20] find a similar finding where maximum oil yield was not obtained by the smallest particle size samples. Smaller particle size samples also have tendency to agglomerate which cause surface contact area between sample and solvent to decrease which results in decrease of oil yield [22]. Similar findings was recorded where the oil yield percentage decreased when the particle size of the sample were increased as there are less surface area contact with the solvent during the extraction process [10, 23]. This shows that the decreasing in particle size of sample in an extraction process helps in improving solid–fluid contact during the extraction process which in return increase the yield of oil [24].

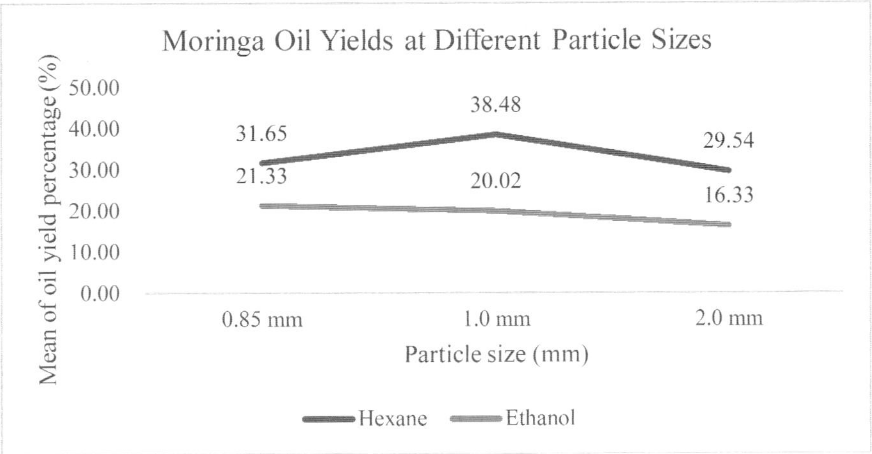

Figure 4. Percentage of Moringa oil extracted at different particle sizes.

Figure 5 illustrate the data obtained from the second parameter of the extraction process which was extraction time and its effect in terms of average oil yield obtained for two different solvent. The highest oil yield obtained for extraction using hexane was 36.7 % from 3 hours extraction time, followed by 33.24 % from 2 hours extraction time and 29.72 % from 1 hour extraction time. The highest oil yield obtained for extraction using ethanol was 24.9 % from 3 hours extraction time, followed by 18.88 % from 2 hours extraction time and 13.9 % from 1 hour extraction time. The same trend was observed in both extraction data where oil yield increased when the extraction time was increased. The increase in the oil yield over time is due to the increase in the diffusivities of oil and solvent over time [10]. Similar findings was recorded by several researchers which used the same extraction time [20, 25–27].

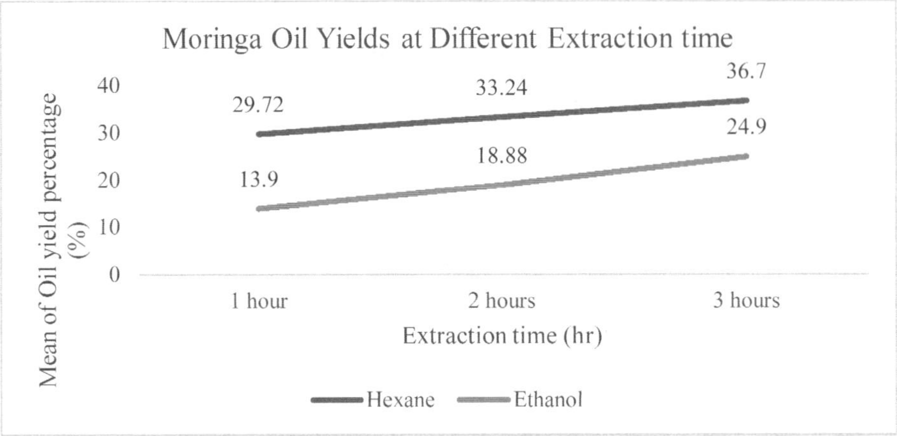

Figure 5. Percentage of Moringa oil extracted at different extraction time.

Figure 6 display data on the effect of type of solvent used in terms of average oil yield. The average oil yield obtained when using hexane as extraction solvent was 33.22 % while the average oil yield obtained when using ethanol as extraction solvent was 19.23 %. The average oil yield obtained when hexane was used as extraction solvent was relatively higher compared to ethanol. The difference in terms of average oil yield recorded between the two types of solvent was high. This might be explained from the characteristic of hexane which is a non-polar solvent compared to ethanol which is a polar solvent [28]. The difference in the oil yield between the two solvent could also be explained by the presence of different compound in the sample used [29]. Alkane compounds are non-polar and can easily be extracted by hexane while the polar compounds can easily be extracted by ethanol since they have O-H ends which enable them to react with hydrogen bond presence in the ethanol [29]. Similar finding was recorded by Zhao & Zhang [29] who used both type of solvents in extraction of Eucalyptus leaves oil. Since Moringa oil contain fatty acids which are non-polar, hexane is more suitable to be used as extraction solvent compared to ethanol [28].

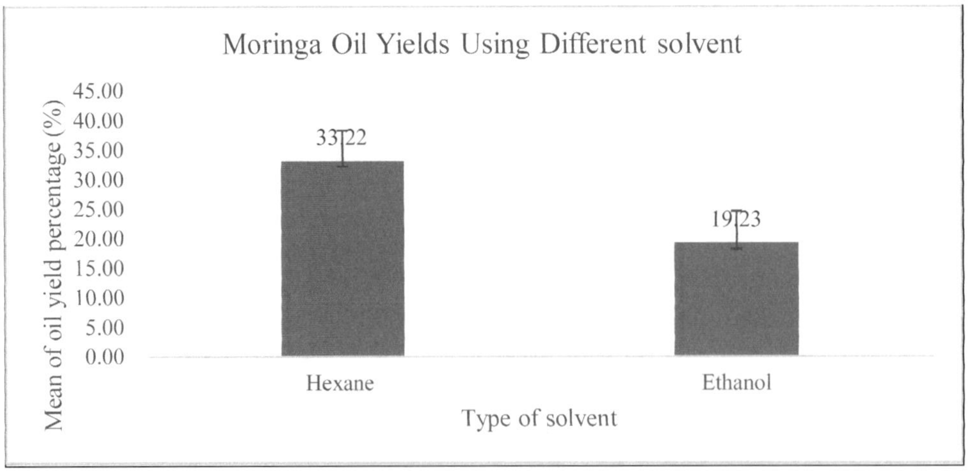

Figure 6. Percentage of Moringa oil extracted using different solvent.

3.3. Optimization of Extraction (Response Surface Method)

Response surface plot for extraction of Moringa oil using hexane are shown in figure 7 and figure 8 below. The relationship of the two extraction parameters which are particle size and extraction time with the oil yield is shown. The highest oil yield from hexane extraction was obtained from the highest extraction time which was 3 hours and the middle particle size used which was 1.0 mm. The highest oil yield was not obtained from the lowest particle size used in the extraction, which was 0.85 mm. Figure 8 shows that the highest oil yield from ethanol extraction was obtained from the highest extraction time s well which was 3 hours and the smallest particle size used which was 0.85 mm. The difference compared to extraction using hexane can be seen in which the highest oil yield from ethanol extraction was obtained was from the lowest particle size used in the extraction which was 0.85 mm. This can be explained by a few reasons. According to Yeop *et al.* [21] smallest particle size samples sometimes tend to float in the solvent of extraction which cause the contact of the samples with the solvent to be reduced and affect the diffusivity of the solvent into the samples to extract oil. Small particle size samples also tend to clump together and cause the surface area in contact with the solvent to be reduced which in return reduce the amount of oil extracted [22]. Similar finding also was recorded where the maximum oil yield from the smallest particle size used in the extraction [20]. The increase in the oil yield percentage alongside the decrease in particle size of the sample was also observed in few studies [10, 23]. Another set of extraction was performed to verify the optimize condition provided by the RSM simulation where 1.4 mm particle size sample, 3 hours of extraction time and hexane as extraction solvent were used. The average Moringa oil yield obtained from the extraction was 39.75 % by weight. This display a significant improvement compared to previous extraction where the average Moringa oil yield using hexane was to be at 33.22 % by weight.

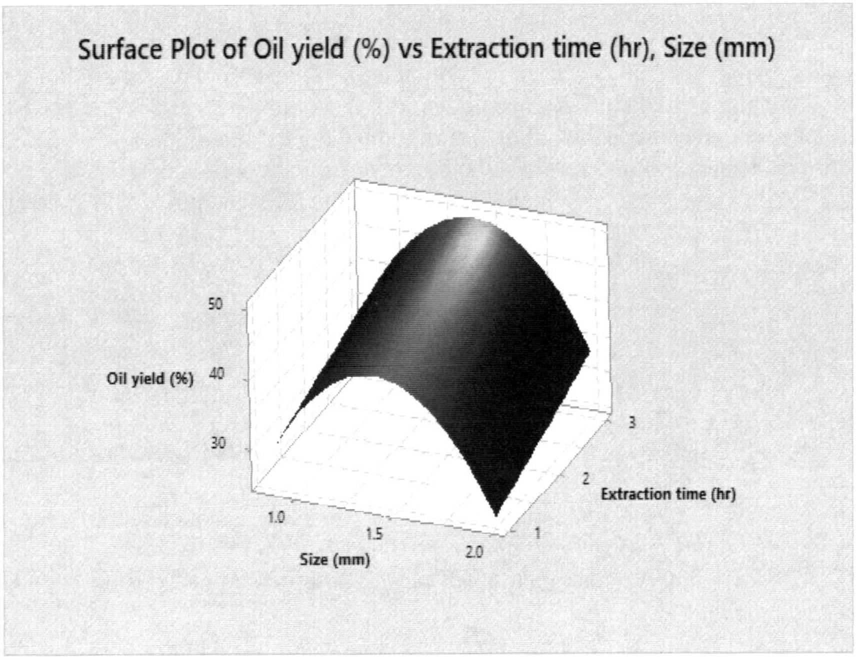

Figure 7. Response surface plot of oil yield at different particle size and extraction time using hexane.

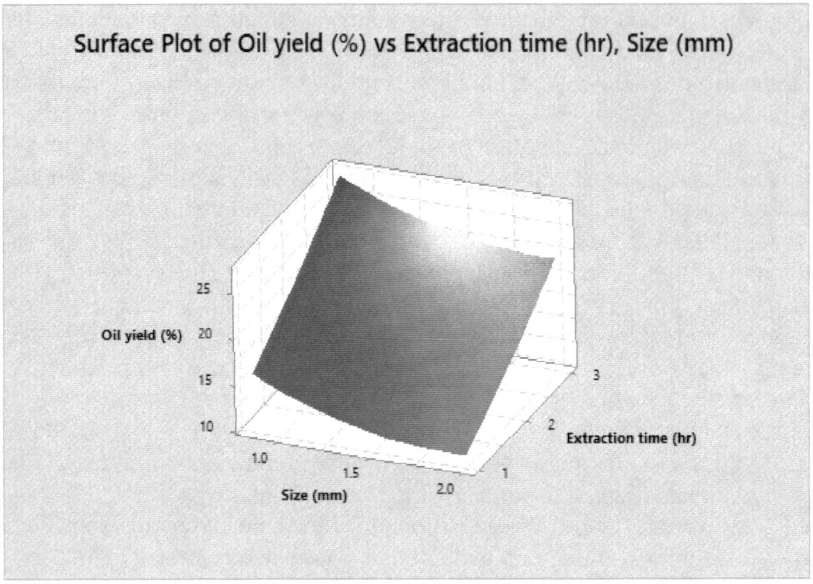

Figure 8. Response surface plot of oil yield at different particle size and extraction time using ethanol.

4. Conclusions

For the conclusion of this study, it could be summarized that hexane produce more extraction yield compared to ethanol signifies that Moringa oil contained more non-polar compound. The extraction output illustrate that smaller particle size sample tend to clump together and cause the surface area in contact with the solvent to be reduced which in return reduce the amount of oil extracted. The optimized extraction conditions obtained based on the whole extraction data are using 1.36 mm particle size of sample, 3 hours of extraction time and hexane as extraction solvent which could give an oil yield of 42.50 %. Verification extraction was made using the condition provided by RSM displayed an average Moringa oil yield of 39.75 % which highly improved compared to previous extraction at an average of 33.22 % by weight.

Reference

[1] Rashid U, Anwar F, Moser B R, and Knothe G 2008 Moringa oleifera oil: A possible source of biodiesel *Bioresour. Technol.* **99** 8175–9

[2] Issariyakul T and Dalai A K 2014 Biodiesel from vegetable oils *Renew Sustain Energy Rev.* **31** 446–71

[3] Ali E N and Tay C I 2013 Characterization of biodiesel produced from palm oil via base catalyzed transesterification *Procedia Eng.* **53** 7–12

[4] Blasio C D 2019 Biodiesel: Fundamentals of biofuels engineering and technology *Green Energy and Technology.* 253–6

[5] Talha N S and Sulaiman S 2016 Overview of catalysts in biodiesel production *ARPN J. Eng. Appl. Sci.* **11** 439–42

[6] Azad A K, Rasul M G, Khan M M K, Sharma S C and Islam R 2015 Prospect of Moringa seed oil as a sustainable biodiesel fuel in Australia: A review *Procedia Eng.* **105** 601–6

[7] Sagona W C J , Chirwa P W and Sajidu S M 2020 The miracle mix of Moringa: Status of Moringa research and development in Malawi *South African J. Bot.* **129** 138–45

[8] Machado S D I, Gastélum J A N, Moreno C R, Wong B R and Cervantes J L 2010 Nutritional quality of edible parts of Moringa oleifera *Food Anal. Methods.* **3** 175–80

[9] Daba M 2016 Miracle tree: A review on multi-purposes of Moringa oleifera and its implication for climate change mitigation *J. Earth Sci. Clim. Change.* **7** 336

[10] Mani S, Jaya S and Vadivambal R 2007 Optimization of solvent extraction of Moringa oleifera

seed kernel oil using response surface methodology *Food Bioprod. Process.* **85** 328–35
[11] Bhutada P R , Jadhav A J, Pinjari D V, Nemade P R and Jain R D 2016 Solvent assisted extraction of oil from Moringa oleifera Lam. seeds *Ind. Crops Prod.* **82** 74–80
[12] Ndabigengesere A and Narasiah K S 1998 Quality of water treated by coagulation using Moringa oleifera seeds *Water Res.* **32** 781–91
[13] Palafox J O, Navarrete A, Rivero J C S, Atoche C R, P.Escoffie P A and Uribe J A R 2012 Extraction and characterization of oil from Moringa oleifera using supercritical co2 and traditional solvents *Am. J. Anal. Chem.* **3** 946–9
[14] Zeković Z, Filip S, Vidović S, Jokić S and Svilović S 2014 Mathematical modeling of ocimum basilicum L. supercritical CO2 extraction *Chem. Eng. Technol.* **37** 2123–8
[15] Ramluckan K, Moodley K G and Bux F 2014 An evaluation of the efficacy of using selected solvents for the extraction of lipids from algal biomass by the soxhlet extraction method *Fuel* **116** 103–8
[16] Cabral M R.P, Santos S A L D, Stropa J M, Silva R C L, Cardoso C A L, Oliveira L C S, Scharf D R, Simionatto E L, Santiago E F and Simionatto E 2016 Chemical composition and thermal properties of methyl and ethyl esters prepared from Aleurites moluccanus (L.) Willd (Euphorbiaceae) nut oil *Ind. Crops Prod.* **85** 109–16
[17] Guzman A 1962 Solubility of triglycerides in aqueous ethanol *IOWA State Cap Dig Repo*
[18] Harris D C 2015 Solvent Polarity Index *Quantitative Chemical Analysis.* **9** 2015
[19] Calvo B, Collado I and Cepeda E A 2009 Solubilities of palmitic acid in pure solvents and its mixtures *J. Chem. Eng. Data.* **54** 64–8
[20] Masime J O, Ogur E, Mbatia B, Aluoch A O and Otieno G 2017 Optimization and thermodynamics of the extraction of yellow oleander seed oil using soxhlet extractor *Optimization.* **2** 43–50
[21] Yeop A, Sandanasam J, Pan S F, Abdulla S, Yusoff M M and Gimbun J 2017 The effect of particle size and solvent type on the gallic acid yield obtained from Labisia pumila by ultrasonic extraction *MATEC Web Conf.* **111** 1–5
[22] Shigidi I and Elkhaleefa A 2015 Original article parameters optimization, modelling and kinetics of balanites aegyptiaca kernel oil extraction *Int. J. Chem. Eng. Appl. Sci.* **5** 1–4
[23] Reshad A S, Tiwari P and Goud V V 2015 Extraction of oil from rubber seeds for biodiesel application : Optimization of parameters *Fuel.* **150** 636–44
[24] Nguyen H N, Gaspillo P D, Maridable J B, Malaluan R M, Hinode H, Salim C and Huynh H K P 2011 Extraction of oil from Moringa oleifera kernels using supercritical carbon dioxide with ethanol for pretreatment: Optimization of the extraction process *Chem. Eng. Process. Process Intensif.* **50** 1207–13
[25] Alara O R , Abdurahman N H and Ukaegbu C I 2018 Soxhlet extraction of phenolic compounds from Vernonia cinerea leaves and its antioxidant activity *J. Appl. Res. Med. Aromat. Plants.* **11** 12–7
[26] Gupta S P, Singh M, Ahmed I, Sharma D N and Kumar P 2018 Effect of temperature and time intervals on the solvent extraction of essential oil from azadirachta indica (neem) leaf powder by using soxhlet extraction method *World. J. Pharma. Resch.* **8** 939–46
[27] Tesfaye B and Tefera T 2017 Extraction of essential oil from neem seed by using soxhlet extraction methods *Int. J. Adv. Eng. Manag. Sci.* **3** 646–50
[28] Ghazali Q and Yasin N H M 2016 The effect of organic solvent, temperature and mixing time on the production of oil from Moringa oleifera seeds *IOP Conf. Ser. Earth Environ. Sci* **36** 0–7
[29] Zhao S and Zhang D 2014 Supercritical CO2 extraction of Eucalyptus leaves oil and comparison with Soxhlet extraction and hydro-distillation methods *Sep. Purif. Technol.* **133** 443–51

Acknowledgement
This work was supported by Short Term Research Grant (STRG) scheme No: STR19035- 65320

Fabrication and characterization of chitosan-titanium oxide nanotubes scaffolds reinforced with tiger milk mushroom

N Rosli[1], H-S Loh[2], C L Chiang[3] and S S Lim[1]

[1] Department of Chemical and Environmental Engineering, Faculty of Science and Engineering, University of Nottingham Malaysia, Jalan Broga, 43500 Semenyih Selangor.
[2] School of Bioscience, Faculty of Science and Engineering, University of Nottingham Malaysia, Jalan Broga, 43500 Semenyih Selangor.
[3] Foundation of Engineering, Faculty of Science and Engineering, University of Nottingham Malaysia, Jalan Broga, 43500 Semenyih Selangor.

Email: ebxnrl@nottingham.edu.my

Abstract. Chitosan-based scaffolds have been reported to promote cellular activities but lack mechanical strength which is much sought after for bone regeneration. The current research work aided to reinforce chitosan-based scaffolds with tiger milk mushroom (TMM) powder, a naturally occurring polysaccharide. Scaffolds of chitosan-titanium oxide nanotubes (TNTs) reinforced with tiger milk mushroom (TMM-CTNTs) were fabricated via direct-blending and freeze-drying methods. Prior to that, TNTs were hydrothermally synthesized and blended with chitosan solution and TMM powder at 1-5 weight percent (wt %). The pore size, microstructure, porosity, swelling, degradation, compressive modulus and functional groups of resultant scaffolds were characterized. These cylindrical scaffolds of TMM-CTNTs showed pore size of 48 – 68 μm. The addition of TMM from 3 wt% to 5 wt% in scaffolds reduced the porosity from 81.7 % to 79.9 %. The compressive modulus of 3 wt%-5 wt% TMM-CTNTs scaffolds increased %from 0.013 MPa – 0.038 MPa. The incorporation of TMM influenced the swelling property of scaffolds. The swelling percentage of TMM-CTNTs reduced from 400% to 373% as TMM powder was introduced from 1 wt% to 5 wt%. The degradation ratio increased from 0.959% to 2.385 % as TMM powder was introduced from 1 wt% to 5 wt%. The Fourier-Transform Infrared (FTIR) spectra of TMM-CTNTs scaffolds revealed the presence of β-glucan which verified that the processing methods in this study preserved the medicinal property of TMM. A preliminary *in vitro* test, MTT assay, was used to study proliferation rate of MG63 (osteoblast-like cells) cultured on TMM-CTNTs scaffolds with different weight percent of TMM. Notably, the cells proliferation of MG63 showed high biocompatibility at 3 days of culture.

Keywords: titanium oxide nanotubes, tiger milk mushroom, chitosan

1. Introduction

The evolution in biomaterial research field has been tremendously become a major attention to researchers in the development of orthopedic tissue engineering. Manufacturing of artificial composite scaffolds using natural bioactive polymer is in high demand compared to synthetic polymer. The physical and chemical characteristic of natural bioactive polymer are similar to the structure of extra cellular matrix (ECM) in human tissues. The designed scaffolds should be capable to mimic ECM functions as a transporter of nutrient, oxygen and metabolic by product at the defect sites within a few weeks of implantation. It is essential for scaffolds to be nontoxic and safe to be used in order to promote normal cellular activities [1].

Chitosan, a natural abundant resource, is a semi-synthetic polymer which is derived from chitin [2]. This linear polysaccharide composed of glucosamine and N-acetyl glucosamine units which can commonly be found in skeletal of fungi, crustaceans and insects. Chitosan has excellent biodegradability, anti-bacterial activity, biocompatibility and non-antigenicity that make it suitable for tissue engineering. Ideal scaffolds should be designed with porous three-dimensional (3D) network, high mechanical strength and controllable degradation rate for renewal and replacement of new bone tissue. However, the fabrication of chitosan-based scaffolds has been not much attractive to researchers due to its insufficient mechanical strength for load bearing usage [3], uncontrollable degradation rate [4] and high acidity and concentration solution [5]. Up to now, there are only a few studies reported on designing scaffolds using chitosan as a based material in orthopedic tissue engineering [6-7]. Chitosan as stand-alone material for scaffolds fabrication could not fulfill ideal requirement as temporary three-dimensional structure for cell growth. To overcome these challenges, reinforcement of chitosan-based scaffolds with organic and inorganic compound can produce nanocomposites scaffolds with improved properties.

Designation of scaffolds template in nano range can effectively enhance surface area and permeability of scaffolds. Therefore, introduction of titanium oxide nanotubes (TNTs), inorganic ceramic materials, is the best material selection in designing porous scaffolds with firm structure. Titanium oxide nanotubes possess great unique features of large surface area, good corrosion resistance, excellent biocompatibility and enhanced bioactivity especially in terms of osteoblast adhesion and proliferation [8-9]. Previous study by Lim *et al.* involved the preparation of chitosan-TNTs scaffolds which showed improved compressive modulus and excellent biocompatibility by promoting enhanced adhesion, proliferation and early differentiation of MG63 cells [2].

Lignosus rhinocerusis commonly known as tiger milk mushroom (TMM), is a medical mushroom which is under Polyporaceae family has been used by native communities in treating cough, fever, asthma, joint pain, cancer and food poisoning. The sclerotium part of TMM has been investigated in a few studies and showed tremendous medicinal benefits, biopharmacological properties and functional food potentials. There are a few research works revealed that TMM function as anti-cancer, anti-inflammatory, immune-modulatory and antioxidant activities [10-12]. These fabulous properties naturally exist in TMM due to β-glucan, a bioactive compound presence in tuber part of sclerotium. Current research in tissue engineering have not explored the potential of β-glucan material in scaffolds. Therefore, chitosan, TNTs and TMM are desired compounds since they are safe and have been reported with high *in vitro* efficacy. This spectacular invention may be useful in improving mechanical strength and enhanced scaffolds compatible performance with living tissues during cellular activities for bone regeneration.

In this paper, TMM-CTNTs scaffolds were fabricated using direct blending and freeze-drying method. The pore size, microstructure, porosity, swelling, degradation, compressive modulus and functional groups of resultant scaffolds were characterized. A preliminary *in vitro* test of cell proliferation on TMM-CTNTs scaffolds using MG63 (osteoblast-like cells) was conducted. The effects of TMM weight percent in chitosan scaffolds on the physical properties and proliferation of MG63 were studied.

2. Materials and methods

2.1. Materials

Titanium dioxide (TiO_2) powder, chitosan powder (middle viscous, 80 % deacetylated), sodium bicarbonate, sodium pyruvate, and dimethyl sulphoxide (DMSO) were purchased from Sigma Aldrich (Germany). NaOH pellets were purchased from R&M Chemicals (UK). Acetic acid, hydrochloric acid and ethanol were purchased from Fisher Scientific (USA). Chemicals for the preparation of phosphate buffered saline (PBS) including sodium chloride (NaCl), potassium chloride (KCl), sodium phosphate dibasic, monopotassium and phosphate were purchased from Sigma Aldrich (Germany). Tiger Milk Mushroom (TMM) powders were purchased from LiGNO (Malaysia). 24-well plates and 96-well plates were purchased from Orange Scientific (Belgium). Minimum essential media (MEM), fetal bovine serum (FBS), peniccilline/streptomycin, 0.25 % trypsin and 3-[4,5-dimethylthiazole-2-yl]-2,5-

diphenyltetrazolium bromide (MTT) were purchased from Gibco (USA). Distilled water was used throughout the sample preparation.

2.2. Fabrication of tiger milk mushroom incorporated chitosan-titanium oxide nanotubes (TMM-CTNTS) scaffolds

Titanium oxide nanotubes (TNTs) were hydrothermally synthesized by mixing 2 g of bulk TiO_2 powder in 40 mL of 10 M NaOH in a Teflon container, put in a stainless steel reactor, and heated 150 °C for 72 h [13]. The hydrothermal product was filtered and washed with distilled water and 0.1 M of hydrochloric acid (HCl) until pH 7 was achieved. The white filter cake was dried overnight in the oven (Binder, United Kindom) at 60 °C and finally was sintered at 400 °C for 5 h. The fabrication of tiger milk mushroom incorporated into chitosan-titanium oxide nanotubes (TMM-CTNTs) scaffolds was referred to a method developed by Lim *et al.* [2] The titanium oxide nanotubes (TNTs) and tiger milk mushroom (TMM) powder (LiGNO, Malaysia) at different weight percent (wt%) were mixed with chitosan solution using a hotplate at 700 rpm for 5 h (see table 1). Chitosan solution was prepared by dissolving 1 g of chitosan flakes in 100 mL of 0.2 M acetic solution at room temperature. The solution was constantly stirred at 300 rpm for 3 h. Next, the colloidal suspension was transferred into a 24-well plate and kept frozen at -20 °C for 24 h. The frozen scaffolds were freeze-dried at -40 °C for 24 h. Finally, scaffolds were rehydrated with absolute alcohol for 1 h and then followed by 30 min rehydration of 70 % alcohol and 50 % alcohol respectively. Rehydrated scaffolds were finally dried prior to characterization.

Table 1. Chemical composition of scaffold samples.

Sample identification	Weight percent of titanium oxide natubes (wt%)	Weight percent of Tiger milk mushroom (wt%)	Chitosan solution (mL)
TMM0	16	0	24
TMM1	15	1	24
TMM2	14	2	24
TMM3	13	3	24
TMM4	12	4	24
TMM5	11	5	24

2.3. Microstructural, physical and chemical characterization of tiger milk mushroom incorporated chitosan-titanium oxide nanotubes (TMM-CTNTS) scaffolds

The pore size and surface morphology of TMM0 – TMM5 scaffolds at different weight percent of TMM powder ranging from 0 to 5 % were characterized using Field Emission Scanning Electron Microscope (FESEM) (FEI QUANTA 400F, United States) at 10 – 20 kV. The pore size of scaffolds at different area was measured and the mean values were taken. The surface morphology of scaffolds was viewed under 500x magnification. The porosity of scaffolds was measured in percentage between 0 to 100 % and determined using equation (1):

$$\text{Porosity} = \left[1 - \left(\frac{Vskeletal}{Vbulk}\right)\right] \times 100\ \% \qquad (1)$$

$V_{skeletal}$ defined as the skeletal volume of scaffolds (cm^3) measured using a pycnometer (Duran, Germany). On the other hand, V_{bulk} was the bulk volume of scaffolds (cm^3) determined from formula of

TA.XT Plus Texture Analyzer (Stable Micro System, United Kingdom) was used to determine compressive modulus via the plot of stress against strain. The load cell of 500 N was used to compress scaffolds until 40 % of its original thickness at a crosshead speed of 1 mm/min. Prior to that, the dimension and thickness of scaffolds was measured using the vernier caliper.

The water absorption capacity of the scaffolds was determined by performing a swelling ratio test. The dry weight (W_d) of all scaffolds at 6 different weight percent of TMM powder ranging from 0-5 % was measured and placed in 24-well plate. Two mL of pure minimum essential media (MEM) was pipetted into each well. Scaffolds were fully immersed in MEM and incubated for 24 h. Next, MEM from each well was removed and the wet weight for all scaffolds was measured and recorded. The swelling ratio was determined using equation (2):

$$\text{Swelling ratio (\%)} = \frac{(Ww - Wd)}{Wd} \times 100 \tag{2}$$

where W_w and W_d are the weight of wet and dry scaffolds respectively.

Degradation test was conducted by immersion of scaffolds in 2 mL of MEM in a 24-well plate. Prior to that, the initial weight (W_o) of scaffolds was measured and recorded. The scaffolds at different weight percent of TMM were incubated at 37 °C for 28 days and changed medium every 2 days. After 28 days of immersion, scaffolds were washed with PBS after MEM was removed. After removal of PBS, wet scaffolds were kept frozen at -20 °C for 5 h. The samples were freeze dried at temperature -40 °C for 24 h. The weight (W_f) of freeze dried scaffolds was measured and recorded. The percentage of degradation was determined using equation (3):

$$\text{Degradation rate (\%)} = \left(\frac{\text{Wo} - \text{Wf}}{\text{Wo}}\right) \text{x } 100 \tag{3}$$

Each testing was performed triplicate on all scaffolds at different weight percent of TMM powders. The data were taken from calculated means values. The chemical compound of TMM powders and TMM-CTNTs scaffolds samples were identified using Fourier Transform Infrared Spectroscopy (FTIR) (Perkin Elmer, United States) at 400 – 4000 cm^{-1} wavelength.

2.4. In vitro cell culture

MG63 cells (ATCC, USA) were cultured in a cell culture flask and maintained using pure MEM added with 5 % FBS and 1 % sodium pyruvate. The cells in the culture flask were incubated at 37 °C with 5 % CO_2 level. Every 2 days, media were replaced to ensure cells supplemented with sufficient nutrients. Subculture and seeding of cells were proceeded when 90 – 100 % confluence of cells was achieved. During subculturing or passaging, 0.25 % trypsin was used to detach cells. Later, some of detached cells were transferred from previous culture to a new culture flask. During seeding procedure, trypsinization method was applied where cells were detached from the culture flask. All the detaching cells were transferred to falcon tube for centrifugation. Cells pellet were produced via centrifugation of cells detached with complete medium in falcon tube at 1800 rpm for 5 min. Finally, cells pellet was resuspended for a few times in complete medium. Prior of that, scaffolds samples which were placed in 24-well plate need to be sterilized under ultra violet (UV) irradiation for 15 min before being seeded with cells. The total number of cells were calculated and diluted to 2.8 x 10^5 cells ml^{-1}. Thirty thousand cells were transferred to each scaffold and empty well of polystyrene plate that served as a control. Both scaffolds and controls were top up with 2 mL of complete medium before the plate of scaffolds with positive control was placed in an incubator at 37 °C for 3, 7 and 14 days. Culture medium was changed every 2 to 3 days to maintain sufficient supplement for cells.

2.5. Cell proliferation

3-[4,5-dimethylthiazole-2-yl]-2,5-diphenyltetrazolium bromide (MTT) assay was used to study the proliferation of cells seeded on scaffolds and well (as control) for incubation duration of 3, 7 and 14 days. MTT assay relies on the reduction of MTT-formazon by mitochondrial succinate dehydrogenase of living cells. The formation of formazon as product was observed by colour change from yellow to dark purple crystal. In details, 1 mL of pure medium was pipetted in each well after discarded the cultured medium. Next, 100 μL of MTT solution with concentration of 5 mg mL^{-1} was transferred to each well before the plate containing cells seeded on scaffolds was incubated for 4 h. After 4 h of incubation, 1 mL of dimethyl sulphoxide (DMSO) was pipetted into each well and resuspended well for a few times using a micropipette. Triplicate of 300 μL purple crystal solution which was formazon product was pipetted to a 96 well-plate. The results of MTT test were evaluated in optical density (OD) values and determined by Versa Max Microplate Reader (Molecular Devices, USA) at wavelength of 570 nm. This MTT test at different incubation period was performed 3 times.

2.6. Statistical analysis

Analysis on quantitative results of *in vitro* proliferation was presented as means ± S.D (N=3) and performed via one-way ANOVA (analysis of varience) analysis. This analysis was performed using LSD and Duncan post-hoc test with $p < 0.05$ was measured as statistically significant.

3. Results and discussion

3.1. Physical and chemical characterization

Pore size and microstructures of scaffold samples (TMM0 – TMM5) were examined at magnification of 500x using FESEM. Figure 1 (TMM0 –TMM5) shows the FESEM images of TMM-CTNTs scaffolds at weight percent (wt%) of 0 – 5 % TMM. All images show 3D porous scaffolds with good interconnectivity network. The incorporation of TMM affected the macroporous size and interconnectivity diameter. The pore size of C-TNTs scaffold without TMM (TMM0) was measured averagely around 110.9 μm (see table 2). Addition of TMM into TMM1, TMM2, TMM3, TMM4 and TMM5 scaffolds obviously reduced the pore size to 35.90 μm, 60.86 μm, 67.69 μm, 49.41 μm and 63.60 μm respectively compared to TMM0 scaffold. The microporous size showed increasing trend as addition of TMM powders increased from 1 to 5 wt%. However, the pore size of TMM4 scaffold did not follow this increasing trend as the pore size reduced to 49.41 μm. The FESEM image (see figure 1 (e)) have confirmed the behaviour of TMM4 scaffold. The image showed smaller pore size with thickened interconnectivity walls compared to TMM2, TMM3 and TMM5 scaffolds. It might be due to irregular pore size distribution produced using the freeze-drying method. The nucleation and growth of ice crystals at a certain freezing rate are important in producing desired shape and size of pores by considering the thickness of the scaffold [14].

Previous studies determined the minimum requirement pore size suitable for artificial 3D scaffolds for bone regeneration lies in between 100 – 300 μm [15]. Apart of that, other reports mentioned pore size between 75 – 100 μm was a minimum requirement for bone regeneration with optimum values ranged between 100 – 135 μm [16]. As the results exhibited pore size less than 75 μm, the outcome from this study expected the reduction of pore size to micropores can effectively increase surface area and enhanced permeability of scaffolds. Hence, both positive effects contributed to production of more proteins adsorption sites which help to improve cell bioactivity such as cell attachment, proliferation and especially differentiation which rapidly increase degradation of scaffolds [17].

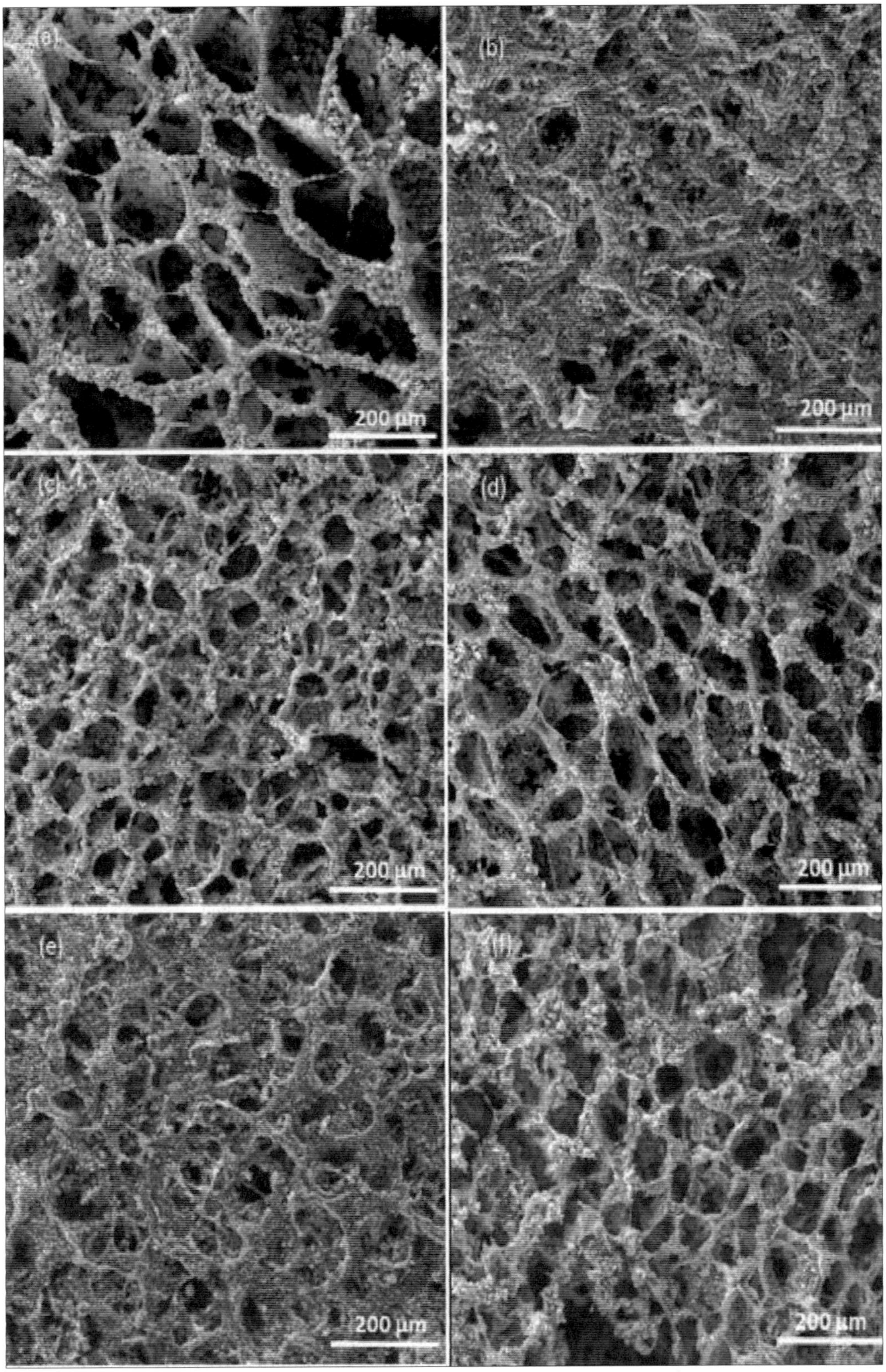

Figure 1. FESEM images of TMM-CTNTs scaffolds (a) TMM0 (b) TMM1 (c) TMM2 (d) TMM3 (e) TMM4 and (f) TMM5.

Table 2. Average pore size of scaffolds at varied wt% TMM.

Scaffold Type	**Weight percent of TMM powder (wt%)**	**Average Pore size (μm)**
TMM0	0	110.9
TMM1	1	35.90
TMM2	2	60.86
TMM3	3	67.69
TMM4	4	49.41
TMM5	5	63.60

The porosity TMM0, TMM1, TMM2, TMM3, TMM4 and TMM5 scaffolds was 83.40 %, 87.10 %, 84.10 %, 81.67 %, 82.97 % and 79.93 % respectively (as shown in table 3). From the result, addition of TMM powders at different wt% did not influence porosity of scaffold samples. The increment of TMM wt% did not show increasing or decreasing trend of porosity. Addition of TMM powders at 5 wt% showed much reduced porosity of 79.93 %. Despite of that, the porosity of TMM-CTNTs scaffolds at varied TMM wt% was in the range of optimum requirement for bone tissue regeneration which was reported in between of 30 – 90 % [18].

Table 3. Porosity of scaffolds at varied wt% TMM.

Scaffold Type	**Weight percent of TMM powder (wt%)**	**Average Porosity (%)**
TMM0	0	83.40
TMM1	1	87.10
TMM2	2	84.10
TMM3	3	81.67
TMM4	4	82.97
TMM5	5	79.93

Fabrication of scaffolds using freeze-drying method generated high porous 3D scaffolds with wide interconnected network of pore structure that works well with degradable and temperature sensitive materials like chitosan as reported in previous studies [19-20]. Freeze drying was proven to preserve the biocompatibility of chitosan scaffolds compared to other conventional drying methods like oven drying [21]. Thus, these scaffolds template were expected to be osteoconductive as it will support osteoblastic cell adhesion, proliferation and mimic the function of extracellular matrix before it degraded and replaced with new regenerated bone tissue.

The compressive modulus of TMM0, TMM1, TMM2, TMM3, TMM4 and TMM5 scaffolds showed 0.012 MPa, 0.009 MPa, 0.029 MPa, 0,013 MPa, 0.01 MPa and 0.038 MPa respectively as shown in figure 2. The incorporation of TMM into CTNTs scaffolds did not tremendously increase the compressive modulus. Overall, the compressive modulus for TMM-CTNTs scaffolds was very low. Besides, the compressive modulus of TMM0 scaffolds did not show any significant difference with TMM1, TMM3 and TMM4. The introduction of TMM in scaffolds changed compressive modulus inconsistently. However, the addition of TMM at the highest wt% (TMM5) obviously enhanced the compressive modulus to 0.038 MPa. Thus, increased wt% of TMM up to 5 %, improved the mechanical properties of scaffolds.

Ideally, increased porosity of scaffolds decreased their mechanical properties and vice versa. Based on table 3, TMM1 scaffolds with the highest porosity showed the lowest compressive modulus among scaffolds. Meanwhile, TMM5 scaffolds that exhibited the lowest porosity showed the highest compressive modulus. Thus, the results of compressive modulus and porosity were in negative correlation to each other. The compressive modulus of TMM2 scaffold exhibited greater increment compared to TMM3 and TMM4. Meanwhile, the porosity of TMM2 showed a higher value compared with TMM3 and TMM4. It was expected that higher in porosity showed lower compressive modulus. This unusual behavior might due to the agglomeration of TMM powder which acts as a filler to strengthen C-TNT scaffolds. Further investigation on the proportional relationship between porosity and compressive modulus of scaffolds could be confirmed by determining the density and linear shrinkage properties due to their unusual mechanical behavior as reported by Kim *et al.* [22].

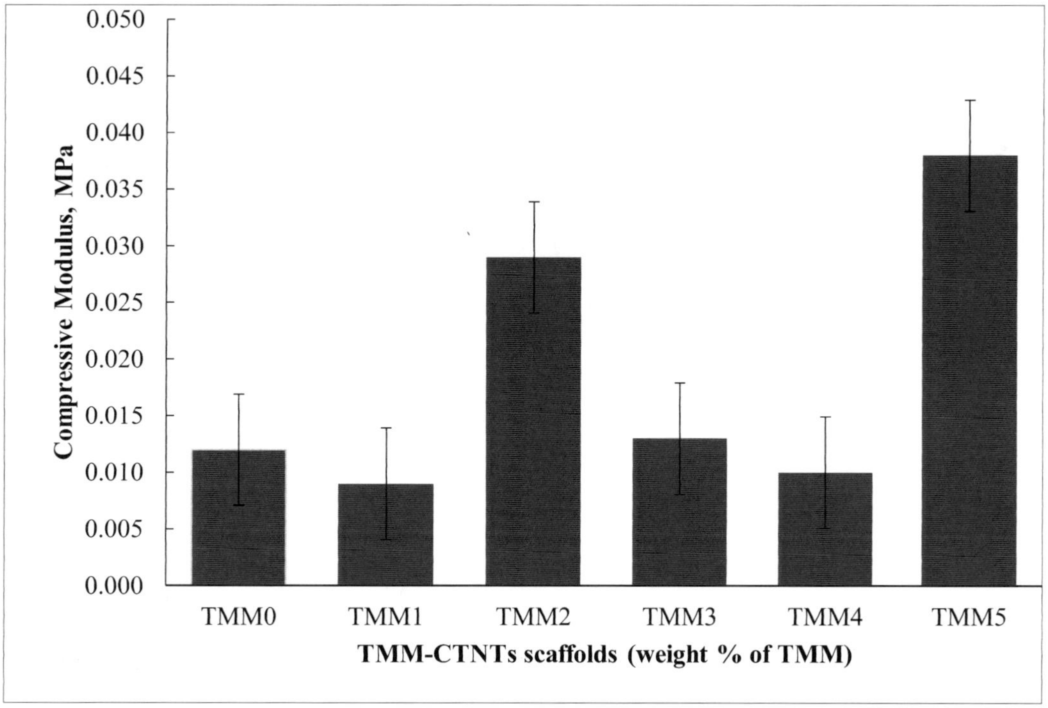

Figure 2. The compressive modulus of TMM-CTNTs scaffolds at varied wt% of TMM.

Figure 3 shows a study of swelling ratio of TMM-CTNTs scaffolds at varied weight percent of TMM. The swelling ratio of TMM0, TMM1, TMM2, TMM3, TMM4 and TMM5 scaffolds was 407.4 %, 373.1 %, 401.8 %, 388.4 %, 386.5% and 371.4 % respectively. In figure 3, introduction of TMM reduced swelling ratio of TMM-CTNTs scaffolds. This was due to the network between chitosan, TNT and TMM forming much thicker pore wall at higher TMM wt%. As a result, the penetration of water into scaffolds was also reduced.

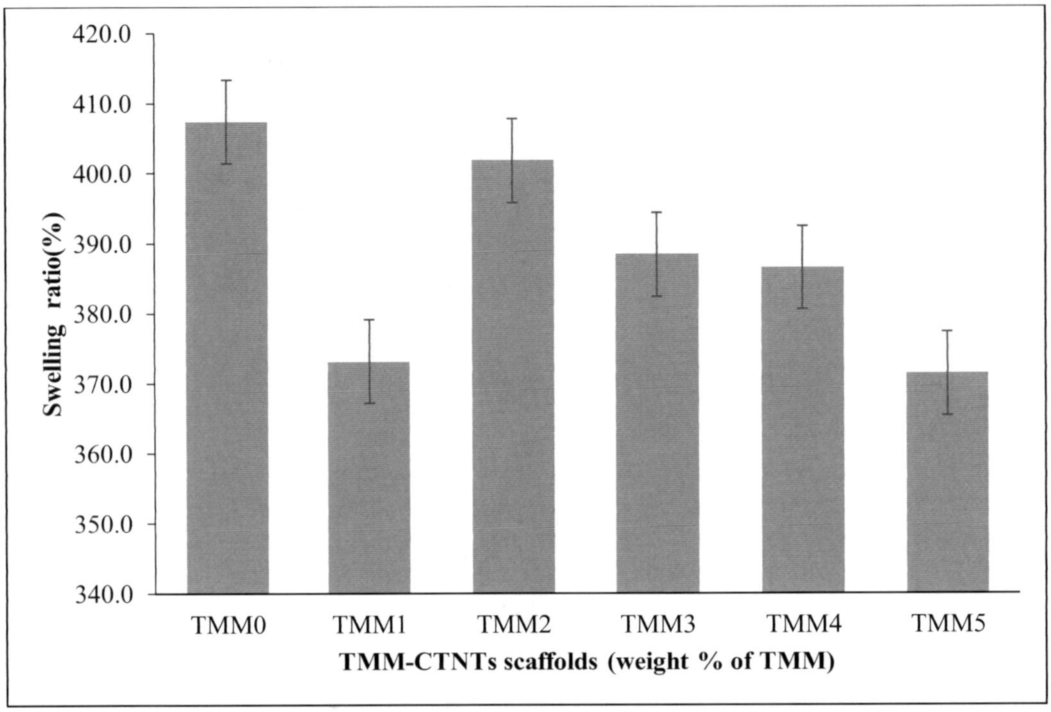

Figure 3. The swelling ratio (%) of TMM-CTNTs scaffolds at varied wt% of TMM.

Hence, there was no correlation between wt% of TMM and swelling ratio even theoretically, scaffolds with larger swelling ratio help cells infusion into the scaffolds for better cell attachment and growth. However, the swelling ratio for all scaffolds showed higher water absorption capacity which was more than 200 % which possessed optimal structure of porous scaffolds for cells to be cultured as reported [23-24]. All scaffolds samples showed high percentage of porosity that revealed their hydrophilic behaviour which allows water to be absorb in scaffolds and offered space for adhesion and proliferation [25].

The degradation performance of TMM-CTNTs scaffolds at different TMM powders weight percent (wt%) after 28 days for each scaffold was evaluated. The degradation ratio (%) of TMM0, TMM2, TMM3, TMM4 and TMM5 scaffolds was 0.959 %, 2.843 %, 1.017 %, 1.390 %, 1.845 % and 2.385 % respectively and illustrated in figure 4. From the results, TMM5 showed highest degradation rate compared to other scaffolds. This showed additional of TMM in scaffolds increased the degradation rate of scaffolds and enhanced the formation of new bone regeneration.

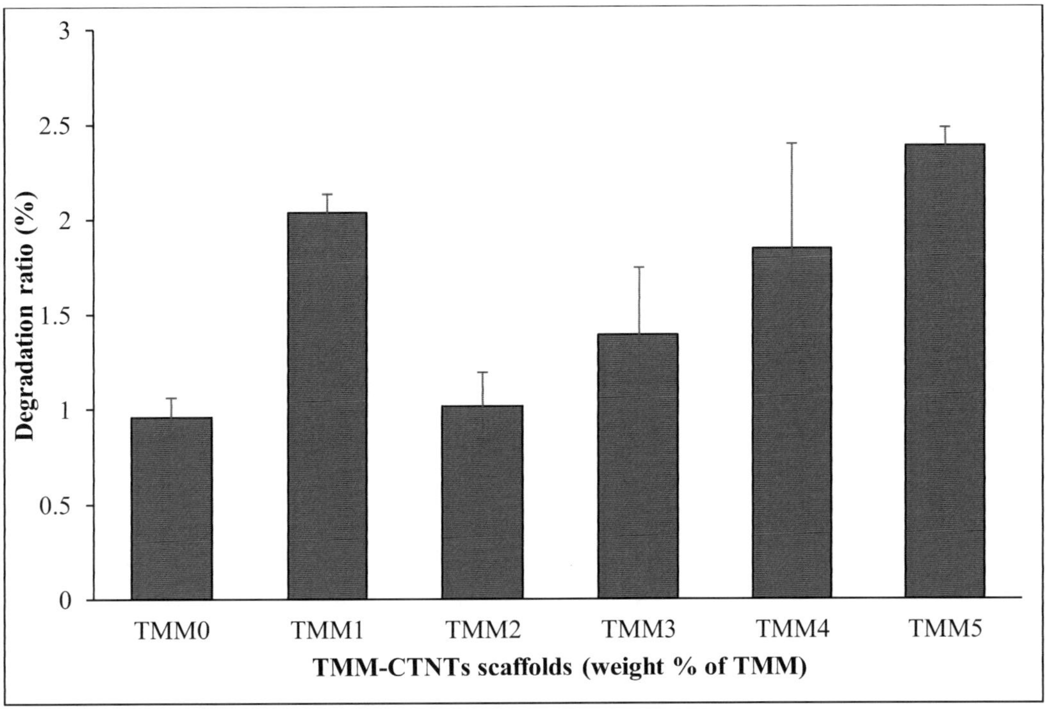

Figure 4. Degradation ratio (%) of TMM-CTNTs scaffolds at varied wt% of TMM.

The FTIR spectra of several important functional groups present in TMM-CTNTs scaffolds at different wt% of TMM are shown in figure 5. TMM was used as a control to confirm the present of β-glucan, the bioactive compound in TMM for inflammation properties in each scaffold. In figure 5, the broad band was visibly seen at 3400 cm^{-1} on the spectra which was attributed to OH stretching vibrations [26]. Another small peak next to –OH group was $-CH_2$ asymmetric stretching group at 2879 cm^{-1}. The peak at 665 to 668 cm^{-1} represented Ti-O-Ti stretching vibration [27]. Another small peak at 1635 to 1640 cm^{-1} showed Ti-OH bending vibration due to the absorption of H_2O [28]. The presence of β-glucan was confirmed in TMM1 to TMM5 scaffolds at spectral range of 1077 to 1080 cm^{-1} under C-O stretching in β-glucan [29]. Moreover all CTNTs-TMM scaffolds showed same peak appearing at spectral range 1022 to 1026 cm^{-1} and 1152 – 1156 cm^{-1} which represented the stretching vibration of C-O α-glycosidic bond and C-O-C asymmetric stretching of glycosidic linkage respectively [30] . This finding confirmed that freeze drying preserved the characteristic compounds of TMM which would later ensure the *in vitro* efficacy of the scaffolds.

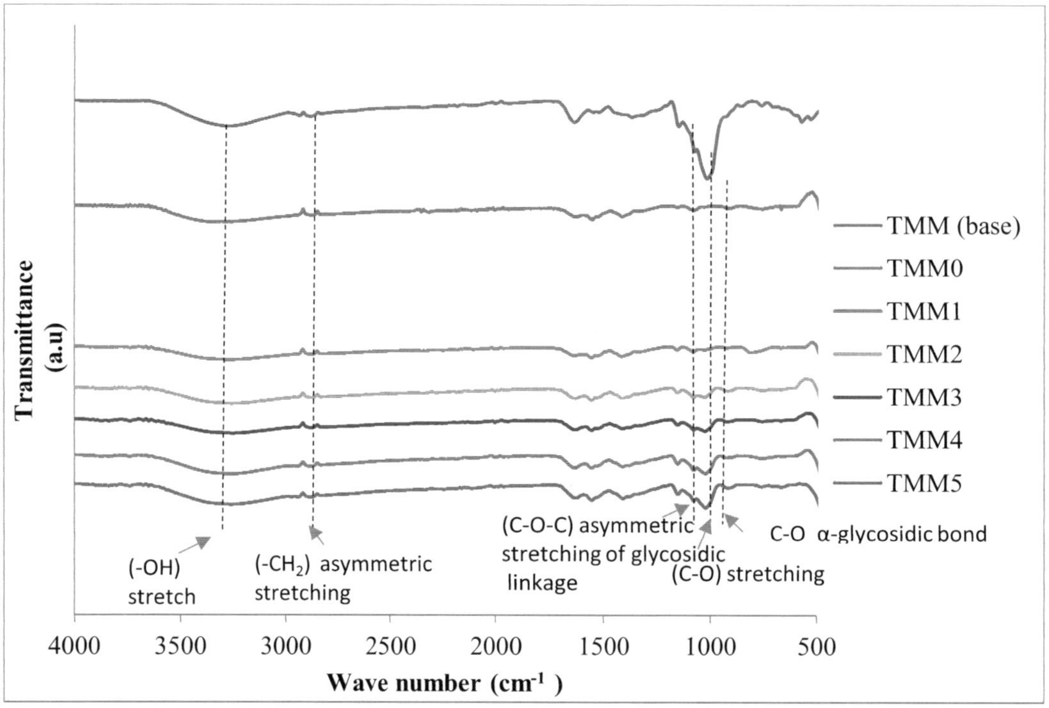

Figure 5. FTIR spectra of TMM-CTNTs scaffolds at varied wt% of TMM.

3.2. Enhanced cell proliferation on TMM-CTNTs scaffolds

Figure 6 showed the cell proliferation of the cells seeded on the scaffolds analysed using MTT assay at 3rd, 7th and 14th days of incubation period. The cell growth of MG63 on CTNTs-TMM scaffolds was measured in terms of optical density (OD) over the positive control. The positive control was referred to cells directly seeded on well without scaffolds. The OD values for all scaffolds showed significantly differences ($p<0.05$) at 3rd, 7th and 14th days of incubation. The values of OD is presented as means $\pm$ S.D (N=3) analyzed using one-way ANOVA analysis.

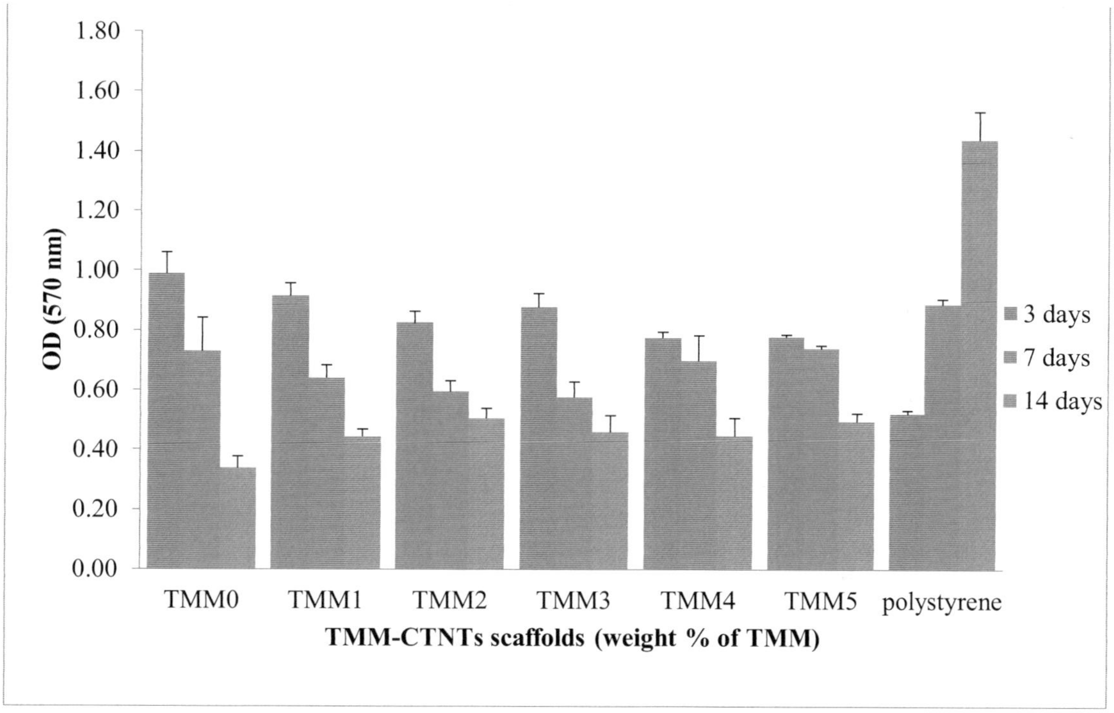

Figure 6. MG63 cells on TMM-CTNTs scaffolds at different weight percent of TMM and polystyrene as positive control evaluated by MTT assays at 3rd, 7th and 14th days of incubation period.

The OD values on 3rd day of incubation revealed TMM0 scaffolds with higher ($p<0.05$) OD values compared to other scaffolds. Furthermore, the proliferation of MG63 showed declined OD values with the wt% of TMM. In addition, the OD values of cells cultured on all TMM scaffolds were significantly higher ($p<0.05$) than OD value of cells on positive control. The proliferation of cells seeded on TMM0 scaffolds was significantly lower ($p<0.05$) compared to OD values of TMM5 scaffolds at 7th days of incubation. TMM5 scaffolds promoted high ($p<0.05$) proliferation of MG63 followed by TMM4 scaffolds. Despite of that, TMM4 and TMM5 scaffolds exhibited slightly small difference in OD values in between 3rd and 7th days of treatment compared to other scaffolds. This pattern of proliferation could not be understood very well. However, introduction of TMM gradually dropped the proliferation compared to scaffolds without additional of TMM powders. Such proliferation pattern also could be observed on TMM2 and TMM3 scaffolds on 14th day of incubation. The proliferation of MG63 on all scaffolds decreased at this incubation period especially TMM0 promoted the lowest ($p<0.05$) proliferation rate. Besides that, the OD values of all TMM scaffolds on 7th and 14th day of incubation were significantly lower ($p<0.05$) compared to cells seeded on positive control.

Based on MTT results, the additions of TMM in CTNTs scaffolds seem to reduce the *in vitro* proliferation of MG63. This can be clearly seen on early stage at 3rd day of incubation. The osteoblastic proliferation became low as the weight percent of TMM increased. Proliferation of cells could be affected by porosity and pore size and these contributed as factors why the proliferation rate became slow at early stages. Porosity result showed decreased pattern which is directly proportional with scaffolds pore size that reduced to less than 100 μm. Scaffolds with higher porosity enhances cell ingrowth, provides more protein sites and promotes nutrient and oxygen exchange [31]. Corresponding to this argument, this could be a reasonable explanation for the proliferation of MG63 on TMM0 scaffolds at the beginning which was higher than cells on other scaffolds containing TMM. TMM0 got higher porosity and pore size which positively affected the cell growth capability. Furthermore, TiO_2 was good filler for biodegradable material with high surface area that enhanced adhesion and proliferation of cell growth [26]. However, the proliferation of MG63 on TMM0 scaffolds drastically reduced on 7th and 14th days of incubation due to rapid migration of cells into TMM0 scaffolds template.

The cells grown on TMM5 scaffolds seemed to need adaptation period on 3rd day of incubation before showed the highest proliferation at 7th days of incubation. This might be due to the addition of TMM at the highest weight percent slowly promotes cells growth which prolonged the duration of proliferation. TMM5 scaffolds presented as the best scaffolds in terms of enhancing cell growth compared with other scaffolds on 7th days of incubation.

4. Conclusion

In conclusion, the reinforcement of TMM in C-TNTs scaffolds improved porous structure, mechanical strength and biocompatibility properties which fulfills the physical requirement of artificial 3D templates for bone regeneration. The cell proliferation through MTT assay test revealed addition of TMM slowly promoted cell fgrowth which prolonged the duration of proliferation. Notably, TMM5 scaffold with 5 wt% of TMM presented as the best scaffolds compared with other scaffolds on 7th days of incubation. Hence, further study on adhesion and differentiation should be carried out to better realize the *in vitro* biocompatibility of TMM-CTNTs scaffolds.

References

[1] Williams D F 2008 *Biomaterials* **29** 2941–53
[2] Lim S S, Chai C Y and Loh H-S 2017 *Mater. Sci. Eng. C.* **76** 144–52
[3] Subramanian A and Lin H 2005 *J. Biomed. Mater. Res.* **75A** 742–53
[4] Tuzlakoglu K and Reis R L 2008 *Handbook of Natural-Based Polymers for Biomedical Application* vol 1, ed R L Reis, N M Neives and J F Mano.(Cambridge:Elsevier) pp 357–73
[5] Levengood S K L and Zhang M 2014 *Eng.* 3161–84
[6]. Madihally S V and Matthew H W T 1999 *Biomaterials* **20** 1133–42
[7] Jana S, Florczyk S J, Leung M and Zhang M 2012 *J. Mater. Chem.* **22** 6291–9
[8] Bjursten L M, Rasmusson L, Oh S, Smith G C, Brammer K S and Jin S 2010 *J. Biomed. Mater. Res. - Part A.* **92** 1218–24
[9] Cheng Y, Yang H, Yang Y, Huang J, Wu K and Chen Z 2018 *J. Mater. Chem. B* **6** 1862–86
[10] Johnathan M, Gan S H, Ezumi M F W, Faezahtul A H, Nurul A A 2016 *BMC Complement. Altern. Med.* **16** 1–13
[11] Eik L F, Naidu M, David P, Wong K H, Tan Y S and Sabaratnam V 2012 *Evidence-based Complement. Altern. Med.* **2012**
[12] Yap Y H, Tan N, Fung S, Aziz AA, Tan C and Ng S 2013 *J. Sci. Food Agric.* **93** 2945–52
[13] Chen Q, Zhou W, Du G H and Peng L M 2002 *Adv. Mater.* **14** 1208–11
[14] Pezeshki M M, Mirzadeh H and Zandi M 2012 *Iran. Polym. J.* **21** 191–200
[15] Bružauskaitė I, Bironaitė D, Bagdonas E and Bernotienė E 2016 *Cytotechnology* **68** 355–69
[16] Murphy C M and O'Brien F J 2010 *Cell Adh. Migr.* **4** 377–81
[17] Zhang K, Fan Y, Dunne N and Li X 2018 *Regen. Biomater.* **5** 115–24
[18] Chen Y, Frith J E, Dehghan-Manshadi A, Attar H, Kent D and Soro N D M 2017 *J. Mech. Behav. Biomed. Mater.***75** 169–74
[19] Sultana N and Wang M 2008 *J. Mater. Sc.* **19** 2555-61
[20] Sun T, Khan T H and Sultana N 2014 *J. Nanomater.* **2014** 194680
[21] Román J, Cabanas M V, Peña J and Vallet-Regí M 2011 *Sci. Technol. Adv. Mater.* **12** 45003
[22] Kim T-R, Kim M-S, Goh T S, Lee J S, Kim Y H and Yoon S-Y 2019 *Appl. Sci.* **9** 1965
[23] Eves P C, Beck A J, Shard A G and Mac-Neil S 2005 *Biomaterials* **26** 7068–81
[24] Wang H M, Chou Y T, Wen Z H, Wang Z R, Chen C H and Ho M-L 2013 *PLoS One* **8** 2–12
[25] Radwan-Pragłowska J, Janus Ł, Piątkowski M, Bogdał D and Matysek D 2020 *Polymers (Basel)* **12**
[26] Kumar P. 2018 *Int. J. Biomater.* **2018** 6576157
[27] Pan H, Wang X D, Xiao S S, Yu L G and Zhang Z J 2013 *Indian J. Eng. Mater. Sci.* **20** 561–7
[28] León A, Reuquen P, Garín C, Segura R, Vargas P and Zapata P 2017 *Appl. Sci.* **7** 1–9
[29] Chen X, Liu X, Sheng D, Huang D, Li W, Wang X 2012 *Spectrochim. Acta Part A Mol. Biomol. Spectrosc.* **97** 667–72
[30] Ma Y, He H, Wu J, Wang C, Chao K and Huang 2018 *Sci. Rep.* **8**
[31] Loh Q L and Choong C 2013 *Tissue Eng. Part B Rev.* **19** 485–502

Acknowledgements
Authors would like to express our sincere gratitude to the Fundamental Research Grant Scheme from the Ministry of Education (FRGS/1/2020/TK0/UNIM/02/9) for funding support in our research work. Special acknowledgement to Ms Khairani Hasuna Jaapar on technical support for FESEM characterization.

Microcrystalline cellulose (MCC) as an adsorbent in copper removal from aqueous solution

N A Khalil[1], N S Abdullah[1], A S A Rahman[1], H A Hamid[2], A N S Fizal[1], M Zulkifli[3], M S Hossain[4] and A N A Yahaya[3]

[1] Universiti Kuala Lumpur, Malaysian Institute of Chemical and BioEngineering Technology, 78000 Alor Gajah, Melaka, Malaysia
[2] Institute of Plantation Studies, Universiti Putra Malaysia, 43400 UPM Serdang, Selangor, Malaysia
[3] Green Chemistry & Sustainable Engineering Technology Cluster, Universiti Kuala Lumpur, Malaysian Institute of Chemical and BioEngineering Technology, 78000 Alor Gajah, Melaka, Malaysia
[4] School of Technology Industry, Universiti Sains Malaysia, 11800 USM Penang, Malaysia

Email: nafifah.khalil@s.unikl.edu.my

Abstract. Many studies have been done on the natural adsorbent, natural/raw cellulose, modified cellulose and modified MCC as media for removing copper. However, the usage of unmodified microcrystalline cellulose (MCC) as an adsorbent to remove heavy metals contaminants such as copper from an aqueous solution is scarcely being explored. Thus, the current study was done to assess the performance of the MCC, without any modification, based on the adsorption capacity and the Cu removal efficiency under varied process parameters. The MCC was successfully used for Cu^{2+} or Cu (II) removal at pH < pH6. The adsorption parameters such as pH (1 to 6), contact time (0.5 to 24 hours), temperature (25 to 70 °C), initial Cu concentrations (1 to 5 mg/L) and MCC dosage (0.05 to 0.5 g) were significantly influenced the adsorption performance of the MCC. This study indicated the adsorption process occurred at pH ranging from pH 2.8 to 6 with 3 to 24 hours of duration were required to achieve the equilibrium condition. Lower temperature ranges (25 to 30 °C) were more favourable for adsorption than higher temperature ranges (40 to 70 °C). The increase in initial Cu concentration enhanced the adsorption capacity of MCC but decreased Cu removal. On the other hand, the increase in the MCC dosage resulted in the decreased adsorption capacity, however, increased Cu removal. MCC dosage of 0.225 g managed to remove Cu (II) with 95% efficiency at the initial Cu concentration of 1 mg/L, pH 5 and temperature 25 °C within 24 hours of contact time with 0.18 mg/g of adsorption capacity. Overall, high Cu removal efficiency (up to 95%) was achieved by the MCC which render its usage as adsorbent.

1. Introduction

Heavy metals such as copper occur naturally as one of the constituents of the earth and are often described as reddish metal [1]. Copper (Cu) is the most precious and widely used metal in diverse industrial applications, for example, electroplating, metal polishing, building, automobiles, manufacturing of electrical machinery, cabling for power and telecommunications and making brass and other similar alloys. Its major application such as in the electrical industries owing to its excellent electrical conductivity and heat conductor properties [2,3]. These industries generate a large volume of wastewaters containing copper as pollutants which can lead to several disorders and diseases if left untreated or inadequately treated [4]. This is because, heavy metals as pollutant such as Cu, has high resistance to biological and chemical degradation, high mobility and the potential to bioaccumulate in the food chain [5]. It has been mentioned that ingesting a certain level of Cu causes temporary gastrointestinal and intestinal problems, kidney or liver injury, and even cancer [4]. This warrants the treatment of Cu contaminated wastewater before discharging to the environment [6]. Cu can be removed either by utilizing adsorption, cementation, membrane filtration, electrodialysis or photocatalysis [3]. Adsorption has become an alternative treatment for those methods due to cleaner process, sustainable, cheap and eco-friendly process [7]. Adsorption is a process where a contaminant, which is Cu, is transferred onto the surface of an adsorbent from the liquid phase, bounded by physical and/ or chemical interactions [6].

Developments of polysaccharides-based adsorbents or biosorbent such as cellulose-based adsorbents have received extensive attention in recent years, due to their advantages such as low cost, renewability and non-toxicity [8]. In order to compete with other adsorbents, the new biosorbent materials must be an effective metal binder and perform well under various process conditions [9]. Sesame husk (SH) was studied as a natural adsorbent for Cu (II) with 2.5 mg/g adsorption capacity and 95.15% of removal at pH 6, 298 K, 1g of SH dosage, 30 mg/L of Cu (II) initial concentration and 300 rpm of agitation speed [10]. The study also indicated that cellulose participated in the adsorption and binding of Cu (II) ions beside other components. The disadvantages of cellulose including low adsorption or ion-exchange capacity and poor physical stability [11]. Therefore, modification of cellulose, for instance, chemical modification, crosslinking or copolymerization can be carried out to increase the adsorption capacity and efficient ion exchange capacity [8,9]. Raw cellulose from refined cotton was modified to cellulose nanofibrils managed to increase the adsorption capacity from 1.24 to 45.66 mg/g (condition: pH 5, temperature 30 °C, 100 mg/L initial Cu (II) concentration and 1 g/L adsorbent dosage) [12]. Cellulose nanofiber isolated from oil palm empty fruit bunch and synthesized into magnetic biosorbent composite with enhancing adsorption capacity of ~0.71 mg/g at 28 °C. 93.25% removal efficiency of Cu (II) by the composite (0.5 g/L dosage) was achieved after 30 mins under experiment conditions of pH 5 and 60 °C [13]. The composite may also easily be removed from an aqueous solution using a magnetic field and it has the potential to be reused. These are among chemical modifications that can give the cellulose new properties such as hydrophilic or hydrophobic character and improved water sorbency, ion-exchange capability elasticity and thermal resistance. The application of magnetic cellulose-based materials as adsorbents is also being explored for an easier separation adsorption process which facilitates regeneration [4].

Many studies have been done on the natural adsorbent, natural/raw cellulose, modified cellulose and modified microcrystalline cellulose as media for removing copper [10,12,13]. However, limited information is available on the adsorption performance of the unmodified microcrystalline cellulose, which is commercially available. The microcrystalline cellulose (MCC) is produced from natural polymeric cellulose of softwood, hardwood and other agricultural waste. The amorphous region of the cellulose was removed by partial depolymerization by enzymes or acids leaving the MCC in the crystalline form [14]. The process making it more suitable for heavy metal adsorption by increasing the surface area considerably [14]. Microcrystalline cellulose consists of separated crystalline aggregates of 1 to 100 μm in sizes with 3.5 to 4.5 m^2/g of the specific surface area [15]. Microcrystalline cellulose is a readily available commercial source and could be used directly without any modification or pre-treatment [8]. Therefore, the objective of this work was to investigate the performance of the unmodified microcrystalline cellulose as an adsorbent for Cu removal. The focus was on the analysis of the parameter's effects such as pH, contact time, temperature, initial Cu concentration and MCC dosage on the adsorption capacity and Cu removal efficiency.

2. Material and method

Microcrystalline cellulose, $C_{14}H_{26}O_{11}$ (C.P., R&M Chemicals), copper (II) sulphate pentahydrate, $CuSO_4.5H_2O$ (C.P., Bendosen) were used as an adsorbent and adsorbate in the batch experiment. 37% hydrochloric acid, HCl (Fisher Scientific) and sodium hydroxide pellets, NaOH (HmbG Chemicals) were used for pH modification. Figure 1 shows the overall flow of the experiment and analysis.

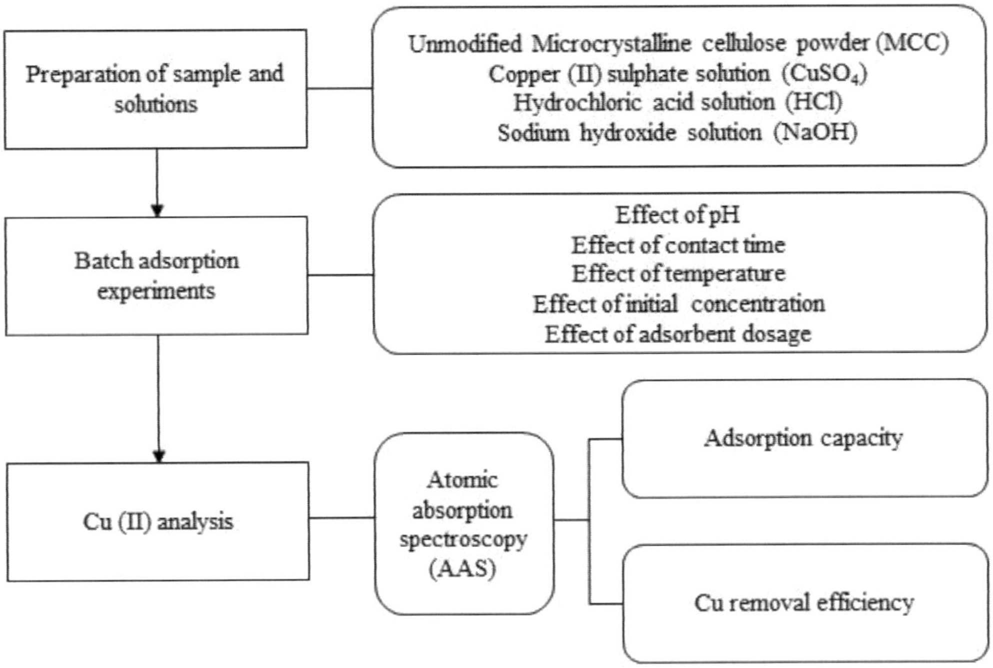

Figure 1. The flow of the experiment and analysis.

2.1. Preparation of solutions

2.1.1. 100 ppm Copper (II) Sulphate solution. 1000 ppm of copper (II) sulphate stock solution by adding 3.93 g of copper (II) sulphate pentahydrate in the 1L volumetric flask. Deionized water was filled until the calibration mark was reached, and the flask was well shaken.

2.1.2. 0.1 M Sodium hydroxide. 0.4 g of sodium hydroxide pellets were weighed, dissolved in deionized water, and stirred using a magnetic stirrer. The solution was poured into a volumetric flask of 100 mL. Deionized water was filled until the calibration mark was reached, and the flask was well shaken.

2.1.3. 0.1 M hydrochloric acid. 0.8 mL of 37% hydrochloric acid was transferred into a 100 mL volumetric flask using a pipette. Deionized water was filled until the calibration mark was reached, and the flask was well shaken.

2.2. Batch adsorption experiments

Microcrystalline cellulose powder was utilized without any modification in the batch adsorption study to assess the adsorption ability of the MCC on Cu. MCC will be used to indicate the unmodified MCC henceforth.

The adsorption studies were performed in a series of batch experiments. The effects of various adsorption parameters were studied, including pH (1 to 7), contact time (0 to 24 hours), temperature

(25 to 70°C), initial concentration of Cu solution (1 to 5 mg/L) and dosage of adsorbent (0.02 to 0.5 g).

Generally, the experiment was conducted by the addition of MCC into Cu solution. The Cu solution of the intended concentration was prepared by diluting the necessary amount of 100 ppm copper (II) sulphate solution in deionized water. Followed by dosing the intended amount of the MCC with the intended dosage was added to the conical flask containing 50 ml of the Cu solution, which was then agitated. Sampling was done by pipetting the mixture using a 5 mL pipette at fixed intervals of 30 mins for the first hour and every 1 hour for the next hours. Each of the samples obtained was filtered before pipetting into the test tube for Cu analysis.

The effect of the pH experiment was conducted by the addition of HCl and NaOH to vary the pH value of the Cu solution. Concurrently, the effect of the contact time experiment was also being conducted by extracting samples at various time intervals. The conical flasks containing Cu solution was immersed in a water bath to obtain the desired temperature to study the effect of temperature. The MCC was added after the temperature of the Cu solution stabilized.

The concentration of Cu was obtained by analyzing the sample extracted during the experiment. Atomic Absorption Spectroscopy (AAS) of PerkinElmer was used to analyze the Cu concentration in the sample. The adsorption capacity (q) was obtained based on the following equation [4]:

$$q = (C_o - C_t)\frac{V}{M} \tag{1}$$

where q is the quantity of Cu adsorbed onto the unmodified MCC [mg/g], C_o is the initial Cu concentration [mg/L], C_t is the concentrations of Cu after the adsorption at a time [mg/L], V is the Cu solution volume [L] and M is the MCC mass [g].

The removal efficiency of Cu was be obtained based on the equation as follows [13]:

$$\text{Removal (\%)} = \frac{C_o - C_t}{C_o} \times 100 \tag{2}$$

where C_o is the initial Cu concentration [mg/L] and C_t is the Cu concentrations the adsorption at a time [mg/L].

3. Results and Discussion

3.1. *Effect of pH*

Experiments were conducted by varying the pH of Cu solution from pH 1 to 7 at 25 °C, 24 hours of contact time, 5 mg/L of Cu initial concentration and 0.225 g of MCC dosage.

Figure 2 shows the adsorption capacity of the unmodified MCC in the Cu solution. The Cu adsorption was found constant at pH 1 to 2 (0.01 mg/g) and started to increase exponentially with an increase in pH from pH 2 to 5 (0.4 mg/g and 0.15 mg/g). The adsorption reached the highest value of 0.39 mg/g at pH 5, which was within an acidic region. The adsorption slightly declined at pH 6 (0.37 mg/g) and further declining to the lowest value (0.04 mg/g) at pH 7. This adsorption trend can be discussed by considering the MCC's surface properties and the Cu species at various pH conditions [16]. The surface properties of the MCC can be explained by associating its isoelectric point (IEP) and zeta potential values. The IEP of the MCC particles occurred at a low pH value of pH 2.8 with a negative zeta potential value at pH above this IEP, which indicated the particles was negatively charged [8,15]. On the other hand, the species of Cu varied with pH where the predominant species is Cu^{2+} ions in pH < pH 6. Cu species such as $CuOH^+$, $Cu(OH)_2^{2-}$, $Cu(OH)_3^-$ $Cu(OH)_2^0$ emerged when the pH of the solution was raised above pH 6 [10,17]. Therefore, the adsorption of Cu onto the MCC at pH < pH 6 was due to the interaction of Cu^{2+} ions (Cu (II)) with the binding site on the surface of the MCC particle, which was negatively charged. As a result, the optimal pH for the adsorption process was determined at pH 5, as demonstrated by the highest adsorption at this pH.

The increasing trend at pH 1 to 6 range also can be explained by looking at the available ions in the solution due to pH adjustment. At pH 1 to 4, a lower value of Cu adsorption was observed as the H^+ ions and Cu (II) competing to occupy the possible limited binding site on MCC [18]. The addition of HCl into the solution during pH adjustment resulted in high H^+ ions concentration. The presence of H^+ ions also resulted in repulsive force or electrostatic repulsion, restricting Cu (II) access to the functional groups of MCC [18,19]. As pH solution is being increased to pH 6, less HCl was added, implying that there were fewer H^+ ions in the solution to compete with the Cu (II). As a result, the probability of Cu (II) binding to the MCC's free binding sites was enhanced [20].

Figure 3 shows the removal efficiency of Cu at various pH. The removal efficiency of Cu shows the same trend as adsorption in between pH 1 to 6 and started to decline beyond pH 6. The removal of Cu at pH < pH 6 was via adsorption of Cu (II) by MCC, however, precipitation occurred above pH 6 [17,21]. Maximum removal of 39.3% was achieved and the optimum pH for removal was found at pH 6, by a possible combination of adsorption and precipitation processes.

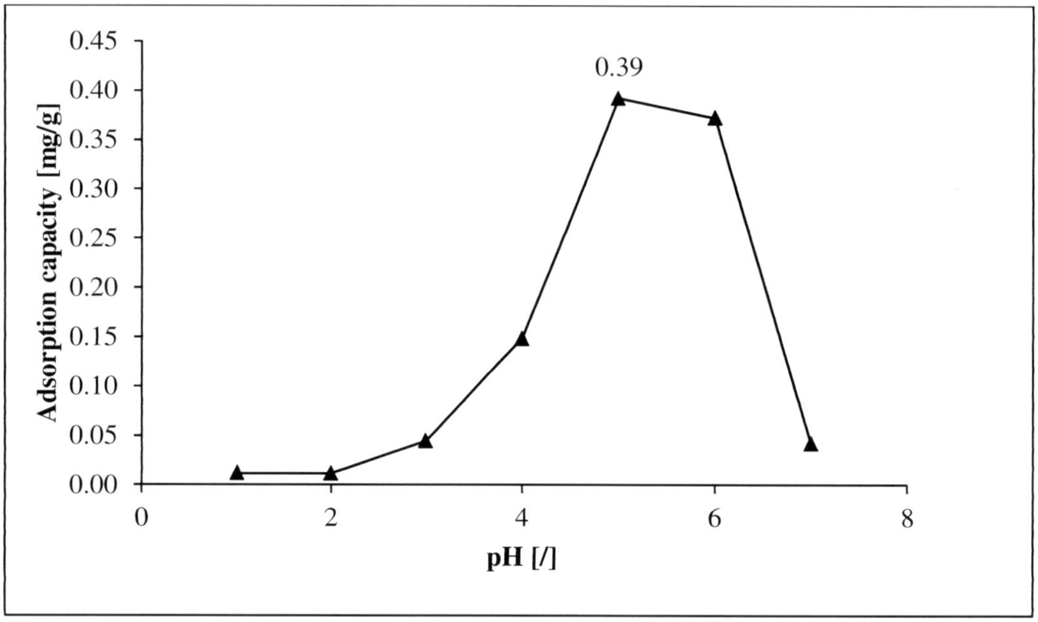

Figure 2. Effect of pH on adsorption capacity (Contact time, 24 hours; Temperature, 25 °C; Initial Cu concentration, 5 mg/L; MCC dosage, 0.225 g).

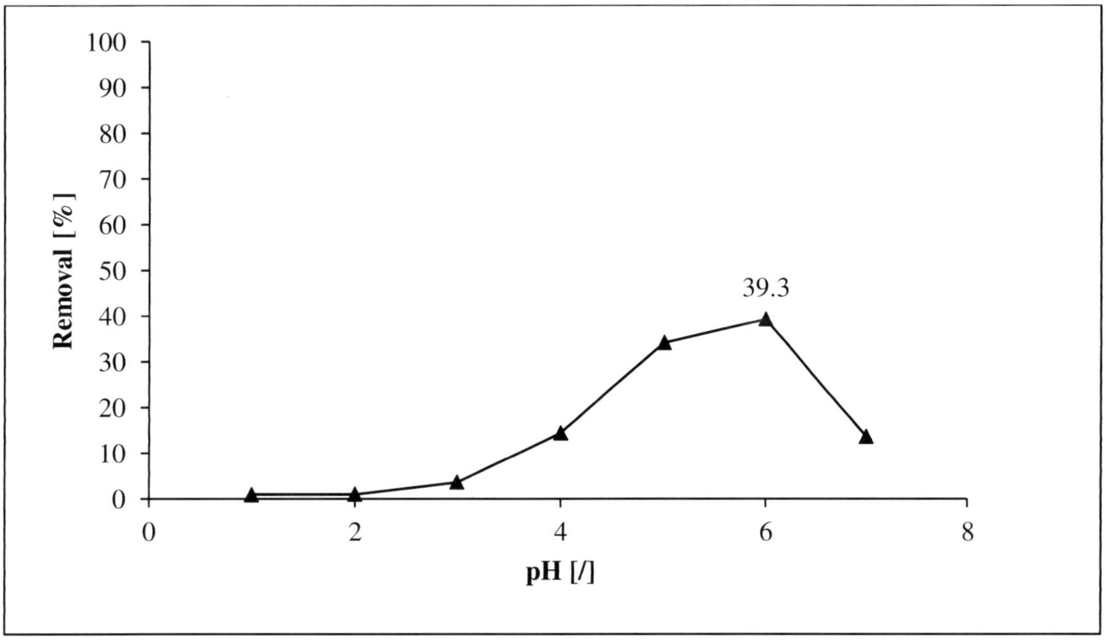

Figure 3. Effect of pH on Cu removal efficiency (Contact time, 24 hours; Temperature, 25 °C; Initial Cu concentration, 5 mg/L; MCC dosage, 0.225 g).

3.2. Effect of contact time

Figure 4 shows the adsorption of Cu onto MCC for different contact time at pH 5 and 25 °C. Rapid adsorption was observed from 0 to 2 hours as the Cu (II) was easily attached to the unoccupied adsorption site of the MCC surface. The adsorption increased slightly in between 2 to 4 hours (0.35 to 0.36 mg/g). The adsorption rate was nearly consistent starting from 3 to 24 hours, indicated that most of the MCC adsorption sites were practically saturated with the Cu (II) [22]. The highest adsorption capacity was obtained at 24 hours, which was 0.39 mg/g. Figure 5 show a similar trend for Cu removal percentage, with the maximum Cu removal of 34.17% at 24 hours of contact time. Based on these analyses, the adsorption has reach equilibrium at 3 hours, with a maximum contact period of 24 hours.

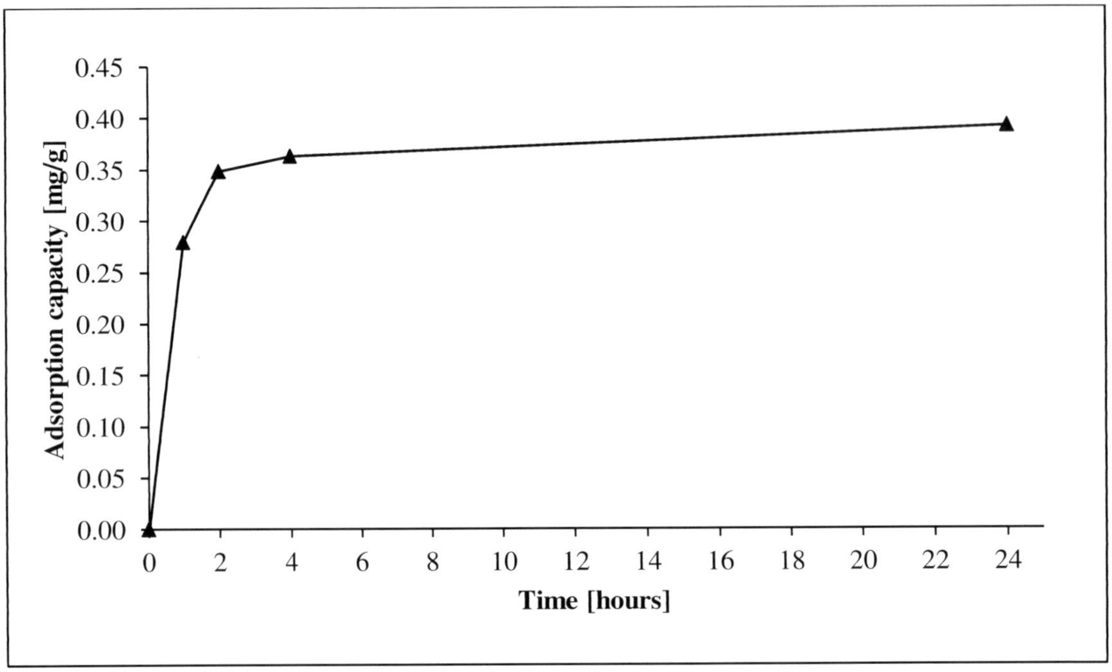

Figure 4. Effect of contact time on adsorption capacity of Cu (pH, pH 5; Temperature, 25 °C; Initial Cu concentration,5 ppm; MCC dosage, 0.225 g).

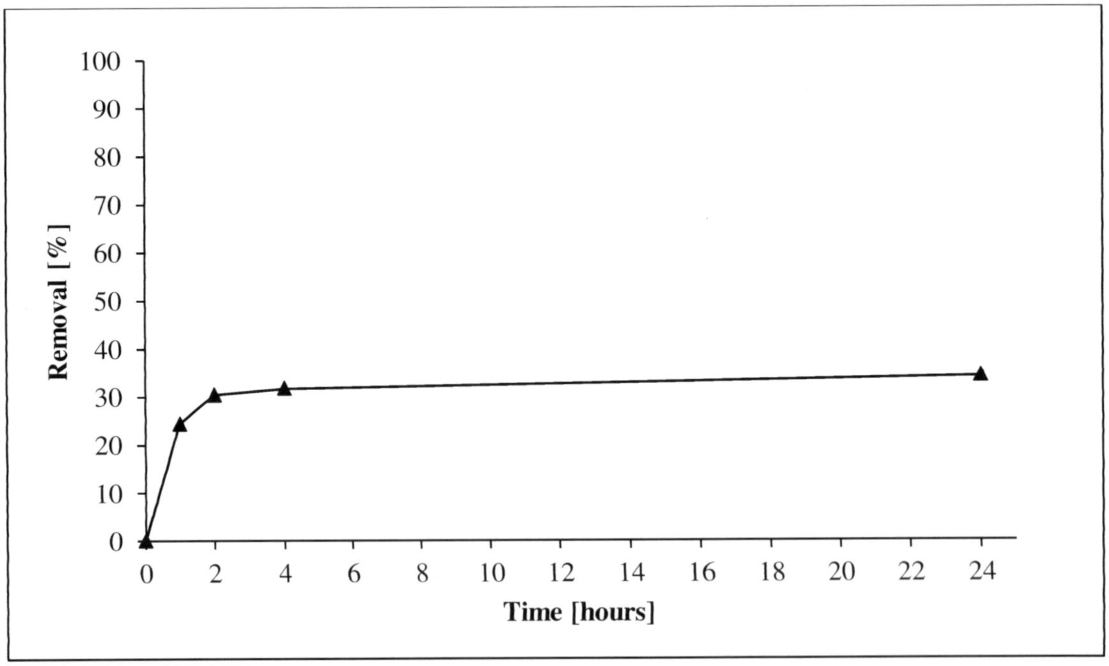

Figure 5. Effect of contact time on Cu removal (pH, pH 5; Temperature, 25°C; Initial Cu concentration, 5 ppm; MCC dosage, 0.225 g).

3.3. *Effect of temperature*

Adsorption capacity was analyzed at various temperatures of 25, 30, 40, 50, 60 and 70 °C for 3 hours of contact time and pH 5 to assess the temperature's effect on adsorption, as shown in figure 6. The adsorption capacity was drastically increased with the increase of temperature from 25 to 30 °C (0.33 to 0.42 mg/g). The increased temperature subsequently increased the mobility of Cu (II) and ensure sufficient energy for the interaction of Cu (II) and the MCC's active sites [23]. The adsorption started to decrease to the lowest value of 0.26 mg/g with the increase in temperature to 50 °C. However, an increasing adsorption trend was observed again from 50, 60 and 70 °C. The further increase in temperature reduced the viscosity of the solution and increase the rate of diffusion of the Cu (II) through the external boundary layer and in the MCC's internal pores, hence, increased adsorption capacity [23]. Higher temperature applied in the adsorption process also caused the internal structure of the MCC to swell which allows more Cu (II) to infiltrate the pores of MCC [23].

The Cu removal efficiency is as shown in figure 7. The trend shows similarity to the adsorption capacity with the highest removal at 30 °C (38%), followed by 70 °C (34%) and the lowest removal happened at 50 °C (24%).

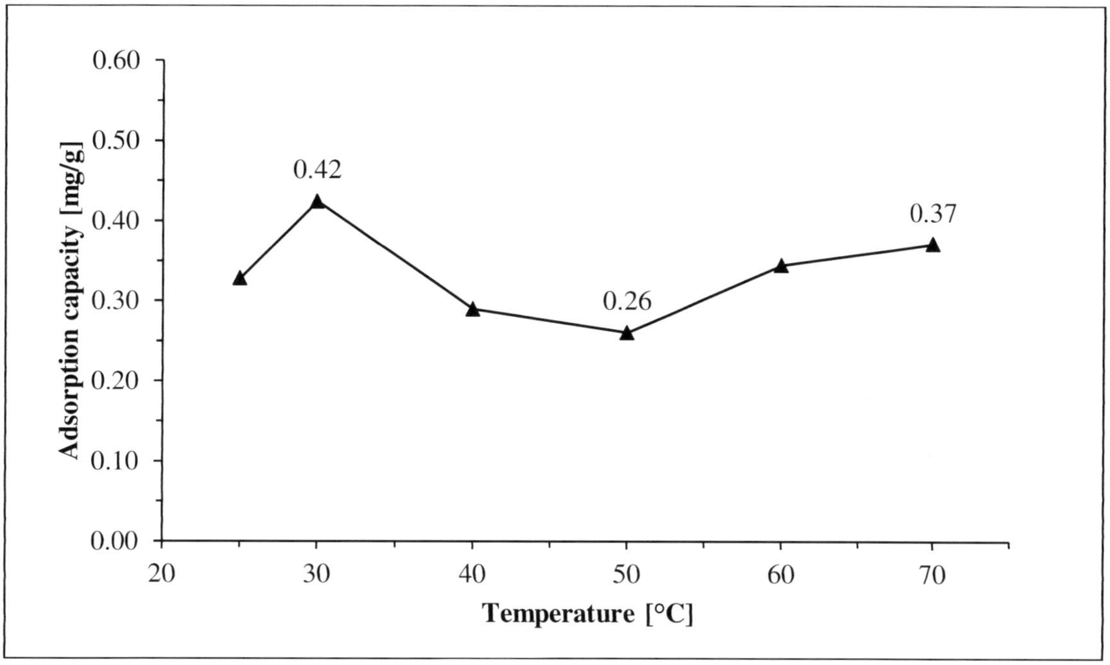

Figure 6. Effect of temperature on adsorption capacity of Cu (pH, pH5; Contact time, 3 hours; Initial Cu concentration, 5 ppm; MCC dosage, 0.225 g).

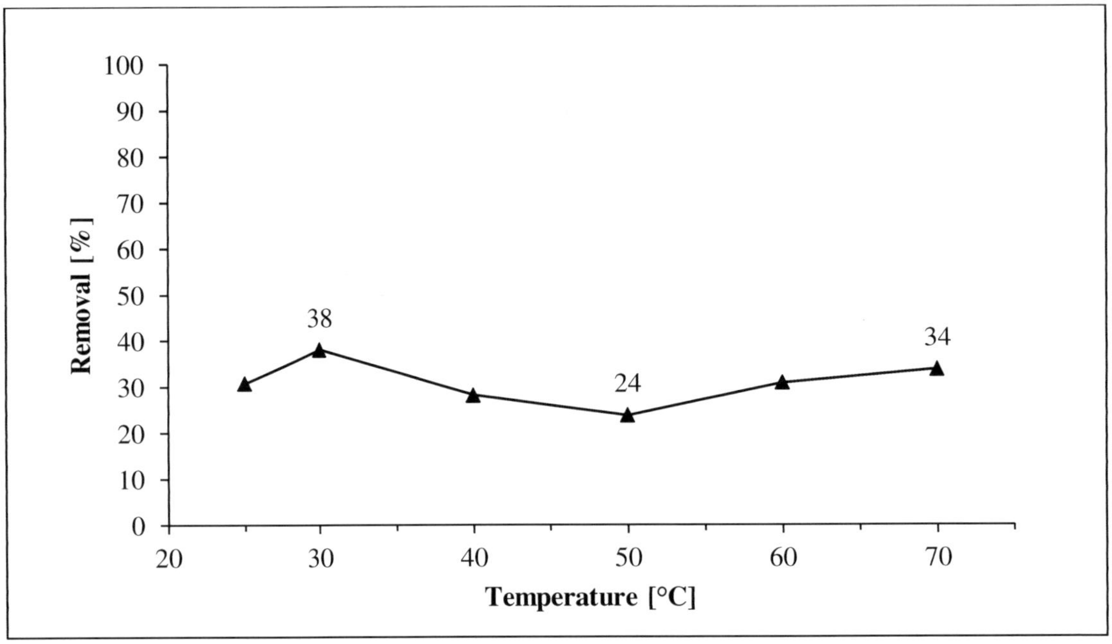

Figure 7. Effect of temperature on Cu removal (pH, pH5; Contact time, 3 hours; Initial Cu concentration, 5 ppm; MCC dosage, 0.225 g).

3.4. Effect of initial copper concentration

Figure 8 shows the adsorption capacity of MCC at varying initial Cu concentrations from 1 ppm to 5 ppm at two different temperatures (25°C and 70°C) and pH 5, respectively. The MCC as adsorbent at 25 °C shows the increasing trend when the Cu initial concentration was increased (1 to 4 mg/L) with the highest adsorption of 0.48 mg/g at 4 mg/L of Cu initial concentration. The increase of adsorption capacity with the increase of Cu initial concentration was due to more number of adsorption sites were available at low metal ion concentration, hence the process was highly effective within this range [24]. The increase in the initial concentration also can cause the mass transfer rate to increase due to the increase of driving force, hence, adsorption capacity increased [25]. A similar trend is shown with adsorption conducted at a higher temperature of 70 °C with a maximum of 0.45 mg/g of adsorption achieved at lower Cu initial concentration (3 mg/L). This trend suggested lesser surface interaction between MCC and Cu at a higher temperature. The high temperature might either tempered the texture of the adsorbent or possibly changed the surface chemistry by altering the amount of MCC's active functional groups for adsorption. This lead to the reduction of sorption ability of the MCC due to degradation or damaging the surface active sites [10,26,27]. Besides, the active adsorption site may be destroyed due to bonds ruptured at higher temperatures leading to heavy metal's desorption [28]. The thickness of the boundary layer of the biomass surface also decreases at higher temperatures causing the metal ions to flee into the solution phase and decreased the adsorption capacity [10]. Further increase of the initial concentration to 5 mg/L for both temperatures (25°C and 70°C) decreased the adsorption capacity to 0.33 mg/L. These declining values can be explained by beyond the said concentration, the adsorbent becomes saturated.

Figure 9 observed the declining trend of Cu removal efficiency for both temperatures (25 °C and 70 °C) with the increase of the initial concentration from 1 to 5 mg/L. When the initial Cu concentration was increased, more metal ions was available in the solution and trying to occupy the same amount of vacant site of MCC. This means lesser available adsorption sites compared to adsorbate, hence, lead to lower removal percentage of Cu (II) [24].

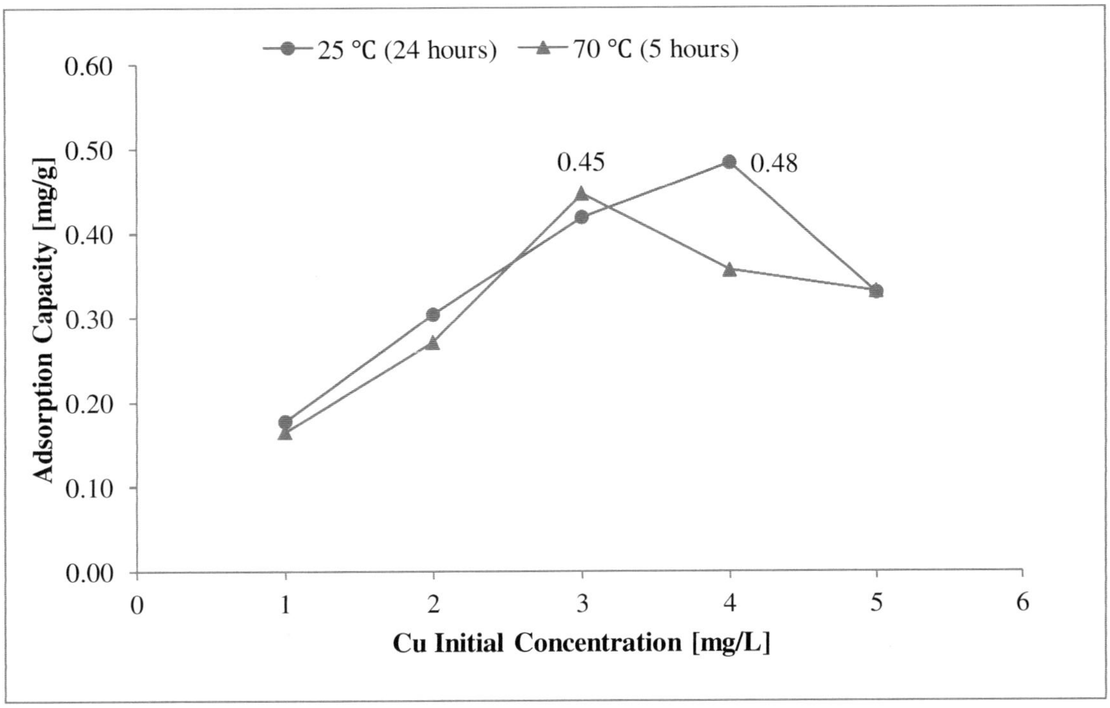

Figure 8. Effect of initial concentration on adsorption capacity (pH, pH5; MCC dosage, 0.225 g).

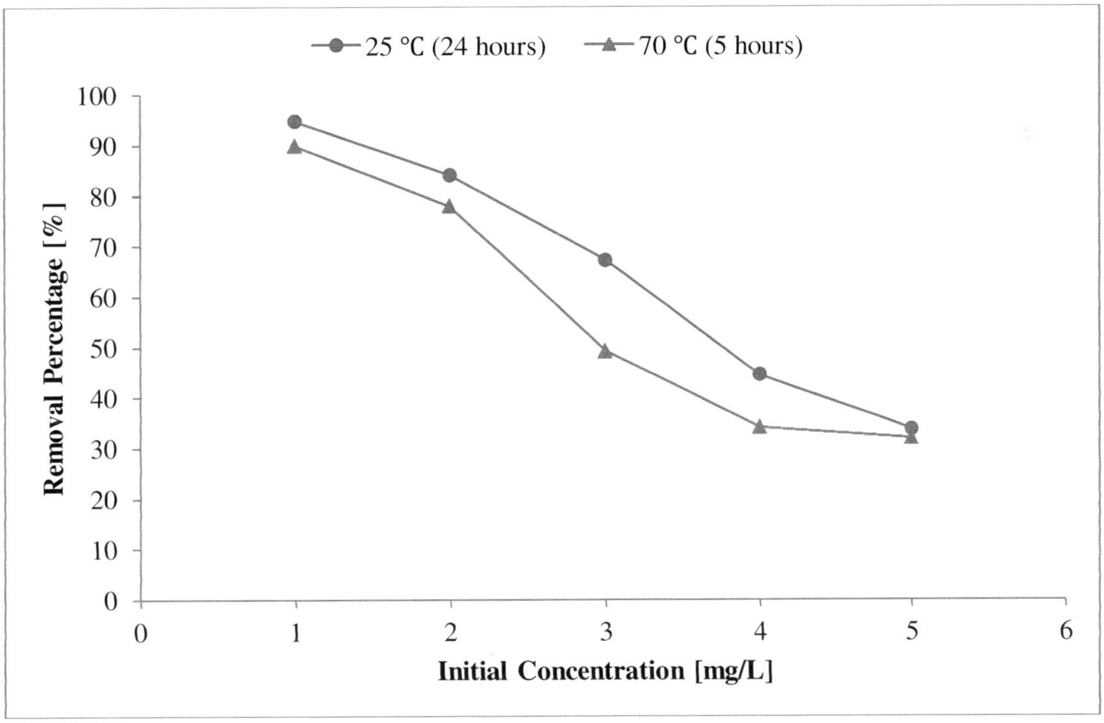

Figure 9. Effect of initial concentration on Cu removal (pH, pH5; MCC dosage, 0.225 g).

3.5. Effect of MCC dosage

The effect of MCC dosage on adsorption was analysed at two different temperature; 25 °C with 24 hours of contact time and 70 °C with 4 hours of contact time (pH 5 and 3 mg/L of initial Cu concentration) as shown in figure 10. At 25 °C, the adsorption capacity achieved the highest value of 0.35 mg/g when 0.1 g of MCC was added and slightly decreased when the dosage was increased to 0.3 g. At the lower dosage of MCC, the available adsorption sites were very low compared to the fixed metal ion concentration, hence, adsorption was easier to happen and achieved its capacity [29]. The adsorption value dropped significantly from 0.33 to 0.22 mg/g with further dosage addition to 0.4 g and remained constant at 0.5 g. In contrast, at higher MCC dosage, adsorbent particles had to compete each other to bind with the same metal ions which resulted in low unit adsorption of metal ions per unit mass of adsorbent [29]. Furthermore, the decreased in adsorption capacity was also due to the aggregation of adsorbent particles that lead to the reduction in total surface area and increased diffusion path length [10]. When the adsorption process was subjected to higher temperature of 70 °C, the adsorption capacity declined rapidly from 0.30 to 0.19 mg/g with the increased of MCC dosage (0.1, 0.225, 0.3, 0.4 and 0.5 mg/L). The higher temperature of 70 °C altered the surface of the MCC due degradation or damaging the active sites, hence the declining trend.

The removal percentage, contrastingly, shows the addition of MCC dosage increased the removal of Cu for both temperature (figure 11). The highest percentage removal was obtained at MCC dosage of 0.5 g/L for both temperatures within the experimental dosage range.The overall surface area and its availability of exchangeable binding sites on the surface for metal ion uptake were increased with the addition of MCC dosage, hence, rising the remval percentage of Cu [10].

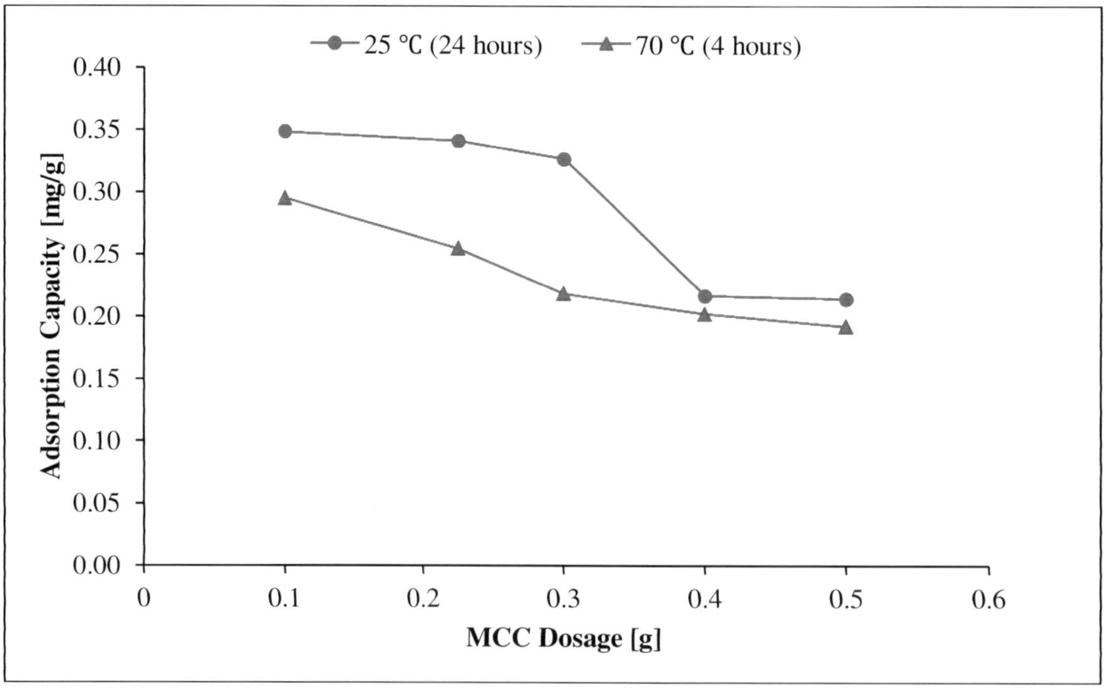

Figure 10. Effect of MCC dosage on adsorption capacity (pH, pH5; Initial Cu concentration, 3 mg/L).

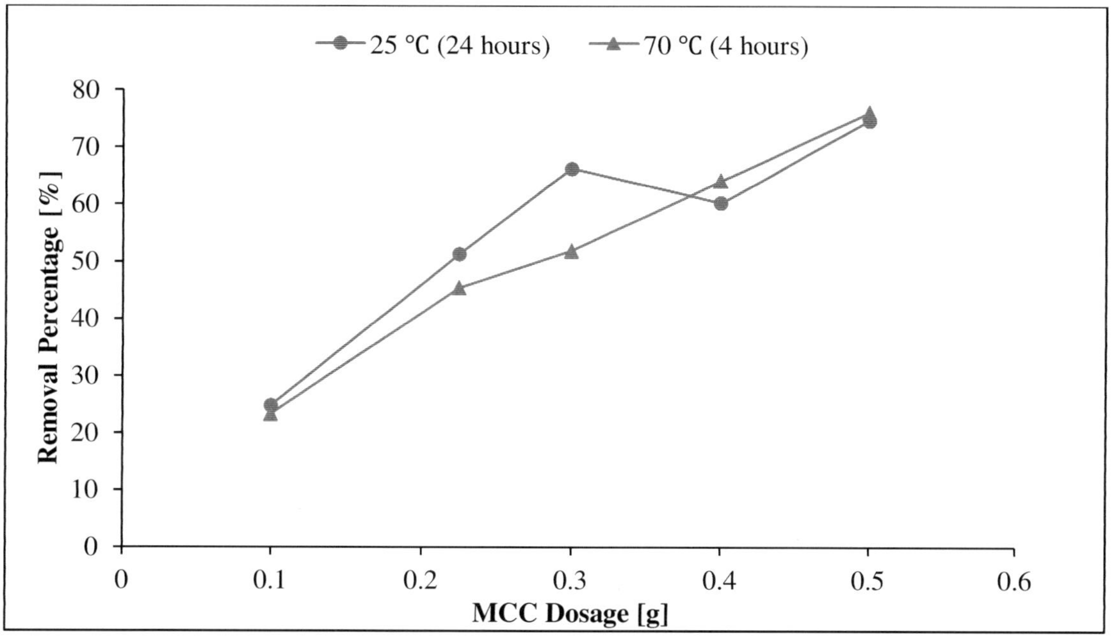

Figure 11. Effect of MCC dosage on Cu removal (pH, pH5; Initial Cu concentration, 3 mg/L).

3.6. Comparison of MMC adsorption with another cellulose-based adsorbent

The adsorption of the MCC (0.44 mg/g) in this study is higher compared to sugarbeet pulp (0.15 mg/g) [10] as specified in table 1. However, the adsorption capacity is lower compared to other plant based adsorbent such as grape stalk wastes (0.78 mg/g), sawdust (1.7 mg/g), baggasse (2.89 mg/g), corn starch (8.57 mg/g) and sesame husk (11.37 mg/g) [10].

In addition, these cellulose-based adsorbents were normally went through modification in order to increase their adsorption capacity, i.e., cellulose nanofibril modified from cellulose of refined cotton managed to increase the adsorption from 1.24 to 45.66 mg/g [8,10]. Some researchers considered the application of magnetic cellulose-based materials as adsorbents due to its simple magnetic separation after the adsorption process which facilitate the regeneration [13,30].

Table 1. Adsorption capacity of various adsorbents.

Type of adsorbent	Adsorption capacity (mg/g)	Reference
Sugar beet pulp	0.15	[10]
Grape stalk wastes	0.78	[10]
Sawdust	1.70	[10]
Bagasse	2.89	[10]
Corn starch	8.57	[10]
Sesame husk (D-R model)	11.37	[10]
Microcrystalline cellulose (unmodified)	0.44	This study
Raw cellulose (from refined cotton)	1.24	[12]
Cellulose nanofibrils (from refined cotton)	45.66	[12]
Magnetic cellulose nanofiber composite	0.71	[13]

D-R = Dubinin-Radushkevich

4. Conclusion

The microcrystalline cellulose (MCC), without any modification, was successfully used as adsorbent for Copper (Cu) removal at pH < pH6 in the form of Cu^{2+} or Cu (II). The adsorption parameters (pH, contact time, temperature, initial Cu concentration and MCC dosage) were significantly influenced the adsorption performance of the MCC. This study indicated that the adsorption process consists of MCC as adsorbent and Cu as adsorbate, occurred at pH ranging from pH 2.8 to 6, with 3 to 24 hours required to achieve the equilibrium condition. Lower temperature ranges (25 to 30 °C) were more favourable for adsorption than higher temperature ranges (40 to 70 °C). MCC dosage of 0.225 g managed to remove Cu (II) with 95% efficiency at 1 mg/L of initial Cu concentration, pH 5 and temperature 25 °C within 24 hours of contact time with 0.18 mg/g of adsorption capacity. The adsorption capacity was increased to 0.48 mg/g when the initial Cu concentration was increased to 4 mg/L, however, the removal efficiency reduced to 44%. The increased of initial Cu concentration increased the adsorption capacity of MCC but decreased the Cu removal. On the other hand, the addition in the MCC dosage resulted in declined adsorption capacity and increased the Cu removal. Overall, high Cu removal efficiency (up to 95%) was achieved by the MCC which render its usage as adsorbent.

Limitation of the unmodified MCC as adsorbent was due to the lower adsorption capacity, which translated into a lower active binding site compared to other adsorbents. Modification of the MCC by chemical modification, copolymerization, crosslinking or grafting can enhance the adsorption capability of the unmodified MCC. The MCC is favourable as a modified adsorbent framework due to its salient features of chemical and mechanical properties as well as the advantages due to its crystalline nature [31].

Acknowledgement

This work was supported by the Ministry of Higher Education (MoHE), Malaysia under Fundamental Research Grant Scheme (FRGS/1/2018/STG07/UNIKL/03/1). The authors would also like to acknowledge Universiti Kuala Lumpur for providing support under UniKL Excellent Research Grant Scheme (UniKL/CoRI/UER20016).

References

[1] Lenntech 2021 Copper - Cu *Lenntech*

[2] Market Research Reports 2021 Copper inductry market research reports, analysis & trends *Market Research Reports*

[3] Al-Saydeh S A, El-Naas M H and Zaidi S J 2017 Copper removal from industrial wastewater: A comprehensive review *J. Ind. Eng. Chem.* **56** 35–44

[4] Peng S, Meng H, Ouyang Y and Chang J 2014 Nanoporous magnetic cellulose-chitosan composite microspheres: Preparation, characterization, and application for Cu(II) adsorption *Ind. Eng. Chem. Res.* **53** 2106–13

[5] Hokkanen S, Bhatnagar A and Sillanpää M 2016 A review on modification methods to cellulose-based adsorbents to improve adsorption capacity *Water Res.* **91** 156–73

[6] Barakat M A 2011 New trends in removing heavy metals from industrial wastewater Arab. *J. Chem.* **4** 361–77

[7] Lichtfouse E 2021 *Green Adsorbents to Remove Metals, Dyes and Boron from Polluted Water* vol 49 (Cham, Switzerland)

[8] Tan K B, Abdullah A Z, Horri B A and Salamatinia B 2016 Adsorption mechanism of microcrystalline cellulose as green adsorbent for the removal of Cationic methylene blue dye *J. Chem. Soc. Pakistan* **38** 651–64

[9] Holan Z R, Volesky B and Prasetyo I 1993 Biosorption of cadmium by biomass of marine algae *Biotechnol. Bioeng.* **42** 548–548

[10] El-Araby H A, Ibrahim A M M A, Mangood A H and Abdel-Rahman A A 2017 Sesame husk as adsorbent for copper(II) ions removal from aqueous solution *J. Geosci. Environ. Prot.* **05**

109–52
[11] Kamel S, Hassan E M and El-Sakhawy M 2006 Preparation and application of acrylonitrile-grafted cyanoethyl cellulose for the removal of copper (II) ions *J. Appl. Polym. Sci.* **100** 329–34
[12] Zhang N, Zang G L, Shi C, Yu H Q and Sheng G P 2016 A novel adsorbent TEMPO-mediated oxidized cellulose nanofibrils modified with PEI: Preparation, characterization, and application for Cu(II) removal *J. Hazard. Mater.* **316** 11–8
[13] Khalid A M, Hossain M S, Ismail N, Khalil N A, Balakrishnan V, Zulkifli M and Yahaya A N A 2021 Isolation and characterization of magnetic oil palm empty fruits bunch cellulose nanofiber composite as a bio-sorbent for Cu(II) and Cr(VI) removal *Polymers (Basel).* **13** 1–22
[14] Nag S and Biswas S 2021 Cellulose-based adsorbents for heavy metal removal *Green Adsorbents to Remove Metals, Dyes and Boron from Polluted Water* ed Inamuddin *et al.* (Cham: Springer International Publishing) pp 113–42
[15] Mosur P M, Chernoberezhskii Y M and Lorentsson A V 2008 Electrosurface properties of microcrystalline cellulose dispersions in aqueous solutions of aluminum chloride, nitrate, and sulfate *Colloid J.* **70** 462–5
[16] Hossain M A, Ngo H H, Guo W S and Nguyen T V 2012 Removal of copper from water by adsorption onto banana peel as bioadsorbent *Int. J. GEOMATE* **2** 227–34
[17] Wang X S and Qin Y 2005 Equilibrium sorption isotherms for of Cu^{2+} on rice bran *Process Biochem.* **40** 677–80
[18] Tan K L and Hameed B H 2017 Insight into the adsorption kinetics models for the removal of contaminants from aqueous solutions *J. Taiwan Inst. Chem. Eng.* **74** 25–48
[19] Cao J, Fei D, Tian X, Zhu Y, Wang S, Zhang Y, Mao Q and Sun M 2017 Novel modified microcrystalline cellulose-based porous material for fast and effective heavy-metal removal from aqueous solution *Cellulose* **24** 5565–77
[20] Reddy D H K and Lee S M 2013 Application of magnetic chitosan composites for the removal of toxic metal and dyes from aqueous solutions *Adv. Colloid Interface Sci.* **201–202** 68–93
[21] Aydin H, Bulut Y and Yerlikaya Ç 2008 Removal of copper (II) from aqueous solution by adsorption onto low-cost adsorbents *J. Environ. Manage.* **87** 37–45
[22] Zango Z U and Shehu Imam S 2018 Evaluation of microcrystalline cellulose from groundnut shell for the removal of Crystal Violet and Methylene Blue *Nanosci. Nanotechnol.* **8** 1–6
[23] Hameed B H and Ahmad A A 2009 Batch adsorption of methylene blue from aqueous solution by garlic peel, an agricultural waste biomass *J. Hazard. Mater.* **164** 870–5
[24] Vijayalakshmi K, Gomathi T, Latha S, Hajeeth T and Sudha P N 2016 Removal of copper(II) from aqueous solution using nanochitosan/sodium alginate/microcrystalline cellulose beads *Int. J. Biol. Macromol.* **82** 440–52
[25] Lin C, Qiao S, Luo W, Liu Y, Liu D, Li X and Liu M 2014 Thermodynamics, kinetics, and regeneration studies for adsorption of Cr(VI) from aqueous solutions using modified cellulose as adsorbent *BioResources* **9** 6998–7017
[26] Shafqat F, Bhatt H N, Hanif M A and Zubair A 2008 Kinetic and equilibrium studies of Cr(III) and Cr(VI) sorption from aqueous solution using Rosa Gruss An Teplitz (Red Rose) waste biomass *J. Chil. Chem. Soc.* **53** 1667–1672
[27] Areco M M and dos Santos Afonso M 2010 Copper, zinc, cadmium and lead biosorption by Gymnogongrus torulosus. Thermodynamics and kinetics studies *Colloids Surfaces B Biointerfaces* **81** 620–8
[28] Ang X W, Sethu V S, Andresen J M and Sivakumar M 2013 Copper(II) ion removal from aqueous solutions using biosorption technology: Thermodynamic and SEM–EDX studies *Clean Technol. Environ. Policy* **15** 401–7
[29] Argun M E, Güclü D and Karatas M 2014 Adsorption of reactive Blue 114 dye by using a new adsorbent: Pomelo peel *J. Ind. Eng. Chem.* **20** 1079–84

[30] Peng S, Meng H, Ouyang Y and Chang J 2014 Nanoporous magnetic cellulose-chitosan composite microspheres: Preparation, characterization, and application for Cu(II) adsorption *Ind. Eng. Chem. Res.* **53** 2106–13

[31] El-Naggar M E, Radwan E K, El-Wakeel S T, Kafafy H, Gad-Allah T A, El-Kalliny A S and Shaheen T I 2018 Synthesis, characterization and adsorption properties of microcrystalline cellulose based nanogel for dyes and heavy metals removal *Int. J. Biol. Macromol.* **113** 248–58

IOP Conf. Series: Materials Science and Engineering 1195 (2021) 012023 doi:10.1088/1757-899X/1195/1/012023

Comparison of chemically surface treated Luffa cylindrica using scanning electron microscopy (SEM)

M J Kamran[1], E Jayamani[1], K H Soon[1] and Y C Wong[2]

[1] Department of Mechanical Engineering, Faculty of Engineering, Computing and Science, Swinburne University of Technology, 93350 Kuching, Sarawak, Malaysia
[2] Senior Lecturer, Faculty of Engineering, Computing and Science, Swinburne University of Technology, 3122, Hawthorn, Melbourne, Australia

Email: mkamran@swinburne.edu.my

Abstract. The world as we probably know is currently facing a big difficulty referred as environmental pollution. Researchers from around the globe have been working on biodegradable materials in order to reduce the overall consequences due to environmental pollution. Furthermore, biodegradable composites also known as green composites made using natural fibers are highly considered over non-green synthetic fiber composites. Additionally, natural fibers are low in cost, have good mechanical properties, biodegradability and require less production energy. Focus of this research paper was on one of the natural fibers, Luffa cylindrica (LC). There are plenty of surface treatments available including chemical and mechanical surface treatments. However, the focus of this paper was on chemical surface treatments. Sodium hydroxide, silane and acetylation chemical surface treatments were utilized. Chemically surface treated LC was compared with untreated LC with the help of scanning electron microscopy (SEM). With the help of SEM, it was observed that all of the chemical surface treatments were effective. Furthermore, it was also noticed that each of the chemical surface treatment affected LC differently. However, sodium hydroxide surface treated LC samples showed the best outcome by removing waxy impurities as well as voids from the surface of LC.

1. Introduction

For decades, composites have been made using synthetic fibers. Some of the commonly used synthetic fibers are glass fiber and carbon fiber. These synthetic fibers have been used due to their mechanical properties thus, their composites are well favoured over steel and aluminium. Although the total emissions are lesser than steel and aluminium, recycling or biodegradability problems arise [1]. Synthetic fibers are not entirely biodegradable. They are flammable and emit huge quantities of smoke and toxic fumes. However, natural fibers on the other hand have decent mechanical properties, good soundproofing, high impact strength and are good fire retardants therefore, natural fibers can be regarded as multipurpose materials [2, 3]. Additionally, natural fibers are abundant, recyclable and biodegradable [4].

Plenty of natural fibers are present however, focus of this research paper is on Luffa cylindrica (LC) which is commonly known as a sponge gourd as seen in figure 1. The LC plant is abundantly available in China, India, and Japan. Furthermore, LC is one of the eight species present in a family of vegetable sponges. The plant itself has a dense outer skin layer which is peeled in order to expose the fibrous mat of LC [5]. As the cells present inside LC expands, which then leads to elongation of cells which then develops a fibrous structure. Dry fibrous structure of LC mainly consists of cellulose, hemicellulose and lignin [6]. As it has been described, LC is fibrous and spongy therefore, LC has great mechanical properties such as high strength, stiffness and energy absorption [7]. Microstructure of LC is mainly made of cellulose, hemicellulose and lignin which make it fall under the category of lignocellulose natural fibers as seen in figure 2. Many researchers have utilized LC to produce composites with various types of polymers, given the mechanical properties of LC [8].

As recorded by other researchers, LC has a high cellulose content of 55–70%, 8–22% hemicellulose, 10–23% lignin, 3.2% extractive and 0.4% ash as seen in Table 1 [7]. LC density is between the range of 0.82–0.92 g/cm^3, which is not so high compared to other widely used natural fibers such as sisal or hemp [10, 11]. Natural fibers such as LC has a higher fiber content than synthetic fiber, and small amounts can provide comparable efficiency, reducing the overall fiber consumption [12].

Figure 1. Luffa fiber [7].

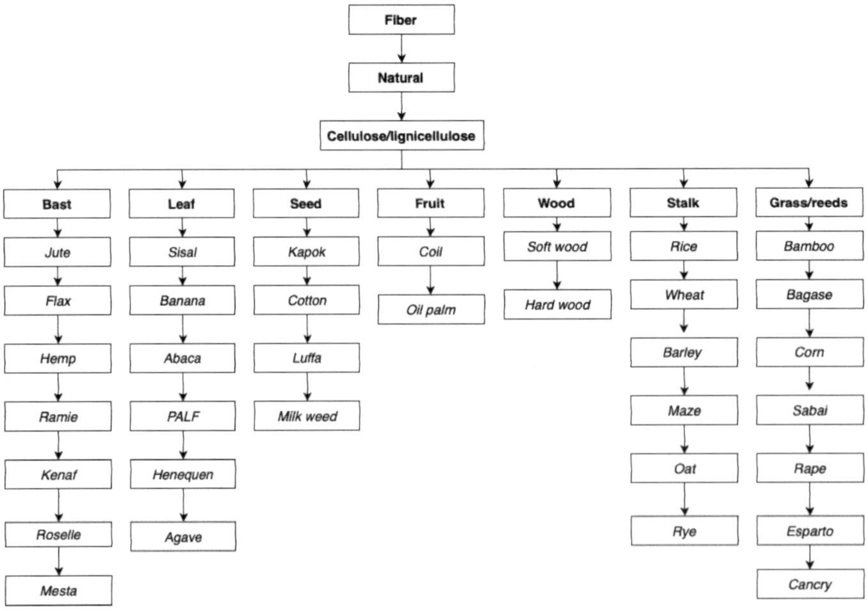

Figure 2. Classification of natural fibers [9].

IOP Conf. Series: Materials Science and Engineering 1195 (2021) 012023 doi:10.1088/1757-899X/1195/1/012023

Table 1. Shows mechanical properties and chemical composition of luffa fiber [7, 13].

Mechanical Properties	Luffa fiber	Chemical Composition	Luffa fiber
Tensile Strength (MPa)	202.3	Cellulose (%)	55–70
Young's Modulus (GPa)	4.5	Hemicellulose (%)	8–22
Elongation at break (%)	2.5	Lignin (%)	10–23
Density (g/cm^3)	1.2	Extractives (%)	3.2

Scanning electron microscopy (SEM) is widely used technique which was first introduced 80 years ago. SEM analysis has been used as a key process for characterization of the sample. With the help of SEM various researchers were able to notice microscopic changes occurring on the surface of the fiber due to chemical surface treatments. Furthermore, this technique of analysing the samples is preferred by researchers due to images being in high resolution with an average magnification range of 10–500,000 times [14]. Previously researchers were able to notice changes in terms of fiber surface, removal of impurities, waxy impurities and voids [15–18].

In this research paper the focus was on analysing the effects on LC surface after chemical surface treatments. Sodium hydroxide, silane and acetylation surface treatments were utilized. Chemically surface treated LC was compared with untreated LC with the help of SEM analysis. Changes were noticed in terms of LC surface, impurities and voids were analysed before and after chemical surface treatments.

2. Methodology

2.1. Preparation of samples

LC utilized in this research was bought from a local dealer. In order to produce composites using natural fiber, surface treatment is a necessary step. In order to enhance the compatibility between LC and any polymer matrix, surface treatments are necessary. In addition to that chemical fiber surface treatments have shown successful removal of naturally occurring waxy impurities and hydroxyl groups which contribute to ineffective bonding between a fiber and a polymer. In this specific research LC fibers were chopped using a chopper and with the help of a sieve, 3mm chopped LC samples were separated from the batch.

Removal of surface impurities like dust is a crucial step before performing any type of chemical surface treatment. LC samples with a length of 3mm were soaked for 2 hours in a solution which was made of detergent and distilled water at 60 °C. This step as mentioned earlier would help in removing any kind of impurities present on the surface such as dust. Furthermore, LC samples were removed from the solution and were rinsed thoroughly using distilled water. Clean and rinsed LC samples were left for drying for 3 days.

2.2. Fiber chemical surface treatment

2.2.1. Sodium hydroxide (NaOH) chemical surface treatment. Dried 3 mm in length LC samples were soaked for an hour at room temperature. Sodium hydroxide (1 N) solution with distilled water was utilized in this research. LC samples were then removed and cleaned thoroughly using distilled water containing a few drops of acetic acid to neutralize the LC samples. Furthermore, LC was rinsed thoroughly under a steady stream of distilled water to ensure sodium hydroxide was eliminated entirely from the surface. Rinsed LC samples were left for drying at room temperature for 24 hours. Furthermore, samples were also dried for 12 hours at 80 °C in the oven. This approach to prepare sodium hydroxide surface treated samples for SEM testing has been standardised by various other researchers in their research using different parameters [19–21].

2.2.2. Silane chemical surface treatment. Dried LC chopped 3 mm samples were soaked for an hour at room temperature in a (60:40) solution made from ethanol and water. Furthermore, 5wt% silane was added in the solution. Silane coupling agents known as Bis-(3-triethoxysilylpropyl) and 3-amino propyltriethoxysilane (APS) were also added in the solution. In order to maintain the pH value at 4, acetic acid was utilized. After an hour of immersion LC samples were removed from the solution and were washed thoroughly with the help of distilled water to remove any remaining solution. Rinsed LC samples were left for drying at room temperature for 24 hours. Furthermore, samples were also dried for 12 hours at 80 °C in the oven. This approach to prepare silane surface treated samples for SEM testing has been standardised by various other researchers in their research using different parameter [21, 22].

2.2.3. Acetylation surface treatment. Dried LC chopped 3 mm samples were soaked for an hour. The solution consisted of acetic acid and acetic anhydride with. Ratio of 1:1 was utilized furthermore, concentrated sulphuric acid (1ml) was added in the solution as well. Concentrated sulphuric acid acted as a catalyst in the solution. LC samples were removed from the solution after an hour of soaking time. Distilled water was utilized to thoroughly clean the samples until the solution was completely removed from the surface of LC. Rinsed LC samples were left for drying at room temperature for 24 hours. Furthermore, samples were also dried for 12 hours at 80 °C in the oven. This technique to perform acetylation surface treatment on the fiber samples for SEM testing has been standardised by various other researchers in their research using different parameter [21, 23].

2.3. Scanning electron microscopy (SEM)
SEM analysis is one of the most commonly used electron microscopies to understand the surface composition of the fibers. Images are produced with the help of electrons from the beam and electrons from the sample combining to create signals. Various researchers have used SEM to study various chemical surface treatments and their effects on the surface of natural fibers. SEM was performed using JOEL (JSM-6300F). Whereas, accelerating voltage was set to 20 KV and working distance of 20mm was used as mentioned in the research. Magnifications of 100x, 500x and 800x were utilized to analyse the surface of the fiber thoroughly. Furthermore, SEM was used to compare treated and untreated fiber samples. Similar standardised approach regarding SEM process has been used by various researchers with different parameters [7, 24, 25].

3. Results and discussion
Process of scanning electron microscopy (SEM) as mentioned above provided images at magnification of 100x, 500x and 800x for untreated and chemically surface treated LC samples. Untreated samples of LC can be seen in figure 3. Smooth fiber surface can be seen due to the waxy impurities present on the surface shown in figure 3 (a). Furthermore, figure 3 (b) confirms the presence of voids on the surface which can be seen as black dots. Furthermore, impurities can be seen in white dots. similar results can also be seen in the research shown by various researchers regarding fiber surface treatments [15, 16]. Due to the presence of waxy impurities, impurities and voids, adhesion between polymer and the LC will be poor, which leads to poor mechanical properties.

Figure 3. Shows untreated luffa fiber samples, (a) 100x magnification and (b) 800x magnification.

Sodium hydroxide surface treated samples can be seen in figure 4 shown underneath. Using the methodology explained earlier, 1 N of sodium hydroxide solution was used to chemically surface treat LC. Rougher fiber surface can be seen which is due to the removal of waxy impurities from the surface of LC. Removal of waxy impurities was achieved due to the impurities being soluble in sodium hydroxide solution [26]. However, some of the impurities can still be seen on the surface of LC as shown in figure 4 (a) and (b). This can be due to the immersion time being only one hour. Furthermore, there is no sign of voids present on the surface of LC. Removal of waxy impurities and voids from the surface of LC can be seen in other research papers as well shown by other various researchers [17, 27]. Better adhesion between LC and polymers can be expected due to surface being rough and without voids. Due to better adhesion, better mechanical properties can be expected from LC and polymer composites.

Figure 4. Shows sodium hydroxide surface treated luffa fiber samples, (a) 100x magnification and (b) 500x magnification.

Silane surface treated samples can be seen in figure 5 shown underneath. Using the methodology explained earlier, 5wt% silane solution with ethanol was used. Bis-(3-triethoxysilylpropyl) and 3-amino propyltriethoxysilane (APS) coupling agents were utilized. It can be seen in figure 5 (a) and (b) that there is not much difference on the fiber surface. Surface of LC is still smooth due to silane sticking on the surface of the fiber [23]. However, figure 5 (b) shows removal of surface impurities but some voids on the surface are still present [28]. This result can be due to immersion time being only one hour. Furthermore, better adhesion can be expected between the fiber and polymer due to removal of waxy impurities and impurities from LC surface. However, due to the voids present on the surface of the fiber, this may affect the overall adhesion between LC and polymer.

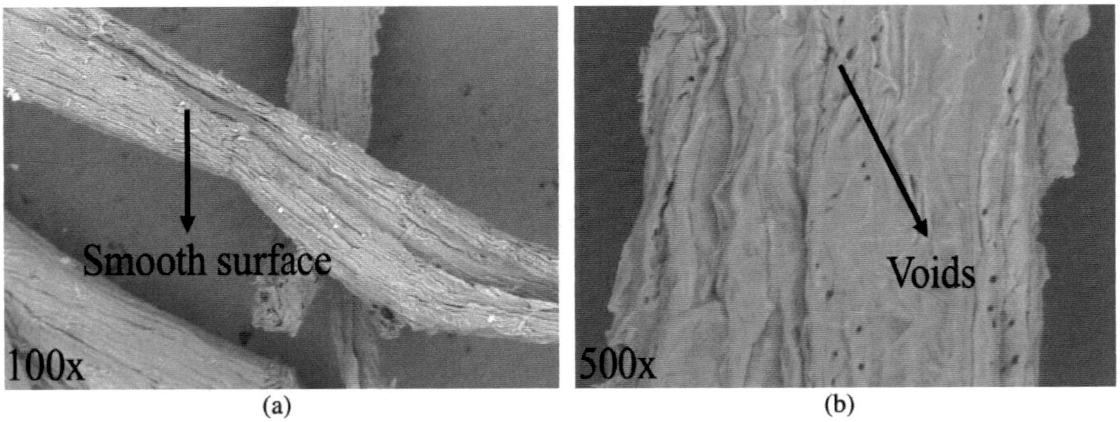

Figure 5. Shows silane surface treated luffa fiber samples, (a) 100x magnification and (b) 500x magnification.

Acetylated surface treated samples can be seen in figure 6 shown underneath. Using the methodology explained earlier, 1:1 mixing was used between acetic acid and acetic anhydride to chemically surface treat LC samples. It can be seen in figure 6 (a) and (b) that fiber surface is still smooth after the surface treatment. This is due to the acetyl functional groups replacing hydroxyl functional groups from the surface of LC [23]. Furthermore, impurities from the surface are removed to some extent. However, figure 6 (b) shows voids present on the surface of luffa fiber samples. This result can be expected due to immersion time being only one hour. Better adhesion can be expected between the fiber surface and polymer due to removal of waxy impurities and impurities to some extent. However, presence of voids may not give the best mechanical properties.

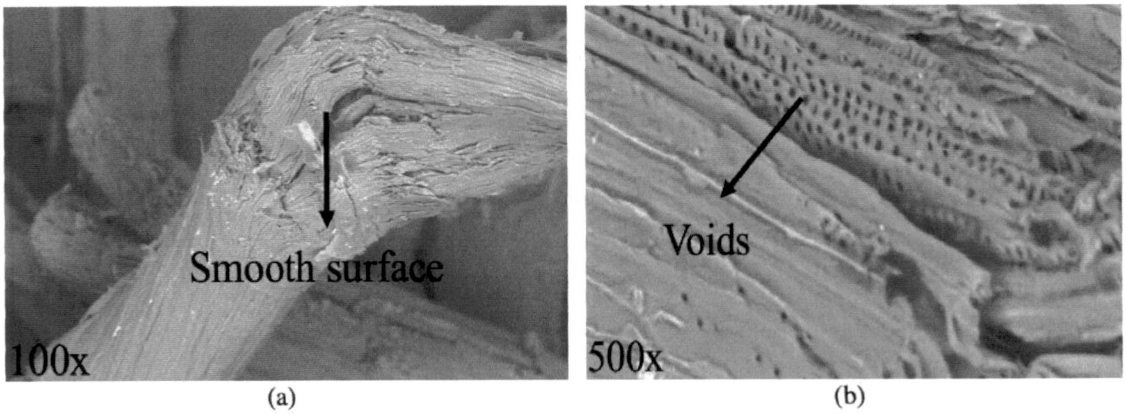

Figure 6. Shows acetylated surface treated luffa fiber samples, (a) 100x magnification and (b) 500x magnification.

4. Conclusion

Focus of this research was on chemically surface treated LC. Sodium hydroxide, silane and acetylation chemical surface treatments were utilized. Untreated LC was compared with chemically surface treated LC with the help of SEM. From the results shown in the previous section, it can be concluded that untreated LC surface consists of waxy impurities which causes smooth surface, other impurities, and voids. All these factors play a huge role in decreasing the adhesion between the fiber and the polymer hence, leading to poor mechanical properties. However, by utilizing chemical surface treatments, these factors can be tackled. Sodium hydroxide surface treatment was able to remove waxy impurities and voids from the surface of LC but, struggled to remove other impurities completely. On the other hand, silane surface treatment showed smoother surface due to silane being deposited on the surface of LC and the impurities were removed significantly. However, voids were not removed completely. Acetylation surface treatment showed smoother surface as well because the acetyl groups replaced the hydroxyl groups in the chemical structure of LC. Furthermore, impurities were removed to some extent; yet voids were still present on the surface after the surface treatment. Overall, with the help of SEM analysis it was predicted that each of the chemical surface treatment utilized in this research according to the methodology was successful. However, each of the chemical surface treatment affected LC differently.

References

[1] AL-Oqla F M 2013 Natural fiber reinforced polymer composites in industrial applications: Feasibility of date palm fibers for sustainable automotive industry *J. Cleaner Prod.* **66** 347–54
[2] Parida C, Das S C and Dash S K 2012 Mechanical analysis of bio nanocomposite prepared from Luffa cylindrica *Procedia Chem.* **4** 53–9
[3] Sanjay M R, Jawaid M, Senthamaraikannan P, Senthil S and Pradeep S 2018 Characterization and properties of natural fiber polymer composites: A comprehensive review *J. Cleaner Prod.* **172** 566–81
[4] Majeed K, Hassan A, Abu Bakar A, Abdul Khalil H P S, Salema A A and Inuwa I 2013 Potential materials for food packaging from nanoclay/natural fibres filled hybrid composites *Mater. Des.* **46** 391–410
[5] Mohanta N 2015 Investigation of mechanical properties of Luffa cylindrica fibre reinforced epoxy hybrid composite *Int. J. Eng. Sci. Technol.* **7** 1–10
[6] Yin S, Li J, Ritchie R, and Xu J 2019 Light but tough bio-inherited materials: Luffa sponge based nickel-plated composites *J. Mech. Behav. Biomed. Mater.* **94** 10–8
[7] Saw S K 2017 Effect of stacking pattern on morphological and mechanical properties of luffa/coir hybrid fiber-reinforced epoxy composite laminates *Hybrid Polymer Composite Materials Structure and Chemistry,* ed V K Thakur *et al.* (Woodhead Publishing) chapter 12 pp 313–33
[8] Su S, Liu J, Zhang H, Li R, Jing X and Wang J 2018 Polyethyleneimine-functionalized Luffa cylindrica for efficient uranium extraction *J. Colloid Interface Sci.* **530** 538–46
[9] Stamboulisa A 2001 Effects of environmental conditions on mechanical and physical properties of flax fibers *Composites, Part A* **32** 1105–15
[10] Joseph K, James B, Thomas S and Carvalho L H 1999 A review on sisal fiber reinforced polymer composites *Brazilian J. Agri. Environ. Eng.* **3** 367–79
[11] Anbukarasi K and Kalaiselvam S 2015 Study of effect of fibre volume and dimension on mechanical, thermal, and water absorption behaviour of luffa reinforced epoxy composites *Mater. Des.* **66** 321–30
[12] Demir H, Atikler U, Balköse D and Tıhmınlıoğlu F 2006 The effect of fiber surface treatments on the tensile and water sorption properties of polypropylene–Luffa fiber composites *Composites, Part A* **37** 447–56
[13] Karthi N, Sathish S, Gokulkumar S, Prabhu L and Vigneshkumar N 2020 An overview: Natural

fiber reinforced hybrid composites, chemical treatments and application areas *Mater. Today: Proc.* **27** 2828–34
[14] Inkson B J 2016 Scanning electron microscopy (SEM) and transmission electron microscopy (TEM) for materials characterization *Materials Characterization Using Nondestructive Evaluation (NDE) Methods*, ed G Hübschen *et al.* (Cambridge: Elsevier Science & Technology) chapter 2 pp 17–43
[15] Boopathi L and Mylsamy K 2012 Investigation of physical, chemical and mechanical properties of raw and alkali treated Borassus fruit fiber *Composites, Part B* **43** 3044–52
[16] Chen Y, Qin S, Canming Y, Kaiting Z, PanPan L, Dingfeng H and Yong G 2020 A novel sound absorbing material comprising discarded luffa scraps and polyester fibers *J. Cleaner Prod.* **245** 1–8
[17] Fiore V and Valenza A 2015 The effect of alkaline treatment on mechanical properties of kenaf fibers and their epoxy composites *Composites, Part B* **68** 14–21
[18] Ghali L, Zidi M and Sakli F 2011 Effects of Fiber weight ratio, structure and fiber modification onto flexural properties of luffa-polyester composites *Adv. Mater. Phys. Chem.* **1** 78–85
[19] Vilay V, Taib M and Todo M 2008 Effect of fiber surface treatment and fiber loading on the properties of bagasse fiber–reinforced unsaturated polyester composites *Compos. Sci. Technol.* **68** 631–8
[20] Paul S A, Ibos L, Candau Y, Joseph K and Thomas S 2008 Effect of fiber loading and chemical treatments on thermophysical properties of banana fiber/polypropylene commingled composite materials *Composites, Part A* **39** 1582–8
[21] Gupta A K, Mohanty S and Nayak S K 2012 Mechanical, Thermal Degradation, and Flammability Studies on Surface Modified Sisal Fiber Reinforced Recycled Polypropylene Composites *Adv. Mech. Eng.* **2012** 1–13
[22] Franco H 2005 A study of the mechanical properties of short natural-fiber reinforced composites *Composites, Part B* **36** 597–608
[23] Hajiha H 2014 Modification and characterization of hemp and sisal fibers *J. Nat. Fibers* **11** 144–68
[24] Sakthivel M and Ramesh S 2014 Production and characterization of luffa/coir reinforced polypropylene composite *Procedia Mater. Sci.* **5** 739–45
[25] Huda M S, Drzal L T, Mohanty A K and Misra M 2007 The effect of silane treated- and untreated-talc on the mechanical and physico-mechanical properties of poly(lactic acid)/newspaper fibers/talc hybrid composites *Composites, Part B* **38** 367–79
[26] Oushabi A, Sair S, Hassani F O, Abboud Y, Tanane O and Bouari A E 2017 The effect of alkali treatment on mechanical, morphological and thermal properties of date palm fibers (DPFs): Study of the interface of DPFePolyurethane composite *S. Afr. J. Chem. Eng.* **23** 116–23
[27] Ghali L, Zidi M, Bendaly H, M'sahli S, and Sakli F 2011 Effect of chemical modification of Luffa cylindrica fibers on the mechanical and hygrothermal behaviours of polyester/luffa composites *Bioresources* **4** 3836–49
[28] Zhou F and Jiang B 2014 Effect of silane treatment on microstructure of sisal fibers *Appl. Surf. Sci.* **292** 806–12

Organic acid-assisted catalytic wet torrefaction of oil palm trunks (OPT)

M Soh, J J Chew and J Sunarso

Research Centre for Sustainable Technologies, Faculty of Engineering, Computing and Science, Swinburne University of Technology Sarawak Campus, Kuching 93350, Sarawak, Malaysia

Email: msoh@swinburne.edu.my

Abstract. Oil palm trunks (OPT) are attractive bio-fuel sources given their abundant availability. Nonetheless, the inherent properties of these biomass can lead to their inefficient use as bio-fuel directly. This work utilizes four organic acids (i.e., acetic, formic, levulinic, and citric acid) as catalyst in wet torrefaction to enhance the fuel properties of OPT hydrochar. In this study, the effects of different catalysts, catalyst concentrations, and residence times on the fuel properties of OPT hydrochar are analyzed. To study the effect of residence time, 0.2M of acid concentration was used for all four acids at 220 °C for 3 hr and 24 hr. Meanwhile, study on the effect of catalyst concentration was performed at 220 °C for 24 hr at 0.2M and 1.0M for all four acids. Increasing the residence time decreased the solid yield of OPT hydrochar treated in deionized water, acetic, formic, and levulinic acid, while wet torrefaction in citric acid results in close solid yield value in both residence time. The energy yield was observed to decrease in all liquid medium with increasing residence time except for formic acid and citric acid. On the other hand, increasing the acid concentration increased the OPT hydrochar solid yield in all acids except formic acid and the highest energy yield of 77.08% was obtained from wet torrefaction in 1.0M of citric acid at 220 °C for 24 hr. In summary, citric acid is an environmentally friendly acid to be used as catalyst to enhance the fuel properties of OPT hydrochar. Further study on the reaction mechanisms that governs such fuel properties enhancement with citric acid is warranted.

1. Introduction

Renewable energy will always be of high demand especially in the near future due to the rapid depletion of fossil fuels owing to the growing industrialization and globalization. Biomass is an attractive option for renewable energy production since they are abundantly available and are carbon-neutral, in which plants take in as much CO_2 as it releases when burned [1]. Malaysia is one of the world's largest palm oil producers coming second after Indonesia [2]. The plantation and processing of palm oil produce various types of biomass including oil palm trunks (OPT), oil palm fronds (OPF), empty fruit bunches (EFB), mesocarp fiber (MF), and palm kernel shell (PKS) [3]. Most studies focus on using EFB, PKS, MF, and OPF as feedstocks for renewable energy production. However, OPT accounts for up to 3.6 million tonnes of unused biomass, which are an attractive option for renewable energy conversion [4].

Direct utilization of biomass in thermochemical conversion processes, i.e., pyrolysis, gasification, and combustion suffer from various drawbacks due to the inherent properties of biomass such as high moisture content, low energy value, high ash content, and hygroscopic nature, which can lead to increased cost for handling, storage, and transport as well as operational problems [5, 6]. Wet torrefaction is an attractive pre-treatment method for biomass prior to their thermochemical conversion to improve the biomass properties since it is performed in liquid medium and thus, pre-drying is not required [7]. Wet torrefaction is typically performed in hot compressed water at temperature between 180 and 260 °C [8, 9]. However, high temperature increases energy consumption and thus increases production cost. Therefore, introduction of catalyst is essential to reduce the temperature required to produce equivalent properties of hydrochar product at lower temperature.

In this preliminary study, four organic acids are selected, i.e., acetic acid, formic acid, levulinic acid, and citric acid to investigate their effects on the wet torrefaction of OPT and its resultant fuel properties. The effects of residence time and acid concentration are the focus of this study.

2. Materials and methods

2.1. Materials

In this study, the oil palm trunks (OPT) were obtained from a local oil palm plantation, i.e., Masranti Plantation at Serian, Kuching. The OPT were first chopped into smaller pieces and dried in a convection oven at 105 °C for 24 hours to remove the moisture. Then, the dried OPT were milled using a rotary miller (Fritsch, Pulverisette 14) and segregated into the desired size range of 0.5-1.0 mm by a mechanical sieve. The processed OPT were then kept in a sealed bag and placed in a desiccator for further use.

Four organic acids, i.e., acetic acid (AA), formic acid (FA), levulinic acid (LA), and citric acid (CA) were used in this study. Glacial acetic acid (R&M Chemicals), reagent grade of ≥ 95% formic acid (Sigma), 98% levulinic acid (Sigma), and citric acid monohydrate (Merck) were purchased from the major chemical suppliers. The four acids were diluted to 0.2M and 1.0M for use in the catalytic wet torrefaction of OPT.

2.2. Wet torrefaction and catalytic wet torrefaction

The wet torrefaction of OPT in deionized water (DI) and the four organic acids were performed with a 200 mL polytetrafluoroethylene (PTFE)-lined acid digestion vessel (Parr Instrument Company, 4748A). The biomass and liquid medium were loaded into the vessel at a 1:25 biomass to liquid ratio by weight. After sealing the vessel tightly, it is placed into the convection oven to be heated to the desired temperature of 220 °C and maintained isothermally until the desired residence time reaches. To investigate the effects of residence time, wet torrefaction was heated at 220 °C for 3 and 24 hours with 0.2M acid concentration. Meanwhile, 0.2 and 1.0M of acid concentrations were used to study the effect of acid concentration and processed at 220 °C for 24 hours for all samples. Once the heating is completed, the acid digestion vessel is taken out from the oven and left to cool down to room temperature. The resultant hydrochar with liquid medium is filtered and rinsed with distilled water until the wash water reaches pH 7 and the solid hydrochar is dried in the convection oven at 105 °C for 24 hours. The dried solid hydrochar is weighed and stored in a plastic sealed bag, which is placed in the desiccator for storage until further analysis. The energy content, i.e., higher heating value (HHV) of the solid hydrochar is measured with a bomb calorimeter (IKA, C200). The solid yield, energy density, and energy yield are calculated following equations 1-3 [10]:

$$Hydrochar\ yield\ (\%) = \frac{Hydrochar\ mass}{Raw\ biomass\ mass} \times 100\% \quad (1)$$

$$Energy\ density, ED = \frac{Hydrochar\ HHV}{Raw\ biomass\ HHV} \quad (2)$$

$$Energy\ yield, EY\ (\%) = Hydrochar\ yield \times ED \quad (3)$$

3. Results and discussion

3.1. Effect of residence time

Figure 1 depicts the effect of two different residence times on the resultant non-catalyzed and catalyzed OPT hydrochar based on their solid yield, energy content presented as HHV, and energy yield. As expected, increasing the residence time from 3 to 24 hours led to decreased solid hydrochar yield from 45.29 to 30.88%, 49.05 to 32.13%, 45.05 to 35.40%, and 55.73 to 35.02% that is treated with deionized water, acetic acid, formic acid, and levulinic acid, respectively (figure 1(a)). Nonetheless, it was observed that increasing the residence time from 3 to 24 hours did not cause significant change in the solid yield when treated in 0.2M of citric acid, where the solid yield was 41.02% in 3 hours residence time and 41.00% for 24 hours residence time.

Lignocellulosic biomasses are generally composed of hemicellulose, cellulose, and lignin. Relative to cellulose and lignin, hemicellulose is the least thermally stable component followed by cellulose and subsequently, lignin. This is since hemicelluloses are amorphous structures of polysaccharides that comprises of various pentose and hexose sugars that can hydrolyze easily in the presence of weak acid or base [11, 12]. The hemicellulose is attached to cellulose microfibrils, which is a strong crystalline structure made of long polymer chain of glucose [11, 12]. Lignin, on the other hand, is a phenylpropane polymer that strongly cross-links the cellulose and hemicelluloses together and has the highest thermal stability [11]. The major cause for the weight loss in OPT during wet torrefaction is due to the decomposition of hemicellulose and parts of the cellulose. Increasing the residence time allows the liquid medium to further hydrolyze the hemicellulose and cellulose. The solid yield of the hydrochar treated in citric acid in both residence times, i.e., 41.00-41.02% is between that of the 3 and 24 hours of the average solid yield of deionized water, acetic acid, formic acid, and levulinic acid (i.e., the average solid yield of the four samples with 3 hours residence time is 48.78%; the average solid yield of the four samples with 24 hours residence time is 33.36%; the average between 33.36 and 48.78% is 41.07%, which is close to that of the solid yield of the hydrochar treated in citric acid). This implies that in citric acid liquid medium, the hydrolyses of hemicellulose and cellulose in OPT can only reach a certain point only, unlike the other liquid mediums that can further hydrolyzes the cellulose component in OPT to a higher extent with increasing reaction time. Thus, further increasing the residence time cannot decrease the solid yield much further in citric acid medium.

The HHV of raw OPT is 16.40 MJ kg^{-1}. Subjecting OPT to wet torrefaction in general, can enhance its energy value content. In addition, increasing the residence time from 3 to 24 hours increases the HHV of the resulting hydrochar from 20.30 to 26.70 MJ kg^{-1}, 18.18 to 26.30 MJ kg^{-1}, 18.25 to 26.37 MJ kg^{-1}, 18.13 to 26.15 MJ kg^{-1}, and 18.64 to 26.32 MJ kg^{-1} that is treated with deionized water, acetic acid, formic acid, and levulinic acid, respectively (figure 1(b)). It is observed that the OPT hydrochars from all acid-catalyzed liquid mediums for 3 hours residence time have lower HHV than those from deionized water. However, the HHV of the OPT hydrochar is relatively same in all liquid mediums when treated at 24 hours. This implies that the change in energy value when acid catalyst was added was highly influenced by the residence time. Wet torrefaction is a densification process whereby low HHV components, i.e., hemicellulose and cellulose are completely or partially decomposed to form a lignin-rich hydrochar, which has a high energy content where lignin is the main contributor to the HHV [13].

The energy yield in figure 1(c) is related to both the solid yield and the energy density, which is calculated as shown in equation (2). Increasing the residence time from 3 to 24 hours decreased the energy yield of the OPT hydrochar from 56.08 to 50.29%, 54.37 to 51.54%, and 61.63 to 55.84% treated in deionized water, acetic acid, and levulinic acid, respectively. The decrease in energy yield in these three samples might be due to the large solid yield loss during wet torrefaction. However, the energy yield increases with increasing residence time from 3 to 24 hours from 50.15 to 56.92% and 46.62 to 65.82% treated in formic acid and citric acid, respectively. This indicates that the formic acid and citric acid catalyzed liquid mediums can cause further reaction with increasing residence time. However, the reactions and mechanism that occur need further investigation.

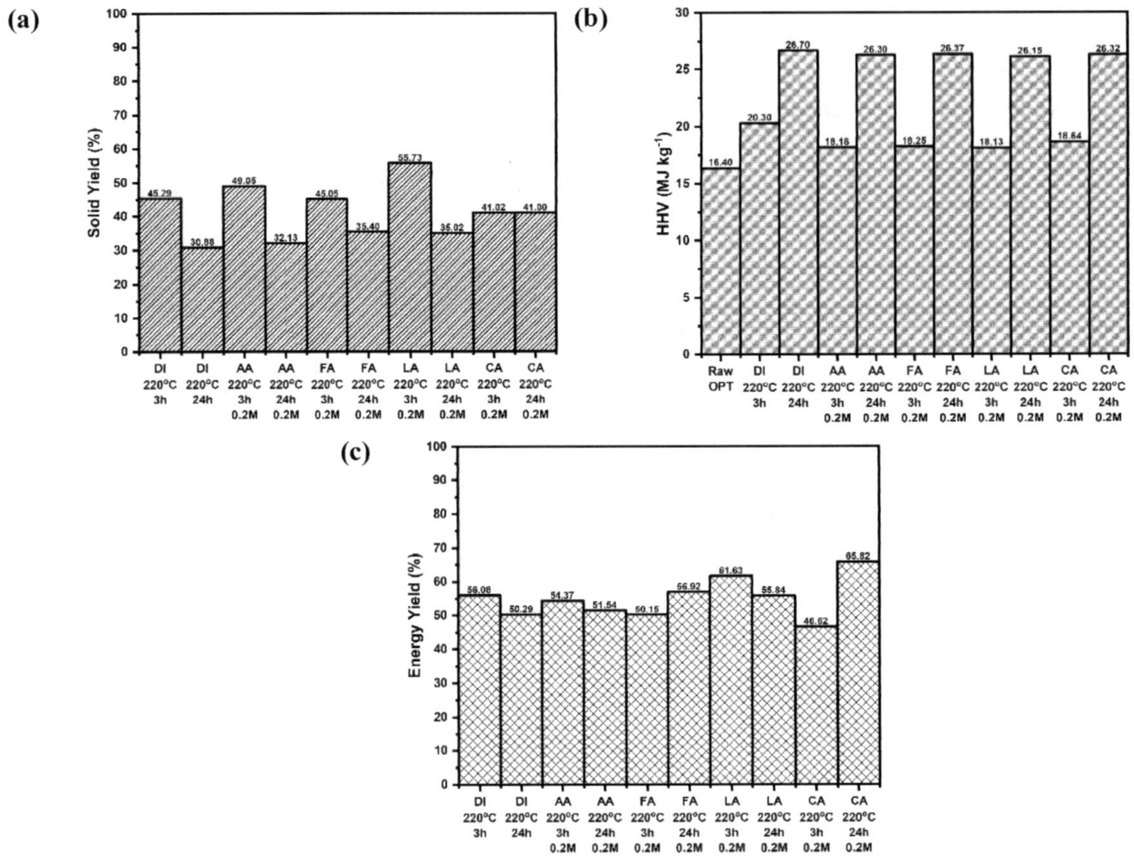

Figure 1. (a) Solid yield, (b) HHV, and (c) energy yield of OPT hydrochar treated at 220 °C for 3 and 24 hours in deionized, acetic acid, formic acid, levulinic acid, and citric acid at 0.2M concentration.

3.2. Effect of acid concentration

The effect of acid concentration on the fuel properties of OPT hydrochar is depicted in figure 2. It is observed in figure 2(a) that increasing the acid concentration from 0.2 to 1.0M increased the solid yield from 32.13 to 33.89%, 35.02 to 40.72%, and 41.00 to 46.67% treated in acetic acid, levulinic acid, and citric acid, respectively, except for that in formic acid, where the solid yield decreased from 35.40 to 24.10% when concentration increased from 0.2 to 1.0M. During wet torrefaction, the hemicellulose and cellulose decompose and hydrolyze into the liquid medium as intermediate hydroxymethylfurfural (5-HMF) [14–16]. The increased solid yield may be due to the increased acid concentration that helps to catalyze the repolymerization reaction of the intermediate 5-HMF components in the liquid medium onto the solid hydrochar. García-Bordejé, Pires [17] studied the wet torrefaction of cellulose in HCl and reported that the presence of 2M HCl promotes cellulose microfibrils fragmentation and formation of carbon spheres, i.e., microspheres. Qi, Song [18] investigated the effects of formic acid and levulinic addition to wet torrefaction of sucrose. They reported that the 5-HMF microspheres contain several reactive groups including hydroxyl groups that can actively react with the carboxyl groups of levulinic acid to promote microspheres formation and growth. They also found that in the case where formic acid was added, the liquid phase had much lower levulinic acid concentration at the initial stage since the addition of formic acid suppressed 5-HMF rehydration to levulinic acid. Thus, it is postulated that the increased in formic acid concentration from 0.2 to 1.0M in this study hinders the formation of levulinic acid during wet torrefaction of OPT that can assist in increasing the solid yield by promoting microspheres formation onto the surface of the hydrochar.

The HHV of raw OPT and OPT hydrochar treated in deionized water was 16.40 and 26.70 MJ kg^{-1}, respectively. Increasing the acid concentration increased the HHV of OPT hydrochar treated in formic acid, levulinic, and citric acid from 26.37 to 26.63 MJ kg^{-1}, 26.15 to 26.99 MJ kg^{-1}, and 26.32 to 27.08 MJ kg^{-1}, respectively, except in acetic acid where the HHV decreases from 26.30 to 25.08 MJ kg^{-1}. The increased acid concentration primarily helped to catalyze hydrolysis reaction and carbonized the OPT to further densify the energy content.

Figure 2(c) depicts the energy yield obtained from the OPT hydrochar at different liquid medium and acid concentration. The energy yield of OPT hydrochar produced with deionized water was 50.29%. Meanwhile, increasing the acid concentration from 0.2 to 1.0M obviously increased the energy yield of OPT hydrochar treated in acetic acid, levulinic acid, and citric acid from 51.54 to 51.85%, 55.84 to 67.02%, and 65.82 to 77.08%, respectively, which is mainly contributed by the increased solid yield as discussed. Nonetheless, the reduced solid yield with increasing formic acid concentration led to decreased energy yield from 56.92 to 39.15%, which is the lowest energy yield obtained in this study. The highest energy yield was obtained from wet torrefaction of OPT in 1.0M of citric acid at 220 °C for 24 hours. Thus, citric acid is the most desired acid to catalyze wet torrefaction of OPT. In actual application nonetheless, optimum energy yield is desired when low acid concentration and residence time is used to minimize cost and downstream effluent processing. Therefore, levulinic acid is also a considerable candidate since a reasonably high energy yield of 61.63% was obtained when treated at 0.2M concentration at 220 °C and 3 hours residence time (figure 1(c)).

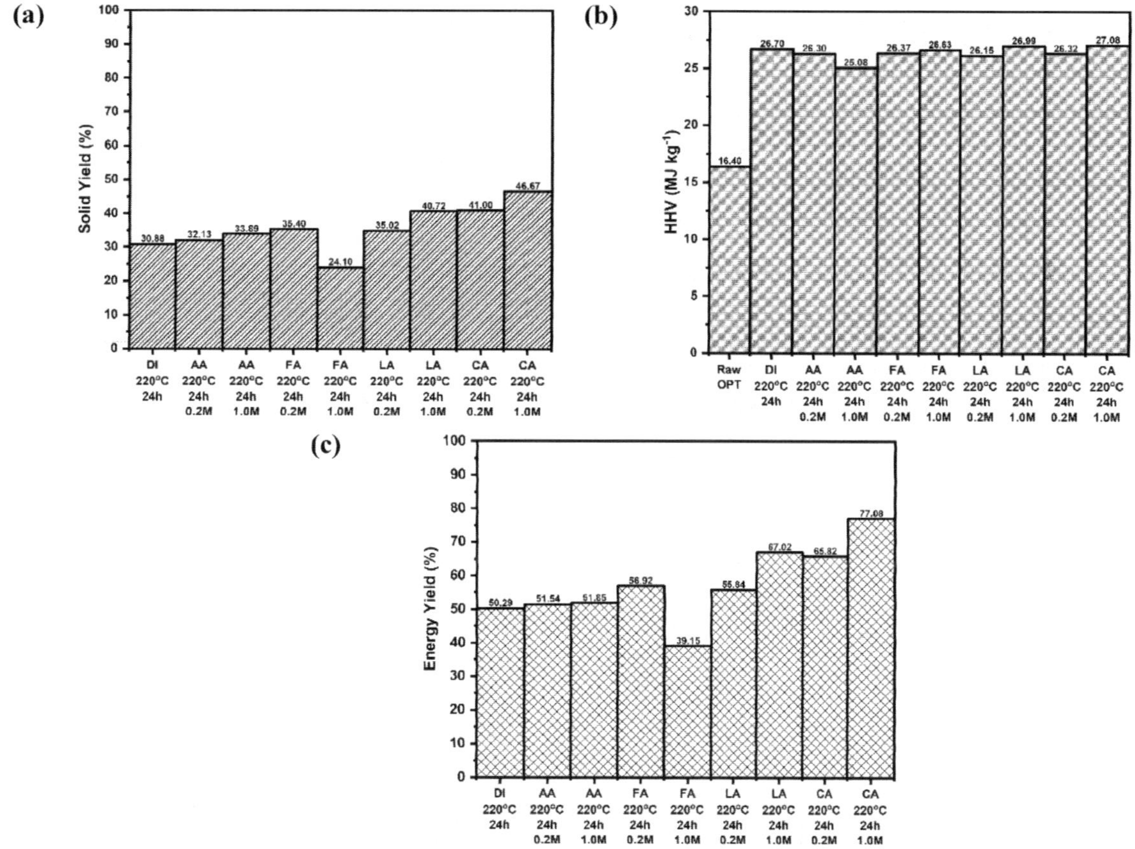

Figure 2. (a) Solid yield, (b) HHV, and (c) energy yield of OPT hydrochar treated at 220 °C for 24 hours in deionized, acetic acid, formic acid, levulinic acid, and citric acid at 0.2M and 1.0M concentration.

4. Conclusion

Organic acid is an excellent medium as catalyst to promote and enhance the fuel properties during wet torrefaction of OPT. Nonetheless, this study concludes that there is a close relationship between various operating parameters, i.e., acids type, residence time, and acid concentration. Therefore, further parametric study and optimization study including response surface methodology (RSM) is necessary to optimize the performance. Citric acid is a good candidate to achieve high energy yield given higher acid concentration and longer residence time. On the other hand, levulinic acid is a good candidate for low concentration and residence time that can help to reduce cost and effluent handling in actual application.

References

[1] Lynam J G, Reza M T, Yan W, Vásquez V R and Coronella C J 2015 Hydrothermal carbonization of various lignocellulosic biomass *Biomass Convers. Biorefin.* **5** 173–81

[2] Lam M K, Tan K T, Lee K T and Mohamed A R 2009 Malaysian palm oil: Surviving the food versus fuel dispute for a sustainable future *Renew. Sust. Energ. Rev.* **13** 1456–64

[3] Loh S K 2017 The potential of the Malaysian oil palm biomass as a renewable energy source *Energy. Conv. Manag.* **141** 285–98

[4] Onoja E, Chandren S, Abdul Razak F I, Mahat N A and Wahab R A 2019 Oil palm (Elaeis guineensis) biomass in Malaysia: The present and future prospects *Waste and Biomass Valorization* **10** 2099–117

[5] Chew J J, Soh M, Sunarso J, Yong S-T, Doshi V and Bhattacharya S 2020 Gasification of torrefied oil palm biomass in a fixed-bed reactor: Effects of gasifying agents on product characteristics *J. Energy Inst.* **93** 711–22

[6] Bach Q-V and Skreiberg Ø 2016 Upgrading biomass fuels via wet torrefaction: A review and comparison with dry torrefaction *Renew. Sust. Energ. Rev.* **54** 665–77

[7] Erlach B, Harder B and Tsatsaronis G 2012 Combined hydrothermal carbonization and gasification of biomass with carbon capture *Energy* **45** 329–38

[8] Kambo H S and Dutta A A 2015 Comparative review of biochar and hydrochar in terms of production, physico-chemical properties and applications *Renew. Sust. Energ. Rev.* **45** 359–78

[9] Bach Q-V, Tran K-Q, Skreiberg Ø, Khalil R A and Phan A N 2014 Effects of wet torrefaction on reactivity and kinetics of wood under air combustion conditions *Fuel* **137** 375–83

[10] Soh M, Khaerudini D S, Chew J J and Sunarso J 2021 Wet torrefaction of empty fruit bunches (EFB) and oil palm trunks (OPT): Effects of process parameters on their physicochemical and structural properties *South African Journal of Chemical Engineering* **35** 126–36

[11] Hu F and Ragauskas A 2012 Pretreatment and lignocellulosic chemistry *BioEnergy Res.* **5** 1043–66

[12] Yang H, Yan R, Chen H, Zheng C, Lee D H and Liang D T 2006 In-depth investigation of biomass pyrolysis based on three major components: Hemicellulose, cellulose and lignin. *Energ. Fuel.* **20** 388–93

[13] Yang W, Shimanouchi T, Iwamura M, Takahashi Y, Mano R, Takashima K, Tanifuji T and Kimura Y 2015 Elevating the fuel properties of Humulus lupulus, Plumeria alba and Calophyllum inophyllum L. through wet torrefaction *Fuel* **146** 88–94

[14] Donar Y O, Çağlar E and Sınağ A 2016 Preparation and characterization of agricultural waste biomass based hydrochars *Fuel* **183** 366–72

[15] Sevilla M and Fuertes A B 2009 The production of carbon materials by hydrothermal carbonization of cellulose *Carbon* **47** 2281–9

[16] Volpe M and Fiori L 2017 From olive waste to solid biofuel through hydrothermal carbonisation: The role of temperature and solid load on secondary char formation and hydrochar energy properties *J. Anal. Appl. Pyrolysis* **124** 63–72

[17] García-Bordejé E, Pires E and Fraile J M 2017 Parametric study of the hydrothermal carbonization of cellulose and effect of acidic conditions *Carbon* **123** 421–32

[18] Qi Y, Song B and Qi Y 2016 The roles of formic acid and levulinic acid on the formation and growth of carbonaceous spheres by hydrothermal carbonization *RSC Advances* **6** 102428–35

Treatment and decolourisation of treated palm oil mill effluent (POME) using oil palm trunk-derived activated carbon as adsorbent

K J Lau, A Lim, J J Chew, L H Ngu and J Sunarso

Research Centre for Sustainable Technologies, Faculty of Engineering, Computing and Science, Swinburne University of Technology Sarawak Campus, Kuching 93350, Sarawak, Malaysia

Email: jchew@swinburne.edu.my

Abstract. Palm oil mill effluent (POME) has become a serious problem for the oil palm industries because of its high organic contents and other contaminant formation that results in dark colour, turbid and unpleasant odour. In the conventional treatment of POME used in Malaysia, treated POME can still pollute receiving water bodies as colour is one of the major contaminants that is not completely removed. Adsorption is a promising technique for addressing this problem, with a large range of adsorbents to choose from. It works by adhering the pollutants on to the high porous surface area of adsorbent. However, the high cost of coal-based AC that is commonly used can be the limiting factor for its wider application in palm oil industry. Therefore, this work looks into resource recovery (i.e., use of waste stream) from plantation as precursor of AC synthesis to treat its own waste from the palm oil mill. Hence, the suitability of oil palm trunk (OPT) as feedstock for AC application in POME colour removal in the mill is investigated. Experimental run at the as-synthesis OPT-derived AC was performed for validation *via* POME adsorption test. The result shows that the OPT-derived AC produced can remove the organic pollutants and colour of POME at the dosage of 15% w v^{-1} within 48 hours.

1. Introduction

Malaysia has been ranked as the second largest palm oil producer in the world, thus it is unquestionable that palm oil industry has significant contribution to the economy in Malaysia. However, this will also result in massive amount of waste being produced during the palm oil milling process (i.e., POME) and in the oil palm plantation (i.e., oil palm trunk (OPT) and oil palm frond (OPF)). POME is a brown oily wastewater that contain approximately 95–96% water, 0.6–0.7% oil and 4–5% total solids including 2–4% suspended solids. Existence of colour in POME hinders the photosynthetic process of aquatic beings and has toxic impact.

Due to the biologically stable pigments present in POME, therefore typical techniques used like biological degradation is inadequate to get rid of the remaining colour and organic compounds from wastewater. As a result, further treatment is required to remove the colour and improve the quality of the POME the conventionally treated POME before it is release into the receiving water bodies. Till date, there are various methods used for reduction of POME colour. Anaerobic digestion has been used commonly by majority palm oil mills for POME treatment [1]. More than 85% of palm oil mills in

Malaysia have adopted the ponding system for the treatment of POME [2] while the others has chosen the open digesting tank method [3]. These aforementioned techniques are known as conventional POME treatment method because it implies lower capital and operating cost but requires long retention time (i.e., 1-2 months) and large treatment areas [4]. Coagulation-flocculation method is also commonly applied for the treatment of POME [5] but is restricted for high-soluble compounds. Thus, the use of low-cost adsorbents for decolourisation of wastewater *via* adsorption process have received much attention. Adsorption is a favourable method for the removal of impurities and colour and has potential application for the treatment of real industrial wastewater. However, commercially available AC are deemed to be expensive [6]. This is because non-renewable and relatively costly are used as the starting material like coal, which is unwarranted in pollution control applications. Thus, in recent years, this matter has driven an increasing research interest in the synthesis of AC using starting materials that are renewable and cheaper which are mostly industrial and agricultural waste, particularly for purpose regarding wastewater treatment.

Potential strategies to reduce the environmental pollution created by palm oil mill industry (i.e., POME) is by utilising its own waste from the plantation and turn it into adsorbents (i.e., AC). Hence, this project proposes to tackle both waste and pollution issue from the palm oil sector using its waste stream. There is limited information regarding the production of AC from OPT despite of its great potential of being the AC precursor due to its high carbon content of around 35-44% and a low inorganic mineral content of around 2-4%. Consequently, OPT is chosen to be investigated on the performance of POME colour reduction in order to fill this knowledge gap. Based on literature, there is only one journal paper by [7] uses OPT as the AC precursor. It uses both physical and chemical activation synthesis method to produce AC with relatively high surface area of 1884 $m^2 g^{-1}$, which is considered a remarkable result.

In this study, efficiencies of OPT-derived AC for the removal of colour, and COD from the treated POME were investigated. Chemical Oxygen Demand (COD) is a common parameter used for evaluating the quality water or wastewater. COD chemical evaluation determines the total amount of all chemicals contained in the liquid, including organic and inorganic materials. It is defined as the equivalent amount of oxygen consumed to the chemical oxidation of organic matter by strong oxidant. The oxidation of matters for polluted water is done using the strong oxidizing agents like potassium dichromate ($K_2Cr_2O_7$), whereas for clean water it is conducted using potassium permanganate ($KMnO_4$) [8]. Water with high COD value has higher chances of causing harm to the environment and the living organisms. The POME COD level arises from the compounds stored in oil palms and the processes to generate palm oil. Generally, raw POME has very high COD value of approximately 50,000 mg L^{-1} [9] and about 1730 mg L^{-1} in bio-treated POME [10]. The discharge limit for COD, based on the Malaysian Department of Environment (DOE) 1986, is 1000 mg L^{-1} [11]. Hence, effective method is required to treat POME and major oil palm industries in Malaysia applied conventional treatment such as ponding system which involves long retaining duration and the usage of large surface area. Colour is another display of water quality that is simple to examine. The dark colour of POME derives from the palm oil mill's sterilising procedure of fresh fruit bunches. The yellowish-brown colour and strong sourish odour are due to the present of complex organic compounds in POME. The degradation of lignocellulosic components (i.e., anthocyanin, lignin, tannin, carotene pigment, polyphenol compounds, polyalcohol, and melanoidin) caused by sterilisation of fresh fruit bunches results in the creation of an oily, viscous texture and the dark brown colour of POME [12-14]. The Malaysian DOE has yet to set a standard regulation for POME colour. Nonetheless, the discharge's murky and dark brown colour raises public concerns about the low water quality.

The aim of this research was to determine whether the optimised synthesis OPT-derived AC from the previous work [15] is a potential adsorbent for the treatment and decolourisation of POME.

2. Materials and methods

2.1. Materials

The OPT used to synthesised AC was collected from an oil palm estate in Bintulu were washed with water to remove dirt from its surface and then dried in an oven at 105 °C for 1 day. Then, the dried OPT were crushed to particle size ranging from 0.3 to 1.18 mm by using a blender and sieved. Raw palm oil mill effluent (POME) was obtained from Felcra Berhad at Kota Samarahan. Then, the raw POME was filtered using filter paper with pore size of 180 μm to obtain filtered POME.

2.2. Sample preparation

The synthesis of AC from OPT was done *via* chemical activation with phosphoric acid as its impregnating agent. The synthesis procedure of OPT-derived AC *via* chemical activation was carried out corresponding to Lim *et al.* [16] with modification as follows: 10 g of dried OPT was mixed with 15.9 mL phosphoric acid (i.e., impregnation ratio of 2.29). To allow the sample to absorb phosphoric acid, distilled water was poured to the OPT until it was completely covered. The sample was then left at room temperature for 24 hours before being dried at 100°C for 1 day to remove excess water and allow the phosphoric acid to absorb entirely.

The resulting samples underwent semi carbonisation at 170 °C in an oven for an hour. Before the activation process, the semi-carbonized sample was allowed to cool to ambient temperature. Then, the sample was put in a muffle furnace for activation at 450 °C for 6 mins. The obtained ACs were washed with distilled water until the wash liquor conductivity level was less than 50 µS. The samples were then dried for 24 hours at 105°C in an oven. After that, the ACs were ground into a fine powder and stored for future use.

2.3. Characterisation

2.3.1. Proximate analysis. Proximate analysis is carried out to determine the physical characteristics of the resulting sample such as the moisture content (%) by mass, volatile matter content (%) by mass, ash content (%) by mass and fixed carbon content (%). It is also used to determine the suitability of OPT as the precursor for the synthesis of AC. The procedures of proximate analysis were carried out according to the [17].

2.3.2. Iodine number. Iodine number denotes the amount (i.e., in milligrams) of iodine that can be adsorbed by 1 g of AC. The purpose of this analysis is to determine the adsorption capacity of the OPT-derived AC. The AC iodine number can be regarded as a quantitative benchmark of the AC porosity, which can be used as a precise forecast of its surface area. Iodine number analysis was carried out in accordance with the standard procedure [18].

2.4. POME adsorption

Batch adsorption studies were carried out in a Memmert shaking water bath at a constant temperature of 30 °C. Around 2 g of AC (dosage of 15 % w v^{-1} with respect to the sample volume) were added to 13.33 mL raw POME in five Erlenmeyer flasks, respectively. Controlled experiment was also conducted where 100 mL of raw POME is added into five Erlenmeyer flasks. In five consecutive days, the parameters of the treated POME (i.e., after adsorption) and degraded raw POME are measured accordingly.

3. Results and discussion

3.1. Iodine number test

The experimental response value (i.e., iodine number from experimental run) and the predicted value is 884.49 mg g^{-1} and 937.94 mg g^{-1} respectively. The results obtained was very close with 5.70%

difference, proving that the model can satisfactorily predict the experimental conditions. The AC prepared using the optimized variable set were then subjected to POME adsorption tests. Iodine number of AC synthesis using the optimum parameter combination has iodine number in the range of 600 to 1100 mg g-1, which is suitable to be used for water treatment [19].

3.2. *Proximate analysis*

According to table 1, raw OPT has high volatile matter and low ash content. According to Rafatullah *et al.* [20], precursor of AC that possess high volatile matter content is preferred as high temperature during carbonisation and activation will cause them to be removed. Consequently, more pores will be formed within the resulting AC. The statement of the removal of volatile matters from AC precursor can be further proven by the decrease in volatile matter content while converting from OPT to AC.

Moreover, low ash content in AC precursor is more advantageous for the production of AC. This is because high ash content represents high inorganic constituents in the resulting carbon, which restricts the pore generation by blocking the pore entrance during the activation stage. Additionally, table 1 also reveals that AC synthesised from OPT fulfil the standard physical characteristics of AC. Thus, based on the proximate analysis results, it can be concluded that OPT is a suitable precursor in the production of AC.

Table 1. Proximate analysis of each sample and the standard acceptance level for AC.

Proximate analysis	Raw OPT	OPT-derived AC	Acceptable level for AC [21, 22]
Moisture content (%)	9.80	0.33	15
Volatile matter (%)	72.43	14.86	13 – 18
Ash (%)	2.95	4.96	2 – 5
Fixed carbon content (%)	24.62	80.18	> 72

3.3. *POME adsorption test*

POME adsorption test was performed to determine the efficiency of OPT-derived AC in removing the organic pollutants and colour of POME. Table 2 shows the initial parameters of the raw POME and filtered POME as well as the percentage reduction of parameters due to filtration through filter paper. Hence, it can be deduced that filter paper can barely remove the organic compound and colour of POME.

Table 2. The parameters measured for raw and filtered POME.

Parameters	Treated POME	Filtered POME	Reduction (%)
COD (mg L^{-1})	1700	1600	6
Colour (Pt Co^{-1})	9200	5500	40

According to table 3, it can be observed that OPT-derived AC can removed most of the pollutants and colour of POME. Therefore, it can be concluded that OPT-derived AC has higher efficiency in treating POME due to its high iodine number and BET surface area. As mentioned earlier, one of the disadvantages of conventional ponding system is that it requires long retention time of approximately 1 to 2 months. Furthermore, even if the POME is biologically treated to meet the government standard discharge limit, it will still bring negative environmental impacts. As for the POME treatment through adsorption by AC, it only requires approximately 2 days to remove most of the organic pollutants and decolourise the POME based on the results obtained from this work.

Table 3. Comparison of POME adsorption test using OPT-derived AC and PKS-derived AC.

Parameters	**Results from This Work**	**Results from [23]**
AC precursor	Oil palm trunk (OPT)	Palm kernel shell (PKS)
BET surface area (m^2 g^{-1})	1605.7	566.27
Iodine number (mg g^{-1})	884.49	-
Dosage of AC (% W V^{-1})	15	15
Optimum removal in COD (%)	99	85
Optimum removal in colour (%)	99	92
Optimum removal in tannin-lignin (%)	99	92
Period treatment to reach maximum pollutant removal (days)	2	3

POME treatment demands an efficient method to deal with the contemporary challenges. Till date, some palm oil mill that apply the current available treatment method, still fail to comply with the standard discharge limit set by the DOE. [24]. Thus, it is understood that the adsorption method can enhance the biological POME treatment method in a more advantageous way such as shorter retention time and smaller surface area (i.e., in terms of land used). In this study, the condition of the POME is not altered and used as it is. The colour of treated POME (i.e., dark brown) was significantly decreased, as more than 99% of its initial value was decreased, and it was observed to be colourless to the naked eye. The removal percentage of this work was 99 % via OPT-derived AC adsorption compared to the 92 % colour removal by Jalani *et al.* [25] after treating POME with PKS-derived AC.

After adsorption by OPT-derived AC, the COD test results indicates a 99 % decrease. The adsorption treatments employing OPT-derived AC greatly reduces the COD value to almost 0 mg L^{-1} from the initial COD value of treated POME (i.e., 1600 mg L^{-1}). It also shows that, the reduction in COD in this work by OPT-derived AC is higher compared to that performed by Jalani *et al.* [25] using PKS-derived AC.

High surface area adsorbent will have high efficiency in treating POME which is the main factor that contributes to the excellent performance of OPT-derived AC in treating POME. Based on the comparison on the maximum removal of pollutants and the duration to reach the maximum pollutant removal shown in table 3, it can be concluded that OPT-derived AC from this work not only has higher efficiency in treating POME, and also shorter duration to reach maximum pollutant removal of POME as compared to the PKS-derived AC by Jalani *et al.* [25].

4. Conclusion
Oil palm trunk have successfully been modified into highly porous AC with high iodine number of 884.49 mg g^{-1}. Based on the proximate analysis results, it can be concluded that OPT has the potential to be the precursor of AC synthesis and the OPT-derived AC fulfils the standard physical characteristics of AC. On the other hand, the performance of OPT-derived AC on POME adsorption was significantly satisfied, specifically measuring parameters in COD, and colour removal in approximately two days with the dosage of 15 %w v^{-1}. In conclusion, this research project has justified that OPT is suitable to be used to produce AC that can treat POME efficiently.

5. Practical Implication
According to the outcome of this work, the production of OPT-derived AC is favourable in large scale production of AC as this will consequently reduce the waste produced in the fill and at the same time able to treat the waste stream from the oil palm mill. Besides, conventional ponding system requires extremely large surface area of land and long retention time of about 1 to 2 months. However, these issues can be solved by using OPT-derived AC from this work as the adsorbent. POME can be treated efficiently in shorter duration which would be an advantage for the palm oil mill manufacturers to handle the substantial amount of POME being produced during the palm oil milling process.

References
[1] Tay J H 1991 Complete reclamation of oil palm wastes *Resour., Conserv. Recycl.* **5** 383–92
[2] Ma A, Cheah S, Chow M and Yeoh B 1993 Current status of palm oil processing waste management, waste management in Malaysia: Current status and prospects for bioremediation 111–36
[3] Yacob S, Hassan M A, Shirai Y, Wakisaka M and Subash S 2005 Baseline study of methane emission from open digesting tanks of palm oil mill effluent treatment *Chemosphere.* **59** 1575–81
[4] Abdurahman N, Rosli Y and Azhari N 2011 Development of a membrane anaerobic system (MAS) for palm oil mill effluent (POME) treatment *Desalination.* **266** 208–12
[5] Bhatia S, Othman Z and Ahmad A L 2007 Pretreatment of palm oil mill effluent (POME) using moringa oleifera seeds as natural coagulant *J. Hazard. Mater.* **145** 120–6
[6] Babel S and Kurniawan T A 2003 Low-cost adsorbents for heavy metals uptake from contaminated water: A review *J. Hazard. Mater.* **97** 219–43
[7] Hussein M Z B, Rahman M B B A, Yahaya A H, Hin T-Y Y and Ahmad N 2001 Oil palm trunk as a raw material for activated carbon production *J. Porous Mat.* **8** 327–34
[8] Ma Y, Tie Z, Zhou M, Wang N, Cao X and Xie Y 2016 Accurate determination of low-level chemical oxygen demand using a multistep chemical oxidation digestion process for treating drinking water samples *Anal Methods.* **8** 3839–46
[9] Adeleke O A, Saphira M R, Daud Z, Ismail N, Ahsan A, Ab Aziz N A, *et al* 2019 Principles and mechanism of adsorption for the effective treatment of palm oil mill effluent for water reuse *Nanotechnol. Water Wastewater Treat.* 1–33
[10] Amosa M K 2016 Sorption of water alkalinity and hardness from high-strength wastewater on bifunctional activated carbon: Process optimization, kinetics and equilibrium studies *Environ. Technol.* **37** 2016–39
[11] Bello M, Nourouzi M, Abdullah LC, Choong TS, Koay Y and Keshani S 2013 POME is treated for removal of color from biologically treated POME in fixed bed column: Applying wavelet neural network (WNN) *J. Hazard Mater.* **262** 106–13
[12] Amat N A, Tan Y, Lau W, Lai G, Ong C, Mokhtar N, *et al* 2015 Tackling colour issue of

anaerobically-treated palm oil mill effluent using membrane technology. *J. Water Process. Eng.* **8** 221–6

[13] Mohammed R R 2013 Decolorisation of biologically treated palm oil mill effluent (POME) using adsorption technique *Int Ref J Eng Sci.* **2** 1–11

[14] Mohammed R R and Chong M F 2014 Treatment and decolorization of biologically treated palm oil mill effluent (POME) using banana peel as novel biosorbent *J. Environ. Manag.* **132** 237–49

[15] Lim A, Chew J J, Ngu L H, Ismadji S, Khaerudini D S and Sunarso J 2020 Synthesis, characterization, adsorption isotherm, and kinetic study of oil palm trunk-derived activated carbon for tannin removal from aqueous solution *ACS Omega* **5** 28673–83

[16] Lim W C, Srinivasakannan C, Balasubramanian N 2010 Activation of palm shells by phosphoric acid impregnation for high yielding activated carbon *J. Anal. Appl. Pyrolysis* **88** 181–6

[17] D1762-84 A 2009 Standard test method for chemical analysis of wood charcoal. American Society for Testing and Materials, Conshohocken, PA. 2009

[18] D4607–94 A 2006 editor Standard test method for determination of iodine number of activated carbon. Standard Test Method for Determination of Iodine Number of Activated Carbon; 2006: ASTM International, West Conshohocken, PA

[19] Gupta T 2017 Carbon: the black, the gray and the transparent *Springer*

[20] Rafatullah M, Ahmad T, Ghazali A, Sulaiman O, Danish M and Hashim R 2013 Oil palm biomass as a precursor of activated carbons: A review *Crit. Rev. Environ. Sci. Technol.* **43** 1117–61

[21] Deshpande R Ultracapacitors: Future of energy storage: McGraw-Hill Education; 2014

[22] Rafatullah M, Ahmad T, Ghazali A, Sulaiman O, Danish M and Hashim R 2013 Oil palm biomass as a precursor of activated carbons: A review *Crit. Rev. Environ. Sci. Technol.* **43** 1117–61

[23] Jalani N F, Aziz A A, Wahab N A, Hassan W H W and Zainal N H 2016 Application of palm kernel shell activated carbon for the removal of pollutant and color in palm oil mill effluent treatment *J. Environ. Health Sci. Eng.* **2** 15

[24] Mohammed R R and Chong M F 2014 Treatment and decolorization of biologically treated palm oil mill effluent (POME) using banana peel as novel biosorbent *J. Environ Manage.* **132** 237–49

[25] Jalani N F, Aziz A A, Wahab N A, Hassan W H W and Zainal N H 2016 Application of palm kernel shell activated carbon for the removal of pollutant and colour in palm oil mill effluent treatment *J. Earth Environ. Health Sci.* **2** 15

Techno-economic evaluation of microalgae-based supply chain: Review on recent approaches

S L Y Lo[1], K G H Kong[1], B S How[1], J Y Lim[2], P L Show[3] and J Sunarso[1]

[1] Research Centre for Sustainable Technologies, Faculty of Engineering, Computing and Science, Swinburne University of Technology, 93350 Kuching, Sarawak, Malaysia
[2] Integrated Engineering, Dept. of Environmental Science and Engineering, College of Engineering, Kyung Hee University, 1732 Deogyeong-daero, Giheung-gu, Yongin-Si, Gyeonggi-do 17104, Republic of Korea
[3] Department of Chemical and Environmental Engineering, Faculty of Science and Engineering, University of Nottingham Malaysia, 43500 Semenyih, Malaysia

Email: bshow@swinburne.edu.my

Abstract. Third generation biomass-derived products such as biofuel has been garnering attention as a viable alternative energy source recently as it does not necessarily require fresh water and vast land for cultivation as compared to first-generation and second-generation biomass. However, extensive studies have to go into the feasibility evaluation for third generation biomass utilization prior to upscaling the process to commercial level. Other than comprehensive technical evaluation such as experimental studies to understand the microalgae productivity, economic evaluation of the utilization of third-generation biomass is also critical specifically in the perspective of supply chain. Therefore, the objective of this review is to lay out an overall picture to the readers the various option of approaches or methods utilized in feasibility evaluation of the microalgae-based supply chain. The outcome of the review paper indicated that approximately 58% of the papers reviewed opted for mathematical modeling with optimization whereas the remaining 42% opted for mathematical modeling without optimization.

1. Introduction

As stated by the International Renewable Energy Agency (IREA), the utilization of renewable energy source has vastly exceeded non-renewable energy source since 2015 [1]. One of the renewable energy sources that has been garnering attention is biomass. Biomass-derived energy source and products have been in the highlights in recent years that in turns slowly phasing out fossil fuels. Biomass can be further categorized into first, second, and third-generation biomass. First generation biomass is crop-like and usually edible biomass such as sugarcane, barley, and corn [2]. Apart from that, second-generation biomass consists of a wide variety of biomass usually inedible lignocellulosic biomass (i.e., wood chips, forest remains, or domestic solid wastes) [2]. Lastly, third-generation biomass is general algal biomass [3]. By definition, unicellular and basic multi-cellular microorganisms are both categorized under prokaryotic microalgae (i.e., cyanobacteria and eukaryotic microalgae) [3]. Researches are now leaning towards third-generation biomass such as microalgae utilization due to multiple reasons. One among these is the high lipid content of certain algal species (i.e., *Chlorella* that has lipid content of approximately 60 to 70%), which allows higher production of biodiesel [4]. Additionally, microalgae can be cultivated in distinct origin of water namely saline water, wastewater, brackish water, and fresh water [5]. The capability of microalgae to be cultivated in wastewater has significant benefits as the discharge of wastewater into the environmental ecosystem can result in the increase of chemical oxygen demand (COD), which further causes the imbalance in ecosystem, contamination towards the environment specifically groundwater, and may pose risk to human health [6]. According to Posadas *et al.* [7], the development of low cost and green treatment for effluents such as wastewater is critical as industries are discharging large volumes of wastewater into the aquatic ecosystem. Although microalgae utilization in the renewable fuels sector is a favorable alternative, Tan [8] stated that the identification of reliable markets for microalgae to venture into has been unsuccessful due to an undesirable return on investment. Furthermore, Andrade *et al.* [9] mentioned that only a few published papers had studied the feasibility of microalgae-based supply chain. Therefore, it is critical to consider the different evaluation approaches available when evaluating the feasibility of the microalgae-based supply chain. As the microalgae-based supply chain is still in its infancy, a more comprehensive study is required to ensure that all aspects have been considered and to avoid any unnecessary losses whether in time or monetary aspects. Furthermore, Madugu and Collu [10] added that careful analysis of the microalgae-based supply chain is critical to for the assurance of the long-term sustainability of microalgae conversion process. According to Chew *et al.* [11], one of the foundation tools that can be adopted to perform feasibility evaluation of microalgae processing is techno-economic analysis.

Therefore, this review paper is equipped with the objective to compile the presently available approaches and methodology utilized in performing techno-economic analysis of microalgae-based supply chain. Subsequently, a list of recent reviews related to microalgae-based supply chain has been listed in table 1. Some of the works listed in this table such as Efroymson *et al.* [12] briefly illustrated a framework for better management practices (BMPs) to achieve sustainability objective that includes techno-economic analysis and provide brief description on the integration of economic indicators with environmental indicators. Nonetheless, they did not provide a discussion on the methods or mathematical models available to be used for techno-economic analysis. Generally, BMP can be obtained from techno-economic analysis and life cycle assessment (LCA) coming hand-in-hand to achieve sustainability objectives such as the reduction of greenhouse gas (GHG) emissions. Apart from that, Doliente *et al.* [13] also reviewed techno-economic analysis of first, second, and third-generation biomass-derived bio-aviation fuel. They however, only provided a cost breakdown for the delivered cost of first-generation and second-generation biomass specifically corn stover, switch grass, oil palm, miscanthus, eucalyptus, forest residues, and sugarcane. However, they did not go on to discuss more comprehensively on the approaches or methods available in performing techno-economic analysis. Aside from that, Rizwan *et al.* [14] discussed the optimization of microalgae-based biorefinery superstructure, which is based on deterministic and stochastic approach alongside the discussions on uncertainties revolving around microalgae-based biorefinery. However, they did not critically review the alternative techno-economic analysis approaches that can be performed without the need for optimization. In contrast, these approaches will be discussed in the current review. Apart from that, the type of microalgae and the corresponding conversion process are outlined in this review as well.

IOP Conf. Series: Materials Science and Engineering 1195 (2021) 012026 doi:10.1088/1757-899X/1195/1/012026

Table 1. A list of current available reviews related to techno-economic analysis of microalgae-based supply chain.

Year	Author	Review Scope
2021	Efroymson *et al.* [12]	• Provided description on algae and algae-derived biofuel production • Discussed the framework for BMPs that relies on LCA, techno-economic assessment, and resource analysis. • Provided examples for BMPs such as the quantity and quality of water
2020	Teng *et al.* [15]	• Discussed the integration of artificial intelligence (AI) in microalgae in different fields such as genetic engineering, experimental analysis, and process integration • Listed the essential databases used for microalgae bioinformatics • Discusses the optimization approaches used with the integration of AI in microalgae-based biorefineries
	Andrade *et al.* [9]	• Focused on Brazilian Microalgae Production Chain (BPMC) • Discussed briefly the studies performed on BPMC • Discussed a few significant elements of BPMC (Shortcomings, threats, challenges, opportunities, efficiency, and demand) • Highlighted a few ways to reduce the BPMC cost
	Doliente *et al.* [13]	• Review the bio-aviation fuel supply chain and their components: microalgae feedstock, conversion pathways, and logistics that include storage and transport • Provide a summary of supply chain studies that include environmental and economic for bio-aviation fuel • Discussed the advantages and limitations of bio-aviation fuel conversion pathways
2019	Rizwan *et al.* [14]	• Discussed microalgae-based biorefinery superstructure optimization • Focus on the description of formulation of optimization models that are mixed-integer linear programming (MILP) and mixed-integer non-linear programming (MINLP)
2018	Deprá *et al.* [16]	• Review microalgae biorefineries including microalgae cultivation and co-products produced during conversion process • Discussed application of process integration and LCA to microalgae biorefineries • Discussed the microalgae biorefineries' bottlenecks
2017	Chew *et al.* [11]	• Discussed briefly the economic viability of microalgae biofuels. However, did not go in-depth towards the method and approach adopted • Emphasized more on microalgae biorefinery (inclusive of their conversion processes), components of microalgae, and potential application of each component of microalgae

The main purpose of this review paper is to provide a compilation of techno-economic studies performed for third-generation biomass utilization. Section 2 illustrates the review methodology adopted in this paper. Subsequently, a brief description of microalgae-based supply chain is included in Section 3. Section 4 then presents the compilation of the methods and approaches covered in this review paper. The concluding remarks and future works are outlined in Section 5.

2. Review methodology and scope of review

The review methodology utilized in this study is illustrated in figure 1. The review began with inputting the desired keyword "*microalgae*" into the database. The database used for this study is Scopus. After obtaining the search outcome from the keyword that includes 29,594 relevant papers, the next step involved refining the search result to the scope of "*techno-economic*" and further refined the search result to the scope of "*supply chain*". The total resulting article was 208 relevant papers that were further refined to consist only journal articles, resulting in 103 total relevant papers. The scope of this review paper is the articles from the recent five years (2017 to 2021). Thus, 74 relevant papers were reviewed in this study. Manual methodology screening was conducted on each paper to determine whether the methodology of study adopted by the papers are relevant to be discussed in this review. After methodology screening, 51 papers (approximately 68% of the total papers obtained from the search results) were found to be less relevant towards the scope of this review paper.

Although specific keywords and article screening have been implemented and the result of the papers obtained are significant and insightful studies but less relevant towards the scope of this review paper (see tables 2A and 2B). For instance, review papers such as Kusmayadi *et al.* [17] found in the search results was screened out. The author briefly addressed the tools that can be used to evaluate the feasibility of production and conversion pathways that are techno-economic analysis and life cycle assessment. However, the paper focuses on the nutritional components of microalgae where the considered application is merely on using as animal and human feed. The largest category of papers is contributed by experimental studies (e.g., Vigor *et al.* [18], Wang *et al.* [19], and etc). Some of the papers performed mathematical modeling as well however did not perform techno-economic analysis. For instance, Andersson *et al.* [20] investigated the potential of process integration (material and heat integration) in microalgae processes (i.e., lipid extraction with transesterification, hydrothermal liquefaction (HTL) with catalytic hydrothermal gasification (CHG) for microalgae, *Nannochloropsis* and macroalgae, *Laminaria saccharina*, respectively). They studied the reuse of heat generated from the co-located oil refinery for algae-derived biofuel production process *via* heat integration. On the other hand, material integration was used to evaluate the potential reuse of hydrogen stream with purity content reaching 90% (categorized as waste stream) from the refinery process. The outcome of the studies was that heat integration and material integration have positive impacts on the biofuel process by reducing the net energy demand and process efficiency. On the other hand, Ryu *et al.* [21] formulated a generic mathematical model to describe the behavior of microalgae under heterotrophic culture conditions. They validated the results using two case studies—lipid production from the microalgae species, *Chlorella protothecoides* and lutein production from *Chlorella protothecoides CS-41*. The last study that performed mathematical modelling without performing techno-economic analysis is the study performed by Goffé and Ferrasse [22]. They studied the impact of stoichiometry ratio (i.e., carbon conversion ratio, energy ratio, and hydrogen conversion ratio) on the process conversion efficiency. Taking hydrogen conversion ratio as an example, it is taken as the amount of hydrogen within the particular biomass that is converted into value-added products.

Although all the listed works are insightful, they do not have direct relevance to the current review that focuses on the techno-economic analysis of microalgae-based supply chain. The brief explanation of microalgae-based supply chain is presented in the subsequent segment of this paper.

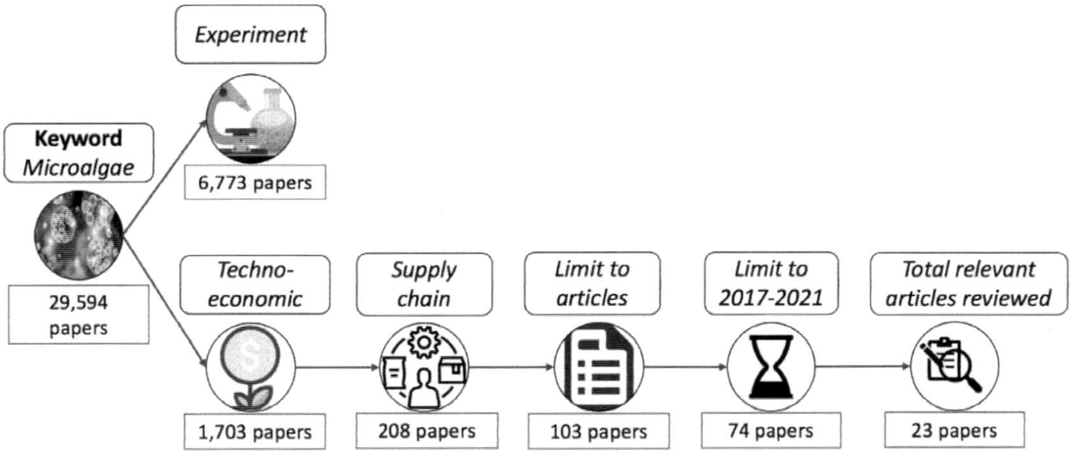

Figure 1. Review methodology.

Table 2A. Categorization of papers not reviewed in current paper. Note: R, Exp, Env, M, D and S denotes review paper, experimental works, environmental studies, mathematical modeling that does not consist of performing techno-economic analysis, decision making method, and simulation studies, respectively.

Year	Author	R	Exp	Env	M	D	S
2021	Kusmayadi *et al.* [17]	✓					
	Chen *et al.* [23]		✓				
	Lu *et al.* [24]	✓					
	Maiolo *et al.* [25]			✓			
	Hu *et al.* [26]		✓				
	Dalheim *et al.* [27]		✓				
	Serrano *et al.* [28]		✓				
	Chauton *et al.* [29]	✓					
2020	Andersson *et al.* [20]				✓		
	Culaba *et al.* [30]			✓			
	Nguyen *et al.* [31]		✓				
	Deprá *et al.* [32]		✓	✓			
	El-Dakar *et al.* [33]		✓				
	Branco-Vieira *et al.* [34]			✓			
	Maiolo *et al.* [35]			✓			
	Vigor *et al.* [18]		✓				
	Hossain *et al.* [36]	✓					
	Morales *et al.* [37]			✓			
	Pankratz *et al.* [38]			✓			
	Wahlen *et al.* [39]		✓				
	Naeini *et al.* [40]					✓	

Table 2B. Categorization of papers not reviewed in current paper. See table 2A caption for the description of abbreviations used here.

Year	Author	R	Exp	Env	M	D	S
2020	Lee *et al.* [41]		✓				
	Ryu *et al.* [20]				✓		
	Desjardins *et al.* [42]	✓					
	Callegari *et al.* [43]	✓					
2019	Rizwan *et al.* [14]	✓					
	Azari *et al.* [44]			✓			
	Morales *et al.* [45]			✓			
	Lee and Sun [46]		✓				
	Zhang and Kendall [47]			✓			
	Song *et al.* [48]		✓				
	Sawant *et al.* [49]						✓
2019	Sun *et al.* [50]			✓			
	Goffé and Ferrasse [21]				✓		
	Bacci di Capaci *et al.* [51]						✓
	Wang *et al.* [19]		✓				
2018	Montero-Lobato *et al.* [52]	✓					
	Pedersen *et al.* [53]		✓				
	Foteinis *et al.* [54]			✓			
	Poddar *et al.* [55]		✓				
	Sturme *et al.* [56]		✓				
	Giraldo-Calderón *et al.* [57]	✓					
	Chen *et al.* [58]		✓				
	Pan *et al.* [59]	✓					
	Phusunti *et al.* [60]		✓				
	Tang *et al.* [61]		✓				
	Chen *et al.* [62]		✓				
2017	Qiu *et al.* [63]		✓				
	Tan *et al.* [64]			✓			
	Sabu *et al.* [65]		✓				
	Chew *et al.* [11]	✓					

3. Microalgae-based supply chain

The microalgae-based supply chain is said to consists of microalgae cultivation followed by harvesting and then the conversion process (inclusive of pre-requirement prior to actual conversion process such as lipid extraction for biofuel production) and lastly, reuse or disposal [12] or transfer to the end users. Microalgae cultivation and harvesting differ from first-generation and second-generation biomass. Doliente *et al.* [13] mentioned that second generation biomass such as waste biomass are co-produced from ago-forestry activities such as lignocellulosic by-products often obtained as a result of post-harvest activities (i.e., wood processing, milling, and etc.). On the other hand, the cultivation of microalgae can be categorized into four types of cultivations (i.e, photoautotrophic, heterotrophic, mixotrophic, and photoheterotrophic) [66], where they can potentially be cultivated in either open ponds, raceways, or sealed photobioreactors (PBR) [12]. Apart from cultivation area, the cultured medium can be specially formulated to contain all required elements (i.e., zinc, copper, calcium, and etc.) [66] on top of the commonly available types that are fresh water, saline water, wastewater, and brackish water [5]. Apart from cultivation methods and medium, there are also various microalgae harvesting methods such as dewatering, wash methods, pumps, drying [12], flocculation, centrifugation, filtration, flotation, sedimentation, and electrolytic process [67]. Furthermore, different components of the microalgae can be extracted for different types of conversion processes. For instance, the carbohydrates and lipids obtained from microalgae can be utilized for biofuel production, while microalgae polysaccharides can be used for cosmetic additives or natural therapeutic agents [11]. Apart from the specific components of microalgae, the microalgae biomass as a whole can undergo thermochemical conversion (i.e., liquefaction, pyrolysis, gasification, and direct combustion), biochemical or biological conversion (i.e., photobiological hydrogen production, anaerobic digestion, and alcoholic fermentation), transesterification (i.e., acid or base catalysis, and supercritical fluid), and photosynthetic microbial fuel cell process to produce value-added products as well [11].

All in all, microalgae-based supply chain has a very wide variety of cultivation and harvesting methods on top of the wide range of conversion processes. Hence, it is critical to understand the current techno-economic evaluation methods available and select the more suitable approach in evaluation to better achieve the desired objective.

4. Techno-economic analysis of microalgae-based supply chain

This section describes the techno-economic analysis methods used for microalgae-based supply chain evaluation. The evaluation methods can be separated into two sections—mathematical modeling without optimization and mathematical modeling with optimization.

4.1. Mathematical modelling without optimization

Recently, Wu *et al.* [68] performed a comparative LCA and economic analysis on four production scenarios (i.e., scenario 1 and 2 that focus on natural gas feedstock; scenario 3: anaerobic co-digestion to produces methanol; scenario 4: electrolysis of salt water that produces hydrogen) with the objective of generating fuel for either hydrogen or methanol fuel cell. In their work, microalgae were fed into the anaerobic digester (scenario 3) where three products, i.e., biogas, liquid digestate, and solid digestate are generated. To efficiently utilize all these products, biogas was proposed to be fed into combined heat and power (CHP) generation system and/or converted into methanol, whereas the liquid and solid digestates were used as the nutrient sources for microalgae cultivation. Based on their cost analysis, the capital expenditure (CAPEX) for scenario 3 is 94% higher than that of the CAPEX for scenario 1 given that approximately 50% of the CAPEX stemmed from conversion of biogas to methanol (i.e., methanol synthesis process, methanol storage, utility, and land). However, the operating expenditure (OPEX) for scenario 3 is slightly lower as compared to scenario 1 due to the higher price of natural gas. As a result, the overall levelized cost of methanol (LCOM) for scenario 3 is 18.6% higher than that of scenario 1. Similarly, Sano Coelho *et al.* [69] performed cost calculation of an economic evaluation indicator, the minimum biodiesel price, to assess the feasibility of utilization of heterotrophic microalgae (*Auxenochlorella protothecoides*) for biodiesel production.

They found that the microalgae biomass production session accounts for more than half of the total equipment cost (i.e., 64% of total equipment cost for fed-batch process and 55.5% for continuous process). On top of that, 95% of the overall equipment cost for the production of microalgae section originates from the bioreactor. Furthermore, a sensitivity analysis was performed to study the influence of various factors towards the minimum biodiesel price ($ 2.51 for fed-batch process and $ 2.27 for continuous process). The results showed that the molasses (substrate generated from the sugar crystallization process that act as substrate for microalgae lipid accumulation) price and bioreactor price are the two most influential factors, which need to be considered by the investors. Additionally, they also benchmarked the feasibility of a stand-alone microalgae plant with another microalgae plant, which is integrated with sugarcane bio-refinery. The result showed that the integrated plant offers a lower gross profit (approximately $ 4.2 million lower) because of the increase in expenses and decrease in revenue as a result of the decrease in ethanol produced. The results obtained shows that there is a need to consider all configuration options for conversion process and thereby, identifying all the bottlenecks that may lead to unfavorable outcomes. By identifying the bottlenecks, potential steps taken that can prevent the bottlenecks can be implemented.

Furthermore, table 3 also reveals that *Chlorella* represents a favorable microalgae type due to its characteristics, e.g., (i) high protein content with substantial vitamins and minerals content as well as a balanced composition of amino acids that make it suitable for human consumption [70]; and (ii) high lipids content making them suitable for biofuel production [4].

Some authors highlighted the concern regarding the price variation. For instance, Schade and Meier [71] investigated the impact of variation in future prices for input materials on the economic performance. Therefore, they performed 1,000 simulation runs with the Geometric Brownian motion to estimate the future prices for input materials then used the mean obtained from the 1,000 simulation results as input data for economic evaluation.

Table 3. Summary of the works adopting mathematical modeling without optimization.

Year	Author	Type of Microalgae	Conversion Process/Product	Mathematical Model Adopted
2021	Wu *et al.* [68]	*Chlorella vulgaris*	Microalgae production	LCA, manual cost calculation
	Sano Coelho *et al.* [69]	*Auxenochlorella protothecoides*	Lipid extraction for biodiesel production	Process simulation, manual cost calculation
	Silva *et al.* [72]	*Chlorella vulgaris*	Effluent treatment, lipids and biogas production	Experiment, mass and energy balance, manual cost calculation
	Schade and Meier [71]	*Nannochloropsis sp.* *Phaeodactylum tricornutum*	Dry biomass or protein-rich biomass and microalgae oil	Manual cost calculation
2020	Nappa *et al.* [73]	-	Microalgae biomass production (closed and open system)	Mass and energy balance, manual cost calculation
	Choudhary and Srivastava [74]	*Chlorella sp.*	Integrated crude oil heating system with anaerobic digestor and solar PV module	Mass and energy balance, manual cost calculation
	Mennella *et al.* [75]	*Chlorella sp.*	Food and biodiesel production	Experimental, manual cost calculation
	Archanaa *et al.* [76]	*Rhodococcus opacus* *Chlorella vulgaris*	Biodiesel production *via* transesterification process	Experimental, manual cost calculation
2019	Tasca *et al.* [77]	*Chlorella vulgaris*	Biomethane production *via* anaerobic digestion	LCA, manual cost calculation
2018	Beal *et al.* [78]	*Desmodesmus sp.*	Fish oil and fishmeal replacement *via* lipid extraction and drying process	Energy and mass balance, energy impact, GHG impact and accounting, manual cost calculation

4.2. Mathematical modelling with optimization

A compilation for the authors adopting mathematical modeling with optimization are presented in tables 4A and 4B. Such model can be used not only to evaluate the performance, but also to determine the set of operational parameters that can provide the best possible performance [79]. Generally, mathematical modeling with optimization can be categorized into deterministic optimization and stochastic optimization. In short, deterministic optimization is an optimization model that does not account for uncertainties surrounding the supply chain [80]. In the case of Kang *et al.* [81], a two-stage MILP model integrated with geographic information system (GIS) is presented to determine an optimal microalgae-based supply chain with an aim to minimize the total supply chain cost. The proposed study considers 13 biorefineries where various conversion technologies (e.g., HTL, fermentation, and protein extraction) are used in their models. The outcome of the study is that the fermentation pathway was not favorable due to the high CAPEX and OPEX. When evaluating from a single biorefinery perspectives, *Nannochloropsis sp.* indicated better economic outcome as compared to *Chlorella sp.* due to the utilization of *Nannochloropsis sp.* in protein extraction process that offset larger revenue from protein sales. However, due to the higher transportation cost required for *Nannochloropsis sp.* as a result of constraints in resource availability and longer distance from supply source, it was only used in six out of the 13 biorefineries. This highlighted the importance of incorporating the supply chain cost to ensure the optimality of the microalgae selection. Nevertheless, they also found that the minimum fuel selling price of the microalgae-derived biodiesel of $ 10.92 per gallon of biodiesel is higher than that of the current market price for biodiesel ($ 3.51 per gallon of biodiesel) in 2019. To address this issue, various strategies (e.g., implementing storage for overproduced microalgae during summer season to be used in winter or autumn season) are proposed to enhance the economic viability. Mathematical modeling with optimization is not limited to merely maximizing total supply chain's economic profitability. In fact, the optimization can be performed to assist in multi-criteria decision analysis (MCDA) to select the best microalgae strain for conversion process. Taking the study performed by Kokkinos *et al.* [82] as an example, they utilized three models that are fuzzy analytic hierarchy process (FAHP), fuzzy technique for order of preference to ideal solution (FTOPSIS), and fuzzy cognitive mapping (FCM) to select an optimal microalgae strain for biofuel production. The outcome from the studies showed that *Chlorella vulgaris sp.* was the best option amongst the four microalgae investigated given the relatively higher daily lipid productivity of *Chlorella vulgaris sp.* (134 mg L^{-1} d^{-1}) amongst the four types of microalgae strain and thus, able to produce greater amount of biofuel. This shows the significance in taking into consideration the microalgae composition (i.e., lipid content) requirement for specific conversion process during techno-economic analysis evaluation.

According to Shabani *et al.* [80], deterministic optimization models are not necessarily enough to reflect the reality of the actual performance of the biomass supply chains specifically when there are uncertain variables involved such as prices and yields. Thus, stochastic models should be applied to capture these uncertainties. Kangas and Kangas [83] mentioned that there can be many definitions or interpretations brought forth for uncertainty, however, it can be defined as the lack of information that can qualitatively or quantitatively describe or predict numerically a system or its characteristics. Fasahati *et al.* [84] highlighted that there are a few uncertainties surrounding biorefinery design such as productivity and residence time of cyanobacteria, supply mode for carbon dioxide, carbon dioxide demand required by cyanobacteria (influenced by potential leakage of carbon dioxide and ratio of cell to product mass), the water and nutrient supply, which is highly dependent on the output from wastewater treatment plant, and etc. Although they did not directly incorporate the uncertainties into their developed mathematical model, they performed sensitivity analysis to evaluate the impact of the uncertainties on the economic performance of the biorefinery.

Apart from performing sensitivity analysis, stochastic optimization model represents another way to incorporate uncertainties directly into the developed economic model. For instance, Beal *et al.* [78] applied Monte Carlo model to study the combined effect of investment cost, energy and material flows and labor cost uncertainty on the NPV, energy return on investment (EROI), and GHG emissions. The microalgae productivity was determined as the most influential factor due the large range of historical data (8.5 to 42 8.5 g/m^2-day). The incorporation of uncertainty has led the NPV value to decrease from $ 26.3 million in the base case to a median NPV of $ 15.8 M. The result obtained emphasized the importance of consideration of uncertainties as the base case NPV is relatively an overestimated value as compared to the median NPV, which may lead to unnecessary losses if not considered more comprehensively. Rizwan *et al.* [14] highlighted in their review that uncertainties tend to be the process parameters (i.e., lipid contents, yield, and productivity, microalgae biomass quality, raw material pricing, utility cost, and consumption). However, as observed in table 4A and table 4B, current studies that perform stochastic optimization to study uncertainties are still scarce. Hence, as a summary of the above discussion, it is critical to consider uncertainties when performing evaluation on the microalgae-based supply chain as a whole and not only consider the microalgae conversion process alone.

Table 4A. Summary of the works adopting mathematical modelling with optimization.

Year	Author	Type of Microalgae	Conversion Process/Product	Mathematical Model Adopted	Model Type[a]
2021	Kokkinos *et al.* [82]	*Chlorella vulgaris sp.* *Schizochytrium limacinum SR21* *Arthrospira (Spirulina) Platensis* *Nannochloropsis sp.*	Biofuel production	FAHP, FTOPSIS, FCM	D
	Correa *et al.* [85]	-	Biofuel production	Integer linear programming (LP)	D
	Correa *et al.* [86]	*Nannochloropsis sp.*	Microalgae cultivation for biofuel production	LP	D
2020	Ahn *et al.* [87]	-	Biodiesel production	Two-stage stochastic model	S
	Sarker *et al.* [88]	*Schizochytrium sp.* *N. oculata*	Fish oil and fishmeal replacement	Hedonic regression analysis	S
	Kang *et al.* [81]	*Chlorella sp.* *Nannochloropsis sp.*	Biofuel with animal feeds and bioethanol produced as by-products	MILP	D
2019	Fasahati *et al.* [84]	Cyanobacteria	Biochemical production	MINLP	D
	Correa *et al.* [89]	-	Biodiesel production	GIS-based MCDA and LP	D

[a] The mathematical model type indicates whether the model is categorized under deterministic (D) or stochastic (S) optimization.

Table 4B. Summary of the works adopting mathematical modelling with optimization.

Year	Author	Type of Microalgae	Conversion Process/Product	Mathematical Model Adopted	Model Type[a]
2019	Shirazi *et al.* [90]	-	Algae-derived syngas *via* supercritical water gasification (SCWG)	Genetic algorithm	
	Thomassen *et al.* [91]	*Dunaliella salina* *Haematococcus pluvialis* *Nannochloropsis sp.*	Superstructure of processes including lipid extraction, thermal and biological process	Multi-objective MINLP, Monte Carlo model	D & S
2018	Beal *et al.* [78]	*Desmodesmus sp.*	Fish oil and fishmeal replacement *via* lipid extraction and drying process	Monte Carlo	S
2017	Gong and You [92]	-	Biodiesel production	Mixed integer nonlinear fractional programming (MINFP)	D
	García Prieto *et al.* [93]	*Haematococcus pluvialis*	Astaxanthin, biodiesel, poly(hydroxybutyrate) (PHB) production	MINLP	D
	Gong and You [94]	-	Poly-3-hydroxybutyrate (PHB) and biodiesel production	Two-stage adaptive robust mixed integer fractional programming (ARMIFP)	S

[a] The mathematical model type indicates whether the model is categorized under deterministic (D) or stochastic (S) optimization.

5. Possible research exploration areas

As microalgae-based feedstock is still in its infancy, modeling approaches on microalgae-based supply chain to evaluate the feasibility of microalgae utilization should consider a wider scope. For instance, the optimization of microalgae-based supply should consider the different types of microalgae, cultivation medium and environment, locations of microalgae cultivation (i.e., on-site of processing site or off-site), and etc. On top of that, stochastic optimization studies should consider multiple uncertainties so tthat the risk associated to the microalgae-based supply chain can be assessed. For example, the variation of microalgae productivity should be considered as it is influenced by multiple environmental conditions such as solar irradiation. This is where artificial intelligence (AI) can be integrated into the study. As an exemplification, an AI model can be developed that is capable of predicting the microalgae productivity based on the fluctuation of environmental conditions.

Microalgae productivity will indirectly impact the production of end-value product. Thus, it can be essential to integrate the prediction of microalgae productivity based on fluctuating environmental conditions into techno-economic evaluation model. Along the years, there has been advancement in government's policy that encourages the exploration of AI. For instance, Malaysia's government had launched a new policy regarding the Fourth Industrial Revolution— National 4IR Policy whereby AI is one of the key technologies that will be focused on [95]. Generally, the implementation of AI can provide few advantages such as acceleration of optimization solving procedures, reduction of uncertainty, and etc [96].

6. Conclusion

The quest for diversification of current energy profile around the world has led to the hunt for potential renewable energy sources. Third-generation biomass is slowly becoming the focus of attention as a potential feedstock for biofuel production. Therefore, comprehensive studies are required to evaluate the feasibility in upscaling the utilization of third-generation biomass to commercial scale. This review presented a compilation of current approaches used in techno-economic evaluation for microalgae-based supply chain. The current findings highlighted that approximately 58% of the studies opted for mathematical modeling with optimization whereas approximately 42% of the papers opted for mathematical modeling without optimization. Amongst the studies adopting mathematical modeling with optimization, deterministic model is deemed as a favorable option. Regarding potential future works, a larger scope of evaluation when performing superstructure analysis that comprises of different cultivation medium and cultivation environment can be substantial due to the wide varieties of cultivation medium and environment that come with their respective costs. Apart from that, it is also critical to consider the regional weather conditions (i.e., solar irradiation, seasonality, rain, and etc.) when selecting a location for the microalgae cultivation or when choosing macroalgae species. Specifically, when performing stochastic techno-economic analysis, it is important to consider the impact of variation of environmental conditions or microalgae species on the overall microalgae productivity of microalgae to attain a more accurate estimation on its economic feasibility. Furthermore, AI model, which is capable of predicting the influence of variation in environmental conditions on the microalgae productivity can be incorporated into the techno-economic analysis model and serves as another promising path to explore.

Acknowledgments

The authors would like to acknowledge the financial support from Swinburne University of Technology Sarawak Campus *via* Research Micro Fund [9-1372 RMFST] and Research Supervision Grant [2-5545 RSG].

References

[1] International Renewable Energy Agency 2019 Renewable Energy Now Accounts for a Third of Global Power Capacity *International Renewable Energy Agency*
[2] Lee R A and Lavoie J M 2013 From first- to third-generation biofuels: Challenges of producing a commodity from a biomass of increasing complexity *Anim. Front.* **3** 6–11
[3] Brennan L and Owende P 2010 Biofuels from microalgae—A review of technologies for production, processing, and extractions of biofuels and co-products *Renew. Sustain. Energy Rev.* **14** 557–77
[4] Liang Y, Sarkany N and Cui Y 2009 Biomass and lipid productivities of Chlorella vulgaris under autotrophic, heterotrophic and mixotrophic growth conditions *Biotechnol. Lett.* **31** 1043–9
[5] Randrianarison G and Ashraf M A 2017 Microalgae: A potential plant for energy production *Geology, Ecology, and Landscapes* **1** 104–20
[6] Drogui P, Asselin M, Brar S K, Benmoussa H and Blais J F 2008 Electrochemical removal of pollutants from agro-industry wastewaters *Sep. Purif. Technol.* **61** 301–10

[7] Posadas E, Bochon S, Coca M, García-González M C, García-Encina P A and Muñoz R 2014 Microalgae-based agro-industrial wastewater treatment: A preliminary screening of biodegradability *J. Appl. Phycol.* **26** 2335–45
[8] Tan W 2018 ROI-Integrated commercialization: An adaptive pathway for microalgae technology *Ind. Biotechnol.* **14** 17–24
[9] Andrade D S, Telles T S and Leite Castro G H 2020 The Brazilian microalgae production chain and alternatives for its consolidation *J. Clean. Prod.* **250** 119526
[10] Madugu F and Collu M 2014 Techno-economic modelling analysis of microalgae cultivation for biofuels and co-products *WIT Trans. Ecol. Environ* **190** 1091–102
[11] Chew K W, Yap J Y, Show P L, Suan N H, Juan J C, Ling T C, Lee D J and Chang J S 2017 Microalgae biorefinery: High value products perspectives *Bioresour. Technol.* **229** 53–62
[12] Efroymson R A, Jager H I, Mandal S, Parish E S and Mathews T J 2021 Better management practices for environmentally sustainable production of microalgae and algal biofuels *J. Clean. Prod.* **289** 125150
[13] Doliente S S, Narayan A, Tapia J F D, Samsatli N J, Zhao Y and Samsatli S 2020 Bio-aviation fuel: A comprehensive review and analysis of the supply chain components *Front. Energy Res.* **8** 110
[14] Rizwan M, Almansoori A and Elkamel A 2019 An overview on synthesis and design of microalgal biorefinery configurations by employing superstructure-based optimization approach *Energy Syst.* **10** 941–66
[15] Teng S Y, Yew G Y, Sukačová K, Show P L, Máša V and Chang J S 2020 Microalgae with artificial intelligence: A digitalized perspective on genetics, systems and products *Biotechnol. Adv.* **44** 107631
[16] Deprá M C, dos Santos A M, Severo I A, Santos A B, Zepka L Q and Jacob Lopes E 2018 Microalgal biorefineries for bioenergy production: Can we move from concept to industrial reality? *BioEnergy Res.* **11** 727–47
[17] Kusmayadi A, Leong Y K, Yen H W, Huang C Y and Chang J S 2021 Microalgae as sustainable food and feed sources for animals and humans – Biotechnological and environmental aspects *Chemosphere* **271** 129800
[18] Vigor C, Oger C, Reversat G, Rocher A, Zhou B, Linares-Maurizi A, Guy A, Bultel-Poncé V, Galano J M, Vercauteren J, Durand T, Potin P, Tonon T and Leblanc C 2020 Isoprostanoid profiling of marine microalgae *Biomolecules* **10** 7
[19] Wang X, Fosse H K, Li K, Chauton M S, Vadstein O and Reitan K I 2019 Influence of nitrogen limitation on lipid accumulation and EPA and DHA content in four marine microalgae for possible use in aquafeed *Front. Mar. Sci.* **6** 95
[20] Andersson V, Heyne S, Harvey S and Berntsson T 2020 Integration of algae-based biofuel production with an oil refinery: Energy and carbon footprint assessment *Int. J. Energy Res.* **44** 10860–77
[21] Ryu K H, Kim B, Heo S, Chang Y K and Lee J H 2020 Mathematical modeling of microalgal internal metabolic behaviors under heterotrophic conditions and its application *Ind. Eng. Chem. Res.* **59** 1631–45
[22] Goffé J and Ferrasse J H 2019 Stoichiometry impact on the optimum efficiency of biomass conversion to biofuels *Energy* **170** 438–58
[23] Chen R, Zhang S, Yang X, Li G, Zhou H, Li Q and Zhang Y 2021 Thermal behaviour and kinetic study of co-pyrolysis of microalgae with different plastics *J. Waste Manag.* **126** 331–9
[24] Lu Q, Li H, Xiao Y and Liu H 2021 A state-of-the-art review on the synthetic mechanisms, production technologies, and practical application of polyunsaturated fatty acids from microalgae *Algal Res.* **55** 102281
[25] Maiolo S, Cristiano S, Gonella F and Pastres R 2021 Ecological sustainability of aquafeed: An emergy assessment of novel or underexploited ingredients *J. Clean. Prod.* **294** 126266

[26] Hu X, Tang X, Bi Z, Zhao Q and Ren L 2021 Adaptive evolution of microalgae Schizochytrium sp. under high temperature for efficient production of docosahexaeonic acid *Algal Res.* **54** 102212
[27] Dalheim L, Svenning J B, Eilertsen H C, Vasskog T and Olsen R L 2021 Stability of lipids during wet storage of the marine diatom Porosira glacialis under semi-preserved conditions at 4 and 20 °C *J. Appl. Phycol.* **33** 385–95
[28] Serrano E, Simpfendorfer R, Medina A, Sandoval C, Martínez A, Morales R and Davies S J 2021 Partially replacing fish oil with microalgae (*Schizochytrium limacinum* and *Nannochloropsis oceanica*) in diets for rainbow trout (*Oncorhynchus mykiss*) reared in saltwater with reference to growth performance, muscle fatty acid composition and liver ultrastructure *Aquac. Res.* **52** 4401–13
[29] Chauton M S, Forbord S, Mäkinen S, Sarno A, Slizyte R, Mozuraityte R, Standal I B and Skjermo J 2021 Sustainable resource production for manufacturing bioactives from micro- and macroalgae: Examples from harvesting and cultivation in the Nordic region *Physiol. Plant.* **173** 495–506
[30] Culaba A B, Ubando A T, Ching P M L, Chen W H and Chang J S 2020 Biofuel from microalgae: Sustainable pathways *Sustainability* **12** 19
[31] Nguyen T H T, Park S, Jeong J, Shin Y S, Sim S J and Jin E 2020 Enhancing lipid productivity by modulating lipid catabolism using the CRISPR-Cas9 system in Chlamydomonas *J. Appl. Phycol.* **32** 2829–40
[32] Deprá M C, Dias R R, Severo I A, de Menezes C R, Zepka L Q and Jacob-Lopes E 2020 Carbon dioxide capture and use in photobioreactors: The role of the carbon dioxide loads in the carbon footprint *Bioresour. Technol.* **314** 123745
[33] El-Dakar M A, Ramzy R R, Ji H and Plath M 2020 Bioaccumulation of residual omega-3 fatty acids from industrial Schizochytrium microalgal waste using black soldier fly (Hermetia illucens) larvae *J. Clean. Prod.* **268** 122288
[34] Branco-Vieira M, Costa D M B, Mata T M, Martins A A, Freitas M A V and Caetano N S 2020 Environmental assessment of industrial production of microalgal biodiesel in central-south Chile *J. Clean. Prod.* **266** 121756
[35] Maiolo S, Parisi G, Biondi N, Lunelli F, Tibaldi E and Pastres R 2020 Fishmeal partial substitution within aquafeed formulations: Life cycle assessment of four alternative protein sources *Int. J. Life Cycle Assess.* **25** 1455–71
[36] Hossain S M Z, Razzak S A, Al-Shater A F, Moniruzzaman M and Hossain M M 2020 Recent advances in enzymatic conversion of microalgal lipids into biodiesel *Energ. Fuels* **34** 6735–50
[37] Morales M, Bonnefond H and Bernard O 2020 Rotating algal biofilm versus planktonic cultivation: LCA perspective *J. Clean. Prod.* **257** 120547
[38] Pankratz S, Kumar M, Oyedun A O, Gemechu E and Kumar A 2020 Environmental performances of diluents and hydrogen production pathways from microalgae in cold climates: Open raceway ponds and photobioreactors coupled with thermochemical conversion *Algal Res.* **47** 101815
[39] Wahlen B D, Wendt L M, Murphy A, Thompson V S, Hartley D S, Dempster T and Gerken H 2020 Preservation of microalgae, lignocellulosic biomass blends by ensiling to enable consistent year-round feedstock supply for thermochemical conversion to biofuels *Front. Bioeng. Biotechnol.* **8** 316
[40] Naeini M A, Zandieh M, Najafi S E and Sajadi S M 2020 Analyzing the development of the third-generation biodiesel production from microalgae by a novel hybrid decision-making method: The case of Iran *Energy* **195** 116895
[41] Lee T, Nam I H, Jung S, Park Y K and Kwon E E 2020 Synthesis of nickel/biochar composite from pyrolysis of Microcystis aeruginosa and its practical use for syngas production *Bioresour. Technol.* **300** 122712

[42] Desjardins S M, Laamanen C A, Basiliko N and Scott J A 2020 Utilization of lipid-extracted biomass (LEB) to improve the economic feasibility of biodiesel production from green microalgae *Environ. Rev.* **28** 325–38
[43] Callegari A, Bolognesi S, Cecconet D and Capodaglio A G 2020 Production technologies, current role, and future prospects of biofuels feedstocks: A state-of-the-art review *Crit. Rev. Environ. Sci. Technol.* **50** 384–436
[44] Azari A, Noorpoor A R and Bozorg-Haddad O 2019 Carbon footprint analyses of microalgae cultivation systems under autotrophic and heterotrophic conditions *Int. J. Environ. Sci. Technol.* **16** 6671–84
[45] Morales M, Hélias A and Bernard O 2019 Optimal integration of microalgae production with photovoltaic panels: Environmental impacts and energy balance *Biotechnol. Biofuels.* **12** 239
[46] Lee Y H and Sun J H 2019 Multifunctional fluorocarbon photobioreactor system: A novel integrated device for CO_2 segregation, O_2 collection, and enhancement of microalgae growth and bioproductions *Bioproc. Biosyst. Eng.* **42** 1591–601
[47] Zhang Y and Kendall A 2019 Consequential analysis of algal biofuels: Benefits to ocean resources *J. Clean. Prod.* **231** 35–42
[48] Song C, Xie M, Qiu Y, Liu Q, Sun L, Wang K and Kansha Y 2019 Integration of CO_2 absorption with biological transformation via using rich ammonia solution as a nutrient source for microalgae cultivation *Energy* **179** 618–27
[49] Sawant S S, Gosavi S N, Khadamkar H P, Mathpati C S, Pandit R and Lali A M 2019 Energy efficient design of high depth raceway pond using computational fluid dynamics *Renew. Energy* **133** 528–37
[50] Sun C H, Fu Q, Liao Q, Xia A, Huang Y, Zhu X, Reungsang A and Chang H X 2019 Life-cycle assessment of biofuel production from microalgae via various bioenergy conversion systems *Energy* **171** 1033–45
[51] Bacci di Capaci R, Tasca A L, Pannocchia G, Scali C, Tognotti L, Brunazzi E, Nicolella C and Puccini M 2019 Biomethane production: Mass and energy balances of alternative supply chains *Ind. Eng. Chem. Res.* **58** 10951–62
[52] Montero-Lobato Z, Vázquez M, Navarro F, Fuentes J L, Bermejo E, Garbayo I, Vílchez C and Cuaresma M 2018 Chemically-induced production of anti-inflammatory molecules in microalgae *Marine drugs* **16** 12
[53] Pedersen T C, Gardner R D, Gerlach R and Peyton B M 2018 Assessment of Nannochloropsis gaditana growth and lipid accumulation with increased inorganic carbon delivery *J. Appl. Phycol.* **30** 2155–66
[54] Foteinis S, Antoniadis-Gavriil A and Tsoutsos T 2018 Life cycle assessment of algae-to-biodiesel shallow pond production systems in the Mediterranean: Influence of species, pond type, by(co)-product valorisation and electricity mix *Biofuel Bioprod. Biorefin.* **12** 542–58
[55] Poddar N, Sen R and Martin G J O 2018 Glycerol and nitrate utilisation by marine microalgae Nannochloropsis salina and Chlorella sp. and associated bacteria during mixotrophic and heterotrophic growth *Algal Res.* **33** 298–309
[56] Sturme M H J, Gong Y, Heinrich J M, Klok A J, Eggink G, Wang D, Xu J and Wijffels R H 2018 Transcriptome analysis reveals the genetic foundation for the dynamics of starch and lipid production in Ettlia oleoabundans *Algal Res.* **33** 142–55
[57] Giraldo-Calderón N D, Romo-Buchelly R J, Arbeláez-Pérez A A, Echeverri-Hincapié D and Atehortúa-Garcés L 2018 Microalgae biorefineries: Applications and emerging technologies *DYNA* **85** 219–33
[58] Chen C Y, Nagarajan D and Cheah W Y 2018 Eicosapentaenoic acid production from Nannochloropsis oceanica CY2 using deep sea water in outdoor plastic-bag type photobioreactors *Bioresour Technol* **253** 1–7
[59] Pan S Y, Chiang P C, Pan W and Kim H 2018 Advances in state-of-art valorization technologies for captured CO_2 toward sustainable carbon cycle *Crit. Rev. Environ. Sci.*

Technol. **48** 471–534
[60] Phusunti N, Phetwarotai W and Tekasakul S 2018 Effects of torrefaction on physical properties, chemical composition and reactivity of microalgae *Korean. J. Chem. Eng.* **35** 503–10
[61] Tang Y, Ma X, Lai Z and Yu Q 2018 Oxy-fuel combustion characteristics and kinetics of microalgae and its mixture with rice straw using thermogravimetric analysis *Int. J. Energy Res.* **42** 532–41
[62] Chen H, Wang J, Zheng Y, Zhan J, He C and Wang Q 2018 Algal biofuel production coupled bioremediation of biomass power plant wastes based on Chlorella sp. C2 cultivation *Appl. Energy* **211** 296–305
[63] Qiu R, Gao S, Lopez P A and Ogden K L 2017 Effects of pH on cell growth, lipid production and CO_2 addition of microalgae Chlorella sorokiniana *Algal Res.* **28** 192–9
[64] Tan J, Tan R R, Aviso K B, Promentilla M A B and Sulaiman N M N 2017 Study of microalgae cultivation systems based on integrated analytic hierarchy process–life cycle optimization *Clean Technol. Envir.* **19** 2075–88
[65] Sabu S, Bright Singh I S and Joseph V 2017 Molecular identification and comparative evaluation of tropical marine microalgae for biodiesel production *Mar. Biotechnol.* **19** 328–44
[66] Daneshvar E, Sik Ok Y, Tavakoli S, Sarkar B, Shaheen S M, Hong H, Luo Y, Rinklebe J, Song H and Bhatnagar A 2021 Insights into upstream processing of microalgae: A review *Bioresour. Technol.* **329** 124870
[67] Morais Junior W G, Gorgich M, Corrêa P S, Martins A A, Mata T M and Caetano N S 2020 Microalgae for biotechnological applications: Cultivation, harvesting and biomass processing *Aquaculture* **528** 735562
[68] Wu W, Pai C T, Viswanathan K and Chang J S 2021 Comparative life cycle assessment and economic analysis of methanol/hydrogen production processes for fuel cell vehicles *J. Clean. Prod.* **300** 126959
[69] Sano Coelho R, Cuellar M C, Franco T T and van der Wielen L A M 2021 Techno-economic assessment of heterotrophic microalgae biodiesel production integrated with a sugarcane bio-refinery *Biofuel Bioprod. Biorefin.* **15** 416–29
[70] Guccione A, Biondi N, Sampietro G, Rodolfi L, Bassi N and Tredici M R 2014 Chlorella for protein and biofuels: From strain selection to outdoor cultivation in a Green Wall Panel photobioreactor *Biotechnol. Biofuels.* **7** 84 1–12
[71] Schade S and Meier T 2021 Techno-economic assessment of microalgae cultivation in a tubular photobioreactor for food in a humid continental climate *Clean Technol. Envir* **23** 1475–92
[72] Silva M I, Gonçalves A L, Vilar V J P and Pires J C M 2021 Experimental and techno-economic study on the use of microalgae for paper industry effluents remediation *Sustainability* **13** 1314
[73] Nappa M, Lienemann M, Tossi C, Blomberg P, Jäntti J, Tittonen I J and Penttilä M 2020 Solar-powered carbon fixation for food and feed production using microorganisms—A comparative techno-economic analysis *ACS Omega* **5** 33242–52
[74] Choudhary P and Srivastava R K 2020 Techno-economic case study: Bio-fixation of industrial emissions at an Indian oil and gas plant *J. Clean. Prod.* **266** 121820
[75] Mennella L, Tosco D, Alberti F, Cembalo L, Crescimanno M, Del Giudice T, Galati A, Moglie M, Scardera A, Schifani G, Solfanelli F and Cicia G 2020 Perspectives and challenges of small scale plant microalgae cultivation. Evidences from Southern Italy *Algal Res.* **45** 101693
[76] Archanaa S, Jose S, Mukherjee A and Suraishkumar G K 2019 Sustainable diesel feedstock: A comparison of oleaginous bacterial and microalgal model systems *BioEnergy Research* **12** 205–16
[77] Tasca A L, Bacci di Capaci R, Tognotti L and Puccini M 2019 Biomethane from short rotation forestry and microalgal open ponds: System modeling and life cycle assessment *Bioresour.*

Technol. **273** 468–77
[78] Beal C M, Gerber L N, Thongrod S, Phromkunthong W, Kiron V, Granados J, Archibald I, Greene C H and Huntley M E 2018 Marine microalgae commercial production improves sustainability of global fisheries and aquaculture *Scientific Reports* **8** 15064
[79] Lo S L Y, How B S, Leong W D, Teng S Y, Rhamdhani M A and Sunarso J 2021 Techno-economic analysis for biomass supply chain: A state-of-the-art review *Renew. Sustain. Energy Rev.* **135** 110164
[80] Shabani N, Akhtari S and Sowlati T 2013 Value chain optimization of forest biomass for bioenergy production: A review *Renew. Sustain. Energy Rev.* **23** 299–311
[81] Kang S, Heo S, Realff M J and Lee J H 2020 Three-stage design of high-resolution microalgae-based biofuel supply chain using geographic information system *Appl. Energy* **265** 114773
[82] Kokkinos K, Karayannis V and Moustakas K 2021 Optimizing microalgal biomass feedstock selection for nanocatalytic conversion into biofuel clean energy, using Fuzzy Multi-Criteria Decision Making processes *Front. Energy Res.* **8** 408
[83] Kangas A S and Kangas J 2004 Probability, possibility and evidence: Approaches to consider risk and uncertainty in forestry decision analysis *For. Policy Econ.* **6** 169–88
[84] Fasahati P, Wu W and Maravelias C T 2019 Process synthesis and economic analysis of cyanobacteria biorefineries: A superstructure-based approach *Appl. Energy* **253** 113625
[85] Correa D F, Beyer H L, Possingham H P, García-Ulloa J, Ghazoul J and Schenk P M 2020 Freeing land from biofuel production through microalgal cultivation in the Neotropical region *Environ. Res. Lett.* **15** 094094
[86] Correa D F, Beyer H L, Possingham H P, Fargione J E, Hill J D and Schenk P M 2021 Microalgal biofuel production at national scales: Reducing conflicts with agricultural lands and biodiversity within countries *Energy* **215** 119033
[87] Ahn Y, Kim J and Han J 2020 Optimization of microalgae-based biodiesel supply chain network under the uncertainty in supplying carbon dioxide *Korean J Chem Eng* **58** 396–407
[88] Sarker P K, Kapuscinski A R, McKuin B, Fitzgerald D S, Nash H M and Greenwood C 2020 Microalgae-blend tilapia feed eliminates fishmeal and fish oil, improves growth, and is cost viable *Sci. Rep.* **10** 19328
[89] Correa D F, Beyer H L, Possingham H P, Thomas-Hall S R and Schenk P M 2019 Global mapping of cost-effective microalgal biofuel production areas with minimal environmental impact *Glob. Change Biol. Bioenergy* **11** 914–29
[90] Shirazi A, Rahbari A, Asselineau C A and Pye J 2019 A solar fuel plant via supercritical water gasification integrated with Fischer–Tropsch synthesis: System-level dynamic simulation and optimisation *Energy Convers. Manag.* **192** 71–87
[91] Thomassen G, Van Dael M, You F and Van Passel S 2019 A multi-objective optimization-extended techno-economic assessment: Exploring the optimal microalgal-based value chain *Green Chem.* **21** 5945–59
[92] Gong J and You F 2017 Consequential life cycle optimization: General conceptual framework and application to algal renewable diesel production *ACS Sustain. Chem. Eng.* **5** 5887–911
[93] García Prieto C V, Ramos F D, Estrada V, Villar M A and Diaz M S 2017 Optimization of an integrated algae-based biorefinery for the production of biodiesel, astaxanthin and PHB *Energy* **139** 1159–72
[94] Gong J and You F 2017 Optimal processing network design under uncertainty for producing fuels and value-added bioproducts from microalgae: Two-stage adaptive robust mixed integer fractional programming model and computationally efficient solution algorithm *AIChE J.* **63** 582–600
[95] Nor Ain Mohamed Radhi 2021 Govt Launches National 4IR Policy *New Straights Time*
[96] Kong K G H, How B S, Teng S Y, Leong W D, Foo D C Y, Tan R R and Sunarso J 2021 Towards data-driven process integration for renewable energy planning *Curr. Opin. Chem. Eng.* **31** 100665

IOP Conf. Series: Materials Science and Engineering 1195 (2021) 012027 doi:10.1088/1757-899X/1195/1/012027

Evaluation and characterisation of candidate prebiotics extracted from coconut husk by ultrasound-assisted extraction technique

S L Tang, C C Koh and S L Hii

Food Technology Programme, School of Engineering and Technology, University College of Technology Sarawak, No.1, Jalan Universiti, 96000 Sibu, Sarawak, Malaysia

Email: hiisl@ucts.edu.my

Abstract. Oligosaccharides are carbohydrates containing between three to ten sugar moieties. Certain oligosaccharides such as inulin and fructo-oligosaccharides are known as prebiotics that promote the growth of beneficial bacteria in the human gastrointestinal tract. This study began by comparing the efficiency of two different solvents (distilled water and 10% w/v sodium hydroxide) in extracting oligosaccharides from the coconut husk by ultrasound-assisted extraction (UAE). Following that, the coconut husk extract (CHE) extract was subjected to a series of prebiotic evaluation tests. The findings indicated that a significantly high extraction yield (40.51 ± 6.00%) could be achieved with 10% w/v NaOH treatment. The in vitro enzymatic digestion study found that there was 43.70 ± 0.15% of hydrolysis at pH 8 after five hours of incubation. For the in vitro gastric juice digestion, 29.21 ± 0.71% of hydrolysis was recorded at pH 1 after four hours of incubation. The extract was able to stimulate the growth of selected beneficial bacterial strains. FTIR and NMR analysis of the CHE revealed that the extract has a similar structure to the well-known prebiotic inulin.

1. Introduction

Coconut (*Cocos nucifera*) is cultivated in 92 countries, with Indonesia producing the most [1]. Malaysia is one of the world's top ten coconut-producing countries. After oil palm, rubber, and rice, coconut is Malaysia's fourth most important industrial crop [2]. Coconut husks, shells, and fronds are the most common agricultural waste generated during the harvesting and processing of coconuts in the industry [3]. Instead of being disposed of directly, these agro-wastes can be converted into value-added products. Coconut husks contain 38% hemicellulose, 28% cellulose, and 32.8% lignin [4]. Due to the high concentration of hemicellulose in coconut husk, it could be a source of oligosaccharides with prebiotic properties.

Prebiotics were originally defined as "non-digestible food ingredients that benefit the host by selectively stimulating the growth and/or activity of one or a limited number of bacteria in the colon, thereby improving host health" [5]. Numerous dietary oligosaccharides and polysaccharides (including dietary fibre) have been claimed to possess prebiotic properties. Oligosaccharides are carbohydrates which made up of 3 to 20 sugar monomer units.

Gibson *et al.* (2004) [6] state that a food ingredient must meet three criteria to be classified as a prebiotic: To begin, a prebiotic must be resistant to digestive processes such as gastric acidity, mammalian enzyme hydrolysis, and gastrointestinal absorption. Second, it must be fermented by the microbiota in the intestine. Thirdly, it must stimulate the growth and activity of beneficial intestinal microbiota in a selective manner.

Extraction of bioactive compounds from natural resources using conventional methods is time consuming and has a low extraction efficiency. Innovative technologies, such as ultrasound-assisted extraction (UAE), have been investigated to overcome these disadvantages. UAE has been used to extract numerous bioactive compounds from a variety of animal, plant, and food matrices [7]. UAE is highly regarded for its straightforward protocols, low investment costs, and high efficiency. UAE's primary advantages are a reduction in extraction time and energy consumption. Reduced operating temperature and processing time are particularly advantageous for heat-labile compound extraction. Apart from expediting extraction, UAE maintains molecular and structural properties [8].

The purpose of this study was to determine the feasibility of extracting prebiotic oligosaccharides from coconut husk using UAE. The UAE was conducted by comparing the effects of two different solvents on the oligosaccharide extract's chemical properties in terms of total carbohydrate (TC), total reducing sugar (TRS), and degree of polymerisation (DP). Prebiotic potency tests were performed on the resulting oligosaccharide extract from coconut husk. Following that, the molecular structure of coconut husk extract (CHE) was compared to that of a well-characterized prebiotic (inulin) using Fourier Transform Infrared (FTIR) and Nuclear Molecular Resonance (NMR) spectroscopy.

2. Materials and Methods

2.1. Sample Preparation

The sample was prepared based on the method proposed by Wichienchot *et al.* (2011) [9] with modification. Coconut husk was cleaned thoroughly under running tap water after collection from the local market. The coconut husk was dried in the oven until constant weight and then grinded into fine powder. The powder was sieved to pass through 250 μm mesh sieve and stored in a clean container at ambient temperature until further analysis.

2.2. Preliminary Study: Selection of Extraction Solvents

The preliminary study was conducted based on the method proposed by Alcántara *et al.* (2020) [10] with modification. There were two solvents used in the preliminary UAE process: (1) distilled water (dH_2O), and (2) 10% w/v sodium hydroxide (NaOH) solution. Approximately two grams of coconut husk powder was mixed with 200 mL of distilled water (solid-to-liquid ratio 1:100). The mixture was then subjected to UAE at two different powers (25 W and 250 W) and time intervals of 5 min and 45 min, respectively.

After UAE, the mixture was centrifuged at 4500 rpm, 4 °C for 10 min. The resulting supernatant was subjected to overnight precipitation with three volumes of cold ethanol (EtOH). After that, the solution was centrifuged at 4500 rpm, 4 °C for 10 min to separate the precipitate. The precipitate that contained the extracted oligosaccharides was collected, dried at 60 °C until constant weight. The total carbohydrates (TC) content, total reducing sugar (TRS) content and degree of polymerisation (DP) of the extracted oligosaccharide was then determined (Section 2.5). All the extraction was conducted at temperature of 30 °C in replicates. The UAE was also performed by replacing the distilled water with 10% w/v NaOH solution.

2.3. Determination of Total Carbohydrates Content

Phenol-sulfuric acid assay was used to determine the total concentration of carbohydrates in the extract by using glucose as standard solution [11]. One millilitre of extract was mixed with 5% phenol solution (1 mL) and concentrated sulphuric acid (5 mL). After vortex, the mixture was incubated in a 30°C water bath for 30 min. After that, the absorbance of the mixture was measured at 490 nm.

2.4. Determination of Total Reducing Sugar (DNSA Analysis)
The concentration of reducing sugar was determined by 3, 5-dinitrosalicylic acid (DNSA) method [12]. One millilitre of the extract was mixed with 1 mL of DNS reagent followed by two drops of 10% sodium hydroxide and boiled for 15 min. The mixture was cooled under running tap water. Next, 10 mL of distilled water was added and vortexed. The mixture was left for 20 min and the absorbance was measured at 540 nm.

2.5. Degree of Polymerisation
The degree of polymerisation of the oligosaccharides was calculated using the equation (1).

$$Degree\ of\ polymerisation = \frac{\text{Total Carbohydrates}}{\text{Total Reducing Sugar}} \quad (1)$$

2.6. In vitro Fermentation Study with CHE as Sole Carbon Source
A single colony of the bacterial cultures (*Lactobacillus acidophilus* DSM 20079, *Lactobacillus casei* DSM 20011 and *Bifidobacterium bifidum* DSM 20239, and *Escherichia coli* DSM 1103) from respective agar plate was transferred into MRS broth or tryptic soy (TS) broth accordingly. The probiotic bacteria were pre-cultured in MRS broth at 37 ± 1 °C for 48 to 72 hours while the *E. coli* in TS broth were cultured for 24 hours. In order to investigate the effect of CHE on the growth of the bacteria, both MRS and TS fermentation media with a series of concentrations of CHE (0.5% w/v to 3.0% w/v) were prepared to replace the glucose.

Next, the medium was inoculated with 1 mL of bacterial culture (approximately 10^4 CFU/mL) and incubated at 37 ± 1 °C for up to 72 hours. Inoculated medium without glucose served as negative control while medium with inulin served as positive control [13]. All incubation, fermentation and enumeration of those probiotic bacteria were performed in an anaerobic system (Oxoid AnaeroJar™ 2.5L, UK), while that of *E. coli* was performed in an aerobic system. The viable count of the cultures (CFU/mL) was determined at predetermined time intervals. The result was expressed as $\log_{10}$ CFU/mL.

The short chain fatty acid (SCFA) namely, lactate, acetate, propionate and butyrate in the fermented culture broth were identified and quantified using High Performance Liquid Chromatography (HPLC) method with minor modifications [14].

2.7. In vitro Enzymatic Digestion
In this analysis, human salivary α-amylase suspension was used to mimic the upper gastrointestinal digestion. The *in vitro* enzymatic digestion was carried out according to the method by Fässler *et al.* (2006) [15].

First, sodium phosphate buffer solution with concentration of 20 mM was prepared. Approximately 3.12 g sodium hydrogen orthophosphate (NaH_2PO_4) and 0.392 g sodium chloride (NaCl) was dissolved with approximately 900 mL distilled water. After that, the pH of the buffer was adjusted to pH 7.0 by using 0.1 N sodium hydroxide (NaOH) solution. The buffer was transferred to a 1 L volumetric flask and made up to volume with distilled water. Human salivary α-amylase was then dissolved in the 20 mM sodium phosphate buffer.

In a clean test tube, 5 mL of the enzyme solution and 5 mL of the extract solution were mixed. The test tube was then incubated in the water bath at 37 ± 1 °C for 5 hours. The mixture was withdrawn in fixed time interval: 0, 1, 2, 3, 4 and 5 hours of post incubation. The total carbohydrates content was measured using phenol-sulphuric acid method whereas the initial and final reducing sugar contents were measured using DNSA assay.

The percentage hydrolysis of the extract was calculated based on the reducing sugar liberated and the total carbohydrates content by using equation (2) [16].

$$\% \, hydrolysis = \frac{\text{Reducing sugar liberated}}{\text{Total carbohydrates content - initial reducing sugar content}} \times 100 \quad (2)$$

where reducing sugar liberated = final reducing sugar content – initial reducing sugar content

2.8. In vitro Artificial Gastric Juice Digestion

The gastric juice digestion analysis was carried out based on the method proposed by Wichienchot, Jatupornpipat and Rastall (2010) [17]. In this analysis, inulin was used as positive control. The samples and inulin were incubated with the artificial gastric juice for predetermined duration.

There were two steps involved for the preparation of artificial gastric juice. First, a phosphate buffer solution was prepared. Then, 5 M HCl was used to adjust the pH of the buffer solution to 1, 2, 3 and 4. The resulting solutions were known as artificial gastric juice.

In the meantime, 1% (w/v) of CHE was prepared. The total carbohydrates content and initial reducing sugar content were measured using phenol sulphuric acid analysis and DNSA assay respectively.

Five millilitres of 1% (w/v) CHE were mixed with 5 mL of HCl buffer solution and incubated in a water bath at 37 ± 1 °C for 5 hours. The mixture solutions were withdrawn for analysis at the time interval of 0, 1, 2, 3, 4 and 5 hours post incubation [17]. The final reducing sugar content were measured using DNSA assay. Percent of hydrolysis was calculated by using equation (2) [16].

2.9. Fourier Transform Infrared (FTIR) Spectroscopy

The structure of extracted oligosaccharides from coconut husk was characterised with Fourier transform infrared spectroscopy (FTIR). The analysis was conducted using Thermo Nicolet iS10 FTIR spectrometer with the resolution of 0.4 cm^{-1} and the spectral range is 7800 cm and 350 cm. Dried CHE were ground into fine powder prior analysis. Spectral scanning was taken between the wavenumber of 4000 cm^{-1} and 650 cm^{-1}. Inulin was used as a standard.

2.10. Nuclear Magnetic Resonance (NMR) Spectroscopy

The structure of oligosaccharides extracted from coconut husk was further identified by using the ^{1}H and ^{13}C NMR spectroscopy [18]. The NMR spectra were recorded using a Varian NMR spectrometer system operating at frequency of 500 MHz. The oligosaccharide was dissolved in deuterated water (D_2O) for both ^{1}H and ^{13}C NMR. Inulin was used as a standard.

3. Results and Discussion

3.1. Preliminary Studies: Selection of Extraction Solvents

As shown in table 1, using NaOH as the extracting solvent results in a higher percentage of recovery than using distilled water. The preliminary study found that extracting oligosaccharides from coconut husk with NaOH solution was more effective than extracting with distilled water.

For total carbohydrates, water extraction at 250 W for 45 min produced 0.72 mg/mL of total carbohydrates. The lowest total carbohydrates content was obtained through water extraction at ultrasonicator power of 250 W for 5 min. This could be attributed to the shorter extraction duration and hence less surface contact between the solvents and sample [19, 20]. In contrast, NaOH extraction produced a significantly higher amount of total carbohydrates.

There was no significant difference in the total reducing sugar content of oligosaccharides extracted from coconut husks using distilled water regardless of the extraction conditions. Nonetheless, when NaOH was used for extraction, a longer extraction time resulted in a higher TRS. The prolonged extraction time results in increased cavitation of microbubbles. The formation of bubbles increases the contact surface area and scours the surfaces of the sample [21].

Table 1. Comparison of two solvents used on extraction of oligosaccharide from coconut husk.

Chemical properties	Type of solvent	Extraction condition			
		5 min, 25 W	**5 min, 250 W**	**45 min, 25 W**	**45 min, 250 W**
Total carbohydrates (TC)	dH_2O	$0.38 \pm 0.35^{ab, A}$	$0.06 \pm 0.03^{a, A}$	$0.17 \pm 0.06^{ab, A}$	$0.72 \pm 0.52^{b, A}$
	NaOH	$120.29 \pm 1.20^{a, B}$	$110.51 \pm 2.99^{a, B}$	$120.07 \pm 20.75^{a, B}$	$108.19 \pm 5.59^{a, B}$
Total reducing sugar (TRS)	dH_2O	$0.005 \pm 0.01^{a, C}$	$0.178 \pm 0.23^{a, C}$	$0.016 \pm 0.02^{a, C}$	$0.076 \pm 0.11^{a, C}$
	NaOH	$22.18 \pm 1.80^{b, D}$	$21.62 \pm 0.06^{ab, D}$	$19.68 \pm 1.52^{a, D}$	$48.17 \pm 0.20^{c, D}$
Degree of polymerisation (DP)	dH_2O	76.0	0.34	10.63	9.47
	NaOH	5.42	5.11	6.10	2.25
% Recovery	dH_2O	$0.555 \pm 0.51^{a, E}$	$0.066 \pm 0.01^{a, E}$	$0.114 \pm 0.01^{a, E}$	$0.435 \pm 0.22^{a, E}$
	NaOH	$40.49 \pm 0.24^{a, F}$	$40.51 \pm 6.00^{a, F}$	$40.37 \pm 5.13^{a, F}$	$34.95\ 16.82^{a, F}$

[a] All the values (except for degree of polymerisation) are expressed as mean ± SD of replicates for the same group of chemical property.
[b] Values within the same rows for same type of solvent followed by different small letters are significantly different ($p < 0.05$).
[c] Values within the same columns for same chemical properties followed by different capital letters are significantly different ($p < 0.05$).

As shown in table 1, for NaOH extraction, a longer extraction time (45 min) combined with a 250W ultrasonicator power resulted in the highest concentration of TRS. As indicated by the lower level of TRS, the lower ultrasonicator power (25 W) may not be strong enough to degrade the carbohydrate derivatives in the sample to a simpler form of sugar.

After the total carbohydrates and total reducing sugar were determined, the degree of polymerisation (DP) was calculated using equation (1). In order to obtain high DP value, high total carbohydrates content and low total reducing sugar content are desirable [22].

3.2. In vitro Fermentation Study

For *in vitro* fermentation study, coconut husk extract (CHE) was used as the sole carbon source for cultivation of selected beneficial bacterial strain, namely *Lactobacillus acidophilus* DSM 20079, *Lactobacillus casei* DSM 20011 and *Bifidobacterium bifidum* DSM 20239, and one available pathogenic strain in the laboratory, i.e., *Escherichia coli* DSM 1103. Inulin was used as the positive control in this study. The colonies forming unit for each strain was determined and expressed as $\log_{10}$ CFU/mL. In addition, pH of the cultivation broth collected at specific time interval was measured. The broth was then centrifuged prior to total carbohydrates and short chain fatty acid (SCFA) determination. The profile of microbes' growth, pH changes and the total carbohydrates consumption were determined. The growth rate (CFU per hour of incubation) and the growth yield (CFU per g of substrate) were calculated.

Figure 1 shows the growth rate (a) and growth yield (b) of respective strains in medium containing different concentrations of CHE (0.5% w/v to 3.0% w/v), in comparison with inulin as sole carbon sources. Through the concentration-dependent analysis, the effect of CHE on the probiotic growth can be evaluated. From the results, all the tested bacteria could utilise CHE. The proliferation effect decreased in the order from *B.bifidum*, followed by *E. coli*, *L. casei* and then *L. acidophilus* in media containing 1.0% w/v of CHE.

3.2.1. L. acidophilus. As shown in the figure 1 (a), the average growth rate of *L. acidophilus* in media using 2.0% w/v and 3.0% w/v of CHE are significantly higher as compared to the 1.0% w/v inulin. Similar growth rate was observed in the media with the substrate concentrations of 0.5%, 1.0% and 1.5% (w/v) CHE. In contrast, there is no significant difference ($p > 0.05$) in the growth rate of *L. casei*, B. bifidum and E. coli cultivated with 1.0% w/v CHE and 1.0% w/v inulin although the growth rate of the three strains in both medium are higher than medium with 0.5%, 1.5%, 2.0% and 3.0% w/v CHE.

Moreover, 1.0% w/v CHE shows mean growth yield of 25.43 ± 3.58 log CFU per mg of substrate on L. acidophilus (figure 1 (b)). The medium with 1.0% w/v inulin gives the lowest growth yield of 3.96 ± 0.91 log CFU per mg of substrate among the other substrates used.

3.2.2. L. casei. In medium with 1% w/v of respective substrate, the average growth rate of *L. casei* was higher compared to other substrates while similar growth rate observed when *L.casei* was grown in medium with 0.5%, 1.5%, 2.0% and 3.0% (w/v) of CHE. In addition, the effect of 1.0% w/v CHE was positive as it stimulates the growth yield of *L. casei* of 20.17 ± 0.97 log CFU per mg of substrate. On the other hand, the cell density of *L. casei* grown in media with 1.0% w/v inulin was the lowest with growth yield of 5.25 ± 1.03 log CFU per mg of substrate.

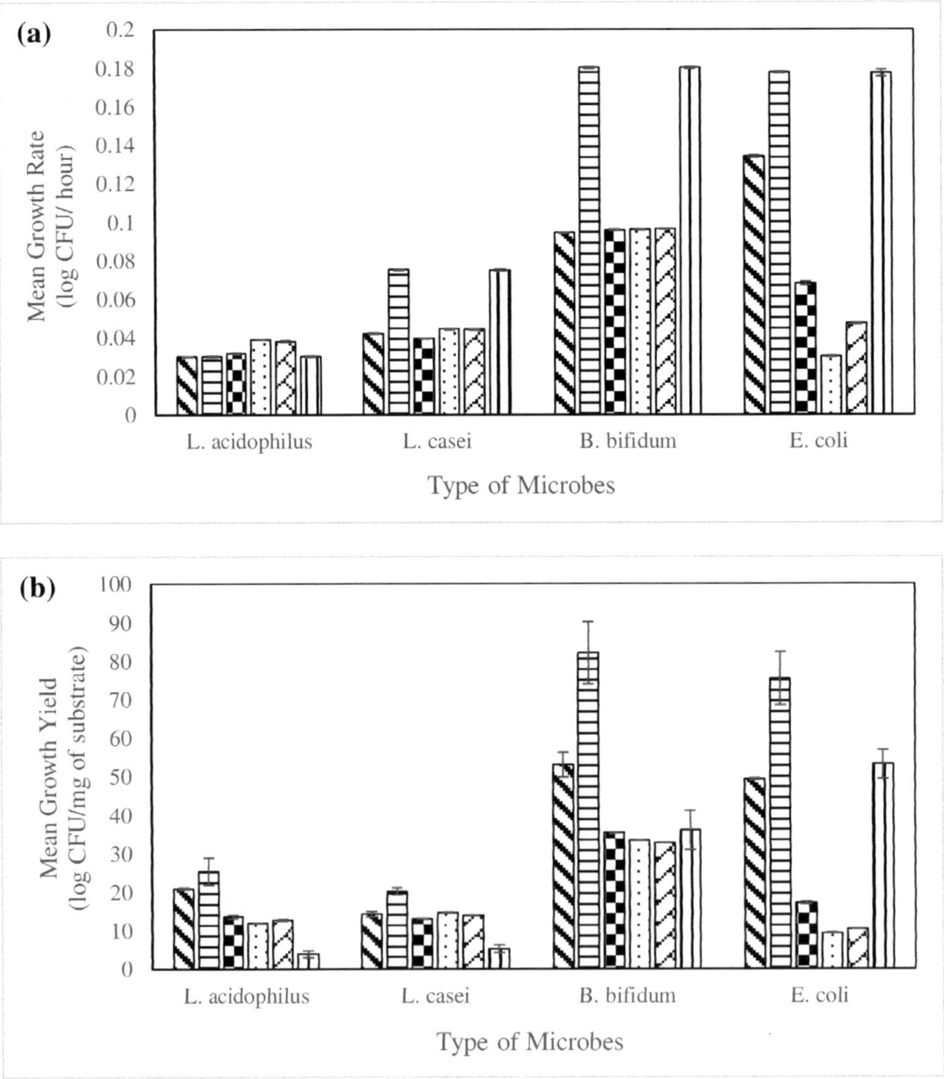

Figure 1. (a) Growth rate and (b) Growth yield of *L. acidophilus, L. casei, B. bifidum* and *E. coli* in medium with different concentration of CHE (% w/v). Symbols: (▧), 0.5%; (▤), 1.0%; (▩), 1.5%; (⬚), 2.0%; (▨), 3.0%; (▥) 1.0% inulin. Error bars indicate the mean ± standard deviation of three replicates. The error bars were smaller than the size of the symbols as regards to the data points without error bars.

3.2.3. B. bifidum. The mean growth rate of *B. bifidum* is high in medium with 1.0% w/v CHE and the growth rate was similar in medium with 1.0% w/v inulin, with growth yield of 94.10 ± 43.97 log CFU per mg of substrate. The growth yield of *B. bifidum* reduced when higher concentration of CHE (above 1.0% w/v) was used as carbon source in the medium. Overall, CHE possess noticeable proliferation effect of the beneficial microorganism with significant growth stimulation effect on *B. bifidum*, followed by *L. casei* and *L. acidophilus*, in comparison with inulin. Similar findings were reported by Lee *et al.* (2002) [23] who were using chitosan oligosaccharides (COS) as the carbon source on the growth of selected strains. Another study using oligosaccharides from jackfruit seed also have been reported to show larger growth stimulatory effect on *lactobacilli* sp. and *bifidobacterium* sp. [24].

3.2.4. E. coli. In addition to beneficial microbe strains mentioned above, the growth profile of a pathogenic microbe – *E. coli*, was also conducted. As shown in figure 1 (a), by using 1.5 % w/v CHE, higher growth rate of E. coli was achieved with the highest average growth yield of 75.48 ± 6.97 log CFU per mg of substrate. According to Fooks and Gibson (2002) [25], the enteropathogenic bacteria, such as *E. coli*, *C. jejuni* and *S. enteriditis*, were able to utilise the different prebiotic as growth substrates, namely FOS, inulin, inulin:FOS mixture (80:20, w/w), XOS, FOS:XOS mixture (50:50, w/w), lactulose, lactitol, starch and dextran, to support their growth.

Of the four species compared with media containing 1.0% w/v CHE, *B. bifidum* grew the fastest with the lowest pH value of fermentation medium recorded. A reduction in pH value indicates the substrates being metabolized and produced organic acid by the strains [11]. Besides, the differences of the growth pattern might due to the chemical composition and molecular conformation of the coconut husk extract [26].

For example, the sugar content of the oligosaccharides could affect the proliferation effect [27]. The proliferation effect of inulin was more significant than CHE probably because of the higher sugar content. Molecular weight is one of the factors that have an effect on the accessibility of oligosaccharides to probiotic strains [27]. Furthermore, the branch degree of the oligosaccharides might influence the prebiotic activity too. Chain length is likely to be one of the contributory factors, since with long chain and multiple branching, more hydrolysis by the organisms are required for the fermentation [25]. According to Stewart, Timm and Slavin (2008) [28], the branched FOS have stronger prebiotic effect than unbranched FOS. Last but not least, water solubility is another important factor. The oligosaccharides with good water solubility could more easily and completely be utilised by the microorganism [29]. Aside from that, there might be other mechanisms that need to be studied and disclosed in the future.

3.2.5. Time Course of Total Short-chain Fatty Acid (SCFA) Production. Figure 2 shows the time course profile of the total amounts of short chain fatty acid (SCFA) produced by selected strains in medium with different concentration of CHE The type of SCFA detected are lactic acid, acetic acid, propionic acid and butyric acid. For *L. acidophilus* strain, the total SCFA produced by using 1.0% w/v CHE and 1.0% w/v inulin are 11.16 ± 0.06 g/L and 11.42 ± 0.13 g/L, respectively after 72 h of cultivation. The results indicate that CHE could produce comparable amount of SCFA as inulin when utilised as carbon source during the fermentation process.

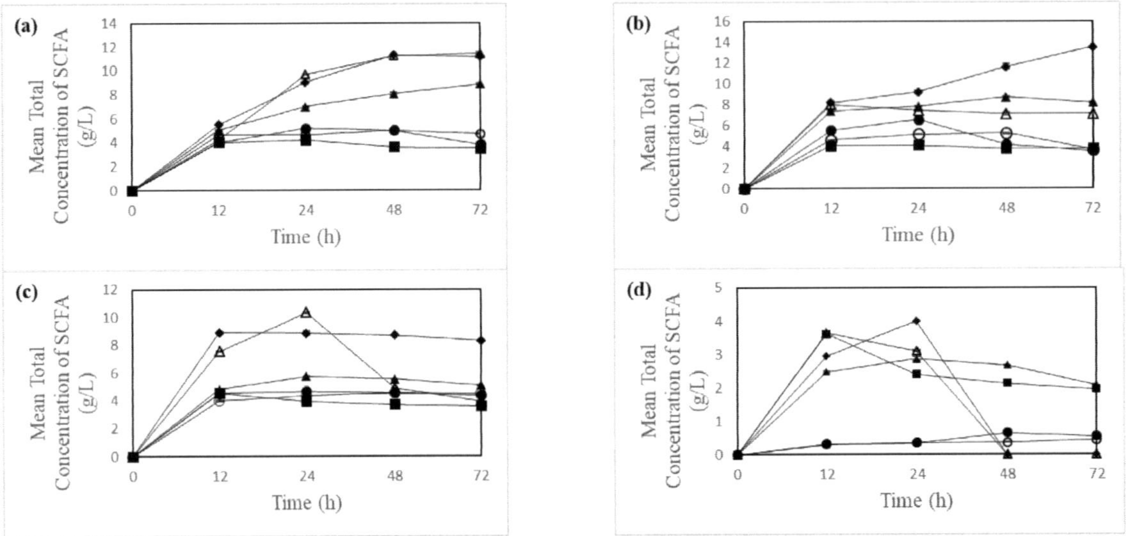

Figure 2. Time course of total short-chain fatty acid (SCFA) production by (a) *L. acidophilus*, (b) *L. casei*, (c) *B. bifidum* and (d) *E. coli* in different concentration of CHE (% w/v). Symbols: (■), 0.5%; (♦), 1.0%; (▲), 1.5%; (●), 2.0%; (○), 3.0%; (Δ) 1.0% inulin. Error bars indicate the mean ± standard deviation of three replicates. The error bars were smaller than the size of the symbols as regards to the data points without error bars.

For *L. casei* strain, 13.50 ± 0.03 g/L of total SCFA was produced when using 1.0% w/v CHE as carbon sources for 72 h of fermentation. In contrast, significantly low level of total SCFA was produced (3.81 ± 0.001 g/L) in medium with 0.5% w/v CHE. When *B. bifidum* was cultivated in 1.0 % w/v CHE, there was 8.97 ± 0.01 g/L of total SCFA produced after 24 h of fermentation. There was a notably decrease of the total SCFA observed after 24 h in medium with 1.0% w/v inulin.

The results of present study indicated that the amount of 1.0% w/v CHE is sufficient for the cultivation of L. *acidophilus, L. casei* and *B. bifidum* strains for SCFA production. The strains utilise the carbon sources to support the growth throughout the cultivation period. The fermentation of the carbon substrate resulted the production of SCFA as the product throughout the fermentation process [30]. There is no SCFA production observed after 24 h of fermentation with 0.5% and 1.0% w/v of CHE on *E. coli* (figure 2 (d)). This might reflect the substrate used do not support the growth of *E. coli* for more than 24 hours. Without the growth of the microbes, no production of total SCFA could be detected. Most of the SCFA produced was lactic acid followed by acetic acid. The anaerobic growth condition was reported to force the microbes in utilising the anaerobic pathway to produce the lactic acid [31].

3.3. *In vitro Enzymatic Digestion*

The enzyme used in this study was human salivary α-amylase. The ability of coconut husk extract (CHE) and inulin to resist enzymatic digestion was determined based on the percentage of hydrolysis (equation (2)). Figure 3 (a) and (b) shows the percentage of hydrolysis of coconut husk extract (CHE) and inulin after enzymatic digestion by salivary α- amylase at pH 6, pH 7 and pH 8.

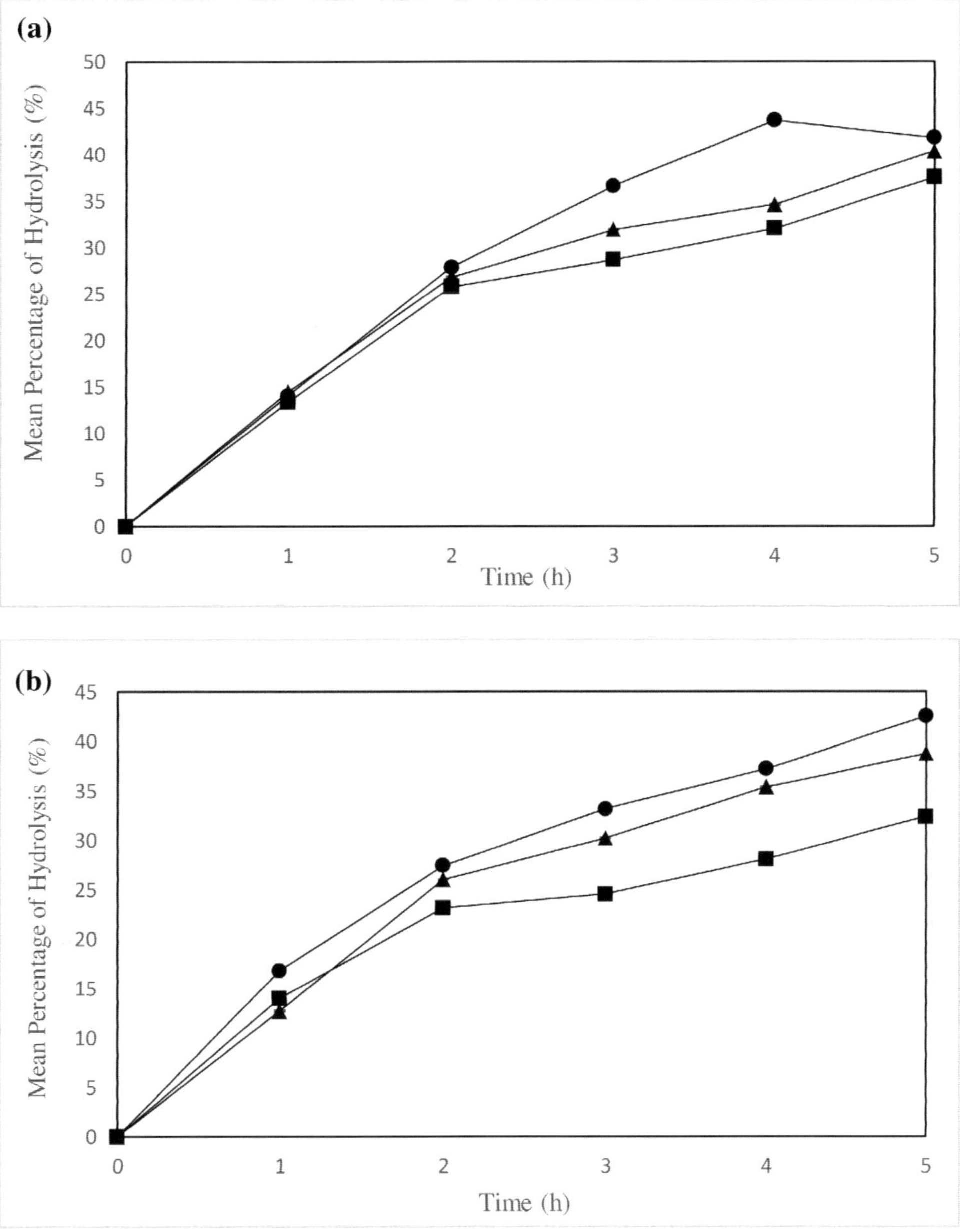

Figure 3. Percentage of hydrolysis (a) coconut husk extract (CHE) and (b) inulin after enzymatic digestion at various pH, incubated at 37 °C for five hours. Symbols: (■), pH 6; (▲), pH 7; (●), pH 8. Error bars indicate the mean ± standard deviation of three replicates. The error bars were smaller than the size of the symbols as regards to the data points without error bars.

The hydrolysis of coconut husk extract increased significantly within first few hours. The maximum percentage of hydrolysis of coconut husk extract with human salivary α-amylase at pH 6, pH 7 and pH 8 was 37.56 ± 0.04%, 40.37 ± 0.04% and 43.70 ± 0.15%, respectively after 5 h of incubation. Meanwhile, the inulin hydrolysis is mainly occurred at pH 6, 7 and 8 with 32.36 ± 0.03%,

38.73 ± 0.03% and 42.49 ± 0.03% recorded, respectively. Based on the findings, alkaline pH conditions gave a significantly higher degree of hydrolysis, indicating the optimal pH for α-amylase activity was in the range of 5.5 to 7.5 [32]. Wang *et al.* (2015) [27] also reported that the digestibility of rapeseed polysaccharides at different pH values were in the order of pH 7> pH 8 > pH 6 > pH 5.

By comparison of the coconut husk extract and inulin, the percentage of hydrolysis of coconut husk extract has significantly differences ($p < 0.05$). From the present study, approximately 30 to 40% of coconut husk extract was expected to reach the colon and utilised by the probiotic bacteria.

3.4. *In vitro Artificial Gastric Juice Digestion*

The coconut husk extract (CHE) was subjected to *in vitro* artificial gastric juice digestion. After five hours of incubation, the content of fermentable sugar in the solution was determined. Throughout the gastric juice digestion test, the reducing sugar content increased after digestion due to hydrolysis of the glycosidic bonds of CHE that would increase the amount of reducing sugars. The increase of the total reducing sugar then further contributed to the increase of the percentage of hydrolysis.

Figure 4 (a) and (b) show the percentage of hydrolysis of coconut husk extract and inulin after gastric juice digestion at pH 1 to 4 for five hours, respectively. The percentage of hydrolysis was calculated using equation (2). The solution was collected at every hour interval immediate after the incubation with artificial gastric juice. The percentage of hydrolysis of coconut husk extract (CHE) increased when highly acidic artificial gastric juice was used. The degree of hydrolysis at pH of 1, 2, 3 and 4 are 24.00 ± 0.70%, 14.90 ± 0.65%, 16.79 ± 0.11% and 9.98 ± 0.52%, respectively after two hours of incubation.

The gastric juice with pH 2 to pH 4 was generally released within 2 h after the consumption of food [17]. At pH 1, the maximum hydrolysis of coconut husk extract at 29.21 ± 0.71% was attained after 4 h of incubation. The maximum hydrolysis of CHE at pH 2 and pH 3 is 19.26 ± 0.55% and 18.18 ± 1.45%, respectively, after 5 h of incubation. From the present study, the coconut husk extract was found to be a potential prebiotic candidate as it shows resistant towards the artificial human gastric juice. There are some other potential candidates that having the similar property are reported in the literature, such as, bamboo shoot crude polysaccharides with 0.41 ± 0.05 to 0.66 ± 0.10% hydrolysis for 6 h of incubation [33], kojiooligosaccharides with 0% hydrolysis for 6 h of incubation, which had 100% resistance to artificial gastric acid [34] or gluco-oligosaccharides with 1.6% hydrolysis for 6 h of incubation produced by *Gluconobacter oxydans* NCIMB 4943 [35].

On the other hand, the inulin hydrolysis of 12.79 ± 1.15%, 1.52 ± 0.03%, 0.91 ± 0.16% and 2.08 ± 0.13% attained at pH of 1, 2, 3 and 4 respectively after two hours of incubation. Although the digestibility of CHE was higher as compared to inulin, there was 76% of the extract was expected to reach the intestine whereas 87.21% of inulin was expected to reach the intestine.

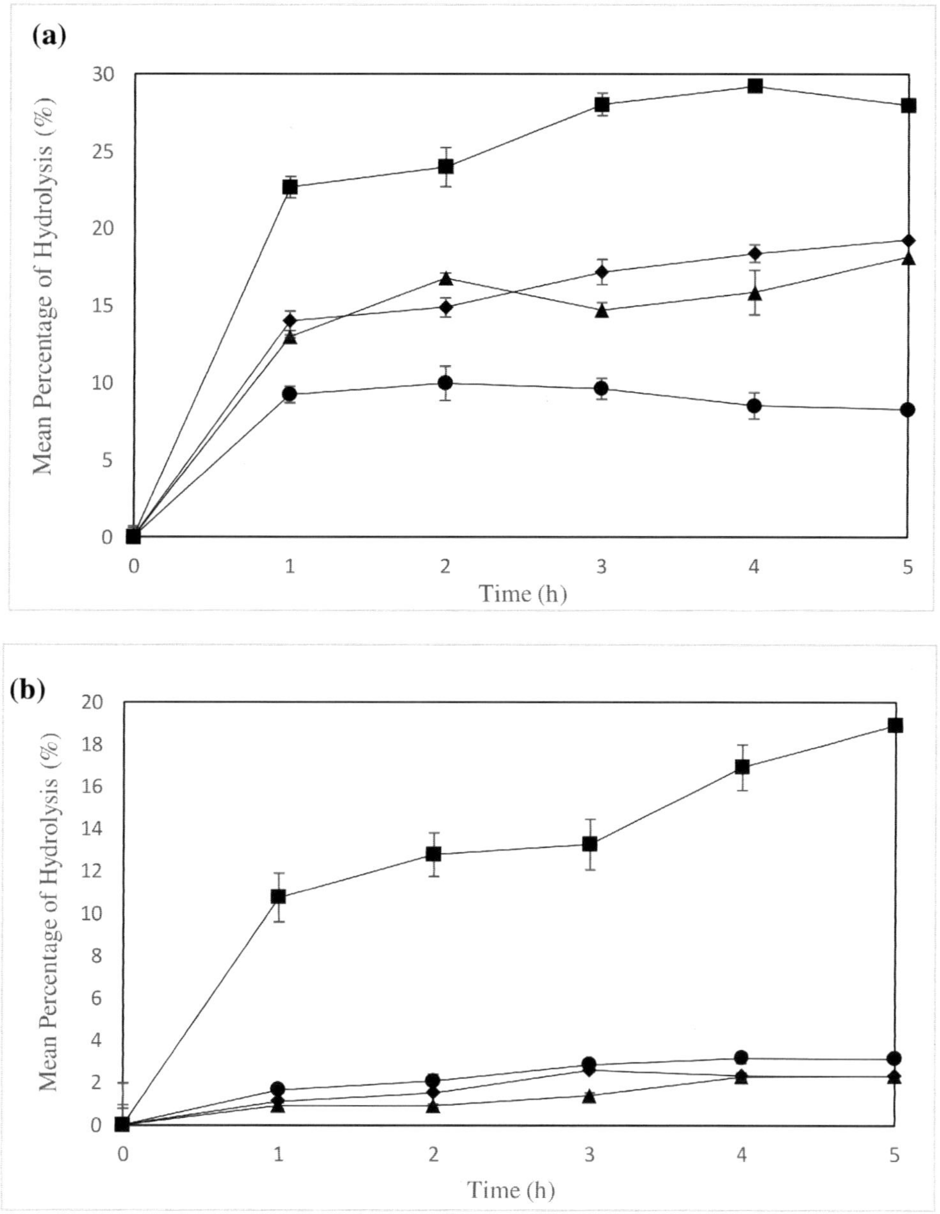

Figure 4. Percentage of hydrolysis (a) coconut husk extract (CHE) and (b) inulin after gastric juice digestion at various pH, Incubated at 37 °C for five hours. Symbols: (■), pH 1; (♦), pH 2; (▲), pH 3; (●) pH 4. Error bars indicate the mean ± standard deviation of three replicates. The error bars were smaller than the size of the symbols as regards to the data points without error bars.

3.5. Fourier Transform Infrared (FTIR) Spectroscopy

Fourier transform infrared (FTIR) spectra of the coconut husk extract (CHE) was performed to identify the main functional groups of extract. Table 2 lists the general assignments of FTIR spectra. The FTIR spectra between 4000 and 650 cm^{-1} are obtained for the analysis of organic functional group in both coconut husk extract (figure 5(a)) and inulin (figure 5 (b)). From the analysis, the FTIR spectrum of coconut husk extract is similar to the FTIR spectrum of inulin, indicating the similar chemical composition of both compounds.

Table 2. General assignment of FTIR spectra of coconut husk extract (CHE) [36].

Wavenumber, cm^{-1}	Experimental FTIR bands, cm^{-1}	Assignment
3200–3400	3278.47	vO-H (OH) H bond
1664–1634	1634.24	Absorption of water
1125–1162	1160.16	vC-C_{ss} (C-O-C) glycosidic bonds

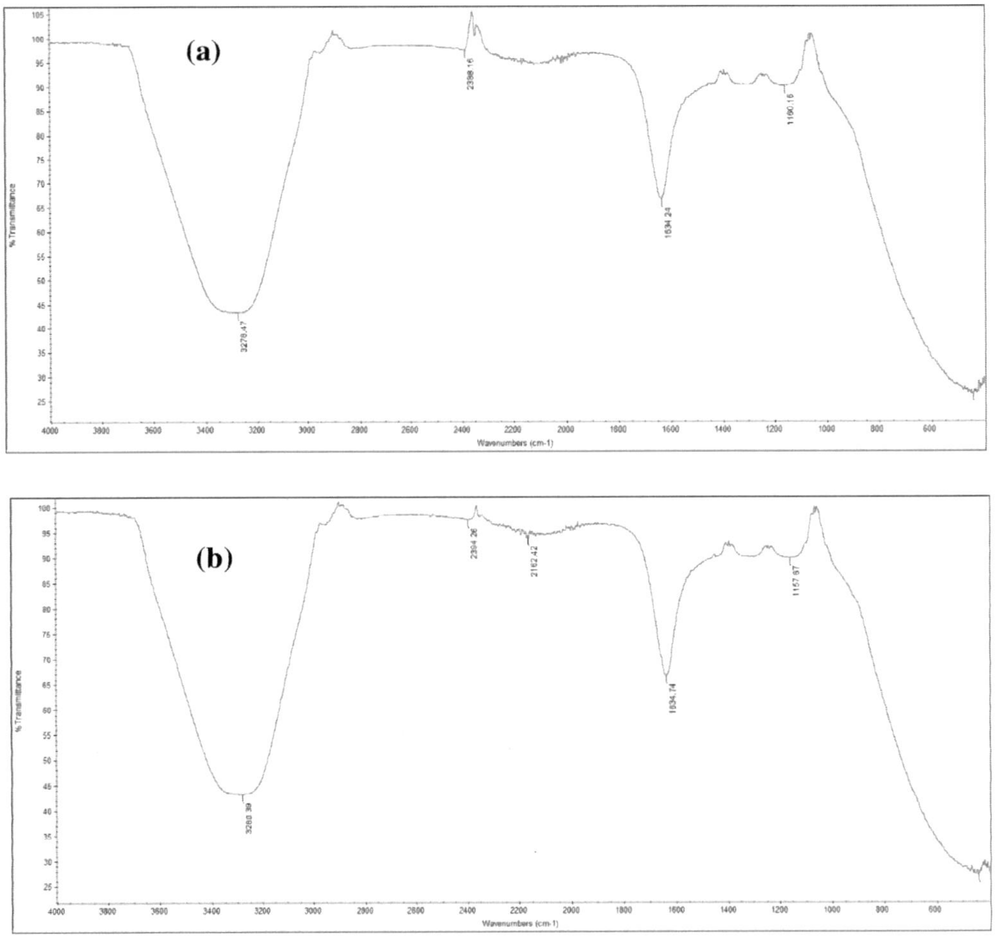

Figure 5. FTIR spectrum of (a) coconut husk extract (CHE) and (b) inulin.

According to figure 5, there are several dominant bands can be observed, such as ~3200 – 3400, ~2200–2400, ~1550–1650 and ~1000–1200 cm^{-1}. A peak with strong and broad band at the absorption between 3200–3400 cm^{-1} is corresponded to the hydroxyl (O-H) bond [33]. Meanwhile, there is another peak with absorption around 1630 cm^{-1} corresponded to the carbonyl group (C=O) bond stretching [37]. This band was assigned with the absorption of water by the substrate [36]. Another peak at around 2400 cm^{-1} indicates C-H stretching of CH_2 group. The band at 1160 cm^{-1} shows the present of C-O-C ring stretching vibration [36].

Overall, the FTIR spectra indicates the coconut husk extract contain hydroxyl group (O-H) group and carbonyl (C=O) group, that are similar to the FTIR spectrum of inulin. The results of present study indicated that both coconut husk extract and inulin contain similar functional groups.

3.6. Nuclear Molecular Resonance (NMR) Spectroscopy

Nuclear molecular resonance (NMR) spectroscopy was conducted to elucidate the structure of coconut husk extract. Figure 6 (a) and (b) show the ^{1}H-NMR spectra of the coconut husk extract and inulin, respectively.

The ^{1}H NMR spectrum of the coconut husk extract (CHE) shows the present of signal in the anomeric region at 5.2 ppm. Meanwhile, the broad peak signal at 3.2 to 4.2 ppm is probably due to the presence of other compounds. On the other hand, the ^{1}H-NMR spectrum of the inulin showed the presence in the anomeric region of two main signals at 5.2 ppm and 5.4 ppm. The signals at 5.2 to 5.4 ppm might due to the presence of vinylic proton [38, 39]. Besides, intense signals were observed at 3.2 to 4.2 ppm. These signals confirmed the identification of the α-Glu unit of free glucose [40].

There are six main signals at 74.04, 94.125, 115.336, 164.299, 165.299 and 189.793 ppm shown in the ^{13}C-NMR spectrum of the CHE (figure 7 (a)). The chemical shift at 74.04 and 94.125 ppm designate the presence of RCH_2O group. Meanwhile, C=C bond is found to be presence in CHE as there is a signal observed at 115.336 ppm. Furthermore, the signals at 164.299 ppm onwards show the presence of C=O bond in CHE.

On the other hand, the ^{13}C-NMR spectrum of the inulin in this study was compared to the frutooligosaccharides from escarole in the literature [41]. There are two major regions observed for ^{13}C-NMR spectrum of inulin – low frequency region at 60.0 to 78.0 ppm and a region between 80.0 ppm and 105.0 ppm, which were believed to be two carbon signals of anomeric monosaccharides [42].

The characteristic signal at 103.57 ppm corresponds to the carbon involved in β-(2→1)-D-fructose intra-chain binding. The signals at 60.875 ppm and 62.18 ppm were assigned to the methylene group (CH_2). Another two signals 76.563 ppm and 75.6 ppm might attribute to the signs of the methinic group (CH) [43, 44].

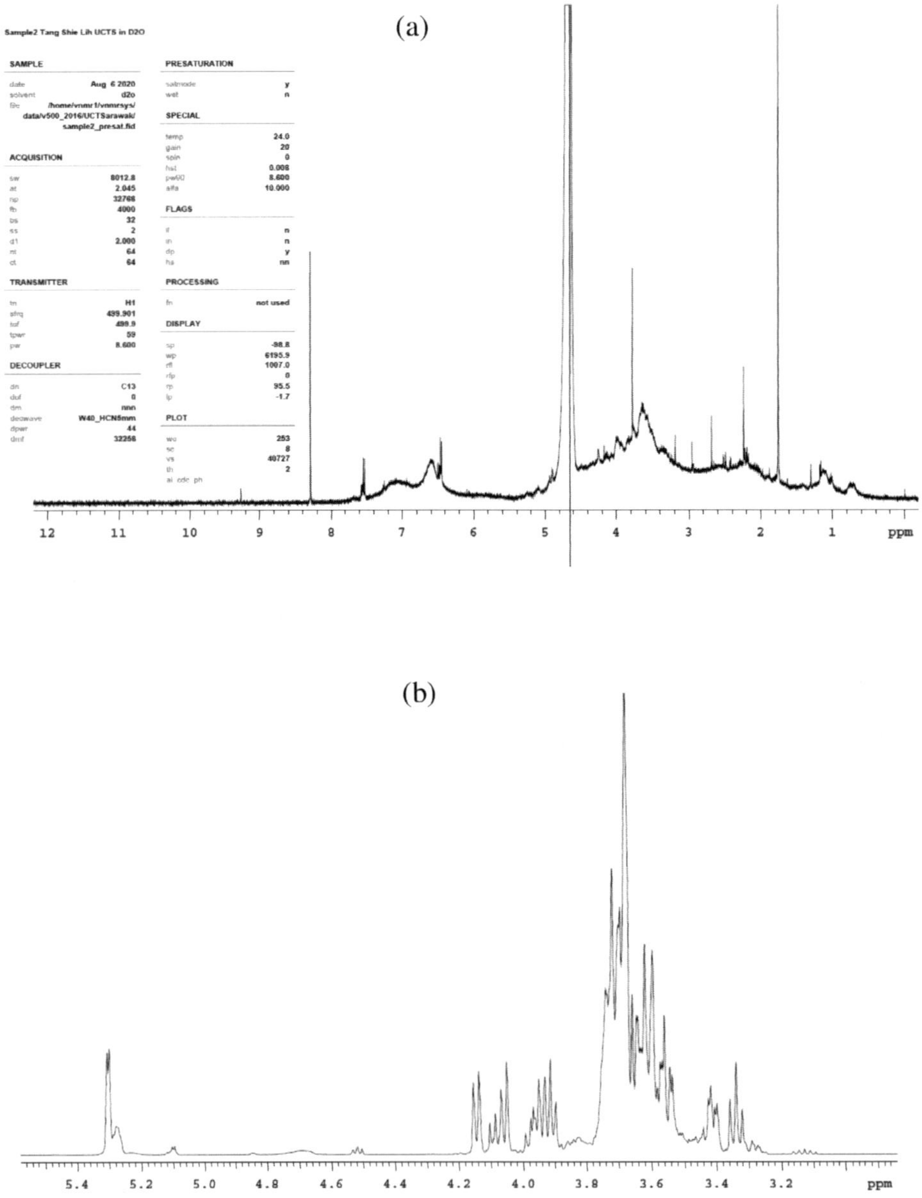

Figure 6. (a) ^{1}H-NMR spectrum of coconut husk extract; (b) ^{1}H-NMR spectrum of inulin.

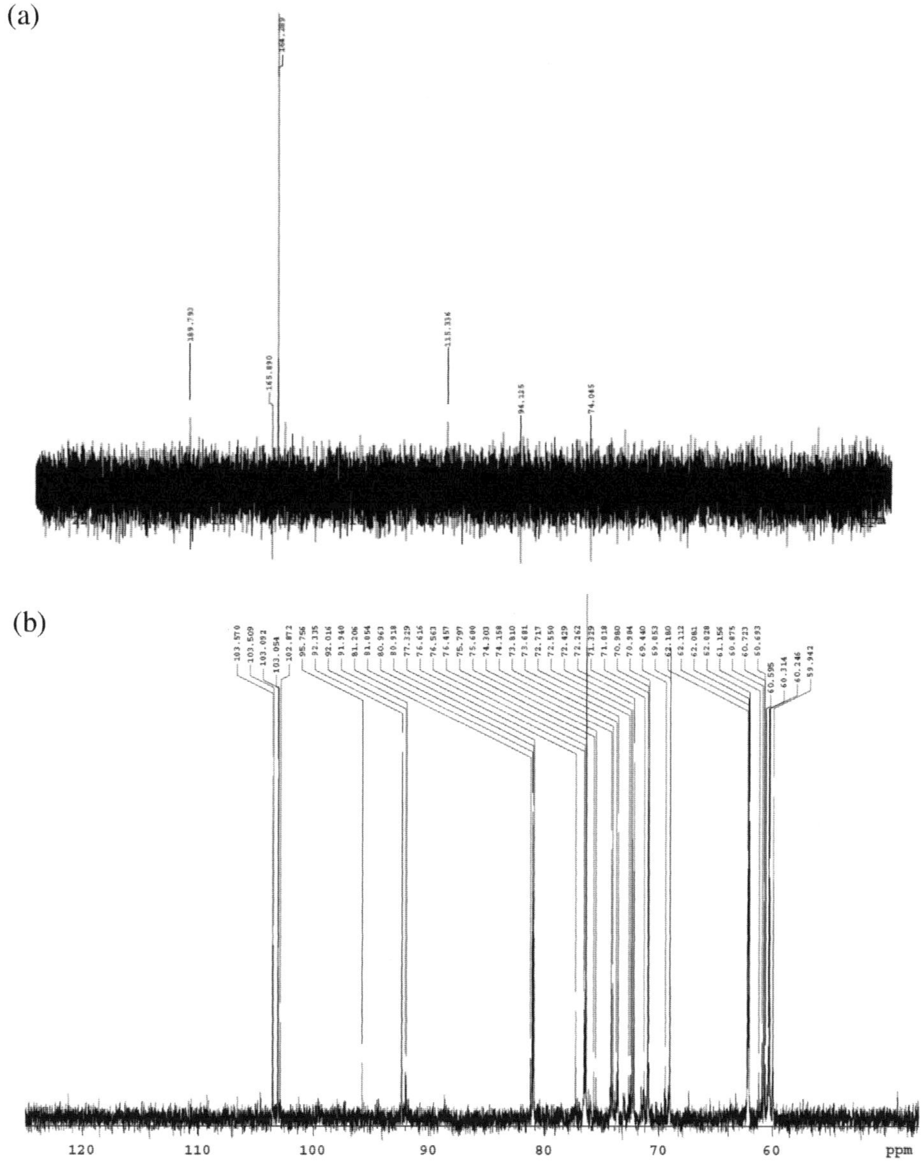

Figure 7. (a) ^{13}C-NMR spectrum of coconut husk extract; (b) ^{13}C-NMR spectrum of inulin.

4. Conclusion

The coconut husk extract (CHE) was subjected to several prebiotic potency tests in this study. The study's findings suggest that the CHE promotes the growth of probiotic bacteria such as *L. acidophilus, L. casei*, and *B. bifidum*. The extract is resistant to *in vitro* enzymatic digestion with a maximum percentage of hydrolysis of 43.70 ± 0.15% at pH 8 after five hours of incubation. The maximum percentage of hydrolysis for *in vitro* gastric juice digestion is 29.21 ± 0.71% at pH 1 after four hours of incubation. These findings suggest that the extract is partially digestible. FTIR spectra indicate that the functional groups, such as the hydroxyl group (O-H) group, carbonyl (C=O) group and C-O-C ring, are available in the CHE. Meanwhile, NMR spectra reveal that CHE has a similar structure to the inulin-type oligosaccharides.

In a nutshell, CHE fulfils the key criteria for being considered as a potential prebiotic. It is resistant to gastric juice and enzymatic digestion, and it promotes the growth of the selected probiotic bacteria. CHE also shows a similar structure to the commercially available inulin-type oligosaccharides. Additionally, the results obtained using CHE are comparable to those obtained using commercially available prebiotics inulin. Thus, CHE may be classified as a prebiotic with the potential to be used as a functional food ingredient.

Acknowledgement
The authors would like to acknowledge the University College of Technology Sarawak (UCTS/RESEARCH/1/2019/03; UCTS/RESEARCH/2/2018/10) for research funding and technical support.

References

[1] Jnawali P, Kumar V, Tanwar B, Hirdyani H and Gupta P 2018 *Waste Biomass Valor.* **9** 1757–66
[2] Sivapragasam A 2008 *Proc. of 2nd Int. Plantation Industry Conf. and Exhibition (IPICEX2008)* (Shah Alam: MARDI) pp 1–9
[3] Vaithanomsat P, Apiwatanapiwat W, Chumchuent N, Kongtud W and Sundhrarajun S 2011 *Kasetsart J. Nat. Sci.* **45** 159–64
[4] Pollard S J T, Fowler G D, Sollars C J and Perry R 1992 *Sci. Total Environ.* **116** 31–52
[5] Gibson G R and Roberfroid M B 1995 *J. Nutr.* **125** 1401–12
[6] Gibson G R, Probert H M, Loo J V, Rastall R A and Roberfroid M B 2004 *Nutr. Res. Rev.* **17** 259–75
[7] Chemat F, Zill-e-Huma and Khan M K 2011 *Ultrason. Sonochem.* **18** 813–35
[8] Jing S, RunCang S, Xiao S and Yin Quan S 2004 *Carbohydr. Res.* **339** 291–300
[9] Wichienchot S, Thammarutwasik P, Jongjareonrak A, Chansuwan W, Hmadhlu P, Hongpattarakere T, Itharat A and Ooraikul B 2011 *Songklanakarin J. Sci. Technol.* **33** 517–23
[10] Alcántara C, Žugčić T, Abdelkebir R, Garcia-Pérez J V, Jambrak A R, Lorenzo J M, Collado M C, Granato D and Barba F J 2020 *Molecules* **25** 1–1
[11] DuBois M, Gilles K A, Hamilton J K, Rebers P A and Smith F 1956 *Anal. Chem.* **28** 350–56
[12] Miller G L 1959 *Anal. Chem.* **31** 426–28
[13] Tan Z N, Rosma A, Napisah H, Karim A A and Liong M T 2015 *J. Food Sci.* **80** 875–82
[14] Cherchi A, Spanedda L, Tuberoso C and Cabras P 1994 *J. Chromatogr. A*, **669**, 59–64
[15] Fässler C, Arrigoni E, Venema K, Hafner V, Brouns F and Amadò R 2006 *Eur.J.Nutr.* **45** 445–53
[16] Korakli M, Gänzle M G and Vogel R F 2002 *J. Appl. Microbiol.* **92** 958–65
[17] Wichienchot S, Jatupornpipat M and Rastall R A 2010 *Food Chem* **120** 850–57
[18] Kaur R, Uppal S K and Sharma P 2018 *Waste Biomass Valor.* **10** 2627–35
[19] Handa S S, Khanuja S P S, Longo G and Rakesh D D 2008 Extraction Technologies for Medicinal and Aromatic Plants *Italy: United Nations Industrial Development Organization and the International Centre for Science and High Technology*
[20] Dhanani T, Shah S, Gajbhiye N A and Kumar S 2017 *Arab. J. Chem.* **10** S1193–99
[21] Vilkhu K, Mawson R, Simons L and Bates D 2008. *Innov Food Sci.Emerg Technol.* **9** 161–69
[22] Astó E, Méndez I, Rodríguez-Prado M, Cuñé J, Espadaler J and Farran-Codina A 2019 *Nutrients* **11** 1–14
[23] Lee H W, Park Y S, Jung J S and Shin W S 2002 *Food Microbiol.* **8** 319–24
[24] Thammarutwasik P, Hongpattarakere T, Chantachum S, Kijroongrojana K, Itharat A, Reanmongkol W, Tewtrakul S and Ooraikul B 2009 *Songklanakarin J. Sci. Technol.* **31** 401–08
[25] Fooks L J and Gibson G R 2002 *FEMS Microbiol. Ecol.* **39** 67–75

[26] Yang Y, Zhao C H, Diao M X, Zhong S N Sun, M C, Sun B, Ye H Q and Zhang T H 2018 *Molecules* **23** 1–14
[27] Wang X, Huang M Y, Yang H F, Sun H J, Zhou X X, Guo Y, Wang X L and Zhang M L 2015 *Carbohydr. Polym.* **125** 232–40
[28] Stewart M L, Timm D A and Slavin J L 2008 *Nutr. Res.* **28** 329–34
[29] Montagne L, Pluske J R and Hampson D J 2003 *Anim. Feed Sci. Technol.* **108** 95–117
[30] Usta-Gorgun B and Yilmaz-Ersan L 2020 *Electron. J. Biotechnol.* **47** 29–35
[31] Macfarlane S and Macfarlane G T 2003 *The 7th International Vahouny Fibre Symp* Royal College of Physicians (Edinburgh. Nutrition Society) pp 67–72
[32] Yadav J K and Prakash V 2011 *Int. J. Food Prop.* **14** 1182–96
[33] Aida F M N A, Shuhaimi M, Dzulkifly M H and Yazid A M 2012 *Molecules* **17** 1635–51
[34] Nakada T, Nishimoto T, Chaen H and Fukuda S 2003 *Oligosaccharides in Food and Agriculture*, ed G Eggleston and G L Côté (Washington, DC: American Chemical Society) pp 104–17
[35] Wichienchot S, Prasertsan S, Hongpattarakere P, Gibson G R and Rastall R A 2006 *Curr Issues Intest Microbiol.* **7** 7–12
[36] Petkova N T, Sherova G and Denev P P 2018 *Int. Food Res. J.* **25** 1876–84
[37] Nordin N A, Sulaiman O, Hashim R and Kassim M H A 2017 *J. Phys. Sci.* **28** 115–26
[38] Starkey L S n.d. ^{1}H NMR Chemical Shift ppm Table *California State Polytechnic University, Pomona*
[39] Mathew S and Abraham E 2005 *Enzyme Microb. Technol.* **36** 565–70
[40] Cérantola S, Kervarec N, Pichon R, Magné C, Bessieres MA and Deslandes E 2004 *Carbohydr. Res.* **339** 2445–49
[41] Mariano T B, Higashi B, Lopes S M S, Carneiro J W P, Almeida R T R, Pilau E J, Gonçalves J E, Gonçalves R A C and Oliveira A J B 2020 *Bioact. Carbohydr. Diet. Fibre.* **24** 100233
[42] Yang Z, Hu J and Zhao M 2011 *Carbohydr. Polym.* **83** 1997–2004
[43] Zhang A Q, Zhang Y, Yang J, Jiang J, Huang F F and Sun P L 2013 *Carbohydr. Res.* **376** 24–28
[44] Lopes S M S, Krausová G, Rada V, Gonçalves J E, Gonçalves R A C and De Oliveira A J B 2015 *Carbohydr. Res* **411** 15–21

IOP Conf. Series: Materials Science and Engineering 1195 (2021) 012028 doi:10.1088/1757-899X/1195/1/012028

Extraction of magnesium from salt pond waste with low lithium grade using sodium silicate reagent

E Sulistiyono[1], S Harjanto[1] and L H Lalasari[2]

[1] Departments of Metallurgical and Materials Engineering, Faculty of Engineering, Universitas Indonesia, Depok16424, Indonesia
[2] Research Center for Metallurgy and Material, Indonesian Institute of Science, Gedung 470, Puspiptek Serpong 15312, Tangerang Selatan, Indonesia

Email: sri.harjanto@ui.ac.id

Abstract. Nowadays, Indonesia's salt production process has shifted from a traditional salt pond process to a salt pond with a geomembrane tarpaulin system. This system has the advantage of producing a faster salting process and a better salt quality and producing a thick yellow liquid waste known as bittern. The waste produced has not been fully utilized, even though the waste contains valuable elements, namely magnesium, lithium, and others. This paper investigates the potential utilization of bittern by extracting magnesium from by-product salt pond using sodium silicate reagent with various concentrations of 125–250 g/l. The bittern used in this experiment had the chemical composition as follows: 33,184 ppm magnesium (Mg), 23,787 ppm sodium (Na), 6.1 ppm lithium (Li), 5,491 ppm potassium (K), 617 ppm calcium (Ca), 152 ppm boron (B). The highest removal of magnesium ion by addition of sodium silicate of 250 g/l was about 96.2%. The results showed that the solid products were magnesium silicate precipitates without containing asbestos compounds. The magnesium silicate precipitate product can be applied as talc raw material for the pharmaceutical industry.

1. Introduction

Indonesia has a traditional salt-works location spread over 44 districts with the total salt production of 2,915,461.17 ton/year by 2016, which can be an abundant magnesium source [1]. Traditional salt production plants are man-made systems exploiting sea water for salt production by wind and solar evaporation [1]. Nowadays, Indonesia's salt production process has shifted from a traditional salt pond process to a salt pond with a geomembrane tarpaulin system. This system has the advantage of producing a faster salting process and a better salt quality as well as producing a thick yellow liquid waste known as bittern. Integrating the "Pugar" and "MP3EI "policies, encouraging the industrialization of community salt is one of the solutions going forward by preparing Madura people's salt to become salt and salt derivatives of a quality that various industries can accept in the Surabaya, Gresik and Sidoarjo regions [2].

Bittern can be used as magnesium sources for magnesium carbonate hydrate products, mineral supplements for drink/food, struvite, supporting fertilizer and cooling system [1]. Sea-water is the main source for $Mg(OH)_2$ due to containing soluble salts of magnesium such as $MgCl_2$ and $MgSO_4$. The concentration of magnesium ion is about 1,272 ppm in sea water and about 30,000 ppm in the end bitterns of NaCl production units from sea-water [3].

Other potential elements, such as lithium in sea water, could be utilized more optimally if the magnesium element is extracted. As an illustration, the mass of the world's oceans is approximately 1.35 x 10^{12} metric ton, making vast amounts of theoretical resources seemingly available announce that more than 2,000,000 metric ton lithium is available from the seas. It is essentially making it an "unlimited" source given its geological abundance [4]. The average lithium concentration in the oceans has been estimated to 0.17 ppm with an Mg/Li ratio of about 7,000 [4]. With a high Mg/Li ratio, there has been no utilization of lithium from sea water for commercial purposes.

The separation process of magnesium from lithium in salt solutions included chemical precipitation processes, electrodialysis membrane, adsorption, and solvent extraction [5]. The separation process using chemical precipitation methods consists of two stages. The first is to carry out the precipitation process of elemental magnesium. The second is precipitating of elemental lithium, or it could be the precipitation of magnesium and lithium simultaneously by forming new complex compounds [6]. The precipitation method was employed by addition of aluminum-based [7], alkaline and alkaline salts [8], phosphate acid, and ammonium phosphate [9], oxalic acid with sodium carbonate [10], and sodium metasilicate [11].

The extraction of magnesium from bittern water, a by-product of salt ponds, aims to obtain valuable magnesium silicate elements. Magnesium silicate in the form of precipitates is a soft and neutral white crystalline that is used as a synthetic talc material. The synthesis talc starts with forming a precipitate obtained from the mixture between a solution of sodium metasilicate and magnesium chloride with a Mg/Si identical to that of talc [12]. The precipitation reaction is equation (1), with m and n as whole integers :

$$4\,(Na_2SiO_3) + 3\,(MgCl_2) + 2\,HCl + m\,H_2O \rightarrow Si_4Mg_3O_{11}\,,\, n\,H_2O + 8\,NaCl + (m-n+1)\,H_2O \quad (1)$$

The mineral talc has the formula $Mg_3Si_4O_{10}(OH)_2$ and theoretical chemical composition, expressed as oxides, of 31.7% by weight (wt) magnesium oxide (MgO), 63.5% silicon dioxide (SiO_2), and 4.8% water (H_2O) [13]. Typical cosmetic talcs have average particle sizes ranging between 4 μm and 15 μm [13]. Talc is a white powder, odorless, stable, and insoluble in water or alcohol [13]. Therefore, natural talc minerals are abandoned in the industrial market and replaced with synthetic minerals with homogeneous composition and grain size [13]. Other applications of talc minerals besides for pharmaceutical and cosmetic needs [13] are ceramic material [14], paper making [15], and a potentiometric sensor for hydrogen ion [16].

Unlike magnesium, which can precipitate into magnesium silicate, lithium compounds cannot react with sodium silicate solution to form lithium silicate. The formation of lithium silicate may occurs at elevated temperatures by solid-state reaction. The products may be lithium metasilicate (Li_2SiO_3) and lithium orthosilicate (Li_2SiO_4). It depends on the temperature and the mole ratio of silica and lithium carbonate [17]. In addition to the solid-state process at high temperatures, lithium silicate could be made by a sol-gel process at a low temperature of 70 °C. The Sol-gel process was carried out by mixing the silica solution with LiOH solution or with lithium methoxide (CH_3OLi) to form lithium metasilicate (Li_2SiO_3) and lithium orthosilicate (Li_2SiO_4) deposits [18].

In this paper, the study aims to extract magnesium from bittern with low lithium grade at room temperature by adding sodium silicate. The bittern is reacted with a sodium silicate solution to form a white precipitate of magnesium silicate. The bittern filtrate from the chemical precipitation process is expected to contain low magnesium because almost all of the magnesium is extracted. While the lithium element is maintained in the filtrate, it caused a decrease in the Mg/Li ratio. Further, the bittern filtrate may be used as raw material to manufacture lithium chemical compounds such as lithium carbonate for lithium battery applications.

2. Methodology

In this study, the material used is a bittern, a by-product from salt processing with a geomembrane tarpaulin system in district Lamongan, East Java, Indonesia. The chemical composition of bitterns which was characterized by using ICP, are listed in table 1.

Table 1. Chemical composition of bittern raw material.

Element	Mg	Na	Li	K	Ca	B	Ratio Mg/Li
Concentration (ppm)	33,184	23,787	6.10	5,491	617	152	5,442

In table 1, it can be seen that the raw material used for bittern contains 33184 ppm magnesium and lithium, 6.10 ppm. Compared to sea water, which contains a lithium of about 0.17 ppm, there is an increase in lithium levels by 36 times. Table 1 shows that the Mg / Li ratio is about 5.442, so it requires many chemical reagents that impact high operating costs to extract lithium from bittern. Therefore, the first step was to take magnesium from the bittern to utilize magnesium from the bittern as an economically valuable magnesium-based product.

2.1. Experimental procedure

In this experiment, 200 ml of the bittern was added with sodium silicate, which varied at 50 g, 60 g, 70 g, 80 g, 90 g, and 100 g. The bittern contains magnesium with a concentration of 33.184 ppm or 33.184 g/l as shown in table 1. Based on stoichiometric calculations, to precipitate magnesium from the bittern requires 168.68 g sodium silicate per liter of the bittern.

The 400 ml of sodium silicate solution in various concentrations, such as 125 g/l, 150 g/l, 175 g/l, 200 g/l, 225 g/l, and 250 g/l were prepared. At first, the sodium silicate solution was cooled to room temperature. Then it was reacted with 200 ml bittern for 2 hours at a stirring speed of 300 rpm in atmospheric pressure. The results of the precipitation process were filtrated to get bitten filtrate, and the solids precipitate. In each precipitation experiment, the filtrate volume and the amount of solids precipitates were also measured. The solid deposits or precipitates were then washed with 1,000 ml aquadest for 30 minutes. The washed solids were dried for 24 hours at a temperature of 110 $^{\circ}$C before further characterization.

2.2. Characterization

The chemical composition of solid magnesium silicate was measured by using XRF (X-Ray Fluorescence, Bruker S 2 Puma). The chemical substances analysis was conducted by using XRD (X-Ray Diffraction, Bentchop X-Ray Diffractometer). The particles' morphology and microstructure were observed by using SEM-EDX (Scanning Electron Microscope – Energy Dispersive X-Ray / Jeol JSM-639OA). The samples' chemical composition was characterized by using ICP-OES (Inductively Coupled Plasma Optical Emission Spectroscopy, Agilent 725 ICP-OES) analysis.

3. Result and discussion

The precipitation experiments have been carried out with various concentrations of sodium silicate solution of 400 ml mixed into 200 ml bittern. In this experiment, the solid and bittern filtrate was separated by a vacuum filtering process.

Table 2. Solid deposit and filtrate in precipitation process.

No	Sample	Sodium silicate concentration (g/l)					
		125	150	175	200	225	250
1	Residual filtrate (ml)	300	271	231	208	213	178
2	Solid mass (g)	60.54	54.88	79.69	111.17	110.34	130.62

Table 2 shows the amount of solid deposit and filtrate which were produced after precipitation process by addition of sodium silicates. The addition of sodium silicate in high concentrations increases the amount of solids precipitate obtained.

3.1. Bittern filtrate

The results of ICP-OES analysis on the bittern filtrate after the precipitation process using a sodium silicate solution at various concentrations could be seen in table 3 as follows. The elements concentration of Sodium (Na), Potassium (K), Calcium (Ca), and Boron (B) are not affected by sodium silicate addition in the filtrate.

Table 3. The chemical composition of bittern filtrate after the precipitation.

Sodium Silicate Concentration (g/litre)	Concentration (ppm)						Ratio
	Mg	Na	Li	K	Ca	B	Mg/Li
125	10,629	1,793	0.854	3,334	60.57	15.12	12,443
150	8,927	2,351	0.432	3,579	36.11	11.61	20,659
175	6,115	2,835	0.579	3,588	27.73	11.45	10,561
200	3,760	3,026	0.0626	3,272	25.09	11.33	60,064
225	2,529	2,995	0.251	2,948	16.65	9.11	10,091
250	1,418	3,307	1.373	3,009	13.07	10.53	1,033

The magnesium and lithium ions removal can be calculated considering the filtrate volume in table 2 after sodium silicate addition that are shown in equation (2).

$$\text{Ion Removal (\%)} = [(V_R \,.\, C_R - V_F \,.\, C_F)/V_R \,.\, C_R] \,.\, 100\ \% \qquad (2)$$

Information :
Ion removal: Ion removal from filtrate (percent)
V_R: Volume from row materials (200 ml)
C_R: Concentration element from row materials (table 1)
V_F: Volume filtrate after precipitation (table 3)
C_R: Concentration filtrate after precipitation (table 3)

Figure 1 shows that the magnesium and lithium ions were removed from the bittern filtrate and precipitated with sodium silicate addition. More than 80% of lithium-ion is removed from filtrate with sodium silicate addition in the concentration range of 125 – 250 g/l. The figure shows that magnesium ions are removed from filtrate and precipitated up to 96.2% by adding 250 g/l sodium silicate. Potential Mg-Li separation may occur with the addition of more than 230 g/l sodium silicate. The addition of 250 g/l sodium silicate gives lithium-ion content of 1.373 ppm, while the Mg/Li ratio is 1,033.

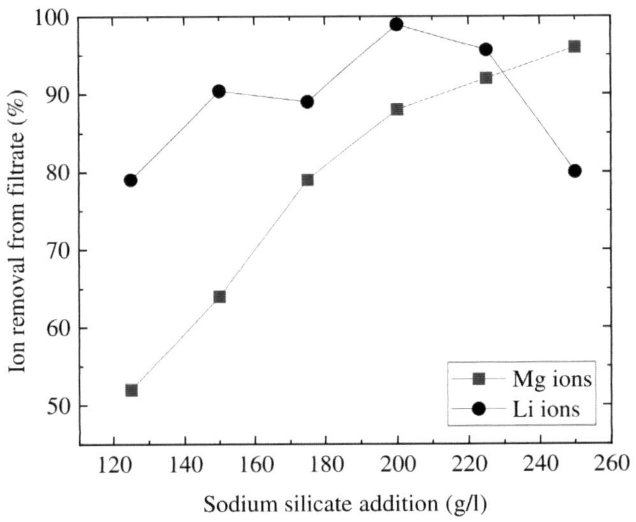

Figure 1. The magnesium and lithium ions removal from the bittern filtrate after sodium silicate addition.

3.2. Precipitate characteristics

3.2.1. The chemical composition of the solid product. Characterization of the solid products, mainly in the form of magnesium silicate product, carried out using XRF and the sodium silicate can be seen in table 4.

Table 4. The chemical composition of magnesium silicate precipitate and raw materials.

Compounds	Concentration (% wt)							
	MgO	Na_2O	SiO_2	P_2O_5	CaO	Fe_2O_3	SO_3	Cl_2
Magnesium silicate product	12.377	nd[a]	70.903	0.673	0.303	0.13	4.258	8.635
Sodium silicate raw material	nd[a]	27.147	71.74	0.466	0.257	0.079	nd[a]	0.199

[a] nd: no detected

It seems that the magnesium silicate precipitates still have high sulfur and chloride impurities. This impurity is due to sulfur and chloride compounds from the raw material of salt pond waste in sulfate and chloride-based salt compounds. Therefore, the magnesium silicate precipitate's washing process needs to be repeated to obtain better purity quality.

3.2.2. The precipitate phase. Figure 2 shows the XRD pattern of the sodium silicate used in experiment (a) and solid product of precipitation reaction (b). The sodium silicate raw material peaks are shown at 16.8°, 24.9°, 29.4°, 34.8°, 37.1°, 48.0°, 52.0°, and 65.69° (figure 2(a)). However, there are other four quartz peaks at 22.7°, 26.7°, 45.5°, and 49.5°, indicating that sodium silicate raw material contains free silica compounds. This XRD analysis results were similar to previous studies that also used crystalline sodium silicate raw material [19]. The presence of quartz (SiO_2) compounds that were insoluble in water caused difficulties in the dissolving of crystalline sodium silicate reagent in the water.

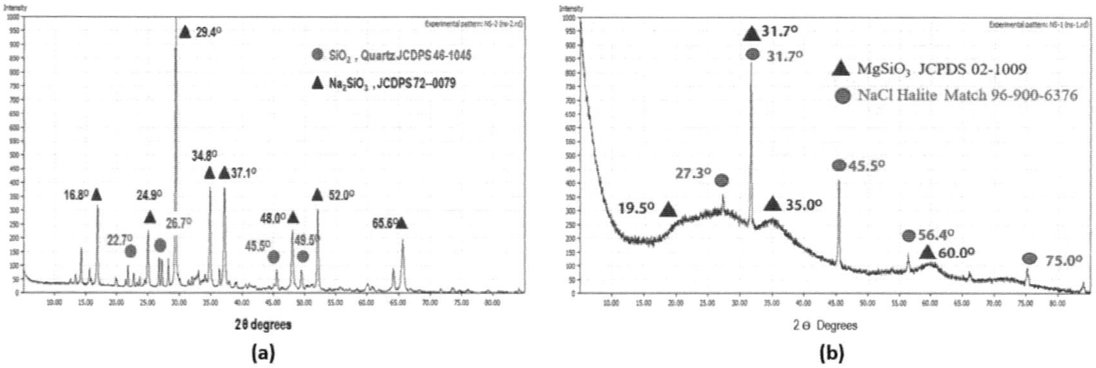

Figure 2. X-ray diffraction pattern of raw material of sodium silicate (a) and solid product (b).

Figure 2 (b) shows four prominent peaks of 27.3°, 31.7°, 45.5°, and 56.4° indicating that solid product contains NaCl halide compounds. The other peaks are observed at 19.5°, 31.7°, 35.0°, and 60.0° as magnesium silicate compound based on the pattern JCPDS 02-1009 [21]. It also looks that there are high intensity overlapping peak showing halite and magnesium silicate compounds at 31.7°. The other peaks such as sodium silicate, quartz and sulfate, and chloride salt compounds are not visible in figure 2 (b).

From previous research explained that commercial talc contained magnesium silicate compounds at 10°, 19.5°, 29° and 37° [24]. According to XRD pattern in figure 2 (b), it can be seen that the the dominat peaks in the solid product are halite compounds. This is due to the raw material used is bittern from salt pond waste that is rich in sodium compound. The pattern of magnesium silicate in solid product was low-intensity crystalline compared to halite compounds. Another benefit of magnesium silica precipitates with a lower crystalline phase was the use of an adsorbent for toxic waste materials such as Rhodamine B [21].

3.2.3. The morphology and microstructure of solid product. Figure 3 shows the morphology of the washed and dried solid precipitates products (a) and commercial magnesium silicate or talc available in the market (b). The morphology of the magnesium silicate crystals that are soft and not sharp like needles is the ultimate goal of making synthetic magnesium silicate. Needle-shaped morphology indicates the presence of compounds such as asbestos that are harmful to health. Free-asbestos magnesium silicate has potential application as talc raw materials since it is safer from a health perspective. Figure 3(b) shows that the morphology of commercial magnesium silicate or talc with needles-like or thread-like granules, which are the characteristic of asbestos minerals [22]. The magnesium silicate precipitate with low crystallinity was advantageous because it can be easily formed into nanoparticles magnesium silicates compound using ultrasonic [23].

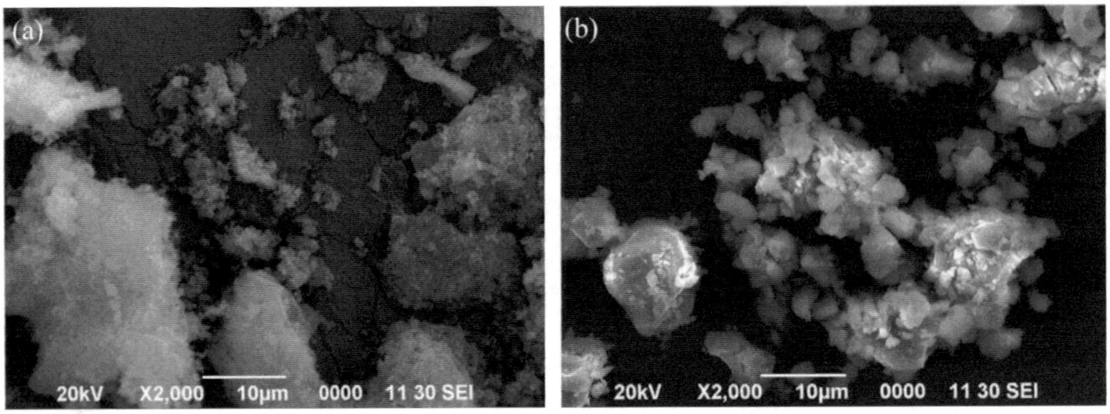

Figure 3. The morphology of samples: (a) solid product (this study); (b) commercial talc.

The morphology of synthetic talc from bittern (figure 3(a)) appeared coarser and dispersed with fine particles than the commercial talc (figure 3(b)). Both of the particle shapes are granules. According to the previous research, to change the rough-looking talc morphology, it is necessary to modify the surface using phthalimide material. One of the surface modifications using Phthalimide material could change the grain structure, physical and optical properties [24].

Figure 4 shows the chemical composition area map of the samples using SEM-EDX. The quantitative chemical composition is listed in table 5. The commercial talc sample contains calcium, which was not identified in the precipitate samples. However, the solid precipitate sample contains sodium and potassium, which are typical ions from sea salts.

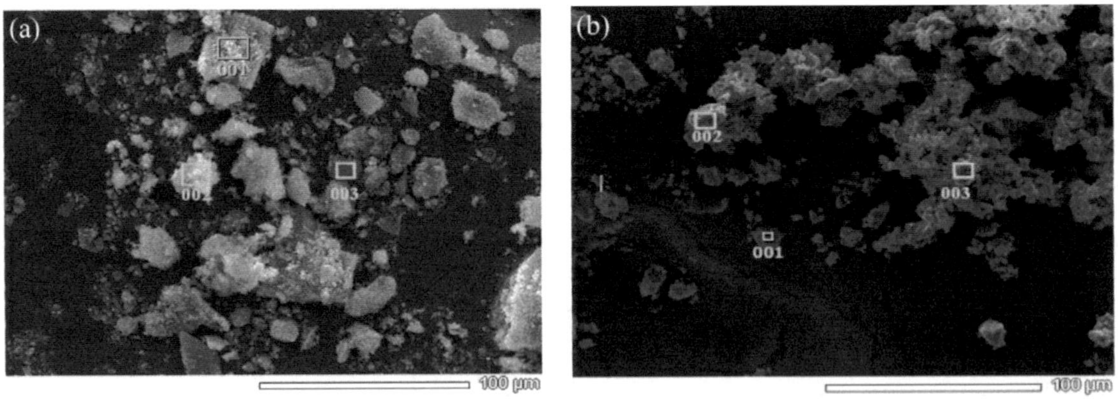

Figure 4. Mapping SEM-EDX: (a) magnesium silicate precipitate; (b) commercial talc.

Table 5. The chemical composition of mapping area by SEM-EDX.

	Magnesium silica precipitate						Commercial talc				
Mapping area	C	O	Na	Mg	Si	K	C	O	Mg	Si	Ca
001	16.32	40.14	4.89	12.05	25.70	0.90	13.55	45.59	0.07	0.28	40.51
002	25.67	42.93	1.86	9.46	19.76	0.32	19.27	43.99	0.11	0.00	36.63
003	33.77	35.62	3.16	7.76	19.06	0.62	31.46	40.17	0.08	0.05	28.24

Based on table 5, the mapping analysis results with SEM-EDX showed that the magnesium silicate residue of this study was different from the talc composition in the market. The granules resulted from the precipitation process were dominated by silica (SiO_2), magnesium silica ($MgSiO_3$), and residual sodium silica (Na_2SiO_3). It was observed a relatively high carbon content indicating the presence of magnesium carbonate compounds. Talc, which is available in the market, is dominated by calcium carbonate ($CaCO_3$). It was characterized by high levels of calcium, oxygen, and carbon. Very low magnesium and silica levels indicated few magnesium and calcium silicate compounds in commercial talc.

4. Conclusion

This research successfully extracted magnesium from bittern using sodium silicate into a magnesium silicate, a valuable product as a raw material for commercial talc. The result of this study shows that potential magnesium-lithium ion separation from the filtrate to obtain a solid magnesium silicate was by adding more than 230 g/l of sodium silicate solution. Compared to commercial talc on the market, magnesium silicate precipitate was free of asbestos that does not contain calcium elements. The experimental results also showed that the Mg/Li ratio reduced from 5,442 to 1,033 in the bittern filtrate.

Acknowledgment
The authors would like to thank Universitas Indonesia for the Hibah Doktor No. NKB-692/UN2.RST/HKP.05.00/2020.

References

[1] Mirna A, Wahyono H and Ali M, 2018 *J. Ecologycal. Eng.* **19** 1–10
[2] Sumarni, S B, Irawan L Y and Fathoni M N 2020 *IOP Conf. Series: Earth Environment Science* **412** 1–9
[3] Alamdari A, Rahimpour M R, Esfandiari N and Nourafkan E 2008 *Chem. Eng. Process* **47** 215–21
[4] Vikström H, Davidsson S and Höök M 2019 *Appl. Energy* **110** 252–66
[5] Liu G, Zhao Z W and Ghahreman A 2019 *Hydrometallurgy* **187** 81–100
[6] Zhang Y, Hu Y H, Wang L and Sun W 2019 *Miner. Eng.* **139** 1–14
[7] Liu X H, Zhong M L, Chen X Y and Zhao Z W 2018 *Hydrometallurgy* **176** 73–7
[8] Hamzaoui A H, M'nif A, Hammi H and Rokbani R 2003 *Desalination* **158** 221–4
[9] He L H, Xu W H, Song Y F, Liu X H and Zhao Z W 2017 *Sep. Purif. Technol.* **187** 214–20
[10] Tran K T, Han K S, Kim S J, Kim M J and Tran T 2016 *Hydrometallurgy* **160** 106–14
[11] Zhang Y, Hu Y H, Sun N, Khoso S A, Wang L and Sun W 2019 *Hydrometallurgy* **187** 125–33
[12] Claverie M, Dumas A, Careme C, Poirier M, Le Roux C, Micoud P, Martin F and Aymonier C 2019 *Chem. Eur. J* 519–42
[13] Fiume M M *et al.* 2015 *Int. J. Toxicol.* **34** 66s–129s
[14] Ghods B, Rezaei M and Meshkani F 2016 *Ceram. Int.* **42** 6883–90
[15] Chauhan V S and Bhardwaj N K 2017 *Arab. J. Chem.* 10 1059–66
[16] Teixeira M S, Cavalheiro C C S, Ramos L A and Neves E A 2003 *Appl. Clay Sci.* **23** 323–8
[17] Tang T, Zhang Z, Meng J B and Luo D L 2019 *Fusion Eng. Des.* **84** 2124–30
[18] Pfeiffer H, Bosch P and Bulbulian S 1998 *J. Nucl. Mater.* **257** 309–17
[19] Kaduku T, Daromala M O, Obazu F O and Iyuke S E 2015 *J. South. Afr. Inst. Min. Metall.* **115** 1175–82
[20] Zhang T T, Vanderperre L J and Cheeseman C R 2014 **65** 8–14
[21] Sun Z W, Duan X H, Srinivakasannan C and Liang J S 2018 *R. Soc. Chem.* **8** 7873–82
[22] Karunaratne P C T and Rohan Fernando G W A 2015 *Proc. of the Int. Conf. on ' Cities, Peoples and Places' - ICCCPP-2015 (*Colombo, Sri Lanka) pp 583–600

[23] Dietemann M, Baillon F, Calvet F E R and Greenhill-Hooper M 2019 *Powder Technol.* **356** 83–96
[24] Marzbani P, Resalati H, Ghasemian A and Shakeri A 2016 *Bioresources* **11** 8720–38

IOP Conf. Series: Materials Science and Engineering 1195 (2021) 012029 doi:10.1088/1757-899X/1195/1/012029

Synthesis and characterization of perovskite-supported CoNi catalyst for CO oxidation via exsolution

G L Lew[1], N Ibrahim[2], S Abdullah[3], W R W Daud[4] and W K W Ramli[1]

[1] Faculty of Chemical Engineering Technology, Universiti Malaysia Perlis (UniMAP), Perlis, Malaysia
[2] Faculty of Civil Engineering Technology, Universiti Malaysia Perlis (UniMAP), Perlis, Malaysia
[3] Faculty of Chemical and Natural Resources Engineering, Universiti Malaysia Pahang (UMP)
[4] Research Centre for Sustainable Process Technology, Universiti Kebangsaan Malaysia (UKM)

Email: wankhairunnisa@unimap.edu.my

Abstract. The introduction of perovskite oxide as catalysts alternative has increased the worldwide interest due to its advantages such as its versatility to accommodate different transition metals. This study set out to evaluate the catalytic activity of CO oxidative perovskite catalysts (LCCNTO), fabricated via solid-state method and reduced under various reducing condition for the exsolution of the active metals, Cobalt-Nickel (CoNi) from the perovskite lattice. The effect of reducing parameter modification towards the catalytic activity of the fabricated LCCNTO was discussed in terms of CO conversion and CO_2 production rate. Through the light-off test, the sample that reduced with the longest deration (S2T10H6-R5H5) showed the highest CO conversion of 45.45% and CO_2 production rate of 0.1409 x 10^{-4} mol $s^{-1}g^{-1}$ at the reaction temperature of 500 °C. Not only that, it was discovered that by controlling the reducing duration, the initiate temperature for the reaction to occur was lowered from 360 °C (S2T10H6-R5H3) enabling the reaction to occur at lower temperature at 280 °C in S2T10H6-R5H5. Under the same reducing temperature, the CO_2 production of sample reduced for 200 minutes (S2T10H6-R5H3) started at 360 °C but as the reducing duration increased to 300 minutes (S2T10H6-R5H5), the CO oxidation initiated at a much lower temperature of 280 °C. Although LCCNTO catalyst still suffer from similar deterioration as the other reported base metal catalyst, but tuning the reducing duration given to a sample, it greatly affects the initiation temperature for the reaction to occur.

1. Introduction

The increasing uses of fossil fuel and motor vehicles had brought the rise in air pollution with their high production of exhaust emission and carbon monoxide (CO) being one of the major atmospheric pollutant emitted from incomplete combustion of carbon source fuel poses a serious threat to human health [1]. Hence the use of oxidative catalyst system is crucial for the removal of exhaust gases such as CO by converting them into a less hazardous gases, carbon dioxide (CO_2) [2]. Supported heterogeneous oxidative catalysts usually consist of high catalytic active metals being uniformly scattered onto the surface of a porous support material which typically fabricated through deposition or precipitation method. Although deposition method has been widely applied in heterogeneous catalyst fabrication but the active metals deposited onto the surface have limited anchorage to the supporting material. Not only they have minimal control over the particle distribution and size during both preparation and catalyst activation stages [3], but also known to be time-consuming and costly [4].

The discovery of the unique structural properties of perovskite oxide which capable of accommodate various base metals and elevating them from within its supporting structure through exsolution process had caught the attention to be a befitting support material for heterogenous catalyst [5]. With the nominal composition of ABO_3, perovskite can couple up two cations, A and B where alkali earth metals with larger ionic radius occupy the A site while B site cations, being a much smaller ionic radius metal can be substituted with different high catalytic elements. Since 90% of the elements in the periodic table could be stable in the perovskite structure, the application of perovskite oxides could be further broadened with partial substitutions of secondary cations (A' and B') to form $A_{1-x}A'_xB_{1-y}B'_yO_3$ [6]. The different sizes of the elements substituted into the perovskite system can cause various degree of structural distortion due to their ionic radii differences in the lattice and deviate from the ideal cubical structure of perovskite. In order for the perovskite oxide with all the substituted formulation to be stable, Goldschmidt's tolerance factor, t is calculated by using the ionic radii of the perovskite structure and the value 0.75 to 1 is required [7]. Figure 1 depicted a simplified illustration of exsolution process.

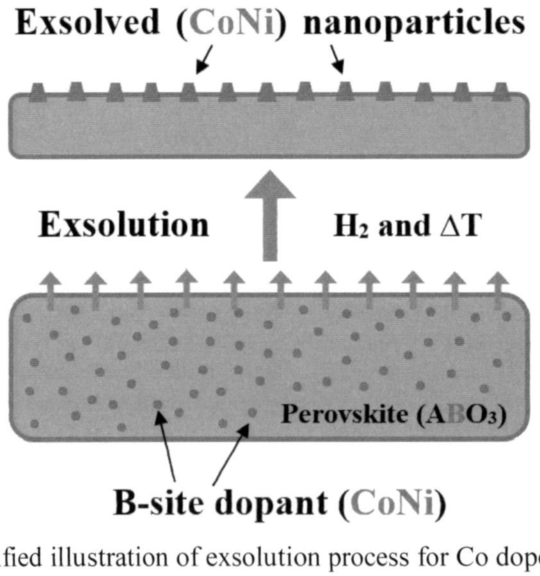

Figure 1. Simplified illustration of exsolution process for Co doped perovskite.

In the place of the high scarcity precious metals such as Pt, Rh, Pd and Au, a lower cost with higher stability substitution is require to fill the uprising demands of oxidative catalyst. Transition metals fits the requirement and deemed to be a more environmentally friendly alternative as heterogeneous catalyst for both their relative affordability and their effective surface redox reactivity [9, 10]. Among the candidates, cobalt oxide and nickel oxide showed their potential in CO oxidation where cobalt oxide had shown impressive reactivity towards CO oxidation at low temperatures [10] and nickel oxide demonstrated astounding catalytic activity in both CO and NO removal [12, 13]. Since both Co and Ni have similar electronic configurations, Co-Ni composite oxides are capable of exhibit synergistic catalytic effects [13] and have a strong adsorption capacity of CO that work splendidly at low temperatures [14].

With the uniqueness of perovskite oxide's structure being a suitable porous supporting material for heterogeneous catalyst, CoNi was chosen as our CO oxidative active metals and were incorporated into the perovskite formulation, LCCNTO to act as a cost-effective alternative. Solid-state reaction was used to fabricate LCCNTO and the active metal CoNi had successfully exsolved onto the surface of LCCNTO for CO oxidation. The structure of LCCNTO along with the single metal system LCCTO and LCNTO were characterized using X-ray diffraction (XRD) to determine their average crystallite size and identify their perovskite structure. The catalytic activity of the fabricated LCCNTO was evaluated in terms of CO conversion and the CO_2 production rate, $r^*_{CO_2}$, obtained from light-off experiments to determine the effects of different reducing duration on the catalytic activity of the samples towards the oxidation of CO.

2. Materials and methods

2.1. Chemical and gases

99.5% Cobalt (III) oxide (Co_3O_4) was purchased from Sigma-Aldrich, United States, while other precursors such as 99% Nickel (II) nitrate hexahydrate ($Ni(NO_3)_2*6H_2O$), 99.9% Ceria oxide (CeO_2), 99.98% Lanthanum oxide (La_2O_3), and Titanium oxide (TiO_2) were purchased from Acros Organics, United States. While the gases used include 30% hydrogen/argon (H_2/Ar) and 99.999% nitrogen (N_2) for the reduction process while 20% oxygen/nitrogen (O_2/N_2) and 20% carbon monoxide/nitrogen (CO/N_2) were used in CO oxidation during light-off test. All the gases mentioned were purchased from Linde group, Ireland. All chemicals and gases for experimental works were used without additional purification.

2.2. Sample preparation

The composite CoNi, Co and Ni incorporated oxidative perovskite catalysts, LCCNTO ($La_{0.7}Ce_{0.1}Co_{0.3}Ni_{0.1}Ti_{0.6}O_3$), LCCTO ($La_{0.7}Ce_{0.1}Co_{0.4}Ti_{0.6}O_3$) and LCNTO ($La_{0.7}Ce_{0.1}Ni_{0.4}Ti_{0.6}O_3$) were fabricated through solid-state reaction which involved the fabrication of perovskite structure through calcination of precisely weighed and well mixed precursors, La_2O_3, CeO_2, Co_3O_4, $Ni(NO_3)_2*6H_2O$ and TiO_2. The mixed precursors were first heated to remove any moisture and volatile substance prior to a calcination process for 12 hours at 1000 °C to produce pure perovskite structure. In order for the metal particles to be exsolved onto the surface, the LCCNTO samples were exposed to a reducing environment with hydrogen gas (H_2) as the reducing agent in a conventional continuous flow reactor at different reducing duration to determine the effect of reducing duration on the catalytic activity of these fabricated LCCNTO.

Both the exsolution and CO oxidation were carried out in a conventional continuous flow reactor consisted of a quartz fixed-packed bed reactor with gases supply connected with mass flow controller (MFC). As-prepared LCCNTO were held in position with two supporting quartz wool to remain at the centre of the heating element prior to the exsolution and CO oxidation. Figure 2 depicted the simplified experimental set-up of the conventional continuous flow reactor. According to the fabrication method mentioned in Wan Ramli *et al.* [15], the reducing temperature between 550 to 860 °C is the suitable range for CoNi exsolution and have direct influence on the particles' size of exsolved CoNi on the perovskite. The exsolved CoNi particles' size reported around 30 nm when reduced at 860 °C but the size of exsolved particles reported to be 10 nm when reduced at around 500 °C, but 10 nm particles size was too small for activation when exposed to CO-rich environment but 30 nm was too big causing agglomeration of particles during the activation. With all their samples were reduced for 5 hours, in order to further investigate the effects of reducing parameter towards the degree of exsolution, reducing temperature of 500 °C and 700 °C with different the reducing duration were chosen to determine the effects of reducing parameters on degree exsolution of CoNi from LCCNTO. The exsolution was done under different reducing condition with a continuous H_2 supply at 20 ml min^{-1} and the reducing condition was as listed in table 1.

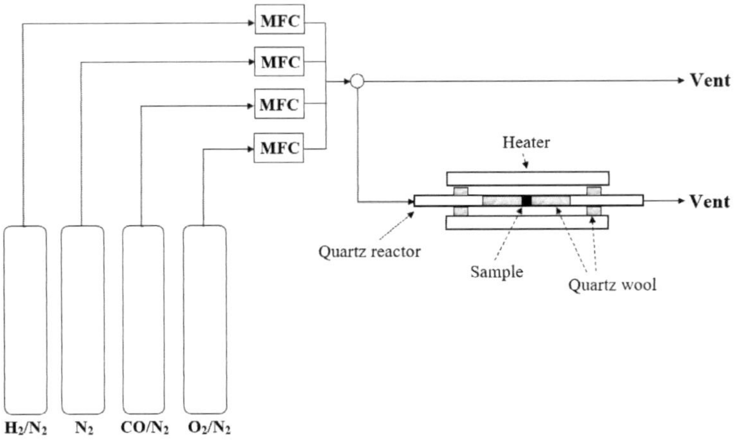

Figure 2. Experimental set up for the exsolution and light-off test.

Table 1. LCCNTO samples with their treatment stages and conditions.

Sample	Stage	Conditions
$S_2T_{10}H_6$	Secondary Calcination	T = 1000 °C, calcined for 360 minutes
$S_2T_{10}H_6$-R_5H_3	Reduction Process	T = 500 °C, reduced for 200 minutes
$S_2T_{10}H_6$-R_5H_4		T = 500 °C, reduced for 250 minutes
$S_2T_{10}H_6$-R_5H_5		T = 500 °C, reduced for 300 minutes
$S_2T_{10}H_6$-R_7H_5		T = 700 °C, reduced for 300 minutes

2.3. Sample characterisation
The crystalline structures of the fabricated samples were analysed using Bruker D2 Phaser benchtop X-ray diffractometer (XRD), equipped with LYNEYE 1D Ultra-fast-solid-state detector with CuKα radiation (λ = 1.54184 Å) to ensure the formation of the desired perovskite structures. LCCTO and LCNTO were fabricated along with LCCNTO to observe the change in structure and average crystallite grain size. The average crystallite grain size of the as-prepared catalysts was estimated by using the major peaks presence from the XRD pattern using Scherrer equation:

$$d = \frac{K\lambda}{\beta\ cos\theta} \tag{1}$$

where d is the is the crystallites size (nm), K represent the Scherrer constant of 0.9, λ depends on the wavelength of the x-ray source (λ = 1.54184 Å), β is the full width at half maximum (FWHM) in radians and θ denote the peak positions in radians.

2.4. Catalytic activity testing
The catalytic activities of reduced LCCNTO ($S_2T_{10}H_6$-$R_5H_{3\text{-}5}$) were tested using the same reactor set up shown in figure 2. The catalyst powder was first diluted with Al_2O_3 powder down to 10 wt% to have a reasonable pressure drop within the reactor [15]. The inlet gases supply to the catalytic testing performed was set according to the stoichiometric reaction of CO oxidation as shown as follow:

$$CO + \frac{1}{2} O_2 \rightarrow CO_2 \tag{2}$$

The total inlet feed gas mixture of 2% CO and 1% O_2 gases, balanced with N_2 gas with the total flowrate F_t = 250 ml min^{-1} given at normal temperatures. The activity of all the reduced catalyst as a function of temperature or also known as the light-off test, was conducted to determine the effectiveness of the catalyst toward CO oxidation with IR analyzer (Testo 340, Germany) to record the CO_2 production at different temperature. In this study, the light-off test was set to obtain the data from room temperature to 520 °C, with a step size of 20 °C. The temperature was held constant at every temperature increment for 40 minutes for the reaction to take place, and also to obtain stable values of CO_2 produced. The CO conversion and reaction rates, $r^*_{CO_2}$, in terms of CO_2 production for all samples were calculated as follow:

$$CO\ conversion, \% = \frac{CO_2\ produced\ (ppm)}{Total\ CO\ supply\ (ppm)}\ x\ 100\% \tag{3}$$

$$r^*_{CO_2}(mol\ (CO_2)s^{-1}g^{-1}) = \frac{y_{CO_2} \times \dot{n}}{w_p} \tag{4}$$

where y_{CO_2} is the CO_2 concentration (ppm) measured by the analyzer at the gas outlet, $\dot{n}$ is the total molar flow rate of the reaction and w_p is the weight of the active catalysts (w_p = 60 mg).

3. Results and discussion

3.1. X-ray diffraction investigation

The analysis of X-ray diffraction pattern consisted of LCCNTO (CoNi), LCCTO (Co) and LCNTO (Ni) was performed to determine the difference in the structure of fabricated perovskite with different active metal. All the samples exhibited the formation of perovskite phase with peaks labelled as shown in figure 3 [16]. Upon the calcination, the formation of the perovskite structure was favoured in all the samples with the peaks corresponded to perovskite phase could be identified with slight difference in intensity. The peaks for each active metals (Co, Ni and CoNi) were also identified indicating the incorporation of active metals in the perovskite structure [17]. Due to the similar electronic configuration of Co and Ni, several of peaks overlapped with each other and the doped LCCNTO and LCCTO have more similarity with difference in intensity and extra peaks from Ni. The peaks indicating the presence of CoNi are around 2θ = 40° and 58°, but with lower intensity. The average crystallite grain size of prepared samples was calculated using Scherrer equation and the results indicate a decrease in crystallite size with the order of LCNTO (25.1280 nm), LCCNTO (25.0408 nm) and LCCTO (18.1051 nm) which suggested that the doping of cobalt with nickel (3:1) has little to none in reducing the crystallite size of LCCNTO. This was in agreement with the work of Yi and colleagues which reported the increase in Co incorporation of NiO had gradually causing the peaks ascribed to NiO becoming wider and weaker even disappeared when content of Co became higher but at the same time, the Ni^{2+} were suspected to be partially substituted Co^{3+} and move into the Co_3O_4 crystal lattice forming Ni-Co spinel structure as they both coexist under a same structure [14].

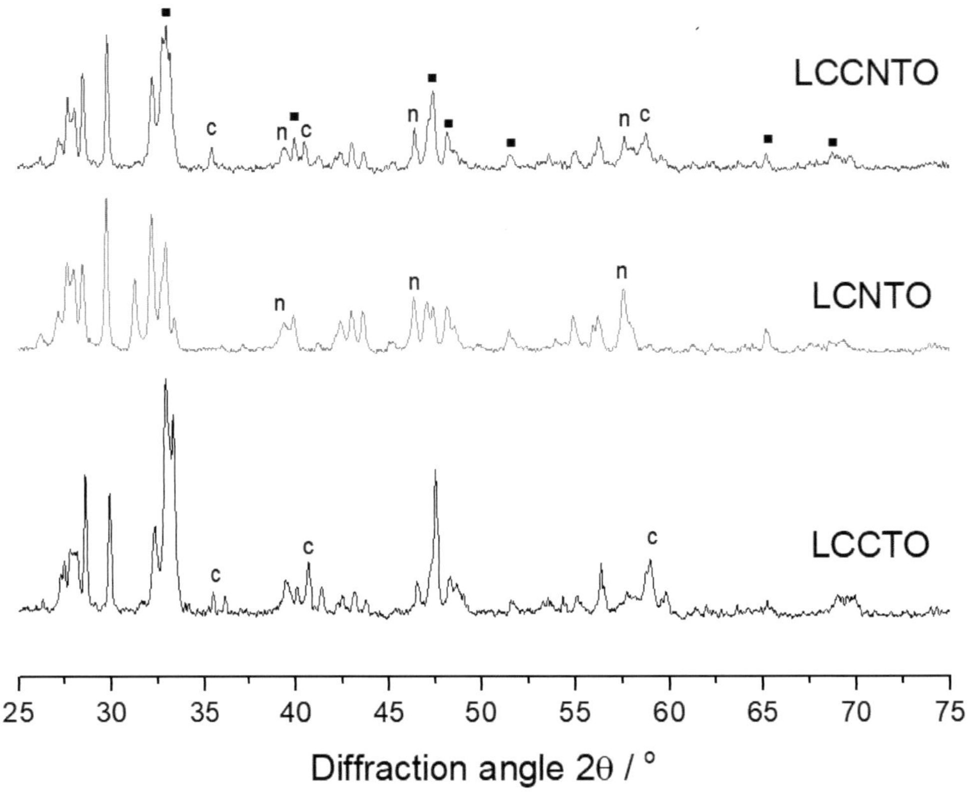

Figure 3. XRD patterns for LCCNTO, LCNTO and LCCTO with ■ denotes the peaks indicating the presence of the perovskite phase while *c* and *n* denotes the peaks for both Co and Ni phases, respectively.

3.2. Catalytic activity testing with light-off test

The catalytic activity of fabricated LCCNTO samples ($S_2T_{10}H_6$-R_xH_y) undergone the light-off test which involved a total flowrate, F_t, of 250 ml min^{-1} under normal temperature and pressure (NTP) with a continuous inlet flow of 2% CO and 1% O_2 gases balanced with N_2 gas supplied during the oxidation process. Figure 4 compares both the CO conversions and the rate of CO_2 production, $r^*_{CO_2}$ for samples with different reducing duration. Unreduced LCCNTO sample ($S_2T_{10}H_6$) was also tested as a control sample but since no CO_2 produced at the temperature range used, proving that, CO oxidation only occurs to reduced samples with their active metals (CoNi) exsolved onto the surface that were accessible for the reaction. All the catalytic activity of all the samples increased with the reaction temperature where $S_2T_{10}H_6$-R_5H_5 marked the highest CO conversion of 45.45 ± 4.5% which corresponded to the CO_2 production rate, $r^*_{CO_2}$ of around 0.1409 x 10^{-4} mol s^{-1}g^{-1} at the temperature of 500 °C, followed by $S_2T_{10}H_6$-R_5H_4 with the highest CO conversion of 24.46 ± 2.45% and $r^*_{CO_2}$ of around 0.0758 x 10^{-4} mol s^{-1}g^{-1} at 500 °C and lastly $S_2T_{10}H_6$-R_5H_3 with the lowest in both highest CO conversion and $r^*_{CO_2}$ achieved at 15.97 ± 1.58% and around 0.0495 x 10^{-4} mol s^{-1}g^{-1} at 420 °C, respectively. S_2T_{10}-H_6-R_5H_5 fabricated exhibited a higher conversion of CO but at much lower CO_2 production rate when compared to the CoNi based exsolved perovskite catalyst reported in Wan Ramli *et al.* [18] with CO conversion rate near 40% but the CO_2 production rate reaching around 0.8 x 10^{-4} mol s^{-1}g^{-1} at the temperature of 520 °C with their CoNi exsolved catalyst even before the activation under CO-rich environment.

An interesting trend was observed during the light-off tests in which the initiation temperature for the CO oxidation to initiate becomes lower by increasing the reducing duration of $S_2T_{10}H_6$. From the initiation temperature recorded at 360 °C for $S_2T_{10}H_6$ that reduced at 500 °C for 200 minutes (R_5H_3) to start the reaction, the initiation temperature had lowered to 320 °C by changing the reducing duration of $S_2T_{10}H_6$ from 200 minutes to 250 minutes (R_5H_4) and the initiation temperature for reaction further dropped as $S_2T_{10}H_6$-R_5H_5 was able to start the CO conversion at the temperature of 280 °C. The improvement in terms of the initiation temperature which allow the oxidation process to start at lower temperature might be caused by the better exsolution of CoNi onto the surface with extended reducing duration. It shows a possible new controlling parameter in the future for manipulating the exsolution of active metal in perovskite by tuning the reducing duration.

A drop in catalytic activity was observed upon reaching the highest CO conversion as both the CO conversion and $r^*_{CO_2}$ started to deteriorate at higher temperature for all samples ($R_5H_{3\text{-}5}$). It was suspected that the active cobalt alloy undergone surface oxidation during the process and formed other inactive phase which deteriorate the catalytic activity of the catalyst as Co have multiple oxidation state that is not reactive towards CO oxidation [19]. Popovic *et al.* reported similar phenomenon with their $La_{0.6}Ca_{0.4}FeO_{3-\delta}$ catalyst where the CO_2 production increased as the reaction temperature reached 300 °C, a sharp drop was observed, as the temperature continue to increase. It was explained to be caused by the oxygen vacancy and the ion mobility of lattice oxygen that act as the reservoir for CO oxidation enabling the CO conversion rate to increase as the reaction temperature increase until it reaches its equilibrium [20]. All of these can only be a speculation of what happened and will be investigated in future studies.

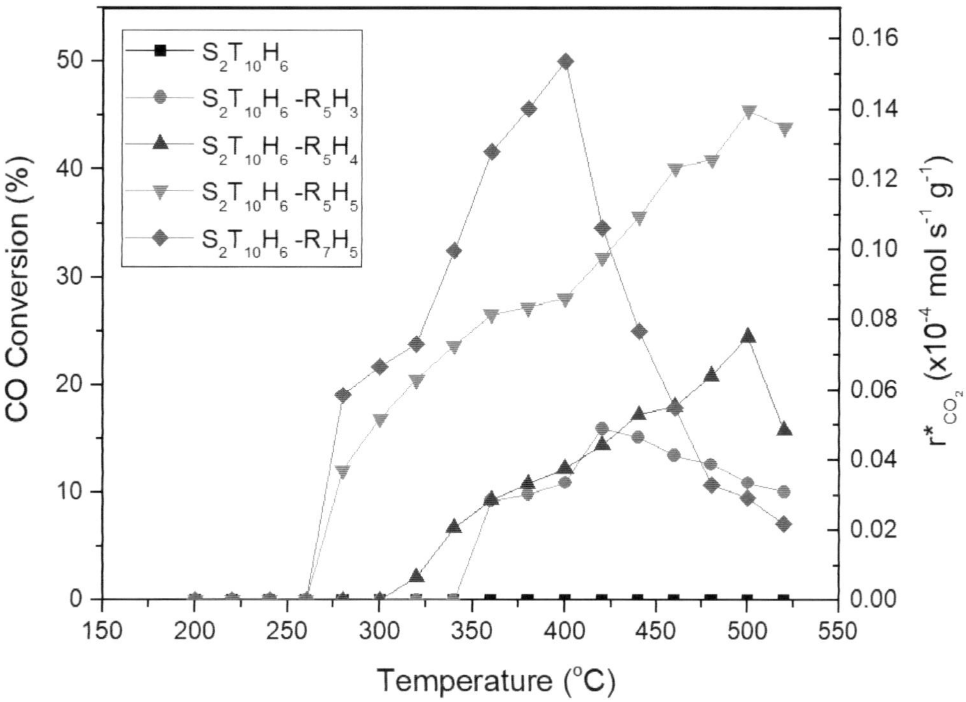

Figure 4. CO conversion and CO_2 production rates, $r^*_{CO_2}$ as a function of temperature (light-off) of the sample $S_2T_{10}H_6$ with different reducing duration during exsolution of CoNi from LCCNTO under CO oxidation process. The inlet feed gas consisted of mixture of 2% of CO and 1% of O_2 balanced with N_2 with total flowrate, F_t, of 250 ml min^{-1} and GHSV = 23575.45 h^{-1}. Powder weight of 60 mg samples had diluted with alumina to obtain the weight of powder bed, W_t of 600 mg. Error of around ±10 % were found during reproducibility test at each measured point.

A sample of S_2T_{10}-H_6, that reduced at a higher reducing temperature of 700 °C (R_7H_5), but under the same reduction duration was also prepared to determine the effect of reducing temperature towards the catalytic activity in comparison to S_2T_{10}-H_6-R_5H_5 and the comparison result is as shown in figure 4. The CO oxidation also initiated at the reaction temperature of 280 °C, similar with the initiating temperature of S_2T_{10}-H_6-R_5H_5, but it reached the highest CO conversion of 50 ± 5% and CO_2 production rate, $r^*_{CO_2}$ of around 0.1550 x 10^{-4} mol $s^{-1}g^{-1}$ at 400 °C, showing a better catalytic activity compared to S_2T_{10}-H_6-R_5H_5, which achieved the highest CO conversion at a higher temperature of 500 °C. However, the conversion started to drop after 400 °C, indicating that S_2T_{10}-H_6-R_7H_5 performs better at lower temperature than S_2T_{10}-H_6-R_5H_5 but less suitable for reaction at higher temperature application. The improvement on CO conversion and CO_2 production rate of S_2T_{10}-H_6-R_7H_5 before 400 °C can be attributed to the better exsolution of active metals with higher reducing temperature causing higher weight percentage and larger particles size of exsolved active metal during reduction [16, 22]. As for the drop in catalytic activity after 400 °C, it might be due to possible sintering effect at higher temperature during CO oxidation that deteriorate the active sites of catalyst as active metals exsolved [22] or the degradation of the active phase to a less active phases as surface oxidation of Co might occurred during the reaction [23]. The experiments were repeated to determine the reproducibility of the results and showed an error around 10% for every measured point. Further analysis such as XPS is required to determine the oxidation state of these active metals on the surface and are currently being investigated.

4. Conclusion

LCCNTO oxidative perovskite catalysts were fabricated through solid-state reaction and the structure phase of perovskite were identified using XRD analysis along with its single metal systems; LCCTO and LCNTO. The average crystallite grain size of this fabricated LCCNTO is determined to be around 25.0408 nm and due to the incorporation of Co, the crystalline size was slightly reduced in comparison to the size of its Ni counterpart (LCNTO) of 25.1280 nm. The Co-counterpart, LCCTO exhibited the smallest average crystallite grain size of 18.1051 nm and this has successfully reduced the crystallite size of LCCNTO in which the CoNi particles become smaller and were able to disperse evenly. The exsolution of active metal CoNi was achieved through continuous reduction at 500 °C with various reducing duration ranging from 200 to 300 minutes, ($R_5H_{3\text{-}5}$). With the reducing temperature of 500 °C, reduced LCCNTO ($S_2T_{10}H_6$-R_5H_5) exhibited the highest CO conversion of 45.45 ± 4.5% and CO_2 production rate of around 0.1409 x 10^{-4} mol $s^{-1}g^{-1}$ at 500 °C. A decreasing trend in the initiation temperature for CO oxidation could be observed as the reducing duration for S_2T_{10}-H_6 increased, where LCCNTO reduced for 200 minutes (S_2T_{10}-H_6-R_5H_3) has its CO conversion started at the temperature of 360 °C but when LCCNTO reduced for 300 minutes ($S_2T_{10}H_6$-R_5H_5), the CO conversion start at 280 °C. Similar initiation temperature of 280 °C could be observed when using LCCNTO reduced with same reducing duration of 300 minutes but higher reducing temperature of 700 °C ($S_2T_{10}H_6$-R_7H_5) which marked a higher CO conversion of 50 ± 5.0% and CO_2 production rate of around 0.1550 x 10^{-4} mol $s^{-1}g^{-1}$ at lower temperature of 400°C. Although all LCCNTO suffers from deterioration in the end, but it opens up the possibility of further tuning a better oxidative perovskite catalyst that can functioned at lower temperature.

Acknowledgments

The authors would like to acknowledge the support from the Fundamental Research Grant Scheme (FRGS) under a grant number of FRGS/1/2018/TK02/UNIMAP/02/8 from the Ministry of Higher Education Malaysia (MOHE). The authors would also like to thank Centre of Excellence for Frontier Materials Research (CFMR), Universiti Malaysia Perlis (UniMAP) for assisting in samples characterizations throughout this work.

References

[1] Duro J A 2016 Intercountry inequality on greenhouse gas emissions and world levels: An

integrated analysis through general distributive sustainability indexes *Ecol. Indic.* **66** 173–9
[2] Park S, Kim Y, Noh Y, Kim T, Han H, Yoon W, Choi J, Yi S H, Lee W J and Kim W B 2019 A sulfur-tolerant cathode catalyst fabricated with in situ exsolved CoNi alloy nanoparticles anchored on a Ruddlesden-Popper support for CO_2 electrolysis *J. Mater. Chem. A* **8** 138– 48
[3] Lindenthal L, Rameshan R, Summerer H, Ruh T, Popovic J, Nenning A, Löffler S, Opitz A K, Blaha P and Rameshan C 2020 Modifying the surface structure of perovskite-based catalysts by nanoparticle exsolution *Catalysts* **10** 1–14
[4] Neagu D, Tsekouras G, Miller D N, Ménard H and Irvine J T D 2013 In situ growth of nanoparticles through control of non-stoichiometry *Nat. Chem.* **5** 916–23
[5] Tsekouras G, Neagu D and Irvine J T S 2013 Step-change in high temperature steam electrolysis performance of perovskite oxide cathodes with exsolution of B-site dopants *Energy Environ. Sci.* **6** 256–66
[6] Atta N F, Galal A and El-Ads E H 2016 Perovskite nanomaterials – Synthesis, characterization, and applications *Perovskite Materials - Synthesis, Characterisation, Properties, and Applications*, ed L Pan and G Zhu (IntechOpen) chapter 4 pp 107–51
[7] Dragan M, Enache S, Varlam M and Petrov K 2019 Perovskite-type lanthanum cobaltite $LaCoO_3$: aspects of processing route toward practical applications *Cobalt Compounds and Applications*, ed Y Yıldız and A Manzak (London: IntechOpen) chapter 6
[8] Faure B and Alphonse P 2016 Co – Mn-oxide spinel catalysts for CO and propane oxidation at mild temperature *Appl. Catal. B Environ.* **180** 716–25
[9] Petitto S C, Marsh E M, Carson G A and Langell M A 2008 Cobalt oxide surface chemistry: the interaction of CoO(1 0 0), Co_3O_4(1 1 0) and Co_3O_4(1 1 1) with oxygen and water *J. Mol. Catal. A Chem.* **281** 49–58
[10] Zhang W, Wu F, Li J and You Z 2017 Dispersion–precipitation synthesis of highly active nanosized Co_3O_4 for catalytic oxidation of carbon monoxide and propane *Appl. Surf. Sci.* **411** 136–43
[11] Zhao B, Ke X K, Bao J H, Wang C L, Dong L, Chen Y W and Chen H L 2009 Synthesis of flower-like NiO and effects of morphology on its catalytic properties *J. Phys. Chem. C* **113** 14440–7
[12] Cheng X, Zhu A, Zhang Y, Wang Y, Au C T and Shi C 2009 A combined DRIFTS and MS study on reaction mechanism of NO reduction by CO over NiO/CeO_2 catalyst *Appl. Catal. B Environ.* **90** 395–404
[13] Gou Y, Liang X and Chen B 2013 Porous Ni-Co bimetal oxides nanosheets and catalytic properties for CO oxidation *J. Alloys Compd.* **574** 181–7
[14] Yi Y, Zhang P, Qin Z, Yu C, Li W, Qin Q, Li B, Fan M, Liang X and Dong L 2018 Low temperature CO oxidation catalysed by flower-like Ni-Co-O: How physicochemical properties influence catalytic performance *RSC Adv.* **8** 7110–22
[15] Wan Ramli W K 2017 Exsolved base metal catalyst systems with anchored nanoparticles for carbon monoxide (CO) and nitric oxides (NO_x) oxidation *Newcastle University*
[16] Lew G L, Ibrahim N, Abdullah S, Daud W R W and Ramli W K W 2021 Exsolution enhancement of metal-support CO oxidation perovskite catalyst with parameter modification *IOP Conf. Series: Earth and Environmental Science* **765** 012078
[17] Gou Y, Liang X and Chen B 2013 Porous Ni – Co bimetal oxides nanosheets and catalytic properties for CO oxidation *J. Alloys Compd.* **574** 181–7
[18] Wan Ramli W K, Papaioannou E, Naegu D and Metcalfe I S 2020 Exsolution of nickel alloys anchored nanoparticles on perovskite oxides for CO oxidation *IOP Conf. Series: Materials Science and Engineering* **778** 012059
[19] Argyle M D and Bartholomew C H 2015 Heterogeneous catalyst deactivation and regeneration: A review *Catalysts* **5** 145–269
[20] Popovic J, Lindenthal L, Rameshan R, Ruh T, Nenning A, Löffler S, Opitz A K and Rameshan C 2020 High temperature water gas shift reactivity of novel perovskite catalysts *Catalysts* **10**

582
[21] Tang C, Kousi K, Neagu D, Portolés J, Papaioannou E I and Metcalfe I S 2019 Towards efficient use of noble metals *via* exsolution exemplified for CO oxidation *Nanoscale* **11** 16935–44
[22] Soliman N K 2019 Factors affecting CO oxidation reaction over nanosized materials: A review *J. Mater. Res. Technol.* **8** 2395–407
[23] Cheng Y, Qiao M and Zong B 2017 Fischer-Tropsch synthesis *Encyclopedia of Sustainable Technologies*, ed M A Abraham (Elsevier) pp 403–10

Simultaneous degumming and deacidification of crude palm oil using mixed matrix PVDF membrane

N H Othman[1], R A Latip[1], A M Noor[1], W J Lau[2], P S Goh[2] and A F Ismail[2]

[1] Department of Processing Technology – Oils & Fats Section (Process & Technology), Sime Darby Plantation Research Sdn Bhd, Lot 2664 Jalan Pulau Carey, 42960 Carey Island, Selangor Darul Ehsan

[2] Advance Membrane Technology Research Centre (AMTEC), Universiti Teknologi Malaysia, 81310 UTM Skudai, Johor Darul Takzim

Email: noor.hidayu.othman@simedarbyplantation.com.

Abstract. The research work aimed to investigate the compatibility of polyvinylidene difluoride (PVDF) membrane embedded with various inorganic adsorbents for producing a mixed matrix membrane (MMM) and to examine the capability of mixed matrix PVDF membrane (PVDF-MMM) for simultaneous degumming and deacidification of crude palm oil (CPO). Four different adsorbent were tested which includes calcium silicate (CaS), magnesium silicate (MagS), ZSM-zeolite (ZSM) and activated carbon (AC). The in-house made PVDF-MMM was fabricated according to the dry-jet wet spinning method. The performances of these PVDF-MMMs were assessed with respect to the removals of phosphorus, FFA and colour. Based on finding showed that the combination of 18PVDF embedded with 3 wt% MagS showed the most compatible polymer-inorganic hybrid MMM to perform pre-treatment of CPO. Increasing the MagS concentration from 3 to 8 wt% in the polymer matrix recorded the highest removals of FFA at 16.51 %, phospholipid at 93.31 % and colour at 18.8 %, respectively. Nevertheless, high amount of MagS added to the polymer matrix affected the spinnability and reproducibility performance of MMM. Future work, surface modifications of inorganic adsorbent can be evaluated to facilitate good dispersion of filler in polymer matrix during membrane fabrication to maintain good membrane reproducibility.

1. Introduction

The production of crude palm oil (CPO) at oil mill processing usually contains variable amounts of minor components such as free fatty acids (FFA), partial acylglycerols, phosphatides, sterols, tocopherols, tocotrienols, hydrocarbons, pigments, traces of pesticides, dioxins, metal ions and oxidation by-products. The CPO must undergo a refining process to remove all unwanted compounds so that the final refined oil is of good quality, stable and safe for human consumption. From an industrial perspective, the goal of refining is to convert the crude oil into a high quality edible oil by reducing unwanted impurities to the desired level as efficiently as possible. This means that, losses of desirable components should be kept to a minimum and cost effective [1]. Two refining methods commonly used at palm oil refinery are the physical refining process for most refined palm oil products and the chemical refining process for more specialized products according to customer requirements.

Degumming and deacidification are very important steps in the palm oil refining process. The main objective of the degumming process is to remove the gums or phosphatides in CPO. The gums must be removed at an early stage of refining to avoid high refining losses as they are emulsifying and thermally unstable that burns at high deodorization temperatures. In addition, the presence of gums contributes to off-flavor and poor oxidation stability as phosphatides coordinate with metal ions such as copper, iron, calcium and magnesium which are catalysts for oil oxidation. There are 2 types of phospholipids (PLs), namely hydratable and non-hydratable. The hydratable phospholipid can be easily removed by adding water and then centrifuging. This process is called water degumming. The non-hydratable phospholipids associated with calcium and magnesium salt need to be converted to hydratable ones by acid treatment with phosphoric acid or citric acid [2].

Deacidification is the process of removing FFA in CPO. FFA is undesirable because it is one of the pro-oxidant compounds that reduce oxidative stability, increase the acidity of the oil and lead to the formation of off-flavors in the refined end products. In the physical refining process, deacidification was carried out at high vacuum-steam distillation and deodorization temperatures between 260 – 265 °C to strip the FFA into volatile odorous compounds and remove them in the form of palm fatty acid distillate. In the chemical refining process, deacidification was carried out with sodium hydroxide (NaOH) to neutralize and remove the FFA in the form of soapstock. The FFA content in CPO determines the amount of NaOH to be added for the neutralization process. Any residual soapstock is removed by adding hot water and then centrifuging. The remaining FFA in the neutralized bleached palm oil is further removed in the final deodorization stage at lower temperatures between 220 – 240 °C [3].

The use of phosphoric acid and caustic soda for the degumming and neutralization process had become an environmental and safety issue due to their disposal. Large amounts of wastewater generated during the deacidification process require costly treatment for soapstock. As a result, both capital and energy costs are high [4]. In physical refining, the elimination of the neutralization step has the advantage of reducing the loss of neutral oil and simplifying process operation. Nevertheless, high operating costs (steam, electricity and water) are required to maintain the operation as steam and operation temperatures are high [5, 6].

Membrane technology is a mature industry that has been successfully applied in various food industries for separation of undesirable compounds, water and waste-water treatments as well as gas separation. It offers a huge potential that worth to explore in palm oil industry to harvest the benefits towards green and mild processing technology to eliminate the use of chemical, low energy consumption as well as retention of nutrients [7]. Köseoglu and Engelgau had reported on membrane applications and research in the edible oil industry since 1990 [8]. Fast forward to the present, one can see the increasing number of relevant research works in refining process in particular in the applications of degumming and deacidification [9–12]. Previous studies have reported that ultrafiltration (UF) membrane can be very effective (depending on the pore size selected) in separating phospholipid from vegetable oils without adding any chemical during degumming process [3,13–15]. Other than pressure-driven filtration membrane, polyphenylsulfone hollow fiber (HF) membrane for CPO deacidification through membrane contactor was reported by Othman *et al.* (2017) to extract the FFA in CPO using NaOH as liquid extractant without soap formation [16,17]. Azmi *et al.* (2016) reported on PVA-crosslinked PVDF membrane to improve the surface affinity toward FFA for direct deacidification of CPO [18].

Our previous studies revealed that membrane-based separation technology using PVDF hollow fiber membranes was a promising alternative for degumming of CPO with phosphorus content in permeate below 2 ppm [3]. However, its effectiveness in removing FFA was not promising due to small differences in molecular weight of major component of CPO (triglycerides ~900 Dalton) and FFA (~300 Dalton) [19]. Mixed matrix membrane (MMM) was one of possible approaches to improve the separation properties of invented PVDF membrane. It is a polymer–inorganic hybrid membrane that may offer a synergistic effects rendered by both polymeric matrix and inorganic fillers such as good selectivity and permeability, superior mechanical strength and high resistant to chemical and temperature [20].

The present research work aimed to investigate the compatibility of in-house made PVDF mixed matrix membrane (MMM) embedded with various inorganic adsorbents and to examine the of PVDF-

MMMs for simultaneous degumming and deacidification of CPO in removing the phospholipids (PLs - measured by phosphorus content) and free fatty acid (FFA).

2. Materials and Methods

2.1. Materials

Crude palm oil (CPO) was obtained from Sime Darby Oils Langat Refinery located at Telok Panglima Garang, Selangor. Table 1 shows the specifications of CPO according to Malaysian Standard MS814:2007. The main membrane raw material of polyvinylidene fluoride (PVDF) pellets (Kynar® 740) was purchased from Arkema Inc., USA. Whereas N-methyl-2-pyrrolidone (NMP) and ethylene glycol (EG) were bought from Merck. The inorganic absorbents to be embedded into PVDF membranes consisted of magnesium silicate or trade name known as Magnesol (MagS) obtained from Dallas Group, activated carbon (AC) obtained from SYSTERM ®, zeolite (ZMS) and calcium silicate (CaS) which both obtained from Sigma Aldrich. All chemicals used in this work were received without further treatment.

Table 1. Guidelines identity characteristic of CPO.

Identity Characteristics	Observed Range	Mean ± Standard Devition
Apparent density, kg/L at 50 °C	0.8889 – 0.8896	0.008892 ± 0.0002
Refractive index at 50 °C	1.4521 – 1.4541	1.4533 ± 0.0003
Saponification value, mg KOH/g oil	194 - 205	199 ± 2.4
Unsaponifiable matter, %	0.19 – 0.44	0.32 ± 0.066
Iodine Value (Wjis)	50.4 – 53.7	52 ± 0.66
Slip Melting Point, °C	33.8 – 39.2	36.7 ± 0.84
Total carotenoid (as ß-carotene), mg/kg	474 - 689	581 ± 45.5

2.2. Formulation of Dope Solution for Mixed Matrix Membrane

PVDF pellet was dried at 100 °C for 5 hours to remove the moisture contents. The dried PVDF was gradually added to a combination of NMP, EG and inorganic absorbent of various properties (MagS, AC, ZSM and CaS) in the reaction flask and mixed thoroughly under constant mechanical stirring at 60 °C. It took at least 1 day to produce a homogenous solution in order to ensure the PVDF-MMM was completely dissolved in the solvent. Finally, the prepared dope solution was degassed to remove any micro-bubbles that might exist before spinning. Table 2 listed all the dope solutions made for fabrication of hollow fiber PVDF-MMMs.

Table 2. Formulation of dope solution for mixed matrix PVDF membranes.

Membrane	Dope Formulation (%)			
	Polymer	Solvent	Additive	Adsorbent
18PVDF+3CaS	18% PVDF	73% NMP	6% EG	3% CaS
18PVDF+3AC	18% PVDF	73% NMP	6% EG	3% AC
18PVDF+3ZSM	18% PVDF	73% NMP	6% EG	3% ZSM
18PVDF+3MagS	18% PVDF	73% NMP	6% EG	3% MagS
18PVDF+5MagS	18% PVDF	71% NMP	6% EG	5% MagS
18PVDF+8MagS	18% PVDF	68% NMP	6% EG	8% MagS

2.3. Fabrication of Hollow Fiber Mixed Matrix Membrane

The hollow fiber membranes were fabricated using a technique known as dry-jet wet spinning. Once the spinning dope and the bore fluid (water) met at the tip of the spinneret with air gap of 10cm, the hollow fiber membrane was produced simultaneously and transferred directly into an external coagulation bath (water). After that, the fibers went through a washing/treatment bath before they being gathered by a wind-up drum. The as-spun membranes were kept in a water bath at room temperature for 1 day, post-treated with ethanol solution to prevent shrinkage and lastly air-dried at room temperature before being used to make a test module. The hollow fiber membranes were potted into bundles of 60 fibers with a length of 25 cm into the U shape test module before ready for testing.

2.4. Mixed Matrix Membranes Performance for Pre-treatment of Crude Palm Oil

The membrane pre-treatment of CPO was set in a simple dead end filtration mode as shown in figure 1. CPO was semi-solid at room temperature. It must be heated at 60 °C to kept it in a liquid state before fed into a MMM processing module. The pre-treatment of CPO was conducted under pressure-driven membrane filtration operated at 2 bar for 1 hour to obtain a sufficient amount of permeate fraction for sample analysis. The performances of these MMMs were evaluated based on the percentage removals of FFA, phosphorus and colour of pre-treated CPO (permeate) as compared to initial CPO (feed).

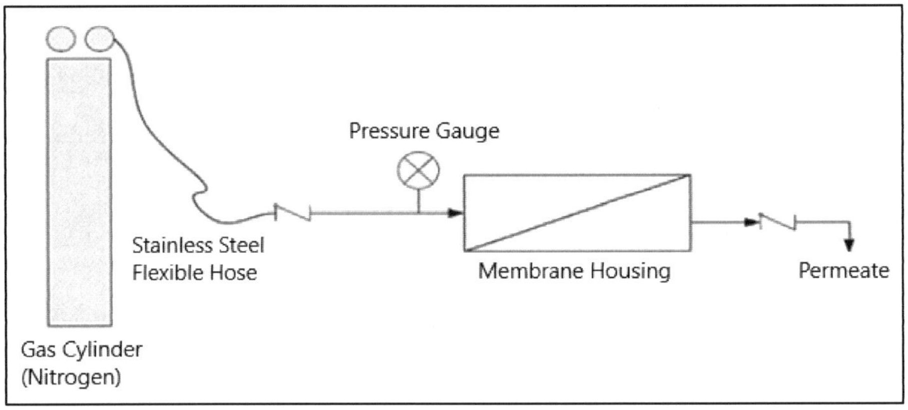

Figure 1. Experimental setup for CPO membrane pre-treatment system.

2.5. Analytical Methods

The analysis of feed CPO and permeate (pre-treated CPO) with respect to phosphorus, FFA and colour were determined in accordance to American Oil Chemists' Society (AOCS) method using AOCS method of Ca 12-55, Ca 5a-40 and Cc 13e-92, respectively.

3. Results and Discussion

Table 3 shows the performances of MMM embedded with four different adsorbents for pre-treatment of CPO. It can be observed that 18PVDF+3MagS MMM exhibited the highest FFA removal among the MMMs tested which recorded at 8.18 %. Besides, it was also able to eliminate more than 90 % phosphorus with 13.3 % colour removal in permeate sample. 18PVDF+3CaS membrane showed slight reduction in FFA (7.8 %), followed by 18PVDF+3AC (6.4 %) and 18PVDF+3ZSM (0 %). Although the efficiency of FFA removal was reduced, the MMMs embedded with CaS, ZSM and AC were able to maintain a promising phosphorus removal (at least 87.3 %), achieving no more than 2 ppm phosphorus content in the permeate sample similar to the conventional phosphoric acid degumming.

PLs in nature had both hydrophilic and hydrophobic functional groups [8,21]. Although it had almost similar molecular weight (700 Dalton) with triglycerides (900 Dalton) [22], it can form a reverse micelle (MW of 20,000 Da) in a non-aqueous environment due to its surfactant properties [8,12]. The large size of the PLs mixed micelles has a synergetic effect during the membrane separation process, causing it to be significantly rejected by the membrane [21]. The significant challenge of using MMM for removing FFA is the possibility of the internal channel blockage and/or active surface coverage of the adsorbent by polymer, reducing the effective area of adsorption which leads to lower FFA rejection

Table 3. The performances of mixed matrix PVDF membranes embedded with different adsorbents for pre-treatment of CPO.

Mixed Matrix Membrane	FFA			Phosphorus			Colour (5R+Y)		
	Feed (%)	Permeate (%)	Rej. (%)	Feed (ppm)	Permeate (ppm)	Rej. (%)	Feed	Permeate	Rej. (%)
18PVDF+3CaS	3.34	3.10	7.46	10.50	1.09	89.62	165	135	18.2
18PVDF+3MagS	3.79	3.48	8.18	14.30	1.39	90.28	158	137	13.3
18PVDF+3ZSM	3.79	3.79	0.00	14.24	1.67	88.27	158	138	12.6
18PVDF+3AC	3.34	3.12	6.31	10.50	1.34	87.33	160	138	13.7

Magnesol in fact very ideal for use as adsorbent in improving the shelf life of frying oil. Official data from the producer have confirmed that Magnesol can remove both solid and dissolved impurities from the oil, owing to its acid and basic adsorptive sites which allow promising removal of polar compounds including FFA. Table 4 compares the performances of the PVDF membrane embedded with different weight content of magnesium silicate (MagS) to enhance the deacidification of CPO. Increasing the amount of MagS in the membrane matrix was able to further increased the FFA removal in the permeate sample from 8.18 % (18PVDF+3MagS) to 16.51 % (18PVDF+8MagS), respectively. It was mainly due to the increased active surface of the adsorbent which improves the adsorption capacity of PVDF-MMM. It must be noted that the promising rejection of phosphorus was not negatively affected with increasing content of MagS in membrane. All PVDF-MMMs modules were reported to remove at least 90 % phosphorus, producing a permeate containing between 1.13 and 1.39 ppm phosphorus.

Nevertheless, it must be pointed out that some issue been faced when high amount of MagS was added into polymer matrix. Severe agglomeration and poor dispersion of MagS was encountered at the bottom of membrane dope solution as a results of incompatibility between PVDF membrane matrix and inorganic filler. As a consequences, it affected the spinnability and reproducibility of MMM performances especially in removing the FFA. ble

Table 4. The performances of mixed matrix PVDF membranes embedded with different amount of MagS for pre-treatment of CPO.

Mixed Matrix Membrane	FFA			Phosphorus			Colour (5R+Y)		
	Feed (%)	Permeate (%)	Rej. (%)	Feed (ppm)	Permeate (ppm)	Rej. (%)	Feed	Permeate	Rej. (%)
18PVDF+3MagS	3.79	3.48	8.18	14.30	1.39	90.28	158	137	13.3
18PVDF+5MagS	3.74	3.42	8.56	14.54	1.18	91.88	158	148	6.3
18PVDF+8MagS	4.18	3.49	16.51	16.90	1.13	93.31	170	138	18.8

4. Conclusion

Mixed matrix membranes embedded with various absorbents successfully fabricated *via* a dry-jet wet spinning method. Among four different adsorbent tested for fabrication of PVDF-MMMs, the combination of 18PVDF+3MagS showed the most compatible to perform a simultaneous degumming and deacdification of CPO with 8.16 % FFA removal, eliminated up to 90.28 % phosphorus with 13.3 % colour has been reduced simultaneously. Increasing the amount of MagS embedded into the PVDF membrane matrix had further improved the removal of FFA at 16.51 %, phosphorus up to 93.31 % and colour up to 18.8 %, respectively as recorded by 18PVDF+8MagS MMM. Unfortunately, high amount of MagS added into polymer matrix unable to sustain the membrane reproducibility due to poor dispersion and severe agglomeration of MagS in the dope solution which affected the filtration performance of CPO. Surface modification of inorganic adsorbent can be further evaluated to improve the dispersion of MagS in the PVDF polymer matrix at high concentration of inorganic filler.

Acknowledgement
The author (N.H. Othman) would like to thanked Sime Darby Plantation Research Sdn. Bhd. for financial support and members of Advanced Membrane Technology Centre (AMTEC), UTM Skudai, Johor for technical support and collaborative works on developing a membrane refining technology specially made for palm oil industry.

References

[1] Gibon V, Greyt W and Kellens M 2007 *Eur. J. Lipid Sci. Tech.* **109** 315–35
[2] Othman N H, Latip R A, Yusoff M S A, Ismail A F, Lau W J, Ng B C, Abdullah M S and Saidin M A R, inventors. Sime Darby Malaysia Berhad, assignee. *A Process for Degumming of Crude Palm Oil.* World Intellectual Property Organization WO 2014/058294 A1. 2014-4-17
[3] Othman N H, Noor A M, Yusoff M S A, Ismail A F, Goh P S and Lau W J· 2016 *Membrane Technology for Water and Wastewater Treatment, Energy and Environment* Vol 3 ed AF Ismail and T Matsuura (London: CRC Press) chapter 5 pp 297–307
[4] Lai L L, Soheili K C and Artz W E 2008 *J. Am. Oil Chem. Soc.* **85**, Issue 2, 189–96
[5] Destaillats F, Craft B D, Sandoz L and Nagy K 2012a *Food Addit. and Contam.* **1** 29–37
[6] Destaillats F, Craft B D, Sandoz L and Nagy K 2012b *Food Chem.* **4** 1391–8
[7] Cui Z F and Muralidhara H S 2010 *Membr. Technol.* **312**
[8] Koseoglu S S and Engelgau D E 1990 *J. Am. Oil Chem. Soc* **4** 239–49
[9] Raman L P, Cheryan M and Rajagopalan N 1996 *J. Am. Oil. Chem. Soc.* **73** 219–24
[10] Pagliero C, Ochoa N, Marchese J and Mattea M 2004 *J. Chem. Technol. Biotechnol.* **79** 148–52
[11] Manjula S and Subramani R 2009 *Sep. Purif. Technol* **66** 223–8
[12] Iyuke S E , Ahmadun F and Majid R A 2004 *J. Food Process Eng.* **27** 476–96
[13] Othman N H, Noor A M, Latip R A, Yusoff M S A, Ibrahim S, Ismail A F, Lau W J, Goh P S, Ng B C, Abdullah M S and Saidin M A R, inventors. Sime Darby Malaysia Berhad, assignee. *A Membrane Pre-treatment System and Process for Producing Refined Oils and Fats.* World Intellectual Property Organization WO 2015/053609 A1. 2015-4-16
[14] Koris E M 2006 *Desalination* **200** 537–9
[15] Kim I C, Kim J H, Lee K H and Tak T M 2002 *J. Membr. Sci.* **205** 113–23
[16] Othman N H, Noor A M, Yusoff M S A, Goh P S and Ismail A F 2017 *J. Membr. Sci. Res.* **3** 291–5
[17] Othman N H, Noor A M, Yusoff M S A, Goh P S, Lau W J, Ng B C and Ismail A F 2017 *Malays. J. Anal. Sci.* **21** 633–42
[18] Azmi R A, Goh P S, Ismail A F, Lau W J, Ng B C, Othman N H, Noor AM and Yusoff M S A 2015 *J. Food Eng.* **166** 165–73
[19] Cheryan M 2005 *Membr. Technol.***2** 5–7
[20] Souza V C and Quadri M G N 2013 *Braz. J. Chem. Eng.* **4** 683–700
[21] Gupta A K S, inventor. Lever Brothers Company, assignee. *Process for Refining Crude Glyceride Oils by Membrane Filtration.* United States Patent 4062882. 1977-12-13
[22] Ochoa N, Pagliero C, Marchese J and Mattea M 2001 *Sep. Purif. Technol.* **22** 417–22

Used lubricating oil recovery process and treatment methods: A review

S M Anisuzzaman[1,2], M H Jumaidi[2] and N N M Nasir[2]

[1] Energy Research Unit (ERU)
[2] Chemical Engineering Programme, Faculty of Engineering,
Universiti Malaysia Sabah, 88400 Kota Kinabalu, Sabah, MALAYSIA

Email: anis_zaman@ums.edu.my

Abstract. Used lubricating oil (ULO) is considered hazardous as it is able to cause pollution and affect the environment. The presence of degraded additives, contaminants, and by-products of degradation render ULO more toxic and harmful to health and environment than virgin base oils. Recovery of ULO generally comprises cleaning, drying, and adsorption in order to eliminate water, sludge, and impurities. As the ULO is one of the hazardous wastes generated in various industries, such as industrial and automotive, it should not be used or disposed of in ways that are harmful for the environment. Recovery of ULO carries out many advantages which includes lower environmental impact, higher energy saving and lower risks. The main objective of this paper was to thoroughly review various recovery process principles and treatment methods for ULO. Importance of ULO recycling and various techniques along with their limitations were also discussed. The significance of this study lies in reviewing the roles of adsorbent and adsorption reclamation processes of ULO and few promising adsorbents were earmarked for further study.

Keywords: used lubricating oil, recycling; treatment, pollutants, contaminants, removal methods, environmental pollution

1. Introduction

Lubricating oil (LO) is petroleum derived product and consists of a mixture of hydrocarbons which are a longer chain of isoalkanes, monocycloalkanes and several short branches on the ring of monoaromatics [1]. LOs are made of two components which are the base oil and the additives and are formed by distillation of petroleum with a temperature range from 300 °C to 400 °C [2]. LO is being classified into five different groups which are industrial oil, grease, automotive oils, metalworking fluids and process oils that are widely being used in industries globally [3]. According to Moura *et al.* [2], LOs are base oils that are mixed with additives to enhance their characteristics which then can be used to minimize wear that are caused by physical friction. During usage, the LO will undergo changes in terms of degradation, oxidation and contamination, which make the LO ineffective for further application and need to be replaced [4]. Literally, LO is used widely in all moving machineries to reduce the friction between two moving parts that are rubbing against each other [5]. In other words, it helps to reduce the build-up of temperature and formation of corrosion within the moving parts that indirectly reduces the efficiency of the machine [1].

Basically, the LOs consist of two components, the base oil and the additives. The main additives include antioxidants, detergents, anti-wear materials, metal deactivators, corrosion inhibitors, rust inhibitors, friction modifiers, pressure-resistant products, anti-foaming agents, viscosity index enhancers, demulsifying or emulsifying agents, stickiness enhancers and so on. However, the performance of the LO will come to an end after a period of time and will reduce the efficiency as the additives become chemically altered and the oil becomes polluted with various harmful toxins and contaminants due to numerous physical and chemical interactions [6]. Degradation of LO also may be due to a mixture of different unwanted matters that contains filings, metal powder, other oils and additives as well. The oxidation of LO produces used lubricating oil (ULO) that has a darker colour, the acid value increases to produce precipitate, bad odour, oil sludge and water content [6,7]. The occurrence of oxidation process will make the LO acidic and these acidic products in the oil will cause corrosion to the internal engine parts, deposits and insoluble oxidation products that literally decrease the performance of the engine. The more the oxidation, the more compounds with higher oxidation states are produced and the acidity of the lubricant increases with the increasing levels of carboxylic acids [8]. Furthermore, at high temperature, the engine oil (EO) will still degrade even with the absence of oxygen. As the temperature increases, the viscosity of the oil increases which is also one of the causes of EO deteriorates [6]. Thus, it could be concluded that by improving thermal stability of EO will also enhance the oxidation stability and viscosity of LO, besides improving the volatility and evaporation rate of EO.

Contaminants containing phenolic compounds, aldehyde, metals, varnish, gums and asphaltic compounds are produced due to the internal combustion of the engine [9]. Emam and Shoaib [10] mentioned that the cause of impurities formation is due to oxidation or thermal degradation of the oil. The further usage of deteriorated ULO in the machineries or equipment's, the contaminants in the oil deposit on the surface of the moving parts and cause malfunctions of the instrument [11].

The lubricant waste has a very high hazard as it contains massive amounts of ash content, carbon residue, asphaltenic materials, metals, water and other dirty materials [12,13]. Various environmental problems can arise if the waste is not disposed correctly [14,15]. Waste oil management is a growing concern to many nations, especially in the industrial and urban areas as this hazardous waste has been disposed of into the environment and caused a lot of complications.

Somehow, the idea of oil recovery is a great effort and it has been a four decades tradition since it was first presented in the 1930s. Initially, the ULO had undergone combustion for energy production then it was re-blended with engine oil (EO) after treatment. The evolution of recycling technology is crucial for re-refining the ULO to remove the physical, mechanical and chemical contaminations through the process such as distillation, hydrogenation, acidic refining, solvent refining, or combinations of the formers [10]. Lately, there are numerous studies conducted on the recovery of ULO using adsorbent through the process called adsorption [16]. Adsorption is a process whereby an adsorbate is used to adsorb the unwanted contaminants from the used oil to improve the properties [14,16–18].

Hence, the main objective of this paper is to thoroughly review various recovery process principles and treatment methods for ULO. This study was based on addressing problems associated with improper application and disposal of ULOs using recovery techniques and re-refining options for protecting the environment and conserving resources. The significance of this study lies in reviewing the roles of adsorbent and adsorption reclamation processes of ULO and few promising adsorbents were earmarked for further study.

2. Cause of LO degradation

According to Rammohan, 2016 [19], if the LO is not frequently changed, contaminants such as the dirt will accumulate in the engine and fail to lubricate the moving part in the engine. Dirty oil can cause serious damage to the moving parts in the engine whereby replacement of the engine might occur in the near upcoming period. Aljabiri [6] found that the vital mechanisms that cause the depletion of LO additives are due to thermal degradation, oxidation, neutralization, shearing, hydrolysis, water washing, particle scrubbing, filtration, contamination, rubbing contact, surface adsorption, rubbing contact, condensation settling and evaporation. There are three major factors that are identified contributing to the degradation of LO, which includes extreme heat, extreme cold and common contaminants [19,20]. However, the growth of contaminants and chemical variation in the oil itself will cause the LO to be degraded and unsuitable for further usage in the engine parts. The factors of degradation also include entrained air, heat, moisture, incompatible gases, process constituent, radiation, internal or external contamination, and inadvertent mixing of fluids.

Water present in the engine due to condensation, can lead to corrosion. Corrosion related damages in the engine would be more critical when water is present in hotter oil due to the severe chemical reaction. Presence of water in LO for a prolonged period of time, can emulsify the oil and a corrosive mixture is formed which will then form sludge that may block the oil passage or filters [6]. Acidic solutions can also be combined with the water within the engine and create an emulsion that can cause problems with oil filters and passageways [21]. The hydrocarbon reacts with oxygen in the initiation phase of the chemical reaction and forms highly reactive peroxide radicals. Finally, hydroperoxides detach and form oxygenated compounds such as aldehydes, ketones, alcohols, and water. Such compounds further react to form high molecular weight organic acids and polymeric materials. Further polymerisation and polycondensation of these products contribute to an insoluble product called sludge, which will precipitate as a thin film to form lacquers or varnishes on cold or hot metal surfaces [22].

High temperature with increased time of usage will degrade the desired properties of LO and thus need to be evacuated and replaced with a new one [23]. ULO contains metals such as lead, zinc, barium, arsenic, chromium and cadmium [24,25]. In terms of chlorinated hydrocarbons, the ULO consists of compounds such as dichlorodifluoromethane and trichlorotrifluoroethane. Moreover, other organic matter is also present in ULO such as benzene, xylene, toluene and naphthalene [25]. The ULO main constituent is the degraded additives, base oil, metallic debris, carbon soot, and oxidation product. A huge number of additives are used to convey the lubricants' performance and during use, these additives lose their characteristic. In addition, the LO picks up a fraction of various metals as the result of wearing of components. Other contaminants such as chlorinated solvents, water, unburned fuel, carbon and dust are also picked up during storage [26].

The air contaminants may contain moisture, dirt, and dust, and even the air itself is a contaminant that could cause the oil to foam. Such contaminants move into the engine *via* the air-cleaner and may consist of minor silicate elements [6]. Meanwhile, the engine contaminants are the metallic particles due to the wear of the engine, carbonaceous particles from the incomplete combustion, metallic oxides from the metal corrosion, water from the cooling system leakage and the product of combustion, and the fuel additive or its by-product which might reach the engine crankcase [20]. In addition to that, since the LO act as an anti-corrosive, cooling and cleaning agent, it picks up a variety of impurities and additional components which include the metal particles such as steel, iron, copper, zinc, lead and other compounds of sulphur, water, barium, carbon, and ash [27]. These contaminants directly interfere with the lubricant's viscosity, lowering its efficiency and performance.

3. Effects of ULO contaminants to human health and environment

The contaminants in ULO render a wide range of problems to human healthiness and environment [28,29]. The presence of degraded contaminants, additives and by-products such as water and sediments will transform the ULO into a toxic one compared to virgin base oil. The improper disposal of ULO can lead to adverse environmental and health problems. For landfills with permeable soils, water - soluble materials in lubricants used can easily penetrate waste and end up contaminating groundwater by runoff leaching and percolation. The waterways and the coastal waters can be affected if the ULO is being disposed of in storm water or waterways. Excessive ULO will cause a threat to plant and animal life, which may lead to economic losses in fishing and recreation industries [24]. Musemić and Bašić, [30] mentioned that ULO has a variety of serious, short and semi-permanent impacts on the environment if disposed of in an uncontrolled manner. The ULO inhibits the penetration of chemical elements such as oxygen into water, therefore degrading the condition of life of several species. ULO contains various poisonous substances, together with polycyclic aromatic hydrocarbons, that are famed to cause cancer if they are inhaled or ingested [31]. Waste oil that has been disposed of in the water will form a thin oil layer that blocks sunlight on the water surface that inhibits the oxygen from dissolving in water [14,31]. This will lead to extinction or death of the aquatic plants and animals in water. Besides, when exposed to air, at high temperature, the ULO can be vaporized to air polluting the fresh air as well. As mentioned, ULO consists of numerous contaminants such as hydrocarbons, heavy metals, halogen compounds, gums, varnish and polychlorinated biphenyls that can harm human beings, aquatic life and environment as well [32]. Not only that, presence of dangerous elements such as polycyclic hydrocarbons (PAHs) and aliphatic hydrocarbons in ULO had a higher flexibility to cause carcinoma [33]. ULO are also very toxic to animals, particularly embryos and new-born animals. Table 1 summarizes the effects of ULO contaminants and impact of oil emissions to human health and environment.

Table 1. Effects of ULO contaminants and impact of oil emissions to human health and environment.

Oil contaminants	Effects	References
Water	Causes rusting of iron or steel Results in the development of water sludge Leads to foaming problems.	[6]
Solid particles	Causes excessive wear, bearing surface score, and potential failure to absorb metal fatigue.	[6]
Sludge and lacquers	Sludge deposits are clogging narrow oil passages and clearances. Causes the valve to stick and resist the continuous operation of the oil pump.	[6]
Liquid contaminants	Unburned fuel from engines dilutes LO and may lessen its viscosity.	[6]
	Impact of oil emissions	
Base oils derived from crude oil	Eco-toxicity, incomplete biodegradability, and highly probable carcinogenicity	[29]
Oil mist	Causing significant health issues such as changes in the lungs, liver, kidneys, adrenal glands, and heart. It results in a negative influence on the respiratory and nervous systems.	[28]
	It is also absorbed by skin and causes skin cancer	[29]

4. Importance of ULO recycling

As previously mentioned ULO is classified as hazardous and constitutes extensive damage to the environment, creating risks of contaminating water, air, and soil with the presence of harmful substances. Consequently, the management of ULO is predominantly vital due to the large amounts generated worldwide, the potential for straight reuse, regenerating, reprocessing, and environmental damaging effects if not handled, treated or disposed of properly. As the number of vehicles increases gradually, the volume of ULO produced is also growing. The disposal and management of the wastage is indeed a rising fear, especially in industrial and urban areas. Many nations are now addressing the predicament of environmental pollution posed by LO wastage [12].

ULO recovery not only guards the environment but likewise preserves a treasured non-renewable resource [25]. ULO is likewise a valuable source of carbon and is considered as one of the important energy sources in the world. ULO may be re-refined into base oil, reprocessed as oil or used as feedstock to supply petroleum-based merchandise or different commercially valuable products through totally different ways. Additionally, the valuable non-renewable resources would be conserved besides solely protecting the environment from ULO [34]. From the point of energy conservation, utilization of ULO expeditiously saves precious, renewable resources. During these days, many companies are investing billions of greenbacks in recovering, exploring and processing ULO into quality LO. Moreover, ULO generated by the customers will be recycled back to helpful merchandise like pure fill oil again and once more, tremendous time and money can be saved as well [34,35]. Presence of refining equipment can reduce cost for hydro treatment, storage of product and pollution can be decreased. On top of that, instead of improper disposal of ULO, it is recycled, re-fined or reused to prevent contamination of soil, groundwater and surface water.

In a nutshell, if ULO is well managed, the scenario of depending on fossil fuels from other countries would be decreased [35]. Proper assortment and regeneration of ULO will be useful to the nation thereby reducing environmental pollution, conserving the mineral resources and saving the foreign exchange for importing the virgin oil [14].

5. ULO recovery and reuse techniques

After being consumed in a period, the petroleum-based LO has become unsuitable for the designed purpose. Due to toxicity through use, the ULO is unable to handle anthropogenic pollutants and become undesirable [36]. However, this wastage can also be considered as a valuable resource in the sense that it can reclaim energy or any profitable material for further use. Thus, recycling or recovery of LO wastage may be an appropriate low-cost alternative and indeed a need of the hour due to its economic, environmental, legal reasons, and public health importance [7]. The disposal management and recovery of ULO have become a crucial issue due to the increasing necessity of environmental protection and strict environmental legislation [37]. One of the disadvantages of recycling ULO is the mixing of different types of waste oil. Besides, there is no clear and specific way of collecting or gathering ULO for recycling. In reality, recovered ULO can be used as an industrial fuel burner, hydraulic oil, integrated into other goods or re-engineered into new LOs [38].

Recovery of ULO can slow down the consumption of natural resources. Recycling aims to a large degree to combat global climate change. By reducing the energy expended on industrial production, recycling helps in minimizing greenhouse gas emissions. By minimizing the energy used, recycling also minimizes the usage of fuel, which in turn decreases the amount of harmful contaminants in the atmosphere [39]. Recovery of ULO also contributes to economic benefits. The recycling process creates employment opportunities for a large number of people involved in the various stages of the process [40].

The reuse of ULO can be approached by three basic methods, namely incineration, reprocessing and re-refining. In incineration, the ULO is treated with the presence of heat. The used oil is thermally destroyed. Diphare *et al.* [26] claimed that, during these contemporary days, incineration is rarely used due to the economic benefits of recycling. Moreover, burning of ULO releases toxic contaminants to the environment. Thus, such a type of recycling is unsafe and impractical. In reprocessing technology, the ULO is heated and processed for energy recovery. It involves water and particulates removal from the oil before burning the fuel to generate heat. Unfortunately, this type of recycling does not provide the best result to recycle the contaminant high ULO. This is because reprocessing methods of ULO may produce by-products that are even more toxic as the secondary pollutant. Furthermore, ULO can be also recycled through a re-refined process. In this process the ULO is re-refined back into lube oil that can be recycled and reused. In this method, the contaminants such as water and dissolved low boiling organic is eliminated. Different treatments are used to recover the lube oil. Re-refining is different from incineration and reprocessing, whereby in this process, the ULO is treated to remove the impurities and increase the properties of the oil, in order to be used as a virgin LO. Re-refining technology of ULO into LO requires less energy than incineration and reprocessing [26]. Thus, it is being preferred as it consumes less energy, produces no damage to the environment and less waste generation. However, the reprocessing of the recovered material is not always free from pollution. Some reprocessing technologies produce residues that are difficult to treat. The acid-clay method for the re-refining of ULO is one of the examples where residual sludge has polluted the soil. Moreover, conventional acid-clay treatments usually require the use of highly expensive adsorbents such as kaolin, alumina, silica gels etc. According to Shakirullah *et al.* [41], existing acid-clay treatment techniques for recovery require endless chemicals and give low yield. Table 2 summarizes the comparison of re-refining methods.

Table 2. Comparison of re-refining methods.

Sl. no.	Method	Type of sample	Outcome	References
1.	Acetic acid-clay treatment	Used car EO	Water and sediment content: 0.013 ml, Carbon residue: 1.65 wt%, Total acid number (TAN): 1.8 mg KOH/g oil, Metal content- Cu:3.56 ppm; Mg:65 ppm; Pb:11.2 ppm; Zn:780 ppm	[42]
2.	Solvent extraction/clay re-refining process	ULO	Sulfur: 0.81 wt%, Water content: Nil; Specific gravity@ 40 °C: 0.6991, Metal content- Fe: Nil; Cu: Nil; Zn: 117.8 ppm	[9]
3.	Acid treatment/clay – percolation re-refining	ULO	Sulfur:0.42wt%, Water content: Nil; Specific gravity@ 40°C: 0.8202, Metal content- Fe: Nil; Cu: 1.048 ppm; Zn:	[9]
4.	Extraction by composite solvent followed by acid treatment	Used EO	Flash point: 150 °C, Specific gravity:0.88, Viscosity @ room temperature: 94 cp, Pour point:-15°C, Ash content (%): 0.09, Iron: 13 ppm	[27]
5.	Ionizing radiation process	Used automotive LO	Mg, Al, Ti, Cr, Mn , Fe ,Ni, Cu, Se, Mo, Nb, Cd, Sn, Ba, Bi and Pb : <10 $\mu g.\,ml^{-1}$	[43]
6.	Solvent extraction process	Used motor EO	The sludge formation increases with increasing solvent 1-butanol produce the best extraction performance in sludge formation	[44]
7.	Microwave pyrolysis reactor	Car EO	Formation of single ring alkyl aromatics such as benzene, toluene, ethyl-benzene and xylene	[45]
8.	Vacuum distillation recycling by continuous extraction with dense propane	ULO	Metal content- Pb: 0.1 ppm; Fe: 0.06 ppm; Zn: 0.01 ppm; Ca: 0.01 ppm	[46]
9	Distillation-clay treatment	LO	Water content: 0.66 v/v Specific gravity: 0.86 Viscosity index: 85.80	[1]

			Flash point: 168 °C	
10	Acid treatment	LO	Water content: 0.60 v/v Specific gravity: 0.86 Viscosity index: 84.40 Flash point: 170 °C	[1]
11	Activated Charcoal-clay treatment	LO	Water content: 0.47 v/v Specific gravity: 0.86 Viscosity index: 86.80 Flash point: 178 °C	[1]
12	Acid treatment	Motor oil	Formic acid specific gravity: 0.915 Flash point: 230 °C Sulphuric acid specific gravity: 0.920 Flash point: 210 °C Phosphoric acid specific gravity: 0.930 Flash point: 200 °C Acetic acid Specific gravity: 0.915 Flash point: 218 °C	[47]

6. Adsorbent and adsorption process in the treatment of ULO

The adsorption process is primarily used for purification or separation purposes. Adsorbent is a micro porous structure that is manufactured to have adsorption properties. The high porous medium offers a large amount of micropores; this results in a large capacity for adsorption. The porous solid material has very small pores in thickness, where adsorbed molecules can find their way to the micropore surface [48].

There are many factors that literally affect the efficiency of the adsorption process such as contact time, temperature and adsorbent dosage [16–18,49]. Few studies have been conducted using activated carbon (AC) in the treatment of the ULO by replacing clay to adsorbents. ULO was recovered by using various types of adsorbents such as date palm kernels powder, bentonite, eggshell powder and so on. According to Riyanto *et al.* [50], the AC is a good material for treatment of ULO, especially to reduce the metal concentrations. The rising amount of AC increases the adsorption potential of the metal in the LO waste. The waste oil treatment using clay or adsorbent can be improved by addition of a catalyst. Bhaskar [51] presented the supported iron oxide catalyst to be used in the conventional clay treatment in producing fuel oil from ULO. The adsorbent has its advantages due to its flexibility and simplicity of design, ease of operation, and insensitivity to toxic pollutants. However, some adsorbents require activation to change the raw material to a more porous material so it can operate at its finest. The main parameters upon selecting the adsorbent are the choice of adsorbent and particle size [52–55].

7. Recommendation of adsorbents for the ULO treatment

Fuller's earth (FE) is a type of sedimentary clay having high magnesium oxide content which is commonly used in bleaching, refining edible oils and clarifying petroleum. FE is preferred due to its high adsorption capacity and low purchasing cost. It decolorizes the oil without variation in chemical properties of oil [56]. Previous studies have reported that it has the ability in removing heavy metals in wastewater and to extract impurities of ULO [56–58]. Recovery of ULO might be done by using different acid solvents paired with FE through acid-clay treatment. River sand (RS) has the ability to remove metal contaminants from aqueous solution was inspired by its wide application in wastewater treatment. It has the ability to remove contaminants and mercaptans from liquid as well. Many characteristics of the RS can make it a very good and cheap adsorbent [59]. Het owever, RS needs to be properly treated in removing contaminants from the ULO.

Bleaching earth are actually bentonite clays or known as montmorillonite that are used as absorbent. Spent bleaching earth (SBE) is known as the by-product that is produced in the palm oil refinery [16,60]. SBE contents lower specific surface and higher pore volume compared to fresh bleaching earth. According to Loh *et al.* [61], the empty pores are being filled by the metals after oil refining. In the palm oil industry bleaching earth is commonly used to adsorb the unwanted colour pigment and undesirable residues. SBE commonly disposes of the landfill upon use which leads to the increased amount of landfill waste. Disposal of SBE in the dumpsite can cause environmental issues as the leaching of the residual fat by rain may pollute groundwater and fire hazards. SBE could be regenerated *via* chemical purification, solvent extraction or combination of these methods [62–63]. After regeneration SBE could be used for the adsorption treatment of ULO [64–66]. However, for the recovery of ULO using adsorbents and optimization of the adsorption process parameters is still lacking in much of the published literature.

8. Conclusion

ULO is a highly hazardous waste that requires very responsible waste management. It can cause detrimental effects not just to the environment but also to humans if it is not properly disposed of, handled or treated. In this literature review, few examples of recycling ULO have been discussed. Besides, the roles of adsorbent in recovering the ULO also have been discussed thoroughly. In order to prove that adsorption can recover the oil, previous studies have been listed and few promising adsorbents were earmarked for further study. However, it is recommended to investigate the optimization of adsorption process parameters which could maximize the efficiency of ULO recovery.

References

[1] Udonne J D 2011 A comparative study of recycling of used lubrication oils using distillation, acid and activated charcoal with clay methods *J. Petroleum Gas Eng.* **2** 12–9

[2] Moura L G M, Assunção Filho J L and Ramos A C S 2010 Recovery of used lubricant oils through adsorption of residues on solid surfaces *Braz. J. Petroleum Gas* **4** 91–102

[3] Mang T and Dresel W 2007 Lubricants and their Market *Lubricants and Lubrication: Second Edition*

[4] Josiah P N and Ikiensikimama S S 2010 Effect of desludging and adsorption ratios on recovery of low pour fuel oil (LPFO) from spent engine oil *Chem. Eng. Res. Bull.* **14** 25–8

[5] Farhanah A N and Bahak M Z 2015 Engine oil wear resistance, *J. Tribol.* **4** 10–20

[6] Aljabiri N A 2018 A comparative study of recycling used lubricating oils using various methods *J. Sci. Eng.* **5** 168–77

[7] Stan C, Andreescu C and Toma M 2018 Some aspects of the regeneration of used motor oil *Procedia Manuf.* **22** 709–13

[8] Khuong L S, Zulkifli N W M, Masjuki H H, Mohamad E N, Arslan A, Mosarof M H and Azham A 2016 A review on the effect of bioethanol dilution on the properties and performance of automotive lubricants in gasoline engines *RSC Adv.* **6** 66847–69

[9] Osman D I, Attia S K and Taman A R 2018 Recycling of used engine oil by different solvent *Egypt. J. Pet.* *27* 221–5

[10] Emam E A and Shoaib A M 2012 Re-refining of used lube oil , II- by solvent/clay and acid/clay-percolation processes ARPN *J. Eng. Appl. Sci.* **2** 1034–41

[11] Hsu Y L and Liu C C 2011 Evaluation and selection of regeneration of waste lubricating oil technology *Environ. Monit. Assess.* **176** 197–212

[12] Shaikh R and Mahanwar P 2018 Reclamation of used engine oil using polymeric flocculants *Int. J. Chem. Sci.* **16** 1–14
[13] Abdel-Jabbar N M, Al Zubaidi E A H and Mehrvar M 2010 Waste lubricating oil treatment by adsorption process using different adsorbents *Int. J. Chem. Biol. Sci.* **3** 70–3
[14] Rahman M M, Siddiquee T A, Samdani S and Kabir K B 2008 Effect of operating variables on regeneration of base-oil from waste oil by conventional acid-clay method *Chem. Eng. Res. Bull.* **12** 24–7
[15] Mohammed R R, Ibrahim I A R, Taha A H and McKay G 2013 Waste lubricating oil treatment by extraction and adsorption *Chem. Eng. J.* **220** 343–51
[16] Anisuzzaman S M, Krishnaiah D, Suali E and Kumaran D 2020 Environmentally friendly recovery and characterization of waste lubricating oil using acid with spent bleaching earth *Malays. J. Chem.* (in press)
[17] Anisuzzaman S M., Abang S, Krishnaiah D and Azlan N A 2019 Removal of used motor oil from water body using modified commercial activated carbon *Malays. J. Chem.* **21** 36–46
[18] Anisuzzaman S M, Krishnaiah D and Alfred D 2018 Adsorption potential of a modified activated carbon for the removal of nitrogen containing compounds from model fuel *AIP Conf. Proc.* **1930** 1–8
[19] Rammohan A 2016 Engine's lubrication oil degradation reasons and detection methods: A review *J. Chem. Pharm. Sci.* **9** 3363–6
[20] Speight J G and Exall D I 2014 Re-refining of used lubricating oils CRC Press Boca Raton
[21] Layzell D 2010 The re-refining of used lubricating oils *VII-Energy-B-Rerefining Used Oils-1* 54–68
[22] Khuong L S, Zulkifli N W M, Masjuki H H, Mohamad E N, Arslan A, Mosarof M H and Azham, A 2016 A review on the effect of bioethanol dilution on the properties and performance of automotive lubricants in gasoline engines *RSC Adv.* **6** 66847–69
[23] Cerón A A, Vilas Boas R N, Biaggio F C and de Castro H F 2018 Synthesis of biolubricant by transesterification of palm kernel oil with simulated fusel oil: Batch and continuous processes *Biomass Bioenerg.* **119** 166–72
[24] Udonne J D and Bakare O A 2013 Recycling of used lubricating oil using three samples of acids and clay as a method of treatment *Inter. Arch. App. Sci.* **4** 8–14
[25] Mekonnen H A 2014 Recycling of used lubricating oil using acid-clay treatment process *M. Sc thesis Addis Ababa University Addis Ababa Institute of Technology (AAiT)* 1–71
[26] Diphare M, Muzenda E, Pilusa T J and Mollagee M 2013 A Comparison of Waste Lubricating Oil Treatment Techniques *2nd International Conference on Environment, Agriculture and Food Sciences* (ICEAFS'2013) Kuala Lumpur, Malaysia 106–9
[27] Abro R, Chen X, Harijan K, Dhakan Z A and Ammar M 2013 A comparative study of recycling of used engine oil using extraction by composite solvent, single solvent and acid treatment methods *ISRN Chem. Eng.* **952589** 1–5
[28] Aluyor E O and Ori-jesu M 2009 Biodegradation of mineral oils-A review *Afr. J. Biotechnol.* **8** 915–20
[29] Nowak P, Kucharska K and Kamiński M 2019 Ecological and health effects of lubricant oils emitted into the environment *Int. J. Environ Res Public Health.* **16** 3002
[30] Musemić R and Bašić A 2011 The waste oil influence and risks on environment in refer to Bosnia and Herzegovina *15th International Research/Expert Conference "Trends in the development of Machinery and Associated Technology"* TMT2011 Prague Czech Republic 12–8
[31] Michael A 2015 Used oil storage and disposal practices in automobile repair garages in Ghana. *Int. J. Sci. Technol. Soc.* **3** 191–201

[32] Ani I, Okafor J, Olutoye M and Akpan U 2015 Effects of process variables and a comparative study of methods for transfer oil production from spent engine oil *Br. J. Appl. Sci. Technol.* **9** 65–76
[33] Wong P K and Wang J 2001 The accumulation of polycyclic aromatic hydrocarbons in lubricating oil over time – a comparison of supercritical fluid and liquid-liquid extraction method *Environ. Pollut.* **112** 407–15
[34] Yang C Y 2008 The feasibility studies on sonochemical processes for treating used oil: Toxin reduction for eliminating recycle interference *Ph.D thesis Faculty of the graduate school University of Southern California* 1–149
[35] Botas J A, Moreno J, Espada J J, Serrano D P and Dufour J 2017 Recycling of used lubricating oil: Evaluation of environmental and energy performance by LCA *Resour. Conserv. Recycl.* **125** 315–23
[36] Owolabi R U, Alabi K A and Oke O B 2013 Reclamation of Spent Automobile Engine Lubricating Oil *Fountain J. Nat. Appl. Sci.* **2** 1–5
[37] Kanokkantapong V, Kiatkittipong W, Panyapinyopol B, Wongsuchoto P and Pavasant P 2009 Used lubricating oil management options based on life cycle thinking. Resources, Conservation and Recycling **53** 294–9
[38] El-Fadel M and Khoury R 2001 Strategies for vehicle waste-oil management: a case study *Resour. Conserv. Recycl.* **33** 75–91
[39] Je-Lueng S, Chen-Yu L and Ching-Yuan C 2003 Recovery of gasoline vapor by a combined process of two-stage dehumidification and condensation *J. Chin. Inst. Chem.* Eng. **34** 605–16
[40] Bartone C R 1990 Economic and policy issues in resource recovery from municipal solid wastes *Resour. Conserv. Recycl.* **4** 7–23
[41] Shakirullah M, Ahmad I, Khan M A, Ishaq M, Rehman H and Saeed M 2006 Spent lubricating oil residues as new precursors for carbon *Fuller. Nanotub. Car. N.* **14** 39–48
[42] Hamawand I, Yusaf T and Rafat S 2013 Recycling of waste engine oils using a new washing agent *Energies* **6** 1023–49
[43] Scapin M A, Duarte C, Sampa M H O and Sato I M 2007 Recycling of the used automotive lubricating oil by ionizing radiation process *Radiat. Phys. Chem.* **76** 1899–902
[44] Omolara A M, Olurotimi A D and Olatunji G O 2015 Regeneration of used lubricating engine oil by solvent extraction process *Int. J. Energ. Env. Res.* **3** 1–12
[45] Abo-Dief H M, Altalhi A A and Mohamed A T 2014 Waste oil recycling using microwave pyrolysis reactors *J. Ind. Intell. Inf.* **2** 314–9
[46] Rincon J, Cañizares P and García M T 2007 Improvement of the waste-oil vacuum-distillation recycling by continuous extraction with dense propane *Ind. Eng. Chem. Res.* **46** 266–72
[47] Abu-Elella R, Ossman M E, Farouq R and Abd-Elfatah M 2015 Used motor oil treatment: turning waste oil into valuable products *Int. J. Chem. Biochem. Sci.* **7** 57–67
[48] Al-Ghouti M A and Da'ana D A 2020 Guidelines for the use and interpretation of adsorption isotherm models: A review *J. Hazard. Mater.* **393** 122383
[49] Ajemba R O and Onukwuli O D 2013 Adsorptive removal of colour pigment from palm oil using acid activated nteje clay, Kinetics, equilibrium and thermodynamics *Physicochem. Probl. MI.* **49** 369–81
[50] Riyanto T A A and Juliantydjawi D P 2018 The effect of treatment with activated carbon on the metal content in reuse of lubricating oil waste *MATEC Web of Conferences* **154** 1–5
[51] Bhaskar T, Uddin A, Muto A, Sakata Y, Omura Y, Kimura K and Kawakami Y 2004 Recycling of waste lubricant oil into chemical feedstock or fuel oil over supported iron oxide catalysts *Fuel* **83** 9–15
[52] Anisuzzaman S M, Joseph C G, Krishnaiah D, Daud W M A W, Suali E and Chee F C 2018

Sorption potential of oil palm shell for the removal of chlorinated phenol from aqueous solution:kinetic investigation *J. Eng. Sci. Technol.* **13** 489–504

[53] Krishnaiah D, Joseph C G, Anisuzzaman S M, Daud W M A W, Sundang M and Leow Y C 2017 Removal of chlorinated phenol from aqueous solution utilizing activated carbon derived from papaya (*Carica Papaya*) seeds *Korean J. Chem. Eng.* **34** 1377–84

[54] Krishnaiah D, Anisuzzaman S M, Bono A and Sarbatly R 2013 Adsorption of 2,4,6-trichlorophenol (TCP) onto activated carbon J. King Saud Univ.-Sci. **25** 251–5

[55] Joseph C G, Anisuzzaman S M, Daud W M A W, Krishnaiah D and Ng K A 2017 Chlorinated phenol removal from aqueous media by tea (*Camellia sinensis)* leaf waste tailored activated carbon *IOP Conf. Ser. Mater. Sci. Eng.* **206** 1–13

[56] Khan A, Hassan S, Naqvi J, Kazmi M A and Ashraf Z 2015 Surface activation of fuller's earth (bentonite clay) using organic acids *Sci. Int. (Lahore)* **27** 329–32

[57] Oubagaranadin J U, Sathyamurthy N and Murthy Z V P 2007 Evaluation of Fuller's earth for the adsorption of mercury from aqueous solutions: A comparative study with activated carbon *J. Hazard. Mater.* **142** 165–74

[58] Malacas M, Balberan M C, Bederi N A J, Ramos C J, Rato M, Salazar A G and Roque E 2019 The removal of copper (II) and lead (II) from aqueous solution using Fuller's earth and Fuller's earth-immobilized nanoscale zero valent iron (FE-NZVI) by adsorption *The 25th Regional Symposium on Chemical Engineering (RSCE 2018) MATEC Web Conf.* **268** 1–6

[59] Yadav S, Srivastava V, Banerjee S, Weng C and Sharma Y C 2013 Adsorption characteristics of modified sand for the removal of hexavalent chromium ions from aqueous solutions: Kinetic, thermodynamic and equilibrium studies *Catena* **100** 120–7

[60] Boey P L, Shangeetha G, Maniam G P and Dafaalla M H A 2011 Ultrasound aided in situ transesterification of crude palm oil adsorbed on spent bleaching clay *Energy Convers. Manag.* **52** 2081–4

[61] Loh S K, Cheong K Y and Salimon J 2017 Surface-active physicochemical characteristics of spent bleaching earth on soil-plant interaction and water-nutrient uptake : A review *Appl. Clay Sci.*, **140** 59–65

[62] Wafti N S A, Yoo C K, Lin S W, Yaw T C S and Abdullah L C 2011 Regeneration and characterization of spent bleaching clay *J. Oil Palm Res.* **23** 999–1004

[63] Sabour M R and Shahi M 2018 Spent bleaching earth recovery of used motor-oil refinery *J. Civ. Eng.* **4** 572–84

[64] Majid R A and Mat C R C 2017 Regenerated spent bleaching earth for the decolourisation and bod reduction of palm oil mill effluent *J. Oil Palm Res.* **29** 579–87

[65] Mana M, Ouali M S, De Menorval L C, Zajac J J and Charnay C 2011 Regeneration of spent bleaching earth by treatment with cethyltrimethylammonium bromide for application in elimination of acid dye *Chem. Eng. J.* **174** 275–80

[66] Wambu E W, Muthakia G K, wa-Thiong'o J K, Paul M and Shiundu P M 2009 Regeneration of spent bleaching earth and its adsorption of copper (II) ions from aqueous solutions *Appl. Clay Sci.* **46** 176–80

Effect of neutralizing agents in the preparation of succinic acid from oil palm trunk

N A Bukhari[1,2], S K Loh[1], A A I Luthfi[2,3], P M Abdul[2,3], S Harun[2,3] and J M Jahim[2,3]

[1] Energy and Environment Unit, Engineering & Processing Research Division, Malaysian Palm Oil Board (MPOB), 6, Persiaran Institusi, Bandar Baru Bangi, 43000 Kajang, Selangor, Malaysia
[2] Department of Chemical and Process Engineering, Faculty of Engineering and Built Environment, Universiti Kebangsaan Malaysia, 43600 Bangi, Selangor, Malaysia
[3] Centre for Sustainable Process Technology (CESPRO), Faculty of Engineering and Built Environment, Universiti Kebangsaan Malaysia, 43600 Bangi, Selangor, Malaysia

Email: adela@mpob.gov.my

Abstract. Neutralization is an important process to control the pH required for enzymatic saccharification of pretreated biomass followed by fermentation for biochemical conversion. In this study, the production of succinic acid as a potential C4 building block was investigated by utilizing lignocellulosic biomass in the form of oil palm trunk (OPT). The effect of different neutralizing agents (NaOH, KOH and NH_4OH) on the enzymatic saccharification of oxalic acid-pretreated OPT and subsequent succinic acid fermentation by *Actinobacillus succinogenes* ATCC 55618 was investigated. The results showed that all neutralizing agents tested were able to assist in the recovery of fermentable sugars with concentrations ranging from 38.1 to 39.6 g/L. However, during succinic acid fermentation, it was found that the soluble NH_4-oxalate salt formed severely inhibited succinic acid fermentation compared to Na and K, thereby decreasing the succinic acid production from 14.0 g/L (using NaOH) to 1.0 g/L (using NH_4OH). In particular, Na- and K-oxalate did not exhibit apparent inhibition for both the saccharification and fermentation processes. Hence, the choice of neutralizing reagent is essential to prevent inhibition in the preparation of succinic acid from lignocellulosic biomass.

1. Introduction

Growing concern over environmental and ecological protection globally has shifted the effort to manufacture industrially important chemicals from renewable resources. Among the promising bio-based chemicals, succinic acid has been identified as one of the most versatile chemicals that can be produced *via* biological transformation. Succinic acid finds a broad range of applications and is a key component to produce various industrially important products, including detergent and surfactant, food additives, biopolymers, fungicides, herbicides, etc. [1].

Conversion of bio-based succinic acid from lignocellulosic agro-industrial residues appears as an attractive approach in promoting green technology [2,3]. Accordingly, *Actinobacillus succinogenes* is the most preferred biocatalyst because of its capability to utilize a wide spectrum of carbon sources including hexose and pentose sugars under anaerobic conditions [3]. The utilization of lignocellulosic feedstocks is sustainable, promising, and eco-friendly approach towards the production of succinic acid since it completely avoids the consumption of food crops [2].

Lignocellulosic material is mainly made up of cellulose, hemicellulose, and lignin, which form a complex structure to withstand the chemical or microbial degradation [4]. In Malaysia, one of the available lignocellulosic materials is oil palm trunk (OPT) amounting to approximately 21.6 million tonnes (dry weight basis) based on a replanted rate of 5 % from 5.9 million ha of oil palm plantations in 2019 [5]. OPT consists of 35 - 40 % cellulose, 25 - 30 % hemicellulose and 20 - 25 % lignin [6,7]. Nevertheless, lignocellulosic OPT currently needs to be pretreated to achieve high sugar yields for the microorganisms to synthesize the succinic acid. Pretreatment is the first step in bioconversion to disrupt the heterogeneous structure of lignocellulose, solubilize hemicellulose and/or lignin, and increase the surface area and porosity of the respective biomass [8]. Dilute acid pretreatment is one of the most widely used processes, which yield promising amounts of digestible substrates for successful enzymatic hydrolysis. This is followed by enzymatic saccharification, where the pretreated lignocellulose is converted into fermentable sugars [7].

Prior to enzymatic saccharification, neutralization step is a prerequisite for adjusting the acidic pretreated hydrolysate or slurry to the required pH favorable for the enzyme reactivity. Generally, alkaline neutralizers such as potassium hydroxide (KOH), sodium acetate ($NaCH_3COO$), sodium chloride (NaCl), sodium hydroxide (NaOH), sodium sulfate (Na_2SO_4), calcium hydroxide (CaH_2O_2), calcium carbonate ($CaCO_3$) and ammonium hydroxide (NH_4OH) are able to neutralize the pretreated acid hydrolysates to the desired pH. However, soluble salts formed during the neutralization step might affect the efficiency of saccharification and fermentation [9]. Several soluble salts, i.e. Na_2SO_4, $NaCH_3COO$ and NaCl are detrimental to microorganisms thus inhibit the cell growth and product formation [9,10].

In this study, different neutralizing agents were used for pH adjustment of acid hydrolysates aiming at exploring the most suitable neutralizer which does not inhibit the subsequent biocatalytic activities. Different alkaline neutralizers were selected i.e. KOH, NaOH and NH_4OH and their influences were investigated on (1) enzymatic saccharification of OPT pretreated with acid and (2) fermentation by *A. succinogenes* to produce succinic acid. In addition, the effects of soluble salts co-exist with complex nutrient and defined mineral salts supplementation during fermentation were also performed to assess the extent of microbial inhibition.

2. Materials and methods

2.1. Oil palm trunk fibers

The OPT sample was ground to achieve a mean particle size of <10 mm using a laboratory stainless steel grinder (Pulverisette 15, Fritsch GmbH, Germany). The OPT fiber, previously characterized, has 56.70 % holocellulose (30.86 % cellulose and 25.84 % hemicellulose), 24.29 % lignin and 2.29 % ash [11].

2.2. Dilute oxalic acid pretreatment

Ground OPT solid at 10 % (w/v) loading were loaded in 250-mL screw-cap Erlenmeyer flasks containing 1 % (w/v) oxalic acid. The samples were further pretreated at 120 °C for 3 h as previously optimized [12].

2.3. Neutralization with different neutralizing agents

In order to compare the effects of different neutralizing agents on saccharification and subsequent fermentation, the whole slurry of acid pretreated samples were tailored to meet the required pH 5.0 by three different bases i.e., 5 M KOH, 5 M NaOH and 30 % (v/v) NH_4OH as detailed in table 1.

Table 1. Bases used during neutralization step of oil palm trunk acid hydrolysate.

Base/Neutralizing agent	Formula	Physical appearance	Molarity/ concentration	Solubility in water	Volume required (mL)
Potassium hydroxide	KOH	White solid, deliquescent	5 M	Soluble	5
Sodium hydroxide	NaOH	White, waxy, opaque crystals	5 M	Soluble	4
Ammonium hydroxide	NH_4OH	Colourless aqueous solution	30% (v/v)	Miscible	2

2.4. Enzymatic saccharification
The neutralized whole slurry of pretreated OPT was enzymatically saccharified using 15 filter paper units (FPU) of commercial cellulase (Cellic CTec2, Novozyme, Denmark) per gram of cellulose, aided by 0.5 % (v/v) Triton X-100 [11]. The samples were incubated at 50 °C, 150 rpm for 72 h. The recovered sugars after enzymatic saccharification were analysed using high performance liquid chromatography (HPLC). The enzymatic hydrolysates were collected and used for subsequent fermentation.

The sugar yield was calculated as follows equation (1):

$$\text{Sugar yield (g g}^{-1}) = \frac{Glucose + xylose\ in\ enzymatic\ hydrolysate}{Holocellulose\ content\ in\ OPT} \tag{1}$$

2.5. Fermentation of OPT hydrolysate
The effect of mineral salts (MS) (per litre of medium: 0.2 g $MgCl_2{\cdot}6H_2O$, 0.2 g $CaCl_2{\cdot}2H_2O$, 3.0 g KH_2PO_4, 1.0 g NaCl) with or without nitrogen source (yeast extract, YE) supplementation (15.0 g per litre) on succinic acid fermentation was also investigated using different acid hydrolysates. Nine types of media were prepared as follows: 1) KOH-oxalate supplemented with YE and MS; 2) KOH-oxalate supplemented with YE without MS; 3) Non-supplemented KOH-oxalate; 4) NaOH-oxalate supplemented with YE and MS; 5) NaOH-oxalate supplemented with YE; 6) Non-supplemented NaOH-oxalate; 7) NH_4OH-oxalate supplemented with YE and MS; 8) NH_4OH-oxalate supplemented with YE and 9) Non-supplemented NH_4OH-oxalate. All samples were supplemeted with 30.0 g $MgCO_3$. The medium was dispensed into a 100-mL serum flask, capped with a butyl rubber stopper and clamped with an aluminum seal. Fermentation was performed anaerobically in a 100-mL serum bottle by adding 10 % (v/v) inoculum of *A. succinogenes* ATCC 55618 into 50 mL medium. The culture was cultivated at 37 °C, 200 rpm for 72 h.

The succinic acid yield was calculated as follows equation (2):

$$\text{Succinic acid yield (g g}^{-1}) = \frac{Succinic\ acid\ in\ fermentation\ broth}{Glucose + xylose\ consumed} \tag{2}$$

2.6. Analytical methods
The sugars and organic acids recovered were quantified using HPLC (Waters 2707, USA) equipped with Phenomenex Rezex™ ROA column (Sunnyvale, USA) as previously described [11]. The column was maintained at 60 °C and the samples were eluted with 0.0025 M H_2SO_4 at 0.6 mL/min. The apparatus was integrated with an auto sampler, refractive index detector (Waters 2414, USA) set at 40 °C and isocratic HPLC pump (Waters 1515, USA).

3. Results and discussion

3.1. Effects of neutralizing agents on enzymatic saccharification of pretreated OPT

The dilute oxalic acid pretreatment of OPT was shown to effectively hydrolyze hemicellulose into monomeric xylose (table 2), releasing 15.17 ± 0.18 g/L xylose in the hydrolysate. Only a small amount of glucose (2.05 ± 0.04 g/L) was detected in the acid hydrolysate derived from the solubilization of hemicellulose. Conversely, acetic acid, one of the derivatives of lignocellulosic component, was also present in the hydrolysate at 4.67 ± 0.10 g/L, mainly due to hydrolysis of acetyl groups of hemicellulose.

Table 2. Concentrations of sugars and inhibitors in the OPT hydrolysate after A) oxalic acid pretreatment; B) enzymatic saccharification of the pretreated whole slurry.

A. After oxalic acid pretreatment

Neutralizing agent	Glucose (g/L)	Xylose (g/L)	Total sugar (g/L)	Acetic acid (g/L)
-	2.05 ± 0.04	15.17 ± 0.18	17.22 ± 0.22	4.67 ± 0.10

B. After enzymatic saccharification

Neutralizing agent	Glucose (g/L)	Xylose (g/L)	Total sugar (g/L)	Acetic acid (g/L)
KOH	20.80 ± 0.25	17.34 ± 0.27	38.14 ± 0.52	4.49 ± 0.06
NaOH	20.62 ± 0.50	17.48 ± 1.16	38.10 ± 1.65	4.44 ± 0.06
NH_4OH	19.66 ± 3.17	19.91 ± 0.86	39.58 ± 4.03	4.53 ± 0.11

The acid hydrolysates were neutralized with different bases to find the most promising neutralizing agent that has no inhibitory effect to enzyme activity and subsequent fermentation. In this study, three bases i.e. KOH, NaOH and NH_4OH that have better solubility in water were investigated. As shown in table 1, the whole slurry of pretreated acid hydrolysate required about 5, 4 and 2 mL of 5 M KOH, 5 M NaOH and 30 % (v/v) NH_4OH, respectively, during pH adjustment to around pH 4.8 - 5.0.

After enzymatic saccharification, the total sugar released from all samples gave similar results in the range of 38.1-39.6 g/L (table 2) with a satisfactory sugar yield (67- 69 %). Cations may influence the catalytic activity of cellulase by acting as cofactor or inhibitor. In previous study, Bin and Hongzhang (2010) found that the K^+, Ca^{2+}, Cu^{2+}, Mn^{2+}, Mg^{2+}, Zn^{2+}, Al^{3+} and Fe^{3+} showed inhibitory effects on cellulase activity during enzymatic saccharification of rice straw at different levels. The higher ash content of rice straw had contributed to an overall higher concentration of these leached cations, which was responsible for the enzymatic inhibitions [13]. In another study using oil palm empty fruit bunch, the tested Ca^{2+}, Co^{2+}, Cu^{2+}, Mn^{2+}, Ni^{2+}, Fe^{2+} and Zn^{2+} all showed otherwise stimulative effects [14]. The result in this study showed that K^+, Na^{2+} and NH^+ did not demonstrate any inhibitory effects on cellulase activity in the whole slurry acid hydrolysate of OPT. This indicated that cellulase might act differently in response to the varying chemical compositions of different types of biomass (substrates).

3.2. Effects of salts from different neutralizing agents on succinic acid fermentation

In common medium consisting of nitrogen source and mineral salts (YE+MS), the highest succinic acid of 13.22 ± 0.22 g/L was achieved from K-oxalate containing hydrolysate. This was followed closely by Na-oxalate containing hydrolysate with 12.8 2 ± 0.19 g/L (figure 1). However, fermentation of NH_4-oxalate containing hydrolysate inhibited succinic acid accumulation by *A. succinogenes* as only 1.22 ± 0.32 g/L was detected in the medium.

The fermentation experiments in OPT hydrolysate containing different neutralization salts with and without supplementation of YE and MS were also performed to investigate their performance on succinic acid production by *A. succinogenes*. As shown in figure 1, the highest succinic acid production of 13.99 ± 0.16 g/L was attained using Na-oxalate hydrolysate supplemented with YE without MS, which increased succinic acid titer by 9 % as compared to the MS supplemented hydrolysate. On the other hand, 10.72 ± 1.56 g/L succinic acid was produced from K-oxalate containing hydrolysate without MS supplementation, showing 23% reduction as compared to that of MS supplementation. Meanwhile, the succinic acid production was strictly inhibited in all NH_4-oxalate containing hydrolysates, producing very poor succinic acid titer (<1.5 g/L) (figure 1).

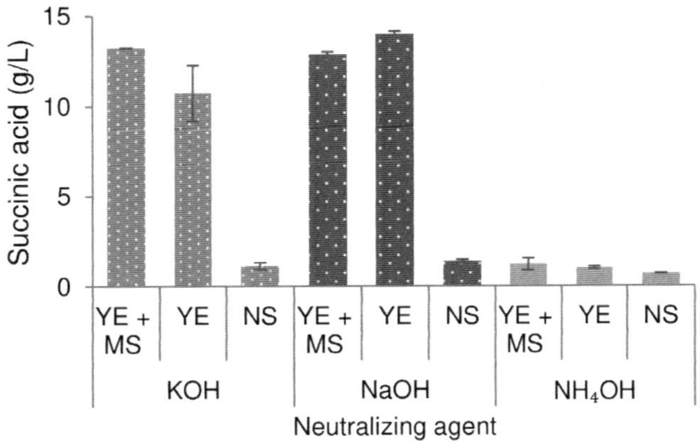

Figure 1. Succinic acid production using oil palm trunk hydrolysates neutralized using different agents. Cells were grown with or without the supplementation of 15 g/L yeast extract (YE) and 4.4 g/L mineral salts (MS) at 37°C, 200 rpm for 60 h. NS: non-supplemented.

Evidently, this showed that fermentation of *A. succinogenes* required YE to synthesis succinic acid while MS supplementation is optional depending on the selection of base (i.e. NaOH) during neutralization. *A. succinogenes* is auxotrophic to several amino acids and vitamins [15], hence, supplementation of complex nutrients such as YE is essential for cell proliferation. It is noteworthy that sufficient MS was required in OPT hydrolysate for bioconversion in this study. Furthermore, the NaOH added during neutralization step of the whole slurry of acid hydrolysates might be sufficient for cellular metabolism to accumulate succinic acid. Owing to this, supplementation of Na could be eliminated, which will benefit the overall process cost besides simplifying medium preparation step [16].

The selection of NH_4OH during neutralization was intentionally made to assess the capability of NH_4^+ to function as nitrogen source to complement the basic requirement of *A. succinogenes* fermentation for succinic acid production. However, the results showed otherwise. The succinic acid yield was negligible in all NH_4OH-containing hydrolysates (table 2). This indicates that inorganic nitrogen source such as NH_4OH was not preferred by *A. succinogenes* in succinic acid fermentation. The poor performance of NH_4 salt was probably related to its toxicity mainly caused by free ammonia (NH_3) to the anaerobic fermentation of *A. succinogenes*. This finding corresponded with Liu *et al.* (2008), reported that inorganic nitrogen source in the form of NH_4Cl inhibited *A. succinogenes* fermentation [17]. Regardless of the presence of the supplemented YE and MS, the soluble salts (NH_4-oxalate) had strictly inhibited the succinic acid fermentation.

Table 2 shows the total sugar consumption during succinic acid production. The highest total sugar consumption was 93.1% using Na-oxalate (YE without MS) containing hydrolysate, followed closely by 92.5 %, 91.3 % and 90.2 % from those of K-oxalate (YE without MS), Na-oxalate (YE + MS) and K-oxalate (YE + MS), respectively. In Na-oxalate (YE without MS) hydrolysate, only 0.9 g/L of glucose and 1.7 g/L of xylose (table 3) remained in the culture from their original base of 20.6 g/L and 17.5 g/L (table 2), respectively, demonstrating that the cells were capable of metabolizing xylose as effectively as glucose. In terms of the formation of by-products, higher formic acid and acetic acid were observed in YE (without MS) hydrolysates compared to those supplemented with MS in both the Na-oxalate and K-oxalate containing hydrolysates. This is possibly because some MS can act as cofactor in *A. succinogenes* metabolic pathway which has led to a higher production of succinic acid and lower by-products formation.

Table 3. Concentrations of sugars and acids in the oil palm trunk hydrolysates after fermentation.

Hydrolysate		Residual glucose (g/L)	Residual xylose (g/L)	Total consumed sugar (%)	Succinic acid yield (g/g)	By-products	
						Formic acid (g/L)	Acetic acid (g/L)
K-oxalate	1) YE + MS[a]	0.73 ± 0.17	3.04 ± 1.03	90.15	0.38	4.19 ± 0.15	9.80 ± 0.42
	2) YE[b]	0.82 ± 0.14	2.04 ± 1.41	92.51	0.30	4.82 ± 0.12	10.36 ± 0.35
	3) NS[c]	11.01 ± 0.97	14.57 ± 1.26	32.46	0.09	0.31 ± 0.14	4.80 ± 0.41
Na-oxalate	4) YE + MS[a]	1.06 ± 0.12	2.10 ± 1.63	91.33	0.39	3.60 ± 0.20	9.72 ± 0.46
	5) YE[b]	0.98 ± 0.07	1.72 ± 0.88	93.14	0.38	4.11 ± 0.13	10.47 ± 0.04
	6) NS[c]	10.72 ± 0.47	15.00 ± 0.73	33.30	0.11	0.31 ± 0.07	4.93 ± 0.16
NH_4-oxalate	7) YE + MS[a]	9.50 ± 0.01	9.33 ± 0.27	54.33	0.05	0.13 ± 0.10	4.72 ± 0.29
	8) YE[b]	5.84 ± 0.84	9.97 ± 1.26	58.33	0.05	0.01 0.01	4.68 ± 0.45
	9) NS[c]	6.79 ± 2.34	8.99 ± 1.46	40.98	0.07	0.18 ± 0.25	4.52 ± 0.33

[a] Yeast extract with mineral salts
[b] Yeast extract without mineral salts
[c] Non-supplemented

The overall performance in terms of sugar and succinic acid yield is summarized in figure 2. Clearly, the enzymatic saccharification was hardly affected by the neutralizing agent. On the other hand, the NH_4-oxalate salt formed during neutralization was detrimental to *A. succinogenes* in the subsequent fermentation. Therefore, the utilization of NH_4OH in succinic acid fermentation by *A. succinogenes* should be avoided.

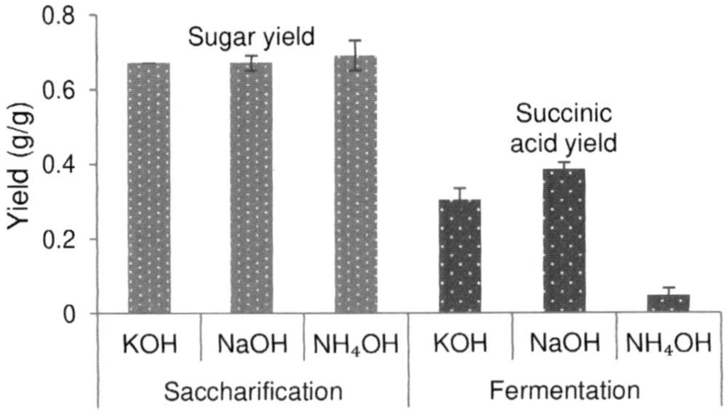

Figure 2. Sugar and succinic acid yields during enzymatic saccharification and fermentation by *A. succinogenes* using oil palm trunk hydrolysates neutralized with different agents.

4. Conclusion

Neutralizing agents such as NaOH, KOH and NH_4OH were shown to affect saccharification and fermentation involving OPT biomass. Based on the findings, NaOH might be the first choice to be considered for designing bioprocess for the preparation of succinic acid from lignocellulosic OPT acid hydrolysates by *A. succinogenes*. Interestingly, supplementation of mineral salts is not required in the subsequent fermentation when incorporating NaOH as the neutralizing agent. A proper selection of suitable neutralizing agent is perhaps wise to reduce any inhibitory effects caused by alkali salts formation during neutralization process. The study provides a scientific basis for adjusting bioprocessing conditions to enhance the feasibility of succinic acid production from lignocellulosic biomass.

Acknowledgments
The authors would like to thank the Director-General of the Malaysian Palm Oil Board (MPOB) for permission to publish this article.

References
[1] Yang Q, Wu M, Dai Z, Xin F, Zhou J, Dong W, Ma J, Jiang M and Zhang W 2020 Comprehensive investigation of succinic acid production by *Actinobacillus succinogenes*: A promising native succinic acid producer *Biofuels, Bioprod. Biorefining* **14** 950–64
[2] Luthfi A A I, Jahim J M, Harun S, Tan J P and Mohammad A W 2017 Potential use of coconut shell activated carbon as an immobilisation carrier for high conversion of succinic acid from oil palm frond hydrolysate *RSC Adv.* 7 49480–9
[3] Dessie W, Xin F, Zhang W, Jiang Y, Wu H, Ma J and Jiang M 2018 Opportunities, challenges, and future perspectives of succinic acid production by *Actinobacillus succinogenes Appl. Microbiol. Biotechnol.* **102** 9893–910
[4] Abdul P M, Md. Jahim J, Harun S, Markom M, Hassan O, Mohammad A W and Asis A J 2013 Biohydrogen production from pentose-rich oil palm empty fruit bunch molasses: A first trial *Int. J. Hydrogen Energy* **38** 15693–9
[5] Parveez G K A, Hishamuddin E, Loh S K, Ong-Abdullah M, Salleh K M, Bidin M N I Z, Sundram S, Hasan Z A A and Idris Z 2020 Oil palm economic performance in Malaysia and R&D progress in 2019 *J. Oil Palm Res.* **32** 159–90
[6] Loh S K 2017 The potential of the Malaysian oil palm biomass as a renewable energy source *Energy Convers. Manag.* **141** 285–98
[7] Bukhari N A, Jahim J M, Loh S K, Bakar N A and Luthfi A A I 2019 Response surface optimisation of enzymatically hydrolysed and dilute acid pretreated oil palm trunk bagasse for succinic acid production *BioResources* **14** 1679–93
[8] Abdul P M, Jahim J M, Harun S, Markom M, Lutpi N A, Hassan O, Balan V, Dale B E and Mohd Nor M T 2016 Effects of changes in chemical and structural characteristic of ammonia fibre expansion (AFEX) pretreated oil palm empty fruit bunch fibre on enzymatic saccharification and fermentability for biohydrogen *Bioresour. Technol.* **211** 200–8
[9] Yang M, Wang J, Nan Y, Zhang J, Li L, Liu G, Vepsäläinen J, Kuittinen S and Pappinen A 2019 Effect of salts formed by neutralization for the enzymatic hydrolysis of cellulose and acetone-butanol-ethanol fermentation *RSC Adv.* **9** 33755–60
[10] Qureshi N, Ezeji T C, Ebener J, Dien B S, Cotta M A and Blaschek H P 2008 Butanol production by *Clostridium beijerinckii*. Part I: Use of acid and enzyme hydrolyzed corn fiber *Bioresour. Technol.* **99** 5915–22
[11] Bukhari N A, Jahim J M, Loh S K, Nasrin A B, Harun S and Abdul P M 2020 Organic acid pretreatment of oil palm trunk biomass for succinic acid production *Waste and Biomass Valorization* **11** 5549–59
[12] Bukhari N A, Loh S K, Luthfi A A I, Nasrin A B, Abdul P M, Harun S and Jahim J M 2021 Oil palm trunk biomass pretreatment with oxalic acid and its effect on enzymatic digestibility and fermentability *Mater. Today Proc.* **42** 119–23
[13] Bin Y and Hongzhang C 2010 Effect of the ash on enzymatic hydrolysis of steam-exploded rice straw *Bioresour. Technol.* **101** 9114–9
[14] Nurul Adela B, Loh S K and Nasrin A B 2015 The improvement on enzymatic hydrolysis of oil palm (*Elaeis guineensis*) empty fruit bunch lignocellulose *Malaysian Appl. Biol.* **44** 95–100
[15] McKinlay J B, Laivenieks M, Schindler B D, McKinlay A A, Siddaramappa S, Challacombe J F, Lowry S R, Clum A, Lapidus A L, Burkhart K B, Harkins V and Vieille C 2010 A genomic

perspective on the potential of *Actinobacillus succinogenes* for industrial succinate production *BMC Genomics* **11** 680

[16] Bukhari N A, Loh S K, Nasrin A B, Luthfi A A I, Harun S, Abdul P M and Jahim J M 2019 Compatibility of utilising nitrogen-rich oil palm trunk sap for succinic acid fermentation by *Actinobacillus succinogenes* 130Z *Bioresour. Technol.* **293** 122085

[17] Liu Y P, Zheng P, Sun Z H, Ni Y, Dong J J and Zhu L L 2008 Economical succinic acid production from cane molasses by *Actinobacillus succinogenes Bioresour. Technol.* **99** 1736–42

Modelling the kinetics of microwave drying of shallot (*Allium cepa*) slices

C L Hii[1], C Govind[1], C L Chiang[2] and D Mohammad[3]

[1] Future Food Malaysia, Department of Chemical and Environmental Engineering, Faculty of Science and Engineering, University of Nottingham Malaysia, Jalan Broga, 43500 Semenyih, Selangor Darul Ehsan, Malaysia
[2] Foundation in Engineering, Faculty of Science and Engineering, University of Nottingham Malaysia, Jalan Broga, 43500 Semenyih, Selangor Darul Ehsan, Malaysia
[3] Department of Chemical Engineering, Diponegoro University, Jl. Prof. Soedarto, SH., Tembalang, Semarang, Indonesia

Email: Ching-Lik.Hii@nottingham.edu.my

Abstract. Convective drying is typically used to dry shallot (*Allium cepa*) commercially. However, a long drying time with a relatively low efficiency has led to the pursuit of new and improved drying methods. Microwave drying was chosen to be used due to its numerous advantages such as improved drying time, high drying efficiency and better product quality. In this research, three microwave power (180 W, 300 W, 450 W) and convective drying at 100°C were used. Results showed that drying kinetics (moisture content and drying rates) decreased the fastest at higher microwave power and the slowest using convective drying. In order to determine the best model to describe the thin-layer drying kinetics, four semi-empirical models were used namely Newton, Page, Logarithmic and Two-term models. Page model was found to be the best in describing the thin-layer microwave drying kinetics. Effective diffusivity values increased with higher microwave power and were found to be in the range of 6.62×10^{-6} m^2/s to 3.69×10^{-5} m^2/s with convective drying being the lowest (6.62×10^{-6} m^2/s) and 450W being the highest (3.69×10^{-5} m^2/s). Microwave drying is therefore able to improve drying kinetics compared to convective drying.

1. Introduction

Drying is an age-old process that has high importance in many different food industries and it is used in food preservation to provide a longer shelf-life and safe storage for the food materials. Conventional methods such as sun drying and convective hot air drying are often chosen due to their relatively low cost as well as their ability to heat regardless of the size, structure and shape of the food item [1]. However, several disadvantages are associated with these drying methods namely relatively long drying time (sun drying), inconsistent quality, unattractive colour/appearance, off flavour/aroma and so on. Therefore, this has ultimately led to the pursuit of other drying methods that are more efficient and one that has been getting a lot of attention recently is microwave assisted drying. Compared to convective drying, there are numerous advantages such as improved drying rates, better drying/energy efficiency and improved product quality [2–4].

Shallot (*Allium cepa*) is a member of the onion family, which is smaller in size with dimensions typically around 2.5 cm (diameter) and 3 cm (height) [5]. Shallots have been widely used as food seasoning due to its nutritional and aromatic nature. In terms of dried shallots, it is typically produced in the forms of powdered, chopped and as well as flaked. These dried shallots are often used as flavor additives in various food dishes including processed food products. Malaysia imported 129.21 metric tonnes of onions and shallots (fresh or chilled) in 2020 which amounting to 55.41 million USD [6]. Most of the onions and shallots are imported from Pakistan, New Zealand and Greece. This trend is expected to increase and grow steadily in the coming years. Hence, there is a need to find an alternative drying method to dry shallots for preservation purpose and diversify the product range to add value to the raw materials. Hence, microwave drying is selected for this purpose due to the various advantages as reported in literature, e.g more efficient heating and better product quality.

The main aim of this research is to dry shallot using a non-conventional method of drying and investigate its drying kinetics as compared to convective drying. Therefore, research works had been carried out by drying shallot samples using a laboratory microwave oven with the following objectives:

- To determine the drying kinetics of the microwave drying process.
- To model the drying process using thin layer drying models.
- To determine the effective diffusivities during drying.

2. Materials and Methods

2.1. Drying experiment

Fresh shallots (*Allium cepa*) were purchased from a local supermarket in Putrajaya, Malaysia. Prior to conducting the experiment, the shallots were peeled and cut into thin circular shape using a sharp knife. The weight of every shallot slice was 9.00 ± 0.5 g with thickness of 0.65 ± 0.05 cm and diameter of 4.8 ± 0.1cm.

Both the drying experiments (convective and microwave) were conducted using a programmable microwave oven (Model C-103FL, Samsung), which has a cavity with a dimension of 0.34 m × 0.34 m × 0.22 m. The samples were placed on the rotating glass turntable. Microwave drying was conducted at three power settings (180 W, 300 W and 450 W) namely low, medium and high microwave heating. Convective drying was conducted at 100 °C. The experiments were conducted for at least 30 minutes and it was dried until equilibrium moisture content (constant weights). However, convective drying was only conducted for only 40 minutes as the intention was to use it as a benchmark for comparison against microwave drying within the similar time range (30 - 40 minutes). All the experiments were replicated twice to obtain the mean values.

2.2. Drying kinetics

The weights of the samples were measured by taking out the samples from the microwave oven at time interval of every 5 minutes. An analytical balance (Metler Toledo, USA) was used to weigh the samples and thereafter the samples were returned to the oven for drying. Moisture content (M_t) was determined according to equation (1) using the oven method [7].

$$\textit{Moisture content: } M_t = \frac{\textit{Mass of sample (g)- Bone dry mass (g)}}{\textit{Bone dry mass (g)}} \tag{1}$$

Bone dry mass was determined by subjecting the final dried samples in an air ventilated oven (Memmert, Germany) at 105 °C for at least 24 hours.

Drying rates (g water/g dry solid.hr) were determined by taking the gradient at each data point along the moisture content profile.

2.3. Mathematical modelling

Modelling using thin layer drying models were carried out using the Newton, Page, Logarithmic and Two-term models [8] as shown in table 1. The justification of using these models were to evaluate the fitting accuracy based on number of constant in the models namely one constant ('k' in Newton), two constants ('k' and 'n' in Page), three constants ('a', 'k' and 'c' in Logarithmic) and four constants ('a', 'b', 'k' and 'g' in Two-term).

Table 1. Semi-empirical thin layer drying models.

Model	Equation	
Newton	$MR=exp^{-kt}$	(2)
Page	$MR=exp^{-kt^n}$	(3)
Logarithmic	$MR = a.exp^{-kt}+c$	(4)
Two-term	$MR = a.exp^{-kt}+b.exp^{-gt}$	(5)

where MR = moisture ratio, M_i = initial moisture content and M_e = equilibrium moisture content as shown in equation (6).

$$Moisture\ ratio: MR = \frac{M_t - M_e}{M_i - M_e} \quad (6)$$

In order to evaluate and determine the constants (a, b, c, g, k and n) of the selected models, non-linear regression analyses were performed by using Excel SOLVER (MS office, USA) by minimizing the sum of the square of the residuals (SSR) as shown in equation (7). The analyses were performed for the mean moisture ratio values determined from the experimental moisture contents. The models were evaluated for root mean square error (RMSE), coefficient of determination (R^2) and Chi square (χ^2) as shown in equation (8) – (10). The best model selected having the lowest RMSE and χ^2 but with the highest R^2 values.

$$SSR= \sum_{i=1}^{N} (MR_{pre}- MR_{exp})^2 \quad (7)$$

$$R^2=1-\{[\sum_{i=1}^{N}(MR_{pre}- MR_{exp})^2]/[\sum_{i=1}^{N}(\overline{MR}_{pre}- MR_{exp})^2]\} \quad (8)$$

$$\chi^2= [\sum_{i=1}^{N}(MR_{exp}- MR_{pre})^2]/(N-z) \quad (9)$$

$$RMSE= \left[\left(\frac{1}{N}\right) \sum_{i=1}^{N} (MR_{pre}- MR_{exp})^2\right]^{1/2} \quad (10)$$

Where subscripts exp and pre denote experimental and predicted values at each drying time, respectively. N and z denote the number of observations and constants, respectively.

2.4. Effective diffusivity

In order to determine the effective moisture diffusivity, Fick's second law equation was used as shown in equation (11) by assuming an infinite slab geometry [9].

$$MR= \frac{8}{\pi^2}\sum_{n=0}^{\infty}\frac{1}{(2n+1)^2} exp\left(\frac{-(2n+1)^2\pi^2 D_e t}{4L^2}\right) \quad (11)$$

where L is the half thickness of sample (m) and D_e is the effective diffusivity (m^2/s).

The equation can be simplified by using only the first term (n = 0) and applying natural log at both sides of the equation as follows [10].

$$ln\ MR=ln\frac{8}{\pi^2} - \frac{\pi^2 D_e t}{4L^2} \quad (12)$$

The slope of the liner graph can be used for estimation of D_e values (equation (13)).

$$Slope = \frac{\pi^2 D_e t}{4L^2} \quad (13)$$

3. Results and Discussion

3.1. Drying kinetics

Figure 1 shows the changes in the moisture ratio of the shallot slices during microwave (MW) and convective (CONV) drying. It could be observed that over time, the moisture ratio decreased exponentially and the drying time was recorded within the range of 30 - 40 minutes. Demiray *et al.* [2] reported similar observation where higher MW power (557 W) recorded shorter drying time (40 minutes) and vice versa. Drying at 450 W MW power took the shortest time to dry the shallot slices as compared to those at 300 W and 180 W. This is mainly due to the greater excitement of water molecules under the action of higher MW power. Also, a faster heating rate promotes better heat transfer within the sample and vaporizes the internal moisture to the surface [11]. Eventually this enhances mass transfer and results in a shorter drying time. Meanwhile, convective drying shows the slowest reduction in moisture content owing to the heating mechanism which is dependent on the driving force between the hot air and sample.

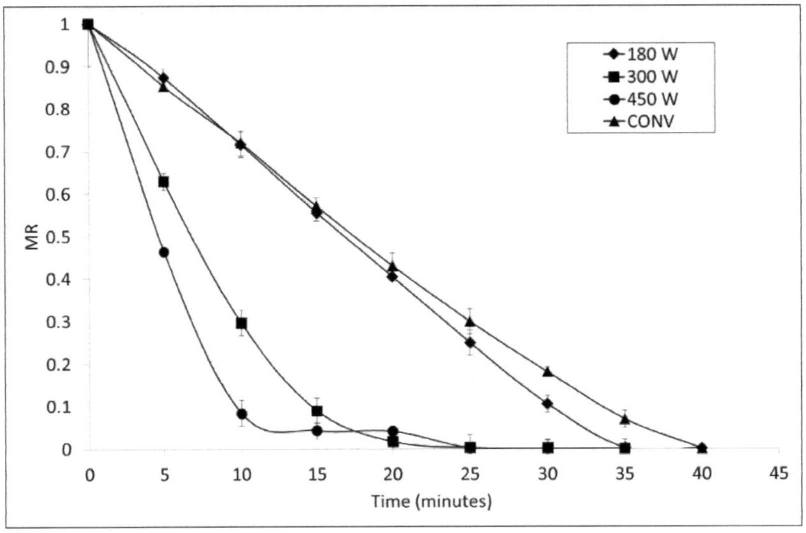

Figure 1. Variation of moisture ratios with drying time for microwave and convective drying.

Figure 2 shows the changes in drying rates with moisture ratio during microwave and convective drying. Microwave drying at 450 W and 300 W showed a more distinct falling rate period as compared to that at 180 W and convective drying. However, initial warming up period was reported by Demiray *et al.* [2] for microwave drying at 328 - 557 W which could be due to the different quantity of samples used (70 g) as compared to current works (45 g). The higher the sample quantity the greater amount of moisture need to be heated/evaporated which could require significant amount of initial warming up/heating. It can be seen that the overall drying rate was higher at higher MW power and the lowest in convective drying. Initial drying rates were -0.934 kg water/kg dry solid.min, -0.628 kg water/kg dry solid.min and -0.188 kg water/kg dry solid.min for microwave drying at 450 W, 300 W, 180 W, respectively, and -0.088 kg water/kg dry solid.min for convective drying. This shows that microwave drying is able to improve the drying rates especially at the early stage of drying when the product moisture content is higher and able to absorb more microwave energy to facilitate drying.

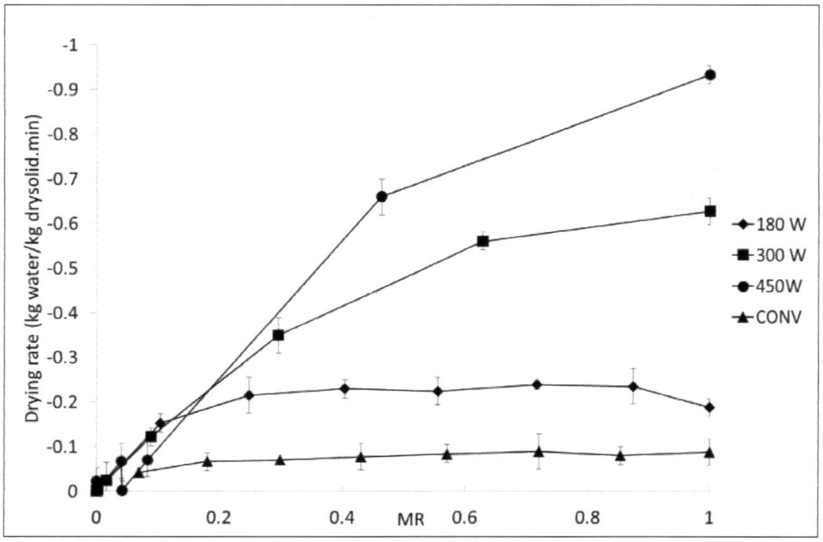

Figure 2. Variation of drying rates with moisture ratios for microwave and convective drying.

3.2. Mathematical Modelling

Table 2 – 5 show the drying constants as well as the statistical parameters (R^2, χ^2 and RMSE) determined from regression analyses. The R^2 values for all the models were in the range of 0.94 to 0.9995, χ^2 values were in the range of 7.28×10^{-5} to 1.49×10^{-2} and the RMSE were in the range of 0.00752 to 0.1. Based on the criteria, the Page model showed the highest R^2 values and the lowest χ^2 and RMSE values in all drying experiments. This implies that the Page model outperformed other models and it is the best in describing the thin layer drying kinetics of the shallot slices.

Table 2. Results of mathematical modelling (MW power =180 W).

Model	Constant	R^2	χ^2	RMSE
Newton	k = 0.053	0.9516	1.12×10^{-2}	1.0×10^{-1}
Page	k = 0.005, n = 1.783	**0.9916**	$\mathbf{1.50 \times 10^{-3}}$	$\mathbf{3.40 \times 10^{-2}}$
Logarithmic	a = 1.099, k = 0.058, c = 0	0.9413	1.20×10^{-2}	9.10×10^{-2}
Two-term	a = 0.549, k = 0.058, b = 0.549, g = 0.058	0.9413	1.50×10^{-2}	9.10×10^{-2}

Table 3. Results of mathematical modelling (MW power = 300 W).

Model	Constant	R^2	χ^2	RMSE
Newton	k = 0.126	0.9861	2.40×10^{-2}	4.60×10^{-2}
Page	k = 0.041, n = 1.491	**0.9995**	$\mathbf{7.28 \times 10^{-5}}$	$\mathbf{7.53 \times 10^{-3}}$
Logarithmic	a = 1.035, k = 0.129, c = 0	0.9851	3.00×10^{-3}	4.47×10^{-2}
Two-term	a = 0.517, k = 0.129, b = 0.517, g = 0.129	0.9851	3.59×10^{-3}	4.47×10^{-2}

Table 4. Results of mathematical modelling (MW power = 450 W).

Model	Constant	R^2	χ^2	RMSE
Newton	k = 0.178	0.9889	1.40×10^{-3}	3.50×10^{-2}
Page	k = 0.072, n = 1.471	**0.9964**	$\mathbf{7.62 \times 10^{-4}}$	$\mathbf{2.39 \times 10^{-2}}$
Logarithmic	a = 1.011, k = 0.180, c = 0	0.9889	1.94×10^{-3}	3.48×10^{-2}
Two-term	a = 0.506, k = 0.180, b = 0.506, g = 0.180	0.9889	2.42×10^{-3}	3.48×10^{-2}

Table 5. Results of mathematical modelling (convective).

Model	Constant	R^2	χ^2	RMSE
Newton	k = 0.048	0.9607	7.75×10^{-3}	8.3×10^{-2}
Page	k = 0.007, n = 1.602	**0.9906**	$\mathbf{1.46 \times 10^{-3}}$	$\mathbf{3.37 \times 10^{-2}}$
Logarithmic	a = 1.078, k = 0.052, c= 0	0.9528	8.71×10^{-3}	7.62×10^{-2}
Two-term	a = 0.539, k = 0.052, b = 0.539, g = 0.052	0.9528	1.04×10^{-2}	7.62×10^{-2}

According to Hii and Ogugo [8], the correction factor 'n' in the Page model has no theoretical meaning but it is merely a correction factor to improve the fitting accuracy. The page model also has the inherent advantage that a good fitting can be obtained with only a few constants (two constants k and n) that need to be determined. figure 3 (a - d) shows fitting of the experimental data to the various thin layer models used in this study.

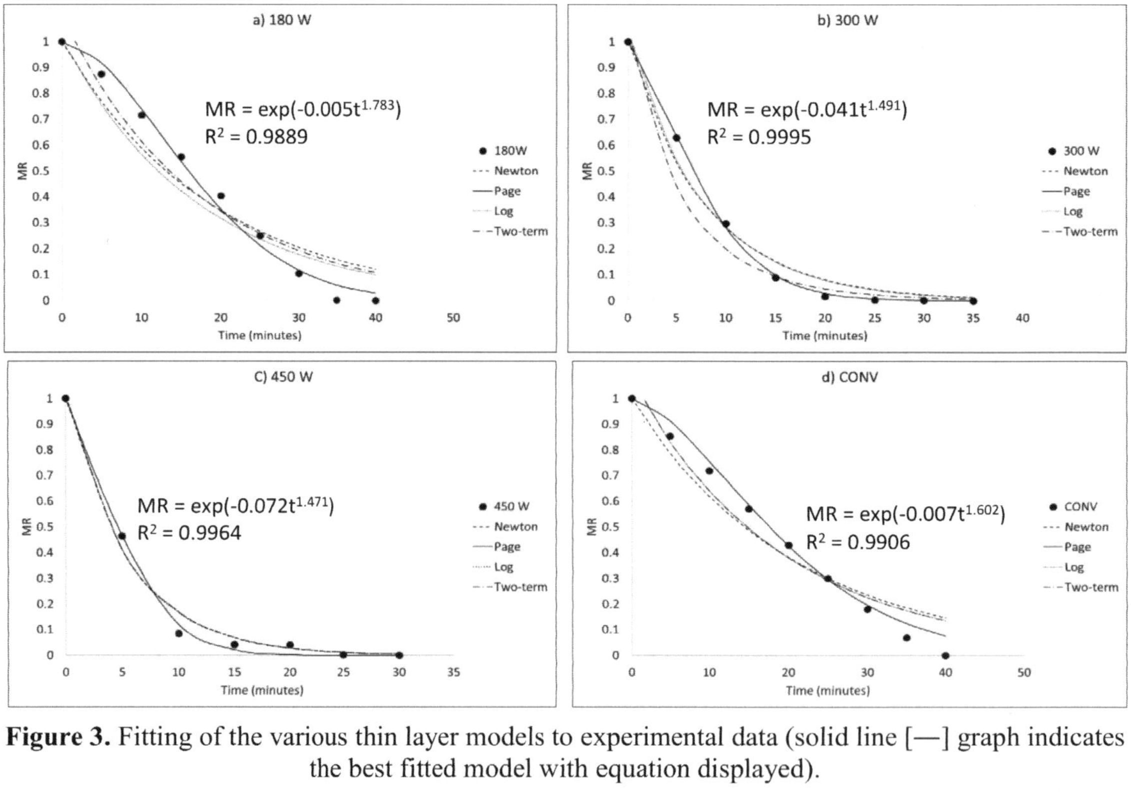

Figure 3. Fitting of the various thin layer models to experimental data (solid line [—] graph indicates the best fitted model with equation displayed).

3.3. Effective diffusivity
Table 6 shows the effective diffusivity (D_e) values determined from the drying experiments. The D_e values ranged from 0.66×10^{-5} to 3.69×10^{-5} m^2/s which are well within the range reported in the literature for food products [12]. The D_e values also increased as the MW power output increased.

Table 6. Effective diffusivity (D_e) values.

Experiment	D_e ($\times 10^{-5}$ m^2/s)
Convective	0.66±0.02
Microwave (180 W)	0.70±0.01
Microwave (300 W)	2.85±0.04
Microwave (450 W)	3.69±0.07

On average, the D_e values of microwave drying is at a higher order of magnitude (10^{-5} m^2/s) compared to convective drying (10^{-6} m^2/s). This is also in agreement with the trend reported by Demiray *et al.* [2] for similar product within the range of 2.59×10^{-7} to 5.08×10^{-8} m^2/s for microwave drying as compared to 3.49×10^{-8} to 9.44×10^{-8} m^2/s for convective drying while Asiah *et al.* [13] reported D_e values at a much higher order of magnitude (10^{-9} m^2/s) for convective drying. However, Khan *et al.* [14] reported a lower D_e value at 1.27×10^{-11} m^2/s which could be due to the lower MW power used (100 W). Furthermore, microwave energy is normally transmitted intermittently during drying and at low power output this resulted in a slower diffusion process at this setting. This shows that microwave drying is able to improve the drying kinetics of drying of shallot slices due to the better heat transfer mechanism attributed MW energy.

4. Conclusions
Drying experiments were conducted using two different methods namely microwave (180 W, 300 W and 450 W) and convective drying (100 °C). Drying rates were found to be the highest in the 450W MW power output followed by 300 W, 180 W and lastly the convective drying. The greater MW power could provide a better heating mechanism by exciting the water molecules through microwave within the inner vicinity of the sample. Mathematical modelling showed that the Page model was the best in describing the thin layer drying characteristics of shallot slices. The effective diffusivity (D_e) values were found ranging from 0.66×10^{-5} to 3.69×10^{-5} m^2/s with convective drying being the lowest value obtained and microwave drying at 450 W being the highest. The effective moisture diffusivity values increased as the MW power increased from 180 W to 450 W. On average, the D_e values of microwave drying is at a higher order of magnitude (10^{-5} m^2/s) compared to convective drying (10^{-6} m^2/s).

Acknowledgments
The project was carried out as part of the MSc research and supported by the Department of Chemical and Environmental Engineering, University of Nottingham Malaysia.

References
[1] Litzler M 2007 Understanding drying *Food Technology* **61** 97–100
[2] Demiray E, Seker A and Tulek Y 2017 Drying kinetics of onion (*Allium cepa* L.) slices with convective and microwave drying *Heat and Mass Transfer/Waerme- und Stoffuebertragung* **53** 1817–27
[3] Onwude D I, Hashim N, Janius R, Abdan K, Chen G and Oladejo A O 2017 Non-thermal hybrid drying of fruits and vegetables: A review of current technologies *Innov. Food Sci. Emerg.* **43** 223–38
[4] Feng L, Zhang M and Adhikari B 2014 Effect of water on the quality of dehydrated products: A review of novel characterization methods and hybrid drying technologies *Dry. Technol.* **32** 1872–84

[5] Gupta K *et al.* 2003 Salad crops: Root, bulb and tuber crops' *Encyclopedia of Food Sciences and Nutrition* 5060–73
[6] Anon 2021 www. tridge.com (accessed 2 April 2021)
[7] Hii C L, Law C L and Cloke M 2009 Modeling using a new thin layer drying model and product quality of cocoa *J. Food Eng.* **90** 191–8
[8] Hii C L and Ogugo J F 2014 Effect of pre-treatment on the drying kinetics and product quality of star fruit slices *J. Eng. Sci. Technol.* **9** 122–34
[9] Crank J 1975 *The Mathematics of Diffusion* second ed. Clearendon Press, Oxford, UK
[10] Yap J Y, Hii C L, Ong S P, Lim K H, Abas F and Pin K Y 2020 Effects of drying on total polyphenols content and antioxidant properties of *Carica papaya* leaves *J. Sci. Food Agric.* **100** 2932–7
[11] Chandrasekaran S, Ramanathan S and Basak T 2013 Microwave food processing – A review *Food Res. Int.* **52** 243–61
[12] Zogzas N P, Maroulis Z B and Marinos-Kouris D 1996 Moisture diffusivity data compilation in foodstuffs *Dry Technol.* **14** 2225–53
[13] Asiah N, Djaeni M and Hii C L 2017 Moisture transport mechanism and drying kinetic of fresh harvested red onion bulbs under dehumidified air *Int. J. Food Eng.* **13**
[14] Khan M K I, Maan A A, Aadil R M, Nazir A, Butt M S, Rashid M I and Afzal M I 2019 Modelling and kinetic study of microwave assisted drying of ginger and onion with simultaneous extraction of bioactive compounds *Food Sci Biotechnol.* **29** 513–9

The effect of process parameters on catalytic direct CO_2 hydrogenation to methanol

M K Koh[1], Y J Wong[2,3] and A R Mohamed[2]

[1]School of Energy and Chemical Engineering, Xiamen University Malaysia, Jalan Sunsuria, Bandar Sunsuria, 43900 Sepang, Selangor, Malaysia
[2]Low Carbon Economy (LCE) Research Group, School of Chemical Engineering, Engineering Campus, Universiti Sains Malaysia, 14300 Nibong Tebal, Penang, Malaysia
[3]Department of Precision Engineering, Graduate School of Engineering, Osaka University, 2-1 Yamadaoka, Suita, Osaka 565-0871, Japan

Email: meikee.koh@xmu.edu.my

Abstract. The direct CO_2 hydrogenation to methanol is an attractive route to actively remove CO_2 and to promote sustainable development. Herein, the performance of Cu-Zn-Mn catalyst supported on mesoporous silica KIT-6 (hereafter, CZM/KIT-6) for methanol synthesis by direct CO_2 hydrogenation reaction was investigated by varying the process parameters, which included the weight-hourly space velocity, reaction temperature and reaction pressure. The CO_2 conversion was found to decrease with the increase of WHSV. On the other hand, CO_2 conversion increased with reaction temperature and pressure. Meanwhile, the methanol selectivity increased with WHSV and reaction pressure but decreased with the increase of reaction temperature. The apparent activation energy of methanol production at low reaction temperature (160 - 220 °C) was 10 kcal/mol. Non-Arrhenius behaviour of methanol formation was observed at high reaction temperature (220 - 260 °C). The performance of CZM/KIT-6 was maintained at high level, with the average methanol yield of 24.4 %, throughout the stability experiment (120-hour time-on-stream). In post-reaction XRD analysis, the copper crystallite growth was found to be 53.5 %, thus, resulting in 35.3 % loss of copper surface area.

1. Introduction

With the growing awareness on environmental protection and energy security issues, methanol production *via* direct CO_2 hydrogenation has been recognized as an advantageous approach to actively remove CO_2 and to promote sustainable development. The reaction is shown in equation (1).

$$CO_2 + 3H_2 \leftrightarrow CH_3OH + H_2O \quad \Delta H^{\circ}_{298} = -49.5\ kJ/mol \tag{1}$$

The approach reduces our dependence on fossil-based resources. It also provides an alternative to CO_2 sequestration, which requires suitable sequestration sites and continuous monitoring [1]. More recently, a new perspective on methanol production from direct CO_2 hydrogenation has emerged, wherein the process is considered as a way to store surplus electrical energy produced from renewable energy sources (RES) such as wind, hydraulic and solar energy [2,3]. The integration of RES and methanol production from CO_2 hydrogenation is also known as power-to-methanol (PtMeOH) process.

Efficient methanol production from CO_2 hydrogenation using inexpensive electricity generated from RES is important for abating global warming and to satisfy the world demand for renewable energy [4].

Despite relentless efforts, the reaction is yet to be operated efficiently due to the presence of the competing reverse water-gas shift reaction (RWGS), which often leads to low methanol production. The RWGS is shown in equation (2).

$$CO_2 + H_2 \leftrightarrow CO + H_2O \quad \Delta H_{298}^{\circ} = 41.2\ kJ/mol \tag{2}$$

At high reaction temperature (≥220 °C), most of the CO_2 are converted to CO through RWGS and/or methanol decomposition [5]. Hence, the methanol selectivity and yield are lowered at high reaction temperature. For instance, Xiao *et al.* [6] reported the methanol selectivity and yield obtained over Cu-ZnO-TiO_2 catalyst at 220 °C were approximately 50 % and 7.5 %, respectively. With the increase of reaction temperature to 260 °C, the methanol selectivity and yield decreased to approximately 33 % and 6.3 %, respectively.

Thermodynamically, the RWGS can be suppressed by low reaction temperature (≤200 °C) [7]. Yet, the CO_2 conversion at ≤200 °C is usually low and therefore limits the methanol yield. The CO_2 molecule is thermodynamically stable due to its negative electron affinity (-0.6 eV) and large ionization potential (13.73 eV) [8]. Besides, the Gibbs energy of formation (ΔG_{298K}°) of the molecule has a large negative value, -394.6 kJ/mol [9]. Consequently, a large input of energy is required to activate CO_2 molecules. Reaction temperatures of ≥220 °C are commonly reported to be the optimum to achieve reasonable CO_2 conversion but at the expense of lower methanol selectivity [10,11]. The approach to facilitate CO_2 conversion by increasing reaction temperature inevitably enhances the RWGS.

In order to address the aforementioned limitations, our research group has been dedicated to explore the reaction at low temperature, in which the kinetic limitation of CO_2 conversion is tackled through the development of Cu-Zn-Mn catalysts. The development of the catalysts has been presented in our earlier contributions [12,13]. Besides catalysts, it is equally important to apply suitable process conditions for achieving high methanol yield since the reaction is reversible and strongly competed by RWGS. In this sense, we investigated the effect of process parameters on direct CO_2 hydrogenation to methanol over Cu-Zn-Mn catalyst supported on mesoporous silica KIT-6. The weight-hourly space velocity (WHSV, 8 - 120 $L/g_{cat}\cdot h$), reaction temperature (160 - 260 °C) and reaction pressure (1.0 - 5.0 MPa) were varied in the study. The range of these process parameters were selected by considering the relevant literature and the equipment limitation. The apparent activation energies of methanol and CO formation were also calculated. Finally, the stability of the catalyst in 120-hour time-on-stream was assessed.

2. Methods

The synthesis of CZM/KIT-6 has been reported elsewhere [13]. The catalyst was tested in a continuous flow tubular micro-reactor. The CO_2:H_2 mole ratio of the feed gas was fixed at 1:3. Other process conditions, i.e. WHSV, reaction temperature and pressure were investigated using change one separate factor at a time (COST) approach. The process parameters investigated are independent of each other, thus, suitable to be examined using COST approach. Traditionally, WHSV has been defined as the weight of feed flowing per unit weight of catalyst per hour. However, to ease reporting of results, the weight of feed flowing is now commonly substituted by the volume of feed flowing [14,15]. Therefore, it should be noted that the WHSV in this study was calculated based on the volume of feed flowing.

In a typical experiment, the catalyst was pre-reduced in hydrogen (60 ml/min) for 2 hours at 300 °C and atmospheric pressure. After cooling, reactant gases were fed into the reactor. Effluent products from reactor were analysed by on-line gas chromatograph (GC, Agilent 7890A) equipped with thermal conductivity detector (TCD) at 20-minute interval. After 6-hour time-on-stream, the conversion of CO_2 and the selectivity of methanol were recorded.

X-ray diffraction (XRD) analysis of the catalysts was conducted using diffractometer PANalytical X'Pert PRO MPD, equipped with Cu Kα radiation. Using Scherrer equation, the Cu crystallite size was calculated from the full width at half-maximum (FWHM) of diffraction peak at 2Θ value of 43.3°. Cu surface area was estimated by equation (3).

$$S_{Cu}(m^2/g) = \frac{6000 \times X_{Cu}}{8.96 \times d_{Cu}} \quad (3)$$

where X_{cu}= mass fraction of Cu and d_{Cu}= Cu crystallite size from XRD analysis (nm) [16].

3. Results and Discussion

3.1. Effect of WHSV

The effect of WHSV on the performance of catalyst was investigated in the range of 8 - 120 L/g_{cat}·h. Meanwhile, the reaction temperature and pressure were held constant at 180 °C and 4.0 MPa, respectively. The CO_2 conversion (C_{CO_2}), product selectivity (S_i) and methanol yield (Y_{CH_3OH}) are shown in figure 1. The CO_2 conversion decreased from 24.2 % to 8.2 % with the increased of WHSV from 8 L/g_{cat}·h to 120 L/g_{cat}·h. The observation could be ascribed to the shorter contact time between reactants and catalytic active sites when WHSV was increased [17,18].

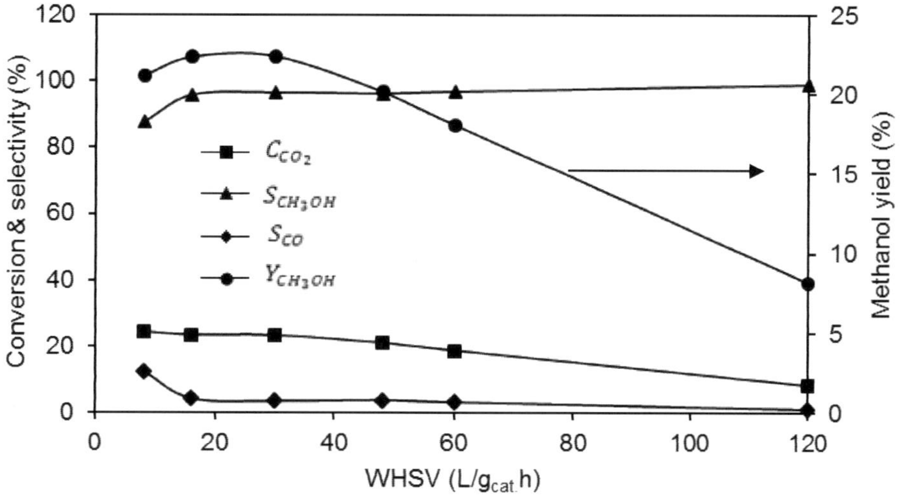

Figure 1. Effect of WHSV on the performance of CZM/KIT-6 at constant reaction temperature (180 °C) and pressure (4.0 MPa).

On stark contrast, the methanol selectivity was increased from 87.7 % to 99.0 % when WHSV was increased from 8 L/g_{cat}·h to 120 L/g_{cat}·h. The increasing trend of methanol selectivity contradicts the findings reported by Ren *et al.* [17], in which the ratio of activated CO_2 to formate intermediates was speculated to decrease with the increase of space velocity (shorter contact time) and thereby lowering the methanol selectivity. The correlation observed between WHSV and methanol selectivity could also be related to the weakening of RWGS with the increased of WHSV. A similar trend has been reported by Razali [19]. Interestingly, it was found the methanol yield increased first with WHSV to a maximum of 22.4 % at 30 L/g_{cat}·h and then decreased with further increase of WHSV. As mentioned earlier, CO_2 conversion and methanol selectivity responded in opposite manner with the increase of WHSV. Thus, it can be concluded that a balance between CO_2 conversion and methanol selectivity was achieved at WHSV of 30 L/g_{cat}·h and therefore resulting in the high methanol yield (22.4 %).

3.2. *Effect of reaction temperature*

The effect of reaction temperature on the performance of catalyst was investigated in the range of 160-260 °C. Meanwhile, the WHSV and pressure were held constant at 60 L/g_{cat}·h and 4.0 MPa, respectively. The catalyst performance is shown in figure 2.

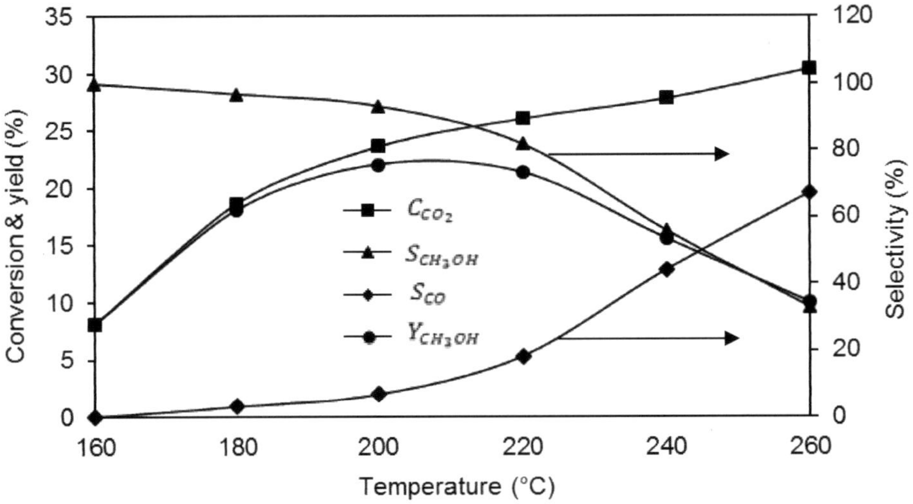

Figure 2. Effect of reaction temperature on the performance of CZM/KIT-6 at constant WHSV (60 L/g_{cat}·h) and reaction pressure (4.0 MPa).

As expected, the CO_2 conversion increased with the reaction temperature. This is because more external energy was supplied for CO_2 activation when reaction temperature was increased. The growth of CO_2 conversion was rapid at 160-200 °C and then became slow at 220 - 260 °C. The deceleration of CO_2 conversion points to the possibility of catalyst deactivation due to the loss of copper surface area when the reaction temperature was increased beyond 220 °C. Such effect of copper surface area on CO_2 conversion has been reported in other studies [20–22]. Using FWHM values from XRD analysis, the average copper crystallite size and copper surface area of CZM/KIT-6 spent in the experiments conducted at 180 °C, 220 °C and 260 °C were calculated and are shown in table 1. It was found that copper crystallite growth intensifies with the increase of reaction temperature, hence, resulting in the loss of copper surface area. The XRD results provide explanation to the deceleration of CO_2 conversion at reaction temperature above 220 °C.

The investigated temperature also showed significant effect on product selectivity. As the reaction temperature increased from 160 °C to 260 °C, the methanol selectivity decreased rapidly from >99.0 % to 33.0 %. On the contrary, the CO selectivity increased from <1.0 % to 67.0 %. These results corroborate well with the thermodynamics of CO_2 hydrogenation discussed earlier. The reaction of CO_2 conversion, either to produce methanol or CO, is highly dependent on the reaction temperature. The reaction of CO_2 hydrogenation to methanol is exothermic, hence, it is favoured at low reaction temperature. Meanwhile, RWGS is an endothermic reaction, thus, it is favoured at high reaction temperature.

Table 1. Average copper crystallite size and copper surface area of catalyst spent in the experiments conducted at 180 °C, 220 °C and 260 °C.

Reaction temperature (°C)	Cu crystallite size (nm)	Cu surface area ($m^2/g_{cat.}$)
180	10.6	19.5
220	11.3	18.3
260	14.1	14.6

At temperature range 160 - 200 °C, the methanol yield increased with temperature due to the rapid increase of CO_2 conversion compared to the decrease of methanol selectivity. However, the methanol yield was decreased with further increase of temperature (220 - 260 °C) due to the drastic decrease of methanol selectivity compared to the gradual raise of CO_2 conversion. A balance between CO_2 conversion and methanol selectivity was attained at 200 °C, and the highest methanol yield (22.0 %) was recorded.

Frequently, the apparent activation energies (*E*a) of methanol and CO formation were estimated from Arrhenius plot, in which the logarithms of formation rates are plotted against the reciprocal of absolute temperatures [23,24]. The Arrhenius plot of methanol formation in this study is depicted in figure 3. It is intriguing to note the Arrhenius plot of methanol formation in the overall temperature range (160 - 260 °C), figure 3(a), severely deviates from the theoretical linear plot, and a concave plot was obtained instead. The non-linearity of the Arrhenius plot is related to the decreased of methanol formation rate with the increase of reaction temperature from 220 - 260 °C. The corresponding apparent activation energy calculated for the high temperature region has a negative value of -10 kcal/mol. It should be noted that it is possible for a chemical reaction to have negative apparent activation energy though rare [25]. The deviation from Arrhenius law is rarely reported but very informative.

In current study, the deviation from Arrhenius law could be related to three factors. Firstly, the increase dominance of RWGS (endothermic) over the reaction of CO_2 hydrogenation to methanol (exothermic) when the reaction temperature was risen beyond 200 °C. The formation rates of methanol and CO in the investigated temperature range are shown in figure 4. It is observable that the formation rates of methanol decreased while CO increased at high temperature range, 220 - 260 °C. Secondly, the reduction of methanol formation rate at high reaction temperature due to catalyst deactivation. As seen in table 1, the increase of reaction temperature was accompanied by the growth of copper crystallites and the loss of copper surface area. The dependence of methanol production rate on the crystallite size and surface area of copper has been widely recognized [25,26]. As such, catalyst deactivation could have contributed to the reduced methanol formation rate at high reaction temperature. Thirdly, there could be a change in the reaction coordinate such as the increase of reverse reaction at high reaction temperature, thus, resulting in the overall negative apparent activation energy. This has been elaborated well by Revell and Williamson [25]. In this regard, Arena *et al.* [28] postulated the formation of CO *via* the decomposition of methanol at reaction temperature above 200 °C. This consumption of the formed methanol could have lowered the apparent methanol formation rate. Despite the non-Arrhenius behaviour of methanol formation at high reaction temperature, the apparent activation energy of methanol formation can be estimated from the low reaction temperature range, 160 - 200 °C. The apparent activation energy of methanol production at low reaction temperature was 10 kcal/mol.

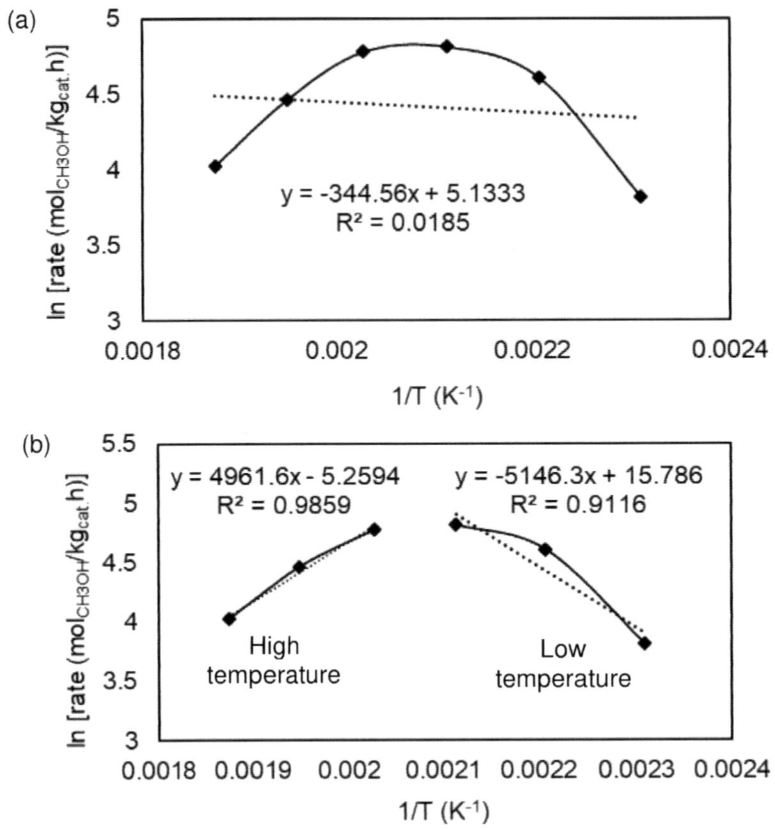

Figure 3. Arrhenius plot for methanol formation in (a) overall temperature range and (b) according to high- and low-temperature range.

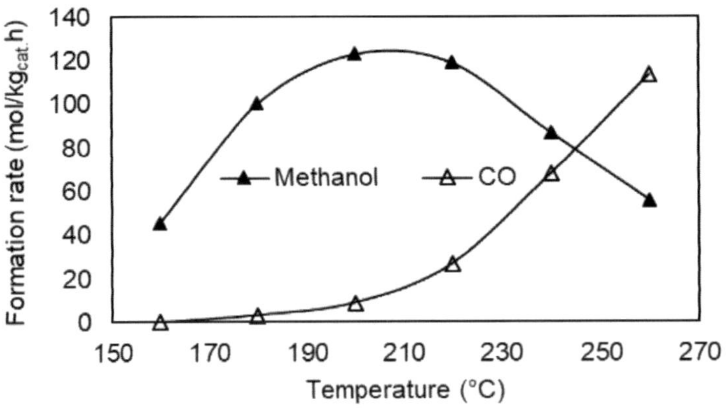

Figure 4. Methanol and CO production rate at reaction temperature of 160 – 260 °C.

The Arrhenius plot of CO formation is shown in figure 5. It should be noted that the formation rate of CO at 160 °C was not included in the Arrhenius plot because virtually no CO was detected at that reaction temperature. The CO production obeys the Arrhenius law, in which the formation rate increased with reaction temperature. The apparent activation energy for CO formation was 22 kcal/mol. A comparison of apparent activation energies achieved in this study and the values reported in literature are shown in table 2. Generally, the apparent activation energies obtained in current study are consistent with other studies. According to table 2, the apparent activation energy of CO formation is generally higher than methanol by 2 - 3 folds. This explains the dominance of methanol formation at low reaction temperature while CO formation grows steeply at high reaction temperature. The experimental data collected in current investigation clearly reflects this trend.

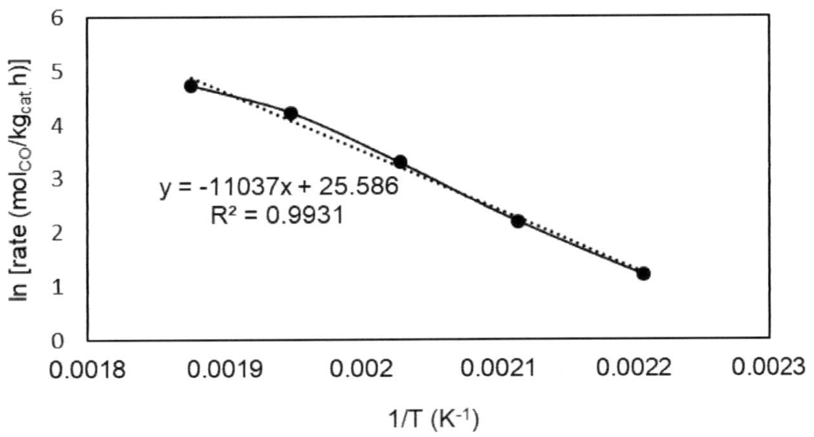

Figure 5. Arrhenius plot for CO formation.

Table 2. Comparison of the apparent activation energies (E_a) for methanol and CO formation attained in current study and in the literature.

Catalyst	Apparent activation energies (kcal/mol)		Reference
	Methanol	CO	
CZM/KIT-6	10	22	This study
$Cu/ZnO/Al_2O_3$	13	28	[29]
$La/Cu/ZrO_2$	9	24	[20]
$Ga-Cu/ZnO/ZrO_2$	7	27	[23]
$Cu/t-ZrO_2$	9	25	[24]
$CuCeTiO_x$	9	18	[21]
Ni/In_2O_3	19	-	[30]

3.3. Effect of reaction pressure
The effect of reaction pressure on the performance of catalyst in CO_2 hydrogenation to methanol was investigated in the range of 1.0 - 5.0 MPa. Meanwhile, the WHSV and temperature were held constant at 60 L/g_{cat}·h and 220 °C, respectively. The catalyst performance is shown in figure 6. The presented results show the reaction pressure has marked influence on CO_2 conversion, in which the conversion increased steeply from 4.1 % to 26.1 % when the reaction pressure was increased from 1.0 MPa to 4.0 MPa. Further increase of reaction pressure to 5.0 MPa marginally increased CO_2 conversion to a maximum of 27.5 %. Under kinetic-controlled regime, the increased of reaction pressure could enhance the adsorptions of CO_2 and hydrogen on the catalyst, thus, improving CO_2 conversion [18].

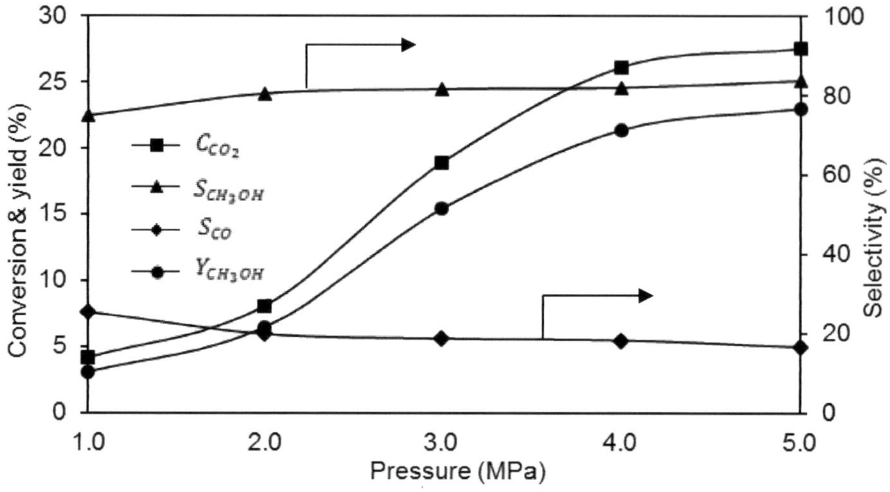

Figure 6. Effect of reaction pressure on the performance of CZM/KIT-6 at constant WHSV (60 L/g_{cat}·h) and reaction temperature (220 °C).

From thermodynamic viewpoint, CO production from the RWGS is pressure-independent [31]. On the contrary, methanol synthesis from CO_2 hydrogenation is sensitive to reaction pressure since the reaction proceeds with the reduction of 4 moles of reactants to 2 moles of products. This is in accord with the Le Chatelier's principle. The data collected shows the methanol selectivity steadily increased (74.6 - 83.5 %) with reaction pressure while CO selectivity decreased. The increased of methanol selectivity and the decreased of CO selectivity were slow when reaction pressure was increased beyond 2.0 MPa. An abrupt growth of methanol yield was observed when reaction pressure was increased from 1.0 MPa to 4.0 MPa (3.1 - 21.3 %), and then slow down with further increase to 5.0 MPa (23.0 %). Based on the observations, it can be concluded that the increase of reaction pressure has positive effect on methanol yield because it remarkably improves CO_2 conversion and to a lesser extend increases methanol selectivity.

3.4. Catalyst stability

The stability of CZM/KIT-6 was evaluated at 200 °C, with WHSV of 30 L/g_{cat}·h, and H_2: CO_2 mole ratio of 3:1 under a reaction pressure of 5.0 MPa. The stability experiment was conducted under high methanol yield conditions for 120 hours. The process conditions applied were selected based on the investigations in Section 3.1-3.3. It should be noted that the effluent stream from the reactor was analysed at hourly interval. The catalyst performance is illustrated in figure 7.

Methanol and carbon monoxide (CO) were the only detectable carbon-containing products during the experiment. On the average, 27.6 % of CO_2 conversion and 88.3 % of methanol selectivity were achieved, leading to 24.4 % of methanol yield. More importantly, the catalytic performance of CZM/KIT-6 was maintained at high level throughout the 120-hour experiment.

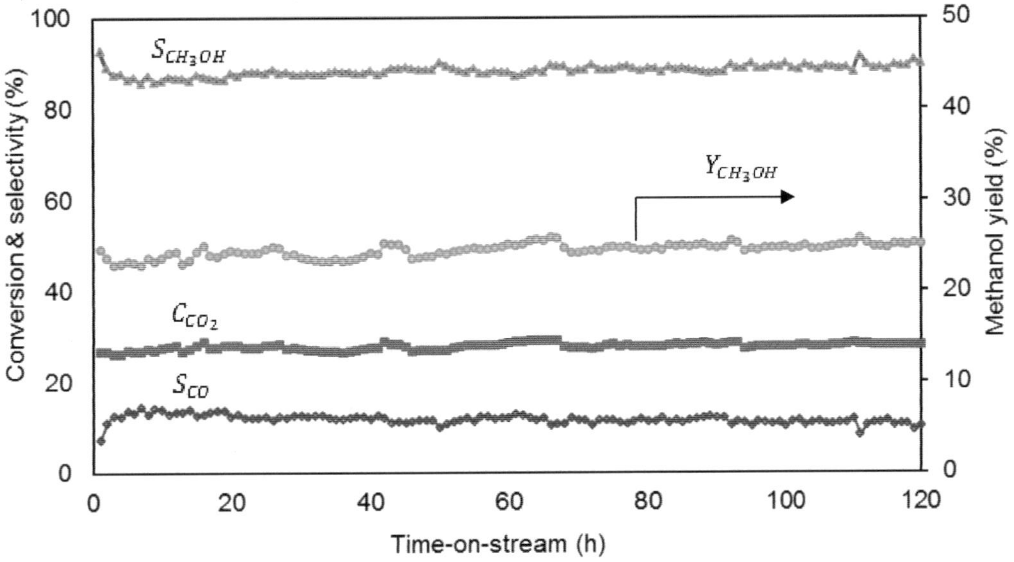

Figure 7. Catalytic performance of CZM/KIT-6 in 120-hour experiment.

The crystalline phase of CZM/KIT-6 after pre-reduction and after stability experiment were analysed by XRD and the results are shown in figure 8. The broad diffraction peak observed in both catalysts at 22° can be attributed to amorphous silica. Furthermore, the diffraction peaks of metallic Cu phase (JCPDS 04-0836) were observed at 43.3°, 50.4° and 74.1°.

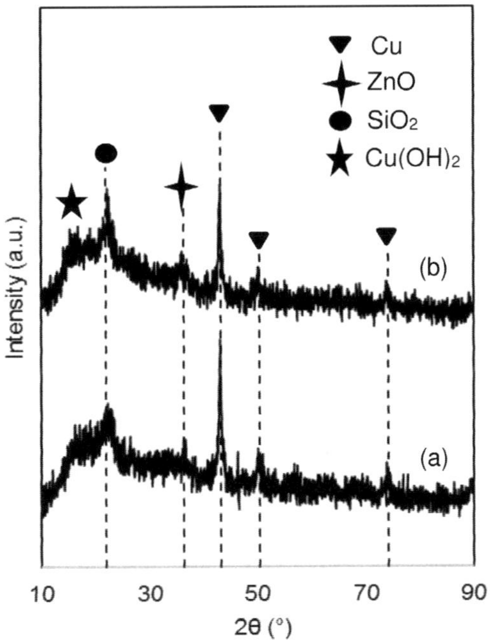

Figure 8. XRD diffraction patterns of (a) pre-reduced CZM/KIT-6 and (b) CZM/KIT-6 spent in stability experiment (after 120 h time-on-stream).

The diffraction peak of ZnO phase (JCPDS 36-1451) was seen at 36.4°. In the spent CZM/KIT-6, a weak diffraction peak corresponds to copper hydroxide ($Cu(OH)_2$) (JCPDS 13-0420) was found at around 16°. Meanwhile, the diffraction peaks of Cu_2O and/or CuO were not detectable, thus, suggesting the re-oxidation of metallic Cu in CZM/KIT-6 was trivial even after 120-hour time-on-stream under high methanol yield conditions.

The average copper crystallite size and copper surface area in pre-reduced and spent catalysts are gathered in table 3. Compared to the pre-reduced catalyst, the copper crystallites in spent CZM/KIT-6 underwent 53.5 % growth. Accordingly, 35.3 % of copper surface area loss was estimated. Although substantial, the crystallite growth and the surface area loss had no significant effect on the catalyst performance in the 120-hour experiment.

Table 3. Average copper crystallite size and copper surface area of pre-reduced catalyst and spent catalyst (after 120 h time-on-stream).

Catalyst treatment	CZM/KIT-6	
	Cu crystallite size (nm)	Cu surface area ($m^2/g_{cat.}$)
Pre-reduced	7.1	29.2
Spent	10.9	18.9

A comparison between the performance of CZM/KIT-6 and other catalysts reported in literature during stability experiment is shown in table 4. In general, the stability of the performance of CZM/KIT-6 is on par with the catalysts reported in literature. Additionally, the high methanol yield achieved using CZM/KIT-6 indicates the catalyst was more active than the others in catalysing direct CO_2 hydrogenation to methanol. The performance demonstrated by CZM/KIT-6 in 120-hour time-on-stream infers the catalyst has the potential for industrial application.

Table 4. Comparison of the performance of CZM/KIT-6 and other catalysts reported in literature during stability experiment.

Catalyst	CO_2 conversion (%)	Methanol		Experiment duration (h)	Reference
		Selectivity (%)	Yield (%)		
CZM/KIT-6	27.6	88.3	24.4	120	This work
$Cu-ZnO-ZrO_2$	12.0	71.0	8.5	100	[32]
Cu/SiO_2	28.0	21.3	6.0	120	[33]
Cu/ZnO	16.1	77.8	12.5	100	[34]
$Cu/ZnO/Al_2O_3$	15.8	63.4	10.0	100	[35]
$Cu/ZnO/Al_2O_3/ZrO_2$	13.6	61.7	8.4	170	[36]

4. Conclusion

The performance of CZM/KIT-6 in catalysing direct CO_2 hydrogenation to methanol has been further assessed by considering the effect of process conditions (WHSV, reaction temperature and pressure) and the stability of the catalyst under high methanol yield conditions. The conversion of CO_2 and the selectivity towards methanol responded to the variation of process parameters according to the reaction thermodynamics and these have substantial effect on the methanol yield. Therefore, the effectiveness of methanol production relies greatly on the selection of operating conditions. The non-Arrhenius behaviour of methanol formation observed at high reaction temperature, 220 – 260 °C, are possibly the results of the increased rate of RWGS, catalyst deactivation and/or a change in the reaction coordinate such as the decomposition of methanol. The CZM/KIT-6 catalyst maintained high performance, with average methanol yield of 24.4 %, throughout the stability experiment. The copper crystallite growth and the loss of copper surface area were 53.5 % and 35.3 %, respectively, after 120-hour time-on-stream under high methanol yield conditions.

Acknowledgments
We acknowledge the financial support given by Universiti Sains Malaysia-NanoMITE (203/PJKIMIA/6720009) and XMUM Research Fund (XMUMRF/2021-C7/IENG/0034).

References

[1] Yuan Z, Eden M R and Gani R 2016 Toward the development and deployment of large-scale carbon dioxide capture and conversion processes *Ind. Eng. Chem. Res.* **55** 3383–419
[2] Rivarolo M, Bellotti D, Magistri L and Massardo A F 2016 Feasibility study of methanol production from different renewable sources and thermo-economic analysis *Int. J. Hydrogen Energy* **41** 2105–16
[3] González-Aparicio I, Pérez-Fortes M, Zucker A and Tzimas E 2017 Opportunities of integrating CO_2 utilization with RES-E: A Power-to-Methanol business model with wind power generation *Energy Procedia* **114** 6905–18
[4] Atsonios K, Panopoulos K D and Kakaras E 2016 Investigation of technical and economic aspects for methanol production through CO_2 hydrogenation *Int. J. Hydrogen Energy* **41** 2202–14
[5] Arena F, Mezzatesta G, Zafarana G, Trunfio G, Frusteri F and Spadaro L 2013 Effects of oxide carriers on surface functionality and process performance of the Cu–ZnO system in the synthesis of methanol *via* CO_2 hydrogenation *J. Catal.* **300** 141–51
[6] Xiao J, Mao D, Guo X and Yu J 2015 Methanol synthesis from CO_2 hydrogenation over CuO – ZnO – TiO_2 catalysts: the influence of TiO_2 Content *Energy Technol.* **3** 32–9
[7] Saeidi S, Amin N A S and Rahimpour M R 2014 Hydrogenation of CO_2 to value-added products—A review and potential future developments *J. CO_2 Util.* **5** 66–81
[8] Viswanathan B 2013 Electro-catalytic reduction of carbon dioxide *New and Future Developments in Catalysis: Activation of Carbon Dioxide* pp 275–95
[9] Havran V, Dudukovi M P and Lo C S 2011 Conversion of methane and carbon dioxide to higher value products *Ind. Eng. Chem. Res.* **50** 7089–100
[10] Porosoff M D, Yan B and Chen J G 2016 Catalytic reduction of CO_2 by H_2 for synthesis of CO, methanol and hydrocarbons: Challenges and opportunities *Energy Environ. Sci.* **9** 62–73
[11] Liu L, Jiang H, Liu H and Li H 2013 Recent advances on the catalysts for activation of CO_2 in several typical processes *New and future developments in catalysis: Activation of carbon dioxide* pp 189–222
[12] Koh M K, Wong Y J, Chai S P and Mohamed A R 2018 Carbon dioxide hydrogenation to methanol over multi-functional catalyst : Effects of reactants adsorption and metal-oxide(s) interfacial area *J. Ind. Eng. Chem.* **62** 156–65
[13] Koh M K, Khavarian M, Chai S P and Mohamed A R 2018 The morphological impact of siliceous porous carriers on copper-catalysts for selective direct CO_2 hydrogenation to methanol *Int. J. Hydrogen Energy* **43** 9334–42
[14] Wang G, Chen L, Sun Y, Wu J, Fu M and Ye D 2015 Carbon dioxide hydrogenation to methanol over Cu/ZrO_2/CNTs: effect of carbon surface chemistry *Rsc Adv.* **5** 45320–30
[15] Yang H, Gao P, Zhang C, Zhong L, Li X, Wang S, Wang H, Wei W and Sun Y 2016 Core-shell structured Cu@m-SiO_2 and Cu/ZnO@m-SiO_2 catalysts for methanol synthesis from CO_2 hydrogenation *Catal. Commun.* **84** 56–60
[16] Ahouari H, Soualah A, Le Valant A, Pinard L, Magnoux P and Pouilloux Y 2013 Methanol synthesis from CO_2 hydrogenation over copper based catalysts *React. Kinet. Mech. Catal.* **110** 131–45

[17] Ren H, Xu C, Zhao H, Wang Y, Liu J and Liu J 2015 Methanol synthesis from CO_2 hydrogenation over Cu/γ-Al_2O_3 catalysts modified by ZnO, ZrO_2 and MgO *J. Ind. Eng. Chem.* **28** 261–7

[18] Portha J F, Parkhomenko K, Kobl K, Roger A C, Arab S, Commenge J M and Falk L 2017 Kinetics of methanol synthesis from carbon dioxide hydrogenation over copper-zinc oxide catalysts *Ind. Eng. Chem. Res.* **56** 13133–45

[19] Razali N A M 2014 *Synthesis, characterization and catalytic performance of Cu/ZnO/SBA-15 for hydrogenation of carbon dioxide to methanol* (Universiti Sains Malaysia)

[20] Guo X, Mao D, Lu G, Wang S and Wu G 2011 The influence of la doping on the catalytic behavior of Cu/ZrO_2 for methanol synthesis from CO_2 hydrogenation *J. Mol. Catal. A Chem.* **345** 60–8

[21] Chang K, Wang T and Chen J G 2017 Hydrogenation of CO_2 to methanol over CuCeTiOx catalysts *Appl. Catal. B Environ.* **206** 704–11

[22] Li F, Zhan H, Zhao N and Xiao F 2017 CO_2 hydrogenation to methanol over La-Mn-Cu-Zn-O based catalysts derived from perovskite precursors *Int. J. Hydrogen Energy* **42** 20649–57

[23] Ladera R, Pérez-Alonso F J, González-Carballo J M, Ojeda M, Rojas S and Fierro J L G 2013 Catalytic valorization of CO_2 *via* methanol synthesis with Ga-promoted Cu-ZnO-ZrO_2 catalysts *Appl. Catal. B Environ.* **142–143** 241–8

[24] Witoon T, Chalorngtham J, Dumrongbunditkul P, Chareonpanich M and Limtrakul J 2016 CO_2 hydrogenation to methanol over Cu/ZrO_2 catalysts: effects of zirconia phases *Chem. Eng. J.* **293** 327–36

[25] Revell L E and Williamson B E 2013 Why are some reactions slower at higher temperatures? *J. Chem. Educ.* **90** 1024–7

[26] Natesakhawat S, Lekse J W, Baltrus J P, Ohodnicki P R, Howard B H, Deng X and Matranga C 2012 Active sites and structure–activity relationships of copper-based catalysts for carbon dioxide hydrogenation to methanol *ACS Catal.* **2** 1667–76

[27] Xiao S, Zhang Y, Gao P, Zhong L, Li X, Zhang Z, Wang H, Wei W and Sun Y 2017 Highly efficient Cu-based catalysts *via* hydrotalcite-like precursors for CO_2 hydrogenation to methanol *Catal. Today* **281** 327–36

[28] Arena F, Mezzatesta G, Zafarana G, Trunfio G, Frusteri F and Spadaro L 2013 How oxide carriers control the catalytic functionality of the Cu – ZnO system in the hydrogenation of CO_2 to methanol *Catal. Today* **210** 39–46

[29] Choi Y, Futagami K, Fujitani T and Nakamura J 2001 Role of ZnO in Cu/ZnO methanol synthesis catalysts - morphology effect or active site model? *Appl. Catal. A Gen.* **208** 163–7

[30] Jia X, Sun K, Wang J, Shen C and Liu C jun 2020 Selective hydrogenation of CO_2 to methanol over Ni/In_2O_3 catalyst *J. Energy Chem.* **50** 409–15

[31] Huš M, Dasireddy V D B C, Strah Štefančič N and Likozar B 2017 Mechanism, kinetics and thermodynamics of carbon dioxide hydrogenation to methanol on Cu/$ZnAl_2O_4$ spinel-type heterogeneous catalysts *Appl. Catal. B Environ.* **207** 267–78

[32] Guo X, Mao D, Lu G, Wang S and Wu G 2010 Glycine-nitrate combustion synthesis of CuO-ZnO-ZrO_2 catalysts for methanol synthesis from CO_2 hydrogenation *J. Catal.* **271** 178–85

[33] Wang Z-Q, Xu Z-N, Peng S-Y, Zhang M-J, Lu G, Chen Q-S, Chen Y and Guo G-C 2015 High-performance and long-lived Cu/SiO_2 nanocatalyst for CO_2 hydrogenation *ACS Catal.* **5** 4255–9

[34] Lei H, Nie R, Wu G and Hou Z 2015 Hydrogenation of CO_2 to CH_3OH over Cu/ZnO catalysts with different ZnO morphology *Fuel* **154** 161–6

[35] Lei H, Hou Z and Xie J 2016 Hydrogenation of CO_2 to CH_3OH over CuO/ZnO/Al_2O_3 catalysts prepared *via* a solvent-free routine *Fuel* **164** 191–8

[36] Zhang Y, Zhong L, Wang H, Gao P, Li X, Xiao S, Ding G and Sun Y 2016 Catalytic performance of spray-dried Cu/ZnO/Al_2O_3/ZrO_2 catalysts for slurry methanol synthesis from CO_2 hydrogenation *J. CO_2 Util.* **15** 72–82

Aloe vera as a natural flocculant for palm oil mill effluent (POME) treatment – characterisation and optimisation studies

R L Chu[1], Vasanthi S[1] and Anurita S[2]

[1] Department of Chemical and Environmental Engineering, Faculty of Science and Engineering, University of Nottingham Malaysia, Jalan Broga, 43500 Semenyih, Selangor

[2] Department of Civil Engineering, Faculty of Science and Engineering, University of Nottingham Malaysia, Jalan Broga, 43500 Semenyih, Selangor

Email: Vasanthi.Sethu@nottingham.edu.my

Abstract. In the present study, fenugreek and aloe vera were investigated for the removal of turbidity (TUR), total suspended solids (TSS) and chemical oxygen demand (COD) from POME by using a central composite design (CCD) in the Design Expert software. The effects of three factors such as pH, coagulant dosage and flocculant dosage were analysed using jar test experiment and optimised using response surface methodology (RSM). The optimum results obtained from process optimisation analysis were pH 4, 24.13 g of coagulant dosage and 20 ml of flocculant dosage that are sufficient to remove 82.78 % of TUR, 83.40 % of TSS and 32.95 % of COD. The maximum error between the optimised values and the experimental values (82.78 % for TUR, 83.08 % for TSS and 33.76 % for COD) were below 4 %, indicating that satisfactory agreement was achieved. This showed that modelling and optimisation of the coagulation-flocculation process can be achieved by RSM approach. From analytical studies, it was found that the interactions between coagulant-flocculant and colloidal particles involve the mechanisms of charge neutralisation, adsorption and bridging, due to the active components such as amine (N-H) and hydroxyl (O-H) groups contained in the fenugreek and aloe vera.

Keywords: Aloe vera, Central Composite Design, Flocculation, Palm oil mill effluent, Response Surface Methodology

1. Introduction

Malaysia is one of the world's biggest producers and exporters of palm oil and palm oil products [1]. It covers 28 % world palm oil production and 33 % world exports [52]. Besides, Malaysia also produces oils and fats that accounts for 11 % of the world's production and 27 % of export trade [54]. Hence, Malaysia plays a crucial role in fulling the increasing demand for oils and fats. Oil palm cultivation in Malaysia covers 4.49 million hectares of land which produce 17.73 million tonnes of palm oil from the fruit flesh and 2.13 tonnes of palm kernel oil from the seeds [53]. It is estimated that 1 tonne of palm kernel oil is produced from every 10 tonnes of palm oil [2]. During the palm oil processing, palm oil mill effluent (POME) is mainly generated from oil extraction, washing and cleaning [58]. For example, about 1000 kg of fresh fruit bunches (FFB) are used for oil extraction using high temperature steam and water, and consequently produce 600-700 kg of POME [3]. Moreover, every 10-35 tonnes of FFB is generated from 1 hectare of oil palm, with more than 70 % by weight of FFB is discarded as waste [4].

These indicate that high amount of polluted wastewater is generated, and POME treatment is essential to prevent environmental pollution.

Raw POME is a highly viscous liquid waste that is brown in colour and comes with unpleasant odour. It is non-hazardous due to absence of chemicals addition in oil extraction process, naturally high in organic content which include mixture of cellulosic wastes, carbohydrates and oils, which are highly polluting to the environment [5]. It is generated from palm oil processing which comprises of oil palm trunks (OPT), oil palm fronds (OPF), empty fruit bunches (EFB), palm pressed fibers (PPF), palm kernel shells, palm kernel cake and liquid sludge discharge [6]. POME is a hundred times more polluting than the municipal sewage, resulting from high concentration of oil and grease (4,000 mg/L), COD (50,000 mg/L), BOD (25,000 mg/L) and total solids (40,500 mg/L) [2]. Furthermore, it is highly toxic to aquatic life as well as possesses significant pollution impact to the surrounding environment. Therefore, POME must be treated according to the discharge standard limit set by the Department of Environment (DOE) to prevent deterioration of the environment quality. The discharge limit in Malaysia is a minimum of 50 mg/L and 20 mg/L for BOD, depending on the location of palm oil mill [7].

In Malaysia, more than 85 % of oil palm mills utilise ponding system that include a series of anaerobic, facultative and aerobic to treat POME because is it more cost effective and much easier to operate. This system requires large surface area of land with long retention time, producing unpleasant odour and greenhouse gas emissions [8]. In fact, such system is unable to consistently meet the standard industrial discharge limit for POME [9]. Thus, an economically and environmentally friendly technology using coagulation and flocculation process is proposed for the POME treatment to complement the ponding system, in order to speed up the treatment process and achieve the discharge limits. This technology is one of the most commonly and widely used in industries to remove colloidal particles in water and wastewater system due to effectiveness and simplicity of the operation [10]. Coagulation process involves the charge neutralisation that produce small flocs, whereas flocculation encourages agglomeration of large flocs under the gentle stirring. Based on Brownian movement in water, the surface charges on the colloidal particles are mainly negative, causing repulsion that inhibit the agglomeration between the particles.

In order to overcome the aforementioned drawback, chemical coagulant and flocculant are added to enhance the destabilisation and agglomeration process, forming larger flocs that can be separated by sedimentation or filtration Sethu *et al.,* [11]. This is due to the various mechanism involved such as charge neutralisation and bridging. Both coagulant and flocculant which is mainly positively charged neutralise the negative charged of the colloidal particles. The most commonly used chemical as coagulant-flocculant is alum (Aluminium sulphate). However, it will cause negative impact on health and the environment. By using it in coagulation-flocculation process will lead to high procurement costs, ineffectiveness in low temperature, variation of pH value and large amount of chemical sludge produced in treated water [12]. In addition, prolonged exposure to water consumption with high level of aluminium residuals will cause Alzheimer's diseases which resulted from the penetration and accumulation of the residuals into the human body and the brain respectively [13].

Studies have been carried out by researchers to find alternative ways to suppress the use of chemical coagulant-flocculant by proposing a cost-effective and environmentally friendly treatment system for POME. It involves the use of toxic free, natural abundance and cost effective of natural plant-based materials extracted from the seeds, leaves, fruits or peels of a plant for the coagulation and flocculation process. The sludge produced from this treatment has fertilizing properties because of the high concentration of degradable organic matter [5]. A study reported by Fatah and Wahab [12] on utilising dragon fruit foliage as natural coagulant for the treatment of the three-phase decanter POME has resulted in removal percentage of 99.2 % for TUR, 98.8 % for TSS and 48.7 % for COD. Another study which uses eggplant seed as natural coagulant was conducted by Oubrayme *et al.* [14] to treat the produced water in petroleum industry, removing 99.42 % of TUR and 92.18 % of COD. Other than that, fenugreek and aloe vera have been found to have good potential as natural coagulant and flocculant in wastewater treatment, but limited studies were conducted by researchers. For example, fenugreek seeds can remove 76.8 % of TSS and 77.6 % of TUR for POME treatment at dosage and pH of 800mg/L and 4 respectively

[11]; Esther-Irma *et al.* [15] reported that the maximum removal efficiency achieved by aloe vera in surface water clarification for TUR, TSS and colour are 72 %, 91 % and 15 % respectively; for textile wastewater clarification, aloe vera removes 92.3 % of TUR, 76.8 % of COD, 83.5 % of BOD and 57.9 % of TSS [16]; and aloe vera with calcium chloride remove 88 % of fluoride in drinking water treatment [17].

Fenugreek, scientifically known as *Trigonella foenum-graecum* which belongs the Leguminosae family and mainly grown as multipurpose crop in Asia, Europe, Africa and Australia [18]. It is well known for its food flavouring and medicinal uses for appetite stimulation, anti-inflammatory, antioxidant, antidiabetic and anticancer [19]. According to Wani and Kumar [20], the active compounds exist in fenugreek consists of 23-26 % protein, 6-7 % fat and 58 % carbohydrates with 25 % of it is dietary fiber, that is effective in enhancing the coagulation process. In order to improve the quality of wastewater, aloe vera as natural flocculant is used to enhance the interparticle interaction to form larger flocs which will be separated through sedimentation or filtration. Aloe vera is a perennial succulent plant that comes from the family of *Xanthorrhoeaceae* and subfamily of *Asphodeloideae* as per the Angiosperm Phylogeny Group III System [21]. It grows especially in the tropical and subtropical area, and is widely used in the medical, pharmaceutical, cosmetic and nutritional purposes [21]. Aloe vera plant has thick, fleshy, sharp leaves that are joined together at the stem, resembling a rosette pattern [22]. Each aloe vera leaf consists of three main parts such as epidermis, latex layer and parenchyma. In this research, the focus will be on the parenchyma where the mucilaginous gel can be found. It includes 98.5 wt. % water and 1.5 wt. % of bioactive compounds such as carbohydrates, proteins, enzymes, vitamins, minerals, organic acids, phenolic substances and phytosterols [21]. However, these chemical compositions in the parenchyma will vary according to the separation techniques, variation of climate conditions and cultivation methods [23]. The existence of monomers, oligomers and polymers in carbohydrates enhance the bridging mechanism in flocculation process. The mucilaginous gel also contains polysaccharides that have strong affinity towards water, leading to intramolecular interactions between polysaccharides molecules and water molecules [24]. As a result, aggregation is formed during the coagulation-flocculation process.

The aim of this research was to study the effectiveness of fenugreek seed and aloe vera as natural organic coagulant and flocculant for the treatment of POME. Experiments were conducted using jar test to study the coagulation-flocculation process. Parameters such as pH, coagulant dosage and flocculant dosage were varied to determine the optimum removal efficiency of turbidity (TUR), total suspended solids (TSS) and chemical oxygen demand (COD) from treated POME. Response surface methodology (RSM) is applied in this study to design the experiment with minimum number of experimental trials and justify the output response towards the nominal and target study requirements. After finalising the optimum conditions, the experiment was repeated at the proposed condition to examine the outcomes from RSM prior conducting analytical analysis such as zeta potential analysis and Fourier transform infrared spectroscopy (FTIR) for further verification of the potential of fenugreek seed and aloe vera in treating the POME.

2. Materials and Methods

2.1. Palm Oil Mill Effluent (POME) Collection

POME samples were obtained from the final pond of the facultative lagoon at KLK Tuan Mee Oil Palm Mill, Sungai Buloh, Selangor Darul Ehsan, Malaysia. The sample collected was the effluent from the oil palm mill at the final pond prior discharging into the environment. It was collected in an airtight plastic container, covered with black plastic bag to avoid direct light exposure from biodegrading the sample due to microbial decomposition. The POME was stored for 12 hours in the laboratory refrigerator.

2.2. Fenugreek Seed Powder Preparation

Fenugreek seeds were obtained from a nearby local market in Semenyih. The seeds were crushed into coarse powder using pestle and mortar. Then, the coarse powder was ground into fine powder of size range 80 – 100 µm using the ultra-centrifugal mill. The fenugreek powder was collected in sample bottle and then stored at the desiccator to prevent re-humidification and mould formation.

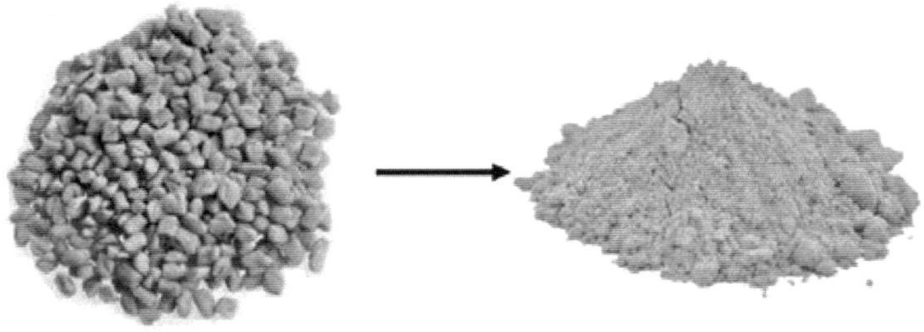

Figure 1. Fenugreek seeds ground into fenugreek seeds powder.

2.3. Aloe Vera Solution Preparation

Fresh aloe vera leaves were obtained from aloe vera plants. The leaves were first rinsed with distilled water to remove impurities. Then, the gel (inner part of the leaves) was obtained by removing the green epidermis of the leaves. The gel was cut into smaller pieces and blended with juice blender to get the aloe vera solution. The solution was collected in the beaker and covered with parafilm. It was then stored in the refrigerator to prevent spoilage in a short term. Moreover, it minimised the growth rate of microorganisms.

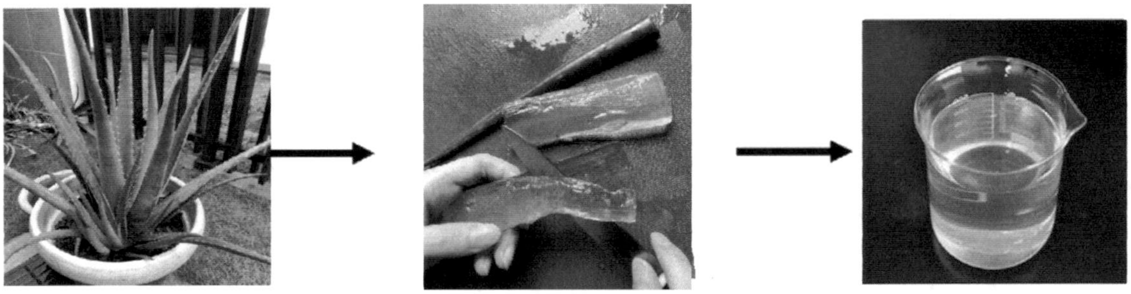

Figure 2. Removal of aloe vera's epidermis to obtain the aloe vera gel solution.

2.4. Jar Test Experiment

A jar test was conducted using the Phipps and Bird flocculator that consists of six impellers as shown below. This flocculator was located at Engineering Research Building (Block N) in University of Nottingham Malaysia.

(a) (b)

Figure 3. (a) Phipps and Bird flocculator; and (b) Phipps and Bird flocculator with beakers covered with parafilm for sedimentation process.

The POME sample was poured into a bucket prior transferring 500 ml each into six 1 L beakers to ease the control of the amount of sample required and minimise spillage. The pH of the sample in each beaker was adjusted to the desired value, ranging from 4 to 9. The initial sample readings for TUR, TSS and COD were measured and recorded. Then, the beakers were put under the agitators, making sure that the impellers were immersed in the POME sample. After that, the dosage used for fenugreek powder and aloe vera solution were varied and added during the rapid mixing and slow mixing stage respectively. Rapid mixing was applied for 5 minutes at 200 rpm for coagulation process, followed by slow mixing for 20 minutes at 30 rpm for flocculation process. This is the typical time range used in a jar test for coagulation and flocculation studies [11]. After that, the beakers were set aside for 4 hours to allow the settling of solids before withdrawing 15 ml of supernatant each into six conical centrifuge tubes from each beaker. The supernatant was then measured and recorded to obtain the final readings of TUR, TSS and COD.

Table 1. Summary of jar test variables.

Variables	Speed (rpm)	Time (min)
Rapid Mixing	200	5
Slow Mixing	30	30
Sedimentation	-	240 (4 hrs)

(a) (b)

Figure 4. (a) Untreated POME; (b) Treated POME at optimal condition.

2.5. Chemical Analysis

The pH of the POME sample was manipulated using hydrochloric acid (HCl) and sodium hydroxide (NaOH), followed by measuring using pH meter. HCl and NaOH were diluted to concentration of 9 M and 10 M respectively using equation (1). The molar mass of each solute was taken from the labelled container.

$$Molarity = \frac{Moles\ of\ Solute}{Litres\ of\ Solution} = \frac{Grams\ of\ Solute}{Molar\ Mass\ Solute \times Litres\ of\ Solution} \quad (1)$$

As the removal efficiency of the POME using fenugreek-aloe vera was evaluated by calculating the removal percentage of TUR, TSS and COD as presented in equation (2). TUR and TSS readings of treated POME were measured and recorded by using turbidity meter and HACH DR2800 spectrophotometer respectively. Before analysing the COD, 2 ml of the treated POME from each beaker was extracted using pipette into 50 ml volumetric flask to dilute the concentration of treated POME. This dilution step was required due to the high concentration of the treated POME that was out of the range for spectrophotometer detection. Then, 1 ml of the mixed solution each was homogenised with high range plus (HR+) Hach COD vials separately. After that, the mixture in vials were heated in a HACH DRB200 Digestor at 150 °C for 2 hours. The COD readings were measured and recorded by using HACH DR2800 spectrophotometer after heated and cooled, where the readings are TUR, TSS and COD of the treated POME.

$$Removal\ Efficiency, \% = \frac{Initial\ Reading - Final\ Reading}{Initial\ Reading} \times 100\% \quad (2)$$

2.6. Experimental Design and Data Analysis

Design Expert version 11.1.2.0 software (Sate Ease, Minneapolis, USA) was used for regression and graphical analysis, statistical experiments design and data analysis. Response surface methodology (RSM) is a collection of mathematical and statistical techniques useful for modelling and analysis of responses affected by several factors [25]. It can be classified into two groups that are Box-Behnken design and central composite design (CCD). However, CCD is used in this research due to its advantages for allowing the experimental designer to understand the effect of factors on responses based on the levels selected for each factor [26]. Both CCD and RSM were applied to optimise the operating variables such as pH, coagulant dosage and flocculant dosage.

A 2^3 full-factorial design for three independent factors, consisting of 8 factorial points coded to the usual ± notation, 6 axial points and 6 replicates at centre points were conducted, given a total of 20 experimental runs for each response. Mathematically, formula in equation (3) was used to calculate the number of experimental runs to be conducted [25].

$$N = 2^n + (2 \times n) + n_c \quad (3)$$

where N is the number of experimental runs; n is the number of factors; and n_c is the number of centre points set between 2 and 6. Assuming that 6 points were set for the calculation, the range and levels of the experimental design were presented in table 2. In order to obtain the lowest and highest level of axial points, alpha (α) value was first calculated using the formula of $2^{n/4}$, where n is the number of factors. Then, the resulted α values was substituted into equation (4) [14].

$$X = \mu \pm \alpha \left(\frac{High\ Level - Low\ Level}{2}\right) \quad (4)$$

where X is the lowest/highest axial point; and μ is the mean of the high level and low level.

Table 2. Range and levels of the variables tested in the CCD design.

Variables	Symbol	Range	Lowest Level (- α)	Low Level (-1)	Zero Level (0)	High Level (+1)	Highest Level (+α)
pH	A	4-9	2.29552	4	6.5	9	10.7045
Coagulant Dosage (mg/L)	B	10-30	3.18207	10	20	30	36.8179
Flocculant Dosage (mg/L)	C	10-20	6.59104	10	15	20	23.409

A second order polynomial model was used in this study to optimise the variables such as pH, coagulant dosage and flocculant dosage in coagulation-flocculation process. The responses such as TUR, TSS and COD were used to develop an empirical model that correlated with the respective variables using the second-degree polynomial equation as shown in equation (5) [27].

$$Y = \beta_o + \sum_{i=1}^{k} \beta_i X_i + \sum_{i=1}^{k} \beta_{ii} {X_i}^2 + \sum_{i<j}^{k} \beta_{ij} X_i X_j + \varepsilon \tag{5}$$

where Y is the predicted response; β_o is the constant coefficient; β_i is the linear coefficients; β_{ii} is the quadratic coefficients, β_{ij} is the interaction coefficient; X_i and X_j are the coded variables values; n is the number of the independent test variables; and ε is the random error. The results obtained were analysed by the analysis of variance (ANOVA). The experimental factors and interaction between the variables and responses can be explained by the determination coefficient, R^2 that expressed the best model fit. The R^2 values measured the variability of the observed response values through analysis of Fischer's 'F' test and P-value (probability). The P-value was evaluated by using 95 % confidence level. Finally, counter plots and regression equation were analysed and solved separately to obtain the optimal values of the variables.

3.Results and Discussion

3.1. Regression Analysis and Model Fitting

For the current study, more experiments are required to obtain the optimal removal efficiencies due to the 'multivariable system' that will consume large amount of time and high operating cost. This is because jar test only allowed one factor at a time to be evaluated and manipulated, with the other factor remains as constant, without considering the interaction between factors. It is important to include the interaction between different factors to obtain the optimal removal efficiencies with optimum condition of the combined factors that derived from the interactions. Therefore, RSM is implied to evaluate the relative significance of several factors to overcome the complex condition.

The relationship between the factors (pH, coagulant, and flocculant dosage) and the responses (TUR, TSS and COD) for the coagulation-flocculation process is analysed by RSM. Table 3 shows 20 sets of the experimental runs from CCD where the experimental values (EV) of the responses are obtained from the coagulation and flocculation process as well as their corresponding coded level.

Table 3. CCD design results for the three experimental variables in coded and actual values.

Run	Factors						Responses		
	A		B		C		1	2	3
	pH		Coagulant Dosage (g)		Flocculant Dosage (ml)		TUR (%)	TSS (%)	COD (%)
	EV	Coded	EV	Coded	EV	Coded			
1	6.5	0	20	0	15	0	66.04	61.54	27.47
2	4	- 1	30	+1	20	+1	82.78	83.08	32.06
3	6.5	0	20	0	15	0	67.65	65.05	23.73
4	9	+1	10	-1	20	+1	32.35	31.21	8.55
5	4	- 1	10	-1	20	+1	75.80	81.54	33.76
6	6.5	0	20	0	15	0	64.17	63.08	28.22
7	6.5	0	20	0	15	0	65.51	64.40	25.18
8	4	- 1	10	-1	10	- 1	75.24	81.10	28.73
9	6.5	0	20	0	15	0	66.58	61.32	26.71
10	2.2955 2	- α	20	0	15	0	72.46	77.80	22.45
11	6.5	0	20	0	6.59104	- α	62.03	58.46	20.16
12	6.5	0	36.8179	+α	15	0	73.80	72.53	11.69
13	9	+1	30	+1	20	+1	37.70	38.68	22.65
14	9	+1	10	- 1	10	- 1	30.21	30.11	18.84
15	6.5	0	20	0	15	0	66.04	64.40	23.57
16	10.704 5	+α	20	0	15	0	20.05	19.34	8.02
17	4	- 1	30	+1	10	- 1	81.18	82.20	30.06
18	6.5	0	3.18207	- α	15	0	30.75	32.09	12.37
19	6.5	0	20	0	23.409	+α	69.25	67.25	27.82
20	9	+1	30	+1	10	- 1	35.83	36.92	9.94

The results obtained for each response was analysed by ANOVA, including Fischer's 'F' test and P-value as shown in table 4. The analysis shown that each response for all the models is significant at 95 % confidence level by using quadratic model, implying that the F-statistic is significant. Besides, the model F-value and lack of fit f-value have a low percentage chance that both of these large values could occur due to noise. If the model P-value and lack of fit P-value are less than 0.05 (P-values of regression ≤ 0.05), these indicate that the model terms are also statistically significant. Thus, the model is correlated with the factors and responses. The lack of fit determines whether a regression model is a poor fit for data [55]. It may due to the poor choice of variables, poor experimental design or important terms are not included in the model. So, model is considered lack of fit when there are large residuals or errors occurred when fitting the model [57]. In addition, the coefficient terms such as pH (A), coagulant dosage (B), flocculant dosage (C), the interactive terms (AB, AC, BC) and square terms (A^2, B^2, C^2) with ≤ 0.05 are significant to the response while terms with ≥ 0.05 are considered to have no effect on the removal of TUR, TSS and COD.

Table 4. Analysis of variance (ANOVA) results for the three responses (turbidity, TSS and COD).

Source	**Sum of Square**	**DOF**	**Mean Square**	**F-Value**	**P-Value**	**Remarks**
Response 1: TUR						
Model	6759.91	9	751.10	14.96	0.0001	significant
A	5222.08	1	5222.08	103.98	< 0.0001	
B	678.93	1	678.93	13.52	0.0043	
C	24.56	1	24.56	0.4889	0.5003	
AB	0.4753	1	0.4753	0.0095	0.9244	
AC	0.4278	1	0.4278	0.0085	0.9283	
BC	0.0741	1	0.0741	0.0015	0.9701	
A^2	606.08	1	606.08	12.07	0.0060	
B^2	273.54	1	273.54	5.45	0.0418	
C^2	1.96	1	1.96	0.0390	0.8474	
Residual	502.21	10	50.22			
Lack of Fit	495.56	5	99.11	74.51	0.0001	significant
Pure Error	6.65	5	1.33			
Cor Total	7262.13	19				
Response 2: TSS						
Model	7099.45	9	788.83	14.48	0.0001	significant
A	6129.13	1	6129.13	112.53	< 0.0001	
B	528.19	1	528.19	9.70	0.0110	
C	26.33	1	26.33	0.4834	0.5027	
AB	16.94	1	16.94	0.3110	0.5894	
AC	0.2964	1	0.2964	0.0054	0.9426	
BC	0.1512	1	0.1512	0.0028	0.9590	
A^2	272.88	1	272.88	5.01	0.0491	
B^2	132.23	1	132.23	2.43	0.1503	
C^2	7.04	1	7.04	0.1293	0.7266	
Residual	544.64	10	54.46			
Lack of Fit	532.09	5	106.42	42.40	0.0004	significant
Pure Error	12.55	5	2.51			
Cor Total	7644.09	19				
Response 3: COD						
Model	931.41	9	103.49	3.53	0.0311	significant
A	578.68	1	578.68	19.74	0.0012	
B	0.9951	1	0.9951	0.0339	0.8575	
C	36.52	1	36.52	1.25	0.2905	
AB	3.88	1	3.88	0.1323	0.7237	
AC	2.66	1	2.66	0.0906	0.7696	
BC	49.85	1	49.85	1.70	0.2215	
A^2	80.57	1	80.57	2.75	0.1284	
B^2	176.30	1	176.30	6.01	0.0341	
C^2	7.70	1	7.70	0.2625	0.6195	
Residual	293.20	10	29.32			
Lack of Fit	274.09	5	54.82	14.34	0.0055	significant
Pure Error	19.11	5	3.82			
Cor Total	1224.61	19				
Notes: A - pH; B - Coagulant Dosage; and C - Flocculant Dosage.						

The empirical model in term of coded factor from ANOVA are shown in table 5. Based on Oubrayme *et al.* [14], the positive sign of the quadratic equation implies synergistic factor effect while negative sign implies antagonistic factor effect. Hence, the results show that the equation for each response is significant and satisfactory.

Table 5. Empirical model in term of coded factors for each response.

Responses	Final Equation in Terms of Coded Factors
TUR	$65.92 - 19.55A + 7.05B + 1.34C - 0.2437AB + 0.2313AC + 0.0963BC - 6.49A^2 - 4.36B^2 + 0.368C^2$
TSS	$63.17 - 21.18A + 6.22B + 1.39C + 1.46AB + 0.1925AC + 0.1375BC - 4.35A^2 - 3.03B^2 + 0.6991C^2$
COD	$25.60 - 6.51A + 0.2699B + 1.64C + 0.6963AB - 0.5762AC + 2.50BC - 2.36A^2 - 3.50B^2 + 0.7309C^2$
Notes: A - pH; B - Coagulant Dosage; and C - Flocculant Dosage.	

Table 6 illustrates other statistical parameters. The coefficient of R^2 provides the proportion of the total response variation predicted by the model, indicating ratio of sum of squares caused by regression (SSR) to total sum of squares (SST) [27]. It measures the global fit of the model [14]. The R^2 coefficient value that close to 1 is desirable and reasonable agreement with adjusted R^2 is required to ensure a satisfactory adjustment of the model to the experimental data [27]. The adjusted coefficient of R^2 (Adj. R^2) compares the models with different number of independent variables [14]. It is observed that the R^2 coefficient for removal of TUR, TSS and COD are 0.9308, 0.9287 and 0.7606 respectively. The values for removal of TUR and TSS are close to 1 and have a satisfactory adjustment of the quadratic model to the experimental data than the removal of COD. Non-significant model terms appear when the predicted R^2 (Pred. R^2) and Adj. R^2 differ dramatically. Based on table 6, the value of Pred. R^2 and Adj. R^2 are not significantly different which confirm the accuracy of the model.

Adequate precision (AP) compares the range between predicted values at the design points and average prediction error. AP ratio greater than 4 shows that the predicted models can be used to navigate CCD design space. The ratio for TUR, TSS and COD are 13.7318, 13.6548 and 6.0840 respectively as shown in table 6, indicating an adequate signal to navigate the design space. The predicted versus the actual values plots of responses removal are presented in figure 5 which it provides the justification for model satisfactoriness. These plots imply an adequate agreement between the experimental data and model data. Based on figure 5, the actual values for TUR, TSS and COD are distributed relatively near to the straight line, indicating that the model data fits well with the predicted values.

The coefficient of variance (CV) is the ratio between standard error of estimate and mean value of the observed response that determine the model reproducibility.

Table 6. Fit statistic for each response.

Responses	R^2	Adj. R^2	Pred. R^2	AP	SD	Mean	CV %
TUR	0.9308	0.8686	0.4804	13.7318	7.09	58.77	12.06
TSS	0.9287	0.8646	0.4692	13.6548	7.38	58.61	12.59
COD	0.7606	0.5451	-0.8260	6.084	5.41	22.10	24.50

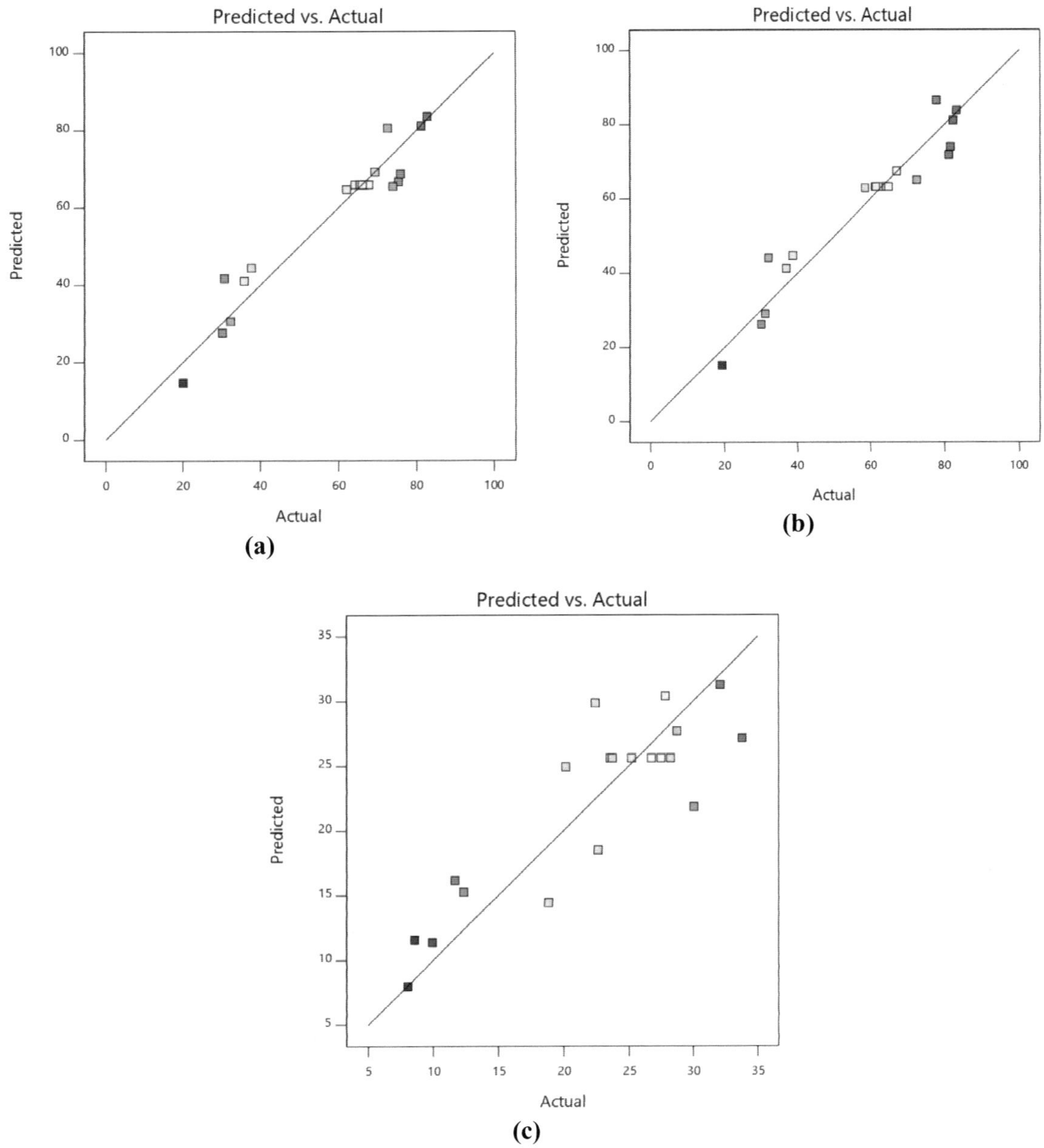

Figure 5. Predicted vs actual values plot for (a) turbidity removal; (b) TSS removal; (c) COD removal.

3.2. Process Analysis
The three-dimensional (3D) surface plots and its corresponding two-dimensional (2D) contour plots for the coagulation-flocculation process are shown in figure 6 to figure 8 as a function of two factors at a time, holding the third factor at fixed levels. Several factors such as pH, coagulant dosage and flocculant dosage are observed to optimise the treatment performance by using the fenugreek and aloe vera.

3.2.1. Effects of pH. pH must be controlled at the beginning of the experiment due to its important role in establishing the optimum condition for coagulation-flocculation process. The pH of the POME was adjusted from 4 to 9 to investigate the performance of the treatment process using fenugreek seed and aloe vera as natural coagulant-flocculant. It is observed that the variation of pH affects the removal efficiency of TUR, TSS and COD to some extents as per the (a), (b), (c) and (d) in figure 6, figure 7 and figure 8. For example, at low pH value yielded high removal efficiency of TUR, TSS and COD. This implies that there is a reduction in negatively charged colloidal particles and increases the number of hydrogen ions (H^+) on its surface. Hence, it generates larger flocs from agglomerating with the positive charged coagulant-flocculant. From zeta analysis, the zeta potential for fenugreek seed and aloe vera are in negative values. However, the zeta potential of fenugreek seed is higher than aloe vera by 13.212 mV. This shows that a higher dosage of fenugreek seed than aloe vera is needed to become positively charged. In addition, hydrogen bonding is much stronger for fenugreek than aloe vera due to the higher density charged [28]. RSM shows a good result in removing the contaminants, indicating that there are interactions between the positively and negatively charged particles. This can be explained by the adsorption of fenugreek with the H^+ ions contained in the POME at low pH, given that HCl is added at the beginning of the experiment. At the same time, it generates positively charges on its surface by neutralising the negatives of colloidal particles. Therefore, high dosage of fenugreek is required to neutralise and agglomerate the colloidal particles. Eventually, the adsorption of active sites on the fenugreek will decrease. More active sites are necessary to enhance the removal efficiency by adding aloe vera into the POME to further promote the agglomeration process.

Besides, the interactions involve the negatively charged of coagulant-flocculant with Na^+ from NaOH added at the beginning of the experiments. The same scenario happened as mentioned earlier. However, the higher the pH value, the lower the removal efficiencies for TUR, TSS and COD. This may due to the fenugreek and aloe vera that are more selectivity to H^+ than Na^+ [29]. As a result, the lowest pH value of 4 over the applied range of pH, is required to obtain such optimum removal. Similar trends conducted by Obiora-okafo [30] explained that high removal efficiency at low pH values are predominant in removing organic contaminants from acid dyes. Another example that shows that natural coagulant are effective at low dosage, conducted by Daud, Ghazi and Ahamad [31] that uses wheat germ as natural coagulant in treating POME reveals high removal percentage of TUR, TSS, COD and colour at pH 2.

3.2.2. Effects of Coagulant Dosage. Coagulant dosage is one of the important factors that affect the performance of the coagulation-flocculation process. The amount of coagulant dosage required is highly dependent on the concentration of contaminants in wastewater [32]. It is added with a constant rapid mixing rate of 200 rpm for 5 mins. The coagulant dosage is varied from 10 to 30 g to determine the highest removal for TUR, TSS and COD. According to figure 6 and figure 7, it is observed that the higher the coagulant dosage, the higher the removal efficiency for TUR and TSS, indicating the effectiveness of fenugreek seed as natural coagulant in treating POME. This high removal efficiencies at high dosage are associated with adsorption mechanisms that increase the electrostatic forces between the positively and negatively charged particles, generating small and fragile flocs (also known as micro-flocs) [33]. The coagulant serves as condensation nuclei and the colloidal particles agglomerate in the form of residuals [30].

Besides, the COD removal efficiency increased as the dosage increases from 10 to 20 g and then decreases from 20 to 30 g. This proves that a positive impact of COD removal is observed at lower dosage and a negative impact at higher dosage. This negative impact can be explained by the presence of soluble organic compounds in fenugreek. The effectiveness of fenugreek seed in removing COD from POME is mainly due to the charge neutralisation and adsorption mechanism (also known as bridging mechanism) [30]. Fenugreek seed is positively charged after reacting with HCl, allowing strong adsorption affinity and charge neutralisation to occur on the negatively charged colloidal particles in POME. The high concentration of coagulant will minimise the optimum dosage. This is caused by overdosing of coagulant in the wastewater which deteriorates the supernatant quality in POME. In this case, 're-stabilisation' of colloidal particles happened that restrict further adsorption and bridging mechanism due to the saturation of active sites.

After eliminating the negative effects and obtain the highest removal for all three responses (82.78 % for TUR, 83.08 % for TSS and 33.76 % for COD), an optimum coagulant dosage is chosen at 24.13 g which agrees with the outcome from the process optimisation of RSM. The effectiveness of fenugreek seed as natural coagulant can be further verified by the studies conducted by Mohd Najib *et al.* [34] that have removed 94.97 % TUR, 92.7 % TSS and 63.11 % COD. Fenugreek seed was also applied in studied conducted by Sethu *et al.* [11] which removed 77.6 % TUR, and 76.8 % TSS.

3.2.3. Effects of Flocculant Dosage. The interaction between the small floc formation after the coagulation process and the remaining colloidal particles in POME generate bigger flocs formation during the flocculation process. In order to improve the quality of supernatant in POME, aloe vera is added at the slow mixing stage of 30 rpm for half an hour. During the slow mixing stage, it facilitates the contacting process from micro-flocs to larger and heavier flocs formation [33]. This phenomenon occurred due to the increasing bonds between the micro-floc particles. This decreases the concentration of colloidal particles in POME while increases the aggregates growth. Also, the size of the particles continues to increase that lead to macro-flocs formation that can be further separated by sedimentation, flotation or filtration. It can be seen from figure 6 to figure 8 that the higher the flocculant dosage, the higher the removal efficiency for TUR, TSS and COD. The usage of aloe vera as natural flocculant can be further verified by Yéwêgnon *et al.* [35] that reported the removal efficiency for aloe vera was found to be 72 % of TUR and 91 % of TSS. Hence, an optimum flocculant dosage is chosen at 20 ml which agrees with the outcome from the process optimisation of RSM.

A similar trend of observations can be observed from Kumar [36] and Esther-Irma *et al.* [15] that wastewater treatment using aloe vera significantly decreases TUR and TSS in treated wastewater. However, there is a very small change on the apparent colour of treated water. Therefore, it is concluded that aloe vera does not supressed the organic matter presence in the water that accountable for the visibility of colour. This study has been proven as shown in figure 4, where the colour of POME becomes clearer at the end of the experiments. With the addition of aloe vera in wastewater, pH decreases indicated that aloe vera is acidic in nature. Another observation on the conductivity analysis carried out, it is deduced that the amount of aloe vera added influences the conductivity of treated water. The conductivity of treated water reduces as dosage of aloe vera increases. On top of that, the colour removal for untreated water is lower than TUR and TSS, indicating that aloe vera removes suspended particles more than colloidal particles. This further verified the effectiveness of aloe vera as natural flocculant in removing high concentration of suspended particles from water and wastewater treatment.

The performance of aloe vera as effective natural flocculant is attributed to the compounds present such as tannins, saponins, flavonoids and quinones that responsible in enhancing the flocculation process. Based on the physico-chemical analysis conducted by Adugna and Gebresilasie [16], the fiber contained in aloe vera is able to adsorb the fat present in wastewater, and consequently reducing the highly loaded organic matter. In addition, a study on aloe vera composition reveals the existence of polysaccharides [23]. This fact is also proven in Fourier Transform Infrared (FTIR) Spectroscopy Spectral Analysis. The solubility of polysaccharides in water determines the interaction between molecules which promote the aggregation process. Polysaccharides contain multi-hydroxyl groups that have strong affinity towards water as well as strong interaction between polysaccharides molecules through hydrogen bonding. Guo *et al.* [24] demonstrated the interaction between polysaccharides molecules and water molecules are energetically favourable for soluble polysaccharides; while intramolecular interactions between polysaccharides molecules are much stronger for polysaccharides with poor solubility of water, causing aggregation of molecules and eventually precipitation. The formation of precipitation in coagulation-flocculation process can be clearly seen from figure 4.

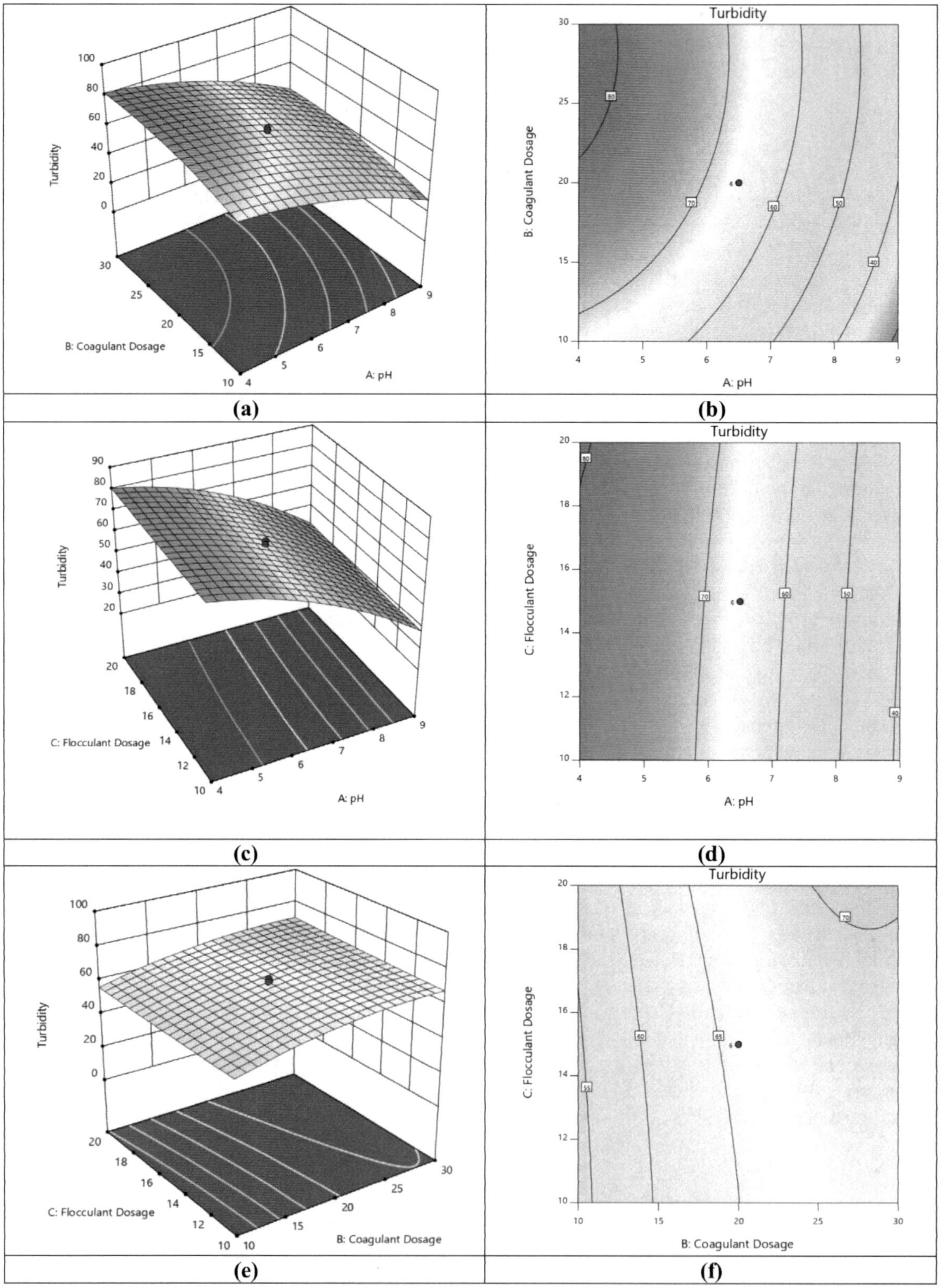

Figure 6. Response surface plot (a), (c), (e); and contour plot (b), (d), (f) for TUR removal.

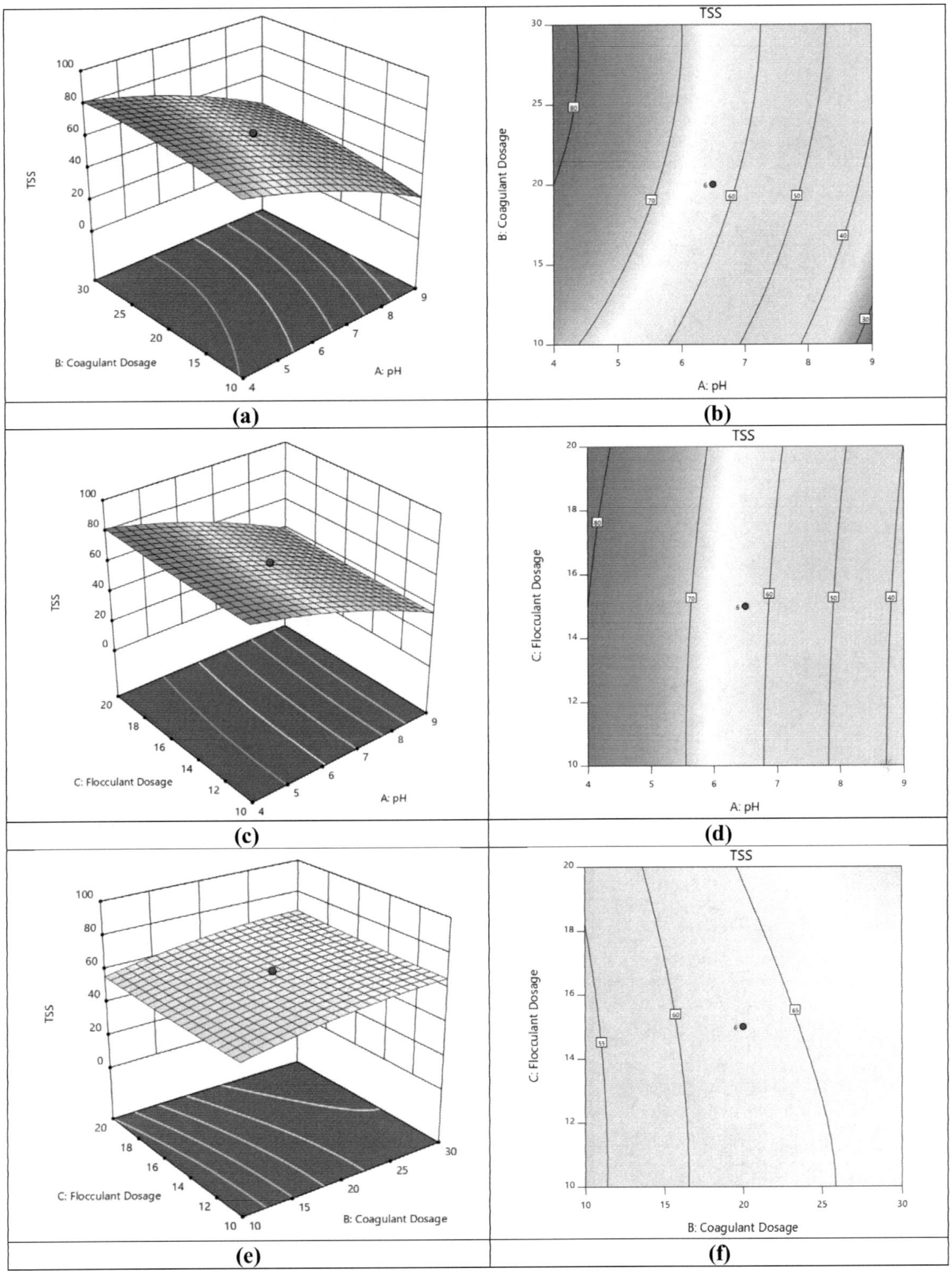

Figure 7. Response surface plot (a), (c), (e); and contour plot (b), (d), (f) for TSS removal.

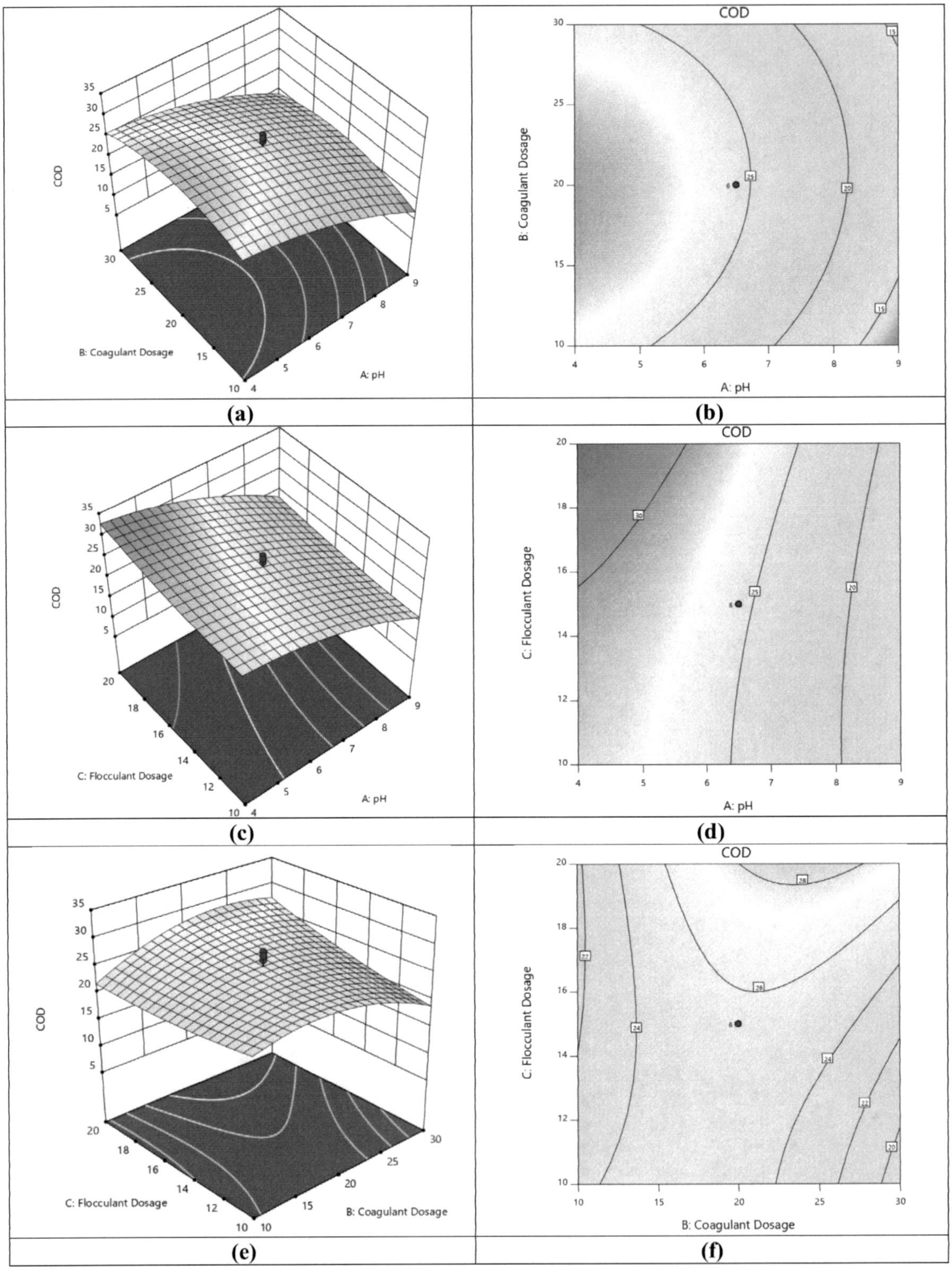

Figure 8. Response surface plot (a), (c), (e); and contour plot (b), (d), (f) for COD removal.

Inefficient performance of flocculation process will occur due to overdosing of either coagulant or flocculant. Therefore, it is important to determine the appropriate dosage required to achieve the targeted performance for wastewater treatment. By comparing the relationship between coagulant dosage and flocculant dosage based on the response surface plots, the higher the coagulant dosage, the lower the flocculant dosage required in the treatment process. This is due to the higher charge density of fenugreek particles that have removed most of the contaminants before aloe vera is added. But, for the removal of COD, it can be seen from figure 8(e) that the coagulant dosage increases and then decreases at flocculant dosage of 20 ml. This may be caused by the soluble organic compounds of fenugreek powder that prevent the actives sites from adsorption. Based on the discussions made, it can be concluded that the require dosages of coagulant and flocculant are 24.13 g and 20 ml respectively, to achieve high removal of contaminants.

3.2.4. Zeta Potential Analysis. Zeta potential is one of the controlling parameters that affects the stability of colloid particles during the process of coagulation-flocculation. It measures the electrochemical equilibrium at the particle-liquid interface as well as the magnitude of electrostatic attraction or repulsion between particles [51]. In accordance to Chemistry LibreTexts [51], it is also known as the electrical potential in the interfacial double layer of surface particles, which is found at the slipping plane. The magnitude of zeta potential ranges from +100 mV to -100 mV. High degrees of stability occur outside the range of +30 mV to -30 mV, while magnitude between +25 mV to -25 mV promote agglomeration of particles in wastewater due to the intraparticle interactions such as van der Waals, hydrophobic interactions and hydrogen bonding [37].

The particles in colloidal suspension such as POME are usually negatively charged which have proven from the results of zeta potential analysis as per table 7. The treated POME and raw POME from the facultative pond are in the negative region and gives a negative net charge on the particles surface. However, there is a reduction in magnitude from 15.5 mV to 0.882 mV, showing that the negative net charges decrease the effectiveness of aloe vera and fenugreek as well. This led to charge neutralisation which is required in the coagulation-flocculation process. This scenario happened due to the positive charge of fenugreek and aloe vera that attract the opposite charge of colloidal particles in POME that reduce the electrostatic repulsion between particles and consequently increase the formation of flocs.

Based on table 7, the resulted analysis for both fenugreek and aloe vera are in negatively charged. These negatively charges neutralise the high acidity of the POME which it contains high concentration of hydrogen ions (H^+), given that HCl is added and remained in the POME. Generally, zeta potential for fenugreek and aloe vera become more positive as it reacts with H^+, simultaneously increasing the pH. Hence, the particles of fenugreek and aloe vera are in positively charged. As a result, fenugreek and aloe vera are more favourable to treat wastewater at lower pH as discussed earlier. On top of that, it is observed that the magnitudes of materials are within the range of +25 mV to -25 mV which has mentioned earlier, proving that the agglomeration of particles occurs due to intraparticle interactions. As a result, the positive charge of fenugreek and aloe vera have further verified the effectiveness as natural coagulant and flocculant respectively in treating the POME.

Table 7. Zeta potential for each material.

Material	**Zeta Potential (mV)**
Fenugreek	-13.700
Aloe Vera	-0.488
Treated POME (from final pond of oil palm mill)	-0.882
Raw POME	-15.500

3.2.5. Fourier Transform Infrared (FTIR) Spectroscopy Spectral Analysis. FTIR is implied in this research to study the functional groups present in the fenugreek seeds (coagulant) and aloe vera (flocculant). Based on figure 9, the spectrums range between a wavelength of 400-4000 cm^{-1} and show a distinct different between the compounds present in the fenugreek seeds and aloe vera. Typically, aloe vera showed fewer and broader bands as compared to fenugreek. The obvious peak patterns observed in figure 9(b) shown three spectral regions: 3700-2800 cm^{-1}, 1800-1500 cm^{-1} and 900-400 cm^{-1}. The broad adsorption spectrum observed between 3700-2800 cm^{-1} and the peak at 3259.13cm^{-1} indicates the stretching of amine (N-H) group and hydroxyl (O-H) group, showing the presence of carbohydrate monomers such as mannose and uronic acid [21]. Furthermore, the strong and broad band centered within 1800-1500 cm^{-1} at 1636.6 cm^{-1} is associated with the asymmetric and symmetric -COO that signify the presence of carboxylate compounds in aloe vera [21]. The adsorption peak at 600.94 cm^{-1} between the wavelength range 900-400 cm^{-1} is due to the C-H out-of-plane deformation of carbohydrate monomers [21].

The adsorption spectrum of fenugreek seeds is shown in figure 9(a). The peak at 3278.99 cm^{-1} with adsorption band of 3500-3000 cm^{-1} indicates the N-H and O-H stretching vibrations that consist of protein amide A and starch fiber respectively [38]. The existence of shoulder band at 3009 cm^{-1} corresponds to the secondary amide, leading to Fermi resonance that resulted from N-H stretching with either of the overtone of amide II band in trans-amides or combination band of C=O stretching and N-H in plane binding in cis-amides. The peaks at 2924.81 cm^{-1} and 2855.07 cm^{-1} between 3000-2500 cm^{-1} specify the asymmetric and symmetric C-H stretching vibrations. The appearance of C-H group is related with aromatic rings, namely CH_2 groups in fatty acids [39]. It is observed from figure 9(a) that many oscillations of peaks occur after 1800 cm^{-1}. The C=O stretching vibration of lipid occurred at medium intensity of 1743.15 cm^{-1}. The appearance of strong bands at 1638 cm^{-1}, 1545.42 cm^{-1} and 1235.97cm^{-1} shows the stretching of C=O (amide I), N-H (amide II) bending and N-H (secondary amide III) bending respectively. The presence of amide in these bands correlated with protein in the solid state. Additionally, band at 1235.97 cm^{-1} provides O-H in plane, CH, CH_2 bending and aromatic alcohols. The bands at 1440.83 cm^{-1} and 1380.82 cm^{-1} are caused by deformation of δ(C-H) bending and COO^- bending groups from primary amide respectively. In the area of 1200-900 cm^{-1} such as the bands at 1141.4 cm^{-1} and 1025.37 cm^{-1} indicate the presence of starch. Studies carried out by El-Bahy [38] deduces that fenugreek mostly include protein, lipids, ash, fibers and soluble sugar.

From the above analysis, the similarities for fenugreek seeds and aloe vera can be concluded from the existence of N-H and O-H groups that indicate the presence of proteins, fatty acids, carbohydrates and lignin components [39]. The treatment of POME using coagulation-flocculation process involves the hydrogen bonding between O-H groups of polysaccharide and functional group of contaminants. In addition, the concentration and solubility of protein are correlated in removing the TUR and TSS from wastewater. For example, the higher the protein concentration and solubility, the higher the removal efficiencies of TUR and TSS.

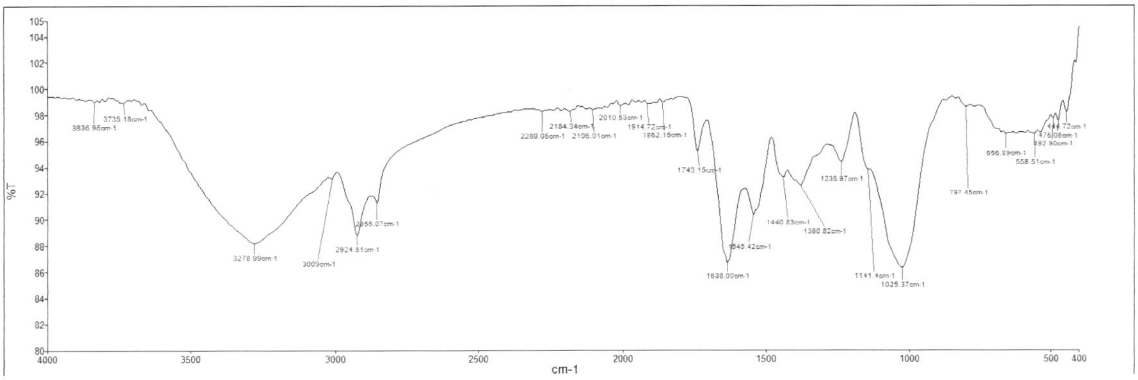

(a)

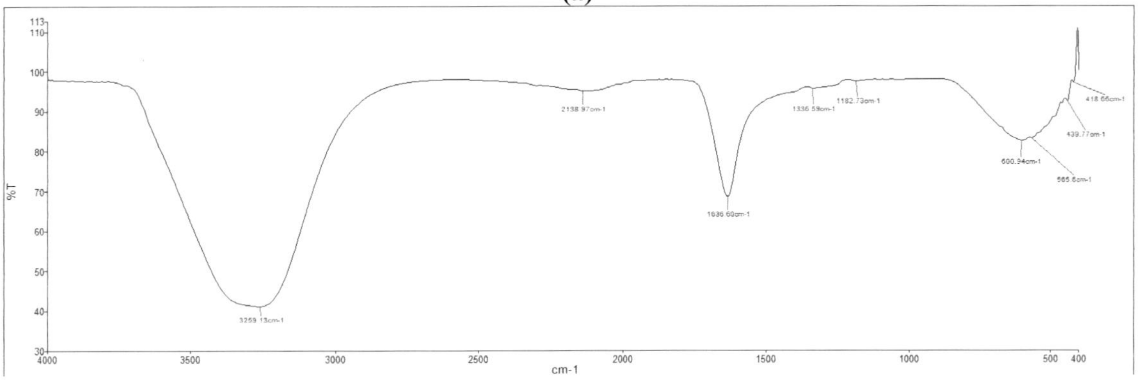

(b)

Figure 9. FTIR spectrum of (a) fenugreek seeds powder; and (b) aloe vera.

3.3. Process Optimisation Analysis

Process optimisation analysis using Design Expert software was conducted to maximise the removal efficiency of TUR, TSS and COD. It explores the combination of factor levels and at the same time fulfil the criteria set on each factor and response [56]. Numerical optimisation was utilised where the goals were set 'in range' for each factor and 'maximise' for each response. This is to obtain the highest removal efficiency for each response within the desired range. The goals combined into an overall desirability function which is to maximise the effectiveness of the function [30].The highest desirability value of 0.993 is obtained for the optimum factors and responses, given the desirability range from 0 to 1. The results obtained from this optimisation were compared with the optimum experimental values as presented in table 8. Obiora-okafo [30] reported that the maximum percentage error between the predicted and experimental values which is less than 4 shows that the model can predict the experimental results well. It is observed that the percentage errors are below 1, showing a good agreement between the predicted value and experimental value. This indicates that implementation of RSM in the coagulation and flocculation process are very effective in obtaining the optimum removal efficiencies that is almost similar as the results obtained from the experiments.

Table 8. Optimisation and comparison between the predicted and experimental optimum values.

Factor			**Response**					
pH	**Coagulant Dosage (g)**	**Flocculant Dosage (ml)**	**TUR (NTU)**		**TSS (mg/L)**		**COD (mg/L)**	
			%PV	%EV	%PV	%EV	%PV	%EV
4	24.13	20	82.78	82.78	83.40	83.08	32.95	33.76
		% Error	0		0.32		0.81	
Notes: %PV = Predicted value in percentage from model; %EV = Experimental value in percentage								

Desirability = **0.993**

3.4. Process Performance Evaluation

The performance of using fenugreek seeds and aloe vera as natural coagulant and flocculant were evaluated. The results obtained from the process optimisation were compared with other types of natural flocculants. The optimum results of TUR, TSS and COD for each respective material used in wastewater at optimum pH and dosage were summarised in table 9.

Table 9. Summary of optimum conditions for each material used in wastewater treatment.

Materials	**pH**	**Dosage**	**TUR Removal %**	**TSS Removal %**	**COD Removal %**	**Effluent Type**
Aloe Vera	4	20 ml	82.78	83.40	32.95	POME
Banana Pith	4	0.1 kg/m3	98.50	96.03	54.30	Polluted River Water [40]
Banana Stem Juice	7	90 ml	98.50	88.60	80.10	Spent Coolant Wastewater [41]
Cassia Obtusifolia Seed Gum	7.2	0.17 g/L	-	89.60	55.40	Raw Pulp and Paper Mill Effluent [42]
Chickpea	6.69	2.60 g/L	85.50	88.50	50.80	POME [43]
Chitosan	6	370 mg/L	97.70	91.70	42.70	POME [44]
Dragon Fruit Foliage	2	300 mg/L	99.20	98.8	48.7	POME from three-phase decanter [12]
Moringa oleifera	6	70 mg/L	63.70	62.05	38.60	Oil Refinery Wastewater [45]
Okra	6	3.20 mg/L	97.24	-	85.69	Textile Wastewater [46]
Opuntia Ficus-indica	5	2.6 mg/L	91.26	-	64.77	Textile Effluent [47]
Rice Starch	3	2 g/L	-	84.10	17.40	POME [48]
Tapioca Starch	12	0.1 g/L	99.00	-	87.00	Semiconductor Waste [49]

It can be deduced that the performance of aloe vera is acceptable and comparable with other natural flocculants. It can be observed that the removal efficiencies of TUR and TSS for natural flocculants are quite high with low removal efficiency of COD. This is because the high solubility of organic compounds is generated from natural flocculants after adding directly into the flocculation process. This high content of organic compounds increases the microbial activity, and consequently increases the COD content in wastewater [50]. Therefore, the effectiveness in removing COD is significantly lower than TUR and TSS. In order to improve the removal efficiency of COD, the active components exist in the natural flocculant must be extracted and tested its properties separately from the coagulation-flocculation process.

4. Conclusion

RSM with CCD was utilised to study the various effects such as pH, coagulant dosage and flocculant dosage on TUR, TSS and COD removal percentage from POME. The optimum combination of the operating parameters that yields the optimum TUR, TSS and COD removal was determined. The results clearly showed that the removal efficiencies are highly influenced by the operating parameters. Based on the process optimisation analysis, the maximum removal efficiency for TUR, TSS and COD are 82.78 %, 83.40 % and 32.95 % respectively. An optimal condition of pH 4, 24.13 g of coagulant dosage and 20 ml of flocculant dosage were obtained from the compromise of the three desirable responses. The maximum error is below 4 % which indicate that the predicted value in agreement with experimental value. This proves that modelling and optimisation of the coagulation-flocculation process can be achieved by RSM approach. It is an economical way of obtaining the maximum amount of information in a short period of time with minimum number of experimental trials.

References

[1] Voora V, Larrea C, Bermudez S and Baliño S 2019 Global market report: Palm oil *Int Inst Sustain Dev*

[2] Kamyab H, Chelliapan S, Din M F M, Rezania S, Khademi T and Kumar A 2018 Palm oil mill effluent as an environmental pollutant *Palm Oil*

[3] Igwe J C and Onyegbado C C 2007 A review of palm oil mill effluent (POME) water treatment *Glob. Environ. Res.* **1** 54–62

[4] Bala J D, Lalung J and Ismail N 2014 Palm oil mill effluent (POME) treatment 'Microbial communities in an anaerobic digester': A review *Int. J. Sci. Res. Publ.* **4** 1–24

[5] Shafaghat A, Tin C, Ali L, Zaimi M and Majid A 2016 Optimum lipid production using agro-industrial wastewater treated microalgae as biofuel substrate *Clean Technol. Environ. Policy.* **18** 2513–23

[6] Abdullah N and Sulaiman F 2013 *Biomass Now-Sustainable Growth and Use.* pp 75–100

[7] Zainal N H, Jalani N F, Mamat R and Astimar A A 2017 A review on the development of palm oil mill effluent (POME) final discharge polishing treatments *J. Oil Palm Res.* **29** 528–40

[8] Silalahi M, Mawengkang H and Syahputri N I 2017 Technical problems of wastewater treatment plant in crude palm oil industry A case study in PT Socfin Indonesia-Kebun Sungai Liput *IOP Conf. Ser. Earth Environ. Sci.* **65** 012048

[9] Sayuti SC and Azoddein A A M 2015 Treatment of palm oil mill effluent (POME) by using electrocoagulation as an alternative method *Malaysian J Anal Sci.* **19** 663–8

[10] Maya S R M R, Tun H O M U, MtNanyan N, Abdul R N, Mariam I K N A and Hashim M K A 2014 Colour removal of reactive dye from textile industrial wastewater using different types of voagulants *Asian J. Appl. Sci.* **2** 650–7

[11] Sethu V, Mendis A, Rajiv R, Chimbayo S and Vejayan S 2015 Fenugreek seeds for the treatment of palm oil mill effluent (POME) *Int. J. Chem. Environ. Eng.* **6** 126–30

[12] Som A M and Abd Wahab A F 2018 Performance study of dragon fruit foliage as a plant- based coagulant for treatment of palm oil mill efluent from three-phase decanters *BioResources.* **13** 4290–300

[13] Beyene H D, Hailegebrial T D and Dirersa W B 2017 Investigation of coagulation activity of cactus powder in water treatment *J. Appl. Chem.* **2016**

[14] Oubrayme H, Souabi S, Bouhria M, Tahiri M, younssi S A and Albizane A 2016 Optimization of the coagulation/flocculation of wastewater from oil refineries use of response surface methodology *J. Mater. Environ. Sci.* **7** 4299–310

[15] Esther-Irma N Y A, Philippe S, Abdoukarim A, Alassane Y A K, Agbangnan C P and Mama D

2015 Evaluation of aloe vera leaf gel as a natural flocculant: Phytochemical screening and turbidity removal trials of water by coagulation flocculation *Res. J. Recent Sci.* **4** 1–9
[16] Adugna A T and Gebresilasie N M 2018 Aloe steudneri gel as natural flocculant for textile wastewater treatment *Water Pract. Technol.* **13** 495–504
[17] Kalwania G S 2014 Removal of fluorides in drinking water by aloe vera and calcium chloride *Chem. Sci. Trans.* **3** 29–36
[18] Mehrafarin A, Rezazadeh S, Naghdi Badi H, Noormohammadi G, Zand E and Qaderi A 2011 A review on biology, cultivation and biotechnology of fenugreek (Trigonella foenum-graecum L.) as a valuable medicinal plant and multipurpose *J. Med. Plants.* **10** 6–24
[19] Ramamurthy C, Maheswari M U, Selvaganabathy N, Kumar M S, Sujatha V and Thirunavukkarasu C 2012 Evaluation of eco-friendly coagulant from *Trigonella foenum-graecum* seed. *Adv. Biol. Chem.* **2** 58–63
[20] Wani S A and Kumar P 2018 Fenugreek: A review on its nutraceutical properties and utilization in various food products *J. Saudi Soc. Agric. Sci.* **17** 97–106
[21] Lim Z X and Cheong K Y 2015 Switching characteristics of natural aloe vera-based memory devices effects of drying temperature and ethanol *Phys. Chem. Chem. Phys.* **17** 26833–53
[22] Giannakoudakis D A, Hosseini-Bandegharaei A, Tsafrakidou P, Triantafyllidis K S, Kornaros M and Anastopoulos I 2018 Aloe vera waste biomass-based adsorbents for the removal of aquatic pollutants: A review *J. Environ. Manage.* **227** 354–64
[23] Hamman J H 2008 Composition and applications of Aloe vera leaf gel *Molecules* **13** 1599–616
[24] Guo M Q, Hu X, Wang C and Ai L 2017 *Solubility of Polysaccharides*
[25] Olawoye B 2016 *Design of Experiment.* pp 1–47
[26] Rakić T, Kasagić-Vujanović I, Jovanović M, Jančić-Stojanović B and Ivanović D 2014 Comparison of full factorial design, central composite design, and Box-behnken design in chromatographic method development for the determination of fluconazole and its impurities *Anal. Lett.* **47** 1334–47
[27] Ghafari S, Aziz H A, Isa M H and Zinatizadeh A A 2009 Application of response surface methodology (RSM) to optimize coagulation-flocculation treatment of leachate using poly-aluminum chloride (PAC) and alum *J. Hazard Mater.* **163** 650–6
[28] Munthali M W, Elsheikh M A, Johan E and Matsue N 2014 Proton adsorption selectivity of zeolites in aqueous media: Effect of Si/Al ratio of zeolites *Molecules.* **19** 20468–81
[29] Krah A, Pogoryelov D, Langer J D, Bond P J, Meier T and Faraldo-Gómez J D 2010 Structural and energetic basis for H+ versus Na+ binding selectivity in ATP synthase F_o rotors *Biochim Biophys. Acta - Bioenerg* **1797** 763–72
[30] Obiora-okafo I A and Onukwuli O D 2017 Optimization of coagulation-flocculation process for colour removal from Azo dye using natural polymers: Response surface methodological approach *Niger. J. Technol.* **36** 482
[31] Daud N S, Ghazi T I M and Ahamad I 2014 Wheat germ as natural coagulant for treatment of palm oil mill effluent (POME) *Int. J. Chem. Environ. Eng.* **5**
[32] Pui K, Shak Y and Wu T Y 2014 Coagulation – flocculation treatment of high-strength agro-industrial wastewater using natural Cassia obtusifolia seed gum: Treatment efficiencies and flocs characterization *Chem. Eng. J.* **256** 293–305
[33] Trösch W 2009 Water treatment *Technol Guid Princ - Appl - Trends.* **33** 394–7
[34] Mohd Najib N A, Sethu V, Arumugasamy S K and Selvarajoo A 2020 Artificial neural network (ANN) modelling of palm oil mill effluent (POME) treatment with natural bio-coagulants *Environ. Process.* **7** 509–35
[35] Yéwêgnon N, Irma A E, Philippe S, Abdoukarim A, Abdou Y and Alassane K 2016. Evaluation of aloe vera leaf gel as a natural flocculant: Phytochemical screening and turbidity removal trials of water by coagulation flocculation *Res. J. Recent Sci.* **5** 9–15
[36] Kumar S S 2016 Application of affluent afresco plants as an affordable aid in the treatment of acquainted effluents. **2** 205–10

[37] Kumar A, Dixit C K and States U 2017 Methods for characterization of nanoparticles *Adv. Naonomed. Delivery Ther. Nucleic Acids*
[38] El-Bahy G M S 2005 (Trigonella foenum graecum L.) SEEDS *J. Appl. Spectrosc.* **72** 106–10
[39] Brian Lee CL, Peter AP, Kevin Hwang QC, Ragu P, Sethu, V, Selvarajoo A and Arumugasamy S K 2018 Treatment of palm oil mill effluent (POME) using chickpea (*Cicer arietinum*) as a natural cogulant and flocculant: Evaluation, process optimization and characterization of chickpea powder *J. Environ. Chem. Eng.* **6** 6243–55
[40] Kakoi B, Kaluli J W, Ndiba P and Thiong'o G 2016 Banana pith as a natural coagulant for polluted river water *Ecol. Eng.* **95** 699–705
[41] Alwi H, Idris J, Musa M, Halim K and Hamid K 2013 A preliminary study of banana stem iuice as a plant-based coagulant for treatment of spent coolant wastewater *J. Chem*
[42] Subramonian W, Wu T Y and Chai S P 2015 An application of response surface methodology for optimizing coagulation process of raw industrial effluent using Cassia obtusifolia seed gum together with alum *Ind. Crops Prod.* **70** 107–15
[43] Brian LCL, Peter AP, Kevin HQC, Ragu P, Sethu, V, Selvarajoo A and Arumugasamy S K 2018 Treatment of palm oil mill effluent (POME) using chickpea (*Cicer arietinum*) as a natural cogulant and flocculant: Evaluation, process optimization and characterization of chickpea powder *J. Environ. Chem. Eng.* **6** 6243–55
[44] Saifuddin N and Dinara S 2011 Pretreatment of palm oil mill effluent (POME) using magnetic chitosan *J. Chem.* **8**
[45] Dehghani M and Alizadeh M H 2016 The effects of the natural coagulant Moringa oleifera and alum in wastewater treatment at the Bandar Abbas Oil Refinery *Environ. Heal. Eng. Manag.* **3** 225–30
[46] Freitas T K F S, Oliveira V M, de Souza M T F, Geraldino H C L, Almeida V C and Fávaro S L 2015 Optimization of coagulation-flocculation process for treatment of industrial textile wastewater using okra (A. esculentus) mucilage as natural coagulant *Ind. Crops. Prod.* **76** 538–44
[47] De Souza M T F, Ambrosio E, De Almeida C A, De Souza F T K F, Santos L B, De Cinque and Almeida V 2014 The use of a natural coagulant (Opuntia ficus-indica) in the removal for organic materials of textile effluents *Environ. Monit Assess.* **186** 5261–71
[48] Teh C Y, Wu T Y and Juan J C 2014 Potential use of rice starch in coagulation-flocculation process of agro-industrial wastewater: Treatment performance and flocs characterization *Ecol. Eng.* **71** 509–19
[49] Fatehah M O, Hossain S and Teng T T 2013 Semiconductor wastewater treatment using tapioca starch as a natural coagulant *J. Water Resour. Prot.* **5** 1018–26
[50] Idris J, Som AM, Musa M, Ku Hamid K H, Husen R and Muhd RMN 2013 Dragon fruit foliage plant-based coagulant for treatment of concentrated latex effluent: Comparison of treatment with ferric sulfate *J. Chem*
[51] Pavan M V R and Andrew R B 2016 2.5: Zeta Potential Analysis *Chemistry LibreTexts*
[52] Macleod A 2017 Indonesian and Malaysian palm oil at a crossroads *Global Risk Insights*
[53] MPOC 2020 About palm oil *Malaysian Palm Oil Council*
[54] MIDA 2020 Circular economy: The way forward for palm-based industries *Malaysian Investment Development Authority*
[55] StatEase 2020 Fit summary (RSM/MIX model selection) *StatEase*
[56] StatEase 2020 Optimization Overview *StatEase*
[57] Stephanie 2019 Lack of fit: Simple definition, examples *Statistics How To*
[58] Veolia Water Technologies 2017 Water treatment for palm oil mill effluents (POME) industry wastewater solutions *Veolia Water Technologies*

Influence of pH on corrosion resistance of electrodeposited Mg-Fe LDH composite films on Mg alloys WE43

J K E Tan[1], **P Balan**[1], **N A N Raj**[2] **and S Nayak**[2]

[1] School of Engineering, Monash University Malaysia, Jalan Lagoon Selatan, 47500, Bandar Sunway, Selangor, Malaysia
[2] Center for Biomaterials, Cellular and Molecular Theranostics (CBCMT), VIT University, Vellore 632014, Tamil Nadu, India

Email: poovarasi.balan@monash.edu

Abstract. Layered double hydroxide (LDH) are widely studied as corrosion inhibition coatings for magnesium (Mg) alloys due to their nano-lamellar structure and anion-exchange ability. The two common approaches in synthesising LDH films on Mg alloys are co-precipitation and hydrothermal treatment. Recently, electrodeposition has drawn more interest as a potential synthesis approach due to the low cost, ease of control, one-step method, and ability to create a compact film. In this work, Mg-Fe LDH film was synthesised on magnesium hydroxide, $Mg(OH)_2$ layer formed on Mg alloy WE43 by electrodeposition approach with the as-prepared iron solution at different pH values. The effect of pH of the solution (3.0, 5.0 and 7.0) on the formation of LDH films was investigated using Field Emission Scanning Electron Microscopy (FESEM). Additionally, the surface chemical components of the film were obtained by Fourier Transform Infrared Spectrometer (FTIR). All the LDH composite films synthesised at different pH values undergo electrochemical tests in Hank's Balanced salt solution (HBSS). All samples promote corrosion resistance of WE43 with the LDH film synthesised at pH 3.0 provided the highest inhibition efficiency (I.E) of 94.48 % and resistance polarisation value of 3559. 07 Ωcm^2.

Keywords. Layered double hydroxide, magnesium alloys, surface modification, corrosion, coatings

1. Introduction

Magnesium (Mg) alloys received much recognition as biomaterials for implant applications due to their biocompatibility and biodegradability [1–3]. For instance, the similarity in density and mechanical properties of Mg alloys to human bone makes them a suitable biomaterial for orthopaedic implant applications [4, 5]. Secondary surgery can be avoided with the degradability properties, which reduces the patients' pain and cost [6]. However, the high electrochemical activity of Mg alloys in physiological environments limits their clinical applications. The rise in local alkalinity and hydrogen evolution hampers the post-surgery healing process [7, 8]. Thus, continuous efforts have been communicated to elevate the corrosion resistance of Mg alloys.

Generally, past literature has highlighted various surface modifications that could delay the fast corrosion activity of Mg alloys, such as conversion coatings [9, 10], plasma electrolytic oxidation (PEO) coatings [11, 12], and layer-by-layer coatings [13–15]. Nevertheless, these coatings can only provide a passive barrier protection in which the protective ability will be at risk once their coating integrity is damaged. Thus, layered double hydroxide (LDH), an active coating system, has gained attention as a potential coating material for Mg alloys as LDH can react to physical and/or chemical damage [2, 16, 17]. As an active coating system, LDH can simultaneously entrap corrosive anions while releasing corrosion inhibitors from the interlayer space during physical and/or chemical damage.

LDH is an anion clay or hydrotalcite-like material stacked with alternative positive charged layers, with an exchangeable anionic layer in the interlayer space, allowing the entrapment of corrosive chlorine anions [18, 19]. Recently, LDH has been reported as a corrosion inhibition coating for Mg alloys due to the ion-exchange ability and flexible structure [19, 20]. Peng *et al.* [21] synthesised Mg-Al LDH film loaded with 5-Fluorouracil on AZ31 alloy. The LDH film constructively elevate the corrosion resistance of the substrate due to their ion-exchange ability. Similarly, Chen *et al.* [22] reported an excellent self-healing ability with improved corrosion resistance of the hydrothermally synthesised Mg-Al LDH film intercalated with aspartic acid.

Looking at the numerous reports on the enhanced corrosion resistance ability of LDH films on Mg alloys, various synthesis approaches have been explored to synthesise LDH on Mg alloys [23, 24]. In particular, *in-situ* growth method, such as co-precipitation [25], hydrothermal process [26, 27], and electrodeposition method [28, 29] is a common approach for forming LDH films on Mg alloys. Despite multiple works which looked into *in-situ* growth method as the synthesis approach for LDH films, the corrosion protection ability and surface morphology of LDH films based on synthesis parameters has not been thoroughly investigated. Based on the crystallography theories, the crystallization rate of LDH and the overall microstructure are affected by the reaction conditions [30, 31]. Kamiyama *et al.* [32] reported that the thickness of the Mg-Al LDH film through steam coating increased when the treatment time increase from 1 h to 9 h. The LDH film synthesised at 9 h showed excellent corrosion protection for the substrate, attributed to the excellent barrier effect.

Additionally, Chen *et al.* [33] studied the effect of different process parameters, including immersion time, pH of the solution, and treatment temperature, on the film properties on the corrosion inhibition ability of the film. They concluded that the LDH film achieved the best quality with excellent corrosion resistance when the film was synthesised at pH 10.5 at 80 °C for 1.5 h. Generally, even though past researchers have studied the effects of process parameters, however, the studies on the impact of pH on the microstructure of LDH film through the electrodeposition method is currently limited based on the author's knowledge.

In this work, Mg-Fe LDH film was prepared successfully by electrodeposition method on as-prepared $Mg(OH)_2$ layer on WE43 alloy. The effect of pH of 3.0, 5.0, and 7.0 on the microstructure of Mg-Fe LDH films was studied. The selection of different solution pH assesses the impact on the surface morphology of Mg-Fe LDH films synthesised via electrodeposition. Additionally, the active protective function of LDH films is evaluated by electrochemical methods.

2. Experimental methods

2.1. Materials

Magnesium alloys WE43 with a composition of Y 3.56 %, Nd 2.20 %, Zr 0.47 % and Mg – balanced (all in wt. %) was used in this work. The WE43 alloys with a 10 x 10 x 6 mm measurement were ground using silicon carbide (SiC) paper of 240 grit to 1500 grit. Afterwards, the samples were ultrasonically cleaned in ethanol for 10 mins.

2.2. Preparation of precursor $Mg(OH)_2$ layer and Mg-Fe LDH films at various pH

Synthesis of an $Mg(OH)_2$ layer on WE43 alloy was carried out through hydrothermal process using DI water at 160 °C for 3 h, as previously reported by Zhu *et al.* [34]. The synthesis of Mg-Fe LDH film was adapted from Lin *et al.* [23], with some modification done for this work. Firstly, an aqueous solution containing Fe^{3+} ions was prepared. Next, the filtered Fe^{3+} solution was adjusted to pH 3.0 with 1.0 M of nitric acid (HNO_3) solution before the electrodeposition of LDH film. Afterwards, Mg-Fe LDH film was electrodeposited through a three-electrode setup at an applied potential of – 1.4 V_{SCE} for 300 s at room temperature (25 °C) without stirring. After electrodeposition, DI water was used to rinse the sample and dried in the oven before any characterisation. Additionally, two other Fe^{3+} solutions were prepared at pH 5.0 and pH 7.0 with 1.0 M of HNO_3 solution and 2.0 M of sodium hydroxide (NaOH) solution, respectively, before electrodeposition. The as-prepared samples were designed as LDH 3, LDH 5, and LDH 7.

2.3. Characterisation

Field Emission Scanning Electron Microscopy (FE-SEM, Hitachi SU8010) was selected to determine the structure and surface morphology of Mg-Fe LDH films. The surface functional groups of the Mg-Fe LDH films were evaluated by Fourier Transform Infrared Spectrometer (FTIR, Thermo Fisher Scientific Nicolet iS10) under attenuated total reflectance (ATR) mode. All coated substrates were scanned from 4000 – 600 cm^{-1} with 64 scans at 4 cm^{-1} resolutions.

Electrochemical studies were conducted via a Gamry FAS2 with a PCI4 Controller Board. The selected medium is Hank's Balanced Salt Solution (HBSS), whereby the temperature was maintained at 37 ± 0.1 °C through a thermostatic water bath. The open-circuit potential (OCP) was conducted for 1 h to stabilise the substrates. Potentiodynamic polarization (PDP) measurements were performed at a scan rate of 1 mV/s in the range of – 300 mV to + 500 mV vs OCP. The inhibition efficiency of the samples was calculated using: [35]

$$I.E = [1 - ({i'_{corr}}/{i_{corr}})] \times 100 \quad (1)$$

where i'_{corr} and i_{corr} are the corrosion current density of the LDH film and uncoated WE43 alloy.

3. Results and discussion

3.1. Physical and chemical characterisation of samples

The structures and surface functional groups of the electrodeposited Mg-Fe LDH films at different pH were examined through FTIR and FESEM.

The functional groups in the as-prepared $Mg(OH)_2$ layer and LDH films were examined through FTIR spectroscopy in wavelengths of 4000 – 600 cm^{-1}. The peaks distribution of the FTIR spectrum for all samples are shown in figure 1 below. Based on figure 1, all samples show a band at 3692 cm^{-1} with LDH 3 showed a distinguish sharp and strong band, describing the stretching vibration of the Mg-OH group [36]. The appearance of a sharper absorption peak at 3692 cm^{-1} for the LDH 3 sample indicates the participation of the $Mg(OH)_2$ layer in the synthesis of LDH film. Asides from the absorption peak at 3692 cm^{-1}, LDH 3 also presented peaks at 1640 cm^{-1} and 1393 cm^{-1}. These peaks correspond to the bending vibration of the hydroxyl group in water molecules and asymmetric stretching of carbonate anions, respectively. Undoubtedly, with 1640 cm^{-1} and 1393 cm^{-1} peaks, it is clear that LDH film is successfully synthesised at a pH of 3.0. On the contrary, LDH 5 and LDH 7 samples presented weak absorption peaks compared to LDH 3, which suggests that the solution's pH at 5.0 and 7.0 are not ideal for forming LDH film via the electrodeposition method.

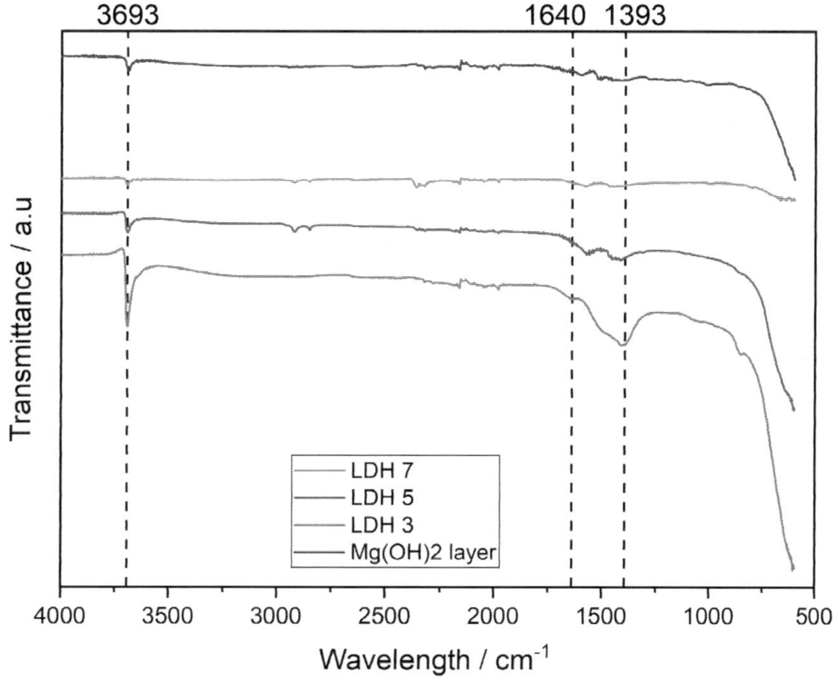

Figure 1. FTIR spectrum for (a) $Mg(OH)_2$ layer; (b) LDH 3; (c) LDH 5; (d) LDH 7.

The FESEM micrographs of the $Mg(OH)_2$ and Mg-Fe LDH samples electrodeposited at various pH values are demonstrated in figure 2. Figure 2 (a,a1) shows the presence of curled nano-plates microstructures, indicating the successful synthesis of $Mg(OH)_2$ layer on the surface of WE43 alloy. Nevertheless, it can be seen that the formation of the $Mg(OH)_2$ layer is not uniform as the Mg alloy's polished surface is still visible. The $Mg(OH)_2$ morphology reported here is similar to the surface morphologies reported by past researchers [37, 38]. Thus, the $Mg(OH)_2$ layer cannot provide excellent corrosion protection for the substrate as the non-uniform layer will accelerate the penetration of corrosive anions to the substrate surface. The surface microstructure of the electrodeposited LDH films at different pH values is shown in figure 2 (b-d). LDH 3 represents the nano-blade structure, eviting the crystalline nature of the LDH film. However, smaller-sized flowers appear at pH 7.0, creating compact and more dense surface morphologies. The formation of flower-like morphology could be due to the decrease in crystallinity.

Nevertheless, the LDH film synthesised at pH 5 contains pores visible in the FESEM micrographs. Additionally, it can be observed from the figure (d, d4) that small needle-like morphology with the presence of cracks is seen in the LDH 7 sample. The presence of cracks and small needle-like morphology could form due to the selection of pH at 7.0 restrict the crystallography growth of the LDH film. These results agree with the FTIR spectrum of the samples. Thus, solution pH of 3.0 is regarded as the most suitable pH to grow a compact and dense Mg-Fe LDH film through electrodeposition.

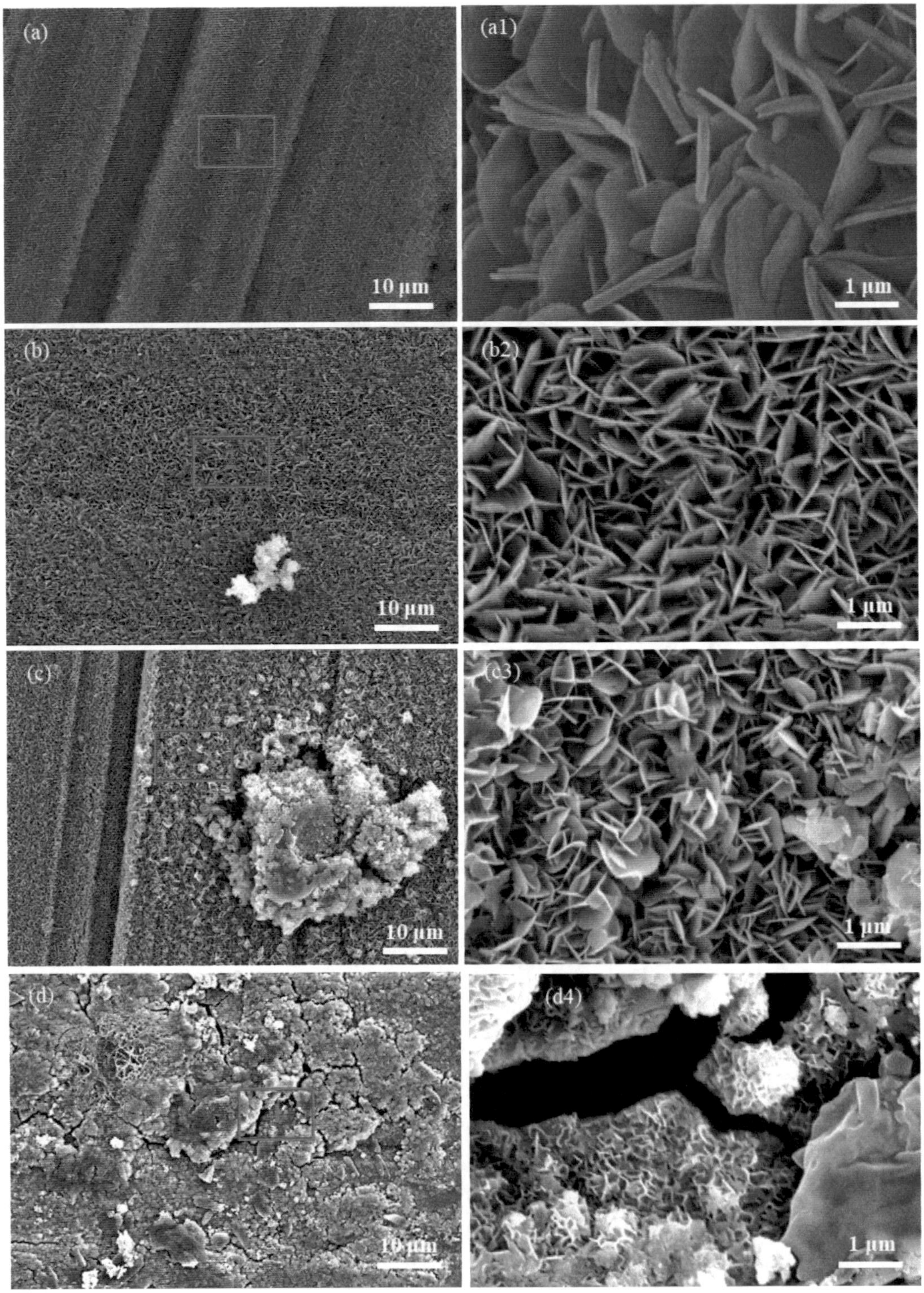

Figure 2. FESEM micrographs of (a, a1) Mg(OH)2 layer, (b, b2) LDH 3, (c, c3) LDH 5, (d, d4) LDH 7.

IOP Conf. Series: Materials Science and Engineering 1195 (2021) 012036 doi:10.1088/1757-899X/1195/1/012036

3.2. Corrosion inhibition of Mg-Fe LDH films at various pH values.

Potentiodynamic polarisation (PDP) technique was selected to assess the electrochemical performance of the Mg-Fe LDH films. PDP curves of Mg-Fe LDH films prepared at different pH values in Hank's balanced salt solution (HBSS) are illustrated in figure 3. Polarization measurements were performed to obtain the values of corrosion potential (E_{corr}), corrosion current density (i_{corr}), polarization resistance (R_p), and inhibition efficiency (I.E), which are presented in Table 1. Based on the E_{corr} and i_{corr} values, it can be seen that both values for the LDH films at pH 3, pH 5, and pH 7 shifted to more positive and lower values, indicating that the LDH films showed improved corrosion resistance in terms of kinetics and thermodynamics as compared to WE43 alloy in HBSS. These results portrait the reliability of the protective ability of the LDH films on WE43 alloy against corrosion attacks in the physiological environment. Above all, these results suggested that LDH 3 sample exhibited the best corrosion resistance, with I.E and R_p values of 94.48 % and 3559.07 Ωcm^2, respectively. Generally, the higher value of I.E and R_p indicated that the sample is less prone to corrosion attacks.

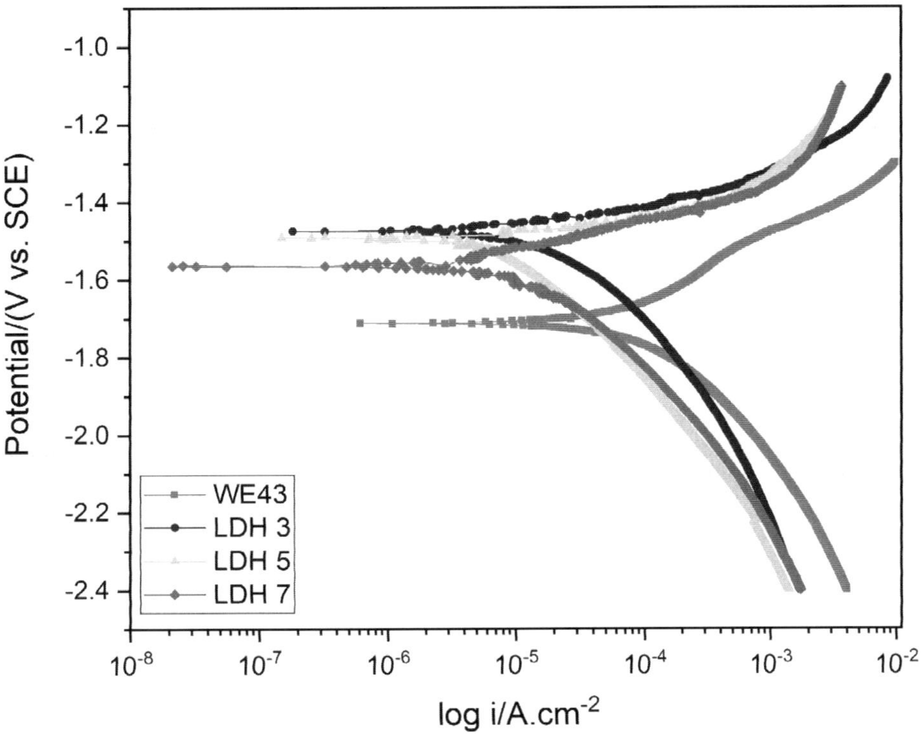

Figure 3. PDP curves of WE43 alloy and Mg-Fe LDH films synthesised at different pH values.

Table 1. Tafel plot values for Mg-Fe LDH samples in HBSS with pH of 7.4 at $37\,^{o}C$.

	β_a (mVdec^{-1})	- β_c (mVdec^{-1})	E_{corr} (V vs. SCE)	i_{corr} (μA.cm^{-2})	I.E (%)	R_p (Ωcm^2)
WE43	0.3331	0.4063	- 1.71	0.000143	-	556.52
LDH 3	0.0804	0.3305	- 1.40	0.0000079	94.48	3559.07
LDH 5	0.0798	0.3644	- 1.49	0.0000090	93.71	3162.51
LDH 7	0.067	0.1165	- 1.56	0.0000098	93.14	1887.17

4. Conclusion

Mg-Fe LDH films were successfully synthesised on WE43 alloy with $Mg(OH)_2$ layer as the precursor layer via electrodeposition. The FTIR spectrum confirmed the formation of LDH film on the WE43 alloy. It is found that the pH value of 3.0 encourages the crystalline growth of LDH, which leads to the formation of a dense and compact nano-blades film. The obtained Mg-Fe LDH film at a pH value of 3.0 shows improved corrosion resistance for the underlying WE43 alloy in HBSS at 37 ^{o}C, rendering its potential for biomedical implant applications.

Reference

[1] Yin Z Z, Qi W C, Zeng R C, Chen X B, Gu C D and Guan S K 2020 Advances in coatings on biodegradable magnesium alloys. *J. Magnesium Alloys.* 42–65

[2] Tan J K E, Balan P and Birbilis N 2021 Advances in LDH coatings on Mg alloys for biomedical applications: A corrosion perspective. *Appl. Clay Sci.* **202** 105948

[3] Liu J, Bian D, Zheng Y, Chu X, Lin Y, and Wang M 2020 Comparative in vitro study on binary Mg-RE (Sc, Y, La, Ce, Pr, Nd, Sm, Eu, Gd, Tb, Dy, Ho, Er, Tm, Yb and Lu) alloy systems. *Acta Biomaterial* **102** 508–28

[4] Chen J, Tan L, Yu X, Etim I P, Ibrahim M and Yang K 2018 Mechanical properties of magnesium alloys for medical application: A review *J. Mecha Behav. Biomed. Mater.* **87** 68–79

[5] Ding W 2016 Opportunities and challenges for the biodegradable magnesium alloys as next-generation biomaterials *Regen Biomater* **3** 79–86

[6] Mao L, Shen L, Niu J, Zhang J, Ding W and Wu Y 2013 Nanophasic biodegradation enhances the durability and biocompatibility of magnesium alloys for the next-generation vascular stents *Nanoscale.* **5** 9517–22

[7] Kuhlmann J, Bartsch I, Willbold E, Schuchardt S, Holz O and Hort N 2013 Fast escape of hydrogen from gas cavities around corroding magnesium implants *Acta Biomaterialia.* **9** 8714–21

[8] Mraied H, Wang W and Cai W 2019 Influence of chemical heterogeneity and microstructure on the corrosion resistance of biodegradable WE43 magnesium alloys *J. Mater Chem B* **7** 6399–411

[9] Yan T, Tan L, Zhang B and Yang K 2014 Fluoride Conversion Coating on Biodegradable AZ31B Magnesium Alloy *J. Mater Sci. & Technol.* **30** 666–74

[10] Abatti G P, Nunes Pires A T, Spinelli A, Scharnagl N and da Conceição T F 2018 Conversion coating on magnesium alloy sheet (AZ31) by vanillic acid treatment: Preparation, characterization and corrosion behavior *J. Alloys Compd.* **738** 224–32

[11] Bakhsheshi-Rad H R, Hamzah E, Ismail A F, Aziz M, Najafinezhad A and Daroonparvar M 2019 Synthesis and in-vitro performance of nanostructured monticellite coating on magnesium alloy for biomedical applications *J. Alloys Compd.* **773** 180–93

[12] Zhang G, Wu L, Tang A, Ma Y, Song G L and Zheng D 2018 Active corrosion protection by a smart coating based on a MgAl-layered double hydroxide on a cerium-modified plasma electrolytic oxidation coating on Mg alloy AZ31 *Corros. Sci.* **139** 370–82

[13] Cui L Y, Xu J, Lu N, Zeng R C, Zou Y H and Li S Q 2017 In vitro corrosion resistance and antibacterial properties of layer-by-layer assembled chitosan/poly-L-glutamic acid coating on AZ31 magnesium alloys *Trans. Nonferrous Met. Soc. China.* **27** 1081–6

[14] Zhao Y B, Liu H P, Li C Y, Chen Y, Li S Q and Zeng R C 2018 Corrosion resistance and adhesion strength of a spin-assisted layer-by-layer assembled coating on AZ31 magnesium alloy *Appl. Surf. Sci.* **434** 787–95

[15] Cui L Y, Cheng S C, Liang L X, Zhang J C, Li S Q and Wang Z L 2020 In vitro corrosion resistance of layer-by-layer assembled polyacrylic acid multilayers induced Ca–P coating on magnesium alloy AZ31 *Bioact. Mater.* **5** 153–63

[16] Wen T, Yan R, Wang N, Li Y, Chen T and Ma H 2020 PPA-containing layered double hydroxide (LDH) films for corrosion protection of a magnesium alloy. *Surf. Coat. Technol.* **383** 1252–55
[17] Wu L, Ding X, Zheng Z, Ma Y, Atrens A and Chen X 2019 Fabrication and characterization of an actively protective Mg-Al LDHs/Al2O3 composite coating on magnesium alloy AZ31 *Appl Surf Sci.* **487** 558–68
[18] Wu L, Yang D, Zhang G, Zhang Z, Zhang S and Tang A 2018 Fabrication and characterization of Mg-M layered double hydroxide films on anodized magnesium alloy AZ31 *Appl. Surf. Sci.* **431** 177–86
[19] Li L X, Xie Z H, Fernandez C, Wu L, Cheng D and Jiang X H 2020 Development of a thiophene derivative modified LDH coating for Mg alloy corrosion protection *Electrochimica Acta.* **330** 1351–86
[20] Shulha T N, Serdechnova M, Lamaka S V, Wieland D C F, Lapko K N and Zheludkevich M L 2018 Chelating agent-assisted in situ LDH growth on the surface of magnesium alloy *Sci. Rep.* **8** 1640–9
[21] Peng F, Wang D, Cao H and Liu X 2018 Loading 5-Fluorouracil into calcined Mg/Al layered double hydroxide on AZ31 via memory effect *Mater. Lett.* **213** 383–6
[22] Chen J, Fang L, Wu F, Xie J, Hu J and Jiang B 2019 Corrosion resistance of a self-healing rose-like MgAl-LDH coating intercalated with aspartic acid on AZ31 Mg alloy *Prog. Org. Coat.* **136** 1052–34
[23] Lin J K, Uan J Y, Wu C P and Huang H H 2011 Direct growth of oriented Mg–Fe layered double hydroxide (LDH) on pure Mg substrates and in vitro corrosion and cell adhesion testing of LDH-coated Mg samples *J. Mater Chem.* **21** 5011–20
[24] Guo L, Wu W, Zhou Y, Zhang F, Zeng R and Zeng J 2018 Layered double hydroxide coatings on magnesium alloys: A review *J. of Mater. Sci. & Technol.* **34** 1455–66
[25] Patil A S, Gunjakar J L, Lokhande C D, Patil U M, Sadavar S V and Padalkar N S 2020 Nanocrystalline copper-chromium-layered double hydroxide with tunable interlayer anions for electrochemical capacitor application *Synth. Met.* **264** 1163–71
[26] Cao K, Yu Z, Zhu L, Yin D, Chen L and Jiang Y 2021 Fabrication of superhydrophobic layered double hydroxide composites to enhance the corrosion-resistant performances of epoxy coatings on Mg alloy *Surf. Coat. Technol.* **407** 1263–67
[27] Cao Y, Zheng D, Li X, Lin J, Wang C and Dong S 2018 Enhanced corrosion resistance of superhydrophobic layered double hydroxide films with long-term stability on Al substrate *ACS Appl. Mater. Interfaces* **10** 15150–62
[28] Shamsayei M, Yamini Y, Asiabi H and Khataei M M 2020 Electrodeposition of layered double hydroxide intercalated with 2,3-dimercaptopropane sulfonate on carbon cloth and application for effective uptake of heavy metals *Appl. Clay Sci.* **196** 105747
[29] Wu F, Liang J, Peng Z and Liu B 2014 Electrochemical deposition and characterization of Zn-Al layered double hydroxides (LDHs) films on magnesium alloy *Appl. Surf. Sci.* **313** 834–40
[30] Wu L, Zheng Z, Pan F-S, Tang A, Zhang G and Liu L 2017 Influence of reaction temperature on the controlled growth of Mg-Al LDH film *Int. J. of Electrochem. Sci.* **12** 6352–64
[31] Hoshino K, Furuya S and Buchheit R G 2019 Effect of Solution pH on Layered Double Hydroxide Formation on Electrogalvanized Steel Sheets *J. Mater. Eng. Perform.* **28** 2237–44
[32] Kamiyama N, Panomsuwan G, Yamamoto E, Sudare T, Saito N and Ishizaki T 2016 Effect of treatment time in the Mg(OH)2/Mg–Al LDH composite film formed on Mg alloy AZ31 by steam coating on the corrosion resistance *Surf. and Coat. Technol.* **286** 172–7
[33] Chen J, Song Y, Shan D and Han E H 2016 In situ growth process of Mg–Al Hydrotalcite conversion film on AZ31 Mg Alloy *J. Mater. Sci. Technol.* **31** 384–90
[34] Zhu Y, Zhao Q, Zhang Y H and Wu G 2012 Hydrothermal synthesis of protective coating on magnesium alloy using de-ionized water *Surf. and Coat. Technol.* **26** 2961–6

[35] Zulfareen N, Kannan K, Venugopal T and Gnanavel S 2016 Synthesis, characterization and corrosion inhibition efficiency of N-(4-(Morpholinomethyl Carbamoyl Phenyl) Furan-2-Carboxamide for brass in HCl medium *Arabian J. Chem.* **9** 121–35
[36] Zhang F, Zhang C, Zeng R, Song L, Guo L and Huang X 2016 Corrosion resistance of the superhydrophobic Mg(OH)2/Mg-Al layered double hydroxide coatings on Magnesium alloys *Met.* **6** 85
[37] Jeong H and Yoo Y 2015 Synthesis and characterization of thin films on magnesium alloy using a hydrothermal method *Surf. and Coat. Technol.* **284** 26–30
[38] Peng F, Li H, Wang D, Tian P, Tian Y and Yuan G 2016 Enhanced corrosion resistance and biocompatibility of Magnesium alloy by Mg–Al-Layered double hydroxide *ACS Appl. Mater. Interfaces*. **8** 35033–44

Acknowledgements
The authors are grateful and would like to thank the Fundamental Research Grant Scheme (FRGS/1/2018/TK05/MUSM/03/2) provided by the Ministry of Higher Education, Malaysia, for the financial support to conduct this research.

Modelling, optimization and control of continuous two-stage Cephalosporin C production

J F Y Chin[1] and J Nandong[1,2]

[1] Department of Chemical Engineering, Curtin University Malaysia, 98009 Miri, Sarawak, Malaysia

[2] Intelligent Systems, Design and Control (ISDCON) Cluster, Curtin University Malaysia, 98009 Miri, Sarawak, Malaysia

Email: jobrun.n@curtin.edu.my

Abstract. Cephalosporin is one of the most consumed antibiotics for its effectiveness against a wide variety of infections. Most cephalosporin products are the semi-derivatives of Cephalosporin C (CPC), a metabolite of the fungus *Acremonium chrysogenum*. Since naturally the desired metabolite is not produced in a large amount by the fungus, an innovative operational strategy is required to increase its yield for the production of the antibiotic to be economically feasible. One way to increase the cephalosporin productivity is by increasing the concentration of thin hyphae cell in the bioreactor, but this will lead to a higher blower power requirement for providing adequate availability of oxygen in the fermentation broth. Lack of oxygen will retard the growth rate and reduce the productivity. Conversely, excessive aeration of the fermentation broth will lead to high shear stress that can kill the cells. The present work investigates through dynamic simulation the effectiveness of a continuous two-stage aerobic fermentation for the CPC production. The operating conditions are optimized to determine an optimal trade-off between the cephalosporin productivity and blower power. An increase of the dissolved oxygen in the first bioreactor from 10 % to 20 % can increase CPC productivity by 75.5 % from 24.42 mg/L.hr to 42.86 mg/L.hr.

1. Introduction

Antibiotics are an important type of drugs that can kill or retard the growth of bacteria and help prevent bacterial infections. The demand for antibiotics is growing as bacterial infections are one of the continuous health problems that plague society. Among the antibiotics available on market today, cephalosporin is one of the types that are the most sold and consumed [1, 2]. Note that the fifth generation of the cephalosporin class antibiotics has been found capable of fighting off resistance strains of bacteria, for example, the methicillin-resistant *Staphylococcus aureus* (MRSA) [3, 4]. Historically, the third generation of the cephalosporin was often used to treat community-acquired pneumonia.

Generally, the production of cephalosporin starts from the fermentation using a suitable fungus, for example, the most common one is *A. chrysogenum* which can produce CPC. The CPC will then be treated *via* chemical or enzymatic hydrolysis which transforms the CPC into 7-aminocephalosporanic acid (7-ACA), which can subsequently be further processed into an antibiotic drug [5]. The exact type of a cephalosporin-based antibiotic that will be synthesized depends on the side chains attached to the 7-ACA. Figure 1 shows the difference in the molecular structures of these compounds as the CPC is turned into 7-ACA, and then from 7-ACA to a few different cephalosporin-based antibiotics.

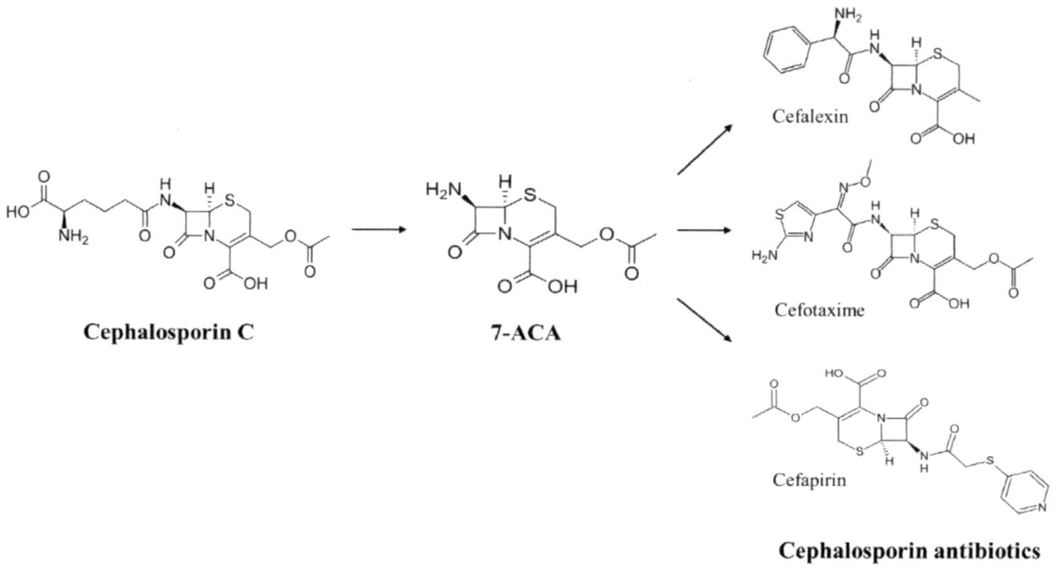

Figure 1. Molecular structure of CPC, 7-ACA and Cephalosporin antibiotics.

The cost of antibiotic production at a commercial scale critically depends on the fermentation process step which requires an antibiotic-producing organism, usually a bacterium or fungus. In the fermentation process, the antibiotic molecules are secreted by the organism involved as its secondary metabolites. The antibiotics will then be separated from the fermentation broth before its subsequent use in antibiotic drug production. For an antibiotic drug to be economically feasible, manufacturers must be able to obtain a high yield of antibiotic from the fermentation process and be able to isolate them from the broth effectively. While antibiotics are produced naturally by microorganisms as their secondary metabolites, the desired metabolites are not often available in the quantities sufficient for large-scale production of antibiotic drugs. Therefore, it often requires an intensive effort to harvest enough amount of these secondary metabolites for mass antibiotic production. Bear in mind that some of the metabolites are produced in the late growth phase of the microorganism involved. This delayed production of desired secondary metabolites is due to inherent regulatory mechanisms of the given microorganism, i.e., the metabolites are suppressed in the logarithmic phase and only generated in low quantities in the stationary phase [6]. One of the reasons for this phenomenon is that the microorganism only needs a small amount of these secondary metabolites for their survival innately, as such the microorganism does not need to produce the metabolites in large quantity [7].

Although that may be true, with the right conditions the production of secondary metabolites can increase substantially. For example, the study by [8] has found that cultivating *A. chrysogenum* with different carbon sources will result in different growth rates and CPC production rates. Furthermore, the availability of sufficient oxygen in the growth environment is equally important for *A. chrysogenum* to produce CPC. Lack of oxygenation will lead to the production of penicillin N instead of CPC [9]. Thus, the determination of optimum conditions for the fermentation process is of prime importance for improving the antibiotic yield. Currently, the pharmaceutical industry is mainly using batch or fed-batch fermentation mode for the production of antibiotics. Nevertheless, the industry is slowly going through a transition towards a continuous production process because the benefits of continuous cultivation over batch fermentation can be realized through suitable operational strategies [10]. The advantages of the continuous process of antibiotic production are higher productivity, lower manufacturing cost, lower waste generation, higher efficiency, lesser energy needs and reduced human errors.

But the transition mentioned above will not be easy as there are still several challenges and research gaps to be resolved [11, 12]. These include the challenge of maintaining the stability of the fermentation operation and biological system over a long period and the challenge of keeping the whole system sterilized as continuous fermentation is prone to contamination even in a closed system. A continuous fermentation process will also require more complicated equipment and process design. In addition, there is a gap in terms of insufficient research on the process dynamics and control of the continuous antibiotics production system. Moreover, the diauxic behaviour that *A. chrysogenum* possesses during fermentation with inverted sugar might pose a difficulty to control using the conventional control system.

Several studies have been reported in the open literature to increase CPC productivity. For example, Liu *et al.* [13] showed that the CPC productivity in *A. chrysogenum* can be substantially increased by using a recombinant strain of the fungus. In another study, Shahidzadeh *et al.* [14] showed that the addition of magnetite nanoparticles into the fermentation broth can increase the CPC production by 60 %. Also, it has been reported that the fermentation production of the CPC can be increased two-fold using algal sugars from the third-generation biomass and glycerol [15]. Zafira and Nandong [16] demonstrated that the feeding strategy using two sugars (glucose and sucrose) in a fed-batch fermentation can be optimized and resulted in the maximum CPC productivity of 65.36 mg/hr. In this study, the idea behind the optimal operating strategy is to maximize the production of the thin cells by adjusting the glucose and sucrose flow rates into the bioreactor.

This study aims to modify a fed-batch model of CPC production into a continuous model, and then creating a MATLAB/Simulink simulation from the continuous model, in which a practical control strategy will be applied. Afterwards, Response Surface Methodology (RSM) is performed to determine the relationship between the various parameters and various factors that might affect the CPC productivity and blower power for the system. Being adopted previously to improve the production of antibiotics successfully in several *Streptomyces* species, RSM has proven to be a useful tool to optimize the parameters of microbial fermentation through mathematics and statistics [17]. Besides that, the Weighted Sum Scalarization optimization approach is also used to combine the two objectives for maximizing CPC production and minimizing blower power requirement. Note that, so far, there has been very little research conducted on the production of CPC by *A. chrysogenum* in a continuous two-stage fermentation process. The idea behind this production strategy is to maximize thin cell concentration through decoupling the thin and thick cell productions in separate but connected bioreactors.

2. Bioreactor modelling

2.1. Bioreactor model

Please note that the fermentation kinetic model used in the present work is obtained from the work of [18]. This kinetic model was derived based on a mixture of glucose and substrate where the former is first consumed by the fungus to produce thin filamentous hyphal cells. After the depletion of the glucose, the thin cells undergo a morphological transformation to thick cells while consuming the sucrose and producing the CPC. In the work of [18], the effect of oxygen limitation is not studied as the kinetic model ignores the oxygen limitation. Therefore, in the present work, the aforementioned kinetic model is further modified by including the dissolved oxygen (DO) term in the model. The kinetic data related to this dissolved oxygen (DO) term is obtained from the work of [19].

Figure 1 shows the schematic diagram of the two-stage continuous bioreactors used to produce CPC by *A. chrysogenum*. The major purpose of the first bioreactor is to produce hyphae cells and to maximize the cell concentration before sending it to the second bioreactor. To repress the formation of thick cells, only glucose is fed into the first bioreactor. The hyphae cells from the first bioreactor will undergo morphological differentiation to thick cells and concurrently the CPC is generated and released into the fermentation broth. To maintain the growth of the thick cells while not repressing the CPC production, only sucrose is fed into the second bioreactor.

The mass balance for the liquid volume in the bioreactor 1:

$$\frac{d[V_1(1-\gamma_1)]}{dt} = F_a - F_b \tag{1}$$

where the fraction of bioreactor holdup volume occupied by the cells is given by:

$$\gamma_1 = C_T/\rho_{cell} \tag{2}$$

The total cell concentration, $C_T = C_h + C_t$ and $\rho_{cell} = 390.0$ g/L.

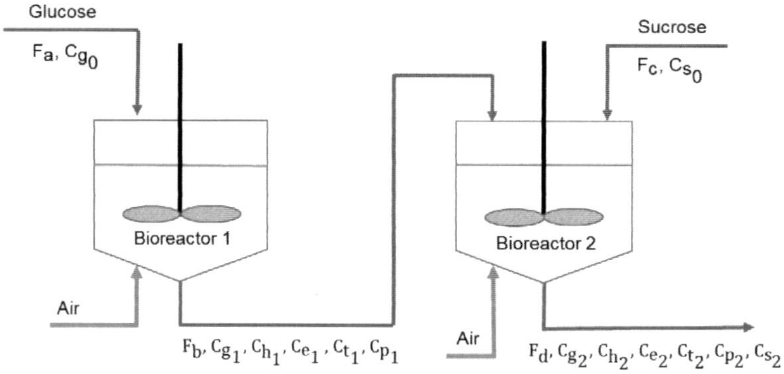

Figure 2. Schematic of aerated two-stage continuous fermentation for CPC production.

Hyphae cells:

$$\frac{d[V_1(1-\gamma_1)C_{h_1}]}{dt} = -F_b C_{h_1} + R_{h_1} V_1(1-\gamma_1) \tag{3}$$

Thick cells:

$$\frac{d[V_1(1-\gamma_1)C_{t_1}]}{dt} = -F_b C_{t_1} + R_{t_1} V_1(1-\gamma_1) \tag{4}$$

Glucose:

$$\frac{d[V_1(1-\gamma_1)C_{g_1}]}{dt} = F_a C_{g_0} - F_b C_{g_1} + R_{g_1} V_1(1-\gamma_1) \tag{5}$$

Enzyme:

$$\frac{d[V_1(1-\gamma_1)C_{e_1}]}{dt} = -F_b C_{e_1} + R_{e_1} V_1(1-\gamma_1) \tag{6}$$

Cephalosporin C:

$$\frac{d[V_1(1-\gamma_1)C_{p_1}]}{dt} = -F_b C_{p_1} + R_{p_1} V_1 \tag{7}$$

Notations: C_{h_1}, C_{e_1} and C_{g_1} denote the concentrations of hyphae cells, enzyme and glucose in the bioreactor 1; V_1 is the bioreactor liquid volume, C_{g_0} inlet glucose concentration; R_{h_1}, R_{g_1} and R_{e_1} denote the rates of hyphae cell growth, glucose consumption and enzyme production in the bioreactor 1. The mass balances for the oxygen in the liquid and gas phases are written as follows:

$$\frac{d[V_1(1-\gamma_1)C_{o_1}]}{dt} = -F_b C_{o_1} + V_1(1-\gamma_1)(R_{otr_1} - R_{our_1}) \tag{8}$$

$$\frac{dy_{o_1}}{dt} = F_{air_1} y_{o_0}/N_{t_1} - R_{otr_1} V_1(1-\gamma_1)/N_{t_1} \tag{9}$$

Where R_{otr_1}, R_{our_1}, C_{o_1}, F_{air_1}, y_{o_1} and N_{t_1} denote the rate of oxygen transfer from gas to liquid phase, rate of oxygen uptake by the cells, dissolved oxygen concentration, air flow rate, mass fraction of oxygen in the headspace and total number of moles of air in the headspace respectively. Note that for the second bioreactor, the above mass balances are repeated with the streams as shown in figure 2.

2.2. *Rate equations*

The rate equations are obtained from [18] and briefly shown as follows. Note that the rates have been modified by considering the limitation imposed by the dissolved oxygen concentration.

Rate of glucose consumption (R_g):

$$R_g = \left(\frac{\mu_g}{Y_g}\right) C_T \quad (10)$$

$$\mu_g = \mu_g^{max}\left(\frac{C_g}{K_{s1} + C_g}\right)\left(\frac{C_o}{k_o + C_o}\right) \quad (11)$$

Rate of sucrose consumption (R_s):

$$R_s = \left(\frac{\mu_s}{Y_s} + m_s\right) C_T \quad (12)$$

$$\mu_s = \theta_d\left(\frac{\mu_s^{max}\, C_s}{K_{s2} + C_s}\right)\left(\frac{C_o}{k_o + C_o}\right) \quad (13)$$

$$\theta_d = \begin{Bmatrix} 0 & \text{if } C_g > \epsilon \\ 1 & \text{if } C_g \leq \epsilon \end{Bmatrix} \quad (14)$$

Rate of hyphae cell growth (R_h):

$$R_h = \mu_h C_T - (\mu_t + \delta_h) C_h \quad (15)$$

$$\mu_h = \begin{cases} \mu_g, & \text{if } C_g > \epsilon \\ \mu_s, & \text{if } C_g \leq \epsilon \end{cases} \quad (16)$$

$$\mu_t = \frac{\mu_t^{max}}{1 + I_2 C_g} \quad (17)$$

$$\delta_h = \frac{\delta_h^{max}}{1 + I_1 C_g} \quad (18)$$

In the current simulation the threshold glucose concentration is taken to be $\varepsilon = 0.01$.

Rate of thick cell growth (R_t):

$$R_t = \mu_t C_h - \delta_t C_t \quad (19)$$

Rate of enzyme production (R_e):

$$R_e = \alpha\mu_t C_h - \beta C_e \quad (20)$$

Rate of Cephalosporin C production (R_p):

$$R_p = C_e C_t - \sigma C_p \quad (21)$$

$$R_{otr} = k_L a\left(C_{o_{eq}} - C_o\right) \quad (22)$$

$$R_{our} = R_{max}\left(\frac{C_o}{k_o + C_o}\right) C_T \quad (23)$$

$$C_{o_{eq}} = y_o P / H_o \quad (24)$$

where $C_{o_{eq}}$, P and H_o denote the oxygen solubility in water, total pressure in the headspace and Henry constant for oxygen respectively. Note that the notations used in the above rate equations are the same as the notations used in [18]. Table 1 shows the parameter values used in the current modelling and simulation.

Table 1. Parameters used in the modelling and simulation obtained from [16,17].

Parameters	Values
α	9.0
β	2.18787
σ	0.01076
H_o $(\frac{\text{Pa.L}}{\text{mol}})$	7.468 x 10^7
$k_L a$ $(\frac{1}{\text{hr}})$	162.0
K_o, $(\frac{\text{mmol}-\text{O}_2}{\text{L}})$	1.132 x 10^{-4}
R_{max}, $(\frac{\text{mmol}-\text{O}_2}{\text{g}-\text{cell.hr}})$	1.31
Y_g $(\frac{\text{g}-\text{cell}}{\text{g}-\text{glucose}})$	0.46188
Y_s, $(\frac{\text{g}-\text{cell}}{\text{g}-\text{sucrose}})$	0.4
I_1	20.0
I_2	300
δ_h^{max}	0.00668
δ_t	0.00441
μ_t^{max}	0.04526
μ_g^{max}	0.042
μ_s^{max}	0.021
m_s	0.02267
K_{s1}	0.1
K_{s2}	10.199

2.3. Blower power equation

The blower power requirement is given by [20]:

$$H_P = 0.01542\left(\frac{Q\,P_i}{\eta}\right)\left(\frac{P_d}{P_i}\right)^{0.283} \tag{25}$$

where H_p the brake horsepower (bhp), Q the blower inlet volumetric flow rate (ft^3/min), P_i the blower inlet pressure (psia), P_d the blower discharge pressure (psia), η the blower efficiency and X the blower adiabatic factor.

The total discharge pressure P_{tot} (psig) from the blower is calculated as follows

$$P_{tot} = 0.433h_w + \Delta P \tag{26}$$

where h_w the depth of water at top of diffuser (ft).

The blower discharge pressure is calculated as follows

$$P_d = P_{tot} + P_i \tag{27}$$

It is assumed that the average pressure drop is $\Delta P = 5$ kPa = 7.252 psi, across the liquid column in the bioreactor. Table 2 shows the parameter values used for the bioreactor design.

Table 2. Parameters of the bioreactor.

Parameters	Values
Bioreactor volume V_R (L)	1500
Bioreactor diameter D (m)	1.0839
Bioreactor height H (m)	1.6258
Pressure drop across liquid height ΔP (psi)	7.2519
Blower inlet pressure P_i (psi)	14.6959

3. Bioreactor control strategy

The CPC fermentation in the two-stage continuous bioreactors exhibits unstable dynamics. For this reason, it is important to control both bioreactors involved before one can optimize the CPC production performance. A simulation study reveals that to achieve stable continuous operation, it is important to control the liquid levels and dissolved oxygen concentrations in both bioreactors. In this process, controlling the liquid levels is the most important factor for the process stability based on the simulation study in this work; a drop or an increase in the liquid level in either bioreactor will severely affect the substrate conversion and hence the CPC productivity. Meanwhile, adequate availability of dissolved oxygen in the fermentation broth is crucial to the fungus growth rate.

When the dissolved oxygen concentration is too low, the growth rate will be severely inhibited and resulting in rapid cell decay; these combined effects will dramatically reduce the CPC productivity. On the other hand, if the dissolved oxygen concentration is too high, then this will cause a dramatic increase in the blower power requirement and hence increasing the cost of operation. Thus, there must be an optimum value of the dissolved oxygen concentration which ensure a sufficiently high growth rate while keeping the blower power requirement at an economic level. In addition to controlling the liquid levels and dissolved oxygen concentrations, it is recommended to control the glucose concentration in the first bioreactor and sucrose concentration in the second bioreactor to enhance the CPC productivity and minimize the unconverted substrates, i.e., saving the cost on substrates. In total, there are 6 proportional-integral (PI) control loops employed in the process (3 control loops for each bioreactor). Table 3 shows the controlled variables (CV), manipulated variables (MV) and their nominal values.

Table 3. Controlled variables, manipulated variables and the nominal values.

MV	CV	Nominal MV value	Nominal CV Setpoint
F_{air_1}	C_{o_1}	302.1	5 x 10^{-4}
F_{air_2}	C_{o_2}	215.8	5 x 10^{-4}
F_b	V_1	6.8079	1100
F_d	V_2	10.247	1100
F_a	C_{g_1}	6.8079	3.0
F_c	C_{s_1}	3.4393	3.0

The mathematical model of the continuous two-stage *A. chrysogenum* fermentation process presented in Section 2 earlier is implemented into MATLAB/Simulink, which is a software that provides a flexible dynamic simulation environment for analyzing dynamic and discrete systems. It is used widely within the industry and academia to represent process behaviour and control systems. The Simulink model, as shown in figure 3 below, consists several input and output signals into and out from the bioreactor blocks. Several PI controllers are used where the brief summary of the controllers have been discussed above.

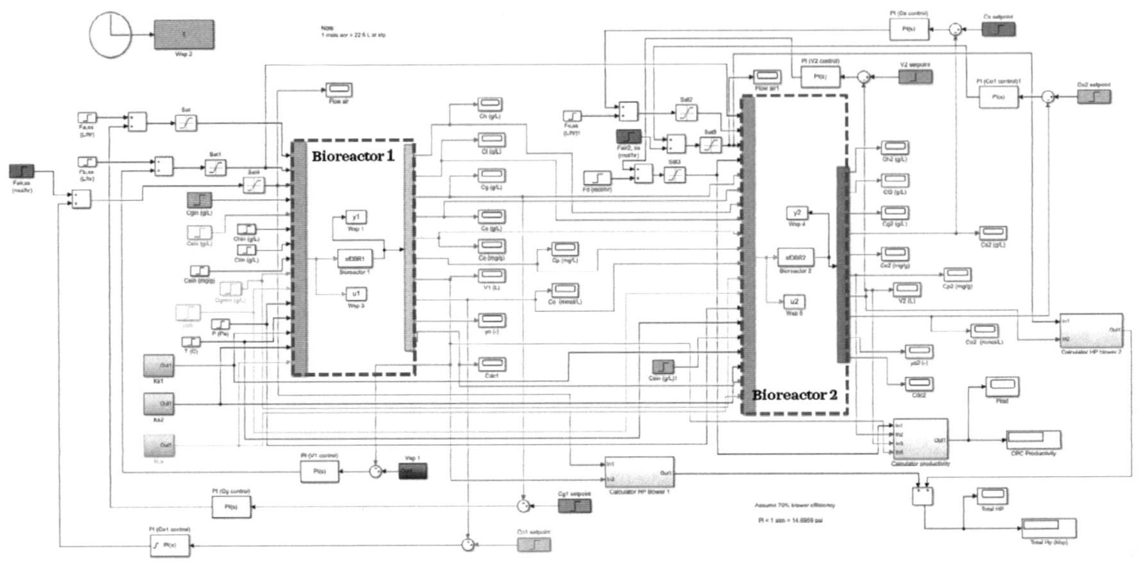

Figure 3. Simulink model for the simulation of the two-stage fermentation process.

4. Results and discussion

4.1. Impact of bioreactor glucose and dissolved oxygen concentrations

The concentration of oxygen in the first bioreactor ($C_{O1\text{-}sp}$) was found to be the limiting factor for the production of CPC. A graph of CPC productivity against $C_{O1\text{-}sp}$ is shown in figure 4 below. The point that is labelled as 10 % oxygen saturation is equivalent to 0.422 x 10^{-4} mmol/L, whereas the point labelled as 20 % oxygen saturation is 0.844 x 10^{-4} mmol/L and 30 % oxygen saturation is 1.265 x 10^{-4} mmol/L. Fairly similar to the results obtained in the study of [21], after the system surpassed the critical oxygen saturation level, the improvement of productivity due to the increase of oxygen slowly plateaued. In this case, the critical oxygen level is around 30 %. The critical oxygen saturation level is defined as the onset beyond which an increase of dissolved oxygen level will lead to reduced rate of productivity.

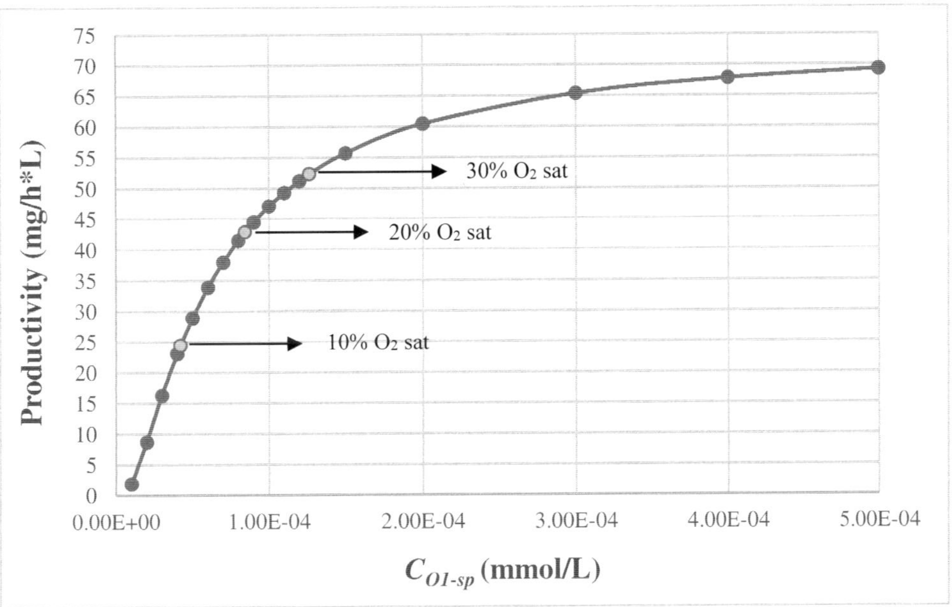

Figure 4. Productivity against $C_{O1\text{-}sp}$.

From the simulation results (figures are not shown) obtained in MATLAB/Simulink, at a low oxygen saturation level, i.e., 10 %, the concentration in the first bioreactor of hyphae cells (C_{h1}) and thick cells (C_{t1}) is around 84.3 g/L and 4.13 g/L respectively while in the second bioreactor, the concentration in the first bioreactor of hyphae cells (C_{h2}) and thick cells (C_{t2}) is at 15.5 g/L and 43 g/L respectively. The CPC concentration in the first bioreactor (C_{p1}) is at 49.25 mg/L whereas in the second bioreactor (C_{p2}), it is at 5150 mg/L. At a moderate oxygen saturation level, i.e., 20 %, the C_{h1} increased to 86.67 g/L whereas the C_{t1} decreased to around 3 g/L. A lesser amount of CPC is also produced in the first bioreactor with C_{p1} at only 28 mg/L. With fewer thick cells being formed prematurely, concentration for both type of cells along with CPC is increased in the second bioreactor. Specifically, the concentration of hyphae cells, thick cells and CPC are at 22.8 g/L, 43.86 g/L and 5870 mg/L respectively. Furthermore, there is more CPC biosynthesis enzyme in both bioreactors.

Given a higher dissolved oxygen saturation level, i.e., 30 %, the C_{h1} increased to 87.48 g/L while C_{t1} further decreased to 2.53 g/L. In the meantime, the C_{p1} dropped to 20.93 mg/L. Similar to the condition in the aforementioned system with different oxygen situation, the hyphae cells in the second bioreactor are also increased while the thick cells are decreased. In this case, the values (C_{h2} and C_{t2}) are 27 g/L and 43 g/L respectively. On the other hand, the value for C_{p2} is 5855 mg/L. Nonetheless, the concentration of CPC biosynthesis enzyme here in both bioreactors (C_{e1} and C_{e2}) are higher than that of the bioreactors at 20 % of oxygen level. Although the CPC concentration (C_{p1} and C_{p2}) is lower in both bioreactors, the productivity of the system is still higher when compared with the system with 20 % of $C_{O1\text{-}sp}$, i.e., dissolved oxygen concentration setpoint in the first bioreactor.

Meanwhile, the concentration of oxygen in the second bioreactor ($C_{O2\text{-}sp}$) has a lower impact on the overall productivity of the system. This is because the thick cells in the second bioreactor are originated from the filamentous hyphal cells in the first bioreactor. When the $C_{O2\text{-}sp}$ is increased from 0.36 x 10^{-4} to 1.2 x 10^{-4} mmol/L with the oxygen level is being set at 10 %, 20 % and 30 % of oxygen saturation, the productivity will increase by an average of merely 1.43 mg/hr.L. Particularly, at 10% oxygen saturation, the volumetric productivity is raised from 24.42 to 26.11 mg/h.L (an increase of 1.69 mg/hr.L); at 20 % oxygen saturation, the volumetric productivity is raised from 42.86 to 44.19 mg/hr.L (increase of 1.33 mg/hr.L); and at 30 % oxygen saturation, the productivity is raised from 52.32 to 53.59 mg/hr.L (an increase of 1.27 mg/hr.L). In addition, it was found that the concentration of oxygen also affects the optimal glucose concentration setpoint in the first bioreactor ($C_{g1\text{-}sp}$). As the concentration of oxygen is increased, the optimal glucose concentration setpoint ($C_{g1\text{-}sp}$) is lower (figure 5). For illustration, as shown in figures 5, 6 and 7, the optimal $C_{g\text{-}sp}$ for the first bioreactor with 10 % oxygen saturation is around 5.5 g/L, while it is around 3.5 g/L for 20% oxygen saturation and 2.5 g/L for 30 % saturation. As the oxygen saturation level is higher, the effect that arises from the changes in $C_{g\text{-}sp}$ also become more apparent.

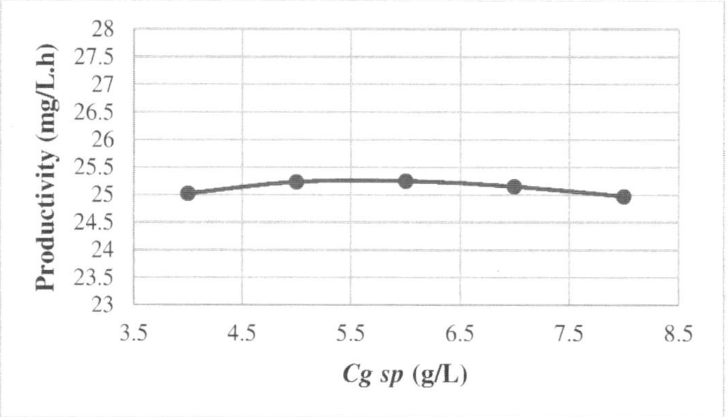

Figure 5. Productivity against $C_{g\text{-}sp}$ under 10 % oxygen saturation.

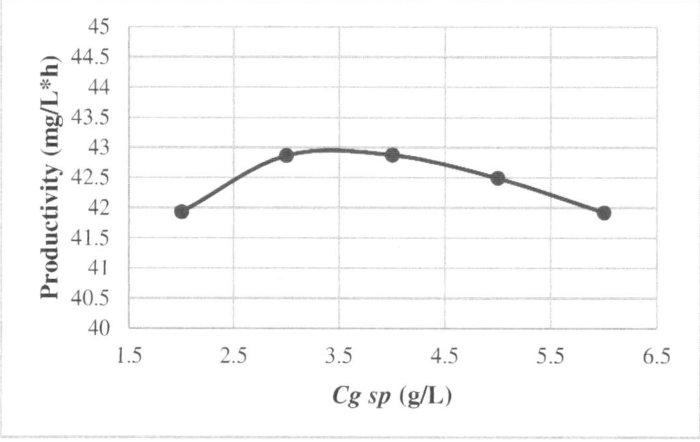

Figure 6. Productivity against $C_{g\text{-}sp}$ under 20 % oxygen saturation.

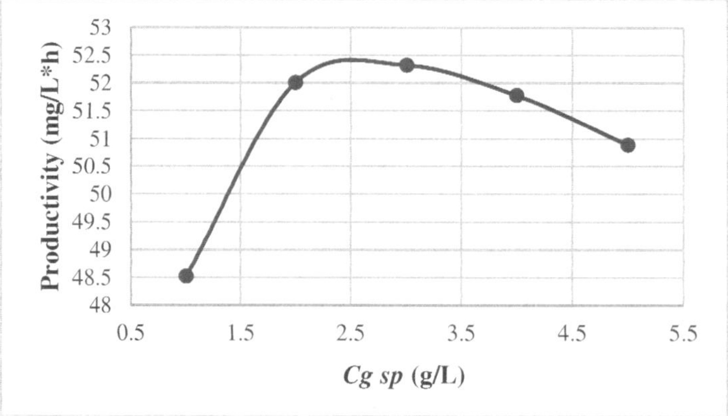

Figure 7. Productivity against $C_{g\text{-}sp}$ under 30 % oxygen saturation.

On the other hand, the effect of changing the sucrose concentration setpoint ($C_{s\text{-}sp}$) is minimal, and it is dependent on a variable like the $C_{O1\text{-}sp}$. When $C_{O1\text{-}sp}$ is at 10 % saturation, productivity will increase following the increment of $C_{s\text{-}sp}$ from 3 g/L up to 9 g/L. Any additional increase of $C_{s\text{-}sp}$ after 9 g/L will result in the reduction of productivity from its peak value of 24.47 mg/hr.L. Contrarily when $C_{O1\text{-}sp}$ is at 20 % and 30 % saturation, productivity will decrease when $C_{s\text{-}sp}$ is increased. In this case, a lower $C_{s\text{-}sp}$ will result in higher productivity. For a 20 % saturation $C_{O1\text{-}sp}$, the peak productivity value of 43.07 mg/hr.L occurs at $C_{s\text{-}sp}$ of 0.5 g/L. Below that value, i.e. 0.1 or 0.25 g/L, the productivity is dropped to 43.06 mg/hr.L. For $C_{O1\text{-}sp}$ of 30% saturation, the peak productivity value of 53.43 mg/hr.L is achieved with $C_{s\text{-}sp}$ of 0.02 g/L and below. After that, in any case that the value of $C_{s\text{-}sp}$ is higher, the productivity will decrease.

4.2. Response surface models

Two response surface model were developed in which one is for the system at 20 % oxygen saturation and another is for the system at 30 % oxygen saturation. A total of 9 variables ($F_{air\text{-}1}$, $F_{air\text{-}2}$, $C_{O2\text{-}sp}$, $V_{1\text{-}sp}$, $V_{2\text{-}sp}$, $C_{g\text{-}in}$, $C_{g\text{-}sp}$, $C_{s\text{-}in}$ and $C_{s\text{-}sp}$) were first screened to eliminate the variables that have insignificant effects on the system. Since the effects of all the variables are the same for both the system at different oxygen saturation level, the same variables ($C_{O2\text{-}sp}$, $V_{1\text{-}sp}$, $V_{2\text{-}sp}$, $C_{g\text{-}in}$ and $C_{s\text{-}in}$) were included to develop the model. The factorial experimental design generated by the Design-Expert 12 software includes all independent variables as the factor with three levels; where 0.9 is 90 % of the baseline value, 1 is the baseline value, and 1.1 is 110 % of the baseline value. The experimental range and coded level of variables are given in table 4.

Table 4. Experimental independent variables and their coded levels.

Independent variable	Symbol	Levels of coded variables		
		Low	Medium	High
		-1	0	+1
$C_{g\text{-}in}$ (g/L)	A	180	200	220
$V_{1\text{-}sp}$ (L)	B	990	1100	1210
$C_{s\text{-}in}$ (g/L)	C	153	170	187
$V_{2\text{-}sp}$ (L)	D	990	1100	1210
$C_{O2\text{-}sp}$ (10^{-5} x mmol/L)	E	3.24	3.60	3.96

After the simulation was run and the response model is generated by Design-Expert 12, regression analysis and analysis of variance (ANOVA) were conducted for the model of productivity and blower power. The R^2 values for both ANOVA analyses are very close to 1, where the adjusted and predicted R^2 value for both models are in good agreement with each other, with a difference of less than 0.0002 between each other. The final regression equation obtained from the software in terms of coded factors for both models is shown below.

At 20 % oxygen saturation the CPC productivity (P_r) and blower power (P_b) are given by:

$$\begin{aligned} P_r = \{&42.8 + 7.104A + 1.986B + 0.457C - 1.975D + 0.158E + 0.356AB \\ &+0.094AC - 0.362AD - 0.019AE + 0.024BC + 0.338BD \\ &+0.016BE - 0.028CD + 0.010CE - 0.007DE + 0.107A^2 \\ &-0.269B^2 - 0.017C^2 - 0.074D^2 + 0.003E^2\} \end{aligned} \tag{28}$$

$$P_b = \{360.9 + 30.89A + 28.88B + 0.664C + 7.892D + 7.972E + 0.060AC + 0.513AD + 0.598AE - 0.007BC + 0.356BD + 0.201BE + 0.107CD + 0.095CE + 0.623DE + 2.596AB - 0.610A^2 - 0.070B^2 - 0.065C^2 - 0.151D^2 - 0.179E^2\} \quad (29)$$

At 30% oxygen saturation:

$$P_r = \{52.3 + 8.761A + 1.542B + 0.541C - 1.533D + 0.186E + 0.320AB + 0.131AC - 0.013AE - 0.319AD - 0.008BC + 0.510BD - 0.004BE + 0.085CE - 0.003CD - 0.008DE + 0.159A^2 - 0.279B^2 - 0.071C^2 - 0.184D^2 - 0.034E^2\} \quad (30)$$

$$P_b = \{444.7 + 40.49A + 36.72B + 0.625C + 8.589D + 8.424E + 3.537AB + 0.094AC + 0.603AD + 0.640AE + 0.013BC + 0.328BD + 0.188BE + 0.089CD + 0.109CE + 0.660DE - 0.492A^2 - 0.094B^2 - 0.085C^2 - 0.141D^2 - 0.184E^2\} \quad (31)$$

The most influential factor, i.e., see the coefficients of A in equations (28) and (30), is the inlet concentration of glucose in the first bioreactor ($C_{g\text{-}in}$), as according to figure 8 and 9, which shows that the slope for $C_{g\text{-}in}$ are the steepest. This is because the increment of sugar in the bioreactor will promote the growth and propagation of cells. Moreover, swollen hyphae cells that produce CPC are derived from the filamentous hyphal cells, which are grown in the first bioreactor from being fed with glucose. With this being said, the enzyme that can stimulate the production of CPC will also be produced in a larger amount since there are more cells. Thus, increasing the inlet concentration of glucose is more effective than increasing sucrose. While glucose is used to build up the cell mass of the fungi, the role of sucrose is to let them attain the physiological state of CPC production and increase the CPC productivity with the amount of cell mass that is present [22]. For this reason, the concentration of swollen hyphae cells in the bioreactor will increase with the increment of $C_{s\text{-}in}$. Small quantities of filamentous hyphal cells are also grown because of this. However, because of the low activity of the enzyme that hydrolyses sucrose, the sugar is harder to be assimilated and there is a limit to its consumption rate [23]. Therefore, sucrose is added in a smaller amount than glucose where its influence on the CPC productivity is also lower than that of glucose. Nevertheless, one thing to be concerned about is the solubility of the substrates in water, as the quantity of sugar added to the reactor must not exceed its solubility limit. Note that at 30 °C, 1 L of water can dissolve about 1200 g of glucose or 2195 g of sucrose. Figures 8 and 9 show the response surface plots of the productivity and blower power against several key parameters: liquid volume setpoint of the first and second bioreactors ($V_{1\text{-}sp}$ and $V_{2\text{-}sp}$), glucose inlet concentration ($C_{g\text{-}in}$) and sucrose inlet concentration ($C_{s\text{-}in}$).

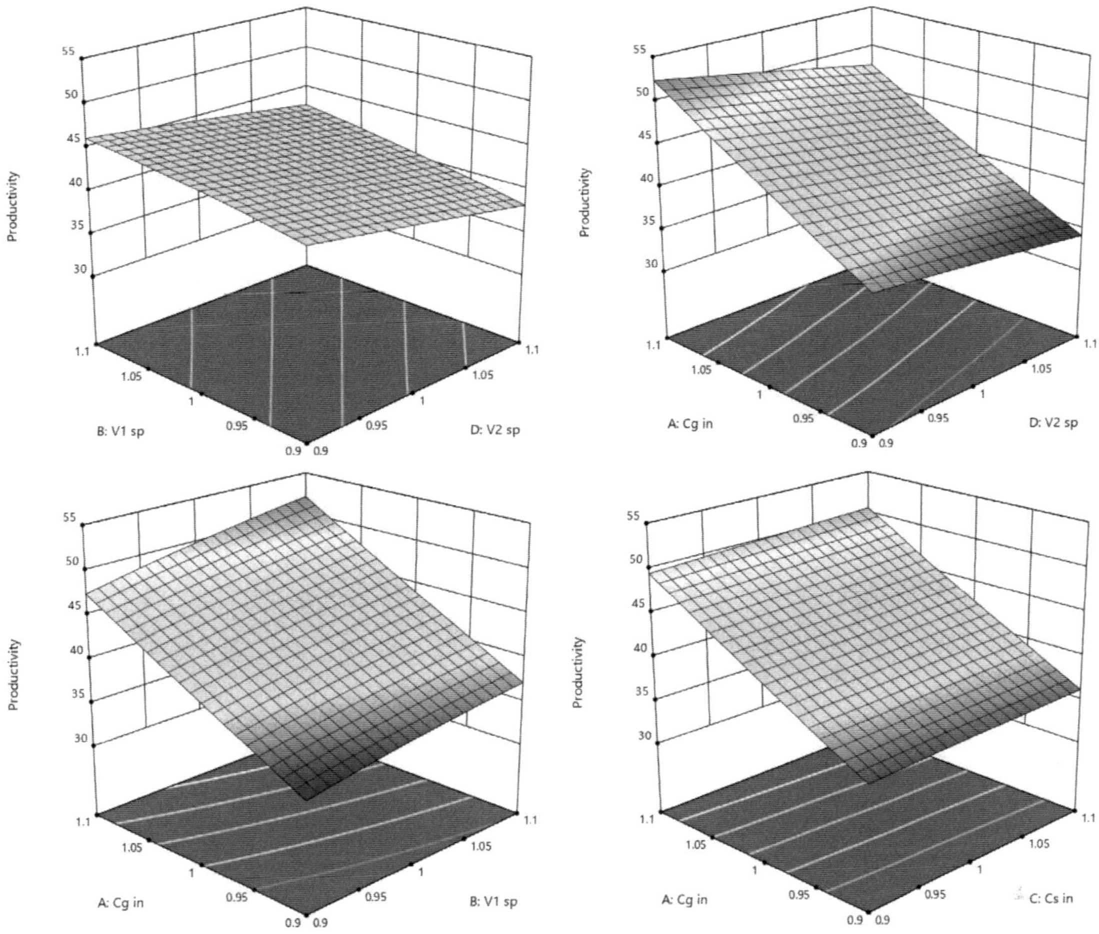

Figure 8. Response surface plots showing the effects of the interaction terms on productivity (Top left to bottom right: Productivity against $V_{1\text{-}sp}$ and $V_{2\text{-}sp}$; $C_{g\text{-}in}$ and $V_{2\text{-}sp}$; $C_{g\text{-}in}$ and $V_{1\text{-}sp}$; $C_{g\text{-}in}$ and $C_{s\text{-}in}$).

At the same time, a high amount of glucose in the reactor might lead to excessive cells growth. This will prevent the cells from taking in enough oxygen as there will be a mass transfer limitation problem due to the filling up of bioreactor by the cells. Since the fermentation of *A. chrysogenum* is highly aerobic, it is necessary to solve this problem by inducing airflow using the blower. For this reason, the requirement for blower power is increased as well when the inlet glucose concentration is increased. However, this situation is undesirable as the increase of blower power will mean the increase in impeller agitation that can cause shear stress to the fungi and causes a breakage of the hyphae cells [24]. Following, the requirement for blower power will be increased as the working volume of the bioreactors get larger, so that aeration will be provided sufficiently for the extended area of the bioreactor. In the meantime, increasing the size of the first bioreactor can benefit the system, as more hyphal cells grow are allowed to grow, with fewer swollen hyphae cells and dead cells. Needless to say, more CPC biosynthesis enzyme is produced as well. Specifically, with the increase of the working volume, the dilution rate is slightly lower, and the residence time is longer. Conversely, a slightly higher dilution rate is needed in the second bioreactor as the productivity is improved when the working volume is decreased for the second bioreactor.

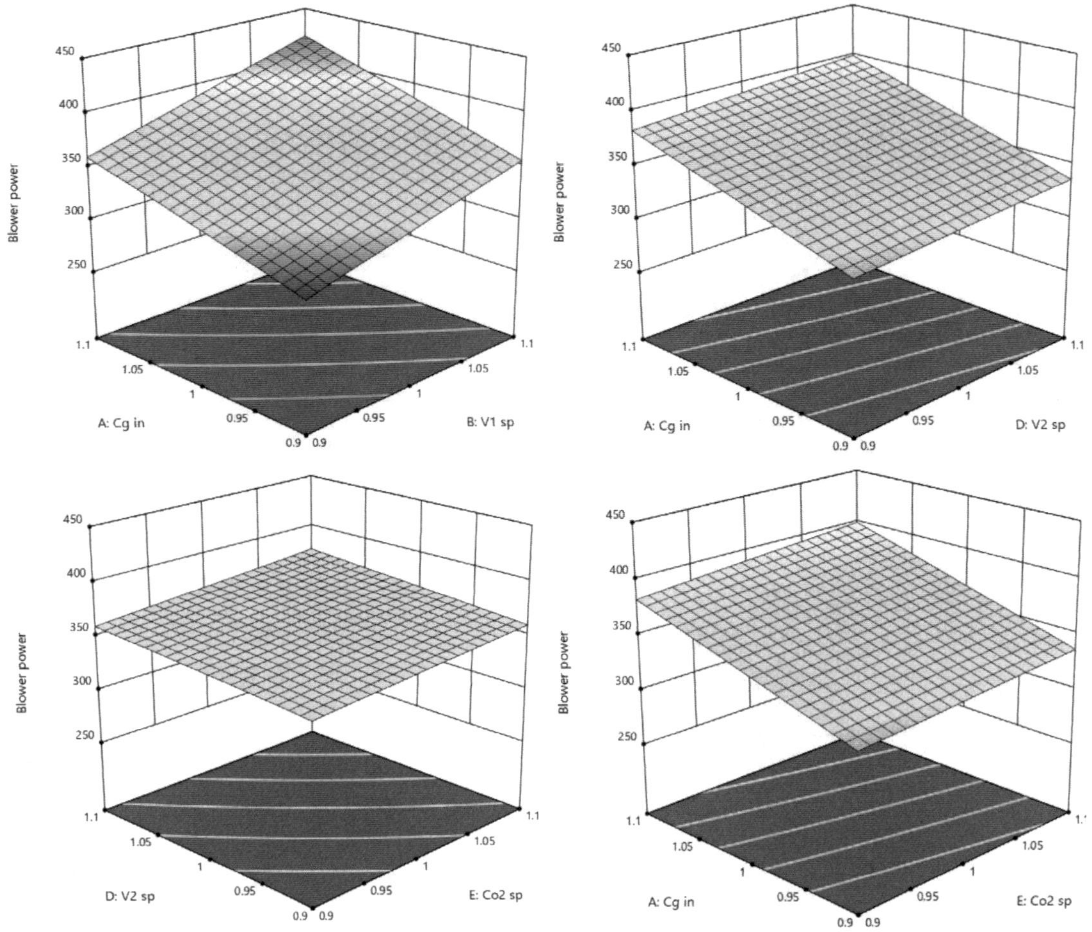

Figure 9. Three-dimensional graph that shows the effects of the interaction terms on requirement of blower power (Top left to bottom right: Blower power requirement against C_{g-in} and V_{1-sp}; C_{g-in} and V_{2-sp}; V_{2-sp} and C_{O2-sp}; C_{g-in} and C_{O2-sp}).

In the MATLAB/Simulink simulation, when the working volume is decreased in the second bioreactor, more filamentous hyphal cells are grown instead of the swollen hyphae cells. Simultaneously, the CPC concentration is decreased. However, more CPC biosynthesis enzyme was produced, and the overall CPC productivity of the system is slightly increased; this is expected from equations (28) and (30) where the coefficients of D in the equations are much smaller than that for A. Not to mention that the concentration of dead cells was also decreased. Thus, although the swollen hyphal cells are known to produce a larger amount of CPC, it was deduced that continuing to grow more filamentous hyphal cells in the second bioreactor may be beneficial for the system. On a side note, when the working volume for both bioreactors is increased or decreased by the same factor at the same time, the productivity will remain unchanged. This is because the beneficial effects of increasing the volume for the first bioreactor are being neutralized by the increment of volume for the second bioreactor and vice versa. The productivity of the system will only increase when the working volume for the first bioreactor is larger than the one for the second bioreactor.

Subsequently, the effects of changing the oxygen concentration in the second bioreactor on the system are significantly lesser than that of raising the oxygen in the first bioreactor. The impact is larger for changing the variable in the first bioreactor in this case since its effect is carried over to the second bioreactor. Thenceforth, it is observed from the simulation that increasing the oxygen can indeed increase productivity, but only by a small amount. This shows that the stage for the production of CPC requires less oxygen than the stage of growing filamentous hyphal cells. Increasing the oxygen will decrease the concentration of dead cells in the bioreactor and increase the amount of CPC biosynthesis enzyme in the bioreactor whereas decreasing the oxygen will do the opposite. All in all, it can be inferred that the CPC biosynthesis enzyme plays a critical role in the productivity of the system, since the fermentation of *A. chrysogenum* is an aerobic process with all rate-limiting enzymes [25].

4.3. Weighted objective scalarization for the multi-objective problem

Given that the optimization for this system is multi-objective, the weighted objective scalarization method will be used in addition to RSM. This approach transforms the multi-objective function into a scalar fitness function by multiplying each objective function with a weighting factor and summing them up altogether [26]:

$$f(x) = \sum_{i=1}^{k} w_i f_i^{norm}(x), \quad \sum_{i=1}^{k} w_i = 1, \quad w_i \geq 0, \quad i = 1, \dots, k \tag{32}$$

where w_i denotes the weighting factor (weight) that represent the relative importance of each normalised objective function. The higher value that the fitness function has, the more desirable is the solution.

As mentioned by [26], the normalisation of objective functions is required to make them scaled and dimensionless. In this case, the normalisation is done by using the following equations when the objectives are to be maximized or minimized:

$$f_i^{norm} = 1 - \frac{\max(f_i) - f_i}{\max(f_i) - \min(f_i)} \tag{33}$$

for objectives to be maximized.

$$f_i^{norm} = \frac{\max(f_i) - f_i}{\max(f_i) - \min(f_i)} \tag{34}$$

for objectives to be minimized.

The bounds of the objective function f_i, $\max(f_i)$ and $\min(f_i)$, is taken as the maximum and minimum value of the responses obtained from the simulation runs. Using the equations above, the value of the scalar fitness function is calculated by applying different sets of the weighting factor, from 0:1 to 1:0, with a difference of 0.1 between each combination of weights, to the productivity and blower power objective function respectively. Productivity will be the objective function to be maximized while blower power will be the objective function to be minimized. When the weighting factor is set to 0:1, it means that the solution is focused on reducing the requirement for blower power with little care given to the yield of the product. Vice versa, when the weighting factor is set as 1:0, it means that the solution is focused on increasing the yield of the product while giving less attention to the requirement of blower power for the process.

The values of fitness functions that are calculated from using different sets of the weighting factor, are inserted as responses into the response surface model that was generated by the Design-Expert 12 software, to perform optimization. The optimized operating conditions in coded factors for a different set of weighting factors along with their predicted responses are tabulated in table 5 and 6 for the system with 20 % oxygen saturation and 30 % oxygen saturation respectively.

Table 5. Optimal operating conditions of the system with 20 % oxygen saturation.

Weightage	$C_{g\text{-}in}$	$V_{1\text{-}sp}$	$C_{s\text{-}in}$	$V_{2\text{-}sp}$	$C_{O2\text{-}sp}$	Productivity (mg/L.hr)	Blower Power (hp)
0:1	0.900	0.900	0.901	0.900	0.900	35.295	288.659
0.1:0.9	0.900	0.900	0.930	0.900	0.900	35.409	288.846
0.2:0.8	0.900	0.900	1.100	0.900	0.900	36.007	289.590
0.3:0.7	0.908	0.900	1.081	0.901	0.909	36.538	292.575
0.4:0.6	1.086	0.900	1.065	0.901	0.900	49.225	340.271
0.5:0.5	1.095	0.900	1.100	0.900	0.900	50.057	342.620
0.6:0.4	1.100	0.900	1.100	0.900	0.900	50.452	343.957
0.7:0.3	1.100	1.011	1.100	0.900	0.900	52.942	378.211
0.8:0.2	1.100	1.085	1.099	0.904	0.980	54.272	408.003
0.9:0.1	1.100	1.100	1.100	0.900	1.045	54.708	417.790
1:0	1.100	1.100	1.099	0.900	1.099	54.810	422.174

Table 6. Optimal operating conditions of the system with 30 % oxygen saturation.

Weightage	$C_{g\ in}$	$V_{1\ sp}$	$C_{s\ in}$	$V_{2\ sp}$	$C_{O2\ sp}$	Productivity (mg/L.hr)	Blower Power (hp)
0:1	0.900	0.900	0.900	0.900	0.900	43.106	355.158
0.1:0.9	0.900	0.900	0.990	0.900	0.900	43.478	355.514
0.2:0.8	0.900	0.900	1.100	0.900	0.900	43.774	355.775
0.3:0.7	0.905	0.901	1.079	0.909	0.906	44.103	359.153
0.4:0.6	1.055	0.902	1.071	0.905	0.902	57.245	412.740
0.5:0.5	1.098	0.901	0.923	0.900	0.903	60.361	426.223
0.6:0.4	1.100	0.900	1.100	0.900	0.900	61.585	427.396
0.7:0.3	1.100	0.902	1.099	0.901	0.914	61.610	429.460
0.8:0.2	1.100	1.007	1.100	0.900	0.977	63.543	476.810
0.9:0.1	1.100	1.097	1.100	0.900	1.100	64.772	522.932
1:0	1.100	1.100	1.100	0.900	1.100	64.803	524.242

5. Conclusion

In this research, a mathematical model for the continuous two-stage fermentation of *A. chrysogenum* to produce CPC has been modified from a fed-batch fermentation reported in the literature, and a simulation of the two-stage process has been conducted in MATLAB/Simulink based on the modified mathematical model. For the control strategy, six PI controllers have been applied in the simulation. In addition, the operating conditions of the continuous fermentation process have been optimized using the Weighted Sum Scalarization and Response Surface Methodology (RSM). Using an equal weighting factor of the CPC productivity and blower power in the optimal operating conditions (refer to table 5 and 6), the optimal productivity increased by 17.1 % when the dissolved oxygen concentration was increased from 20 % to 30 %. This led to an increase in the blower power by 24.4 %. As the weighting factor placed is higher on the productivity and lower on the blower power (refer to table 5 and 6), the optimal productivity and blower power would increase. When weighting factor is all placed on the productivity (i.e., only the productivity was optimized), the productivity only increased by 7.4 % from the value obtained when equal weighting factor were placed for the 30 % dissolved oxygen concentration scenario in table 6. Meanwhile, the blower power increased by 23 % which is a larger increase than that of the productivity. This result shows that it is important to consider both the productivity and blower power when optimizing the CPC production process. It was found in this study that as the dissolved oxygen concentration increased, the sensitivity of the fermentation performance towards mass transfer rate decreased. Without an oxygen saturation limit, the maximum productivity can be achieved with a low mass-transfer rate, but the blower power requirement would be impractically high. Also, it was found that changing the glucose concentration gave a higher impact than changing sucrose concentration on the CPC productivity and blower power requirement. It can be concluded that the application of continuous two-stage fermentation can increase the CPC productivity from that in batch fermentation.

Acknowledgments
Authors wishing to acknowledge assistance or encouragement from colleagues, special work by technical staff or financial support from Curtin University Malaysia.

References

[1] Khan N T 2017 Cephalosporin C production from Acremonium chrysogenum *Enzyme Eng.* **6**
[2] Chaudhry S B, Veve M P and Wagner J L 2019 Cephalosporins: A focus on side chains and β-Lactam cross-reactivity *Pharm.* **7** 103
[3] Duplessis C and Crum-Cianflone N 2011 Ceftaroline: A new cephalosporin with activity against methicillin-resistant Staphylococcus aureus (MRSA) *Clin. Med. Ther.* **3** 1–17
[4] Dauner D G, Nelson R E and Taketa D C 2010 Ceftobiprole: a novel, broad-spectrum cephalosporin with activity against methicillin-resistant Staphylococcus aureus *Am. J. Health-Syst. Pharm.* **67** 83–993
[5] Vardanyan R and Hruby V 2016 *Synthesis of Best-Seller Drugs* (Elsevier)
[6] Thirumurugan D, Cholarajan A, Raja S S and Vijayakumar R 2018 An introductory chapter: Secondary metabolites *Cholarajan A. Secondary Metabolites - Sources and Applications* (IntechOpen)
[7] Demain L and Adrio J L 2008 Strain improvement for production of pharmaceuticals and other microbial metabolites by fermentation *Prog. Drug Res.* **65** 252–89
[8] Kennel Y M and Demain A L 1978 Effect of carbon sources on β-lactam antibiotic formation by Cephalosporium acremonium *Exp. Mycol.* **2** 234–8
[9] Chakravarty I, Kundu K, Ojha S and Kundu S 2017 Development of various processing strategies for new generation antibiotics using different modes of bioreactors *JSM Biotechnol. Bioeng.* **4** 1073
[10] Gernaey V, Cervera-Padrell V A and Woodley J M 2012 Development of continuous

pharmaceutical production processes supported by process systems engineering methods and tools *Future Med. Chem.* **4** 1371–4

[11] Li T, Chen X B, Chen J C, Wu Q and Chen G Q 2014 Open and continuous fermentation: products, conditions and bioprocess economy *Biotechnol. J.* **9** 1503–11

[12] Hausner D, Lustri J and Brodbeck P Continuing progress in continuous manufacturing. 1 March 2016. [Online]. Available: https://www.pharmaceuticalprocessingworld.com/continuing-progress-in-continuous-manufacturing/

[13] Liu Y, Gong G, Xie L, Yuan N, Zhu C, Zhu B and Hu Y 2010 Improvement of cephalosporin C production by recombinant DNA integration in Acremonium chrysogenum *Mol. Biotechnol.* **44** 101–9

[14] Shahidzadeh H, Labbeiki G and Attar H 2017 Enhanced fermentation production of cephalosporin C by magnetite nanoparticles in culture of Acremonium chrysogenum *IET Nanobiotechnol.* **11** 644–9

[15] Lee H, Yoo H Y, Yang X, Kim D S, Lee J H, Lee S K, Han S O and Kim S W 2017 Utilization of algal sugars and glycerol for enhanced cephalosporin C production by Acremonium chrysogenum M35 *Lett. Appl. Microbiol.* **64** 66–72

[16] Zafira I Z and Nandong J 2019 Optimal feeding strategy of cephalosporin C fermentation *IOP Conf. Ser. Mater. Sci. Eng.*

[17] Yun T Y, Feng R J, Zhou D B, Pan Y Y, Chen Y F, Wang F, Yin L Y, Zhang Y D and Xie J H 2018 Optimization of fermentation conditions through response surface methodology for enhanced antibacterial metabolite production by Streptomyces sp. 1–14 from cassava rhizosphere. PLoS One **13**

[18] Basak S, Velayudhan A and Ladisch M R 1995 Simulation of diauxic production of cephalosporin C by Cephalosporium acremonium: Lag model for fed-batch fermentation *Biotechnol. Prog.* **11** 626–31

[19] Araujo L G C, Oliveira R P, Giordano R C and Hokka C O 1996 Comparative studies on cephalosporin C production process with free and immobilized cells of Cephalosporium acremonium *Chem. Eng. Sci.* **51** 2835–40

[20] Jenkins T E 2007 Evaluating and comparing aeration blower power requirements *Proc. Water Environ. Fed.*

[21] Zhou W, Holzhauer-Rieger K, Dors M and Schügerl K 1992 Influence of dissolved oxygen concentration on the biosynthesis of cephalosporin C *Enzyme Microb. Technol.* **14** 848–54

[22] Hassanein S H, Nasr M I, El-aziz R M A, Elbaz A F and Malash M N 2009 Induction of cephalosporin C in Cephalosporium acremonium using U.V. light and selection of a mutant that showed highest fermentation capability *Minufiya J. Agric. Res.* **34** 35–55

[23] Srivastava P, Mishra P and Kundu S 2006 Process strategies for cephalosporin C fermentation J. Sci. Ind. Res. **65** 599–602

[24] Cai M, Zhang Y, Hu W, Shen W, Yu Z, Zhou W, Jiang T, Zhou X and Zhang Y 2014 Genetically shaping morphology of the filamentous fungus Aspergillus glaucus for production of antitumor polyketide aspergiolide *A. Microbial Cell Factories* **13** 73

[25] Hu Y J and Zhu B Q 2016 Study on genetic engineering of Acremonium chrysogenum, the cephalosporin C producer *Synth. Syst. Biotechnol.* **1** 143–9

[26] Jakob W and Blume C 2014 Pareto optimization or cascaded weighted sum: a comparison of concepts *Algorithms* **7** 166–85

IOP Conf. Series: Materials Science and Engineering 1195 (2021) 012038 doi:10.1088/1757-899X/1195/1/012038

Towards zero carbon dioxide concentration in sweet natural gas product from amine sweetening plant

Abdulqader Bin Sahl, Tharindu Siyambalapitiya, Ahmed Mahmoud and Jaka Sunarso

Research Centre for Sustainable Technologies, Faculty of Engineering, Computing and Science, Swinburne University of Technology, Jalan Simpang Tiga, 93350 Kuching, Sarawak, Malaysia

Email: melhadi@swinburne.edu.my

Abstract. This work presents a two-step method to reduce CO_2 concentration of sweet natural gas product from amine sweetening plant *via* amine blending (Step 1) followed by minor process modification (Step 2). In Step 1, an industrial natural gas sweetening plant was simulated using Aspen HYSYS and the simulation results were validated against the plant data. Afterwards, different blends of methyl diethanolamine and monoethanolamine (MDEA-MEA) and methyl diethanolamine and diethanolamine (MDEA-DEA) were investigated. Then the optimum amine blend of 28 wt.% MDEA and 10 wt.% MEA was reported. The optimum amine blend achieved a significant reduction in CO_2 concentration of sweet natural gas of 99.9% compared to the base case (plant data). In Step 2, two types of amine stream splits, i.e., lean amine stream split and semi-lean amine stream split were studied. The study covered split stream amount, absorber recycle stage, and regenerator stage withdrawal. Both types of stream splits attained a significant reduction in CO_2 concentration of sweet natural gas product and amine circulation rate compared to Step 1. However, the semi-lean amine stream split was superior to lean amine split with 69.1% and 63.6% reduction in CO_2 concentration of sweet natural gas and lean amine circulation rate, respectively.

1. Introduction

Natural gas has an immense impact on today's economy given its much-increased demand over the last decade [1]. In comparison to other conventional fossil fuels such as coal and oil, natural gas provides cleaner combustion as it produces 71% and 56% less carbon dioxide (CO_2) emissions, respectively [2]. Nonetheless, raw natural gas usually contains unwanted inorganic gas components that are often referred to as sour gases given their acidic characteristics [3]. Such inorganic components particularly CO_2 and hydrogen sulphide (H_2S) are to be removed from natural gas to meet the allowable limit for gas pipelines regulations, which is 2 mol.% and 4 ppm, respectively [4]. Carbon dioxide is responsible for the corrosion of transmission pipelines and other equipment including pumps, turbines, and compressors by its reaction with water vapour to form carbonic acid. Moreover, the presence of CO_2 reduces natural gas heating value [5]. Many methods have been used to remove acid gases from natural gas including absorption, adsorption, membrane separation, cryogenic distillation, and chemical conversion [6]. From these methods, absorption-based gas sweetening technology attracts considerable attention due to its high acid gas removal percentage [7]. Absorption can be achieved by means of chemical solvents, physical solvents, and their mixtures [8]. However, the commercialized absorption technology adopts aqueous alkanolamines solutions to remove the acid gas components, i.e., amine gas sweetening due to their low operating cost and flexibility of modifying the solvent composition to fit the sour gas properties [9]. Alkanolamines are organic bases formed from ammonia (NH_3) where one, two, or three alkyl groups substitute ammonia's hydrogen resulting in primary, secondary, and tertiary amines [10]. Monoethanolamine (MEA) and diethanolamine (DEA), which are examples of primary and secondary amines, respectively, have a high absorption rate due to their highly reactive behaviour whereas tertiary amine such as methyl diethanolamine (MDEA) has a relatively slower absorption rate given its lower reactivity. However, this attribute of MDEA translates to a lower regeneration energy requirement compared to DEA and MEA, with MEA having the highest regeneration requirements [11]. Thus, to achieve higher removal of acid components whilst maintaining low regeneration cost, amine blending has become an attractive option. Amine blending involves the combination of a tertiary amine (such as MDEA) with a primary or secondary amine (MEA and DEA, respectively). Other techniques to reduce stripper regeneration energy including alternative configurations like stream-split, vapour recompression, and their combinations have been recently investigated. Results approved the possibility of decreasing the stripper reboiler duty significantly with the different configurations [12]. To this end, this work aims to reduce CO_2 concentration in a sweet gas product far beyond the standard pipeline specification (2 mol.%), i.e., towards zero concentration of CO_2 in sweet gas product to maintain the sustainable operation of natural gas facilities by eliminating pipeline and equipment corrosion as well as maintaining higher heating value of natural gas. Hence, it is proposed to achieve this aim *via* employing amine blends followed by a stream splitting technique.

2. Amine Sweetening Process Description

Figure 1 shows a conventional amine sweetening process flow diagram [13]. A sour gas feed is introduced directly into the bottom of the absorption column (absorber) where it is counter-currently contacted with the CO_2-lean alkanolamines solution (lean amine), which flows into the column from the top stage of the absorber. The lean amine solution reacts with CO_2 in the feed gas on the column trays in an exothermic reaction, which causes the absorption of CO_2 into the lean amine solution. The CO_2-deprived gas (sweet gas product) exits overhead from the column while the remaining CO_2-rich amine solution (rich amine) exits the bottom of the absorber entering a flash separator, where it is throttled and flashed to remove the light hydrocarbons. The rich amine solution is then heated as it flows through a lean/rich amine heat exchanger to attain the required stripping column (stripper) temperature. The stripper receives hot rich amine flowing downwards from the top stage contacting rising steam, which is supplied by a reboiler, in a counter-current manner. The cross-counter between the rich amine and steam strips the CO_2 from the rich amine, with the remaining CO_2-rich gas leaving at the top of the column while the stripped-amine solution comes out at the bottom as lean amine. The lean amine solution's flow rate and concentration are maintained by mixing it with a makeup stream to compensate for any solvent loss throughout the process. Finally, the lean amine solution is pumped and cooled by an air cooler before being recycled back to the top of the absorber.

3. Methodology

3.1. Base case

An amine-based sweetening process for the removal of CO_2 from natural gas was simulated in Aspen HYSYS version 11 (figure 2). The sour natural gas composition, process parameters, and process design specification are adopted from a real plant located in Argentina (figure 1) [14]. The thermodynamic fluid package used for the simulation was Acid Gas package designed by Aspen Technology for amine gas sweetening process. The accuracy of the findings and the reliability of the base case simulation model are essential factors to carry out parametric studies. Thus, to verify the reliability of the model, the simulation results were validated against the plant data where the percentage error was found to be less than 10% (table 2). This minor discrepancy validates the accuracy of the simulation. It is worth

IOP Conf. Series: Materials Science and Engineering 1195 (2021) 012038 doi:10.1088/1757-899X/1195/1/012038

noting that the presented simulation results in table 2 have been accomplished at a lean amine circulation rate of 15.4 $m^3\ hr^{-1}$, reboiler duty of 0.8309 MW, and sweet gas product CO_2 concentration of 0.3418 mol.%.

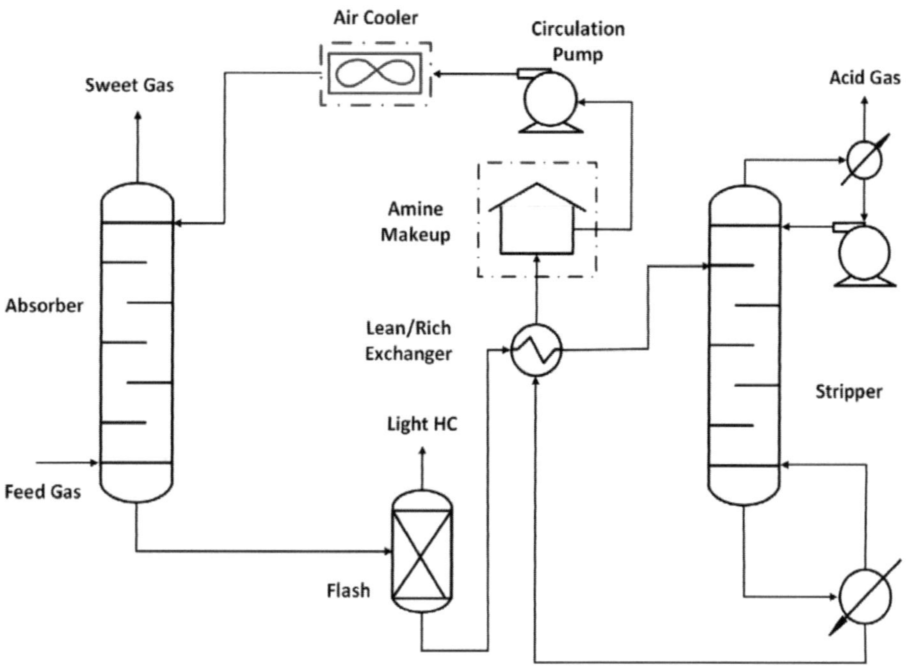

Figure 1. A basic amine sweetening process flow diagram [13].

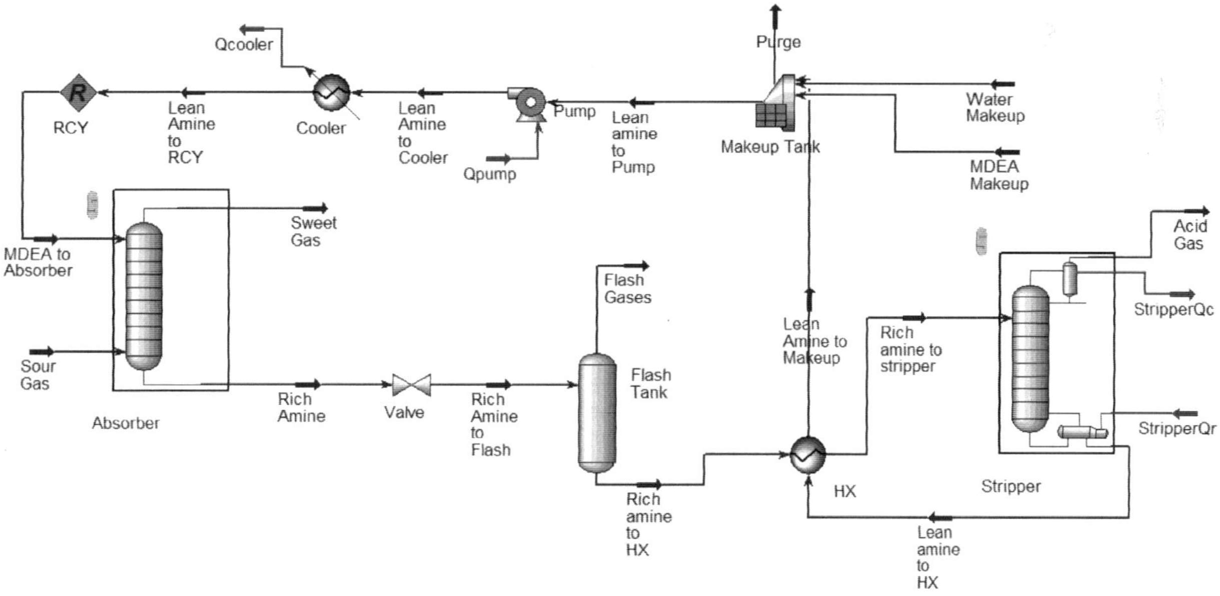

Figure 2. Base case aspen HYSYS v11 configuration (Note: HX – Heat Exchanger, RCY – Recycle).

Table 1. Input data for Aspen HYSYS v11 base case simulation.

Parameters	Operating Data	Simulation Data	Error (%)
Sweet Gas Temperature (°C)	40	41.3	3.25
Sweet Gas Pressure (kPa)	6000	6000	0
Sweet Gas Molar Flow (m^3/d)	240000	239470	0.22
Sweet Gas CO_2 (mol.%)	0.35	0.3418	2.31
Rich Amine Pressure (kPa)	6000	6000	0
Acid Gas Mole Flow (m^3/d)	9755	10320	5.79
Acid Gas Pressure (Pa)	112	112	0
Rich Amine to Exchanger Temperature (°C)	50	54.79	9.58
Stripper Bottoms Temperature	110	109.4	0.55
Stripper Bottoms Pressure (kPa)	132	132	0

Table 2. Plant data validation.

Parameter	Value
Feed Gas Temperature	40 °C
Feed Gas Pressure	6000 kPa
Feed Gas Flow	250 Mm^3/d
CO_2 in Feed Gas	4 mol.%
CH_4 in Feed Gas	93 mol.%
Lean Amine Temperature	42 °C
Lean Amine Pressure	9610 kPa
MDEA in Lean Amine	38 wt.%
Number of stages in Absorber	24
Number of stages in Stripper Tower	20
Stripper Feed Temperature	90 °C
Top of the Stripper Temperature	87.78 °C
Reflux Ratio	1.1

3.2. Parametric Analysis

In this work, the parametric analysis was carried out in a two-step method with the aim of achieving the lowest CO_2 concentration in sweet gas product at the minimum levels of lean amine circulation rate and stripper reboiler duty, which are considered as indicators for minimising operating cost. In Step 1, different blends of MDEA-MEA and MDEA-DEA were studied to find the best blending ratio, while in Step 2, two types of amine stream split configurations; lean amine split stream configuration (figure 3) and semi-lean split stream configuration (figure 4) were investigated to further improve the process performance in terms of stripper reboiler duty and CO_2 concentration in sweet gas product. In the lean amine split stream configuration (figure 3), the lean amine leaving the stripper is split into two portions where one portion is recycled back into some tray below the top of the absorber and the remaining is recycled to the top tray. In the semi-lean split stream configuration (figure 4), however, a partly regenerated (semi-lean) amine solution drawn from a certain stage in the stripper is recycled back into some tray below the top of the absorber, while the completely regenerated (ultra-lean) amine is recycled to the top. The partly regenerated amine draw from the stripper contains residual CO_2, which allows for bulk CO_2 capture in the absorption column without extra regeneration costs [12,15–17]. Apart from the base case parameters that affect the process performance, the amount recycled (lean amine split-ratio for figure 3 and stripper semi-lean side draw amount for figure 4) to the absorption column along with its feed stage-number as well as the stripper draw stage-number (for figure 4) need to be adjusted to enhance the overall performance of the plant [17].

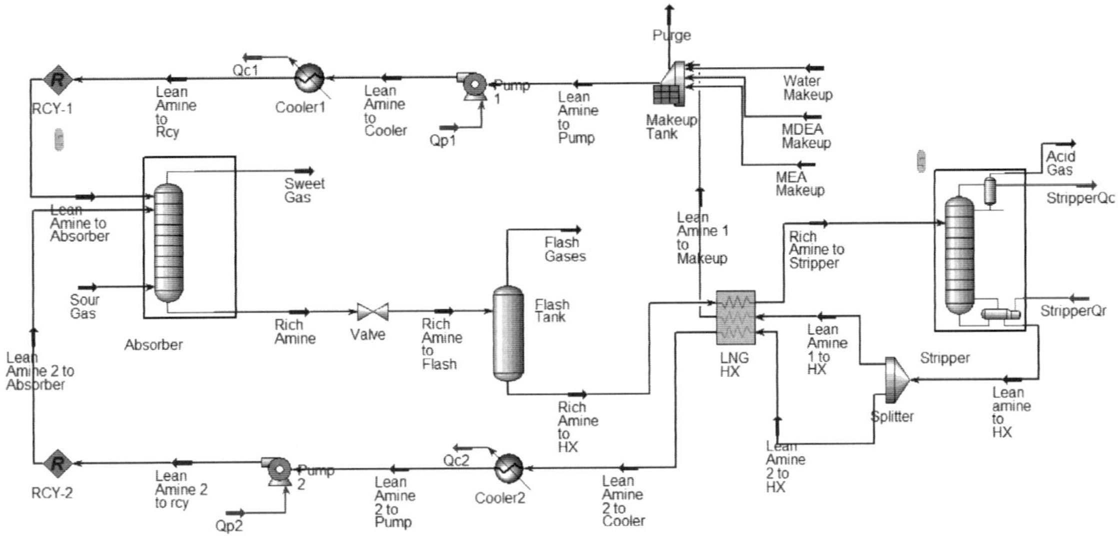

Figure 3. Lean amine split stream configuration (Note: HX – Heat Exchanger, RCY – Recycle).

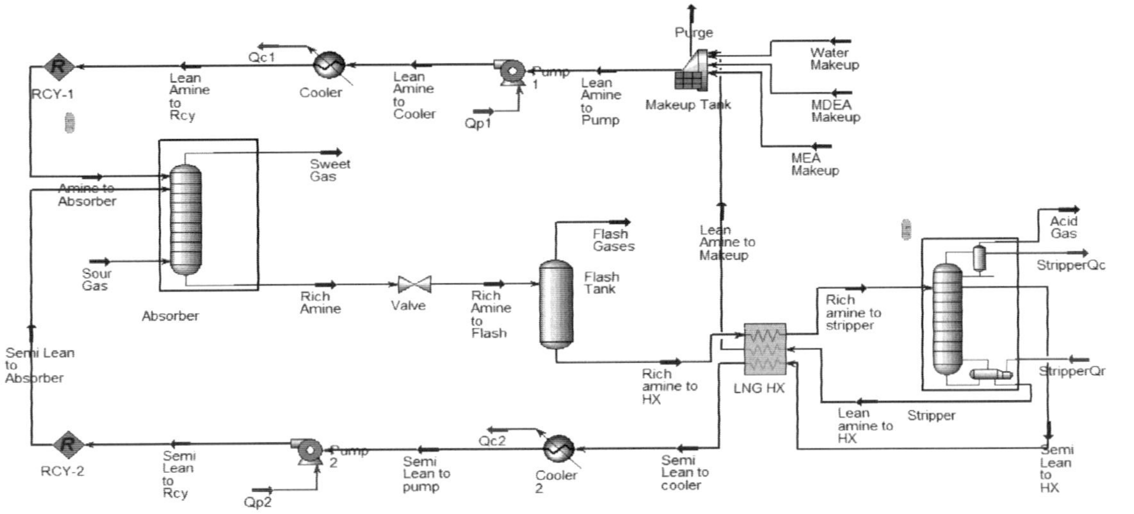

Figure 4. Semi-lean amine split stream configuration (Note: HX – Heat Exchanger, RCY – Recycle).

Figure 5 shows the parametric study framework proposed in this work. The first section focuses on selecting the best amine blend between (28 wt.% MDEA & 10 wt.% MDEA) and (28 wt.% MDEA & 10 wt.% DEA), which were benchmarked against the base case (38 wt.% MDEA). The chosen weight percentage for MEA and DEA are both 10 wt. % as the secondary amine is generally less than 25% of the total amine on a molar basis [8]. To this end, the chosen amine blend was subjected to further investigation in Section 2 to select the best blending ratio with five different ratios of secondary amine A (which is either MEA or DEA) ranging from 0 wt.% to 20 wt.% with the remaining MDEA, respectively. In Section 3, a minor process modification through two different types of split streams is carried out (lean amine and semi-lean amine split stream configuration).

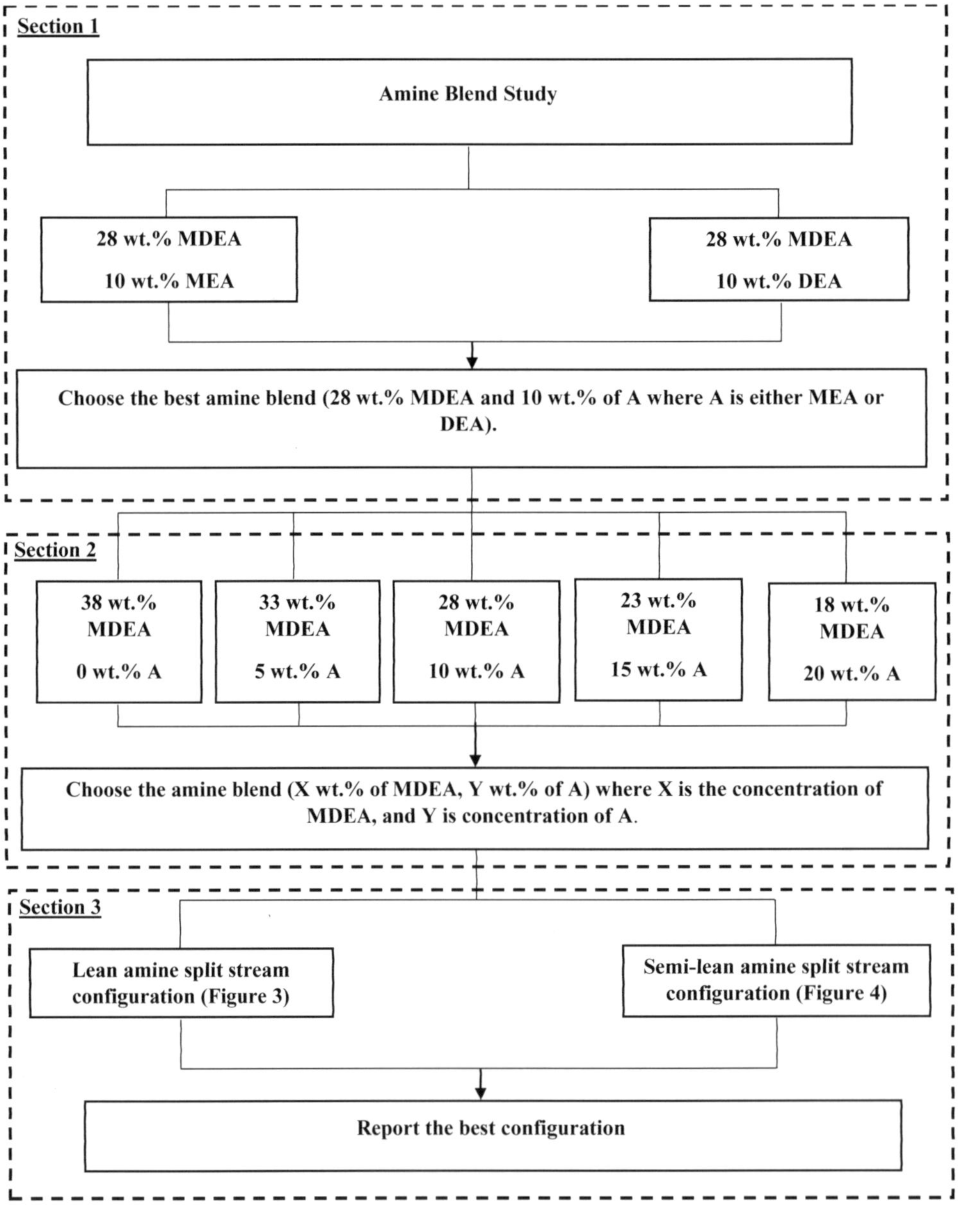

Figure 5. Parametric study framework.

IOP Conf. Series: Materials Science and Engineering 1195 (2021) 012038 doi:10.1088/1757-899X/1195/1/012038

4. Results & Discussion

The parametric study framework depicted in figure 5 is demonstrated section by section as follow.

4.1. Section 1

The effect of the lean amine circulation rate on the reboiler duty and the sweet gas product CO_2 concentration for different amine blends is demonstrated in figure 6. The trends in figure 6 indicate that as the lean amine circulation increases, the CO_2 concentration in sweet gas product decreases for all the different blends. On the other hand, the reboiler duty increases with increasing lean amine circulation rate due to the increased rich amine fed to the stripper. Figure 6 also shows a significant decrease in the sweet gas product CO_2 concentration for the MDEA-DEA and MDEA-MEA blends compared to MDEA alone. Nonetheless, MDEA provides the lowest reboiler duty compared to the amine blends given its lower heat of reaction. However, the minimum sweet gas product CO_2 concentration is achieved at 11 $m^3\ hr^{-1}$ in the case of MDEA-MEA blend followed by the 12 $m^3\ hr^{-1}$ for the MDEA-DEA blend. There is a negligible decrease in CO_2 concentration after this point for both blends, but a constant increase in reboiler duty. Thus, MDEA-MEA blend is viewed as the best amine blend and will be subjected to further analysis to determine the best blending ratio that leads to better process performance.

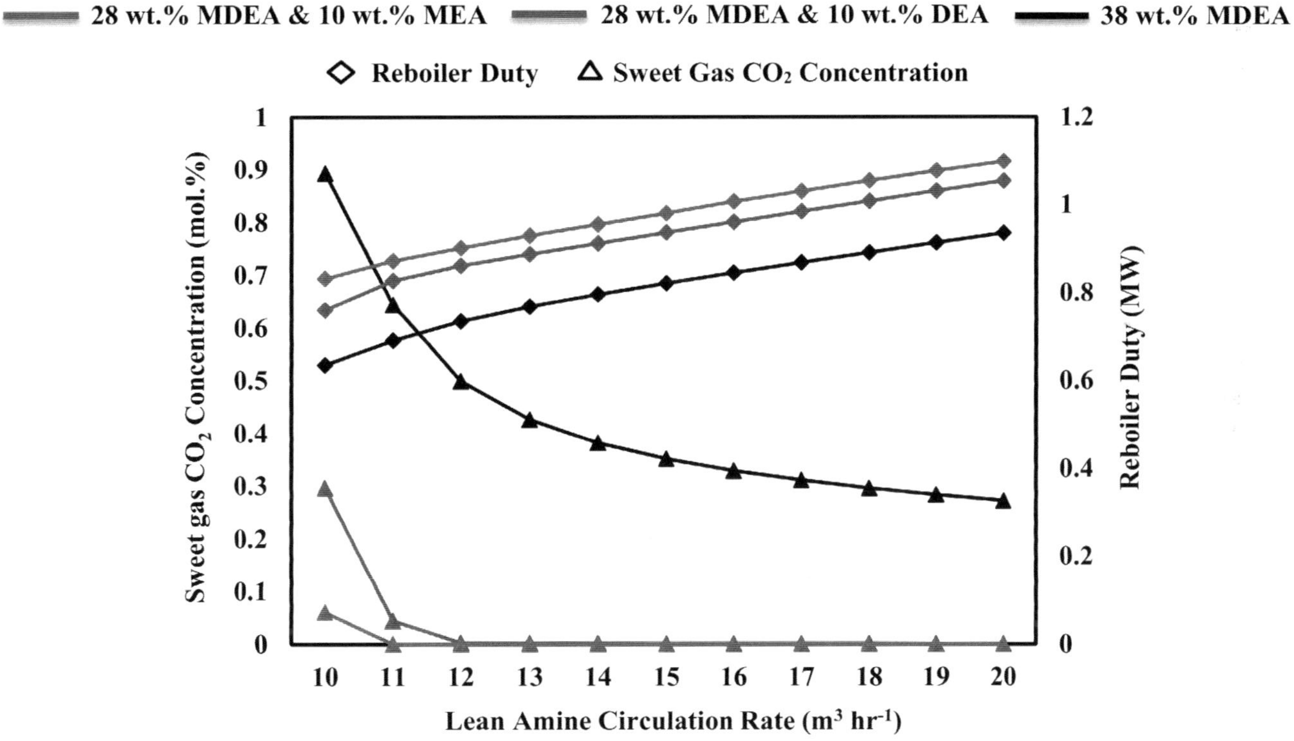

Figure 6. Effect of lean amine circulation rate on sweet gas product CO_2 concentration and reboiler duty at different amine blends.

4.2. Section 2

Figure 7 and 8 show the effect of the lean amine circulation rate on the reboiler duty and the sweet gas product CO_2 concentration for different MDEA-MEA blend ratios. The sweet gas product CO_2 concentration decreases with increasing lean amine's circulation rate (figure 7). However, the lowest sweet product CO_2 gas concentration is achieved at blended amine of 28 wt.% MDEA and 10 wt.% MEA. Figure 8, on the other hand, reveals a different trend as sweet product CO_2 gas concentration increases with the lean amine circulation rate due to the mass transfer limitations that occur at MEA ratios of 15 wt. % and higher. As expected, the reboiler duty is shown to increase as the lean amine circulation rate increases for amine blends (figure 7 and figure 8). Out of the 5 MDEA-MEA blends presented in figure 7 and 8, the 28 wt.% MDEA and 10 wt.% MEA achieved the lowest sweet gas product CO_2 concentration at a lean amine circulation rate of 11 m^3 hr^{-1}. However, increasing lean amine circulation rate beyond 11 m^3 hr^{-1} provides a negligible improvement in sweet gas product CO_2 concentration while adversely increases the reboiler duty.

4.3. Section 3

The effect of different split lean amine ratio on the sweet gas product CO_2 concentration and the reboiler duty is depicted in figure 9. The sweet gas product CO_2 concentration decreases with increasing split lean amine ratio due to extra lean amine flow to some trays below the top of the absorber, allowing for better mass transfer between the amine and the contacting sour gas. This enhances the absorption of CO_2, which is in agreement with semi-lean amine technique that picks up CO_2 when fed into some tray below the top of the absorber [15, 16]. As anticipated, the reboiler duty increases as the split lean amine ratio is increased because of a greater regeneration requirement to account for a larger CO_2 loading. The best split lean amine ratio is 0.5 as any further increase provides no significant improvement in the sweet gas product CO_2 concentration but increases the reboiler duty (figure 9).

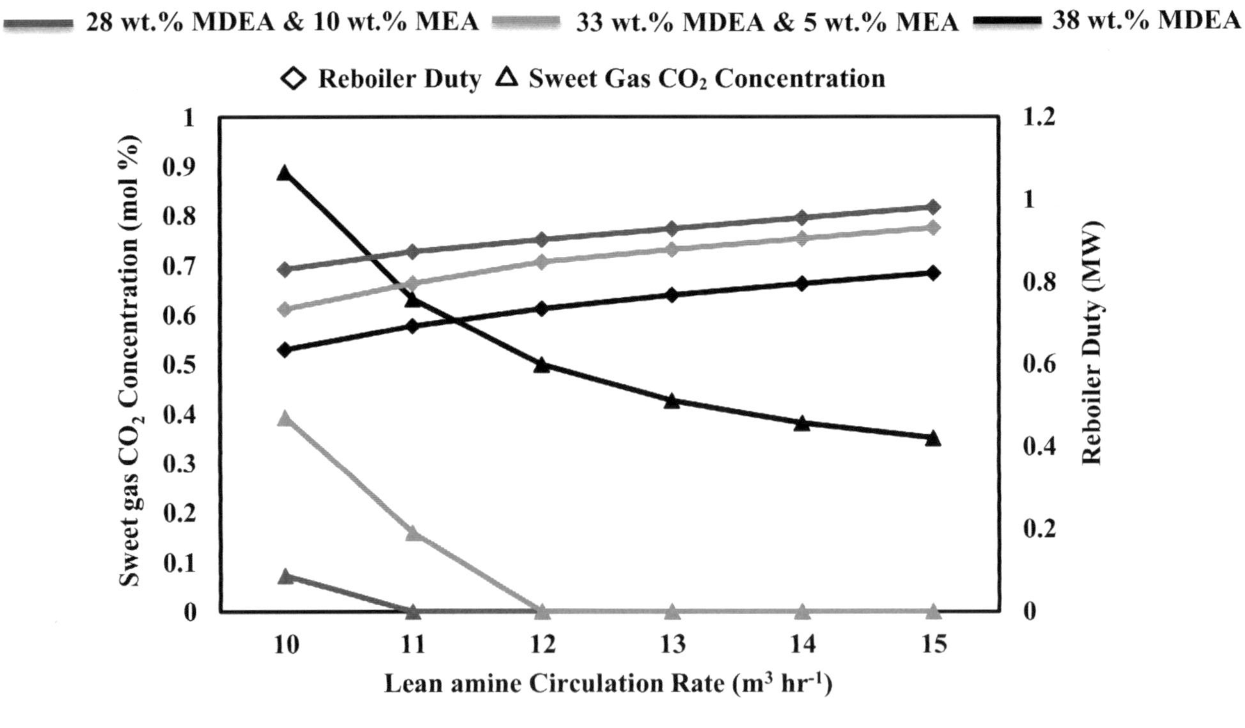

Figure 7. The effect of lean amine circulation rate on sweet gas product CO_2 concentration and the reboiler duty at different MEA blend ratios.

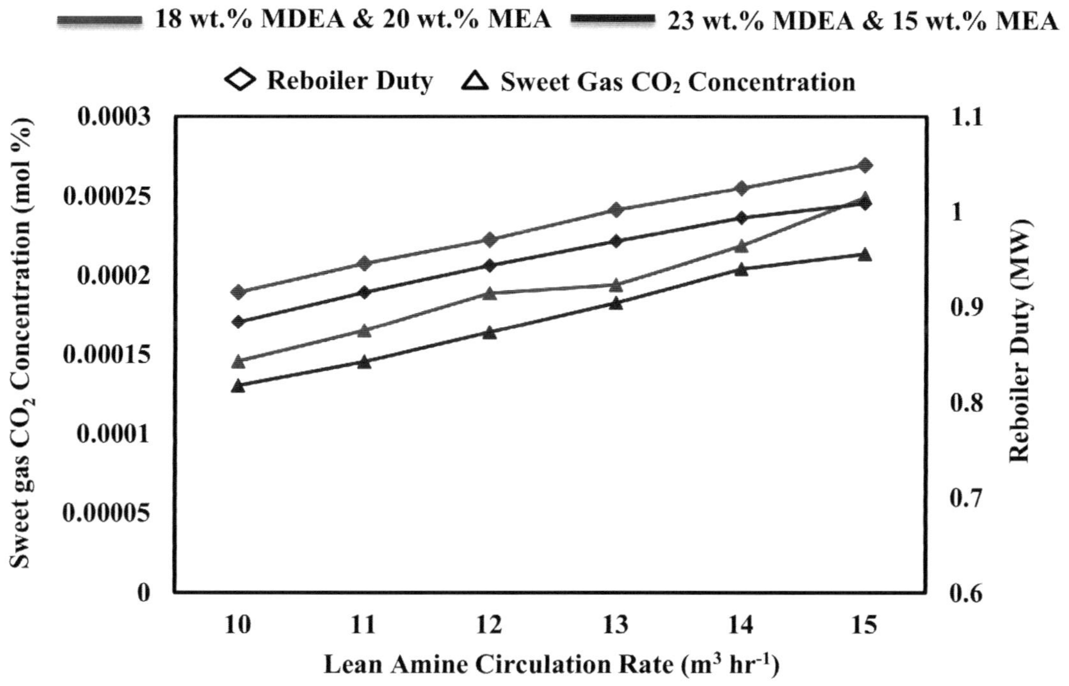

Figure 8. The effect of lean amine circulation rate on sweet gas product CO_2 concentration and the reboiler duty at different MEA blend ratios (2).

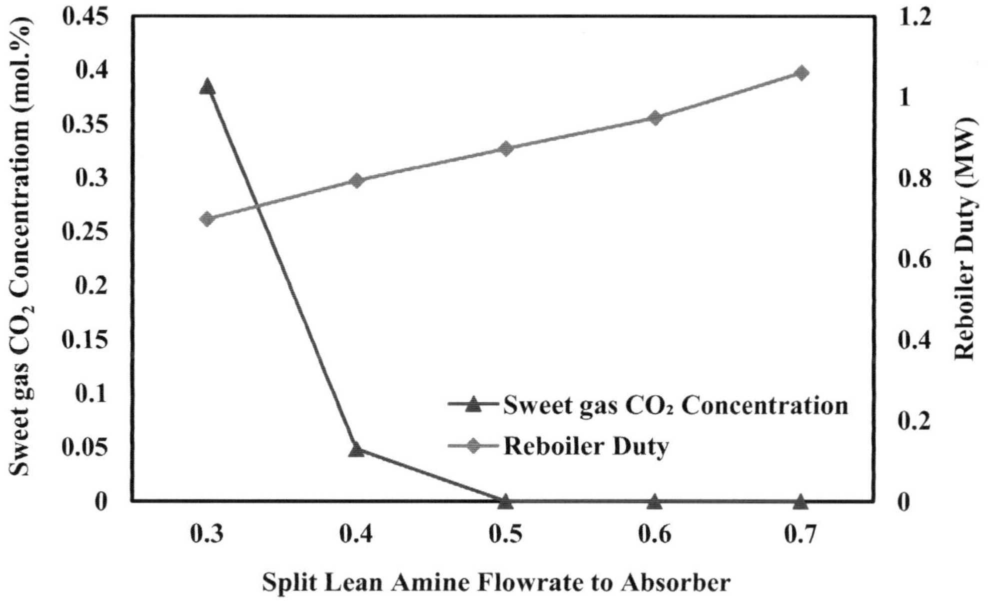

Figure 9. The effect of different split lean amine ratio on the sweet gas product CO_2 concentration and the reboiler duty.

Figure 10 shows the effect of the absorber feed stage-number for the split lean amine on the sweet gas product CO_2 concentration and the reboiler duty. The results show an increase in sweet gas product CO_2 concentration with an increase in feed stage-number (from top to bottom). At the same time, the reboiler duty tends to decrease between stage 6 and 10, then increases between stage 10 and 14 followed by another slight decrease in stage 16. Stage 10 is chosen as the best split lean amine feed stage for its minimum reboiler duty compared to the other stages. Figure 11 shows the effect of the semi-lean amine molar flow rate fed to the absorber on the sweet gas product CO_2 concentration and reboiler duty. The results display an overall decrease in the sweet gas product CO_2 concentration as the semi-lean amine flow rate increases, facing a significant decrease between 630 kgmol hr^{-1} and 650 kgmol hr^{-1}, which gives the minimum concentration. On the other hand, the reboiler duty is shown to increase linearly as the semi-lean flow rate is increased. Therefore the best semi-lean amine flow rate is 650 kgmol hr^{-1} drawn from stage 10 (from top to bottom) of the stripper (figure 11). Figure 12 shows the effect of the absorber feed stage-number for semi-lean amine on the sweet gas product CO_2 concentration and the reboiler duty. The sweet gas product CO_2 concentration increases with the number of absorber semi-lean amine feed stages (from top to bottom) up to stage 16, where it faces a significant decrease till its minimum at stage 19 with a slight increase at stage 20. It appears that from stages 16 to 19, the semi-lean is enabled for efficient contact with sour gas improving absorption performance, while it is limited after stage 19. As for the reboiler duty, it shows a minimal increase as the number of semi-lean absorber feed stage is increased. Therefore, the best absorber stage for semi-lean feed is stage 19, which achieves the best sweet gas product CO_2 concentration.

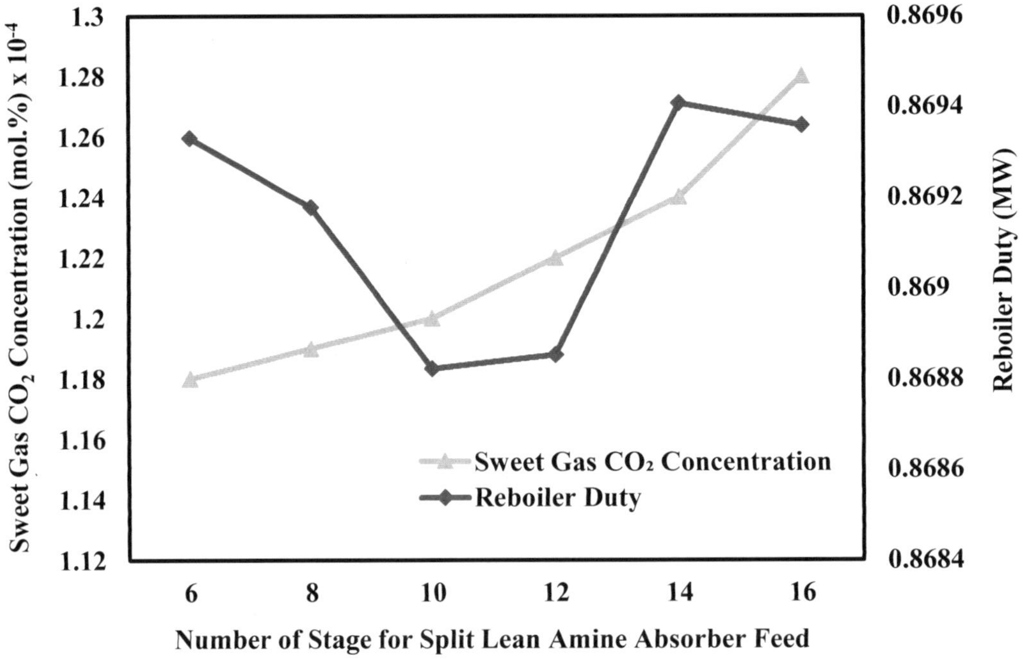

Figure 10. Effect of the number of stages for split lean amine absorber feed on the sweet gas product CO_2 concentration and reboiler duty.

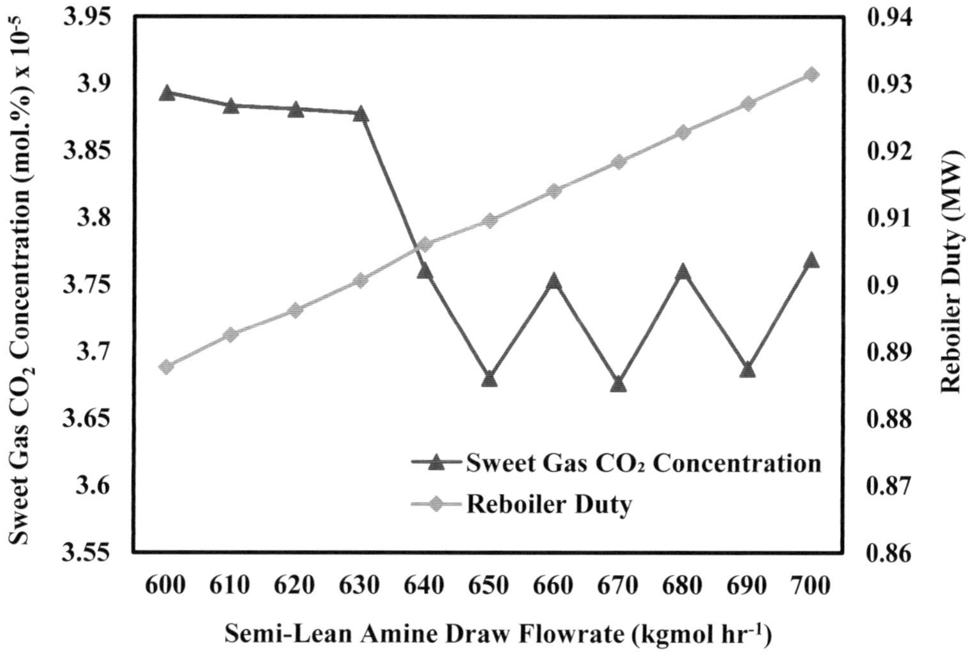

Figure 11. The effect of the semi-lean amine molar flow rate fed to the absorber on the sweet gas product CO_2 concentration and reboiler duty.

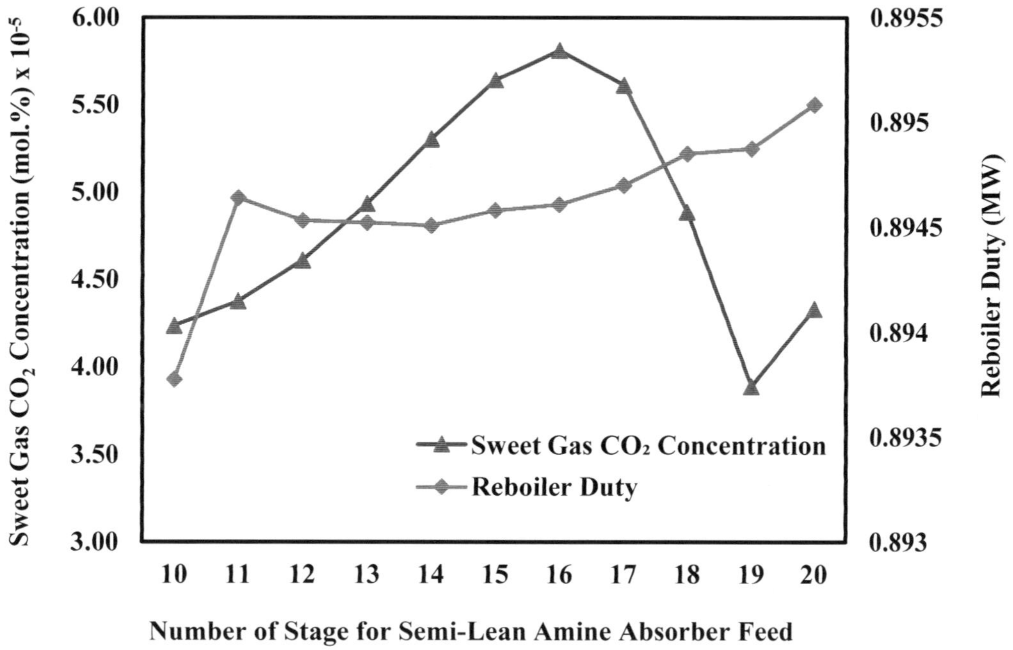

Figure 12. The effect of the absorber feed stage-number for semi-lean amine on the sweet gas product CO_2 concentration and the reboiler duty.

Figure 13 shows the effect of the stage-number of semi-lean amine drawn from the stripper on the sweet gas product CO_2 concentration and reboiler duty. The sweet gas product CO_2 concentration decreases with stage-number of semi-lean amine drawn from the stripper until stage 10 (from top to bottom) followed by significant increase in the sweet gas product CO_2 concentration after stage 10. In contrast, the reboiler duty increases with the stage-number of semi-lean amine drawn from the stripper (from top to bottom). Thus, the stage-number 10 from the top of the stripper is the best stage as it possesses the minimum sweet gas product CO_2 concentration.

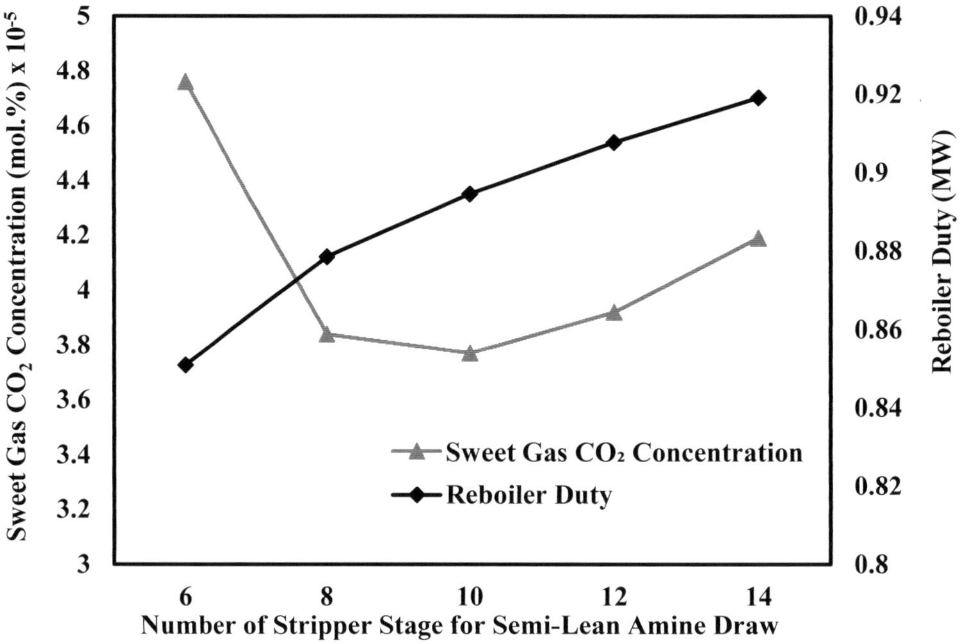

Figure 13. The effect of the stage-number of semi-lean amine drawn from the stripper on the sweet gas product CO_2 concentration and reboiler duty.

Table 3 summarises the overall results of the proposed methodology, which comprises two steps; Step 1 (amine blending) followed by Step 2 (lean/semi-lean amine split). The results reveal that both steps of the proposed methodology achieved significant reduction in sweet gas CO_2 concentration and amine circulation rate at the expense of increase in the reboiler duty compared to the base case. Thus, since both the lean amine split and the semi-lean amine split configurations are combined with Step 1's outcomes (choosing the best amine blend type and ratio), it is obvious that the semi-lean amine split configuration is the best option for its lower CO_2 sweet gas product concentration and lean amine circulation rate compared to lean amine split configuration.

Table 3. Overall results summary.

Process configuration	Lean amine circulation rate ($m^3\ hr^{-1}$)	Sweet gas CO_2 conc. (mol.%)	Reboiler duty (MW)
Base case	15.4	0.3418	0.8309
Step 1	11	1.26×10^{-4}	0.8730
Step 2 (Lean amine split)	5.5	1.204×10^{-4}	0.8692
Step 2 (Semi-lean amine split)	2	3.721×10^{-5}	0.9093

5. Conclusion

In this work, a two-step method consist of amine blending (Step 1) followed by minor process modification (Step 2) has been presented. In Step 1, different blends of MDEA-MEA and MDEA-DEA were examined and the best amine blend of 28 wt.% MDEA and 10 wt.% MEA was selected. The best amine blend achieved a sweet gas CO_2 concentration of 1.26×10^{-4} mol.% with a lean amine circulation rate of 11 $m^3\ hr^{-1}$ and a reboiler duty of 0.8730 MW. This translates into a significant reduction of 99.9% in CO_2 concentration of sweet natural gas compared to the base case (plant data). In Step 2, two types of amine stream split, namely lean amine stream split and semi-lean amine stream split were considered. Several process parameters including the split stream's ratio and amount, absorber's recycle stage-number and stripper's draw stage-number were studied. Compared to the lean amine configuration, the semi-lean amine configuration achieved a lower CO_2 concentration of sweet natural gas and a lean amine circulation rate of 69.1% and 63.6%, respectively. The application of semi-lean amine technique can be extended to Benfield process to investigate the effect of recycling semi-lean potassium carbonate on the performance of natural gas sweetening process.

References

[1] Abdulrahman R K and Sebastine I M 2013 Natural gas sweetening process simulation optimization: A case study of khurmala field in Iraqi Kurdistan region *J. Nat. Gas Sci. Eng.* **14** 116–20

[2] Zahid U 2020 Simulation of an acid gas removal unit using a DGA and MDEA blend instead of a single amine *Chem. Prod. Process Model.* **15** 20190044

[3] Mokhatab S, Poe W and Mak J 2018 *Handbook of Natural Gas Transmission and Processing*, 4th ed

[4] Stewart M and Arnold K 2011 *Gas Sweetening and Processing Field Manual* (Amsterdam: Gulf Professional)

[5] El-Maghraby R, Salah A and Shoaib A 2019 Carbon dioxide capturing from natural gas using di-glycol amine and piperazine — A new solvent mixture *Int. J. Recent Technol. Eng.* **8** 11378–83.

[6] Ngu L, Mahmoud A and Sunarso J 2020 Aspen Plus simulation-based parametric study of benfield process using hot potassium carbonate promoted by diethanolamine *IOP Conf. Ser.:Mater. Sci. Eng.* **778** 012058

[7] Shoukat U and Knuutila H 2020 Effect of various parameters on the thermal stability and corrosion of CO_2-loaded tertiary amine blends *Energies* **13** 2626

[8] Yusuf M 2017 Anisotropic selection of amine in natural gas sweetening process for acid gases removal: A review of recent studies *Pet. Petro. Chem. Eng. J.* **1** 000114

[9] Abbas T, Ghauri M, Shahid Z and Rashid M 2011 Dynamic simulation of sweetening process of natural gas *Can. J. Chem. Eng. Technol.* **2** 156–61

[10] Gutierrez J, Benitez L, Ale Ruiz E and Erdmann E 2016 A sensitivity analysis and a comparison of two simulators performance for the process of natural gas sweetening *J. Nat. Gas Sci. Eng.* **31** 800–7

[11] Reza M, Hamlehdar O, Boor H, Monnavar A and Rostami S 2011 Mixed amines application in

gas sweetening plants *Chem. Eng. Trans.* **24** 265–70
[12] Øi L, Bråthen T, Berg C, Brekne S, Flatin M, Johnsen R, Moen I and Thomassen E 2014 Optimization of configurations for amine based CO_2 absorption using aspen hysys *Energy Procedia* **51** 224–33
[13] Khalaf A 2020 Influence of using different mixtures of amines on the performance of natural gas sweetening process at Iraqi north gas company *Pet. Coal* **62** 735–42
[14] Gutierrez J, Ale Ruiz E and Erdmann E 2017 Energy requirements, GHG emissions and investment costs in natural gas sweetening processes *J. Nat. Gas Sci. Eng.* **38** 187–94
[15] Bae H, Kim S and Lee B 2011 Simulation of CO_2 removal in a split-flow gas sweetening process *Korean J. Chem. Eng.* **28** 643–48
[16] Polasek J, Bullin J and Donnelly S T 2006 Alternative flow schemes to reduce capital and operating costs of amine sweetening units *Bryan Research and Engineering, Inc.*
[17] Al-Lagtah N, Al-Habsi S and Onaizi S 2015 Optimization and performance improvement of Lekhwair natural gas sweetening plant using aspen hysys *J. Nat. Gas Sci. Eng.* **26** 367–81

Fractionation of phycocyanin and carbohydrate from *Spirulina platensis* using ionic liquid-based aqueous two-phase system

Y H Choi, S K Mah, Y S Ng and S Y Lee

Department of Chemical Engineering, Lee Kong Chian Faculty of Engineering and Science, University Tunku Abdul Rahman, 43000, Kajang, Selangor, Malaysia

Email: choiyuhui@gmail.com

Abstract. Microalgae have gained considerable attention due to their high-value biomolecules and potential applications in the pharmaceutical, food, medical and cosmeceutical field. However, the conventional biorefinery process of microalgae are costly, energy-intensive, and time-consuming. Aqueous two-phase system (ATPS) has emerged as a potential technique for the separation and fractionation of biomolecules in the biorefinery field. This study proposed the application of ionic liquid (IL) based ATPS for the fractionation of high-value phycocyanin and carbohydrates from a crude extract of *Spirulina platensis*. The biomass was first lysed by a high-speed homogenization with a solid to liquid ratio (S/L) of 1:10 and an operation time of 5 min to achieve optimum yields of multiple products. Next, the feasibility of several IL-based ATPS with phosphate/citrate buffer was evaluated by investigating the effects of cation, anion, and alkyl chain's length of ILs. Among the IL-based ATPS, the system comprising of 1-butyl-1-methylpyrrolidinium dicyanamide and citrate buffer demonstrated the optimum phycocyanin extraction efficiency of 83.26 $\pm$ 0.05% at the top phase and a concurrent recovery of 73.89$\pm$0.06% carbohydrate at the bottom phase. In this work, the IL-based ATPS performed better than conventional polymer-based ATPS. This work suggests that IL-based ATPS allows efficient fractionation of phycocyanin and carbohydrates.

1. Introduction

Recently, microalgae have been gaining substantial attention as a potential source of high-value biochemical compounds owing to their capability to accumulate proteins, antioxidants, antibiotics, vitamins, pigments, carbohydrates and lipids [1]. The blue-green microalgae *Spirulina* belongs to the Arthrospira genus and can be classified into two common species, namely, *Spirulina platensis* and *Spirulina maxima*. They have been marketed as healthy food ingredients and supplements in capsules, tablets, powders, and liquids form due to their high nutritional value. *Spirulina sp.* is a multi-filamentous photosynthetic cyanobacterium that is intensively investigated in the fields of medicine and food science due to its wide market potentials [2–3]. It is rich in proteins, vitamins, essential fatty acids, polysaccharides and pigments such as chlorophylls, β-carotene, xanthophylls and phycobiliproteins [3].

Phycobiliproteins are photosynthetic pigments that are accumulated in the cells which are attached to the thylakoid membranes of chloroplasts. They are categorized into three dominant classes which are phycocyanin, phycoerythrin and allophycocyanin [4–5]. *Spirulina sp.* is found to potentially produce high content of phycocyanin with a small quantity of allophycocyanin [6]. Phycocyanin is a water-soluble, non-toxic and blue-color pigment-protein complex with plenty of promising bioactivities. The phycocyanin is comprised of α and β subunits with the molecular weights of 140 to 210 kDa. The number of α and β subunits that exist in the phycocyanin depends on the sources of species [7]. It is widely utilized as natural colour additives in food and cosmetic products, fluorescent markers, as well as probes and tracers in medical diagnosis, immunology, biological engineering, and other research fields. Moreover, a few studies reported the use of phycocyanin in pharmaceutical and medical applications, which serves as stimulants of the immune system, antioxidant, anti-inflammatory, anti-tumor, prevention, and treatment of some neurological disorders and neurodegenerative diseases such as Alzheimer's and Parkinson's [8]. Other than phycocyanin, S*pirulina sp.* also composed of 8 to 14% of carbohydrate on dry weight basis. However, the carbohydrate content in spirulina can be increased by varying the culture medium during the cultivation of microalgae. Carbohydrates in microalgae can be converted into biofuel such as bioethanol *via* fermentation processes [9–11].

Typically, biorefinery processes involve a series of intensive steps such as cell disruption, product extraction and purification. Cell disruption methods such as freezing and thawing, mechanical method, and sonication, followed by the purification approaches such as ammonium sulfate precipitation, ultrafiltration and chromatographic techniques (i.e., ion exchange, hydrophobic interaction and gel filtration) are generally used for the recovery of phycocyanin from biomass. However, the biorefinery of microalgae encounters challenges such as energy-intensive, time-consuming, costly processes and the difficulty of scaling up in the industry [12]. In addition to that, degradation of purity and loss of targeted products often occur in these tedious processing steps. Despite the high-value of microalgae as biochemical sources, the economic feasibility of production of phycocyanin derived from *Spirulina sp.* is still limited. In the interest of sustainability and cost competition in extraction and purification of phycocyanin from *Spirulina sp.*, there is an urgent need to exploit a reliable, mild and scalable biorefinery process to eliminate the energy-intensive multiprocessing steps. At this juncture, a product recovery platform which is environmentally beneficial and effective in term of high yield or purity is preferable.

Aqueous two-phase system (ATPS) has been explored as an alternative to isolate and fractionate the high-value compounds from microalgae [11,13]. ATPS has benefits over conventional methods such as short processing time, energy efficient as well as easier operation and scaling up in industry. Nevertheless, several authors reported that the conventional ATPS consisting of polymer and salt was constrained by the limited polarity range between two phases and resulted in unfavourable selectivity of desired biomolecules. In contrast, ionic liquid (IL) has emerged as a promising phase former to replace conventional polymer in ATPS. The tunable properties of IL can establish an ATPS with higher flexibility in designing the polarity range between two phases by pairing with appropriate cation and anion. IL-based ATPS (IL-TPS) can provide a mild environment and achieve higher selectivity which reduce the intensive purification steps, thus minimizing the degradation of phycocyanin purity. To date, only a few reports on employment of IL-ATPS in extraction and fractionation or purification of high-value compounds from microalgae are available [2,14]. Santos *et al.* [15] and Suarez Ruiz *et al.* [16] investigated the selective and mild separation of protein and carbohydrate from the crude extract of microalgae. These findings showed that the migration of protein to the top phase while carbohydrates to the bottom phase, proving the successful application of IL-ATPS in fractionation of compounds from crude extract of microalgae.

To the best of the authors' knowledge, numerous studies have reported the use of conventional polymer-based ATPS to fractionate and purify phycocyanin. However, only one study reported the use of IL containing bromide anion as phase forming components for the isolation of phycocyanin from *Spirulina sp.*. Aiming at validating the concept of multiproduct biorefinery, this study implements various combination of ILs and inorganic salts to investigate the feasibility of IL-ATPS for fractionation of phycocyanin and carbohydrate as well as phycocyanin purification from *Spirulina sp.* crude extract. The effects of type of salts, ILs cation alkyl chain length and different ILs anions on the biomolecule fractionation and purification of phycocyanin were discussed. Prior to fractionation and purification, biomolecules extraction operating conditions such as solid biomass to liquid ratio and the homogenization time were evaluated to attain optimum biomolecules yield and phycocyanin purity ratio

Lastly, the result of IL-based ATPS with best performance was compared with conventional polymer-based ATPS.

2. Methodology

2.1. Materials

Spray-dried *Spirulina sp.* was obtained from the Algae International Sdn. Bhd. located in Negeri Sembilan, Malaysia. ILs used in this work including 1-ethyl-3-methylimidazolium bromide, $[C_2min]Br$ (purity=99%), 1-butyl-3-methylimidazolium bromide, $[C_4min]Br$ (purity=99%), 1-ethyl-3-methylimidazolium dicyanamide, $[C_2min]DCN$ (purity>98%), 1-butyl-3-methylimidazolium dicyanamide, $[C_4min]DCN$ (purity>98%), 1-ethyl-3-methylimidazolium thiocyanate, $[C_2min]SCN$ (purity>98%), and 1-butyl-1-methylpyrrolidinium dicyanamide, $[C_4mpy]DCN$ (purity>98%) were purchased from Sigma-Aldrich. Potassium citrate tribasic monohydrate, $C_6H_5O_7K_3$ (purity >99%), citric acid, $C_6H_8O_7$ (purity>99%), potassium phosphate dibasic, K_2HPO_4 (purity >99%) and monopotassium phosphate, KH_2PO_4 (purity>99%), polyethylene glycol 4000 (PEG4000), Brilliant Blue G-250, ethanol (purity >95%), phosphoric acid, H_3PO_4 (purity=85wt%), sulfuric acid, H_2SO_4 (purity>99%), glucose standard, bovine serum albumin (BSA), acetonitrile, ACN (HPLC grade), trifluoroacetic acid, TFA, phycocyanin from *Spirulina sp.* were also procured from Sigma Aldrich. Bicinchoninic acid assay (BCA) test kit was sourced from Thermo Fisher Scientific.

2.2. Preparation of crude extract from microalgae

Extraction of biomolecules was performed via homogenization method as demonstrated in previous literature with modifications [17,18]. Spray-dried *Spirulina sp.* powder was mechanically disrupted using ultra-turrax homogenizer (IKA dispersion S18N-19G) at 12000 rpm in the presence of distilled water as extraction solvent. The experiment was conducted in a water bath to maintain the temperature of crude extract at 25±1 °C under dark condition since phycocyanin is photosensitive and thermo-sensitive biomolecules. The resulting supernatant was centrifuged at 10000 rpm for 15 min at 4 °C to remove cell debris. The operating parameters, which were solid biomass to liquid (S:L) ratio and the homogenization time were investigated using one factor at-a-time (OFAT) method. The S:L ratio varied at 0.200, 0.100, 0.067 0.050, 0.040 g/ml while the homogenization time was manipulated from 0 to 20 min by collecting the samples at the interval time of 5 min. The concentration of phycocyanin, total protein and carbohydrate were analysed and the supernatant was stored at 4 °C for further use.

2.3. Morphology and optical analysis of Spirulina sp.

The dried cell debris from homogenization and *Spirulina sp.* powder were immersed in 1% glutaraldehyde phosphate buffer solution for 3 h. Then, dehydration of samples was conducted using a series of ethanol solutions with concentrations of 10, 30, 50, 70, 90 and 100% for 5 min for each concentration and the resulting samples were dried under room condition [19]. The samples were analysed using Scanning electron microscopy (SEM) and microscope to study the effect of homogenization on the optical and morphology changes of samples.

2.4. Construction of phase diagram

Phase diagram comprising of different ILs and salts buffer such as phosphate buffer and citrate buffer were constructed using cloud point method at room temperature and atmospheric pressure. The aqueous solutions of ILs and salts with known concentration were prepared gravimetrically ($\pm 10^{-4}$ g) and were used for the determination of saturation curves. Salt buffer solution was added dropwise into the IL solution under constant stirring and the addition of salt solution was stopped once the solution turned cloudy, which indicated the formation of biphasic region. To detect the formation of monophasic regime, distilled water was added until the solution turned clear. The weight of salt solution and distilled water added were measured and recorded. The weight composition of ILs and salts in ATPS were calculated and binodal curves of IL-ATPS were plotted [20]. The experimental steps were repeated to attain a number of cloud points that enable the formation of binodal curve which correlates to Merchuk equation as in equation (1):

$$w_T = A\exp\left[B\, w_B^{0.5} - C w_B^{3}\right] \quad (1)$$

where w refer to weight fraction and subscript T and B represent top phase (ILs) and bottom phase (salts) while A, B, C are constant of Merchuk parameter [21]. The mixture points with volume ratio of 1 were selected based on the plotted phase diagram and to be used in further ATPS step.

2.5. *Fractionation of biomolecules using ATPS*

Phosphate buffer consisting of K_2HPO_4 and KH_2PO_4 and citrate buffer consisting of $C_6H_5O_7K_3$ and $C_6H_8O_7$ were prepared at pH 7 and pH 6, respectively, by mixing the salts at appropriate ratios. Different combinations of ATPS comprising of ILs, namely[C_2min]Br, [C_4min]Br, [C_2min]DCN, [C_4min]DCN, [C_2min]SCN and [C_4mpy]DCN, and the salt buffers were prepared gravimetrically by adding an appropriate amount of IL, salts and crude extract into the centrifuge tube. Distilled water was added into the mixture to bring the final weight of the system to 1 g. To allow the contact between phase components and crude extract, the mixture was stirred gently in a vortex mixer and was centrifuged at 2000 rpm for 5 min to facilitate the phases separation. The top and bottom phases were separated carefully using a micropipette. To compare IL-based ATPS with conventional polymer-based ATPS, a system composing of PEG4000 and salt buffer were also prepared. To avoid interference, the reference system for each trial was prepared without the addition of crude extract. The volume of the top and bottom phase were measured while phycocyanin and carbohydrate presence at the top and bottom phase were quantified [22]. For all systems, the volume ratio was fixed at 1 ± 0.1.

2.6. *Spectrophotometric estimation of phycobiliproteins*

The concentrations of phycocyanin in crude extract, top phase and bottom phase were determined by measuring the absorbance spectrum at 620 and 652 nm using UV-vis spectrophotometer in accordance with the equation (2) [23].

$$C_{PC}\left(\frac{mg}{mL}\right) = \frac{[(A620 - 0.70\,(A652))]}{7.38} \quad (2)$$

The corrective equations developed by were implemented in the cases which exhibited the presence of chlorophyll as shown in equation (3) and (4) [24].

$$A620 = 1.012(A620) - 0.215\,(A675) \quad (3)$$

$$A652 = 1.038(A652) - 0.256(\,A675) \quad (4)$$

where C_{PC} represents the phycocyanin concentration, A620 is the peak absorbance of phycocyanin, A652 is the maximum absorbance of allophycocyanin and A675 indicates the presence of chlorophyll. The purity ratio of phycocyanin, P was calculated using equation (5):

$$P = \frac{C_{PC}}{C_P} \quad (5)$$

where C_P is the concentration of total protein and C_{PC}.

2.7. *Total protein content determination*

Total protein content was quantified by BCA method. BSA was used as protein standard for plotting of calibration curve. In a 96-well microplate, 10 µL of sample was mixed with 200 µL of BCA working reagent. The microplate was shaken for 30 s and incubated for 15 min at 37 °C. After the incubation period, the mixture in the microplate was cooled for 5 min and their absorbance was measured at 562 nm using a microplate reader and the total protein concentration in crude extract, top phase and bottom phase were determined [15].

2.8. *Total carbohydrates content determination*

Total carbohydrate contents were analysed by phenol sulfuric acid method adapted from Santos et al [15]. Firstly, 80 μL of sample, 150 μL of 5% (w/v) of an aqueous phenol solution and 1 mL of concentrated H_2SO_4 were pipetted into a long test tube. The test tube was stirred under dark environment for 10 min in 100 ºC water bath. Quantification was carried out by measuring the absorbance of the yellow-orange colour sample at 490 nm using the spectrophotometer. A calibration curve (0.05 – 0.3 mg/mL) was established using glucose as the predominant sugar standard in the *Spirulina sp.*.

2.9. *Calculation and statistical analysis*

Extraction efficiency, EE of each biomolecule at the top and bottom phase was evaluated based on equation (6) and (7) [15].

$$EE_{i,T} = \frac{C_{i,T}V_T}{C_oV_o} \times 100\% \tag{6}$$

$$EE_{i,B} = \frac{C_{i,B}V_B}{C_oV_o} \times 100\% \tag{7}$$

where, $C_{i,T}$ and $C_{i,B}$ represent the concentration of biomolecules while subscript i represents phycocyanin, total proteins and carbohydrates at the top and bottom phase, respectively. The purification factor, P_f of each ATPS system was calculated using equation (8)

$$P_f = \frac{P_A}{P_B} \tag{8}$$

where P_A denotes phycocyanin purity ratio after ATPS and P_B represents the phycocyanin purity ratio before ATPS. The volume ratio of the top and bottom phases can be calculated based on equation (9):

$$V_R = \frac{V_T}{V_B} \tag{9}$$

where V_R indicates the volume ratio, V_T is the volume of top phase while V_B is the volume of bottom phase. Triplicate independent runs were conducted for each extraction parameters study and fractionation runs. The mean values and standard deviation were evaluated.

3. Results

3.1. *Extraction of biomolecules*

In this work, rotor-stator-based homogenization method was employed to disrupt the cell wall of biomass by rapidly moving the inner rotor. The strong shear stress applied on the cells attributed to the disruption of the rigid cell wall [25]. Eventually, phycocyanin, carbohydrate and total protein could be discharged from cell into the extraction solvent. Upon preliminary test of crude extraction, a blue green color extract was observed instead of typical blue color, which signifying the release of chlorophylls from the cells. Similar observation was revealed by Syrpas *et al.* [17] This would induce interference and deviation in phycocyanin content estimation with spectrophotometric analysis due to the overlapping of absorbance bands of phycobiliproteins and chlorophyll *a* (Chl*a*). The presence of Chl*a* could be further verified by the existence of peak absorbance around 675 nm. Lauceri *et al.* [24] claimed that the contamination of phycobiliprotein by Chl*a* was possible when the extraction method was not sufficiently mild. The extreme extraction process triggered the release of undesired biochemical compounds from the cytoplasm or other cell organelles. In fact, equation (2) proposed by Bennett and Bogorad [23] did not consider the presence of Chl*a* in the crude extract and the interference was generally neglected in other studies. In order to compute a more accurate phycocyanin quantification, the corrected equations (3) and (4) that were developed by Lauceri *et al.* [24] were included in this study. Additionally, allophycocyanin was not detected in the crude extract due to its small quantity as compared to phycocyanin [26]. To obtain optimum biomolecules yield, the effect of extraction conditions such as S:L ratio and homogenization time were investigated at room temperature. Although it was reported that higher stability of phycocyanin could be achieved at 4 °C, it was suggested to conduct the experiment at room temperature with regard to the perspective of economic viability and industrial scale-up [27].

Figure 1 displays the effect of solid biomass to solvent ratio on yield of phycocyanin, carbohydrate, total protein and phycocyanin purity ratio. The lower S:L ratio indicates the presence of more solvent in the extraction process. As the S:L ratio decreased from 0.200 to 0.100 g/ml, the trend demonstrates a drastic increase in phycocyanin yield from 39.85±3.43 mg/g to 101.20±3.48 mg/g. The phycocyanin yield gradually increased at ratio of 0.067 g/ml and remained unchanged when the S:L ratio was further decreased to 0.04 g/ml. At highest S:L ratio, the solvent volume available for contact between solvent and biomass became saturated, in which their interaction became limited and caused an incomplete extraction of phycocyanin from dry biomass, leading lower phycocyanin yield [28]. It can also be observed that the suspension was highly concentrated due to the presence of high amount of biomass at highest S:L ratio. With decreasing ratio, higher amount of solvent available for the contact between solvent and biomass thus increasing the driving force for mass transfer and diffusion of solvent into cells, thereby, the desorption of phycocyanin from the cells into the solvent was enhanced [29]. The finding was in agreement with literature which reported that the presence of more solvent resulted in better extraction of phycocyanin from *Spirulina sp.* [18,30]. Nevertheless, the purity ratio of phycocyanin shows a decreasing trend from 0.611±0.03 to 0.396±0.02 with a decreasing S:L ratio. In this study, the phycocyanin purity ratio is defined as the ratio of concentration of phycocyanin to concentration of total protein. The decline in purity ratio can be explained by the increase in availability of solvent for extraction of total protein from the biomass [18]. The result can be evidenced by increasing total protein yield trend with decreasing S:L ratio as described in figure 1. For carbohydrate extraction, a maximum yield of 103.55±1.02 mg/g of carbohydrate was obtained at S:L ratio of 0.100 g/ml and further decrease in the ratio did not result in significant change in the yield. The result was in good accordance with the works of literature which stated that the polysaccharides extraction yield increased when the solvent available increased, and was remained unchanged beyond a certain ratio [29,31]. The results described that the effect of S:L ratio is an essential parameter to take into consideration. Considering the phycocyanin yield and purity, the S:L ratio of 0.100 g/ml was chosen for further study due to its maximum phycocyanin yield and phycocyanin purity ratio as well as comparable total protein and carbohydrate yield.

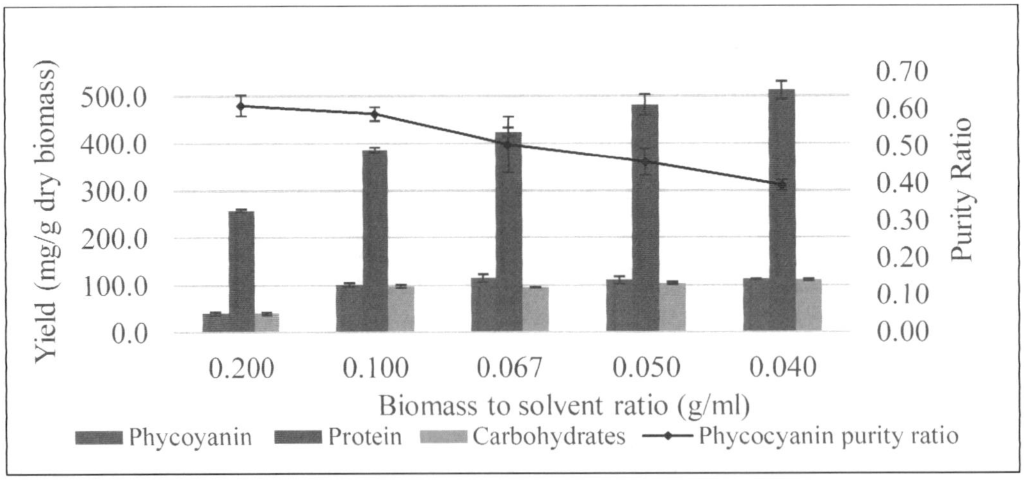

Figure 1. Effect of solid biomass to liquid ratio on biomolecules yield and phycocyanin purity ratio at constant homogenization time.

The effect of homogenization time on biomolecules yield and phycocyanin purity ratio is illustrated in figure 2. As depicted in figure 2, all biomolecules yield show a drastic increase with the increasing of homogenization time from 0 min to 5 min. The sharp increment in the yield at initial time was due to large concentration gradient of biomolecules between the solvent, which caused more phycocyanin, carbohydrates and total protein to diffuse out from the cells into the exterior solvent. Nevertheless, the further extension of homogenization time did not trigger significance change in phycocyanin and carbohydrate yield. This phenomena indicated that there was mass transfer limitation between the cell and the solvent when the solvent became saturated with biomolecules due to continuous exposure of biomass to solvent [28]. However, the trend for total protein yield was differed from phycocyanin and carbohydrate yielded. It can be seen that total protein yield was increased gradually with prolonged homogenization time. The high protein content in *Spirulina sp.* caused the yield to increase when the biomass was continuously exposed with the solvent. Summing up, an optimum yield of 110.30±1.07 mg/g of phycocyanin, 289.00±1.70 mg/g of total protein, 30.74±1.02 mg/g of carbohydrate with phycocyanin purity ratio of 0.656±0.01 were attained at S:L ratio of 1:10 and homogenization time of 5 min. These conditions were chosen for preparation of crude extract in further ATPS study.

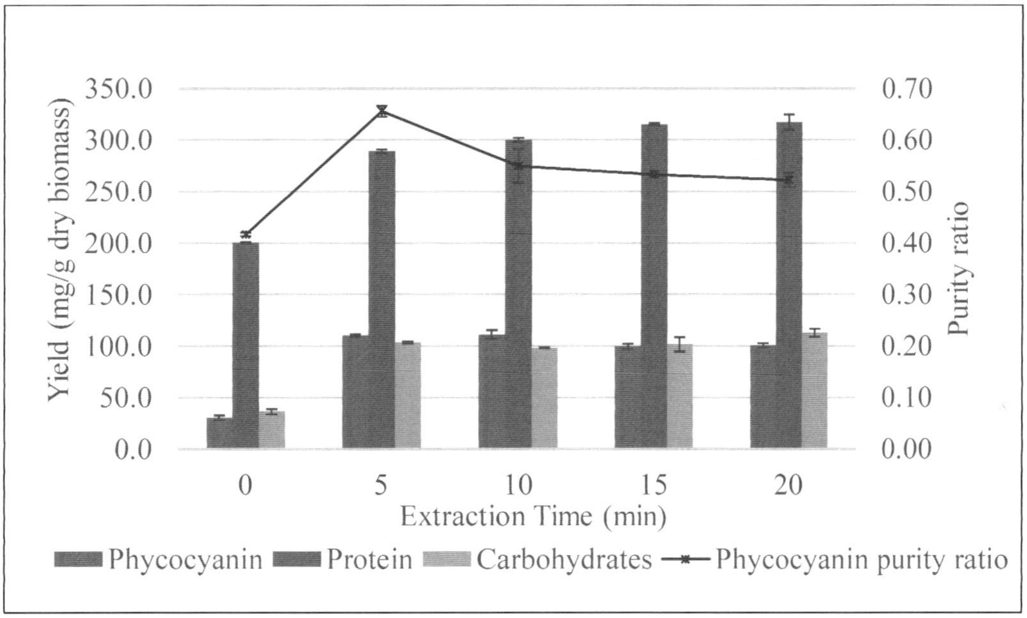

Figure 2. Effect of homogenization time on biomolecules yield and phycocyanin purity ratio at constant solid to liquid ratio.

Figure 3 and 4 depict the optical structure and morphology features of *Spirulina sp.* before and after homogenization. From figure 3a and 3b, it can be observed that the helical shape filament of *Spirulina sp.* is broken down into smaller fragments after subjecting to high shear force during homogenization. As portraited in figure 4a, the morphology feature of *Spirulina sp.* is characterized by intact and smooth cell wall surface before the homogenization. After homogenization, the cell surface is found to be altered, and rough cell surface is observed in figure 4b. This indicated that homogenization method was an effective cell disruption method which allowed the release of biomolecules from the phycobilisomes of the *Spirulina sp.*. In the study of Ilter *et al.* [30], they concluded that homogenization was the most appropriate method in comparison to ultrasonication and microwave-assisted extraction. By using homogenization method, a maximum phycocyanin yield of 74.74 mg/g was achieved in their work at S:L ratio of 0.32% (w/w), speed of 7000 rpm and homogenization time of 18.41 min. Previously, Tavanandi *et al.* [18] achieved phycocyanin yield of 109.03 mg/g and purity of 0.8 by a combination of ultrasonication and freezing and thawing method (2 cycles) at S:L ratio of 1:6 and ultrasonication time of 2.5 min. Tavanandi *et al.* [32] also achieved the highest phycocyanin yield of 98.24 mg/g and purity ratio of 1.09 by a combination of ultrasonication and enzyme-assisted extraction method. The use of ultrasound-assisted extraction by Bachchhav *et al.* [33] on the extraction of phycocyanin followed by B-carotene also showed a phycocyanin yield of 67 mg/g at 80% amplitude and duty cycle of 66%. Although a lower phycocyanin purity ratio was attained, it can be concluded that the homogenization method used in the present study obtained a phycocyanin yield with 10.79 to 32.23% higher than that was obtained in other studies [18, 32–33].

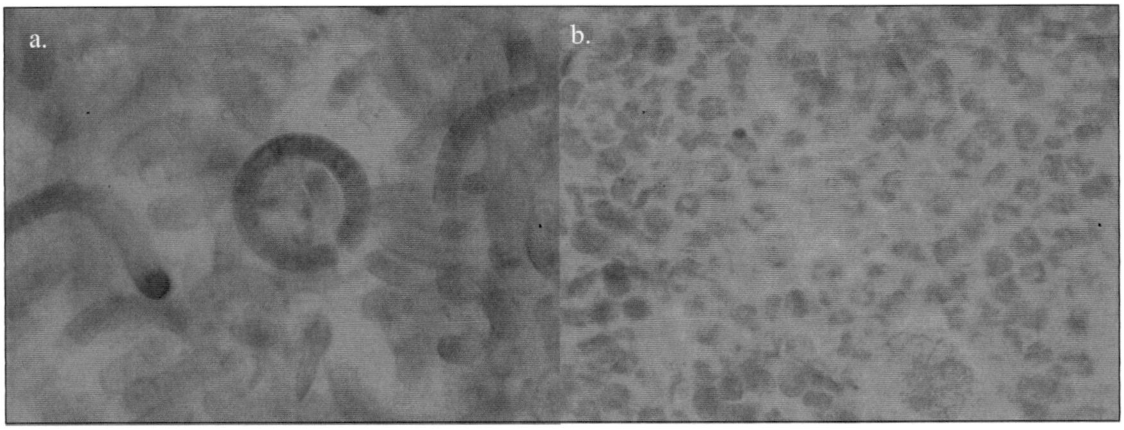

Figure 3. Morphology feature of Spirulina sp, at 400 × magnification (a) Before homogenization (b) After homogenization.

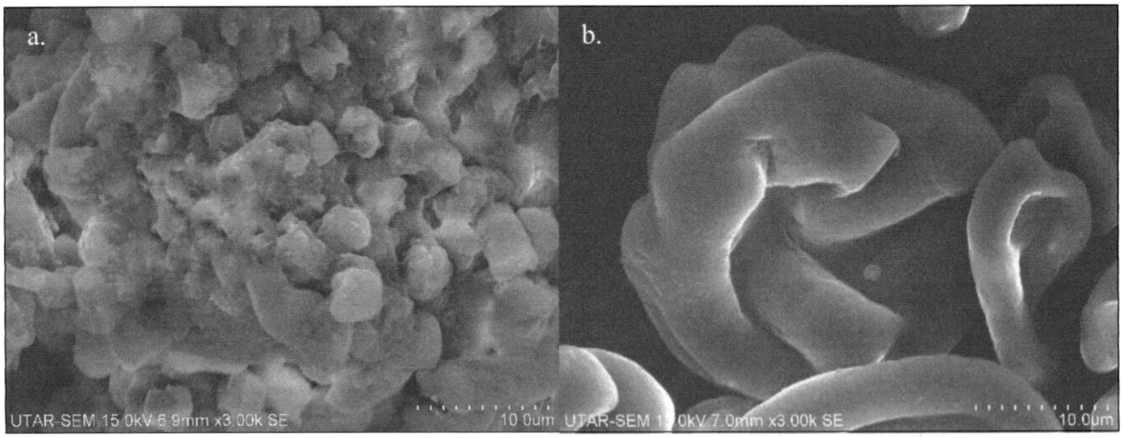

Figure 4. Optical image of Spirulina sp. at 3000 × magnification, (a) Before homogenization (b) After homogenization.

3.2. Fractionation of phycocyanin and carbohydrate using IL-based ATPS

In this study, kosmotropic buffer salts such as phosphate buffer at pH 7 and citrate buffer at pH 6 were utilized as phase-forming components to maintain the pH of the system during partitioning of biomolecules. Kosmotropic salts are designated as water-structuring salts which can stabilise and precipitate protein as well as promote salting-out effect. The addition of kosmotropic salts into the IL solution leads to entropic effect which is contributed by formation of water-kosmotropic ions complexes. The high affinity of kosmotropic ions towards water causes dehydration of hydrophilic IL and a decrease in its solubility in aqueous solution [34]. When the salt and IL reach certain concentration, hydrophilic IL containing anions with less water structuring characteristic and hydrophobic cation are separated and salted out [35–36].

In an ATPS, the partitioning behaviours of biomolecules are complex and could be driven by several interactions such as hydrophobic interaction, electrostatic interaction, hydrogen bonding interaction, van der Waals force and steric effects, depending on types of biomolecule [37–39]. The properties of protein such as isoelectric charges, hydrophobicity, solubility and size could influence the type of interaction involved during the partitioning [16]. A few studies have explored the domination of hydrophobic interaction between the phase former in polymer-based ATPS [40]. Nevertheless, the partitioning behaviour of biomolecules in IL-based ATPS seems to be more complicated due to the presence of cation and anion of IL which contributes to a wide range of polarity. Figure 5 shows the extraction efficiency of phycocyanin (top phase), total protein and carbohydrate (bottom phase) as well as the purification factor of phycocyanin using ATPS composing of ILs and salts. Purification factor is an important parameter for the measurement of the effectiveness of purification platform by comparing the purity ratio of phycocyanin before and after ATPS process.

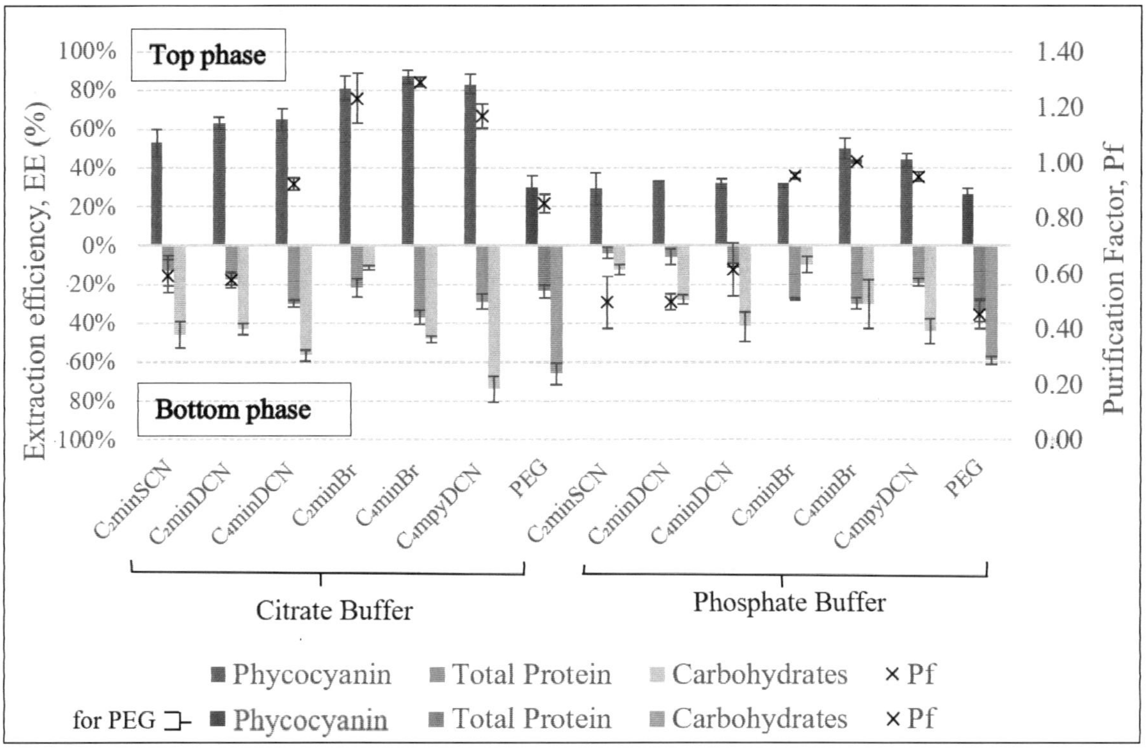

Figure 5. Fractionation of phycocyanin and carbohydrate using various ILs and salt buffers.

3.2.1. Effect of inorganic salts. The effect of inorganic salts on the partitioning behaviour of phycocyanin and carbohydrates are depicted and compared in figure 5. Total protein at bottom phase were also estimated to study the migration of contaminated protein which could affect the purity ratio of phycocyanin. The result indicated that the ATPS comprising of citrate buffer resulted in higher phycocyanin extraction efficiency and *purification factor* at top phase. The good partitioning behaviour and purification of phycocyanin can be explained based on the pH associated with each ATPS, its biocompatibility and salting-out effect. The pH of the system played a major role in partitioning behaviour of biomolecules, especially phycocyanin and total protein since pH could greatly influence their net charge, charge distribution, zeta potential and conformation structure change of phycocyanin and protein [41]. In current study, it was portrayed that higher phycocyanin extraction efficiency was found when using citrate buffer at pH 6. However, the result was in disagreement with another study which indicated that the optimum pH of 7 for extraction of phycocyanin with extraction efficiency and partition coefficient of 97.8% and 20.46, respectively, using ATPS consisting of K_2HPO_4 and $[C_8min]Br$ [2]. Chew *et al.* [41] also disclosed that working pH of 6 triggered unfavourable phycocyanin purity ratio and yield when compared with working pH 7 in ATPS formulated with PEG and K_2HPO_4. They claimed that phycocyanin becomes more negatively charge and being partitioned more effectively as the pH of the system increases, thereby the positively charge cation of IL tends to undergo electrostatic interaction with the negative charge of phycocyanin [41]. In this context, it is worth mentioning that the isoelectric point (PI) of phycocyanin varies between 4.6 to 5.2 [42]. When the pH above its PI, phycocyanin becomes negatively charges and interacts electrostatically with the cation of ILs [41]. At pH 6, there was insignificant charge on phycocyanin surface and the protein was in neutral state. For this reason, it was assumed that the surface properties or hydrophobic effect are the crucial effect on the partitioning behaviour of phycocyanin instead of electrostatic interaction in this study. The result was conformed with the disclosure of Patil *et al.* [43], which reported a greater purity ratio of phycocyanin at pH 6 using PEG4000 and potassium phosphate. The authors also claimed that the interaction that happened for partitioning behaviour of phycocyanin at this working pH was due to surface properties instead of electrostatic charge. Apart from that, the pH of ATPS also affects the stability of phycocyanin since it is heat, light and pH-sensitive biomolecules. Several studies revealed that a higher solubility of phycocyanin at neutral working pH. It might suffer from structural conformation change and degradation at acidic and alkaline pH [41,44–45]. According to Chentir *et al.* [46], phycocyanin tends to dissociate and aggregate in the form of various oligomer structures such as monomers, trimers, hexamers and higher degree of aggregation, depending on the working pH ionic strength and concentration. At working pH range between 4.5 to 6, phycocyanin was predominantly retained their hexameric structure, thus entraining their higher spectrum property and maximum stability [46]. At higher pH, phycocyanin undergoes denaturation as they dissociate from hexamers to trimers [47]. This was reported by Adam and Kao [48–49] whom revealed that the existence of phycocyanin in hexamer form was 77% at pH 6 and 18% only at pH 7. Several authors postulated that maximum phycocyanin stability were achieved at pH of 6 [46, 47]. Moreover, citrate salt is well-known for its environmental friendly characteristic and biocompatible with biomolecules, thus offering a conducive environment for phycocyanin [16]. It was inferred that the presence of citric acid in citrate buffer aided in maintenance of stability since citric acid has been touted as good preservative for food-grade phycocyanin [45]. By referring to extraction efficiency of total protein, it was deduced that ATPS comprising of citrate buffer has higher ability to partition total protein to bottom phase, thus giving rise to higher purification factor of phycocyanin at top phase.

On the other hand, the results presented in figure 5 shows that all ATPSs composing of citrate buffer exhibited better carbohydrate extraction efficiency at salt-rich bottom phase. The higher salting-out effect of citrate buffer plays a crucial role in migration of carbohydrate to bottom phase and phycocyanin to the IL-rich top phase [22]. The citrate buffer with trivalent anion ($C_6H_5O_7^{3-}$) reflects a greater salting-out effect than phosphate buffer consisting of monovalent and bivalent anions ($H_2PO_4^-/HPO_4^{2-}$). The high charge of salt anion contributes to stronger electric field and causes alignment of dipole of water molecules around the surrounding of anion, leading to hydration of salt over ILs [35]. At the same time, the hydrogen bond interaction was established between carbohydrate and water since carbohydrates have high polarity due to presence of numerous hydroxyl groups. This caused the exclusion of carbohydrates to the bottom phase and hence less free water molecules at the bottom phase causing the phycocyanin to migrate to the IL-rich top phase [2, 39].

3.2.2. Effect of Ionic liquids nature. The effects of ionic liquid nature which were anion, cation and cation alkyl chain length of ILs on the partitioning behaviour of biomolecules are as shown in figure 5. From figure 5, the extraction efficiency of phycocyanin for all ATPS with citrate buffer in descent order are $[C_4min]Br > [C_4mpy]DCN \geq [C_2min]Br > [C_4min]DCN \geq [C_2min]DCN > [C_2min]SCN$ while the purification factor showed decreasing trend of purity ratio of $[C_4min]Br > [C_2min]Br > [C_4mpy]DCN$

> [C_4min]DCN > [C_2min]DCN ≥ [C_2min]SCN. The ATPS with citrate buffer and phosphate buffer exhibited similar phycocyanin extraction efficiency and P_f trends as well as carbohydrate extraction efficiency with descent order of[C_4mpy]DCN > [C_4min]DCN > C_4min]Br > [C_2min]SCN > [C_2min]DCN > [C_2min]Br.

From the performance trend, the effect of anions with increasing extraction efficiency of phycocyanin was [SCN]$^-$ < [DCN]$^-$ < [Br]$^-$. The finding suggests that the effect of anions on extraction performance is highly dependent on their hydrophilicity. In this sense, the hydrophilic [Br]$^-$ anion has highest hydrogen bond acceptor ability than [DCN]$^-$ and [SCN]$^-$, thereby, [Br]$^-$ possessed a stronger interaction with the surrounding aqueous solution. Hence, the ILs with [Br]$^-$ anion resulted in superior phycocyanin extraction and purification performance as a consequence of stronger interaction IL and water-soluble phycocyanin molecules. This result was agreed with Martins *et al.* [50], which proved that the use of hydrophilic ILs was effective in extraction of phycobiliproteins such as phycoerythrin, phycocyanin and allophycocyanin. On the other hand, with an increasing number of cyanamide groups from [SCN]$^-$ to [DCN]$^-$, the hydrogen bonding interaction with water was enhanced and thus increasing IL hydrophilicity [51]. The finding was in close agreement with the solvatochromic parameter β reported by Batista *et al.* [52], whom found that the value of β of [C_4min]DCN was higher than [C_4min]SCN. This indicated that the hydrogen bond acceptor ability of anion [DCN]$^-$ was stronger than [SCN]$^-$ thus leading to greater phycocyanin extraction efficiency [52]. In contrast, the carbohydrate extraction efficiency at bottom phase showed an opposite trend. As the hydrophilicity of ILs anion increased, the ability of carbohydrate to form hydrogen bond with ILs at the top phase increased. This could lead to lower carbohydrate extraction efficiency at the bottom phase [39]. The hydrophobic ILs anion which were [SCN]$^-$ and [DCN]$^-$ lead to higher enrichment of total carbohydrate to bottom phase due to their higher propensity migration toward more hydrated phase. Santos *et al.* [15] reported similar opposite trend against protein partitioning.

In term of purification performance, both [DCN]$^-$ and [SCN]$^-$ exhibited close value of purification factor with non-ideal performance. The hydrophobicity of ILs which correlates to their toxicity can affect the stability of phycocyanin in ILs. With increasing hydrophobicity of ILs, the toxicity increases and this could cause degradation of phycocyanin during the partitioning [53]. The hydrophobic character of ILs with [SCN]$^-$ and [DCN]$^-$ anions had negatively affected phycocyanin purity ratio owing to their high toxicity. The purification factor of phycocyanin not only accounted for its partitioning behaviour but also enrichment of total protein at bottom phase. In ATPS composing of citrate buffer, it was found that the total protein extraction efficiency at bottom phase followed an descending order of [C_4min]Br > [C_4min]DCN >[C_4mpy]DCN ≥ [C_2min]Br > [C_2min]DCN > [C_2min]SCN. It can be seen that IL containing [Br]$^-$ anion tended to partition more total protein to bottom phase when compared with [DCN]$^-$ and [SCN]$^-$.

Moreover, the effect of cation which were imidazolium and pyrrolidinium on the fractionation of biomolecules was also studied. It was found that [C_4bmpy]DCN shows a higher extraction efficiency of all biomolecules and phycocyanin purification than [C_4min]DCN. It could be inferred that the good result of [C_4bmpy]DCN was associated with good stability of phycocyanin in pyrrolidinium-based ionic liquid. Bui-le *et al.* [54] proved that protein retains its secondary structure and tertiary structure in pyrrolidinium-based ionic liquid which was verified by presence of negative band at particular wavelength and maximum absorbance associated with the protein chromophore using circular dichroism spectrum and UV-vis spectrum analysis.

In addition, the cation alkyl chain length of ILs with anions [DCN]$^-$ and [Br]$^-$ were varied from n = 2 to 4 to evaluate their effects. By referring to figure 5, it can be seen that the carbohydrate and phycocyanin extraction efficiency and its purification factor increase with IL cation alkyl chain length. The finding was consistent with the study of Chang *et al.* [2] which reported an increase in extraction efficiency of phycocyanin as the cation alkyl chain length increased. The author claimed that the partitioning behaviour of phycocyanin accounted for both hydrophobic interaction and electrostatic interaction. As aforementioned, the driving force for the fractionation of phycocyanin is mainly contributed by hydrophobic interaction or surface properties. The increase in cation alkyl chain length enhanced the hydrophobic interaction between the IL cations and the hydrophobic residues on the surface of phycocyanin. This caused an increase in phycocyanin extraction efficiency at top phase. Furthermore, the ILs with longer alkyl chain length has higher ability to promote the migration of carbohydrate and total protein to the bottom phase as depicted in figure 5, thus resulting in higher purity ratio of phycocyanin. As the hydrophobicity of IL increased with cation alkyl chain length, the hydrogen interaction between carbohydrate and water in salt-rich phase was intensified.

By looking at the overall performance of each ATPS, [C_4mpy]DCN and citrate salt was chosen as the best ATPS constituents for the fractionation of phycocyanin and carbohydrate to top and bottom phase, separately. In this ATPS, optimum extraction efficiency of 83.26$\pm$0.05% phycocyanin and 73.89$\pm$0.06% of carbohydrate were achieved. Despite [C_4min]Br exhibited higher phycocyanin and purification performance, this system only recovered 48.23$\pm$0.02% of carbohydrate at the bottom phase. In ATPS comprising of [C_4mpy]DCN and citrate salt, simultaneous fractionation of phycocyanin and carbohydrate to opposite phase can be achieved. When comparing the IL-based ATPS and conventional polymer-based ATPS as displayed in figure 5, it is reflected that IL-based ATPS performed more effectively for fractionation of multiple products.

4. Conclusion

In this work, the implementation of various ILs-based ATPS for fractionation of biomolecules from *Spirulina* crude extract as well as purification of phycocyanin were investigated. The results suggested that the use of biocompatible citrate buffer at pH 6 provided a more conductive environment for phycocyanin and resulted in high salting-out effect in contrast to phosphate buffer which was generally used in other studies. This work postulated that ILs-based ATPS are not touted as good purification platform for phycocyanin since the non-ideal purification factor of 1.29 was achieved during purification. However, it can be concluded that ILs-based ATPS is a promising method for the fractionation of biomolecules which can achieve the multiproduct biorefinery concept. Moreover, [C_4mpy]DCN and citrate buffer were inferred as best phase forming components due to its high phycocyanin and carbohydrate recovery of 83.26$\pm$0.05% and 73.89$\pm$0.06% at the top and bottom phase, respectively. Lastly, the phycocyanin recovery of this system is higher than those obtained in conventional polymer based ATPS. Therefore, the application of ILs-based ATPS is substantially envisaged in biorefinery field.

References

[1] Barkia I, Saari N and Manning S R 2019 Microalgae for high-value products towards human health and nutrition *Mar. Drugs* **17** 1–29
[2] Chang Y K, Show P L, Lan J C W, Tsai J C and Huang C R 2018 Isolation of C-phycocyanin from Spirulina platensis microalga using ionic liquid based aqueous two-phase system *Bioresour. Technol.* **270** 320–7
[3] Saranraj P and Sivasakthi S 2014 Spirulina platensis food for future: A review *J. Pharm. Sci. Technol.* 26–33
[4] Pan-utai W and Iamtham S 2019 Physical extraction and extrusion entrapment of C-phycocyanin from Arthrospira platensis *J. King Saud Univ. Sci.* **31** 1535–42
[5] Kumar D, Dhar D W, Pabbi S, Kumar N and Walia S 2014 Extraction and purification of C-phycocyanin from Spirulina platensis (CCC540) *Indian J. Plant Physiol.* **19** 184–8
[6] Stanic-Vucinic D, Minic S, Nikolic M R and Velickovic T C 2018 Spirulina Phycobiliproteins as food components and complements *Microalgal Biotechnol.*
[7] Park W S, Kim H J, Li M, Lim D H, Kim J, Kwak S S, Kang C M, Ferruzzi M G and Ahn M J 2018 Two classes of pigments, carotenoids and c-phycocyanin, in spirulina powder and their antioxidant activities *Molecules* **23** 1–11
[8] Prabakaran G, Sampathkumar P, Kavisri M and Moovendhan M 2020 Extraction and characterization of phycocyanin from Spirulina platensis and evaluation of its anticancer,

antidiabetic and antiinflammatory effect *Int. J. Biol. Macromol.* **153** 256–63
[9] Demirbas A 2010 Use of algae as biofuel sources *Energy Convers. Manag.* **51** 2738– 49
[10] Costa J A V, Freitas B C B, Rosa G M, Moraes L, Morais M G and Mitchell B G 2019 Operational and economic aspects of Spirulina-based biorefinery *Bioresour. Technol.* **292** 121946
[11] Zhao L, Peng Y L, Gao J M and Cai W M 2014 Bioprocess intensification: An aqueous two-phase process for the purification of C-phycocyanin from dry Spirulina platensis *Eur. Food Res. Technol.* **238** 451–7
[12] Luo X, Smith P, Raston C L and Zhang W 2016 Vortex fluidic device-intensified aqueous two phase extraction of C-phycocyanin from spirulina maxima *ACS Sustain. Chem. Eng.* **4** 3905–11
[13] Patil G, Chethana S, Madhusudhan M C and Raghavarao K S M S 2008 Fractionation and purification of the phycobiliproteins from Spirulina platensis *Bioresour. Technol.* **99** 7393–6
[14] Zhang X, Zhang F, Luo G, Yang S and Wang D 2014 Extraction and Separation of Phycocyanin from *Spirulina* using aqueous two-phase systems of ionic liquid and salt *J. Food Nutr. Res.* **3** 15–9
[15] Santos H P M 2018 Fractionation of Isochrysis galbana proteins, arabinans, and glucans using ionic-liquid-based aqueous biphasic systems′ *ACS Sustain. Chem. Eng.* **6** 14042–53
[16] Suarez Ruiz C A, van den Berg C, Wijffels R H and Eppink M H M 2018 Rubisco separation using biocompatible aqueous two-phase systems *Sep. Purif. Technol.* **196** 254–61
[17] Syrpas M, Bukauskaitė J, Ramanauskienė K, Karosienė J R, Majienė D, Bašinskienė L and Venskutonis P R 2019 Ultrasound-assisted extraction and assessment of biological activity of phycobiliprotein-rich aqueous extracts from wild Cyanobacteria (Aphanizomenon flos-aquae) *J. Agric. Food Chem.*
[18] Tavanandi H A, Mittal R, Chandrasekhar J and Raghavarao K S M S 2018 Simple and efficient method for extraction of C-Phycocyanin from dry biomass of Arthospira platensis *Algal Res.* **31** 239–51
[19] Chen W, Xu J, Yu Q, Yuan Z, Kong X Y, Sun Y M, Wang Z M, Zhuang X S, Zhang Y and Guo Y 2020 Structural insights reveal the effective Spirulina platensis cell wall dissociation methods for multi-output recovery *Bioresour. Technol.* **300** 122628
[20] Gómez E, Domínguez I, Domínguez Á and Macedo E A 2018 Ionic liquids-based aqueous biphasic systems with citrate biodegradable salts *J. Chem. Eng. Data* **63** 1103–8
[21] Jose C M , Barbara A and Andrews J A A 1998 Aqueous two-phase systems for protein separation studies on phase inversion *J. Chromatogr. B* **711** 285–93
[22] Suarez Garcia E, Suarez Ruiz C A, Tilaye T, Eppink M H M, Wijffels R H and van den Berg C 2018 Fractionation of proteins and carbohydrates from crude microalgae extracts using an ionic liquid based-aqueous two phase system *Sep. Purif. Technol.* **204** 56–65
[23] Bennett A and Bogobad L 1973 Complementary chromatic adaptation in a filamentous blue-green alga *J. Cell Biol.* **58** 419–35
[24] Lauceri R, Bresciani M, Lami A and Morabito G 2018 Chlorophyll a interference in phycocyanin and allophycocyanin spectrophotometric quantification *J. Limnol.* **77** 169–77
[25] Wang D, Li Y, Hu X, Su W and Zhong M 2015 Combined enzymatic and mechanical cell disruption and lipid extraction of green alga Neochloris oleoabundans *Int. J. Mol. Sci.* **16** 7707–22
[26] Yoshikawa N and Belay A 2008 Single-laboratory validation of a method for the determination of c-phycocyanin and allophycocyanin in spirulina (Arthrospira) supplements and raw materials by spectrophotometry *J. AOAC Int.* **91** 524–9
[27] Choi W Y and Lee H Y 2018 Effect of ultrasonic extraction on production and structural changes of C-phycocyanin from marine Spirulina maxima *Int. J. Mol. Sci.* **19**
[28] Dey S and Rathod V K 2013 Ultrasound assisted extraction of β-carotene from Spirulina platensis *Ultrason. Sonochem.* **20** 271–6
[29] Bendahou A, Dufresne A, Kaddami H and Habibi Y 2007 Isolation and structural characterization

of hemicelluloses from palm of Phoenix dactylifera L. *Carbohydr. Polym.* **68** 601–8
[30] İlter I, Akyıl S, Demirel Z, Koç M, Conk-Dalay M and Kaymak-Ertekin F 2018 Optimization of phycocyanin extraction from Spirulina platensis using different techniques *J. Food Compos. Anal.* **70** 78–88
[31] Samavati V 2013 Polysaccharide extraction from Abelmoschus esculentus: Optimization by response surface methodology *Carbohydr. Polym.* **95** 588–97
[32] Tavanandi H A and Raghavarao K S M S 2020 Ultrasound-assisted enzymatic extraction of natural food colorant C-Phycocyanin from dry biomass of Arthrospira platensis *LWT - Food Sci. Technol.* **118** 108802
[33] Bachchhav M B, Kulkarni M V and Ingale A G 2020 Process-intensified extraction of phycocyanin followed by β-carotene from Spirulina platensis using ultrasound-assisted extraction *Sep. Sci. Technol. (Philadelphia)* **55** 932–44
[34] Passos H, Trindade M P, Vaz T S M, Da Costa L P, Freire M G and Coutinho J A P 2013 The impact of self-aggregation on the extraction of biomolecules in ionic-liquid-based aqueous two-phase systems *Sep. Sci. Technol.* **108** 174–80
[35] Freire M G, Cláudio A F M, Araújo J M M, Coutinho J A P, Marrucho I M, Canongia Lopes J N and Rebelo L P N 2012 Aqueous biphasic systems: A boost brought about by using ionic liquids *Chem. Soc. Rev.* **41** 4966–95
[36] Gutowski K E, Broker G A, Willauer H D, Huddleston J G, Swatloski R P, Holbrey J D and Rogers R D 2003 Controlling the aqueous miscibility of ionic liquids: Aqueous biphasic systems of water-miscible ionic liquids and water-structuring salts for recycle, metathesis, and separations *J. Am. Chem. Soc.* **125** 6632–3
[37] Hachem F, Andrews B A and Asenjo J A 1996 Hydrophobic partitioning of proteins in aqueous two-phase systems *Enzyme Microb. Technol.* **19** 507–17
[38] Andrews B A, Schmidt A S and Asenjo J A 2005 Correlation for the partition behavior of proteins in aqueous two-phase systems: Effect of surface hydrophobicity and charge *Biotechnol. Bioeng.* **90** 380–90
[39] Suarez Ruiz C A, Kwaijtaal J, Peinado O C, Van Den Berg C, Wijffels R H and Eppink M H M 2020 Multistep fractionation of microalgal biomolecules using selective aqueous two-phase systems *ACS Sustain. Chem. Eng.*
[40] Suarez Ruiz CA, Emmery D P, Wijffels R H, Eppink M H M and van den Berg C 2018 Selective and mild fractionation of microalgal proteins and pigments using aqueous two-phase systems *J. Chem. Technol. Biotechnol.* **93** 2774–83
[41] Chew K W, Chia S R, Krishnamoorthy R, Tao Y, Chu D T and Show P L 2019 Liquid biphasic flotation for the purification of C-phycocyanin from Spirulina platensis microalga *Bioresour. Technol.* **288**
[42] Sala L, Figueira F S, Cerveira G P, Moraes C C and Kalil S J 2014 Kinetics and adsorption isotherm of C-phycocyanin from Spirulina platensis on ion-exchange resins *Brazilian Chem. Eng. J.* **31** 1013–22
[43] Patil G and Raghavarao K S M S 2007 Aqueous two phase extraction for purification of C-phycocyanin *Biochem. Eng. J.* **34** 156–64
[44] Rito-Palomares M, Nuez L and Amador D 2001 Practical application of aqueous two-phase systems for the development of a prototype process for c-phycocyanin recovery from Spirulina maxima *J. Chem. Technol. Biotechnol.* **76** 1273–80
[45] Mishra S K, Shrivastav A and Mishra S 2008 Effect of preservatives for food grade C-PC from Spirulina platensis *Process Biochem.* **43** 339–45
[46] Chentir I, Hamdi M, Li S, Doumandji A and Markou G 2018 Stability , bio-functionality and bio-activity of crude phycocyanin from a two-phase cultured Saharian Arthrospira sp . strain *Algal Res.* **35** 395–406
[47] Chaiklahan R, Chirasuwan N and Bunnag B 2012 Stability of phycocyanin extracted from Spirulina sp .: Influence of temperature , pH and preservatives *Process Biochem.* **47** 659–64

[48] Adams S M, Kao O H W and Berns D S 1979 Psychrophile C-Phycocyanin 525–7
[49] Kao O H W, Edwards M R and Berns D S 1975 Physical chemical properties of C phycocyanin isolated from an acido thermophilic eukaryote, Cyanidium caldarium *Biochem. J.* **147** 63–70
[50] Martins M, Vieira F A, Correia I, Ferreira R A S, Abreu H, Coutinho J A P and Ventura S P M 2016 Recovery of phycobiliproteins from the red macroalga: Gracilaria sp. using ionic liquid aqueous solutions *Green Chem.* **18** 4287–96
[51] Vataščin E and Dohnal V 2017 Thermodynamic properties of aqueous solutions of [EMIM] thiocyanate and [EMIM] dicyanamide *J. Chem. Thermodyn.* **106** 262–75
[52] Batista M L S, Kurnia K A, Pinho S P, Gomes J R B and Coutinho J A P 2015 Computational and experimental study of the behavior of cyano-based ionic liquids in aqueous solution *J. Phys. Chem. B* **119** 1567–78
[53] Studzińska S and Buszewski B 2009 Study of toxicity of imidazolium ionic liquids to watercress (Lepidium sativum L.) *Anal. Bioanal. Chem.* **393** 983–90
[54] Bui-Le L, Clarke CJ, Bröhl A, Brogan A P S, Arpino J A J, Polizzi K M and Hallett J P 2020 Revealing the complexity of ionic liquid–protein interactions through a multi-technique investigation *Commun. Chem.* **3** 1–9

Acknowledgments
This work was supported by Universiti Tunku Abdul Rahman Research Fund (UTARRF), (IPSR/RMC/UTARRF/2019-C1/L05).

IOP Conf. Series: Materials Science and Engineering 1195 (2021) 012040 doi:10.1088/1757-899X/1195/1/012040

Performance enhancement of a baffle-type solar heat collector through CFD simulation study

A S T Tan[1], J Janaun[1], H J Tham[1], N J Siambun[1] and A Abdullah[2]

[1] Industrial Dryer and Palm Oil Recovery Research Group, Universiti Malaysia Sabah, Jalan UMS, 88400 Kota Kinabalu, Sabah, Malaysia
[2] Chancellery Office, Universiti Malaysia Kelantan, 16300 Bachok, Kelantan, Malaysia

Email: jidon@ums.edu.my; arham.a@umk.edu.my

Abstract. The application of solar energy conversion has been extensively utilized as an alternative energy source to generate heat. This approach would be a step towards sustainable energy development particularly in the manufacturing industry with energy-intensive process. In this paper, thermal enhancement on the key component of a solar energy device – solar heat collector (SHC), has been evaluated by proposing a baffle-type SHC with various geometric configuration in the air passage namely longitudinal baffle and transversal baffle. The performance of SHC is evaluated in term of efficiency, temperature distribution, airflow pattern and pressure drop across the collector outlet through Computational Fluid Dynamic (CFD) investigation. It was observed that maximum collector efficiency was achieved in the Longitudinal-SHC (L-SHC), with a value of 46.2 % followed by Transversal-SHC (T-SHC) and without baffles. Maximum drying temperature at the collector outlet was 332.43 K for L-SHC, showing temperature rise of 0.35 % and 4.21 % from T-SHC and without baffles, respectively. The velocity vector indicated that turbulence flow was created in the T-SHC which consequently improved the heat transfer. Whereas in L-SHC, enhancement was achieved through the prolonged heating time in the passage. Considering the thermo-hydraulic performance factor evaluated, these enhancement features had diminished the effect of pressure drop.

1. Introduction

In recent years, the application of solar energy has been extensively utilized in the industry for drying, space heating, water heating and other industrial application [1–3]. Solar energy is a clean, sustainable and universally accessible energy source. Hence, numerous research have been performed to study the potential of applying solar energy in rural areas. According to Borhanazad *et al.* [4], electricity coverage of rural areas in Malaysia is about 79 % compared to Peninsular Malaysia with 99.62 %. Since the major source of income in these areas is agricultural production, implementation of least expensive energy solution is essential. Utilization of solar energy not only offers a cheap alternative energy source, but also resolves the issues of limited accessibility to electricity in rural areas. Ultimately, this could boost the local economic opportunities.

As a key component to harvest solar energy and convert into useful heat, many researchers have introduced several thermal enhancement techniques for solar heat collector (SHC). Wide research has been performed to improve the convective heat transfer in the SHC by means of modifying the flow passage as reported by Arunkumar *et al.* [5]. The two major enhancement mechanism in the air passage that are widely discussed in the literature are number of flow pass and attachment of tabulator. Satcunanathan *et al.* [6] first introduced the concept of double pass. It was reported that the efficiency was improved 10 to 15 % higher than that of single-pass SHC. Ho *et al.* [7] confirmed that the enhancement was achieved due to the increase heat transfer area and time hence, reduce the heat loss from the absorber plate. Jia *et al.* [8] proposed a flow path modification where air is forced to flow through in a spiral path. An average efficiency ranged from 58 % to 62 % was recorded under irradiance above 700 Wm^{-2} and wind velocity of 0 – 3 m/s.

Introduction of baffles into the air passage improve the heat transfer in solar heat collector by increasing the heating time between the fluid and absorber (residence time), to create turbulence for intense mixing and to avoid 'hot spot' which consequently reduces the heat loss [9]. These enhancement were reported by Romdhabe [10]. However, findings indicate that the addition of baffles in the air passage increases the pressure drop. Hu *et al.* [11] studied the pressure drop in a solar heat collector with baffles positioned perpendicularly towards the air direction. It was found that the variation of pressure drop across the passage is low with increasing baffle number. In Hu *et al.* [12] later study, it was reported that reducing the width at the first chamber air duct shows significant improvement on the thermal efficiency but less influence on the pressure drop. Moummi *et al.* [13] developed a solar heat collector with baffles attached perpendicular to the flow to create high turbulence flow. The thermal improvement was remarkable compared to that without baffles. Bensaci *et al.* [14] performed experimental testing to study the effect of transversal baffle arrangement on the heat transfer coefficient and thermal efficiency of solar heat collector. Highest performance was found in solar heat collector with baffles attached all over the air passage. Although the enhancement on thermal performance was significant with the presence of baffles, several researchers analysed the overall performance through the evaluation of thermo-hydraulic performance, which consider the heat transfer enhancement with respect to the pressure drop established throughout the passage [15□–□16].

According to the literature review of research in the field of thermal enhancement of solar heat collector through the modification of flow configuration, it is evident that limited study on solar heat collector with baffles positioned parallel to the air direction was reported. It was confirmed by Potgieter *et al.* [17] that counter-/parallel flow created by this baffle arrangement could improves the turbulence and convection heat transfer in the collector. In addition, less pressure drop is expected as the baffles arranged in parallel does not provokes high pressure drop along the passage. In this present study, a comparative study of the Longitudinal Solar Heat Collector and Transversal Solar Heat Collector is performed through Computational Fluid Dynamics (CFD) simulation to identify the enhancement mechanism occurs in the air passage with different baffle arrangement.

2. Methodology

This study aims to evaluate the thermal performance enhancement on an air-based solar heat collector with different baffles arrangement, namely Longitudinal Solar Heat Collector (L-SHC) and Transversal Solar Heat Collector (T-SHC), using Computational Fluid Dynamics (CFD) simulation. Computational modelling adopted in this study was developed based on the assumptions as follow:

- Three-dimensional fluid flow and heat transfer.
- Steady-state and incompressible flow condition.
- Negligible heat loss from the bottom and side wall of the solar heat collector frame.
- Negligible heat loss from the inlet and outlet of solar heat collector.
- Thermo-physical properties of fluid (air) and solid (aluminium frame and baffle plate) remain constant throughout the process.

Firstly, the computational model adopted in this present study was extracted from previous work that studied the performance of solar heat collector integrated to UMS Eco-Solar Dryer and validated with the experimental results [18]. Experimental testing performed previously was based on solar heat collector with absorber made from recycled aluminium cans, enclosed with plywood frame and acrylic cover at the top. In this present study, recycled aluminium cans absorber was replaced with baffle plate arranged in longitudinal and transversal with regard to the collector. The computational model was developed based on the geometry and dimension as of the previous work. Quality of mesh generated was then analysed through grid independence study. After the validation of CFD model developed, it was adopted to study the influence of baffles arrangement on the performance of solar heat collector. The enhancement mechanism in solar heat collector was evaluated in respect with the temperature rise of fluid across the collector air passage, velocity vector and pressure drop established approaching to the outlet air vent. In order to analyse the overall performance of solar heat collector, parameters such as Nusselt number, local heat transfer coefficient and friction factor across the air passage was calculated to identify the thermo-hydraulic performance of solar heat collector.

Two different cases were considered in this study depending on the baffle arrangement in the air passage – L-SHC and T-SHC. A solar heat collector without baffle in the air passage was simulated for comparison purpose. For each case, simulation was performed according to the boundary condition specified under natural convection with air velocity of 0.12 m/s. The results were then compared in the aspect of heat transfer characteristics, fluid flow condition, thermal efficiency and thermo-hydraulic performance.

2.1. Description of solar heat collectors

The 3D geometry of solar heat collectors covered in this study are as illustrated in figure 1. The solar heat collectors consist of five main parts: transparent cover, absorber plate, baffles, collector frame and air passage between the absorber plate and bottom plate where heated air circulates. The dimensions of solar heat collector were 2112 mm length, 910 mm width, and 100 mm height according to the prototype – UMS Eco-Solar Dryer, developed in Faculty of Engineering, Universiti Malaysia Sabah (UMS). The cover was a transparent acrylic with 5 mm thickness. Aluminium absorber with thickness of 3 mm was fixed below the cover to absorb solar radiation pass through it. An air gap with 25 mm thickness was left in between the cover and the absorber to ensure greenhouse heating and to prevent heat loss through the top cover. In addition, collector frame made from aluminium was well-insulated to ensure minimal heat loss to the surrounding. Baffles were attached on the bottom plate of the collector in a way to cover 80% of the total height of the air passage to achieve maximum heat transfer between the drying air and absorber wall [19].

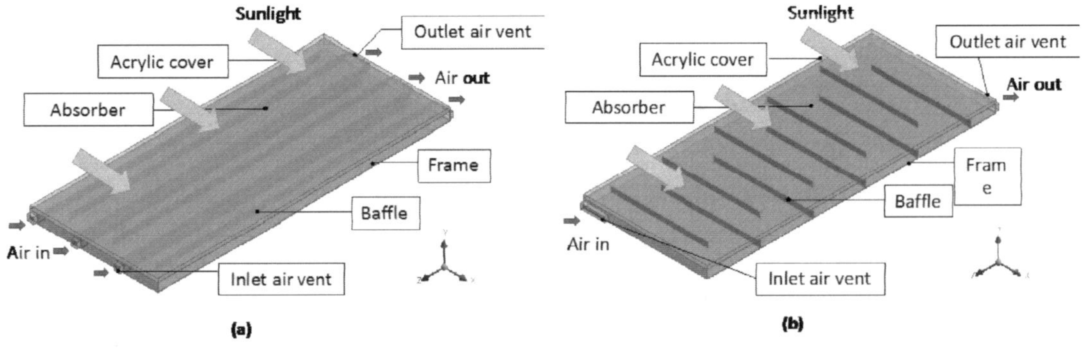

Figure 1. Three-dimensional geometry of solar heat collectors used for CFD model – (a) Longitudinal Solar Heat Collector (L-SHC) and (b) Transversal Solar Heat Collector (T-SHC).

In L-SHC, baffles were arranged parallel with regard to the air direction. In this design, fluid is forced to flow through three time the length of collector while approaching to the outlet air vent. As shown in figure 1 (a), the longitudinal baffle arrangement adopted in this study creates three sets of this particular air channel. In an attempt to minimize the pressure drop across the collector, the width of air duct between two baffles was increased from 50, 100 to 150 mm from the first to third air duct in a single air channel. In T-SHC, baffles were arranged perpendicularly occupying 80 % of the total width of solar heat collector to create a zig-zag air passage across the collector. The width of air duct between baffles was fixed at 182 mm in the collector, which is equivalent to the width of the inlet air vent.

3. Computational model

The CFD model of solar heat collector used in the present study was made up with a rectangular solid geometry comprises seven major domain namely top wall (cover), side wall (collector frame), absorber wall, baffle plates, fluid domain flow through air channel between the top and absorber wall, fluid domain flow through air channel between absorber and bottom wall, air inlet and air outlet. The L-SHC consists of three sets of air channel, each with an air inlet vent and a corresponding outlet vent, whereas the T-SHC consists only one inlet and outlet air vent respectively at each end of the solar heat collector. The simulation was done using ANSYS 2021 R1 Fluent package. The flow in solar heat collector is considered steady-state, three dimensional and incompressible. In addition, since this study aimed to investigate the performance of solar heat collector under natural convection, gravity is considered. All fluid and solid material properties remain constant throughout the process at average air temperature of 300.15 K.

3.1. Boundary condition

The radiation model utilized in this study was Rosseland model incorporated with solar ray tracing algorithm. Realizable k-ε turbulent model with enhanced wall treatment was selected for the calculation of turbulence flow created in the air passage due to the presence of baffles. All wall interfaces were defined as no-slip condition. The top wall (cover) was defined as semi-transparent acrylic with 0.8 transmissivity and participated in solar ray tracing. The side wall (collector frame) was aluminium material defined as adiabatic wall (heat flux = 0) which does not participate in solar ray tracing. Absorber plate positioned below the cover was assigned as aluminium and subject to solar radiation. A mass flow rate condition was assigned at the inlet section whereas the outlet was modelled as pressure outlet.

Coupled algorithm was selected to use with the pressure-based solver to solve the momentum equation and pressure-based continuity equations together [20]. Residual value of 10^{-6} and 10^{-3} was selected to monitor the convergence of solution for energy and velocity equation respectively.

3.2. Modelling

Performance parameters used to evaluate the enhancement mechanism in the solar heat collector are the outlet temperature, velocity and pressure drop across the air passage, which contour plot was extracted from CFD post-processing software. Dimensionless parameters such as Nusselt number, Nu and friction factor, f that define the thermo-hydraulic performance of solar heat collector can be estimated from the following equations [21]:

$$Nu = \frac{hD}{k} \tag{1}$$

where h is the heat transfer coefficient (W/m^2-K) obtained directly from Fluent simulation and k is the thermal conductivity (W/m-K) of fluid.

$$f = \frac{\Delta PD}{2\rho LV^2} \tag{2}$$

where ΔP is the pressure drop (Pa), ρ is the fluid density (kg/m^3), L is the total length of air passage (m) and V is the velocity (m/s). D is the hydraulic diameter of the air duct (m), calculated by multiplying A, the cross-sectional area (m^2), of the air duct by 2 and dividing by the sum of width, W (m) and height, H (m) of the air duct as follow:

$$D = \frac{2A}{W + H} \tag{3}$$

Thermal efficiency, η of the collector is calculated by the equation as follows [22]:

$$\eta = \frac{Q_u}{G \times A} \tag{4}$$

where G is the global solar radiation (W/m^2) and A is the area of the collector (m^2). The useful heat gain, Q_u (W) in the solar heat collector is expressed as:

$$Q_u = m \times C_p \times (T_{out} - T_{in}) \tag{5}$$

where m is the mass flow rate (kg/s), C_p is the specific heat (J/kg-K) of the fluid, T_{out} is the outlet temperature (K) and T_{in} is the inlet temperature (K).

4. Validation of model

The validation of computational model proposed in Section 3 was performed in the aspect of grid independency analysis and by experimental results. The latter approach was reported in previous work, which identified the computational model with a percentage deviation of 1.76 % and 8.60 % compared to the measured value for single-pass solar heat collector and multiple-pass solar heat collector respectively [18]. The basic structure of L-SHC and T-SHC discussed in this study is very similar to the model reported in previous work, which used recycled aluminium cans to modify the air flow configuration in the air passage. Hence, it is safe to assume that the experimental validation reported is applicable to the present study.

4.1. Grid independence test

The computational grid for the domains was generated using unstructured mesh with uniform growth rate of 1.2. During meshing, it is crucial to identify the number of nodes generated in the domain is sufficient prior to obtain an accurate numerical result [23]. Hence, it is necessary to perform grid independence test to evaluate the mesh quality at different grid size. Five different mesh size was used in this study – 0.014, 0.015, 0.018, 0.020, and 0.025 mm. The number of cells were increased from 101,991 to 443,928 using the built-in meshing software in ANSYS Fluent. Each case was carried out under turbulent flow with Re = 7580 (velocity = 2 m/s). The quality of mesh was evaluated based on the temperature of drying air at the outlet vent as this performance parameter is well established in the literature to study the performance of solar heat collector. Results of the mesh independence test for L-SHC and T-SHC are shown in table 1 and table 2 respectively. Based on the results presented, coarse mesh method provides less accurate solution for both model with floating error indicated in L-SHC and large percentage error in T-SHC. Hence, very fine element size was selected for both case due to less percentage deviation (less than 1%) was recorded.

Table 1. Grid independence test results for Longitudinal Solar Heat Collector (L-SHC).

Trial	Mesh size	Cell number	Outlet temperature (K)	Percentage deviation (%)
1	Extremely coarse	111,164	Floating error	-
2	Less coarse	183,968	298.870	-
3	Normal	260,788	301.625	0.920
4	Fine	397,750	301.648	0.008
5	Very fine	457,284	301.660	0.004

Table 2. Grid independence test results for Transversal Solar Heat Collector (T-SHC).

Trial	Mesh size	Cell number	Outlet temperature (K)	Percentage deviation (%)
1	Extremely coarse	101,991	80.798	-
2	Less coarse	177,498	284.217	251.8
3	Normal	241,879	292.441	2.894
4	Fine	382,380	301.581	3.125
5	Very fine	443,928	301.872	0.096

5. Results and discussion

In this section, the predicted heat transfer and fluid flow characteristics across the solar heat collector are presented in term of contour plot extracted from CFD. Results presented were simulated under natural convection with Reynold number of 1303. The effect of different baffle arrangement on the performance of solar heat collector is discussed in the aspect of thermal efficiency, Nusselt number, heat transfer coefficient and friction factor. Lastly, the thermo-hydraulic performance of each solar heat collector is studied by also taking account of the pressure drop together with the heat transfer enhancement achieved in the air passage.

5.1. Temperature distribution and thermal efficiency

The predicted temperature across the length of the smooth and baffled solar heat collectors (L-SHC and T-SHC) are shown in figure 2. The results were simulated under the condition with inlet velocity of 0.12 m/s and solar radiation of 438.578 W/m^2. Based on the plotted graph, results simulated in CFD indicate an increase in fluid temperature while approaching to the outlet air vent of the solar heat collector. Both baffled solar heat collector offers more uniform distribution along the length, indicating the reduction of dead zone due to the presence of baffles in the air passage. In addition, the performance of baffled solar heat collector is significantly improved in term of outlet temperature as compared to the smooth solar heat collector, which indicates a maximum fluid temperature of 319.00 K at the collector outlet. For L-SHC and T-SHC, the highest predicted outlet temperature was 332.43 K and 331.26 K respectively. From the plotted graph, it is noticeable that in the L-SHC, a sudden drop of drying temperature is observed at length of 1.2 m. This can be inferred as the reattachment of fluid flow at the turning edge results higher heat transfer rate between the absorber and the fluid hence, higher fluid temperature was observed in the turning region compared to the middle zone (1.2 m).

Temperature uniformity for L-SHC and T-SHC is apparent in figure 3. The temperature contour indicates larger temperature gradient in the T-SHC compared to L-SHC, inferring vortex was formed in the air passage of T-SHC, particularly at the high temperature zone observed near the outlet vent (figure 3b). In this region, heat transfer between the fluid and the absorber is improved due to the separation and reattachment of airflow that would occurs at the downstream of baffle, allowing the air to be constantly heated by the absorber hence, resulting higher temperature at this region. Hu *et al.* [11] reported similar trend in a four-baffles solar heat collector. The study reveals that in a serpentine flow channel, the fluid temperature attained across the air channel is exactly correspond to the flow pattern. For L-SHC, high temperature zone was observed in the turning region (figure 3a). At each turn, an increment in temperature was observed, followed by a slight reduction until the next turning edge. Similarly, it can be inferred that reattachment of flow occurs in the turning edge resulting high heat transfer rate at this region. In addition, a more uniform temperature distribution was established in L-SHC compared to T-SHC. Potgieter *et al.* [17] revealed that the presence of fins that allow fluid to flow in parallel and counter pattern exhibits uniform temperature at each consecutive pass.

The predicted thermal efficiency for smooth, L-SHC and T-SHC is 26.6, 46.2, and 43.4%, respectively. It can be concluded that the installation of baffles in solar heat collector significantly improve the efficiency. Despite the fact that vortex zone formed in the air passage enhance the heat transfer between the fluid and absorber, the flow configuration in baffled solar heat collectors increases the residence time of fluid in the air passage. Consequently, heat on the absorber plate is fully utilized by the fluid due to prolonged heating time thus, improve the efficiency of collector. A slightly higher thermal enhancement was observed in the L-SHC, which can be explained due to the parallel and counter flow pattern produced in the air passage.

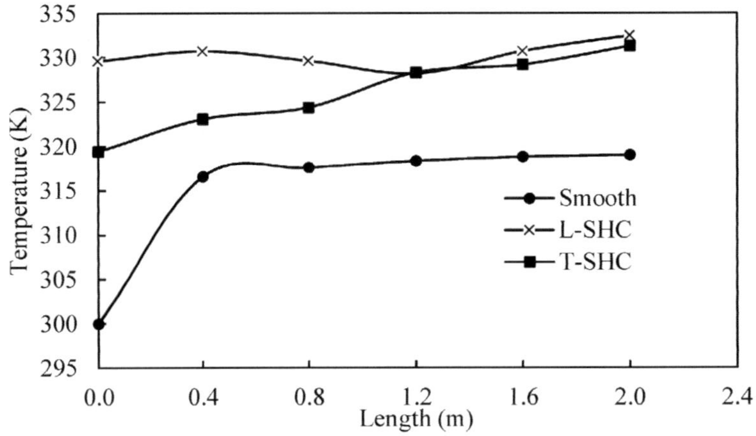

Figure 2. The predicted fluid temperature at the middle region (y = 0.05 m) of the solar heat collector across the length of air passage (m = 0.012 kg/s).

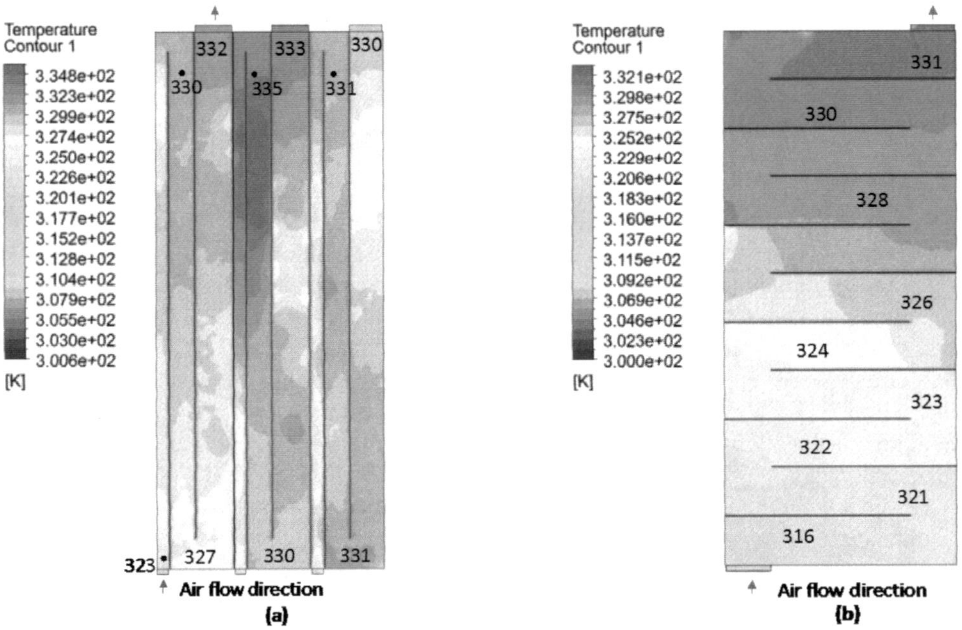

Figure 3. Temperature distribution of fluid at the middle region (y = 0.05 m) of air passage in (a) L-SHC and (b) T-SHC.

5.2. *Heat transfer characteristics*

In this section, heat transfer characteristics between the fluid and the absorber within the air passage is studied in terms of the local convection heat transfer coefficient (h_x) obtained directly from Fluent and Nusselt number (Nu) calculated as in equation (6). The variation of predicted local convective heat transfer coefficient along the length of air passage for each solar heat collector is presented in figure 4. As illustrated in the graph plotted, both baffled solar heat collector exhibits higher h_x compared to the smooth one, with maximum value of 4.15 W/m^2-K recorded along the air passage. As a comparison, highest heat transfer coefficient recorded in L-SHC and T-SHC was 16.68 W/m^2-K and 18.81 W/m^2-K respectively. This finding indicates that the heat transfer in solar heat collector can be significantly improved by using baffles. Mahanand and Senapati [24] reported in their previous study that the convective heat transfer in solar heat collector attached with ribs is improved due to the breaking of boundary layer in the air passage, promoting rapid heat transfer between the fluid and absorber plate.

The effect of baffles arrangement on the local convective heat transfer coefficient can also be identified in figure 4. As observed, T-SHC exhibits higher h_x along the air passage compared to L-SHC, inferring larger turbulence was created in T-SHC. As baffles were arranged transversely across the passage, fluid flow will be constantly interrupted by the baffles while approaching towards the outlet vent. This disturbance created in the air passage then results rapid fluid mixing, leading to the improvement of convective heat transfer. Conversely in L-SHC, flow modification occurs only at the turning edge. This explained the lower h_x observed across the air duct.

The Nusselt number calculated using the average convective heat transfer coefficient across the smooth passage, L-SHC and T-SHC is 27.63, 110.93, and 125.12, respectively. As expected, T-SHC has a higher Nusselt number value, inferring fewer dead zones among the tested solar heat collector. Ozgen *et al.* [25] and Abdullah *et al.* [26] confirmed the reduction of dead zone corresponding to the turbulence regime. In the L-SHC, less turbulence region is expected hence, yields slightly lower Nu value. Significantly, the heat transfer enhancement in a transversal baffle type solar heat collector is a result of convective heat transfer improvement correspond to the strong vortex flow produced in the air passage. Whereas the major enhancement mechanism occurs in a longitudinal baffle type solar heat collector is the result of prolonged heating time between the fluid and absorber plate.

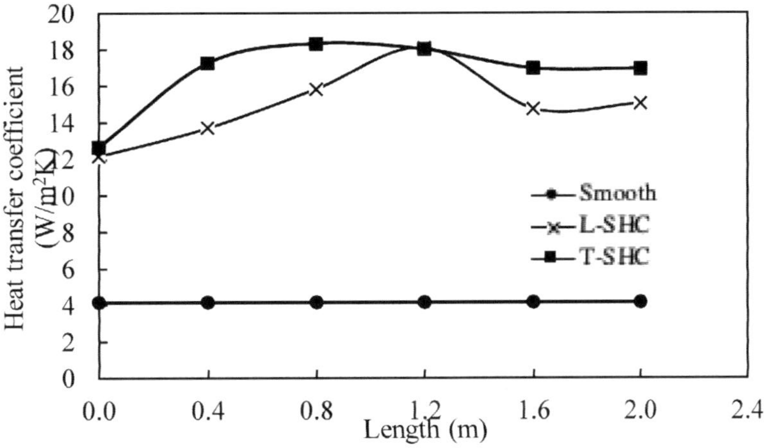

Figure 4. Variation of predicted local convective heat transfer coefficient at the middle region (y = 0.05 m) of air passage in L-SHC and T-SHC (m = 0.012 kg/s).

5.3. Fluid flow characteristics

This section presents the internal flow of both L-SHC and T-SHC at the middle region (y = 0.05 m) of the air passage. The velocity vectors are displayed in figure 5. It was found that the flow pattern in L-SHC is less turbulence than that of the T-SHC with significant vortex indicated along the passage. This finding revealed that the thermal enhancement in solar heat collector through the separation and reattachment of flow in longitudinal baffles arrangement is less significant than the transverse baffles which are attached perpendicular to the airflow direction. It was reported that the flow pattern created by the presence of transversal baffles could increase the air velocity, resulting turbulence flow across the passage [27]. In L-SHC, no significant increment in the variation of air velocity was observed, indicating the heat transfer in this collector occurs completely under laminar flow regime. Hence, explained T-SHC exhibits more efficient convective heat transfer.

Despite to ensure maximum heat transfer between the fluid and absorber occurs in the passage, it is equally important to reduce heat loss from the absorber to the external surrounding through radiation heat transfer. Therefore, it is vital to ensure that convective heat transfer dominates the heat transfer process in the solar heat collector. Temperature distribution across the absorber for L-SHC and T-SHC is as illustrated in figure 6. As expected, L-SHC exhibits higher temperature zone on the absorber, indicating more convective heat loss. Abo-Elfadl *et al.* [28] confirmed that the heat loss from solar heat collector depends on the absorber temperature. In spite of the fact that the convective heat transfer in L-SHC is less efficient, this flaw is compensated by the increasing residence time of fluid over the absorber. Consequently, the efficiency can be improved through the prolonged heating time in the air passage. Therefore, it may be argued that the drawback of heat loss in L-SHC can be diminished.

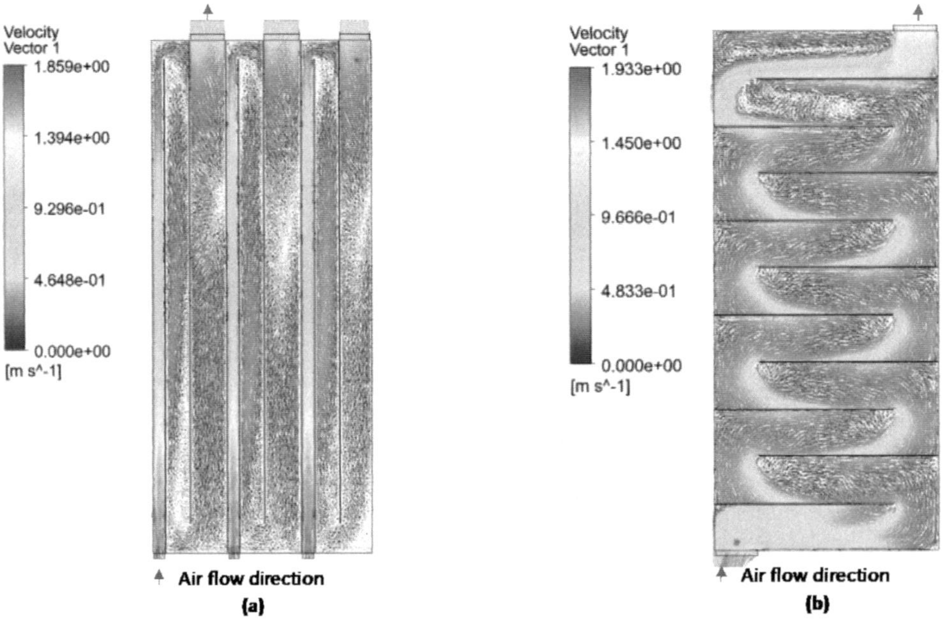

Figure 5. Velocity vector of fluid at the middle region (y = 0.05 m) of air passage in (a) L-SHC and (b) T-SHC.

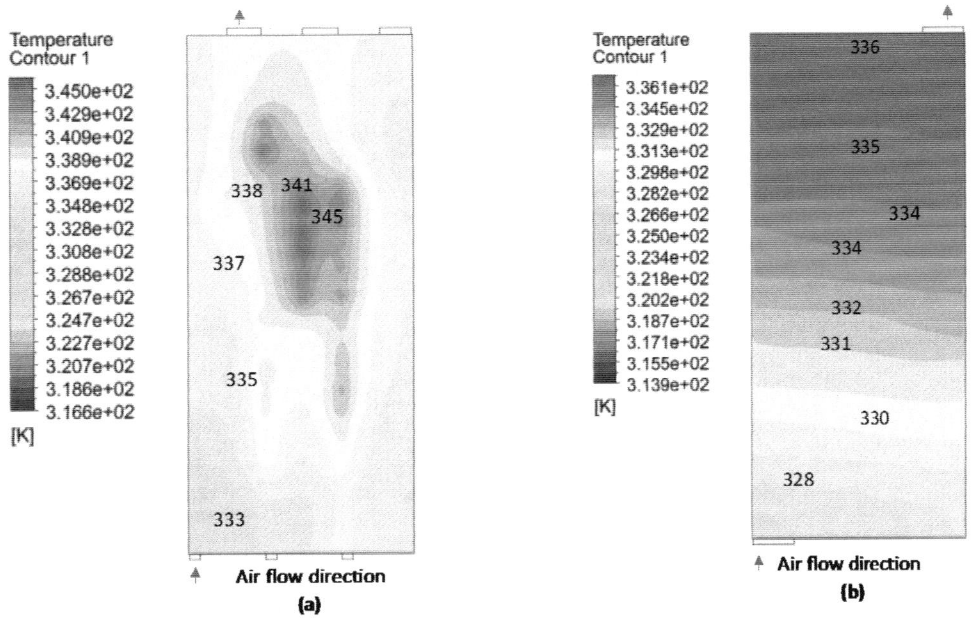

Figure 6. Temperature distribution across the absorber in (a) L-SHC and (b) T-SHC.

5.4. Thermo-hydraulic performance

Thermo-hydraulic performance of a solar heat collector concerns with the enhancement of heat transfer and also the pressure drop established across the collector towards the outlet vent. Pressure contour for both baffled solar heat collector is shown in figure 7. As in the smooth air passage, the predicted pressure drop was 0.038 Pa. From figure 7, it was found that both L-SHC and T-SHC shows higher pressure drop with maximum value of 0.27 and 6.84 Pa respectively recorded in the air passage. This indicates that the position of baffles has significant effect on flow resistance. As calculated using equation (7), the friction factor for L-SHC and T-SHC is 0.58 and 14.66 respectively. The corresponding thermo-hydraulic performance factor (THPF) is 2.1 and 0.8 for L-SHC and T-SHC respectively. Lower THPF was found in the T-SHC, inferring that the flow resistance (friction factor) outweighed the enhancement of heat transfer in the collector. However, the friction factor is expected to reduce at increasing Reynold number [29]. Hence, pressure drop might not have a sensible effect under forced convection. The results predicted by CFD in this study is within the range reported in the literature with Reynolds number ranging from 1200 to 20 000 [30].

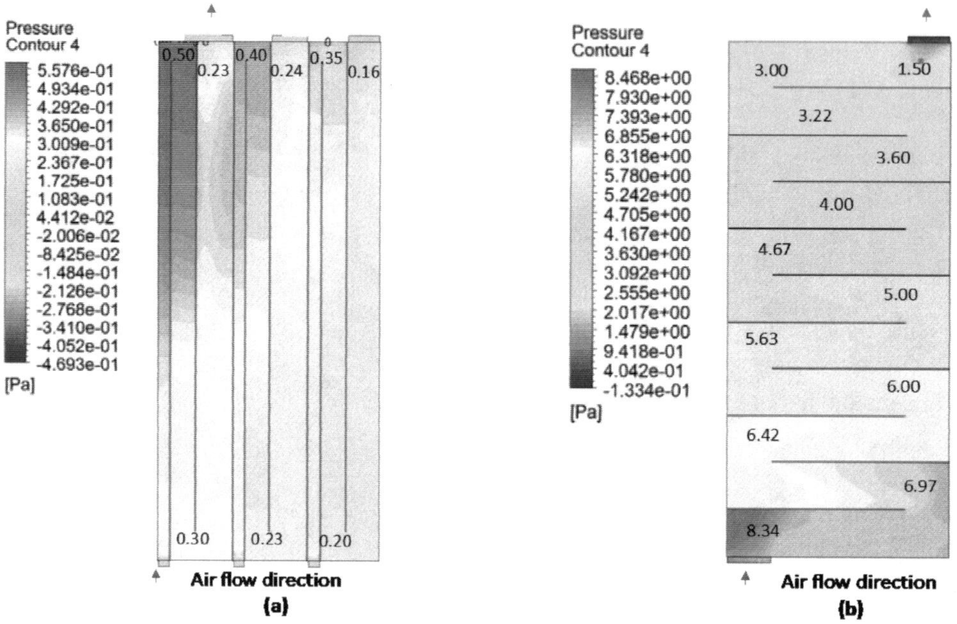

Figure 7. Pressure drop at the middle region (y = 0.05 m) of air passage in (a) L-SHC and (b) T-SHC.

6. Conclusion

The effect of different baffle arrangement on the heat transfer characteristics, fluid flow characteristics and pressure drop across the solar heat collector has been studied. The main objective of this study is to identify the enhancement mechanism in the proposed arrangement – baffles fixed parallel towards the air direction (L-SHC) and baffles fixed perpendicularly towards the air direction (T-SHC). The major findings from this research are listed as follow:

1. The L-SHC exhibits higher thermal efficiency compared to the T-SHC due to the longer residence time in the air passage. This indicates that fluid pass through the L-SHC fully utilized heat on the absorber through the prolonged heating time thereby attained higher fluid temperature at the collector outlet.
2. Highest local heat transfer coefficient was identified in the T-SHC with a maximum value of 18.81 Wm^{-2}, indicating the presence of greater turbulence flow compared to L-SHC. The Nusselt number was found as 125.12. This finding implied that the enhancement in T-SHC was achieved through rapid heat transfer.
3. Velocity contour extracted from CFD shows that fluid in T-SHC has more intense mixing, inferring convective heat transfer dominates the heat transfer process in the collector and that the heat loss can be minimized.
4. Despite the fact that the convective heat transfer in L-SHC is less efficient, the high temperature zone on the absorber plate is insignificant indicating lower potential of heat loss occurs. This can be explained as the prolonged heating time could compensate this drawback.
5. Pressure drop in the T-SHC is significantly higher than that of L-SHC. The corresponding thermo-hydraulic performance factor (THPF) is 2.1 and 0.8 for L-SHC and T-SHC respectively.

References

[1] Chwieduk D A 2016 8—Active solar space heating *Advances in Solar Heating and Cooling.* (Woodhead Publishing); pp 151–202
[2] Chand T K, Mohanty M K and Mohanty R C 2015 An overview of solar energy and its application

in solar dryers with brief concept of energy and energy analysis *Int. J. Res.* **2** 870–7
[3] Singh I and Singh S 2018 A review of artificial roughness geometries employed in solar air heaters *Renew. Sustain Energy Rev.* **92** 405–25
[4] Borhanazad H, Mekhilef S, Saidur R, and Boroumandjazi G 2013 Potential application of renewable energy for rural electrification in Malaysia *Renew. Energy* **59** 210–9
[5] Arunkumar H S, Vasudeva Karanth K and Kumar S 2020 Review on the design modifications of a solar air heater for improvement in the thermal performance *Sustain. Energy Technol. Assessments* **39** 100685
[6] Satcunanathan S and Deonarine S 1973 A two-pass solar air heater *Sol. Energy* **15** 41–9
[7] Ho C D, Yeh H M and Chen T C 2011 Collector efficiency of upward-type double-pass solar air heaters with fins attached *Int. Commun. Heat Mass Transf.* **38** 49–56
[8] Jia B, Liu F and Wang D 2019 Experimental study on the performance of spiral solar air heater *Sol. Energy* **182** 16–21
[9] Hu J and Zhang G 2019 Performance improvement of solar air collector based on airflow reorganization: A review *Appl. Therm. Eng.* **155** 592–611
[10] Romdhane B S 1987 *Contribution to the study and the development of pumps and solar air collectors* (University of Valenciennes, France).
[11] Hu J, Sun X, Xu J and Li Z 2013 Numerical analysis of mechanical ventilation solar air collector with internal baffle *Energy Build* **62** 230–8
[12] Hu J, Liu K, Guo M, Zhang G, Chu Z and Wang M 2019 Performance improvement of baffle-type solar air collector based on first chamber narrowing *Renew. Energy* **135** 701–10
[13] Moummi N, Youcef-Ali S, Moummi A and Desmons J Y 2004 Energy analysis of a solar air collector with rows of fins. *Renew. Energy* **29** 2053–64
[14] Bensaci C E, Moummi A and Labed A 2020 Experimental investigation on heat transfer coefficient and thermal efficiency of solar air heaters having different baffles *Green Energy and Technology* (Springer International Publishing) pp 309–32
[15] Kumar A and Kim M-H 2016 Thermal hydraulic Performance in a solar air heater channel with Multi V-Type Perforated Baffles *Energies* **9** 564
[16] Kumar A and Layek A 2018 Thermo-hydraulic performance of solar air heater having twisted rib over the absorber plate *Int. J. Therm. Sci.* **133** 181–95
[17] Potgieter M S W, Bester C R and Bhamjee M 2020 Experimental and CFD investigation of a hybrid solar air heater *Sol. Energy* **195** 413–28
[18] Tan A S T, Janaun J, Tham H J, Siambun N J and Abdullah A 2021 Performance analysis of a solar heat collector through experimental and CFD investigation
[19] Romdhane B S 2007 The air solar collectors: Comparative study, introduction of baffles to favor the heat transfer *Sol. Energy* **81** 139–49
[20] ANSYS 2009 ANSYS FLUENT 12.0 Theory Guide *ANSYS Inc.*
[21] Bhushan B and Singh R 2011 Nusselt number and friction factor correlations for solar air heater duct having artificially roughened absorber plate *Sol. Energy* **85** 1109–18
[22] Chabane F, Grira F, Moummi N and Brima A 2019 Experimental study of a solar air heater by adding an arrangement of transverse rectangular baffles perpendicular to the air stream. *Int. J. Green Energy* **16** 1264–77
[23] Lee M, Park G, Park and Kim C 2020 Improvement of grid independence test for computational fluid dynamics model of building based on grid resolution *Adv. Civ. Eng.* **2020**
[24] Mahanand Y and Senapati J R 2021 Thermo-hydraulic performance analysis of a solar air heater (SAH) with quarter-circular ribs on the absorber plate: A comparative study *Int. J. Therm. Sci.* **161** 106747
[25] Ozgen F, Esen M and Esen H 2009 Experimental investigation of thermal performance of a double-flow solar air heater having aluminium cans *Renew. Energy* **34** 2391–8
[26] Abdullah A S, Abou Al-sood M M, Omara Z M, Bek M A and Kabeel A E 2018 Performance evaluation of a new counter flow double pass solar air heater with turbulators *Sol. Energy* **173**

398–406
[27] El-khawajah M F, Aldabbagh L B Y and Egelioglu F 2011 The effect of using transverse fins on a double pass flow solar air heater using wire mesh as an absorber *Sol. Energy* **85** 1479–87
[28] Abo-Elfadl S, Hassan H and El-Dosoky M F 2020 Study of the performance of double pass solar air heater of a new designed absorber: An experimental work. *Sol. Energy* **198** 479–89
[29] Bensaci C E, Moummi A, Sanchez de la Flor F J, Rodriguez Jara E A, Rincon-Casado A and Ruiz-Pardo A 2020 Numerical and experimental study of the heat transfer and hydraulic performance of solar air heaters with different baffle positions *Renew. Energy* **155** 1231–44
[30] Chamarthi S and Singh S 2021 A comprehensive review of experimental investigation procedures and thermal performance enhancement techniques of solar air heaters *Int. J. Energy Res.* **45** 5098–164

Acknowledgments
This study was supported by Ministry of Finance Malaysia (MOF) Social Enterprise Project (Project Code: R/MOF/B0108/01759A/001/2020/0072), in collaboration between Universiti Malaysia Sabah and Universiti Malaysia Kelantan (UMS Project Code: LPK2001).

Chitosan-TiO_2 nanotubes scaffolds for proliferation and early differentiation of MG63 by functionalization with fetal bovine serum

S S Lim[1], H M Zu[2] and H S Loh[2]

[1] Department of Chemical and Environmental Engineering, Faculty of Science and Engineering, University of Nottingham Malaysia, Jalan Broga 43500 Semenyih Selangor.
[2] School of Biosciences, Faculty of Science and Engineering, University of Nottingham Malaysia, Jalan Broga 43500 Semenyih Selangor.

Email: SiewShee.Lim@nottingham.edu.my

Abstract. Scaffolds have been used as alternative biomaterials to overcome physiological bone disorders. Production of scaffolds has been challenging to fulfil the following criteria: biodegradability, mechanical sustainability, and biocompatibility. For cellular interaction, protein adsorbed on scaffold surface is important for osteoblastic activities. This study aimed to functionalize chitosan-TiO_2 nanotubes scaffolds with fetal bovine serum and investigate in vitro efficacy of such scaffolds with fetal bovine serum. Chitosan-TiO_2 nanotubes scaffolds were prepared via direct blending and lyophilization. They were then functionalized with fetal bovine serum via adsorption for 4, 8, 12 and 24 h. The in vitro efficacy of the functionalized scaffolds was evaluated using MG63 cells. The adsorption of fetal bovine serum onto the scaffolds was complex where saturation of adsorption was hardly attained. The in vitro efficacy of scaffolds with adsorbed fetal bovine serum was higher than that of those without fetal bovine serum by promoting better osteoblastic functions. Notably, the scaffolds functionalized for 4 h enhanced cell adhesion and proliferation on 7 day suggesting good regulation of osteoblastic binding and proliferation. ALP protein was expressed on 26 day in all functionalized scaffolds. Chitosan-TiO_2 nanotubes scaffolds with adsorbed fetal bovine serum can be a potential regenerative material for bone regeneration.

1. Introduction

A huge number of people suffered from bone diseases like osteomalacia, osteogenesis imperfecta and Paget's disease. In most cases, the removal of defective bones is inevitable and results in bone void or loss. This greatly affects the functionality of the patients. The current replacement of bones is very troublesome of several revisions of surgeries which are time consuming and cause finance burden to patients. Therefore, bone tissue engineering as a therapeutic method has gradually gained attention in treating bone diseases and disorders. To achieve guided bone regeneration, three-dimensional (3D) scaffolds serve as temporary support to stimulate bone growth. These scaffolds have often been fabricated with improved physicochemical properties, biocompatibility and mechanical properties.

Scaffolds with bioinspired features are very critical for cellular functions like proliferation, differentiation and mineralization. The ultimate faith of scaffolds is to get replaced by new bone tissue.

Ideally, the degradation rate of scaffolds matches the bone growth rate. To make scaffolds degradable, natural polymers like chitosan have been extensively researched in biomedical aspects [1-2]. Chitosan possesses high hydrophilicity which is responsible for its exceptionally high swelling behaviour. This feature facilitates the anchorage of bone cells on the scaffolds which is pre-requisite of subsequent osteoblastic activities [3]. However, chitosan cannot be used alone in guiding bone growth due to its poor mechanical strength and low bone-bonding ability [4]. Hence, the reinforcement of chitosan with fillers especially nanoparticles becomes an imperative step.

In recent decades, nanoparticles evidently promoted enhanced osteoblastic functions due to their close resemblance to the building blocks of bone [5–6]. Notably, nanohydroxyapatite and TiO_2 nanotubes showed better interactions with the host cells after their incorporation into chitosan-based scaffolds [4,7,8]. These studies affirmed the excellent osteoconductivity and mechanical stability of nanoparticles by promoting cellular adhesion and proliferation via enhanced adsorption of proteins and metal ions [4,9]. In particular, the high adsorption affinity of chitosan-TiO_2 nanotubes scaffolds towards calcium ions was proven to elevate the *in vitro* efficacy of the scaffolds [4].

Upon the implantation of scaffolds in host tissue, this is quickly followed by plasma proteins adsorption to the scaffolds [10]. The deposition of plasma proteins on scaffolds regulates the efficacy of the biomaterials. This can also be perceived as an in-situ functionalization of the scaffolds. Protein adsorption is a dynamic process, as serum proteins arrive and leave the surface of biomaterials in a time-dependent manner [11]. Besides, the adsorption behaviour of plasma proteins depends on the type of scaffolds [12]. With high adsorption affinity towards calcium ions, chitosan-TiO_2 nanotubes scaffolds were expected to facilitate sites for the adsorption of plasma proteins. This study aimed to develop bioinspired scaffolds by functionalizing chitosan-TiO_2 nanotubes scaffolds with fetal bovine serum (FBS) via adsorption over 24 h. Subsequently, the *in vitro* efficacy of functionalized scaffolds was verified using MG63 osteoblast-like cell for adhesion, proliferation, and early differentiation. Finally, the expression of housekeeping protein in osteoblast differentiation was examined by using a dot-blot semi-quantitative analysis. The correlation between the adsorption duration of proteins and *in vitro* efficacy of the new scaffolds would then be elucidated.

2. Methods

2.1. *Fabrication of Chitosan-TiO_2 nanotubes scaffolds*

Synthesis of the TiO_2 nanotubes was attempted by employing the method developed by Chen *et al.* (2002). Two grams of TiO_2 (Sigma Aldrich, USA) was mixed with 40 mL of 10 M NaOH (Sigma Aldrich, USA) in a Teflon container for 5 min. The Teflon container containing mixture of TiO_2 and NaOH was screwed tightly in a stainless-steel reactor and heated to 150 °C for 72 h. Under hydrothermal condition (at 150 °C), the corrosiveness of NaOH was elevated. As a result, NaOH delaminated the TiO_2 in micron into nanosheets. All these nanosheets were subsequently rolled into nanotubes by the movement of NaOH. After 72 h, white powder was collected from the reactor and filtered with distilled water and 0.1 M HCl (Merck, Germany). The washing was completed when a neutral pH of the powder was achieved. Finally, the powder was dried at 50 °C for 6 h prior to the grinding and 5-h of calcination at 400 °C to increase the crystallinity of the powder.

Subsequently, the chitosan powder was dissolved in 0.2 M of acetic acid (Merck, Germany) and stirred homogenously for 2 h. Approximately 16 weight percent (wt%) of TiO_2 nanotubes which was previously synthesized and chitosan solution were agitated using a magnetic stirrer for 5 h. One mL of the mixture was transferred to each well of a 24-well plate. The same plate was frozen at -20 °C for 24 h. It was then freeze dried at -40 °C and 0.12 mbar for 24 h. Freeze-dried scaffolds were rehydrated with different concentration of ethanol starting with 1-h immersion in absolute ethanol. This was then followed by immersion in 70% ethanol and 50% ethanol for 30 min respectively. This rehydration ensured sterilization of the scaffolds and also gradually introduced water into the scaffolds. After rehydration, scaffolds were dehydrated in a desiccator prior to protein adsorption and *in vitro* tests.

2.2. *Functionalization of Scaffolds via Adsorption of FBS*
The mass of the scaffolds was measured. Then, scaffolds were completely immersed in phosphate buffered saline (PBS) overnight before the functionalization with FBS. After 24 h of incubation with PBS at room temperature, the solution was removed. In the functionalization, chitosan-TiO_2 nanotubes scaffolds were immersed in 2 mL of FBS at different time intervals (4, 8, 12 and 24 h). The incubation was carried out at 37 °C. After each time interval of incubation, 1 mL of FBS was taken to measure the concentration of the protein remnants after the adsorption by the scaffolds. The remaining FBS was removed and scaffolds were gently washed with PBS for three times to remove free and loosely bound proteins. The Bradford assay was used to quantify the amount of adsorbed proteins and the initial concentration of protein in FBS was also determined. The remaining FBS was diluted by 4000 times with PBS and the diluted solution was mixed with Bradford reagent (Bio-Rad Laboratories) at 1:1 ratio in a 96-well plate. The absorbance of the diluted FBS solution was measured at 595 nm using Varioskan Flash multimode reader (Thermo Scientific, USA). The amount of FBS adsorbed (q_e) was calculated using equation (1).

$$q_e = \frac{(c_o - c_e)V}{w} \quad (1)$$

Where q_e was the amount of protein adsorbed per gram of scaffold (mg/g), C_o was initial concentration of protein in FBS, C_e was the concentration of FBS after adsorption (mg/mL) at different time intervals, V was the volume of FBS in each well (2 mL) and w is the mass of scaffolds (g).

2.3. *Cell Culture*
MG63 cells (ATCC, non-USA) were cultured in a 75-mL cell flask with minimum essential medium (MEM) supplemented with 10% FBS, 100 mM of sodium pyruvate, 1% penicillin-streptomycin, and 1% sodium bicarbonate. The cell flask was incubated at 37 °C with 5% CO_2. The culture medium was changed every 3 days and harvested prior seeding on the scaffolds.

2.4. *Fluorescein Diacetate (FDA) and Propidium Iodide (PI) Staining*
MG63 cells ($3x10^4$ cells/well) were seeded on functionalized the chitosan-TiO_2 nanotubes scaffolds and incubated for 4 h. After 4 h, 5 μg/ml of FDA solution and 1 μg/ml of PI solution were added to observe the survival and death of cells under a florescence microscope (Nikon, Japan).

2.5. *Proliferation Assay*
MG63 cells ($3x10^4$ cells/well) were seeded on the functionalized scaffolds and incubated in complete media for 3, 5, and 7 days. MG63 cultured directly on well served as control. After each incubation period, complete media was discarded and replaced with pure MEM in each well. Subsequently, 5 mg/mL of MTT solution (3-(4,5-dimethylthiazol-2-yl)-2,5-diphenyltetrazolium bromide) was added to each well and incubated for 4 h at 37 °C until the reduction of the yellow tetrazole (MTT solution) to purple formazan by the living cells. After 4 h, dimethyl sulfoxide (DMSO, 1.0 mL/well) was added and mixed with purple crystal homogeneously. The absorbance of DMSO-formazan solution was measured and recorded at 570 nm using Varioskan Flash multimode reader. The reading was recorded three times for each scaffold. The viability of MG63 was determined by using equation (2).

$$Cell\ viability\ (\%) = \frac{OD_{scaffold}}{OD_{control}} x\ 100\% \quad (2)$$

Where $OD_{scaffold}$ was the optical density of MG63 cultured on the scaffolds functionalized for different time intervals and $OD_{control}$ was the optical density of MG63 on cultured on well.

2.6. *Alkaline Phosphatase (ALP) Activity Assay*
Early differentiation of MG63 was analysed by using alkaline phosphatase (ALP) at 7, 14, 21, and 26 days. The cells and scaffolds were washed with PBS for three times and homogenized with cell lysis solution (Pro-Prep, protein extraction solution). The mixtures were centrifuged at 10,000 rpm at 4 °C

for 10 min. Approximately 100 μL of supernatant was collected and incubated with *P*-nitrophenyl-phosphate liquid substrate at 1:1 volume ratio in 96-well plate for 2 h. The incubation was carried out at 37 °C. In the presence of alkaline phosphatase, the liquid substrate changes its color from transparent to pale yellow. After 2 h of incubation, the absorbance of the solution was measured at 405 nm using Varioskan Flash multimode reader. The remaining supernatants were stored at -80 °C for dot blot analysis.

2.7. Expression of Housekeeping Proteins (GAPDH and ALP)
The supernatant from ALP assay was used for the expression of housekeeping protein such as GAPDH (Glyceraldehyde 3-phosphate dehydrogenase) and ALP (Alkaline Phosphatase). Polyvinylidene fluoride (PVDF) membrane was used for the dot blot analysis. The grid lines were drawn on the membrane for the supernatant to load. Approximately 10 μg (10 μL) of supernatant was loaded on the membrane and air dried. Then, the membrane was soaked with blocking buffer (PBS-Tween containing 4% milk) in a petri dish and shaken at 60 rpm horizontally at room temperature for 1 h. The blots were incubated in 1 μg/mL of mouse monoclonal anti-GAPDH (Abcam, UK) or 0.5 mg/mL of anti-alkaline phosphatase antibody (ABcam, UK) at 1:10000 dilution in blocking buffer and shaken horizontally at 60-70 rpm for 2 h. Then, the membranes were washed with 10 mL of PBS-Tween 3 times for 5 min. About 2 mg/mL of rabbit polyclonal anti-mouse IgG Horseradish Peroxidase conjugated antibody (ABcam, UK) was diluted at 1:10000 in PBS-Tween and used as secondary antibody incubation for 2 h. After washing with PBS-Tween for 4 times, 1 mL of tetramethylbenzidine (TMB) substrate was added to observe the band. The relative protein bands were analysed with QualityOne GS-800 (Biorad, United States).

2.8. Statistical Analysis
Two-way ANOVA was used to analyse MTT and ALP assays. The dot blot analysis and ALP assay were analysed with Pearson correlation for comparison of relationship.

3. Results

3.1. High Adsorption Affinity of Chitosan-TiO_2 Nanotubes Scaffolds towards FBS
Chitosan-TiO_2 nanotubes scaffolds were incubated in FBS at different time intervals and the amount of adsorbed proteins by the scaffolds was quantified. As shown in figure 1, the amount of adsorbed proteins (q_e) after 4 and 8 h of incubation was approximately 300 mg/g of scaffold. However, q_e reduced slightly to 277 mg/g after the incubation was increased to 12 h. This was explained by the replacement of heavier proteins like fibronectin (FN) by lighter proteins like albumin at the incubation period of 12 h. Ultimately, 24 h of incubation marked the highest amount of protein adsorbed by the scaffold at 442 mg. There was no sign of saturation was observed in these scaffolds due to the dynamic nature of protein adsorption and desorption. The high affinity of chitosan-TiO_2 nanotubes scaffolds towards FBS proteins was evidently demonstrated in this study.

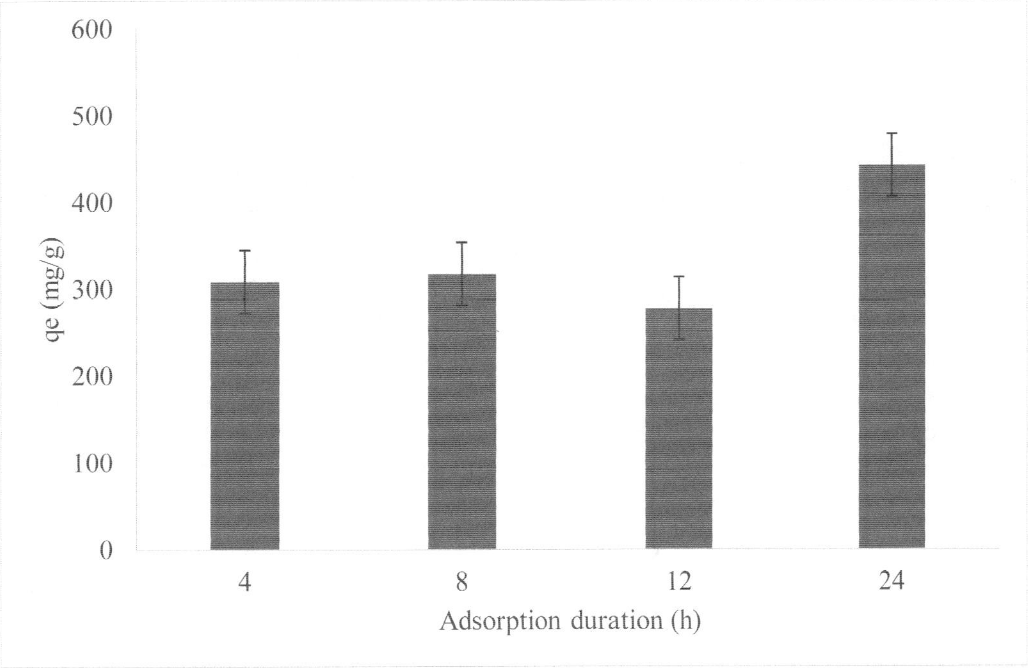

Figure 1. The amount of adsorbed proteins (q_e) by chitosan-TiO_2 nanotubes scaffolds at different time intervals.

3.2. Cell Adhesion on Chitosan-TiO2 nanotubes Scaffolds

FDA and PI reagents were used to observe the viability and death of MG63 on the functionalized scaffolds respectively after 4 h of incubation. FDA is converted by intracellular esterases into fluorescein with green fluorescence which represents for living cells. On the other hand, PI interacts with nucleotides to emit red fluorescence which represents for dead cells. More FDA stained cells were observed in chitosan-TiO_2 nanotubes scaffolds functionalized with FBS for 4 h as shown in figure 2(a). This promising result was also coupled with less PI stained cells on the same scaffold (in figure 2(b)). This suggested that the adsorbed proteins at this time interval were likely to be FN (cell adhesive protein) which facilitated the adhesion of bone cells. Similar outcome was observed in cells cultured directly on well (in figure 2(e)). However, more PI stained cells were detected in scaffolds functionalised for 12 h. The amount of adsorbed protein at 12 h was slightly less than that of at 4 h but the proteins adsorbed at 12 h caused more death of MG63. The scaffolds without adsorbed FBS also caused cell death by showing more PI stained cells.

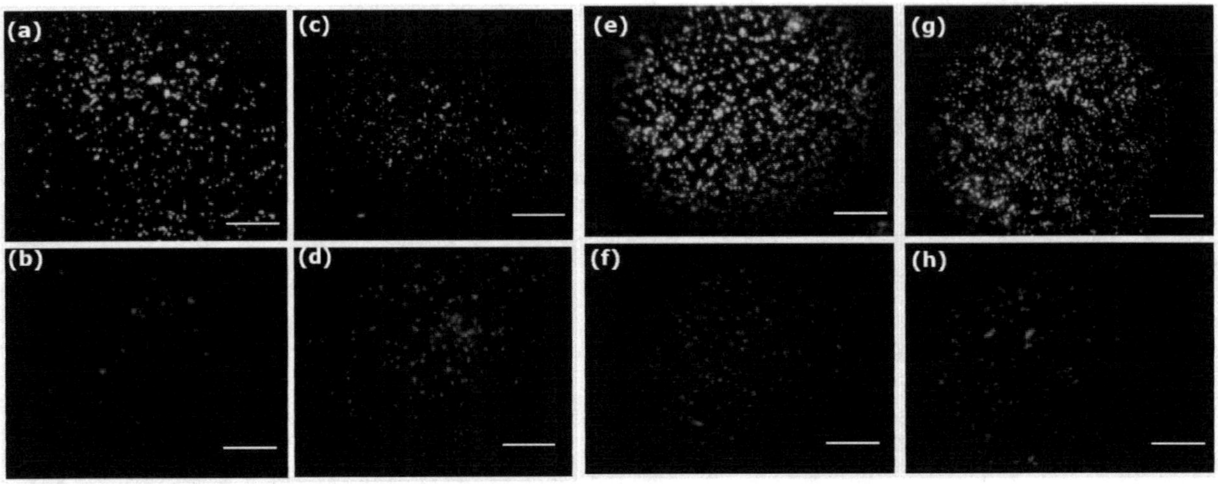

Figure 2. MG63 stained with (a) FDA and (b) PI after 4 h of FBS functionalization (c) FDA and (d) PI after 12 h of FBS functionalization (e) FDA and (f) PI cultured directly on well (g) FDA and (h) PI without FBS functionalization. Scale bars represent 1 mm.

3.3. High cell viability on FBS-functionalize scaffolds

MTT is a colorimetric assay for quantifying cell viability via the reduction of MTT dye (tetrazolium) into formazan (purple colour) by cellular oxidoreductase enzymes in living cells. The cell viability of MG63 cultured on the functionalized chitosan-TiO_2 nanotubes scaffolds after 3, 5 and 7 days was evaluated and MG63 cultured directly on well was control. In figure 3, the cell viability of MG63 cultured on the functionalised scaffolds ranged roughly from 120% to 180% which was higher than the control. This suggested that the adsorbed proteins promoted cell proliferation of MG63 more effectively. In general, the proliferation of MG63 cultured on the functionalized scaffolds increased with incubation periods. The scaffolds functionalized with FBS for 4 h gave the highest proliferation at day 7 compared to the remaining scaffolds as well as the controls (*$p<0.05$).

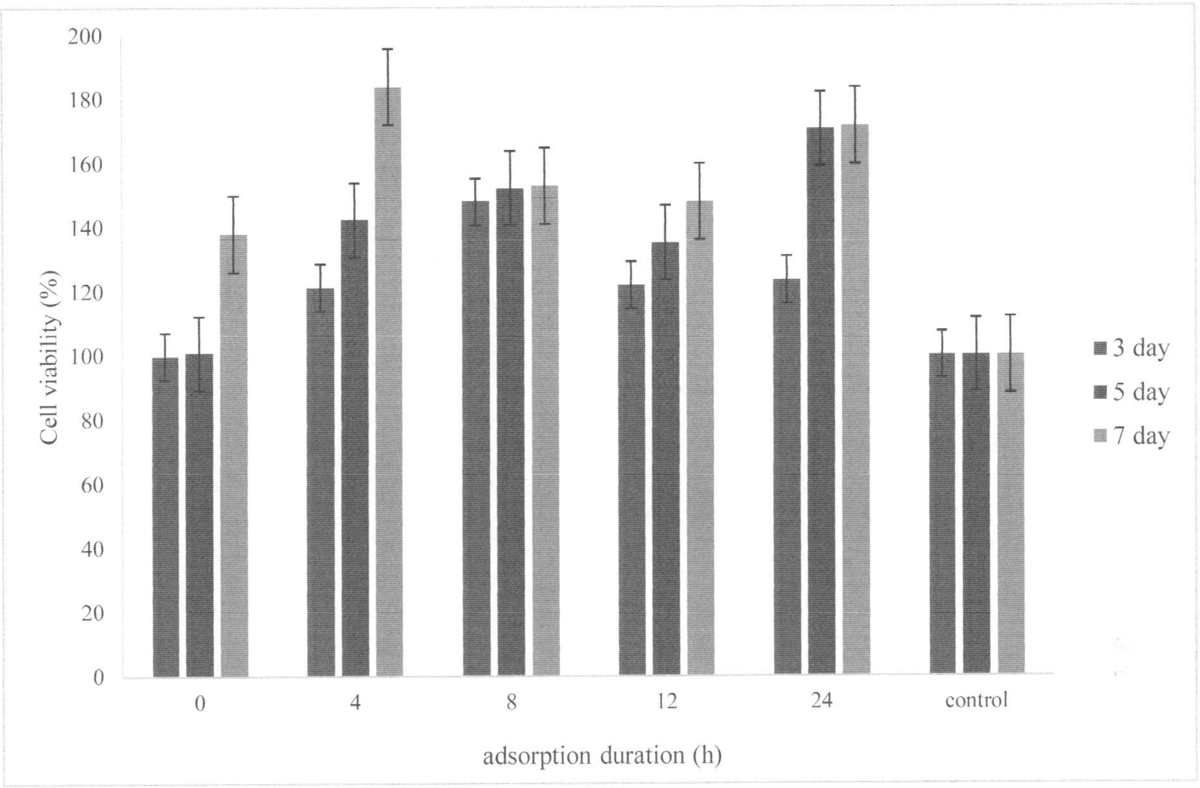

Figure 3. Histogram showing proliferation of MG63 cells seeded on functionalized chitosan- TiO_2 nanotubes scaffolds after 3, 5, and 7 days of incubation in terms of cell viability. The error bars represent mean ± SD. These experiments were repeated in triplicate (*$p<0.05$).

3.4. High ALP Activity on Day 26

ALP was used as an early detection of cellular differentiation markers on MG63 cells and was measured with *P*-nitrophenyl-phosphate liquid substrate on 7, 14, 21 and 26 days. As shown in figure 4, ALP activity of MG63 cells grown on chitosan- TiO_2 nanotubes scaffolds gave the highest significant on 26 day than the controls (**p<0.01).

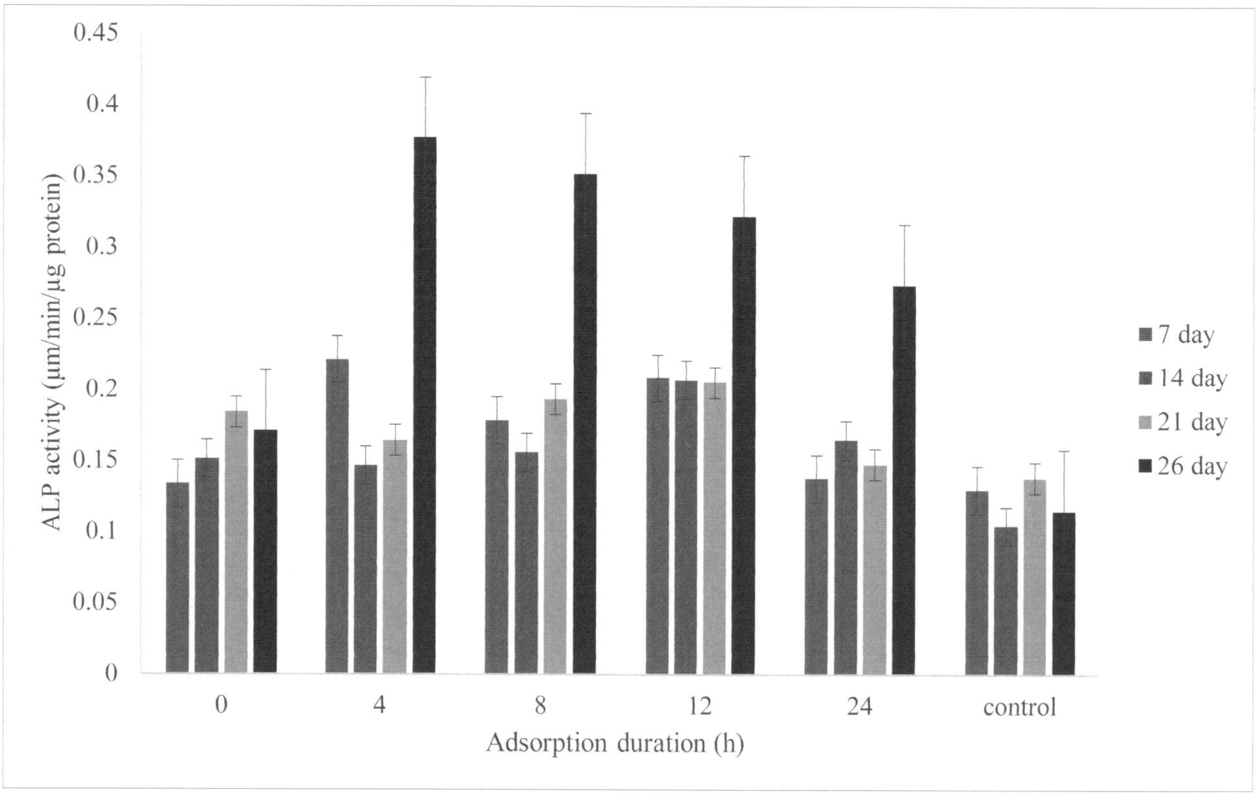

Figure 4. Histogram showing ALP activity of MG63 cells on functionalized chitosan-TiO_2 nanotubes scaffolds at 4 different time intervals after 7, 14, 21 and 26 day of incubation period (**p<0.01).

3.5. *Expression of Housekeeping Proteins of GAPDH and ALP*

GAPDH is one of the housekeeping proteins expressed in all types of cells and thus it can be used as a control for the expression of ALP protein. ALP protein plays an important role in bone differentiation. The functionalized chitosan-TiO_2 nanotubes scaffolds gave differentiation on 26 day and it may be due to ALP protein expression. Therefore, the dot blot analysis was performed to confirm the level of ALP expression on 26 day. It was found that there was positive correlation between ALP enzyme assay and dot blot analysis (figure 5, $p<0.05$).

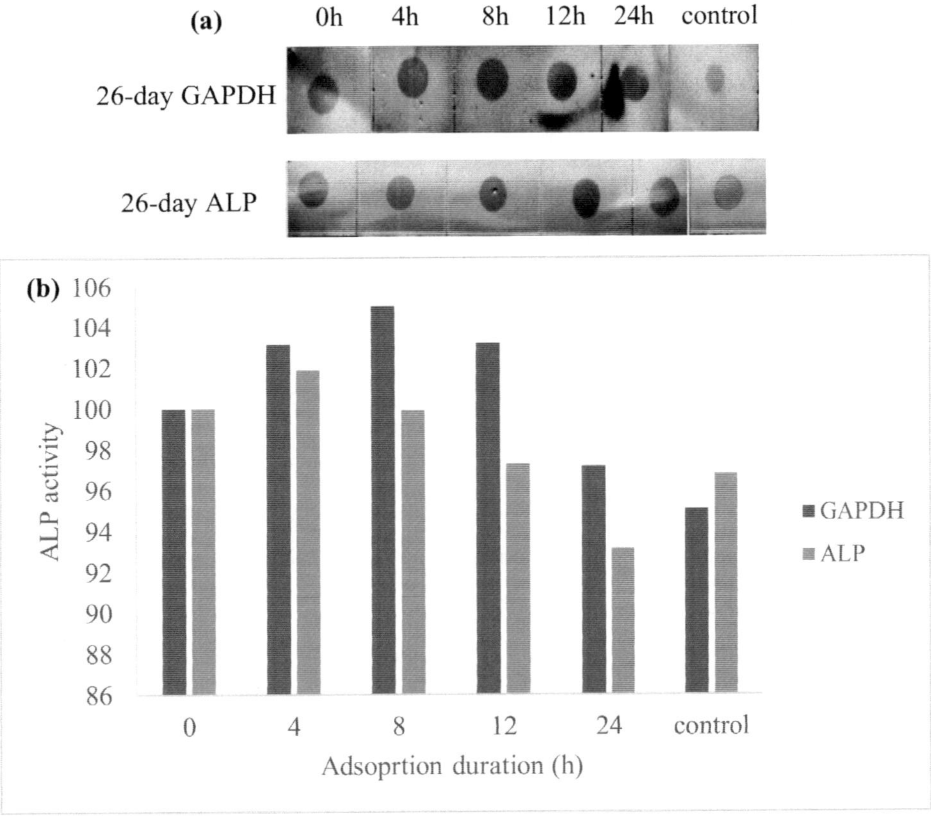

Figure 5. (a) Dot blot analysis of housekeeping proteins and (b) percentage of ALP activity comparison with GAPDH expression on 26 day of incubation. ($p<0.05$, correlation with ALP assay on 26 day).

IOP Conf. Series: Materials Science and Engineering 1195 (2021) 012041 doi:10.1088/1757-899X/1195/1/012041

4. Discussion

In this study, adsorption is a dynamic process between nanoscale surface of the scaffolds and FBS proteins. This is associated with attractive and repulsive forces by local changes of the scaffold surface which can cause changes in density and orientation of adsorbed proteins [13]. It is clear to notice that the amount of adsorbed proteins increases with the increase of adsorption duration except the adsorption for 12 h. The proteins undergo orientation for more contact with the surface of scaffolds by removing some of existing binding proteins [14], which causes the reduction of protein adsorption after 12 h as shown figure 1. The chitosan-TiO_2 nanotubes scaffolds are still not saturated with FBS proteins even after 24 h of adsorption, as there is no plateau observed in figure 1. A plateau is an indicative of scaffolds can no longer uptake any protein. The structural rearrangement of FBS proteins is responsible for accommodating more sites for the subsequently adsorbed proteins [14]. This explains the high adsorption affinity of the scaffolds towards FBS proteins.

In the FDA staining, scaffolds functionalised with FBS for 4 h gave the highest cellular adhesion compared to the other functionalized ones. These results highlighted the role of adsorbed proteins on the modulation of cellular adhesion. This suggests that majority of the adsorbed protein is FN after 4 h of adsorption. FN as a cell adhesive protein regulates the adhesion of bone cells like MG63. As the adsorption duration is prolonged to another 4-8 h, FN which is initially adsorbed by the scaffolds is likely to be displaced by other FBS proteins like albumin [15]. The displacement of FN by other FBS proteins contributed to lower adhesion of MG63 on scaffolds functionalised for 8 and 12 h. Even though the scaffolds without functionalization also show high adhesion of MG63 in figure 2(g), it is also followed by high cell death as shown in figure 2(h). This again affirms the importance of adsorbed proteins in facilitating osteoblastic adhesion. As an adherent type of cell line, MG63 seeded directly on well is also stained with FDA the most. The type of protein bind to the scaffolds should be analysed by using Western blot analysis or recombinant fragments of FN [16]. This would help to identify the actual type of adsorbed protein by the scaffolds.

In MTT assay, protein adsorbed for 4 h gave the highest proliferation on 7 day (figure 3). In general, the cell viability of MG63 on the functionalized scaffolds is higher than that of control. Since cell adhesion is dynamic process with initial cell-protein contact prior to long term adhesion for cellular responses [17]. From the response of MG63, FN was probably activated after being adsorbed on the surface of scaffolds for cellular regulation [15]. Other scaffolds functionalized for more than 4 h exhibited slightly lower cell viability due to the desorption process of FN [17]. It can also be due to albumin that can co-adsorb with FN for competition and replacing FN and resulted in low cellular response [15]. Further extension of MTT assay should be done to examine more proliferation profile to compare with the events of ALP assay.

ALP activity was significantly expressed on 26 day (figure 4). In chitosan-based scaffolds, ALP activity was seen on 14-21 days for osteosarcoma cell [3]. Interestingly, all functionalized scaffolds gave differentiation on 26 day and was even higher than the control which has no change (figure 4). This not only signified the presence of adsorbed proteins as effective biological cues for the scaffolds, but also 3D structure of the scaffolds is the key feature of promoting the early differentiation of MG63. This finding is corroborated with a previous study conducted by Lim and co-workers (2017). They also demonstrated the significance of 3D structure of scaffolds in promoting early differentiation of MG63. Moreover, dot blot analysis confirmed that it was caused by ALP protein expression (figure 5(a)-(b)). Therefore, chitosan-TiO_2 nanotubes scaffolds functionalized with FBS were able to promote MG63 cell adhesion, proliferation as well as differentiation.

5. Conclusion

In conclusion, the chitosan-TiO_2 nanotubes scaffolds showed high adsorption affinity toward FBS over 24 h. Chitosan-TiO_2 nanotubes scaffolds functionalized with FBS for 4 h promoted the highest osteoblastic adhesion which was then coupled with the highest proliferation at day 7. This suggested that the type of FBS protein adsorbed by the scaffolds functionalized for 4 h was mainly FN, the cell adhesive protein. Surprisingly, high ALP activity and ALP protein were demonstrated on 26 day in

MG63 cultured on functionalized scaffolds. Hence, chitosan-TiO_2 nanotubes scaffolds with adsorbed FBS proteins can be an alternative regenerative material in bone tissue engineering.

Acknowledgments
Authors would like to acknowledge University of Nottingham Malaysia for providing finance support in this research work.

References

[1] Costa-Pinto A R, Reis R L and Neves N M 2011 Scaffolds based bone tissue engineering: The role of chitosan *Tissue Eng.* **17** 1–15
[2] Martino A D, Sitinger M and Risbud M V 2005 Chitosan: A versatile biopolymer for orthopaedic tissue-engineering *Biomaterials.* **26** 5983–90
[3] Ozdemir E, Sendemir-Urkmez A and Yesil-Celiktas O 2013 Supercritical CO_2 processing of a chitosan- based scaffold: Can implantation of osteoblastic cells be enhanced *J Supercrit Fluids.* **75** 120–7
[4] Lim S S, Chai C Y and Loh H S 2017 In vitro evaluation of osteoblast adhesion, proliferation and differentiation on chitosan-TiO2 nanotubes scaffolds with Ca^{2+} ions *Mater. Sci. Eng.C.* **76** 144–52
[5] Webster T J, Schadler L S, Siegel R W and Bizios R 2001 Mechanisms of enhanced osteoblast adhesion on nanophase alumina involve vitronectin *Tissue Eng.* **7** 291–301
[6] Webster T J 2007 Nanotechnology for the regeneration of hard and soft tissues *River Edge. NJ, USA*: *Word Scientific.* 3
[7] Brammer K S, Frandsen C and Jin S 2012 TiO_2 nanotubes for bone regeneration *Trends Biotechnol.* **30** 315–22
[8] Oh S H, Finones R R, Daraio C, Chen L H and Jin S 2005 Growth of nanoscale hydroxyapatite using chemically treated titanium oxide nanotubes *Biomaterials.* **26** 4938–43
[9] Kommiredde D S, Sriram S M, Lvov Y M and Mills D K 2006 Stem cell attachment to layer-by-layer assembled TiO_2 nanoparticle thin films *Biomaterials.* **27** 4296–303
[10] Dietscheweiler C 2012 Protein adsorption at solid surface *Swiss Federal Institute of Technology (ETH)*, Zurich
[11] Robert A 2005 Biomaterials: protein-surface interactions *Ency biomater.biomed. eng.* **28** 1–15
[12] Brash J L and Horbett T A 1995 Proteins at interfaces: an overview *ACS Symp. Ser.* **602** 1–25
[13] Roach P, Eglin D, Rohde K, Perry C C, Runt J and Donahue H J 2007 Modern biomaterials: a review – bulk properties and implications of surface modifications *J Mater. Sci.:Mater. Med.* **18** 1263–77
[14] Rabe M, Verdes D and Seeger S 2011 Understanding protein adsorption phenomena at solid surfaces *Adv. Colloid Interface Sci.* **162** 87–106
[15] Horbett T A 1994 The role of adsorbed proteins in animal cell adhesion *Colloids Surf. B.* **2** 225–40
[16] Michael K E, Keslowsky B G and Vernekar V N 2003 Adsorption-induced conformational changes in fibronectin due to interactions with well-defined surface chemistries *Langmuir.* **19** 8033–40
[17] Lord M S, Foss M and Besenbacher F 2010 Influence of nanoscale surface topography on protein adsorption and cellular response *Nano Today.* **5** 66–78

IOP Conf. Series: Materials Science and Engineering 1195 (2021) 012042 doi:10.1088/1757-899X/1195/1/012042

Effect of open pore and pore interconnectivity in the Ni-SDC cermet anode microstructure on the performance of solid oxide fuel cells

S A Muhammed Ali[1], A Muchtar[1,2,*], J C W Mah[1], J Raharjo[3] and D S Khaerudini[4]

[1] Fuel Cell Institute, Universiti Kebangsaan Malaysia, 43600, UKM Bangi, Selangor, Malaysia
[2] Department of Mechanical and Manufacturing Engineering, Faculty of Engineering and Built Environment, Universiti Kebangsaan Malaysia, 43600, UKM Bangi, Selangor, Malaysia
[3] Centre of Materials Technology, Agency for the Assessment and Application of Technology (BPPT), Jl. M.H. Thamrin 8, Jakarta Pusat, 10340, Indonesia
[4] Research Centre for Physics, Indonesian Institute of Sciences (LIPI), Gd.440-442, Kawasan Puspiptek Serpong, Tangerang Selatan, Banten 15314, Indonesia

Email: muchtar@ukm.edu.my

Abstract. In nickel-samarium-doped ceria (Ni-SDC) cermet anode layers, the open pores and interconnected pores in the microstructure are the main factors that affect the mechanical and electrical properties. In this work, porous Ni-SDC cermet anode layers are fabricated using various quantities of potato starch (0 to 25 wt.%) as a pore forming in the anode powders. The properties of the Ni-SDC cermet anode layers were characterised by FESEM-BSE microscopy, Archimedes method for density measurement, Vickers hardness, flexural strength, and DC four-point electrical conductivity. The findings revealed that the different content of potato starch greatly affected the percentage of porosity and pore interconnectivity in the microstructure and consequently altered the mechanical and electrical properties of the Ni-SDC cermet anode. The degree of shrinkage, relative density, mechanical strength and electrical conductivity of the Ni-SDC cermet anodes decreased as their pore former content increased. Furthermore, the research shows that the large porosity (> 40%) in the Ni-SDC cermet anode microstructure affected the continuity of Ni-Ni, SDC and Ni-SDC phases and thereby affected the mechanical and electrical properties. The Ni-SDC cermet anode with 10 wt.% exhibited sufficient porosity, Vickers hardness, flexural strength and electrical conductivity of 34%, 48 MPa, 72 MPa and 2028 S/cm (at 800 °C), respectively. Therefore, optimisation of porosity in the Ni-SDC cermet anode microstructure strongly contributes to the well-connected pore channels for the rapid diffusion of hydrogen for oxidation and mechanical strength.

1. Introduction

Solid oxide fuel cells (SOFCs) are extremely efficient at converting chemical energy to electrical energy [1]. The key benefits that make SOFCs suitable for both stationary and portable applications are the high energy conversion, fuel flexibility, use of inexpensive catalysts, and recycling utilisation of exhaust heat

[2]. A cutting-edge technology SOFCs are made up of three layers: the anode and cathode electrodes, as well as an intermediate layer that serves as an ion-conducting electrolyte [3]. The anode is a critical component of the SOFC because it offers active sites for hydrogen oxidation by interacting with oxide ions in the solid electrolyte, hence enabling fuel access and product removal [4]. Nickel-samarium-doped ceria (Ni-SDC) cermet anodes have attracted attention as a SOFC anode material due to their outstanding hydrogen-catalytic capabilities and internal reforming of hydrocarbon fuels [5]. This is because the combination of Ni and SDC significantly enhanced the anode's catalytic activity toward hydrogen fuel oxidation and also promoted efficient hydrocarbon fuel internal reforming [6].

The hydrogen adsorption/diffusion process on Ni surfaces and charge transfer process in SDC electrolyte can be significantly tailored by changing their composition, microstructure, and fabrication conditions. Many studies have sought to improve microstructure of the Ni-SDC anode to facilitate mass transport of the reactants and increase its mechanical strength and electrical conductivity. In these studies, the following processing parameters were examined: (1) Ni-SDC ratio [7], (2) combination of coarse and fine Ni and doped ceria powders [8], (3) effect of Ni powder synthesis on the performance of the Ni-SDC [9], (4) effect of different synthesis techniques on the preparation of Ni-SDC [10], (5) influence of calcination temperature [11], (6) effect of the reduction conditions of Ni on its mechanical and electrical properties [12], and (7) effect of pore former addition and sintering temperature on the microstructure [13]. These studies demonstrated that the performance of the Ni-SDC cermet anode can be tailored substantially by a change in its microstructure and porosity. In this regards, the anode structure should possess excellent permeability; better connectivity of the Ni-SDC, Ni-Ni, and SDC-SDC particles in the cermet [14]; and adequate pores in the structure (30 - 40%) [15]. Additionally, a continuous connection of pores is required for the rapid diffusion of reactants to and from the triple phase boundary (TPB) [16].

The porosity of the microstructure of the anode has an effect on the anode's performance [17]. Reducing the pore size or porosity from the anode microstructure can significantly limit gas diffusion and subsequently increase interfacial polarization resistance [18]. Conversely, high porosity will decrease the TPB area available for reaction [19]. Typically, the anode structure has a porosity of 30 - 40% [20]. This level of porosity can be achieved by optimising the microstructural properties of NiO-based composite anodes. Porous Ni-SDC cermet anodes are often fabricated by addition of organic pore formers or reduction using hydrogen gas.

Among the pore formers, starch is a commonly used pore former in the preparation of porous materials because it is cheap, renewable nature, widely available, and environmental friendliness [21]. Potato starch is a commonly used pore-forming agent in ceramic manufacture [22]. This natural biomaterial is completely consumed during the firing process, leaving no remnant in the final ceramic body. This is because of its chemical composition, which is predominantly a polysaccharide that is composed of just carbon (C), hydrogen (H), and oxygen (O). However, addition of organic pore formers to the initial powders and reduction of the nickel oxide (NiO)-SDC composite anode can cause drastic changes in its porosity, density, and physical and mechanical properties. In this study, potato starch is used as a pore former to produce porous Ni-SDC cermet anodes, and the influence of the pore former on the final product are investigated.

In the present communication, attention was mainly focused on elucidating the influence of porosity, particle distribution and particle to particle connectivity on mechanical and electrical properties of Ni-SDC cermet anode. In this work, a systematic approach was established to examine the effects of porosity on the mechanical and electrical properties of Ni-SDC cermet anodes to which different weight percentages of a pore former, in order to understand the microstructure-mechanical-electrical performance relationship.

2. Methodology

2.1. NiO-SDC powder preparation

The nickel oxide (NiO)-SDC composite anode powder (60 wt.% of NiO and 40 wt.% of SDC) were prepared by mechanical mixing using high speed ball milling technique. Both the NiO powders of average particle size of 5 nm and ($Sm_{0.2}Ce_{0.8}O_{1.9}$) SDC powders with average particle size of 500 nm were used as raw materials. The true theoretical density of NiO and SDC powders were 6.67 g/cm^3 and 7.1417 g/cm^3, respectively. All of the raw materials were purchased from Sigma Aldrich, Malaysia. The commercial powders of NiO and SDC were mixed together with ethanol and ball-milled for 24 h. The slurry was then dried in an oven at 120 °C for 12 h. The dried powder was calcined at 700 °C for 2 h to obtain the desired NiO-SDC composite anode powder. To investigate the effects of amount of pore former on the microstructure and electrical properties of Ni-SDC cermet anode, the NiO-SDC composite anode powder was ball milled with potato starch with different wt.% from 0 to 20 wt.%. The mixed powders were designated as NS0, NS5, NS10, NS15 and NS20, respectively. Potato starch was purchased from Sigma Aldrich, Malaysia. Potato starch has a moisture content of 18 - 21% and an average diameter of 48 μm [23].

The prepared anode composite powder was pressed at 50 MPa to fabricate a disk-shaped sample. The as-prepared green pellets were subsequently sintered at 1350 °C for 5 h in air. The sintered pellet of NiO-SDC composite anode was reduced to metallic Ni-SDC cermet anode with mixture of wet hydrogen and wet nitrogen gas (10% H_2 and 90% N_2) at 800 °C for 5h.

2.2. Characterization

The phase composition of the prepared powder was conducted by X-ray Diffraction (XRD) with CuKα (λ = 0.15418 nm) using an X-ray diffractometer (Shimadzu XRD-6000, Bruker, Karlsruhe, Germany). The powder morphology of the prepared powders was observed through transmission electron microscopy (TEM, HitachiHT7700, Tokyo, Japan) at 120 kV. The density and porosity of the Ni-SDC cermet samples were performed by Archimedes method with deionised (DI) water at room temperature. The degree of thermal shrinkage was also calculated for the Ni-SDC cermet. Field emission scanning electron microscopy (FESEM), combined with back scattering electron (BSE) microscopy (Hitachi FESEM SU5000, Tokyo, Japan) was utilised to examine the morphology of the Ni-SDC cermet pellets. The Vickers hardness tester (ZHV 30-m, Zwick Roell Indentee, Stourbridge, UK) was used to measure the Ni-SDC cermet's hardness. The flexural strength of the Ni-SDC cermet was determined using the piston-on-three-ball method in accordance with the ISO 6872:2008 standard. Using the van der Pauw technique, a DC four-point electrical conductivity of the Ni-SDC cermet was measured in humidified hydrogen at 300 to 800 °C.

3. Results and Discussion

3.1. Powder analysis

From figure 1(a), it is obvious that no phase shifts or chemical reactions happened during the mixing and calcination of NiO and SDC powders. There were no secondary peaks of impurities observed in this sample, indicating that the composite anode powder was chemically compatible after mixing and calcination. Cubic NiO (JCPDS #96-432- 9324) and face-centred cubic fluorite SDC (JCPDS #96-434-3152) have been identified as the two principal phase structures [24]. The NiO–SDC composite anode showed no structural change and was composed entirely of two distinct phases with steep and strong peaks. The theoretical densities of the NiO-SDC composite powder were 6.85 g/cm^3. These findings corresponded to previously published values in the literature [25].

Figure 1(b) shows a TEM image at 5000x magnification of NiO-SDC composite anode powder. The TEM results showed a consistent distribution of small spherical particles and a non-spherical distribution of macro-sized particles. SDC particles are small, spherical particles and NiO particles are larger, non-spherical particles. A similar observation was made during the synthesis of NiO–SDC composite anode at a 60% by weight NiO concentration [26]. Furthermore, homogeneously dispersed SDC particles operate as an inhibitor, limiting NiO particle grain growth. As a result, the macro-sized non-spherical sized particles in the NS-24 sample can be identified as NiO particles [27].

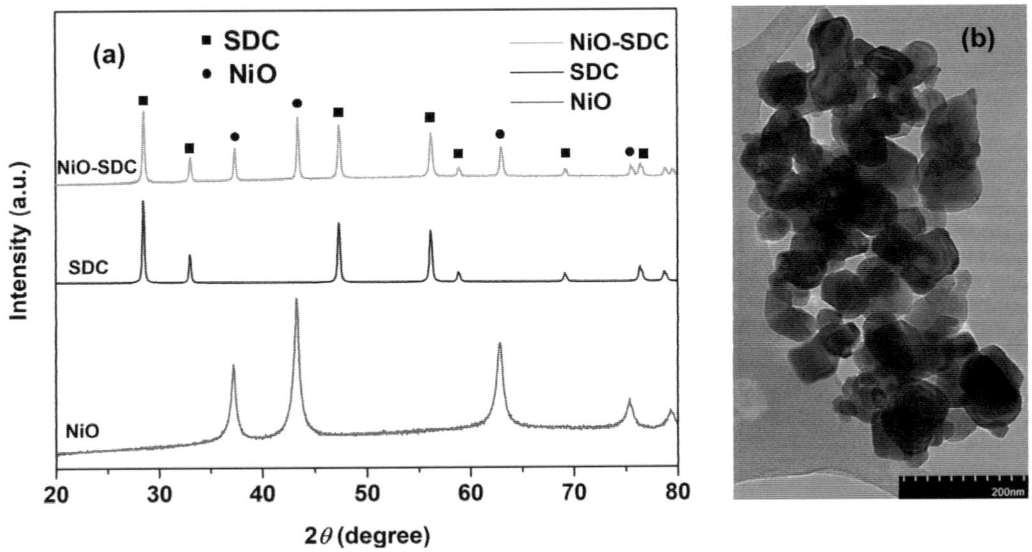

Figure 1. NiO-SDC and composite anode powders (a) XRD patterns and (b) TEM images at 5000x.

3.2. Sintered pellet properties

Prior to microstructural and electrical analysis of the sintered pellets of NiO-SDC composite anode, all the samples were reduced to metallic Ni-SDC cermet anode with mixture of wet hydrogen and wet nitrogen gas (10% H_2 and 90% N_2) at 800 °C for 5h. The impact of potato starch addition and consequently the degree of thermal shrinkage can be reflected on the sintering behaviour of the pellets. It is clearly found that the thermal shrinkage decreased with increasing the amount of potato starch (table 1). This means that the percentage of thermal shrinkage decreases corresponding to the increasing of the potato starch addition amount, indicating the percentage of the pores increases with the increasing of the potato starch addition. In addition, increasing potato starch loading hinders the degree of thermal or sintering shrinkage because it eliminates the influence of pellet densification behaviour. The relationship between the potato starch addition amount and the relative densities of Ni-SDC cermet is listed in table 1. When the amount of potato starch is increased, the relative density of the Ni-SDC cermet anode drops proportionately. For instance, the NS0 show a higher relative density of 83% than the NS5 cermet with 70% relative density. Densities were reduced as a result of the enhanced porosity created by the potato starch addition and reduction process [13]. In general, the percentage of porosity increases with increasing potato starch content.

The reduced pellets of Ni-SDC cermet anode are depicted in cross-sectional images in figure 2 (a-e). FESEM imaging revealed no significant differences in particle shape and size. Three distinct and equally distributed interfaces were identified in the microstructure. SDC, Ni, and pores are symbolised by the white, grey, and dark grey colours, respectively. Increasing potato starch wt.% resulted in higher porous morphological structure. It appears that the size of the grains of Ni-SDC pellets had no correlation with the addition of potato starch, however, the density and porosity of the pellets was dependent on the starch content. The pore distribution was uniform with lots of fine pores and fine-grained microstructure was evident. The potato starch addition and reduction process has a stronger influence on the connection between Ni-SDC, Ni-Ni and SDC-SDC particles and distribution of Ni and SDC within the anode microstructure [15]. It is notable that the grain-grain contact regarded as the greater importance to facilitate better oxidation of H_2 from Ni contact and mechanical support for Ni from SDC contact. Ni–Ni and SDC–SDC had a loose agglomeration in the NS10 sample, which produced a highly porous structure. When starch addition exceeded 10 wt.%, there was weak connection between particles, and in the reduced sample, a more agglomerated morphology was observed (figure 2(e)). The adequate level of porosity was induced with 10 wt.% and 15 wt.% potato starch are evident in the micrographs and consistent with the porosity measurement. Therefore, the NiO-SDC composite anode with 5 wt.% and 10 wt.% of potato starch as pore former can satisfy the functional demands of anode SOFCs.

Table 1. Thermal shrinkage, relative density, porosity, flexural strength and Vickers hardness properties of Ni-SDC cermet anode with different potato starch content.

Sample	Shrinkage (%)	Relative Density (%)	Porosity (%)	Mechanical Properties	
				Flexural Strength (MPa)	Vickers Hardness (MPa)
NS0	22.7 ± 0.3	83 ± 0.7	17 ± 0.7	169 ± 2.5	165 ± 1
NS5	20 ± 0.2	70 ± 0.6	30 ± 0.6	98 ± 4	72 ± 1.5
NS10	20 ± 0.4	65 ± 1.4	35 ± 1.4	73 ± 3	48 ± 2.3
NS15	19.5 ± 0.1	56 ± 1.7	44 ± 1.7	59 ± 2	33 ± 1.7
NS20	18.6 ± 0.1	57 ± 1.2	43 ± 1.2	35 ± 3	14 ± 2.8

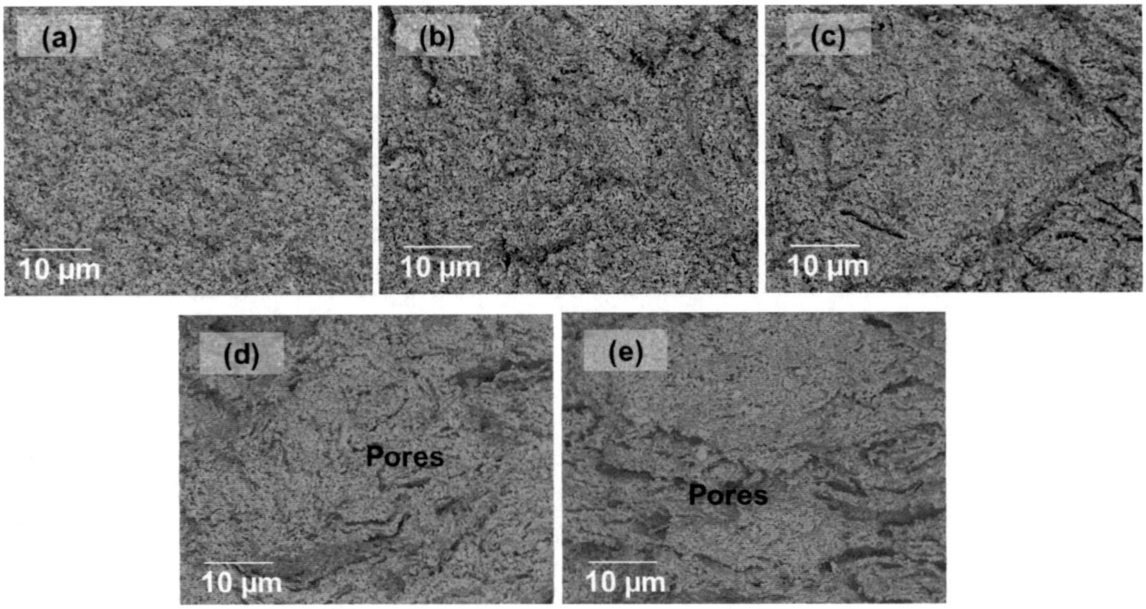

Figure 2. FESEM images of Ni-SDC cermet anode (a) NS0, (b) NS5, (c) NS10, (d) NS15, and (e) NS20.

The flexural strength and Vickers hardness of the Ni-SDC cermet composites with different potato starch addition were listed in table 1. The results show that the flexural strength and Vickers hardness value decreased with increasing potato starch content from 0 to 20 wt.%. It seems that the difference in mechanical properties is mainly attributed by a difference in porosity, which is in good agreement with their relative density from Archimedes method. In general, it is essential to highlight that the degradation of the mechanical strength with increasing the potato starch addition is consequence of increase in porosity and change in microstructure. Increase in porosity not only show adverse effect on the mechanical characteristics of Ni-SDC cermet anode but also on the catalytic activity of the Ni-SDC cermet anode materials.

3.3. Electrical properties

The electrical conductivity of Ni-SDC cermet anode is strongly dependent on the potato starch addition amount as shown in figure 3. When the potato starch addition amount increases from 5 to 20 wt.%, the DC electrical conductivity of the Ni-SDC cermet anode decreased from 2163 to 1022 S/cm, which is significantly lower than that of NS0 (3642 S/cm) at 800 °C. This decrease is mainly attributed to the porous microstructure and particle connectivity which disrupts the electronic pathways in the cermet anode, thereby can lose or have diminished electrical conductivity [28]. It is undeniable that the addition of potato starch has a significant impact on the amount of pores present. It can be observed that the volume of open pores is increasing in correlation with the quantity of pore forming used (figure 2). In contract, the volume of closed pores decreased with the increasing potato starch amount. This is because the increase in potato starch addition helps to extend the pores. Moreover, the adequate addition of pore forming agent will facilitate presences of macro-sized pores and uniform pore size distribution in the anode microstructure, thus increasing the cell performance.

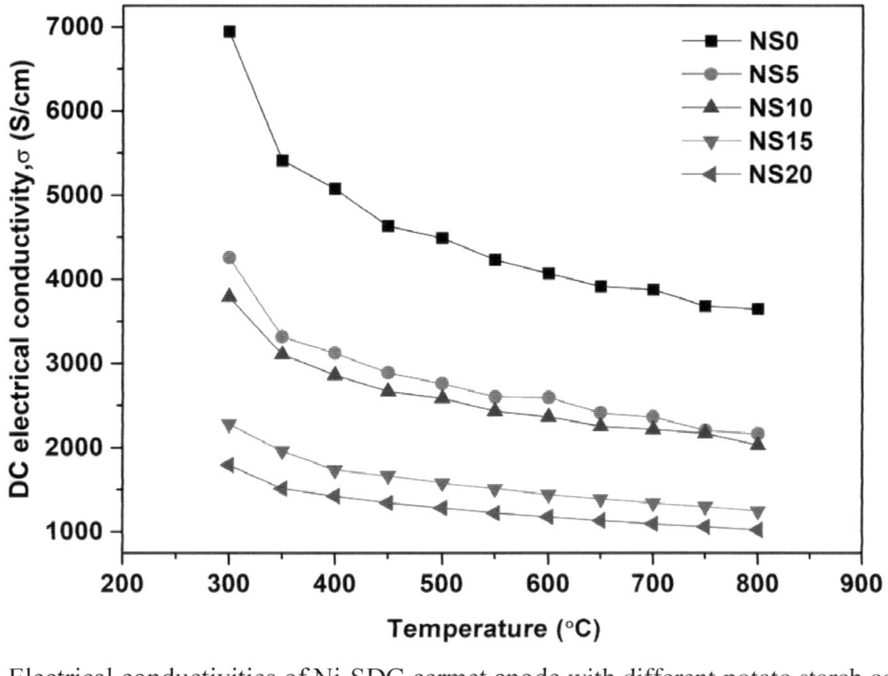

Figure 3. Electrical conductivities of Ni-SDC cermet anode with different potato starch content measured between 300 °C and 800 °C.

4. Conclusion

In this study, effects of amount of pore former on the microstructural and electrical properties of nickel-samarium doped ceria cermet anode were investigated in detailed. As electrical and mechanical studies demonstrate that a continuous network structure comprising Ni-Ni, SDC-SDC, and Ni-SDC particles in cermet, along with sufficient porosity, meets the functional requirements of a solid oxide fuel cell anode. The porosity, mechanical and electrical properties of NS5 and NS10 is sufficient enough to meet the requirement for an SOFC anode. Therefore, in order to maintain an acceptable performance of anode-supported SOFC, the amount of potato starch addition should be in the range between 5 to 10 wt.% for Ni-SDC cermet anode prepared by ball milling technique.

Acknowledgments
This work was supported by the Universiti Kebangsaan Malaysia (UKM) through the research grant: GUP-2019-076. The authors would also like to thank the Centre for Research and Instrumentation Management, UKM for the use of their excellent testing equipment.

References

[1] Mah J C W, Muchtar A, Somalu M R and Ghazali M J 2017 Metallic interconnects for solid oxide fuel cell: A review on protective coating and deposition techniques *Int. J. Hydrogen Energy* **42** 9219–2
[2] Mahmud L S, Muchtar A and Somalu M R 2017 Challenges in fabricating planar solid oxide fuel cells: a review *Renew. Sustain. Energy Rev.* **72** 105–16
[3] Jiang Y, Chen F and Xia C 2021 A review on cathode processes and materials for electro-reduction of carbon dioxide in solid oxide electrolysis cells *J. Power Sources* **493** 229713
[4] Chien A C and Ye N J 2021 Effect of preparation method and particle size of Ni/SDC catalyst on methane oxidation *Catal. Commun.* **154** 106312
[5] Skalar T, Zupan K and Marinšek M 2019 Microstructure tailoring of combustion-derived Ni-GDC and Ni-SDC composites as anode materials for intermediate temperature solid oxide fuel cells *J. Aust. Ceram. Soc.* **55** 123–33
[6] Ideris A, Croiset E and Pritzker M 2017 Ni-samaria-doped ceria (Ni-SDC) anode-supported solid oxide fuel cell (SOFC) operating with CO *Int. J. Hydrogen Energy* **42** 9180–7
[7] Chen M, Kim B H, Xu Q, Ahn B G and Huang D P 2010 Effect of Ni content on the microstructure and electrochemical properties of Ni-SDC anodes for IT-SOFC *Solid State Ionics* **181** 1119–24
[8] Zha S, Rauch W and Liu M 2004 Ni-$Ce_{0.9}Gd_{0.1}O_{1.95}$ anode for GDC electrolyte-based low temperature SOFCs *Solid State Ionics* **166** 241–50
[9] Li Y, Luo Z, Yu C, Luo D, Xu Z and Cen K 2005 The impact of NiO on microstructure and electrical property of solid oxide fuel cell anode *J. Zhejiang Univ. Sci.* **6B** 1124–9
[10] Liu Q, Dong X, Yang C, Ma S and Chen F 2010 Self-rising synthesis of Ni-SDC cermets as anodes for solid oxide fuel cells *J. Power Sources* **195** 1543–50
[11] Chen M, Kim B H, Xu Q and Ahn B G 2009 Preparation and electrochemical properties of Ni-SDC thin films for IT-SOFC anode *J. Memb. Sci.* **334** 138–47
[12] Brodnikovskyi Y, Vasyliv B, Podhurska V, Andrzejczuk M, McDonald N, Kyrpa O, Ostash O, Steinberger-Wilckens R and Lewandowska M 2016 Influence of reduction conditions of NiO on its mechanical and electrical properties *J. Electrochem. Sci. Eng.* **6** 113
[13] Drozdz E, Stachura M, Wyrwa J and Rękas M 2015 Effect of the addition of pore former: graphite and ammonium bicarbonate on the properties of Ni/Al_2O_3–3YSZ composite composite materials *J. Therm. Anal. Calorim.* **122** 157–66
[14] Wang H, Liu X, Bi H, Yu S, Han F, Sun J, Zhu L, Yu H and Pei L 2016 Effects of NiO on the conductivity of $Ce_{0.85}Sm_{0.15}O_{1.925}$ and on electrochemical properties of the cathode/electrolyte interface *J. Power Sources* **320** 86–93
[15] Sarikaya A, Petrovsky V and Dogan F 2013 Development of the anode pore structure and its effects on the performance of solid oxide fuel cells *Int. J. Hydrogen Energy* **38** 10081–91
[16] Ahsan M, Irshad M, Fu P F, Siraj K, Raza R and Javed F 2020 The effect of calcination temperature on the properties of Ni-SDC cermet anode *Ceram. Int.* **46** 2780–5
[17] Somalu M R, Muchtar A, Daud W R W and Brandon N P 2017 Screen-printing inks for the fabrication of solid oxide fuel cell films: a review *Renew. Sustain. Energy Rev.* **75** 426–39
[18] Choi Y G, Park J Y, Song H, Kim H R, Son J W, Lee J H, Je H J, Kim B K, Lee H W and Yoon K J 2013 Microstructure-polarization relations in nickel/gadolinia-doped ceria anode for intermediate-temperature solid oxide fuel cells *Ceram. Int.* **39** 4713–8
[19] Muhammed Ali S A, Anwar M, Baharuddin N A, Somalu M R and Muchtar A 2018 Enhanced

electrochemical performance of LSCF cathode through selection of optimum fabrication parameters *J. Solid State Electrochem.* **22** 263–73
[20] Ho T X, Kosinski P, Hoffmann A C and Vik A 2009 Numerical analysis of a planar anode-supported SOFC with composite electrodes *Int. J. Hydrogen Energy* **34** 3488–99
[21] Mohamad H and Lim Y M 2015 Effect of corn starch and potato starch as the pore forming agent to the structure of porous cordierite *Mater. Sci. Forum* **819** 146–51
[22] Garrido L B, Albano M P, Genova L A and Plucknett K P 2011 Influence of starch type on characteristics of porous 3Y-ZrO_2 prepared from a direct consolidation casting method *Mater. Res.* **14** 39–45
[23] Larrea-Wachtendorff D, Tabilo-Munizaga G and Ferrari G 2019 Potato starch hydrogels produced by high hydrostatic pressure (HHP): a first approach *Polymers (Basel).* **11** 1673
[24] Soltani N, Arcos L H, Bahrami A and Carvayar J C 2019 Structural changes in NiO-$Ce_{0.8}Sm_{0.2}O_{2-x}$ anode under reducing atmosphere *Mater. Charact.* **150** 8–12
[25] Skalar T, Zupan K, Marinšek M, Novosel B and Maček J 2014 Microstructure evaluation of Ni-SDC synthesized with an innovative method and Ni-SDC/SDC bi-layer construction *J. Eur. Ceram. Soc.* **34** 347–54
[26] Ding C and Hashida T 2011 Synthesis and evaluation of NiO-$Ce_{0.8}Sm_{0.2}O_{1.9}$ nanocomposite powders for low-temperature solid oxide fuel cells *Int. J. Hydrogen Energy* **36** 5567–73
[27] Muhammed Ali S A, Yahaya A Z, Anwar M, Soo M T, Muchtar A and Kovrugin V M 2021 Effect of synthesis method of nickel-samarium-doped ceria anode on distribution of triple-phase boundary and electrochemical performance *Crystals* **11** 523
[28] Haslam J J, Pham A Q, Chung B W, DiCarlo J F and Glass R S 2005 Effects of the use of pore formers on performance of an anode supported solid oxide fuel cell *J. Am. Ceram. Soc.* **88** 513–8

Eggshell waste as a catalyst for biodiesel production: A preliminary study

N A Ali[1], N Khairuddin[1,2] and B M Siddique[3]

[1] Department of Science and Technology, Faculty of Humanities, Management and Science, Universiti Putra Malaysia Bintulu Sarawak Campus, 97008 Bintulu, Sarawak, Malaysia
[2] Institut Ekosains Borneo, Universiti Putra Malaysia Bintulu Sarawak Campus, 97008 Bintulu, Sarawak, Malaysia
[3] Faculty of Engineering, Computing and Science, Swinburne University of Technology Sarawak, 93350 Kuching, Sarawak, Malaysia

Email: nozieana@upm.edu.my

Graphical abstract

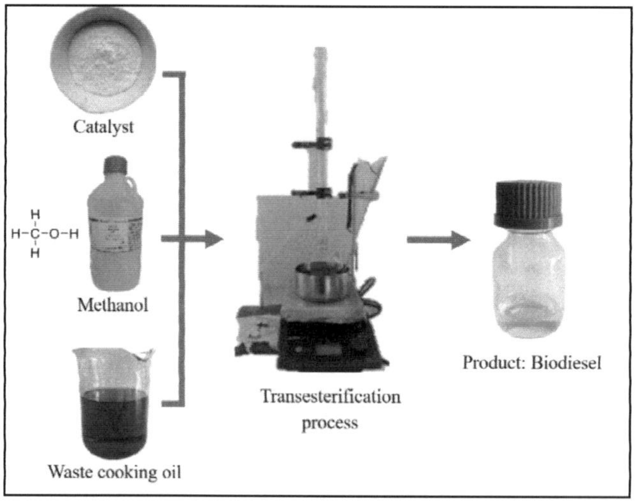

Abstract. This research aims to observe the usability of waste eggshells as a heterogeneous CaO-based catalyst to produce biodiesel *via* the transesterification process. The waste eggshell contains CaO and serves as a heterogeneous catalyst that can be extracted using a simple heat treatment technique. The catalyst loading used in this study were 1, 2, 3, 4, and 5 wt.%. The biodiesel yield affects the variables of the reaction, including the amount of catalyst used. According to the findings in this work, the low-weight catalyst produced more biodiesel (64% of yield) than other catalysts in the studies. Using the same CaO catalyst with a 1 wt.% catalyst, previous studies showed a biodiesel yield of 56.64%. Outcomes revealed that waste eggshells could be used as an effective catalyst to transform waste cooking oil into biodiesel, thereby significantly lowering biodiesel costs and improving product yield and fuel properties.

1. Introduction

Along with the rise in human growth, industrial globalisation, and transportation needs, energy demand is constantly rising [1]. Consequent to the growing global population, as well as the development of the economy and industry, petroleum consumption rises every year [2], where there is a great concern on the toxic exhaust emissions from cars that significantly impact the human health and the environment, in addition to the ever-rising prices of petroleum-based fuels [3]. Due to the current strong reliance on petroleum as a main source of fuel for transportation and electricity production, petroleum prices have recently reached new highs [4]. Figure 1 illustrates that the consumption of non-renewable energy keeps increasing. Biodiesel fuel is widely used in most nations since it is eco-friendly, harmless, and environmentally benign, emitting fewer greenhouse gases (GHGs) [5].

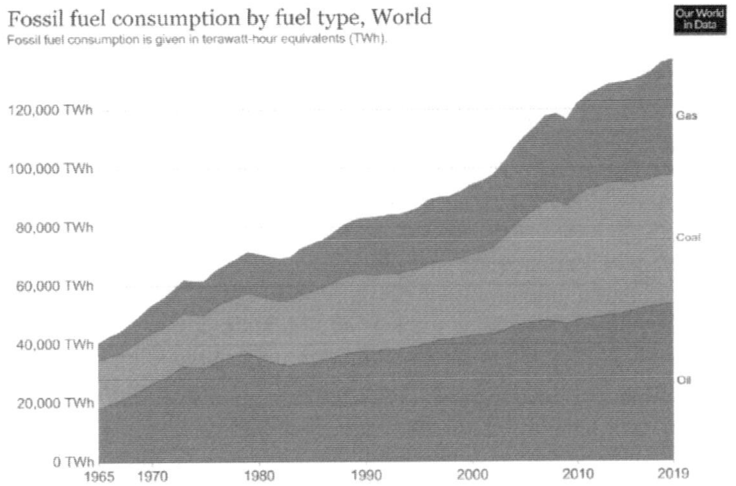

Figure 1. Transitions in energy: Global and national viewpoints & BP world energy's statistical analysis [6].

Efforts to reduce diesel engine emissions without engine modification have accelerated rapid growth and the increased output of biodiesel, which has also been driven by the urgent need to reduce diesel engine emissions to minimise global warming drastically. Biodiesel has become an enticing and sustainable green fuel, recognised as Fatty Acid Methyl Ester (FAME) with a global recognition as an alternate energy source [2]. Biodiesel is typically generated through the transesterification of vegetable oils or tallow with methanol using a suitable catalyst, where it is a healthier option to diesel, whereby the former consists primarily of mono-alkyl esters of unsaturated fatty acids and has similar physical properties to that of diesel, with the added favour of being green and non-toxic biodegradable [7]. In the early 1990s, biodiesel, also known as mono-alkyl ester, became widely available, and production has steadily increased since then [8]. Waste lipids must be used to achieve higher biodiesel yields without compromising the atmosphere or food sources [9]. Transesterification generates FAME while resulting in saponification or soap formation [5]. Biodiesel has been called the most favourable alternative to conventional petroleum diesel because it is renewable and eco-friendly. The downside of biodiesel marketing is the combined high cost of raw materials and processing attributes to the high cost of catalysts and biodiesel feedstocks. Therefore, waste cooking oil (WCO) may be used successfully in biodiesel processing to reduce the expense of raw materials. It also helps to reduce the disposal of WCO, which may cause environmental pollution. Figure 2 shows that the production of cooking oil in Malaysia has a steady increase, where it would increase the amount of WCO required in the present and in the future, therefore, polluting the environment.

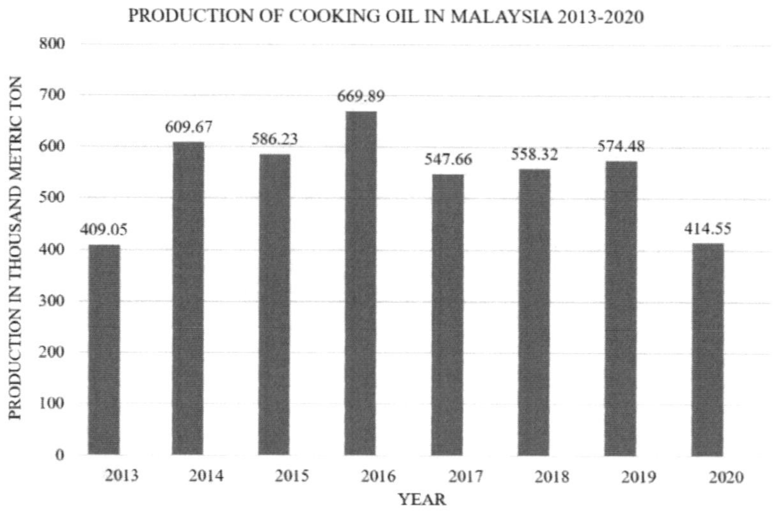

Figure 2. Malaysian cooking oil production 2013–2019 [10].

Catalyst has a crucial role in biodiesel development. Various homogeneous and heterogeneous catalysts have been used in the manufacture of biodiesel. Furthermore, the use of waste material catalysts would significantly incur an additional rate of biodiesel output. Homogeneous catalysts, such as potassium hydroxide (KOH), sodium hydroxide (NaOH), potassium methoxide (CH_3OK), and sodium methoxide (CH_3ONa), are commonly used during the transesterification process [5]. After transesterification on heterogeneous catalysts, a greater grade of FAME can be produced [5]. Calcium oxide (CaO) is among the most extensively used heterogeneous catalysts contained in waste eggshells, which can easily be converted into CaO by high-temperature calcination with a large amount of calcium compounds ($CaCO_3$) [2]. The CaO catalyst from chicken eggshell can be used for successful transesterification [11]. An average egg is estimated to weigh between 60.0–60.2 g, whereas an empty shell weighs between 6.6–7.3 g [12]. Calcium carbonate in the form of calcite accounts for 94–97 % [13–14] of the chemical composition of eggshells, with organic matter accounting for 3–4.5 % [12]. This shows that chicken eggshells are reliable sources of CaO. The CaO catalysts are easy to obtain, as the production of chicken eggs showed higher increment every year. Figure 3 demonstrates that the production of chicken eggs in Peninsular Malaysia have increased year by year, where the eggshell waste containing valuable sources of CaO can be easily obtained from restaurants, among other places [15].

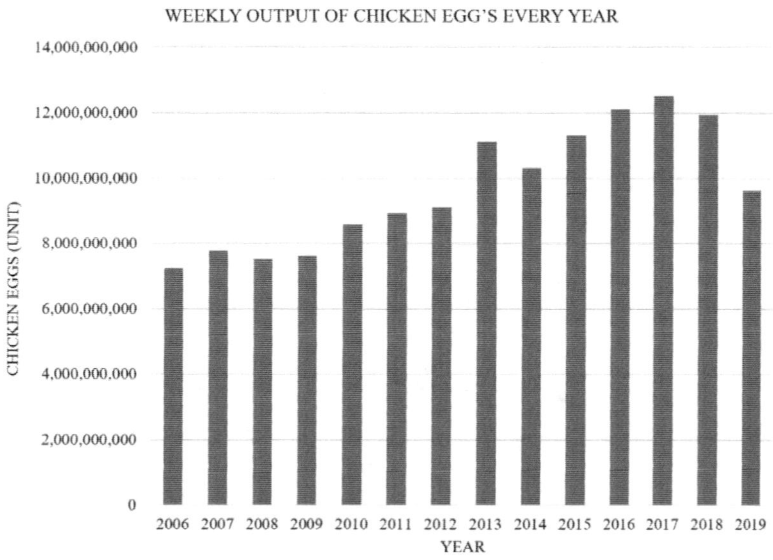

Figure 3. Weekly output of chicken eggs every year [16].

A substantial increase of 91% in biodiesel yield for the catalyst compared with a 78% yield for the unsupported CaO catalyst was obtained [17].

This research aims to study the development of biodiesel using a low-cost, heterogeneous CaO-based catalyst from waste eggshells. The development of biodiesel through the transesterification method using WCO was carried out using the established catalyst. In addition, the influence of eggshell catalyst loading on biodiesel yield was observed.

2. Methodology

2.1. *Materials*

In Malaysia, the wide availability of WCO in all eateries has led to its use as raw material in this study. The WCO used in this experiment was acquired from the restaurant in Bintulu, Sarawak. Methanol was purchased from HmbG Chemicals (Hamburg, Germany) to produce high-quality biodiesel fuel. The chicken eggshells were collected from bakery and restaurant in Bintulu, Sarawak.

2.2. *Catalyst preparation*

To minimise impurities present on the surface and inside of the eggshells, the waste chicken eggshells collected were rinsed with distilled water several times. The eggshells were then washed and dried in a 100 °C oven for 24 hours. Then, using a 0.08-mm sieve wire, the waste eggshell was crushed into powder and sieved. Finally, the powdered eggshell experienced a calcination process at 900 °C for 3 hours.

The following equation (1) demonstrates the conversion of chicken eggshells into calcium oxide (CaO) under calcination process:

$$Chicken\ eggshells\ (CaCO_3) \xrightarrow[900°C]{} CaO + CO_2 \uparrow \tag{1}$$

2.3. *Reaction of transesterification*
The following reaction of transesterification procedure described by [2] shows a slight modification. The transesterification reaction was performed using a 100-mL three-necked round-bottom glass flask fitted with a reflux condenser, magnetic stirrer, and thermometer. To regulate the temperature, a water bath at a constant temperature was also used. Molar ratio: methanol/oil using in the experiment carried out was 6:1. Time taken to carry out the process of transesterification was 1 hour. At a stirring rate of 600 rpm for the length of the reaction, the catalyst was dispelled in methanol at 100 °C and subsequently add on oil. The reaction mixture was separated using the separating funnel. The latter occurred in the separation between the two phases, where the upper layer consisted of biodiesel, and the lower layer is the glycerol. With a mass yield of 10 %, the glycerol obtained as a by-product of biodiesel processing is referred to as crude glycerol because it contains free fatty acids (FFA) [18]. Figure 4 below shows the transesterification reaction, where biodiesel is made from triglycerides in the presence of alcohol and a catalyst.

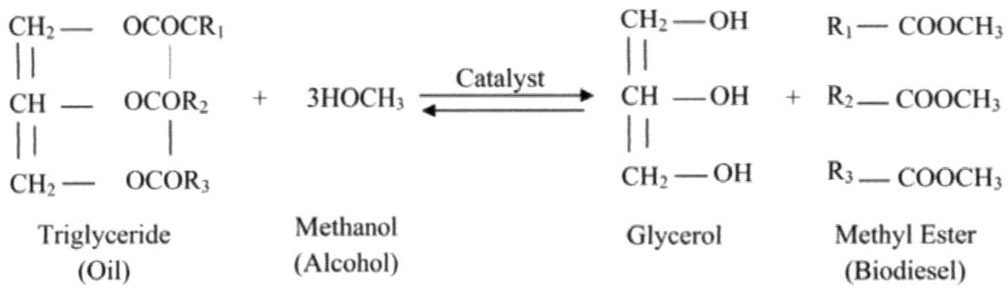

Figure 4. The reaction equation of process conversion from triglycerides to biodiesel in the presence of alcohol with a catalyst through the transesterification process [19].

2.4. *Post - treatment process*
To extract unreacted oil and impurities, biodiesel with excess methanol and unreacted oil was washed with hot distilled water at 90.3 °C. Next, the final mixture was heated for 30 min at 100 °C to eliminate excess moisture content and obtain pure biodiesel.

2.5. *Fourier-Transform Infrared Spectroscopy (FTIR) analysis of biodiesel and WCO*
The biodiesel content C=O was determined using the Shimadzu IRAffinity-1 equipment of the Fourier-Transform Infrared Spectroscopy (FTIR) for the biodiesel characterisation study of the functional group and composition of the biodiesel, where the infrared absorption spectra were obtained in the range of 600 to 4000 cm^{-1}.

3. Results and discussion

3.1. *Effect of calcination temperature*
The temperature for calcination was one of the parameters of nanocatalytic synthesis, where it demonstrates improvement in the catalytic surface and improves the action of a catalyst [20–22]. Table 1 shows that the biodiesel production yield is inconsistent when the loading catalyst increases. Logically, it demonstrates that 1 wt.% of catalyst can produce high yields for others. According to previous studies conducted by Goli and Sahu [23], a biodiesel yield of 63.2% is obtained at 2 wt.% of catalyst loading and at a low reaction duration of 1 hour; yet, at the same reaction time, the catalyst loading of 12 wt.% had reduced the biodiesel yield by 55.2%. The 2 wt.% catalyst, on the other hand, shows that it produced the second highest biodiesel yield. Results showed a somewhat similar outcome to that of the previous study, where a high catalyst was used, showing a reduction rate of biodiesel compared with that using a low catalyst. For every catalyst loading weight percentage, the biodiesel yield obtained was distinct. Maximum biodiesel yield of 64% was achieved at 100 °C. The biodiesel

yield, as shown in table 1, showed a decrease when 4 wt.% of catalyst loading was used. As shown in table 1, the yield of biodiesel was quite low (34.0%) at 100 °C. High oil viscosity causes a decreased diffusion rate and irreversible transesterification process conversion on the catalyst surface, which can be attributed to this [24]. The excess catalyst was shown to make the reaction mixture more viscous, resulting in poor diffusion of reactants in the methanol–oil–catalyst systems and a reduction in the yield and conversion of the reactions [25–30]. The yield reduced as the catalyst quantity increased for two possible reasons, which are, firstly, at increasing catalyst concentrations, diffusion in methanol-oil-catalyst systems is inhibited [25]. Whilst, secondly, greater catalyst concentrations might adsorb more products, lowering yield [31–32].

Table 1. Biodiesel yield production by weight percentage of catalyst loading.

Catalyst loading, (wt.%)	Biodiesel yield, (% ± S.E.)
1	64.0 ± 0.00
2	56.0 ± 1.00
3	53.4 ± 0.65
4	45.0 ± 2.75
5	52.0 ± 1.00

3.2. Characterisation of biodiesel

Figure 5 illustrates the FTIR biodiesel output and WCO spectra with different catalyst loading percentage and mid-infrared between 600 and 4000 cm^{-1}. The FTIR study was conducted to analyse the percentage of biodiesel in the biodiesel transesterification reaction, depending on the catalyst. Chemically, biodiesel is a blend of mono-alkyl esters of long-chain fatty acids (C_{16}–C_{18}) derived from vegetable oils and animal fats [33].

The methyl ester peaks showed strong C=O and O–CH_3 group bands, demonstrating the occurrence of transesterification process. It is vital to identify the O–CH_3 group, as it plays a significant role in ensuring the transesterification had occurred. For biodiesel and WCO, there were no noticeable peaks in the bands 1800–2700 cm^{-1} and 3000–4000 cm^{-1}. Furthermore, the difference is seen in the area between band 3400–2600 cm^{-1}, where the C–H stretch is indicated at different absorption spectra. Therefore, all of the spectra in figure 5 displayed high peaks at precise bands of 2922.16 and 2852.72 cm^{-1} prior to C–H vibrations.

The O–CH_3 group is present in biodiesel but absent in WCO, which indicates the WCO successfully converted to biodiesel where can be found between band 1188–1200 cm^{-1}. According to a prior study by Goli and Sahu [23], peak of O–CH_3 can be found at 1170.16 cm^{-1} with 19.5% for eggshell based FAME product.

The peaks that show in figure 5, the presence of C–O group can be found precisely at bands 1159.22 cm^{-1}, and some of the bands at 1234.44, 1159.22, 1114.86, and 1097.50 cm^{-1} perchance attributed towards the category of (C–O) ester group stretching vibrations. The cis C=C bond produced a small band with a wavelength between 1750–1500 cm^{-1}.

The bands of CH_2 and CH_3 aliphatic groups resulted at band 1460.11 cm^{-1}. Other biodiesel diagnostic peaks were found at positions 1743.65 cm^{-1} for the C=O group of glycerides. The greater the amount of free fatty acid, the less likely it is to convert into fatty acid methyl ester. Because of the variety of bending and stretching within the molecule, the area between 600–1500 cm^{-1} is usually a very complicated absorption sequence. Previous research by Goli and Sahu [23] confirmed the eggshell-based FAME obtain was 21.5%, where the –CH_3 asymmetric bending vibration at 1458.19 cm^{-1}. This study refers to a band spectra that summarise the functional groups assigned to the main and minor absorption bands from previous studies [34]. Both the biodiesel and WCO did not show a significant difference. In the FTIR, the only difference is that glycerine had been removed from the hydrocarbon chain and replaced with methanol, where the difference can be found between 600–1500 cm^{-1}. Some are recognisable in a specific band, such as 1425–1447 cm^{-1} and 1188–1200 cm^{-1}, where the biodiesel spectrum differs slightly [35–37]. In the biodiesel spectrum, these peaks are not present in the oil spectrum [35].

Figure 5 illustrates the probable mechanistic feature of the transesterification reaction in the presence of the calcined eggshells. The calcinated eggshell containing mainly calcium oxide (CaO) acts as a catalyst in the transesterification reaction by drawing protons from the reacting media to the solid catalyst surface, forming an extremely fundamental and active methoxide anion(I) [15]. As a strong nucleophile, methoxide anion targets the oil's carbonyl carbon atom to form a tetrahedral intermediate. CaO proton attacks the intermediate species with simultaneous tetrahedral rearrangement to generate biodiesel or methyl ester (FAME) and glycerol fatty acids [38–39].

According to Zaman *et al.* [40], CaO calcined at 800 °C had a higher degree of crystallinity than others, implying that this is the optimal temperature for the synthesis of CaO. Besides, a prior study conducted by Habte *et al.* [41] also indicated that the absence of a greater absorption at 1415.52 cm^{-1} in the FTIR analysis implies that the $CaCO_3$ as an essential element of the eggshell was no longer present because it has been transformed to CaO. According to the XRD analysis performed by Kirubakaran and Selvan [42], calcium hydroxide $Ca(OH)_2$ was generated during the decarbonation of $CaCO_3$ into CaO due to reactive moisture adsorption displays at $2\theta = 47.1^o$ and $2\theta = 18^o$ and 47.2^o, respectively, in contrast, calcium carbonate was noticed at $2\theta = 28.7^o$ due to incomplete conversion.

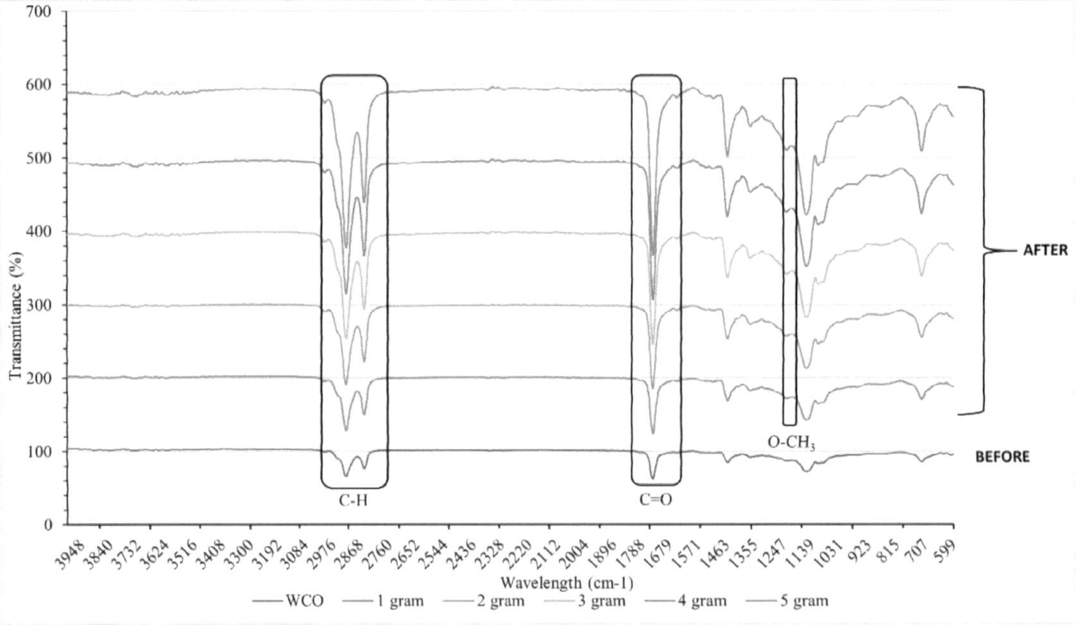

Figure 5. FT-IR spectrum of biodiesel and WCO.

3.3. *Acid value of biodiesel*

The acid value of the biodiesel is different for each catalyst loading. It was found that 3 wt.% of catalyst loading was found to be 2.02 mg of titration oil CaO/g corresponding to the 1.01% FFA value. Extensive research was done on the yield of biodiesel in terms of FFA content, where it was learnt that transesterification will not occur if the FFA content in the oil is more than 2 wt.%; hence, the FFA value must be reduced to below 1% [43–44]. A pre-treatment step was necessary for oils with acid value greater than 2 mg KOH/g, [45]. A high acid value may cause rubber seals and hoses in the engine to lose their solvency, resulting in early failure causing deposits that can clog the fuel filter and lower fuel pressure. [46–47].

Table 2. Free fatty acid biodiesel oil and acid value.

Catalyst loading, (wt%)	Acid value, (mg)	Free fatty acid, (%)
1	1.36	0.68
2	1.67	0.84
3	2.02	1.01
4	1.60	0.80
5	1.39	0.70

4. Conclusion

This study demonstrates that the different weight percentages of CaO catalyst used had produced different yield of biodiesel production, demonstrating the highest biodiesel production at 64%. The biodiesel yield varied depending on the amount of catalyst used. It appears that a low-cost, heterogeneous CaO-based catalyst made from waste eggshells was used to produce biodiesel by using powdered eggshell that was calcined at 900 °C, which majorly helped to ensure the success of the transesterification conducted. The presence of CaO with high basicity in eggshell after calcination is directly linked to these findings. Eggshell waste served as a catalyst give an outstanding contribution as a heterogeneous catalyst to synthesis biodiesel production. The development of biodiesel through the transesterification method using WCO was carried out using the established catalyst. Therefore, the compatibility of the CaO catalyst and WCO as a potential candidate for biodiesel production in future fuel regimes are found to be promising. Further investigation needs to be done on more parameters such as reaction time and temperature and catalyst loading to observe the suitability of waste eggshell as a catalyst for biodiesel production. Meanwhile, the experimental approach used in this study can be expanded. Thus, additional catalyst reusability experiments might be conducted to determine the biodiesel yield production at which the catalyst could be further tuned. Thus, it may assist biodiesel manufacturing by enhancing product quality and significantly reducing purification procedures and ancillary wastes.

Acknowledgment

The authors are grateful to the Universiti Putra Malaysia Bintulu Sarawak Campus, Malaysia, and Ministry of Higher Education (MOHE). This study was supported by (GP-IPB vot no: 9671301) and Fundamental Research Grant Scheme (Ref.: FRGS/1/2018/TK05/UPM/02/8) that enabled the success of this work.

References

[1] Putra M D, Ristianingsih Y, Jelita R, Irawan C and Nata I F 2017 *RSC Adv.* **7** 55547–54
[2] Borah M J, Das A, Das V, Bhuyan N and Deka D 2019 *Fuel* **242** 345–54
[3] Mansir N, Teo S H, Rabiu I and Taufiq-Yap Y H 2018 *Chem. Eng. J.* **347** 137–44
[4] Sahar, Sadaf S, Iqbal J, Ullah I, Bhatti H N, Nouren S, ur Rehman H, Nisar J and Iqbal M 2018 *Sustain. Cities Soc.* **41** 220–6
[5] Hamzah N H C, Khairuddin N, Siddique B M and Hassan M A 2020 *Processes* **8** 1–11
[6] Ritchie H and Roser M 2017 Published online at OurWorldInData.org
[7] Savaliya M L, Bhakhar M S and Dholakiya B Z 2016 *Catal. Letters* **146** 2313–23
[8] Mishra V K and Goswami R 2018 *Biofuels* **9** 273–89
[9] Jamal Y, Shah I H and Park H S 2019 *Biofuels* **0** 1–7

[10] Hirschmann R 2019 Production of blended/deodorized cooking oil in Malaysia from 2013 to 2020 *Statista*
[11] Chung Z L, Tan Y H, Chan Y S, Kansedo J, Mubarak N M, Ghasemi M and Abdullah M O 2019 *Biocatal. Agric. Biotechnol.* **21** 101317
[12] John-Jaja S A, Udoh U H and Nwokolo S C 2016 *Beni-Suef Univ. J. Basic Appl. Sci.* **5** 389–94
[13] Mittal A, Teotia M, Soni R K and Mittal J 2016 *J. Mol. Liq.* **223** 376–87
[14] Mohan T P and Kanny K 2018 *J. Compos. Mater.* **52** 3989–4000
[15] Kavitha V, Geetha V and Jacqueline P J 2019 *Process Saf. Environ. Prot.* **125** 279–87
[16] Federation of Livestock Farmers' Associations of Malaysia 2020 Weekly output of broiler, chicken eggs and meat duck in Peninsular Malaysia 2006—forecast in 2020 industry statistics *Federation of Livestock Farmers' Associations of Malaysia.*
[17] Yaşar F 2019 *Fuel* **255** 115828
[18] Rodrigues A, Bordado J C and Dos Santos R G 2017 *Energies* **10**
[19] Degfie T A, Mamo T T and Mekonnen Y S 2019 *Sci. Rep.* **9** 1–8
[20] Naveenkumar R and Baskar G 2019 *Bioresour. Technol.* **280** 493–6
[21] Baskar G, Aberna Ebenezer Selvakumari I and Aiswarya R 2018 *Bioresour. Technol.* **250** 793–8
[22] Meng Y L, Wang B Y, Li S F, Tian S J and Zhang M H 2013 *Bioresour. Technol.* **128** 305–9
[23] Goli J and Sahu O 2018 *Renew. Energy* **128** 142–54
[24] Afsharizadeh M and Mohsennia M 2019 *React. Kinet. Mech. Catal.* **128** 443–59
[25] Maneerung T, Kawi S, Dai Y and Wang C H 2016 *Energy Convers. Manag.* **123** 487–97
[26] Birla A, Singh B, Upadhyay S N and Sharma Y C 2012 *Bioresour. Technol.* **106** 95–100
[27] Yan S, Kim M, Salley S O and Ng K Y S 2009 *Appl. Catal. A Gen.* **360** 163–70
[28] Kotwal M S, Niphadkar P S, Deshpande S S, Bokade V V and Joshi P N 2009 *Fuel* **88** 1773–8
[29] Jitputti J, Kitiyanan B, Rangsunvigit P, Bunyakiat K, Attanatho L, and Jenvanitpanjakul P 2006 *Chem. Eng. J.* **116** 61–6
[30] Kim H J, Kang B S, Kim M J, Park Y M, Kim D K, Lee J S and Lee K Y 2004 *Catal. Today* **93–95** 315–20
[31] Khatibi M, Khorasheh F and Larimi A 2021 *Renew. Energy* **163** 1626–36
[32] Mansir N, Teo S H, Rashid U and Taufiq-Yap Y H 2018 *Fuel* **211** 67–75
[33] Da Silva H R G, Quintella C M and Meira M 2017 *J. Braz. Chem. Soc.* **28** 2348–56
[34] Wembabazi E, Mugisha P J, Ratibu A, Wendiro D, Kyambadde J and Vuzi P C 2015 *J. Spectrosc.* **2015**
[35] Rosset M and Perez-Lopez O W 2019 *Vib. Spectrosc.* **105** 102990
[36] Mahamuni N N and Adewuyi Y G 2009 *Energy and Fuels* **23** 3773–82
[37] Siatis N G, Kimbaris A C, Pappas C S, Tarantilis P A and Polissiou M G 2006 *JAOCS, J. Am. Oil Chem. Soc.* **83** 53–7
[38] Dhmees A S, Rashad A M and Abdulla E S 2020 *Egypt. J. Chem.* **63** 1033–44
[39] Boro J, Thakur A J and Deka D 2011 *Fuel Process Technol.* **92** 2061–7
[40] Zaman T, Mostari M S, Mahmood M A Al and Rahman M S 2018 *Cerâmica* **64** 236–41
[41] Habte L, Shiferaw N, Khan M D, Thriveni T and Ahn J W 2020 *Sustain.* **12** 1174
[42] Kirubakaran M and Arul Mozhi Selvan V 2021 *Bioresour. Technol. Reports.* **14** 100658
[43] Idowu I, Wylie S, Teng K H, Kot P, Phipps D and Shaw A 2019 *Renew. Energy* **142** 535–42
[44] Canakci M 2007 *Bioresour. Technol.* **98** 183–90
[45] Bouaid A, Vázquez R, Martinez M and Aracil J 2016 *Fuel* **174** 54–62
[46] Yesilyurt M K 2020 *Fuel* **275** 117893
[47] Hossain A B M S and Mazen M A 2010 *Aust. J. Crop Sci.* **4** 550–5

Numerical simulation and experimental study on hematite production by oxidation of mill scale

M T Aldiramadan[1], J Nabilah[2], M Ridwan[1] and D S Khaerudini[3]

[1] Department of Chemistry, University of Indonesia, Depok, Jawa Barat, 16424, Indonesia
[2] Department of Physics, State University of Jakarta, Jalan Rawamangun Muka, Jakarta Timur, 13220, Indonesia
[3] Research Center for Physics, Indonesian Institute of Sciences (LIPI), Kawasan PUSPIPTEK Serpong, Tangerang Selatan, 15314, Indonesia

Email: deni.shidqi.khaerudini@lipi.go.id

Abstract. Mill scale is a waste product from steel hot rolling processes and containing a high amount of FeO and Fe_3O_4. It is crucial to recycle and reuse this waste for the recovery into a metallic iron or its single oxide derivative. One of the end products of mill scale oxidation is hematite which has multitude of uses in various application. In this present study, mill scale is converted to hematite by roasting a mixture of mill scale at a specific set of conditions at 900°C for a varied time under ambient air environment by addition of pelletized limestone as a heat storage media. In this work, the Dimensionless Degree of Oxidation Prediction Model (DDOPM) has been constructed to gain hematite purity approach in the resulting powder by using Matlab. The Buckingham Pi Theorem is used to find dimensionless parameters considering the effect of different parameters including the geometric factor of tubular horizontal furnace used, characteristic of mill scale, ambient air factor, and thermal characteristics of limestone. The degree of oxidation from experimental result was obtained from XRF analysis and compared to the result of DDOPM to show the correlation between the experimental and numerical.

1. Introduction

Mill scale is a steel-making by-product from steel hot rolling processes and is composed of iron oxides and metallic iron with varying oil and grease contents [1]. It was formed during the continuous casting and rolling mill processes when steel is submitted to thermal gradients in oxidant temperatures, promoting the growth of the iron oxide layer at the surface of steel [2]. Mill scale was separated from the steel through some processes such as flame cleaning, water spray, pickling, and blasting [3]. The oil component in the rolling mill scale makes recycling difficult, and its direct re-use in sintering may lead to environmental pollution [1].

It was estimated that in 2020, the capacity of steel production in Indonesia was reaching 13.67 million tons every year [4]. Around 35-40 kg of mill scale was produced for every ton of hot-rolled product [1]. Therefore, 546.8 tons of mill scale was estimated to be generated every year. Legodi *et al.* [1] reported that the bulk of mill scale waste was dumped in landfills by a steel manufacturing company in the Republic of South Africa. Leaching minor percentages of heavy metals into soil and groundwater is threatening the environment [1]. The mill scale also needs to be pretreated before it can be re-utilized.

Therefore, it is increasingly encouraging to treat these solid wastes more sustainably to be re-utilized as a functional material.

One way to treat mill scale is to oxidize it into its highest level of oxide, i.e., hematite (Fe_2O_3). As the mill scale is the multi-oxide component with the amount of iron oxide varied depends on its location of steel production. Thus, oxidation into pure hematite would be economically beneficial because hematite is a functional material in various applications. Hematite is commonly used in synthetic pigments, such as in paints, ceramics, and porcelain [5]. Hematite nanomaterials have been investigated intensively as a promising anode material for Li-ion batteries due to their advantages including high theoretical capacity, low cost, environmental friendliness, and high resistance to corrosion [6]. It has also been a catalyst for many industrially essential reactions, such as synthesis of NH_3, high-temperature water gas shift reaction, and natural gas desulfurization [7].

Many studies in the literature deal with different methods for preparing hematite [1, 8–11]. Jikar *et al.* [12–13] have successfully prepared hematite from a ferrous mill scale by calcination under a 100 % oxygen environment at a temperature higher than 900°C. Pure oxygen was consumed and wasted in this process. A way to improve this method is to use the oxygen from ambient air. Another problem is that calcination under an ambient air environment needed longer time due to slower oxygen flow rate. This problem can be overcome by either increasing the holding time, which takes more energy (electricity), or employing heat storage media (passive heating) to extend its heating time for oxidation to be occurred even after the furnace (active heating) was turned off.

Heat storage is a technology that stores thermal energy over a particular time and releases the stored energy at a later time [14]. Heat storage by increasing the temperature of the material is known as sensible heat storage. Materials used for an efficient, sensible heat storage system should have high specific heat capacity, low thermal conductivity, long-term stability in terms of thermal cycling, and be compatible with the material where storage takes place [15]. Brick, concrete, granite, sandstone, and limestone are common material used as a heat storage media [16]. Limestone was selected as the heat storage material due to its low conductivity, high specific heat, low cost, and abundantly available in Indonesia.

Thus, the whole oxidation process can be mathematically modelled considering the effect of different parameters by using the Dimensionless Analysis Approach, i.e., Buckingham Pi Theorem [12]. The use of dimensionless analysis has been widely applied in processing metallurgy when more than two parameters influence its behavior [13]. Jikar *et al.* [12] attempted experimentation and mathematical modelling in this process with thirteen dimensionless groups and simplified it to a simple power-law equation with two dimensionless groups.

In this work, hematite was successfully prepared from mill scale. Pelletized limestone was added as heat storage media, and calcination is carried out under an ambient air environment at 900 °C. The attempt has been made to model ideal conditions in the tubular horizontal furnace (THF) and is modelled using the Buckingham Pi Theorem to estimate the degree of oxidation for predicting the output by assuming 20 input parameters. Subsequently, the supporting experimentation is also carried out, and the resulting powder was then characterized using X-ray fluorescence (XRF) to investigate the obtained hematite content. The affecting parameters on the degree of oxidation considered in the dimensionless analysis modelling are listed in table 1.

2. Formulation of Dimensionless Degree of Oxidation Prediction Model (DDOPM)

The degree of oxidation (O) is defined as %Fe_2O_3 content in the oxidized powder obtained by XRF (S2 Puma, Bruker) analysis. In this present study, it was assumed that the degree of oxidation is influenced by the different process parameters, as described in table 1, and its relationship is stated by:

$$X = f(T, t, T_R, L_T, D_T, V, W_M, \rho_M, d_M, S_M, P_O, \rho_O, \eta, W_C, D_C, S_C, \rho_C, C_P, \kappa, P_C) \quad (1)$$

It also considered that ρ_O, η, and P_C are a function of temperature [17, 18].

Table 1. List of variables input of physical and material parameters.

No	Variables	Symbol	SI Unit	Dimensions
	Tubular Horizontal Furnace			
1	Holding Temperature	T	K	$[M^0L^0T^0\theta^1]$
2	Holding time	t	s	$[M^0L^0T^1\theta^0]$
3	Heating rate	T_R	Ks^{-1}	$[M^0L^0T^{-1}\theta^1]$
4	Length of the tube	L_T	m	$[M^0L^1T^0\theta^0]$
5	Diameter of the tube	D_T	m	$[M^0L^1T^0\theta^0]$
6	Crucible volume	V	m^3	$[M^0L^3T^0\theta^0]$
	Mill Scale Parameter			
7	Weight of mill scale	W_M	Kg	$[M^1L^0T^0\theta^0]$
8	Apparent density of mill scale	ρ_M	Kgm^{-3}	$[M^1L^{-3}T^0\theta^0]$
9	Mill scale particle size	d_M	m	$[M^0L^1T^0\theta^0]$
10	Specific surface area of mill scale powder	S_M	m^2Kg^{-1}	$[M^{-1}L^2T^0\theta^0]$
	Ambient Air Parameter			
11	Oxygen purity	P_O		$[M^0L^0T^0\theta^0]$
12	Density of oxygen	ρ_O	Kgm^{-3}	$[M^1L^{-3}T^0\theta^0]$
13	Viscosity of oxygen	η	$Kgm^{-1}s^{-1}$	$[M^1L^{-1}T^{-1}\theta^0]$
	Heat Storage (Limestone) Parameter			
14	Weight of limestone	W_C	Kg	$[M^1L^0T^0\theta^0]$
15	Limestone pellet diameter	D_C	m	$[M^0L^1T^0\theta^0]$
16	Surface area of limestone pellet	S_C	m^2Kg^{-1}	$[M^{-1}L^2T^0\theta^0]$
17	Apparent density of limestone	ρ_C	Kgm^{-3}	$[M^1L^{-3}T^0\theta^0]$
18	Specific heat capacity of limestone	C_P	$JKg^{-1}K^{-1}$	$[M^0L^2T^{-2}\theta^{-1}]$
19	Thermal conductivity	κ	$Wm^{-1}K^{-1}$	$[M^1L^1T^{-3}\theta^{-1}]$
20	Equilibrium pressure of limestone decomposition	P_C	Pa	$[M^1L^{-1}T^{-2}\theta^0]$

Equation (1) was determined using the computed Buckingham's Pi Theorem with Gauss-Jordan elimination MatLab syntax `rref`. This equation involving 20 variables with 4 primary or fundamental dimensions, which are mass, length, time, and temperature. Thus, it can be rewritten in terms of a set of 16 dimensionless Π parameters constructed from the original variables.

The final form of Process Parameter (X) is written as in equation (2) and the detailed calculations are listed in Appendix A.

$$X = \left[\frac{T_R^8 \, \kappa^8 \, T \, t \, L_T \, D_T \, V \, W_M \, \rho_M \, d_M \, P_O \, \rho_O \, W_C \, D_C \, S_C \, \rho_C \, C_P \, P_C}{S_M^{22} \, \eta^{35} \, \}} \right] \quad (2)$$

The equation (2) is determined using constant values of parameters, as listed in table B.1.

The final form of the Dimensionless Degree of Oxidation Prediction Model (DDOPM) is then given as in equation (3). This model depicts the degree of oxidation (O) governed by the process parameters. There are two dimensionless constants of the model that emerged, i.e., ϕ and A, known as characteristic constant. These constants can be obtained by conducting an experiment. To solving equation (3), it was carried out by varying the time and keeping other parameters (as listed in table B1) constant.

$$O = \Phi(X)^A \quad (3)$$

Taking the logarithmic of equation (3) on both sides:

$$\log O = \log \Phi + A \log X \quad (4)$$

This is similar with the general of linear equation of $y = c + mx$.

3. Experimental Procedures

3.1. Material preparation

The mill scale was mechanically grinded into a fine powder and sieved into 400 mesh that provided by PT Krakatau Steel Tbk., Indonesia. It was then heated at the temperature of 200 °C for 1 hour as pretreatment to reduce dirt and lubricant. Chemical analysis was conducted by using XRF Analyzer (S2 Puma, Bruker, Germany).

Limestone (CAS:1317-65-3) technical grade was grounded into a fine powder then sieved into 325 mesh. Limestone powder was prepared as a pellet by pressing 2 grams of powder using Hydraulic Lab Press CMG 100H-15 (Carver, America) at the pressure of 200 MPa for 5 minutes. The pellet diameter is about 2 cm with a density of 0.5127 g cm^{-3}. The conductivity and specific heat capacity of limestone are measured using TCi Thermal Conductivity Analyzer (C-Therm, Canada).

3.2. Mill Scale Oxidation Processes

The applied ceramic boat in this work has the dimension of half-cylinder shape with length of 9.3 cm, width of 3 cm, and height of 1.5 cm, as shown in figure 1. The 6 limestone tablets with each weight of 2 grams placed on the bottom (figure 1(a) and top of mill scale powder (figure 1(c)), and the 20 grams of mill scale powder (figure 1(b) were loaded into the boat for each heating process. The crucible was then placed inside the center of heating zone that covered with the glass tube with length of 1 m and diameter of 5.5 cm (figure 2). The tubular furnace (figure 2) used in this present study was made by the Research Center for Physics, Indonesian Institute of Sciences (LIPI, Indonesia). There was no pure oxygen supply used in this process, and the oxidant used in the calcination process was solely ambient air. Therefore, both end of the glass tube was free exposed into the air. The calcination process was carried out at holding temperature of 900 °C, heating rate of 5 °C/min, at a varying holding time of 1, 2, and 4 hours. The corresponding samples are then abbreviated as HEM1, HEM2, and HEM4; the same approach was also carried out for CON1, CON2, and CON4, as listed in table 2, to showing with and without limestone addition respectively.

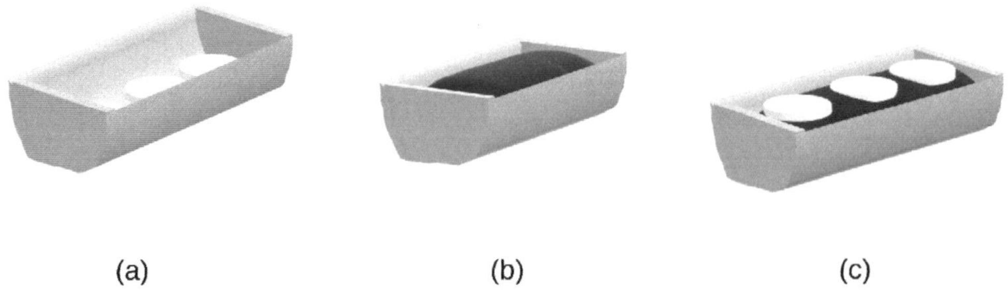

Figure 1. (a) Schematic diagram of limestone pellets on the bottom of the crucible, (b) mill scale powder was placed on top of limestone pellets, and (c) the three limestone pellets were placed on top of mill scale powder.

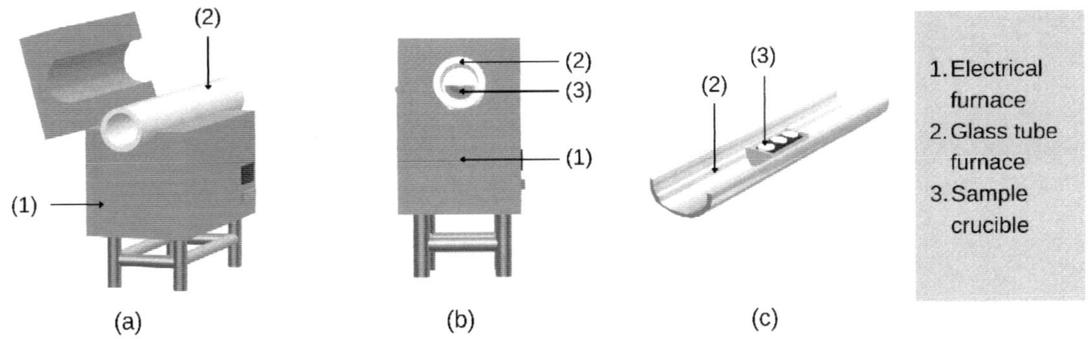

Figure 2. (a) Schematic of furnace used for oxidation of mill scale, (b) side view of furnace showing the crucible inside the tube and (c) explode view of the tube showing the placement of the crucible.

Table 2. Process condition for oxidation of mill scale.

Sample	W_M (g)	W_C (g)	T (°C)	t (h)
HEM1	20	12	900	1
HEM2	20	12	900	2
HEM4	20	12	900	4
CON1	20	0	900	1
CON2	20	0	900	2
CON4	20	0	900	4

The resulting powder was then ball milled at 300 rpm for 2 hours. The ball to powder weight ratio was set at 20:1 wt.% with large to medium to small weight ratio of the ceramic ball used are 4:2:4 wt.%.

4. Results and Discussion

The XRF results of used mill scale in this work mainly consist of wustite and magnetite, as shown in table 3. The oxidation of wustite into magnetite and magnetite into hematite occurs according to the following reaction:

$$6\,FeO + O_2 \rightarrow 2\,Fe_3O_4 \tag{5}$$
$$\Delta H_{900°C} = -641.36\,kJ/mol\,, \Delta G_{900°C} = -316.59\,kJ/mol$$
$$\Delta H_{25°C} = -604.56\,kJ/mol\,, \Delta G_{25°C} = -522.02\,kJ/mol$$

$$4\,Fe_3O_4 + O_2 \rightarrow 6\,Fe_2O_3 \tag{6}$$
$$\Delta H_{900°C} = -454.08\,kJ/mol\,, \Delta G_{900°C} = -141.62\,kJ/mol$$
$$\Delta H_{900°C} = -471.60\,kJ/mol, \Delta G_{900°C} = -392.19\,kJ/mol$$

This reaction was set-up in two-step processes. First step, the wustite converted into magnetite and then the next oxidation step is converted the magnetite into hematite. This means that there are two solid phases can coexist during the reaction either with pure oxygen or oxygen from ambient air, i.e., wustite and magnetite phase or magnetite and hematite phase [19].

Both of the reaction are exothermic, and thermodynamically possible to occur at room temperature; however, the kinetic rate of the oxidation process is extremely slow at low temperature. A study by Monazam *et al.* [20] shows that oxidation of magnetite into hematite starts to occur at a decent rate at

750 °C. It takes 10 to 60 minutes for the oxidation process of wustite to magnetite to occur, and 30 to 240 minutes to convert magnetite into hematite. Both depend on the oxygen supply flow rate.

Table 3. XRF results of mill scale as raw material.

Components	Weight (%)
Fe_T	98.744
FeO	54.070
Fe_3O_4	44.829
MnO_2	0.756
SiO_2	0.171
Cr_2O_3	0.083
TiO_2	0.025
V_2O_5	0.023
CaO	0.021
Others	0.023

The black-colored powder was obtained after the calcination process for both HEM and CON samples. When the sample is held close to a hard magnet (NdFeB), all HEM samples show weak attraction on it, while all CON samples show paramagnetic behavior, as shown in figure 3. HEM and CON turn into bright, red-colored powder and brown powder, respectively, after dry milling processes. This due to some extra oxidations happen during the milling as a consequence of ball impact and speed. The complete XRF results for all samples after the oxidation processes at 900 °C and dry milling for 2 hours are listed in table 4.

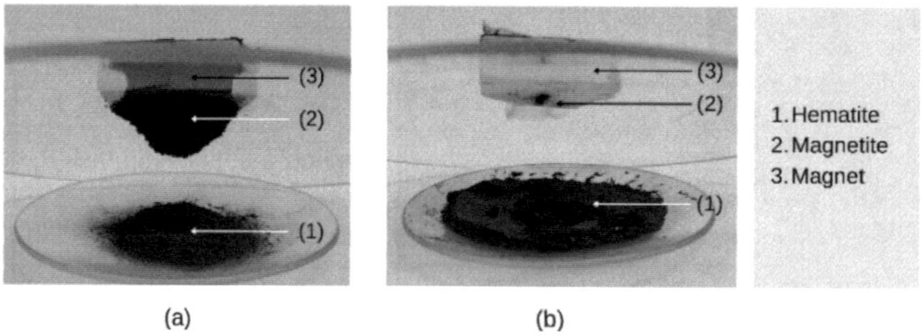

Figure 3. Magnetic interaction of (a) CON1 and (b) HEM1.

Table 4. XRF analysis of oxidized mill scale at 900 °C.

Sample	Weight (%)					
	Fe_2O_3	Fe_3O_4	CaO	SiO_2	MnO_2	Others
HEM1	93.382	5.866	0.093	0.110	0.476	0.072
HEM2	95.487	3.440	0.260	0.147	0.545	0.117
HEM4	96.769	1.871	0.400	0.108	0.482	0.370
CON1	37.053	61.905	0.018	0.120	0.403	0.502
CON2	43.062	56.097	0.011	0.113	0.395	0.321
CON4	49.737	49.944	0.000	0.153	0.166	0.000

The number of Fe_2O_3 content in table 4 shows significant different number between with and without of limestone pellets. This can be explained due to the effect as a heating storage media of limestone during the calcination (roasting). Therefore, it will extend the effect of heat distribution into the mill scale powder during the calcination, as illustrated in figure 4. Then, resulting into better oxidation process and significant effect into the number of resulted hematite. Figure 4 shows that HEM4 undergo a longer time at a temperature region above the minimum limit for the oxidation process to occur than CON4. The same phenomenon occurs between HEM2 & CON2 and HEM1 & CON1. It should be noted that the values given on this graph are rough estimation from the temperature measurements made every 30 minutes during the oxidation process. For the first 3 hours, all HEM and CON samples undergo the temperature rise step from 30 to 900 °C. Oxidation of wustite in mill scale to magnetite starts to occur at this step. The second step is the holding time which varied as shown in table 2. Then it was the temperature drop step, where all CON samples undergo a drop of 2 to 5 $Kmin^{-1}$, whereas all HEM samples undergo a slower temperature drop. This phenomenon happens due to the presence of limestone pellet as heat storage media. The placement of the pellets was arranged in such a way that heat was trapped around the sample. Conductivity and specific heat capacity measurement results are 0.086 $Wm^{-1}K^{-1}$ and 238.564 $Jkg^{-1}K^{-1}$, respectively. Limestone's low thermal conductivity plays a huge role in maintaining temperature drop as slow as possible. Limestone's excessive specific heat capacity and low thermal conductivity allow it to absorb a high amount of heat during the temperature rise step and release it at a slower rate during the cooling temperature drop step [18].

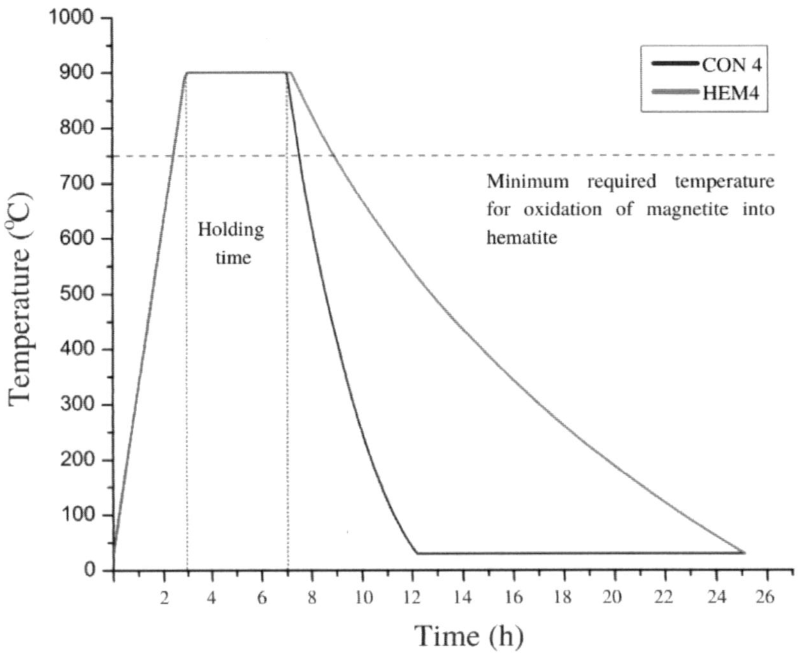

Figure 4. Estimated changes of temperature of CON4 and HEM4 inside the furnace during the oxidation processes.

The forgoing analysis highlights the individual effect of the holding time (t) on the purity of hematite in the resulting powder or the degree of oxidation (O). As given by equation (3), the dimensionless mathematical model has been formulated to establish a logical relationship. A summary of the experimental degree of oxidation (O_E), predicted degree of oxidation (O_P) from DDOPM calculation, % relative error, and logarithmic value of (X) for HEM1, HEM2, and HEM 4 are given in table 5.

From equation 4, a graph of log X vs. log O_E was fitted using Fit Linear Regression Model from MatLab, fitlm. It was found that the linear correlation coefficient of adjustment is 0.959. Conforming to the linearization shown in figure 5, The values of the characteristic constant of A and ϕ are found to be 0.0257 and 0.0128, respectively. Therefore, the DDOPM equation is:

$$O = 0.0128(X)^{0.0257} \tag{7}$$

which gives the value of the predicted degree of oxidation (O_P) as listed in table 5. An infinitesimal relative error shows that the DDOPM calculation successfully predicted the hematite purity in the resulting powder. It should be emphasized that this DDOPM equation, shown in equation (7), can only be used to predict hematite purity based on the variation of time. Further experiment with the more varied variable is required to obtain the individual characteristic constant if other variable relationships will be further measured.

Table 5. Process condition for oxidation of mill scale.

Sample	log (X)	O_P	O_E	% Relative Error
HEM1	150.3283	93.4646	93.382	0.088%
HEM2	150.6293	95.1445	95.487	-0.360%
HEM4	150.9303	96.8545	96.769	-0.088%

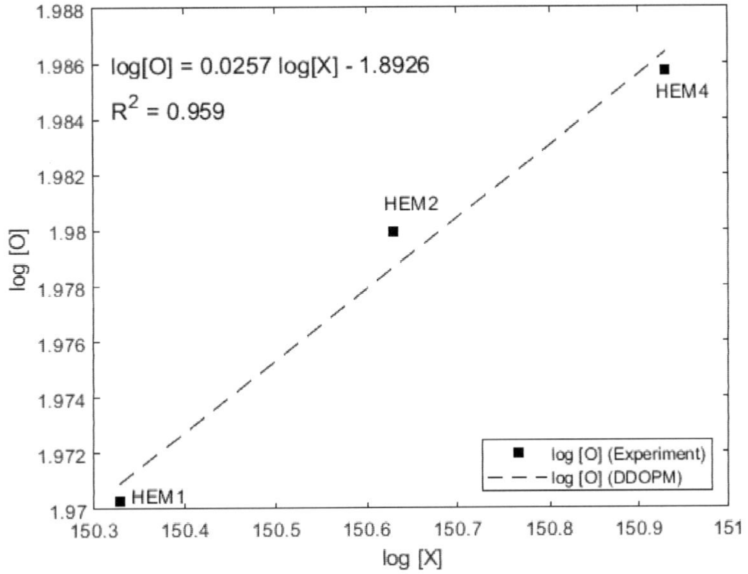

Figure 5. Linear relationship of the logarithmic value of dimensionless degree of oxidation log[O] vs. log[X] to obtain characteristic constant.

The Fe_2O_3 purity shows a power-law relationship with time, as indicated in figure 6. This graph is computed by MatLab curve fit of the 'power order-1'. It shows that the increase rate of Fe_2O_3 purity tends to decelerate after 4 hours of holding time as the gradient of the graph tend to flatten. This imply that the roasting process with holding time above 4 hours is unnecessary and not effectively convert into targeted Fe_2O_3 phase. Other consideration is the discharge energy and time wasted for the furnace to run up to 14 hours.

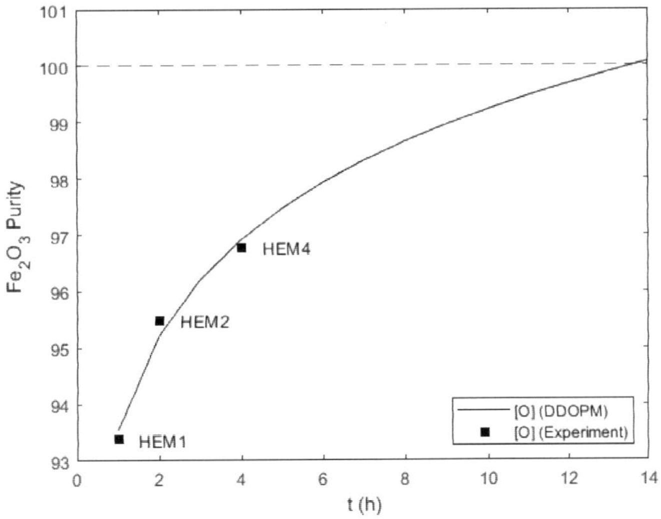

Figure 6. Power-law relationship of Fe_2O_3 purity based on DDOPM calculation vs. time.

5. Conclusion

The Dimensionless Degree of Oxidation Prediction Model (DDOPM) has been obtained using the Buckingham Pi theorem. The characteristic constant derived from this model can be used to predict

Fe_2O_3 purity relationship with time as selected parameters. The dimensionless model for analyzing the degree of oxidation has been established with $R^2 = 0.959$ and characteristic constant of $\phi = 0.0128$ and $A = 0.0257$. O configures the degree of oxidation, and the influencing parameters process are configured by X. A relationship between the experimental degree of oxidation (O_E) and predicted degree of oxidation (O_P) is in close agreement with % relative error less than 0.36%. The XRF analysis results show that the addition of limestone pellets as heat storage media in the proposed experimental design significantly increases the Fe_2O_3 purity in the resulting powder.

References

[1] Legodi M A and de Waal D 2007 *Dyes Pigm.* **74** 161–8
[2] Bagatini M C, Zymla V, Osorio E and Vilela A C F 2011 *ISIJ Int.* **51** 1072–9
[3] Nurjaman F, Prilitasari N M, Prasetyo A E and Nugroho E 2019 *Maj. Metall.* **34** 37–48
[4] Zulhan Z 2013 *Maj.Metall.* **28** 105–20
[5] Cornell R and Schwertmann U 2003 *The Iron Oxides: Structures, Properties, Reaction, Occurences and Uses* (John Wiley & Sons, Ltd)
[6] Zheng X and Li J 2014 *Ion.* **20** 1651–63
[7] Teja A S and Koh P Y 2009 *Prog. Cryst. Growth Charact. Mater.* **55** 22–45
[8] Ismail H, Zaki M, Hussein G and Magar M 1990 *Powder Technol.* **63** 87–96
[9] Ismail H M, Cadenhead D and Zaki M I 1996 *J. Colloid Interface Sci.* **183** 320–8
[10] Bailey J K, Brinker C and Mecartney M L 1993 *J. Colloid Interface Sci.* **157** 1–13
[11] Itoh H and Sugimoto T 2003 *J. Colloid Interface Sci.* **265** 283–95
[12] Jikar P, Sabban R, Tadwalkar C and Dhokey N 2020 *Mater. Today : Proc.* **44** 4013–19
[13] Dhokey N and Jikar P 2020 *J. Sustain. Metall.* **6** 622–30
[14] Wu T and Lei C 2016 *Energy Build.* **116** 198–208
[15] Khare S, Dell'Amico M, Knight C and McGarry S 2013 *Sol. Energy Mater. Sol. Cells* **115** 114-22
[16] Dinker A, Agarwal M and Agarwal G 2017 *J. Energy Inst.* **90** 1–11
[17] Harold W 1953 *Thermodynamic Properties of Molecular Oxygen* (National Bureau of Standards Report)
[18] Khosa A A, Xu T, Xia B, Yan J and Zhao C 2019 *Sol. Energy* **193** 618–36
[19] Mori K, Okada T, Takagi Y, Takada Y and Mizoguchi T 1999 *Jpn. J. Appl. Phys.* **38** 189–91
[20] Monazam E R, Breault R W and Siriwardane R V 2014 *Ind. Eng. Chem. Res.* **53** 13320–8

Acknowledgments

This work was supported by the Lembaga Ilmu Pengetahuan Indonesia (LIPI), Indonesia. The authors would also like to thank the Research Center for Physics of LIPI for allowing the use of their excellent testing equipment.

Authorship contribution statement

M T Aldiramadan: Methodology, Formal analysis, Investigation, Validation, Visualization, Writing - original draft. **J Nabilah**: Formal analysis, Investigation, Validation, Writing - original draft. **M Ridwan**: Validation, Supervision. **D S Khaerudini**: Conceptualization, Methodology, Formal analysis, Investigation, Supervision, Writing - review & editing, Funding acquisition.

Appendix A. Derivation of Process Parameter (X) with Buckingham Pi Theorem
Total Number of Variable: 20
Total number of Fundamental Dimensions: 4
Π Terms Required: 20 – 4 = 16

$$X = f(\Pi_1, \Pi_2, \Pi_3, \Pi_4, \Pi_5, \Pi_6, \Pi_7, \Pi_8, \Pi_9, \Pi_{10}, \Pi_{11}, \Pi_{12}, \Pi_{13}, \Pi_{14}\Pi_{15}\Pi_{16}) \quad \text{(A.1)}$$

Repeating Variables
1] Heating Rate (T_R) [$M^0L^0T^{-1}\theta^1$]
2] Specific Surface Area of Mill Scale Powder (S_M) [$M^{-1}L^2T^0\theta^0$]
3] Viscosity of Oxygen (η) [$M^1L^{-1}T^{-1}\theta^0$]
4] Thermal Conductivity of Calcite (κ) [$M^1L^1T^{-3}\theta^{-1}$]

1. **Π_1 Term Calculation**

Non-repeating variable: holding temperature (T) [$M^0L^0T^0\theta^1$]
$\Pi_1 = T_R^a S_M^b \eta^c \kappa^d T$
$[M^0L^0T^0\theta^0] = [M^0L^0T^{-1}\theta^1]^a[M^{-1}L^2T^0\theta^0]^b[M^1L^{-1}T^{-1}\theta^0]^c[M^1L^1T^{-3}\theta^{-1}]^d[M^0L^0T^0\theta^1]$
For M : $-b + c + d = 0$
For L : $2b - c + d = 0$
For T : $-a - c - 3d = 0$
For θ : $a - d + 1 = 0$
$a = 0; b = -2; c = -3; d = 1$

$$\Pi_1 = \frac{\kappa T}{S_M^2 \eta^3} \quad \text{(A.2)}$$

2. **Π_2 Term Calculation**

Non-repeating variable: holding time (t) [$M^0L^0T^1\theta^0$]
$\Pi_2 = T_R^a S_M^b \eta^c \kappa^d t$
$[M^0L^0T^0\theta^0] = [M^0L^0T^{-1}\theta^1]^a[M^{-1}L^2T^0\theta^0]^b[M^1L^{-1}T^{-1}\theta^0]^c[M^1L^1T^{-3}\theta^{-1}]^d[M^0L^0T^1\theta^0]$
For M : $-b + c + d = 0$
For L : $2b - c + d = 0$
For T : $-a - c - 3d + 1 = 0$
For θ : $a - d = 0$
$a = 1; b = -2; c = -3; d = 1$

$$\Pi_2 = \frac{T_R \kappa\, t}{S_M^2 \eta^3} \quad \text{(A.3)}$$

3. **Π_3 Term Calculation**

Non-repeating variable: Length of the tube (L_T) [$M^0L^1T^0\theta^0$]
$\Pi_3 = T_R^a S_M^b \eta^c \kappa^d L_T$
$[M^0L^0T^0\theta^0] = [M^0L^0T^{-1}\theta^1]^a[M^{-1}L^2T^0\theta^0]^b[M^1L^{-1}T^{-1}\theta^0]^c[M^1L^1T^{-3}\theta^{-1}]^d[M^0L^1T^0\theta^0]$
For M : $-b + c + d = 0$
For L : $2b - c + d + 1 = 0$
For T : $-a - c - 3d = 0$
For θ : $a - d = 0$
$a = 1; b = -3; c = -4; d = 1$

$$\Pi_3 = \frac{T_R \kappa\, L_T}{S_M^3 \eta^4} \quad \text{(A.4)}$$

4. Π_4 Term Calculation

Non-repeating variable: Diameter of the tube (D_T) $[M^0L^1T^0\theta^0]$

$\Pi_4 = T_R^a S_M^b \eta^c \kappa^d D_T$

$[M^0L^0T^0\theta^0] = [M^0L^0T^{-1}\theta^1]^a[M^{-1}L^2T^0\theta^0]^b[M^1L^{-1}T^{-1}\theta^0]^c[M^1L^1T^{-3}\theta^{-1}]^d[M^0L^1T^0\theta^0]$

For M : $-b + c + d = 0$
For L : $2b - c + d + 1 = 0$
For T : $-a - c - 3d = 0$
For θ : $a - d = 0$

$a = 1; b = -3; c = -4; d = 1$

$$\Pi_4 = \frac{T_R \kappa\, D_T}{S_M^3 \eta^4} \tag{A.5}$$

5. Π_5 Term Calculation

Non-repeating variable: Crucible Volume (V) $[M^0L^3T^0\theta^0]$

$\Pi_5 = T_R^a S_M^b \eta^c \kappa^d V$

$[M^0L^0T^0\theta^0] = [M^0L^0T^{-1}\theta^1]^a[M^{-1}L^2T^0\theta^0]^b[M^1L^{-1}T^{-1}\theta^0]^c[M^1L^1T^{-3}\theta^{-1}]^d[M^0L^3T^0\theta^0]$

For M : $-b + c + d = 0$
For L : $2b - c + d + 3 = 0$
For T : $-a - c - 3d = 0$
For θ : $a - d = 0$

$a = 3; b = -9; c = -12; d = 3$

$$\Pi_5 = \frac{T_R^3 \kappa^3\, V}{S_M^9 \eta^{12}} \tag{A.6}$$

6. Π_6 Term Calculation

Non-repeating variable: weight of mill scale (W_M) $[M^1L^0T^0\theta^0]$

$\Pi_6 = T_R^a S_M^b \eta^c \kappa^d W_M$

$[M^0L^0T^0\theta^0] = [M^0L^0T^{-1}\theta^1]^a[M^{-1}L^2T^0\theta^0]^b[M^1L^{-1}T^{-1}\theta^0]^c[M^1L^1T^{-3}\theta^{-1}]^d[M^1L^0T^0\theta^0]$

For M : $-b + c + d + 1 = 0$
For L : $2b - c + d = 0$
For T : $-a - c - 3d = 0$
For θ : $a - d = 0$

$a = 2; b = -5; c = -8; d = 2$

$$\Pi_6 = \frac{T_R^2 \kappa^2 W_M}{S_M^5 \eta^8} \tag{A.7}$$

7. Π_7 Term Calculation

Non-repeating variable: apparent density of mill scale (ρ_M) $[M^1L^{-3}T^0\theta^0]$

$\Pi_7 = T_R^a S_M^b \eta^c \kappa^d \rho_M$

$[M^0L^0T^0\theta^0] = [M^0L^0T^{-1}\theta^1]^a[M^{-1}L^2T^0\theta^0]^b[M^1L^{-1}T^{-1}\theta^0]^c[M^1L^1T^{-3}\theta^{-1}]^d[M^1L^{-3}T^0\theta^0]$

For M : $-b + c + d + 1 = 0$
For L : $2b - c + d - 3 = 0$
For T : $-a - c - 3d = 0$
For θ : $a - d = 0$

$a = -1; b = 4; c = 4; d = -1$

$$\Pi_7 = \frac{S_M^4 \eta^4 \rho_M}{T_R \kappa} \tag{A.8}$$

8. Π_8 Term Calculation

Non-repeating variable: Mill Scale Particle Size (d_M) $[M^0L^1T^0\theta^0]$

$\Pi_8 = T_R^a S_M^b \eta^c \kappa^d d_M$

$[M^0L^0T^0\theta^0] = [M^0L^0T^{-1}\theta^1]^a[M^{-1}L^2T^0\theta^0]^b[M^1L^{-1}T^{-1}\theta^0]^c[M^1L^1T^{-3}\theta^{-1}]^d[M^0L^1T^0\theta^0]$

For M : $-b + c + d = 0$
For L : $2b - c + d + 1 = 0$
For T : $-a - c - 3d = 0$
For θ : $a - d = 0$

a = 1; b = -3; c = -4; d = 1

$$\Pi_8 = \frac{T_R \kappa\, d_M}{S_M^3 \eta^4} \tag{A.9}$$

9. Π_9 Term Calculation

Non-repeating variable: Oxygen purity (P_O) $[M^0L^0T^0\theta^0]$

$\Pi_9 = T_R^a S_M^b \eta^c \kappa^d P_O$

$[M^0L^0T^0\theta^0] = [M^0L^0T^{-1}\theta^1]^a[M^{-1}L^2T^0\theta^0]^b[M^1L^{-1}T^{-1}\theta^0]^c[M^1L^1T^{-3}\theta^{-1}]^d[M^0L^0T^0\theta^0]$

For M : $-b + c + d = 0$
For L : $2b - c + d = 0$
For T : $-a - c - 3d = 0$
For θ : $a - d = 0$

a = 0; b = 0; c = 0; d = 0

$$\Pi_9 = P_O \tag{A.10}$$

10. Π_{10} Term Calculation

Non-repeating variable: Density of oxygen (ρ_O) $[M^1L^{-3}T^0\theta^0]$

$\Pi_{10} = T_R^a S_M^b \eta^c \kappa^d \rho_O$

$[M^0L^0T^0\theta^0] = [M^0L^0T^{-1}\theta^1]^a[M^{-1}L^2T^0\theta^0]^b[M^1L^{-1}T^{-1}\theta^0]^c[M^1L^1T^{-3}\theta^{-1}]^d[M^1L^{-3}T^0\theta^0]$

For M : $-b + c + d + 1 = 0$
For L : $2b - c + d - 3 = 0$
For T : $-a - c - 3d = 0$
For θ : $a - d = 0$

a = -1; b = 4; c = 4; d = -1

$$\Pi_{10} = \frac{S_M^4 \eta^4 \rho_O}{T_R \kappa} \tag{A.11}$$

11. Π_{11} Term Calculation

Non-repeating variable: Weight of Calcite (W_C) $[M^1L^0T^0\theta^0]$

$\Pi_{11} = T_R^a S_M^b \eta^c \kappa^d W_C$

$[M^0L^0T^0\theta^0] = [M^0L^0T^{-1}\theta^1]^a[M^{-1}L^2T^0\theta^0]^b[M^1L^{-1}T^{-1}\theta^0]^c[M^1L^1T^{-3}\theta^{-1}]^d[M^1L^0T^0\theta^0]$

For M : $-b + c + d + 1 = 0$
For L : $2b - c + d = 0$
For T : $-a - c - 3d = 0$
For θ : $a - d = 0$

a = 2; b = -5; c = -8; d = 2

$$\Pi_{11} = \frac{T_R^2 \kappa^2 W_C}{S_M^5 \eta^8} \tag{A.12}$$

12. Π_{12} Term Calculation

Non-repeating variable: Calcite Pellet Diameter (D_C) $[M^0L^1T^0\theta^0]$

$\Pi_{12} = T_R{}^a S_M{}^b \eta^c \kappa^d D_C$

$[M^0L^0T^0\theta^0] = [M^0L^0T^{-1}\theta^1]^a[M^{-1}L^2T^0\theta^0]^b[M^1L^{-1}T^{-1}\theta^0]^c[M^1L^1T^{-3}\theta^{-1}]^d[M^0L^1T^0\theta^0]$

For M : $-b + c + d = 0$
For L : $2b - c + d + 1 = 0$
For T : $-a - c - 3d = 0$
For θ : $a - d = 0$

$a = 1; b = -3; c = -4; d = 1$

$$\Pi_{12} = \frac{T_R\kappa\, D_C}{S_M^3\eta^4} \tag{A.13}$$

13. Π_{13} Term Calculation

Non-repeating variable: Surface Area of Calcite Pellet (S_C) $[M^{-1}L^2T^0\theta^0]$

$\Pi_{13} = T_R{}^a S_M{}^b \eta^c \kappa^d D_T$

$[M^0L^0T^0\theta^0] = [M^0L^0T^{-1}\theta^1]^a[M^{-1}L^2T^0\theta^0]^b[M^1L^{-1}T^{-1}\theta^0]^c[M^1L^1T^{-3}\theta^{-1}]^d[M^{-1}L^2T^0\theta^0]$

For M : $-b + c + d - 1 = 0$
For L : $2b - c + d + 2 = 0$
For T : $-a - c - 3d = 0$
For θ : $a - d = 0$

$a = 0; b = -1; c = 0; d = 0$

$$\Pi_{13} = \frac{S_C}{S_M} \tag{A.14}$$

14. Π_{14} Term Calculation

Non-repeating variable: apparent density of calcite (ρ_C) $[M^1L^{-3}T^0\theta^0]$

$\Pi_{14} = T_R{}^a S_M{}^b \eta^c \kappa^d \rho_C$

$[M^0L^0T^0\theta^0] – [M^0L^0T^{-1}\theta^1]^a[M^{-1}L^2T^0\theta^0]^b[M^1L^{-1}T^{-1}\theta^0]^c[M^1L^1T^{-3}0^{-1}]^d[M^1L^{-3}T^0 0^0]$

For M : $-b + c + d + 1 = 0$
For L : $2b - c + d - 3 = 0$
For T : $-a - c - 3d = 0$
For θ : $a - d = 0$

$a = -1; b = 4; c = 4; d = -1$

$$\Pi_{14} = \frac{S_M^4\eta^4\rho_C}{T_R\kappa} \tag{A.15}$$

15. Π_{15} Term Calculation

Non-repeating variable: Specific Heat Capacity of Calcite (C_P) $[M^0L^2T^{-2}\theta^{-1}]$

$\Pi_{15} = T_R{}^a S_M{}^b \eta^c \kappa^d C_P$

$[M^0L^0T^0\theta^0] = [M^0L^0T^{-1}\theta^1]^a[M^{-1}L^2T^0\theta^0]^b[M^1L^{-1}T^{-1}\theta^0]^c[M^1L^1T^{-3}\theta^{-1}]^d[M^0L^2T^{-2}\theta^{-1}]$

For M : $-b + c + d = 0$
For L : $2b - c + d + 2 = 0$
For T : $-a - c - 3d - 2 = 0$
For θ : $a - d - 1 = 0$

$a = 0; b = 0; c = 1; d = -1$

$$\Pi_{15} = \frac{\eta C_P}{\kappa} \tag{A.16}$$

16. Π_{16} Term Calculation

Non-repeating variable: Equilibrium Pressure of Calcite Decomposition (P_C) $[M^1L^{-1}T^{-2}\theta^0]$

$\Pi_{16} = T_R^a S_M^b \eta^c \kappa^d P_C$

$[M^0L^0T^0\theta^0] = [M^0L^0T^{-1}\theta^1]^a[M^{-1}L^2T^0\theta^0]^b[M^1L^{-1}T^{-1}\theta^0]^c[M^1L^1T^{-3}\theta^{-1}]^d[M^1L^{-1}T^{-2}\theta^0]$

For M : $-b+c+d+1=0$

For L : $2b-c+d-1=0$

For T : $-a-c-3d-2=0$

For θ : $a-d=0$

$a=-1;\ b=2;\ c=2;\ d=-1$

$$\boldsymbol{\Pi_{16} = \frac{S_M^2\eta^2 P_C}{T_R\kappa}} \tag{A.17}$$

$$X = \frac{\kappa\, T}{S_M^2\eta^3}\times\frac{T_R\kappa\, t}{S_M^2\eta^3}\times\frac{T_R\kappa\, L_T}{S_M^3\eta^4}\times\frac{T_R\kappa\, D_T}{S_M^3\eta^4}\times\frac{T_R^3\kappa^3\, V}{S_M^9\eta^{12}}\times\frac{T_R^2\kappa^2W_M}{S_M^5\eta^8}\times\frac{S_M^4\eta^4\rho_M}{T_R\kappa}\times\frac{T_R\kappa\, d_M}{S_M^3\eta^4}\times P_O$$
$$\times\frac{T_R^2\kappa^2W_C}{S_M^5\eta^8}\times\frac{T_R\kappa\, D_C}{S_M^3\eta^4}\times\frac{S_C}{S_M}\times\frac{S_M^4\eta^4\rho_C}{T_R\kappa}\times\frac{\eta C_P}{\kappa}\times\frac{S_M^2\eta^2P_C}{T_R\kappa}\times\frac{S_M^4\eta^4\rho_O}{T_R\kappa} \tag{A.18}$$

$$X = \left[\frac{T_R^8\kappa^8 T t L_T D_T V W_M \rho_M d_M P_O \rho_O W_C D_C S_C \rho_C C_P P_C}{S_M^{22}\eta^{35}}\right] \tag{A.19}$$

IOP Conf. Series: Materials Science and Engineering 1195 (2021) 012044 doi:10.1088/1757-899X/1195/1/012044

Appendix B. Calculation of Process Parameter

Process Parameter X is defined by,

$$X = \left[\frac{T_R^8 \kappa^8 T t L_T D_T V W_M \rho_M d_M P_O \rho_O W_C D_C S_C \rho_C C_P P_C}{S_M^{22} \eta^{35}}\right] \quad (B.1)$$

This equation is solved using constant values of parameters listed in table B.1.

Table B.1. Experimental constant values used to solve Process Parameter (X).

No	Variables	Symbol	SI Unit	Value
1	Holding Temperature	T	K	1173
2	Holding time	t	s	Varied
3	Heating rate	T_R	Ks^{-1}	0.0833
4	Length of the tube	L_T	m	1.0
5	Diameter of the tube	D_T	m	0.055
6	Crucible volume	V	m^3	3.287×10^{-5}
7	Weight of mill scale	W_M	Kg	0.02
8	Apparent density of mill scale	ρ_M	Kgm^{-3}	5106.7
9	Mill scale particle size	d_M	m	3.70×10^{-5}
10	Specific surface area of mill scale powder	S_M	m^2Kg^{-1}	0.2
11	Oxygen purity	P_O		0.21
12	Density of oxygen	ρ_O	Kgm^{-3}	Function of T [17]
13	Viscosity of oxygen	η	$Kgm^{-1}s^{-1}$	Function of T [17]
14	Weight of limestone	W_C	Kg	0.012
15	Limestone pellet diameter	D_C	m	0.02
16	Surface area of limestone pellet	S_C	m^2Kg^{-1}	0.44
17	Apparent density of limestone	ρ_C	Kgm^{-3}	2580
18	Specific heat capacity of limestone	C_P	$JKg^{-1}K^{-1}$	238.564
19	Thermal conductivity	κ	$Wm^{-1}K^{-1}$	0.086
20	Equilibrium pressure of limestone decomposition	P_C	Pa	Function of T [18]

Oxygen density (ρ) as a function of temperature is defined by

$$\rho(T) = 397.65\, T^{-1.003} \quad (B.2)$$

with R-squared value of 1. Equation (B.2) is obtained by curve fitting the data of dimensionless density of molecular Oxygen (ρ/ρ_0) at 1 atm with the range of temperature from 100 K to 2300 K [17]. The tabulated value was converted to the desired unit (Kg/m^3) by multiply the dimensionless density with 1.42904.

η as a function of temperature is defined by

$$\eta(T) = 4 \times 10^{-7} T^{0.7071} \quad (B.3)$$

with R-squared value of 0.9964. Equation (B.3) is obtained by curve fitting the data of dimensionless coefficient of molecular oxygen viscosity (η/η_0) from 100 K to 2000 K [17]. The tabulated value was converted to the desired unit (Kg/ms) by multiply the dimensionless viscosity with 1.9191944 × 10^{-5}.

P_C as a function of temperature is defined by

$$P_C(T) = 101325 \times \exp\left(17.74 - 0.00108\,T + 0.332\ln T - \frac{22020}{T}\right) \quad \text{(B.4)}$$

Equation above defines as equilibrium pressure when Limestone undergo a decomposition process and it was calculated by Azhar et al [18].

Evolution of predictive emissions in PETRONAS

A Rusman, TA Tuan Mat and N Shabudin

Group Technical Solutions, PETRONAS, PETRONAS Tower 3, 50088 Kuala Lumpur, Wilayah Persekutuan, Malaysia

Email: andriny@petronas.com, amansyah@petronas.com, normaya_shabudin@petronas.com

Abstract. The effect of global warming and unprecedented black swan events give rise to innovative solutions to keep manufacturing companies stay afloat. Many countries have started to employ more stringent emissions limits and stricter monitoring enforcements. PETRONAS is not spared from these policies and with the enforcement of Clean Air Regulation Act (CAR) 2014, PETRONAS embarked on its emissions monitoring *via* statistical modelling methods. Traditional analysers are still employed for the emissions monitoring, but the focus of the paper is to spotlight the efforts in utilising soft sensors in ensuring compliance to CAR2014. Soft sensors are predictive models developed to infer primary parameters. The models can be derived *via* statistical tools or based on first principles using chemical engineering knowledge. The use of soft sensors is not only pervasive in the manufacturing industry but is also widely used elsewhere as it provides an indication of the future position of the primary indicators for the specific industry. The paper will also provide some insights on potential next steps in supporting PETRONAS efforts in its Sustainability Agenda. It also tells the story of collaborative efforts in maintaining good positive governance and relationship with its stakeholders.

1. Introduction on emissions monitoring in Malaysia

All continuous emission monitoring method compliance and acceptance are subjected to regulatory framework within a particular country. Worldwide, the major regulatory framework on environment protection on air emission is defined by the U.S. Environment Protection Agency and associated standards issued by the European Union. Most countries in the world align their domestic regulations with these standards.

In the middle of 2014, the Malaysian government issued a new law called the Clean Air Regulation (CAR) 2014, making it mandatory for industries to monitor stack emissions continuously. It was a significant move from the Environmental Quality (Clean Air Regulations) 1978 where industries are required to self-regulate their own emissions with continuous reporting to the Malaysian Department of Environment (DOE). PETRONAS utilised this opportunity to work with the Malaysian Department of Environment (DOE) and a PETRONAS subsidiary and implemented its first PETRONAS-Predictive Emissions Monitoring System (or P-PEMS) in 2015. In December 2015, DOE provided certification of this application and ever since then, PETRONAS began to roll-out this application to all its facilities.

2. The Malaysian regulatory framework of CAR 2014

The Environmental Quality (Clean Air Regulations) 2014 (CAR 2014) was gazetted on June 4, 2014 and came into force on June 5, 2014. The aim of the regulation is to regulate emissions of air pollutants from industrial activities including power plants, waste fuel plants as well as Oil and Gas industries as specified in First Schedule of CAR 2014 document. The affected activities and industries shall comply with the limit value and technical standards as specified in Second and Third Schedules.

2.1. Compliance to CAR2014

Under this regulation, premises shall comply with the limit value and technical standards specified in the regulations. For existing premises, five (5) years of grace period is given for opacity and limit compliance as specified in regulations 12 and 13.

2.1.1. Continuous emissions monitoring. For continuous emission monitoring as specified in regulations 16, owner of the premise shall carry out continuous emission monitoring as specified in Second and Third Schedule in the regulation. As stipulated in regulation 17, pollutants' limits value are complied to, if the following are met:

a) No daily average of emission reading exceeds the emission standard within any one calendar year
b) No half hour average of emission reading exceeds the emission standard more than two times within any one calendar year.

3. Emissions monitoring system

There are 2 different types of technology in emissions monitoring system defined under CAR 2014: Continuous Emissions Monitoring System (CEMS) and Predictive Emissions Monitoring System (PEMS). Each has its own merits and needs to follow the CAR2014 requirements for certification.

3.1. Continuous emissions monitoring system

In emissions monitoring, the common system utilized is the Continuous Emissions Monitoring System (CEMS). Continuous Emission Monitoring System (CEMS) is a hardware-based analyser system that comprises sampling system, transport line and analyser for online continuous emission monitoring. CEMS is the traditional, default solution for emissions monitoring. However, CEMS may face issues with reliability, mainly related to the sample conditioning system and complex installation especially for high moisture, high temperature systems as well as systems with long sampling lines. CEMS also requires preventive maintenance practices and parts replacement, to ensure it continues to operate effectively and under the conditions at which it was certified.

3.2. Predictive emissions monitoring system

An alternative to CEMS is Predictive Emissions Monitoring System (PEMS), a model-based prediction system to continuously monitor emissions using readily available plant process data. PEMS model/s is derived using various inferential techniques based on historical process data pattern, cause and effect relationship between process measurements and emission properties. PEMS can provide equal accuracy and quality of data comparable to CEMS. Typically, PEMS are installed due to lower operational and maintenance cost and higher reliability.

4. Background of PETRONAS PEMS (P-PEMS)

PETRONAS has been using predicted properties of process for over 20 years to control processes to meet its product specifications. Typically, properties prediction is developed in PETRONAS for the purpose of accurate and repeatable control of complex process systems and to achieve high levels of process optimisation. The prediction of properties is used to cater to issues of time delay in process analyser and unavailability of process analyser. The same proven technology used in process properties prediction is used in PETRONAS PEMS development.

4.1. PEMS technology

PETRONAS chose Statistical Regression method to for its emissions monitoring programme. In regression method, a relationship is developed between a target variable 'Y' against known 'X' values using statistical techniques. The relationship takes the form of a mathematical equation that can then be used to estimate 'Y' given a series of 'X' values. This modelling technique is used to describe relation between independent and dependent variables in scatter diagram or parity plot.

Simple linear regression is summary of relationship between one dependent and one independent variable. When there are more than one independent variables, Multiple Linear Regression (or MLR) is utilized to develop the relationship:

$$Y = a + b_1 *X_1 + b_2 *X_2 + ... + b_n *X_n \quad (1)$$

Where Y is the primary variable to be inferred (example, a component composition, e.g. CO_2, H_2S), and X is the input variables used to predict the Y (e.g. pressure, temperature or flow of the process). n indicates the multiple number of input variables that is utilised in the equation (1).

In the statistical methods, the goal is to find an equation that best summarizes the test data and predicts emissions to within an acceptable level of confidence. Linear regression statistical techniques have been used widely to develop PEMS

Other modeling techniques, e.g. First principle or combination of statistical and first principle, may be applied for a specific application to yield better models. This statistical/regression modelling technology is supplemented with several additional features to comply with regulatory requirements.

4.2. P-PEMS technology

PETRONAS first developed a P-PEMS proof-of-concept in 2013. Through collaboration with the regulatory body, the Malaysian Department of Environment (DoE) and with two trial sites, PETRONAS was able to achieve success in delivering two functional and regulatory approved applications by 2015. A typical P-PEMS architecture is as shown in figure 1:

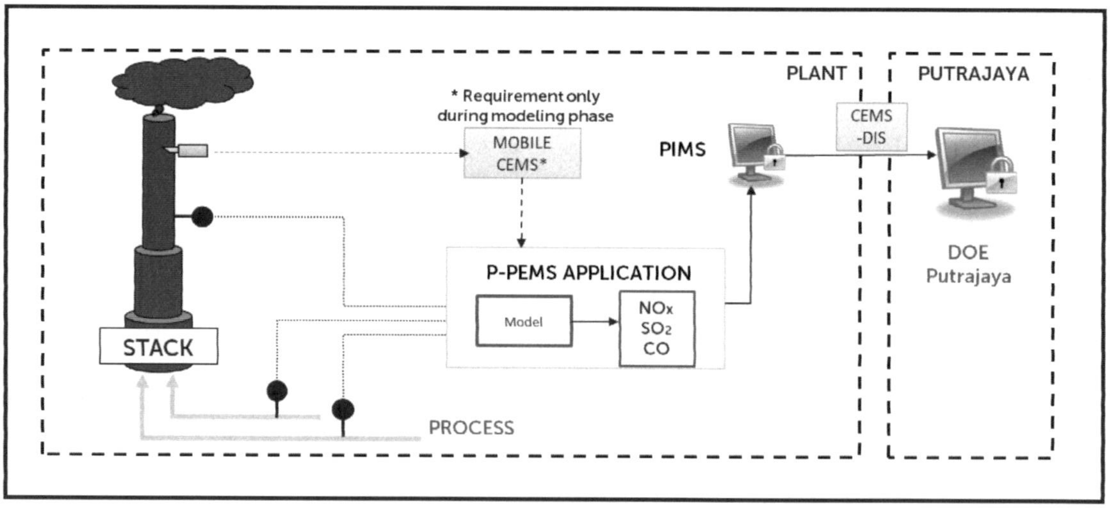

Figure 1. P-PEMS architecture.

The above figure 1 indicates how the process data from the stack is used in the P-PEMS application (modelling the emitted NO_x, SO_x and/or CO). The predicted emission values are then sent to the Plant Historian (PIMS) server. The emission data is then sent to DoE Putrajaya office *via* the CEMS-DIS server which belongs to PETRONAS. During audit/modelling period, a mobile CEMS will be hooked up to the stack to measure actual emission readings to enable comparison between the prediction and actual values.

P-PEMS complies with US EPA 40 CFR Part 60 and Part 75 as a tool for continuous emission monitoring system. P-PEMS application runs in an operating unit's highly secure server environment to protect the model, as well managing DoE's confidence on compliance.

The P-PEMS server is a system that is made for the sole intent to provide continuous transmission of emissions to DoE, as per CAR 2014. Due to this critical activity, the server is made secure, with tight access control. In addition, to ensure that there is no single point of failure, it is equipped with redundancy and Uninterrupted Power Supply (or UPS) systems which allow automatic switching of servers and power in case of system failure to provide high system availability.

P-PEMS incorporates features such as input and output validation system to ensure prediction reliability and accuracy. A dedicated graphic is configured for the users to allow continuous monitoring of emission for a particular equipment. The application will trigger alarms to alert operation personnel should there be any abnormalities either to the system or to the emission prediction. A sample of the display is as below:

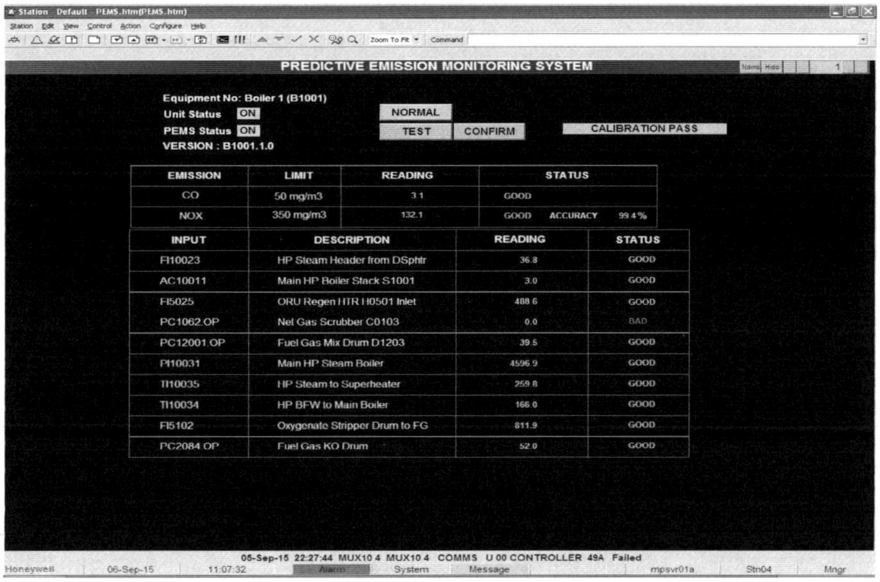

Figure 2. Sample of P-PEMS DCS display.

5. Methodology

PETRONAS developed a unique approach to modelling for Emissions Prediction. This included utilizing a third party for providing accurate stack measurements to support model accuracy, developing a rigorous approach to regression modelling with an improved understanding of model prediction limits.

5.1. *Initial phase: PEMS feasibility assessment*

To assess the applicability of PEMS, a Feasibility Assessment was developed in the initial phase. This structured assessment assesses the suitability of CEMS or PEMS for a particular stack and incorporates targeted areas such as the sufficiency of modelling inputs for PEMS and the physical modifications for CEMS. The assessment team looks at type of equipment, complexity of installation, physical dimensions, available sample results, operating modes, existing instrumentations (including reliability) and available infrastructure to support both technologies. In addition, the assessment may call for rectification to existing instrumentation, additional instrumentation, enhancement of data information systems, etc.

5.2. Implementation phase: PEMS implementation and certification

PEMS implementation phase is subjected to results obtained from initial assessment results. If initial assessment indicates PEMS is suitable technology, approval by DoE must be obtained prior to PEMS installation. The implementation of PEMS involved the coordination of activities by several stakeholders coupled with the technical aspects of the development itself. Its implementation consists of several critical stages as explained in figure 2 and the paragraph below:

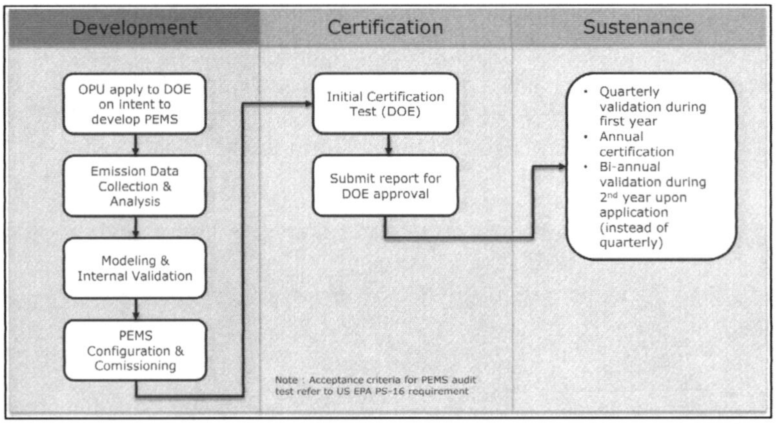

Figure 3. PEMS development and maintenance workflow.

a. Emission Data Collection and Analysis
   Both emission data from reference measurement or mobile CEMS and relevant plant process measurement are extracted on daily basis to ensure reliability and validity of data.
b. Modelling & Internal Validation
   Identification of inputs to P-PEMS model and correlation to emission is carried out at this step. Input selection to the PEMS model is based on process and operation knowledge and offline validation is performed using different sets of data to ensure model robustness.
c. Configuration and commissioning
   During the system configuration and testing, all relevant personnel including Instrument, Process Control and Operation shall be on site to ensure the system loading is executed smoothly and verify the testing. P-PEMS model will be deployed and emission prediction model will be validated against actual measurement from mobile CEMS.
d. Certification
   PEMS application shall undergo the initial certification test as stated in Section 2.1 of PS-16 before reporting any PEMS data as quality-assured Performance Specification 16 (PS-16) is used to determine whether PEMS is acceptable for use and compliance with applicable requirement. Full report on PEMS implementation shall be submitted to DoE for approval before PEMS is accepted as emission monitoring method for a particular stack.

6. Challenges

PETRONAS, being the first company in Malaysia to develop its own PEMS technology, required an excellent understanding of Malaysia's Clean Air Regulation and US EPA specifically for PEMS. To overcome this, several engagements were organized with DoE and data verification with CEMS vendor.

The PEMS server application involved the development of numerous unique algorithms that were required to comply with specific sections of the US EPA PS-16 requirements. Each of these algorithms required a development, testing and documentation phase in order to ensure compliance and acceptance by regulatory body. Multiple engagement with development team to ensure modules execute concurrently and continuously to ensure input parameters and emission predictions are reliable and accurate.

PETRONAS-PEMS represents the first Malaysian-made predictive emissions application that is in full compliance to the Clean Air Regulations (CAR) 2014, leveraging on skills by Malaysian engineers to that venture into developing own product they believe has immense potential to grow and has positive impact to environment. From the first pilot implementation, a comprehensive Project Plan and Organization Chart was developed to achieve a successful project within the allotted timeframe. Through extensive collaboration between several organizations such as the premise owner, the Solution Provider, a third party CEMS vendor that provides accurate measurements for modelling, and DoE, PETRONAS has been approved as PEMS consultant by DoE.

In terms of application technology, P-PEMS initially was implemented in Plant Distributed Control System (DCS) taking into consideration of system security, internal capability of DCS coding by PETRONAS' Automation Engineer, available features in DCS to satisfy with requirements in US EPA and less network communication requirements. However, upon increment in the number of P-PEMS implementation which covers different DCS brand (Honeywell, Yokogawa, Foxboro), housing the applications in DCS were no longer effective as the coding needs to be specific for different DCS brand. In addition, system troubleshooting will be dictated by DCS brand which is cumbersome for the personnel.

7. Sustainability

Upon successful implementation of P-PEMS and officially approved by DoE, it is mandatory for the application to be maintained accordingly to ensure the prediction accuracy and validity of the emission being reported. Malaysian DoE has outlined to all industries utilising PEMS as a monitoring and reporting tool on emission to follow US EPA 40 CFR Part 60 specifically under PS-16 for application audits and verifications. An overview of the mandatory actions required is as summarised in figure 4 below:

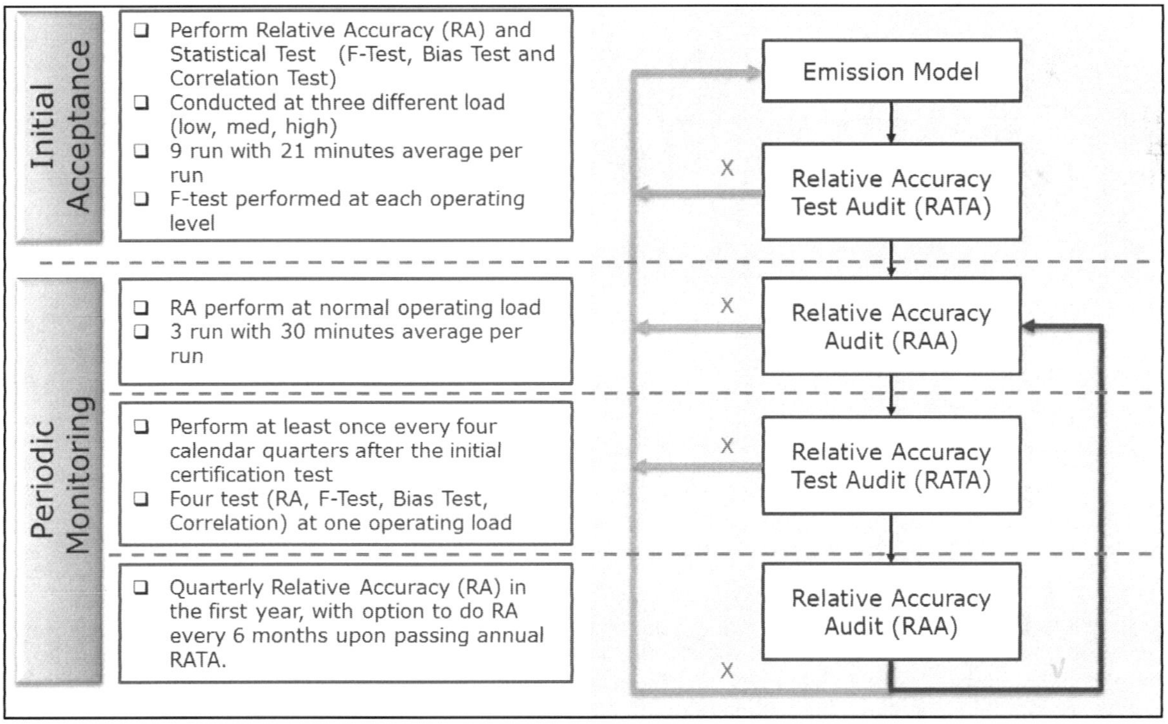

Figure 4. Sustainability requirements as required under US EPA 40 CFR part 60.

Due to this, PETRONAS has developed a sustenance program to comply to these requirements and most importantly keeping the applications continuously verified and validated by DoE. This sustenance program among others (not limited to), consist of the following:

a. To conduct a quarterly RAA
b. To conduct Yearly RATA
c. P-PEMS maintenance log/record
d. Sensors' maintenance log/record
e. P-PEMS alarm journal
f. Capability development

The most important facet to ensure P-PEMS sustainability over the long run is to secure a long-term partnership with local Mobile CEMS services provider. PETRONAS has established a contract with one of the local Mobile CEMS service provider to conduct P-PEMS verification audit on quarterly basis (either RAA or Annual RATA). The activities are planned accordingly with inputs from sites on equipment/stack availability as well as any issues or constraints to be handled. Based on this Master Plan, the sustenance program is executed in a coordinated manner.

7.1. Example of quarterly RAA & annual RATA conducted

P-PEMS will be subjected to verification every quarter either for Relative Accuracy Audit (RAA) or Annual Relative Accuracy Test Audit (RATA). The frequency is 3 times RAA and 1-time Annual RATA for 1-year cycle. The verified audit result will be submitted to DoE for approval and the cycles continues. Figure 5 and figure 6 are samples of Quarterly RAA and Annual RATA result, indicating the number of runs, period of runs and results whether it passes the test:

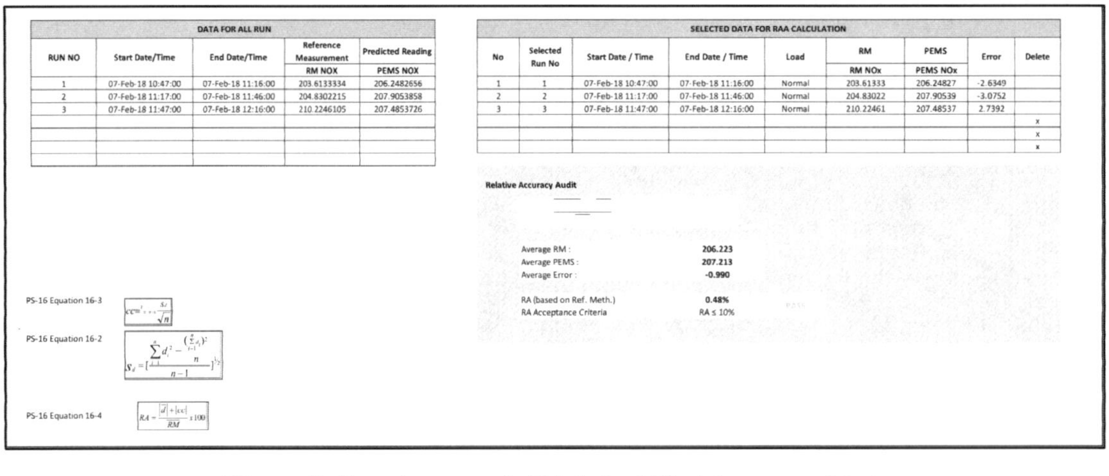

DATA FOR ALL RUN				
RUN NO	Start Date/Time	End Date/Time	Reference Measurement RM NOX	Predicted Reading PEMS NOX
1	07-Feb-18 10:47:00	07-Feb-18 11:16:00	203.6133334	206.2482656
2	07-Feb-18 11:17:00	07-Feb-18 11:46:00	204.8302215	207.9053858
3	07-Feb-18 11:47:00	07-Feb-18 12:16:00	210.2246105	207.4853726

SELECTED DATA FOR RAA CALCULATION								
No	Selected Run No	Start Date / Time	End Date / Time	Load	RM RM NOx	PEMS PEMS NOx	Error	Delete
1	1	07-Feb-18 10:47:00	07-Feb-18 11:16:00	Normal	203.61333	206.24827	-2.6349	
2	2	07-Feb-18 11:17:00	07-Feb-18 11:46:00	Normal	204.83022	207.90539	-3.0752	
3	3	07-Feb-18 11:47:00	07-Feb-18 12:16:00	Normal	210.22461	207.48537	2.7392	
								x
								x
								x

Figure 5. Sample quarterly RAA for NO_x (nitrogen oxides).

DATA FOR ALL RUN					
LOAD	RUN	Start Date/Time	End Date/Time	Reference Measurement	Predicted Reading
				RM NOX	PEMS NOX
Normal	1	07-Jun-18 14:03:00	07-Jun-18 14:23:00	200.343	204.358
Normal	2	07-Jun-18 14:24:00	07-Jun-18 14:44:00	200.050	204.106
Normal	3	07-Jun-18 14:45:00	07-Jun-18 15:05:00	200.164	202.808
Normal	4	07-Jun-18 15:06:00	07-Jun-18 15:26:00	201.559	203.106
Normal	5	07-Jun-18 15:27:00	07-Jun-18 15:47:00	204.207	203.223
Normal	6	07-Jun-18 15:48:00	07-Jun-18 16:08:00	205.619	204.120
Normal	7	07-Jun-18 16:09:00	07-Jun-18 16:29:00	205.675	204.511
Normal	8	07-Jun-18 16:30:00	07-Jun-18 16:50:00	206.697	204.337
Normal	9	07-Jun-18 16:51:00	07-Jun-18 17:11:00	206.190	204.379

SELECTED DATA FOR RATA CALCULATION								
No	Selected Run No	Start Date / Time	End Date / Time	Load	RM	PEMS	Error	Delete
					RM NOX	PEMS NOX		
1	1	07-Jun-18 14:03:00	07-Jun-18 14:23:00	Normal	200.343	204.358	-4.02	
2	2	07-Jun-18 14:24:00	07-Jun-18 14:44:00	Normal	200.050	204.106	-4.06	
3	3	07-Jun-18 14:45:00	07-Jun-18 15:05:00	Normal	200.164	202.808	-2.64	
4	4	07-Jun-18 15:06:00	07-Jun-18 15:26:00	Normal	201.559	203.106	-1.55	
5	5	07-Jun-18 15:27:00	07-Jun-18 15:47:00	Normal	204.207	203.223	0.98	
6	6	07-Jun-18 15:48:00	07-Jun-18 16:08:00	Normal	205.619	204.120	1.50	
7	7	07-Jun-18 16:09:00	07-Jun-18 16:29:00	Normal	205.675	204.511	1.16	
8	8	07-Jun-18 16:30:00	07-Jun-18 16:50:00	Normal	206.697	204.337	2.36	
9	9	07-Jun-18 16:51:00	07-Jun-18 17:11:00	Normal	206.190	204.379	1.81	

1. Relative Accuracy (RA)

Average RM :	**203.389**	n-1 =	8
Average PEMS :	**203.883**	t =	2.0560
Average Error :	**-0.494**		
Standard Deviation Error :	**2.578**		
Confidence Coefficient =	1.8742		
RA (based on Ref. Meth.) (%)	1.16%		
Absolute Difference	**0.49**		

Figure 6. Sample annual RATA for NO_x.

8. Future

P-PEMS has undergone various transformations since its pilot inception in 2016. From just a handful, now a whole section has the capability to implement and maintain P-PEMS. In looking ahead, the challenges in Section 6 must be addressed to improve P-PEMS, as well as innovating the solution to assist PETRONAS in addressing our own Sustainability Agenda.

PETRONAS has embarked on consolidating all P-PEMS in one platform, housing the application in PETRONAS Centralised Data Centre and utilising Microsoft platform for programming. Latest, the applications have been migrated to the Cloud and it has been integrated with CEMS monitoring as part of PETRONAS Digital Transformation Journey. This system is now accessible *via* web with proper access granted to different viewer. This is depicted in figure 7 below.

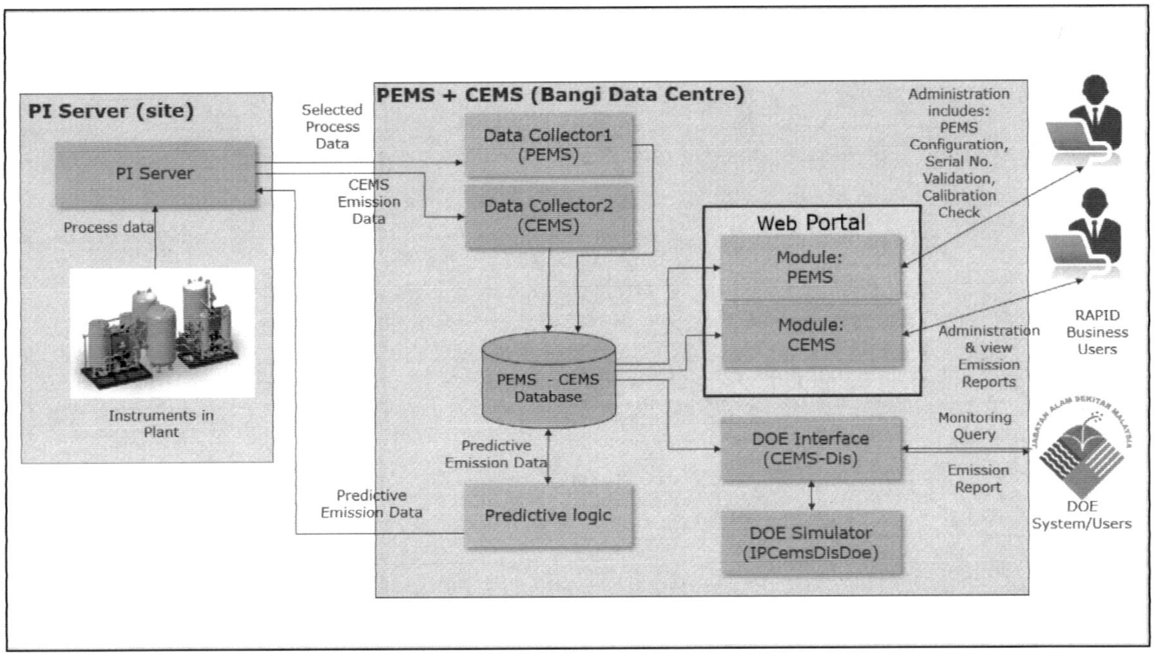

Figure 7. Overview of P-PEMS current system.

As the P-PEMS application number increases drastically, it is important for PETRONAS to improve on the overall P-PEMS Work Processes in order to be more efficient, systematic and well documented. There is an on-going effort to streamline RAA/RATA execution template, audit planning and tracking, issues to be resolved and communication with DoE.

Almost all P-PEMS implemented so far are for the existing facilities, i.e. equipment already in operation prior to CAR2014. PETRONAS has successfully piloted P-PEMS for new boiler project recently and this has capped a major milestone in P-PEMS future growth and evolution.

Moving forward, the team is in the midst of improving the application one step further with the proposal of Emission Monitoring and Intervention Initiative (EMI). This is a 3-phase plan where in the 1st phase, it is to improve existing P-PEMS emission model (prediction) utilising more advanced and sophisticated modelling techniques from currently being used. Once the new techniques have been selected, tested and validated, the new model will be implemented as an advisory to Operations personnel to adjust combustion equipment handles to optimise energy required vs emission released. Later in Phase 3, a control strategy will be devised to automatically optimise energy (fuel gas) while still meeting the emission limit.

Acknowledgments

This work was supported by PETRONAS GTS Management to further advance PETRONAS' Engineering Solutions.

References

[1] Department of Environment (DoE) Malaysia 2019 Guidelines for the installation & maintenance of continuous emission monitoring systems (CEMS) for industrial premises/ facilities *Department of Environment (DoE) Malaysia*

[2] Department of Environment (DoE) Malaysia 2014 Environmental Quality (Clean Air) Regulations, *Department of Environment (DoE) Malaysia*

[3] Evans S 1994 Predictive monitoring: The way things ought to be *Air & Waste Management Association Meeting, Cincinnati, OH*

Assessment of air pollution control technologies to reduce SO_x emission from thermal oxidizer for oil and gas industry

N H B Haji Nawawi and M N Jaafar

Gas & LNG Section, Process Department, Group Technical Solution, Project Delivery & Technology (PD&T), PETRONAS, KLCC Tower 3, 50450 Kuala Lumpur, Malaysia

Email: noradnin@petronas.com.my, nasrullah.jaafar@petronas.com.my

Abstract. Many countries have put in place, various legislations that govern air emission limits/pollutants from the industries. The common pollutants being monitored are Sulphur Oxides (SO_X), Nitrogen Oxides (NO_X), Carbon Monoxide (CO), Carbon Dioxide (CO_2), Volatile Organic Compounds (VOCs), particulate matters and dioxins. In Malaysia, the regulatory requirement aims to regulate emissions of air pollutants from industrial activities including oil and gas, power plants, waste fuel plants and asphalt mixing plants. One of the emission limits under Clean Air Regulation (CAR2014) is emission level for SO_X should be less than 600 mg/m^3 (reference condition at 3 % of O_2, 273 K, 101.3 kPa) whereby sum of SO_2 and SO_3 expressed as SO_X. Excessive SO_X emission can affect both health and the environment. Aligning with the regulation requirement, Group Technical Solution (GTS) under PETRONAS has embarked on assessment of technology solutions to meet the emission limit on SO_X emission limit for thermal oxidizers which cover new and existing facilities. This paper describes on the work methodology and approach adopted during the assessment. The objective of the assessment is to determine the suitable process technology to reduce SO_X emission in order to achieve the desired emission limit for flue gas at outlet stream of thermal oxidizer. Thorough evaluation was carried out based on proposal submission from various technology providers and Vendors. The selection criteria was developed and established. For existing thermal oxidizers, the assessment is more complex taking into consideration the nature of brownfield project and to ensure the proposed modification has minor impact to operability and maintainability of existing facilities. This study has successfully enabled identification of feasible process technologies such as Caustic Scrubber, Seawater Flue Gas Desulfurization and Ammonia based Desulfurization to meet the desired emission limit at thermal oxidizer outlet for Oil and Gas Industry and supporting environmental protection. The selected technology is varies based on plant/project specific requirement. Among main considerations are the by-product management, consumable and utility consumption as well as compatibility of the technology with existing plant on shutdown requirement.

1. Introduction

Many countries have put in place, various legislations that govern air emission limits/pollutants from the industries. The common pollutants being monitored are Sulphur Oxides (SO_X), Nitrogen Oxides (NO_X), Carbon Monoxide (CO), Carbon Dioxide (CO_2), Volatile Organic Compounds (VOCs), particulate matters and dioxins. In Malaysia, the regulatory requirement aims to regulate emissions of air pollutants from industrial activities including oil and gas, power plants, waste fuel plants and asphalt mixing plants. One of the flue gas emission limits under Clean Air Regulation (CAR2014) is emission level at the thermal oxidizer exhaust stack for SO_X should be less than 600 mg/m^3 (reference condition at 3 % of O_2, 273 K, 101.3 kPa), whereby sum of SO_2 and SO_3 expressed as SO_X.

The function of the thermal oxidizer is to dispose-off acid gas from upstream unit to the atmosphere in an environmentally acceptable manner. This is achieved by combusting a mixture of fuel gas and the acid gas at high temperature of about 900 °C to decompose H_2S and hydrocarbons including aromatics such as benzene, toluene and xylene. The flue gas emission shall meet the requirements of the Malaysian Laws as well as the requirements of Department of Environment (DOE).

SO_2 is colourless gas with suffocating odour. It is very toxic and corrosive to the respiratory tract. Excessive SO_X emission can affect both health and the environment. Short-term exposures to SO_2 can harm the human respiratory system and make breathing difficult. People with asthma, especially children, are sensitive to these effects of SO_2. At high concentrations, gaseous SO_X can harm trees and plants by damaging vegetation and decreasing growth. SO_2 and other sulfur oxides can contribute to acid rain which can harm sensitive ecosystems.

2. Background

Aligning with the regulation requirement, Group Technical Solution (GTS) under PETRONAS has embarked on assessment of technology solutions to meet the emission limit on SO_X emission for thermal oxidizers which cover new and existing facilities. The objective of the assessment is to determine the suitable process technology and cost-effective solution to reduce SO_X emission to achieve the desired emission limit for flue gas at outlet stream of new and existing thermal oxidizer.

3. Methodology

The methodology adopted for the study involved the following steps such as establishing the problem statement, reaching-out to the technology providers, technology selection, risk assessment and finally conclusion and recommendations.

The initial step was to establish the problem statement and the design requirement for the oxidizer including the emission limit. Then, the team reached out to the technology provider/vendors worldwide to identify the available process technologies. The technology provider refers to the inventor and provider of technology whereas vendors refer to the supplier of equipment product/service. The received technical proposals were consolidated and reviewed. Further questions were being clarified to ensure the proposal meeting the problem statement and design requirement.

Technology selection is performed using evaluation matrix based on technology selection criteria. Subsequently, Technology Risk Assessment was conducted to assess potential risk of adopting and implementing the technology. Finally based on the outcome of evaluation, team came-up with the conclusion and recommendations for the way forward. The methodology of work process for the study is illustrated in figure 1.

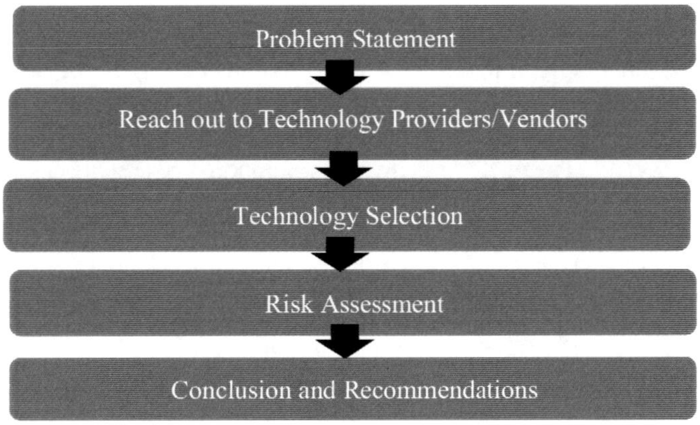

Figure 1. Methodology for the study.

4. Findings

The available SO_X Removal Technologies were divided into two categories which are pre-treatment and post-treatment technology to the thermal oxidizer. Pre-treatment is referring to reduction of H_2S in sour gas prior entering the thermal oxidizer meanwhile post-treatment is referring to recover SO_2 in the flue gas from thermal oxidizer stack. Regardless of the category, the objective is the same which is to achieve flue gas emission within the specified regulatory limit at the exhaust stack.

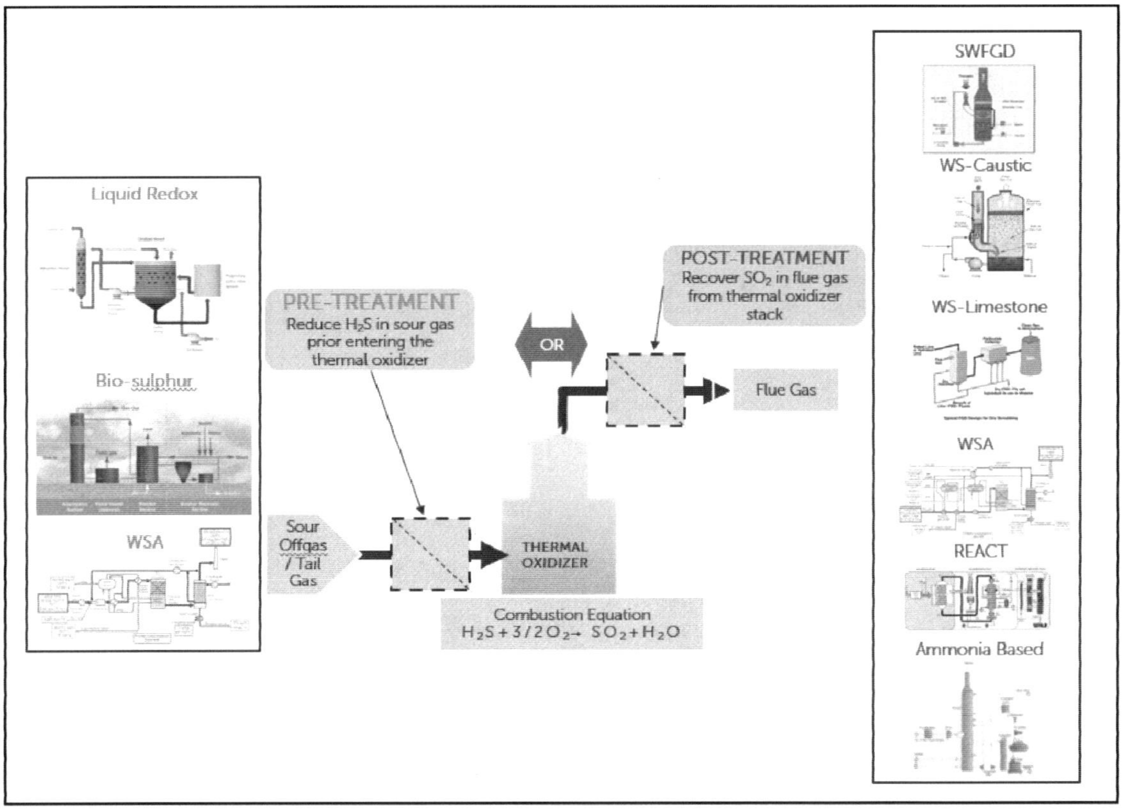

Figure 2. Categorization of available SO_X removal technologies.

5. Technology Selection

The selection criteria were established based on the problem statement and objectives of the study. The following are among others the important categories considered in the technology screening and selection criteria.

Table 1. Categories for technology screening.

Categories	Description of Selection Criteria
Technology Performance	• Ability to meet emission limit • Utility consumption within plant limit • Within provided plot space
Coft-Effective	• Capital Expenditure (CAPEX) • Operational Expenditure (OPEX) • Life Cycle Cost (LCC)
Technology Track Record	• Proven reference plant
Technology Operability and Maintainability	• Number of stream days • Turndown requirement • Turnaround interval requirement
Impact on Health Safety and Environment (HSE)	• Compliance of effluent discharge to regulatory limit • Product handling/hazards

Based on the consolidated inputs of technologies, scoring was included into comparison table taking into account the assigned weightage for each criteria. The weightage was assigned based on the importance on the categories and its selection criteria elements. The scoring was given based on relative comparison between each technology against the minimum requirement which has been determined by the team, i.e, the required plot have to meet available plot size of 3800 m^2.

From the evaluation, ranking is done based on total score to select the most technical and economical feasible technology to be further studied in the next phase. Based on the study for acid gas treating plant, the identified suitable process technologies are summarized below.

Table 2. Summary of feasible desulfurization technologies.

Technologies	Description
Caustic Scrubber	• Flue gas enters Caustic Scrubber unit and contact with circulating caustic solution (NaOH). • Effluent is cooled down and sent to Effluent Treatment System.
Seawater Flue Gas Desulfurization	• Scrubbing of SO_2 in flue gas to SO_3^{2-} by seawater and disengaging of Treated Gas and SO_3^{2-} in Absorber • Oxidation of SO_3^{2-} to SO_4^{2-} at Oxidation Basin and seawater is discharged to sea.
Ammonia based Desulfurization	• SO_2 in flue gas is absorbed by ammonia and converted into ammonium sulfate. • The ammonium sulfate solution is further processed to obtain ammonium sulfate product as fertilizer.

6. Risk Assessment

Technology Risk Assessment (TRA) was then carried out for the selected technology which met the minimum scoring requirement. The objectives of TRA are as to identify potential risks and operability issues related to the selection of the technology and to ensure adequate control measures are in place. The categories considered under TRA are:

- Performance
- Operability/Maintainability
- Readiness
- Hazards
- Implementation

Table 3. Example of TRA worksheet.

Guide Word	Cause	Consequence	Recommendations
Feed gas quality	1. Varying feed gas composition (i.e., Hydrocarbon). 2. Presence of contamination in Ammonia.	1. Potential slowdown and shutdown, resulting in SO_2 not meeting DOE specification. 2. Product not meeting specification. 3. Potential damage to equipment.	1. To consult Technology Licensor about potential contaminants that will disrupt operation of the SO_2 removal technology unit. 2. To include the potential feed gas quality variation in the Basis of Design (BOD).
Hazard studies, HSE case, Hazard and effect register	Lack of familiarization and awareness to manage ammonia leak/release due to newly introduced substance in existing facility	Potential LOPC, hazards to people and environment and asset damage.	1. To conduct thorough HSE studies relevant to ammonia handling including the safeguard (e.g., people protection, emergency response etc.). 2. To include ammonia in relevant process safety and HSE documents (e.g., Hazard Effect Register)

Based on the TRA review, recommendations were then provided to mitigate the risks of the technology under study. One of the identified concerns of applying ammonia based desulfurization technology in the natural gas liquefaction plant is the variation of feed gas composition (i.e., hydrocarbon) and presence of contaminant in ammonia solution. This can cause potential slowdown and shutdown of the unit and resulted the SO_2 may be exceed the limit. The recommendation is to include the potential flue gas quality range in the Basis of Design (BOD).

In addition, the other concern is lack of familiarization with handling, storing and Health Safety and Environment (HSE) management of newly introduced substances such as ammonia to existing facility. As part of the recommendation, it is suggested to conduct thorough HSE studies relevant to ammonia handling including the mitigation plans (e.g., people protection, emergency response etc.) during the next study phase.

7. Conclusion

Thorough evaluation was carried out based on proposal submission from various technology providers and vendors. The evaluation was performed based on the technology selection criteria and followed by technology risk assessment review. This evaluation study has successfully enabled identification of feasible process technologies such as Caustic Scrubber, Seawater Flue Gas Desulfurization and Ammonia based Desulfurization to meet the desired emission limit at thermal oxidizer outlet for Oil and Gas Industry and supporting environmental protection. The selected technology is varies based on plant/project specific requirement. Among main considerations are the by-product management limitation, consumable, utility availability and compatibility of the technology with existing plant on shutdown requirement.

References

[1] Gary J H, Handwerk J H, Kaiser M J and Geddes D 2007 *Petroleum Refining Technology and Economics* (Boca Raton: CRC Press) chapter 13 pp 283–91
[2] Sorrels J L 2017 Incinerator and Oxidizer (Research Triangle Institute)
[3] Sulfur Dioxide (SO_2) Pollution 2019 [Available from: https://www.epa.gov › so2-pollution]
[4] Nava J Proven Methods for Reducing Sulfur Dioxide (SO_2) Emissions 2019 [Available from: https://www.bba.ca/publication/proven-methods-for-reducing-sulfur-dioxide-so2-emissions/]

IOP Conf. Series: Materials Science and Engineering 1195 (2021) 012047 doi:10.1088/1757-899X/1195/1/012047

Synthesis and characterization of (3-aminopropyl) triethoxysilane (APTES) functionalized zeolite AlPO-18

L D Anbealagan[1,2], T L Chew[1,2], Y F Yeong[1,2], Z A Jawad[3] and C D Ho[4]

[1] Department of Chemical Engineering, Faculty of Engineering, University Teknologi PETRONAS, 32610 Seri Iskandar, Perak, Malaysia
[2] CO_2 Research Centre (CO_2RES), Institute Contaminant Management, University Teknologi PETRONAS, 32610 Seri Iskandar, Perak, Malaysia
[3] Department of Chemical Engineering, College of Engineering, Qatar University, P.O. Box: 2713, Doha, Qatar
[4] Department of Chemical and Materials Engineering, Tamkang University, New Taipei City 25137, Taiwan

Email: lanisha_19000945@utp.edu.my, thiamleng.chew@utp.edu.my

Abstract. Over the years, functionalization of zeolite is gaining popularity among researchers to further modify the properties of the zeolite for wide applications. The procedure of functionalization is crucial to ensure that the framework and structure of the zeolite would not be destroyed by the functionalization process. In this work, zeolite AlPO-18 was synthesized via hydrothermal synthesis method and functionalized by (3-Aminopropyl) triethoxysilane (APTES). The effect of the APTES functionalization on zeolite AlPO-18 was investigated in this work. Both unfunctionalized and silane-functionalized zeolite AlPO-18 were characterized using Fourier-transform infrared spectroscopy (FT-IR), X-ray diffraction (XRD), and Thermogravimetric analysis (TGA) for their properties. The morphology and the composition of the elements present in zeolite AlPO-18 and zeolite NH_2-AlPO-18 were examined using Field Emission Scanning Electron Microscopy (FESEM) and Energy-Dispersive spectroscopy (EDX) respectively. The XRD pattern of NH_2-AlPO-18 was similar to that of zeolite AlPO-18, however, the intensity of the peaks was lower compared to zeolite AlPO-18. Based on the FTIR spectra, the presence of N-H stretching and bending vibration band of aminosilane were observed in the NH_2-AlPO-18 sample. According to FESEM images, the morphology of NH_2-AlPO-18 was comparable to that of zeolite AlPO-18 even after functionalization, proving that functionalization of aminosilane on zeolite does not affect on the zeolite structure. Besides that, EDX proves the presence of 3.02 % of element N in the NH_2-AlPO-18 sample which is absent in the zeolite AlPO-18 sample. All of the characterizations evinced the presence of aminosilane, APTES in the NH_2-AlPO-18 sample.

1. Introduction

Zeolites are crystalline aluminosilicates with a well-defined pore framework made from interlinked tetrahedra of SiO_4 and AlO_4 [1]. Currently, there are more than 200 types of zeolites have been documented, such as zeolite T, SAPO-34, DDR, SSZ-13, Si-CHA, and AlPO-18 [2]. Zeolites can be differentiated based on their crystal structure and framework. It exhibits a well-defined and uniform pore size with a large surface area and high porosity. Zeolites can be categorized based on the pore size such as small pore (6-, 8- and 9- membered ring), medium pore (10- membered ring), and large pore (12- membered ring) [3]. Zeolites have been used in various industrial applications such as adsorption, catalysis, and gas separations due to their intrinsic characteristics such as high chemical, thermal and mechanical stabilities [3].

Zeolite AlPO-18 is a type of zeolite that is gaining favor among researchers because of its small pore structure. Zeolite AlPO-18 is a type of zeolite that is made up of aluminum, phosphorus, and oxygen and consists of 3D structured pores possessing 8-membered rings with a pore size of 3.8 Å. The framework density of AlPO-18 of 15.1 T/nm^3 is the lowest among the aluminophosphate family. Zeolite AlPO-18 also has a low hydrophilicity nature. The low framework density, as well as the low hydrophilicity properties, are the predominant feature of AlPO [4]. It can be synthesized at a moderate synthesis temperature and a short synthesis duration. Zeolite AlPO-18 has been researched for applications including gas separation. It is remained an interest among researchers to explore approaches in modifying or improving the properties of zeolite AlPO-18 to be suitably applied at different applications.

Functionalization of zeolites using different types of silane coupling agents is among the effective technique used to modify the zeolite properties for wide applications [5–9]. The most commonly used silane coupling agents are (γ-aminopropyl)-triethoxysilane (APTES), 3-aminopropylmethyldiethoxysilane (APMDES) and (γ-aminopropyl)-diethoxymethylsilane (APDEMS) [10-15]. It contains two types of reactive functional groups including organic and inorganic groups. $R - (CH_2)_n - Si - X_{4-n}$ is the typical structure of silane coupling agents, where R represents the organo-functional group (amine, methacryloxy or, epoxy) and X represents the hydrolysable group (methoxy, ethoxy, or acetoxy) [16, 17]. During the functionalization process, the hydroxyl group on the inorganic filler surface reacts with the silane [18, 19].

In the present work, APTES has been chosen to be used as the silane coupling agent as it has been extensively used for a wide range of applications and it could also be considered as one of the less expensive silane coupling agents compared to the others [20–28]. APTES consists of an amine group (NH_2) and three ethoxy groups that can be adhered to the zeolite surface [22, 29]. Figure 1 depicts the schematic illustration for zeolite AlPO-18 functionalization using an APTES silane coupling agent. The novelty of the work is the functionalization of zeolite AlPO-18 using APTES and investigating the effect of the functionalization on the zeolite. Both unfunctionalized and silane-functionalized zeolite AlPO-18 were characterized using XRD, FT-IR, TGA, and FESEM. To date, there are no research works reported on the functionalization of zeolite AlPO-18 by using APTES.

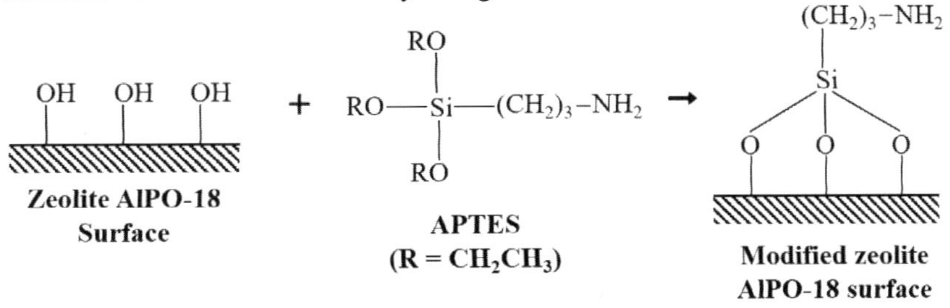

Figure 1. Schematic illustration for aminosilane functionalization of zeolite AlPO-18.

2. Materials and experimental procedures

2.1. Materials

Aluminium isopropoxide (98 %, Sigma Aldrich), tetramethylammonium hydroxide (TEAOH) (35 % in water, Sigma Aldrich), phosphoric acid (H_3PO_4) (85 wt % aqueous solution, Sigma Aldrich), and deionized water (DI water) were used for zeolite AlPO-18 synthesis. The silane coupling agent, (3-aminopropyl)-triethoxysilane (APTES) (99 %, Sigma Aldrich) was used to modify zeolite. Toluene (> 99.9 %) and ethanol (> 99.9 %) were supplied by Merck Co. All the chemicals were used as received.

2.2. *Synthesis of zeolite AlPO-18*
Zeolite AlPO-18 was synthesized via hydrothermal synthesis method by following the literature reported [30, 31]. A precursor solution with a molar composition of 1.0 Al_2O_3: 3.16 P_2O_5: 6.32 TEAOH: 186 H_2O was prepared. Aluminium isopropoxide, TEAOH, and DI water were mixed and stirred for an hour at room temperature to form a homogeneous mixture. Phosphoric acid was added dropwise into the stirring solution. The resulting precursor solution was stirred again for 2 hours before heated hydrothermally at 150°C for 20 hours. The synthesized particles were centrifuged at 6000 rpm for 10 minutes to collect the seeds and washed with DI water. The resulting crystal was dried at 50 °C for overnight.

2.3. *Functionalization of zeolite AlPO-18*
The synthesized zeolite AlPO-18 powder was dried overnight at 50 °C before functionalization. For functionalization, 2 g of zeolite AlPO-18 powder was uniformly dispersed into 50 mL of toluene. Then, 4 mL of APTES was added dropwise to the resultant mixture and refluxed at 110 °C for 4 hours. After the reflux process, the mixture was filtered and rinsed with toluene and ethanol absolute in order to remove the unreacted APTES. The washed particles were dried at 50 °C overnight. AlPO-18 that was functionalized by the silane group was denoted as NH_2-AlPO-18 in the current project.

2.4. *Characterization of zeolite AlPO-18 and NH_2-AlPO-18*
XRD (X'Pert^3 Powder & Empyrean, PANalytical) was used to study the crystallinity of zeolite AlPO-18 and NH_2-AlPO-18. The analysis was carried out at an accelerating voltage of 40 kV and current of 40 mA and by using Cu Kα radiation at 2θ in the range of 5° - 45° with a step size of 0.05°. Fourier Transform Infrared-Attenuated Total Reflection, FTIR-ATR (Perkin Elmer, Frontier) was used to identify the functional groups and chemical bondings presence in the zeolite AlPO-18 and NH_2-AlPO-18. The analysis was carried out with a wavelength of 4000 cm^{-1} to 400 cm^{-1}. Thermogravimetric Analysis, TGA (Perkin Elmer, STA 6000) was used to study the thermal stability of zeolite AlPO-18 and NH_2-AlPO-18 based on the weight loss of the sample due to change in temperature over time. The samples were heated from 30 °C to 800 °C at a constant heating rate of 10 °C/min under N_2 atmosphere. Field Emission Scanning Electron Microscopy, FESEM (Zeiss Supra 55VP) was used to study the morphology of the zeolite AlPO-18 and NH_2-AlPO-18. Energy-Dispersive spectroscopy (EDX) was used to identify the elemental compositions of zeolite AlPO-18 and NH_2-AlPO-18.

3. Results and discussions

3.1. *Crystallinity analysis of zeolite AlPO-18 and NH_2-AlPO-18*
The XRD pattern of the calcined zeolite AlPO-18 and NH_2-AlPO-18 are illustrated in figure 2. From figure 2(a), it can be observed that the XRD pattern display the peaks at two theta of 9.6°, 12.8°, 16.8°, 21°, 23.6°, 26.2°, and 32.0°. All the XRD peaks of the synthesized zeolite AlPO-18 were similar to the XRD patterns reported in the literature previously [32–35]. The results obtained prove that zeolite AlPO-18 was successfully synthesized. According to Carreon *et al.* [35], the XRD pattern exhibit broad and less intense peaks due to the presence of amorphous regions and/or there is a high degree of structural disorder in the synthesized zeolite AlPO-18 framework.

From figure 2(b), it can be seen that the XRD pattern of NH_2-AlPO-18 was almost similar to the zeolite AlPO-18 peaks. This implies the zeolite structure was not affected by the functionalization of aminosilane on the zeolite surface. Despite that, the intensity of NH_2-AlPO-18 peak was lower compared to the zeolite AlPO-18 patterns. This could be due to the pore filling effect, where the pore surface of zeolite AlPO-18 is probably covered by APTES groups. Thus, it affects the crystallinity of the sample by causing a slight decrement in the peak crystallinity [36, 37].

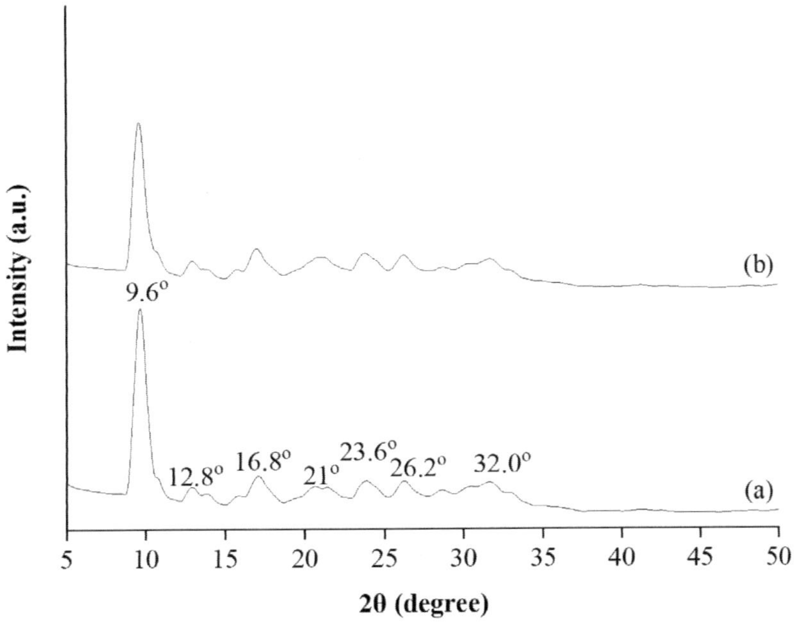

Figure 2. XRD patterns of (a) zeolite AlPO-18 and (b) NH_2-AlPO-18.

3.2. Spectroscopic analysis of zeolite AlPO-18 and NH_2-AlPO-18

Figure 3 and figure 4 shows the FTIR spectra of calcined zeolite AlPO-18 and NH_2-AlPO-18 in the region of 4000-400 cm^{-1}. FTIR was used to identify the functional group and type of bonding present in zeolite AlPO-18 and zeolite NH_2-AlPO-18. From figure 3(a), it can be seen that the FTIR spectra exhibits the bands around 3500 cm^{-1}, 1640 cm^{-1}, 1100 cm^{-1}, 600 cm^{-1} and 500 cm^{-1}. All these bands are the typical zeolite AlPO-18 peaks that are in good agreement with the literature reported previously [34, 38, 39].

The broad absorption band around 3800-3200 cm^{-1} was attributed to H- bonded Al-OH and P-OH groups [34, 40–42]. Besides that, the band around 1700-1600 cm^{-1} attributed to the interlayer bending vibration of physically adsorbed water molecules [43–45]. P-O stretching and bending were ascribed to the bands observed in the region of 1200-1000 cm^{-1} and about 500 cm^{-1}, respectively [43, 46]. The presence of double 8-rings framework in the zeolite AlPO-18 was linked to a small intensity band formed around 650 cm^{-1} [42, 47, 48].

From figure 3(b) and figure 4(b), it can be seen that the FTIR spectra of NH_2-AlPO-18 was almost similar to the zeolite AlPO-18 spectra with a few additional bands around 1500 ~ 1300 cm^{-1} and 1560 cm^{-1}. Compared to zeolite AlPO-18, the zeolite NH_2 - AlPO-18 showed a broader band at the frequency of 3800-3200 cm^{-1}, where the band assigned to N-H stretching of a primary amine overlaps with the O-H stretching of hydroxyl group, indicating the presence of aminosilane in the sample [15, 49]. Besides that, Si-CH_2 and Si-CH_3 stretching vibrations of the aminosilane were also apparent around 1500-1300 cm^{-1} in NH_2-AlPO-18 [12, 40, 50]. Moreover, the presence of an absorption peak at 1560 cm^{-1}, which indicates N-H bending of the amine, was also absent in zeolite AlPO-18 [15].

The type of bonding represented by each of the peak presence in zeolite AlPO-18 and NH_2-AlPO-18 spectra were summarised in table 1.

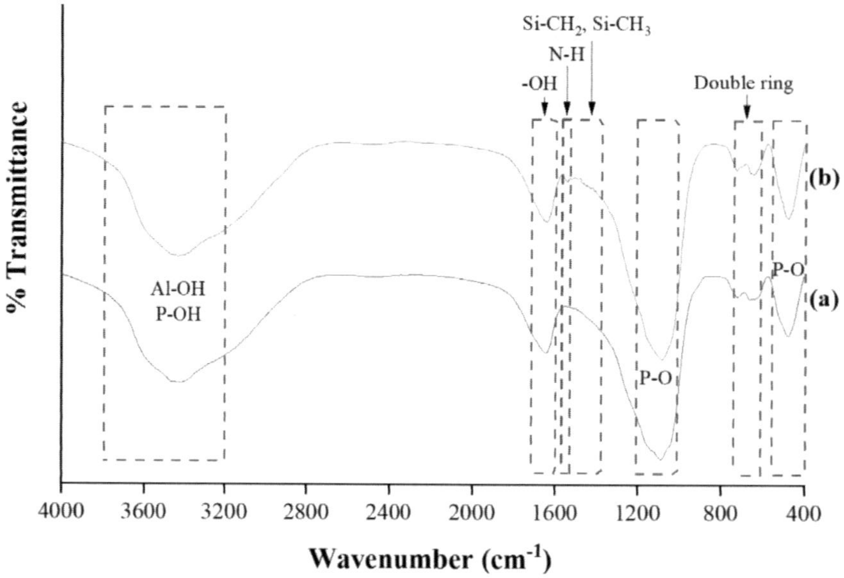

Figure 3. FT-IR spectra of (a) zeolite AlPO-18 and (b) NH_2-AlPO-18.

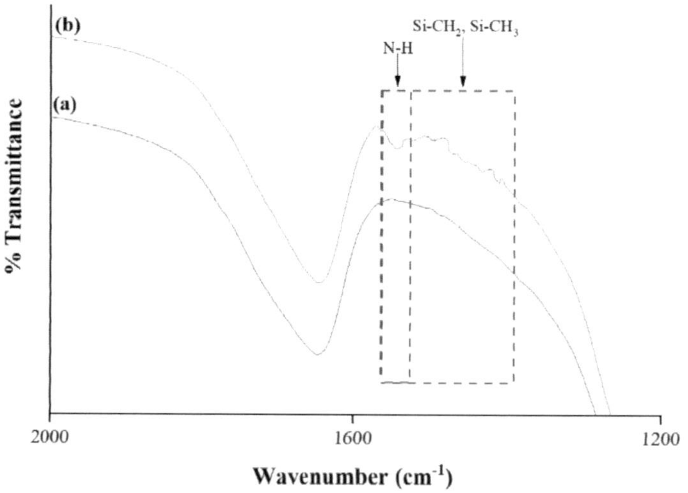

Figure 4. FT-IR spectra of (a) zeolite AlPO-18 and (b) NH_2-AlPO-18 in the range 2000-1200 cm^{-1}.

Table 1. Type of bonding shown in the FTIR spectra of zeolite AlPO-18 and NH_2-AlPO-18.

Wavenumber (cm^{-1})	Type of bonding
3800 ~ 3200	–OH stretching of hydroxyl group
3434#	N–H stretching of primary amine
1640	–OH bending of hydroxyl group
1560#	N–H bending of primary amine
1500 ~ 1300#	Si–CH_2 and Si–CH_3 stretching of aminosilane
1200 ~ 1000	P–O stretching
650 ~ 500	Double 8-rings framework
~ 500	P–O bending

#*Attributed to silane groups*

IOP Conf. Series: Materials Science and Engineering 1195 (2021) 012047 doi:10.1088/1757-899X/1195/1/012047

3.3. *Thermal analysis of zeolite AlPO-18 and NH_2-AlPO-18*

Figure 5 illustrates the TGA diagram of zeolite AlPO-18 and NH_2-AlPO-18. TGA profile shows the percentage of weight loss of the zeolite AlPO-18 and NH_2-AlPO-18 as a function of time.

From the figure 5, it can be observed that zeolite AlPO-18 exhibits a single-stage decomposition process in the temperature range of 30-200 °C. Whereas, NH_2-AlPO-18 shows a two-stage decomposition at the temperature ranged from 30-200 °C and 200-800 °C. In figure 5(a), the TGA profile of zeolite AlPO-18 shows a significant weight loss of 24.72% below 200 °C, which was almost similar to the TGA profile reported in the literature [51, 52]. This weight loss was ascribed to the removal of physiosorbed water molecules within the zeolite pores [51, 53, 54]. After 200 °C, there is no obvious mass variation of zeolite AlPO-18 up to 800 °C, which clarifies that zeolite AlPO-18 has good thermal stability.

The TGA profile of NH_2-AlPO-18 in figure 5(b) illustrates the first-stage decomposition with a weight loss of 19.52% at the temperature range of 30-200 °C. The thermal decomposition rate of NH_2-AlPO-18 was lower compared to zeolite AlPO-18. This justifies the attachment of APTES silane groups on the hydroxyl groups (-OH) of zeolite AlPO-18 during the grafting process [55–57]. Thus, it can be concluded that the small weight loss in NH_2-AlPO-18 is because of less -OH group presence in the NH_2-AlPO-18 sample. The second-stage decomposition of NH_2-AlPO-18 with a weight loss of 3.94 % occurs at the temperature range 200-800 °C. This weight loss was attributed to gradual organic volatilization-decomposition of propyl chain in APTES molecule [22, 58–61]. The thermal properties of zeolite AlPO-18 and NH_2-AlPO-18 are summarized in table 2.

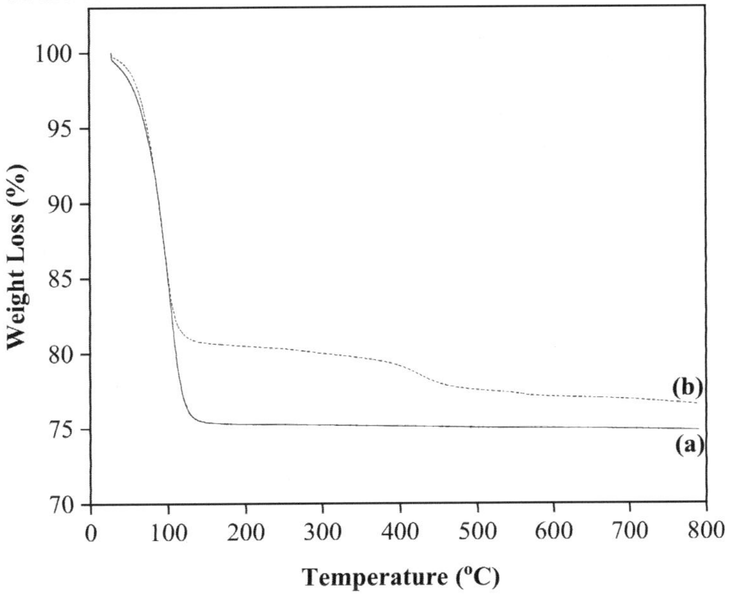

Figure 5. TGA diagram of (a) zeolite AlPO-18 and (b) NH_2-AlPO-18.

Table 2. Weight loss of zeolite AlPO-18 and NH_2-AlPO-18 at different temperatures.

Stages of weight loss	Weight loss (%)	
	Zeolite AlPO-18	NH_2-AlPO-18
Desorption of adsorbed water (30 – 200°C)	24.72	19.52
Decomposition of APTES (200 – 800°C)	-	3.94
Total	24.72	23.46

3.4. Morphological analysis of zeolite AlPO-18 and NH_2-AlPO-18

Figure 6(a) and (b) shows the FESEM images of zeolite AlPO-18 and NH_2-AlPO-18 at a magnification of 20kX respectively. FESEM was used to examine the morphology of the zeolite AlPO-18 and NH_2-AlPO-18.

The FESEM images in figure 6(a) show a thin elongated plate-like structure of zeolite AlPO-18. The morphology obtained is the typical zeolite AlPO-18 structure that looks similar to the literature reported previously [30, 51]. However, the morphology of the samples displays particle aggregates. Besides that, from figure 6(b), it can be observed that the morphology of NH_2-AlPO-18 was similar to that of zeolite AlPO-18 even after functionalization on the surface of zeolite AlPO-18. This shows that the functionalization of aminosilane on the zeolite surface had no effect on the zeolite structure.

Moreover, EDX spectroscopy was used to identify the composition of the element presence in zeolite AlPO-18 and NH_2-AlPO-18. To ensure the consistency of the composition, five trials of the analysis were carried out on the samples. The elemental composition of the samples is summarized in table 3. Aluminium (Al), Phosphorus (P), and Oxygen (O) are the main elements of zeolite AlPO-18 whereas Nitrogen (N) is the main element in APTES. Thus, the presence of N in the EDX of NH_2-AlPO-18 sample proves the presence of APTES in the sample.

Figure 6. FESEM images of (a) zeolite AlPO-18 and (b) NH_2-AlPO-18.

Table 3. EDX of zeolite AlPO-18 and NH_2-AlPO-18.

Samples	Weight composition (Weight %)				
	Al	P	O	N	Total
Zeolite AlPO-18	19.42	25.52	55.06	Not detected	100.00
NH_2-AlPO-18	18.39	23.75	54.84	3.02	100.00

4. Conclusion

In the current project, the effect of aminosilane functionalization on zeolite AlPO-18 was investigated. The XRD pattern of NH_2-AlPO-18 was comparable to that of zeolite AlPO-18, although the peak intensity was lower than that of zeolite AlPO-18. The existence of N-H stretching and bending vibration bands of aminosilane was observed in the FTIR spectra of NH_2-AlPO-18 sample. According to FESEM analysis, even after functionalization, the morphology of NH_2-AlPO-18 was similar to that of zeolite AlPO-18, demonstrating that aminosilane functionalization has no effect on zeolite structure. Furthermore, EDX confirms the existence of 3.02 percent element N in the NH_2-AlPO-18 sample. All of the characterizations revealed the presence of APTES in the NH_2-AlPO-18 sample. The obtained NH_2-AlPO-18 can be used widely in industrial applications such as molecular separation, adsorption, and catalysis. For future research, the NH_2-AlPO-18 can be potentially used as filler for CO_2/CH_4 gas separation. Besides that, the effect of various types of silane coupling agents such as (3-aminopropyl) dimethylethoxysilane (APDMES) and (3-aminopropyl) methyldiethoxysilane (APMDES) on the zeolite properties can also be investigated.

Acknowledgments
The authors would like to express gratitude to financial support from Fundamental Research Grant Scheme (FRGS) Ref: FRGS/1/2020/TK0/UTP/02/28 (Cost center: 015MA0-123) from the Ministry of Higher Education Malaysia (MOHE). This work was also supported by Universiti Teknologi PETRONAS, Institute Contaminant Management UTP, CO_2 Research Centre (CO_2RES) UTP.

References

[1] Prech J, Pizarro P, Serrano D P and Cejka J 2018 From 3d to 2d zeolite catalytic materials *Chem. Soc. Rev.* **47** 8263–306
[2] Wu T, Tanaka K, Chen X, Kumakiri I and Kita H 2018 Synthesis and gas permeation properties of aei zeolite membranes by dipea as a template *Membrane* **43** 67–73
[3] Feng C, Khulbe K C, Matsuura T, Farnood R and Ismail A F 2015 Recent progress in zeolite/zeotype membranes *J. Membr. Sci. Res.* **1** 49–72
[4] Database of zeolite structures [Internet] Available from: http://www.iza-structure.org/databases/
[5] Huang L, Xiao H and Ni Y 2006 Cationic-modified microporous zeolites/anionic polymer system for simultaneous removal of dissolved and colloidal substances from wastewater *Sep. Purif. Technol.* **49** 264–70
[6] Kuwahara Y, Kamegawa T, Mori K, Matsumura Y and Yamashita H 2009 Fabrication of hydrophobic zeolites using triethoxyfluorosilane and their application for photocatalytic degradation of acetaldehyde *Top. Catal.* **52** 643–8
[7] Smaïhi M, Gavilan E, Durand J O and Valtchev V P 2004 Colloidal functionalized calcined zeolite nanocrystals *J. Mater. Chem.* **14** 1347–51
[8] Zhang X, Lai E S M, Martin-Aranda R and Yeung K L 2004 An investigation of knoevenagel condensation reaction in microreactors using a new zeolite catalyst *Appl. Catal. A-Gen.* **261** 109–18
[9] Zhang X Y, Wang Q C, Zhang S Q, Sun X J and Zhang Z S 2009 Stabilization/solidification (s/s) of mercury-contaminated hazardous wastes using thiol-functionalized zeolite and portland cement *J. Hazard. Mater.* **168** 1575–80
[10] Chen X Y, Nik O G, Rodrigue D and Kaliaguine S 2012 Mixed matrix membranes of aminosilanes grafted fau/emt zeolite and cross-linked polyimide for co_2/ch_4 separation *Polymer* **53** 3269–80
[11] Chua L K, Jusoh N and Yeong Y K 2015 Fabrication of sapo-34 and silane-modified sapo-34/polyimide mixed matrix membranes for co_2/ch_4 separation *J. Appl. Sci. Agric.* **10** 215–21
[12] Amooghin A E, Omidkhah M and Kargari A 2015 The effects of aminosilane grafting on nay zeolite–matrimid®5218 mixed matrix membranes for co_2/ch_4 separation *J. Membr. Sci.* **490** 364–79

[13] Nik O G, Chen X Y and Kaliaguine S 2011 Amine-functionalized zeolite fau/emt-polyimide mixed matrix membranes for co_2/ch_4 separation *J. Membr. Sci.* **379** 468–78
[14] Pechar T W, Tsapatsis M, Marand E and Davis R 2002 Preparation and characterization of a glassy fluorinated polyimide zeolite-mixed matrix membrane *Desalination* **146** 3–9
[15] Sanaeepur H, Kargari A and Nasernejad B 2014 Aminosilane-functionalization of a nanoporous y-type zeolite for application in a cellulose acetate based mixed matrix membrane for co_2 separation *RSC Adv.* **4** 63966–76
[16] Junaidi M U M, Khoo C P, Leo C P and Ahmad A L 2014 The effects of solvents on the modification of sapo-34 zeolite using 3-aminopropyl trimethoxy silane for the preparation of asymmetric polysulfone mixed matrix membrane in the application of co_2 separation *Microporous Mesoporous Mater.* **192** 52–9
[17] Kumudinie C 2001 Polymer-ceramic nanocomposites: Interfacial bonding agents *Encyclopedia of Materials: Science and Technology* 7574–7
[18] Ahmad N N R, Mukhtar H, Mohshim D F, Nasir R and Man Z 2016 Surface modification in inorganic filler of mixed matrix membrane for enhancing the gas separation performance *Rev. Chem. Eng.* **32** 1–20
[19] Chung T S, Jiang L Y, Li Y and Kulprathipanja S 2007 Mixed matrix membranes (mmms) comprising organic polymers with dispersed inorganic fillers for gas separation *Prog. Polym. Sci.* **32** 483–507
[20] Liu X M, Thomason J L and Jones F R: Taylor & Francis Group; 2009 Xps and afm study of the structure of hydrolysed aminosilane on e-glass surfaces In: Mittal KL, editor *Silanes and other coupling agents* 5 Boston p 39–50
[21] Smith E A C and Wei 2008 How to prevent the loss of surface functionality derived from aminosilanes *Langmuir* **2008** 12405–9
[22] Berktas I, Ghafar A N, Fontana P, Caputcu A, Menceloglu Y and Okan B S 2020 Facile synthesis of graphene from waste tire/silica hybrid additives and optimization study for the fabrication of thermally enhanced cement grouts *Molecules* **25** 1–15
[23] Liu Y, Zhang S, He Y, Chen C, Zhang C, Xie P, Zhong F, Li H, Chen J and Li Z 2021 Aptes modification of molybdenum disulfide to improve the corrosion resistance of waterborne epoxy coating *Coatings* **11** 1–13
[24] Lo X C, Li J Y, Lee M T and Yao D J 2020 Frequency shift of a sh-saw biosensor with glutaraldehyde and 3-aminopropyltriethoxysilane functionalized films for detection of epidermal growth factor *Biosensors (Basel)* **10** 1–12
[25] Wang J 2020 A simple, rapid and low-cost 3-aminopropyltriethoxysilane (aptes) based surface plasmon resonance sensor for tnt explosive detection *Anal. Sci.* 1–17
[26] Yang J, Liao M, Hong G, Dai S, Shen J, Xie H and Chen C 2020 Effect of aptes- or mpts-conditioned nanozirconia fillers on mechanical properties of bis-gma-based resin composites *ACS Omega* **5** 3254–50
[27] Zhang B, Wang Y, Zhang J, Qiao S, Fan Z, Wan J and Chen K 2020 Well-defined 3-aminopropyltriethoxysilane functionalized magnetite nanoparticles and their adsorption performance for partially hydrolyzed polyacrylamide from aqueous solution *Colloids and Surfaces A: Physicochemical and Engineering Aspects* **586** 1–31
[28] Soria S, Berneschi S, Barucci A, Cosci A, Farnesi D, Conti G N, Pelli S and Righinia G C: Jenny Stanford; 2019 Biomedical sensing applications of microspherical resonators In: Righini GC, editor *Glass microand nanospheres : Physics and applications* USA p 165–201
[29] Rao X, Tatoulian M, Guyon C, Ognier S, Chu C and Abou Hassan A 2019 A comparison study of functional groups (amine vs. Thiol) for immobilizing aunps on zeolite surface *Nanomaterials (Basel)* **9** 1–14
[30] Wang B, Gao F, Zhang F, Xing W and Zhou R 2019 Highly permeable and oriented alpo-18 membranes prepared using directly synthesized nanosheets for co_2/ch_4 separation *J. Mater. Chem. A.* **7** 13164–72

[31] Zong Z, Elsaidi S K, Thallapally P K and Carreon M A 2017 Highly permeable alpo-18 membranes for n_2/ch_4 separation *Ind. Eng. Chem. Res.* **56** 4113–8
[32] Vilaseca M, Mintova S, Valtchev V, Metzger T H and Bein T 2003 Synthesis of colloidal alpo4-18 crystals and their use for supported film growth *J. Mater. Chem.* **13** 1526–8
[33] Wu T, Wang B, Lu Z, Zhou R and Chen X 2014 Alumina-supported alpo-18 membranes for co_2/ch_4 separation *J. Membr. Sci.* **471** 338–46
[34] Chen J, Thomas J M, Wright P A and Townsend R P 1994 Silicoaluminophosphate number eighteen (sapo- 18): A new microporous solid acid catalyst *Catal. Lett.* **28** 241–8
[35] Carreon M L, Li S and Carreon M A 2012 Alpo-18 membranes for co2/ch4 separation *Chem Commun (Camb)* **48** 2310–2
[36] Babaei M, Anbia M and Kazemipour M 2017 Improving co_2 adsorption with new amine-functionalized y-type zeolite *J. Adv. Environ. Health. Res.* **5** 70–7
[37] Kim S, Ida J, Guliants V V and Lin J Y S 2005 Tailoring pore properties of mcm-48 silica for selective adsorption of co_2 *J. Phys. Chem. B* **109** 6287–93
[38] Peng L, Li J, Yu J, Li G, Fang Q and Xu R 2005 [h3n(ch2)4nh3]2[al4(c2o4)(h2po4)2(po4)4]·4[h2o]: A new layered aluminum phosphate-oxalate *Journal of Solid State Chemistry* **178** 2686–91
[39] Preethi M E L, Umasankari A, C.H. R, Palanichamy M, Sivakumar T and Pandurangan A 2018 Selective oxidation of cyclohexane to ka oil over ce-alpo-18 molecular sieves *Int. J. Eng. Technol.* **7** 352–4
[40] Gianotti E, Oliveira E C, Dellarocca V, Coluccia S, Pastore H O and Marchese L 2002 Meso-alpo prepared by thermal decomposition of the organic-inorganic composite: A ftir study *Nanoporous materials iii, proceedings of the 3rd international symposium on nanoporous materials* Studies in surface science and catalysis 417–22
[41] Martis M In situ and ex situ characterization studies of transition metal containing nanoporous catalysts: University College London; 2011
[42] Byarappa K and Kumar S B V 2007 Characterization of zeolites by infrared spectroscopy *Asian J. Chem.* **19** 4933–5
[43] Boonchom B, Youngme S, Srithanratana T and Danvirutai C 2008 Synthesis of $alpo_4$ and kinetics of thermal decomposition of $alpo_4 \cdot h_2o$-h_4 precursor *J. Therm. Anal. Calorim.* **91** 511–6
[44] Popescu S C, Thomson S and Howe R F 2001 Microspectroscopic studies of template interactions in $alpo_4$-5 and sapo-5 crystals *Phys. Chem. Chem. Phys.* **3** 111–8
[45] Ng E P, Awala H, Komaty S and Mintova S 2019 Microwave-green synthesis of alpo-n and sapo-n (n = 5 and 18) nanosized crystals and their assembly in layers *Microporous Mesoporous Mater.* **280** 256–63
[46] Mustafa S, Javid M and Zaman M I 2006 Effect of activation on the sorption properties of $alpo_4$ *Sep. Sci. Technol.* **41** 3467–84
[47] Venkatathri N 2005 Synthesis and characterization of alpo4-21 from aqueous and non-aqueous systems *Bulletin of the Catalysis Society of India* **4** 18–27
[48] Zhao D, Zhang Y, Li Z, Wang Y and Yu J 2017 Synthesis of sapo-18/34 intergrowth zeolites and their enhanced stability for dimethyl ether to olefins *RSC Adv.* **7** 939–46
[49] Cheng F, Sajedin S M, Kelly S M, Lee A F and Kornherr A 2015 Uv-stable paper coated with aptes-modified p_{25} tio_2 nanoparticles *Elsevier* 1–18
[50] Ismail A, Kusworo T and Mustafa A 2008 Enhanced gas permeation performance of polyethersulfone mixed matrix hollow fiber membranes using novel dynasylan ameo silane agent *J. Membr. Sci.* **319** 306–12
[51] van Heyden H, Mintova S and Bein T 2006 Alpo-18 nanocrystals synthesized under microwave irradiation *J. Mater. Chem.* **16** 514–8
[52] Tosheva L N, Ng E P, Mintova S, Holzl M, Metzger T H and Doyle A M 2008 Alpo4-18 seed layers and films by secondary growth *Chem. Mater.* **2008** 5721–6

[53] Chen C, Feng B, Hu S, Zhang Y, Li S, Gao L, Zhang X and Yu K 2018 Control of aluminum phosphate coating on mullite fibers by surface modification with polyethylenimine *Ceramics International* **44** 216–24
[54] Karka S, Kodukula S, Nandury S V and Pal U 2019 Polyethylenimine-modified zeolite 13x for co_2 capture: Adsorption and kinetic studies *ACS Omega* **4** 16441–9
[55] Hanim S A M, Malek N A N N and Ibrahim Z 2016 Amine-functionalized, silver-exchanged zeolite nay: Preparation, characterization and antibacterial activity *Appl. Surf. Sci.* **360** 121–30
[56] Li P, Zhang Y, Zuo Y, Lu J, Yuan G and Wu Y 2019 Comparative study of organic and inorganic modification of chinese fir wood based on the respiratory impregnation method
[57] Wang J, Stevens L A, Drage T C and Wood J 2012 Preparation and co2 adsorption of amine modified mg–al ldh via exfoliation route *Chemical Engineering Science* **68** 424–31
[58] Abdi J, Vossoughi M, Mahmoodi N M and Alemzadeh I 2017 Synthesis of amine-modified zeolitic imidazolate framework-8, ultrasound-assisted dye removal and modeling *Ultrason. Sonochem.* **39** 550–64
[59] Bakdash R S, Aljundi I H, Basheer C and Abdulazeez I 2020 Rice husk derived aminated silica for the efficient adsorption of different gases *Sci Rep* **10** 19526
[60] Maity D, Chandrasekharan P, Feng S S and Jun D 2010 Synthesis and studies of aptes functionalized magnetite nanoparticles International Conference on Nanoscience and Nanotechnology (ICONN) 94–7
[61] Sehlleier Y H, Abdali A, Schnurre S M, Wiggers H and Schulz C 2014 Surface functionalization of microwave plasma-synthesized silica nanoparticles for enhancing the stability of dispersions *Journal of Nanoparticle Research* **16** 1–11

Modified kinetic models for Cr (VI) adsorption in polymer inclusion membranes

A Amador, R V Canaria, N Santos and M Gomez

School of Chemical, Biological, and Materials Engineering and Sciences, Mapua University, Intramuros, Manila, Philippines 1002

Email: mcgomez@mapua.edu.ph

Abstract. Hexavalent chromium is a highly toxic environmental inorganic pollutant. To eliminate toxic Cr (VI) ions in natural waters, polymer inclusion membranes (PIMs) have been developed for highly selective metal ion transport applications. The investigation of the effectiveness of Cr (VI) recovery in aqueous solutions using PIMs with varying amounts of plasticizer was studied. The pseudo-first order (PFO) kinetic model was modified to describe the amount of Cr (VI) ions that have accumulated onto the PIMs at a specific time and to evaluate the performance of the PIMs. A quantitative analysis of the modified PFO a model based on their non-linear representation and using the coefficient of determination indicates that the adsorptive properties of the PIMs are best described by the modified non-linear pseudo-first-order kinetic model (R^2 > 0.9748), suggesting that the sorption process is physisorption. To show the applicability of the modified model to other transport studies, modified PFO was fitted into the experimental data that studies the transport of Zn (II) ions onto PIM (R^2> 0.95).

1. Introduction

Heavy metals occur in the environment naturally, and it has a density greater than that of the water which means its components are much more held together [1]. A common environmental pollutant is chromium, which is widely used in various industrial processes, from leather treatment to manufacture of steel and paint. As an environmental pollutant, it has two oxidation states, Cr (III) and Cr (VI), of which the latter is significantly more toxic [2]. Accordingly, there is a need for the environmental chromium concentrations to be reduced to acceptable levels because chromium concentrations that exceed 0.25 mg/L are considered harmful [3].

Other Due to the high-risk effects on health of this heavy metal ion, other conventional methods in extracting hexavalent chromium ions were also reported from different studies where membrane extraction was utilized, including ion exchange resins, ultrafiltration, and liquid membranes. Different liquid membrane modalities are being utilized through the years, supported liquid membranes (SLMs) have been used widely to separate different pollutants from aqueous solutions [4]. The disadvantage of using SLMs in separating pollutants was the loss of either the carrier or the membranes solvent, or both, on the aqueous phase which resulted in its inability to be exploited industrially [5].

On the other hand, polymer inclusion membrane (PIM) was introduced by Sugiura in 1980 as an alternative for SLMs. It is a thin film cast from an ion carrier, a base polymer and a membrane plasticizer. The extraction of Cr (VI) ions and other metal ions using PIMs was reported in different studies that involve equilibrium and kinetic data. There are several adsorption models developed to study the process of physisorption and chemisorption, including Lagergen's Pseudo-First Order Model and Elovich Model [6]. However, these models only describe the adsorption capacity as a function of time, and not the more experimentally observed parameter of accumulation factor with contact time.

In this paper, the Pseudo First Order Model was modified to fit the experimental data that describes the relationship of accumulation factor with time. This paper also interpreted the prime cause of the gradual decrease of Cr (VI) ions that accumulates in the membrane as time increases using the initial conditions: pH=1, 10 ppm of Cr (VI) ions in HCl in the feed phase; 0.1 M NaOH in the strip phase.

The ability of the modified model to describe the relationship of accumulation factor with time will significantly aid not only the study of membrane adsorption but industrial operations as well since the more directly observable parameter was described here. Also, this model can predict the relationship between pH and time which is known to gradually change because of the effect of the HCl- media used in the feed phase. Since the modification of kinetic models that describe the relationship of accumulation factor concerning time was not yet reported, this can be very useful to have a better understanding of the effect of saturation on heavy metal ion transport caused by the pH and stripping phase.

The modified model that showed a good fit on the experimental data of Cr (VI) transport in PIMs was also fitted in the experimental data on Zn (II) transport that uses a different composition of PIMs to show the versatility of the developed model. This paper did not dwell more on the study of the PIMs physicochemical composition but focused on its ability to recover Cr (VI) ions as well as the factors that affect it during the sorption process.

2. Methodology

2.1. Experimental Data

All experimental data needed were collected from previous studies that involve Polymer Inclusion Membranes (PIMs) having different CTA/PBAT blend with Aliquat 336 [7] and PIMs having 2-hydroxy-5-nonylacetophenone oxime (LIX 84I) and di(2-Ethylhexyl) phosphoric acid (D 2 EHPA) as their composition [8].

2.2. Transport Studies

The adsorption capacity of metal ions per mass unit of adsorbent (Q_t) at time t is calculated by:

$$Q_t = \frac{(C_0 - C_t)V}{m} \tag{1}$$

where C_o is the initial concentration of metal ions (mg/L), C_t is the concentration of a metal ion in the liquid phase at time t (mg/L), respectively. V is the volume of the aqueous phase (L) and m is the mass of the adsorbent (g).

The accumulation factor (AF) was used to determine the efficiency of PIMs in the transport of Cr (VI) using the equation (2):

$$AF = \frac{C_{mem}}{C_0} \times 100\% \tag{2}$$

where C_{mem} (mg/L) is the concentration of the metal ion that accumulates in the membrane phase at time t (hr) and is equal to $C_{mem} = C_o - C_t$.

2.3. Pseudo-first-order (PFO) kinetics model modification

Hexavalent chromium adsorption kinetics can be modelled by the PFO (Largeren) equation based on the capacity to uptake the Cr (VI) ions. PFO model usually assumes that the sorption of metal ions into the sorption sites of the membrane and the initial surface coverage of sorbent is zero [9]:

$$S_{solid} + M^{+2}_{aqueous} \rightarrow SM^{+2}_{adsorbed\ phase} \tag{3}$$

Lagergren initially introduced PFO which is generally described as [10]:

$$\frac{dQ_t}{dt} = k_1 (Q_e - Q_t) \tag{4}$$

after integrating equation (4) with boundary conditions of $Q_t = 0$ to $Q_t = Q_t$ and $t = 0$ to $t = t$ the linearized equation will be:

$$\log(Q_e - Q_t) = \log Q_e - \frac{K_1}{2.303} t \tag{5}$$

Lastly, its analytical equation can be expressed as:

$$Q_t = Q_e \left(1 - e^{-K_1 t}\right) \tag{6}$$

Since the PFO model is known to only describe the relationship of adsorption capacities and contact time, the analytical equation of the PFO kinetic model was modified to be expressed in terms of accumulation factor (AF), initial concentration (C_0) and equilibrium concentration (C_e):

$$Q_t = \frac{C_0 AF\, V}{100\,\% m} \tag{7}$$

$$Q_e = \frac{(C_0 - C_e) V}{m} \tag{8}$$

$$\frac{C_0 AF\, V}{100\, m} = \frac{(C_0 - C_e) V}{m} \left(1 - e^{-K_1 t}\right) \tag{9}$$

simplifying the equation will result in the accumulation factor as a function of time:

$$AF = \frac{(C_0 - C_e)\left(1 - e^{-k_1 t}\right)}{C_0} \times 100\,\% \tag{10}$$

Since the relationship of pH with time can be expressed as [11]:

$$pH = \log\left(a t^b\right) \tag{11}$$

and the membrane adsorbent concentration as [12]:

$$C_m = \frac{C_e}{1 + \alpha} \tag{12}$$

$$\log \alpha = pH - pk_a \tag{13}$$

$$C_e = C_m \left(1 + 10^{\log at^b - pk_a}\right) \quad (14)$$

The final form of the modified PFO used in this study is:

$$AF = \frac{C_0 - C_m \left(1 + 10^{\log at^b - pK_a}\right)}{C_0} \left(1 - e^{-K_1 t}\right) \times 100\% \quad (15)$$

2.4. Parameters estimation and model fitting
Python 3.8 Matplotlib package was used to for the experimental data to the modified PFO model [13]. Initial parameters were first graphically estimated and then further optimized using the nonlinear optimization function curve fit in the SciPy package [14]. The goodness of fit of the modified PFO model was evaluated using sums of squares method.

Table 1. Constants used in the study.

Symbol	Description
AF	Accumulation Factor of Cr (VI) in the membrane
a	Numerical constant from equation (11)
b	Numerical constant from equation (11)
C_e	Equilibrium concentration of Cr (VI) in strip phase
C_m	Maximum concentration of Cr (VI) in strip phase
C_{mem}	Concentration of Cr (VI) in the membrane phase
C_o	Initial concentration of Cr (VI) in the feed phase
C_t	Concentration of Cr (VI) in the feed phase
K_1	Rate constant of adsorption process using PFO kinetic model
Q_e	Adsorption capacity of adsorbent at equilibrium
Q_t	Adsorption capacity of the adsorbent
α	Dissociation constant at equilibrium
t	Contact time

3. Results and Discussion

3.1. Adsorption kinetics models
Adsorption kinetics is described as the uptake ore release of solute as a function of time. These models are also used to determine the mechanisms which may influence the process which includes the chemical reaction and the mass transfers that occur [15]. Several adsorption models were tried, including PFO and Elovich models; however, only the PFO showed promising fit to the experimental data obtained from Sellami *et al.* and Wang *et al.*

3.1.1. PFO Model for Cr (VI). The modified PFO model fitted to the transport of Cr (VI) is shown in figure 1. The calculated parameters of the modified non-linear model are shown in table 2, as observed the rate constant K1 increased with plasticizer content. This can be due to the plasticizers on the membrane that improves its permeability to enhance the PIM's capacity for carrier movement. Also, the polarity and viscosity of the membrane plasticizer has a substantial effect on Cr (VI) transport [16]. The results indicated that the modified PFO model has a good fit with the data published by Sellami *et al.* [7]. A good correlation coefficient (R^2) was obtained for the modified PFO model, which showed that the transport of the hexavalent chromium process followed the rate expression of pseudo-first order. The goodness of fit of the modified PFO to the experimental data is further proven by the values of the maximum concentration (C_m) of the Cr (VI) ions in the liquid phase that is predicted from the model, as those have positive values and is less than the initial concentration (C_0). The high quantity of carriers and plasticizers also influences transport properties positively [7] as it can be adjusted to achieve a high recovery factor [17].

The good fit of the Sellami *et al.* data to the modified PFO model indicates that the hexavalent chromium adsorption onto the PIM that uses Aliquat 336 as the carrier is more inclined towards physisorption. Physisorption is mainly associated with forces that are formed because of the physical changes happening during the process such as the pore volume, surface area, and the functional groups along the surface of the PIM [18], these physical changes are because of the saturation that is influenced by the stripping phase and the change of pH concerning the contact time.

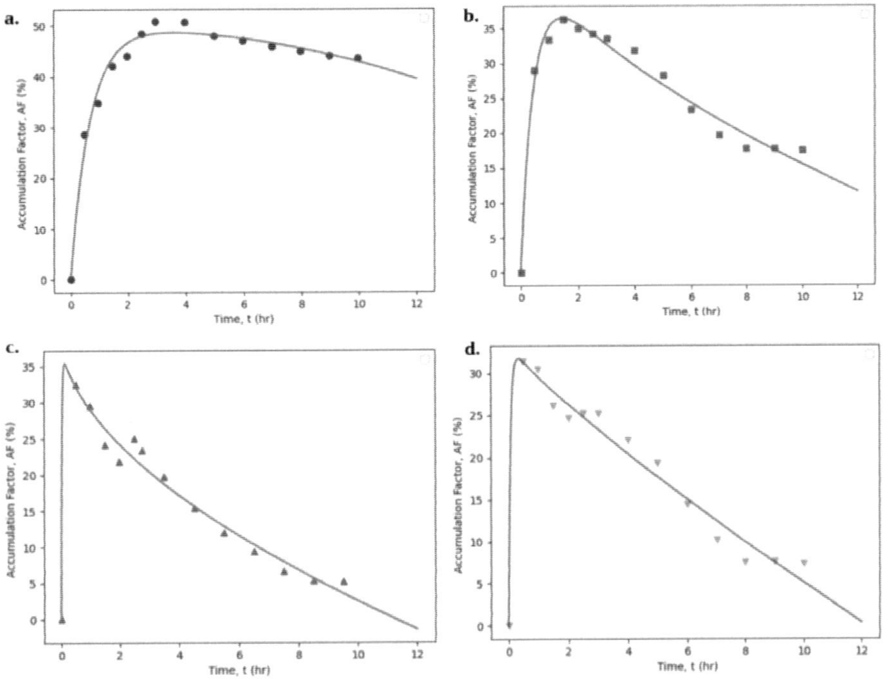

Figure 1. (a) Kinetic model plot using modified PFO for the Cr (VI) transport using 70CTA/30Aliquat 33, R^2=0.9813; (b) Kinetic model plot using modified PFO for the Cr (VI) transport using 52.5CTA/17.5PBAT/30Aliquat 336, R^2 =0.9796; (c) Kinetic model plot using modified PFO for the Cr (VI) transport using 35CTA/35PBAT/30Aliquat 336, R^2=0.9768; (d) Kinetic model plot using modified PFO for the Cr (VI) transport using 17.5CTA/52.5PBAT/30Aliquat 336, R^2=0.9748.

Table 2. Modified pseudo-first-order kinetics parameters for Cr (VI) ions.

PIM	R^2	C_m	a	b	K_l
70CTA-30Aliquat 336	0.9813	5.0424	0.0157	2.2203	1.433
52.5CTA-17.5PBAT-30Aliquat 336	0.9796	5.0651	1.2139	0.5920	1.995
35CTA-35PBAT-30Aliquat 336	0.9768	6.0785	0.5620	0.5553	13.765
17.5CTA-52.5PBAT-30Aliquat 336	0.9748	6.6258	0.8560	0.8351	31.953

Physical changes due to the stripping phase are constituted with the strength of the chosen stripping medium. In this case, NaOH was chosen to remove the extracted Cr (VI) ions. The strength of sodium hydroxide solution has substantially affected the Cr (VI) transport because the driving force for the diffusion through the polymer inclusion membrane has become saturated, due to the concentration of the metal complexes at the interface of the membrane phase and strip phase [19]. On the other hand, in conditions of initial pH of the solution, it is expected to change gradually because of the reverse transport of HCl therefore as shown in table 2, the constants for pH change vary with the composition of the PIM. As shown, the PIM without plasticizer on its composition will observe a faster change of pH since plasticizers are known to help the membrane be more flexible and be a better medium for carrier movement.

3.1.2. PFO Model for Zn (II). The To support the fitting of the modified PFO model to the study of accumulation factor as a function of time, the recent study that transported Zn (II) ions onto PIM was also fitted to the non-linear model (figure 2). The experiment was conducted using the following experimental data: Feed solution: C_0 =32.69 mg/L of Zn (II) ions, pH=4, and Stripping solution: H2SO4. As shown in table 3, the non-linear modified PFO model fitted the experimental data published by Wang *et al.* [8] produced high values of correlation coefficient (R^2 > 0.95). The membranes with different compositions where the PIM that fitted well in the model is composed of 45% D 2 EHPA and 55% of PVC [8].

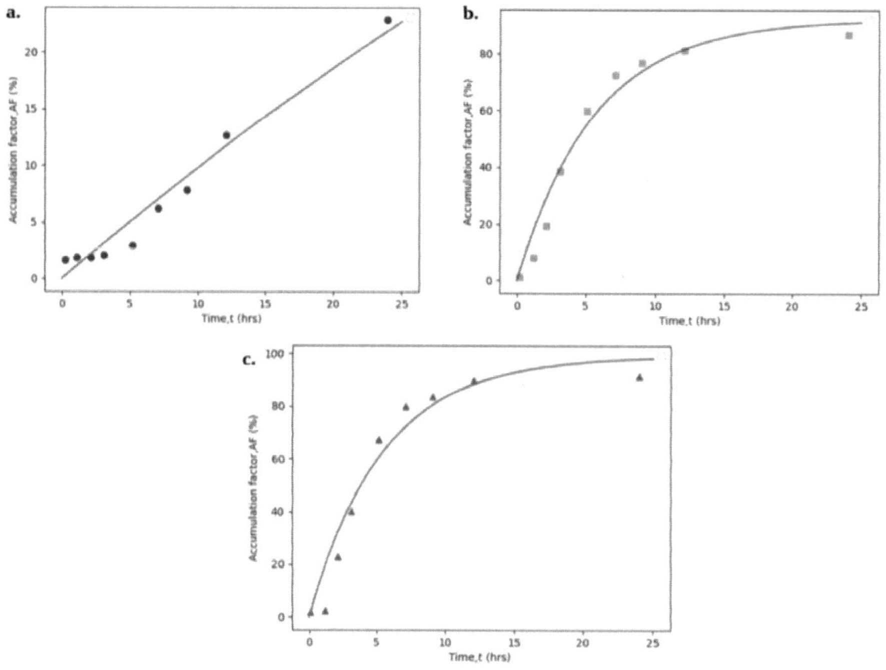

Figure 2. (a) K Kinetic model plot for the transport of Zn (II) ions using D2EHPA:PVC: NPOE =30:40:30, R^2=0.96928; (b) Kinetic model plot for the transport of Zn (II) ions using D2EHPA: PVC =45:55, R^2=0.96908; (c) Kinetic model plot for the transport of Zn (II) ions using D2EHPA:PVC =55:45, R^2=0.95079.

Table 3. Modified pseudo-first-order kinetics parameters for Zn (II) ions.

PIM	R^2	C_m	a	b	K_1
M1, D2EHPA: PVC: NPOE =30:40:30	0.96928	0.9035	$3.31x10^{-8}$	1.0002	0.01064
M2, D2EHPA: PVC =45:55	0.96908	2.5724	0.05	1.0000	0.17929
M3, D2EHPA: PVC =55:45	0.95079	0.2554	1.00	1.0000	0.18385

The arbitrary constants calculated from fitting the modified model with the experimental data could shed light in the effects of pH and time in the physical adsorption of ions and solutes in the membrane. However, further experimental data using various metal ions and polymer blends with well-characterized molecular and supramolecular structures and properties would be needed to fully describe the adsorption behavior of solutes in polymer membranes.

4. Conclusion

The composition of the membrane, specifically the amount of plasticizer, increases the ability of the metal to be recovered by improving the permeability of the membrane. Cr (VI) transport of hexavalent chromium was also affected by pH. As the maximum adsorption rate is achieved, pH also changed. Which caused a gradual decrease of in Cr (VI) ions accumulated in the membrane. To better evaluate adsorption mechanisms, models showing the relationship between accumulation factor and contact time are developed. Non-linear modified pseudo-first order (PFO) is used in modelling the kinetics of Cr (VI) adsorption in PIM with different CTA/PBAT blends. Correlation coefficient (R^2) values for the modified PFO model ranges from ~ 0.91 to 0.98 which indicates that the Cr (VI) adsorption follows pseudo-first order rate expression. The Cr (VI) maximum concentration predicted from the model also has values less than the initial concentration. The effect of plasticizer with Cr (VI) transport was also observed as the rate constant value (K) increases with an increase unproportionally with the PBAT amount for each PIM used. A good correlation coefficient was also obtained from experimental data collected from a Zn (II) transport study with D2EHPA-based PIM using the modified PFO model. The experimental data fit the modified pseudo-first-order (PFO) model is an indication that the Cr (VI) adsorption onto PIMs is inclined toward physisorption.

Acknowledgments

Highest honor and praise to the LORD for the wisdom, knowledge, and understanding He has given us and for the motivation and strength to pursue and accomplish this thesis. Thank you for delivering us through the challenges and trials that we faced and for directing us to the following persons who have been a channel of blessing in the completion of our study.

The assistance provided by Professor. Nanette D. Santos, the Thesis adviser, is greatly appreciated for her guidance every step of the way as she shares all her knowledge and wisdom while motivating us to keep our eyes on the goal.

We also wish to show our appreciation to Professor. Marineil C. Gomez, Thesis adviser, for her dedication and patience as she imparted new ideas and concepts that made a big impact as we complete our study.

We would like to take this opportunity to give our thanks to Professor. Flordeliza C. De Vera, Course adviser, and a panel member, for her never-ending guidance and support to the completion of this thesis.

We are extremely grateful to Professors Kristopher Ray S. Pamintuan and Medarlo B. De Jesus, our panel members, for contributing their time and expertise in helping us complete this study. Through their continued support, we were able to improve our thesis.

We also want to extend our gratitude to Dr. Kateryna Fatyeyeva, Professor (Assistant), Université de Rouen, who permitted us to access valuable data from their work that influenced this study.

To our family, our sincerest thanks for all the support, prayers, and sacrifices that you have done for us. Thank you for being our inspiration. You have given us hope amidst trying times. Special thanks to our friends, that reminded us to continue believing that through perseverance and hard work we can achieve our dreams.

Truly with GOD nothing is impossible. To Him be all the honor, all the glory, and all the praise.

References

[1] Decena S C P, Arguelles M S and Robel L L 2018 Assessing heavy metal contamination in surface sediments in an urban river in the Philippines *Pol. J. Environ. Stud.* **27** 1983–95

[2] Agarwal S, Tyagi I, Gupta V K, Dehghani M H, Jaafari J, Balarak D and Asif M 2016 Rapid removal of noxious nickel (II) using novel γ-alumina nanoparticles and multiwalled carbon nanotubes: Kinetic and isotherm studies *J. Mol. Liq.* **224** 618–23

[3] Jahanmahin O, Mehdi M, Rahmati M, Mohammadi T, Babaee J and Khosravi A 2016 Cr (VI) ion removal from artificial waste water using supported liquid membrane *Chem. Zvesti* **70** 913–25

[4] Jung C, Heo J, Han J, Her N, Lee S J, Oh J, Ryu J and Yoon Y 2013 Hexavalent chromium removal by various adsorbents: Powdered activated carbon, chitosan, and single/multi-walled carbon nanotubes *Sep. Purif. Technol.* **106** 63–71

[5] Neplenbroek A M, Bargeman D and Smolders C A 1992 Supported liquid membranes:instability effects *J. Memb. Sci.* **67** 121–32

[6] Chien S H and Clayton W R 1980 Application of Elovich Equation to the kinetics of phosphate release and sorption in soils *Soil Sci. Soc. Am. J.* **44** 265–8

[7] Sellami F, Kebiche-Senhadji O, Marais S, Couvrat N and Fatyeyeva K 2019 Polymer inclusion membranes based on CTA/PBAT blend containing Aliquat 336 as extractant for removal of Cr(VI): Efficiency, stability and selectivity *React. Funct. Polym.* **139** 120–32

[8] Wang D, Hu J, Liu D, Chen Q and Li J 2017 Selective transport and simultaneous separation of Cu(II), Zn(II) and Mg(II) using dual polymer inclusion membrane system *J. Memb. Sci.* **524** 205–13

[9] Guo X and Wang J 2019 A general kinetic model for adsorption : Theoretical analysis and modeling *J. Mol. Liq.* **288** 111100

[10] Matouq M. Jildeh N, Qtaishat M, Hindiyeh M and Al M Q 2015 The adsorption kinetics and modeling for heavy metals removal from wastewater by *Moringa*pods *Biochem. Pharmacol.* **3** 775–84

[11] Ahmed D N, Faisal A A H, Jassam S H, Naji L A and Naushad M 2020 Kinetic model for pH variation resulted from interaction of aqueous solution contaminated with nickel ions and cement kiln dust *J. Chem.* **2020** 11

[12] Nouri S 2003 Effect of functional groups and pH on the affinity and adsorption capacity of activated carbon : Comparison of homogeneous and binary Langmuir *Adsorp Sci Technol* **21** 511–24

[13] Hunter J D 2007 Matplotlib: A 2D graphics environment *Comput. Sci & Eng.* **9** 90–5

[14] Virtanen P *et al.* 2020 SciPy 1.0: Fundamental algorithms for scientific computing in Python *Nat. Methods* **17** 261–72

[15] Senthil Kumar P, Vincent C, Kirthika K and Sathish Kumar K 2010 Kinetics and equilibrium studies of Pb^{2+} ion removal from aqueous solutions by use of nanosilversol-coated activated carbon *Brazilian J. Chem. Eng.* **27** 339–46

[16] Kebiche-sendhajie O, Tingry S, Seta P and Benamor M 2010 Selective extraction of Cr (VI) over metallic species by polymer inclusion membrane (PIM) using anion (Aliquat 336) as carrier *DES* **258** 59–65

[17] Almeida M I G S, Cattrall R W and Kolev S D 2012 Recent trends in extraction and transport of

metal ions using polymer inclusion membranes (PIMs) *J. Memb. Sci.* **415–416** 9–23
[18] Essandoh M, Kunwar B, Pittman C U, Mohan D and Mlsna T 2015 Sorptive removal of salicylic acid and ibuprofen from aqueous solutions using pine wood fast pyrolysis biochar *Chem. Eng J.* **265** 219–27
[19] Vithanage M, Mayakaduwa S S, Herath I, Ok Y S and Mohan D 2016 Kinetics, thermodynamics and mechanistic studies of carbofuran removal using biochars from tea waste and rice husks *Chemosphere* **150** 781–9

IOP Conf. Series: Materials Science and Engineering 1195 (2021) 012049 doi:10.1088/1757-899X/1195/1/012049

Development of optimisation model for black soldier fly-based aquaculture feed supply chains in Malaysia

C A Ng[1], C W Chan[2], V Andiappan[1], L Y Ng[1] and D K S Ng[1]

[1] School of Engineering and Physical Sciences, Heriot-Watt University Malaysia, 62200, Putrajaya, Wilayah Persekutuan Putrajaya, Malaysia
[2] Agridon Technologies, Lot 2745-D, Jalan Industri 12, 47000 Kampung Baru Sungai Buloh

Email: Denny.Ng@hw.ac.uk

Abstract. Aquaculture is identified as one of the critical food supplies in Malaysia. Due to the increasing demand for aquaculture products, the demand for protein sources for fish feed is also increased accordingly. Black soldier fly larvae is identified as one of the main protein sources that can be used in fish feed. Such larvae can be grown using different types of organic materials, such as food waste, agriculture waste, etc. As Malaysia is the second-largest palm oil producer in the world, therefore, a large number of agricultural wastes, also known as palm-based biomass (e.g., empty fruit bunches, mesocarp fibre, decanter cake, etc.) are generated annually. Based on the current industry practise, palm-based biomass can be converted into value-added products. However, using palm-based biomass as feedback to grow black soldier fly larvae is a relatively recent discovery. Thus, a viable supply chain model has yet to be established. In this work, a mathematical optimisation model is developed via commercial optimisation software (Lingo v.16) to synthesise an optimum black soldier fly-based aquaculture feed supply chain that utilised palm–based biomass as the feedstock. Based on the optimised result, the annual operating cost of the aquaculture feed supply chain is estimated as RM 5.2 million.

1. Introduction

The aquaculture industry involves cultivation of aquatic lifeforms for sustenance and food source. In Malaysia, the aquaculture industry has been identified as a critical food supply since the 7th Malaysia Plan [1]. As the population of Malaysia increases, it is estimated that the demand for fish will also increase. From 1983 to 2019, the production of aquaculture products has increased from 44,000 tonnes per annum to 220,000 tonnes per annum [1].

In Malaysia, brackish water fish culture, freshwater fish culture, and marine fish culture are the common type of aquaculture system. Brackish water fish culture is aquaculture farms that are built on water with salinity between freshwater and seawater; freshwater fish water culture is aquaculture farms that are built on freshwaters such as river and lake; and marine fish culture are aquaculture farms that are built offshore. Note that brackish water fish culture is the dominant practice in Malaysia, with 70 % of the aquaculture farms in Malaysia [1]. Besides, freshwater and marine aquaculture encompass the remaining 30 %. Cobia fish is the most common fish that rears in brackish fish culture [1]. Since Cobia fish are carnivorous, the feed must have sufficient digestible materials [2].

According to the literature [3], larvae from larvae of black soldier fly (*Hermatia ilucens*) can be used as a raw material to produce fish feed for carnivorous fish. Black soldier fly is a type of fly in the

Hermatia genus. The average size of this fly larvae is approximate 24 mm long and 8 mm wide. The life cycle of a black soldier fly last approximate 45 days: Eggs (4 days), Larvae stage (18 days), Pupae stage (14 days), and adult stage (9 days) [4]. Note that the larvae is suitable to be used as fish feed because of the high nutrient and protein contents (16 %), as shown in table 1.

Table 1. Proximate analysis of black soldier fly larvae [5].

Parameters	% dry weight
Moisture	62.0
Lipids	7.0
Protein	16.0
Ash	3.0
Crude Fibre	3.0
Carbohydrate	9.0

Note that black soldier fly larvae contain a high level of protein which is important to enhance the health of the fish. Black soldier fly larvae grow well in agricultural wastes such as rotten vegetables, spent rice grains and palm-based biomass [6]. Therefore, growing larvae via palm-based biomass can be used as a strategy for sustainable waste management.

Palm oil products is ranked the top ten Malaysia's export goods [7]. In 2020, Malaysia exported 17.3 million tonnes of palm oil products [7]. Therefore, it is important to ensure a sustainable production of palm oil products. During the extraction of crude palm oil from fresh fruit bunches, various types of palm-based biomass such as empty fruit bunches, palm kernel cake, decanter cake, palm oil mill effluent, etc. are produced [8] Due to the large quantity of biomass being produced from palm oil mills, various technologies have been developed to convert palm-based biomass into value-added products [9]. For example, empty fruit bunches and palm kernel shell are used as fuel in the boilers to generate steam and electricity [10], whereas palm kernel cake can be used as animal feeds [11].

As the demand of fish feed increased, such palm-based biomass can be used as raw material for breeding black soldier fly for larvae production [12]. Palm-based biomass such as palm kernel cake is suitable to be used as food for black soldier fly larvae as it consists of sufficient protein content that is needed for insect rearing [13]. In order to sustainable production of black soldier fly larvae as aquaculture feed, an aquaculture supply chain needs to be developed. Such supply chain is a complex network of material flow through the aquaculture industry that involves different materials such as aquaculture feed, fish egg, fish products and intermediate products. Due to the complexity of the entire aquaculture supply chain, the aquaculture supply chain can be subcategorised into multiple sections which are aquaculture feed production supply chain, fish farm management supply chain and end-user fish product supply chain. In this work, aquaculture feed production supply chain is to be analysed.

To optimise the aquaculture feed supply chain, a mathematical model needs to be developed. However, there are limited research works focus on aquaculture supply chain optimisation. Aquaculture supply chain optimisation covers the optimisation of configuration and material flow network in the aquaculture supply to maximise profitability, minimise operating cost and maximise the productivity in the aquaculture supply. Gagalyuk *et al.* [14] presented a supply chain network for the fish supply chain to maximise the profitability of fish retailers in Germany and tested its validity empirically using Partial Least Squares Structural Equation Modelling based on surveys from fish retailers in Germany. Bakhrankova *et al.* [15] developed an integral stochastic model that includes raw material supplies and downstream product flow to improve capacity utilisation, operational efficiency, and profitability. Bakhrankova *et al.* [15] also accounted for fish quality deterioration and shelf life restrictions when developing the model. Malindretos *et al.* [16] used value stream mapping methodology to improve cost savings and lead time from production to consumption of fish in Greece. Abedi and Zhu [17] used mixed-integer linear programming (MILP) to optimise the production of trout farm based on spawn purchase quantity, the best time to harvest and the farming period. Table 2 summarises the optimisation objective of the previous works.

Table 2. Summary of literature involving aquaculture supply chain optimisation.

Literature	Optimisation objective
Gagalyuk *et al.* (2010)	Maximise the profitability of fish retailers in Germany
Bakhrankova *et al.* (2014)	Maximise capacity utilisation, operational efficiency and profitability.
Malindretos *et al.* (2016)	Minimum lead time from production to consumption of fish
Abedi and Zhu (2017)	Maximum profitability of trout sold

It is noted that the abovementioned works [14–17] are focusing on maximising the profitability of the end-user aquaculture products. Those works do not consider the cost of aquaculture feed in their approaches when optimising the aquaculture feed supply chain. As shown in literatures [14–17], the previous works focused mainly on the supply chain downstream from the aquaculture fish farm such as the production efficiency of aquaculture farms, the benefits of different aquaculture farms cooperating and the flow of product from fish retailers to consumers. Limited works are presented to optimise the upstream of aquaculture fish farms such as minimising the cost of fish feed to aquaculture farm. Furthermore, none of the work utilises palm-based biomass as feed in an aquaculture feed supply chain. Therefore, this work aims to integrate palm-based biomass into the development of aquaculture feed supply chain that produce fish feed via bioconversion of black soldier larvae.

As reported in the previous works [12,18,19], different agricultural wastes have been used to grow black soldier fly larvae. Leong *et al.* [18] analysed the proximate analysis of black soldier fly larvae fed with sewage sludge, palm decanter and fruit waste. Bokau and Witoko [12] presented the effect of probiotics concentration mixed with palm kernel cake towards crude protein content in the black soldier fly larvae. Mujahid *et al.* [19] analysed the growth rate of black soldier fly larvae fed with empty fruit bunches.

Based on the literature information, a mathematical optimisation model of an aquaculture feed supply chain that product fish feed from black soldier fly larvae is developed in this work. In this supply chain, source of palm-based biomass, quality of the biomass, location of the black soldier fly farm and operating cost of production are considered.

2. Problem Statement

A superstructure of the aquaculture feed supply chain is first developed in this work. Given palm-based biomass $\boldsymbol{k} \in \boldsymbol{K}$ (Palm kernel cake, palm oil mill effluent, empty fruit bunch etc.) are collected from palm oil mills $\boldsymbol{i} \in \boldsymbol{I}$ and then be transported to black soldier fly farm $\boldsymbol{j} \in \boldsymbol{J}$ using transportation modes $\boldsymbol{z} \in \boldsymbol{Z}$. For cases where black soldier fly farm j located in palm oil mill i, the distance will be given as 0. The mathematical model also considers the possibility of separating black soldier fly rearing processes and feed processing facilities into different locations. For cases where black soldier fly farm and feed processing facilities are located together, the distance will be given as 0. In the black soldier fly farm $\boldsymbol{j} \in \boldsymbol{J}$, processing technologies $\boldsymbol{w} \in \boldsymbol{W}$ are employed to convert palm-based biomass k into suitable composition for bioconversion into black soldier fly larvae. Black soldier fly larvae are then further processed into fish feed using various processing technologies $\boldsymbol{w'} \in \boldsymbol{W'}$. Technologies considered in the intermediate stages are processes such as drying, pelleting and pre-treatment process. The final fish feed is then transported to the port $\boldsymbol{l} \in \boldsymbol{L}$ via transportation modes $\boldsymbol{z'} \in \boldsymbol{Z'}$. Figure 1 illustrates the superstructure aquaculture feed supply chain.

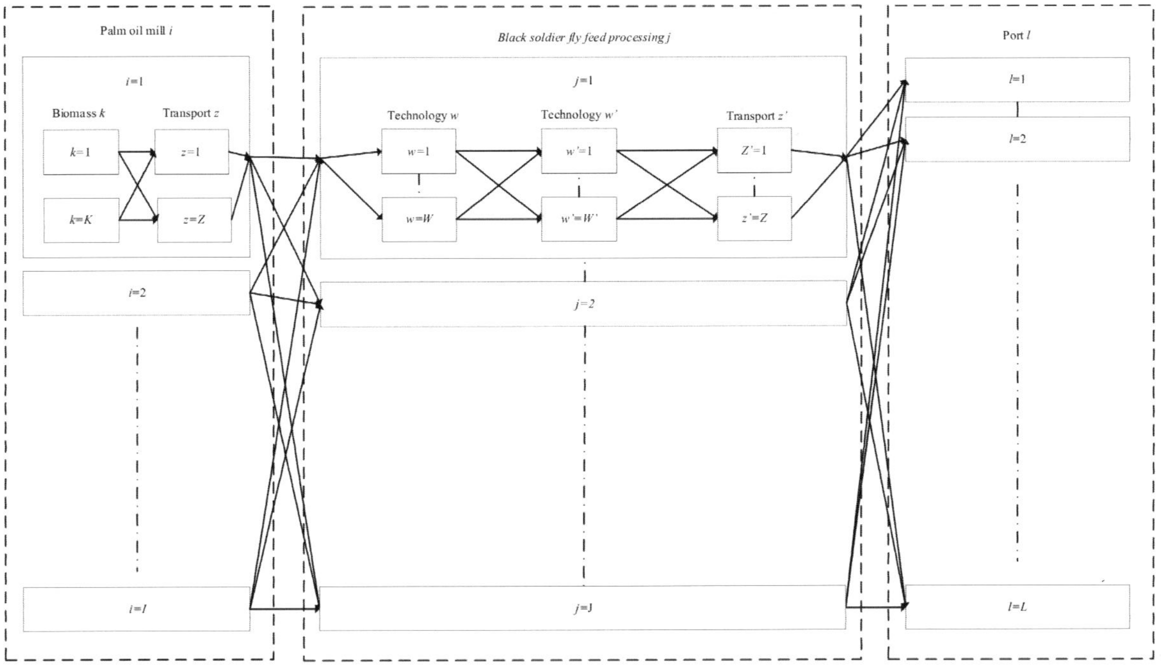

Figure 1. Aquaculture feed supply chain.

3. Mathematical Formulation

To synthesize an optimum aquaculture feed supply chain with minimum operating cost, the following mathematical model, which presented in the following sub-sections, is formulated using the commercial optimisation software.

3.1. Material Balance

Based on the superstructure in figure 1, the availability of biomass k per batch ($F_k^{\text{biomass,batch}}$) is given in this work. The distribution of palm-based biomass k from palm oil mills i to farm j per batch ($F_{k,j}^{\text{batch}}$) can be to determine via equation (1).

$$F_{k,j}^{\text{biomass,batch}} = \sum_{i=1}^{i=I} F_{i,j,k}^{\text{biomass,batch}} \quad \forall j \forall k \tag{1}$$

The production of larvae per batch from biomass k ($F_{k,j}^{\text{larvae,batch}}$) can be determined based on the given conversion of biomass k to larvae, X_k^{larvae}.

$$F_{k,j}^{\text{larvae,batch}} = F_{k,j}^{\text{biomass,batch}} X_{k,j}^{\text{larvae}} \quad \forall j \forall k \tag{2}$$

To convert the production of larvae from per batch basis to annual basis, $F_{k,j}^{\text{larvae}}$. Production of larvae per annum from biomass k $F_{k,j}^{\text{larvae}}$ can be determined by on the number of larvae batches produced from biomass k per annum, $n_{k,j}^{\text{batch}}$.

$$F_{k,j}^{\text{larvae}} = F_{k,j}^{\text{larvae,batch}} n_{k,j}^{\text{batch}} \quad \forall j \forall k \tag{3}$$

$$F_{k,j}^{\text{biomass}} = F_{k,j}^{\text{biomass,batch}} n_{k,j}^{\text{batch}} \quad \forall j \forall k \tag{4}$$

As larvae growing is a batch process, different biomass requires a different amount of time t_k^{batch} to grow the larvae. Therefore, $F_{k,j}^{\text{larvae}}$ must not exceed the maximum amount of larvae that can be produced using biomass k $F_{k,j}^{\text{max,larvae}}$ as shown in equation (5). $F_{k,j}^{\text{max,larvae}}$ is determine using equation (6).

$$F_{k,j}^{\text{larvae}} \leq F_{k,j}^{\text{max,larvae}} \quad \forall j \; \forall k \tag{5}$$

$$F_{k,j}^{max,larvae}=\frac{F_{k,j}^{larvae,batch}}{t_{k,j}^{batch}}\times t_{max}\ \forall j \forall k \tag{6}$$

where t_{max} is the maximum available time to produce the larvae per year. The output of larvae from farm j F_j^{larvae} is the summation of larvae output from each biomass k. Besides, the mass of dried larvae is the mass of larvae F_j^{larvae} multiplied with dry fraction.

$$F_j^{larvae}=\sum_{k=1}^{K} F_{k,j}^{larvae}\ \forall j \tag{7}$$

$$F_j^{dried}=\left(1\text{-}x^{moisture}\right)F_j^{larvae}\ \forall j \tag{8}$$

3.2. *Operating Cost Modelling*

For the operating cost modelling, it is categorised into three different costs, mainly biomass purchasing cost $C^{biomass}$, intermediate processing costs $C^{process}$ and transport cost $C^{transport}$. The biomass cost is modelled as the summation of the flowrate of each biomass k to farm j multiplied by the cost of biomass, C_k, respectively.

$$C_j^{biomass}=\sum_{k=1}^{K}\left(F_{k,j}^{biomass}C_{k,j}\right)\ \forall j \tag{9}$$

The intermediate processing costs $C^{process}$ is the cost of converting palm based biomass into the final product, fish feed. Hence, it is the cost summation of each processing step as shown in equation (10) below.

$$C_j^{process}=\sum_{k=1}^{K}\sum_{w=1}^{W}\left(F_{k,j}C_w\right)+\sum_{k=1}^{K}\sum_{w'=1}^{W'}\left(F_{k,j}^{larvae}C_{w'}\right)\ \forall j \tag{10}$$

The cost of transport $C^{transport}$ is the cost summation of all transportation cost which includes transporting biomass from palm oil mill i to black soldier fly farm j as well as transporting final feed product from black soldier fly farm j to port l as shown in equation (11) below.

$$C_j^{transport}=\sum_{i=1}^{I}\sum_{k=1}^{K}\sum_{z=1}^{Z}\left(D_{i,j,k}\times C_{z,j}\right)+\sum_{l=1}^{L}\sum_{z'=1}^{Z'}\left(D_{l,j}\times C_{z',j}\right)\ \forall j \tag{11}$$

The mass transport of biomass from palm oil mill i to black soldier fly farm j, $D_{i,j,k}$ is the product of mass flowrate of biomass k from palm oil mill i to black soldier fly farm j, $F_{i,j,k}$ and the distance between farm j and palm oil mill i, $Dis_{i,j}$. The distance between the farm j and palm oil mill i was calculated by inputting the coordinates of farm j and palm oil mill i into Google Maps then determining the shortest road distance between the 2 locations.

$$D_{i,j,k}=F_{i,j,k}Dis_{i,j} \tag{12}$$

$$F_{i,j,k}=\sum_{i=1}^{I}\left(F_{i,j,k}^{biomass,batch}\times n_{k,j}^{batch}\right)\ \forall j,\forall k \tag{13}$$

The mass transport of final feed product from black soldier fly farm j to port l, $D_{l,j}$ is the product of mass flowrate of final feed product from farm j to port l, $F_{l,j}^{dried}$ and the distance between farm j and port l, $Dis_{l,j}$. The distance between the farm j and port l was also calculated by inputting the coordinates of farm j and port l into Google Maps then determining the shortest road distance between the 2 locations.

$$D_{l,j}=F_{l,j}^{dried}\times Dis_{l,j} \tag{14}$$

The operating cost at each farm in the supply chain is thus the sum of all the different costs at each farm.

$$C_j^{total}=C_j^{biomass}+C_j^{process}+C_j^{transport}\ \forall j \tag{15}$$

The objective function of this mathematical model is to minimise the operating cost C^{total}. Hence the objective function is given as in equation (16).

$$\text{Minimise } C^{total} \tag{16}$$

4. Case Study

In this supply chain optimisation approach, the objective function is to minimise operating cost C^{total}. In this case study, three Black Soldier Fly (BSF) farms operate to produce 1200 tonnes of pelletised feed per annum in each farm. To convert the biomass into larvae then aquaculture feed, the supply chain will have to go through several unit operations which are pre-treatment, bioconversion, drying and pelletisation. Each process has several technologies that can be applied; thus the selection of technology was considered by the model based on each technology's operating cost. On the other hand, the protein content of the larvae produced was not constrained in this case study.

With the data collected, a mathematical model of the superstructure was developed using LINGO 18.0. The target feed production output was set as 1200 tonnes per annum. This is because a typical black soldier fly farm produces 150 kg per hour and operates 8000 hours per annum (20). The mathematical model was solved with the constraints in mind and the feed production configuration per batch of BSF larvae is illustrated in figure 2 while the optimised supply chain is illustrated in figure 3.

Figure 2 shows that the pre-treatment technology selected is aerobic digestion as the cost of aerobic digestion is lower. The drying technology selected is rotary drying as it has the lowest operating cost. Moreover, the pelletisation technology chosen is the pelletiser rather than single screw extrusion. This is because the pelletiser operating cost is lower than the single screw extrusion.

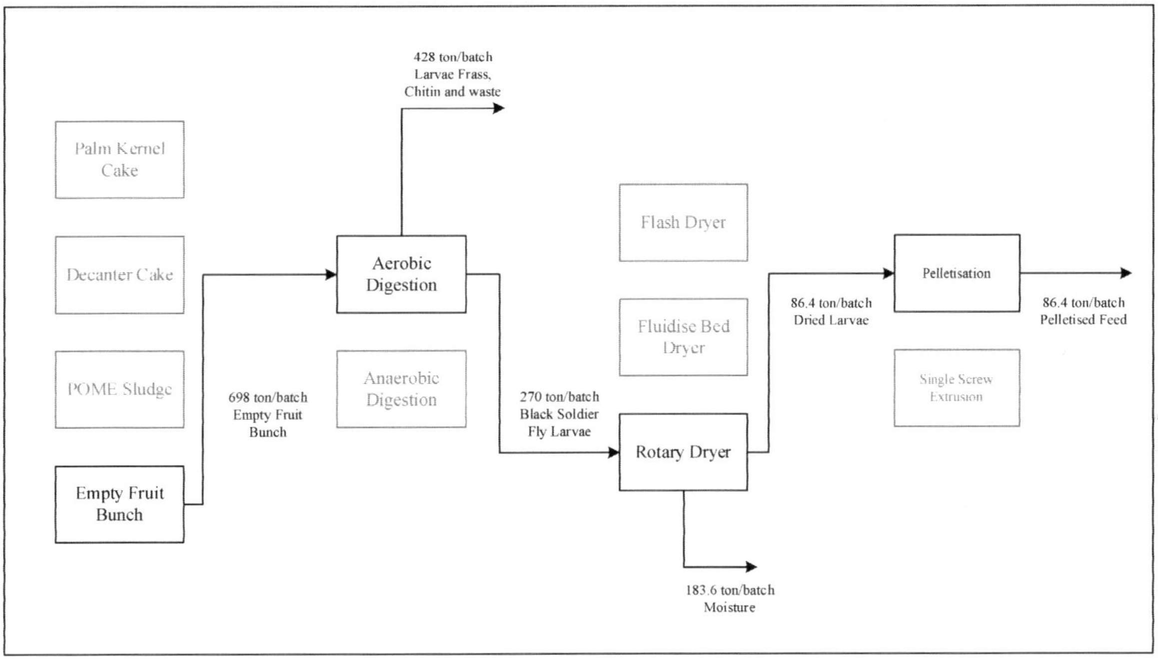

Figure 2. Optimised feed production configuration.

Figure 3 shows that Palm Oil Mill (POM) i_2, i_3 and i_7 is selected to supply the biomass to Farm j_2, j_1, and j_3 respectively. This is because the distance is the shortest between i_2 to j_2, i_3 to j_1 and i_7 to j_3. Besides, that the biomass selected is only Empty Fruit Bunches (EFB). This is because the EFB bioconversion rate is the highest and thus less EFB is needed to be fed to the larvae. Furthermore, the price of EFB is the lowest at RM 32 per ton. Thus, EFB is selected as the biomass in the model. The biomass was then transported using the 10 tons truck. Although the 10 tons truck cost more per kilometre transported, the truck's capacity is higher therefore a smaller number of trips between the farm and the mill is needed. Furthermore, feed produced from farm j_1 and j_2 is transported to Port Klang for storage and feed produced from farm j_3 is transported to Port Dickson for storage. This is because farm j_1 and j_2 are closer to Port Klang and farm j_3 is closer to Port Dickson.

Additionally, it is noted that the model decided against separating the black soldier fly farm from the feed processing facilities. This is because it would cost more to transport the larvae to a separate facility to process larvae into feed thus it would increase the transport cost and operating cost. The operating cost of each BSF farm is illustrated in table 3. Overall, the operating cost is RM 5.2 million per annum for BSF farm j_1, RM 5.3 million per annum for BSF farm j_2 and RM 5.4 million per annum for BSF farm j_3. The difference in the operating cost is attributed to the transportation cost to transport biomass to the farm and to transport the pelletised feed to the port.

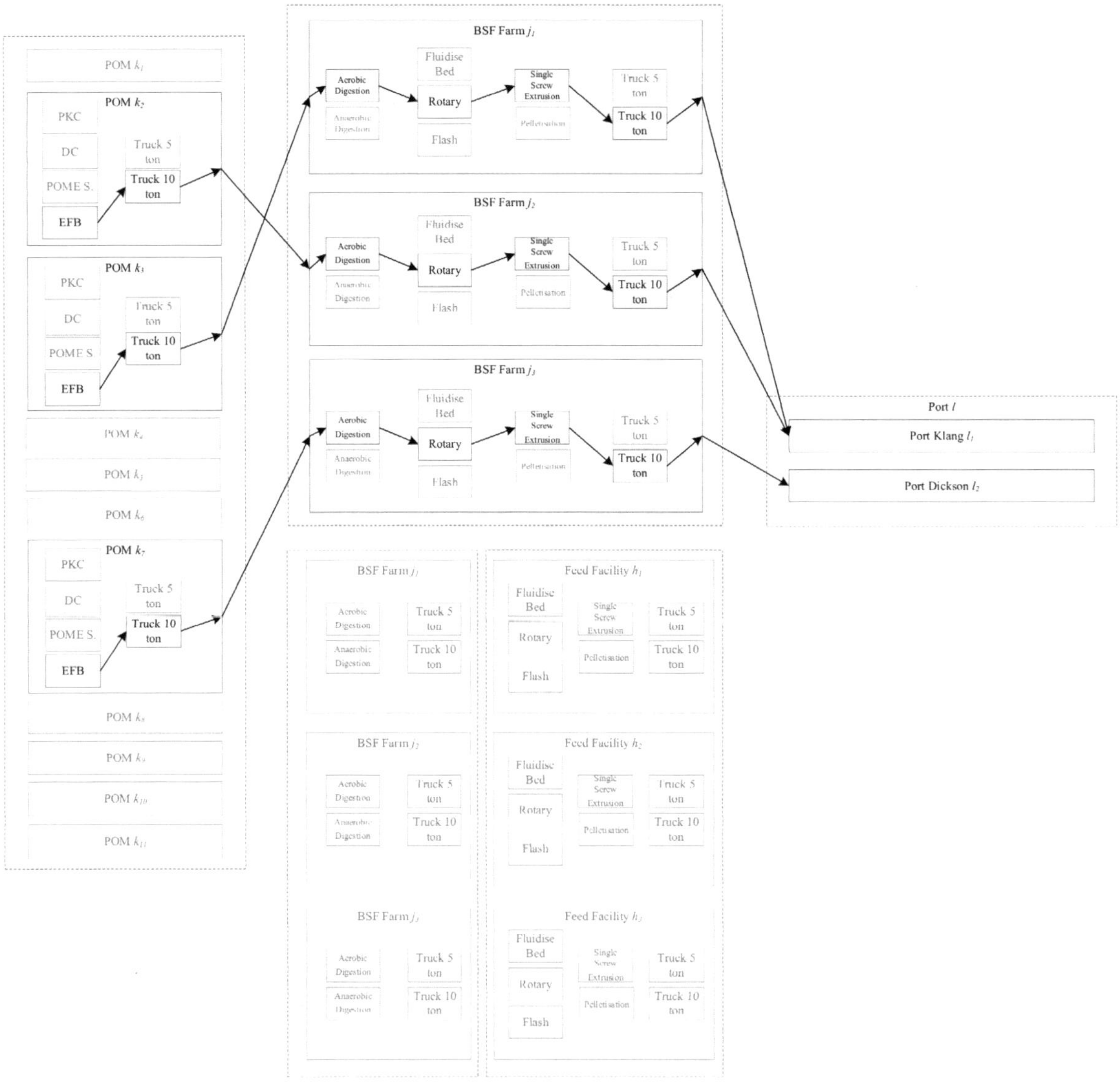

Figure 3. Optimised feed production superstructure.

Table 3. Optimised BSF feed production cost.

		BSF Farm j_1		BSF Farm j_2		BSF Farm j_3	
Biomass Cost, $C^{biomass}$	(RM/year)	RM	265,000	RM	265,000	RM	265,000
Process Cost, $C^{process}$	(RM/year)	RM	4,871,000	RM	4,871,000	RM	4,871,000
Transport Cost, $C^{transport}$	(RM/year)	RM	63,000	RM	159,000	RM	277,000
Total Operating Cost, C^{total}	(RM/year)	RM	5,200,000	RM	5,300,000	RM	5,400,000
Larvae crude protein content	(% wt)		13.49		13.49		13.49

5. Conclusion

In conclusion, a mathematical model for the aquaculture feed supply chain is synthesised in this work. The developed mathematical model considers the mass balance of the supply chian, distance between palm oil mill and black soldier fly farm, protein content of the biomass, bioconversion rate and technology cost when selecting the biomass and technology. The optimised model shows that the optimised feed production cost is lower compared to commercial feed prices and thus the results show that it is more cost-effective for aquaculture companies to manufacture feed compared to purchasing commercial feeds in the market. Future works that can be done to improve the model is to introduce the GHG emissions of each unit operations as a constraint for the model to consider in the technology and biomass selection for each unit operations. Besides, the capital cost of setting up the feed manufacturing process was not considered and can be a prospect for future works. It can be noted that the variability of biomass prices, biomass quality and technology cost can be easily updated and remodelled based on updated data.

Acknowledgments

The authors are grateful to the School of Engineering and Physical Science, Heriot-Watt University Malaysia and Triple Vs Venture for supporting this research work. The contributions of Bluestream Mariculture Sdn Bhd for sharing data on the aquaculture feed supply chain is also acknowledged and appreciated.

References

[1] FAO 2019 Malaysia National Aquaculture Sector Overview *FAO Ctry notes* **1380** 1–16
[2] Rombenso A N and Trushenski J T 2021 Is a fish oil-free cobia feed possible? *Insights gained from lipids research at Southern Illinois* 1–7
[3] Swinscoe I, Oliver D M, Gilburn A S, Lunestad B, Lock E J and Ørnsrud R 2019 Seaweed-fed black soldier fly (Hermetia illucens) larvae as feed for salmon aquaculture: Assessing the risks of pathogen transfer *J Insects as Food Feed.* **5** 15–27
[4] Ferrarezi R and Bailey D S 2016 UVI/AES Annual Report 2016 *Alternative Sources of Food for Aquaponics in the U.S. Virgin Islands: A Case Study with Black Soldier Flies*
[5] Shumo M, Osuga I M, Khamis F M, Tanga C M, Fiaboe K K M and Subramanian S 2019 The nutritive value of black soldier fly larvae reared on common organic waste streams in Kenya *Sci Rep* **9** 1–13
[6] Nyakeri E M, Ogola H J, Ayieko M A and Amimo F A 2017 An open system for farming black soldier fly larvae as a source of proteins for smallscale poultry and fish production *J Insects as Food Feed* **3** 51–6
[7] Malaysian Palm Oil Board 2020 Overview of the malaysian oil palm industry *Malaysian Palm Oil Board* 1–4
[8] Ng R T L, Ng D K S, Tan R R and El-Halwagi M M 2014 Disjunctive fuzzy optimisation for planning and synthesis of bioenergy-based industrial symbiosis system *J Environ Chem Eng.* **2** 652–64
[9] Ng D K S and Ng R T L 2013 Applications of process system engineering in palm-based biomass processing industry *Curr Opin Chem Eng* **2** 448–54
[10] Husain Z, Zainal Z A and Abdullah M Z 2003 Analysis of biomass-residue-based cogeneration

system in palm oil mills *Biomass and Bioenergy* **24** 117–24

[11] Aspar H M 2001 Malaysian palm kernel cake as animal feed *Palm Oil Dev* **34** 4–6

[12] Bokau R J and Witoko P 2017 Optimalization of bioconversion prosess of palm kernel cake for production maggot Hermetia illucens as a source of animal protein in fish farming *Aquac Indones.* **18** 20

[13] Hem S, Toure S, Sagbla C and Legendre M 2008 Bioconversion of palm kernel meal for aquaculture: Experiences from the forest region (Republic of Guinea) *African J Biotechnol.* **7** 1192–8

[14] Gagalyuk T, Hanf J and Herzlieb C 2010 Managing supply chains successfully: An empirical testing of success of supply chain networks in the German fish sector *Food Econ - Acta Agric Scand Sect C* **7** 139–50

[15] Bakhrankova K, Midthun K T and Uggen K T 2014 Stochastic optimization of operational production planning for fisheries *Fish Res* **157** 147–53

[16] Malindretos G, Vlachos I, Manikas I and Chatzimanolakis M 2016 Future prospects of sustainable aquaculture supply chain practices *Smart Innov Syst Technol.* **52** 487–97

[17] Abedi A and Zhu W 2017 An optimisation model for purchase, production and distribution in fish supply chain–a case study *Int J Prod Res.* **55** 3451–64

[18] Leong S Y, Kutty S R M, Malakahmad A and Tan C K 2016 Feasibility study of biodiesel production using lipids of Hermetia illucens larva fed with organic waste *Waste Manag.* **47** 84–90

[19] Mujahid M, Amin A A, Hariyadi H, Fahmi M R 2017 Oil palm empty bunches bioconversion using Trichoderma sp. and Black Soldier fly larvae as poultry feed composition. *J of Animal Prod & Proc Technol.* **05** 5–10

[20] Bullock N, Chapin E, Evans A, Elder B, Givens M, Jeffay N, Pierce B and Robinson W 2013 *The Black Soldier Fly How-to-Guide*

Optimising of vacuum distillation units using surrogate models

S X H'ng, L Y Ng, D K S Ng and V Andiappan

School of Engineering and Physical Sciences, Heriot-Watt University Malaysia, 62200, Putrajaya, Wilayah Persekutuan Putrajaya, Malaysia

Email: v.murugappan@hw.ac.uk

Abstract. Crude oil blending is an important step for the operation of crude distillation systems in the refinery to improve the yield and profitability of the products. The product's yield and quality are strongly dependent on the properties of the crude oil. However, the products of crude distillation units, especially the vacuum distillation unit (VDU) need to satisfy the yield and quality requirements of the downstream process units in the refinery. Otherwise, the performance of downstream processes will be affected, and loss of profitability in the refinery. Hence, it is important to optimise the performance of the VDU to ensure the optimum operation of VDU. This work covers the process simulation of VDU, surrogate modelling and mathematical optimisation model. The objective of the developed optimisation model is to determine an optimal for maximum high vacuum gas oil (HVGO) yield and minimum total annualised cost (TAC) respectively. To do this, crude oil blending ratio, column temperature, column pressure, stripping steam flowrate, pump-around flowrate in the VDU are optimised. Based on the optimised result, the heavy-light crude blend achieves higher HVGO yield and lower TAC as compared to the heavy-medium crude blend and heavy-medium-light crude blend. The optimised results can provide insight into the optimal process conditions of VDU for the refiners. With this insight, effective operating strategies can be developed to overcome the limitations present in real VDU operations.

1. Introduction

Vacuum distillation unit (VDU) is part of a typical refinery process that helps to further recover the higher boiling gas oil from the atmospheric residue of an atmospheric distillation unit. The major products of the VDU include light vacuum gas oil and heavy vacuum gas oil (HVGO). In the VDU, the separation of atmospheric residue into products involves many complicated relationships between operating conditions. The operating conditions of VDU need to be adjusted from time to time due to this variation of crude oil properties. These adjustments may consequently impact the product yield and quality. Nevertheless, it is critical for the products of the VDU to meet the targeted yield and quality specifications. This is because the products of VDU are the feedstocks to the downstream process units such as hydrocracker, fluid catalytic cracker or lube oil facility to produce commercially valuable products [1]. If the product yield or product quality specifications cannot be met, the downstream process units of the refinery will suffer. It may also cause an unplanned shutdown of the units and eventually incur a financial loss. Hence, all these sequences raise the need to develop a systematic optimisation method for determining the blending ratio of different kinds of crude oil and the optimal operating conditions of the VDU. By doing so, VDU can achieve high production throughput, satisfactory product quality and excellent economic potential.

Mathematical programming has been widely used in the synthesis, design and optimisation of crude distillation systems. The implementation of the mathematical optimisation technique helps to determine the optimal solutions of the system based on the problems of the mathematical descriptions [2]. For instance, Seo *et al.* [3] implemented mixed integer nonlinear programming to determine the optimal feed location and operating conditions of a crude distillation unit. With that, the proposed framework reduced the operating and capital costs of the existing crude distillation unit by 86 %. Next, Basak *et al.* [4] developed a non-linear, steady-state crude distillation unit to maximise the net profit while satisfying the product properties. A noticeable increase in profitability was observed with this optimisation model. Furthermore, Gu *et al.* [5] combined exergy analysis and mathematical programming for the energy optimisation of the multistage crude oil distillation units. The results showed a considerable reduction of 2.79 % in energy consumption which proved that this approach was implemented effectively. Besides, Inamdar *et al.* [6] introduced elitist non-dominated sorting genetic algorithms to address the multi-objective optimisation of the crude distillation systems. From this work, better optimal conditions and higher profit of the crude distillation system were reported. Similarly, Xiaoqiao *et al.* [7] also employed a multi-objective optimisation model to investigate the trade-off between economic benefit, furnace energy consumptions and CO_2 emissions of the crude distillation unit. Meanwhile, the effect of binary feed composition impact on the crude distillation unit had investigated as well. The proposed model provided a set of Pareto-optimal solutions that can solve the limitations of the refinery. Based on the literature cited above, a lot of researchers have considered the mathematical programming techniques in the optimisation work of crude distillation unit. Nevertheless, the complexity and nonlinearities associated with the distillation units may restrict the use of the mathematical optimisation model.

This has brought the attention of the researchers to employ the surrogate modelling technique into the distillation optimisation process, particularly for the crude distillation units in the refinery. A surrogate model is a sophisticated analytic model that mimics the complex behaviour of the simulation environment based on the regression of statistical input variables and output responses [8,9]. For example, López *et al.* [10] implemented surrogate models in the non-linear programming model to investigate the effect of crude composition and operating conditions of the crude distillation unit. An increase of 13 % in profit was observed from the proposed optimisation model. Besides, Yao and Chu [11] developed a surrogate model with the support vector regression and improved design of experiment method to optimise the operating conditions of the atmospheric distillation unit. A remarkable increase in profitability was also reported in the proposed approach. Apart from that, Ochoa-estopier *et al.* [12] introduced an artificial neural network model to represent the distillation columns. The proposed model associated with the heat exchanger networks was incorporated in the optimisation framework to determine the optimal operating conditions that give maximum process economy. Moreover, a cut-point temperature surrogate modelling technique was proposed by Gut *et al.* [13] to determine the product yield and properties of the crude distillation units. The surrogate models were proven to have accurate estimations and the models were able to be implemented into the planning, control and scheduling applications of the crude oil distillation units. Based on the literature surveys above, most of the optimisation works have been devoted to the overall crude distillation systems and atmospheric distillation units, rather less attention has been paid to the VDUs.

The performance of the VDU is significant as it needs to satisfy the feedstocks quality and rate target of the downstream process units. Therefore, this raised the focus of some researchers on the optimisation work of the VDU. With the aid of Aspen HYSYS simulation, Mittal *et al.* [14] developed an optimisation framework for crude oil blending and processing with the simultaneous considerations of the furnace energy consumptions, CO_2 emissions and economic potential of the VDU. Next, Gu *et al.* [15] proposed a methodology to analyse and evaluate three crude oil vacuum distillation processes. Recovery energy and exergy efficiencies, economical potential and product yield were also taken into considerations in the proposed methodology. Consequently, the results provided insight for engineers to determine a suitable process and outlet temperature of the VDU. Also, it is worth noting that Li *et al.* [16] introduced a dividing wall column in a lubricant type VDU to enhance the product yield and quality. The proposed configuration showed that the boiling point range of the lube cut was reduced significantly which improved the product yield and quality.

A critical observation of the literature review presented above infers that the relationship between the product yield and total annualised cost (TAC) of VDU has not been studied for crude oil blending. The refiners have been looking for the opportunity in processing lower-cost heavy crude oil due to the operational and feed availability constraints. By blending the crude oil, the diversity of crude oil processing as feedstock has increased which maximise the profitability of refiners. Therefore, crude oil blending is taken into consideration in the optimisation work of the VDU. Meanwhile, another observation that can be drawn from the literature cited above is that no work has combined the process simulation, surrogate modelling and mathematical programming in the distillation system optimisation works. Process simulation is used to simulate and provide basic data of the distillation process. Nevertheless, a lot of challenges occur in the optimisation process using process simulation due to the high degree of freedom and nonlinearities associated with the distillation units. The process simulation tool does not reflect the optimal solution to run a system. It can only give the best possible solution based on a set of predefined scenarios determined by the decision-makers. To address the issues of process simulation, surrogate modelling and mathematical programming are introduced. Surrogate modelling technique is applied to mimic the behaviour of the distillation system due to their simplicity and satisfactory accuracy. Surrogate modelling techniques are employed to capture the relationship between the independent X variables and dependent Y variables of the distillation system. The independent X variables are the variables that can be controlled (i.e., crude oil blending ratio and operational variables), whereas dependent Y variables are the variables that are measured and dependent on the independent variables (i.e, HVGO yield, pump-around cooler flowrate, column diameter, etc.). The surrogate model is constructed by regressing against a set of statistical data. The data is later converted to a fairly representative equation that can describe the operation of the distillation system. Also, mathematical programming can be used to analyse and provide the optimal solution of a given distillation system. Therefore, the combination of the individual strengths of these three methods can reveal the results that are not possible to be determined when these three methods are used separately. The combination of these methods is essential for decreasing the overall process costs, predicting the process behaviour, improving the utilisation of resources and providing detailed information on how the system operates. Hence, this work considers the combination of both process simulation, surrogate modelling and mathematical programming in the optimisation work of the VDU.

In order to fulfil the research gaps discussed above, this paper develops an optimisation model with the combination of process simulation, surrogate modelling and mathematical programming. The proposed model is used to investigate the impact of crude blending composition and the operating conditions of the VDU on the HVGO yield. Besides, upon considering the economic profit of refineries, the total annualised cost (TAC) of VDU is investigated as well. This study can provide economical and operational benefits to the refiners as it can overcome the tighter supply-demand constraints in the refinery. The objectives of this research are:

- To determine the optimal VDU operating performance and conditions that give maximum HVGO.
- To determine the optimal VDU operating performance and conditions that give minimum TAC.

The remainder of this paper is organised as follows. Section 2 presents the methodology for the optimisation of VDU, based on the process simulation, surrogate modelling and mathematical programming. Section 3 illustrates the application of the proposed methodology in a case study. In the case study, different crude blending ratios are compared and analysed to identify the optimal crude blending ratio with the corresponding operating conditions that give maximum HVGO yield and minimum TAC respectively. Moreover, the development of surrogate models is illustrated in this section as well. Section 4 presents the optimised results of the maximum HVGO yield and the minimum TAC of VDU respectively. Finally, the conclusion and future work are given in Section 5.

2. Methodology

In this research study, the methodology is conducted in three stages: process simulation, surrogate modelling and mathematical optimisation.

2.1. Process Simulation

Two types of models are mainly used to design the crude distillation unit, namely shortcut model and rigorous model. A rigorous model can simulate mass balance, energy balance and equilibrium relations for every stage of the distillation unit. Therefore, the rigorous model is chosen in this research study as it provides more accurate predictions. Commercial rigorous simulation software (i.e., Aspen HYSYS, Aspen Plus, Pro/II of SimSci-Esscor, etc.) are commonly used for modelling crude distillation units in the refinery. In this work, Aspen HYSYS version 10.0 is selected to develop the crude oil blending model and rigorous distillation model of the vacuum distillation unit (VDU). This is because Aspen HYSYS contains existing routines specific to solving the distillation columns, unlike other software.

2.1.1. Crude Oil Characterisation and Blending. For the simulation, three crude assays are considered as different choices of crude blends. Heavy crude is considered as one of the crude assays. Meanwhile, higher API gravity crude assays, namely light and medium assays are considered for the remaining crudes. As mentioned in the research gap, crude oil blending plays an important role to maximise the profitability of the refinery. Consequently, a crude oil blending model is considered here to mix the heavy crude with other crudes to be processed as feedstock in the distillation operation. Thus, the assay data of the crude oil is first added and calculated by Aspen HYSYS. By doing so, the working curves such as internal true boiling point, molecular weight, density and viscosity curves are generated. Then, a set of hypothetical pseudo-components representing each crude oil is created from the working curves. By using the oil product cut option, different types of crude oil are blended in a designated ratio in the oil environment of Aspen HYSYS. Once the blending is completed, the crude oil is installed in the simulation environment and the rigorous distillation model of the crude distillation unit is built and discussed in Section 2.1.2.

2.1.2. Simulation Development. The crude distillation unit is simulated based on the operating parameters and column specifications listed in Aspen HYSYS version 10 [17], which has been validated. Peng-robinson equation-of-state is selected to simulate the simulation model as it is the most suitable thermodynamic fluid package to predict the volumetric properties and phase behaviour of the selected crude oil. The simulation model of crude distillation unit included preheating trains, atmospheric distillation unit and VDU, as shown in figure 1. The atmospheric distillation unit consisted of 50 trays, 1 total condenser, 3 side strippers and 3 pump-arounds. The products of the atmospheric distillation unit are naphtha, kerosene, diesel, atmospheric gas oil and atmospheric residue. The atmospheric residue of the atmospheric distillation unit is fed into VDU which consists of 14 trays and 2 pump-arounds. After that, VDU allowed fractionation of the atmospheric residue into off-gas, light vacuum gas oil, heavy vacuum gas oil, slop wax and vacuum residue under vacuum pressure.

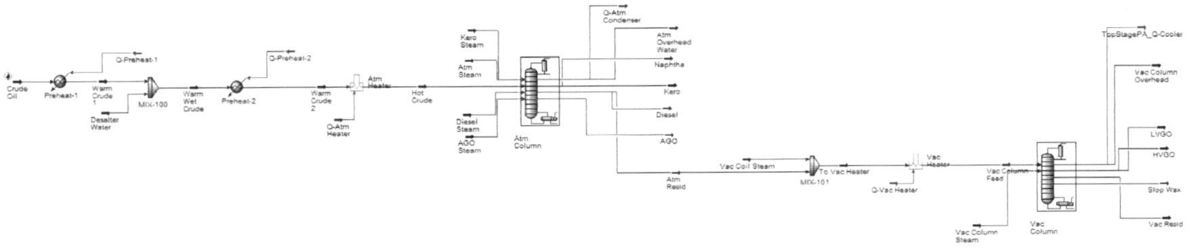

Figure 1. Simulation model of crude distillation unit.

It is worth noting that the number of stages for the distillation columns was determined using a series of procedures. The procedures consist of various levels of simulation rigour. Firstly, a component splitter was used to determine the removal ratios and operating conditions required to achieve the required yield. With the removal ratios and operating conditions, a shortcut distillation simulation was performed to determine the number of stages required to achieve the said yield. The number of stages here is then used as input data for the rigorous simulation model to establish the baseline performance of the distillation column. This baseline simulation model is later used to develop sampling points for the regression analysis. Thus, it is assumed that the number of stages in the column is fixed.

2.2. *Surrogate Modelling*

In the second stage, the surrogate modelling technique is employed to represent the simulation environment of the vacuum distillation unit (VDU). The surrogate model is built around the data generated from the simulation model with a random selection of input variables and their corresponding bound limits. The data is then regressed and converted into equations that can be used to represent the operation of the distillation system. One advantage of the surrogate model is that it allows a rapid calculation of the model output responses by applying equations to relate the input variables to the output responses. There are two types of surrogate models that have been reported in many publications such as response surface method and artificial neural network. As compared to the response surface method, an artificial neural network in the distillation column is more complicated, consists of more intensive steps and requires more investigation for the data regression [8]. On the other hand, the major advantage of the response surface method is that it requires very short computational time to study the relationship between the factors. The behaviour between the output responses and the important influencing variables can be captured easily using regression analysis [18]. Besides, the response surface method can produce a mathematical expression based on the data supplied and this mathematical expression essentially helps to represent the performance of the system mathematically. The mathematical expression from RSM is allowed to be integrated into a mathematical optimisation model to enhance and optimise a process or a system. It can be used in a mathematical model to obtain optimal responses with minimum variance and a small number of controlled parameters. These combinations allow the decision-makers to make more informed decisions about the operation of the system. Hence, the response surface method is selected to be used in this work. A general first-order polynomial function of a response surface equation is expressed as the following form:

$$Y = \mathrm{c} + \sum_{n=1}^{N} \beta_{\mathrm{n}} X_n \tag{1}$$

where Y represents the dependent variables, X represents the independent variables, c is the intercept regression coefficient, β is the linear term regression coefficient and N is the number of factors.

Due to the complexity and non-linearity of VDU operations, the surrogate modelling technique is applied in this work to represent the relationship between the input and output responses. The surrogate models of VDU are built based on the aforementioned response surface method. Before the construction of VDU surrogate models, the dependent variables and independent variables are determined. In order to achieve the objectives of this research study, 4 dependent variables are considered such as HVGO yield, HVGO pump-around cooler utility flowrate, LVGO pump-around cooler utility flowrate and column diameter. Meanwhile, crude oil blending ratio and operating conditions of VDU are selected as the independent variables for all the dependent variables. In order to determine the coefficient values that make up the response surface equations above, a large number of sample points are generated from the simulation model developed in the previous stage. The number of sample points is determined based on the amount of data available for the equipment considered, the amount of data available after data pre-processing and the quality of the regressions generated using the available data. In this work, the sample points are generated by varying the crude oil blending ratio and operating conditions of VDU. Statistical analysis software (i.e., JMP, Python, etc.) is typically used to regress the generated sample points in response surface equations [8]. In this work, JMP version 15.0 is selected to regress the response surface equations because it can easily assess the data from various sources which allows user to build their model rapidly. Besides, JMP can link the statistical data to interactive graphics which helps the user to explore and visualise their data better. After the regression analysis, validation analysis is carried out to evaluate the accuracy of surrogate models. An evaluation of R-square, P-value and leverage plots are considered. R-square of the equations must be greater than 0.90 to indicate a good fit of data, while the P-value of each variable must be smaller than 0.05 to show a significant impact on the dependent variable. The impacts of each independent variable on the dependent variables are further identified using the leverage plots. By performing the validation analysis, those independent variables that have a weak impact on the dependent variables are excluded from the development of the surrogate models. Finally, the surrogate models are built and implemented in the optimisation model, as discussed in Section 2.3.

2.3. Mathematical Optimisation
Large numbers of mathematical programming tools are available for the development of the distillation optimisation model. In this work, LINGO version 18.0 is used to develop an optimisation model to maximise the HVGO yield and minimise the TAC of the VDU respectively. The optimisation model is developed based on the response surface equations of the surrogate models, as discussed in Section 2.2.

3. Case Study
In this section, a case study is presented to illustrate the methodology that is discussed in Section 2. The case study aims to determine the optimal crude blending composition with their respective operating conditions that give maximum heavy vacuum gas oil (HVGO) yield and minimum total annualised cost (TAC) respectively.

The surrogate models are built to represent the relationship between the independent variables and dependent variables of the vacuum distillation unit (VDU). As discussed in the methodology section, HVGO yield, HVGO pump-around cooler utility flowrate, light vacuum gas oil (LVGO) pump-around cooler utility flowrate and column diameter are chosen as the dependent variables of the surrogate models. Besides, the development of the VDU surrogate model considered the crude oil blending ratio and operating conditions as the independent variables. Three different types of crude oil blending are conducted based on different crude blending ratios of light crude, medium crude and heavy crude. The bound limits of each crude oil ratio are given in table 1.

Table 1. Bounds of each crude ratio.

Crude Blending Ratio	Lower	Upper
Light crude ratio	0.2	0.8
Medium crude ratio	0.2	0.8
Heavy crude ratio	0.2	0.8

The operating conditions of the case study are varied based on the crude blending ratio. 7 input variables, related to the operating conditions are selected. Those variables are furnace outlet temperature, flash zone temperature, column top pressure, column bottom pressure, stripping steam flowrate, HVGO pump-around flowrate and LVGO pump-around flowrate. Each variable with the respective lower and upper bound limits is summarised in table 2.

Table 2. Bounds of each operating condition.

Operating Conditions	Lower	Upper	References
Furnace outlet temperature (°F)	720	780	[19]
Flash zone temperature (°F)	650	750	[19]
Top pressure (mmHg)	40	60	[20]
Bottom pressure (mmHg)	65	85	[19]
Stripping steam flowrate (lb/hr)	15,000	30,000	[21]
HVGO pump-around flowrate (bbl/day)	10,000	45,000	-
LVGO pump-around flowrate (bbl/day)	10,000	35,000	-

The lower and upper limits of each operating condition are determined by performing a literature review. Note that the lower and upper limits of HVGO and LVGO pump-around flowrate are not reported in the literature review. Hence, case studies are conducted in the simulation model to fine-tune the limits of HVGO and LVGO pump-around flowrate. Besides, the process constraints are also determined, as given in table 3.

Table 3. Process constraints.

Constraints	Specifications
Feedstock (bbl/day)	99,000
Column Pressure Drop (mmHg)	25
LVGO ASTM D86 95 % recovery (°F)	915
HVGO ASTM D86 95 % recovery (°F)	1,050

The total crude blending flowrate that fed into the distillation model is held constant at the value specified in the Aspen HYSYS listed conditions. On the other hand, the pressure drop of VDU is fixed at the maximum value of 25 mmHg, suggested by Jones [19]. In order to meet the composition specification of the products, ASTM D86 95 % cut points for both HVGO and LVGO are set as constant. Consequently, the product flowrate of VDU is allowed to be varied.

After determining the independent variables and process constraints, the surrogate models for each independent variable are constructed. 654 sets of sampling points are generated from the Aspen HYSYS simulation model for the development of VDU surrogate models. For each crude oil blending, 4 response surface equations are generated based on the above-mentioned dependent variables. In the case of heavy-light crude, the sampling points are generated based on the different ratios of the heavy-light crude blends with their respective operating conditions. Then, JMP is used to regress the corresponding data in the response surface equations for the heavy-light crude blend. The following discussions use one of the dependent variables, HVGO yield as an example. The regression results of the HVGO yield in the heavy-light crude blend surrogate model are demonstrated in figure 2, figure 3 and table 4.

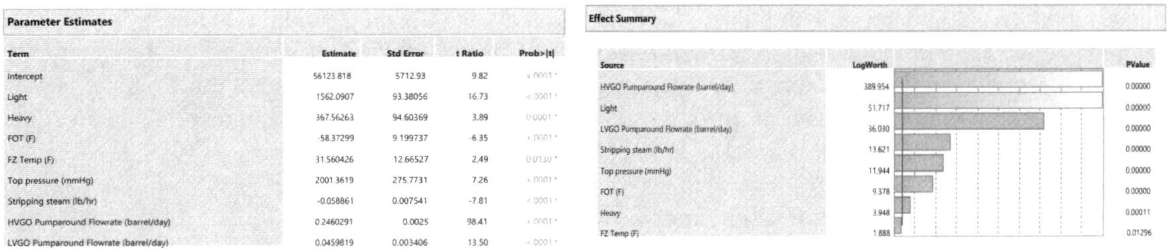

Parameter Estimates

Term	Estimate	Std Error	t Ratio	Prob>|t|
Intercept	56123.818	5712.93	9.82	<.0001*
Light	1562.0907	93.38056	16.73	<.0001*
Heavy	367.56263	94.60369	3.89	0.0001*
FOT (F)	-58.37299	9.199737	-6.35	<.0001*
FZ Temp (F)	31.560426	12.66527	2.49	0.0130*
Top pressure (mmHg)	2001.3619	275.7731	7.26	<.0001*
Stripping steam (lb/hr)	-0.058861	0.007541	-7.81	<.0001*
HVGO Pumparound Flowrate (barrel/day)	0.2460291	0.0025	98.41	<.0001*
LVGO Pumparound Flowrate (barrel/day)	0.0459819	0.003406	13.50	<.0001*

Effect Summary

Source	LogWorth	PValue
HVGO Pumparound Flowrate (barrel/day)	389.954	0.00000
Light	51.717	0.00000
LVGO Pumparound Flowrate (barrel/day)	36.030	0.00000
Stripping steam (lb/hr)	13.621	0.00000
Top pressure (mmHg)	11.944	0.00000
FOT (F)	9.378	0.00000
Heavy	3.948	0.00011
FZ Temp (F)	1.888	0.01296

Figure 2. Regression results of HVGO yield for the heavy-light crude blend that is taken from JMP.

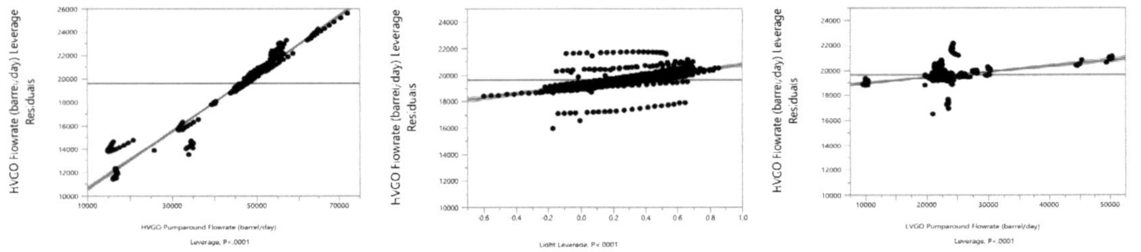

Figure 3. Leverage plots of HVGO yield for the heavy-light crude blend that is taken from JMP.

Table 4. Summary of fit of HVGO yield for the heavy-light crude that is taken from JMP.

Summary of Fit	Value
R-squared	0.9573
Adjusted R-squared	0.9568
Root Mean Square Error	636.5115
Mean of Response	19,630.41
Observations	654

The regression coefficients of the crude blending ratio and 7 operating variables that are calculated by JMP version 15 are given in figure 2. As observed, the HVGO pump-around flowrate has influenced the HVGO yield the most as HVGO pump-around flowrate has the highest log worth value. Log worth value is defined as $-\log_{10}(\text{p-value})$. The lower the P-value, the higher the log worth value. This is then followed by the light crude ratio, LVGO pump-around flowrate and so on. On the other hand, it can be observed that heavy crude ratio and flash zone temperature have the least but still significant impacts on the HVGO yield. Apart from that, the leverage plots can help to further investigate the effect of the independent variable on the HVGO yield. The leverage plots of the three independent variables that have the most significant impacts on HVGO yield are illustrated in figure 3. The effect is significant at the 5 % level when the confidence curve (red coloured line) crosses the hypothesis line (blue coloured horizontal line). In contrast, if the confidence curve does not cross the hypothesis line, the effect is considered not significant. With that, the leverage plots suggested that these three independent variables have a significant impact on the HVGO yield. To further validate the model, the R-square of the model must be higher than 0.9. As observed from table 4, the R-square of the HVGO yield surrogate model is 0.957 and this has confirmed that the model has high accuracy. Hence, all the selected independent variables are included in the development of the surrogate model. The constructed surrogate models are then integrated into the optimisation model.

An optimisation model of VDU is formulated using LINGO version 18.0 to determine the maximum HVGO yield and minimum TAC respectively. The optimisation work of the VDU has led to mixed-integer non-linear programming containing linear, non-linear and integer variables as well as constraints. The model had a total of 169 variables, 12 non-linear variables, 3 integer variables, 196 constraints and 12 non-linear constraints. Based on the previous surrogate model development, 4 response surface equations of the surrogate model are generated for heavy-light crude blend, heavy medium crude blend and heavy-medium-light crude blend, respectively. The HVGO yield surrogate model is integrated into the optimisation model to address the objective function of maximising HVGO yield. Meanwhile, the other 3 surrogate models are incorporated in the optimisation model to address the objective function of minimising TAC. In order to achieve the objective functions, all the surrogate model equations in three crude blendings are incorporated in the optimisation model. In this model, two stages of optimisation procedures are being considered, covering (a) maximise HVGO yield under varying the crude oil blending ratio and operating conditions, (b) minimise TAC under varying the crude oil blending ratio and operating conditions.

4. Results and Discussions

In this section, the optimal solution of maximising heavy vacuum gas oil (HVGO) yield and minimising total annualised cost (TAC) are presented, as given in table 5.

Table 5. Optimised results of maximising HVGO yield and minimising TAC.

Objective Function	Maximise HVGO yield	Minimise TAC
Selected Crude Blending	Heavy - Light Crude	Heavy - Light Crude
HVGO Yield (bbl/day)	23,788	14,028
TAC ($ MM)	1.75	1.70
Light crude fraction	0.8	0.8
Medium crude fraction	0	0
Heavy crude fraction	0.2	0.2
Furnace outlet temperature (°F)	720	720
Flash zone temperature (°F)	750	750
Top pressure (mmHg)	-13.54	-13.54
Bottom pressure (mmHg)	-13.05	-13.05
Stripping steam flowrate (lb/hr)	15,000	15,000
HVGO pump-around flowrate (bbl/day)	45,000	10,000
LVGO pump-around flowrate (bbl/day)	35,000	10,000

The purpose of these two objectives is to provide a comparative analysis of the resulting configurations. Based on the results presented above, the crude blending has considered a light crude ratio of 0.8 and a heavy crude ratio of 0.2 in both objectives. The light crude contains more distillate and it is easier to refine as compared to medium and heavy crude. It can be proven that the heavy-light crude achieves maximum HVGO yield and minimum TAC as compared to the heavy-medium crude blend and heavy-medium-light crude blend. Moreover, other decision variables of VDU such as the operating conditions have significant impacts on the HVGO yield and TAC as well. Maximum HVGO yield is achieved by operating the VDU at the recommended operating conditions, as shown in table 5. On the other hand, it is important to note that only the HVGO pump-around flowrate and LVGO pump-around flowrate are optimised and stayed at the lowest bound of the variable in order to minimise the TAC of the VDU. Hence, it can be concluded that the optimisation results offer the possibilities of the different configurations based on the objective functions. The results can be used to provide insights into the optimal operating conditions of the VDU. By doing so, it can help the refiners to develop an effective strategic planning application for the VDU operation [22]. The developed insight can also help to overcome the current limitations such as tighter crude demand present in the refineries. Apart from that, the optimisation model can be used to perform computational experiments which can help to avoid costly mistakes in the building pilot-scale plant. In terms of flexibility, the proposed surrogate-assisted mathematical optimisation model can be extended to other equipment or industries. This can be done by incorporating the proposed methodology to develop the surrogate models based on the statistical data of the equipment or industry. Then, the next thing is to develop a new optimisation model based on the surrogate models of the equipment or industry.

5. Conclusions and Future Works

In this work, a surrogate-assisted methodology is developed to capture the relationship between input and output responses of the vacuum distillation unit (VDU), and thus predicted the complex behaviour of the VDU. The constructed surrogate models are then integrated into the optimisation model to obtain an optimal solution for maximising heavy vacuum gas oil (HVGO) yield and minimising total annualised cost (TAC) respectively. The proposed methodology is implemented in a case study where different crude oil blending ratios are compared and analysed. As a result, the heavy-light crude blend provided a higher HVGO yield and lower TAC as compared to other types of crude oil blending. As concluded, the proposed model provided the optimal operating condition of VDU which can be used to provide insight for the decision-makers upon deciding a preferred solution. With this insight, the refiners can easily develop strategic planning procedures to tackle the issues present in the real VDU operation in the refinery.

Future work will focus on the development of a multi-objective optimisation model to optimise the HVGO yield and TAC of VDU simultaneously. This is because these two objectives are contradicting in nature, which means that improvement in HVGO yield may bring a negative impact on TAC or vice versa. Therefore, a multi-objective optimisation model can be used to determine the trade-off between these two objectives. Additionally, further studies will be carried out to analyse the impact of the number of stages for the distillation columns on the HVGO yield and TAC. Lastly, future work will also focus on using the real-time data from the refinery to improve the accuracy of the proposed model.

Acknowledgments

This work has supported by the School of Engineering and Physical Sciences at Heriot-Watt University Malaysia.

References

[1] Martin G R and Nigg J M 2001 Vacuum pressure control : Impact on profitability *Pet. Technol. Q.* **6** 73–81

[2] Pintarič Z N and Kravanja Z 2006 Selection of the economic objective function for the optimisation of process flow sheets *Ind. Eng. Chem. Res.* **45** 4222–32

[3] Seo J W, Oh M and Lee T H 2000 Design optimisation of crude oil distillation *Chem. Eng. Technol.* **23** 157–64

[4] Basak K, Abhilash K S, Ganguly S and Saraf D N 2002 On-line optimisation of a crude distillation unit with constraints on product properties *Ind. Eng. Chem. Res.* **41** 1557–68

[5] Gu W, Wang K, Huang Y, Zhang B, Chen Q and Hui C W 2015 Energy optimisation for a multistage crude oil distillation process *Chem. Eng. Technol.* **38** 1243–53

[6] Inamdar S V, Gupta S K and Saraf D N 2004 Multi-objective optimisation of an industrial crude distillation unit using the elitish non-dominated sorting genetic algorithm *Chem. Eng. Res. Des.* **82** 611–23

[7] Huang X, Zhao T, Li N, Ma Z, Song L and Li J 2017 Emission-reductive and multi-objective coordinative optimisation of binary feed for atmospheric and vacuum distillation unit *China Pet. Process Petrochemical Technol.* **19** 101–12

[8] Loper M L 2015 *Modeling and Simulation in the Systems Engineering Life Cycle: Core Concepts and Accompanying Lectures* (Springer)

[9] Denimal E, Nechak L, Sinou J and Nacivet S 2016 Kriging surrogate models for predicting the complex eigenvalues of mechanical systems subjected to friction-induced vibration *Shock Vib.* **2016** 1–22

[10] López D C, Hoyos L J, Mahecha C A, Arellano-Garcia H and Wozny G 2013 Optimisation model of crude oil distillation units for optimal crude oil blending and operating conditions *Ind. Eng. Chem. Res.* **52** 12993–3005

[11] Yao H and Chu J 2012 Operational optimisation of a simulated atmospheric distillation column using support vector regression models and information analysis *Chem. Eng. Res. Des.* **90** 2247–61

[12] Ochoa-estopier L M, Jobson M and Smith R 2013 Operational optimisation of crude oil distillation systems using artificial neural networks *Comput. Chem. Eng.* **59** 178–85

[13] Gut J A W, Franzoi R E, Menezes B C, Kelly J D and Grossmann I E 2020 Cutpoint temperature surrogate modeling for distillation yields and properties *Ind. Eng. Chem. Res.* **59** 18616–28

[14] Mittal V, Zhang J, Yang X and Xu Q 2011 E3 Analysis for crude and vacuum distillation system *Chem. Eng. Technol.* **34** 1854–63

[15] Gu W, Huang Y, Wang K, Zhang B, Chen Q and Hui C W 2014 Comparative analysis and evaluation of three crude oil vacuum distillation processes for process selection *Energy* **76** 559–71

[16] Li X, Cui C and Sun J 2018 Enhanced product quality in lubricant type vacuum distillation unit by implementing dividing wall column *Chem. Eng. Process Process Intensif.* **123** 1–11

[17] Aspen HYSYS 2017 *Crude Tower Simulation - HYSYS v10* MA Aspen Technology Inc

[18] Aydar A Y 2018 Utilisation of response surface methodology in optimisation of extraction plant materials *Stat. Approaches With Emphas. Des. Exp. Appl. to Chem. Process.* 157–69

[19] Jones D S J 2015 *Handbook of Petroleum Processing vol. 1* (Springer Science & Business Media)

[20] Lousdad S 2020 What is vacuum distillation 1–5. https://www.arab-oil-naturalgas.com/what-is-vaccum-distillation/ (accessed December 20, 2020)

[21] More R K, Bulasara V K, Uppaluri R and Banjara V R 2010 Optimisation of crude distillation system using Aspen Plus: Effect of binary feed selection on grass-root design *Chem. Eng. Res. Des.* **88** 121–34

[22] Geoffrion A M 1976 The purpose of mathematical programming is insight, not numbers *Interfaces (Providence)* **7** 81–92

Functionalization of polylactic acid thin films via polydopamine-assisted chelation of copper (II) ions for antibacterial applications

R B Leron[1], L L Tayo[1] and R R Aquino[2]

[1]School of Chemical, Biological, and Materials Engineering and Sciences, Mapua University, Intramuros, Manila 1002, Philippines
[2]General Education Department, Colegio de Muntinlupa, Sucat Muntinlupa City 1770, Metro Manila, Philippines

Email: rbleron@mapua.edu.ph

Abstract. Polylactic acid (PLA) is a biodegradable polymer, which has been widely investigated for use in biomedical and packaging applications due to its excellent biodegradability, biocompatibility, non-toxicity, low cost, good stability, and thermal processability. In this work, PLA was functionalized to improve the membrane's hydrophilicity and impart antimicrobial activity by simultaneously depositing polydopamine (PDA) and chelating Cu^{2+} metal ions on the membrane surface. Pristine PLA films were modified via one-pot dip coating method using dopamine-copper (II) solution at different coating times (6, 12, 24 h). FTIR analysis confirmed the deposition of PDA on the modified membranes (PLA/PDA/Cu) as indicated by the presence of catechol and amine moieties on the samples. TGA results revealed the degradation of the same functional groups on PLA/PDA/Cu. The hydropilicity of PLA was significantly reduced upon coating with PDA as indicated by the decrease in the membrane's contact angle from 96.5 ± 5.3° to 56.2 ± 4.7°. SEM images and EDS results clearly showed that copper particles were deposited on the PLA/PDA/Cu membranes (atomic % ~ up to 0.88) and coating with PDA did not alter the porous structure of the pristine PLA film. Results also demonstrated that the concentration of copper immobilized on the modified membranes increased with longer coating; thus, offering a way of tailoring the metal concentration on the membrane for its specific use. PLA/PDA/Cu membranes showed antibacterial property against the *B. subtilis*, which could be attributed to the chelation of Cu^{2+} ions with the catechol moiety of the PDA coating.

1. Introduction

One of the best-known biopolymers is polylactic acid (PLA), a biodegradable aliphatic polyester that can be derived from lactic acid and can be produced from fermentation of renewable sources. It has been widely investigated for practical medical applications—scaffolds in tissue engineering [1], surgical sutures and dressing materials [2], and wound healing dermal patch [3]—and as green alternative for food packaging applications [4]. Such considerable interest has been given to PLA and its copolymers because of their excellent biocompatibility and biodegradability. The biopolymer degrades, by random hydrolysis of its ester bonds, to form lactic acid or its naturally occurring metabolites that are normally present in the human body. In addition, it exhibits non-toxicity, low cost, stability, and thermal processability [5]. But despite its many advantages, PLA also exhibit several limitations including high hydropobicity and lack of antibacterial activity; hence, it requires some modifications to increase its desirability and further its use in certain applications.

One of the promising approaches to modify and impart specific functions to polymeric materials including PLA, is the use of mussel-inspired polydopamine (PDA) coating [1,6,7]. Marine mussels are well known to adhere on different substrates under moist conditions. The adhesion is due to *Mytilusedlis* foot protein 5 (Mefp-5), which has the reactive catechol-containing compound 3,4-dihydroxyphenyl-L-alanine (DOPA) and lysine, distributed at the interface between the protein and the substrate. The ortho-dihydroxyphenyl group of the catechol is responsible for the excellent adhesive property of the organism to a wide variety of organic and inorganic surfaces [8]. Lee and co-workers [9] identified dopamine (DA)—a small-molecule compound that contains both catechol (DOPA) and amine (lysine) functional groups—as a simple structural mimic of Mefp-5, and showed that such is a powerful building block for the spontaneous deposition of thin polymer films on virtually any bulk material surface and that the films are adapted for a wide variety of functional uses. Inspired by this phenomenon, they introduced a simple method to form multifunctional polymer coatings by simple immersion of a material in an aqueous solution of DA under alkaline conditions where the latter undergo self-polymerization to form PDA. The most interesting property of the PDA coating is its adhesive function such that it can strongly adhere to a variety of materials including metals, oxides, semiconductors, polymers, and ceramics [9]. It also promotes cell adhesion and immobilization of bioactive molecules and is amenable to post-functionalization via catechol chemistry [8]. Another advantage of PDA is it is non-toxic and produces no harmful by-products during mild formation process. Nonetheless, despite the excellent adhesive property and biocompatibility of PDA, it has been reported that its antibacterial effect—an important requirement for medical biopolymers—is only moderate [10–12].

In this work, Cu (II) metal ions were used to impart antibacterial function to PLA films with the use of PDA surface coating. Metal nanoparticles are well known to have antibacterial properties and have been extensively studied for the development of antibacterial surface coatings. In particular, silver nanoparticles (AgNPs) have been found highly effective in imparting antibacterial function when utilized in titanium implants, medical devices, and organic polymers [13–16]. However, its use may be limited due to its high cost and relative toxicity. Another metal with excellent antibacterial property, yet less expensive than silver, is copper. It has been reported copper has less toxicity compared to other heavy metals, including silver [16,17]. Several studies suggest that copper is a promising metal ion for antibacterial application not only because of its lower toxicity but also its higher cytocompatibility. Xing *et al.* [18] successfully constructed an antimicrobial polyethersulfone (PES) membrane via PDA-assisted immobilization of copper ions on the polymer surface. They showed that the prepared PES-PDA-Cu membranes exhibited highly effective antibacterial activity with a relatively safe release amount of copper ions. Also, it has been demonstrated that upon incorporation in titania (TiO_2) coating, Cu^{2+} would have the best compromise between antibacterial effectiveness and cytotoxicity as compared to Ag^+, Hg^{2+}, Al^{3+}, Co^{2+}, and Zn^{2+} ions [19]. In addition, copper is naturally present in the body and it undergoes metabolization, while silver does not which results in increased body silver serum level [17,20]. Although copper is a well-recognized antibacterial metal, its use as an antibacterial agent on biodegradable PLA systems has not yet been fully investigated to date. Thus, this study aimed to modify the surface of PLA thin film by coating with PDA and impart antibacterial function through deposition of Cu (II) metal ions on the polymer surface. It also provided an extensive characterization of the functionalized PLA/PDA/Cu membrane composite.

2. Experimental Section

2.1. Materials

Polylactic acid (PLA polymer 6202D, pellets, MW = 140 kDa; 98 % L-Lactide, 2 % D-Lactide) was obtained from Natureworks. *N*-methyl-2-pyrollidone (NMP, ≥ 99.8 %) was supplied by RCI Labscan Ltd. while dopamine hydrochloride (3-hydroxytyramine hydrochloride, 98 %) and tris(2-amino-2-

hydroxymethyl-propane-1,3-diol) (Tris-HCl, 99 %) were purchased from Sigma Aldrich (USA). Copper sulfate pentahydrate and hydrochloric acid was obtained from Merck KGaA (Germany). All chemicals were used as received. Type 1 deionized water was used in the preparation of solutions.

2.2. Method

Polylactic acid (PLA) films were prepared via liquid induced phase separation (LIPS) technique. First, PLA polymer solution was prepared by dissolving a specific amount of PLA (13 wt. %) pellets in NMP while stirring at 60 °C for at least 6 hours. The solution was then cast on a glass plate using a casting knife, exposed to air for 10 s, and immersed into a nonsolvent (water) bath at room temperature. Both plate and knife were heated in an oven at 60 °C before use. The membrane was allowed to form, then soaked into and rinsed thoroughly with deionized water several times to remove the remaining solvent (NMP). Finally, the membrane was dried in air and stored at room temperature.

The surface of prepared PLA film was modified using a one-pot dip coating procedure, according to method proposed by Liu *et al.* [21]. Copper sulfate solution (100mM) was prepared by dissolving $CuSO_4{\cdot}5H_2O$ in Tris-HCl buffer solution (50 mM, pH = 8.5). The pH of the solution was further adjusted to 4.5 by adding a small amount of 1 M HCl. The pH was measured using a pH meter (Sartorius PB-10). The coating solution (dopamine-Cu^{2+}) was then prepared by adding 100 mg of dopamine HCl to 50 mL (2 mg dopamine/mL) of the prepared $CuSO_4$ solutions. The PLA membrane (2 × 3 in) was soaked in the freshly prepared dopamine-Cu^{2+} solution in a plastic pan with mild heating at 30 °C using an incubator shaker (Fine PCR BAE07-H300) for 6 h. The solution was kept agitated at 50 rpm to minimize aggregation of PDA and Cu^{2+} particles on the surface of the membrane. After which, the modified PLA membrane was gently rinsed with deionized water several times to wash loosely bounded particles from the membrane surface before drying in air. The dried membrane composite (PLA-/PDA/Cu) was stored at room temperature until further characterization. The experiment was done at varying soaking times (6, 12 and 24 h) to investigate the effect of coating time on the deposition of PDA and Cu^{2+} on the PLA membrane surface. The samples were identified as PLA-/PDA/Cu-6h, PLA-/PDA/Cu-12h, and PLA-/PDA/Cu-24h, respectively.

A Fourier transform infrared spectroscopy (FTIR, Perkin Elmer) was performed in attenuated total reflection (ATR) mode, in the wavenumber range 4000 – 400 cm^{-1}, to confirm the deposition of PDA on PLA membrane surface. The thermogravimetric analysis (TGA) was done using a simultaneous thermogravimetric analyser (DSC/TGA, Hitachi NEXTA STA 200RV) using nitrogen as purge gas. The sample (approximately 5 mg) was heated from 25 to 500 °C at a heating rate of 10 °C/min. Furthermore, the hydrophilicity of PLA and modified membrane samples was determined using an optical tensiometer (Theta Lite). The morphology of each membrane sample was characterized using a scanning electron microscope (SEM, Hitachi SU3500) using an accelerating voltage of 15 kV. The equipment was coupled with an energy dispersive X-ray spectroscope (EDS, Bruker X Flash 6/60), which was used to determine the concentration and distribution of copper on the surface of the membrane by area and spot scanning.

Antibacterial function of the prepared PLA and modified membrane samples was tested against the gram-positive bacteria *Bacillus subtilis* using the disc diffusion method. The bacteria were spread in a sterilized agar plate using metal loop spreader, then the membrane samples were placed atop of it. One-inch discs of membrane samples, which were cut into quarters, were used in the experiment. The plate was then left to incubate in an incubator oven at 37 °C for 24 h with monitoring at (3, 6, 12, 24) h. Tests were done for samples coated in different concentrations of dopamine-$CuSO_4$ solutions (50, 100, 150 mM).

3. Results and Discussion

3.1. FTIR Analysis

FTIR spectra were determined for pristine PLA and PLA/PDA/Cu films to key out the functional groups in the fabricated polymer and confirm the deposition of polydopamine on the PLA surface.

FTIR results for pristine PLA and PLA/PDA/Cu composites are shown in figure 1. Characteristic peaks at 2995 and 2946 cm^{-1} were observed in all samples, which belong to the symmetric and asymmetric C–H stretching of $-CH_3$ in PLA, respectively. PLA also showed stretching frequencies for C=O at 1751 cm^{-1} and for C–O at 1182 and 1084 cm^{-1}. Bands identified at 1452 and 1359 cm^{-1} can be assigned to the bending of $-CH_3$ groups in PLA. Such observations exhibit close similarity with those reported in the literature [22, 23]. The identified peaks in pristine PLA were also present in the spectra of PLA/PDA/Cu composites. For the said composites, broad peaks were observed at 3198 cm^{-1} and 3228 cm^{-1} for PLA/PDA/Cu-6h and PLA/PDA/Cu-24h, respectively. These absorbances correspond to both the catechol O–H and N–H stretching vibrations present in the PDA layer. Furthermore, some new absorption peaks were observed at 1617 and 1509 cm^{-1}. The peaks at 1617 cm^{-1} can be attributed to the overlapping of the C=C stretching vibration in the aromatic ring and N–H bending vibration in PDA. These can be identified with the FTIR spectra of PDA reported by Zhao *et al.* [24] as shown in figure 2a. The slight shift in the bending bands from 1623 (for PDA) to 1617 cm^{-1} in the case of the PLA/PDA/Cu composites can be due to the molecular interactions of the amine group of dopamine on the surface of the PLA. On the other hand, the peaks at 1509 cm^{-1} can be ascribed to the N–H shearing vibration of PDA. Such results are consistent with those reported by Liu *et al.* [21] and provide strong evidences for the successful coating of PDA on the PLA substrate.

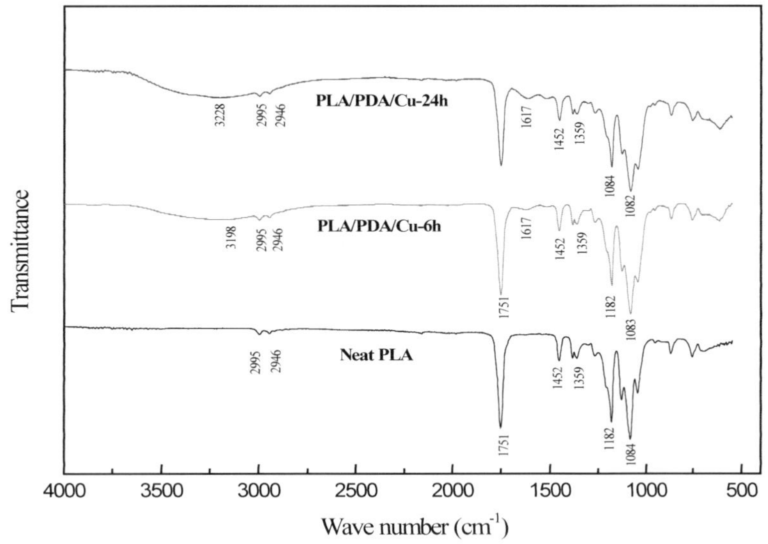

Figure 1. FTIR spectra of (a) pristine PLA; (b) PLA/PDA/Cu-6h; (c) PLA/PDA/Cu-24h.

It can be hypothesized that PDA adsorbed on PLA since catechol—one of the main functional groups in PDA—interact effectively with almost any kind of surface. In this study, the binding of PDA on PLA can be attributed partially to hydrogen bonding between the hydroxyl and/or amine groups in the catecholamine and the polar carbonyl and hydroxyl groups in the PLA substrate. Such nature of interaction was also reported for some organic and polymeric substrates bearing polar groups including poly(allylamine), polyethylene glycol (PEG) and polyethylenimine (PEI). However, it has to be noted that the contributions from covalent bonding cannot be completely discounted [25, 26]. Zhao *et al.* [24] reported similar observations for PDA deposited on PLA pellets where N–H/O–H stretching bands were identified between 3680 and 3050 cm^{-1} (figure 2b). It is also worth noting that the spectra of PLA/PDA/Cu-6h and PLA/PDA/Cu-24h films were greatly similar. However, the intensity of the peak at 3223 cm^{-1} (PLA/PDA/Cu-24h) significantly increased compared with that at 3198 cm^{-1}

(PLA/PDA/Cu-6h) suggesting that the amount of polydopamine deposited on the PLA surface increased with longer coating time.

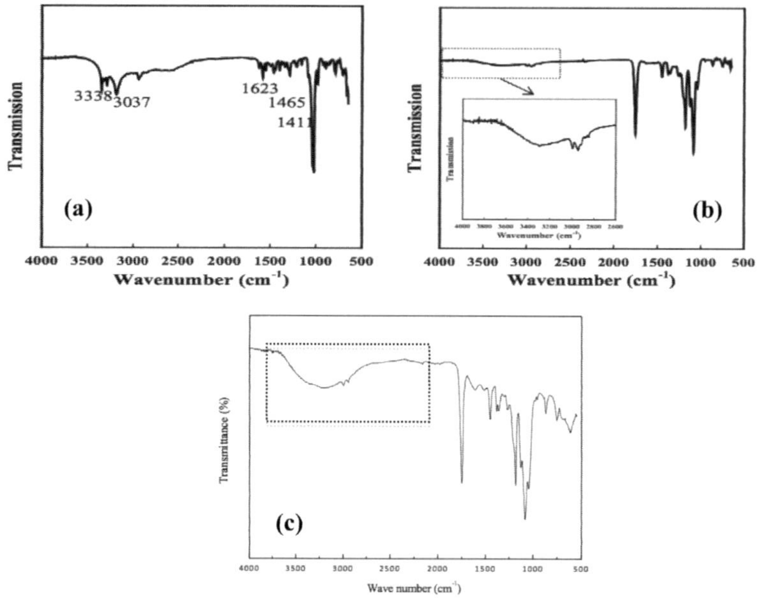

Figure 2. FTIR spectra of (a) PDA [24]; (b) PLA/PDA [24]; (c) PLA/PDA/Cu-24h (this work).

3.2. Contact Angle Measurements

To further validate the deposition of PDA on the substrate, static contact angle measurements were performed on the membrane surfaces. As shown in figure 3, the water contact angle of pristine PLA was measured to be 96.5 ± 5.3° indicating the inherent hydrophobicity of the membrane. After coating the PLA sample with PDA and copper for 24 h, the water contact angle decreased remarkably to 56.2 ± 4.7°. It is to be noted that the reported the contact angle of PDA on glass substrate was 27.8 ± 4.3° indicating that it is highly hydrophilic [24]. Hence, the relative hydrophilicity of PLA/PDA/Cu-24h membrane sample was a good indication that the PLA membrane surface was coated with PDA. Such improvement of the hydrophilicity of the membrane surface was expected because of the presence of hydrophilic hydroxyl and amino groups in the PDA molecule and would be advantageous in promoting cell adhesion [21].

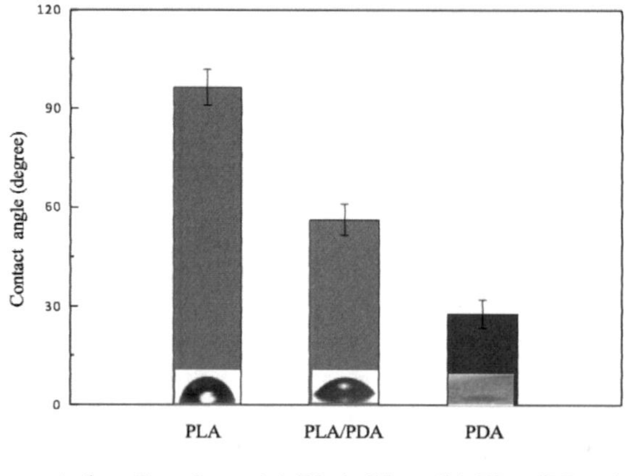

Figure 3. Water contact angle of surface: (a) PLA film; (b) PLA/PDA/Cu-24h; (c) PDA [24].

3.3. *TGA Analysis*
Figure 4 shows the TGA curves for pristine PLA and coated PLA/PDA/Cu-24h from this work and for pure DA obtained by De Guzman and co-workers [27]. PLA showed a one-step degradation process similar to that reported by Dai *et al.* [28] where the polymer started to degrade at 250 °C until 360 °C indicating the decomposition of oxygen-containing functional groups [29]. After heating the sample to about 500 °C, 1.37 % residual remained, which could be expected as PLA completely degrades above 900 °C [30]. De Guzman *et al.* [27] reported the decomposition of DA, starting at 180 °C to 220 °C, which was attributed to the evaporation of adsorb water in the monomer. Continuous decomposition occurred at 210 °C to 325 °C, which is due to the degradation of the catechol moiety while the amine functional groups decomposed from 325 °C to 425 °C (third decomposition). It can be observed that the thermal decomposition behaviour of the coated PLA/PDA/Cu-24h sample was similar with that of PLA. However, the initial degradation temperature of the coated sample was lower than that of the pristine polymer. As shown, the first observable weight loss of about 3.14 % occurred at 50 °C to 200 °C indicating the removal of adsorb water on the sample. Such result can be attributed to the low starting degradation temperature of PDA layer deposited on the PLA substrate. Further weight loss to up to 300 °C corresponded to the decomposition of the hydroxyl group in PDA. Above 300 °C, the simultaneous degradation of the amine moiety in PDA (at 300 °C to 335 °C) and oxygen-containing functional groups in PLA may have caused the shift in the degradation curve of PLA. It can also be noted that the coated PLA/PDA/Cu-24h exhibited a slower degradation compared to pristine PLA, which indicated the relative stability of the coated sample at higher temperatures. Such could be attributed to the presence of copper particles on the sample surface, which may have prevented the burning. This finding was further supported by the increased residual weight obtained for PLA/PDA/Cu-24h of approximately 5.90 %.

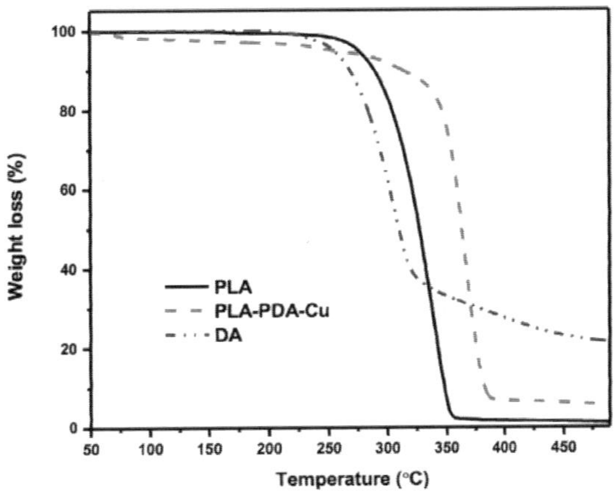

Figure 4. TGA curves for PLA and PLA/PDA/Cu-24h films (this work), and dopamine (DA) nanoparticle [27].

3.4. *Morphology*
Scanning electron microscopy (SEM) was used to observe the morphologies of the membrane samples before and after deposition of PDA and copper. Figure 5a shows the SEM image of pristine PLA. It was observed that the membrane was highly porous and characterized by a microporous structure with pore sizes ranging from 0.8 to 2.5 μm. Figure 5 also shows a comparison of the pristine PLA sample with the coated PLA/PDA/Cu composites. It is important to note that the even after deposition of PDA on the substrate's surface, the highly porous structure of PLA was preserved. This attribute is greatly advantageous in the utilization of the biopolymer in numerous applications. Moreover, the composites were characterized by the presence of microcracks on the sample surfaces, which could be identified as the PDA film coated on the substrate surface. This observation can be more clearly seen in figure 6

with the SEM images of PLA/PDA/Cu-6h and PLA/PDA/Cu-24h. The occurrence of such microsized cracks was also observed by Zhang *et al.* [31] on PDA films deposited on polydimethylsiloxane (PDMS), and was attributed to the difference in elastic modulus between the rigid PDA film and its soft substrate. Drying and storage of the membrane samples in air resulted in the formation of these microcracks. The cracking of PDA films on poly(L-lactide) and silicon oxide substrates were also reported in other studies [32, 33]. Furthermore, the microcracks found in PLA/PDA/Cu-24h were more prominent than those in PLA/PDA/Cu-6h sample indicating the formation of a thicker layer of PDA on the substrate that was coated for 24 h. This finding supports the results obtained from the FTIR analysis where the concentration of amine groups (or PDA) on the sample increased with longer coating time.

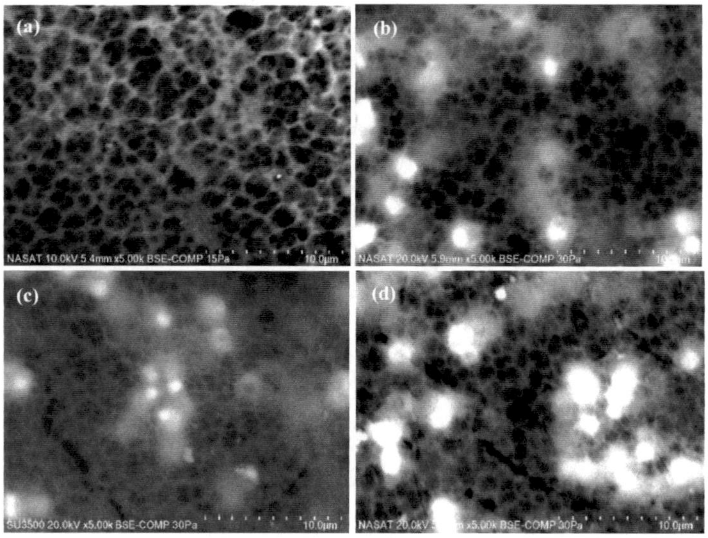

Figure 5. SEM images of (a) pristine PLA; (b) PLA/PDA/Cu-6h; (c) PLA/PDA/Cu-12h; (d) PLA/PDA/Cu-24h.

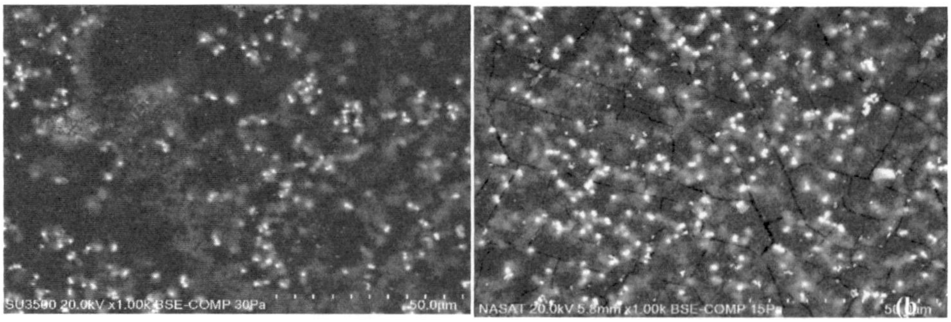

Figure 6. SEM images of (a) PLA/PDA/Cu-6h and (b) PLA/PDA/Cu-24h.

SEM images in figures 5 and 6 also display the presence of copper particles (light points) deposited on the surface of the coated samples (PLA/PDA/Cu-6h, PLA/PDA/Cu-12h and PLA/PDA/Cu-24h). Such finding was confirmed by an EDS point analysis on the particles, which revealed that they were mainly consisted of copper. According to Liu *et al.* [21], who characterized the valence states of copper present during the polymerization of dopamine in the presence of $CuSO_4$ in solution, copper ions were present as Cu^+ and Cu^{2+}. As it is well known that PDA binds with metals via chelation, it can be said that the immobilization of copper on the PDA coating can be attributed to the chelation of Cu^{2+}or Cu^{+} ions. In particular, copper ions are proposed to exhibit interaction with the catechol moiety of the catecholamine [21, 32, 34]. It could also be noted that the hydrophilicity of PDA may have played a role in the deposition of the metal ions. It is worth noting that copper played an important role in the formation of PDA layer on the coated membrane because the polymerization of dopamine was initiated by Cu^{2+} metal ion in the solution. Although dopamine is known to self-polymerize only under alkaline environment, it has been reported in recent works [21, 32] that Cu^{2+}, in the presence of air, initiates the polymerization of dopamine in acidic conditions (pH = 4.5). Such can be indicated by the rapid color change of the coating solution from clear to dark brown observed in this work. Moreover, formation of dopamine-Cu^{2+}-oxygen complex may have possibly enhanced the rate of polymerization of dopamine. Although the exact structure of PDA has not yet been fully elucidated, literature suggests that the configuration of PDA becomes crosslinked at the phenyl sites upon oxidation of dopamine [35]. Nonetheless, since the exact structure of PDA remains under debate further investigations would be necessary to fully understand and explain the mechanism of reaction of Cu^{2+} in the polymerization of dopamine. It is clearly shown in figure 6 that copper particles were distributed on sample surfaces where a more even distribution obtained in PLA/PDA/Cu-24h. The particle sizes ranged from 300 – 900 nm. However, it seemed that agglomeration of the particles during the coating process was inevitable that some form into bigger clusters (up to ~1.5 μm). Nevertheless, the results evidently show that the PLA substrate was successfully coated with a layer of PDA and copper.

3.5. EDS Analysis

Energy-dispersive X-ray spectroscopy (EDS) area scanning was done to estimate the distribution and atomic percentage of copper for different coating times and concentrations of $CuSO_4$ in the soaking solution. Table 1 shows the EDS analyses of PLA/PDA/Cu samples for different coating times. Results indicated that a significant amount of Cu element was present on the sample surface with increasing atomic percentage as coating time in the dopamine-Cu^{2+} solution increased. A similar observation can be seen for the atomic percentage of nitrogen element. Furthermore, a high percentage of elemental nitrogen gives a clear indication that the PDA was abundantly present on the coated membrane. These results suggest that the concentration of copper immobilized on the modified membrane may be controlled for specific applications by adjusting the coating time. Such will offer a way of tailoring the concentration of copper in the membrane to levels that would provide antibacterial effect and pose limited toxicity risks when leached. Although leaching of copper was not addressed in the present work, a comparison with results obtained by Xing *et al.* [18] for PES-PDA-Cu membrane revealed that given the amounts of copper immobilized in the studied membrane, possible leaching of copper would be below the maximum contaminant limit in the human body (below 1 ppm).

Table 1. Energy-dispersive X-ray spectroscopy (EDS) analysis of PLA/PDA/Cu samples for different coating times.

Time (h)	Atom (%)			
	C	O	N	Cu
6	41.00	50.50	7.96	0.53
12	41.42	49.48	8.36	0.75
24	39.68	49.79	9.68	0.86

3.6. Antibacterial Test

Pristine PLA and coated PLA/PDA/Cu membrane samples were subjected to antimicrobial test against the gram-negative bacteria *Bacillus subtilis* at different incubation times. The results are summarized in table 2. Results showed that bacterial growth started on the edges of pristine PLA as early as 3 h of incubation. This was the case for the coated PLA/PDA/Cu50 sample after 6 h incubation time, and for the modified PLA/PDA/Cu100 and PLA/PDA/Cu150 only after 12 h. More importantly, no visible bacterial growth was observed on the surface of all modified samples even after 24 h, while significant growth was seen covering the entire surface of the control sample (PLA) after 12 h. As indicated in table 2, for samples PLA/PDA/Cu50, PLA/PDA/Cu100 and PLA/PDA/Cu150, bacterial growth was absent in the entire duration of the test on the whole sample surface where PDA and Cu^{2+} ions were available. These results suggest that the coated samples exhibit antimicrobial protection against *B. subtilis*. The antimicrobial property of the studied PLA/PDA/Cu membrane composites may be attributed to the presence of copper ions, mainly Cu^{2+}, on the samples' surfaces. It is known that Cu^{2+} has strong antimicrobial activity, as it is capable of penetrating bacterial cell walls causing oxidation of the core cell, which eventually leads to death [17, 19, 36–37]. In this study, it was evident that copper was immobilized on the surface of all modified membrane samples, and their antimicrobial effect on a representative bacterial strain (*B. subtilis*) was successfully demonstrated. These results would suggest that the prepared PLA/PDA/Cu membrane could be effective in preventing bacterial infection.

Table 2. Antibacterial test results for coated PLA/PDA/Cu membrane samples *vs. Bacillus subtilis* at different incubation times.

Sample	Time (h)			
	3	6	12	24
PLA	+	++	+++	+++
PLA/PDA/Cu50	−	+	++	++
PLA/PDA/Cu100	−	−	++	++
PLA/PDA/Cu150	−	−	++	++

Legend:
− No visible growth along sample edges
\+ Visible growth along sample edges
++ Visible growth along sample edges but none on sample surface
+++ Visible growth along sample edges and surface

4. Conclusions

PLA films were successfully functionalized by simultaneous deposition of PDA membrane surface and chelation of Cu^{2+} metal ions on the PDA coating using a one-pot dipping method. FTIR and TGA results indicated that the PLA surface was coated with PDA as confirmed by the presence and degradation of −OH and −NH functionalities. The modified membrane was shown to be hydrophilic by contact angle measurements, while SEM-EDS analyses provided clear evidence of copper particles deposited on the membrane surface. The fabricated PLA/PDA/Cu membrane showed decent antibacterial property against the *B. subtilis* due to Cu^{2+} ions chelated on the membrane surface. Thus, the prepared membrane can be potentially be used for antimicrobial applications.

References

[1] Lin C C and Fu S J 2016 *Mater. Sci. Eng.* C **58** 254–63
[2] Li W Q, Wang Z, Hao S, He H, Wan Y, Zhu C, Sun L P, Cheng G and Zheng S Y 2017 *ACS Appl. Mater. Inter.* **9** 16793–802
[3] Pandey K A and Dwivedi A K 2019 *Int. J. Res. Pharm. Sci.* **10** 2572–7
[4] Ahmed J and Varshney S K 2011 *Int. J. Food Prop.* **14** 37–58
[5] Pawar R, Tekale S U, Shisodia S, Totre J T and Domb A J 2014 *Recent Pat. Regener. Med.* **4** 40–51
[6] Kao C T, Lin C C, Chen Y W, Yeh C H, Fang H Y and Shie M Y 2015 *Mater. Sci. Eng.* C **56** 165–73
[7] Yang Z, Si J, Cui Z, Ye J, Wang X, Wang Q, Peng K, Chen W and Chen S C 2017 *Carbohydr. Polym.* **174** 750–9

[8] Lu M and Yu J 2018 *In Novel Biomaterials for Regenerative Medicine*, ed Chun H J, Park K, Kim C-H and Khang G (Singapore: Springer) p 451
[9] Lee H, Dellatore S M, Miller W M and Messersmith P B 2007 *Science* **318** 426–30
[10] Bettinger C J, Bruggeman J P, Misra A, Borenstein J T and Langer R 2009 *Biomater.* **30** 3050–7
[11] Fullenkamp D E, Rivera J G, Gong Y-K, Lau K H A, He L, Varshney R and Messersmith P B 2012 *Biomater.* **33** 3783–91
[12] Karkhanechi H, Takagi R and Matsuyama H 2014 *Desalination* **336** 87–96
[13] Harrasser N, Jüssen S, Obermeir A, Kmeth R, Stritzker B, Gollwitzer H and Burgkart R 2016 *Biomater.* Res. **20** 17
[14] He T, Zhu W, Wang X, Yu P, Wang S, Tan G and Ning C 2015 *Mater. Technol.* **30** B68–72
[15] Song Y, Jiang H, Wang B, Kong Y and Chen J 2018 *ACS Appl. Mater. Inter.* **10** 1792–801
[16] Zhang W, Luo Y, Wang H, Pu S and Chu P K 2009 *Surf. Coat. Technol.* **203** 2550–3
[17] Bergemann C, Zaatreh S, Wegner K, Arndt K, Podbielski A, Bader R, Prinz C, Lembke U and Nebe J B 2017 *World J. Transplantation* **7** 193–202
[18] Xing J, Wang Q, He T, Zhou Z, Chen D, Yi X, Wang Z, Wang R, Tan G, Yu P and Ning C 2018 *ACS Appl. Bio Mater.* **1** 1236–1243
[19] Heidenau F, Mittelmeier W, Detsch R, Haenle M, Stenzel F, Ziegler G and Gollwitzer H 2005 *J. Mater. Sci. - Mater. Med.* **16** 883–8
[20] Gritsch L, Lovell C, Goldmann W H and Boccaccini A R 2018 *Carbohydr. Polym.* **179** 370–8
[21] Liu Y, Xu C, Gu Y, Shen X, Zhang Y, Li B and Chen L 2019 *RSC Adv.* **9** 11722–36
[22] Chieng B W, Ibrahim N, Yunus W and Hussein M 2013 *Polymers.* **6** 93–104
[23] Singh S K, Anthony P and Chowdhury A 2018 Orient J Chem **34** 1984–90
[24] Zhao X G, Hwang K-J, Lee D, Kim T and Kim N 2018 *Appl. Surf. Sci.* **441** 381–7
[25] Chou S C, Chung W A, Fan T L, Dordi Y, Koike J and Wu P W 2020 *J. Electrochem. Soc.* **167** 042507
[26] Saiz-Poseu J, Mancebo-Aracil J, Nador F, Busqué F and Ruiz-Molina D 2019 *Angew Chem Int Ed* **58** 696–714
[27] De Guzman M R, Andra C K A, Ang M B M Y, Dizon G V C, Caparanga A R, Huang S H, Lee K R 2021 *J. Mem. Sci.* **620** 118881
[28] Dai X, Li X and Wang X 2018 *Chem. Eng. J.* **338** 82–91
[29] Niu M, Wang H, Li J, Chen H, Li L, Yang H, Liu X, Chen Z, Liu H and Chen J 2020 *Royal Soc. Open Sci.* **7** 192154
[30] Yang S H, Kang S M, Lee K-B, Chung T D, Lee H and Choi I S 2011 *J. Am. Chem. Soc.* **133** 2795–7
[31] Zhang W, Yang F K, Han Y, Gaikwad R, Leonenko Z and Zhao B 2013 *Biomacromolecules* **14** 394–405
[32] Bernsmann F, Ball V, Addiego F, Ponche A, Michel M, Gracio J J d A, Toniazzo V and Ruch D 2011 *Langmuir* **27** 2819–25
[33] Xie J, Michael P L, Zhong S, Ma B, MacEwan M R and Lim C T 2012 *J. Bio. Mater. Res. Part A* **100A** 929–38
[34] Salomäki M, Marttila L, Kivelä H, Ouvinen T and Lukkari J 2018 *J. Phys. Chem. B* **122** 6314–27
[35] Hofman A H, van Hees I A, Yang J and Kamperman M 2018 *Adv. Mater.* **30** 1704640
[36] Badaraev A D, Nemoykina A L, Bolbasov E N and Tverdokhlebov S I 2017 *Resour.Eff. Technol.* **3** 204–11
[37] Zhang M, He X, Chen L and Zhang Y 2010 *J. Mater. Chem.* **20** 10696–704

Acknowledgments
This work was supported by a grant from the Engineering Research Development for Technology consortium of the Department of Science and Technology, Philippines. We would also like to thank

the Nanotech Analytical Services and Training Corp. (NASAT Labs) which provided the analytical tools used in the experiments.

Effect of magnetite on alginate-based hydrogel beads composite bio-sorbent for copper removal

A S A Rahman[1], A N A Yahaya[2], N A Khalil[1], A N S Fizal[1], M A S A'zim[1], M S Hossain[3] and M Zulkifli[2]

[1] Universiti Kuala Lumpur, Branch Campus Malaysian Institute of Chemical and Bioengineering Technology, 78000 Alor Gajah, Melaka, Malaysia
[2] Green Chemistry and Sustainability Cluster, Universiti Kuala Lumpur, Branch Campus Malaysian Institute of Chemical and Bioengineering Technology, 78000 Alor Gajah, Melaka, Malaysia
[3] School of Technology Industry, Universiti Sains Malaysia, Penang, Malaysia

Email: muzafar@unikl.edu.my

Abstract. A composite magnetite alginate-based bio-sorbent in hydrogel beads form as adsorbent for copper ion removal was prepared through this work. Two types of composite bio-sorbents which are cellulose-magnetite-alginate (CeMA) and chitosan-magnetite-alginate (CMA) hydrogel beads were synthesized by the physical cross-linking method. Ratios of magnetite iron oxide 0, 0.1, and 1.0 were used during the synthesis of bio-sorbents to observe the effect of magnetite ratios on copper ion removals. Based on the performance of bio-sorbents on copper removals, 24.6% of the highest percentage copper removal was achieved by CMA with a magnetite ratio of 0.1 at an adsorption equilibrium time of 24 hours and initial concentration of 100 mg/L. In addition, through this work, magnetite embedded bio-sorbent with the simple synthesized method was done by utilizing the capability of alginate to instantaneously form hydrogel beads upon addition into calcium chloride (Ca^{2+}). Therefore, this work proves the potential of magnetite embedded in alginate-based composite bio-sorbent hydrogel beads for heavy metal industrial wastewater.

1. Introduction

Contamination of water sources resulting from industrial activities had recently been in the spotlight due to multiple health risks to humans and the environment [1]. Industrial wastewater contained common heavy metals such as copper, lead and chromium are usually associated with the manufacturing industry, such as the electroplating industry. Heavy metals are known as toxic and harmful to humans and living beings even if ingested or exposed at low levels which can initiate disorders and even diseases due to accumulation in living tissues [2, 3]. Ingestion of heavy metal copper over the permissible amount in the human body leads to kidney, liver, stomach, and intestines damage and failure. Therefore, the National Water Quality Standards of Malaysia (NWQS) outlined the permissible limit for copper discharge into the watercourse at 0.02 mg/L. Presently, synthetic polymers are being commercialized for heavy metals removal, such as polyacrylamide (PAAM) as it is efficient in heavy metal removal through a coagulation-flocculation process [3]. However, the incorporation of PAAM in wastewater treatment creates another issue, which is environmental pollution due to its disposal [4]. Hence, the application of renewable resources in industries and daily lives was seen as a way towards environmentally friendly industrialization [5]. In recent years, researchers had commonly used processes such as membrane filtration, adsorption, coagulation/flocculation, ion exchange, and catalytic oxidation for heavy metal removals [6]. However, out of all processes, adsorption process has gained massive interest due to its simplicity and efficiency using the force of attractions between bodies during the removal of a certain substance from wastewater [2].

Currently, the development of natural polysaccharides such as cellulose, chitosan, and alginate as adsorbents had been in the centre of focus among researchers [1, 3, 4, 6–9]. Cellulose is a naturally occurred polysaccharides consist of building units of β-1,4-linked anhydro-D-glucose that can be found from the cell wall of plants [4]. Aside from that, cellulose has multiple beneficial properties such as high tensile strength, flexibility, thermal stability and high surface area that is admirable for researchers to develop an environmentally friendly composite sorbent [10]. Whereas chitosan is a linear polysaccharide consist of poly[β-(1-4)-linked-2-amino-2-deoxy-D-glucose] in which is high in functional groups amine (NH_2) and hydroxyl (OH). These functional groups were proven to have a high affinity towards heavy metals cation due to different particle charges that are attracted towards one another [1]. Other than that, alginate consists of α-L-guluronic acid and 1-4 linked β-D-mannuronic acid which is abundant in carboxyl functional groups (COOH) which also proven to be attracted towards heavy metal cation [1]. Besides, alginate also possessed a great ability to bind with divalent ions such as calcium (Ca^{2+}) instantaneously in an aqueous medium through the cross-linking process [11]. Recently, the addition of magnetite in adsorbents had been sought after by researchers to produce adsorbents with an easier separation process from an aqueous medium. Conventional separation techniques such as centrifugation and filtration had been the limitation in the development of adsorbents [6]. Hence, the addition of magnetite will solve this limitation.

Previously, a study made by Yi *et al.* 2018 had focused on the removal of heavy metals copper and uranium by composite alginate chitosan hydrogel beads. However, up to our knowledge, very limited research was found on studying the effect of magnetite ratios in hydrogel bead adsorbents. Therefore, through this work, a composite alginate-based bio-sorbent in hydrogel beads form was synthesized and the effect of magnetite ratios (0, 0.1, and 1.0) in the bio-sorbents on copper removal performance was studied. The addition of magnetite in the hydrogel bead was proposed for easier separation of adsorbents from an aqueous medium. Besides, an easy sol-gel and cross-linking process using alginate to form adsorbent in hydrogel beads form are found to be desirable due to its temperature-independent formation process. Cross-linking of alginate with other natural polysaccharides, which is cellulose, and chitosan was also made to overcome the limitations of individual components, which are their thermal and mechanical properties. Two types of composite bio-sorbents, which are cellulose-magnetite-alginate (CeMA) and chitosan-magnetite-alginate (CMA) was synthesized and compared throughout this work. The effect of pH and contact time on copper removal were investigated for all composite bio-sorbent hydrogel beads synthesized in this work. Other than that, SEM analysis was also done in this work to characterize the morphological surface of synthesized bio-sorbents in this work.

2. Experimental section

2.1. Materials

Microcrystalline cellulose (MCC) C.P, copper (II) sulphate anhydrous and sodium alginate C.P was purchased from R&M Chemicals. Chitosan (De-acetylation degree > 90%) is purchased from Bio Basic. Magnetite iron (II, III) oxide powder with a particle size of 50 μm was purchased from Sigma Aldrich. Calcium chloride anhydrous powder is purchased from Merck. All chemical reagents purchased are of analytical grade and acquired from commercial sources in Malaysia.

2.2. Synthesis of CeMA hydrogel beads

Sodium alginate solution was prepared by mixing 4 g sodium alginate into 200mL ultrapure water. The solution then was stirred for 1 hour. Then, 2 g microcrystalline cellulose powder and magnetite iron oxide with different ratios, CeMA x: y was added into the solution to study the effect of magnetite ratios on copper removal performance. In which, x = 1 representing mass of microcrystalline cellulose powder, whereas y = 0, 0.1, and 1.0 representing mass ratios of magnetite iron oxide to cellulose added into the formulation. Then, for the formation of hydrogel beads, the well-mixed solution was dropped into 100 mL of 0.2 M calcium chloride solution using a burette at a constant speed with approximately 10 cm distance from the solution after 3 hours of mixing. The hydrogel beads formed was left to cure in the calcium chloride solution for 24 hours at room temperature. The collected hydrogel beads undergo multiple rinsing using deionized water and stored for further study.

2.3. Synthesis of CMA hydrogel beads

Sodium alginate solution was prepared by mixing 4g sodium alginate into 200mL ultrapure water. The solution then was stirred for 1 hour. Then, 2 g chitosan powder and magnetite iron oxide with different ratios, CMA x: y was added into the solution to study the effect of magnetite ratios on copper removal performance. In which, x = 1 representing mass of chitosan powder, whereas y = 0, 0.1, and 1.0 representing mass ratios of magnetite iron oxide to chitosan added into the formulation. Then, for the formation of hydrogel beads, the well-mixed solution was dropped into 100 mL of 0.2 M calcium chloride solution using a burette at a constant speed with approximately 10 cm distance from the solution after 3 hours of mixing. The hydrogel beads formed was left to cure in the calcium chloride solution for 24 hours at room temperature. The collected hydrogel beads undergo multiple rinsing using deionized water and stored for further study.

2.4. Characterization of hydrogel beads bio-sorbent

Morphological surfaces of hydrogel beads bio-sorbent were analyzed through scanning electron microscopy (SEM, Quanta FEG 650). All samples were dried before the analysis.

2.5. Copper removal performance

The performance of each hydrogel bead on copper removal was performed in a 250 mL conical flask containing 100 mL copper solution with a concentration of 100 mg/L. 0.4 g of hydrogel beads were poured into each flask and undergoes constant shaking at 150 rpm in a thermoshaker incubator. An adequate amount of samples were taken and filtered through a 0.45 micron filter every 10 min, 20 min, 30 min, 60 min, 120 min, 240 min, 24 hours, and 48 hours before subjected to AAS analysis to observe the effect of contact time on copper removal. The effect of pH on the performance of copper removal was investigated in pH 3 and 5 with 0.4 g hydrogel beads weight at room temperature for 48 hours. All flasks were sealed to reduce the evaporation of the solution. All data were taken in triplicate. A blank solution containing copper solution without any hydrogel beads were also prepared to ensure no adsorption of heavy metal ion to flask wall occurred.

3. Results and discussion

3.1. Characterization of hydrogel beads bio-sorbent

3.1.1. SEM analysis. The morphological surface of CeMA and CMA was observed by SEM. Figure 1(a), (b), (c), and (d) shows SEM images of CeMA and CMA of magnetite ratios 0.1 and 1.0 to compare the effect of magnetite on the morphological surface of bio-sorbents. Based on figures 1(a), (b), (c), and (d), the morphological surface of the bio-sorbents was observed to be rough and uneven. This could be contributed by the particle aggregation that occurred between magnetite iron oxide with cellulose (in figure 1(a) and 1(b)) and chitosan (in figure 1(c) and 1(d)) [9]. Besides that, agglomeration of magnetite iron oxide particles on the surface of bio-sorbents occurred due to the existence of strong inter-particle Van Der Waals force and magnetic attraction [12]. Other than that, porous structures could be seen in figures 1(a), (b), (c), and (d). However, based on figures 1(b) and (d), porosity structures on surface of CeMA 1.0 and CMA 1.0 are less compared to CeMA 0.1 and CMA 0.1. This might be due to magnetite particles filling in the pores on the surface of the adsorbent as the magnetite ratio increase [6]. As reported by previous studies, high porosity structure in adsorbent will assist in high adsorption or removal of heavy metals [9, 13]. Hence, the porosity structures possess by the hydrogel beads bio-sorbent in this work proves that there are potentials as an efficient adsorbent for heavy metal removal.

Figure 1. SEM images of (a) CeMA 0.1, (b) CeMA 1.0, (c) CMA 0.1, (d) CMA 1.0.

3.2. *Copper removal performance*

3.2.1. Effect of pH. pH is one of the important parameters that affect metal adsorption by adsorbents due to the influence of ions that exist in the metal solution and functional groups on the surface of adsorbents. The removal of copper ions from copper solution as a function of pH is portrayed in figure 2. The removal of copper increases with the increase in pH. The low copper removal percentage at pH 3 is possibly due to the existence of hydronium ion, H^+ that is competing with copper ion Cu^{2+} for active sites on the surface of adsorbents [1]. This is causing less interaction between copper ions and active sites on the adsorbent, hence reducing the percentage of copper removal from copper solution. Other than that, as pH range increase, strong electrostatic interaction between copper ions, Cu^{2+} and active sites in functional groups -$COOH^-$ in alginate, -NH_2^- in chitosan, and -OH^- in both chitosan and cellulose is bound to happen, thus leading to higher copper removal percentage from copper solution. This is due to deprotonation occurs on the surface of the adsorbent as the pH range increase, leaving vacant active sites for interactions with copper ions [3]. Therefore, pH 5 was chosen for other effect studies for the performance of copper removal by hydrogel beads in this study.

Theoretically, the adsorption mechanism by heavy metal ion and adsorbent is ion exchange and/or electrostatic interaction due to different charges that exist [14]. Therefore, the negatively charged functional groups -$COOH^-$ in alginate, -NH_2^- in chitosan, and -OH^- in both cellulose and chitosan that exist on the surface of the bio-sorbent hydrogel beads were possibly responsible for the adsorption of the positively charged copper ions, Cu^{2+}. This is also supported by the findings from Yu *et al.* 2013, Wang *et al.* 2019, and Hu *et al.* 2017, in which the metal ion forms complex with the functional groups on the surface of the composite chitosan and composite cellulose adsorbents [3, 14–15]. Figure 3 shows a schematic diagram of the possible adsorption mechanism through electrostatic interaction between copper ions with CeMA and CMA.

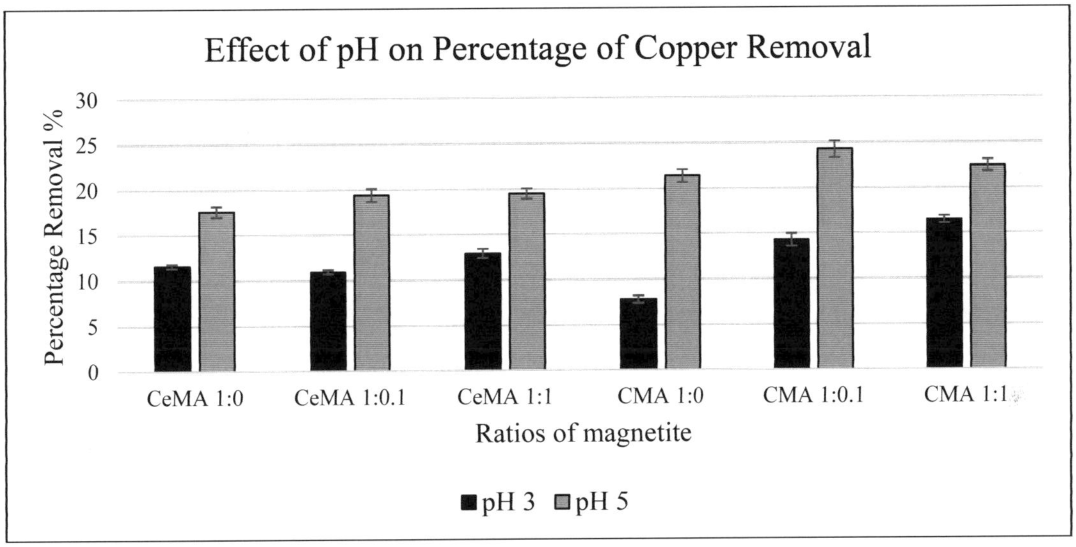

Figure 2. Effect of pH on copper removal at the initial concentration of copper solution at 100 mg/L and contact time at 48 hours.

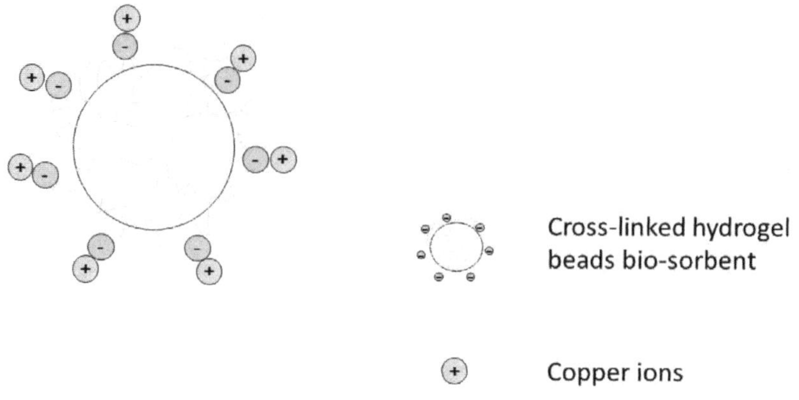

Figure 3. Schematic diagram of the possible adsorption mechanism by electrostatic interaction.

3.2.2. Effect of contact time. The removal of copper ions against time is presented in figure 4. It can be seen from figure 4 that the removal of copper increased until they reached equilibrium at 24 hours. The same trends can be observed in both CeMA and CMA hydrogel beads. Therefore, it can be concluded that the bio-sorbent hydrogel beads synthesized in this work reach their equilibrium in 24 hours. This is due to vacant active sites on the surface of adsorbent during the initial stage in the adsorption process, causing rapid adsorption of copper ions. However, as time increases, vacant active sites on the surface of the adsorbent gradually filled with copper ions until no more vacant active sites left [3]. This leads to slower and constant percentage removal of copper ions as all vacant active sites in adsorbent are filled, thus indicating that concentration equilibrium between adsorbate solution and adsorbent has been reached. Moreover, the smooth curve increasing up to the equilibrium point suggesting that the adsorption of copper ions onto adsorbent is by surface monolayer adsorption [16].

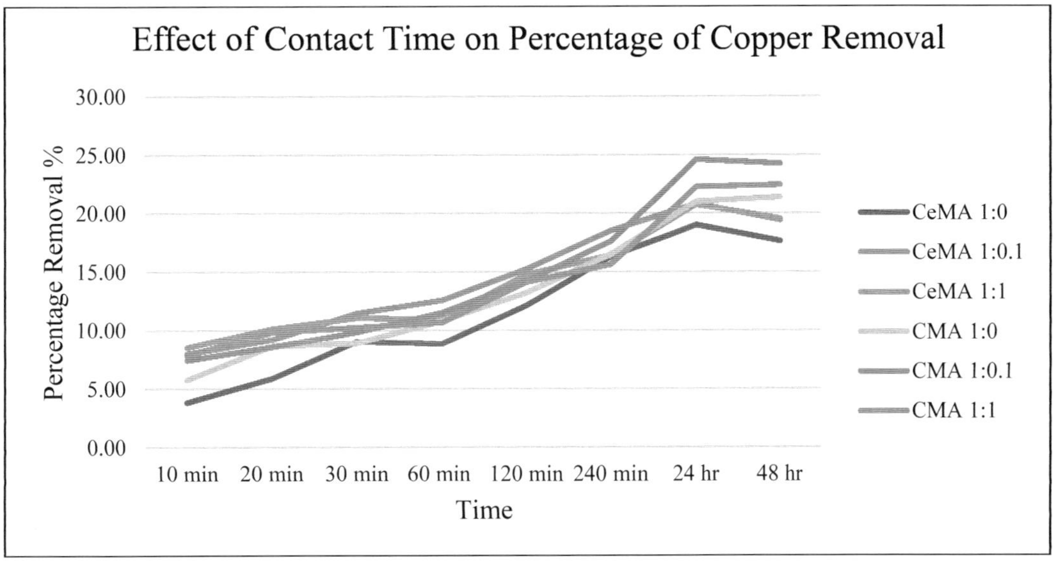

Figure 4. Effect of contact time on copper removal at the initial concentration of copper solution at 100 mg/L.

3.2.3. Effect of magnetite ratios on copper removal. To investigate the effect of magnetite ratios on the performance of hydrogel beads in copper removal, different magnetite ratios of 0, 0.1, and 1.0 were formulated during the synthesis of CeMA and CMA composite hydrogel beads bio-sorbent. Based on figure 5 and figure 6, the magnetite ratio of 0.1 shows the best copper removal in CeMA and CMA compared to magnetite ratios 0 and 1.0. This implied that the addition of magnetite in the composite hydrogel bead improves the percentage of copper removal. Besides, it is also proven that magnetite plays an important role in enhancing the removal of copper from the solution. However, at a magnetite ratio of 1.0, the removal of copper ions is slightly reduced. The same trends can be seen in findings by Zhang *et al.* 2017 which may be explained by magnetite filling the active sites on the adsorbent, hence reducing the interactions between adsorbent and copper ion [6].

Furthermore, the percentage of copper removal was also seen to be higher in CMA compared to CeMA. This could be contributed by the 3D porous network composite in CMA that is high in multiple active functional groups hydroxyl and amine with high affinity towards metal ions on the surface of bio-sorbent [1, 13]. Therefore, through this work, it can be concluded that the presence of magnetite in CMA exhibited the highest percentage of copper removal compared to CeMA.

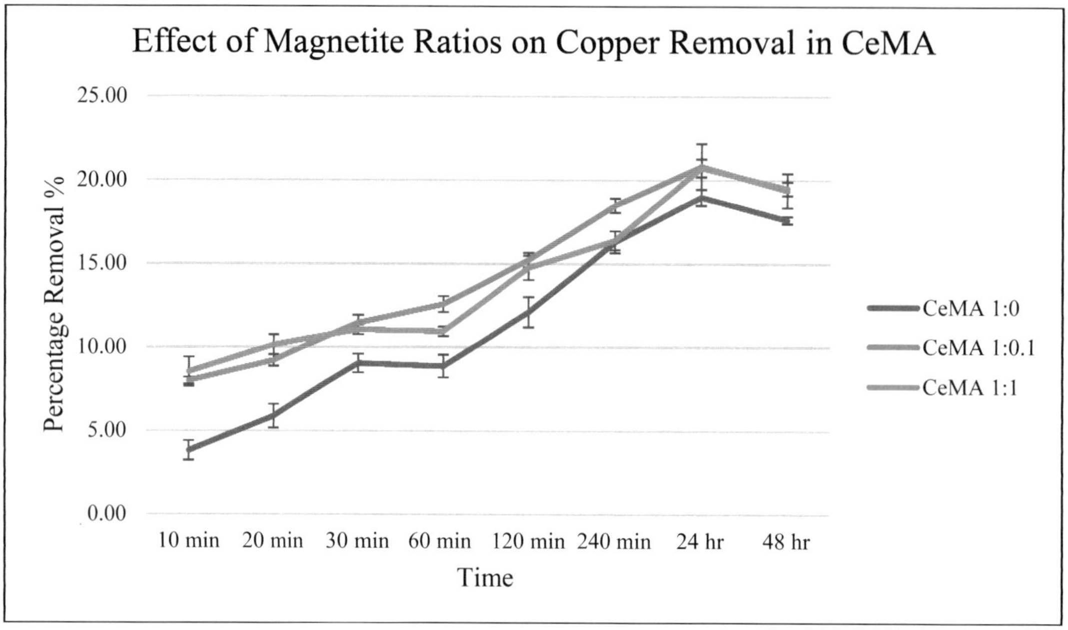

Figure 5. Effect of magnetite ratios on copper removal in CeMA.

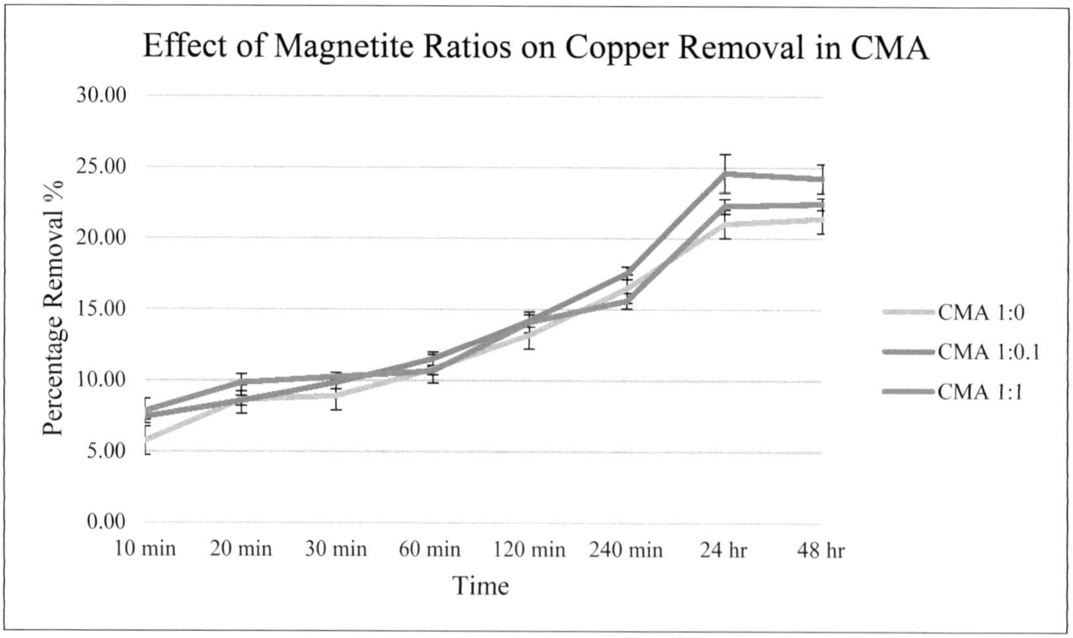

Figure 6. Effect of magnetite ratios on copper removal in CMA.

4. Conclusions

Cellulose-magnetite-alginate (CeMA) and chitosan-magnetite-alginate (CMA) composite bio-sorbent in hydrogel beads form were successfully synthesized using the sol-gel method in this work. The addition of magnetite in the composite bio-sorbent hydrogel beads was also proven to enhance the removal of copper from copper solution. Besides, through this work, the different magnetite ratios in composite bio-sorbent hydrogel beads were proven to be affecting the percentage of copper removal from the solution. The highest percentage of copper removal was achieved at 24.60% by CMA at a magnetite ratio of 0.1 in 24 hours. Therefore, it can be concluded that the magnetite composite bio-sorbent in hydrogel beads form synthesized through the sol-gel method in this work have the potential to be further studied for the development of efficient heavy metal bio-sorbents.

Acknowledgement
Authors would like to acknowledge to Ministry of Higher Education (MoHE), Malaysia for providing Fundamental Research Grant Scheme (FRGS/1/2018/STG07/UNIKL/03/1), as a financial support.

References

[1] Yi X, He J, Guo Y, Han Z, Yang M, Jin J, Gu J, Ou M and Xu X 2018 Encapsulating Fe_3O_4 into calcium alginate coated chitosan hydrochloride hydrogel beads for removal of Cu (II) and U (VI) from aqueous solutions *Ecotoxicol. Environ. Saf.* **147** 699–707
[2] Peng S, Meng H, Ouyang Y and Chang J 2014 Nanoporous magnetic cellulose-chitosan composite microspheres: Preparation, characterization, and application for Cu(II) adsorption *Ind. Eng. Chem. Res.* **53** 2106–13
[3] Wang X, Jiang S, Cui S, Tang Y, Pei Z and Duan H 2019 Magnetic-controlled aerogels from carboxylated cellulose and $MnFe_2O_4$ as a novel adsorbent for removal of Cu(II) *Cellulose* **26** 5051–63
[4] Mautner A 2020 Nanocellulose water treatment membranes and filters: A review *Polym. Int.* **69** 741–51
[5] Phanthong P, Reubroycharoen P, Hao X, Xu G, Abudula A and Guan G 2018 Nanocellulose: Extraction and application *Carbon Resour. Convers.* **1** 32–43
[6] Zhang H, Luan Q, Tang H, Huang F H, Zheng M M, Deng Q C, Xiang X, Yang C, Shi J, Zheng C and Zhou Q 2017 Removal of methyl orange from aqueous solutions by adsorption on cellulose hydrogel assisted with Fe_2O_3 nanoparticles *Cellulose* **24** 903–14
[7] Abdul Khalil H P S, Saurabh C K, Adnan A S, Nurul Fazita M R, Syakir M I, Davoudpour Y, Rafatullah M, Abdullah C K, Haafiz M K M and Dungani R 2016 A review on chitosan-cellulose blends and nanocellulose reinforced chitosan biocomposites: Properties and their applications *Carbohydr. Polym.* **150** 216–26
[8] An B, Lee H, Lee S, Lee S H and Choi J W 2015 Determining the selectivity of divalent metal cations for the carboxyl group of alginate hydrogel beads during competitive sorption *J. Hazard. Mater.* **298** 11–8
[9] Khalid A M, Hossain M S, Ismail N, Khalil N A, Balakrishnan V, Zulkifli M and Ahmad Yahaya A N 2021 Isolation and characterization of magnetic oil palm empty fruits bunch cellulose nanofiber composite as a bio-sorbent for Cu(II) and Cr(VI) removal *Polymers (Basel)* **13** 1–22
[10] A Fatah I Y, Abdul Khalil H P S, Hossain M S, Aziz A A, Davoudpour Y, Dungani R and Bhat A 2014 Exploration of a chemo-mechanical technique for the isolation of nanofibrillated cellulosic fiber from oil palm empty fruit bunch as a reinforcing agent in composites materials *Polymers (Basel)* **6** 2611–24
[11] Hecht H and Srebnik S 2016 Structural characterization of sodium alginate and calcium alginate *Biomacromolecules* **17** 2160–67
[12] Takai Z I, Mustafa M K, Asman S and Sekak K A 2019 Preparation and characterization of magnetite (Fe_3O_4) nanoparticles by sol-gel method *Int. J. Nanoelectron. Mater.* **12** 37–46
[13] Li B, Zhang Q, Pan Y, Li Y, Huang Z, Li M and Xiao H 2020 Functionalized porous magnetic cellulose/Fe_3O_4 beads prepared from ionic liquid for removal of dyes from aqueous solution *Int. J. Biol. Macromol.* **163** 309–16
[14] Yu Z, Zhang X and Huang Y 2013 Magnetic chitosan-iron(III) hydrogel as a fast and reusable adsorbent for chromium(VI) removal *Ind. Eng. Chem. Res.* **52** 11956–66
[15] Hu Z H, Omer A M, Ouyang X K and Yu D 2018 Fabrication of carboxylated cellulose nanocrystal/sodium alginate hydrogel beads for adsorption of Pb(II) from aqueous solution

Int. J. Biol. Macromol. **108** 149–57
[16] Tang H, Zhou W, Lu A and Zhang L 2014 Characterization of new sorbent constructed from Fe_3O_4/chitin magnetic beads for the dynamic adsorption of Cd^{2+} ions *J. Mater. Sci.* **49** 123–33
[17] Peng S, Liu Y, Xue Z, Yin W, Liang X, Li M and Chang J 2017 Modified nanoporous magnetic cellulose–chitosan microspheres for efficient removal of Pb(II) and methylene blue from aqueous solution *Cellulose* **24** 4793–806

Photocatalytic degradation of anthracene by biochar-based graphitic carbon nitride

S N Q Syed Abd Halim[1], N A M Nazri[1] and N A H M Nordin[2]

[1] Section of Chemical Engineering Universiti Kuala Lumpur Kampus Malaysian Institute of Chemical and Bioengineering Technology (UniKL MICET), Lot 1988, Kawasan Perindustrian Bandar Vendor, Taboh Naning, 78000, Alor Gajah, Malacca, Malaysia
[2] Chemical Engineering Department, Universiti Teknologi Petronas, Seri Iskandar, 32610, Perak, Malaysia

Email: syarifah.halim19@s.unikl.edu.my

Abstract. Polycyclic aromatic hydrocarbons (PAHs) are persistent and toxic to living organisms, that can be classified as carcinogenic, mutagenic and teratogenic pollutants. In this study, a green photocatalyst of biochar-based graphitic carbon nitride (BC/g-C_3N_4-M) is derived from sugarcane bagasse (SB) and melamine were developed as a potential material for the degradation of PAHs. BC(SB)/g-C_3N_4-M prepared with varied ratios of melamine to SB and different synthesis temperature were characterized by FTIR, BET and UV-DRS. The efficiency of the catalyst for the degradation of anthracene were investigated further in terms of its efficiency at various pH mediums, catalyst dosage and anthracene initial concentration. Experimental results revealed that g-C_3N_4 showed better degradation efficiency to anthracene than BC(SB)/g-C_3N_4-M. Additionally, the best degradation efficiency of anthracene by g-C_3N_4 and BC/g-C_3N_4-M75% composites were found at pH 3 with 1.0 g/L dosage at 2 ppm and 1 ppm initial concentration, respectively. The catalysts were also discovered to be reusable for 5 cycles with a slight decrease in photocatalytic degradation. The use of agricultural biomass waste in this study could provide a promising approach to water and wastewater treatment.

Keywords: Biochar; graphitic carbon nitride; thermal polycondensation; photocatalytic degradation; melamine.

1. Introduction

Polycyclic aromatic hydrocarbons (PAHs) can be regarded as one of the priority pollutants in water system, affecting not only the living and growing environment of plants and animals, but also human health. This is due to a number of factors such as hydrophobicity, bioaccumulation tendency, acutely toxicity and persistence in the environment, all of which contribute to global pollution [1]. Natural and anthropogenic sources of PAHs in riverine environments and water systems include petroleum spills, industrial effluent, municipal waste water discharge, chemical refineries and volcanoes [1, 2]. In the past, several conventional process like physical techniques (i.e., coagulation, flocculation, activated carbon adsorption and photo degradation) [3, 4], chemical and biological method (i.e., chemical oxidation, aerobic and anaerobic treatment) [5], have been used for the removal of pollutants (i.e., dyes, antibiotics, pesticides and herbicides) from water system. However, the conventional used methods are associated with the formation of secondary waste, generation of toxic by-products (i.e., bromine, hydrogen peroxide) [6, 7] and also inefficient due to strict control of conditions, time consuming and also costly. Currently, heterogenous photocatalysis which is one of the advanced oxidation processes (AOP) has received growing interest in photodegradation of various toxic compounds because of its advantages such as non-toxic, mild reaction conditions [8], reduced or no formation of secondary pollution [4] using catalyst that can be regenerated after process.

Under UV light irradiation, semiconductor photocatalyst (i.e., TiO_2, CdS, ZnO, WO_3 and g-C_3N_4) [6, 9] can degrade a wide range of organic pollutants such as herbicides, dyes, pesticides, phenolic compounds, tetracycline, and others [4]. Recently, a metal free material, graphitic carbon nitride (g-C_3N_4), that consist of carbon and nitrogen compounds has been widely utilized as photocatalyst due its high oxidation efficiency, non-toxic nature, significant chemical and thermal resilience [10,11]. Unlike TiO_2, which can only be activated in UV regions (3.2 eV), g-C_3N_4 has a 2.7 eV band gap and a good visible light response up to 460 nm [12]. Furthermore, using inexpensive nitrogen rich precursors such as melamine, dicyanamide, urea and thiourea, g-C_3N_4 can be easily synthesized by thermal polycondensation [13]. However, the use of g-C_3N_4 is limited due to a number of disadvantages such as low electrical conductivity [14], a small specific surface area (<10 $m^2.g^{-1}$) due to the graphitic layered structure [15, 16], low charge separation [17] and high recombination rate of photogenerated charge carriers [18], all of which reduced photocatalytic efficiency and hindering large scale applications.

To overcome this limitation, various modifications of g-C_3N_4 has been explored in order to enhance its photocatalytic efficiency for the remediate the environmental pollutants such as metal (i.e., Ze, Fe^3, Na, Ag) and non-metal doping (i.e., S, P, C, O), hybridization with carbon nanomaterials (i.e., carbon nanotubes, carbon nanospheres, graphene oxide) and coupling with other semiconductors. Introduction of biochar to g-C_3N_4 is also favourable to further improve the features of BC/g-C_3N_4, and consequently its performance for degradation of various pollutants as conductive carbon materials in the biochar can serve as an efficient electron-transfer channel and acceptors to allowing photogenerated electron-hole pairs to be separated more effectively [19,20]. Organic materials with a high carbon content, primarily biomass waste from agriculture, forestry and food processing can be used as feedstock materials for the production of biochar [21]. The main compositions of biomass known as insoluble fibre include cellulose, hemicellulose and lignin which highly indigestible by animals and has low biodegradability [22]. Biomass conversions into value-added carbon materials considered as an economically and environmentally viable solution rather than disposal in a landfill. To date, biomass waste has received significant attentions in photocatalytic treatment as it can be used as a carbon source of biochar for the preparation of g-C_3N_4 due to its distinctive properties, easily tunable functional groups, chemical stability and high conductivity [7]. Moreover, biochar has a remarkable adsorption capacity [21, 22] and high surface area that can improve the photocatalytic degradation.

The objective of this study is to synthesize BC(SB)/g-C_3N_4-M by thermal polycondensation method with improves adsorption and photocatalytic performance with high surface area and high photoresponsiveness under visible light. The present study also discusses the roles of biochar in BC(SB)/g-C_3N_4-M for photocatalytic degradation derived from SB. This works evaluates the effect of important parameters to identify the optimum conditions for high efficiency photocatalytic degradation of anthracene.

2. Methodology

2.1. Materials and chemicals

Sugarcane bagasse wastes as biochar feedstock was examined in this study. Melamine (AR, 99.0%), absolute denatured ethanol (AR, 99.5%), buffer solution (pH 7 and pH 4), sodium hydroxide (NaOH) and hydrochloric acid (HCl) were purchased from DNS Saintifik Sdn Bhd (Malaysia). Anthracene

(AR, 97.0%) was purchased from Polyscientific Enterprise Sdn Bhd (Malaysia). All the chemicals were used exactly as they were received. Throughout the experiments, deionized water was used.

2.2. *Preliminary thermal gravimetric analysis*

The thermal analysis of g-C_3N_4 were performed by using Thermal Gravimetric Analysis thermal analyzer (TA instruments) under constant static air atmosphere ramping of 5 °C.min^{-1} from 30 to 800 °C.

2.3. *Synthesis of BC(SB)/g-C_3N_4-M*

Waste biomass was washed, dried at 100 °C for 8 hours, crushed with a grinder (12,000 rpm), and sieved through a 30 mesh sieve (0.6 mm). Thermal polycondensation was used to produced BC(SB)/g-C_3N_4-M composites. In detail, melamine as a precursor was first ground in mortar with biomass at different melamine to biomass ratios indicated as BC(SB)/g-C_3N_4-MX (X = 25%, 50% and 75% of melamine). For 3 hours, samples were placed in a closed alumina crucible and heated in a muffle furnace at various synthesis temperatures (450 °C, 500 °C, 550 °C and 600 °C), at a constant rate of 5 °C.min^{-1}. Pure g-C_3N_4 is synthesized from melamine, and biochar is synthesized from SB under the same reaction conditions. The prepared samples were collected and kept in a desiccator at 25 °C.

2.4. *Characterization and analysis*

The average sample yield was calculated as the weight of thermal polycondensation product divided by the weight of the original materials. The surface area was determined by running N_2 sorption isotherms on an automated surface area using the Brunauer-Emmett-Teller (BET) methods (4 hour degassed). The UV-vis diffuse reflection spectra (UV-DRS) for dry-pressed disc samples were obtained using a Scan UV-vis spectrophotometer (Perkin Elmer Lambda 35) equipped with a 60 mm integrating sphere assembly and $BaSO_4$ as the reflectance sample. Tauc plot methods were used to calculate the optical band gap of the samples. Fourier-transform infrared (FTIR) spectroscopy (Thermo Scientific, US) was used with dried potassium bromide (KBr) in the 400 to 4500 cm^{-1} range with a resolution of 4 cm^{-1} and 8 accumulation modes.

2.5. *Photocatalytic activity of anthracene*

The photocatalytic activity of the catalyst was evaluated by anthracene degradation. The experiments were conducted in a batch mode of photocatalytic reactor. A magnetic stirrer was used to agitate the reaction mixture. The pH of the anthracene solution can be adjusted by adding of 0.1 M NaOH or 0.1 M HCl. The suspension was stirred in the dark until the anthracene concentration was horizontal to achieve adsorption equilibrium. The photocatalytic degradation experiments were carried out for 240 minutes while being illuminated by a 24 W LED light with a UV filter to maintain the main wavelength. With air circulation, the experimental temperature was kept at room temperature. At 1 hour intervals, 3 mL of suspension was syringed and filtered through a 0.45 μm microfilter. Anthracene concentration was determined using high performance liquid chromatography (HPLC)(Perkin Elmer) on an Ace C-18 column (150 mm x 4.6 mm) at a wavelength of 254 nm. Acetonitrile and deionized water (75%:25%) were used as the mobile phase, with a flowrate of 1.0 mL.min^{-1} for 12 minutes. The catalyst`s reusability was determined by reusing it for 5 consecutive cycles in the same photocatalytic degradation process. After each cycle, the catalyst was filtered, ethanol washed, and oven dried before being reused. The effluent was disposed of as laboratory waste.

3. Result and discussion

3.1. *Thermal gravimetric analysis*

The thermal polycondensation of melamine had been recognized as a facile synthesis pathway to generate the g-C_3N_4 polymer [25]. Thermal gravimetric analysis (TGA) has been widely implemented to obtain information on the processes taking place for physical and chemical transformation from melamine to g-C_3N_4. In this set of experiments, air at constant flow rate of 20 mL.min^{-1} was used with constant heating rate of 5 °C.min^{-1} ramping from 30 to 800 °C. The thermal gravimetric analysis (TGA) in figure 1 showed that g-C_3N_4 had small weight loss below 250 °C resulted by removal of bound moisture [26]. Above 250 °C, the significant weight loss possibly due to the rapid volatilization of moisture and deammonia condensation of melamine to form melam [27]. When the temperature increased from 360 °C, further condensation process of melam to form melem, which has the tri-s-triazine structure unit. Then, the structure unit was condensed into based compounds such as melon and formation of polymeric N-bridge tri-s-triazine graphitic carbon nitride started by 540 °C. The g-C_3N_4 decomposition process started from 360 to 630 °C with less than 10% residual content. The thermal analysis of melamine indicates two temperature regions to analyses in this research: (a) 360 to 540 °C and (b) 540 to 630 °C. The catalyst synthesized in muffle furnace via thermal polycondensation at different synthesis temperature within the range at 450 °C, 500 °C, 550 °C and 600 °C.

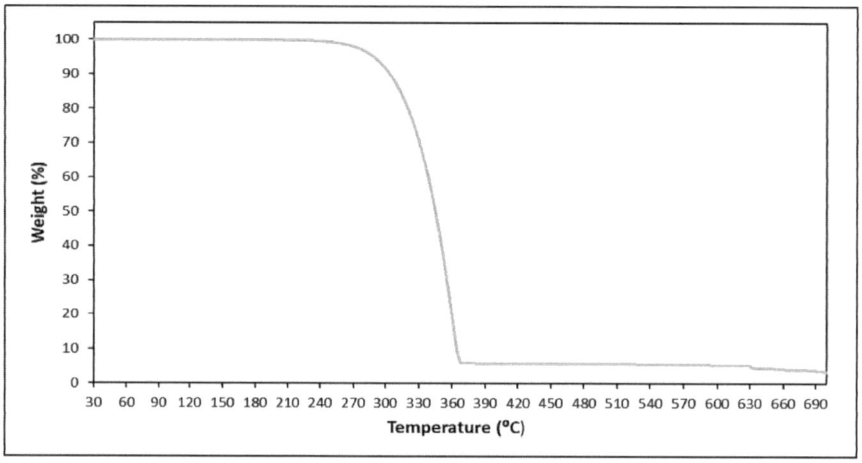

Figure 1. TGA graph g-C_3N_4.

3.2. Characterization and analysis

The FTIR spectra of g-C_3N_4 samples obtained at various thermal polycondensation synthesis temperatures are shown in figure 2. The stretching vibrational modes of N-H in primary (3510 to 3330 cm^{-1}), secondary (3330 to 3130 cm^{-1}) amides and their intermolecular hydrogen bonds are responsible for the broad band at around peak 3000 to 3700 cm^{-1}. The intensity of these peaks decreases with the increase of synthesis temperatures indicating the reduction of N-H or amide groups. The series of peak in the region of 1270 to 1630 cm^{-1} are related to the typical stretching mode of aromatic C-N heterocycles [19]. With higher synthesis temperatures, the intensity of these peaks became more intense, implying that the degree of polycondensation increased. The breathing mode of tri-s-triazine structure unit was assigned to the peaks around 810 to 830 cm^{-1} [28]. A weak band observed at 2360 cm^{-1} are attributed to C-N, implying that some molecules were not entirely react during the thermal polycondensation process [29].

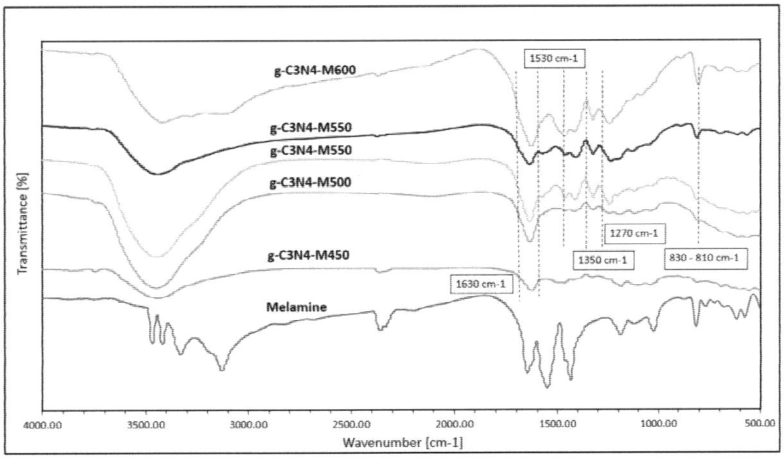

Figure 2. FTIR spectra of g-C_3N_4 synthesized at various temperatures.

As shown in figure 3, the FTIR spectra of BC(SB), BC(SB)/g-C_3N_4-M composites and g-C_3N_4 that is derived at 600 °C synthesis thermal polycondensation temperature. The commonly functional groups on biochar can be observed from BC(SB) sample. The BC(SB) showed bands 3700 to 780 cm^{-1}, which assigned to the O-H stretching mode of hydroxyl groups due to the absorption of water molecules (3700 to 3400 cm^{-1}), aliphatic CH_2 and CH_3 groups (2970 cm^{-1}), N-H stretching vibration (2300 cm^{-1}), N-H groups in amide (1550 cm^{-1}), aliphatic C-H and C=C stretching vibration (1400 to 1480 cm^{-1}), C-N stretching and C-O aromatic ester (1290 to 1293 cm^{-1}) and aromatic C-H bending (780 to 880 cm^{-1}), respectively [30]. The BC(SB)/g-C_3N_4-M composites varied in the intensities when the weight percentage melamine to SB changes, however, the composites show no obvious g-C_3N_4 characteristic peaks, implying that the BC(SB)/g-C_3N_4-M structure more likely as BC(SB). In contrast to the BC/g-C_3N_4-M composites produced by Pi *et al.* [19], the g-C_3N_4 characteristic in FTIR peaks can be observed suggested that biochar comprised the composites without change the polymeric reaction of melamine.

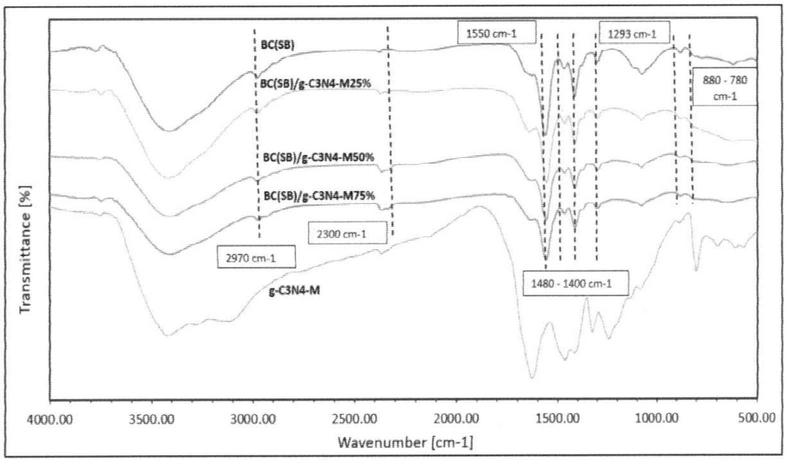

Figure 3. FTIR spectra of BC(SB), BC(SB)/g-C_3N_4-M and g-C_3N_4.

Table 1 shows the average yield, surface area (S_{BET}), and bandgap (E_g) of the catalyst. The average catalyst yield was calculated by dividing the synthesized product by the original materials used in thermal polycondensation. The average yield produced decreases as the amount of SB in the samples increases. This is because there is a significant amount of residual plant material present, which increases the volatile matter volatilization and organic material loss at high temperature thermal polycondensation temperatures, resulting in a more porous structure [31].

The S_{BET} for g-C_3N_4 was 14.6 $m^2.g^{-1}$, which was nearly 29 times less than the S_{BET} for BC(SB), which was 425.1 $m^2.g^{-1}$. As shown in table 1, the different percentage melamine to SB ratios have a significant effect on the S_{BET} of BC(SB)/g-C_3N_4-M composites. As the melamine ratios increase, the S_{BET} decreases, resulting in g-C_3N_4 aggregation clogging the biochar porous structure [32]. The addition of biochar can act as an attachment, preventing g-C_3N_4 reunion and stacking, as well as exhibiting a more developed nanopore structure, which provides more reaction sites to improve adsorption and photocatalytic degradation [33]. Furthermore, while a high surface area of the catalyst may increase adsorption capacity, it may reduce the rate of photocatalytic degradation. The availability of active sites on the surface decreases as more pollutant molecules adsorb onto the catalyst, limiting active radical generation and thus the pollutants ability to be attacked [29, 34].

The Tauc plot method was used to calculate the indirect E_g. The g-C_3N_4 exhibits strong visible light absorptions at 480 nm, which can be attributed to a narrow bandgap with E_g value of 2.60 eV, which is comparable to previous reported [35]. As for the BC(SB) and BC(SB)/g-C_3N_4-M composites, they have a significant shift towards shorter wavelengths at 340 nm and 400 nm, respectively, with a wide bandgap of 3.28 to 3.27 eV, which slightly higher than that of the common photocatalyst, TiO_2 semiconductor (3.05 to 3.26 eV) [36]. The findings suggest that g-C_3N_4 can be used as a visible light driven catalyst, whereas BC/g-C_3N_4-M is beneficial in the UV region.

Table 1. Average yield, S_{BET}, and E_g of BC(SB), BC(SB)/g-C_3N_4-M and g-C_3N_4.

Catalyst	Average Yield (%)	S_{BET} ($m^2.g^{-1}$)	E_g (eV)
BC(SB)	1.5	425.1	3.28
BC(SB)/g-C_3N_4-M25	1.8	360.6	3.28
BC(SB)/g-C_3N_4-M50	2.0	327.5	3.28
BC(SB)/g-C_3N_4-M75	2.1	188.3	3.27
g-C_3N_4	10.4	14.6	2.60

3.3. Photocatalytic degradation of anthracene

3.3.1. Effect of thermal polycondensation synthesis temperature. Adsorption in the dark and degradation of anthracene under visible light irradiation were used to evaluate the photocatalytic degradation of g-C_3N_4 derived at various thermal polycondensation synthesis temperatures. When exposed to visible light, anthracene self-degrades in the absence of a catalyst, demonstrating that it has a self-degradation process. Surprisingly, the thermal polycondensation synthesis temperature of g-C_3N_4 had an effect on photocatalytic degradation towards anthracene. Adsorption removals of g-C_3N_4-M450, g-C_3N_4-M500, g-C_3N_4-M550 and g-C_3N_4-M600 for 2 ppm anthracene in dark were approximately 6.2%, 8.8%, 8.4% and 9.7%, respectively, with constant catalyst dosage (1.0 g/L) and original pH anthracene (pH 9). Figure 4 shows that photocatalytic degradation increased as the synthesis g-C_3N_4 temperature increased from 450 to 600 °C, with anthracene degradation of 62.4%, 64.8%, 82.2% and 100%. The highest degradation efficiency of g-C_3N_4-M600 could be assigned to its higher visible light absorption efficiency, which could lead to better photogenerated electrons generation, more efficient separation charge and thus improved photocatalytic activity [37]. However, the lowest degradation efficiency by g-C_3N_4-M450 could be possibly due to the insufficient reaction of melamine during thermal polycondensation process to produce g-C_3N_4, as demonstrated by the FTIR spectra in figure 2, resulting in fewer reaction sites to produce photogenerated electrons to react with radicals for anthracene degradation [33, 13].

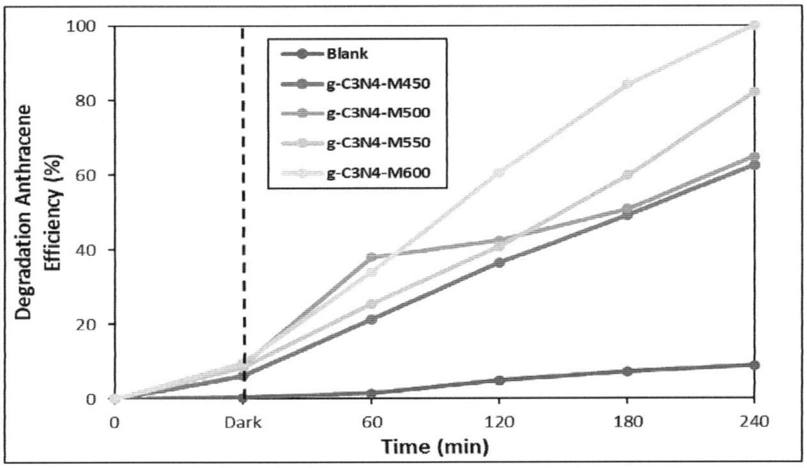

Figure 4. Photocatalytic degradation by g-C_3N_4 synthesized at different synthesis temperature.

3.3.2. Effect pH of anthracene. The pH of the solution influences the photocatalytic degradation of anthracene, as shown in figure 5. The pH of the anthracene affects the surface charge of g-C_3N_4 via electrostatic attraction [37], thereby shifting the potential of photocatalytic degradation reactions. The surface of g-C_3N_4 can be protonated (positively charged) or deprotonated (negatively charged) under acidic or alkaline conditions, respectively. By varying the initial pH of the anthracene solution, the effect of pH on the photocatalytic degradation of g-C_3N_4 was investigated with constant catalyst dosage (1.0 g/L) and initial concentration of anthracene at 2 ppm. Figure 5 shows that the degradation efficiency of g-C_3N_4 achieved complete removals of 100% for both pH 3 and pH 9 anthracene solutions. However, the anthracene solution at pH 3 demonstrated higher adsorption (15.3 %) and a faster rate of anthracene degradation than the pH 9 solution. The photocatalytic degradation of anthracene under high alkaline conditions at pH 12 is not favoured by 61.1% efficiency, presumably due to competitive adsorption by hydroxyl groups onto the surface of g-C_3N_4 [4].

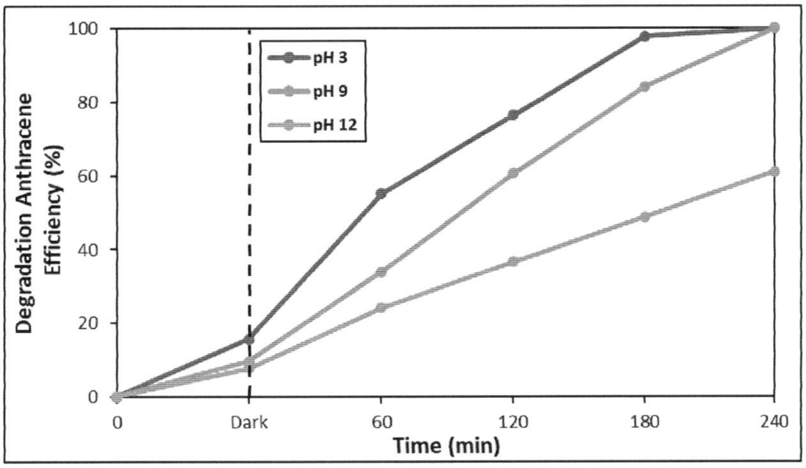

Figure 5. Photocatalytic degradation by g-C_3N_4 at different pH medium.

3.3.3. Effect of BC(SB)/g-C_3N_4-M ratios. Melamine ratios in BC(SB)/g-C_3N_4-M have a great impact on the performance of the catalyst for anthracene degradation. Different initial ratios of melamine and SB were investigated during photocatalytic degradation of 2 ppm anthracene at pH 3 with a catalyst dosage of 1.0 g/L. Figure 6 clearly shows that the BC(SB)/g-C_3N_4-M composites with the highest degradation efficiency are BC(SB)/g-C_3N_4-M75%, which reached 72.8% degradation, while anthracene degradation efficiency was only 47.6% for BC(SB)/g-C_3N_4-M25%. All of the BC(SB)/g-C_3N_4-M composites, on the other hand, had lower anthracene degradation than g-C_3N_4, which had a 100% degradation efficiency after 240 minutes of light exposure. This is because the biochar content on the g-C_3N_4 surface may have reduced the light transmission opportunity and thus the photocatalytic degradation performance to anthracene. When the BC(SB) sample was saturated with anthracene molecules, horizontal anthracene removals revealed that it lost its decontamination ability. Due to the integration of g-C_3N_4 materials that are light reactive, all BC(SB)/g-C_3N_4-M exhibits different capability towards anthracene. Even though it is already saturated during adsorption in dark conditions, the material can degrade the molecules when exposed to light. Contrary to the findings of Pi *et al.* [19], where BC/g-C_3N_4 composites derived from chestnut leaves and melamine demonstrated that increasing biochar content could improve photocatalytic degradation of methylene blue (MB) and p-nitrophenol by implying the adsorption process played more crucial role in photocatalytic activity.

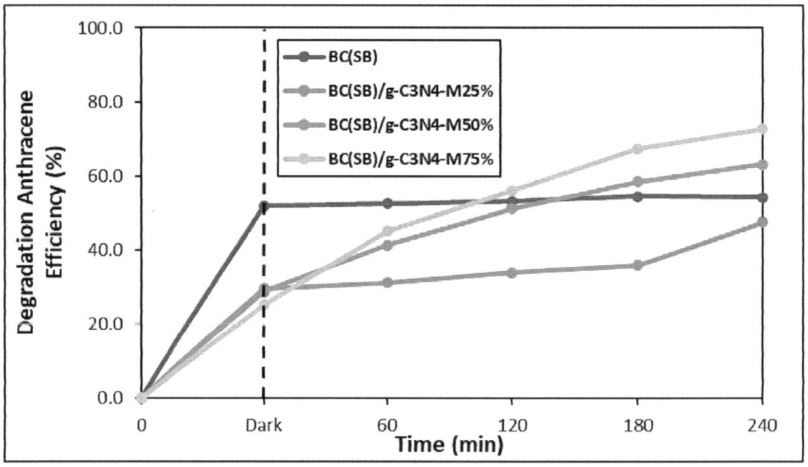

Figure 6. Photocatalytic degradation by BC(SB) and BC(SB)/g-C_3N_4-M composites.

3.3.4. Effect of catalyst dosage. In photocatalytic degradation with 2 ppm anthracene and pH 3, the dosage of g-C_3N_4 and BC(SB)/g-C_3N_4-M75% was varied from 0.5 to 2.0 g/L to analyses the effect on anthracene degradation, as shown in figures 7 and 8, respectively. When the dosage of both g-C_3N_4 and BC(SB)/g-C_3N_4-M75% was increased from 0.5 to 1.0 g/L, photocatalytic degradation was found to be increasing. This could be due to high availability of active sites on the catalyst surface to receive light photons, resulting in an increase in the formation number of active radicals that participate in anthracene degradation. Increases in g-C_3N_4 and BC(SB)/g-C_3N_4-M75% dosage above 1.0 g/L may reduce degradation due to anthracene molecules light penetration, which could be due to the light scattered domination at higher heterogeneous photocatalyst dosage [37]. The generation of electrons and holes was reduced as a result of the reduced light intensity, and thus the photocatalytic degradation of anthracene was also reduced. As a result, the optimum catalyst dosage for both g-C_3N_4 and BC(SB)/g-C_3N_4-M75% was determined to be 1.0 g/L.

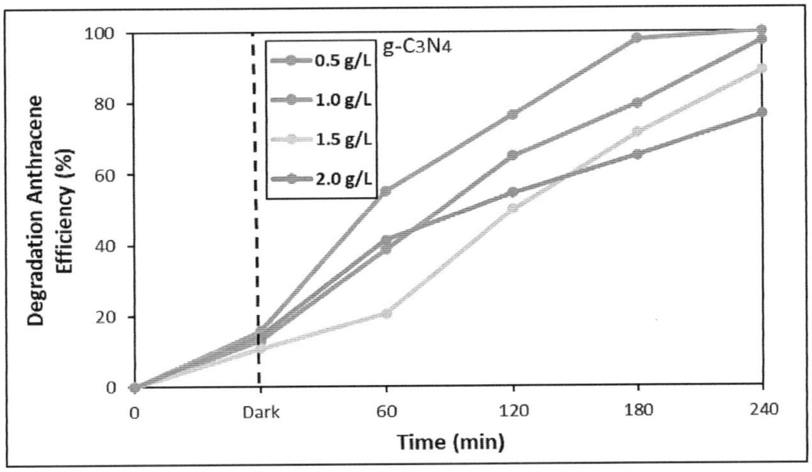

Figure 7. Photocatalytic degradation by g-C_3N_4 for different catalyst dosage.

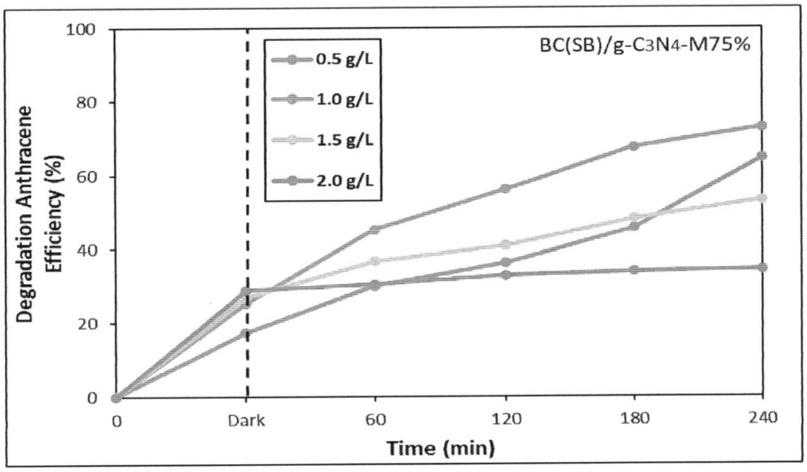

Figure 8. Photocatalytic degradation by BC(SB)/g-C_3N_4-M75% for different catalyst dosage.

3.3.5. Effect initial concentration of anthracene. Figure 9 shows that increasing the initial anthracene concentration from 1 to 2 ppm increased the degradation efficiency from 97.2 to 100%. This is explained by the fact that at low anthracene concentrations, fewer anthracene molecules approach the g-C_3N_4 surface, where the active radicals present may be deactivated. Increased anthracene initial concentrations above 2 ppm decreased photocatalytic degradation anthracene possibly due to anthracene molecules saturating the active sites of g-C_3N_4 by adsorption, resulting in a reduction in the total number of photons that can reach the catalyst surface, resulting in fewer active radicals formed and thus lower degradation efficiency [39]. Figure 10 shows that increasing the initial anthracene concentration from 1 to 4 ppm with 1.0 g/L dosage of BC(SB)/g-C_3N_4-M75% at pH 3 results in a significant decrease in anthracene degradation efficiency from 75.2 to 51.1%. The main reason was that as anthracene concentration increased, so did the number of anthracene molecules that needed to be degraded, but the catalyst dosage, the amount of electron-hole pairs generated, light intensity, and irradiation time remained constant during the photocatalytic degradation process [34].

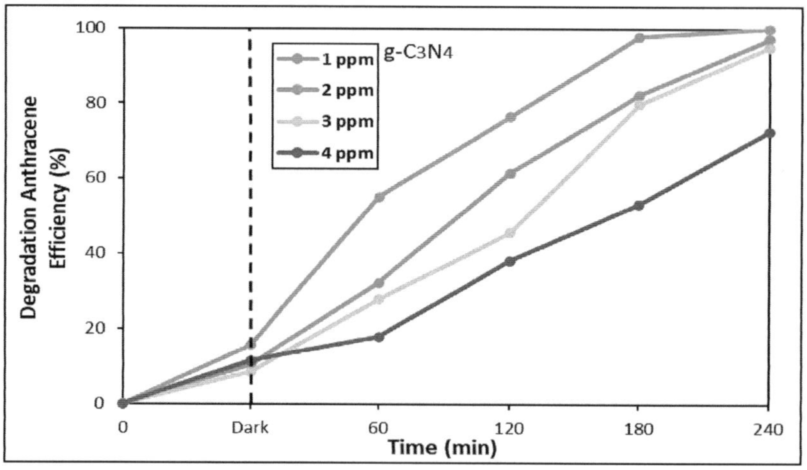

Figure 9. Photocatalytic degradation by g-C_3N_4 for different anthracene initial concentration.

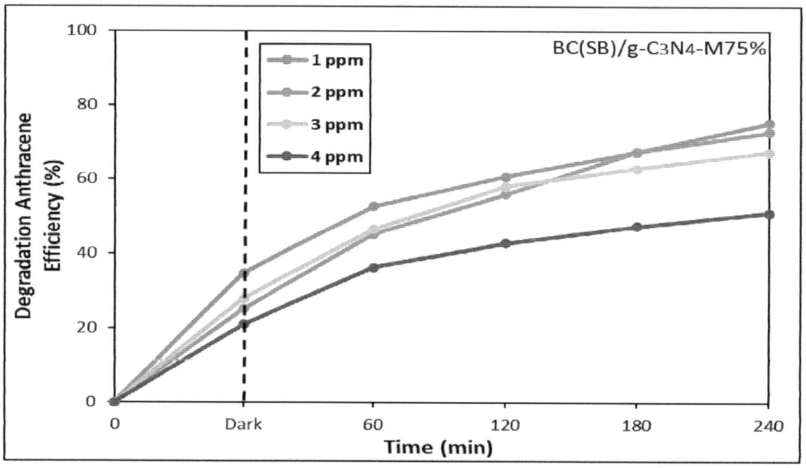

Figure 10. Photocatalytic degradation by BC(SB)/g-C_3N_4-M75% for different anthracene initial concentration.

3.4. Reusability of catalyst

Photocatalyst reusability is a method of determining the stability and actual cost of anthracene photocatalytic degradation. The reusability of g-C_3N_4 and BC(SB)/g-C_3N_4-M75% was investigated using anthracene degradation at 2 ppm solution with pH 3 and 1.0 g/L catalyst dosage over 5 cycles, as shown in figure 11 and figure 12, respectively. From the 1st to the 5th cycle, the reusability of g-C_3N_4 and BC(SB)/g-C_3N_4-M75% both decreased by 11.6% and 17.4%, respectively. This could be due to the used catalyst is lost during each sampling [39] or the catalyst surface could be clogged with anthracene molecules that the ethanol washed process could not effectively remove.

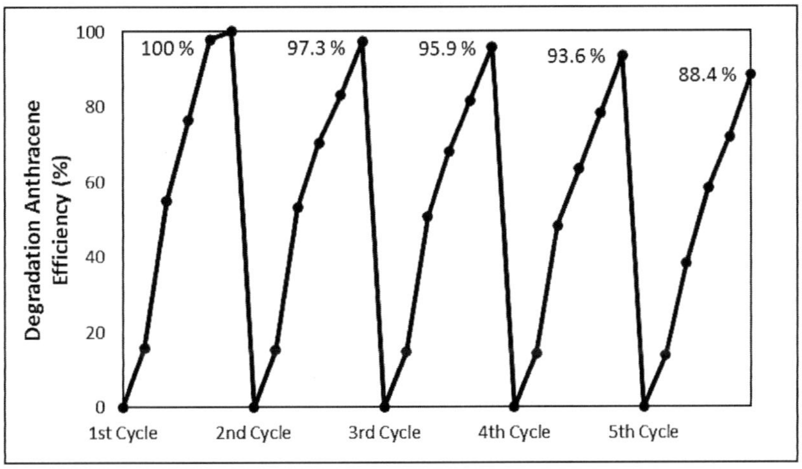

Figure 11. Reusability of $g-C_3N_4$ for anthracene photocatalytic degradation.

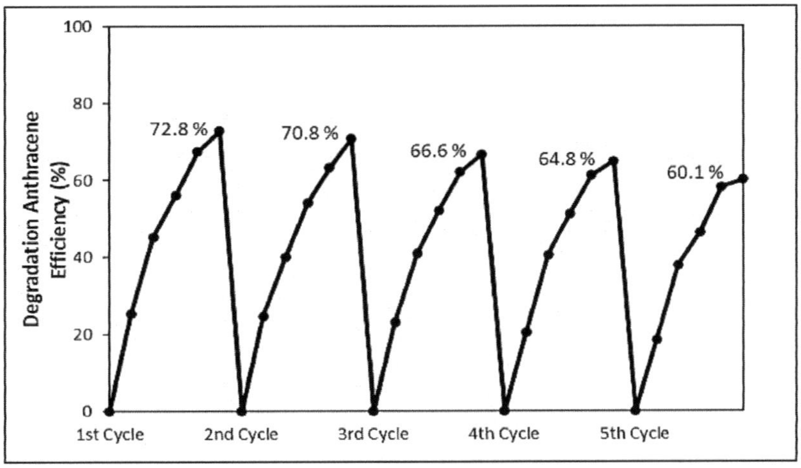

Figure 12. Reusability of BC(SB)/$g-C_3N_4$-M75% for anthracene photocatalytic degradation.

4. Conclusions

In summary, the average yield produced decreased with increasing of SB ratios due to higher developed of porous structure that resulting in improved the $g-C_3N_4$ surface area, which biochar can act as an attachment to reduce the $g-C_3N_4$ reunion and stacking. However, the agglomeration of biochar onto the surface of BC(SB)/$g-C_3N_4$-M has shifted the light absorption within UV range with larger bandgap while narrow bandgap $g-C_3N_4$ able to absorb light photon in visible light regions. The $g-C_3N_4$ that is derived at 600 °C thermal polycondensation synthesis temperature has the highest photocatalytic degradation of anthracene (100%) at pH 3 with 1.0 g/L catalyst dosage and 2 ppm anthracene initial concentration. The BC(SB)/$g-C_3N_4$-M75% have the highest photocatalytic degradation among BC(SB)/$g-C_3N_4$-M composites at same pH medium and catalyst dosage as $g-C_3N_4$ but in the optimum 1 ppm anthracene initial concentration with 75.2% anthracene degradation. The efficiency of the $g-C_3N_4$ and BC(SB)/$g-C_3N_4$-M75% slightly reduced in photocatalytic degradation of anthracene after 5 complete cycles that proved the reusability of the catalyst in pollutants degrading. To conclude, BC(SB)/$g-C_3N_4$-M that is derived from agricultural waste, SB successfully prepared by thermal polycondensation but the challenges to improve the photocatalytic degradation performance and catalyst features required further research. It should be noted this research provide a new sight in utilization of agricultural waste biomass as a visible light driven photocatalyst in degrading the organic pollutants.

Acknowledgements
The authors gratefully acknowledge financial support by the Ministry of Higher Education (Fundamental Research Grant Scheme (FRGS) (Ref: FRGS/1/2019/STG07/UNIKL/02/4)). The author would also like to acknowledge HICOE AMTEC (A.J090301.5300.07092) for their collaborative research work undertaken at Advanced Membrane Technology Research Centre (AMTEC), Universiti Teknologi Malaysia.

References
[1] Adeniji A O, Okoh O O and Okoh A I 2018 Analytical methods for polycyclic aromatic hydrocarbons and their global trend of distribution in water and sediment: A review *Recent Insights in Petroleum Science and Engineering*, ed M Zoveidavianpoor (Croatia: InTech) chapter 19 pp 393–413
[2] Abdel-Shafy H I and Mansour M S M 2016 A review on polycyclic aromatic hydrocarbons: source, environmental impact, effect on human health and remediation *Egypt. J. Pet.* **25** 107–23
[3] Sun Y, Niu W, Hu X, Ma X, Sun Y and Wen Y 2020 Oxidative degradation of polycyclic aromatic hydrocarbons in contaminated industrial soil using chlorine dioxide *Chem. Eng. J.* **394** 1–12
[4] Kumar A and Pandey G 2017 A review on the factors affecting the photocatalytic degradation of hazardous materials *Mater. Sci. Eng. Int. J.* **1** 106–14
[5] Kadri T, Rouissi T, Kaur B S, Cledon M, Sarma S and Verma M 2017 Biodegradation of polycyclic aromatic hydrocarbons (PAHs) by fungal enzymes: A review *J. Environ. Sci.* **51** 52–74
[6] Machulek A *et al.* 2013 Application of different advanced oxidation processes for the degradation of organic pollutants *Organic Pollutants – Monitoring, Risk and Treatment*, ed M N Rashed (Croatia: InTech) chapter 6 141–62
[7] Mian M M and Liu G 2018 Recent progress in biochar-supported photocatalysts: Synthesis, role of biochar, and applications *RSC Adv.* **8** 14237–48
[8] Gomes J F *et al.* 2017 Photocatalytic ozonation using doped TiO_2 catalysts for the removal of parabens in water *Sci. Total Environ.* **609** 329–40
[9] Darkwah W K and Ao Y 2018 Mini review on the structure and properties (photocatalysis), and preparation techniques of graphitic carbon nitride nano-based particle, and its applications *Nanoscale Res. Lett.* **13** 1–15
[10] Jeon P, Lee M E and Baek K 2017 Adsorption and photocatalytic activity of biochar with graphitic carbon nitride (g-C_3N_4) *J. Taiwan Inst. Chem. Eng.* **77** 244–9
[11] Ong W, Tan L, Ng Y H, Yong S and Chai S 2016 Graphitic carbon nitride (g-C_3N_4)-based photocatalysts for artificial photosynthesis and environmental remediation: Are we a step closer to achieving sustainability? *Chem. Rev.* **116** 7159–329
[12] Jiang L *et al.* 2017 Doping of graphitic carbon nitride for photocatalysis: A review *Appl. Catalysis B: Environ.* **217** 388–406
[13] Papailias I, Giannakopoulou T, Todorova N, Demotikali D, Vaimakis T and Trapalis C 2015 Effect of processing temperature on structure and photocatalytic properties of g-C_3N_4 *Appl. Surf. Sci.* 1–24
[14] Mamba G and Mishra A K 2016 Graphitic carbon nitride (g-C_3N_4) nanocomposites: A new and exciting generation of visible light driven photocatalysts for environmental pollution remediation *Appl. Catalysis B: Environ.* **198** 347–77
[15] Jin Z, Zhang Q, Yuan S and Ohno T 2015 Synthesis high specific surface area nanotube g-C_3N_4 with two-step condensation treatment of melamine to enhance photocatalysis properties *RSC*

Adv. **5** 4026–9
[16] Zhao Z, Fan J, Xue Y, Chang H, Masubuchi Y and Yin S 2018 Synthesis of graphitic carbon nitride from different precursors by fractional thermal polymerization method and their visible light induced photocatalytic activities *J. Alloys Compounds* **735** 1297–305
[17] Yuan D, Ding J, Zhou J, Wang L, Wan H, Dai W and Guan G 2018 Graphite carbon nitride nanosheets decorated with ZIF-8 nanoparticles: Effects of the preparation method and their special hybrid structures on the photocatalytic performance *J. Alloys Compounds* **762** 98–108
[18] Zhu J, Xiao P, Li H and Carabineiro S A C 2014 Graphitic carbon nitride: Synthesis, properties, and applications in catalysis *ACS Appl. Materials and Interfaces* **6** 16449–65
[19] Pi L, Jiang R, Zhou W, Zhu H, Xiao W, Wang D and Mao X 2015 g-C_3N_4 modified biochar as an adsorptive and photocatalytic material for decontamination of aqueous organic pollutants *Appl. Surf. Sci.* **358** 231–9
[20] Huang T, Fu Y, Peng Q, Yu C, Zhu J, Yu A and Wang X 2019 Catalytic hydrogenation of p-nitrophenol using a metal-free catalyst of porous crimped graphitic carbon nitride *Appl. Surf. Sci.* **480** 888–95
[21] Saleem J, Shahid U, Hijab M, Mackey H and McKay G 2019 Production and applications of activated carbons as adsorbents from olive stones *Biomass Conv. Bioref.* **9** 775–802
[22] Xu H, Zhan M, Cai P, Ji L, Chen T and Li X 2019 Adsorption characteristics of polycyclic aromatic hydrocarbons by biomass-activated carbon in flue gas *Energy and Fuels* **33** 11477–85
[23] Ji X, Li B, Yuan B and Guo M 2017 Preparation and characterizations of a chitosan-based medium-density fiberboard adhesive with high bonding strength and water resistance *Carbohydr. Polym.* **176** 273–80
[24] Samsudin M F R, Bacho N and Sufian S 2019 Recent development of graphitic carbon nitride-based photocatalyst for environmental pollution remediation *Nanocatalysts*, ed I Simha and M Shukla (London: IntechOpen) chapter 3 pp 1–11
[25] Zheng Y, Lin L, Wang B and Wang X 2015 Graphitic carbon nitride polymers toward sustainable photoredox catalysis *Angew. Chem. Int. Ed.* **54** 12868–84
[26] Varma A K and Mondal P 2016 Physicochemical characterization and pyrolysis kinetic study of sugarcane bagasse using thermogravimetric analysis *J. Energy Resour. Technol.* **138** 1–9
[27] Liu X, Hao J and Gaan S 2016 Recent studies on the decomposition and strategies of smoke and toxicity suppression for polyurethane-based materials *RSC Advances* **6** 74742–56
[28] Devthade V, Kulhari D and Umare S 2018 Role of precursors on photocatalytic behavior of graphitic carbon nitride *Mater. Today Proc.* **5** 9203–10
[29] Xu H, Wu L, Zhao H, Jin L and Qi S 2015 Synergic effect between adsorption and photocatalysis of metal-free g-C_3N_4 derived from different precursors *PLOS One* **10** 1–20
[30] Hue N 2020 Biochar for maintaining soil health *Soil Health* (*Soil Biology* vol 59) ed B Giri and A Varma (Springer) pp 21–46
[31] Ding Z, Zhang F, Sun N and Chi J 2019 Sorption of polycyclic aromatic hydrocarbons by biochars of wheat straw with different pyrolysis temperatures *IOP Conf. Series: Earth and Environmentalence* **237** 1–4
[32] Zheng Y, Yang Y, Zhang Y, Zou W, Luo Y, Dong L and Gao B 2019 Facile one-step synthesis of graphitic carbon nitride-modified biochar for the removal of reactive red 120 through adsorption and photocatalytic degradation *Biochar* **1** 89–96
[33] Zhu Z, Fan W, Liu Z, Yu Y, Dong H, Huo P and Yan Y 2018 Fabrication of the metal-free biochar-based graphitic carbon nitride for improved 2-Mercaptobenzothiazole degradation activity *J. Photochem. Photobiol.: A Chem.* **358** 284–93
[34] Luan J, Shen Y, Zhang L and Guo N 2016 Property characterization and photocatalytic activity evaluation of $BiGdO_3$ nanoparticles under visible light irradiation *Int. J. Mol. Sci.* **17** 1–26
[35] Zhang W, Zhang Q, Dong F and Zhao Z 2013 The multiple effects of precursors on the

properties of polymeric carbon nitride *Int. J. Photoenergy* **2013** 1–8

[36] Fonseca-Cervantes O R, Pérez-Larios A, Arellano V H R, Sulbaran-Rangel B and González C A G 2020 Effects in band gap for photocatalysis in TiO_2 support by adding gold and ruthenium *Processes* **8** 1–15

[37] Paul D R, Sharma R, Nehra S P and Sharma A 2019 Effect of calcination temperature, pH and catalyst loading on photodegradation efficiency of urea derived graphitic carbon nitride towards methylene blue dye solution *RSC Adv.* **9** 15381–91

[38] Lan Y, Li Z, Li D, Yan G, Yang Z and Guo S 2019 Graphitic carbon nitride synthesized at different temperatures for enhanced visible-light photodegradation of 2-naphthol *Appl. Surf. Sci.* **467–468** 411–22

[39] Parvizi E, Tayebee R, Koushki E, Abdizadeh M F, Maleki B, Audebert P and Galmiche L 2019 Photocatalytic efficacy of supported tetrazine on MgZnO nanoparticles for the heterogeneous photodegradation of methylene blue and ciprofloxacin *RSC Adv.* **9** 23818–31

Utilization of supercritical carbon dioxide (SC-CO_2) in lipids extraction from sewage sludge cake: A preliminary study

N A Khalil[1], H A Hamid[2], A N S Fizal[1], M Zulkifli[3], M S Hossain[4] and A N A Yahaya[3]

[1]Universiti Kuala Lumpur, Malaysian Institute of Chemical and BioEngineering Technology, 78000 Alor Gajah, Melaka, Malaysia
[2]Institute of Plantation Studies, Universiti Putra Malaysia, 43400 UPM Serdang, Selangor, Malaysia
[3] Green Chemistry & Sustainable Engineering Technology Cluster, Universiti Kuala Lumpur, Malaysian Institute of Chemical and BioEngineering Technology, 78000 Alor Gajah, Melaka, Malaysia
[4]School of Technology Industry, Universiti Sains Malaysia, 11800 USM Penang, Malaysia

Email: nafifah.khalil@s.unikl.edu.my; ahmadnaim@unikl.edu.my

Abstract. Sewage sludge containing a large number of lipids that can be recovered and utilised as a promising raw material in the production of biodiesel. Studies have been conducted to extract lipids from sludge using conventional solvent methods. However, all these conventional methods have some limitations such as extensive product separation and long extraction time (between 4 to 8 hours), which lead to high energy consumption. Supercritical carbon dioxide extraction (SFE) which utilises carbon dioxide (CO_2) gas at its critical condition as solvent has been studied extensively in various fields for oil extraction especially for plant and vegetative. This is due to the shorter extraction time and the lipids can be easily separated from the extraction system. The present research has undertaken a comparison study of supercritical carbon dioxide (SC-CO_2) utilisation in the extraction of lipids from sewage sludge against conventional soxhlet extraction of methanol and ethanol as solvent. The extraction of lipids from sewage sludge utilising SC-CO_2 extraction was successfully being conducted with lipids yield of 0.69 % within 0.5 hours at the operating temperature of 50 °C and pressure of 20 MPa. The lipids were easily separated subsequently from the SFE system when CO_2 is being released in gas form through the outlet valve during lipids collection. Whilst soxhlet extraction using methanol and ethanol as solvent (sludge: solvent ratio of 1:10) managed to extract 1.95 % and 2.81 % within 4 hours of extraction time at 60 °C, with the additional time needed to separate the lipids from solvent by evaporation.

1. Introduction

Municipal wastewater treatments plants generate a massive amount of sewage sludge each year as residual from wastewater treatment originated from household and light industrial sectors. The generation of annual sewage sludge per capita is estimated within the range of 0.1 to 30.8 kg/population.year [1–3]. Massive economic development and the population increase led to rapid urbanization resulting in the generation of a huge volume of sewage sludge, which posing significant environmental challenges for management and disposal. Many countries regulate the usage of sludge for land application such as fertilizer due to poor odour and the presence of heavy metals, toxic organic compounds, and pathogenic microorganism, while sludge incineration produces emissions comprising heavy metals and dioxins [4, 7]. Sewage sludge comprises a lipids fraction, which is a composite organic matrix containing oils, greases, fats and long-chain fatty acid originated from domestic and industrial wastes [4]. The volume of sewage sludge could be reduced by extracting the lipids fraction and converting it into a value-added product as a viable alternative in sustainable sludge management and disposal [5–9].

Lipids are extracted from sewage sludge mostly using solvent extraction (solid-liquid extraction) by organic solvents (non-polar or polar solvent) or supercritical carbon dioxide (SC-CO_2) [4, 5, 10]. The lipids extraction and its yield are influenced by the type of solvent used due to its unique characteristics and also the extraction parameters such as temperature, extraction time and the sludge-to-solvent ratio [10, 11].

The extraction of lipids using SC-CO_2 is a sustainable alternative compared to other solvents because carbon dioxide (CO_2) is inexpensive, readily available, non-toxic, and non-flammable [12]. SC-CO_2 utilize its solvating properties when the temperature and pressure are raised beyond the critical point which are 30.9 °C and 7.39 MPa [14, 16]. The SC-CO_2 has desirable properties such as high diffusivity, low surface tension and viscosity, density, heat capacity, thermal conductivity and ease of high solute recovery, which make it suitable for challenging extraction processes [5, 12, 13]. Supercritical fluid extraction technology with supercritical carbon dioxide (SFE-CO_2) is an effective extraction and sterilization technology and SC-CO_2 is an excellent solvent for the extraction of thermally sensitive compounds and materials [14, 15]. The SC-CO_2 technology has been proven as promising, advantageous and environmentally friendly over other conventional methods for extraction and fractionation of oil and its valuable components especially for plant and vegetative [14, 17]. Unfortunately, limited study has been conducted on the extraction of lipids from sewage sludge by using SC-CO_2. Therefore, this study aims to compare the lipids extraction from sewage sludge cake utilizing SC-CO_2, methanol, and ethanol as solvent.

2. Material and method

2.1. Sludge sample collection, preparation, and characterization

The sewage sludge cake was gathered from sludge storage yard using a shovel at 5 different points with similar weight. The sludge cake was well mixed to ensure the sample is consistent. The sludge cake was kept in a zipper bag and transported in a cool box to the laboratory.

Upon reaching the laboratory, the sludge cake was analyzed for moisture content (MC), volatile matter, crude ash, fixed carbon and elemental composition. Heavy metal analysis was also conducted for iron (Fe), copper (Cu), zinc (Zn) and lead (Pb).

The remaining sludge cake was kept in the refrigerator at 4 °C until further processing and testing. The sludge cake was oven-dried at temperature 105 °C to obtain <5 % MC. Then, the dried sample was ground using pastel & mortar and sieved through a 100 mesh before lipids extraction [18].

2.2. Lipids extraction experiments

Two extraction processes used for this study: Supercritical fluid extraction (SFE) using SC-CO_2 and soxhlet extraction using methanol and ethanol as the solvent. All the chemicals used were of analytical grade and were stored in a dry cabinet to avoid any changes due to humidity.

2.2.1. Extraction by SC-CO_2. Lipids extraction by the SC-CO_2 extraction system was conducted in static mode. Briefly, the desired amount of sludge sample was placed inside the high-pressure extractor vessel and the heating system is switched on. Carbon dioxide was pumped into the vessel at the desired temperature until the pressure reaches its operating value. The sludge sample was left in contact with the CO_2 fluid for the intended extraction time. After extraction, the outlet valve was opened slowly to collect the extracted lipids.

2.2.2. Extraction by soxhlet extractor using solvent. Lipids extraction by Soxhlet extractor was conducted using two different solvents: methanol and ethanol. The sludge sample was placed in a thimble and put into the extraction chamber. The solvent was filled up into the round bottom flask and placed on the heating mantle. All parts (round bottom flask, extraction chamber and condenser) were assembled accordingly and the extraction process was started at the desired temperature and extraction time. After extraction, the resultant mixture was filtered and further concentrated using a rotary evaporator at a temperature of 80 °C and under vacuum (50 mbar). The resultant lipids were stored in a desiccator and weighed the day after.

2.2.3. Lipids yield. The extracted lipids yield by both extraction methods was calculated using the following equation [19].

$$Lipid\ Yield\ [\%] = \frac{Residual\ mass\ [g]}{Sludge\ solid\ mass\ [g]}\ x\ 100\% \quad (1)$$

3. Results and discussion

3.1. The site and sample collection

The sewage sludge was collected from the sludge storage yard of the Centralized Sludge Treatment Facility (CSTF), situated in Central Malacca District, Malacca, Malaysia. The facility is being managed and operates by the Indah Water Konsortium Sdn. Bhd (IWK Malacca), where it is designed to cater 190 m^3/day of wastewater or 300,000 population equivalents [20]. The facility received an average of 63.86 m^3/day of wet sludge to be treated and produced an average of 1.2 ton/day of sludge cake by centrifuge dewatering unit [20].

Figure 1 shows the flow of the sludge treatment process within the CSTF [20]. The sewage sludge generated from individual septic tanks (IST), sewage treatment plants (STP) and communal septic tanks (CST) is transported by a tanker. Upon reaching the facility, the sludge is passed through a coarse screen into a scum collection tank to remove oil and grease. The screened sludge is temporarily stored in a sludge holding tank. The sludge is mixed with a polymer for thickening and subsequently dewatered by centrifugation to ensure the sludge is at least 20 % dried. The dewatered sludge as sludge cake is transported using sludge hoppers to the sludge cake storage yard (figure 2) and keep until disposal.

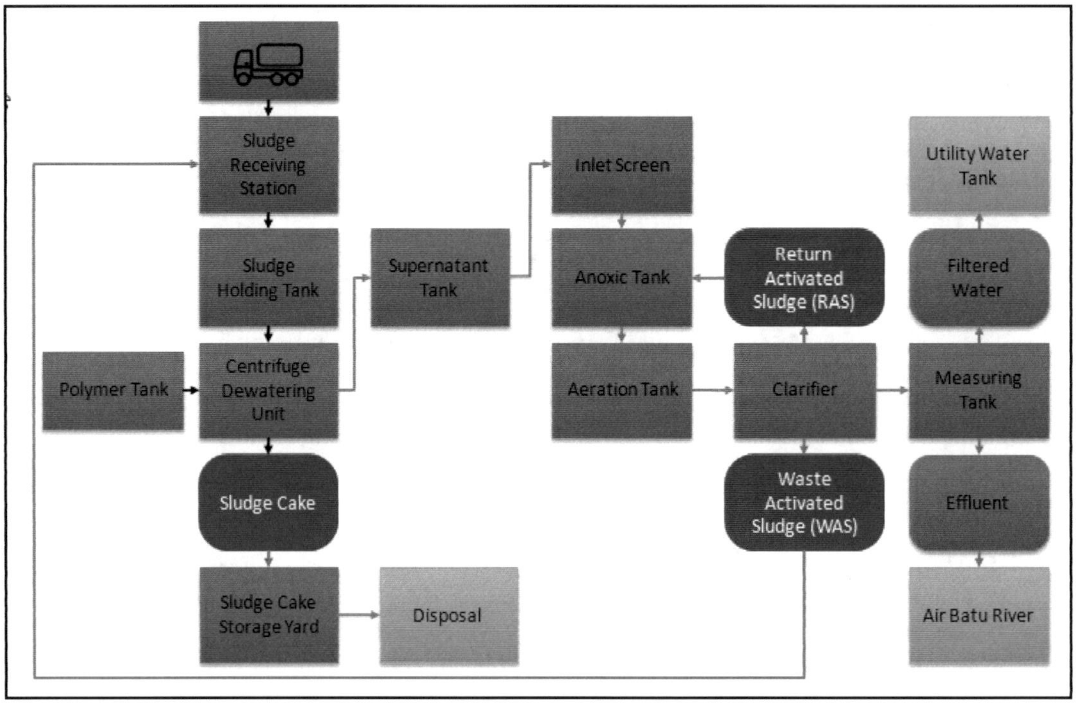

Figure 1. The sludge treatment process flow in the CSTF [20].

Figure 2. Sludge cake in the sludge cake storage yard in the CSTF.

3.2. Characteristics of the sludge cake

Preliminary analysis (physical and chemical analysis) was conducted on the sludge cake obtained from the CSTF. Properties of the sludge cake such as moisture content, elemental analysis, ash content, volatile matters and fixed carbon are shown in table 1.

The collected sludge sample from CSTF had an initial moisture content of 66.68 wt. % and reduced subsequently to 2.36 wt. % following oven drying for 6 hours at 105 °C as shown in figure 3. Oven drying at 105 °C shows the highest extracted lipids compared to other drying methods such as the addition of $MgSO_4.H_20$, freeze-drying, fume hood-drying and sun-drying [7]. The final moisture content of the sludge is a significant factor affecting the extraction of lipids because water can surround the sludge particles and hinder efficient solvent penetration into the sludge [7]. The sewage sludge also had high volatile matter (35.98 %) and fixed carbon (55.60 %) and low ash content (0.85 %). Heavy metal analysis conducted shows high Fe content, followed by Zn, Cu and Pb.

Table 1. Characteristics of sewage sludge cake.

Properties	**Item**	**Value**
Moisture content	Sludge Cake	66.68 ± 1.67 %
	Oven dried sludge (105 °C, 6 hr)	2.36 ± 0.34 %
Elemental Analysis	Carbon	0.26 ± 0.07 %
	Hydrogen	0.9± 0.00 %
	Oxygen	4.63 ± 0.02 %
Ash Content		0.85 ± 0.06 %
Volatile Matters		35.98 ± 7.95 %
Fixed Carbon		55.60 ± 7.89 %
Heavy metal	Fe	141.28 ± 12.69 mg/L
	Cu	5.03 ± 0.88 mg/L
	Zn	30.95 ± 1.63 mg/L
	Pb	0.81 ± 0.06 mg/L

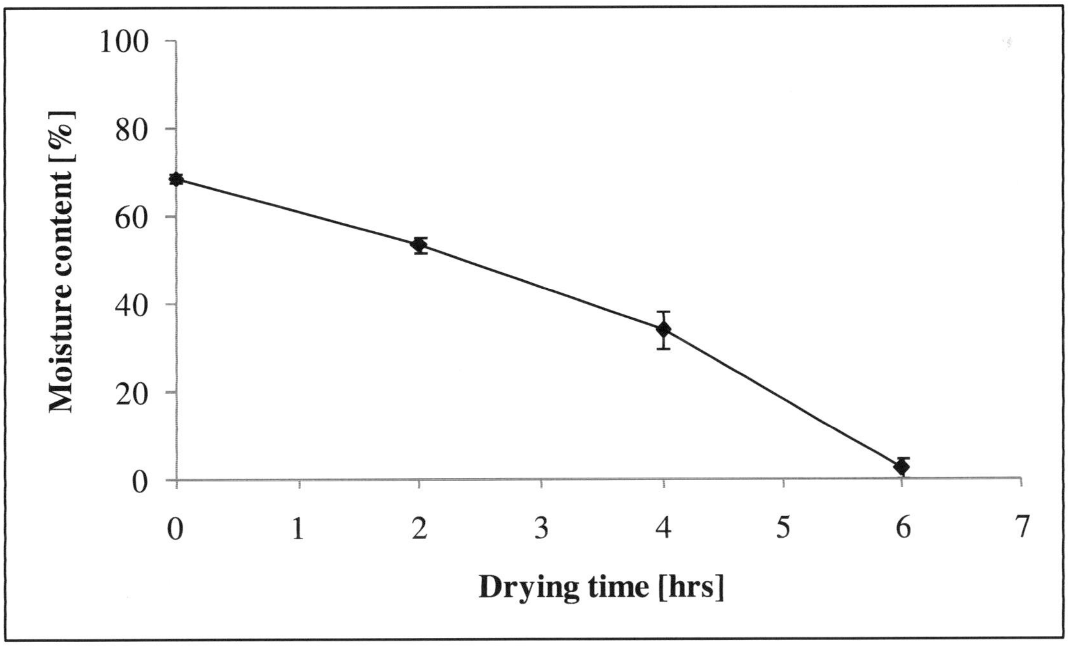

Figure 3. Moisture content for oven-dried sludge.

3.3. Sludge particle size distribution after oven drying.

Particle size analysis was conducted on the oven-dried sludge sample used in the SC-CO_2 and soxhlet extraction. Figure 4 shows the largest fraction of size lies in range 0.425≤ x < 2 mm, followed by <0.425 mm, 2 mm≤ x <3.35 mm and x≥3.35 mm.

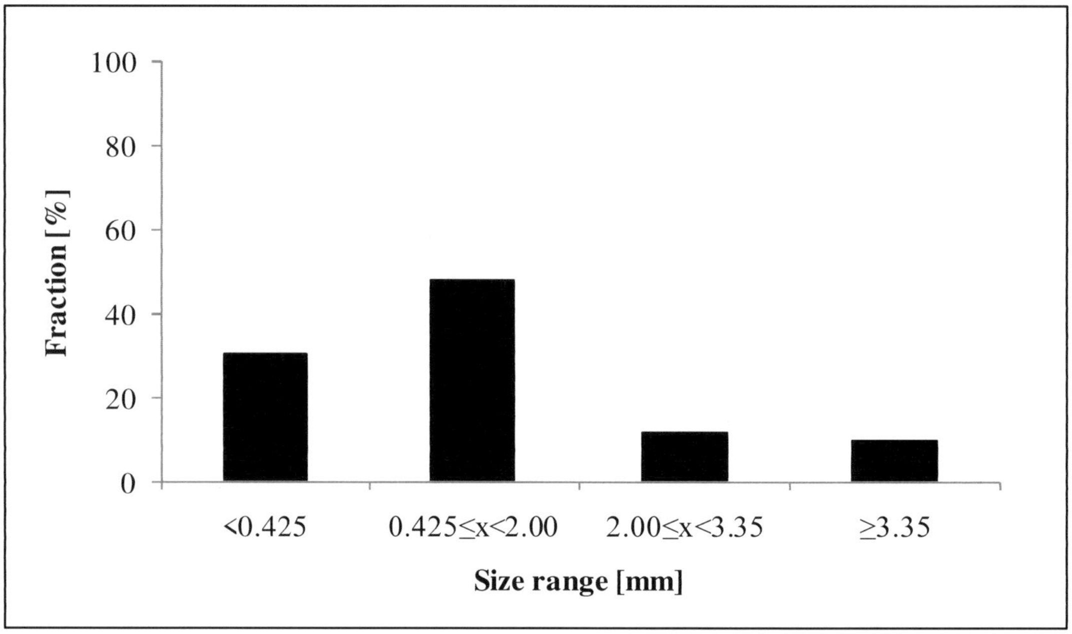

Figure 4. Particle size range for sludge sample.

3.4. Lipids yield for SC-CO_2 and soxhlet extraction.

Figure 5a shows the lipids collection from the outlet valve after extraction by SC-CO_2. A batch SC-CO_2 extraction process was conducted under the constant operating parameter of 20 MPa of pressure and 50 °C of temperature for 0.5 hour and yielded 0.65 % of lipids. The recovered lipids were high in viscosity and yellowish in appearance as shown in figure 5b and figure 5c.

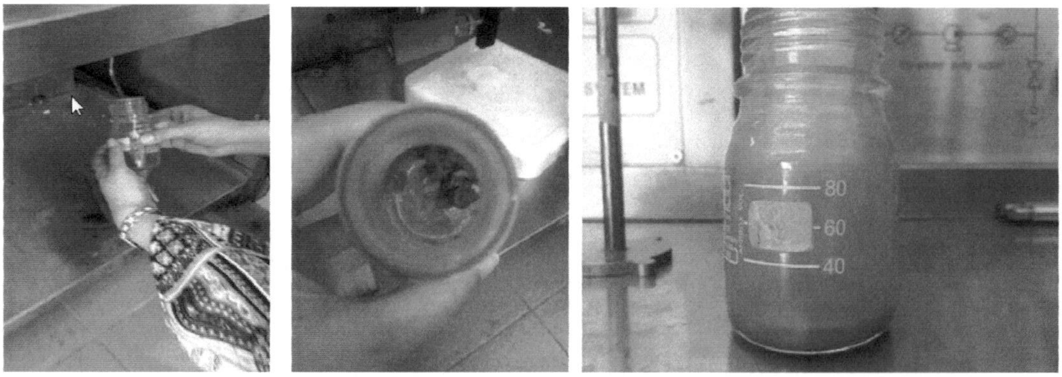

(a) (b) (c)

Figure 5. Lipids collection activity after SFE-CO_2 extraction.

Figures 6 and 7 show the lipids yield from soxhlet extraction by methanol and ethanol as solvent (sludge: solvent ratio of 1:10). The yield of lipids significantly increased ($R^2 > 0.9$) with the increased extraction time (4, 5 and 6 hours) for both solvents. The extraction temperature of 80 °C gives the highest lipids yield for both solvents. The extraction using ethanol yielded more lipids at all temperature (60, 70 and 80 °C) and extraction time (4, 5 and 6 hours) with the highest lipids yield (5.16 %) compared to methanol (4.05 %).

Extraction of lipids by SC-CO_2 yielded the lowest lipids followed by methanol and ethanol (SC-CO_2 < methanol < ethanol). It is because the SC-CO_2 has a lower polarity index compared to ethanol and methanol [5, 21, 22]. Even though the lipids extracted from SC-CO_2 is lower compared to soxhlet extraction, a shorter extraction time (0.5 hour) needed for this process. Furthermore, lipids were immediately separated from the CO_2 and no residual traces of solvent once CO_2 becomes gaseous after being released from the SC-CO_2 system due to depressurization [14, 16].

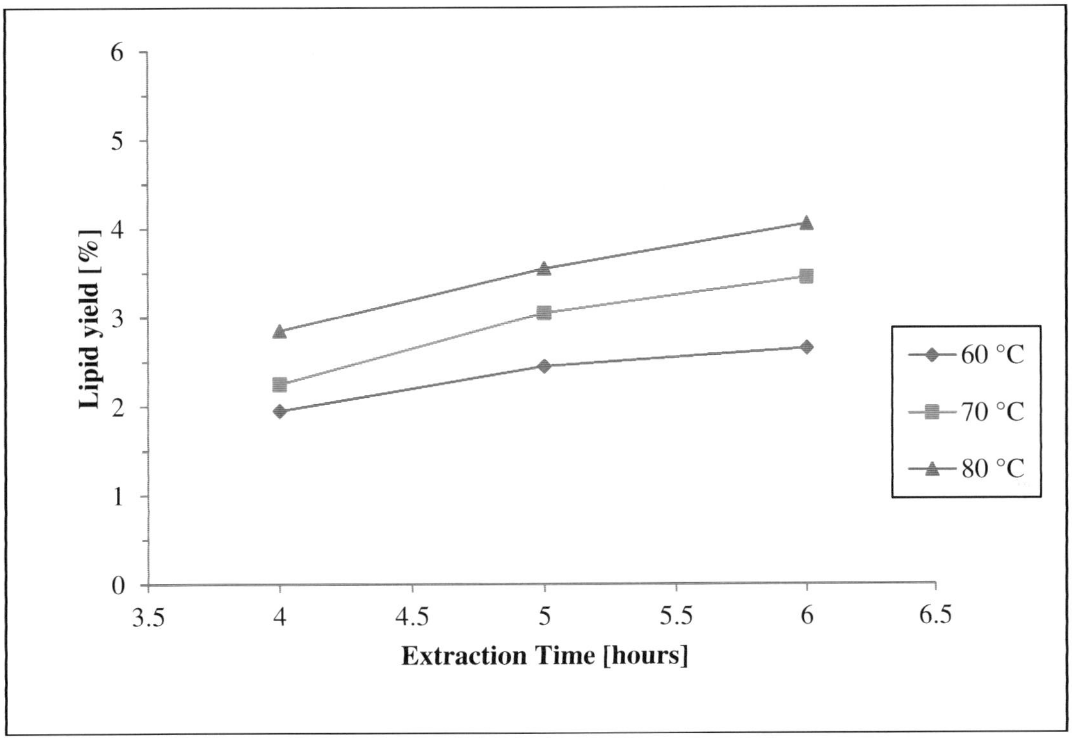

Figure 6. Lipids yield for extraction using soxhlet extraction and methanol as solvent.

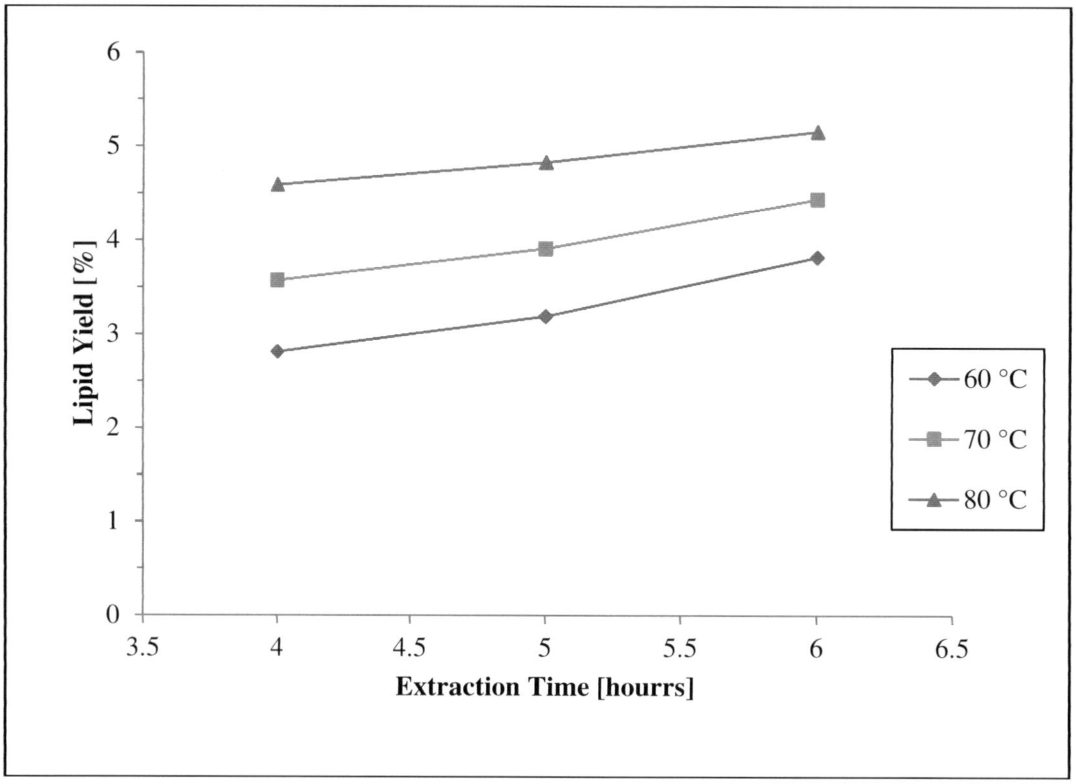

Figure 7. Lipids yield for extraction using soxhlet extraction and ethanol as solvent.

4. Conclusion

The extraction of lipids from sewage sludge utilizing SC-CO_2 extraction was successfully being conducted with lipids yield of 0.69 % within 0.5 hours at the operating temperature and pressure of 50 °C and 20 MPa. Lipids were easily separated subsequently from the SC-CO_2 system when CO_2 is being released in gas form through the outlet valve during lipids collection. Whilst soxhlet extraction using methanol and ethanol as solvent (sludge: solvent ratio of 1:10) managed to extract 1.95 % and 2.81 % within 4 hours of extraction time at 60 °C, with the additional time needed to separate the lipids from solvent by evaporation.

Limitation on the usage of SFE-CO_2 in the industrial process is always associated with low lipids yield, high initial investment costs as a stand-alone technology and the end-product, which is not at of high-value which hinder the further development of this technology especially in lipids extraction from sewage sludge. Hence, further exploration is needed to be conducted in SC-CO_2 lipids extraction from sewage sludge specially to increase the lipids yield such as the variation of SC-CO_2 techniques (static, dynamic and combination mode), co-solvent usage and optimization of extraction parameters (temperature, pressure, extraction time and CO_2 flow rate) and the quality to be used for the high-value product. Techno-economic and life cycle assessment of SFE-CO_2 in this application also can be conducted as a whole scheme (wastewater treatment, sludge management and disposal, lipids extraction and end-product refinery) rather than a stand-alone technology to assess the sustainability of integrating SFE-CO_2 in the overall processes.

Acknowledgement

This work was supported by UniKL Excellent Research Grant Scheme (UniKL/CoRI/UER20016). The authors would like to acknowledge Indah Water Konsortium Sdn. Bhd. (IWK Malacca) for the sewage sludge provided. The authors also would like to thank Mr. Charles Bidi anak Amban and Mr. Muhamad Alif Bin Adam for their contribution to this project.

References

[1] Peccia J and Westerhoff P 2015 We Should Expect More out of Our Sewage Sludge *Environ. Sci. Technol.*

[2] Syed-Hassan S S A, Wang Y, Hu S, Su S and Xiang J 2017 Thermochemical processing of sewage sludge to energy and fuel: Fundamentals, challenges and considerations *Renew. Sustain. Energy Rev.* **80** 888–913

[3] Yang G, Zhang G and Wang H 2015 Current state of sludge production, management, treatment and disposal in China *Water Res.* **78** 60–73

[4] Siddiquee M N and Rohani S 2011 Lipid extraction and biodiesel production from municipal sewage sludges: A review *Renew. Sustain. Energy Rev.* **15** 1067–72

[5] Dufreche S, Hernandez R, French T, Sparks D, Zappi M and Alley E 2007 Extraction of Lipids from Municipal Wastewater Plant Microorganisms for Production of Biodiesel *J. Am. Oil Chem. Soc.* **84** 181–7

[6] Kargbo D M 2010 Biodiesel Production from Municipal Sewage Sludges *Energy & Fuels* **24** 2791–4

[7] Olkiewicz M, Caporgno M P, Fortuny A, Stüber F, Fabregat A, Font J and Bengoa C 2014 Direct liquid-liquid extraction of lipid from municipal sewage sludge for biodiesel production *Fuel Process. Technol.* **128** 331–8

[8] Demirbas A, Bamufleh H S, Edris G and Al-Sasi B O 2017 Biodiesel production from lipids of municipal sewage sludge by direct methanol transesterification *Energy Sources, Part A Recover. Util. Environ. Eff.* **39** 800–5

[9] Liu S, Luo T, Liu G hua, Xu X, Shao Y, Qi L and Wang H 2020 Characterization and reutilization potential of lipids in sludges from wastewater treatment processes *Sci. Rep.* **10** 1–10

[10] Aiman N, Mustapha H, Sing W and Rahman R A 2017 Optimisation of Lipid Extraction from Primary Sludge by Soxhlet Extraction *Chem. Eng. Trans.* **56** 1321–6

[11] Cea M, Sangaletti-Gerhard N, Acuña P, Fuentes I, Jorquera M, Godoy K, Osses F and Navia R 2015 Screening transesterifiable lipid accumulating bacteria from sewage sludge for biodiesel production *Biotechnol. Reports* **8** 116–23

[12] Hyatt J A 1985 Liquid and Supercritical Carbon Dioxide as Organic Solvents *J. Org. Chem.* **50** 3246

[13] Tutek K, Masek A, Kosmalska A and Cichosz S 2021 Application of fluids in supercritical conditions in the polymer industry *Polymers (Basel).* **13** 1–17

[14] Pourmortazavi S M and Hajimirsadeghi S S 2007 Supercritical fluid extraction in plant essential and volatile oil analysis *J. Chromatogr. A* **1163** 2–24

[15] Hossain M S, Nik Ab Rahman N N, Balakrishnan V, F M Alkarkhi A, Ahmad Rajion Z and Ab Kadir M O 2015 Optimizing supercritical carbon dioxide in the inactivation of bacteria in clinical solid waste by using response surface methodology *Waste Manag.* **38** 462–73

[16] Chandra R, Vishal G, Sánchez C E G and Uribe J A G 2020 *Bioreactor for algae cultivation and biodiesel production* (INC)

[17] Wen D, Jiang H and Zhang K 2009 Supercritical fluids technology for clean biofuel production *Prog. Nat. Sci.* **19** 273–84

[18] Choi O K, Song J S, Cha D K and Lee J W 2014 Biodiesel production from wet municipal sludge: Evaluation of in situ transesterification using xylene as a cosolvent *Bioresour. Technol.* **166** 51–6

[19] Hag Ibrahim S N and Hamza E A 2017 Optimization of Lipid Extraction from Municipal Scum Sludge for Biodiesel Production Using Statistical Approach *Int. J. Renew. Energy Dev.* **6** 171–9

[20] Rahmat S, Kusin F M and Wan Johari W L 2016 Performance assessment of centrifuge dewatering unit using multivariate statistical approach: a case study of a centralized sludge treatment facility (CSTF) in Malacca, Malaysia *Desalin. Water Treat.* **57** 3904–15

[21] Abarca-Vargas R, Peña Malacara C F and Petricevich V L 2016 Characterization of chemical compounds with antioxidant and cytotoxic activities in bougainvillea x buttiana holttum and standl, (Var. rose) extracts *Antioxidants* **5**

[22] Bøwadt S and Hawthorne S B 1995 Supercritical fluid extraction in environmental analysis *J. Chromatogr. A* **703** 549–71

IOP Conf. Series: Materials Science and Engineering 1195 (2021) 012055 doi:10.1088/1757-899X/1195/1/012055

Fundamental study of colloidal stability and dispersion of novel nanosized conductive polyaniline (PANI)/prevulcanised latex (PVL) film for antimicrobial applications

M F Banjar[1], H D Suphi[1], M I Sarizan[1], A N A Yahaya[1], N A Khalil[1], M Singh[3] and M Zulkifli[2]

[1] Universiti Kuala Lumpur, Branch Campus Malaysian Institute of Chemical and BioEngineering Technology, 78000 Alor Gajah, Melaka, Malaysia
[2] Green Chemistry and Sustainability Cluster, University Kuala Lumpur, Branch Campus Malaysian Institute of Chemical and Bioengineering Technology, Taboh Naning, 78000 Alor Gajah, Melaka, Malaysia
[3] Malaysian Rubber Board, RRIM Research Station, 47000 Sungai Buloh, Selangor, Malaysia

Email: faizar.banjar16@s.unikl.edu.my;muzafar@unikl.edu.my

Abstract. A smart material possessed enhanced conductivity integrated in prevulcanized latex (PVL) film produced throughout this work. Also recognizing the synthesis route of PANI was vast and vary, choosing suitable method was great importance corresponding to the aim of study. PANI was prepared through chemical oxidative polymerization of aniline carried out in aqueous solution which aniline dissolved in strong acidic solution (1 M HCl) with the presence of Ammonium Persulphate (APS) as the oxidizing agent and Sodium Dodecyl Sulphate (SDS) as surfactant. However, PANI was readily in acidic condition while PVL in basic and consequently causes a state of immiscibility upon mixing. Hence, PANI formed then mixed with 0.1 – 0.5 % KOH *via* homogenizer to increase the pH and maintain the homogeneity as well as dispersion to be combined with PVL. Various studies on PANI synthesis and incorporation with latex had been reported but very limited in focusing the colloidal and dispersion stability of the mixture. Zeta potential measurements revealed an effective dispersion and the colloidal stability as the pH of PANI increases. Analysis of mechanical performance using Universal testing Machine revealed that addition of PANI improves greatly in novel film tensile strength and Young's Modulus by 109 % and 230 %, respectively.

1. Introduction

Polyaniline (PANI) had gained a lot of attention among researchers due to the reversible doping/de-doping character, pH switching properties, good environmental stability and modifiable electrical conductivity [1]. This conducting polymer has been studied for potential applications including anticorrosion coatings, lightweight battery electrodes and sensors. However, PANI on its own do have few drawbacks such as poor mechanical properties, low infusibility and low solubility in organic solvents limiting its usage. Introduction of secondary component into PANI further optimizes its functionality and enhanced performance. Usage of conductive polymers such as PANI and polypyrrole (PPy) combined with natural rubber latex had been reported broadly [2, 3]. This combination create a beneficial values especially in developing surgical gloves with enhanced conductivity that acts as antimicrobial defense agent and low percolation threshold. Current surgical procedures do involve hard practices and creates surgical site infections (SSI) which arise due to the failure of these gloves to protect the transmission of skin flora from surgeon's hand to the surgical site. SSI is the most costly healthcare-associated infection type and associated with huge amount of additional inpatient-days annually [4].

Increasing the thickness of gloves appear to be inefficient since more cost will be required for the manufacturing processes or raw material. In this work, a smart material presented as solution and putting together the best method to develop a nanosized conductive polyaniline (PANI) with ammonium persulfate (APS) as initiator as well as surfactant, sodium dodecyl sulfate (SDS). PANI synthesized through one-step redox reactions where chemical oxidative polymerization and formation of nanoparticle take places. Production of nanostructures creates advantages where new physio-chemical properties and new functional materials could be yield [5]. PANI nanofibers could be obtained by suppressing secondary growth of irregular shaped polyaniline. Simple approaches taken by rapidly mixing the initiator solution into monomer solution.

However, limited papers were focusing on the colloidal and dispersion stability which are very vital characteristic in determining the stability of latex mixture [6]. Therefore, this work also concern on this aspect and carried out zeta potential analysis that reflect the stability of colloidal dispersions. Challenge was there as PANI was readily in acidic condition, attributed by the synthesizing method which mainly involve hydrochloric acid while PVL was base in nature. Different pH of substances to be mixed tend to create problem in dispersion, consequently prevent a homogenized mixture [7]. Realizing this particular issues, addition with KOH increased the pH and improve the homogeneity. The mechanical performances also evaluated in order to ensure further effectiveness of PANI upon addition. Manipulation of PANI loading in PVL was studied throughout this work.

2. Experimental section

2.1. Materials

Aniline monomers with density of 1.02 g/ml and molecular weight of 93.13 g/mol and ammonium persulfate (APS) were purchased from Bendosen. Sodium dodecyl sulfate (SDS) obtained from R&M Chemicals Sdn Bhd. Hydrochloric acid (HCl) purchased from Sigma Aldrich while PV (pre-vulcanized) latex obtained from Getahindus (M) Sdn Bhd., and used as received without further purification.

2.2. Synthesis of PANI

PANI was prepared *via* chemical oxidative polymerization. The molar ratio of aniline monomers to oxidizing agents or initiator (SDS) is 1:1. The work began with mixing 36.6 mL of aniline into 2 L of 1.0 M HCl solution inside 5 L beaker and stirred for 15 minutes with 600 rpm speed. Then, 60 g of SDS added into the solution and further stirred for 45 minutes. On the other hand, preparation of APS solution carried out by mixing 91.2 g APS into 2 L of 1.0 M HCl solution. Next, the monomer and initiator solutions (in 1.0 M HCl) were rapidly mixed together all at once under ice bath condition. Using mechanical stirrer with fixed stirring speed (600 rpm), the solution stirred for 5 hours to disperse the monomer and initiator molecules evenly before polymerization. Polymerization of aniline occurred as the colour changes from colourless to dark green. Lastly, the sample sealed with aluminum foil and stored in refrigerator for 24 hours.

2.3. Synthesis of PANI/PVL film

After being stored in refrigerator for 24 hours, purification of PANI done by filtration in Buchner setup. The sample then heated using hot plate at temperature 80 °C to remove HCl and soluble oligomers *via* evaporation process, and becomes precipitated solution. With consistent stirring, 30 mL of 1.0 M potassium hydroxide (KOH) solution was added drop by drop till the pH reached 7. While reaching neutral condition, the sample was taken to be analyzed further in term of zeta potentials and pH correlations. Different PANI loading were incorporated into PVL, 3 phr, 5 phr, 10 phr, 20 phr and 40 phr. After weighted specifically, the samples were added into PVL and mixed using glass rod and casted into 50 mm × 50 mm × 10 mm casting plate. The film kept dry at ambient temperature for 24 hours.

2.4. Mechanical performance
Mechanical performance of each sample were performed using Universal Testing Machine characterizing in term of tensile strength (TS), Young's modulus (YM) and elongation at break (EB). The prepared samples evaluated for tensile test according to ASTM D3039 with strain rate of 500 mm/min. The films cut into standardized size with constant rectangular section, 2.6 mm wide, 100 mm long and 10 mm gauge length.

2.5. Colloidal and dispersion stability
The colloidal and dispersion stability examined using zeta potential analyser as a function of pH. The instrument works by using electrophoretic light scattering (ELS) and laser Doppler velocimetry (LDV) to determine the electrophoretic mobility, μ.

3. Results and discussion

3.1. Mechanical performance

3.1.1. Tensile strength. The tensile strength of PANI/PVL sample with different PANI content was observed and displayed in figure 1. With control sample, pure PVL with tensile strength of 17.39 MPa,

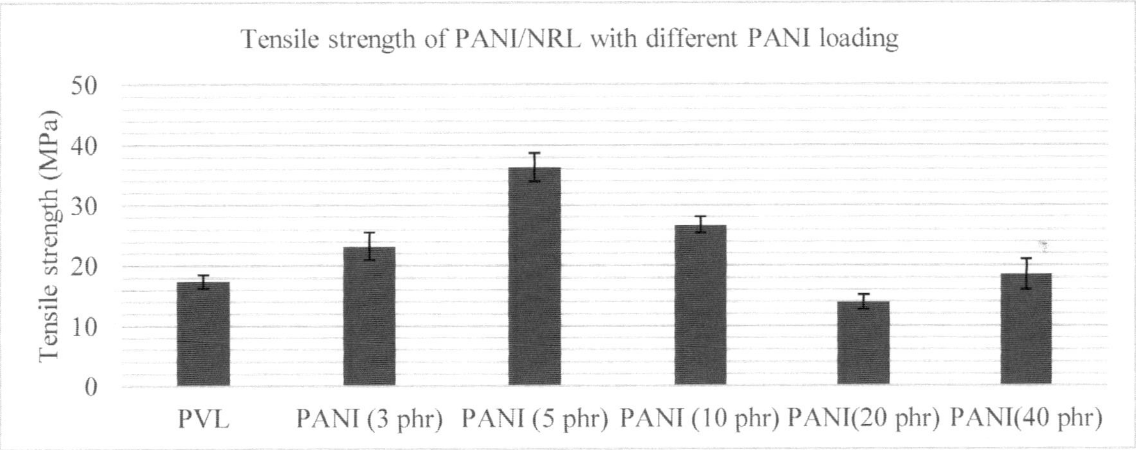

Figure 1. Tensile strength of PANI/PVL with different PANI loading.

comparable results obtained. The tensile strength of PANI/PVL composites depends on the effectiveness of stress transfer between matrix and fillers. The applied stress from tensile test able to be transferred effectively to the well-bonded PANI particles and improves the film strength significantly. However, further addition of particles could lead to strength reduction [8]. From the graph it was obvious that addition of PANI into PVL increase the tensile strength than pure PVL films except for 20 phr loading with 19.89 % decrement. The decrease in tensile strength at higher PANI loading is due to particle-particle aggregation and consequently poor interaction with rubber matrix [8]. Optimum reinforcement effect offered by 5 phr PANI/PVL film with 109.14 % increment, which attributed to the rigidity of PANI [9].

3.1.2. Young's modulus. Young's Modulus (YM) expressed as the ratio of stress to strain in elastic linear region or measurement of stiffness (MPa). The blank sample, PVL hold YM of 1.34 MPa at figure 2, incorporation of PANI into PVL shows significant increase in Young's modulus as might be expected

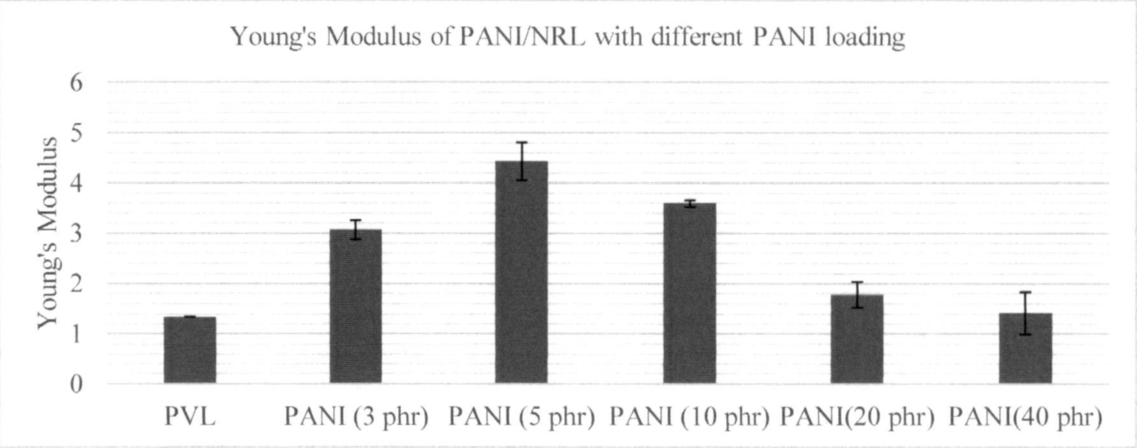

Figure 2. Young's modulus of PANI/PVL with different PANI loading.

for an elastomer matrix filled with particulate filler. Adding up the loading to 5 phr, increasing trend of modulus obtained and further decrease as more loading applied. Such optimum improvement up to 230.56 % is due to the fact that PANI act like crosslinking points and restricting the movement of PVL chains [10].

3.1.3. Elongation at break. Elongation at break (EB) indicates the highest extension of sample could achieved under tension. Elongation at break for PANI/PVL film with different loading shown in figure 3. The EB depends on polymer-filler interaction. EB greatly reduced as PANI incorporated into PVL at

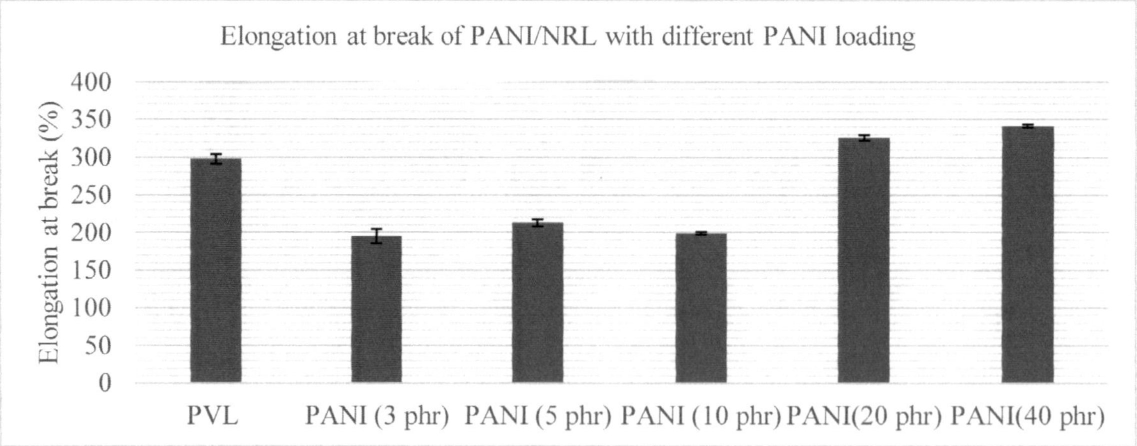

Figure 3. Elongation at break of PANI/PVL with different PANI loading.

3 to 10 phr loading, and increase later at higher amount of loading. The reduction in elongation at break at lower PANI loading is apparently due to the increase in Young's modulus as stiffer materials are expected to not elongate as much as soft materials [11].

3.2. Colloidal and dispersions stability

In this characterization, zeta (ζ) potential as a function of pH was carried out for samples of PANI during pH adjustment. The zeta potential is an important indicator of the stability of colloidal dispersions. The magnitude itself reflects on the degree of electrostatic repulsion between similar charged particles in a dispersion. Referring to figure 4, note that the ζ is mostly negative for the whole pH, and it increase in

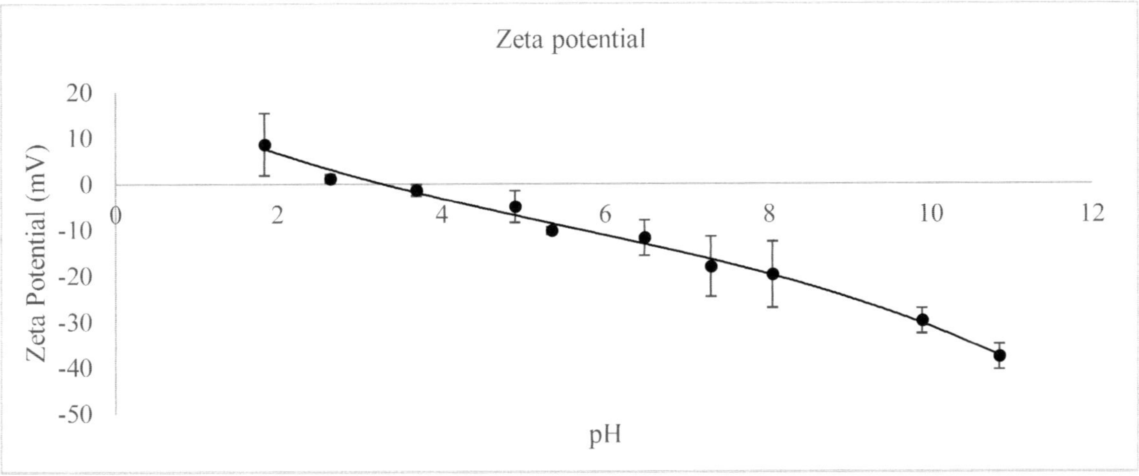

Figure 4. Zeta potential of PANI as function of pH.

absolute value till up to a pH close to electroneutrality. The chemical species involved is responsible for the generation of negative surface charge corresponding to dissociated sodium dodecyl Sulphate (SDS) end molecules. This is practicable, since it can likely described that part of the SDS used in synthesis adsorbed on surface of PANI even after cleaning process. Moreover, these results could qualitatively explain the stability. At acidic condition (pH below 7), the zeta potential value is low (10 to -10 mV), consequently formation of aggregates can occur at faster rate. Hence, the average particle-particle distance will be rather large. In contrast, at pH 7 and above, the Electrokinetic potential of particles is high (-18 to -38 mV) and this will encourage strong electrostatic repulsion between PANI particles. The film can be considered colloidally stable and the average particle-particle distance will be small, and particles will stay at primary minimum of their interaction potential [12].

4. Conclusions

Prevulcanized latex (PVL) films can be modified with conducting polymer, polyaniline. The colloidal stability is maintained only if the pH of aniline adjusted to neutral state to prevent coagulation during film synthesis. The mechanical properties of PANI/PVL films are sufficient for gloves or antimicrobial applications as PANI loading increased up to 5 phr with optimum TS of 36.37 MPa and YM of 4.43 MPa. However, PANI incorporation have poor capability at low loading when prioritizing EB performance but slightly improved as more PANI used up (20 to 40 phr).

Acknowledgments

This work was supported by Ministry of Higher Education (MoHE), Malaysia under Fundamental Research Grant Scheme (FRGS/1/2020/TK0/UNIKL/02/4).

References

[1] Sen T, Shimpi N G and Mishra S 2016 Synthesis and sensing applications of polyaniline *RSC Adv.* **6** 42196–222

[2] Mahmood W A K and Azarian M H 2015 Thermal, surface, nanomechanical and electrical properties of epoxidized natural rubber (ENR-50)/ polyaniline composite films *Curr. Appl. Phys.* **15** 599–607

[3] Al-Dulaimi A A and Wanrosli W D 2016 Preparation of colloidal properties and water dispersible conductive polypyrrole nanocomposite of nanocrystalline cellulose *Polym. Compos.* **24** 695–701

[4] National Healthcare Safety Network 2021 Surgical site infection event (SSI) procedure-associated module 2021 pp 1–39

[5] Skotheim T A and Reynolds J 2007 *Handbook of Conducting Polymers Third Edition* ed Skotheim T A and Reynolds J (London: CRC Press) chapter 7 pp 1–39
[6] Sarfraz M, ur Rehman Z and Ba-Shammakh M 2019 Pursuit of electroconducting thermoplastic vulcanizates: activated charcoal-filled polypropylene/ethylene–propylene–diene monomer blends with upgraded electrical, mechanical and thermal properties *Polym. Bull.* **76** 2005–20
[7] Hu W, Chen S, Yang Z, Liu L and Wang H 2011 Flexible electrically conductive nanocomposite membrane based on bacterial cellulose and polyaniline *J. Phys. Chem.* B. **115** 8453–7
[8] Fu S Y, Feng X Q, Lauke B and Mai Y W 2008 Effects of particle size, particle/matrix interface adhesion and particle loading on mechanical properties of particulate–polymer composites *Compos. B. Eng.* **39** 933–61
[9] Silva M J, Sanches A O, Malmonge L F and Malmonge J A 2014 Electrical, mechanical, and thermal analysis of natural rubber/polyaniline-Dbsa composite *Mater. Res.* **7** 59–63
[10] Kalasad M N, Gadyal M A, Hiremath R K, Mohamed Ikram I, Mulimani B G, Khazi L M, Anantha Krishnan S K and Rabinal M K 2008 Synthesis and characterization of polyaniline rubber composites *Compos. Sci. Technol.* **68** 1787–93
[11] Anand K, Varglese S and Kurian T 2015 Effect of micro and nano zinc oxide on the properties of pre-vulcanized natural rubber latex films *Prog. Rubber Plast. Recycl. Technol.* **31** 145–56
[12] Shin J S, Kim J H and Cheong I W 2005 Colloidal and thermal stability of polyaniline-coated multi-core shell polystyrene latexes prepared using sulfonated N-hydroxyethyl aniline *Synth. Met.* **151** 246–55

Membrane pervaporation performance applied for brackish water prepared by vacuum impregnation method

N Huda[1,2], E Lulu Atika Rampun[2], R Ayu Lestari[1], Y Raharjo[3], D Heri Yuli Yanto[4] and M Elma[1,2]

[1] Chemical Engineering Department, Faculty of Engineering, Lambung Mangkurat University, Jl A. Yani KM 36, Banjarbaru, South Kalimantan, Indonesia 70714
[2] Materials and Membranes Research Group (M^2ReG), Lambung Mangkurat University, Jalan A. Yani KM 36, Banjarbaru, South Kalimantan, Indonesia 70714
[3] Membrane Science and Technology Research Group, Chemistry Department, Faculty of Science and Technology, Universitas Airlangga, 60115 Surabaya, Indonesia
[4] Research Center for Biomaterials, Indonesian Institute of Sciences, Jl. Raya Bogor Km.46, Cibinong Science Center, Cibinong, Bogor 16911, Indonesia

Email: melma@ulm.ac.id

Abstract. Coating method and number of membrane layer are crucial factors on membrane performance. Through a vacuum impregnation method allows a sol solution uniformly fill into membrane support and it is required only less solution. The aim of this study is to apply vacuum impregnation method through vacuum calcination and air calcination during fabrication of silica membranes and to investigate the effect of layer variations on silica membranes performance to apply for brackish water. The sol solution was made from TEOS as silane precursor, ethanol and dual catalysts (citric acid + ammonia). Alumina membrane support was coated by vacuum impregnation method and calcined the membrane under air and vacuum condition. From the FTIR result, it indicates that silica membranes calcined in air and vacuum calcination have Si-O-Si and Si-OH. The vacuum impregnation obtained smoother surface membranes. The silica membrane calcined *via* vacuum calcination performs excellent water fluxes and salt rejection of 22.01 $kg.m^{-2}.h^{-1}$ and 98.98 %. If compare to silica membranes calcined in air, the water flux (19.11 $kg.m^{-2}.h^{-1}$) and salt rejection (98.75 %). It also found the two layers silica membrane is better than three layers for the membrane performance result.

1. Introduction

WHO which stand for the global organization concerning about world health states up to 15 % of human population in the world has difficulty to get clean water. The biggest issue around the world is water scarcity. Especially people who live in wetland areas, river is an important water source for their daily activities. The water becomes salty due to seawater intrusion into river [1], especially during rainy season [2]. So that, wetland saline water has an abundant amount so that it can be used as a source of clean water [3]. Currently, pervaporation is used for treat saline water into clean water through desalination. This high water salt content may impact the environment if it may not be treated [4].

Membrane technology is widely chosen to solve the problem of access to clean water because it has advantages that are eco-friendly, energy efficient, high efficiency and easy to operate [5]. Zeolite and silica are types of membranes that are often used in desalination [6, 7]. Membranes used are generally composed of a porous material-thin-film-medium that functions to separate particles with a specific molecular size from the solution [2, 8, 9].

Silica membrane has several advantages, namely high molecular sieving when compared to zeolite membrane [6], resistant with high temperatures [10] and have a good performance [11]. This unique properties can be applied to remove the salt particles in the water [11, 12]. In addition, it has a pore size ranging from 3-5 Å according to the size of the water kinetics diameter (dk = 2.6 Å) [6]. However, it can block the hydrated salt ions (Na+ dk = 7.2Å) and Cl in the desalination process [13].

Many coating methods on membranes have been introduced such as dip-coating [14–20], electrospinning [21] and vacuum impregnation [22, 23]. Non-destructive method for presenting a solution of a certain composition into a porous matrix it is called vacuum impregnation. Mechanically induced difference pressure can transpires the mass transfer in vacuum impregnation [24]. Currently, the vacuum impregnation method is getting attention by researchers because it can minimize the possibility of physical damage to the membrane and can save membrane coating time [22, 25]. The structural formation of membrane can be easily control through vacuum impregnation [23].

Other parameters that can also affect the membrane performance are the number of membrane layers and calcination process [26]. Therefore, the aim of this research is to compare a vacuum calcination and air calcination in a silica membrane fabrication through vacuum impregnation and to investigate the effect of layer variations (2 layers and 3 layers) on the performance of silica membranes to 0.3 % NaCl (brackish water) as feed.

2. Methodology

2.1. Preparation and characterization of pure silica xerogel

Silica sol synthesis ware carried out by sol-gel method employing tetraethyl orthosilicate (TEOS, Merck), ethanol 96 % (EtOH), the dilute nitric acid (0.00078 M, HNO_3, Merck) and ammonia (NH_3, Merck). solution diluted with ethanol. First, TEOS was added drop-wise into ethanol and stirred for 5 min in an ice bath subsequently added HNO_3 by drop it slowly within 15 min. Then the solution was refluxed for 1 hour at 50 °C on a hot plate. After that, add the base catalyst in the form of ammonia which has been diluted first by drop-wise and reflux continued for 2 hours at 50 °C. At the end of 3 h refluxing the pH of the sol is measured to show pH 5 or pH 6. Then, dry the sol-gel at 60 °C for 24 hours to obtain xerogel and then grind it t`o a powder and grind it into a powder. Xerogel was then calcined at 600 °C for 4 hours. Xerogel characterization using FTIR test (Fourier transform infrared spectrometer).

2.2. Membrane preparation and characterisation

The pure sols were firstly put into quartz tube. Then, thin film membrane was coated on membrane support alumina-based tubular (α-Al_2O_3 ϕE100 nm), using the pure sols *via* a vacuum impregnation method with a dwell time of 2 min. After the coating process on each layer of the membrane, it is then dried by air and vacuum calcination for 1 h. This step was conducted for 2, 3, and 4 time to get varied number of thin film layers by vacuum impregnation method. The final membrane were evaluate by pervaporation to desalination of saline water. Vacuum impregnation set-up presented in figure 1 below:

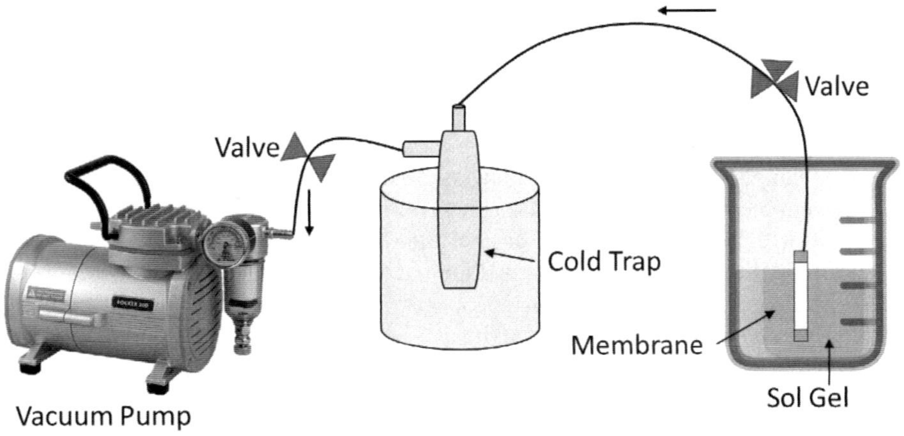

Figure 1. Set-up impregnation method.

The pervaporation process in this research is presented in figure 2. The membrane was installed by dead end system which the one side connected to a cold trap (immersed in liquid nitrogen) and another side connected to vacuum pump operating at 1.5 kPa. The feed solution used was made of sodium chloride (NaCl, Sigma-Aldrich) dissolved in deionized water at a concentration of 0.3 %wt and the temperature used in this research is room temperature (□25 °C). The feed solution was stirring constantly at 500 rpm and recirculation to avoid the concentration polarization on the contact membrane side. The water flux, F (kg m^{-2} h^{-1}), can be calculated using the equation $F= m/(A\Delta t)$, where m is the mass of permeate (kg) left in the cold trap, A is the surface-active area (m^2) of the membrane and Δt is the time measurement (h). The salt rejection, R (%), can be calculated using the equation $R=(C_f-C_p)/C_f$ x 100 %, where C_f and C_p are the feed and permeate concentrations of salt (wt%). The concentration of salt has a correlation with the conductivity of the solution, the conductivity is measured using a conductivity meter.

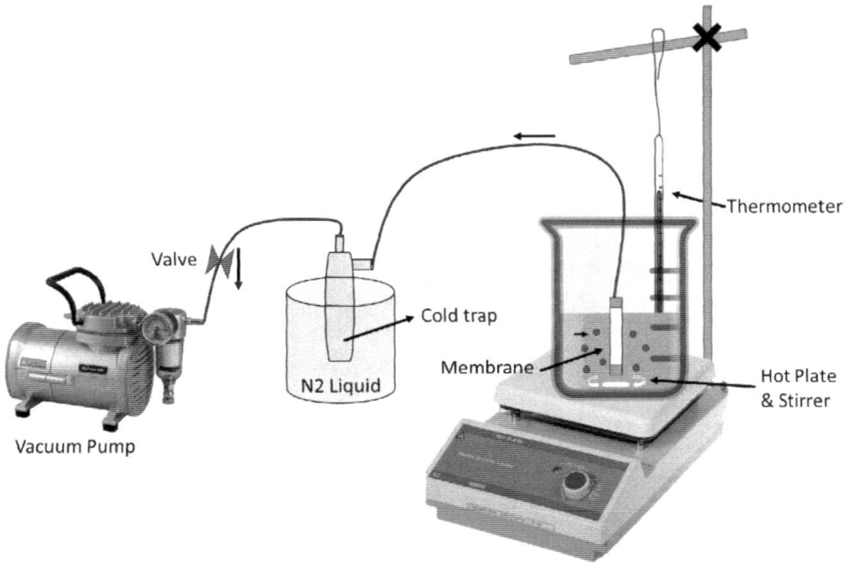

Figure 2. Set-up for membrane pervaporation.

3. Result and Discussion

This study, the sol-gel method in the process has chosen because it is the most effective, simplest, cheapest when compared to the other methods [27]. It can also easily modify the pore size of the membrane and its morphological structure. The addition of a two-step acid-base in a sol-gel solution results in a relatively combination of small and large structures called bottleneck pores [4]. Firstly, the sol gel process is done under acidic conditions that lead to the formation of many silanol groups. The final silica sol pH is 6, which the pH is above the isoelectric point limit (pH 1-3) of the silica bridges [28, 29]. In this pH, the siloxane bridges formed high concentration due to the silanol group is to be deprotonated participating in the polycondensation reaction [13].

The FTIR (Fourier Transform Infrared) spectrum in the region of 1400-750 cm^{-1} of the calcined xerogel sample at temperature of 600 °C through air calcination shows in Figure 3. The peak wavelength of 1080 cm^{-1} represents siloxane (Si-O-Si). While the initial peak at 973 cm^{-1} shows vibrations of the silanol (Si-OH) group. Previous study has reported that silanol and siloxane appeared in silica membranes calcined in air and vacuum calcination [13]. Silica material has a high resistance temperature up to 800 °C [30].

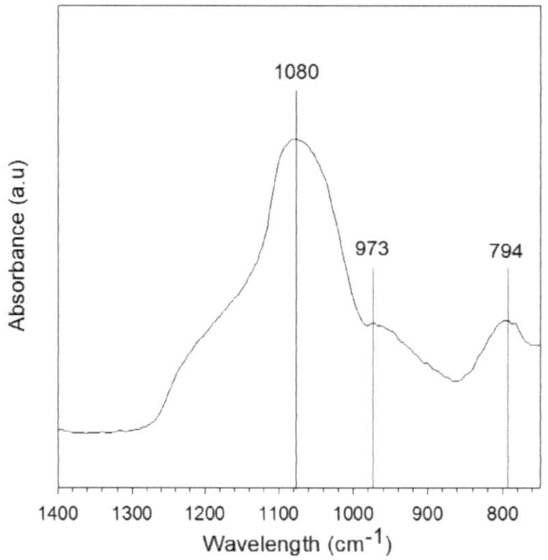

Figure 3. Spectrum FTIR xerogel pure silica.

Based on figure 3, FTIR spectrum of the pure silica xerogel has siloxane bonds [Si-O-Si] and silanol bonds [Si-OH]. Siloxane group is seen at wavelength of 1080 cm^{-1} while the silanol group is seen at a wavelength of 973 cm^{-1}. According to Elma, Yacou, Diniz da Costa and Wang [6], a low silanol concentration and a high siloxane concentration appeared in pH 6 of sol. In addition, the sol gel with pH 6 showed a tendency to form combination of micropores and mesopores. This pore size pattern could hinder small particles in micropore, yet allow the water passed through the membrane because of mesopores presence. A high salt rejection and water flux might be achieved [14]. Silanol are hydrophilic so that it reacts with water and forms fouling. Thus, silica network is designed to configure siloxane (Si-O-Si) and silanol (Si-OH) group [19].

Pure silica membrane fabricated by vacuum impregnation both calcined under air and vacuum condition have a clean surface, shining surface and no damage as shown in figure 4. This is because porous alumina domains provide the ideal conditions to hindered the thermal stresses induced while sintering process [13]. Therefore, there is no defect and crack on top layer of the membrane. Both of calcination types resulting in white color's membrane because it utilized the alumina as support. In our study, the vacuum impregnation only required less volume of solution than dip-coating process. The silica thin film attached on alumina support resulting in shining surface.

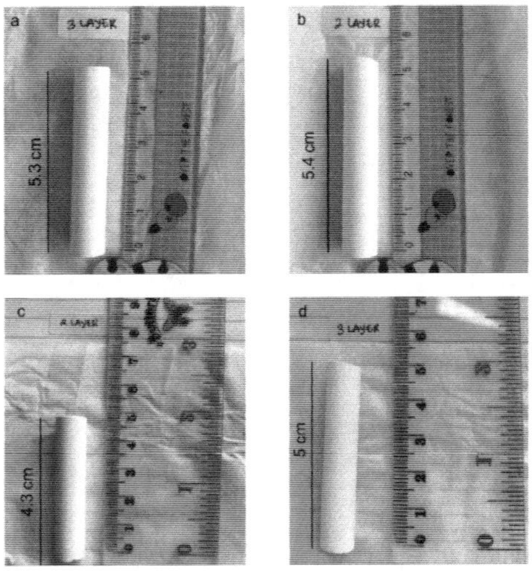

Figure 4. Picture of membrane pure silica (a) air calcination 2 layers (b) air calcination 3 layers (c) vacuum calcination 2 layers and (d) vacuum calcination 3 layers.

Figure 5 illustrates all pure silica membranes calcined with vacuum or air have relatively high rejection and water fluxes. The result shows the membranes water fluxes calcined under vacuum in variations of 2 layers and 3 layers are 22.01; 16.72 kg.m^{-2}.h^{-1}, respectively. While membranes calcined with air in variations of 2 layers and 3 layers are 19.11 and 12.46 kg.m^{-2}.h^{-1}, respectively. The highest water fluxes are found in membrane with 2 layers for both vacuum and air calcination. The salt rejection of membranes *via* vacuum calcination that have been tested with NaCl 0.3 % are 98.75 % for 2 layers and 99.94 % for 3 layers. On the other hand, the membranes made by air calcination at 2 layers and 3 layers are 98.98 % and 89.49 %, respectively. It happened because of the compound of silanol and siloxane presence in silica matrix. These results are in line with previous studies that reported a silica-based membranes have a good rejection ability and salt rejection >99 % [31].

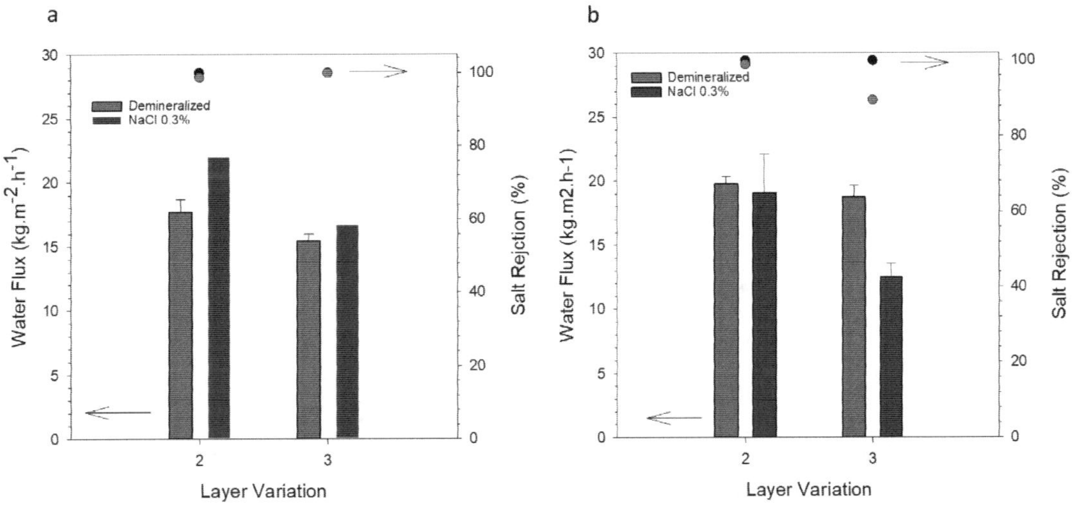

Figure 5. Membrane performance of (a) vacuum calcination and (b) air calcination.

4. Conclusions

Based on the results, pure silica membranes calcined in air and vacuum calcination have siloxane and silanol groups. All the silica membranes prepared through the vacuum impregnation obtained a shining surface because of the thin film attached on alumina support. The 2 layers of silica membranes *via* vacuum calcination (22.01 $kg.m^{-2}.h^{-1}$; 98.98 %) and air calcination (19.11 $kg.m^{-2}.h^{-1}$; 98.75 %) resulting in better performance than 3 layers for NaCl 0.3 % (brackish water). The highest water flux achieved from silica membrane calcined in vacuum and less membrane thickness.

Acknowledgments

Authors wishing to acknowledge assistance or encouragement and facility from materials and membranes research groups (M^2ReG) Lambung Mangkurat University as well as colleagues, special work by technical staff or financial support from Applied Research of Universities Grant 2021-2023, Basic Research Grant 2021-2022 and World Class Research Grant 2021-2023 Deputy of Research and Development National Research and Innovation Agency, Directorate General of Higher Education, The Ministry of Cultural and Education Republic of Indonesia.

References

[1] Elma M and Assyaifi Z L 2018 Desalination process *via* pervaporation of wetland saline water. In: *IOP Conf. Series: Earth Environ. Sci.*: IOP Publishing) 012009

[2] Elma M, Ayu R, Rampun E L, Annahdliyah S, Suparsih D R, Sari N L and Pratomo D A 2019 Fabrication of interlayer-free silica-based membranes–effect of low calcination temperature using an organo-catalyst *Membr. Technol.* **2019** 6–10

[3] Rahma A, Elma M, Mahmud M, Irawan C, Pratiwi A E and Rampun E L A 2019 Penyisihan bahan organik alami pada desalinasi air rawa asin menggunakan proses koagulasi-Pervaporasi *Jurnal Kimia Sains dan Aplikasi* **22** 85–92

[4] Elma M, Sari N, Pratomo D, Annadliyah S, Rampun E, Rahma A and Pratiwi A 2020 Organo-silica membrane for brine water pervaporation. In: *IOP Conf. Series: Earth Environ. Sci*: IOP Publishing) 012129

[5] Elma M and Riskawati N 2018 Silica membranes for wetland saline water desalination: performance and long term stability *E&ES* **175** 012006

[6] Elma M, Yacou C, Diniz da Costa J C and Wang D K 2013 Performance and long term stability of mesoporous silica membranes for desalination *Membr.* **3** 136–50

[7] Tsuru T, Igi R, Kanezashi M, Yoshioka T, Fujisaki S and Iwamoto Y 2011 Permeation properties of hydrogen and water vapor through porous silica membranes at high temperatures *AIChE J.* **57** 618–29

[8] Pradhana E, Elma M, Othman M, Huda N, Ul-haq M, Rampun E L and Rahma A 2021 The functionalization study of PVDF/TiO2 hollow fibre membranes under vacuum calcination exposure. In: *J. Phys.: Conf. Series*: IOP Publishing) 012035

[9] Assyaifi Z L, Elma M, Syauqiah I, Rampun E L, Rahma A, Sumardi A, Lestari A E, Suryani L, Mustalifah F R and Huda N 2021 Photocatalytic–pervaporation using membranes based on organo-silica for wetland saline water desalination *Membr. Technol.* **2021** 7–11

[10] Elma M, Mahmud M, Akhbar A, Suryani L, Mustalifah F R, Rahma A, Rahmah D and Baity N 2020 Aplikasi membran silika-pektin untuk desalinasi air payau *Jukung (Jurnal Teknik Lingkungan)* **6**

[11] Rahman S K, Rahma A, Syauqiah I and Elma M 2020 Functionalization of hybrid organosilica based membranes for water desalination–Preparation using ethyl silicate 40 and P123 *Mater. Today: Proc.*

[12] Elma M and Hairullah A Z 2018 Desalination process *via* pervaporation of wetland saline water. In: *IOP Conf. Series: Earth Environ. Sci.* pp 1–6

[13] Elma M, Wang D K, Yacou C and da Costa J C D 2015 Interlayer-free P123 carbonised template silica membranes for desalination with reduced salt concentration polarisation *J. Membr. Sci.*

475 376–83
[14] Elma M 2018 Pembuatan silica thin film sebagai pelapis membran dari precursor TEOS (tetra ethyl orto silicate) *Quantum: Jurnal Inovasi Pendidikan Sains* **8** 78–82
[15] Elma M, Mahmud M, Rahmah A, Handayani N and Putri V S 2019 Pengaruh penambahan pektin sebagai pelapis membran interlayer-free silika-pektin melalui proses desalinasi air rawa asin. In: *Prosiding Seminar Nasional Lingkungan Lahan Basah,* pp 274–8
[16] Zhang T, Elma M, Xie F, Motuzas J, Zhang X and Wang D K 2018 Rapid thermally processed hierarchical titania-based hollow fibres with tunable physicochemical and photocatalytic properties *Sep. Purif. Technol.* **206** 99–106
[17] Yang H, Elma M, Wang D K, Motuzas J and da Costa J C D 2017 Interlayer-free hybrid carbon-silica membranes for processing brackish to brine salt solutions by pervaporation *J. Membr. Sci.* **523** 197–204
[18] Elma M 2020 Interlayer-free hybrid carbon-silica membranes for processing brackish to brine salt solutions by pervaporation *J. Membr. Sci.*
[19] Elma M 2020 Functionalization of Si-C using TEOS (tetra ethyl ortho silica) as precursor and organic catalyst
[20] Sumardi A, Elma M, Lestari A E, Assyaifi Z L, Darmawan A, Syauqiah I, Rampun E L A, Mawaddah Y and Wati L S 2021 Deconvolution of TEOS/TEVS xerogel by single or dual organic catalyst addition *Jurnal Kimia Valensi* **6** 208–14
[21] Krisnandika V E 2017 Produksi nanofiber dan aplikasinya dalam pengolahan air
[22] Song Y 2014 Carbon molecular sieve membranes for desalination
[23] Abd Jalil S N, Wang D K, Yacou C, Motuzas J, Smart S and da Costa J C D 2017 Vacuum-assisted tailoring of pore structures of phenolic resin derived carbon membranes *J. Membr. Sci.* **525** 240–8
[24] Radziejewska-Kubzdela E, Biegańska-Marecik R and Kidoń M 2014 Applicability of vacuum impregnation to modify physico-chemical, sensory and nutritive characteristics of plant origin products—A review *Int. J. Mol. Sci.* **15** 16577–610
[25] Song Y, Wang D K, Birkett G, Smart S and da Costa J C D 2017 Vacuum film etching effect of carbon alumina mixed matrix membranes *J. Membr. Sci.* **541** 53–61
[26] Ishigami T, Amano K, Fujii A, Ohmukai Y, Kamio E, Maruyama T and Matsuyama H 2012 Fouling reduction of reverse osmosis membrane by surface modification *via* layer-by-layer assembly *Sep. Purif. Technol.* **99** 1–7
[27] Purnomo P 2013 Sintesis α-Fe2O3 menggunakan metode sol gel *Institut Pertanian Bogor: Bogor*
[28] Brinker C J and Scherer G W 2013 *Sol-gel science: The physics and chemistry of sol-gel processing*: Academic press)
[29] Elma M, Sumardi A, Paramita A, Rahma A, Lestari A E, Yanto D H Y, Hadi S, Assyaifi Z L and Raharjo Y 2021 Physicochemical properties of mesoporous organo–silica xerogels fabricated through organo catalyst *Membr.* **11** 607
[30] Wang L J, Zhao S Y and Yang M 2009 Structural characteristics and thermal conductivity of ambient pressure dried silica aerogels with one-step solvent exchange/surface modification *Mater. Chem. Phys.* **113** 485–90
[31] Elma M, Pratiwi A E, Rahma A, Rampun E L A and Handayani N 2020 The performance of membranes interlayer-free silica-pectin templated for seawater desalination *via* pervaporation operated at high temperature of feed solution. In: *Mater. Sci. Forum*: Trans Tech Publ) pp 349–55

IOP Conf. Series: Materials Science and Engineering 1195 (2021) 012057 doi:10.1088/1757-899X/1195/1/012057

Chemical cleaning to evaluate the performance of silica-pectin membrane on acid mine drainage desalination

F R Mustalifah[1,3], A Rahma[3], Mahmud[2], Sunardi[4] and M Elma[1,3]

[1] Chemical Engineering Department, Engineering Faculty, Lambung Mangkurat University (ULM), Jalan A. Yani KM 36, Banjarbaru, South Kalimantan, Indonesia 70714
[2] Environmental Engineering Department, Engineering Faculty, Lambung Mangkurat University (ULM), Jalan A. Yani KM 36, Banjarbaru, South Kalimantan, Indonesia 70714
[3] Materials and Membranes Research Group (M^2ReG), Lambung Mangkurat University, Jalan A. Yani KM 36, Banjarbaru, South Kalimantan, Indonesia70714
[4] Chemistry Department, Earth and Science Faculty, Lambung Mangkurat University (ULM), Jalan A. Yani KM 36, Banjarbaru, South Kalimantan, Indonesia 70714

Email: melma@ulm.ac.id

Abstract. Pervaporation process is an excellent and potential way applied for desalting acid mine drainage water. Nevertheless, the water flux was reduced gradually due to the issue of membrane fouling. To resolve this problem, cleaning process was chosen to maintain the water flux of silica-pectin membranes. This study aims to recover the water flux and salt rejection of the silica-pectin membranes *via* chemical cleaning process applied for acid mine drainage water desalination with various temperature of feed water (25-60 °C). Silica-pectin membrane was formulated by employing TEOS functioning as silica precursor and pectin as carbon template from banana peels. Chemical cleaning of the membrane carried out by employing TiO_2 solution + UV light radiation for an hour. Performance of the silica-pectin membrane was evaluated *via* pervaporation process under dead-end system. The performance of silica-pectin banana peels membrane found flux recovery from 10.6 $kg.m^{-2}.h^{-1}$ and flux recovery of 17.54 $kg.m^{-2}.h^{-1}$. It shows that flux recovery higher than before backwashing process. Also, silica-pectin membrane results in all of the salt rejection <99 %. It is concluded that the chemical backwashing process is important to apply to recover the water flux of membrane, also, this process considers to save and reduce the operational costs.

1. Introduction

South Kalimantan, Indonesia has potential of natural mineral resources, it is found that the mining activities is still more important for economy region. Generally, mining activities in South Kalimantan, Indonesia using open pit mining method and generated Acid Mine Drainage (AMD) [1]. The AMD has negative effect to the environmental [2] . The characteristics of AMD on production activities/pit area have pH ± 4.28, high content of Fe reaching 9.1 mg/L and Mn content of 6.4 mg/L [3]. Modernise, product AMD has been processed using membrane technology with desalination process to reduce fouling or pollutants AMD [4]. Pervaporation in membrane desalination has preferred technique to convert to be portable water [5–9]. The pervaporation process has advantage of the difference partial pressure from the pump as the driving force in the feed water and the permeate [5, 9–11]. Silica membrane with Tetraethyl orthosilicate (TEOS) as main precursor in the polymerisation process have a longer life time for desalination water [12]. Nonetheless, silica still has limitation due to low hydro-stability properties [13], consequently a new innovation by addiction of pectin into the silica membrane can increasing hydro-stability properties of membrane [14, 15] The carbon content in pectin includes neutral sugar chains derived from complex polysaccharide compounds [16] strengthen the pore structure of the membrane and makes the membrane pore resistance stronger and has high salt rejection [4, 17, 18]. Hence, the large of pollutants will decrease water flux and performance of membrane, resulting in membrane fouling [19, 20].

Generally, fouling membrane type are reversible and irreversible. Particulates are deposited on the membrane such as colloids, macromolecules, salts and microorganisms. In other research there are various types of particulates that have an impact on membrane surface fouling, including inorganic particulates, dissolved organic matter, and dissolved solids [21, 22]. Impurity particles on the membrane surface impact on decreasing the permeate flux of the membrane, the difficulty of operating performance and the short life time and make decrease productivity of the system [23, 24]. Fouling membrane may reduce using backwashing membrane or chemical cleaning membrane [25]. There are type of backwashing membrane process, such as: a physical backwashing and chemical cleaning, but pollutants stronger adhering to the membrane surface, needed chemical materials to optimally cleaning. Especially photocatalytic and catalyst titanium dioxide (TiO_2) is a material that has advantages to removed fouling membrane [26]. The other advantages of using photocatalytic and TiO_2 such as catalyst in membrane separation can be recycled for reuse and increase flux recovery in the membrane [27].

Photocatalytic is a method that had founded with various applications in the reduction of persistent organic micro pollutants [28]. Concept of photocatalytic + TiO_2 as catalyst utilizes the concept of generating energy bandgaps in the excitation of photons which have semiconductor materials [29]. Photocatalytic-membrane application rehabilitated performance of the membrane [30]. In addition, TiO_2 as well as environmentally friendly catalyst [31, 32], and TiO_2 catalyst can reduce the activation energy which accelerates the reaction and pollutants are oxidised to carbon dioxide and water [33]. Utilising photocatalytic + catalyst TiO_2 methods in reducing pollutants, this work is aimed at water flux and salt rejection recovery of silica-pectin membrane by applying chemical backwash process evaluated on acid mine drainage water desalination with varous feed water (25 - 60 °C).

2. Methodology

2.1. Chemicals and materials

Acid Mine Drainage (AMD) taken from South of Kalimantan-Indonesia specially at the Kintap region. Membrane support in this study from a macro-porous alumina tubular with pore size ~100 nm. Membrane. Silica-pectin sols is fabricated from Tetraethyl Orthosilicate (TEOS, 99 %, Sigma-Aldrich), ethanol (EtOH, 96 %), ammonia (NH_3, 0.0003 M, Merck) and citric acid (HNO_3, 0.0078 M, Merck), pectin from banana peels (concentration 0.1 % and 0.5 %), glycerol (85 %, Merck). Chemical cleaning membrane in this study using liquid titanium dioxide (TiO_2, Merck) with photocatalytic UV (type C Philip 18 Watt).

2.2. Fabrication and characterisation silica-pectin membrane

Sol-gel membrane silica-pectin was method to product thin film membrane. pH sols at final process adjusted to pH ± 6. In other side, preparing carbon template from pectin which extract from banana peels. Prepared glycerol at 50 °C for 90 times to mix pectin-glycerol (0.1 % and 0.5 % pectin concentration). Final molar ratio was conducted from our previous work which become EtOH : TEOS :H_2O :HNO_3 :NH_3 : pectin, measure 38: 1: 5: 0.0008: 0.0003: x, where x was variation of pectin from banana peels concentration with

0.0063 grams to concentration 0.1 % and 0.032 grams to concentration 0.5 % [4]. Finishing process, the silica-pectin sols was coating in tubular membrane 1 hour for 1 layer until 4 layers. Membrane was characterisation preparing with dried sols into oven for 24 hours to product xerogel. Xerogel was calcined at 300 °C and 400 °C temperature using Rapid Thermal Processing (RTP). Finally, xerogel was ready to be characterised using FTIR spectra.

2.3. Membrane pervaporation

Performance and evaluation of membrane silica-pectin determine using pervaporation process. AMD water is used to determine the resistance and ability of silica-pectin membrane with measure of water flux, water flux recovery, and salt rejection. Feed water temperature was variations were carried out at room temperature (25 °C), 40 °C and 60 °C. Pervaporation was carried out for 20 minutes. Set up pervaporation was design as shown in figure 1.

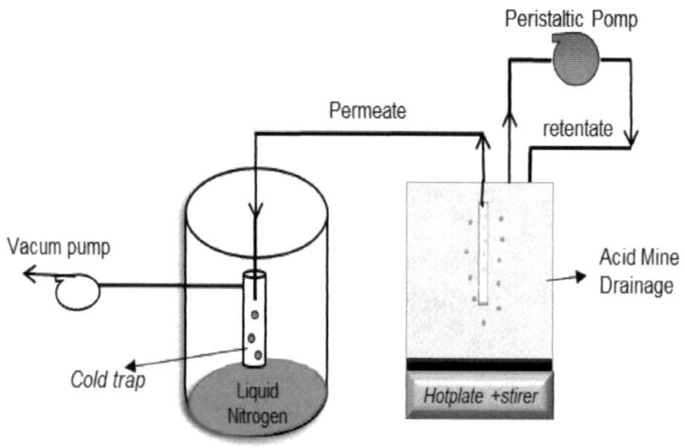

Figure 1. Set up pervaporation process.

Measuring of the conductivity of the permeate solution can use a conductivity meter [13]. Water flux of silica-pectin membrane was measured based equation:

$$F = \left(\frac{M}{A.dt}\right) \tag{1}$$

Wherein, mass of flux permeate (kg) collected in the clod trap (M), A is an activated-area on surface silica-pectin membrane (m^2) and operational times was simbol of dt (h). Salt rejection in silica-pectin banana peels membrane measured with the equations (2):

$$R = \left(\frac{CF-CP}{CF}\right) X\ 100\ \% \tag{2}$$

Wherein, *CF* is a permeate solution and *CP* is a retentate solution. Salinity of solution can measure using conductivity meter (OHAUS SF300C-G)

2.4. Chemical cleaning (Backwashing membrane)

Chemical cleaning to recovery performance of membrane silica-pectin prepared with hired submersion of membrane silica-pectin using titanium dioxide solid and accompanied by irradiation photocatalytic UV for an hour. Set up of chemical cleaning membrane as shown in figure 2.

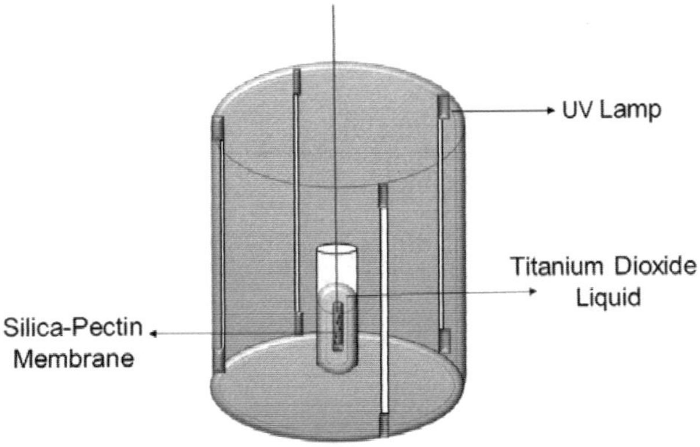

Figure 2. Set up chemical cleaning silica-pectin membrane.

3. Result and Discussion

3.1. Characteristic of Acid Mine Drainage
Acid Mine Drainage (AMD) as samples taken from Kintap, South Kalimantan-Indonesia was treated using pervaporation of membrane silica-pectin. Firstly, AMD was characterized with parameters which shown in table 1 and integrated with Indonesian regulatory standards regarding AMD for hygiene sanitation:

Table 1. Characterisation of acid mine drainage.

No.	Parameters	Unit	Value	Standard Indonesia's Ministry of Health No. 32 2017th
1.	Fe	mg/L	3.69	1
2.	Total Dissolved Solid	mg/L	652	-
3.	Conductivity	µS/cm	1400	-
4.	pH	-	4	6.5-8.5

3.2. Morphology and Characterisation of Silica-pectin Membrane
Acid Mine Drainage (AMD) as samples taken from Kintap, South Kalimantan-Indonesia was treated using pervaporation of membrane silica-pectin. Firstly, AMD was characterized with parameters which shown in table 1 and integrated with Indonesian regulatory standards regarding AMD for hygiene sanitation:

3.2.1. Scanning Electro-Microscopy Membrane. The structure of the silica-pectin membrane was analysed using Scanning Electron Microscopy (SEM) which shows on figure 3. Analysed SEM aims to determine the surface morphological structure of the membrane. The surface area of the membrane affected the performance of membrane infiltration process [13, 34].

Figure 3. SEM of membrane silica-pectin: (a) Cross section of membrane and (b) surface area of membrane.

The thickness of membrane silica-pectin was shown in figure 3(a) the thickness of membrane in this study ~2 µm. The membrane with thin film was thinner result higher of water flux [7, 35–37]. In this study, the silica-pectin membrane using interlayer free membrane to reduce the thickness of membrane and to low cost in production. Therefore, in this work dip-coating of membrane silica-pectin was carried out 4 times to result 4 layers. Figure 3(b) shows the surface area of membrane silica-pectin. It shown there are no cracks on surface area membrane. The membrane silica-pectin looks asymmetrical due the calcination technique. In this work, calcination process using RTP method. There is an instantaneous temperature increasing during the calcined process [38, 39]. This technique has the advantage of speeding up the manufacture of membranes with still good results.

3.2.2. Fourier Transform Infra-Red. Silica-pectin membrane was analysed using FTIR-ATR technique. Figure 4(a). Shows region the FTIR-ATR spectra calculated peaks at 1300-700 cm^{-1}. Peaks of siloxane (Si-O-Si) region of silica-pectin membrane measure on region 1080 cm^{-1}, silanol (Si-OH) on region 976 cm^{-1} and pectin on region 790 cm^{-1}. Region ± 970 cm^{-1} on the deconvolution of silanol peaks identified as the spin out mode of silica-carbon and shaking mode of carbon couple to Silica (Si) [15, 40–43]. Peaks at region 1080 cm^{-1} are delegated to a symmetric bond spin out vibration of Si-O-Si network [44]. In region 2001-2029 cm^{-1} there are the carbon bond vibration. The deconvolution of the band on region 976 cm^{-1} and region 1080 cm^{-1} to investigated of pectin with systematic incorporation measured using the peaks area ratio of Si-OH group vs Si-O-Si group was shown in figure 4(b). Fityk software was used to evaluate the amounts of areas on the silica-pectin membrane. The lowest peak area ratio between silanol and siloxane at xerogel sample on the membrane with concentration pectin 0.1 % calcined in air 400 °C. High concentration of siloxane region with an area of silanol region representing mesoporous or micropores [12, 13]. Actually, silica-carbon group contributed to determines the pore on the membrane silica-pectin. carbon chain clusters precedence to the formation of small pores but strengthen Si-OH bonds. This work results, the membrane concentration 0.5 % has high peak area ratio between silanol and carbon chains [45].

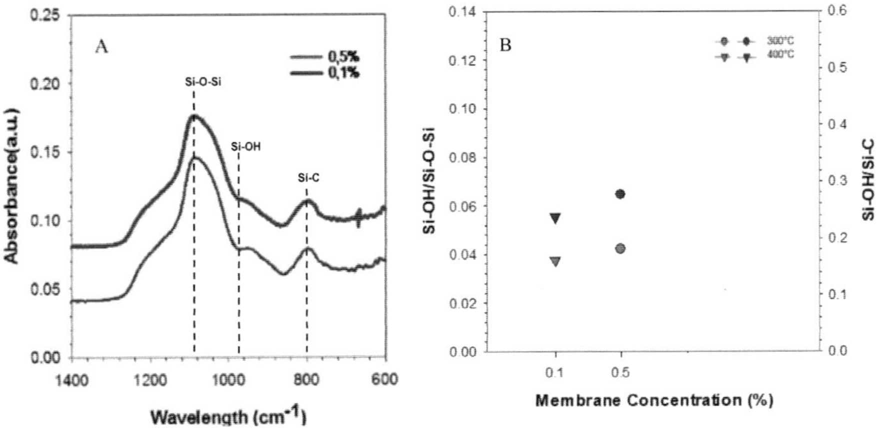

Figure 4. (A) FTIR spectra (Fourier-transform Infrared Spectroscopy) xerogel Silica-pectin Membrane and (B) peaks area ratio Si-OH (silanol) vs Si-O-Si (siloxane).

3.3. Performance of membrane after chemical cleaning process
The performance of silica-pectin membrane after chemical cleaning process of the membrane was shown in figure 5.

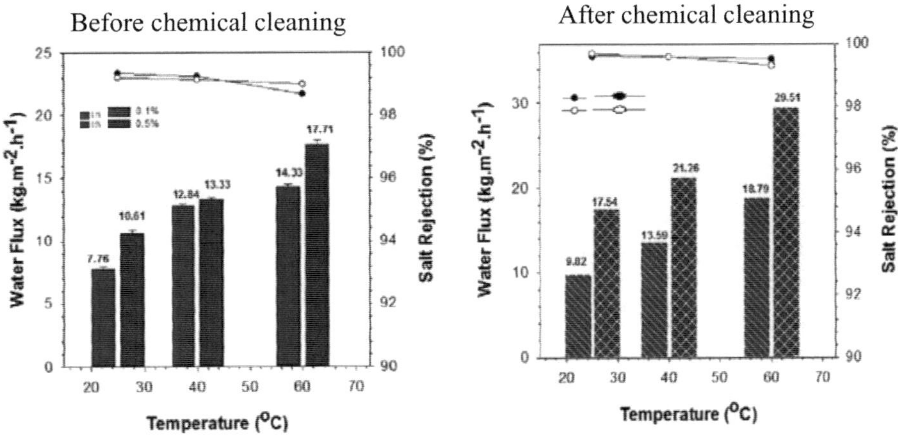

Figure 5. Evaluated performance of silica-pectin membrane.

Generally, the membrane was cleaned due to membrane fouling which caused particles adhering to the membrane surface. Irreversible type of membrane fouling can be removed by physical washing [46]. However, particles fouling too attached the membrane surface, physical cleaning not optimal. Hence, chemical cleaning method of membranes can cleaning the membrane optimally and improve the flux and life time the membrane longer period [35, 47]. Therefore, this work using titanium dioxide such as chemical material in chemical cleaning membrane. Figure 5 was shown the performance of silica-pectin membrane with variations temperature of AMD (25 - 60 °C). Result of this work showed performance of silica-pectin membrane after chemical cleaning using titanium dioxide is excellent. The flux recovery after chemical cleaning is higher than before cleaning process. Flux recovery from silica-pectin membrane concentration 0.1 % result 9.82 kg.m^{-2}.h^{-1} with salt rejection 99.61 %. Then, flux recovery on silica-pectin membrane was 27.20 % higher before cleaning. The best result in this work is still flux recovery from membrane with concentration of pectin 0.5 %. The value of flux recovery during the pervaporation process 20 minutes is 17.54 kg.m^{-2}.h^{-1} with salt rejection <99 % at room temperature. This flux showed that flux recovery of silica-pectin membrane recovery 65.32 % higher before cleaning. In other results, at temperature variations also showed that backwashing membrane with chemical cleaning was able to recovery water flux membrane. On feed temperature of 40 °C because of the driving force of the feed water temperature increase a high flux recovery. The flux value come to 13.59 kg.m^{-2}.h^{-1} with salt rejection reaching 99.60 % at 0.1 % pectin concentration. Meanwhile, membranes with a pectin concentration of 0.5 % were able to produce a flux recovery of 21.26 kg.m^{-2}.h^{-1} with a salt rejection of 99.60 %. The feed water temperature to evaluate the membrane performance was also carried out at on operating temperature of 60 °C. The resulting flux recovery increased at pectin concentrations of 0.1 % and 0.5 % consecutive 18.79 kg.m^{-2}.h^{-1} and 29.51 kg.m^{-2}.h^{-1} with the ability of salt rejection results still 99.52 % and 99.31 %. This proves that the silica pectin membrane that has been backwashed with chemical cleaning process exactly may improves the performance of the silica pectin membrane, also, the membrane is able to maintain salt rejection <99 %. The results of high salt rejection after chemical cleaning also prove that photocatalytic irradiation + TiO_2 catalyst able to change the compounds/ pollutants attached to the membrane surface can be converted into simple compounds which reduce fouling on the membrane as evidenced by the high salt rejection in the pervaporation process using a silica-pectin membrane.

4. Conclusion

Chemical cleaning of silica-pectin membranes using a photocatalytic membrane + TiO_2 catalyst shows that the evaluation results with pervaporation on various temperature are very excellent. Result shown in this work produce flux recovery reached 27.20 % and 65.32 % higher than the water flux before backwashing process. It is at an operating temperature of 25 °C. The increasing of flux recovery feed water temperature variations resulted in an increase with the ability of salt rejection <99 % in all variations in the temperature of the feed water.

Acknowledgments

Muthia thanks to the Advanced Membrane Technology Research Centre (AMTEC), University Teknologi Malaysia (UTM) and Materials and Membranes Research Group (M2ReG), Lambung Mangkurat University for the facilities. Muthis thanks to Applied Research of Universities Grant 2021-2023, Basic Research Grant 2021-2022 and World Class Research Grant 2021-2023 Deputy of Research and Development National Research and Innovation Agency, Directorate General of Higher Education, The Ministry of Cultural and Education Republic of Indonesia.

References

[1] Wahyudin I, Widodo S and Nurwaskito A 2018 Analysis of coal mine acid water handling *Geomine J.* **6**

[2] Hidayat L 2017 Environmental management of coal mining area:(Case study of acid mining drainage at PT. Bhumi Rantau Energi Tapin Regency South Kalimantan) *Adhum: J. Res. Dev. Administrative Sci. Humanit.* **7** 44–52

[3] Ferdian I 2020 Analysis of successful handling of mine acid water based on pH, TSS, Fe and Mn parameters at KPL AL 01 PT Bukit Asam, Tbk. In: *National Seminar on Suboptimal Land*

[4] Elma M, Mustalifah F R, Suryani L, Rampun E L A and Rahma A 2020 Wetland saline water and acid mine drainage desalination by interlayefree silica pectin membrane from banan peels *Nusantara Sci. Technol. Proc.* 271–9
[5] Elma M, Pratiwi A E, Rahma A, Rampun E L A and Handayani N 2020 The performance of membranes interlayer-free silica-pectin templated for seawater desalination *via* pervaporation operated at high temperature of feed solution. In: *Mater. Sci. Forum*: Trans Tech Publ) pp 349–55
[6] Elma M and Saputro G S 2020 Performance of cobalt-silica membranes through pervaporation process with different feed solution concentrations. In: *Mater. Sci. Forum*: Trans Tech Publ) pp 342–8
[7] Elma M and Assyaifi Z L 2018 Desalination process *via* pervaporation of wetland saline water. In: *IOP Conf. Series: Earth Environ. Sci.*: IOP Publishing) p 012009
[8] Elma M, Rakhman A and Hidayati R 2018 Silica P123 membranes for desalination of wetland saline water in South Kalimantan. In: *IOP Conf. Series: Earth Environ. Sci.*: IOP Publishing) p 012007
[9] Elma M and Riskawati N 2018 Silica membranes for wetland saline water desalination: performance and long term stability. In: *IOP Conf. Series: Earth Environ. Sci.*: IOP Publishing) p 012006
[10] Wang Q, Li N, Bolto B, Hoang M and Xie Z 2016 Desalination by pervaporation: A review *Desalin.* **387** 46–60
[11] Ayu Lestari R, Elma M, Rahma A, Suparsih D, Anadhliyah S, Ledyana Sari N, Ari Pratomo D, Sumardi A, Lestari A and Assyaifi Z L 2020 Organo silica membranes for wetland saline water desalination: effect of membranes calcination temperatures. In: *E3S Web Conf.* p 07006
[12] Elma M, Yacou C, Diniz da Costa J C and Wang D K 2013 Performance and long term stability of mesoporous silica membranes for desalination *Membr.* **3** 136–50
[13] Elma M, Yacou C, Wang D K, Smart S and Diniz da Costa J C 2012 Microporous silica based membranes for desalination *Water* **4** 629–49
[14] Rahman S K, Rahma A, Syauqiah I and Elma M 2020 Functionalization of hybrid organosilica based membranes for water desalination–Preparation using ethyl silicate 40 and P123 *Mater. Today: Proc.* **31** 60–4
[15] Elma M, Rampun E L, Rahma A, Assyaifi Z L, Sumardi A, Lestari A E, Saputro G S, Bilad M R and Darmawan A 2020 Carbon templated strategies of mesoporous silica applied for water desalination: A review *J. Water Process Eng.* **38** 101520
[16] Tuhuloula A, Budiyarti L and Fitriana E N 2013 Characterization of pectin by utilizing banana peel waste using conversion extraction method **2** 21–7
[17] Elma S and Wang Y's research influenced the addition of PECTIN as an interlayer-free silica-pectin membrane coating through the salt marsh water desalination process.
[18] Elma M, Mahmud M, Handayani N, Putri V S and Rahmah A 2019 Performance of interlayer-free silica-pectin membranes in seawater desalination with temperature variations of 25 °C and 40°C with membrane calcination temperature of 300°C. In: *Proced. Natl. Semin. Wetland Environ.* pp 279–8
[19] Elma M, Mahmud M, Handayani N, Putri V S and Rahmah A 2019 Performance of interlayer-free silica-pectin membranes in seawater desalination with temperature variations of 25°C and 40°C with 300°C temperature membrane calcination. In: *Proc. Natl. Wetland Environ. Semin.* pp 279–84
[20] Waqas S, Bilad M R, Aqsha A, Harun N Y, Ayoub M, Wirzal M D H, Jaafar J, Mulyati S and Elma M 2021 Effect of membrane properties in a membrane rotating biological contactor for wastewater treatment *J. Environ. Chem. Eng.* **9** 104869
[21] Beyer M, Lohrengel B and Nghiem L D 2010 Membrane fouling and chemical cleaning in water

recycling applications *Desalin.* **250** 977–81

[22] Waqas S, Bilad M R, Man Z B, Suleman H, Nordin N A H, Jaafar J, Othman M H D and Elma M 2021 An energy-efficient membrane rotating biological contactor for wastewater treatment *J. Cleaner Prod.* **282** 124544

[23] Razak N N A N, Rahmawati R, Bilad M R, Pratiwi A E, Elma M, Nawi N I M, Jaafar J and Lam M K 2020 Finned spacer for enhancing the impact of air bubbles for membrane fouling control in Chlorella vulgaris filtration *Bioresour. Technol. Rep.* **11** 100429

[24] Elma M 2017 Optimum ratio between waste cooking oil and coconut oilas raw material for biodiesel production *International J. Adv. Sci. Eng. Inf. Technol.*

[25] Sulistyani E and Fitrianingtyas M 2010 Ultrafiltration membrane fouling control with automatic backwash and membrane washing system. Department of Chemical Engineering Faculty of Engineering UNDIP)

[26] Khuzwayo Z and Chirwa E M 2017 Analysis of catalyst photo-oxidation selectivity in the degradation of polyorganochlorinated pollutants in batch systems using UV and UV/TiO2 *S. Arf. J. Chem. Eng.* **23** 17–25

[27] Wang D K, Elma M, Motuzas J, Hou W C, Schmeda-Lopez D R, Zhang T and Zhang X 2016 Physicochemical and photocatalytic properties of carbonaceous char and titania composite hollow fibers for wastewater treatment *Carbon* **109** 182–91

[28] Leyva-Díaz J, López-López C, Martín-Pascual J, Muñío M and Poyatos J 2015 Kinetic study of the combined processes of a membrane bioreactor and a hybrid moving bed biofilm reactor-membrane bioreactor with advanced oxidation processes as a post-treatment stage for wastewater treatment *Chem. Eng. Proc.: Process Intensification* **91** 57–66

[29] Chun H, Yizhong W and Hongxiao T 2000 Destruction of phenol aqueous solution by photocatalysis or direct photolysis *Chemosphere* **41** 1205–9

[30] Wang D K, Elma M, Motuzas J, Hou W C, Xie F and Zhang X 2017 Rational design and synthesis of molecular-sieving, photocatalytic, hollow fiber membranes for advanced water treatment applications *J. Membr. Sci.* **524** 163–73

[31] Molinari R, Argurio P and Palmisano L 2015 *Advances in Membrane Technologies for Water Treatment,* ed A Basile, *et al.* (Oxford: Woodhead Publishing) pp 205–38

[32] Assyaifi Z L, Elma M, Syauqiah I, Rampun E L, Rahma A, Sumardi A, Lestari A E, Suryani L, Mustalifah F R and Huda N 2021 Photocatalytic–pervaporation using membranes based on organo-silica for wetland saline water desalination *Membr. Technol.* **2021** 7–11

[33] Ramadhani S U, Destiarti L and Syahbanu I 2017 Degradation of organic matter in peat water with photocatalyst TiO2 thin layer *J. Equat. Chem.* **6**

[34] Elma M, Ayu R, Rampun E L A, Annahdliyah S, Suparsih D R, Sari N L and Pratomo D A 2019 Fabrication of interlayer-free silica-based membranes – Effect of low calcination temperature using an organo-catalyst *Membr. Technol.* **2019** 6–10

[35] Rampun E L A, Elma M, Syauqiah I, Putra M D, Rahma A and Pratiwi A E 2019 Interlayer-free silica pectin membrane for pervaporation of salt marsh water Journal of Chemical Science and Applications **22** 99–104

[36] Rampun E L, Elma M, Rahma A and Pratiwi A E 2019 Interlayer-free silica–pectin membrane for sea-water desalination *Membr. Technol.* **2019** 5–9

[37] Elma M, Setyawan H, Rahma A, Pratiwi A and Rampun E L A 2019 Fabrication of interlayer-free P123 caronised template silica membranes for water desalination: Conventional versus rapid thermal processing (CTP vs RTP) techniques. In: *IOP Conf. Series: Earth Environ. Sci.*: IOP Publishing) p 012076

[38] Elma M 2020 Rapid thermal processing and long term stability of interlayer-free silica-P123

membranes for wetland saline water desalination

[39] Rahma A, Elma M, Rampun E L A, Pratiwi A E, Rakhman A and Fitriani F 2021 Rapid thermal processing and long term stability of interlayer-free silica-P123 membranes for wetland saline water desalination *J. Adv. Res. Fluid Mech. Therm. Sci.* **71** 1–9

[40] Wieder H 1979 M. Cardona and CR Guarnieri *Phys. Status Solidi B* **92** 99

[41] Lestari R A, Elma M, Rampun E L A, Sumardi A, Paramitha A, Lestari A E, Rabiah S, Assyaifi Z L and Satriaji G 2020 Functionalization of Si-C using TEOS (tetra ethyl ortho silica) as precursor and organic catalyst *E3S Web Conf.* **148** 07008

[42] Pratiwi A E, Elma M, Rahma A, Rampun E L A and Saputro G S 2019 Deconvolution of pectin carbonised template silica thin-film: synthesis and characterisation *Membr. Technol.* **2019** 5–8

[43] Elma M and Setyawan H 2018 Synthesis of silica xerogels obtained in organic catalyst *via* sol gel route. In: *IOP Conf. Series: Earth Environ. Sci.*: IOP Publishing) p 012008

[44] Elma M, Wang D K, Yacou C and Diniz da Costa J C 2015 Interlayer-free P123 carbonised template silica membranes for desalination with reduced salt concentration polarisation *J. Membr. Sci.* **475** 376–83

[45] Elma M, Wang D K, Yacou C, Motuzas J and Diniz da Costa J C 2015 High performance interlayer-free mesoporous cobalt oxide silica membranes for desalination applications *Desalin.* **365** 308–15

[46] Jayanti R D and Widiasa I N 2016 Fouling and Cleaning Of Low Pressure Reverse Osmosis Membranes for Domestic Wastewater Recycling Applications. In: *National Semin. Chem. Eng.* p 2

[47] Arnal J M, Garcia-Fayos B and Sancho M 2011 *Expanding issues in desalination*: IntechOpen)

IOP Conf. Series: Materials Science and Engineering 1195 (2021) 012058 doi:10.1088/1757-899X/1195/1/012058

Effect of external heat source on temperature and moisture variation for composting of food waste

J Rose[1], J C Lai[1], Y L Then[2] and C K Vithanawasam[2]

[1] School of Chemical Engineering and Science, Faculty of Engineering, Computing and Science, Swinburne University of Technology, 93350 Kuching, Sarawak, Malaysia

[2] School of Engineering, Faculty of Engineering, Computing and Science, Swinburne University of Technology, 93350 Kuching, Sarawak, Malaysia

Email: jlai@swinburne.edu.my

Abstract. This paper investigates the outcome of having an external heat source on temperature and moisture variations in the food waste composting process. Food waste accumulation is a growing concern in many countries. Converting food waste into usable compost is a more desirable tactic than dumping to crowded landfill sites. Closed composting was applied in this work, which relies on a controlled but uninterrupted airflow during the organic material degradation process. However, undesirable odour released at low aeration rate due to low temperature and high moisture content found in the compost. Finding the ideal aeration rate with the least possible loss of moisture is needed, which was discussed in this paper. The vegetable-fruit waste used in the experiment was given an aeration rate of 0.3 L/min at a moisture setting of 60% and 70%. For 15 mins/day, the forced aeration was carried out at 3-day intervals. Results showed that 0.3 L/min with 60% and 70% moisture content attained best temperature peaks of 32.4°C and 31.6°C, respectively at day 13 for 28 days composting. A strong odour continued to exist with the compost and was mitigated by using an external additional heat source (light bulb). The light bulb also helped to provide a higher temperature for the compost of 41.5°C by day 1 for 10 days composting.

1. Introduction

A very large amount of total solid waste in Malaysia (around 50% to 60%) will comprise municipal solid waste, which is mostly food waste. This creates numerous concerns in the local context and internationally as the problem is the same in most countries. The food waste originates from daily markets (24.5%), restaurants (23.4%) and from domestic household (38.2%) environments [23]. These can then be categorized into three types, food losses, unavoidable food waste and avoidable waste. It should also be noted that this waste will continuously increase with the population, with Malaysia's 32.45 million people producing food waste parallel to the population growth [1].

The bulk of this waste makes its way into landfills, as approximately 95% of the total municipal solid waste, which amounts to around 10.5 million tons of waste, will be disposed of in operating landfills [2]. This is expected to increase by around 31,000 tons of waste by 2020 [3]. Despite the relative ease with which this waste can be disposed of, it emits greenhouse gases. High odor emission is also a problem, along with the space constraints. With a population increase of 2.4%, the landfills are piling up very fast and there will be a shortage of landfills, with large costs incurred in finding new landfill sites [4]. The treatment of municipal waste has not been investigated extensively, and therefore landfill operation continues as one of the accepted methods [5].

Composting is seen as one acceptable method of getting rid of degradable waste, however, it is not seen as a successful method. The low value of compost products is one reason, along with the pollution that occurs from gas emissions, despite changes made to reduce such emissions [6]. The factors taken into consideration include the moisture content, temperature, aeration rate, porosity, and the Carbon-to-Nitrogen (C:N) ratio. The composting material can vary significantly and therefore there will be differences in the quality of the compost, with poor quality compost being a major concern [7].

The temperature of the food waste compost is observed to show which phase the compost is currently at. The microbial activity inside the compost can also be observed from the temperature changes that occur. It has been concluded from former research that composting temperature reflects on the degradation rate of the compost, where a higher temperature of 50°C to 70°C reflects a higher degradation rate [8]. Alternate research, however, indicated that 70°C will inhibit microbial activity [9]. Instances where quick rapid external heating was used did not benefit microorganism growth. Their optimum temperature range was observed to be 45°C to 70°C for a thermophilic phase within 3 to 4 days for the total elimination of pathogens. Ranges like this are observed for the thermophilic phase (figure 1). The mesophilic phase was also observed to help in elimination where mesophilic bacteria initiates the microbial activity [10]. Achieving the desirable temperature is a challenge, as the compost pile in most cases obtains the highest possible temperature at a maximum of 38°C, which means there is insufficient elimination of pathogens. The low temperature observed indicates that the compost did not reach the status of thermophilic phase, rather it remains composting at the mesophilic phase. This can be concluded as an effect of high moisture contents ranging from 92% to 94%. Low heating values are observed for the vegetable waste as well. Convection heat loss can be experienced when performed in the laboratory reactor [11].

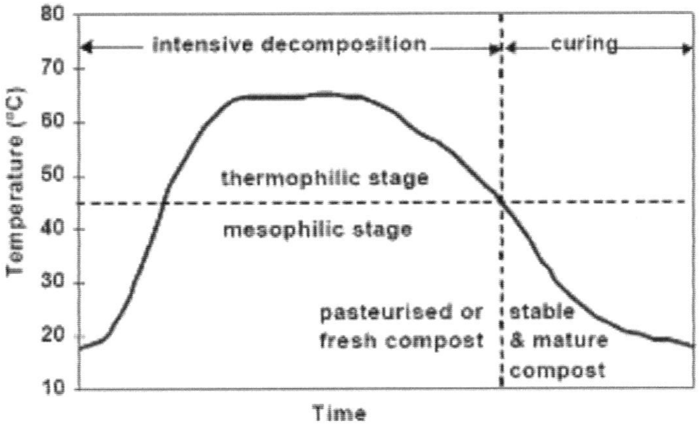

Figure 1. Range of composting temperature at different phases.

Another experiment used carton and vegetable waste composting, which was able to conclude that the compost will achieve peak temperatures of 45°C to 50°C. This can be observed due to low isolation of the composting vessel and the smaller volume [12]. It resulted in a period of composting of 308.33 days, which was longer. Self-regulating heat in a sample will allow the temperature to slowly increase and drop under anaerobic conditions. A suitable growing temperature is acquired by the microbes and the remaining organic matter will mature in the short thermophilic phase [9]. It can therefore be concluded that the compost temperature and thermophilic phase are highly correlated in order to obtain a high-quality compost amount, while remaining at the thermophilic phase for a longer period.

Much like the thermophilic phase, moisture plays a crucial role in the composting process. Moisture, created by the addition of water, will carry the necessary nutrients for the physiological and metabolic activity of the microorganism. Previous research has indicated that the moisture content is a dominant factor in an experiment with compost which takes a toll on the microbial activity [13]. The experimental work was justified with the need for 50% moisture content, which indicated that an increment in the temperature can be realized with increased moisture content alone. This does, however, require further research. Food waste will, in most instances, have very high moisture contents, which will cause a longer composting period for most experiments. It is also a clear indication of the low degradation efficiency when it comes to composting [9]. Previous research suggests that the ideal moisture content for the bulk of composting materials will be in the range of 50% to 70% [14]. Another experiment showed impressive results by finding green vegetables with the combination food waste co-composting will provide a standalone moisture content of 60% [10].

The moisture content and the ratio of the bulking agent used will be related. This was evident from a previous experiment when large ratios of bulking agent were used on leftover food waste at a ratio of 70:30 of bulking agent to food waste which showed no leachate formation for home composting [15]. When the food waste was higher than the bulking agent, for example, with a ratio of 30:70 of bulking agent to food waste, there was gas exchange, unpleasant odour and an inclination toward the anaerobic process. Reasoning behind this may be due to the high moisture exceeding the absorption capacity of the bulking agent used. Therefore, the moisture content will need to be controlled effectively along with the bulking agent to food waste ratio in order to get the ideal temperatures required and the best composting process.

Improper selection of aeration rate and mode could slow down the breakdown process of organic waste, resulting in a thermophilic phase that could not be achieved in a laboratory scale of composting reactor. In past research, composting mixture aerated in continuous mode at high rate could achieve C:N ratio below than 20 to show a good usability of the final product [17]. Higher aeration is typically chosen to achieve good composting performance as well as to obtain a good quality of final product. However, usage of high aeration and continuous aeration mode could induce high operating cost. Besides, high aeration rate could result in extensive moisture and heat loss, which would cause a longer time to reach maturity of stage in the composting process. In contrast to high aeration rate, the anaerobic condition would be developed at different pile regions if low or not enough air is pumped into the compost pile [19]. This is unfavourable as anaerobic condition would cause odour problems and lengthen the composting process to attain maturity stage. Thus, intermittent aeration mode is proposed to pump oxygen into compost pile within the intervals with the appropriate pumping time, resting time and aeration rate. By adopting intermittent aeration mode, moisture and heat loss as well odour problem could be reduced since there would not be excessive pumping of air into the compost pile. Pumping air into the compost pile and resting within the period where the oxygen saturation falls would cause conservation of energy and enhancement of composting quality [20]. Intermittent aeration could also reduce the total greenhouse gases (GHG) emission by 17.8% compared to continuous aeration mode [20]. Though intermittent aeration sounds to be more economical and efficient than continuous aeration mode, substantial finding is required to justify the usage of intermittent aeration for composting of food waste, which is limited at present and arguable.

Households are responsible for most of the food waste in Malaysia. The main food waste from these households include rice, vegetables and meat [21]. It is generally grouped into 60% carbohydrates, 20% protein and 10% fats. Fruits and vegetables are definite sources of carbohydrates, thereby making them the predominant wastes from Malaysian households. Though meat is also a part of household food waste, it will not be applied in the composting process due to risk of pest infestation if not handled adequately [22]. This process is likely not favoured by the general public. Therefore, this study will utilize fruits and vegetables as the main substrate in the mixture. Besides, most research findings found that the thermophilic phase could not be achieved in a small scale of composting system. With the aim to improvise home composting on food waste, a light bulb is proposed to attach inside the composting vessel as an affordable source to assist temperature rise during the composting process. Self-regulating heat is typically slow and not able to be maintained in a small-scale composting system, thus an external heat source is required to compensate for the heat loss in this system.

Taking all these factors into consideration, this paper will investigate numerous factors, including the effect of additional heat sources on temperature and moisture changes during composting of food waste with intermittent aeration technique. The work will investigate the potential similarities between the aeration rate and high peak temperatures which allows the compost pile to mature in a shorter time period.

2. Materials and methods

2.1. *Raw materials*

In order to create a mix of food waste, the following items were used: vegetable waste (raw uneaten green vegetables waste), fruit waste (left over fruit peels), and spent coffee grounds. These were collected from the local market, a coffee shop and a juice stall. The fruit along with the vegetable waste were cut into pieces ranging from 1.5 to 2.5 cm in size to promote degradation rate [18], and then mixed in with the rice husks and spent in coffee grounds. Rice husk is the bulking material in this experiment, and it was added as a source of carbon to regulate the carbon to nitrogen ratio of the food waste mixture, along with its moisture content. The ground coffee, acquired from a coffee shop as a byproduct of its brewing process, was added as a nitrogen source to balance the carbon to nitrogen ratio. The containers used were 5 L capacity water dispenser units with a dimension of 29.0 x 19.7 x 10.1 cm. These had holes cut into them to perform intermittent aeration. The vessels had a mixture of 300 g vegetable waste, 200 g fruit waste, 450 g coffee ground, and 50 g rice husk. Carbon to nitrogen (C:N) ratio was maintained at 23, which was ideal for the initial composting process. The type of green vegetable waste and fruit peels were fixed in each collection to ensure uniformity of the mixture in each experiment. Sampling was done in each preparation to make sure the initial C:N ratio of the mixture had fulfilled the requirement.

2.2. *Experimental setup and composting procedure*

The container that was described in Section 2.1 was used with the necessary additions carried out to it. Initially, the system was designed to investigate the optimum aeration rate which will assist in providing a temperature that is high for the longest possible time period. In the meantime, a fixed moisture content was retained without a source of heat. The experiment then proceeded to introduce a heat source and identify the peak temperatures as well. The external heat source used was a light bulb, which generated substantial amounts of heat. This is so that the compost will be able to climb to a higher temperature profile, to eliminate the odor, and to cut down the composting period due to rapid degradation rate. The differences in results between the experimental part with an external heat source and without an external heat source will be useful in understanding the usefulness of an external heat source.

The experiment required five important components for each vessel: the water dispenser, heat insulator, pump and tube, moisture sensors, and temperature sensors. These were connected to a microcontroller which will display the data at the location itself and upload the data online to ThingSpeak, an open source IoT platform, which will help remotely log the data. The water dispenser used was a high-density polyethylene 5 L capacity water dispenser (figure 2 (a)), made from a high quality PP5 material. Containing polypropylene, which has a melting point of 167°C, this vessel displays good heat resistivity and is therefore suitable for the job. The heat shield used was a Xinfrax Aluminium Heat Shield (figure 2 (b)). This was to help establish the thermophilic phase, which was a technique used for a closed composting system which can allow the self-regulating heat of the microbial to take place. Minimum heat loss subsequently occurs due to this insulation. The air pump used was the variety normally utilized in aquariums (figure 2 (c)). It has a maximum capacity of 3.0 L/min and is utilized in order to supply air flow during the composting period. It provided aeration rate supply ranging from 0.1 to 1.0 L/min for the various vessels. A silicon tube was used to penetrate the compost in the vessel, and there were holes cut into it to evenly distribute the distilled water or air. The tube size was approximately 2.5 to 3.0 cm.

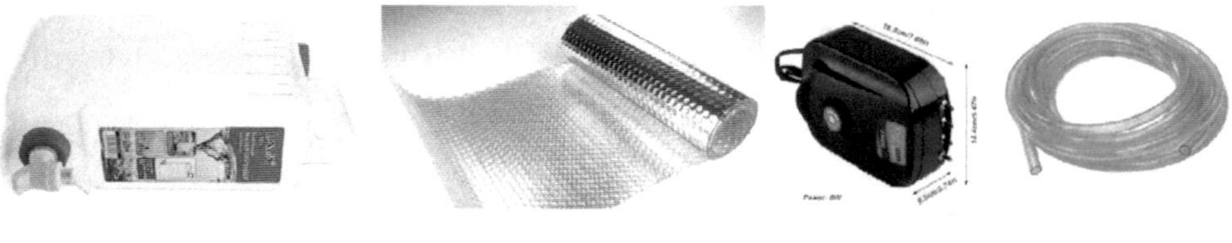

(a) (b) (c)

Figure 2. Water dispenser (a), heat insulator (b), and air pump (c).

The sensors were attached toward the center of the vessel using drilled holes, with the hole for the air pump being the vessel's premade openings (figure 3). Each vessel used had its own respective temperature sensors and moisture sensors, along with a temperature sensor placed outside the system to measure ambient temperature. The temperature sensor used was the DS18B20 waterproof system, and the moisture sensor was a low-cost YL-69 hydrometer sensor, which has previously been calibrated and tested for use with various soil samples [16].

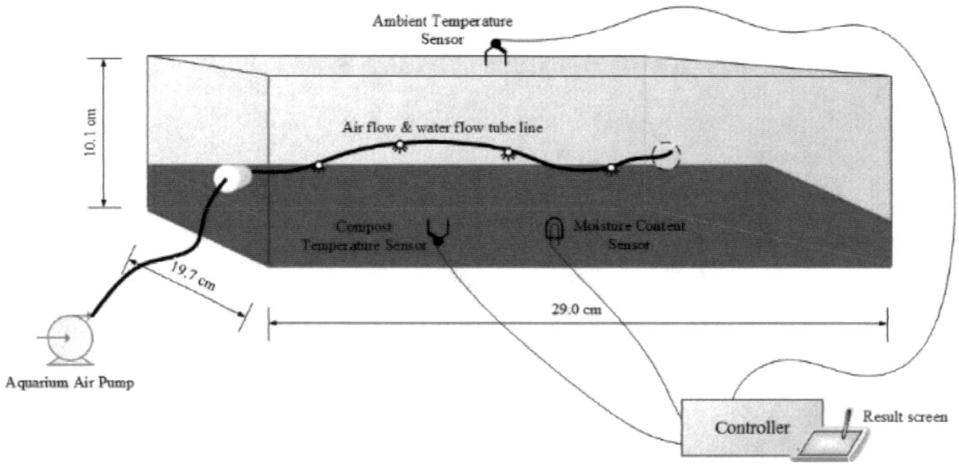

Figure 3. Illustration of the closed composting system used for the project.

For the initial experiment, where no external heat source was used, the vessels were placed on tables with closed pipes placed for air pump fixing when needed (figure 4 (a) and (b)). The microcontroller setup was kept at the center of the set of vessels, where the YL-69 moisture sensors were connected to a microcontroller with internet connecting capabilities, the Node-MCU unit (figure 5). Since there were multiple sets of moisture sensors and the microcontrollers have limited analog inputs, multiplexer modules were used to measure numerous sensors at the same time. This was not needed for the temperature sensors as each sensor had a unique ID and multiple sensors could be read through one digital GPIO pin. Since continuous current flow through the moisture sensors would cause them to rust, relay modules were placed which were controlled by the microcontrollers, which would switch on the moisture sensors for the time duration when they were measured.

(a) (b)

Figure 4. Illustration of the closed composting system used for the project.

The moisture sensors were calibrated by inserting them into samples with known moisture content levels and then using the model with the compost system. The equation obtained in Section 2.3 was used to adjust the moisture content of compost to the set point of 50, 60 and 70%. The moisture sensors and temperature sensors were first installed using hot glue and then the compost was carefully inserted into the vessel. Nine vessels were prepared and given the aeration rate and moisture content in table 1. This aeration was supplied using the aquarium pump and was supplied every Monday and Thursday at the same time for 15 minutes. The system was then gently shaken to evenly distribute the compost and air. Distilled water would be added into the vessel when moisture content is reduced. The composting was carried out for 28 days without any additional heat source.

Table 1. Aeration rate with the moisture content for the nine composting vessels.

Vessel	Aeration Rate, AR (L/min)	Moisture Content, MC (%)
A1	0.3	50
A2	0.6	50
A3	0.9	50
A4	0.3	60
A5	0.6	60
A6	0.9	60
A7	0.3	70
A8	0.6	70
A9	0.9	70

IOP Conf. Series: Materials Science and Engineering 1195 (2021) 012058 doi:10.1088/1757-899X/1195/1/012058

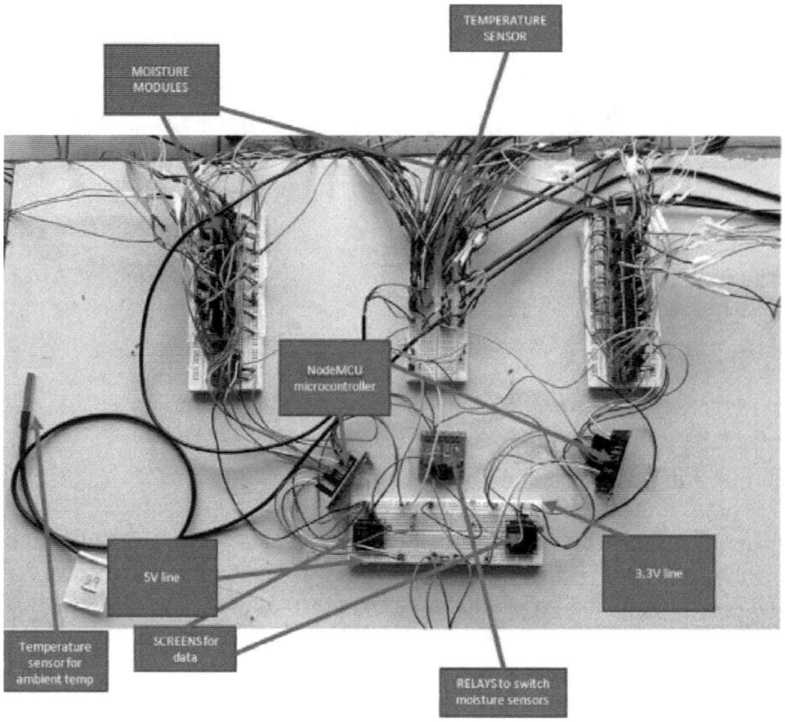

Figure 5. Microcontroller setup.

The second portion of the experiment used the external heat source in order to achieve a higher temperature profile. A 58-Watt tungsten incandescent candle bulb with screw base was chosen because this light bulb could emit more heat than a fluorescent lamp. The light bulb used as a heat source performed the additional task of odor elimination. A faster degradation rate was also achieved with the heat source. The three highest peak temperatures found using the first portion of the experiment were the ones repeated for the second section using the additional heat source. Figure 6 shows the composting vessels with the desired aeration rate and moisture content based on the results from the first section. Yellow was the least desired, red was moderately desired, and green was highly desired.

Figure 6. Composting vessel with light bulb installation. (Yellow — least desired aeration rate, red — medium desired aeration rate, green — most desired aeration rate).

Vessel openings were sealed using hot glue in this instance as well. Similar to the first portion of the experiment, the aeration and moisture content were kept controlled. The food waste used was then allowed to decompose for 10 days provided the compost had reached a state of maturity. Samples were tested with the same parameters after the experiment was carried out.

2.3. Monitoring and analysis

The YL-69 moisture sensor was calibrated using two methods. The two methods are compared to ensure a certain degree of accuracy when using the system. Firstly, the drying oven method was used where oven dried samples with known moisture contents can be measured using the YL-69 sensors (figure 7 (a)). The other method was the standard soil gravimetric method where a food waste sample had water mixed into it to calculate the moisture value (figure 7 (b)).

Both methods had obtained a 4th degree polynomial equation and linear regression analysis was used to determine the accuracy of the moisture sensors. The equations were then used to measure the moisture of the compost samples. The initial values were collected in bits, as the analog signal received from the moisture sensor gave a value between 0 and 1024. These bit values, along with the calibrated moisture percentages were uploaded online to the ThingSpeak server, where they can be downloaded when required.

Sampling was done to determine properties of final compost. For pH measurement, the sample was used to prepare aqueous extract at a 1:10 compost: water ratio (w/v) and mechanically shaken for 1 hour. The pH was then measured with a pH meter [17]. Same sample was used for electrical conductivity (EC) measurement, where the EC was measured with an EC meter. For C/N ratio analysis, 100 g of sample was sent to the Nabbir Laboratory Sdn Bhd (Kuching) for Total Organic Carbon (TOC) and Total Kjeldahl Nitrogen (TKN) analysis. The C/N ratio was then calculated by dividing TOC with TKN.

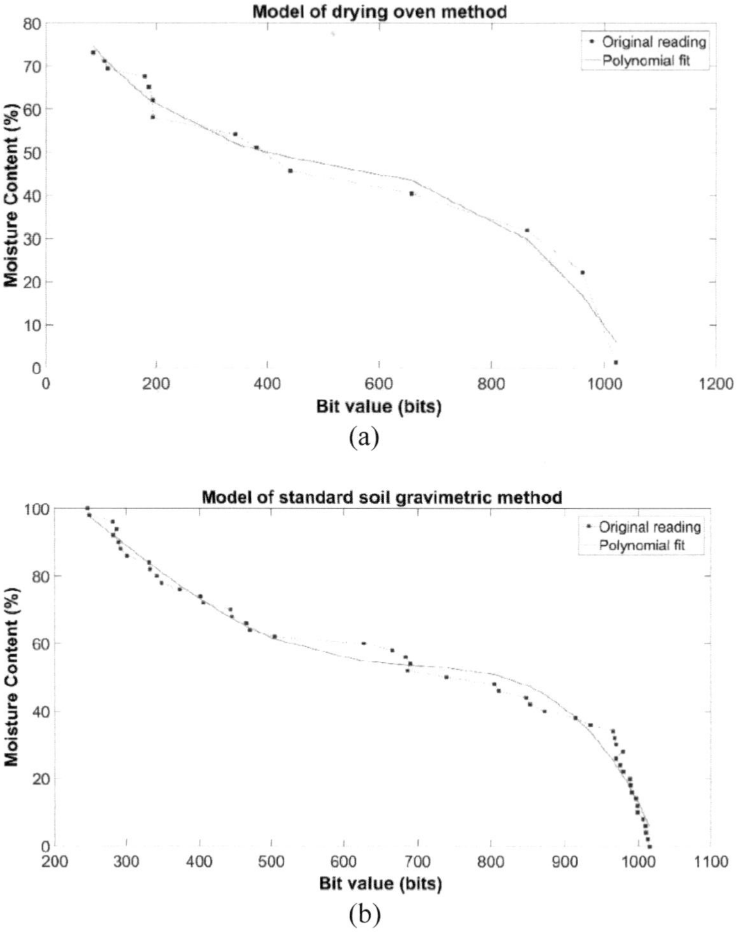

Figure 7. Bit Value recorded for YL-69 Sensor vs Moisture Content (%) for (a) drying oven method and (b) gravimetric method.

3. Results and Discussion

Table 2 displays the maximum temperatures observed during the experiment without the use of light bulbs. Forced aeration was carried out as mentioned previously. The tabulated moisture values were given by moisture sensors, which were calibrated by standard soil gravimetric method with 5.02% error. It shows a more promising accuracy in predicting moisture content of food waste compost as compared to the drying oven method with 10.06% error.

Vessel A4 is the vessel with the highest peak temperature of 32.38 °C in a 28-day composting period. This is the highest value which came about a low aeration rate of 0.3 L/min. Vessel A7, which also contained a 0.3 L/min aeration rate, has a significantly high temperature. Vessel A9 is the notable exception, with a 0.9 L/min aeration rate and a high temperature. It is therefore possible to conclude that if a medium and low aeration rate can achieve higher peak temperatures, it is unnecessary to use higher aeration rate for the composting procedure to reach a higher peak temperature. It can also be observed that the aeration rate impacts the temperature profile, as the aeration rate has a relationship with the oxygen supply for microbial activity. It should be noted however that the temperatures do not exceed 32.38 °C for the experiment, and there it is not considered as a good food waste composting stage as the mesophilic phase has not been reached. This phase would be where the mesophilic bacteria start killing pathogens at temperatures of 35 °C and above.

Table 2. Maximum temperature achieved during 28 days of composting.

Vessels	AR (L/min)	Peak T (°C)
A1	0.3	31.13
A2	0.6	31.81
A3	0.9	31.75
A4	0.3	32.38
A5	0.6	31.25
A6	0.9	31.57
A7	0.3	31.63
A8	0.6	31.38
A9	0.9	31.81

Figure 8, 9 and 10 indicate the points at which the moisture content for the best vessels was observed at varying aeration rates. Vessel A4 had a moisture content of 50.05% to reach the peak temperature in 13 days. Within 16 days, for the medium aeration rate, the most attractive aeration rate was achieved by vessel A2 at 31.81 °C with a moisture content of 60.05%. The use of a high aeration rate 0.9 L/min achieved temperature of 31.81 °C happened to show a moisture content at 80 % by day 3. This shows that a high aeration rate will provide a high moisture content inside the compost. The 0.6 L/min aeration rate and the 0.9 L/min one, showed similar temperature peaks. This did differ greatly in its moisture content. High aeration rates can be rejected as a choice which justifies this better. A large enough moisture content can reduce the degradation rate, resulting in a longer period of composting. Considerably low aeration rates can provide a high temperature profile at 50.05 % moisture content. This shows that the moisture content was enough to transport the dissolved nutrients, necessary for microbial activity. Vessel A9 most likely experienced an excessive amount of cooling with high levels of moisture content. This certainly created a difficulty in obtaining higher peak temperatures even with the use of high aeration rates. Water vapour was observed on the walls of the composting vessel as the experiment was being carried out, which shows that evaporation has occurred inside the vessel. Water vapour could be observed dropping back into the compost which certainly led to increased moisture content levels. It is likely that this phenomenon occurred in vessel A2 and A4, even though it had achieved a higher temperature peak. It should be noted that it did not achieve a mesophilic phase. Moisture content can be seen to heavily influence the low temperature profiles.

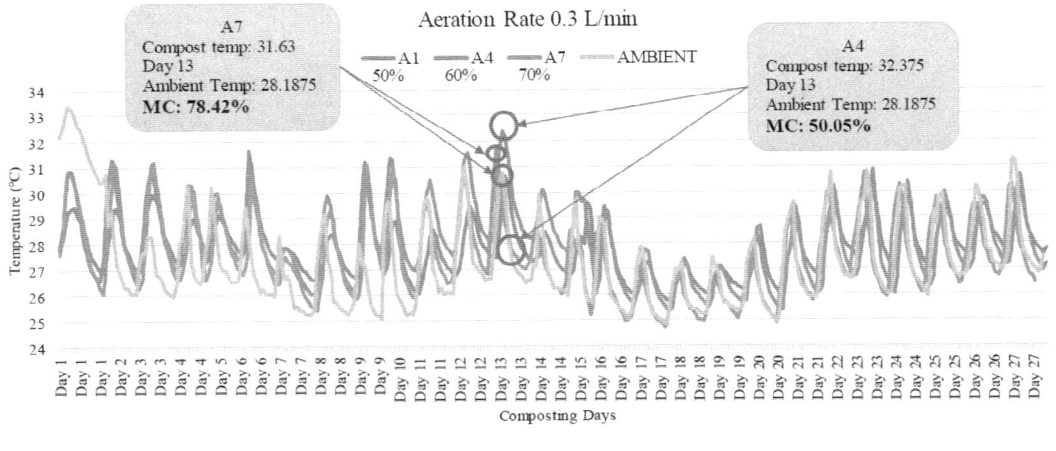

Figure 8. Peak temperature achieved at a low aeration rate of 0.3 L/min in a 28-day composting period.

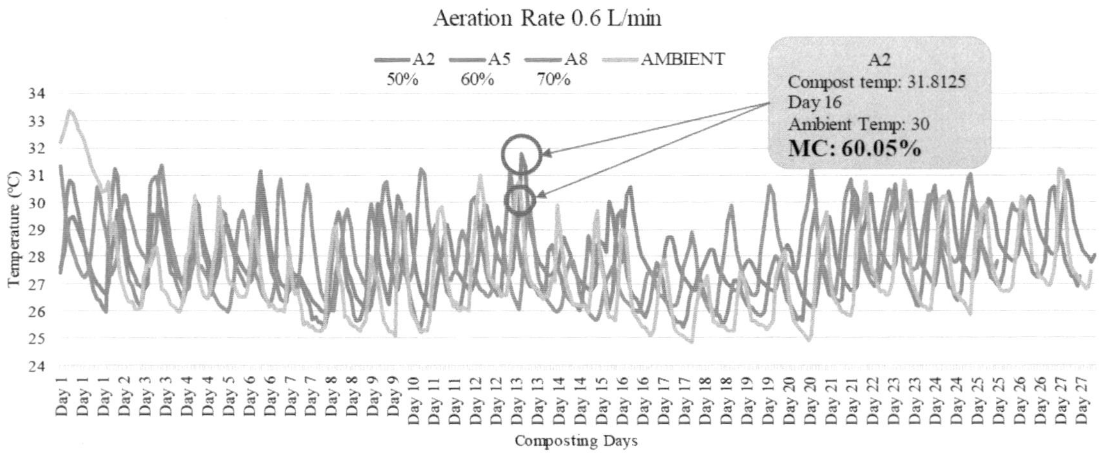

Figure 9. Peak temperature achieved at medium aeration rate of 0.6 L/min in a 28-day composting period.

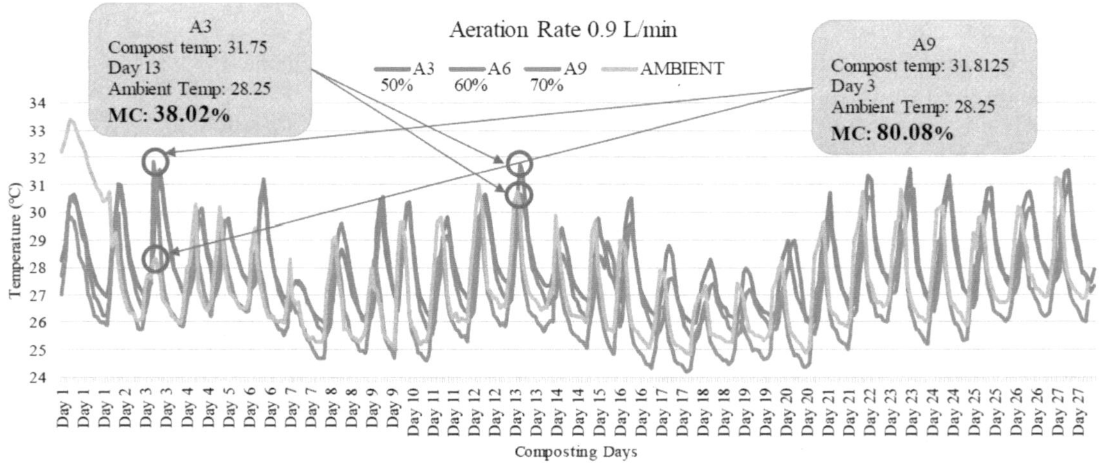

Figure 10. Peak temperature achieved at a high aeration rate of 0.9 L/min in a 28-day composting period.

Figure 11 shows the ranking of the top 5 vessels with the highest peak temperatures. This helped to pick the best three vessels for use with the heat source in the next part of the experiment. The first vessel selected was vessel A4. It used a low aeration rate and moisture content set point of 60 % to achieve 32.38°C. A2 and A9 were the next vessels chosen. Vessel A2's reason for being selected is due to the temperature of 31.81 °C being achieved with medium aeration rate without the need for a high aeration rate. Vessel A9 is then not selected due to its use of high aeration rate and a high moisture content in relation (%). One of the vessels from A3 and A7 needed to be picked as well and they recorded peak temperatures of 31.75 °C and 31.63 °C respectively. In this case, a higher peak temperature is reached by aeration rate of 0.9 L/min compared to lower aeration rate. However, A7 is selected as the dominant first vessel, as it displayed the highest peak achieved from the vessels with a low aeration rate. A7 is chosen instead of A3 due to its aeration rate and optimum moisture content.

The three selected vessels, A2, A4, and A7 were then experimented with for a 10-day period. A 58-Watt light bulb is used for providing an additional heat source in each of the vessels. The resulting peak temperatures are displayed in table 3.

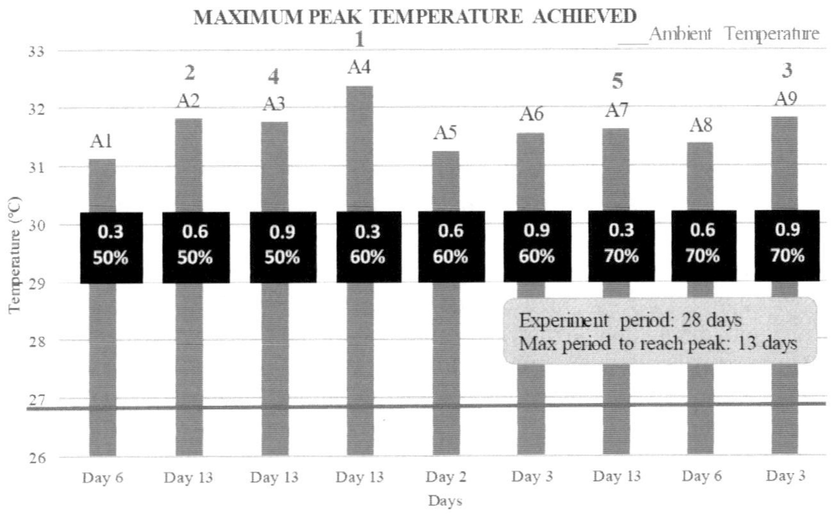

Figure 11. Peak temperature achieved by the top 5 vessels.

Table 3. Temperature measurements for selected 3 vessels during 10 days of composting.

Vessels	AR (L/min)	T (°C)	Period to peak T
A2	0.6	39.69	Day 1
A4	0.3	41.50	Day 1
A7	0.3	39.94	Day 3

The 58-Watt light bulb is introduced as the extra heat source. With this new heat source, vessel A4 can be seen to show a high peak temperature recorded at 41.5 °C for a 10-day period of composting. The aeration rate used is low, which is at 0.3 L/min. The moisture content set point thus varies the three vessels composting condition as shown in figure 12 below.

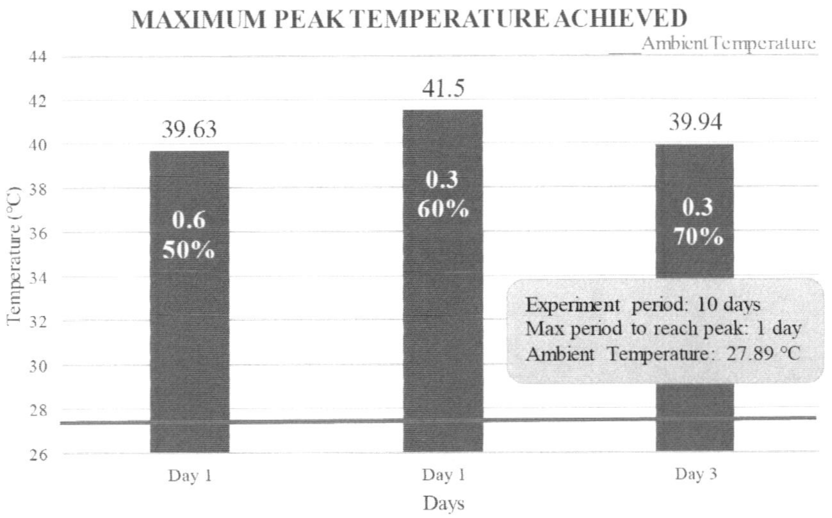

Figure 12. Maximum peak temperature against days to reach peak temperature.

It may be seen that with the extra heat source the composting process has accelerated the degradation rate of the microbial movement. From figure 12, the peak temperature was reached within 24 hours for all 3 vessels in a 10-day period. This is roughly 12 days earlier than without the use of the external heat source. Degradation rate is precisely indicated as the increase of temperature from its ambient of approximately 27.89 °C is observed. A difference of 13.7 °C was noted for vessel A4 with a low aeration rate of 0.3 L/min and 50% moisture content. Medium aeration's temperature peak for 0.6 L/min with 50% moisture content was 39.69 °C. Apart from a noticeable rise in temperature, the rise occurred significantly fast in this case as well.

Figure 13 shows that the aeration rate of 0.6 L/min which was able to reach 39.7 °C within 24 hours contains a moisture content of 53%. This can be considered relatively optimum as the set point for moisture content of the compost is 50%. Meanwhile, figure 14 indicates that the aeration rate of 0.3 L/min can obtain the highest peak temperature of 41.5 °C. This was also achieved within 24 hours and had a moisture content of 58%. The moisture content set point is at 60%. Figure 15 shows 0.3 L/min with a moisture content of 70 % as its set point. This last vessel has a notable difference, where, at the period of peak of 48 °C, it had a certain degree of moisture. A substantial reduction in moisture was observed at this point where a temperature of 39.94 °C is recorded, thus explaining that the vessel cannot climb to a higher peak temperature. It can be explained as being due to the exhaustion of nutrient supply for the microbial activity. When comparing vessels A2 and A4, the moisture content was maintained. This shows that enough nutrient and oxygen is available for medium and low aeration rate vessels to obtain a high peak temperature.

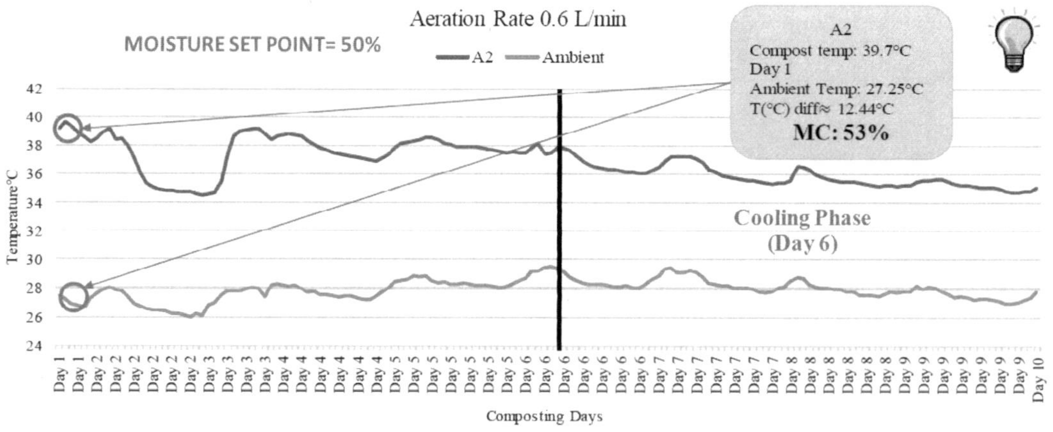

Figure 13. Moisture Content (%) and high peak temperature of A2 with medium aeration rate of 0.6 L/min.

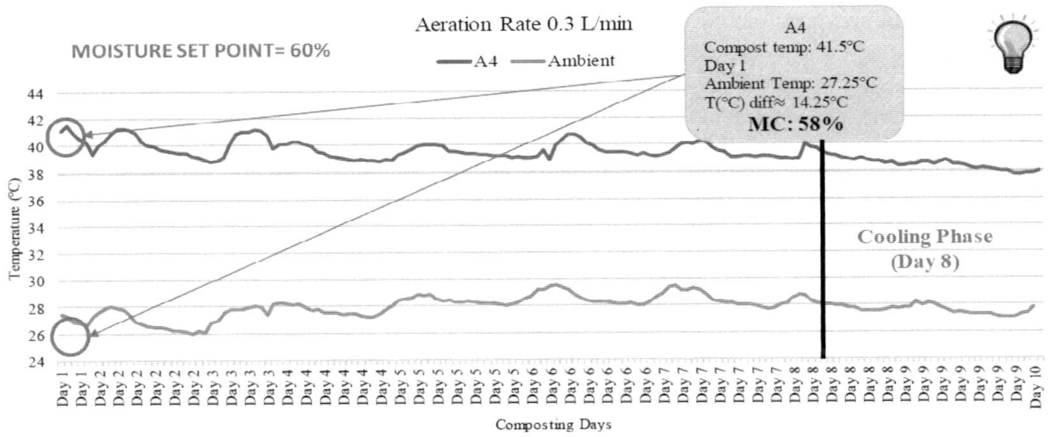

Figure 14. Moisture Content (%) and high peak temperature of A4 with medium aeration rate of 0.6 L/min.

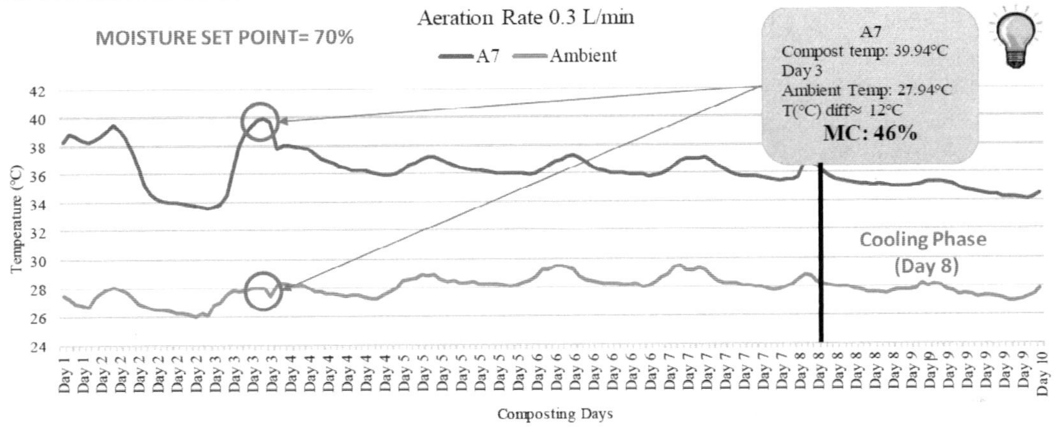

Figure 15. Moisture Content (%) and high peak temperature of A7 with medium aeration rate of 0.6 L/min.

Table 4. Final physicochemical parameters for the 3 selected vessels after 28 days composting.

Vessel	AR (L/min)	C:N	pH	EC (μS/cm)	Peak T (°C)
A2	0.6	19.11	7.31	578	31.81
A4	0.3	18.59	5.21	702	32.38
A7	0.3	20.07	5.58	819	31.63

The final physicochemical properties of the 3 best vessels without the heat source is shown in table 4. A pH of 4.17 was recorded from all the vessels used. This can be understandable and is likely from the ingredients of the food waste mixture. The mixture consisted of banana peels, green vegetables, orange skin and coffee grounds. The acidity can be notably due to the orange skin, as oranges are known for their acidic properties. The pH value, in the beginning, most likely has been impacted by external sources. This could most likely be the acidity of the food waste type. All 3 of the vessels that were selected, A2, A4 and A7 gave a notable increment in the pH value during the 28 days of composting. The degradation phase's occurrence is apparent due to these observed readings. Vessel A2 is the only one that fell within the acceptable range of 7 to 8, for the final pH value. Vessels A4 and A7 recorded slightly lower pH values. In terms of a final product indicator, this is considered quite poor, possibly due to the degree of aeration being insufficient or an insufficient supply of oxygen. The pH values can most certainly naturally vary. This could be resulting from the waste type used.

A carbon to nitrogen ratio of 22.46 was recorded for the total set of vessels in the beginning. The final C:N ratio for the 3 vessels in table 4 above showed ratios below 21. In detail, vessel A2 had a C:N of 19.11 and A4 had a C:N of 18.59. A7, on the other hand, had a value that is somewhat higher, at 20.07. The initial composting was planned for 40 days, but it was achieved in 28 days, as seen by the results above. This can be justified by referring to figure 8, 9 and 10, where, after a time duration of 22 days, the graph trends indicated that the stability of the compost is at a temperature range that is similar. The maturity of the compost is confirmed by this. It is therefore unnecessary to extend the composting period which was done within a short time period of 28 days. The result shown here also means that the experiment started off with a good C:N ratio. The somewhat high C:N ratio in Vessel A7 can be explained as being due to a moisture content that is elevated. This most likely made the composting period longer. It can also be concluded that A7 will perhaps need a composting duration that is more than 28 days in order to achieve a C:N ratio of 20.

The electrical conductivity results that were given above give information on the salt content. This content is mainly in the form of magnesium and calcium which is found inside the compost. A substantial difference is not observed among the vessels as compared to the initial EC value of 739 μS/cm. It is evident from the research that the vessels will all be under the acceptable range for the EC (< 30,000 μS/cm).

There was a strong odour emission produced throughout the experimental procedure which is not surprising as the temperature profiles are all noted to be less than 35 °C. This creates the notion that there is no active phase of bacteria killing.

A rough coarse kind of texture was observed with the food waste mix. Initially it was roughly 2.0 cm to 2.5 cm. After 28 days of the composting procedure, a clear conversion of the food waste into compost was observed, which met the requirement of usability. A darker shade was observed for the product and it could now be termed as a sort of fertilizer. Smaller particle sizes were observed, and it resembles a rough kind soil texture (figure 16 (a) and (b)).

Figure 16. Initial (a) and final (b) stage of compost.

Table 5. Final physicochemical parameters for the 3 selected vessels for 10 days composting.

Vessel	AR (L/min)	C:N	pH	EC (μS/cm)	Peak T (◦C)
A2	0.6	N/A	7.96	987	39.69
A4	0.3	18.59	7.79	1075	41.50
A7	0.3	N/A	7.54	693	39.94

In the event of the vessels with the heat source attached, the readings are as indicated in table 5. An acidic pH property of 4.17 was recorded for all the vessels. This was like the food waste mixture which consisted of banana peels, orange skin, coffee ground, and green vegetables. Vessel A2, A4 and A7 have shown an increment in the pH value for a duration of 10 days. This increased pH value shows that there is degradation taking place, which makes the product very appealing for use as a source of compost.

An initial C:N ratio of 22.46 was recorded for all the vessels and vessel A4 had attained the requirement of C:N ratio below than 20 to indicate good usability of the final product. Vessel A2 and A7 were not selected for C:N ratio measurement due to their peak temperature being below 40, which did not fulfil the purpose of adding heat source to the composting vessel. This study aimed to raise the temperature of the compost pile by using affordable sources such as attaching light bulbs in the second experiment. The electrical conductivity of all the vessels was under an acceptable EC of 30,000 µS/cm and below. No emission of odour was reported for this second experiment, which is not surprising as the temperature profile was recorded to be higher than 40 °C which would show that bacteria killing has begun. The physical changes appear to be like the first part of the experiment (figure 17 (a) and (b). A very dark shade was observed for the final product. It can be said to resemble the look of a common fertilizer. The usability of this product was justified clearly in relation with its C:N, pH and EC.

Figure 17. Initial (a) and final (b) stage of compost.

4. Conclusions

Acknowledging the importance of composting food waste, the effects of external heat sources on the temperature and moisture variations in food waste composting has been provided in this paper. The effect of these additional heat sources has been investigated in order to observe the temperature and moisture changes during composting of food waste with the intermittent aeration technique. The experiments explained the possibility of shortening the composting period from 28 to 10 days with the addition of a heat source. Minimal loss of moisture has occurred with the use of an intermittent aeration rate instead of a continuous aeration rate. In the 28 days of composting, intermittent aeration at rate of 0.6 L/min with 50 % moisture content and 0.3 L/min with 60% and 70% moisture content had showed the highest temperature peak of 31.8°C, 32.4°C and 31.6°C respectively by day 16 compared to other vessels. However, all vessels had strong odour emission. Hence, an additional heat source (light bulb) was utilized to minimize the release of unpleasant odour. A higher temperature profile between 39.7°C and 41.5°C was attained within 1 day in the second part of the study with 10 days composting. The forced aeration with intermittent mode and low aeration rate has shown a good composting process. Further optimization results showed that intermittent aeration with aeration rate between 0.2 to 0.3 L/min can achieve a high temperature profile with peak temperature of approximate 42°C achieved within 24 hours. In addition, these low aeration rates will also create less odour and reduce the loss of moisture content. The final product has properties of C:N ratio below 20, pH in range of 7–8.5, EC below 30,000 µS/cm and moisture content between 40–60% that make it a great fertilizer. These research findings act as a groundwork for future experimentations on utilizing intermittent aeration at low rate to compost the food waste. The high temperature is achievable with use of low aeration rate and composting process can be sped up to days or hours to avoid long accumulation of waste. With this, a simple automated home composting system can be built with an improvising to this groundwork.

References

[1] Worldometers 2017 World population prospects *Worldmeters*
[2] Muhd Yunus M N and Abd Kadir K Z 2006 The development of municipal solid waste treatment technology based on refuse derived fuel and bio-gasification integration *Proc. of the International Symposium on Renewable Energy: Environ. Protection and Energy Solution* 762
[3] Sreenivasan J, Govindan M, Chinnasami M and Kadiresu I 2012 Solid waste management in Malaysia — A move towards sustainability *Waste Manag. An Integr. Visions* 55–70
[4] Kathirvale S, Muhd Yunus M N, Sopian K and Samsuddin A H 2004 Energy potential from municipal solid waste in Malaysia *Renewable Energy* **29** 559–67
[5] Wan A and Kadir W R 2001 A comparative analysis of Malaysian and the UK *London England Addison-Wesley Publishing Company Inc.*
[6] Sakawi Z 2010 Municipal solid waste management in Malaysia: Solution for sustainable waste management *J. Appl. Sci. Environ. Sanit.* **6** 29–38
[7] Cerda A, Artola A, Font X, Barrena R, Gea T and Sanchez A 2018 Composting of food wastes: Status and challenges *Bioresour. Technol.* **248** 57–67
[8] Guo W Y,Zhou Y, Zhu N W ,Hu H G, Shen W H, Huang X X, Zhang T P, Wu P X and Li Z B 2018 On site composting of food waste: A pilot scale case study in China *Resour. Conserv. Recycl.* **132** 130–8
[9] Li Z T, Lu H W, Ren L X and He L 2013 Experimental and modeling approaches for food waste composting: A review *Chemosphere* **93** 1247–57
[10] Kumar M, Ou Y L, and Lin J G 2010 Co-composting of green waste and food waste at low C/N ratio *Waste Manag.* **30** 602–9
[11] Chang J I, Tsai J J and Wu K H 2006 Composting of vegetable waste *Waste Manag. Res.* **24** 354–62
[12] Rawoteea S A, Mudhoo A and Kumar S 2017 Co-composting of vegetable wastes and carton: Effect of carton composition and parameter variations *Bioresour. Technol.* **227** 171–8
[13] Liang C, Das K C and McClendon R W 2003 The influence of temperature and moisture contents regimes on the aerobic microbial activity of a biosolids composting blend *Bioresour. Technol.* **86** 131–7
[14] Iqbal M K, Shafiq T and Ahmed K 2010 Characterization of bulking agents and its effects on physical properties of compost *Bioresour. Technol.* **101** 913–9
[15] Guidoni L L C, Marques R V, Moncks R B, Botelho F T, da Paz M F, Correa L B and Correa E K 2018 Home composting using different ratios of bulking agent to food waste *J. Environ. Manag.* **207** 141–50
[16] Tan W Y, Then Y L, Lew Y L and Tay F S 2019 Newly calibrated analytical models for soil moisture content and pH value by low-cost YL-69 hygrometer sensor *Measurement* **134** 166–78
[17] Gao M, Li B, Yu A, Liang F, Yang L and Sun Y 2010 The effect of aeration rate on forced-aeration composting of chicken manure and sawdust *Bioresour. Technol.* **101** 1899–903
[18] Jolanun B, Towprayoon S and Chiemchaisri C 2008 Aeration improvement in fed batch composting of vegetable and fruit wastes *Environ. Progress* **27** 250–6
[19] Trautmann N 1996 Compost physics Cornell Waste Management Institute
[20] Tao J, Li G X, Tang Q, Ma X G, Wang G and Schuchardt F 2015 Effects of aeration method and aeration rate on greenhouse gas emissions during composting of pig feces in pilot scale *J. Environ. Sci.* **31** 124–32
[21] Hafid H S, Rahman N A A, Shah U K M, Baharuddin A S and Ariff A B 2017 Feasibility of using kitchen waste as future substrate for bioethanol production: A review *Renew. Sustain. Energy Rev.* **74** 671–86

[22] Storino F, Arizmendiarrieta J S, Irigoyen I , Muro J and Aparicio-Tejo P M 2016 Meat waste as feedstock for home composting: Effects on the process and quality of compost *Waste Manag.* **56** 53–62

[23] Abd Razak S, Abd Ghafar S W, Mohd Padzil N A, Kamaruddin A, Mat Zin N, Saim M and Suhaimi A H M S 2018 Household food wastage prevention in Malaysia: An issue processes model perspective *Economic Technol. Manag. Rev.* **13** 51–62

Acknowledgments
This research project acknowledges the support from the Faculty of Engineering, Computing and Science and Chemical Engineering Research lab for continuous accommodating the needs of this project.

Study of forced aeration system for fruit and vegetable waste composting

Y H Then[1], J C Lai[1] and Y L Then[2]

[1] School of Chemical Engineering and Science, Faculty of Engineering, Computing and Science, Swinburne University of Technology, 93350 Kuching, Sarawak, Malaysia
[2] School of Engineering, Faculty of Engineering, Computing and Science, Swinburne University of Technology, 93350 Kuching, Sarawak, Malaysia

Email: jlai@swinburne.edu.my

Abstract. Composting of fruit and vegetable waste (FVW) was investigated in a forced aeration system to determine the effects of aeration rate (AR) and aeration method (AM) on compost quality. Rice husk was mixed with FVW as bulking agent to compost for seven days and aerated in continuous and intermittent at rates of 0.3, 0.6, and 0.9 L min^{-1}. Intermittent aeration at rate of 0.3 L min^{-1} was observed to have higher initial temperature increasing rate and peak temperature compared to continuous aeration. Moisture loss at rate of 0.3 L min^{-1} was found smaller than other two rates regardless of AM. With comparison made for similar AR, intermittent aeration had experienced less moisture loss with an average of 17.15 %. The pH of compost was found to vary in aeration method where higher peak pH was found in intermittent aeration at low rate, corresponding to their temperature profile. Largest carbon-to-nitrogen (C/N) ratio reduction was attained by compost treated in intermittent aeration at rate of 0.3 L min^{-1}. Further optimization was performed at a rate of 0.1 L min^{-1}, 0.2 L min^{-1}, and 0.4 L min^{-1}. The result showed an improvement at rate of 0.2 L min^{-1} in intermittent aeration.

1. Introduction

The year 2018 shows a total amount of 16 metric kilotons of food is wasted daily in Malaysia [1]. Statistics show that 20–50 % of fruits and vegetables are discarded along its production chain in the year 2018 due to poor management and failure to achieve required specifications [2]. Such high loading of waste has a considerable role in environmental pollution if it is not treated correctly. Conventionally, waste management strategies, such as landfilling and incineration, are used to treat this waste. However, this led to a tremendous amount of gas emission and caused related environmental issues. Composting was explored as a more sustainable and effective solution to manage food waste to overcome the drawbacks. It helps to recover biodegradable waste materials and minimize gas emissions. Composting is the process of turning organic waste matter into a humus-like substance that can greatly improve soil conditions due to its adequately dense structure and nutrient contents [3]. A successful composting will produce a final product that is sufficiently stable for application to land without adversely affecting the environment. The waste volume, weight and water content subsequently can be reduced, and the valuable nutrients contained inside the waste can be returned to soil for enhancement of soil quality.

Composting is a biological degradation and stabilization process of organic substrates to carbon dioxide (CO_2) and water under aerobic conditions. The biologically generated heat is sufficient to develop thermophilic temperatures inside the composting system [4]. Generally, the composting process and compost quality are governed by temperature, moisture content, oxygen (O_2) level, pH, and carbon-to-nitrogen (C/N) ratio. These factors, however, are dependent upon each other, and hence a proper control of operating conditions is required [5].

Past researchers investigated the effects of various aeration rate (AR) and aeration method (AM) in forced aeration system on composting a wide variety of waste. Shen *et al.* [6] composted a mixture of chicken manure, straw, and dry grasses at different ARs of 0.01, 0.1, and 0.2 m^3 min^{-1} m^{-3}. Rasapoor *et al.* [7] studied the effects of AR (0.4, 0.6, and 0.9 L min^{-1} kg^{-1}) on composting of municipal waste. Li *et al.* [8] compared the composting performance under continuous aeration and intermittent aeration. Although considerable research has been conducted on AR and AM's impact on effective composting, little is known about the effect of AR and AM on the composting process of fruit and vegetable waste (FVW) in forced aeration system. As has been discovered, the households are responsible for the majority of the food waste in Malaysia. The main food waste from these households include rice, vegetables and meat [9]. It was further mentioned that the waste is generally grouped into 60% carbohydrates, 20 % protein and 10 % fats. Fruits and vegetables are definite sources of carbohydrates, thereby making them the predominant wastes from Malaysian households. On the other hand, while meat is also considered a part of household food waste, it will not be applied in the composting process. This is because the composting of meat products can result in significant odour problems if not handled adequately, thus increasing the risk of pest infestations [10]. This process is likely not favoured by the general public as well. Therefore, this study will utilize fruits and vegetables as the main substrate in the mixture.

The composting process is a process that decomposes organic matters in the presence of air, or more specifically oxygen [11]. Aeration is classified into two categories, passive aeration, and forced aeration. Passive aeration is where diffusion and natural air movement occur in a system, whereas forced aeration depends on fans or pumps to move the air through the composting pile's mass. For forced aeration mode, there are two types of forced aeration used in a composting process: continuous aeration mode and intermittent aeration mode.

In continuous aeration mode, the air is continuously pumped into the vessel containing food waste. Studies show various disadvantages of adopting continuous aeration mode into the composting system. One of the disadvantages of this conventional method as it excessively pumps air into the system, which may lead to ample moisture and heat losses and the release of undesired gas [12]. Extensive moisture and heat losses in a composting process are not favorable as it would slow down the decomposition process occurring within the composting pile [13]. However, if not enough air is pumped into the system, the system develops anaerobic conditions at different pile regions [13]. This is unfavorable as anaerobic conditions would cause odour problems and a longer time to reach the maturity stage compared to aerobic conditions [14].

Intermittent aeration mode supplies oxygen within the intervals with the appropriate pumping time, resting time, and aeration rate. By adopting intermittent aeration mode, there would not be excessive pumping of the air into the system. Therefore, reducing moisture and heat loss as well as the release of undesired gas. By reducing moisture and heat loss, the time taken for composting to reach maturity under these ideal conditions would be shortened compared to continuous aeration mode. Hence, this would be a significant factor when promoting this in-vessel composting system into households. Besides, intermittent aeration mode has been used in the current years in order to reduce energy consumption [8]. Based on previous studies, it can be observed that the oxygen content in the center of the pile may fall from saturation in approximately 25–50 minutes at a rate of 5 %. By adopting intermittent aeration mode, pumping air into the system and resting within the period where the oxygen saturation falls would cause conservation of energy and enhancement of composting quality [8]. Li *et al.* [8], intermittent aeration could increase the oxygen supply efficiency and reduce the total greenhouse gases (GHG) emission by 17.8 % compared to continuous aeration mode. This reduction would then increase if the composting process were ended after approximately 36 days. In another article, intermittent aeration mode has proved to be the best compared to continuous aeration mode. When utilizing the same aeration rate for these two methods, intermittent aeration was shown to be more efficient and economical than continuous aeration mode [12].

Hence, this study aimed to determine the effect of AR and AM on forced aeration composting of FVW to supply an optimum aeration condition for a successful composting process.

2. Materials and methods

2.1. Experimental design and setup

Twelve plastic containers that have a volume of 5 L with a dimension of 29.0 (L) x 19.7 (W) x 10.1 (H) cm were utilized as a composting vessel. The top cap was used to facilitate its filling and emptying of compost material. A small hole (X) was drilled on the top to facilitate the inflow of air. The airflow rate was adjusted by manipulating the valve openings, and it is measured by using an anemometer. Several small holes were also drilled at the vessel's bottom surface to allow the outflow of leachate. The composting vessel was placed inside a polystyrene box, which acts as a heat insulation layer to minimize heat loss to the surroundings. Besides, a light bulb was placed inside the polystyrene box to provide a heat source to maintain the ambient temperature at around 30 °C. The composting setup is shown in figure 1.

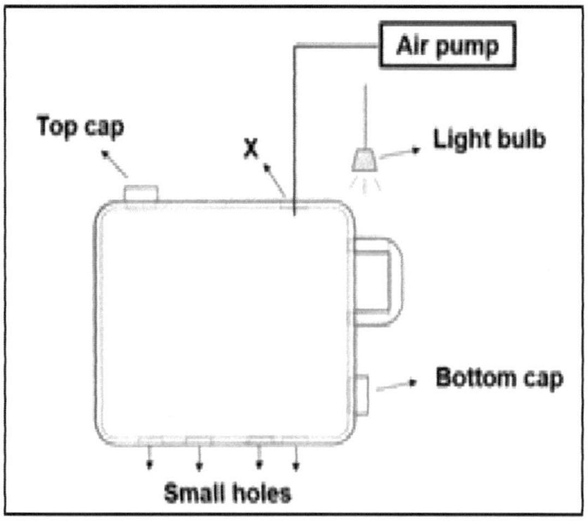

Figure 1. Composting setup.

2.2. Composting material

Vegetable waste (raw uneaten green vegetable waste) and fruit waste (leftover orange peels) was collected from the local market and juice stall. This FVW was cut into small pieces, then mixed in with rice husk to enhance the composting process. Rice husk as a source of carbon was employed as a bulking agent to adjust the C/N ratio of the composting materials as well as its moisture content. Rice husk is considered as one of the major by-products from the rice production process. This agricultural residue could absorb extra moisture from compost material. Controlling the moisture content of compost material is important because fruits and vegetables waste are usually high in moisture. Besides, with addition of rice husk, oxygen transfer would be increased as the porosity of the compost materials were increased [33]. Each container had a mixture of 480 g vegetable waste, 320 g fruit waste and 200 g rice husk (ratio of 12:8:5). The waste mixture had filled up approximately 90 % volume of the container with remaining left for ventilation purposes.

2.3. Composting process

Continuous aeration with AR of 0 (Control), 0.3 (C1), 0.6 (C2) and 0.9 (C3) L min^{-1} were used in the first part of the experiment. The aeration rate was selected based on literature research from past experiments, which were widely used for composting of various wastes except fruit and vegetable waste. It was repeated with intermittent aeration with an aeration interval of 1 day based on 20 minutes' aeration (before sampling). I1, I2 represented these aeration conditions, and I3, corresponding to 0.3, 0.6, and 0.9 L min-1 AR. The material was composted for seven days, and each treatment was performed in duplicate to ensure the experiment's accuracy. The composting material inside the vessel was manually mixed to have a homogeneous mass before sampling at 9 a.m. each day. Table 1 shows the aeration condition for all composting vessels used in the first part of the experiment. The second part of the experiment was conducted with an aeration method that had shown good performance in the first part of the experiment. The peak temperature, moisture loss, final pH and final C/N ratio were the criteria used to assess performance of the composting process.

Table 1. Aeration rate and mode for seven composting vessels.

Abbreviation	Aeration rate (L min^{-1})	Air velocity (m/s)	Aeration method
Control	0	0	-
C1	0.3	0.4	Continuous
C2	0.6	0.8	Continuous
C3	0.9	1.1	Continuous
I1	0.3	0.4	Intermittent
I2	0.6	0.8	Intermittent
I3	0.9	1.1	Intermittent

2.4. Sampling and analysis

Approximately 5 g of sample was collected from the top cap each day after the mixing activity. 3 g of this sample was immediately used to determine the moisture content placed in the oven for 24 hours under 105 °C [15]. For pH measurement, the remaining 2 g of sample was used to prepare aqueous extract at a 1:10 compost: water ratio (w/v) and mechanically shaken for 1 hour. The pH was then measured with a pH meter [16]. A total amount of 35 g sample was extracted over the period of 7-day composting, which was only around 3.5% of the smallest total compost weight of 1 kg. The total amount of sample was sufficiently small in quantities to keep the majority of the compost remaining in the vessel so as not to interfere with overall composting. The temperature of composting material was measured using a stainless-steel probed digital thermometer each day at morning and evening during resting period without aeration. The temperature was measured in duplicates at the center of compost and the average value was taken for that day. For C/N ratio analysis, 100 g of sample was sent to the Nabbir Laboratory Sdn Bhd (Kuching) for Total Organic Carbon (TOC) and Total Kjeldahl Nitrogen (TKN) analysis. The C/N ratio was then calculated by dividing TOC with TKN.

2.5. Statistical analysis

Experimental data were analyzed by Duncan's Test and analysis of variance (ANOVA) with the least significant difference (LSD-t). The difference between duplicate data was compared at the level of 1 % and 5 %.

3. Results and discussion

3.1. Physical characteristics

Throughout the composting process, mould was observed on Day 2 and 4 in I1 and I2 treatments. Figure 2 (a) shows the mould was white in colour and had a cotton-like texture. The presence of mould signifies full decomposition, and it is often seen on dead matter such as compost. After seven days of composting, it was observed that only rice husk was leftover in the finished compost. This implies that FVW decomposes entirely throughout the process. The finished compost had a light brown colour despite Control or I1 as shown in figure 2 (b) and 2 (c), respectively. The others compost using continuous and intermittent aeration mode at different rates also had the same colour and appearance as shown in figure 2 (b) and 2 (c). This indicates that the compost had low humic substances content and was considered immature to be utilized in agricultural activity [17].

Besides, the odour was produced as composting progress, and it was maintained in the finished compost. This was due to the anaerobic degradation process resulting from insufficient diffusion of O_2 into a deeper compost material area in the early stage of composting. As a result, anaerobic bacteria took a foothold and produced odour through its metabolic process. Moreover, a 100 % aerobic degradation was hard to accomplish due to small anaerobic voids inside the vessels. However, mixing activity was performed daily to allow equal distribution of airflow.

(a) (b) (c)

Figure 2. Mould formation (a), finished compost appearance of Control (b), and I1 (c).

3.2. Temperature

In the composting process, the temperature tends to change in response to microbial activity [18]. The temperature can classify the composting stages, namely the mesophilic phase, thermophilic phase, cooling phase, and maturation phase [19]. Besides, the temperature must be maintained above 55 °C for days to eliminate the pathogens in the compost material [20]. The changes in temperature under different aeration conditions is shown in figure 3. Duncan's Test showed no significant difference between the two trials ($p<0.01$). This is supported by ANOVA, which showed no significance between the trials ($p<0.05$). Based on figure 3, the initial temperature in all treatments was approximately 31 °C. The highest temperature was found in I1, which was 42 °C, and this indicates that the thermophilic phase was not achieved in this experiment.

Besides, it is also investigated that low AR treatments could achieve a higher initial temperature increasing rate and peak temperature. Under similar AR, the initial temperature increasing rate and peak temperature in intermittent aeration were higher than continuous aeration. This implies that excessive aeration from high AR and continuous aeration could cool down the compost and restrict the vessel's temperature rise. A similar trend was also found in the study conducted by Lu *et al.* [21] and Arslan *et al.* [15]. In contrast with this finding, Shen *et al.* [6] and Rasapoor *et al.* [7] reported that high AR was able to provide more O_2 for degradation and the heat generated surpassed the cooling rate caused by high AR. However, this does not apply in this case as different compost material was used in the experiment, and the volume was relatively small compared to their studies. Hence, high AR could quickly cool down the compost in this experiment.

Based on the temperature curve, it increased upon the initiation of the experiment. This explained that O_2 was utilized by the microorganisms to break down the degradable organic material. Based on the temperature curve of Control, it is found that the absence of forced aeration had reduced the O_2 level, which slowed down microbial activity and caused less generation of heat.

The temperature began to decrease upon reaching its peak due to reducing organic matter degradation rate. The cooling phase occurred on Day 5–7, corresponding to the depletion of organic matter by mesophilic microorganisms.

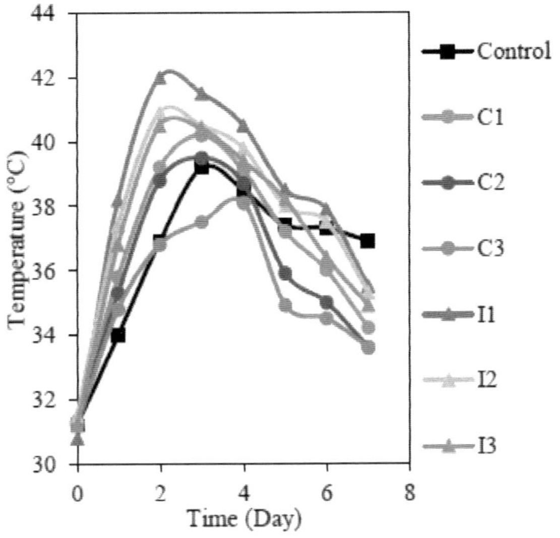

Figure 3. Changes in temperature under differential aeration condition.

Maturation had not been achieved in this work due to the shorter composting period. The result showed that the Control set resulted in the highest temperature at the end of the experiment. This was because the degradation was not completed in the vessel owing to low O_2. Under similar AM, the temperature decreased at a higher rate in high AR conditions due to a more significant cooling effect. Similarly, under similar AR, the cooling effect under continuous aeration was greater compared to intermittent aeration.

3.3. Moisture content

An optimum moisture content level is necessary throughout the process to allow the growth of microorganisms and improve microbial activity [22]. If the moisture content is higher than 60%, it will decrease airflow supply efficiency and adversely affect the aerobic degradation process [23]. On the other hand, if the moisture content is lower than 40%, the microbial activity will slow down and become dormant. The ideal moisture content of compost is between 40–60 % [24]. Figure 4 shows the moisture content changes under different aeration conditions, and figure 5 shows the total moisture loss based on the initial moisture content. Both Duncan's Test and ANOVA showed no significant difference between the trials ($p<0.01$ & $p<0.05$).

Based on figure 4, all the composting experiments were initiated with an approximately similar moisture content of 62 %, and this was slightly higher than the recommended value. However, all treatments' moisture content was then maintained between 40–60 % on Day 1 onward, which were acceptable. There was a slight increase in moisture content of vessel I3, C1 and C2 between day 5 to 7. This could be due to water vapor at the wall dropping back into compost material during the cooling phase that occurred during this period. The microbial activity starts to slow down and less heat is generated to evaporate the excess moisture from compost material.

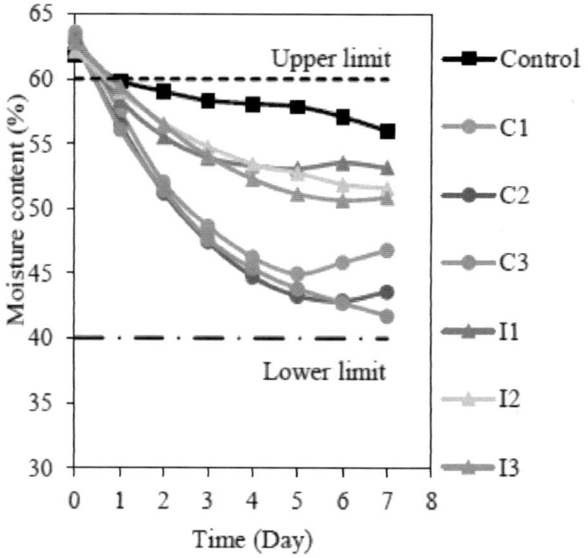

Figure 4. Changes in moisture content under different aeration conditions.

In the composting process, moisture loss resulted from leachate formation as well as evaporation of water. Additionally, there are two sources of evaporation of water whereby it can be either resulting from heat formation inside the vessel or the drying from aeration. The moisture content in all treatments except Control showed a sharp decrease upon the experiment's initiation, corresponding to the development to the mesophilic phase. This explained that the presence of forced aeration dried the compost material and, at the same time, generated heat that evaporated the water content in the compost material. Apart from that, it was determined that high AR evaporated more water and reduced the moisture content significantly when continuous aeration was used.

Based on figure 5, I1 had the highest initial rate of moisture loss on Day 1, corresponding to its highest heat formation and highest temperature difference. This implies that the heat generated from degradation activity evaporated the moisture at a higher rate compared to drying from aeration. Because of the highest evaporation rate, C3 produced a compost with lowest moisture content such that 42.63% was achieved. This finding was supported by other researchers where the researchers found out that the high AR produced a low moisture content compost [15]. Comparing the AM, it shows that intermittent aeration was able to retain the moisture in the compost material. Although the heat formation was higher in intermittent aeration, the excess airflow in continuous aeration had a more significant impact in evaporating the moisture, which then produced compost that had lower moisture content compared to intermittent aeration. This is agreed by Jiang *et al.* [32], the researchers reported a higher decreasing rate of moisture content when continuous aeration was used, and its final moisture content was significantly lower compared to intermittent aeration.

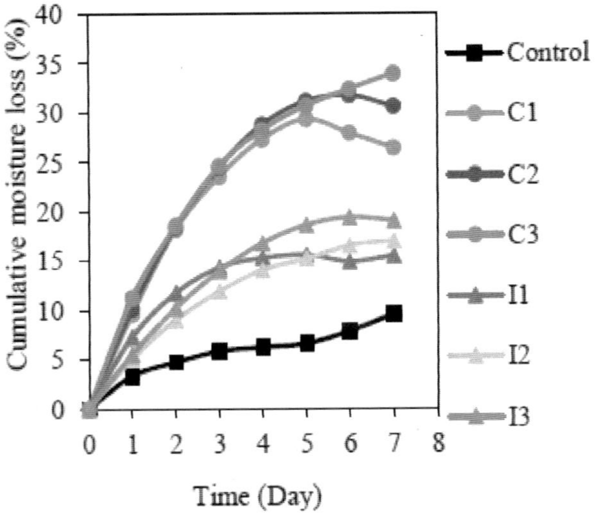

Figure 5. Total moisture loss based on initial moisture content.

3.4. pH

The optimal pH range for bacteria and fungi to operate is between 6–7.5 and 5.8–8, respectively [25]. Besides, the minimum limit of composting mixture pH is 6.5, in which any pH lower than this value will inhibit the composting process [26]. Furthermore, the final product should have a neutral pH range from 7–8.5 to ensure the compost's safety application [27]. Figure 5 shows the changes in pH under different aeration conditions. Both Duncan's Test and ANOVA showed no significant difference between the trials ($p<0.01$ & $p<0.05$).

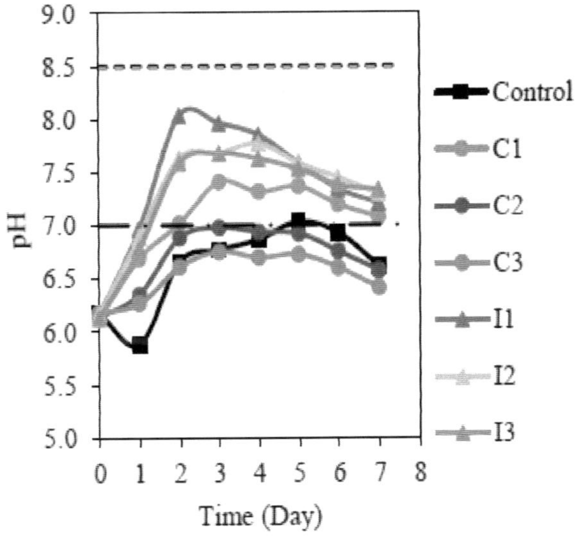

Figure 6. Changes in pH under different aeration conditions.

Based on figure 6, the composting experiments began with an initial pH of approximately 6.1. This was slightly lower than the optimal pH of 6.5, but it was still within the bacteria and fungi operating pH. In the early stage of composting, the pH increment was contributed by the formation of ammonium ions from organic matter mineralization. However, this experiment's ammonium ion production is low as it is highly produced in the thermophilic phase. This can be observed from the highest pH achieved was only 8.01 compared to 9 achieved by Arslan *et al.* [15].

Figure 6 also shows that low AR achieved a higher peak pH owing to its higher degradation rate. Lu *et al.* [21] reported a similar trend when composting dog food. Moreover, under similar AR, intermittent aeration achieved a higher peak pH, which the highest peak pH of 8.01 was found in I1.

The pH began to stabilize at the latter stage of composting. This was because the release of H^+ ion from the volatilization of ammonium ion and nitrification process inhibited pH from increasing. The variation of AR in continuous aeration showed a significant impact on produced compost pH where only C1 returned to a neutral pH of 7.08 while C2 and C3 produced a slightly acidic compost. This could be due to the high AR aerated in continuous mode causing the slow breakdown of soluble compounds into organic acids. The breakdown process could be completed if the composting period is prolonged. On the other hand, AR's effect on the final pH of compost under intermittent aeration was minimal. This was further supported by results obtained in Section 3.6 with compost treated at different AR in intermittent aeration mode. The final pH stabilized between 7.21 and 7.34, within the safety range (7–8.5).

3.5. C/N ratio

Table 2 shows the C/N ratio and TKN of Control, I1, I2, C2, and C3. Kumar *et al.* [28] stated that the ideal initial C/N ratio is between 25–30. However, due to the inconsistent supply of FVW (different type and origin of FVW), the initial C/N ratio of compost material varied at each experiment run. In the composting process, the C/N ratio variation depends on the degradation of organic carbon and nitrogen. The result shows that the C/N ratio of Control decreased slightly from 38.15 to 36.65 after seven days of composting, indicating that the absence of forced aeration reduced the organic degradation rate.

Comparing the C/N ratio of I1 and C1 clearly showed that AM had a significant impact on the C/N ratio. Since more organic matter was degraded in I1, the C/N ratio of I1 was significantly larger than C1. Similarly, Jiang *et al.* [32] reported higher C/N ratio changes when intermittent aeration was applied. Based on the final C/N ratio obtained, only I1 and I2 could produce a mature compost where its final C/N ratio falls within 12–20 [29]. Furthermore, it is also investigated that TKN increased in all treatments. The activity of nitrogen-fixing bacteria can explain this during the composting process [30]. Besides, the increase in organic nitrogen can also be attributed to a concentration effect whereby organic carbon degrades at a high rate [31].

Table 2. C/N ratio and TKN of control, I1, I2, C1 and C3.

Parameter	Control	I1	I2	C1	C3
C/N ratio (I)^{i}	38.15	32.36	27.19	38.15	39.56
C/N ratio (F)^{ii}	36.65	19.67	18.96	30.69	36.34
C/N ratio (D)^{iii}	1.50	12.69	8.23	7.45	3.22
TKN (I)^{iv}	0.27	0.36	0.42	0.27	0.32
TKN (F)^{v}	0.34	0.64	0.70	0.36	0.38
TKN (D)^{vi}	0.07	0.28	0.28	0.09	0.06

i = initial C/N ratio; ii = final C/N ratio; iii = changes in C/N ratio; iv = initial TKN; v = final TKN; vi = changes in TKN.

3.6. *Optimization of aeration condition*
Based on the result obtained in the first part of the experiment, it shows that intermittent aeration could attain higher peak temperature than compost in continuous aeration. In addition, the compost aerated with intermittent aeration method had experienced less moisture loss, which kept degradation proceed until compost attained final C/N ratio below 20. The compost obtained from the 7-day composting period falls within the safety range of pH value. Under this AM, I1 and I2 performed better compared to the others. Hence, optimization was conducted around these two ARs with AR of 0.1, 0.2, and 0.4 L min^{-1}. Their abbreviations are O1, O2, and O3, respectively. Figure 7, 8, and 9 show the changes in temperature, moisture content, and pH in optimization treatments.

Based on the result obtained, O1 was able to retain the moisture content throughout the process. However, it developed a low-temperature profile. In contrast, I1 and O2 developed a better temperature profile but resulted in a slightly higher moisture loss. Additionally, it is also found that O2 achieved the highest initial temperature increasing rate and peak temperature compared to I1, which indicates 0.2 L min^{-1} resulted in significantly increased microbial activity to degrade organic matter in vessel O2. Corresponding to that, the peak pH of O2 was also higher compared to I1, and the final pH in I1 was closer to the lower boundary of the safety limit (7). This indicates that I1 has a higher possibility of producing acidic compost. Based on the result, it is determined that O2 is the optimum aeration that can carry out composting efficiently.

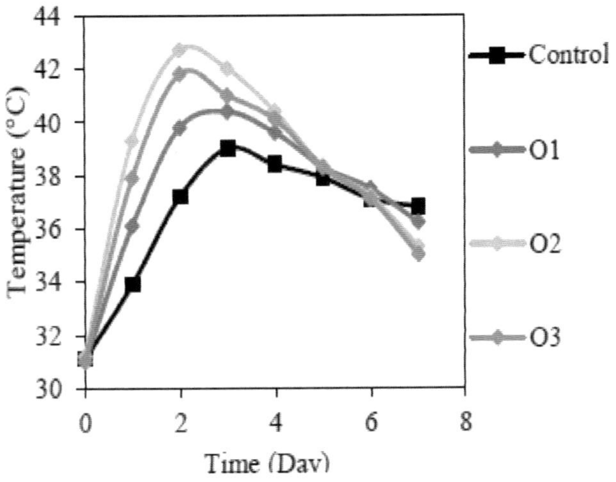

Figure 7. Changes in temperature in optimization treatments.

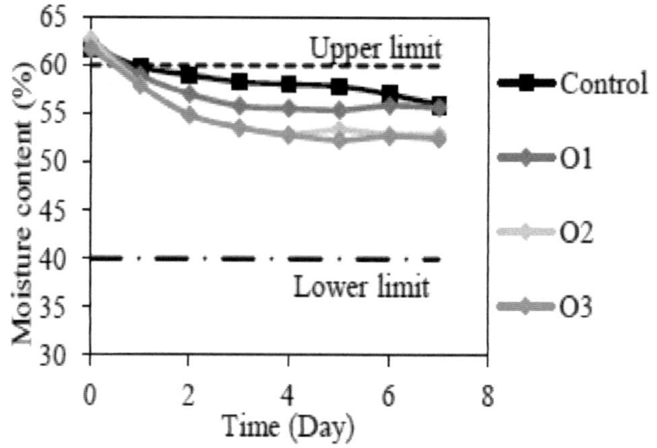

Figure 8. Changes in moisture content in optimization treatments.

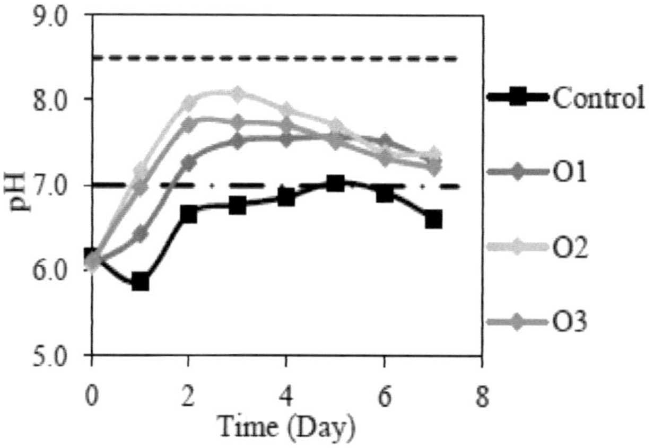

Figure 9. Changes in pH in optimization treatments.

4. Conclusion

In conclusion, it is determined that both AR and AM had a significant impact on the FVW composting process. The initial temperature increasing rate and peak temperature were generally higher in the low AR treatments. This reflected a higher rate of microbial activity when low AR was used. Intermittent aeration further promoted the degradation of organic matter and generated more heat during the FVW composting process. Apart from that, the temperature gradually decreased upon reaching its peak. It occurred at a higher rate when high AR and continuous aeration were used, owing to the cooling effect. Compared to high AR, low AR resulted in minimal moisture loss throughout the process. Additionally, intermittent aeration was able to retain the moisture in the compost material effectively compared to continuous aeration. Nevertheless, all finished compost's moisture content falls within 4–60 %, which is acceptable. The result showed that low AR also resulted in higher peak pH because of the higher production of ammonium ions from mineralization. However, the variation of obtained compost pH with AR was not significant as they were close to each other when aerated with different AR. Since intermittent aeration enhanced the degradation, it allowed higher peak pH achievement during the composting process. Furthermore, the changes in the C/N ratio were governed by both AR and AM. The obtained compost C/N ratio was between 18.96–36.65. The highest C/N reduction was found in 0.3 L min^{-1} with intermittent aeration, which its C/N ratio reduced from 32.36 to 19.67. The present study showed that AR of 0.3 L min^{-1} and 0.6 L min^{-1} with intermittent aeration mode provided better composting performance. Further optimization results showed that intermittent aeration with AR of 0.2 L min^{-1} was generally better than the other aeration conditions. This aeration mode was able to achieve the highest initial temperature increasing rate and peak temperature. However, the moisture reduction rate was the highest compared to the others. In this research study, the FVW composting process progressed well with the intermittent aeration mode. The compost treated in intermittent aeration could attain C/N ratio below 20, pH in range of 7–8.5, moisture content between 40–60 % and reach peak temperature higher than others. The conventional composting process requires the achievement of a thermophilic phase to degrade organic matter. Based on these research findings, although the thermophilic phase was not achieved, the compost produced under the mesophilic phase had an excellent fertilizer condition. Hence, this finding shows the potential to carry out composting with only the mesophilic phase. This helps to reduce gas emission as high gas emission only occurs during the thermophilic phase. The only downside of this process is that it could not achieve pathogen elimination and humic substances content might be lower than compost treated with thermophilic phase.

References

[1] Newsflash R A G E 2018 Food wastage in Malaysia *The Star Online*
[2] The Star Online 2018 Malaysians throwing away food at alarming rate 2018 *The Star Online*
[3] Kadir A A, Jamaludin S N and Azhari N W 2016 An overview of composting based on variable feedstock material *MATEC Web Conf. EDP Sci.* **47** 05016
[4] Wang W D, Wang X F, Liu J B, Ishii M, Igarashi Y and Cui Z J 2007 Effect of oxygen concentration on the composting process and maturity *Compost Sci.Util.* **15** 184–90
[5] Diaz L F and Savage G M 2007 Chapter 4 factors that affect the process *Waste Mang.* **8** 49–65
[6] Shen Y J, Ren L M, Li G X, Chen T B and Guo G 2011 Influence of aeration on CH_4, N_2O and NH_3 emissions during aerobic composting of a chicken manure and high C/N waste mixture *Waste Manag.* **31** 33–8
[7] Rasapoor M, Nasrabadi T, Kamali M and Hoveidi H 2009 The effects of aeration rate on generated compost quality using aerated static pile method *Waste Manag.* **29** 570–73
[8] Tao J, Li G X, Tang Q, Ma X G, Wang G and Schuchardt F 2015 Effects of aeration method and aeration rate on greenhouse gas emissions during composting of pig feces in pilot scale *J. Environ. Sci.* **31** 124–32
[9] Hafid H S, Rahman N A A, Shah U K M, Baharuddin A S and Ariff A B 2017 Feasibility of using kitchen waste as future substrate for bioethanol production: A review *Renew. Sustain. Energy Rev.* **74** 671–86
[10] Storino F, Arizmendiarrieta J S, Irigoyen I, Muro J and Aparicio-Tejo P M 2016 Meat waste as feedstock for home composting: Effects on the process and quality of compost *Waste Manag.* **56** 53–62
[11] Lin L, Xu F Q, Ge X M and Li Y B 2018 Improving the sustainability of organic waste management practices in the food-energy-water nexus: A comparative review of anaerobic digestion and composting *Renew. Sustain. Energy Rev.* **89** 151–67
[12] Lin L, Xu F, Ge X and Li Y 2019 Chapter four - biological treatment of organic materials for energy and nutrients production — Anaerobic digestion and composting *Adv. Bioenergy* **4** 121–81
[13] Trautmann N 1996 Compost physics *Cornell Waste Management Institute*
[14] Washington State University 2021 Aerobic decomposition *Washington State University*
[15] Arslan Topal E I, Ünlü A and Topal M 2011 Determination of the effect of aeration rate on composting of vegetable–fruit wastes *Clean-Soil Air Water* **39** 1014–21
[16] Gao M, Li B, Yu A, Liang F, Yang L and Sun Y 2010 The effect of aeration rate on forced-

aeration composting of chicken manure and sawdust *Bioresour. Technol.* **101** 1899–903
[17] Guo X X, Liu H T and Wu S B 2019 Humic substances developed during organic waste composting: Formation mechanisms, structural properties, and agronomic functions *Sci. Total Environ.* **662** 501–10
[18] Tiquia-Arashiro S and Tam N F Y 2002 Characterization and composting of poultry litter in forced aeration piles *Process Biochem.* **37** 869–80
[19] Sánchez Ó J, Ospina D A and Montoya S 2017 Compost supplementation with nutrients and microorganisms in composting process *Waste Manag.* **69** 136–53
[20] United States Environmental Protection Agency 2000 Biosolids technology fact sheet: in-vessel composting of *biosolids United States Environmental Protection Agency*
[21] Lu S, Imai T, Li H, Ukita M, Sekine M and T Higuchi T 2010 Effect of enforced aeration on in-vessel food waste composting *Environ. Technol.* **22** 1177–82
[22] Sesay A A, Lasaridi K E and Stentiford E I 1998 Aerated static pile composting of municipal solid waste (MSW): A comparison of positive pressure aeration with hybrid positive and negative aeration *Waste Manag. Res.* **16** 264–72
[23] Madejón E, Díaz Blanco M, López R and Cabrera F 2002 New approaches to establish optimum moisture content for compostable materials *Bioresour. Technol.* **85** 73–8
[24] El-Haggar S 2007 Sustainable industrial design and waste management: cradle-to-cradle for sustainable development *London England Elsevier Academic Press* p 190
[25] Chen L, de Haro Marti M E, Moore A and Falen C 2011 Dairy compost production and use in Idaho: The composting process University of Idaho
[26] Sundberg C, Smårs S and Jönsson H 2004 Low pH as an inhibiting factor in the transition from mesophilic to thermophilic phase in composting *Bioresour. Technol.* **95** 145–50
[27] Masó M and Bonmatí A 2008 Evaluation of composting as a strategy for managing organic wastes from a municipal market in Nicaragua *Bioresour. Technol.* **99** 5120–4
[28] Kumar M, Ou Y L and Lin J G 2010 Co-composting of green waste and food waste at low C/N ratio *Waste Manag.* **30** 602–9
[29] Elango D, Thinakaran N, Perumal P and Sivanesan S 2009 Thermophilic composting of municipal solid waste *Appl. Energy* **86** 663–8
[30] Kamolmanit N and Reungsang A 2006 Effect of carbon to nitrogen ratio on the composting of cassava pulp with swine manure *J. Water Environ. Technol.* **4** 33–50
[31] Huang G F, Wong J W C, Wu Q T and Nagar B B 2004 Effect of C/N on composting of pig manure with sawdust *Waste Manag.* **24** 805–13
[32] Jiang T, Li G X, Tang Q, Ma X G, Wang G and Schuchardt F 2015 Effects of aeration method and aeration rate on greenhouse gas emissions during composting of pig feces in pilot scale *J. Environ. Sci.* **31** 124–32
[33] Li M X, He X S, Tang J, Li X, Zhao R, Tao Y Q, Wang C and Qiu Z P 2020 Influence of moisture content on chicken manure stabilization during microbial agent-enhanced composting *Chemosphere* **264** 128549

Acknowledgments
This research project acknowledges the support from the Faculty of Engineering, Computing and Science and Chemical Engineering Research lab for continuous accommodating the needs of this project.

Pure oxygen separation from air using dual-phase SDC-SCFZ disc membrane: A modelling approach

J Chua, C Li and J Sunarso

Research Centre for Sustainable Technologies, Faculty of Engineering, Computing and Science, Swinburne University of Technology Sarawak Campus, Kuching 93350, Sarawak, Malaysia

Email: cli@swinburne.edu.my

Abstract. Novel $Ce_{0.8}Sm_{0.2}O_{1.9}$-$SrCo_{0.4}Fe_{0.55}Zr_{0.05}O_{3-\delta}$ (SDC-SCFZ) disc membranes consist of 25 wt.% SDC fluorite ionic conducting phase and 75 wt.% SCFZ perovskite mixed conducting phase, which is more promising than perovskite oxide SCFZ single-phase membrane in terms of the oxygen permeation flux. This work features a modelling approach to simulate the oxygen permeation fluxes of the SDC-SCFZ membrane. Simplified model equations from the Zhu model and Xu-Thomson model based on the limiting cases of surface exchange reactions and bulk diffusion are compared. The Zhu model is found to be more applicable for the membranes with overall good correlation and low sum of squared error. Furthermore, modelling studies revealed that the oxygen transport is limited by surface exchange reactions from 700 to 850 °C and a mixture of both limiting cases above 850 up to 950 °C. It is concluded that the membranes exhibit high oxygen permeation flux of up to 2×10^{-6} $mol\ s^{-1}\ cm^{-2}$ at 950 °C with P_{air} of 5 atm and $P^{II}_{O_2}$ of 0.005 atm. The optimum range of operating conditions of the membrane are found to be at 950 °C with minimum P_{air} of 1 atm and $P^{II}_{O_2}$ lower than 0.025 atm.

1. Introduction

Mixed ionic-electronic conducting (MIEC) membrane is extensively utilised as an oxygen separation pathway alternative to pressure swing adsorption and cryogenic distillation [1]. This is due to its numerous advantages such as lower capital and operational costs, lower energy consumption, simpler process engineering and better scalability [1]. The membrane can attain up to 99.99 % oxygen product purity and 100 % oxygen selectivity when operated at elevated temperature (above 700 °C). The central theme of research activities for oxygen separation using MIEC membranes is primarily centred on achieving high oxygen permeation flux and good chemical stability of materials as well as exploring engineering methods and designs to improve the performance of the membrane. Different techniques such as engineering approaches, which are exemplified by mixing of ionic conducting phase and electronic conducting phase, geometry modification, internal and external short circuits, and modification of membrane surface or materials development approaches, which are represented by chemical doping have been widely investigated to boost the oxygen permeation flux of the membrane [2].

$SrCo_{0.4}Fe_{0.55}Zr_{0.05}O_{3-\delta}$ (SCFZ) is a perovskite-type oxygen permeable membrane, which is prepared by doping of fluorite-related structure ZrO_2 into perovskite-related structure $SrCo_{0.8}Fe_{0.2}O_{3-\delta}$ (SCF) [3]. The doping of Zr into SCF enhances the structural stability in the presence of CO_2 and H_2O-containing atmosphere, making it possible to retain its perovskite structure even after operating at 700 °C for 10 hours [4]. This is ascribed to the reduction of the oxygen vacancy concentration by the Zr^{4+} metal ion substitution [5]. Since the SCFZ membrane can retain its structure in the presence of CO_2 and H_2O, ambient air can be directly utilised as feed. Although the doping of Zr into SCFZ membrane may cause the oxygen flux of the membrane to be reduced, the flux is found to be more stable than SCF membrane over long time operation [6]. The doping of 9 wt.% of ZrO_2 in SCFZ membrane has produced $4.04\times10^{-7}\,mol\,s^{-1}\,cm^{-2}$ of stable oxygen permeation flux at 900 °C over a 240 hours period [7].

$Ce_{0.8}Sm_{0.2}O_{1.9}$-$SrCo_{0.4}Fe_{0.55}Zr_{0.05}O_{3-\delta}$ (SDC-SCFZ) disc membrane, which consists of 25 wt.% $Ce_{0.8}Sm_{0.2}O_{1.9}$ (SDC) fluorite ionic conducting phase and 75 wt.% $SrCo_{0.4}Fe_{0.55}Zr_{0.05}O_{3-\delta}$ (SCFZ) perovskite mixed conducting phase, is found to be more promising than perovskite oxide SCFZ single-phase membrane in terms of oxygen permeation flux and structural stability. The high ionic diffusivity and structural stability of SDC fluorite (as the ionic conducting phase) [1] and the stable and good oxygen permeation in SCFZ perovskite (as the mixed conducting phase) [3] are expected to endow SDC-SCFZ dual-phase membranes with good oxygen permeation performance [7].

Modelling plays a significant role towards describing oxygen permeation behaviour. Using small sets of experimental data, modelling provides insights on dominant variables as well as predicts close to actual experimental results and behaviour from the simulation [8]. Thus, modelling can contribute better knowledge and understanding towards development of membranes. Thus far, numerous oxygen permeation models had been developed so that the diffusion kinetics, oxygen exchange kinetics and degradation mechanism for MIEC membranes can be well understood [9]. Among the available oxygen permeation models, the Zhu and Xu-Thomson models have been widely applied to various MIEC membranes. The Zhu model is a simple permeation model that was introduced by Zhu *et al.* [10] based on the theoretical analysis of electrochemical processes within the membrane interfaces as well as within the membrane bulk. Another oxygen permeation model by Xu and Thomson [11] was introduced based on the surface reaction model from Lin, Wang and Han [12], whereby the oxygen permeation is driven by the oxygen chemical potential gradient and the surface reactions are not influenced by the membrane interface thickness.

Thus far, a suitable oxygen permeation model that conforms well to the permeation data of SDC-SCFZ disc membrane and can determine the oxygen permeation behaviour has yet to be identified. Therefore, this work features the determination of a suitable oxygen permeation model for SDC-SCFZ disc membrane through a modelling approach using the Zhu and Xu-Thomson models. This paper determines the limiting case of SDC-SCFZ disc membrane in each temperature range using simplified model equations based on the surface exchange or bulk diffusion limited cases, which constitutes a simple approach to replace the entire model for simulation while still attaining accurate results. Furthermore, this paper provides new insights on the performance of SDC-SCFZ disc membrane through a parametric study.

2. Methodology

2.1 Source of experimental data

SCFZ powders were synthesised through solid-state reaction, as reported elsewhere [14]. $SrCO_3$ (99.0%), Co_2O_3 (99.0%), Fe_2O_3 (99.0%), and ZrO_2 (99.0%) were stoichiometrically mixed and milled for 5 hours followed by drying for 24 hours. Then the mixed powders were calcinated in air at 950 °C for 4 hours with ramping and cooling rates of 3 °C min^{-1}. The calcined SCFZ powders were mixed with $Sm_{0.2}Ce_{0.8}O_{1.925}$ (SDC, fuelcellmaterials, SDC20-TX) powders using SDC:SCFZ weight ratio of 25:75 in an agate mortar. The weight ratio of 25:75 was selected in this work based on the highest oxygen permeation fluxes reported among other SDC-SCFZ dual-phase membranes [14]. The ground powder mixture was ball milled in ethanol for 5 hours followed by drying for at least 24 hours. The powders were then pressed into disks of 0.95 cm diameter using a mould before they were sintered in air at 1200 °C for 5 hours to obtain dense membranes with a thickness of 0.06 cm.

The oxygen permeation performance of the membrane was evaluated from 950 to 700 °C at decreasing intervals of 50 °C. At each temperature, feed side oxygen partial pressure ($P^{I}_{O_2}$) was varied from 0.7 to 0.2 atm at decreasing intervals of 0.1 atm. Helium gas was introduced to sweep the permeated oxygen and the permeate side oxygen partial pressure ($P_{O_2}^{II}$) was recorded. The oxygen permeation flux was determined from the measured oxygen concentration and outlet flow rate.

2.2 Modelling for oxygen permeation tests

The Zhu model and Xu-Thomson model equations are simplified according to the surface exchange reactions and bulk diffusion limited cases to simplify the flux simulation of SDC-SCFZ disc membrane. equation 1 and 2 represent the oxygen fluxes calculated using the Zhu model while equation 3 and 4 represent oxygen fluxes calculated using the Xu-Thomson model. $J_{O_2,ex}$ ($mol\, s^{-1}\, cm^{-2}$) represents the fluxes calculated based on surface exchange reaction limited case while $J_{O_2,diff}$ ($mol\, s^{-1}\, cm^{-2}$) represents the fluxes calculated based on bulk diffusion limited case. All the simplified model equations are adapted from Shubnikova *et al.* [13].

$$J_{O_2,ex} = \gamma_{(T),surface\ ex} \frac{(P_{O_2}^{I} P_{O_2}^{II})^{\frac{1}{2}}}{\left(\frac{k_1}{k_2} P_{O_2}^{I}\right)^{\frac{1}{2}} + \left(\frac{k_2}{k_1} P_{O_2}^{II}\right)^{\frac{1}{2}}} \ln\left(\frac{P_{O_2}^{I}}{P_{O_2}^{II}}\right) \qquad \gamma_{(T),surface\ ex} = \frac{1}{4}(k_1 k_2)^{\frac{1}{2}} \tag{1}$$

$$J_{O_2,diff} = \gamma_{(T),bulk} \ln\left(\frac{P_{O_2}^{I}}{P_{O_2}^{II}}\right) \qquad \gamma_{(T),bulk} = \frac{CD}{4h} \tag{2}$$

$$J_{O_2,ex} = \gamma_{(T),surface\ ex} \frac{{P_{O_2}^{I}}^{\frac{1}{2}} - {P_{O_2}^{II}}^{\frac{1}{2}}}{{P_{O_2}^{I}}^{\frac{1}{2}} + {P_{O_2}^{II}}^{\frac{1}{2}}} \qquad \gamma_{(T),surface\ ex} = k_r \tag{3}$$

$$J_{O_2,diff} = \gamma_{(T),bulk} \frac{{P_{O_2}^{I}}^{\frac{1}{2}} - {P_{O_2}^{II}}^{\frac{1}{2}}}{\left(P_{O_2}^{I} P_{O_2}^{II}\right)^{\frac{1}{2}}} \qquad \gamma_{(T),bulk} = \frac{k_r}{k_f}\frac{D}{2h} \tag{4}$$

k_1 ($mol\, s^{-1}\, cm^{-2}$) is the rate constant corresponding to surface transport on the feed side on the membrane whereas k_2 ($mol\, s^{-1}\, cm^{-2}$) is the rate constant corresponding to surface transport on the permeate side on the membrane. C is the oxygen ion concentration ($mol\, cm^{-3}$), D is the bulk diffusion coefficient ($cm^2\, s^{-1}$) while h is the membrane thickness (cm). k_r and k_f are reverse (desorption) surface exchange rate constants ($mol\, s^{-1}\, cm^{-2}$) and forward (adsorption) surface exchange rate constants ($cm\, atm^{-0.5}\, s^{-1}$), respectively. $P_{O_2}^{I}$ and $P_{O_2}^{II}$, each with a unit of atm, are the feed and permeate side oxygen partial pressure, respectively. $\gamma_{(T)}$ is a temperature-dependent parameter that governs the oxygen transport depending on the limiting case of either bulk diffusion or surface exchange limited.

3. Results and discussion

3.1. Preliminary results

The model parameters in each limiting case of the Zhu and Xu-Thomson models are determined using MATLAB, the results of which are displayed in tables 1 to 4 along with the corresponding resultant coefficient of determination (R^2) and sum of squared error (SSE) at each temperature point.

Table 1. Regressed oxygen permeation parameters for surface exchange reaction limited case using Zhu model.

Temperature (°C)	Model parameter k_1 $(mol\ s^{-1}\ cm^{-2})$	Model parameter k_2 $(mol\ s^{-1}\ cm^{-2})$	R^2	SSE
950	3.800×10^{12}	9.262×10^{-6}	0.845	1.526×10^{-14}
900	5.059×10^{10}	7.169×10^{-6}	0.988	9.907×10^{-16}
850	1.879×10^{-6}	8.452×10^{-6}	0.998	1.042×10^{-16}
800	6.457×10^{-7}	9.149×10^{-6}	0.980	3.662×10^{-16}
750	3.803×10^{-7}	1.067×10^{-5}	0.990	1.023×10^{-16}
700	2.526×10^{-7}	1.888×10^{10}	0.875	8.707×10^{-16}

Table 2. Regressed oxygen permeation parameters for bulk diffusion limited case using Zhu model.

Temperature (°C)	Model parameter $\gamma_{(T),bulk}$ $(mol\ s^{-1}\ cm^{-2})$	R^2	SSE
950	3.495×10^{-7}	0.882	1.163×10^{-14}
900	2.427×10^{-7}	0.703	2.454×10^{-14}
850	1.354×10^{-7}	0.398	3.316×10^{-14}
800	7.518×10^{-8}	0.295	1.282×10^{-14}
750	5.147×10^{-8}	0.161	8.252×10^{-15}
700	4.240×10^{-8}	0.081	6.376×10^{-15}

Table 3. Regressed oxygen permeation parameters for surface exchange reaction limited case using Xu-Thomson model.

Temperature (°C)	Model parameter $\gamma_{(T),surface\ ex}$ $(mol\ s^{-1}\ cm^{-2})$	R^2	SSE
950	1.641×10^{-6}	0.748	2.476×10^{-14}
900	1.164×10^{-6}	0.533	3.865×10^{-14}
850	6.741×10^{-7}	0.263	4.058×10^{-14}
800	3.762×10^{-7}	0.190	1.475×10^{-14}
750	2.604×10^{-7}	0.100	8.859×10^{-15}
700	2.133×10^{-7}	0.050	6.592×10^{-15}

Table 4. Regressed oxygen permeation parameters for bulk diffusion limited case using Xu-Thomson model.

Temperature (°C)	Model parameter $\gamma_{(T),bulk}$ $(mol\ atm^{0.5}\ cm^{-2}\ s^{-1})$	R^2	SSE
950	1.982×10^{-7}	-0.773	1.744×10^{-13}
900	1.257×10^{-7}	-1.151	1.779×10^{-13}
850	6.055×10^{-8}	-1.454	1.352×10^{-13}
800	3.253×10^{-8}	-1.858	5.200×10^{-14}
750	2.115×10^{-8}	-2.190	3.140×10^{-14}
700	1.747×10^{-8}	-2.394	2.355×10^{-14}

3.2. Determination of SDC-SCFZ disc membrane's oxygen permeation model

Based on the R^2 from preliminary results, the Zhu model exhibits good correlation between the modelled and experimental results for the surface exchange limited case given that the R^2 is greater than 0.8 between 700 and 950 °C (table 1), which exhibits relatively better correlation than the Xu-Thomson model (table 3). For the bulk diffusion limited case, the Zhu model only shows good correlation at 900 and 950 °C (table 2) while the Xu-Thomson model exhibits poor model behaviour given the negative R^2 which is attained from constraining the regressed model parameters to produce only positive values (table 4). Therefore, the oxygen transport according to the Zhu model is assumed to be limited by surface exchange reactions between 700 and 850 °C. The oxygen transport above 850 up to 950 °C is assumed to be limited by a mixture of both limiting cases as R^2 for the bulk diffusion case is relatively high compared to the other temperatures. However, the distribution of the two limiting cases cannot be accurately determined. To incorporate both limiting cases, a single equation (equation 5) was developed based on the aforementioned equations 1 and 2 used in this work. For the Xu-Thomson model, oxygen transport is assumed to be limited by surface exchange reactions from 700 to 950 °C. Therefore, equation 3 and the parameters from table 3 will be utilised for the Xu-Thomson model.

The model parameters of the Zhu model as well as R^2 and SSE for 900 and 950 °C are regressed from equation 5 based on the assumptions for mixed limiting case whereas the parameters between 700 and 850 °C are adapted from table 1 based on the assumptions of surface exchange limited case. The final Zhu model parameters used in this work are displayed in table 5. The large value of k_2 at above 900 °C in table 5 can be attributed to the larger influence of bulk diffusion.

Figure 1 exhibits a comparison of the linearised fluxes modelled using the Zhu and Xu-Thomson models. The modelled oxygen permeation fluxes calculated from the Zhu model have a good correlation with the experimental data based on figure 1(a) and also according to the high R^2 and low SSE in Table 5. From figure 1(b), the modelled oxygen permeation fluxes calculated using the Xu-Thomson model exhibit a bad correlation with experimental results. This together with low R^2 and high SSE attained in table 3 as compared to table 5 indicate a bad fitting and high error with the experimental data. Overall, the Zhu model produces better correlation than the Xu-Thomson model for dual-phase SDC-SCFZ disc membrane and will be utilised in the subsequent sections.

$$J = \gamma_{(T),surface\ ex} \frac{4 {P_{O2}^{I}}^{0.5} {P_{O_2}^{II}}^{0.5} \gamma_{(T),bulk}}{4\gamma_{(T),bulk} \frac{k_2^{0.5}}{k_1^{0.5}} {P_{O_2}^{II}}^{0.5} + k_1^{0.5} k_2^{0.5} {P_{O_2}^{I}}^{0.5} {P_{O_2}^{II}}^{0.5} + 4\gamma_{(T),bulk} \frac{k_1^{0.5}}{k_2^{0.5}} {P_{O_2}^{I}}^{0.5}} \ln \frac{P_{O_2}^{I}}{P_{O_2}^{II}}$$

$$\gamma_{(T),surface\ ex} = \frac{1}{4}(k_1 k_2)^{\frac{1}{2}} \qquad \gamma_{(T),bulk} = \frac{CD}{4h} \tag{5}$$

Table 5. Regressed oxygen permeation parameter for Zhu model.

Temperature (°C)	k_1 $(mol\, s^{-1}\, cm^{-2})$	k_2 $(mol\, s^{-1}\, cm^{-2})$	$\gamma_{(T),bulk}$ $(mol\, s^{-1}\, cm^{-2})$	R^2	SSE
950	1.308×10^{-4}	1.492×10^{11}	3.495×10^{-7}	0.880	1.182×10^{-14}
900	3.832×10^{-5}	1.581×10^{11}	2.427×10^{-7}	0.696	2.515×10^{-14}
850	1.879×10^{-6}	8.452×10^{-6}	-	0.998	1.042×10^{-16}
800	6.457×10^{-7}	9.149×10^{-6}	-	0.980	3.662×10^{-16}
750	3.803×10^{-7}	1.067×10^{-5}	-	0.990	1.023×10^{-16}
700	2.526×10^{-7}	1.888×10^{10}	-	0.875	8.707×10^{-16}

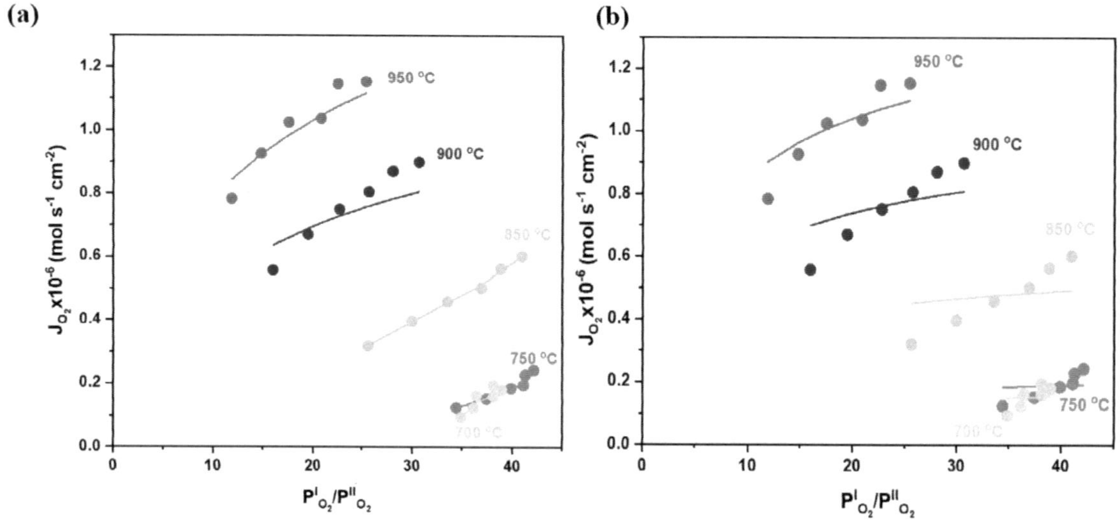

Figure 1. Linearisation of oxygen permeation fluxes of SDC-SCFZ disc membrane between 700 and 950 °C. Symbols perform experimental results, whereas lines perform values calculated from (a) Zhu model (b) Xu-Thomson model.

3.3. Distribution of resistances

The distribution of permeation resistances is determined as a product of temperature between 700 and 950 °C at two distinct permeate side oxygen partial pressures, $P_{O_2}^{II}$ of 0.02 and 0.005 atm whereby the feed side oxygen partial pressure, $P_{O_2}^{I}$ is fixed at 0.21 atm (figure 2). At $P_{O_2}^{II}$ of 0.02 atm, an increase in r'' is observed between 700 and 850 °C. Such increase is compensated by the equal decrease in the distribution of resistance from r'. Above 850 °C, the distribution of resistance from surface exchange reaction reduces significantly whereas distribution of resistance from bulk diffusion, r^b increases and accounts for approximately 95% of the total resistance at 900 °C. When $P_{O_2}^{II}$ is decreased to 0.005 atm, the distribution of resistance from r'' increases significantly relative to $P_{O_2}^{II}$ of 0.02 atm. The same trend in $P_{O_2}^{II}$ of 0.02 atm is observed when temperature rises above 850 °C and r^b is involved. This indicates the limiting case changed from surface exchange reactions to mixture of surface exchange reactions and bulk diffusion at 850 °C. Furthermore, bulk diffusion becomes the main resistance contributor as compared to surface exchange reactions when temperature rises above 900 °C.

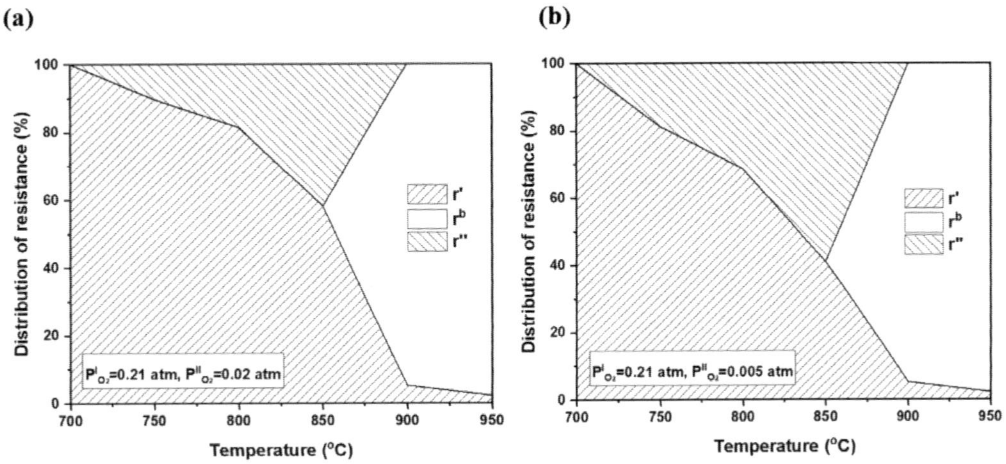

Figure 2. Distribution of permeation resistance between 700 and 950 °C at $P_{O_2}^{I}$ of 0.21 atm and $P_{O_2}^{II}$ of (a) 0.02 atm (b) 0.005 atm.

3.4. Activation energy

The activation energies, E_a of fluxes calculated based on r' and r'', r^b, and r^{tot}, are determined at feed side partial pressure of 0.21 atm and permeate side partial pressure of 0.02 atm (figure 3). Two different E_a for both limiting cases and r^{tot} are observed, which indicate the change in the limiting case at 850 °C. An E_a of 117.8 ± 12.8 $kJ\ mol^{-1}$ was attained for the flux based on r^{tot} at high temperature region (850–950 °C). In this region, the activation energies for fluxes based on surface exchange resistances and bulk diffusion resistances are similar, which suggests that the oxygen transport is limited by a mixture of surface exchange reactions and bulk diffusion. At low temperature region (700–850 °C), an E_a of 71.9 ± 13.4 $kJ\ mol^{-1}$ was attained for the fluxes based on r^{tot}, which is comparable to the E_a calculated from the regression of fluxes based on surface exchange resistances (69.4 ± 13.4 $kJ\ mol^{-1}$). This tallies with the earlier assumption that the fluxes are limited by surface exchange reactions between 700 and 850 °C.

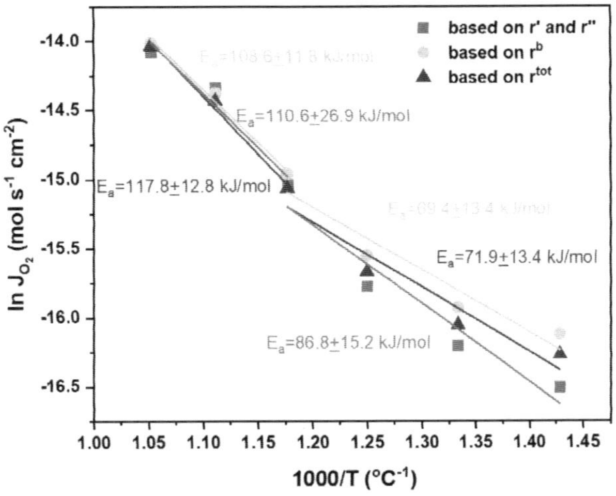

Figure 3. Arrhenius plot of the oxygen permeation fluxes determined based on r' and r'', r^b, and r^{tot} under oxygen partial pressure gradient of 0.21/0.02 atm (feed/permeate).

3.5. Influence of oxygen partial pressure variation

Figure 4(a) illustrates the oxygen permeation fluxes as a product of feed side pressure, P_{air} at 850 and 950 °C, at two distinct permeate side oxygen partial pressures, $P_{O_2}^{II}$ of 0.1 and 0.02 atm. At each individual temperature and permeate side oxygen partial pressure, the increase in P_{air} brings about an increase of oxygen permeation flux. At $P_{O_2}^{II}$ of 0.1 atm, the oxygen permeation fluxes at 850 °C is higher than at 950 °C above the crosssover point, P_{air} of 12 atm. This may be attributed to the additional denominator term $({k_1}^{0.5}{k_2}^{0.5}{P_{O_2}^{I}}^{0.5}{P_{O_2}^{II}}^{0.5})$ in equation 5 that may contribute to lower oxygen permeation fluxes at 950 °C as compared to 850 °C with increasing P_{air} and $P_{O_2}^{II}$ of 0.1 atm. At $P_{O_2}^{II}$ of 0.02 atm, the additional term in the denominator of equation 5 is smaller in value, and the flux is not reduced as much as that at $P_{O_2}^{II}$ of 0.1 atm.

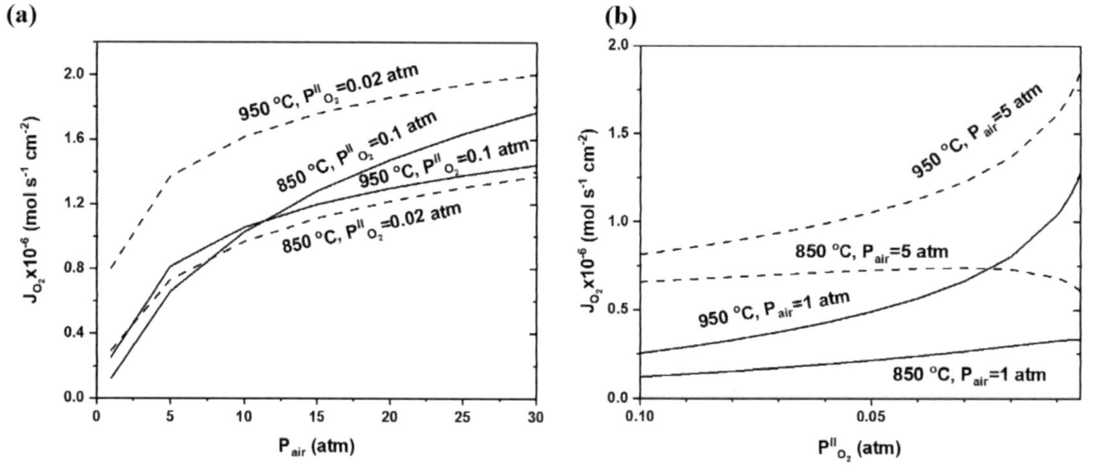

Figure 4. Influence of variation in (a) feed side pressure, P_{air} (b) permeate side oxygen partial pressure, $P_{O_2}^{II}$, on the oxygen permeation fluxes at 850 and 950 °C.

Figure 4(b) illustrates the oxygen permeation fluxes as a product of $P_{O_2}^{II}$ at 850 and 950 °C, and at two distinct P_{air} of 1 and 5 atm. A decrease in $P_{O_2}^{II}$ brings about an increasing trend in oxygen permeation flux at 950 °C and a decreasing trend at 850 °C. This is attributed to the increase in r'' at 850 °C along with the decrease in $P_{O_2}^{II}$ (figure 4), which brings about a decrease in oxygen permeation flux. In contrast, the r'' at 950 °C remains unchanged and thus leads to an increase in flux. Increasing the feed side pressure from 1 to 5 atm is expected to boost the oxygen permeation fluxes by an average of 5.65×10^{-7} and 5.14×10^{-7} $mol\ s^{-1}\ cm^{-2}$ at 850 and 950 °C, respectively. However, the increase of oxygen permeation fluxes from 1 to 5 atm at 950 °C is found to be more stable than at 850 °C. Therefore, operating the membrane at higher temperature will exhibit more stable and higher oxygen permeation flux.

4. Conclusion

In conclusion, the Zhu model is more applicable for the simulation of oxygen permeation fluxes of dual-phase SDC-SCFZ disc membrane as compared to the Xu-Thomson model given the overall good correlation and low sum of squared error. The oxygen transport is found to be limited by surface exchange reactions between 700 and 850 °C and a mixture of surface exchange reactions and bulk diffusion above 850 up to 950 °C. Bulk diffusion is revealed as the main resistance contributor above 900 °C. Through a parametric study, the increase in feed side pressure, P_{air} from 1 to 30 atm increases the oxygen permeation flux. Decreasing the permeate side oxygen partial pressure, $P_{O_2}^{II}$ from 0.1 to 0.01

atm increases the oxygen permeation flux for 950 °C but decreases for 850 °C. As such, SDC-SCFZ disc membranes exhibit high oxygen flux up to 2×10^{-6} $mol\ s^{-1}\ cm^{-2}$ at 950 °C with P_{air} of 5 atm but $P_{O_2}^{II}$ of 0.005 atm. Thus, SDC-SDCZ disc membrane indicates higher oxygen flux than other dual-phase membranes mixed with ionic and electronic conductor. The optimum range of operating conditions to maximise the oxygen permeation flux for SDC-SCFZ disc membrane are at 950 °C with minimum P_{air} of 1 atm and $P_{O_2}^{II}$ lower than 0.025 atm. This work is significant in providing insights of SDC-SCFZ disc membrane and further improvement of similar membrane compositions can be developed based on optimum operating conditions found in this work. To further extend this work, it is recommended to conduct the simulation using different membrane compositions to determine the extent of the applicability of the Zhu model.

References

[1] Zhu Y, Li W, Cai L, Zhu X and Yang W 2020 Universally applicable kinetic model for mixed ionic-electronic conducting membranes *Chem. Eng. Sci.* **215** 115455

[2] Li C, Chew J J, Mahmoud A, Liu S and Sunarso J 2018 Modelling of oxygen transport through mixed ionic-electronic conducting (MIEC) ceramic-based membranes: An overview *J. Membr. Sci.* **567** 228–60

[3] Lv X, Chen H, Zhou W, Li S-D and Shao Z 2020 A CO_2-tolerant $SrCo_{0.8}Fe_{0.15}Zr_{0.05}O_{3-\delta}$ cathode for proton-conducting solid oxide fuel cells *J. Mater. Chem. A.* **8** 11292–301

[4] Shao L, Siok F, Fu X-Z and Luo J-L 2018 Stable $SrCo_{0.7}Fe_{0.2}Zr_{0.1}O_{3-\delta}$ cathode material for proton conducting solid oxide fuel cell reactors *Int. J. Hydrogen Energy* **43** 7511–4

[5] He Y J, Chang X F, Dong X L, Zhang C, Jin W Q and Xu N P 2008 Effect of dopant valence on the oxygen desorption and oxygen permeability of $SrCo_{0.4}Fe_{0.5}M_{0.1}O_{3-\delta}$ (M=Ni, Al and Zr) mixed-conducting oxides *Chin. Chem. Lett.* **19** 725–9

[6] Yang L, Gu X, Tan L, Zhang L, Wang C and Xu N 2003 Role of ZrO_2 addition on oxygen transport and stability of ZrO_2-promoted $SrCo_{0.4}Fe_{0.6}O_{3-\delta}$ *Sep. Purif. Technol.* **32** 301–6

[7] Yang L, Gu X, Tan L, Jin W, Zhang L and Xu N 2002 Oxygen transport properties and stability of mixed-conducting ZrO_2-promoted $SrCo_{0.4}Fe_{0.6}O_{3-\delta}$ oxides *Ind. Eng. Chem. Res.* **41** 4273–80

[8] Jin Y, Meng X, Yang N, Meng B, Sunarso J and Liu S 2017 Modeling study of oxygen permeation through an electronically short-circuited YSZ-based asymmetric hollow fiber membrane *AIChE J.* **63** 3491–500

[9] Zhu Y, Li W, Liu Y, Zhu X and Yang W 2017 Selection of oxygen permeation models for different mixed ionic-electronic conducting membranes *AIChE J.* **63** 4043–53

[10] Zhu X, Liu H, Cong Y and Yang W 2012 Permeation model and experimental investigation of mixed conducting membranes *AIChE J.* **58** 1744–54

[11] Xu S J and Thomson W J 1999 Oxygen permeation rates through ion-conducting perovskite membranes *Chem. Eng. Sci.* **54** 3839–50

[12] Lin Y S, Wang W and Han J 1994 Oxygen permeation through thin mixed-conducting solid oxide membranes *AIChE J.* **40** 786–98

[13] Shubnikova E, Popov M, Bychkov S, Chizhik S and Nemudry A 2019 The modeling of oxygen transport in MIEC oxide hollow fiber membranes *Chem. Eng. J.* **372** 251–9

[14] Li C, Song J, Zhang S, Tan X, Meng X, Sunarso J and Liu S 2021 SDC-SCFZ dual-phase ceramics: structure, electrical conductivity, thermal expansion, and O_2 permeability *J. Am. Ceram. Soc.* **104** 2268–84

IOP Conf. Series: Materials Science and Engineering 1195 (2021) 012061 doi:10.1088/1757-899X/1195/1/012061

Engineering solutions for the palm mill of the future: Increasing extraction rate and sustainability through biotechnology

Hemavathi Silvamany[1], Nik Suhaimi Mat Hassan[1], Nik Mohd Farid Mat Yasin[1], Muliadi Mustaner[1], Amirul Al Hafiz Abdul Hamid[1], Muhammad Izhar Khairi[1], Muhammad Zaidy Arnan[1], Mohammed Faisal Mohammed Yunus[1], Harikrishna Kulaveerasingam[1], Christopher J. R.Dunning[2], Karthini Nagalingam[3] and Chien Lye Chew[1]

[1]Sime Darby Plantation Research, R&D Centre – Carey Island, Lot 2664 Jalan Pulau Carey, 42960 Pulau Carey, Selangor, Malaysia
[2]Novozymes Innovation Campus, Novozymes A/S, 2800 Kgs., Lyngby, Denmark
[3]Novozymes Malaysia Sdn Bhd, Jalan Inovasi 1 Technology Park Malaysia Bukit Jalil 57000 Kuala Lumpur Malaysia

Email: hemavathi.s@simedarbyplantation.com

Abstract. The edible oil industry involving extraction and processing of oils and fats from vegetable sources has witnessed a tremendous growth over the decade accounting for rapid expansion of urbanization and industrialization. Consequently, concerted efforts have been undertaken to increase the palm oil extraction rate (OER) at processing mills. This paper explores the commercial application of enzymatic treatment of sterilised palm fruitlets and diluted crude oil (DCO) stream in a 60 t/hr oil palm mill. Incubation of sterilised palm fruitlets mixed with enzymes in a pre-digester vessel at 70 °C and retention time of 30 minutes resulted in an increment of 1.15 % OER. Parallelly, dosing of enzyme into DCO with simpler system yielded in OER increment up to 1.19 %. This study offers insight into the potential oil left in pressed fibre using different methods, such as prolonged Soxhlet extraction, ultrasonic treatment, and supercritical fluid extraction. Additionally, confocal microscopy images of treated samples showed disintegration and thinning of cell walls resulting in higher free oil droplets.

1. Introduction

The global oils and fats consumption is forecasted to increase to 250 million metric tonnes to fulfil the growing population of over 8.5 billion by 2050 [1]. Palm oil acts as the most relevant and crucial component of this food supply chain and accountable for nearly half of the global edible oil consumption. Oil palm is by far the most efficient biological source of oil-based hydrocarbons, outperforming any other commercial oil crop [2]. In short, of 300 million hectares for global oil crop cultivation, palm oil only accounts for 6 % of this land use with a significant contribution of 36 % global consumption [2, 3]. However, with limited land allowable to be planted under the RSPO requirement, with weather changes that has negatively impacting the yield and with very long development of higher yielding planting material, it is essential for palm oil producers to strengthen the role and address the substantial increase in demand. Therefore, ways are needed to increase the sustainability aspects of palm oil and productivity by limiting the expansion of agriculture areas that could jeopardize these biodiversity-rich lands [4].

The palm oil milling industry that is based on conventional mechanical extraction has almost reaching the optimum extraction efficiency rate with the extraction efficiency of approximately between 92 to 94 %. Palm oil mills use oil extraction rate (OER) to measure its efficiency and is calculated based on the mass ratio of the crude palm oil (CPO) extracted to the fresh fruit brunches (FFB) processed. For the past 40 years, Malaysia's average palm OER has remained stagnant between 19 and 21 %. Globally, for every 1 % rise in the OER, nearly 3 million tonnes of palm oil can be produced [5].

$$\text{Oil extraction rate, OER (\%)} = \frac{\text{total oil produced}}{\text{total fresh fruit bunches processed}} \text{ x } 100\% \quad (1)$$

With sustainable biotechnology solutions, this greater amount of oil may be generated without the requirement of additional land, water, energy, or fertilizer for crop production. Solvent extraction that supposedly will provide the highest means of extraction has not been acceptably recognized for palm oil extraction by the consumer, therefore will have to compromise with the conventional mechanical means of extraction. This will require palm oil milling to modify the process by either altering the process condition or using process catalyst [6–8].

Some of the developmental works are in the area of sterilization whereby sterilization at higher pressure and temperature will hydrolyze the hemi-cellulose that is part of the cell wall protecting the oil globules in oil palm fruit mesocarp, thus it will improve the oil availability for the extraction [9]. Another team of researcher has been exploring for the separation of oil palm fruit mesocarp from the nut that has improved the digestion of the oil-bearing cell, giving a potential increment of OER of 0.5 to 0.7 % [10]. As a result, alternative techniques to the aforementioned solvent extraction methods are being investigated to meet the increased need for product purity, nonpolluting, and energy-efficient processes, with the added benefit of requiring less solvent. Extraction assisted by pulsed electric field, microwave extraction, instant controlled pressure drop technology, microwave hydro diffusion and gravity, ultrasound assisted extraction, ionic liquid extraction, subcritical water extraction, high pressure assisted extraction, aqueous two phase extraction, and enzyme assisted aqueous extraction are some of the most recent extraction techniques developed in the quest to create commercially viable, efficient, energy-saving, safe, compact, and sustainable extraction processes [6].

Enzymatic extraction of palm oil which involves addition of hydrolytic enzymes during maceration process, has gained increased attention in deciphering the nation's OER stagnation issue [11]. The synergetic hydrolysis of complex constituents of cellulose, hemicellulose, lignin, and pectin, together with the digestion and screw pressing steps could efficiently results in liberation of higher oil and reduction in oil losses in pressed fibre. The enzymatic extraction has been widely explored with various oil-bearing materials such as soybean, grape seed, sunflower, peanut, and olive, and proven to improve oil extraction rates by more than 90 % and promises superior quality [12–16].

The research and development of enzymatic technology for extraction in palm oil milling began in year 1990 with an increment of 6.6 % extraction rate that was obtained by treating palm mesocarp with pectinase followed by hydraulic press [17]. Teixeira *et al.* (2013) have reported highest oil extraction efficiency of 90 - 93 % total oil utilizing 4 % of enzyme cocktail of tannase, cellulase and pectinase, pH 4, ratio of solution to pulp of 2:1 and 30 minutes incubation at 50 °C. Recently, Javvadi *et al.* (2012) presented an invention for extracting oil from the mesocarp of oil palm fruit using enzymes that includes exocellulolytic, pectinolytic, mannanolytic and glucanolytic activity with improved efficiency and increased yield of 90 %. In another study by Eshtiaghi *et al.* (2015), samples treated with 0.15 % cellulase resulted in distinctively 95 % oil extraction yield and reduction of remaining press pulp by 5 %.

Therefore, in this study, we have conducted laboratory evaluation of potential oil extractable and with positive outcome, the plant trials were carried out. During the mill trial and subsequent commercial implementation, we evaluated the application of enzyme to sterilized oil palm fruitlets as well as diluted crude oil (DCO) with the objective of enhancing the palm oil extraction.

2. Materials and Method

2.1. Enzymatic treatment
PALMORA OER, a carbohydrase cocktail of xylanase and cellulase was obtained from Novozymes Malaysia Sdn Bhd. The Palmora OER is a thermotolerant enzyme, suitable to be used at processing temperature between 80 to 90 °C. Cellulases and xylanases catalyze the hydrolysis of the 1,4-beta-D-glycosidic bonds in cellulose, hemicellulose, lignin and cereal beta-D-glucans to break down the cellulose found in plants cells.

2.2. Enzymatic Treatment
Approximately 20 grams of sterilized palm mesocarp pulp was dissolved in enzyme solution (enzyme diluted with water) and extraction was carried out for 1 hour in a 100 mL conical flask placed in a water bath shaker operating at 80 to 90 ºC and constant shaking of 200 rpm, followed by enzyme deactivation [11]. Three times serial centrifugation at 6500 rpm for 30 minutes was used to separate the oil, liquid, and solid fractions from the mixture, followed by hexane washing and filtration. The samples were dried overnight at 103 °C and oil content was quantified using Soxhlet extraction method.

2.3. Soxhlet Extraction
Total amount of oil in samples was determined by solvent extraction procedure to compare the extraction efficiency of palm mesocarp and pressed fibre treated with different methods. 10 g sample was extracted using n-hexane in a Soxhlet apparatus of 100 mL capacity between 4 to 16 hours. The extract was then filtrated, and rotary evaporator was used to remove the n-hexane contained in the filtrate.

2.4. Ultrasonic treatment
20 g palm pressed fibre was weighed into a 250 mL conical flask, 100 mL of distilled water was added, and the mixture was placed inside the Brasonic® Ultrasonic Cleaner Model 5510 (Emerson, USA) for ultrasonic pre-treatment at 485 W with operating hours ranging from 0 to 5 hours. Three times serial centrifugation at 6500 rpm for 30 minutes was used to separate the oil, liquid, and solid fractions from the mixture, followed by hexane washing and filtration. The samples were dried overnight at 103 °C and oil content was quantified using Soxhlet extraction method of 4 hours.

2.5. Supercritical Fluid Extraction (SFE)
This study was conducted using Supercritical fluid extraction (SFE) facility in UKM based on the method by Mohamed *et al.* (2018) with minor modifications. The SFE system comprise of a carbon dioxide pump (PU-2080, JASCO Corporation, Japan), BP 1580-81 model back pressure regulator (BPR, JASCO Corporation, Japan), series 111 solvent pump (Lab Alliance, USA), pressure transmitter (model 682-8, Dwyer Instrument, USA), extractor vessel enclosed in a FX2-2 model air circulating oven (Sheldon Manufacturing, USA) and sample collector. Before the extraction process began, a chiller (Protech Electronic, Malaysia) was utilized to keep the liquefied carbon dioxide (CO_2) in a liquid form at - 4 °C. The extraction was done in a controlled environment. The liquid CO_2 flow rates were set at 4 mL/minutes, 60 °C, and 200 barg. 4 g palm fibre samples (at 10 % moisture content) were inserted in the extractor vessel, and the extraction was begun using the dynamic extraction mode, with each fraction collected every 30 minutes. Each portion was weighed after drying in a 45 °C oven.

2.6. Plant Scale trial
The enzyme dosing rate has been optimized during the laboratory trials was used during the plant trial. Two different dosing strategies have been studied in 60 metric t/hr palm oil mills (figure 1). A methodology of alternate dosing for two (2) weeks for a period of six (6) cycles were carried out. Process performances were evaluated based on Delta OER as described in equation (2);

$$\text{Delta OER (\%)} = \text{OER Enzyme} - \text{OER Control} \quad (2)$$

In milling process, sterilized FFB are transferred to rotating threshing drums to remove the fruitlets from the bunches. These loose fruitlets are then passed to a digester and screw press for oil extraction, which is known as mass passing to digester (MPD). The first strategy was dosing of enzymes to sterilized palm loose fruitlets, a patented process developed by Sime Darby Plantation Research [22]. The enzyme was dosed into the MPD from closed enzyme spraying system to ensure longer contact time between enzymes and the fruitlets and a more homogenous mixing to enable penetration of enzymes into the mesocarp. Thereafter, the mixture of MPD and enzymes were fed into a Pre-Digester vessel to obtain an additional retention time of 30 minutes and followed by conventional maceration in digester. Due to mixing limitation, second strategy of dosing of enzymes to diluted crude oil (DCO) was evaluated. DCO is liquid stream post digestion and press, comprising of oil, water, non-oily solid. The enzyme was dosed into the DCO pipeline, a patented process developed by Sime Darby Plantation Research [23]. A much-simplified dosing system at lower capital cost was installed for this method.

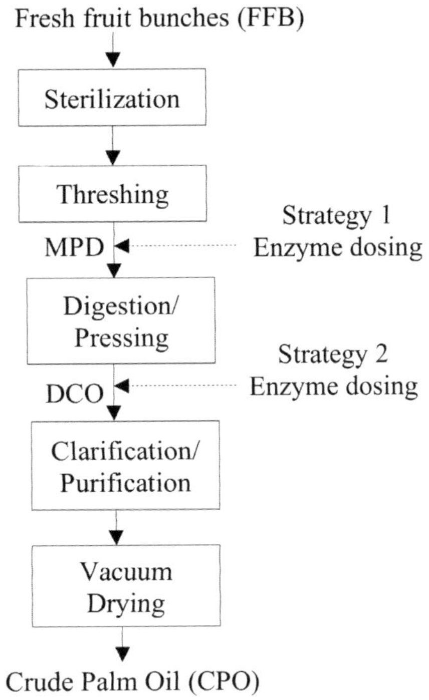

Figure 1. Plant scale trial enzyme dosing strategy.

2.7. *Reducing sugar content*

Reducing sugars (monosaccharide) in the clarifier were analysed using HPLC (Agilent Technologies, USA) equipped with refractive index (RI) detector and Rezex column (ROA Organic acids H+ (8 %) 4E, 7.8 mm × 300 mm) (Phenomenex, USA). The HPLC was run at 60 °C using a mobile phase of 0.005 N sulphuric acid with 0.6 mL/minutes flowrate [24].

2.8. *Confocal Microscopic Imaging*

Control and enzyme treated DCO were skimmed of oil and incubated at 90 °C for 1 hr. Oil and cells were subsequently labelled by the addition of an equal volume of 3.2 μM Calcofluor white (Sigma Aldrich) and 20 mM BODIPY493/503 (Invitrogen), incubated at room temperature for 10 minutes prior to imaging in 96 well imaging plates (Corning). Images were captured on an Olympus FV3000 confocal laser scanning microscope (CLSM) (Olympus, model FV3000 Japan) with a 20x objective. Image processing and visualization was performed with Imaris 9.6.

2.9. Statistical Analysis

The lab experiments were repeated at least three times. The data was statistically analysed by computing mean and standard deviation values for various parameters. The mean data was further analysed using one-way ANOVA at a 5 % significance level to see if there were any significant differences between extraction strategies.

3. Results and Discussion

3.1. Potential oil in Palm pressed fibre and palm mesocarp

In traditional palm oil milling, an average of 20 % of the oil is obtained from fresh fruit bunches (OER equivalent) while also producing 23 % empty fruit bunch (EFB), 15 % fibre, and 12 % nut. Oil is lost in a variety of by-products, including the fibre that remains after a screw press extracts the mesocarp oil. Oil content of pressed fibre ranges from 5 to 11 % w/w oil, however the fibre is usually burned as fuel to power the mill. Several studies were conducted using different methods such as prolonged soxhlet extraction, ultrasonic treatment, and supercritical fluid extraction to determine the potential oil left in pressed fibre as shown in figure 2. Results represent the mean values and error bars indicate the standard deviation of the mean (n = 3). The baseline for control comparison is 5.44 % w/w oil content by wet basis in pressed fibre sample. Extending the duration of soxhlet extraction, about 27 % of increment of oil content is obtained as compared to the usual 4 hours. Generally, increase of extraction duration and temperature is proportionate to the increase of oil yield. Solvent extraction using non-polar solvent, namely hexane has been widely utilized in various vegetable oil extractions such as rapeseed and soybean. In palm oil milling sector, soxhlet extraction with hexane is used as oil content analysis for product (mesocarp) and by-product (fibre, EFB, raw effluent) as process monitoring.

Ultrasound-assisted extraction is a new simple technique for the recovery of oil and bioactive compounds from different sources [25, 26]. Ultrasound intensity presented a positive effect on the extraction, indicating that its exposure overtime can lead to higher yields. The highest oil content, 8.3 % w/w was obtained at the prolonged ultrasound treatment of 4 hours, an increment potential of 27 %. Similar extraction trend was observed for supercritical fluid extraction. Supercritical fluid technology using carbon dioxide (CO2) has been applied in extraction, purification and fractionation of crude palm oil. An increment of 50 % higher and proves the existence of more oil than solvent extracted method (figure 2c). By comparing to the control sample, SFE treatment of 5 hours yields additional oil content of 2.77 % w/w. All our findings regarding potential palm oil extraction with emerging green technology are in similar accordance to review by Chew *et al.* (2021b).

Extended treatment with higher diffusivity and mass transfer of oil from complex plant matrices shows that additional oil as high as 3 % w/w are extractable, however left unextracted and contained as intact oil bodies due to the limitation of traditional thermo-mechanical extraction process. These oil bodies are either trapped in an insoluble extracellular denatured protein matrix or coalesced oil droplets that are too big to pass through the partially disrupted cellular matrix [27]. Alternatively, this additional potential oil can be extracted by deploying feasible biotechnology means, i.e., enzymatic technology due to enzymes' ability to successfully break down plant cell walls. As a result, enzymes could be the palm oil milling industry's next great operational breakthrough.

According to Silvamany *et al.* (2015), cellulose and hemicellulose make up the majority of the polysaccharides constituents in the palm mesocarp cell wall, followed by soluble lignin. The lignin layer on the mesocarp fibre functions as a physical barrier, preventing cellulose and hemicellulose from being accessed.

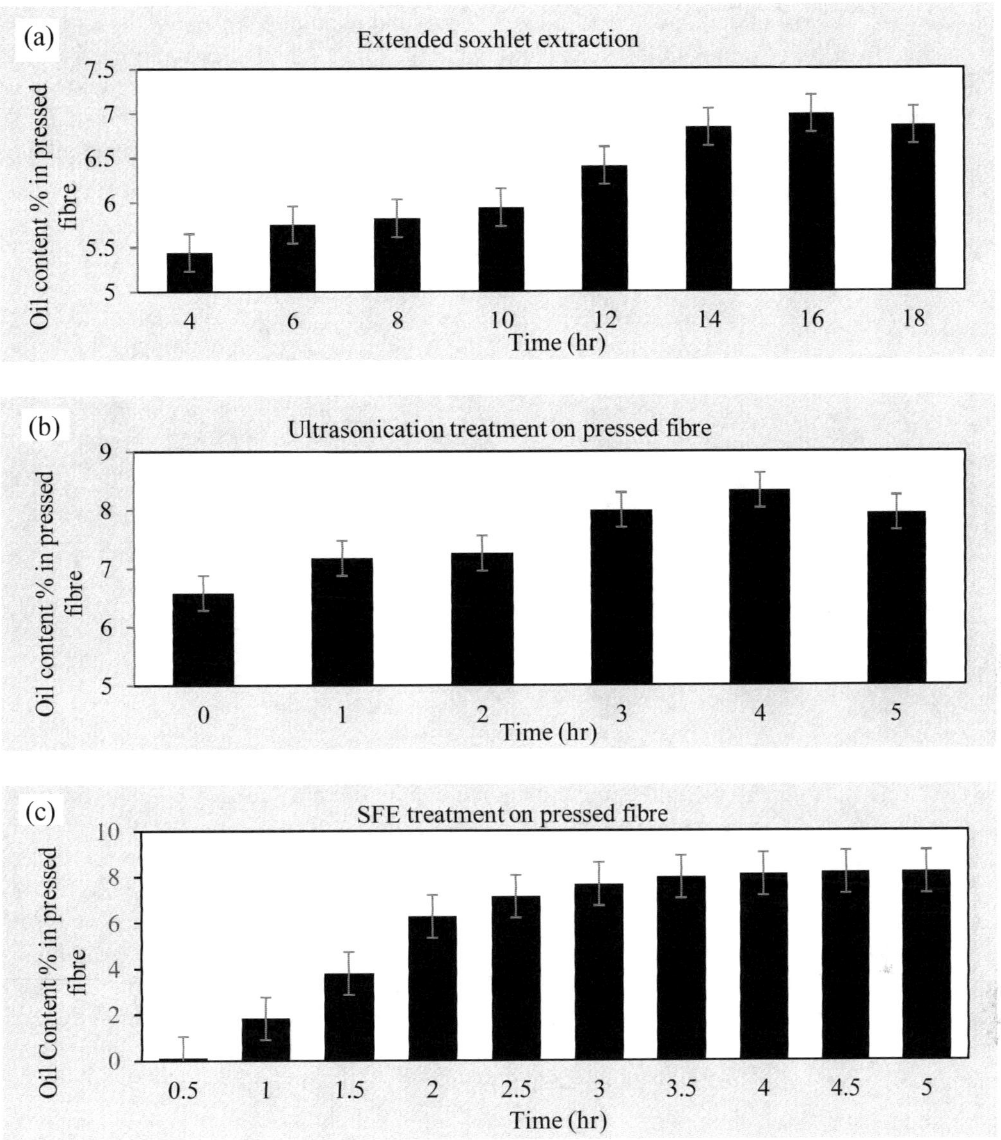

Figure 2. Potential oil left in pressed fibre by treatments of (a) extended soxhlet extraction, (b) ultrasonication and (c) supercritical fluid extraction.

Hence, to enable the release of oil from vacuoles and cytoplasmic membranes, the cellular wall of the fruit mesocarp must be degraded and ruptured. Palm mesocarp samples were treated with enzymes and subjected to extended soxhlet extraction alongside control samples as shown in figure 3. Results represent the mean values and error bars indicates the standard deviation of the mean (n = 3). Enzymatic treatment followed by the standard 4 hours soxhlet extraction showed only extra 1 % w/w oil content. Nevertheless, extending the solvent extraction of treated palm mesocarp to 16 hours yielded an additional 10.7 % w/w oil content, while omitting the solvent extraction contribution, enzyme alone could give 6 % w/w additional oil (1.2 % OER equivalent).

The findings are supported by the basics where enzyme acts as a pre-treatment step before the extraction procedure. During soxhlet treatment, the addition of the enzyme clearly breaks the cells and enables the escape of intact oil. Perez *et al.* 2013 reported a substantial oil recovery of 110.85 % from sunflower seeds using pectinase assisted solvent extraction in comparison to solvent extraction alone.

Similarly, enzyme cocktail of cellulase, xylanase, protease, and pectinase were used as a pre-treatment before solvent extraction of grape seed oil. Solvent extraction for 24 hours and 120 hours showed oil yield recovery of 106 % and 163 %, respectively. The increased oil extraction efficiency was attributable to the additional breaking of the cotyledon cells, which released the oils bodies and made the cell structure more porous, allowing for more efficient hexane percolation and hence improved oil liberation [13].

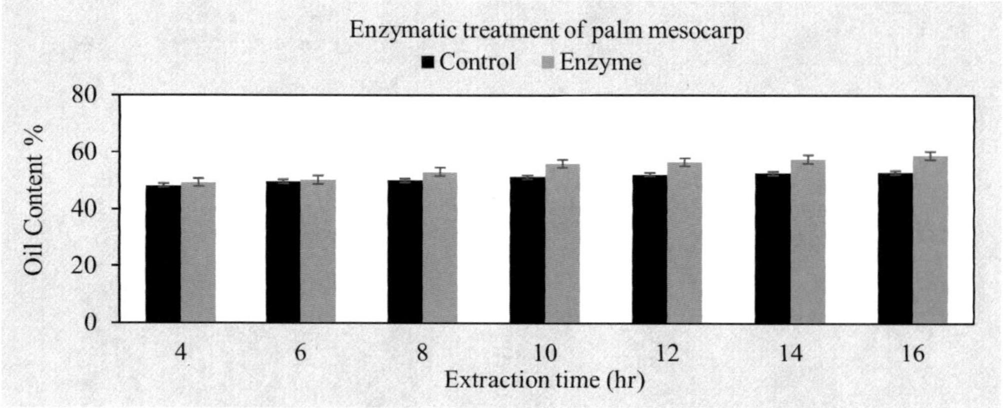

Figure 3. Enzymatic treatment of palm mesocarp followed by extended soxhlet extraction.

3.2. Plant scale trials

Commercial trial of 6-months period with alternating sequences of enzyme cycles and control cycles conducted in Sime Darby palm oil mill achieved an average increment OER of 1.15 % (figure 4). The conclusion is based on mean OER on enzyme vs. control runs, with a rigorous statistical analysis to confirm 95 % statistical degree of confidence. The nature of enzymatic hydrolysis requires incubation time of at least 30 minutes to enable effective reaction, thus, a pre-digester was installed to hold the MPD before digester and screw pressing. It was observed that additional maceration time of MPD with predigester for 40 minutes yielded delta OER of 0.32 %, while addition of enzyme effect to the processing resulted in 0.83 % delta OER. The synergetic hydrolysis of complex constituents of cellulose, hemicellulose, lignin and pectin, together with the digestion and screw pressing steps has resulted in more oil being liberated when dosing the enzyme into the MPD. It is evident that enzyme integration into mechanical processing effectively breaks down plant cell walls to facilitate release of trapped oils.

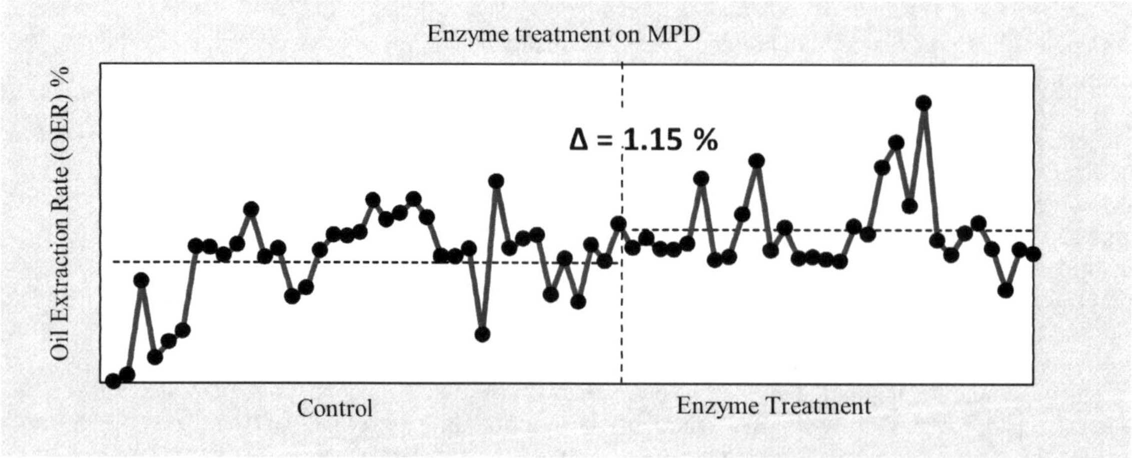

Figure 4. Oil extraction rate increment by enzyme treatment on MPD in plant.

To further minimize mechanical limitation of enzyme technology in palm oil milling, another method of dosing the enzyme mixture into DCO was explored. Enzyme dosing to DCO is more feasible as it requires a very simplified dosing system of enzyme into liquid stream while eliminating limitation in effective mixing and installation of pre-digester. Some unbroken cells could be still present in DCO, thus addition of enzyme will rupture these cell walls enabling more oil to be released and recovered. The moisture level in DCO assists the hydrolytic reaction by increasing diffusivity, mobility of the enzymes and products, thus shortening the reaction time required to obtain higher oil yields [28].

Two methodologies of single line (mill 1 & 2) and dual line (mill 3 and 4) system monitoring were used for evaluation. Single line approach involves the dosing of enzymes alternately two weeks and control two weeks (1 cycle), with a total of 6 cycles based on normal harvesting interval to ensure both periods will get crops from the similar area. Monitoring was based on OER declared by mill on daily basis. On the other hand, dual line involves set up of two lines with identical equipment to ensure similar efficiency and direct comparison of dosing line to control at minimal variation. Evaluation was based on the accumulated OER which is equivalent to pure oil (PO) produced over DCO generated on daily basis. OER comparison for dual line mill was much simpler as the input materials are the same for both enzyme and control line. Summary of delta OER obtained for 4 mills is shown in figure 5. Results represent the mean values and error bars indicate the standard deviation of the mean (n=40). Delta OER ranged between 0.71 to 1.19 %, with an average of 0.91 %.

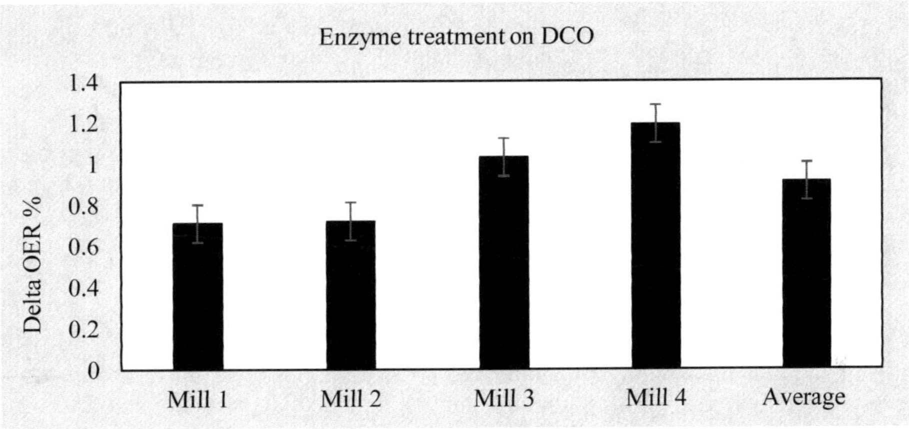

Figure 5. Oil extraction rate increment by enzyme treatment on DCO in plant.

Enzymatic treatment could potentially facilitate coalescence of small oil droplets to bigger oil drops that float up and improve clarification process by reducing viscosity and breaking emulsion [29, 30]. Therefore, an improved and shorter clarification time will be more favourable, given the importance of energy expenditure and time in industrial extraction operations. Reduction in consumption of process water, on the other hand, reduces overall processing costs, eliminates larger reaction and storage systems, and creates a significant waste management challenge [31, 32].

3.3. Reducing sugar content

Reducing sugar analysis is an indication to evaluate the degree of enzymatic hydrolysis. Figure 6 shows the reducing sugar content profile in clarifier underflow. Reducing sugar yield indicates the breakdown of carbohydrate polymer (xylan and glucan). Reducing sugar comprise of monomers and dimers mainly glucose, xylose, fructose, sucrose, mannose and arabinose. Total reducing sugar content in biocatalyst and control samples were 12.29 g/L and 10.13 g/L, respectively. With the presence of enzyme, the difference in reducing sugar is 2 g/L. The increase in reducing sugar is close to 21 %. This increment is considered high considering the available substrate.

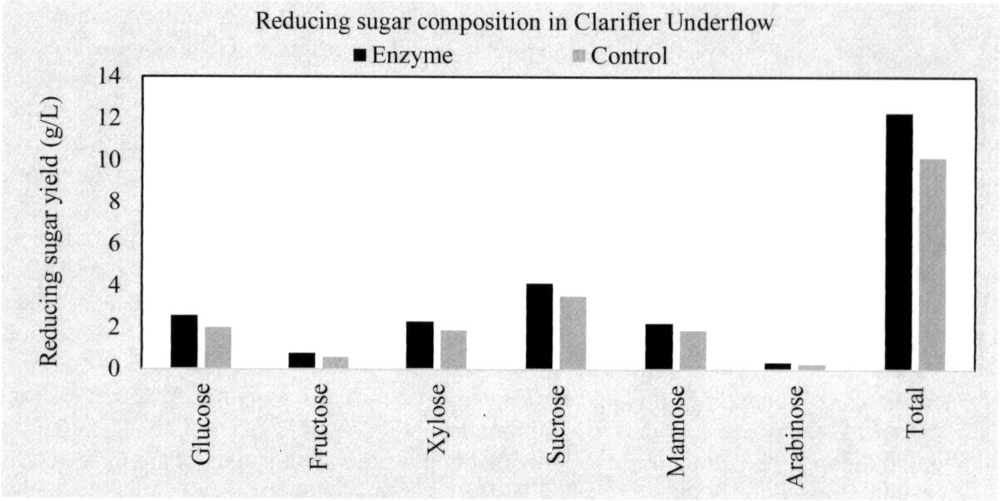

Figure 6. Reducing sugar composition in clarifier underflow.

DCO composition during the sampling were in the range of; moisture content: 52. 47 %, oil content: 40.8 %; NOS (substrate): 6.76 % and structural carbohydrate of 13.8 % [10]. For 1 L of DCO, the substrate available is about 67.6 g of NOS and of that, cellulose and hemicellulose polymer make up about 10 g. Theoretically 1 g of xylan hydrolysed will yield 1.1 g of xylose. Back calculating, for release of 12 g/l monomers, substrate available should be 10.9 g, tallying the findings. Based on this, we can prove the hydrolytic enzymes acted on unbroken cell walls, causing an increase in the release of total reducing sugar content in the reaction media, indicating that the hydrolysis process was efficient during clarification process [33]. However, the value doesn't really represent the complete reaction as oligomers such as cellobiose was not quantified and further testing is required.

3.4. Confocal Microscopy Imaging

Most of the oil bodies are found in the vacuoles of plant cells as free oil, as well as dispersed in the cytoplasm [34]. Confocal microscopy on fluorescently labelled DCO reveals that enzyme treatment results in larger oil bodies free of cells and that remaining cells display weaker Calcofluor fluorescence, suggesting degradation of the cell wall (Fig 7 a, b). This degradation is clearly visible when the surface of the cells are 3D rendered with Imaris (Fig 7 c-f), with predominantly intact cells visible in control samples while the majority of cells in the enzyme treated sample have large areas devoid of any cell wall staining. It can be observed that the cellular architecture of cell wall is disrupted with absolute release of oil globules while leaving only some traces of cell wall. The structural alteration in the cell wall of the palm tissue demonstrated the action of enzymes [11].

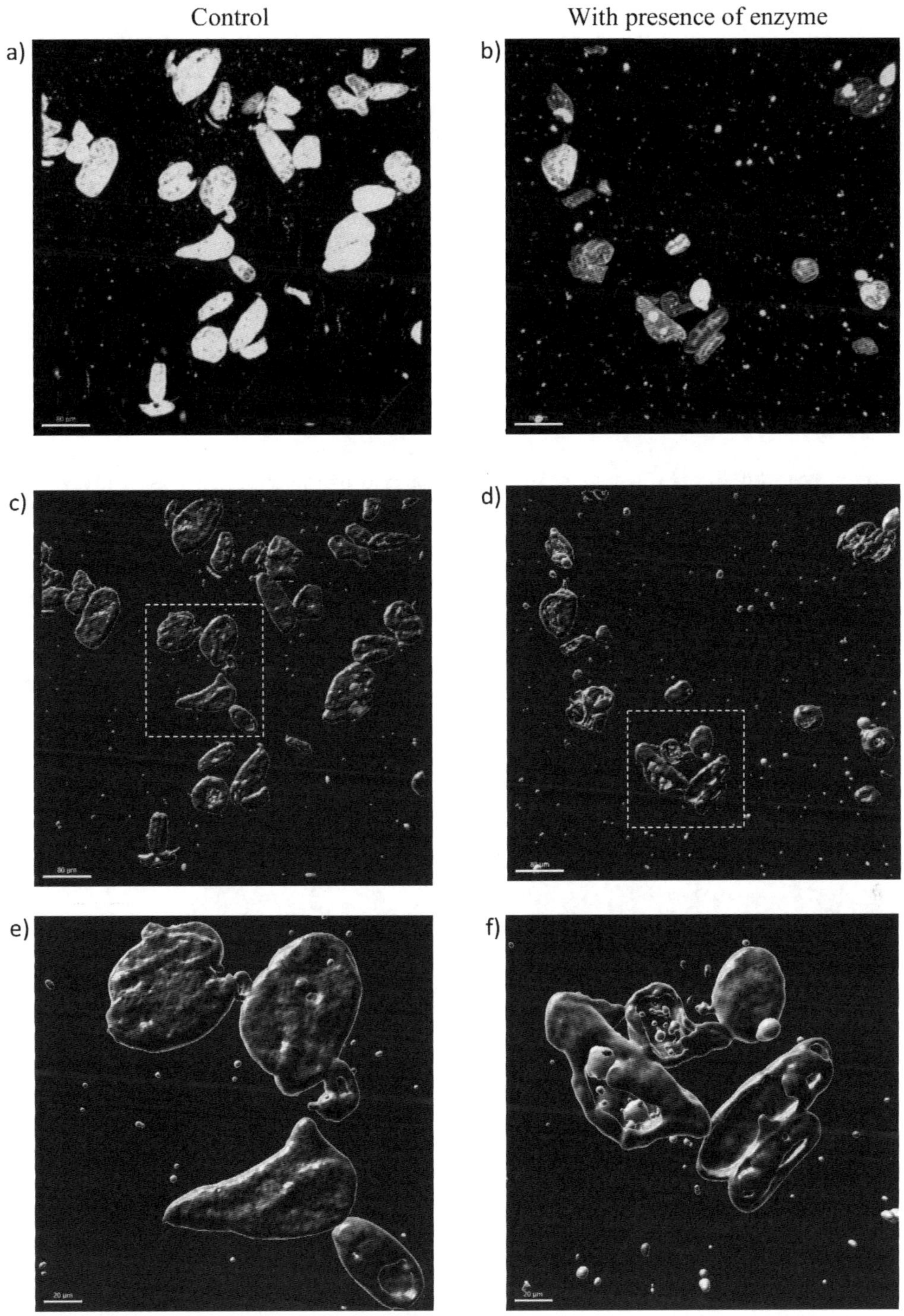

Figure 7. Visualization of oil-bearing cells. (a,b) CLSM fluorescent imaging of DCO stained with 1.6 μM Calcofluor white (Cyan) and 10 mM BODIPY 493/503 (Green) in the absence (a) or presence (b) of enzyme. (c,d) 3D rendering of stained surfaces. (e,f) Zoomed view of indicated area.

4. Conclusion

Deforestation, biodiversity loss, and greenhouse gas (GHG) emissions due to agriculture expansion are all ongoing environmental and sustainability challenges that require urgent attention and solutions. New breeding technologies such as genomics and gene editing, promise to increase yielding cultivars, oil profiles, disease resistance, and crop architecture, but they are time consuming. In terms of yield and delivery of a variety of specialised oils for human use, there are no viable alternatives to palm oil. Rather than demanding bans or boycotts on oil palm goods, it is critical to use appropriate biological extraction methods to boost palm oil processing yield and productivity. A green and novel process is developed to decipher the OER stagnation issue of the palm industry. Dosing of enzyme into MPD yielded in OER increment of 1.15 %. Meanwhile, dosing of enzyme into DCO with a simple system yielded in OER increment of 0.71 to 1.19 %. Cell wall degrading enzymes are used to hydrolyse complex structures of the oil-bearing cells to liberate more oil, reducing losses, and improving downstream processing. A 1.0 % OER increase can potentially generate additional revenue of RM2 billion/year for the palm oil industry in Malaysia.

References

[1] Mielke T 2019 World markets for vegetable oils: Status and prospects. In: Kaltschmitt M (eds) energy from organic materials (biomass) *Encyclopedia of Sustainability Science and Technology Series* Springer New York 261–98

[2] United States Department of Agriculture (USDA) 2021 production, supply and distribution. Available at: https:// https://apps.fas.usda.gov/psdonline/circulars/oilseeds.pdf accessed 7 Sep 2020

[3] James F D and Fitton C 2010 The importance of the global oils and fats supply and the role that palm oil plays in meeting the demand for oils and fats worldwide *J. Am. Coll. Nutr.* **29** 245–52

[4] Dubos B, Baron V, Bonneau X, Dassou O, Flori A and Impens R 2019 Precision agriculture in oil palm plantations: diagnostic tools for sustainable N and K nutrient supply *OCL Oilseeds Fats Crop Lipids* **26** 1–8

[5] Chew C L, Ng C Y, Hong W O, Wu T Y, Lee Y Y, Low L E, Kong P S and Chan E S 2021a Improving sustainability of palm oil production by increasing oil extraction rate: A review *Food Bioproc. Tech.* **14** 573–86

[6] Chew C L, Low L E, Chia W Y, Chew K W, Liew Z K, Chan E S, Chan Y J, Kong P S and Show P L 2021b Prospects of palm fruit extraction technology: Palm oil recovery processes and quality enhancement *Food Rev. Int.* 1–28

[7] Basiron Y 2005 Palm oil processing 6th Bailey's Industrial Oil and Fat Products (John Wiley & Sons Inc) pp 333–429

[8] Dijkstra A D 2007 Production of vegetable oils and fats 3rd Edition the lipid handbook United States of America: CRC Press pp 155–75

[9] Noerhidajat Y R, Zurina Z A, Syafiie S, Ramanaidu V and Rashid U 2016 Effect of high pressurized sterilization on oil palm fruit digestion operation *Food Rev. Int.* **23** 129–34

[10] Zakaria A 2019 MICRONES helps overcome loss of crude palm oiduring extraction https://core.ac.uk/download/pdf/153830942.pdf accessed 3 Oct 2019

[11] Silvamany H and Jahim J 2015 Enhancement of palm oil extraction using cell wall degrading enzyme formulation *Malays. J. Anal. Sci.*, **19** 77–87

[12] Grasso F V, Montoya P A, Camusso C C and Maroto B G 2012. Improvement of soybean oil solvent extraction through enzymatic pretreatment *Int. J. Agron.* 1–7

[13] Passos C P, Yilmaz S, Silva C M and Coimbra M A 2009 Enhancement of grape seed oil extraction using a cell wall degrading enzyme cocktail *Food Chem.* **115** 48–53

[14] Perez E E, Fernandez M B, Nolasco S M and Crapiste G H 2013 Effect of pectinase on the oil solvent extraction from different genotypes of sunflower (Helianthus Annuus L.) *J Food Eng.* **117** 393–8

[15] Jiang L, Hua D, Wang Z and Xu S 2010 Aqueous enzymatic extraction of peanut oil and protein hydrolysates *Food Bioprod. Process* **88** 233–8
[16] Hadj-Taieb N, Grati N, Ayadi M, Attia I, Bensalem H and Gargouri A 2012 Optimisation of olive oil extraction and minor compounds content of tunisian olive oil using enzymatic formulations during malaxation *Biochem. Eng. J.* **62** 79–85
[17] Cheah S C, Augustin M A and L O L C 1990 Enzymatic extraction of palm oil *Palm Oil Res. Bull. Malaysia* **20** 30–6
[18] Teixeira C B, Macedo G A, Macedo J A, Da Silva L H M and Rodrigues A M D C 2017 Efficiency of the use of solvents in vegetable oil extraction at oleaginous crops *Renew. Sust. Energ. Rev.* **80** 121–4
[19] Javvadi S, Pradhan S, Rathi C L and Wani An 2012 enzyme composition and process for extracting oil from oil palm fruits WO2012011130A3 March 15
[20] Eshtiaghi M N, Paoplook K, Yoswathana N and Kuldilloke J 2015 Enhanced oil extraction from palm fruit mesocarp using technical enzymes *Int. J. Adv. Sci. Technol.* **3** 42–5
[21] Mohamed M, Markom M, Daud W R W and Nadirah A 2018 Cosolvent selection for supercritical fluid extraction (SFE) of bioactive compounds from orthosiphon stamineus *Sains Malays.* **47** 1741–7
[22] Hashim N, Asis A J and Abdullah M A 2018 A system for extracting palm oil from plurality of palm fruitlet WO2018135937A1 July 26
[23] Mat Yasin N M F, Hashim N and Zaidy Arnan M 2019 Process for producing crude palm fruit oil (silicate / enzyme application) WO2019209100A1, October 31
[24] Abdul P M, Jahim J M, Harun S, Markom M, Lutpi N A, Hassan O, Balan V, Dale B E and Nor M T 2016 Effects of changes in chemical and structural characteristic of ammonia fibre expansion (AFEX) pretreated oil palm empty fruit bunch fibre on enzymatic saccharification and fermentability for biohydrogen *Bioresour. Technol.* **211** 200–8
[25] Dal Prá V, Lunelli F C, Vendruscolo RG, Martins R, Wagner R, Lazzaretti A P, Freire D M G, Alexandri M, Koutinas M, Mazutti M A and Barcellos M 2017 Ultrasound-assisted extraction of bioactive compounds from palm pressed fiber with high antioxidant and photoprotective activities *Ultrason. Sonochem.* **36** 362–6
[26] Gam D H, Yi Kim S and Kim J W 2020 Optimization of ultrasound-assisted extraction condition for phenolic compounds, antioxidant activity, and epigallocatechin gallate in lipid-extracted microalgae *Molecules* **25** 454
[27] Campbell K A and Glatz C E 2009 Mechanisms of aqueous extraction of soybean oil *J. Agric. Food Chem.* **57** 10904–12
[28] Yang L, Jiang L, Sui X and Wang S 2011 Optimization of the aqueous enzymatic extraction of pine kernel oil by response surface methodology *Procedia Eng.* **15** 4641–52
[29] Rosenthal A, Pyle D L and Niranjan K 1996 Aqueous and enzymatic processes for edible oil extraction *Enzyme Microb.* **19** 402–20
[30] Niu R, Zhou L, Chen F, Zhu T and Duan X 2021 Effect of enzyme on the demulsification of emulsion during aqueous enzymatic extraction and the corresponding mechanism *Cereal Chem.* **98** 594–603
[31] Mwaurah P W, Kumar S, Kumar N, Attkan A R, Panghal A, Singh V K and Garg M K 2019 Novel oil extraction technologies: Process conditions, quality parameters, and optimization *Compr. Rev. Food Sci. Food Saf.* **19** 3–20
[32] Yusoff M M, Gordon M H and Niranjan K 2015 Aqueous enzyme assisted oil extraction from oilseeds and emulsion de-emulsifying methods: A review *Trends Food Sci. Technol.* **41** 60–82
[33] Yuanyuan M, Yali Z, Jiang L and Hongbo S 2009 Roles of plant soluble sugars and their responses to plant cold stress *Afr. J. Biotechnol.* **8** 2004–10
[34] Ho L S, Anusha N, Hirzun M Y, Harikrishna K and Jangi S 2014 Morphometry of lipid bodies in embryo, kernel and mesocarp of oil palm: Its relationship to yield *Am. J. Plant Sci.* **05** 1163–

73

Acknowledgments
The authors are grateful for the financial and technical support from the Sime Darby Plantation Research, SDP Processing & Engineering Unit and Palm Milling Technical team from Novozymes Malaysia.

Extraction of cellulose from oil palm empty fruit bunch using eco-friendly solvents for preparation of transparent cellulose thin film

N S Izzaha[1], A N A Yahaya[2], N Z M Zuhudi[3], N A Khalil[1] and M Zulkifli[1]

[1] Universiti Kuala Lumpur, Branch Campus Malaysian Institute of Chemical and BioEngineering Technology, 78000 Alor Gajah, Melaka, Malaysia
[2] Green Chemistry and Sustainability Cluster, University Kuala Lumpur, Branch Campus Malaysian Institute of Chemical and Bioengineering Technology, Taboh Naning, 78000 Alor Gajah, Melaka, Malaysia
[3] Aero Composite Cluster, University Kuala Lumpur, Branch Campus Malaysian Institute of Aviation Technology, Jalan Jenderam Hulu, 43800 Dengkil, Selangor, Malaysia

Email: syazwana.izzaha@s.unikl.edu.my

Abstract. The purpose of this research is to extract the cellulose using eco-friendly reagents of hydrogen peroxide and formic acid and determine the optimum reaction time for delignification process. The extracted cellulose and characterised using FTIR, TGA and PSA. The percentage yield of extracted cellulose were calculated. The highest yield was found to be 65.78 % at reaction time 120 min. The FTIR spectral studies confirm the removal of lignin from the delignified cellulose at peak 1613 cm^{-1} and the TGA result shows the thermal degradation of extracted cellulose at 329.04, 329.92 and 330.99 °C at reaction time 60, 90 and 120 min. The PSA studies provided the evidence of extracted particle size of the cellulose become finer as the reaction time increase. The particle size observed for delignified cellulose at 60, 90 and 120 min are 68.4, 64.6 and 57.3 μm. The extraction of cellulose and characterization to determine the optimum reaction time was able to obtain. From the result obtained, it can be concluded that the longer the reaction time, the higher the percentage yield of cellulose extracted. Film formation was later carried out using the extracted cellulose from different reaction time.

Keywords: Transparent thin film; Extraction; Eco-friendly reagent

1. Introduction

Cellulose $(C_6H_{10}O_5)_n$ is a natural linear polysaccharide. It is one of the abundant biomaterials on earth prooduced by plants or some bacteria. Plant cell walls are mainly composed by cellulose which help the plant to remain strong and rigid whereas the other component composed together with the cellulose [1]. Cellulose in plant fiber are in amorphous phase that linked with crystalline phase through the intermolecular and intramolecular of hydrogen bonding. The unique properties of the cellulose are low density, good mechanical properties and biodegradable [2].

Cellulose can be extracted by chemical, mechanical or enzymatic technique [3]. The chemicals method can be conducted by using both alkali and acidic chemicals whereas the enzymatic technique involves specific enzyme such as cellulase and cultural medium with the presence of glucose and ethanol for cellulase consumptions in extraction process [4]. The mechanical method usually involves grinding and milling of the fiber with addition of small amount of chemicals as well as thermomechanical process such as steam fiber explosion [5].

The cellulose-based film has several characteristics such as light in weight, renewable, recyclable and compatible to the health and the environment. The casting technique are widely used as the thickness of the film produced will be distributed uniformly, maximum optical purity and low haze. Nowadays, the solvent cast technology are used widely in productions of film as it gives Extremely high quality [6].

Currently, the natural fibers such as lignocellulosic biomass from oil palm waste such as cellulose is one of the major applications in food packaging and film industry. The cellulose obtains from OPEFB via extraction method using eco-friendly reagent is widely explored and well documented in the literature. However, the optimization of reaction time in the extraction process has not been determined. Throughout this study, the optimum parameter specifically the delignification reaction time will be determine to achieve the most optimize and effective extraction process.

Besides, the preparation of transparent cellulose thin film are also widely explored and documented but the production of transparent cellulose thin film from cellulose extracted from OPEFB is limited [7]. The cellulose extracted were used to prepare the film.

The analytical, thermal and mechanical tests were conducted to determine the characterization of each sample. The test were performed by using Fourier-Transform Infrared (FTIR) Spectroscopy, Particle Size Analyzer (PSA), Thermogravimetric Analysis (TGA) and Tensile testing to determine the functional group, particle size, thermal stability and mechanical properties of the sample respectively. By conducting this study, the optimum reaction time of the extraction process were finalised.

2. Experimental section

2.1. Materials

Oil Palm Empty Fruit Bunch (OPEFB) of size 500 μm was purchased from MPOB, Selangor. Ethanol and Hydrogen Peroxide was purchased from Sigma Aldrich. Sodium Hydroxide, Dimethylacetamide (DMAc) and Lithium Chloride was purchased from R&M Chemicals Sdn Bhd. chemicals reagents purchased are of analytical grade and acquired from commercial sources in Malaysia.

2.2. Sample preparation

1000 g of 500 μm OPEFB were soaked with distilled water for 30 min and washed several times until the water observed become clear. The purpose of washing is to remove the dirt and other impurities from raw materials. The fiber was further dried at 50 °C until constant weight was obtained after 3 days.

2.3. Pre-treatment of oil palm empty fruit bunch (OPEFB)

The 30 g of OPEFB were dewaxed with 300 ml of 70 % ethanol using Soxhlet extractor for 3 hours and dried for overnight. 50 g of the dewaxed fibers were suspended in 10 % NaOH and autoclaved for one hours at 121 °C and 1.5 atm. The mixture of supernatant and fibers were separated and washed until clear water was observed. 10 g of treated fiber were suspended in 10 % H_2O_2 and placed into incubator shaker for 1 hour at room temperature for bleaching treatment. The bleached fibers were washed with distilled water and dried overnight.

2.4. OPEFB-cellulose extraction process

15 g of pre-treated fibers were soaked with the mixture of 20 % formic acid and 10 % H_2O_2 at ratio 1:1 (v/v). The soaked fibers were placed in water bath at 85 °C at different reaction times, which is 60 min, 90 min and 120 min each for delignification process. The delignified cellulose were washed with 10 % concentration of formic acid before further rinse with distilled water until pH 7 were obtained. The

yellowish extracted cellulose were then re-suspended in 10 % H_2O_2 each for 90 min at 60 °C for further bleaching. The process was repeated until the fiber turned whitish. The insoluble fraction of cellulose was collected and the percentage of yield (% wt) was calculated. The α-cellulose content was determined using Mohtar *et al.* [8].

Percentage of yield equation:

$$Y_i\ (\%\ wt) = \frac{A}{B} \times 100 \tag{1}$$

Where,
i = Cellulose fraction
A = Weight of extracted cellulose from OPEFB, g
B = Weight of respective OPEFB used the extraction process, g

2.5. Transparent cellulose thin film preparation

2.5.1. Transparent cellulose thin film preparation method 1. According to Acharya *et al.* [9], 3 g of extracted cellulose at reaction time 60, 90 and 120 min were suspended in DMAc with concentrations of 5 % w/v. The suspension were heated for 30 min at 80 °C under continuous stirring condition and 8 g of LiCl were added after 30 min. The mixture continue heated at 80 °C under stirring condition for another 3 hours. The temperature were decreased to 50 °C at the end of heating and the temperature were maintained for overnight.

2.5.2. Transparent cellulose thin film preparation method 2. According to Zailuddin & Husseinsyah [10], 10 g of extracted cellulose from different reaction time were activated in distilled water for 5 hours at room temperature. The activation process was to swell the cellulose fiber before dehydrated in DMAc for 4 hours at room temperature. The prepared cellulose were filtered and dissolved in DMAc for another 20 min. 8 % of lithium chloride were added and the solution were stirred for 40 min continuously until the lithium chloride was completely dissolve in the solution.

2.5.3. Casting method to prepare transparent cellulose thin film. The cellulose solution prepared were casted on casting glass slide and were gently distributed to spread the solution uniformly on the glass slides for film preparation. The glass slides that were spread with the solution were left for gelation at ambient temperature for overnight. The gelation film were immersed in distilled water for generation. The water were exchanged every 5 hours for 3 days to regenerate the film and to remove the solvent. The wet film obtained were dried in freeze-dryer for 4 days. The freeze-dried film then undergo further characterization.

2.6. Characterization

2.6.1. Fourier-transform infra-red spectroscopy (FTIR). FTIR Spectrophotometer model Perkin Elmer Spectrum 400 FT-IR/FT-NIR was used to identify the functional group of the raw OPEFB, treated OPEFB, cellulose extracted and the transparent thin film prepared. The sample and KBr were mixed at ratio 100:1 until it is fully homogenised. FTIR spectra were analysed within wavelength of ranged between 400 and 4000 cm^{-1} at a rate of 0.02 cm^{-1}.

2.6.2. Thermogravimetric analysis (TGA). The thermal stability of raw OPEFB, treated OPEFB and cellulose extracted as well as the transparent thin film prepared were determined using Mettler Toledo TGA machine according to ASTM E1131 method. Nitrogen gas were flowed at 20 ml min^{-1} and heated from room temperature, which is 30 °C to 700 °C with heating rate of 10 °C min^{-1}.

IOP Conf. Series: Materials Science and Engineering 1195 (2021) 012062 doi:10.1088/1757-899X/1195/1/012062

2.6.3. Particle size analyzer (PSA). Particle size analysis were done using Mastersizer, Malvern 3000 model according to ASTM C209-20. The temperature set was at room temperature and the refractive index used was 1.746 with 1.000 absorption index. Water was used as the dispersant medium with the dispersant refractive index of 1.330.

3. Result and discussion

3.1. OPEFB pre-treatment

The primary steps before extracting cellulose from the fiber is pre-treatment. The main purpose of pre-treatment process is to destroy the lignin structure, decrease the cellulose crystallinity and increase the lignocellulosic materials porosity prior to further treatment [11]. The transmittance peak located at 3400 cm^{-1} in both raw and pretreated OPEFB is attributed to the –OH stretching vibration of cellulose [12]. The peak was slightly decreasing in the intensity after pre-treatment due to heating process introduced at high temperature to the fiber during autoclaved. As the lignin was removed along during the pre-treatment, the intensity of -CO stretching band of lignin at peak 1613 cm^{-1} were decreased [13]. The removal of lignin was proven in figure 1. Delignification was done prior to further removal of lignin. The peaks at 1170 cm^{-1} were shown in all sample as it attributes to the deformation of C-H rocking vibration and C-O-C pyranose [14]. To conclude, lignin functional groups presented in untreated OPEFB were diminished along the process.

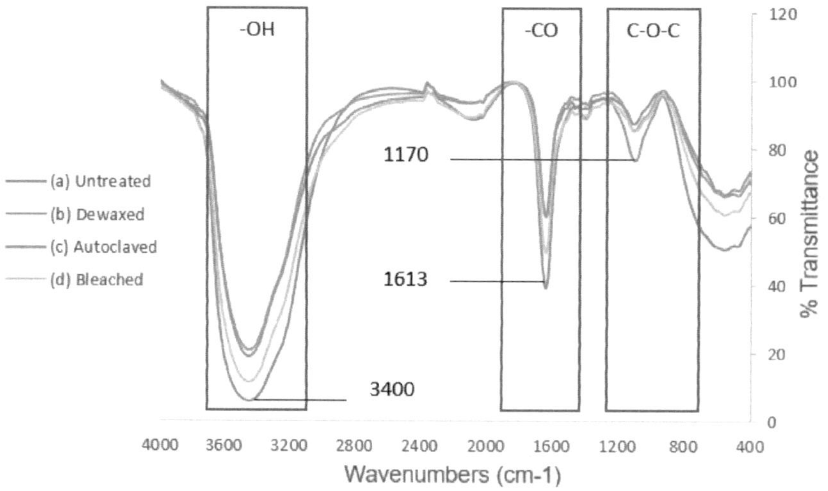

Figure 1. FTIR spectrum of (a) untreated; (b) dewaxed; (c) autoclaved and (d) bleached OPEFB.

3.2. Extraction of cellulose from pre-treated OPEFB

Further delignification of cellulose fiber has effectively removed the lignin fraction and produced chlorine-free cellulose with percentage yield as reported in previous research of Ranganagowda *et al.* [15] and Nazir *et al.* [12] that reported the extracted cellulose from OPEFB at reaction time 120 min is at 64 % and 65 % respectively. The synergistic effect of formic acid will remove lignin by partial oxidation. The lignin were converted into carboxylic acids and removed in the form of water-soluble fraction and leave insoluble cellulose suspended in aqueous medium [16]. The percentage yield of extracted cellulose calculated is computed in the table 1. The highest percentage yield of delignified cellulose is at 120 min reaction time followed by at 90 and 60 min. The content of cellulose are 65.78 %, 64.13 % and 61.65 % respectively. Thus, the optimum reaction time for delignification of cellulose is at 120 min.

Table 1. Percentage yield of extracted cellulose from OPEFB.

Reaction time, min	Initial weight, g	Final weight, g	Yield, % (w/w)
60	15.02	9.26	61.65
90	15.00	9.62	64.13
120	14.99	9.86	65.78

The transmittance peak at 1613 cm^{-1} in figure 2 were assigned to -CO stretching of lignin. At reaction time 120 min, the transmittance peak of lignin functional group shows the least intense. Thus, the presence of lignin was removed form delignification process. The intensity peak of C-O-C of aryl-alkyl ether in lignin (figure 2) decreases as the reaction time increases. The delignified cellulose at 120 min reaction time has the lowest lignin intensity compared to delignification of cellulose at 60 min reaction time. The FTIR Spectra of delignified cellulose at reaction time 60, 90 and 120 min shows the transmittance peak at 2907 cm^{-1}. The peak were attributed to -CH of cellulose groups stretching vibrations was presence after delignification process but become less intense due to damage of cellulose structure during the process. To conclude, the lignin component was successfully removed from the untreated fiber.

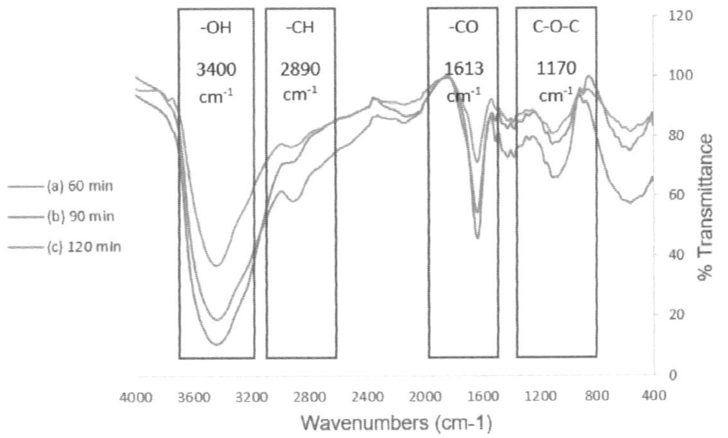

Figure 2. FTIR spectrum for delignified cellulose at reaction time (a) 60 min; (b) 90 min; (c) 120 min.

3.3. Thermogravimetric analysis (TGA)

From the result obtained, the initial weight loss for first decomposition is attributed to loss of water content and volatile substance at temperature for the first 300 °C [17]. The second composition shows the thermal degradation of the cellulose. Dewaxed fiber has the highest inflection point, which is at 345.35 °C. The bleached fiber has the second highest inflection point at 339.50 °C, which then followed by autoclaved at 345.35 °C and untreated fiber at 317.03 °C. The dewaxed fiber shows the highest inflection point compared to other pre-treated and untreated fiber due to the presence of –OH group of cellulose during dewaxed process, which was absence in untreated and other pre-treated fiber. Inflection point for untreated fiber was the least since the fiber consists of wax, pectin, and other lignocellulosic component that has low thermal degradation. Figure 3 shows the overlay thermal degradation curve for untreated OPEFB and pre-treated fiber.

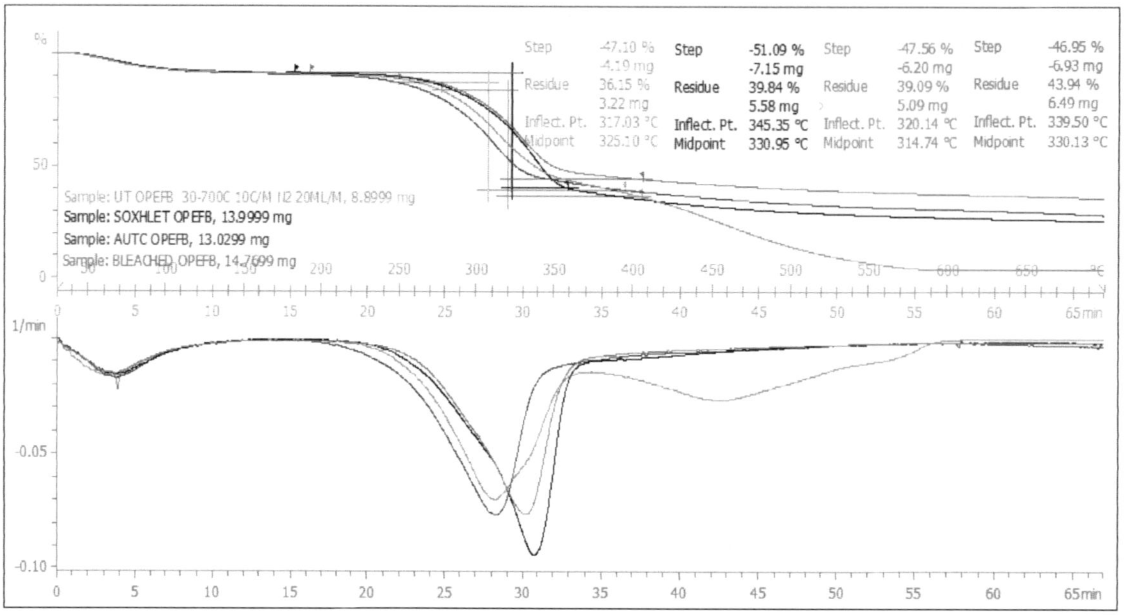

Figure 3. Overlay thermal degradation curve for untreated OPEFB and pre-treated fiber.

The delignified cellulose at 120 min reaction time has the highest inflection point for the second decomposition which is 330.99 °C and followed by delignified cellulose at 90 min and the least is delignified cellulose at 60 min with inflection point of 329.92 °C and 329.04 °C, respectively. The highest cellulose yield was at 120 min delignified cellulose and it has the highest concentration of cellulose as it was delignified at optimum reaction time compared to other delignified cellulose. Thus, more heat is needed to degrade the more concentrated cellulose which makes the inflection point obtained is high. The thermal stability of the samples shows increment along with the process due to the increased in crystallinity and intermolecular hydrogen-bonded after the removal of amorphous content. Figure 4 shows the overlay thermal degradation curve for delignified cellulose at different reaction time.

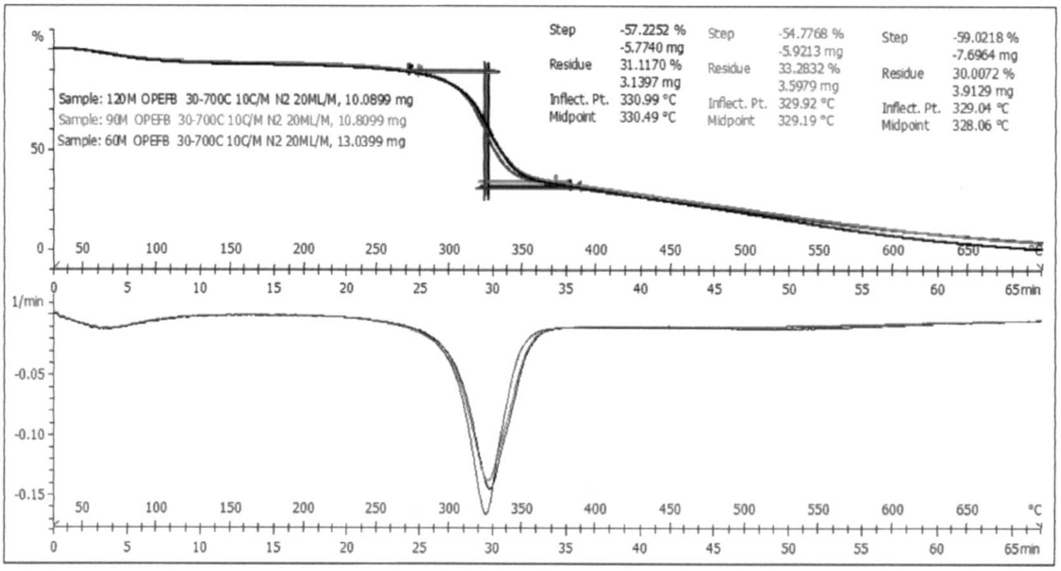

Figure 4. Overlay thermal degradation curve for delignified cellulose at different reaction time.

3.4. Particle size analyzer (PSA)

The result were obtained in the point of distribution of untreated OPEFB at Dv (90) is 207 µm. The size of sample become finer for dewaxed and autoclaved fiber with Dv (90) which are 202 µm and 190 µm, respectively. The smallest particle size was obtained during bleaching process with Dv (90) of 50.8 µm. The particle size in pre-treatment was decreasing along the process. During bleaching process, the fiber were destructed or modified to remove the colour bodies by degrading it into smaller and soluble units [18].

For delignified cellulose, smallest size of particle is 57.3 µm at 120 min delignification reaction time. The delignified cellulose at 90 and 60 min reaction time has the derivative size of 64.6 µm and 68.4 µm. Theoretically, the longer the reaction time, the smaller the particle distribution size until it reach its optimum condition as the delignification process removed amorphous structure of the fiber and the microfibrils were obtained in the process. According to Mohtar *et al.* [8], it is reported that the smaller size of particle were obtained when longer hydrolysis time applied to isolate the nanocellulose from OPEFB. The particle size and distribution gives impact on the properties such as rheology, film gloss, surface area and packing density of film. Table 2 shows the particle size obtained for untreated OPEFB, pre-treated and delignified cellulose.

Table 2. The particle size distribution for untreated OPEFB, pre-treated and delignified cellulose.

Sample	**Size classes, µm**		
	Dv (10)	**Dv (50)**	**Dv (90)**
Untreated OPEFB	18.90	83.70	207.00
Dewaxed	24.50	94.60	202.00
Autoclaved	34.80	78.00	190.00
Bleached	23.40	35.00	50.80
Delignified cellulose at 60 min	33.60	48.60	68.40
Delignified cellulose at 90 min	31.50	45.70	64.60
Delignified cellulose at 120 min	30.26	46.40	57.30

3.5. Transparent cellulose thin film

The cellulose-based transparent thin films were prepared using 2 different methods. The weight of cellulose used is according to method 1 and 2 of Acharya *et al.* [9] and Zailuddin & Husseinsyah [10], respectively. The cellulose film was unable to be prepared using method 1 as the film did not solidified. According to Acharya *et al.* [9], hydrolysed cellulose fiber with size 30 µm [9] were used, meanwhile the extracted cellulose size is 57.3 µm fiber. The cellulose fiber was only dissolved partially in the solvent during cellulose solution preparation which cause less interaction between cellulose and the solvent. This is because the surface area exposed of the fiber to solvent is limited compared to the approached method [19]. Figure 5 indicates the outcome for the film prepared for method 1.

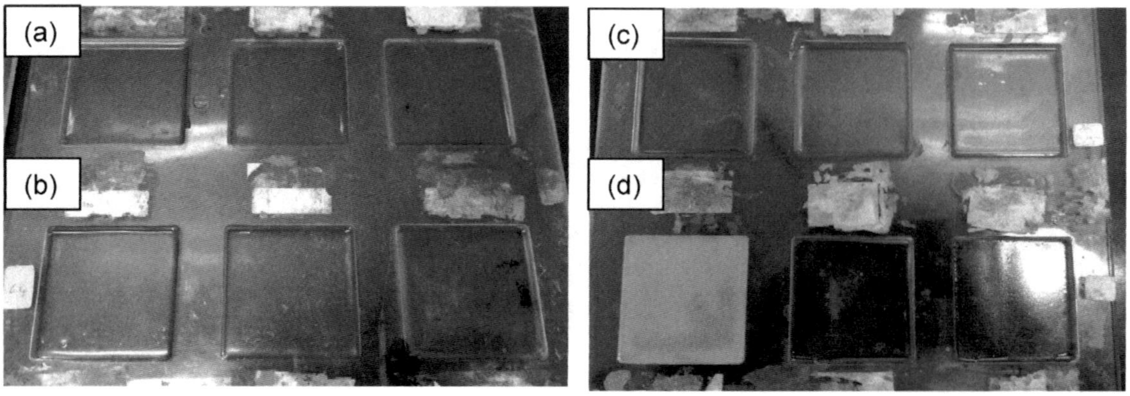

Figure 5. Casted cellulose solution from delignified cellulose (a) 60 min; (b) 90 min using method 1 and delignified cellulose (c) 120 min and (d) MCC using method 1.

The approached method used cellulose fiber with size 50 μm [10]. The film also unable to be prepared but after several repeated preparation the film was successfully formed according to method 2, but no tests has been done to date.

4. Conclusions

To conclude, from this study the cellulose were successfully extracted using eco-friendly reagents which are formic acids and hydrogen peroxide which make the cellulose pulp obtained was chlorine-free. The optimum reaction time of cellulose were also determined by calculating the percentage yield of cellulose at 120 min with the yield of 65.78 %. The result extracted cellulose were characterized using FTIR, TGA and PSA. The FTIR result shows the lowest lignin intensity at 120 min compared to the other with the thermal degradation obtained at 330.99 °C, which is the highest, compared to other reaction time and the finest particle size were obtained at 120 min reaction time which is 57.3 μm. However, the transparent cellulose thin film was not able to be prepared according to method 1, as the extracted cellulose used was larger than the required size according to the respective method meanwhile the film was successfully prepared using method 2. Therefore, the objective to extract cellulose and preparation of transparent cellulose thin film proposed were able to be achieved, but the characterization of the film were unable to be achieved. The cellulose were recommended to undergo acid hydrolysis treatment to produce nanocellulose before film preparation as the size of the cellulose might affect the film produced.

References

[1] Bailey R The structure and function of a cell wall [web page on the Internet]. 2019 [cited 2021 May 31]. Available from https://www.thoughtco.com/cell-wall-373613
[2] Wulandari W T, Rochliadi A and Arcana I M 2016 Nanocellulose prepared by acid hydrolysis of isolated cellulose from sugarcane *bagasse IOP Conf. Ser.: Mater. Sci. Eng.* **107**
[3] Zhang B X, Azuma J I and Uyama H 2015 Preparation and characterization of a transparent amorphous cellulose film *RSC Adv.* **5** 2900–7
[4] Binod P, Janu K U, Sindhu R and Pandey A 2011 Hydrolysis of lignocellulosic biomass for bioethanol production (*Biofuels* 1st edn)
[5] Shahi N, Min B, Sapkota B and Rangari V K 2020 Eco-friendly cellulose nanofiber extraction from sugarcane bagasse and film fabrication *Sustainability* **12** 1–15
[6] Siemann U 2005 Solvent cast technology — A versatile tool for thin film production *Progr. Colloid Polym. Sci.* **130** 1–14
[7] Hastuti N, Kanomata K and Kitaoka T 2020 High translucent of polymeric membrane reinforced by nanocellulose from oil palm empty fruit bunches *IOP Conf. Ser.: Mater. Sci. Eng.* **935**
[8] Mohtar S S, Tengku Malim Busu T N Z, Md Noor A M, Shaari N and Mat H 2017 An ionic liquid

treatment and fractionation of cellulose, hemicellulose and lignin from oil palm empty fruit bunch *Carbohydr. Polym.* **166** 291–9

[9] Acharya S, Hu Y, Moussa H and Abidi N 2017 Preparation and characterization of transparent cellulose films using an improved cellulose dissolution process *J. Appl. Polym. Sci.* **134** 1–12

[10] Zailuddin N L I and Husseinsyah S 2016 Tensile properties and morphology of oil palm empty fruit bunch regenerated cellulose biocomposite films *Procedia Chem.* **19** 366–72

[11] Rame 2018 Oil palm empty fruit bunches (OPEFB): Existing utilization and current trends biorefinery in Indonesia *E3S Web Conf.* **31** 1–5

[12] Nazir M S, Wahjoedi B A, Yussof A W and Abdullah M A 2013 Eco-friendly extraction and characterization of cellulose from oil palm empty fruit bunches **8** 2161–72

[13] Isroi I and Cifriadi A 2018 Oxidation of cellulose from oil palm empty fruit bunch using hydrogen peroxide in alkaline condition *Jurnal Selulosa* **8** 51

[14] Wong E V 2.4: sugars [web page on the Internet]. 2020 [cited 2021 May 31]. Available from https://bio.libretexts.org/Bookshelves/Cell_and_Molecular_Biology/Book%3A_Cells_-_Molecules_and_Mechanisms_(Wong)/02%3A_Basic_Cell_Chemistry_-_Chemical_Compounds_and_their_Interactions/2.04%3A_Sugars

[15] Ranganagowda R P G, Kamath S S and Bennehalli B 2019 Extraction and characterization of cellulose from natural areca fiber *Mater. Sci. Res. India* **16** 86–93

[16] Li B, Xu W Y, Kronlund D, Määttänen A, Liu J, Smått J H, Peltonen J, Willför S, Mu X D and Xu C L 2015 Cellulose nanocrystals prepared via formic acid hydrolysis followed by TEMPO-mediated oxidation *Carbohydr. Polym.* **133** 605–12

[17] Chen Y, Tshabalala M A, Stark N M, Gao J, Fan Y and Ibach R E 2013 Thermal properties of extracted and delignified wood flour for use in wood-plastic composites 1–10

[18] Alomar A and Manubens E 2015 Bleaching agents *European Handbook of Dermatological Treatments* vol 4 (Berlin: Springer) 1421–6

[19] Zaman A, Huang F, Jiang M, Wei W and Zhou Z 2020 Preparation, properties, and applications of natural cellulosic aerogels: A review EBE 1 60–76

Acknowledgement

This work was supported by Short Term Research Grant (STRG) under Universiti Kuala Lumpur (UniKL/CoRI/str19036).

AUTHOR INDEX